I0789941

Silicon Nitride, Silicon Dioxide, and Emerging Dielectrics 11

Editor:

R. Ekwal Sah
Fraunhofer Institute for Applied Solid State Physics
Freiburg, Germany

Sponsoring Division:

 Dielectric Science & Technology

Published by
The Electrochemical Society

65 South Main Street, Building D
Pennington, NJ 08534-2839, USA

tel   609 737 1902
fax   609 737 2743
www.electrochem.org

ecstransactions ™

Vol. 35, No. 4

Copyright 2011 by The Electrochemical Society.
All rights reserved.

This book has been registered with Copyright Clearance Center.
For further information, please contact the Copyright Clearance Center,
Salem, Massachusetts.

Published by:

The Electrochemical Society
65 South Main Street
Pennington, New Jersey 08534-2839, USA

Telephone 609.737.1902
Fax 609.737.2743
e-mail: ecs@electrochem.org
Web: www.electrochem.org

ISSN 1938-6737 (online)
ISSN 1938-5862 (print)
ISSN 2151-2051 (cd-rom)

ISBN 978-1-56677-865-7 (Hardcover)
ISBN 978-1-60768-215-8 (PDF)

Printed in the United States of America.

Symposium Organizers:

R. E. Sah, Fraunhofer Institute for Applied Solid State Physics, e-mail: sah@iaf.fhg.de;

J. Zhang, Liverpool John Moores University, e-mail: j.f.zhang@ljmu.ac.uk ;

M. J. Deen, McMaster University, e-mail: jamal@mcmaster.ca ;

A. Toriumi, The University of Tokyo, e-mail: toriumi@material.t.u-tokyo.ac.jp ;

D. Bauza, CNRS/IMEP/Minatec-INPG, e-mail: bauza@enserg.fr;

S. W. King, Intel, e-mail: sean.king@intel.com.

Technical Program Committee:

P. Joshi, Sharp Laboratories of America, Inc., e-mail: pjoshi@sharplabs.com;

P. Srinivasan, Texas Instruments, e-mail: psrinivasan@ti.com;

R. Todi, Scmiconductor Research and Development Center, IBM Microelectronics,
e-mail: rmtodi@us.ibm.com.

ECS Transactions, Volume 35, Issue 4
Silicon Nitride, Silicon Dioxide, and Emerging Dielectrics 11

Table of Contents

Preface   *iii*

Chapter 1
Interface Characterization

Fluctuations in Electronic Properties of MOS Interface in Nanoscale MOSFETs   3
   T. Tsuchiya, Y. Morimura, and Y. Mori

Electron States in MOS Systems   19
   O. Engström

Process Engineering and Trap Distribution for Dielectric/Si Interfacial Layer in High-k   39
Gated MOS Devices
   K. Chang-Liao, C. Fu, C. Lu, Y. Chang, Y. Hsu, C. Tsao, T. Wang, D. Heh, Y. Li,
   W. Tsai, C. Ai, F. Hou, and Y. Hsu

Current Understanding of the Transport Behavior of Hydrogen Species in MOS Stacks and   55
Their Relation to Reliability Degradation
   Z. Liu, S. Fujieda, H. Ishigaki, M. Wilde, and K. Fukutani

Impact of Silicon Nitride Gate Dielectric Composition on the Stability of Low Temperature   73
Nanocrystalline Silicon Thin Film Transistors
   M. Esmaeili-Rad, G. Chaji, F. Li, M. Moradi, A. Sazonov, and A. Nathan

Development of a Fast Technique for Characterizing Interface States   81
   L. Lin, Z. Ji, J. Zhang, and W. Zhang

Detailed Analysis of Si-SiO$_2$ Interface Traps in State-of-the-Art MOSFETs Using Charge   95
Pumping
   D. Bauza

Clear Difference between the Chemical Structure of SiO$_2$/Si Interfaces Formed Using   115
Oxygen Radicals versus Oxygen Molecules
   T. Suwa, Y. Kumagai, A. Teramoto, T. Muro, T. Kinoshita, T. Ohmi, and T. Hattori

Chapter 2
Ultra-Thin Film/Reliability

Impact of Twofold Coordinated Nitrogen on the Generation of Deep-Level Hole Traps 125
under Negative-Bias Temperature Stressing
  C. Gu, D. Ang, and Z. Teo

Essential aspects of Negative Bias Temperature Instability (NBTI) 145
  A. E. Islam, S. Mahapatra, S. Deora, V. D. Maheta, and M. Alam

Atomic Imaging of Atomic H Cleaning of InGAs and InP for ALD 175
  W. Melitz, J. Shen, T. Kent, R. Droopad, P. Hurley, and A. C. Kummel

Plasma-Assisted Atomic Layer Deposition of Low Temperature SiO_2 191
  G. Dingemans, C. Van Helvoirt, M. Van de Sanden, and W. M. Kessels

Surface Passivation of InGaAs/InP HBTs Using Atomic Layer Deposited Al_2O_3 205
  R. Driad, F. Benkhelifa, L. Kirste, R. Lösch, M. Mikulla, and O. Ambacher

Nitric Acid Oxidation Method to Form a Gate Oxide Layer in Sub-Micrometer TFT 217
  T. Matsumoto, Y. Kubota, S. Imai, and H. Kobayashi

Effects of Deposition Method of PECVD Silicon Nitride as MIM Capacitor Dielectric 229
for GaAs HBT Technology
  J. Yota

Low Temperature Processing of Si-Based Dielectric Thin Films 241
  P. Joshi, A. Voutsas, and J. Hartzell

Negative Charge in Plasma Oxidized SiO_2 Layers 259
  A. Boogaard, A. Y. Kovalgin, and R. Wolters

Optical and Electrical Properties of Si-Based Multilayer Structures for Solar Cell 273
Applications
  R. Nalini, J. Cardin, K. R. Dey, X. Portier, C. Dufour, and F. Gourbilleau

Context Dependence Effects in Si/SiON Based Advanced CMOS Devices 287
  O. O. Olubuyide

Quantitative Discussion on Electron-Hole Universal Tunnel Mass in Ultrathin Dielectric 303
of Oxide and Oxide-Nitride
  H. Watanabe

Physics-Based Hot-Carrier Degradation Modeling 321
S. E. Tyaginov, I. Starkov, H. Enichlmair, J. Park, C. Jungemann, and T. Grasser

Intrinsic Variability and Reliability in Nano-CMOS 353
J. Velamala, C. Wang, R. Zheng, Y. Ye, and Y. Cao

Bias-Temperature Instabilities and Radiation Effects on SiC MOSFETs 369
*E. Zhang, C. Zhang, D. Fleetwood, R. Schrimpf, S. Dhar, S. Ryu, X. Shen, and
S. T. Pantelides*

Chapter 3
Emerging Dielectrics

Impact of Gate Dielectric Geometry on the Nanowire MOSFETs Performance and Scaling 383
M. Li, W. Cao, D. Huang, C. Shen, S. Cheng, C. Yao, and H. Yu

Role of Oxygen Transfer for High-k/SiO$_2$/Si Stack Structure on Flatband Voltage Shift 403
T. Nabatame, A. Ohi, and T. Chikyow

Charge Trapping and Reliability Properties of MONOS Memory with High-k Blocking 417
Layer
N. Yasuda, S. Fujii, J. Fujiki, and H. Kusai

Dynamic Negative Bias Stress Instability Effects in Hafnium Silicon Oxynitride and 447
Silicon Dioxide
J. Mee, R. Devine, H. Hjalmarson, and K. Kambour

Electrical and Structural Properties of Ternary Rare-Earth Oxides on Si and Higher 461
Mobility Substrates and their Integration as High-k Gate Dielectrics in MOSFET Devices
*J. Lopes, E. Durğun Özben, M. Schnee, R. Luptak, A. Nichau, A. Tiedemann, W. Yu,
Q. Zhao, A. Besmehn, U. Breuer, M. Luysberg, S. Lenk, J. Schubert, and S. Mantl*

High-k Integration and Interface Engineering for III-V MOSFETs 481
H. Oh, A. B. Sumarlina, and S. Lee

Plasma Enhanced Atomic Layer Deposition of ZrO$_2$: A Thermodynamic Approach 497
*E. Blanquet, D. Monnier, I. Nuta, F. Volpi, B. Doisneau, S. Coindeau, J. Roy, B. Detlefs,
Y. Mi, J. Zegenhagen, C. Martinet, C. Wyon, and M. Gros-Jean*

ix

V_T Stability Of High-K/Metal Gate Stacks with Device Scaling in 30nm FDSOI   515
Technology
X. Garros, L. Brunet, M. Cassé, O. Weber, F. Andrieu, D. Lafond, C. Gaumer,
G. Reimbold, and F. Boulanger

Investigation of Electron and Hole Charge Trapping in $LaLuO_3$ Stack MOS Capacitor   531
Using the 3-Pulse CV Technique
N. Sedghi, I. Mitrovic, J. Lopes, J. Schubert, and S. Hall

Inelastic Electron Tunneling Spectroscopy (IETS) Study of Ultra-Thin Gate Dielectrics   545
for Advanced CMOS Technology
T. Ma

High-k Gate Dielectric MOSFETs: Meeting the Challenges of Characterization and   563
Modeling
M. M. De Souza, S. Sicre, and D. Casterman

Universal Set/Reset Characteristics of Metal-Oxide Resistance Switching Memories   581
D. Ielmini

Resistive Switching Behaviors of ReRAM Having $W/CeO_2/Si/TiN$ Structures   597
C. Dou, K. Mukai, K. Kakushima, P. Ahmet, K. Tsutsui, A. Nishiyama, N. Sugii,
K. Natori, T. Hattori, and H. Iwai

Electrically Detected Magnetic Resonance in Dielectric Semiconductor Systems of   605
Current Interest
P. M. Lenahan, C. Cochrane, J. Campbell, and J. Ryan

Synthesis, Pore Morphology, and Dielectric Property of Mesoporous Low-k Material   629
PSMSQ Using a Reactive High-Temperature Porogen, TEPSS
S. Chiu, H. Hsu, M. Che, and J. Leu

Electrical Characteristics Analysis at "Oxide Flat-Band Voltage" for $Al-SiO_2-Si$ Capacitor   639
H. Lu, T. Chen, and J. Hwu

Novel Hardmask for Sub-20nm Copper/Low K Backend Dual Damascene Integration   651
L. Xia, Z. Cui, M. Balseanu, V. Nguyen, K. Zhou, J. Pender, and M. Naik

Study of Porous SiOCH Patterning Using Metallic Hard Mask: Challenges and Solutions   667
N. Posseme, T. David, T. Chevolleau, M. Darnon, F. Bailly, R. Bouyssou, J. Ducote,
H. Chaabouni, M. El Kodadi, C. Licitra, C. Verove, and O. Joubert

Process Challenges for Integration of Copper Interconnects with Low-k Dielectrics   687
J. Gambino

Patterning with Amorphous Carbon Thin Films 701
G. A. Antonelli, S. Reddy, P. Subramonium, J. Henri, J. Sims, J. O'loughlin, N. Shamma,
D. Schlosser, T. Mountsier, W. Guo, and H. Sawin

Ultra Low Dielectric Constant Materials for 22 nm Technology Node and Beyond 717
M. R. Baklanov, E. A. Smirnov, and L. Zhao

Development of Porosimetry Techniques for the Characterization of Plasma-Treated 729
Porous Ultra Low-k Materials
C. Licitra, T. Chevolleau, R. Bouyssou, M. El Kodadi, G. Haberfehlner, J. Hazart,
L. Virot, M. Besacier, N. Posseme, M. Darnon, R. Hurand, P. Schiavone, and F. Bertin

An Electron Paramagnetic Resonance Study of Defects in Interlayer Dielectrics 747
B. C. Bittel, T. Pomorski, P. M. Lenahan, and S. King

Development of Voltammetry-Based Techniques for Characterization of Porous 757
Low-k/Cu Interconnect Integration Reliability
C. Kim, L. Chen, N. Michael, W. Bang, Y. Park, T. Ryan, and S. King

Recent Findings in Electrical Behavior of CMOS High-K Dielectric/Metal Gate Stacks 773
G. Ghibaudo, J. Coignus, M. Charbonnier, J. Mitard, C. Leroux, X. Garros, R. Clerc,
and G. Reimbold

Flatband Voltage Tuning of HfSiON-Based Gate Stacks: Impact of High Temperature 805
Activation Annealing and LaO_x Capping Layers
R. Boujamaa, S. Baudot, E. Martinez, O. Renault, B. Detlefs, J. Zegenhagen, V. Loup,
F. Martin, M. Gros-Jean, F. Bertin, and C. Dubourdieu

Physical and Electrical Effects of the Dep-Anneal-Dep-Anneal (DADA) Process for 815
HfO_2 in High K/Metal Gate Stacks
R. D. Clark, S. Aoyama, S. Consiglio, G. Nakamura, and G. Leusink

Interface Structure and Charge Trapping in Hf-Incorporated Y_2O_3 Gate Dielectrics on 835
Germanium
C. Mahata, S. Mallik, T. Das, M. Hota, and C. Maiti

Chapter 4
Poster Session

Schottky Barrier Height at Dielectric Barrier/Cu Interface in Low-K/Cu Interconnects 849
S. King, M. French, M. Jaehnig, M. Kuhn, and B. French

Global and Local Stress Characterization of SiN/Si(100) Wafers Using Optical Surface 861
Profilometer and Multiwavelength Raman Spectroscopy
W. Yoo, J. Kajiwara, T. Ueda, T. Ishigaki, and K. Kang

Mechanisms of Difficulty to Correlate the Leakage Current of High-k Capacitor Structures 873
with Defect States Detected Spectroscopically by the Thermally Stimulated Current
Technique
W. S. Lau

Degradation Mechanisms of MILC P-Channel Poly-Si TFTs under Dynamic Hot-Carrier 889
Stress Using a Novel Test Structure
C. Lin, W. Hong, T. Lin, H. Lin, and T. Huang

Solution Processed High-k Lanthanide Oxides for Low Voltage Driven Transparent 901
Oxide Semiconductor Thin Film Transistors
S. Choi, B. Park, M. Jang, S. Jeong, J. Lee, B. Ryu, T. Seong, and H. Jung

Reliability Properties and Current Conduction Mechanisms of HfO_2 MIS Capacitor with 909
Dual Plasma Treatment
K. Chang, T. Chang, S. Chen, and I. Deng

A MIM Diode with Ultra Abrupt Switching Process and High On/Off Current Ratio 923
L. Zhang and R. Huang

Author Index 931

Facts about ECS

The Electrochemical Society (ECS) is an international, nonprofit, scientific, educational organization founded for the advancement of the theory and practice of electrochemistry, electrothermics, electronics, and allied subjects. The Society was founded in Philadelphia in 1902 and incorporated in 1930. There are currently over 7,000 scientists and engineers from more than 70 countries who hold individual membership; the Society is also supported by more than 100 corporations through Corporate Memberships.

The technical activities of the Society are carried on by Divisions. Sections of the Society have been organized in a number of cities and regions. Major international meetings of the Society are held in the spring and fall of each year. At these meetings, the Divisions and Groups hold general sessions and sponsor symposia on specialized subjects.

The Society has an active publications program that includes the following.

Journal of The Electrochemical Society — JES is the peer-reviewed leader in the field of electrochemical and solid-state science and technology. Articles are posted online as soon as they become available for publication. This archival journal is also available in a paper edition, published monthly following electronic publication.

Electrochemical and Solid-State Letters — ESL is the first and only rapid-publication electronic journal covering the same technical areas as JES. Articles are posted online as soon as they become available for publication. This peer-reviewed, archival journal is also available in a paper edition, published monthly following electronic publication. It is a joint publication of ECS and the IEEE Electron Devices Society.

Interface — *Interface* is ECS's quarterly news magazine. It provides a forum for the lively exchange of ideas and news among members of ECS and the international scientific community at large. Published online (with free access to all) and in paper, issues highlight special features on the state of electrochemical and solid-state science and technology. The paper edition is automatically sent to all ECS members.

Meeting Abstracts (formerly Extended Abstracts) — Abstracts of the technical papers presented at the spring and fall meetings of the Society are published on CD-ROM.

ECS Transactions — This online database provides access to full-text articles presented at ECS and ECS-sponsored meetings. Content is available through individual articles, or as collections of articles representing entire symposia.

Monograph Volumes — The Society sponsors the publication of hardbound monograph volumes, which provide authoritative accounts of specific topics in electrochemistry, solid-state science, and related disciplines.

For more information on these and other Society activities, visit the ECS website:

www.electrochem.org

CHAPTER 1

INTERFACE CHARACTERIZATION

2

Fluctuations in Electronic Properties of MOS Interface in Nanoscale MOSFETs

Toshiaki Tsuchiya, Yuta Morimura, and Yuki Mori

Interdisciplinary Faculty of Science and Engineering
Shimane University
1060 Nishikawatsu, Matsue 690-8504, Japan

Fluctuations in the number, energy level, and carrier capture rate of Si/SiO_2 interface traps in small-gate-area metal-oxide-semiconductor field-effect transistors (MOSFETs) containing only a few interface traps have been directly observed. This observation is based on an understanding of charge pumping phenomena and a newly observed phenomenon of on(off)-time-dependent charge pumping characteristics in their rising (falling) portion. It is directly shown from experimental results that the fluctuation in the number of interface traps contained in an individual nanoscale MOSFET is fairly large, and that there are various interface traps with individual discrete energy levels and carrier capture rates. These fluctuations may have an impact on the variation in device performance of future digital metal-insulator-semiconductor (MIS) devices.

Introduction

Anomalously large fluctuations in threshold voltage due to random telegraph noise (RTN) in floating-gate memories and static random access memories have recently been reported. It has been suggested that RTN will become one of the greatest reliability issues in scaled-down digital devices [1-6]. RTN is considered to be caused by charge transport fluctuation by the capture and emission of charge at a single trap, which is considered to be located at some distance inside the gate dielectric. Therefore, it will be essential to identify the individual traps at and near the metal-insulator-semiconductor (MIS) interface in order to improve the device performance and reliability, and to reduce the fluctuations in device characteristics in future nanoscale MIS devices.

The charge pumping (CP) technique [7-8] is recognized as a very high-precision method of evaluating interface traps between the gate oxide and the semiconductor surface in metal-oxide-semiconductor field-effect transistors (MOSFETs). However, several studies even on the in-depth exploration of traps near the MIS interface have been reported using the CP method [9-13], and furthermore statistical distributions of individual trap position inside the gate dielectric, trap energy levels, capture/emission time constants, RTN amplitude, and correlation among these properties have been recently reported by characterization of individual traps based on RTN measurements [14].

On the other hand, the observation of single traps at the interface has been demonstrated by the CP method using submicron MOSFETs [15-16], and the trap capture cross sections for electrons and holes ($\sim 10^{-16}$ cm^2) were derived for a particular trap measured [17]. Moreover, the capture cross sections for electrons (σ_n) and holes (σ_p), and the energy level (E_T) of a single interface trap have been examined performing three-voltage level CP method [18] using deep-submicron MOSFETs [19-20]. They obtained $E_T \sim E_V + 0.83$ eV and $\sigma_n \sim 6 \times 10^{-17}$ cm^2 for a particular trap, and $E_T \sim E_V + 0.31$ eV and $\sigma_p \sim 2 \times 10^{-16}$ cm^2 for another particular trap, where E_V is the valence band edge.

For the next stage, it will be required to experimentally verify statistical distributions of the electronic properties of individual interface traps, individual trap position, and so on in actual nanoscale MIS devices, by simpler and easier evaluation methods as much as possible, and then it will be also expected to clarify the influence of individual traps on the device performance and reliability.

In this study, we focus on the fluctuations in electronic properties of MOS interface traps in nanoscale MOSFETs containing only a few interface traps, i.e., fluctuations in their number, energy levels, carrier capture/emission properties. We observed novel transient CP characteristics. Based on an understanding of this transient phenomenon, we then successfully observed the fluctuations in the electronic characteristics of individual interface traps in nanoscale MOSFETs. Moreover, we experimentally demonstrated fluctuations in the number of interface traps contained in individual nanoscale MOSFETs, as well as the energy levels of individual interface traps using the CP method.

Experimental Procedure

The CP measurement configuration for nMOSFETs is shown in Fig. 1(a). A pulsed voltage, as shown in Fig. 1(b), is applied to the gate to alternately form inversion and accumulation layers. Electrons are captured in the interface traps during inversion, and the trapped electrons recombine with holes coming from the substrate during accumulation. The CP current (I_{CP}) is due to such captured and recombined electrons; it is therefore proportional to the density of interface traps (N_{it}), and can simply be described as $I_{CP} = fqA_GN_{it}$, where f is the gate pulse frequency, q is the electron charge, and A_G is the gate area. Here, the CP current flows in the direction opposite to that of the reverse-biased ($V_j = 0.05$ V) source/drain-junction current. All measurements were carried out at room temperature.

We used polycrystalline-silicon-gate nMOSFETs in this study. Their gate pattern lengths (L_p) and gate pattern widths (W_p) are 110–250 nm and 120–300 nm, respectively. Actual polycrystalline-silicon-gate lengths (L_G) measured by SEM (Secondary Electron Microscopy) are approximately $L_G = L_p - 55$ nm, and electrically-measured effective channel widths (W_{eff}) are approximately $W_{eff} = W_p - 40$ nm. The thickness of the gate oxide grown by rapid thermal oxidation is 4 nm.

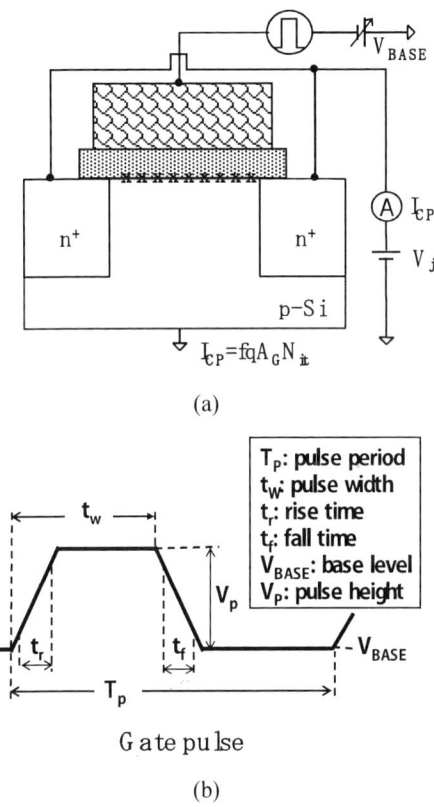

(a)

(b)

Fig. 1 (a) Configuration of charge pumping measurements, and
(b) pulsed voltage applied to gate.

Fluctuation in the Number of Interface Traps

First, we evaluated the fluctuation in the number of MOS interface traps contained in individual small-gate-area MOSFETs using the CP method. An example of the CP characteristics in a small MOSFET is shown in Fig. 2. In this figure, since the gate pulse frequency (f) is 250 kHz, the maximum CP current (I_{CPMAX}) for a single trap (i.e., $A_G N_{it} = 1$) should be $I_{CPMAX} = fqA_G N_{it} = fq = 40$ fA; thus, it is relatively easily found from the maximum CP current that the MOSFET in Fig. 2 contains four interface traps. We measured several MOSFETs of $L_p = 130$ nm and various values of W_p to evaluate the numbers of interface traps contained. The numbers obtained are shown as a function of gate area in Fig. 3, which indicates that the fluctuation in the number of interface traps is fairly large. Here, the average density of interface traps in the measured MOSFETs is approximately 10^{11} cm^{-2}. We confirmed that this fluctuation is not caused by the scatter of the gate size [22].

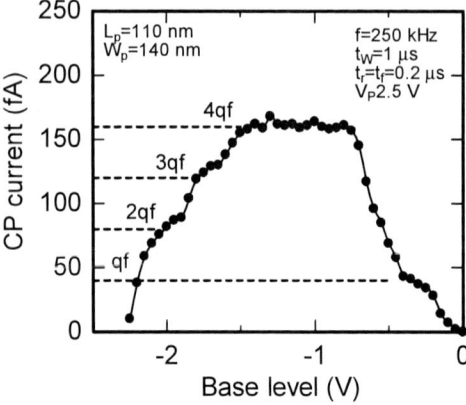

Fig. 2 Charge pumping characteristics in an nMOSFET containing only four interface traps. I_{CP} (one trap) = 40 fA at f = 250 kHz.

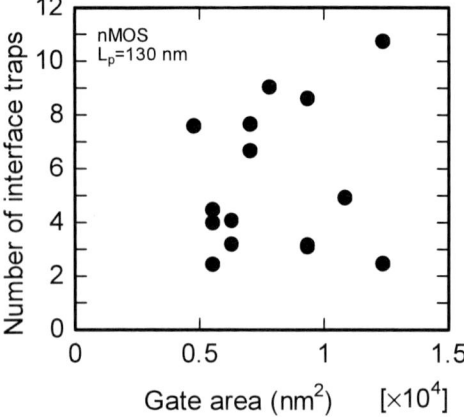

Fig. 3 Fluctuation in the number of interface traps contained in a small nMOSFET.

Fluctuation in the Carrier Capture Rate of Interface Traps

Newly Developed Evaluation Method

It is well known that the maximum CP current (I_{CPMAX}) depends on the rise time (t_r) of the gate pulse, even when sufficient inversion and accumulation conditions are satisfied during CP measurement, and I_{CPMAX} increases with decreasing t_r. This is due to the decrease in the emission current from majority carriers (holes) in the non-steady-state interface traps near the valence band moving to the valence band because of the rapidly rising gate voltage [7], as will be explained in greater detail later.

As shown in Fig. 4, we observed a novel transient phenomenon in the rising portion of the CP characteristics, which depend on the pulse width (t_W), i.e., the on-time of the gate pulse [22-23]. In the rising portion, the CP current (I_{CP}) at a given base level (V_{BASE}) of the gate pulse increases with increasing on-time. The CP current at a constant V_{BASE} as a function of the on-time with V_{BASE} as a parameter is shown in Fig. 5, which

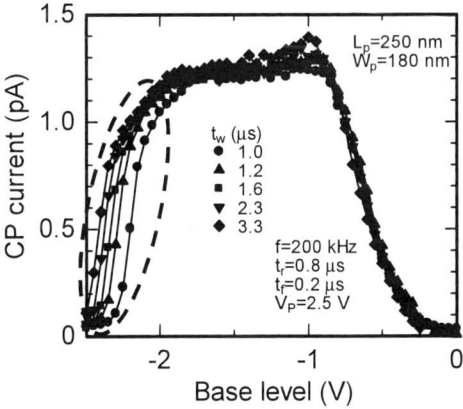

Fig. 4 Novel transient behavior appearing in the rising portion of the charge pumping characteristics, which depends on the width of the gate pulse (t_W), i.e., the on-time of the gate pulse.

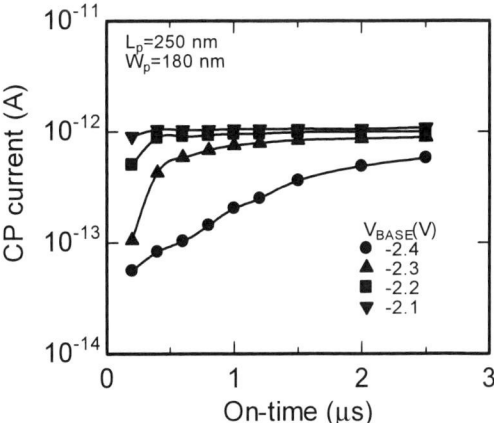

Fig. 5 Saturation behavior of charge pumping current at a constant base level of the gate pulse in the rising portion of the charge pumping characteristics.

indicates saturation behavior. This behavior can be considered to be related to the process of electron capture by interface traps as follows. Changes in the energy band diagram during this process are shown in Fig. 6. Electrons trapped in the interface traps recombine with holes coming from the accumulation layer formed while at the base level (i.e., during accumulation), as shown in Fig. 6(a). Some of the electrons in the inversion layer are captured by interface traps during the on-time (i.e., during inversion), as shown in Figs. 6(b) and 6(b)'. However, in this case, the inversion is less strong because of the rising portion in the CP characteristics, and only a small fraction of the interface traps can capture electrons if the on-time is insufficient (Fig. 6(b)). Therefore, we can observe the process of electron capture in the interface traps using this newly developed method.

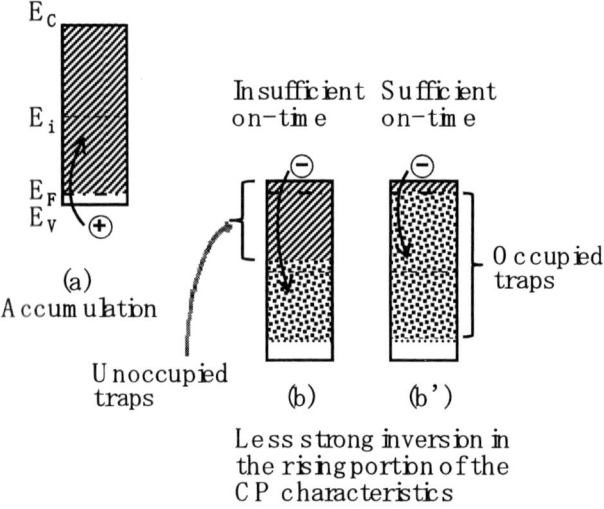

Fig. 6 Energy band diagrams illustrating the mechanism of the novel transient behavior, which is related to the electron capture process.

Thus, the effects of the on-time are related to the electron capture process, and interface traps over a wide range within the energy band gap are related to this process. On the other hand, the effects of the rise time are related to the hole emission process near the valence band edge, and interface traps only near this edge take part in this process. Therefore, the influence of the on-time on the CP current is fairly large compared with that of the rise time.

In the same manner, we confirmed that the process of capturing majority carriers (i.e., holes) in individual interface traps can be observed in the off-time-dependent transient CP characteristics.

Carrier Capture Rates of Individual Interface Traps

We applied the method to nanoscale MOSFETs and directly observed the carrier capture rates of individual interface traps. The t_W-dependent CP characteristics in these very small MOSFETs are shown in Fig. 7, where the MOSFET in Fig. 7(a) contains only one interface trap and that in Fig. 7(b) contains three traps. From these CP characteristics, we obtained individual characteristics for each interface trap using the method shown in Fig. 8(a). The CP current at a given base level in the rising portion for each trap should be

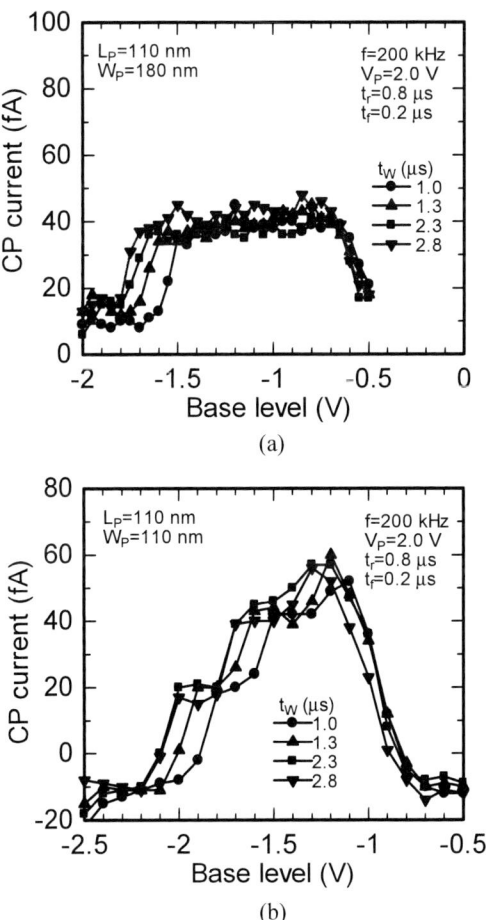

(a)

(b)

Fig. 7 Charge pumping characteristics dependent on the gate-pulse-width (t_W) in very small MOSFETs, containing one (a) and three interface traps (b), respectively. We can observe the electron capture process of individual interface traps by applying this method to very small MOSFETs.

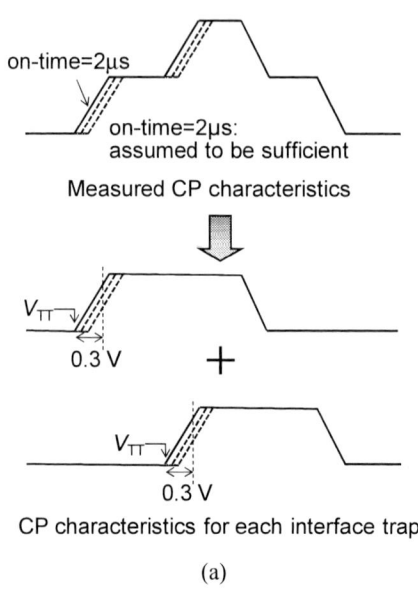

(a)

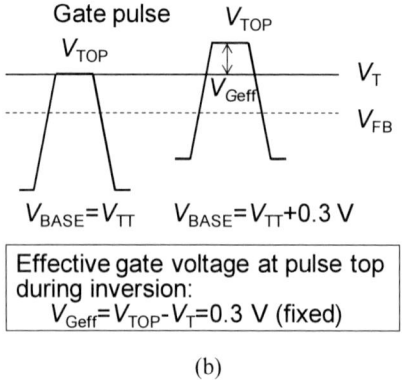

(b)

Fig. 8 (a) Schematic charge pumping characteristics for individual interface traps obtained from the measured characteristics, and definition of threshold voltage V_{TT} for each trap where CP current begins to flow for on-time = 2 μs. (b) Given that the effective gate voltage at pulse top during less strong inversion is fixed at 0.3 V for each trap, the density of electrons in the inversion layer is considered to be identical for each trap.

described by $I_{CP} = I_{CP0}[1 - \exp(-t/\tau_T)]$ with time constant τ_T, where I_{CP0} is the saturated CP current. The capture rate for electrons (c_n) can be described by $c_n = 1/\tau_T = \sigma_n v_{TH} n$, where σ_n is the electron capture cross section, v_{TH} is the free carrier thermal velocity, and n is the density of electrons in the less strong inversion layer. In order to compare c_n for individual traps, τ_T must be measured under a fixed n for each trap. To achieve this purpose, we defined the threshold voltage V_{TT} for each trap as the value of V_{BASE} where the CP current begins to flow for on-time = 2 μs, which is assumed to be a sufficient on-time. We then obtained the CP current at $V_{BASE} = V_{TT} + 0.3$ V for individual traps as a function of on-time, as shown in Fig. 9. In this case, since the effective gate voltage at pulse top during less strong inversion is fixed at 0.3 V for each trap (Fig. 8(b)), the value of n is considered to be identical for each trap. It was found from Fig. 9 that there are various interface traps with individual carrier capture rates, i.e., individual electron capture cross sections.

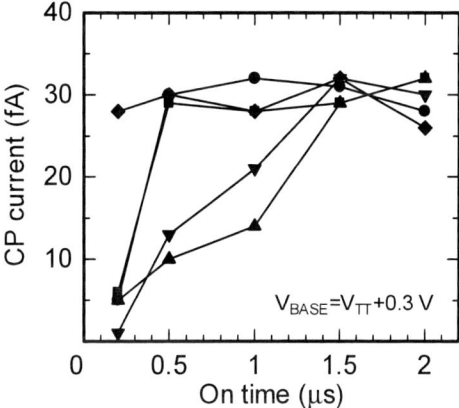

Fig. 9 Charge pumping current at $V_{BASE} = V_{TT} + 0.3$ V for individual interface traps as a function of on-time. This figure shows the fluctuation in the carrier capture rates of individual traps.

Fluctuation in the Energy Level of Interface Traps

We have also sought to evaluate the energy levels of individual interface traps. It is well known that the energy distribution of interface traps can be obtained from the dependences of the rise (t_r) and fall times (t_f) of the gate pulse on the CP characteristics [7].

As described previously, the maximum CP current (I_{CPMAX}) increases with decreasing rise time (t_r) of the gate pulse, even when sufficient inversion and accumulation conditions are satisfied during CP measurement. Similarly, I_{CPMAX} increases with decreasing fall time (t_f) of the gate pulse. The former behavior is related to e change in the process of emitting majority carriers (holes for nMOSFETs) in the non-steady-state interface traps (which depend on the gate signal characteristics) near the lence band, as shown in Fig. 10. During accumulation, holes are captured in the terface traps, as shown in Fig 10(a). That is, electrons trapped in interface traps combine with holes, and electrons are captured in the interface traps during inversion igs. 10(c) and 10(c')). In addition, note that during the rise time, some of the trapped les can be emitted to the valence band, and the number of such holes decreases with creasing rise time (Figs. 10(b) and 10(b')). This is why the number of trapped electrons ring inversion increases with decreasing rise time, and the maximum CP current reases with decreasing rise time. Similarly, the fall-time-dependent I_{CPMAX} is related to change in the emission process of minority carriers (electrons for nMOSFETs) in the n-steady-state interface traps near the conduction band.

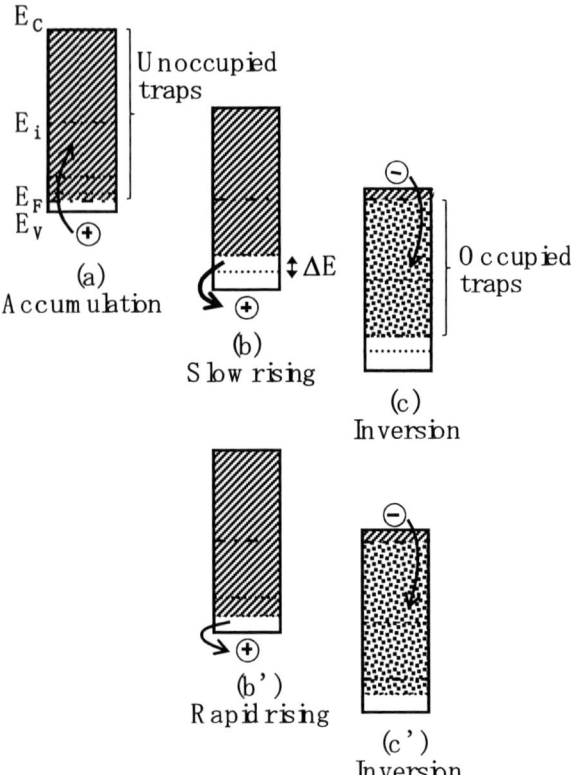

Fig. 10 Energy band diagrams explaining the rise-time-dependent charge pumping characteristics.

Therefore, the change in I_{CPMAX} due to the change in rise time is caused by interface traps present in the energy range of ΔE in the lower part of the band gap (Fig. 10(b)). In addition, the change in I_{CPMAX} due to the change in fall time is caused by interface traps of the corresponding ΔE in the upper part of the energy gap. Thus, the energy distribution of interface traps can be obtained using the following equations [7]:

$$D_{it}(E_1) = -\frac{t_r}{fqA_G kT}\frac{dI_{CP}}{dt_r}, \tag{1}$$

$$E_1 = E_i + kT \ln\left[v_{th}\sigma_n n_i \frac{|V_{FB} - V_T|}{|\Delta V_G|} t_r \right], \tag{2}$$

$$D_{it}(E_2) = -\frac{t_f}{fqA_G kT}\frac{dI_{CP}}{dt_f}, \tag{3}$$

$$E_2 = E_i - kT \ln\left[v_{th}\sigma_n n_i \frac{|V_{FB} - V_T|}{|\Delta V_G|} t_f \right], \tag{4}$$

where $D_{it}(E_1)$ and $D_{it}(E_2)$ are the densities of interface traps at the energy levels of E_1 and E_2, respectively, and v_{th} is the thermal velocity, σ_n the capture cross section for electrons, n_i the intrinsic concentration, and ΔV_G the amplitude of the gate pulse (V_P).

We attempted to apply this method to nanoscale MOSFETs containing a few interface traps. If a discrete interface trap is present in an energy range of ΔE, a change in I_{CPMAX} of fq should be observed by measuring CP current by varying t_r and t_f. The effects of t_r and t_f on the CP characteristics at room temperature are shown in Figs. 11(a) and 11(b), respectively for an MOSFET containing three interface traps (traps A, B, C). In these experiments, we used gate pulse of 200 kHz; therefore, the change in I_{CPMAX} for a single trap should be $fq = 32$ fA. It is found from Fig. 11(a) that a decrease of almost 30 fA in I_{CPMAX} is observed by increasing t_r from 0.1 to 0.8 μs, indicating that one trap (referred to as trap A) is located at near the valence band edge (E_V). Moreover, a decrease of almost 32 fA in I_{CPMAX} can be observed by increasing t_f from 0.1 to 0.5 μs, indicating that another trap (trap B) is located at near the conduction band edge (E_C). The calculated positions of traps A and B are about 0.3 eV below and above the intrinsic Fermi level E_i, respectively, as shown in Fig. 11(c). The other trap (trap C) should be present in the range of $E_i \pm 0.3$ eV.

Another example for an MOSFET containing two traps (referred to as traps D and E) is shown in Fig. 12(a). The CP current in this sample depends neither on t_r nor t_f. This indicates that neither trap is located at near the band edge, and that they should be present in the range of $E_i \pm 0.3$ eV, as shown in Fig. 12(b).

From the results above, we experimentally showed the discrete energy levels of individual interface traps, and demonstrated fluctuation in the energy levels. Using this method, we can identify interface traps with deeper or shallower energy levels by changing the temperature, and more precisely determine the discrete energy levels over a wider range within the energy gap.

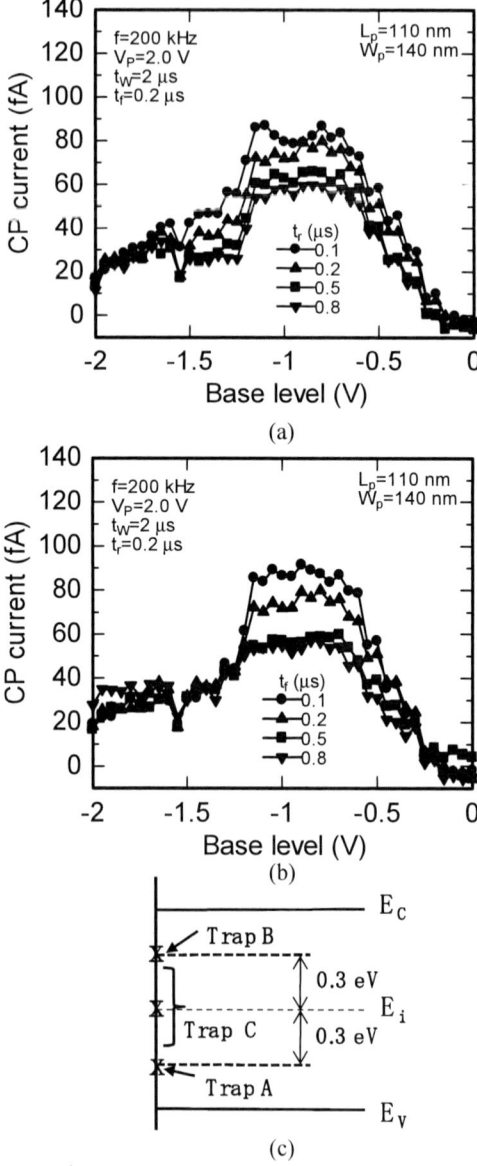

Fig. 11 Effects of (a) the rise time and (b) fall time of the gate pulse on the charge pumping characteristics at room temperature, and (c) the energy levels of the interface traps in the measured MOSFET containing three interface traps (traps A, B, and C).

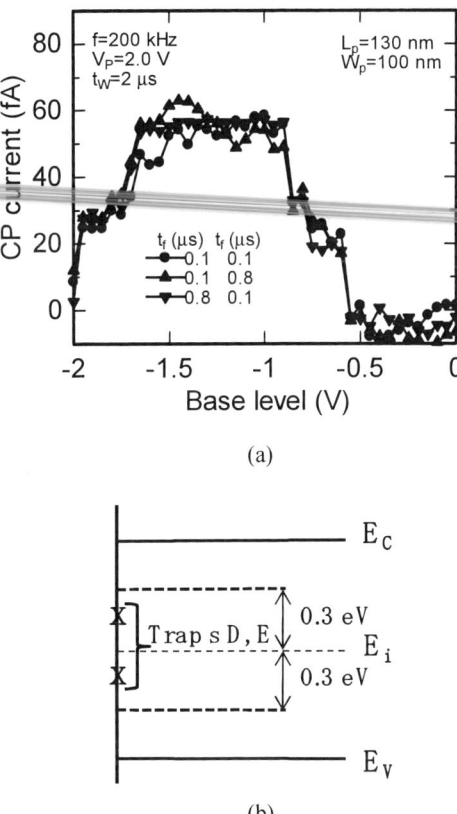

(a)

(b)

Fig. 12 Effects of (a) the rise time and fall time of the gate pulse on the charge pumping characteristics at room temperature, and (b) the energy level region where the interface traps are located in the measured MOSFET containing two traps. It is possible to identify the interface traps with deeper energy levels by lowering the temperature.

Conclusions

We have directly observed fluctuations not only in the number of interface traps, but also in the energy levels and carrier capture rates of individual interface traps in nanoscale MOSFETs containing only a few interface traps. This observation is based on an understanding of charge pumping phenomena and a newly observed phenomenon of the dependence of the rising portion of the charge pumping characteristics on the on(off)-time of the gate pulse. From an understanding of these phenomena, we experimentally verified that the fluctuation in the number of interface traps is fairly large, and that there are various interface traps with individual electronic properties such as energy level and carrier capture rate. These fluctuations in the properties of individual interface traps could have potential impact on variation in the device performance and reliability of future digital MIS devices. The newly developed method will provide a powerful way to characterize individual traps.

Acknowledgments

The authors would like to express thanks to Tohru Mogami of Semiconductor Leading Edge Technologies, Inc. for his cooperation in device fabrication. This work was partially supported by a Grant-in-Aid for Scientific Research on Priority Areas (18063016) from the Ministry of Education, Culture, Sports, Science and Technology.

References

1. N. Tega, H. Miki, T. Osabe, A. Kotabe, K. Otsuga, H. Kurata, S. Kamohara, K. Tokami, Y. Ikeda, and R. Yamada, IEEE IEDM, p. 491 (2006).
2. N. Tega, H. Miki, F. Pagette, D. J. Frank, A. Ray, M. J. Rooks, W. Haensch, and K. Torii, Symp. VLSI Tech., p. 50 (2009).
3. K. Takeuchi, T. Nagumo, S. Yokogawa, K. Imai, and Y. Hayashi, Symp. VLSI Tech., p. 54 (2009).
4. T. Nagumo, K. Takeuchi, S. Yokogawa, K. Imai, and Y. Hayashi, IEEE IEDM, p. 759 (2009).
5. S. Lee, H. J. Cho, Y. Son, D. S. Lee, and H. Shin, IEEE IEDM, p. 763 (2009).
6. N. Tega, H. Miki, Z. Ren, C. P. D'Emic, Y. Zhu, D. J. Frank, J. Cai, M. A. Guillorn, D. G. Park, W. Haensch, and K. Torii, IEEE IEDM, p. 771 (2009).
7. G. Groeseneken, H. E. Maes, N. Beltran, and R. F. DE Keersmaecker, *IEEE Trans. Electron Devices*, **31**, 42 (1984).
8. P. Heremans, J. Witters, G. Groeseneken, and H. E. Maes, *IEEE Trans. Electron Devices*, **36**, 1318 (1989).
9. D. Bauza and Y. Maneglia, *IEEE Trans. Electron Devices*, **44**, 2262 (1997).
10. M. Masuduzzaman, A. E. Islam, and M. A. Alam, *IEEE Trans. Electron Devices*, **55**, 3421 (2008).
11. D. Bauza, *IEEE Trans. Electron Devices*, **56**, 70 (2009).
12. D. Bauza, *IEEE Trans. Electron Devices*, **56**, 78 (2009).
13. M. B. Zahid, R. Degraeve, M. Cho, L. Pantisano, D. R. Aguado, J. Van. Houdt, G. Groeseneken, and M. Jurczak, Proc. 47th Int. Reliability Physics Symp., Montreal, p. 21 (2009).
14. T. Nagumo, K. Takeuchi, T. Hase, and Y. Hayashi, IEEE IEDM, p. 628 (2010).
15. N. S. Saks, G. Groeseneken, and I. DeWolf, *Appl. Phys. Lett.*, **68**, 1383 (1996).
16. G. V. Groeseneken, I. De Wolf, R. Bellens, and H. E. Maes, *IEEE Trans. Electron Devices*, **43**, 940 (1996).
17. N. S. Saks, *Appl. Phys. Lett.*, **70**, 3380 (1997).
18. N. S. Saks and M. G. Ancona, *IEEE Electron Device Lett.*, **11**, 339 (1990).
19. L. Militaru, P. Masson, and G. Guegan, *IEEE Electron Device Lett.*, **23**, 94 (2002).
20. L. Militaru and A. Souifi: *Appl. Phys. Lett.* **83**, 2456 (2003).

21. T. Tsuchiya, Y. Mori, Y. Morimura, T. Mogami, and Y. Ohji, European Solid-State Device Research Conf., p. 387 (2009).
22. T. Tsuchiya, Y. Mori, Y. Morimura, T. Mogami, and Y. Ohji, *Jpn. J. Appl. Phys.*, **49** (6), 064001(2010).
23. T. Tsuchiya, K. Yoshida, M. Sakuraba, and J. Murota, 4th Int. SiGe Tech. and Dev. Meet., p. 64 (2008).

ECS Transactions, 35 (4) 19-38 (2011)
10.1149/1.3572273 ©The Electrochemical Society

Electron States in MOS Systems

O. Engström

Department of Microtechnology and Nanoscoience
Chalmers University of Technology, SE-412 96 Göteborg, Sweden

The properties of charge carrier traps in the oxide bulk, at high-
k/silicon transition regions, at the silicon interface and as dipoles
are discussed from physical and electrical perspectives. In order to
elucidate the charging properties of oxide traps, the statistical
mechanics for occupation is derived based on a constant pressure
ensemble and used to interpret the influence of negative-U states
occurring in high-k oxides. For the transition region close to the
silicon interface, the existence of unstable traps in the continuous
shift of the energy bands between SiO_2 and HfO_2 is pointed out.
The physical background for electrical measurements on interface
states is examined and, finally, dipoles constituted by traps in high-
k dielectrics for regulating threshold voltage of MOS transistors are
considered.

Introduction

Extrinsic electron states occur in solid materials as a result of perturbations of the atomic
arrangements due to the existence of defects and impurities. As MOS structures involve a
metal/insulator junction and an insulator/semiconductor heterojunction, a multifarious
appearance of this kind of states has an important technical impact when MOS structures are
used as transistor gates [1, 2]. Traps positioned in the bulk of the insulator and at the two
interfaces may influence threshold voltages, channel mobilities and gate leakage.
Furthermore, they are more common and severe in high-k oxides than in traditional thermal
SiO_2 [3].

In many cases, the interaction between charge carriers and traps with energy levels in
the band gap of a solid is dependent, not only on potentials governed by pure Coulomb forces,
but also by the local dynamics of the atomic configurations constituting the trap geometry.
The energy supplied or released in processes for emission and capture of charge carriers,
therefore, includes components of both electronic and atomic origin. For example, when an
electron in the conduction band is captured into a state with a narrow electronic potential,
which strongly localizes the carrier, it may influence the local atomic bonding properties in
the trap volume [4]. As a consequence, the local heat, stored through vibrational frequency,
changes. This heat is included in the total energy exchange connected with the transition and
can be considered as originating from a change in entropy [5]. The trap can be represented by
an oscillator potential connected with the electronic potential, where energy states of the latter
may be shifted by the former in relation to the energy bands. As the oscillator potential gives
rise to tightly separate vibrational eigen energies, transitions by local phonon interaction are
probable [6].

This mechanism for capture and emission is named by the prefix "multiphonon" and has
a long history for transitions in alkali halides [7], but also in III-V semiconductors [6] and

more lately for oxides and their interfaces to silicon [8-15]. For traps with multiphonon capture mechanisms one would expect that (i) the capture processes are thermally activated [8-13]; (ii) the values of capture cross sections *increase* with temperature [6, 9, 13]; (iii) the measured values of energy levels would be different and have different thermodynamic meanings depending on whether they are determined from probing by the position of the Fermi-level (e.g. capacitance-voltage technique) or from Arrhenius-plots [8, 10, 14]; (iv) optically determined energy values would be different from values obtained by thermal method [15].

For traps with wider electronic potentials such that the wave functions of captured carriers are extended across a number of atomic distances, a smaller carrier portion takes part in bonding. Therefore, the vibrational frequency does not change in the capture process and (i) the values of capture cross sections *decrease* with temperature [13, 16]; (ii) the measured values of energy levels would differ by small amounts between different measurement methods, due to the influence of electronic degeneracies [14]; (iii) A capture mechanism, suggested originally by Lax [15] for these traps includes phonon interaction while electrons step down in energy along states in the electronic potential.

The most common charge carrier trap in the bulk of high-k oxides is argued to be the oxygen vacancy [3]. Theoretical treatments for a number of materials indicate that large atomic relaxations take place for these traps [17 – 19]. This influences tunneling processes and thus the leakage properties of oxides thin enough for technical interest [20]. For electron states at high-k /silicon interfaces, it has been found that both kinds of traps exist [13]. For shallower traps, with levels close to the silicon conduction band edge, the Lax-mechanism seem probable, while for deeper states, with P_b character, multiphonon processes dominate. Controlled doping of oxide traps close to the metal-insulator junction is utilized for regulating the Schottky-barrier at this position and thus the threshold voltage of transistor gates [21]. Also, there have been suggestions for the existence of dipoles close to high-k – silicon interfaces [22, 23].

This paper will present an overview of the most important oxide traps, the properties of interface states and how to understand the influence on electrical behavior from their physical properties.

Occupation Statistics

Considering the schematic picture in Fig. 1(a), where the trap volume is surrounded by atoms vibrating in breathing modes, one expects that emitting an electron will change the local properties of atomic bonds. This will change the atomic positions and also the atomic vibrational frequencies. The vibrations can be defined as local phonons carrying energy which should be added to the total energy of an electron captured into the trap. The trap captures electrons into an electronic potential as depicted in Fig. 1(b), including eigen states with energies $E_{i,r}$, where r is the number of captured electrons, whereas i denotes the electronic states available. Modeling the atomic potential as a harmonic oscillator, as shown in Fig.1 (c), the energy levels are, $E_{m,r}$, where m counts the vibronic levels and r again is the number of captured electrons. Assuming that the atomic and electronic wave functions are independent (adiabatic), the total energy of the captured electron is the sum $E_{i,r} + E_{m,r}$, where $E_{m,r}$ represents the eigenenergy m of the oscillator for r electrons captured into the trap [2, 24, 25].

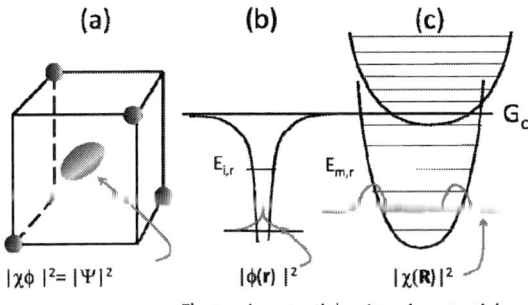

$$|\chi\phi|^2=|\Psi|^2 \qquad\qquad |\phi(\mathbf{r})|^2 \qquad\qquad |\chi(\mathbf{R})|^2$$

Electronic potential   Atomic potential

Figure 1. (a) Trap configuration with captured electron represented by its wave function. (b) Electronic potential for the captured electron. (c) Atomic oscillator potentials for a captured (lower) and emitted (upper) electron. The energy scale in (c) represents the total energy: electronic plus vibrational. Thermal emission occurs by excitation of the atom along the eigenstates until it reaches the crossing point between the oscillator potentials, where the electron goes to the conduction band at G_c and the oscillator relaxes to a potential with lower curvature and tighter eigenstates.

Furthermore, assuming that the charge carriers are in thermal equilibrium with the rest of the material, the average energy per carrier corresponds to the Fermi-energy, μ. Therefore, the capture of r electrons contributes to the trap energy by an amount $E_{m,r} + E_{i,r} - r\mu$. The probability for occupying the trap with r electrons is proportional to a Boltzmann factor including this energy while the total number of possibilities for the event is the sum of Boltzmann factors along r, m and i. Hence, the probability for capturing r electrons when the oscillator is in state m and the electron is in state i is [24, 25]

$$P(r,m,i) = \frac{1}{\Theta}\exp\left(-\frac{E_{m,r} + E_{i,r} - r\mu}{k_B T}\right) \tag{1}$$

where k_B is Boltzmann's constant, T is absolute temperature and Θ is

$$\Theta = \sum_r \sum_m \sum_i \exp\left(-\frac{E_{m,r} + E_{i,r} - r\mu}{k_B T}\right), \tag{2}$$

which is the combined partition function of two ensembles in statistical mechanics, the Canonical for the atoms and the Grand Canonical for the electrons. The former is expressed by

$$\Lambda_r = \sum_m \exp\left(-\frac{E_{m,r}}{k_B T}\right) \equiv \exp\left(-\frac{A_r}{k_B T}\right) \tag{3}$$

and defines a free energy A_r for the oscillator. Similarly for the electronic part in Eq. (2), the latter is

$$\Xi = \sum_r \sum_i \exp\left(-\frac{E_{i,r} - r\mu}{k_B T}\right) \equiv \sum_r \exp\left(-\frac{F_r - r\mu}{k_B T}\right), \tag{4}$$

where we have defined F_r as the free energy of the electronic part, also given by the partition function Z_r:

$$Z_r = \sum_i \exp\left(-\frac{E_{i,r}}{k_B T}\right) = \exp\left(-\frac{F_r}{k_B T}\right). \tag{5}$$

Using Eq. (2) – (4) in Eq. (1), we find the probability $P(r)$ that the trap has captured r electrons for any states m and i.

$$P(r) = \frac{\exp\left(-\dfrac{G_r}{k_B T}\right) \exp\left(\dfrac{r\mu}{k_B T}\right)}{\sum_r \exp\left(-\dfrac{G_r}{k_B T}\right) \exp\left(\dfrac{r\mu}{k_B T}\right)}, \tag{6}$$

where

$$G_r = A_r + F_r, \tag{7}$$

and the energy levels, G_{Tr} in relation to to the Fermi-level is shown to be

$$G_{Tr} = G_r - G_{r-1}. \tag{8}$$

This is a free energy for the combined electronic-atomic trap system. The free energy of the oscillator, A_r, is built by the eigenenergies $E_{m,r}$, of the atomic oscillator, where the wave functions occupy larger space for higher energy values (Fig. 1 (c)). This means that they are related with a volume increase at constant pressure, which motivates to assign A_r to a Gibbs ensemble and mention G_r as a Gibbs free energy [5,14]. From Eq. (6) we can formulate $P(1)$ as

$$P(1) = \frac{1}{1 + \exp\left(\dfrac{G_{T1} - \mu}{k_B T}\right)} \tag{9}$$

where

$$G_{T1} = G_1 - G_0 \tag{10}$$

corresponding to the energy positions in the band gap of an oxide or a semiconductor. We notice from Eqs. (9) that the probability for passing the critical point of $P(1) = \frac{1}{2}$ occurs when the Fermi-level coincides with the Gibbs free energy positions, G_{T1}, of the traps and not with the eigenenergy.

In order to demonstrate the features of Eq. (6), the occupation probabilities of a trap with an arbitrary set of energy levels in the bandgap between 0.1 eV and 1 eV from the conduction band edge are shown in Fig. 2. The trap is assumed to have five charge states and the ability to capture four electrons. When empty, it has a charge +2 for an energy level at 1.0 eV (Fig. 2(b)). The energy level for the first electron added, thus resulting in a charge state of +1, occurs at 0.59 eV. The reason that this energy level is higher is due to a mix between the added negative charge and the change in local atomic configuration. Continued addition of one electron at a time first makes the trap become neutral, followed by transfer to -1 and finally -2 charges at 0.1 eV. For every added electron, the positive energy contribution is often mentioned as a "positive U". If the lattice relaxation is large enough upon adding an electron, this may change the local bonds around the trap in such a way that a "negative U"-trap is formed, which means that adding an electron will give rise to a lower energy level [26]. As will be discussed in the next section, it has been argued that the oxygen vacancy in a number of high-k oxides has such properties.

(a)　　　　　　　　(b)

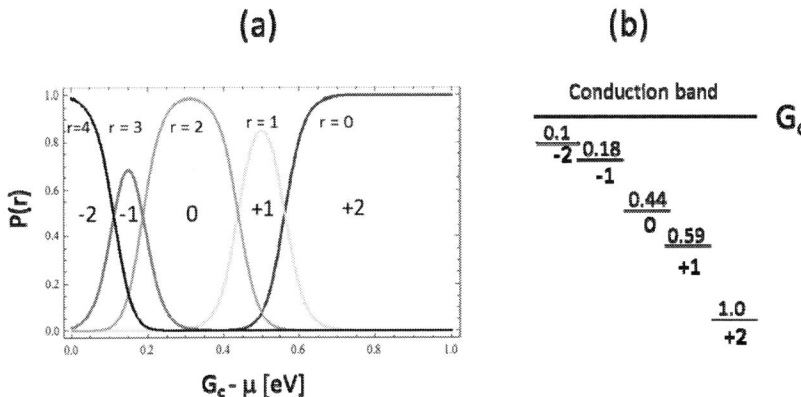

Figure 2. (a) Occupation probabilities for a trap with free energy levels as shown in (b) for different positions of the Fermi-level, μ, in relation to the conduction band edge at G_c. The numbers in (b) denote distances in eV to the conduction band edge.

Considering the graphs in Fig. 2 (a) for the occupation probabilities, we notice that when moving the Fermi-level, μ, from a low energy, the +2 state survives until μ reaches about 0.59 eV, where the +1 state takes over. Due to the close distance to the next energy level, the +1 state never reaches $P(1) = 1$ before the neutral states start to dominate at about 0.44 eV. By the same reason, this repeats for the 0.18 eV level, until the trap approaches $P(4) = 1$ when μ passes the 0.1 eV level. This illustrates the conditions for occupation probabilities based on charge carrier statistics and will be further discussed below in relation with oxide traps.

Oxide Traps

Due to its abundance in high-k oxides based on transition and rare-earth metals, the oxygen vacancy has attracted a considerable attention, especially in theoretical work [17 – 19]. The process for capturing charge carriers into this kind of traps seems more complicated than normally for traps in crystalline silicon or at the oxide/silicon interface. The oxygen vacancy is a common defect also in SiO_2 but because of the stiffer properties of this material and its glassy appearance, it occurs with lower concentration. Taking HfO_2 as an example, the oxygen vacancy has five charge states, +2, +1, 0, -1 and -2 as suggested from theory by Robertson's group [17]. Lacking information about the vibrational properties, we use the energy values found as free energies in the following treatment. In the 0-state it has captured two electrons and is neutral with an energy level at about 3.8 eV above the HfO_2 valence band or about 2.1 eV below the conduction band. The four Hf ions surrounding the vacant oxygen position are vibrating and relaxing when electrons are emitted or captured. Emitting one electron, thus making the trap charge +1, relaxes the surrounding Hf atoms slightly outwards [17]. This changes the electronic potential for the last electron such that its energy level moves up closer to the conduction band edge at a distance of about 1.4 eV. Taking away the last electron further relaxes the atomic positions such that the +2 level is at about 0.7 eV from the conduction band edge. This means that adding an electron to the +2 and the +1 states, gives rise to energy levels with positions deeper into the band gap. Thus, the system has "negative U"-properties and gives rise to unusual behavior of the capture properties as will be discussed below. Adding electrons to the neutral trap in the 0-state gives rise to -1 and -2 charges with energy levels stepping closer to the conduction band. This is the more common behavior for traps in crystalline solids and gives rise to "normal" capture statistics as was demonstrated above in relation to Fig. 2. The difference from the positively charged trap states is, that in this case the relaxation goes inwards [17].

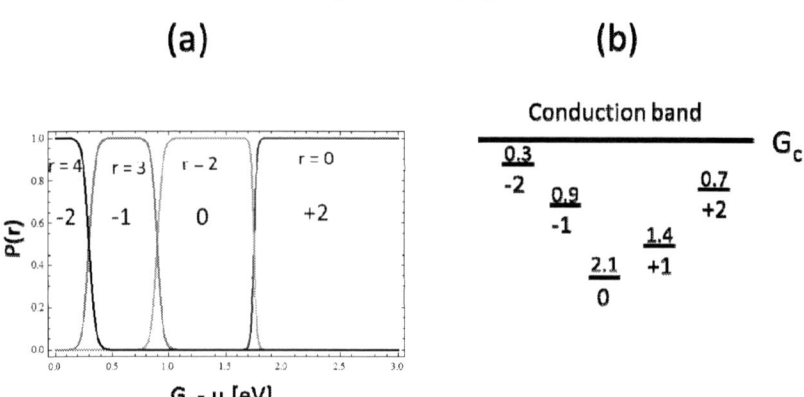

Figure 3. (a) Occupation probabilities for the oxygen vacancy in HfO_2 with free energy levels as shown in (b) for different positions of the Fermi-level, μ, in relation to the conduction band edge at G_c. The numbers in (b) denote distances in eV to the conduction band edge and the charge states of the different levels.

The occupation probabilities for the different charge states of the oxygen vacancy in HfO_2, discussed above are calculated by using Eq. (6) and shown in Fig. 3. In order to demonstrate the influence of the negative-U properties of states +1 and +2, Fig. 3(a) shows the occupation probabilities, where the energy levels approach the conduction band edge as the number of captured electrons decreases. Starting at a situation where the Fermi-level, μ, is below the neutral 2.1 eV level, all traps are empty and thus in the +2 state. Moving the Fermi-level upwards, one would expect from traditional thinking that when it approaches 2.1 eV, the energy level at this position becomes occupied. However, as can be noticed in Fig. 3(a) this is not the case. The 2.1 eV level fills only after μ has reached G_c - 1.75 eV. This is a result of the negative-U properties, which pushes the +1 level to 0.7 eV above the 0 level, and of the occupation number going from 0 to 2. As mentioned above, the Fermi-level is a measure of the average energy per electron in the system, while the energy levels represent the change of adding one electron at a time (Eq. (7)). Therefore, a transfer of the trap from +2 to 0, meaning an increase of two electrons, will occur when $2(G_c - \mu) = 2 \times 2.1$ eV + U, where $U = -0.7$ eV is the correlation energy. This gives $G_c - \mu = 1.75$ eV. Energy level schemes of the standard type shown in Fig. 3(b) are valid for single electron transfers. Due to the negative-U properties in the present case, the neutral trap level switches from +2 to 0, confusing the thinking in single electron schemes. Continued movement of the Fermi-level, adding one electron at a time for the -1 and -2 states follows the standard course as can be seen in Fig. 3(a) and discussed in connection with Fig. 2

The practical consequences of these properties of the oxygen vacancy in HfO_2 are that in experiments, where thermal emission of carriers are studied, for example in Poole-Frenkel transport, mainly the -2 and -1 states at 0.3 and 0.9 eV, respectively, may play roles. As can be noticed from the occupation graphs in Fig. 3(a), after emptying the the -1 state nothing happens until the Fermi-level has reached 1.75 eV. For normal trap properties, the emission rates from that depth would correspond to time constants in the range of many years. However, for non-thermal processes, like tunneling or optical excitations all charge states are expected to contribute.

Additional theoretical investigations to the one for HfO_2, suggest that other high-k oxides, like La_2O_3, Lu_2O_3, $LaLuO_3$ and $HfSiO_4$ [19, 27] have similar energy level schemes with negative-U behavior for the positive states, while Al_2O_3 [28] is not expected to show this property due to smaller relaxation effects.

Traps in High-k/SiO$_x$ Transitions Regions

As a result of the reactivity between high-k oxides and silicon, an interlayer of SiO_x is commonly formed between the high-k film and the semiconductor crystal. This has an advantage in transistor applications because lower interface state densities are achieved this way. However, a drawback is that the effective k-value of the high-k/SiO$_x$/Si stack is lowered. The transition from high-k oxide to SiO_x is not to be considered abrupt. A number of experimental results have demonstrated that within a range of about 1 nm or even wider, the concentrations of all elements, Si, O and oxide metal vary [29, 30]. In this transition region, a high concentration of traps is expected. For the HfO_2/SiO$_x$ stack, it was recently demonstrated that these traps are structurally unstable and communicate with the free carriers in the silicon crystal by a combined thermal-tunneling process [30, 31].

The tendency to form an interlayer depends on the heat of formation between the high-k material and the reaction product, which may be an oxide, a silicide or a silicate. As the reaction is driven by free energy, the balance depends on temperature. Under the right temperature conditions a silicate has been demonstrated to form between silicon and oxides like for example LaLuO$_3$ and Y$_2$O$_3$ [32 – 34]. Because silicates often have higher k-values than for SiO$_x$, such solutions are to be preferred, especially because the qualities of the interlayer/silicon interface states seem to be very similar for both these combinations.

For the HfO$_2$/SiO$_x$ double layer, the concentration of Hf atoms have been found to increase from zero to the stoichiometric value for HfO$_2$ within a range of about 1 nm starting at about 0.5 nanometer from the SiO$_x$/Si interface [35]. The dielectric constant for Hf$_x$Si$_{1-x}$O has been shown to change close to linear with x from the value of $k = 25$ for HfO$_2$ to $k = 3.9$ for SiO$_2$ [36]. Fig. 4 shows the energy of the conduction band edge for a HfO$_2$/SiO$_x$ combination comprising a high-k oxide with a physical thickness of 4 nm. Using data in Refs. 35 and 36, the k-value is assumed to change from about 3.9 at the SiO$_x$/Si interface to 25 at the gate metal/HfO$_2$ interface. The variation in k as a function of the distance gives rise to an oxide capacitance expressed by

$$C_{ox}(x) = \left[\int_0^x \frac{du}{k(u)\varepsilon_0} \right]^{-1}, \qquad (11)$$

where x and u represent the length scale perpendicular to the surface of the stack. From this, the shape of the conduction band edge as a function of depth in the oxide can be calculated. The result is shown for varying oxide voltage together with the graph for $k(x)$ in Fig. 4. We observe that Fowler-Nordheim tunneling into the stoichiometric HfO$_2$ range is possible only for voltages at about 1.6 V. Above this voltage the tunneling process takes place inside the transition region. Here, a large concentration of acceptor traps has been found to exist [31], which gives the opportunity for increased tunneling processes of trap assisted type. Due to the influence of the lower k-values in the transition region, for the present stack the effective k-value is 11.9, corresponding to EOT = 1.3 nm.

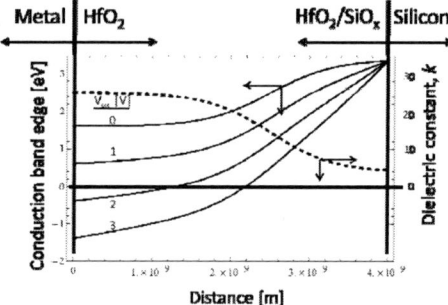

Figure 4. Dashed curve: The dielectric constant, k, calculated as a function of distance between the metal contact and the silicon interface of a metal/HfO$_2$/SiO$_x$/Si structure, based on data in Refs. 35 and 36. Solid curves: The conduction band edge of the HfO$_2$/ SiO$_x$ combination as a function of distance between the metal contact and the silicon interface for oxide voltages at 0, 1, 2 and 3V, is calculated by using the dashed k data. The energy level at 0 eV on the left scale corresponds to the position of the conduction band edge in the silicon crystal.

An interesting observation from Fig. 4 is that for voltages above about 2 V, the fundamental leakage current by tunneling is limited merely by the properties of the oxide within a distance of about 2.5 nm from the SiO_2/Si interface. However, the unstable traps found in this part of the stack certainly may act as sources of trap assisted processes.

From thermally stimulated current (TSC) measurements on metal/$HfO_2/SiO_x/Si$ structures it has been found that traps in the transition region discussed above, communicate with states in the silicon conduction band by a combined thermal-tunneling process [31]. Filling the traps at room temperature with a positive bias on the gate metal, followed by cooling down to $T = 50$ K under continued positive bias, switching to negative bias and heating with a linear temperature ramp from 50 K up to room temperature gave rise to current peaks when electrons were leaving the traps. The schematic diagram in Fig. 5 (a), depicts the energy relations for the trap, where the captured electron is thermally excited and resonates with states from the conduction band of the silicon crystal to perform tunneling. This gives rise to the TSC peaks A and B shown in Fig. 5 (b) from consecutive measurements on the same sample. Repeated measurements gave rise to peak positions at various temperatures, suggesting structural instabilities of these traps thus changing their energy positions and emission properties [31]. The solid TSC curves in this diagram are theoretical while the measured data coincide only up to a certain temperature, where the signal collapses. This indicates that the decreasing negative charge in the transition region lowers the electric field across the transition region until the tunneling process suddenly ceases.

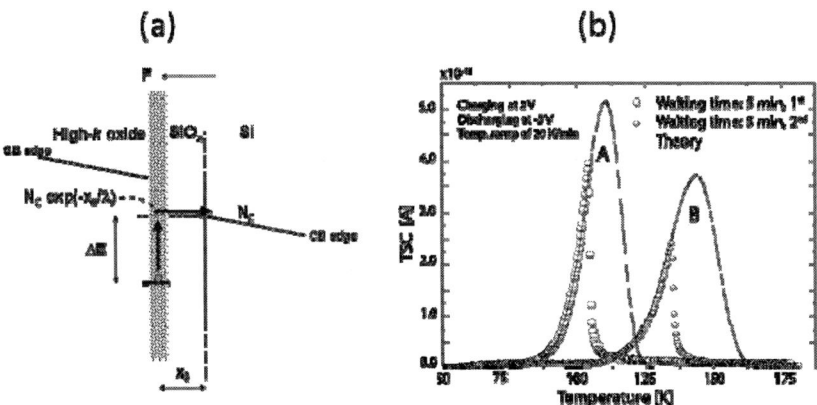

Figure 5. (a) Energy scheme for the emission of electrons from traps in the HfO_2/SiO_x transition region to the silicon conduction band. (b) Data from thermally stimulated current (TSC) measurements (points) compared with theoretical curves (solid) from two consecutive measurement cycles on the same samples. The change in peak positions indicates structural instability. The sudden termination of the TSC signals at 110 and 130 K is a result of decreased tunneling probability when electrons leave the traps and the electric field, F, decreases.

Interface States

The dynamic properties of MOS systems are governed by the existence of charge carrier states at the intimate contact area between the oxide and the silicon crystal. Such states communicate with the silicon band states through thermal emission and capture of carriers, at room temperature delayed by time constants in the ranges of ms down to below μs. For SiO_2/Si interfaces, extensive investigations have been performed to increase the understanding of their occurrence. The main trap discovered was the so called P_b center, which was found to be amphoteric with three charge states, +1, 0 and -1 and with the +/0 level at about 0.3 eV above the silicon valence band edge, while the 0/- level exists at about the same energy distance from the conduction band edge [37]. The center occurs as a result of electron orbitals, *dangling bonds*, pointing out from the crystalline silicon surface into the amorphous oxide. The orientation of these dangling bonds is a little different for (1,1,1) surfaces compared with (1,0,0), but the energy level scheme and positions are similar [38]. It was found from thermal capture measurements and optical investigations that the properties of P_b centers are controlled by atomic and electronic potentials in combination, similar but with less fierce consequences compared with the case of the oxygen vacancy in high-k oxides [8 – 15]. Furthermore, as mentioned above, it was recently demonstrated for HfO_2/SiO_x/Si stacks that electron interface states with energy positions closer the silicon conduction band, are of a different quality [13]. For those states, the atomic vibrational potential does not have a role. The states seem to be of pure electronic character without vibrational properties. This existence of varying sets of states gives rise to more intriguing charge carrier statistics which will be discussed in this section.

The energy band diagram of an MOS-structure is shown in Fig. 6 (a) with a more detailed sketch of the oxide-semiconductor interface in Fig. 6 (b). Here, V_G is the voltage applied on the metal gate, which divides between the oxide voltage V_{ox} and the surface potential, Ψ_s, occurring in the semiconductor. This partition depends on the charge distributed through the structure by bulk oxide charge, Q_{ox}, charge in interface states, Q_{it} and charge, Q_s, collecting in the semiconductor close to the interface as schematically depicted in Fig. 6 (c). From simple electrostatic reasoning it is found that

$$V_G = \Phi_{ms} - \frac{Q_{it} + Q_s + \frac{x_m}{d} Q_{ox}}{C_{ox}} + \Psi_s,$$ (12)

where Φ_{ms} is the difference in work function between the metal and the semiconductor, x_m is the center of gravity for the oxide charge, d is the oxide thickness and C_{ox} is the oxide capacitance. A DC voltage, V_G, will determine the position of the Fermi-level at the interface as shown in Fig. 6 (b) and thus the occupation of interface states as given by Eq. (6).

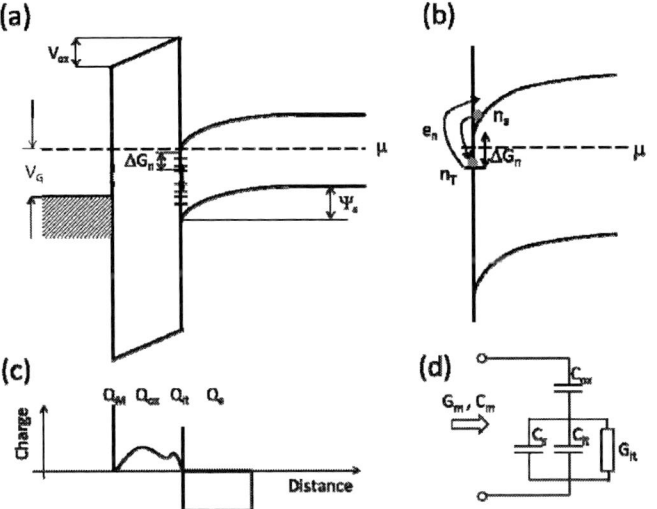

Figure 6. (a) Band diagram of the MOS structure. (b) Detail to illustrate an interface trap position close to the Fermi-level taking part in the emission and capture of electrons for an applied AC small-signal. (c) Charge distribution along the depth from the gate metal into the silicon crystal. (d) Equivalent circuit for calculating measured admittance data.

Assuming that the interface traps can capture only one electron each, the probability for occupation follows Eq. (9). Then, at thermal equilibrium for any interface state, the capture and emission of electrons balance such that

$$e_n P(1) = [1 - P(1)] v_{th} \sigma_n n_s,$$ (13)

where e_n is the thermal emission rate of electrons from the interface state to the conduction band, v_{th} is the average thermal velocity of electrons in the conduction band of the semiconductor, σ_n is the cross section for capturing an electron into the interface state and n_s is the volume concentration of carriers in the semiconductor conduction band close to the interface. Using standard expressions for the quantities in Eq.(13) it is readily shown that

$$e_n = v_{th} \sigma_n N_c \exp\left(-\frac{\Delta G_n}{k_B T}\right),$$ (14)

where N_c is the effective density of states in the semiconductor conduction band and ΔG_n is the Gibbs free energy for releasing an electron from the trap level to the conduction band edge.

The most common method for investigating the electrical properties of interface states is by admittance measurements on the MOS-system. This is performed by applying a DC gate voltage V_G with a small signal AC voltage of about 10 mV amplitude superimposed and by measuring AC current and phase angle. The AC voltage will bring the semiconductor energy bands and the energy levels at the interface to oscillate in relation to the Fermi-level. For electrons occupying the interface states with energy levels close to the Fermi-level, this means that they will be emitted and captured in pace with the frequency of the applied AC voltage and with a probability that depends on their distance to the DC position of the Fermi-level. This probability depends on the slope of the Fermi-function, $f = P(1)$, as expressed in Eq. (9). Knowing the magnitude of the AC voltage, the admittance added by the interface states can be obtained and divided into a capacitive and a conductive part. Taking into consideration that the interface state energy distribution is more or less a continuous function $D_{it}(\Delta G_n)$ and that states with free energy levels beside the Fermi-level also contribute, it has been demonstrated earlier in detail [13, 39] that a conductive, γ_{it}, and a capacitive component, χ_{it}, can be defined for the *admittance density*. This is the contribution of admittance from the interface states close to the Fermi-level position within an energy range determined by its sloping part:

$$\gamma_{it} = \frac{q^2}{k_B T} \frac{D_{it}}{2} \frac{\omega^2 e_n}{4e_n^2 + \omega^2} f(1-f) \tag{15}$$

$$\chi_{it} = \frac{q^2}{k_B T} \frac{D_{it}}{2} \frac{2e_n^2}{4e_n^2 + \omega^2} f(1-f). \tag{16}$$

In these expressions, the quantities, D_{it} and e_n are functions of ΔG_n, while $f = P(1)$ from Eq. (9) is a function of the difference $\Delta\mu - \Delta G_n$, defining $\Delta\mu = G_c - \mu$ and $\Delta G_n = G_c - G_T$. The electron charge is denoted q and ω is the angular frequency of the AC signal used for measurement. This means that for each position of the Fermi-level, which is set by the applied DC voltage, energy states with positions ΔG_n contribute to the admittance with a strength given by $f(1-f)$ depending on their energy distance, $\Delta\mu - \Delta G_n$, to the Fermi-level. The function $f(1-f)$ has a maximum for $\Delta G_n = \mu$ and decays on both sides of $\Delta\mu - \Delta G_n = 0$ with a width at half maximum of about $4 k_B T$ [14]. This means that the conductance and capacitance densities γ_{it} and χ_{it}, respectively, are functions of ΔG_n and $\Delta\mu$ to be considered valid for each specific position of μ. In order to find the total capacitance and conductance contributions from an interface state distribution $D_{it}(\Delta G_n)$ at a specific position of the Fermi-level, the conductance and capacitance densities in Eqs. (15) and (16) need to be integrated across the semiconductor band gap, E_g.

$$G_{it}(\Delta\mu) = \int_0^{E_g} \gamma_{it}(\Delta\mu, \Delta G_n) d(\Delta G_n) \tag{17}$$

$$C_{it}(\Delta\mu) = \int_0^{E_g} \chi_{it}(\Delta\mu, \Delta G_n) d(\Delta G_n) \tag{18}$$

The total admittance of the MOS structure can be modeled as a parallel connection between G_{it}, C_{it} and the semiconductor capacitance C_s, connected in series with the oxide capacitance C_{ox} as demonstrated in Fig. 6 (d). The measured conductance and capacitance values are the equivalent quantities of those measured on the input of the circuit shown in the figure. The measured AC current and its phase may be transferred into a series or a parallel connection between a conductance and a capacitance. For a parallel case the measured quantities, G_m and C_m, are obtained from the components in the circuit shown in Fig. 6 (d) as [13]

$$G_m = \frac{C_{ox}^2 G_{it}}{C_t^2 + G_{it}^2 / \omega^2} \tag{19}$$

and

$$C_m = \frac{C_{ox}\left[C_t(C_{it} + C_s) + G_{it}^2 / \omega^2\right]}{C_t^2 + G_{it}^2 / \omega^2}, \tag{20}$$

where $C_t = C_{ox} + C_s + C_{it}$. Formulas for the reverse calculation, to obtain G_{it} and C_{it} from measured data, can be found in Ref. 40.

It should be noted that the expressions presented for G_{it} and C_{it} are valid for a case where the interface is hosting electron states of one single quality. As shown in Fig. 7, the total capture and emission of carriers may differ depending on whether one set of states is present with capture cross sections changing as a function of energy or, for example, two sets of states with two different capture mechanisms. This is especially significant when the distributions have a considerable overlap as demonstrated in Fig. 7 (b). For such cases, each set of traps, numbered by suffix l, needs to be allotted specific $\gamma_{it,l}$ and $\chi_{it,l}$ functions as given by Eqs. (15) and (16) giving specific $G_{it,l}(\Delta\mu)$ and $C_{it,l}(\Delta\mu)$ functions following Eqs. (17) and (18), which are to be added according to

$$G_{it} = \sum_l G_{it,l} \tag{21}$$

$$C_{it} = \sum_l C_{it,l} \tag{22}$$

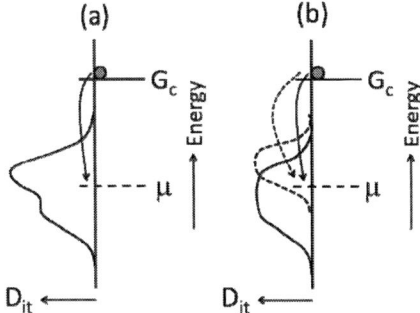

Figure 7. Illustration of the difference in influence from one D_{it} distribution as shown in (a), as compared with two distributions with different capture properties in (b). In (a) the same energy distribution of the capture, $\sigma_n(\Delta G_n)$, is valid for the whole distribution, which requires just single functions of γ_{it} and χ_{it}. In (b) two different functions each of γ_{it} and χ_{it} are required, one for each D_{it} distribution.

Considering a situation where $D_{it}(\Delta G_n)$ has been measured, for example by a capacitance versus voltage (C-V) method, it is hard to decide whether the capacitance features are results of one type of interface states, as shown in Fig. 7 (a) or of a sum of a number of different sets of states as indicated in Fig. 7 (b). In the energy range where the D_{it} distributions overlap, the capture and emission of charge carriers take place at two different types of traps with different dependences on frequency and temperature which would give different results depending on the measurement conditions. An experimental example of such conditions is the different types of capture mechanisms recently discovered for interface states of HfO$_2$/SiO$_x$/Si structures [13]. It was found that energy states close to the conduction band are of the Lax type with negative energy and temperature dependences of the capture cross sections while the deeper states with P_b character follow a multiphonon scheme with opposite dependences of those quantities [13]. The problem is clarified by a couple of theoretical examples from multiparameter admittance spectroscopy (MPAS) as shown below in Figs. 8 and 9. The $D_{it}(\Delta G_n)$ and $\sigma_n(\Delta G_n)$ functions in Fig. 8 (a) and (b) are, in each case, the sum of the of two sub-functions as shown separately in Fig 9 (a) and (b), respectively. For the interface states in Fig. 8 (a), this means that their capture cross sections change gradually as shown in Fig. 8 (b) for varying ΔG_n. On the other hand, in Fig. 9 (b), we assume that two different and independent $\sigma_n(\Delta G_n)$ distributions exist with their sum equal to the feature of Fig. 8 (b). The two MPAS maps in Figs. 8(c) and 9(c) are contour plots of the conductivity G_{it} as a function of the surface potential Ψ_s and log[$1/\omega$]. Sample data for the calculations are given in the figure captions. It is noticed that the results of conductance data for the two cases differ considerably.

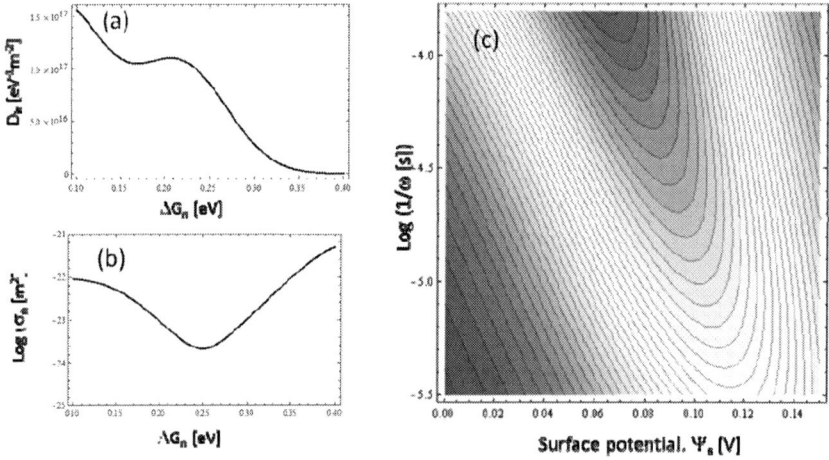

Figure 8. (a) Density of interfaces states, $D_{it}(\Delta G_n)$ and (b) their capture cross section, $\sigma_n(\Delta G_n)$, for a case assuming one single distribution of states constituting the sum of the states in Fig. 9. (c) The corresponding MPAS map. Sample data: EOT = 11.5 nm, silicon doping 4 x 10^{20} m^{-3}, n-type, T = 300 K.

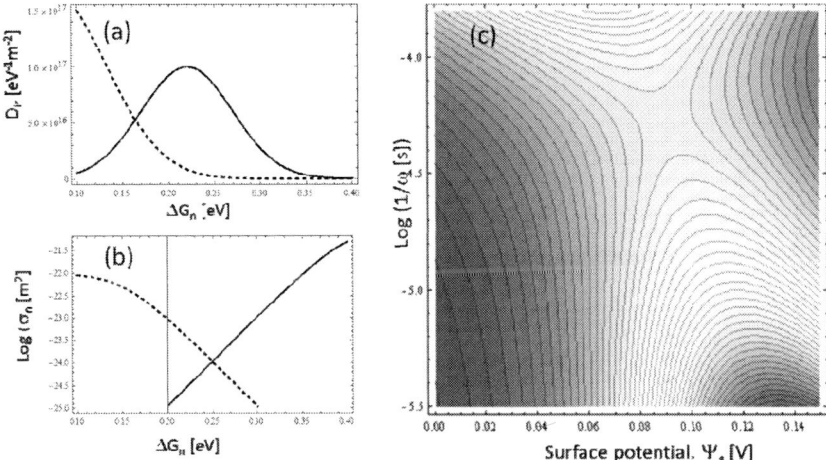

Figure 9. (a) Density of interfaces states, $D_{it}(\Delta G_n)$ and (b) their capture cross section, $\sigma_n(\Delta G_n)$, for a case assuming two separate distribution of states. Dashed curves: "Lax"-states. Solid curves: Multiphonon type (See text). (c) The corresponding MPAS map. Sample data: See Fig. 8.

In an experimental situation, when interpreting conductance data from a measured conductance plot, the use of MPAS is to prefer compared with measurements of single conductance curves as functions of one of the variables in Figs. 8 (c) and 9(c). The MPAS map in the latter figure, where two mountains are separated by a saddle point, can normally not be achieved by fitting one single $\sigma_n(\Delta G_n)$ distribution of the type shown in Fig. 8(b).

High frequency C-V technique is commonly used to obtain fast diagnoses of MOS samples, with the measurements in a frequency range of about 100 Hz to 1 MHz. A theoretical example is shown in Fig. 10, with $D_{it}(\Delta G_n)$ and $\sigma_n(\Delta G_n)$ data shown in Figs. 10 (a) and (b), respectively and for sample data as mentioned in the figure caption. In this case the P_b like D_{it} peak at about 0.28 eV in Fig. 10 (a) is well separated from the rest of the state distribution and two independent $\sigma_n(\Delta G_n)$ distributions are assumed as shown in Fig. 10 (b). The C-V curve, calculated for three different frequencies and using Eq.(12) for the relation between surface potential and gate voltage, is shown in Fig. 10 (c). The graph has a shoulder which decreases with increasing frequency as a result of the relation between ω and e_n appearing in Eq. (16). This feature is a typical indication of the existence of the P_b center and is often found in samples with SiO_2 as well as in those with high-k dielectrics [41]. By choosing appropriate voltages and measuring the frequency dependences of the capacitance on the shoulder as demonstrated in Fig. 10 (d), the capture cross sections can be found as a function of ΔG_n [11, 12].

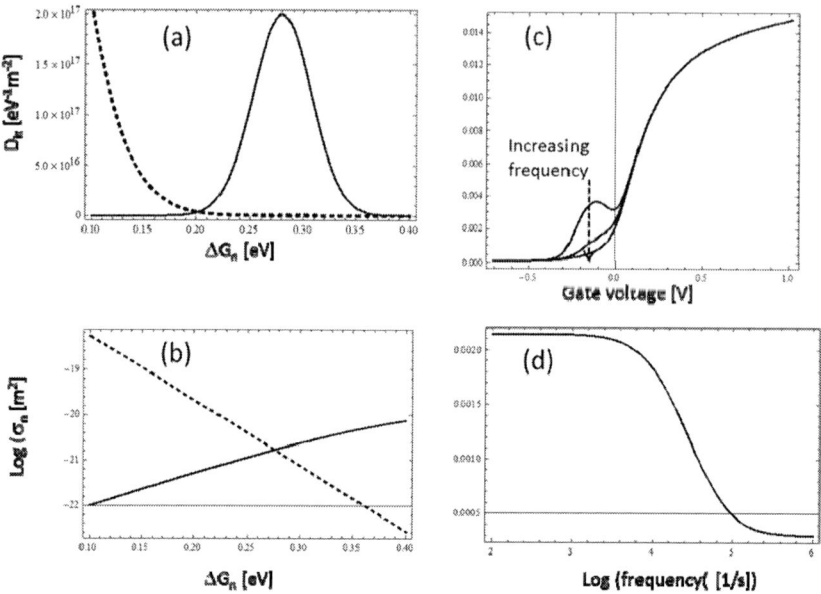

Figure 10. (a) Density of interfaces states, $D_{it}(\Delta G_n)$ and (b) their capture cross section, $\sigma_n(\Delta G_n)$, for a case assuming two separate distribution of states. Dashed curves: "Lax"-states. Solid curves: Multiphonon type (See text). (c) The corresponding capacitance versus voltage curves for frequencies 1 kHz, 10 kHz and 100 kHz. (d) Capacitance as a function of frequency at gate voltage = -1.5 V. Sample data: See Fig. 8.

Dipoles

Recently, dipoles have been pointed out as important features of trap combinations for control of threshold voltage when MOS structures are used as transistor gates. These objects have found roles at the metal/high-k and the high-k/Si interfaces for regulating threshold voltages in MOS transistors.

The energy barrier at metal-insulator interfaces has the same physical origin as that of Schottky-barriers between metals and semiconductors. This structure has a simple geometry but a comprehensive history. Two important works, published in 1965 initiated a tremendous international activity lasting for three decades in order to find a basic theory for the metal-semiconductor problem. In the first case, a phenomenological model was developed by Cowley and Sze [42] to handle the influence of surface/interface charge on the barrier height. The argument was to use a dipole charge close to the metal-semiconductor interface to change the distance between the metal Fermi-level and the energy bands of the semiconductor. This was done by taking into consideration the influence of interface states separated at close distance from the metal by a thin oxide. Then, from electrostatic reasoning, the barrier height

Φ_{Bn} in metal-insulator terms for electrons can be expressed

$$\Phi_{Bn} = \gamma(\Phi_M - \chi_i) + (1 - \gamma)\Phi_0, \tag{23}$$

where Φ_M is the work function of the metal, χ_i is the electron affinity of the insulator and Φ_0 is the "neutral level" of the metal/insulator interface. The latter is defined as the energy level, measured from the insulator conduction band, up to which the interface energy distribution needs to be filled in order to obtain a neutral interface. Finally, the factor γ in Eq. (23) expresses the influence of interface state concentration:

$$\gamma = \frac{1}{1 + q^2 \dfrac{d}{k\varepsilon_0} D_{it}}. \tag{24}$$

Here, d is the distance between the metal plane and the plane of the interface states and k is the dielectric constant of the material included in that separation.

The second work was published by Heine [43] and introduced the possibility of electron states in the metal, penetrating the bandgap of the semiconductor, thus giving rise to a charge separated from the interface and creating a dipole potential. In later refinements of this idea, the term metal induced gap states (MIGS) was introduced for this phenomenon and has survived also in works on gate stacks. This model gave a physical motivation for the existence of the dipole suggested in Ref. 42 and an incentive for using Eqs. (23) and (24) for practical application. Assuming an interface free of charge, i.e. $D_{it} = 0$, makes $\gamma = 1$ and $\Phi_{Bn} = \Phi_M - \chi_i$, whereas high values of D_{it} make γ approach zero and Φ_{Bn} pinned to the neutral level Φ_0. Hence, choosing appropriate energy distributions of traps close to the metal/insulator interface gives opportunities to tune the "gate metal workfunction" as Φ_{Bn} often a little misleadingly is called within the MOS community. A number of different doping elements has been used in recent research to find appropriate tuning possibilities [21].

For the high-k/transition region/Si interfaces, an experiment separating the influence of dipoles at this position from that of single charges would be to measure the flat band voltage, V_{FB}, of MOS structures with different oxide thickness, but otherwise identical. A voltage drop changing V_{FB} as a result of a dipole charge would appear between the positive and negative charge of the dipole such that V_{FB} would be independent on oxide thickness. On the other hand, a single charge in the transition region gives rise to a voltage drop proportional to its distance to the gate metal, which would change V_{FB} in proportion to the oxide thickness. A number of investigations can be found in the literature, where the dipole phenomenon has been investigated. Also, a saturation of V_{FB} with increasing oxide thickness has been demonstrated for HfO$_2$ [22]. However, it has been shown that such effect also may occur due to an expected variation in oxygen vacancies [30]. The interesting ideas for the existence of dipoles in the transition region of high-k/interlayer/Si stacks, therefore, may need further confirmation.

Discussion

As the electrical influence of electron states is decisive for device functions, traditional electrically based methods play an important role for characterizing electron states in MOS structures. Investigations of MOS interface electron states based on C-V and conductance has a history starting in the 1960s when the first successful steps in mastering the SiO_2/Si interface were taken [40, 44]. Later progress in the development of this measurement technique has mainly been done in instrumentation for measurement of electrical quantities together with more advanced methods for data collection. MPAS as described in this report is an example. Alternative methods based on optical interaction are only rarely found in the literature [15, 45]. One reason for this might be the small cross sections for photoionization which seem to exist for interface states, making such detailed data hard to take. The possibility of exciting charge carriers with photons gives new insight into the atomic vibrational properties of traps, especially in combination with electrical measurements at varying temperature. An interesting finding is that the electron state properties for many high-k/silicon interfaces are very similar to those of SiO_2/Si interfaces, even where it has been found that a silicate replaces the SiO_x interlayer [11, 33, 41].

For investigations of bulk oxide traps, the electrical methods offer less detailed results. Coarse data on total oxide charge contents are achieved by C-V from parallel shifts of the capacitance versus voltage graphs along the voltage axis. However, as seen from Eq. (12), this just gives a minimum value of this quantity as long as the center of gravity of the charge distribution is not known. Another problem with published results in this connection is that the waiting times for filling and emptying of traps at the voltage turning points of these characteristics seldom are mentioned. Methods for measuring C-V data by using fast voltage pulses, under development by a number of groups [46 – 48], are interesting for capture investigations, but a unified model for the charge injection processes is still lacking [49, 50]. The complicated energy band relations due to inter-diffusion, expected at high-k/interlayer transition regions (Fig. 4), makes injection studies hard to interpret. In most works published so far, an abrupt interface at this point is assumed, which may give rise to deviations in theoretical estimates. The fascinating structure of oxygen vacancies in high-k materials discussed in this overview, probably will require detailed optical investigations based on photoconductivity and luminescence for confirming the theoretical results in Refs. 17 -19.

Acknowledgment

This work was financed by the European NANOSIL Network of Excellence (Grant No IST 216171)

References

1. O. Engström, I. Z. Mitrovic, S. Hall, P. K. Hurley, K, Cherkaoui, S. Monaghan, H. D. B. Gottlob and M. C. Lemme in *"Nanoscale CMOS"*, p. 23, (Wiley, 2010, Editor: F. Balestra).
2. O. Engström, *"The MOS System"*, Cambridge University Press, to be published 2012.
3. J. Robertson, Rep. Prog. Phys., **69**, 327 (2006).
4. J. A. Van Vechten and C. D. Thurmond, Phys. Rev. B, **14**, 3539 (1976).
5. O. Engström and A. Alm, Solid-State Electron., **21**, 1571 (1978).
6. C. H. Henry and D. V. Lang, Phys. Rev. B, **15**, 989 (1977).
7. N. F. Mott and R. W. Gurney, *"Electronic processes in ionic crystals"* (Oxford University Press, 1946).
8. O. Engström and H. G. Grimmeiss, Semicond. Sci. Technol., **4**, 1106 (1989).
9. A. Ricksand and O. Engström, J. Appl. Phys. **70**, 6927 (1991).
10. O. Engström, T. Gutt and H. M. Przewlocki, J. Telecomm. Inf. Tech., **2**, 86 (2007).
11. B. Raeissi, J. Piscator, O. Engström, S. Hall, O. Buiu, M. C. Lemme, H. D. B. Gottlob, P. K. Hurley, K. Cherkaoui and H. J. Osten, Solid-State Electron., **52**, 1274 (2008).
12. O. Engström, B. Raeissi and J. Piscator, J. Appl. Phys., **103**, 104101 (2008).
13. J. Piscator, B. Raeissi and O. Engström, J. Appl. Phys., **106**, 054510 (2009).
14. O. Engström and A. Alm, J. Appl. Phys., **54**, 5240 (1983).
15. M. O. Andersson and O. Engström, Appl. Surf. Sci., **39**, (1989).
16. M. Lax, Phys. Rev., **119**, 1502 (1960).
17. K. Xiong, J. Robertson, M. C. Gibson and S. J. Clark, Appl. Phys. Lett., **87**, 183505 (2005).
18. L. Lin and J. Robertson, Appl. Phys. Lett., **95**, 012906 (2009).
19. K. Xiong and J. Robertson, Appl. Phys. Lett., **95**, 022903 (2009).
20. W. B. Fowler, J. K. Rudra, M. E. Swanut and F. J. Feigl, Phys. Rev. B, **41** 8313 (1990).
21. X. P. Wang, H. Y. Yu, M. F. Li, C. X. Zhu, S. Biesemans,, A. Chin, Y. Sun, Y. P. Feng, A. Lim, Y. C. Yeo, W. P. Loh and D. L. Kwong, IEEE Electron Dev. Lett., **28**, 258 (2007).
22. K. Kita and A. Toriumi, Appl. Phys. Lett., **94**, 132902 (2009).
23. Xiaolei Wang, Kai Han, Wenwu Wang,_ Shijie Chen, Xueli Ma, Dapeng Chen, Jing Zhang, Jun Du, Yuhua Xiong and Anping Huang, Appl. Phys. Lett., **96**, 152907 (2010).
24. P. T. Landsberg, "Recombination in Semiconductors", (Cambridge University Press, 1991).
25. P. T. Landsberg and O. Engström, *"Handbook on Semiconductors"*, Vol.1, p. 197 (North Holland, 1992, Editor: T. S. Moss).
26. P. W. Anderson, Phys. Rev. Lett., **34**, 953 (1975).
27. K. Xiong, Y. Du, K. Tse and J. Robertson, Appl. Phys. Lett., **101**, 024101 (2007).
28. D. Liu, S. J. Clark and J. Robertson, Appl. Phys. Lett., **96**, 032905 (2010).
29. H. D. B. Gottlob, A. Stefani, M. Schmodt, M. C. Lemme, H. Kurz, I. Z. Mitrovic, M. Werner, W. M. Davey, S. Hall, P. R. Chalker, K. Cherkaoui, P. K. Hurley, J. Piscator, O. Engström and S. B. Newcomb, J. Vac. Sci. Technol. B, **27**, 249 (2009).
30. O. Engström, B. Raeissi, J. Piscator, I. Z. Mitrovic, S. Hall, H. D. B. Gottlob, M. Schmidt, P. K. Hurley, and K. Cherkaoui, J. Telecomm. Inf. Tech., **1**, 10 (2010).
31. B. Raeissi, S. Piscator, Y. Y. Chen and O. Engström, J. Electrochem. Soc., **158**, G63, (2011).
32. J.M.J. Lopes, E. Durgun-Özben, M. Roeckerath, U. Littmark, R Lupták, St. Lenk, A Beshem, U. Breuer, J. Schubert and S. Mantl, Proc. 10th International Conference on Ultimate Integration of Silicon (ULIS), 2009, p.99.

33. E. Durgun-Özben, J.M.J. Lopes, A. Nichau, M. Schnee, S. Lenk, A. Besmehm, K. K. Bourdelle, Q. T. Zhao, J. Schubert and S. Mantl, to appear in IEEE Electron. Dev. Lett., 2011.
34. M. Copel, Appl. Phys. Lett,, **82**, 1580 (2003).
35. M. P. Augustin, L. R. C. Fonseca, J. C. Hooker and S. Stemmer, Appl. Phys. Lett., **87**, 121909 (2005).
36. P. Broquist and A. Pasquarello, Appl. Phys. Lett., **90**, 082907 (2007).
37. P. M. Lenahan and P. V. Dressendorfer J. Appl. Phys. **55**, 3495 (1984).
38. E. H. Poindexter, Semicond. Sci. Technol., **4**, 961 (1989).
39. O. Engström and B. Raeissi, ECS Trans., **33** (3), 257 (2010).
40. E. H. Nicollian and A. Goetzberger, Bell. Syst. Tech. J., **46**, 1055 (1967).
41. P. Hurley, P K. Cherkauoi, E. O'Connor, M. C. Lemme, H. D. B. Gottlob, S. Hall, Y Lu, O. Buiu, B. Raeissi, J. Piscator and O. Engström, J. Electrochem. Soc., **155**, G13, (2008).
42. A. M. Cowley and S. M. J. Sze, Appl. Phys., **36**, 3212 (1965).
43. V. Heine, Phys. Rev., **138**, 1689 (1965).
44. A. S. Grove, E. H. Snow, B. E Deal and C. T. Sah, J. Appl. Phys. **35**, 2458 (1964).
45. H. G. Grimmeiss, W. R. Buchwald, E. H. Poindexter, P. J. Caplan, M. Harmatz and N. M. Johnson, Phys. Rev. B, **39**, 5175 (1989).
46. W. D. Zhang, B. Govoreanu, X. F. Zheng, D. Ruiz Aguado, M. Rosmeulen, P. Blomme, J. F. Zhang, and J. Van Houdt, IEEE Trans. Electron. Dev., **29**, 1043 (2008).
47. R. Rao and F. Irrera, J. Appl. Phys., **107**, 103708 (2010).
48. N. Sedghi, W. Davey, I. Mitrovic, J. M. P. Lopez, J. Schubert and S. Hall, accepted for publication in J. Vac. Sci. Technol. 2011.
49. S. Zafar, A. Callegari, E. Gusev and M. V. Fischetti, J. Appl. Phys. **93**, 9298 (2003).
50. C. Z. Zhao, J. F. Zhang, M. B. Zahid, B. Govoreanu, G. Groeseneken and S. De Gendt, J. Appl. Phys., **100**, 093716 (2006).

ECS Transactions, 35 (4) 39-53 (2011)
10.1149/1.3572274 ©The Electrochemical Society

Process Engineering and Trap Distribution for Dielectric/Si Interfacial Layer in High-k gated MOS Devices

Kuei-Shu Chang-Liao, Chung-Hao Fu, Chun-Chang Lu, Yu-An Chang, Ya-Yin Hsu, Che-Hao Tsao, Tien-Ko Wang, Da-Wei Heh[1], Y.C. Li[2], Wen-Fa Tsai[2], Chi-Fong Ai[2], Fu-Chung Hou[2], and Yao-Tung Hsu[2]

Department of Engineering and System Science, National Tsing Hua University, Hsinchu 30013, Taiwan, R.O.C.
[1]National Nano Device Laboratories, Hsinchu 300, Taiwan, R.O.C.
[2]Institution of Nuclear Energy Research, Taoyuan 32546, Taiwan, R.O.C.

The effects of interfacial layer at high-k dielectric/Si substrate formed by using stress-relieved pre-oxide (SRPO) treatment on electrical characteristics of MOS devices were studied in this work. The equivalent oxide thickness value could be scaled with reducing the thickness of the high quality IL. The reliability in terms of stress-induced leakage and stress-induced V_{fb} shift is clearly improved for MOS device with a SRPO treatment. Besides, the constant-voltage stress-induced interface trap generation in MOSFET was measured by charge-pumping techniques. The influences of stress and recovery on devices with HfO_2 high-k dielectric are also compared. Results show that the stress induced V_{th} shifts can be separated into two stages, namely, trap filling and generation. The trap generation stage is only determined by the stress voltage and temperature.

Introduction

To enhance the performance of MOS device, the thickness of gate dielectric needs to be scaled. However, as the thickness of a traditional SiO_2 dielectric would lie below 2 nm, the gate leakage current would increase obviously and the power consumption would have been too large for practical applications. High-k gate dielectric is one of the key technologies for the nano-scale MOS device in controlling the leakage issue as the equivalent oxide thickness (EOT) is reduced to 1 nm and below [1-3]. For MOS device with high-k dielectric, the interfacial layer would affect many electrical characteristics such as mobility, transconductance, and reliability. For a low EOT value of MOS device, the thickness of interfacial layer (IL) should be as thin as possible. However for IL reduction, gate leakage current is largely increased, and mobility and reliabilities are degraded obviously [4, 5]. Besides, since the thickness of IL would often be increased after high temperature processes, the EOT value is hardly to be scaled [6]. It is an important issue to obtain a thin and high quality IL, which could keep good electrical characteristics for the scaled device. An IL using stress relieved pre-oxide (SRPO) method [7, 8] and post oxidation in-situ annealing in nitrogen is reported to reduce the density of interfacial defects generated during the oxidation process [9]. Thus, the potential application of IL with SRPO method for high-k gated MOS device was studied in this work. The interfacial layers with chemical oxide treatment by a H_2O_2 solution or SRPO on electrical characteristics of MOS devices are compared.

Although Hf based high-k gate dielectric is one of the most promising gate dielectric for high performance MOS devices, it still suffers from the low crystallization temperature [10], low channel mobility [11], and high bias temperature instability (BTI) degradation [12-14]. In this work, the distribution of interface traps of HfO_2 high-k dielectric are extracted by modified charge-pumping (CP) techniques [15-20]. Yet, the gate leakage current is the main problem as the CP technique is applied to measure a thin high-k gated MOS device. Hence, a method in reference [21, 22] is modified in this work to separate the CP current from the parasitic tunneling component in MOS devices with thin high-k gate dielectric. The influences of stress and recovery on nMOSFETs with HfO_2 high-k dielectric are compared. The threshold voltage instability under constant voltage stress (CVS) is observed by power law dependence with respect to stress and recovery time.

Experiment

Process of MOS capacitors

The MOS capacitors were fabricated on (100)-oriented 6-in P-type Si wafer with resistivity of 15-25 Ω.cm. After RCA clean, for SRPO samples, a 7.5 Å chemical oxide layers was formed by H_2O_2 solution and then an annealing treatment was performed at 900 °C in N_2 gas for 30 s. After annealing, the thickness of IL is increased to 14 Å, then it was etched back to 7.5 or 6 Å by dilute HF (D.I. water : Hf = 400 : 1). For chemical oxide samples, 7.5 Å and 6 Å thick chemical oxides (etched by dilute Hf) were prepared.

Next, for all samples, a 3 nm thick HfO_2 was deposited by a metal-organic chemical vapor deposition (MOCVD) at 500 °C. Then, a 50 nm thick TaN film was deposited by sputtering to serve as the metal gate. Subsequently, a post metal annealing (PMA) was carried out at 700 °C in N_2 gas for 30 s. After a 300 nm Al deposition, the Al/TaN was then patterned by a helicon- wave plasma etching. Finally, a sintering was conducted in a N_2/H_2 ambient at 400 °C for 30 min.

Measurement for MOS capacitors

The electrical measurement was performed on MOS capacitors with gate area of 100 $\times$ 100 μm^2 in this work. The HP4145B instrument was used to measure the current-voltage (I-V) characteristics and the high-frequency C-V measurements were carried out at 100 KHz by using HP4284A. The EOT and flat-band voltage (V_{fb}) were extracted from the simulation program considering quantum effect [23]. Besides, stress-induced leakage current (SILC) and stress-induced V_{fb} shift were measured to investigate the reliability of MOS capacitors. Thirty samples were measured for capacitance and leakage current data. The mean values were plotted in the figures and the variation of capacitance value is less than 5%.

Fig. 1 shows the structure and process sequence of MOS capacitors in this work. Tables I and II show various process conditions for MOS capacitors studied in this work. The thickness of oxide after annealing treatment is increased to 14 Å and the etch rate of the dilute HF is about 0.1 Å/s. The oxide thickness could be controlled precisely by the low and stable etch rate of the dilute HF.

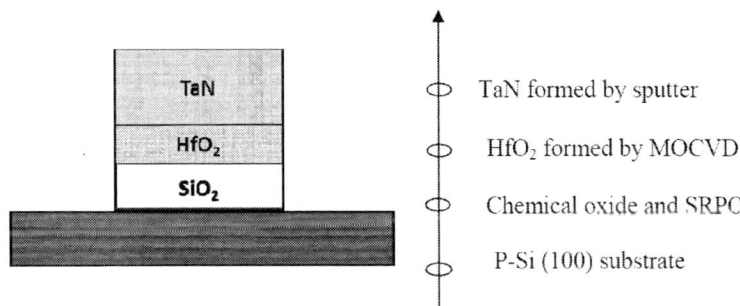

Figure 1. Structure and process sequence of MOS capacitors in this work.

Table I. Annealing and etching time conditions for various interfacial layers.

	Initial thickness	900 °C annealing	Thickness (After annealing)	400:1 Hf etching	Final thickness
Chemical oxide	7.5Å	X	7.5 Å	X	7.5 Å
Chemical oxide	7.5 Å	X	7.5 Å	10s	6 Å
SRPO	7.5 Å	V	14 Å	50s	7.5 Å
SRPO	7.5 Å	V	14 Å	60s	6 Å

Table II. MOS capacitors with various IL process conditions in this work.

Sample	Interfacial layer	HfO$_2$	Gate	PMA (°C)
G1	Chemical oxide 7.5 Å	30 Å	TaN	700
G2	Chemical oxide 6 Å	300 °C	(500 Å)	
G3	SRPO 7.5 Å	90 sec		
G4	SRPO 6 Å			

<u>MOSFETs and parameters for charge pumping measurement</u>

The gate dielectric of MOSFET devices studied in this work is HfO$_2$ high-k formed by an atomic layer deposition (ALD) with SiO$_2$ interfacial layer below 1.0 nm. The physical thickness of deposited HfO$_2$ dielectric is 2.0 nm. The dimension of MOSFET devices measured in this work is W/L = 10/0.5 μm. For the measurement of energy distribution of interface trap density, the rise/fall times of gate pulse are varied from 100 ns to 1 μs for frequency of 500 kHz and voltage swing is fixed at 1.2 V. The CVS is applied with gate voltage overdrive 1.3 V (V_g-V_{th} = 1.3 V), and 0 V (V_g = 0 V) for recovery.

Results and Discussion

Effects of various IL treatments on capacitor-voltage curves

Fig. 2 shows the measured high-frequency and the simulated low-frequency capacitance–voltage (C–V) curves of MOS capacitors with different thicknesses of chemical oxide or SRPO treatment. The measured data match well with the simulated ones, indicating the interface quality of MOS device is good and both kinds of interfacial layers could provide good electrical properties.

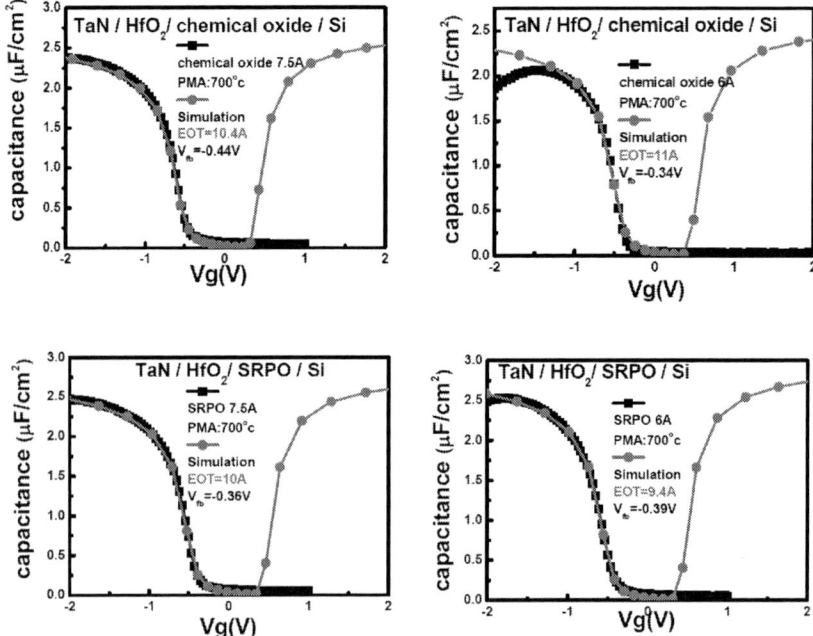

Figure 2. Experimental and simulated C-V curves for MOS devices with different thicknesses of chemical oxide or SRPO treatment.

Fig. 3 shows the high-frequency C-V curves for MOS device with different thicknesses of chemical oxide or SRPO treatment. In the accumulation region, the capacitance values for samples with SRPO are larger than those with only chemical oxide. Especially for 6 Å thick interfacial layer, CV curves of samples with traditional chemical oxides are clearly distorted in the accumulation region. This indicates that more defects may be generated and more leakage paths are formed in the dielectric of sample with 6 Å thick chemical oxide.

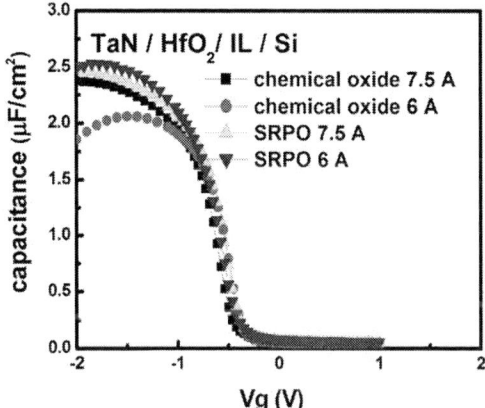

Figure 3. C-V curves for MOS device with different thicknesses of chemical oxide or SRPO treatment.

Fig. 4 shows the comparison of EOT and leakage current density for MOS devices with different thicknesses of chemical oxide or SRPO treatment. After the IL thickness of SRPO reduces from 7.5 to 6 Å, the EOT value of MOS device reduces 0.6 Å. However, for the sample with chemical oxide, the EOT value increases from 10.4 to 11 Å and the leakage current also increases as the thickness of chemical oxide reduces. This may be resulted from the re-formation of IL after PMA treatment because of oxygen diffusing from metal gate and through IL. The quality of additional IL formed by high temperature annealing is poor and not uniform. Since more defects in the IL would be formed during PMA treatment, more leakage current paths would pass through the dielectric. On the other hand, for samples with SRPO, the bonding between IL and Si substrate is stronger and fewer defects exist in the IL. Thus, oxygen is hard to diffuse through the IL for forming additional IL. As reported before [24], the SRPO formed by N_2 annealing would reduce the leakage current.

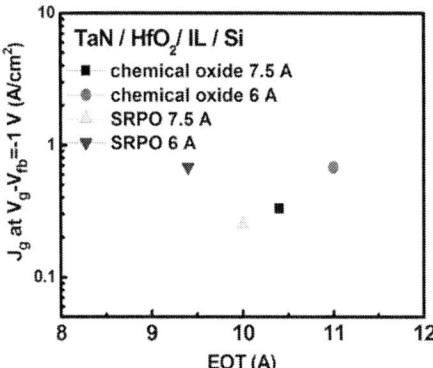

Figure 4. Comparison of EOT and leakage current density for MOS devices with different thicknesses of chemical oxide or SRPO treatment.

Effects of various IL treatments on hysteresis and uniformity

Fig. 5 shows the hysteresis effects of MOS devices with different thicknesses of chemical oxide or SRPO treatment as gate voltage sweeps from 2 to −2 V and then returns to 2 V. Ten samples were measured for hysteresis data. The hysteresis values for samples with SRPO are much smaller than those with chemical oxide. This may be because the interfacial layer after high temperature annealing in N_2 could reduce the defect in oxide and then fewer traps are formed in the HfO_2 deposited on the SRPO [9, 25]. Besides, the defects at SiO_2/HfO_2 interface are fewer for SRPO sample. Thus, it is an important issue to keep the interface and dielectric quality after scaling EOT and IL. Among all samples, MOS device with 6 Å SRPO has the lowest hysteresis value, indicating that for the sample with thin SRPO IL, the high-k dielectric shows good quality and it is suitable for scaling EOT value.

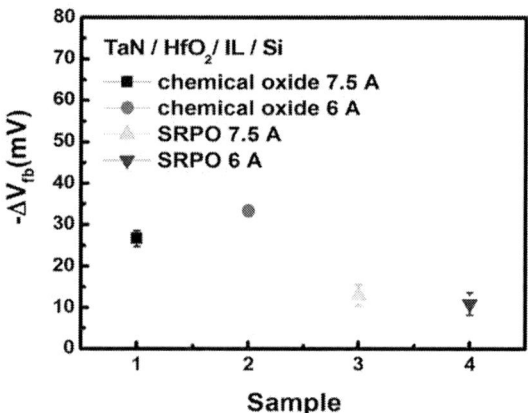

Figure 5. Hysteresis effect of MOS devices with different thicknesses of chemical oxide or SRPO treatment.

Fig. 6 shows the data distribution of leakage current density of MOS devices with different thicknesses of chemical oxide or SRPO treatment. For all samples, the uniformity is good. This is because the IL treatment was performed by a H_2O_2 solution and the thickness of chemical oxide would be saturated at 7.5~8 Å [26]. During the etching back process of the SRPO, the process could be controlled precisely since the etch rate of the dilute HF is small and stable [7]. Especially for the samples with SRPO, the variation of leakage current could be much reduced, indicating that the SRPO treatment could provide more stable electrical characteristics of MOS devices.

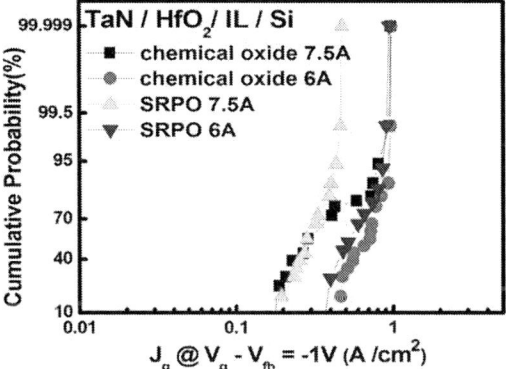

Figure 6. Cumulative probability of leakage current of MOS devices with various thicknesses of chemical oxide or rapid SRPO.

Effects of various IL treatments on reliability

Fig. 7 shows (a) stress-induced flat-band voltage (V_{fb}) shift and (b) stress-induced leakage current (SILC) of MOS devices with stress time under a constant electrical field stress of -14 MV/cm. Stress-induced ΔV_{fb} values for samples with SRPO are smaller than those with chemical oxide. This result suggests that the oxide traps in high-k dielectric could be reduced by a high quality IL and suppressing oxygen vacancy diffusion [27-29]. The SILC values for sample with SRPO treatment are clearly smaller as compared to those with chemical oxide. This is because after high temperature annealing, the atom bonding between IL and Si substrate is stronger and the defect generation rate would be suppressed during the stress [7]. Besides, since oxygen diffusion to the interface is suppressed by SRPO treatment, the generation of poor quality IL is minimized.

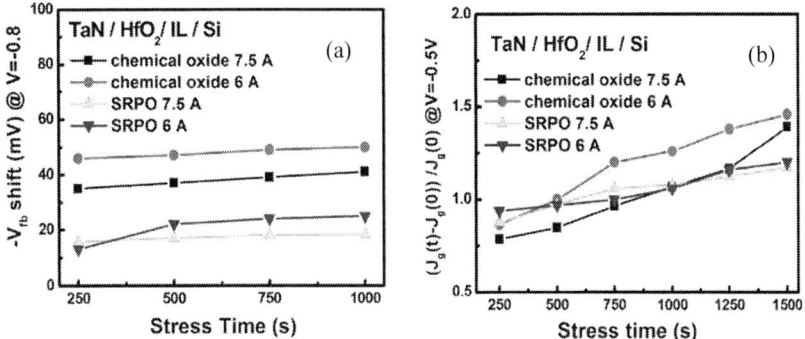

Figure 7. (a) stress-induced flat-band voltage (V_{fb}) shift and (b) stress-induced leakage current (SILC) of MOS devices with stress time under a constant electrical field stress of -14 MV/cm.

Electrical characterization on MOSFETS by charge pumping measurement

The gate dielectric of MOS device studied in this section is 3.0 nm thick HfO_2 formed by a sputter with SiO_2 interfacial layer about 1.0 nm. The gate electrodes are TaN/polysilicon stacks. In Fig. 8 below, the rise/fall times of gate pulse were fixed at 100 ns and voltage swing was fixed at 1.0 V for the CP current measurement. Since the tunneling component dominates CP measurement when the frequency is very low, we first measured CP curves at different gate pulse frequencies. Then, CP current measured at high frequency is subtracted by that at low frequency to directly obtain tunneling corrected CP data [21, 22]. For example, CP currents are measured at 1 MHz and 1 kHz, respectively. Then, CP currents measured at 1 MHz is subtracted by that at 1 kHz and we have the corrected 999 kHz CP data. Fig. 8 shows charge pumping current density (I_{cp}) vs. base level voltage (V_b) for CP w/o and with correction. Mark (□) is measured at 1 kHz without correction. Mark (●) is measured at 1 kHz, which is dominated by tunneling component. Mark (△) is the tunneling corrected CP data. A good correction of CP curve is obtained, indicating the tunneling component is completely eliminated. Although the tunneling component in CP current increases with increasing voltage swing, the well corrected CP currents can be obtained by subtracting low frequency CP current, which is dominated by tunneling component. For different gate voltage swings in the CP measurement, the proposed method shows similar correction effects as illustrated in Fig. 9, indicating it is promising for overcoming the bottleneck of CP measurement.

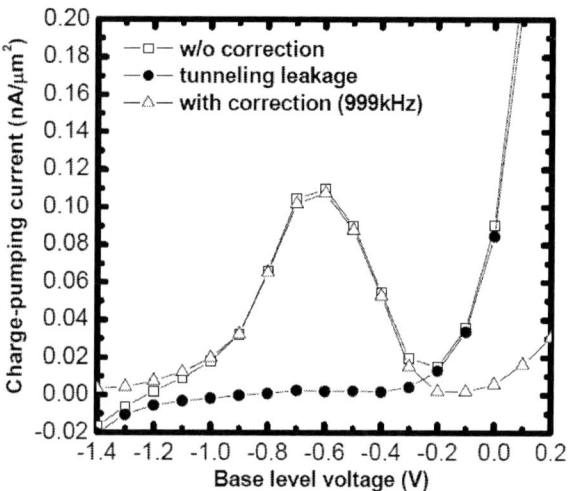

Figure 8. I_{cp} vs. V_b for CP w/o and with correction. (□) is measured at 1 MHz without correction. (●) is measured at 1 kHz, which is dominated by tunneling component. (△) is the tunneling corrected CP data.

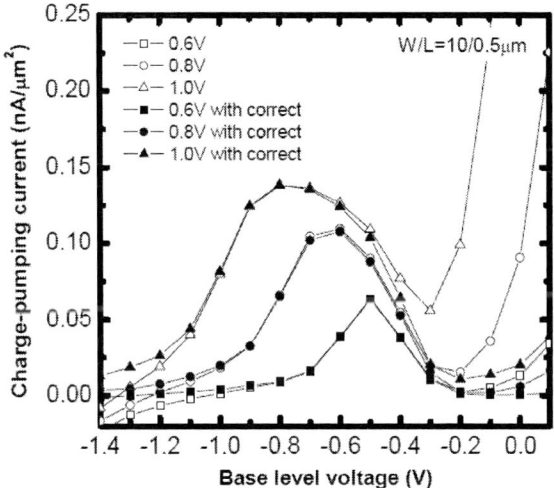

Figure 9. I_{cp} vs. V_b for CP using various gate voltage swings before and after correction.

The gate dielectric of MOS device studied in this section is 2.0 nm thick HfO_2 formed by an atomic layer deposition (ALD) with about 0.6~0.7 nm thick SiO_2 interfacial layer. The gate electrodes are TaN/polysilicon stacks. Halo implantation was used in the device fabrication. The total EOT is about 1.5 nm. Detailed flow for device fabrication can be found elsewhere [30]. In the following experiments, the CVS is applied with gate voltage fixed at $(V_g - V_{th}) = 1.3$ V, and the gate voltage is 0 V for recovery. In Fig. 10 below, the rise/fall times of gate pulse were fixed at 100 ns. Frequency was fixed at 1 MHz and voltage swing was fixed at 1.0 V for the CP current measurement. For the measurement of energy distribution of interface trap density, the rise/fall times of gate pulse were varied from 100 ns to 1 μs for frequency of 500 kHz and voltage swing was fixed at 1.0 V. Fig. 10 shows the CP current curves measured on the original and after 500 sec stressing for HfO_2 high-k gated device. It can be seen that CVS causes obvious increase in the $I_{cp,max}$ (maximum value of CP current). The value is almost double after CVS for 500 sec. It is suggested that there are many interface traps induced by CVS.

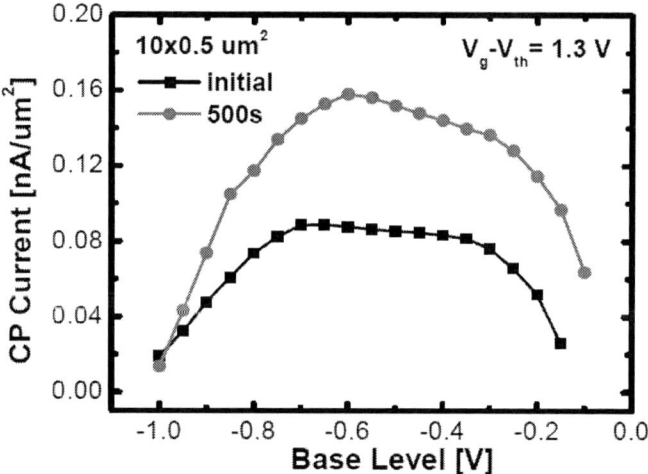

Figure 10. CP current curves measured on the original and after 500 sec stressing for HfO$_2$ high-k gated device.

Fig. 11 shows the energy distribution of interface trap density (D$_{it}$) measured on the original and after 500 sec stressing for HfO$_2$ high-k gated device. Consistent with Fig. 10, the D$_{it}$ increases a lot after 500 sec CVS. Besides, it can be seen from Fig. 11 that the energy distribution of interface trap density seems to move toward the conduction band and valence band, respectively. This phenomenon can be resulted from the variation of capture cross section of electron and hole caused by CVS.

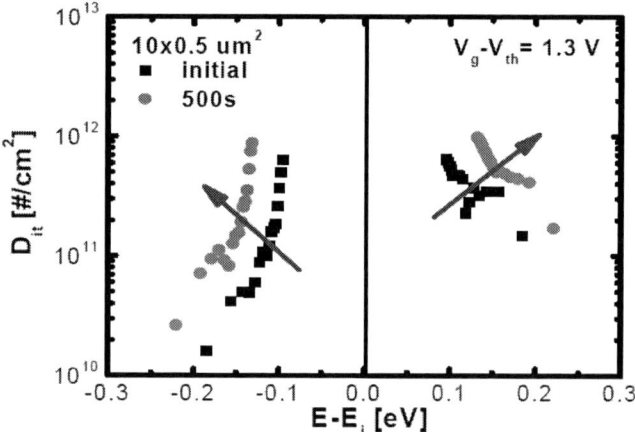

Figure 11. Energy distribution of interface trap density (D$_{it}$) measured on the original and after 500 sec stressing for HfO$_2$ high-k gated device.

Study of trap generation and recovery after stress and relax tests

Fig. 12 shows the ΔV_{th} (total threshold voltage shift), ΔV_{ot} (ΔV_{th} induced by oxide trap charge) and ΔV_{it} (ΔV_{th} induced by interface trap) extracted with respect to stress time in logarithm scale [31-33]. Results in Fig. 12 show that there are two stages observed on the V_{th} shift caused by CVS. The first stage is a fast process, which can be attributed to the fast charge filling of traps. The second stage is a slow process. The moderate voltage shifts is due to the additional trap generation. Also, it can be seen from Fig. 12 that the V_{th} shift resulted from interface traps is very little compared to that from oxide trapped charge.

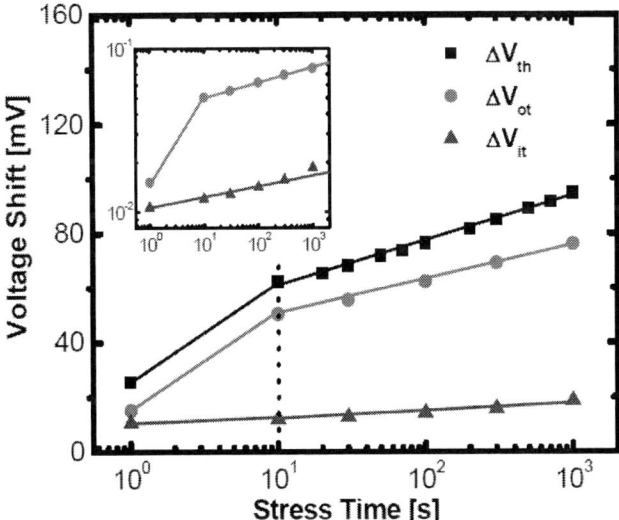

Figure 12. ΔV_{th} (total threshold voltage shift), V_{ot} (ΔV_{th} induced by oxide trap charge) and ΔV_{it} (ΔV_{th} induced by interface trap) extracted with respect to the stress time in logarithm scale.

Fig. 13 shows the time dependence of threshold voltage shifts under CVS and recovery cycle. The V_{th} shift induced by the second CVS seems smaller than that induced by the first CVS, but both of them can be partially recovered by the recovery cycles. Besides, the amounts of V_{th} shift recovered by the first and the second cycle are almost identical.

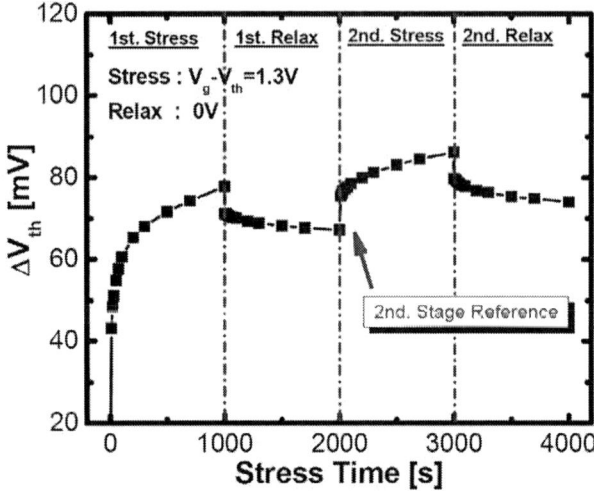

Figure 13. Time dependence of threshold voltage shifts under CVS and recovery cycle.

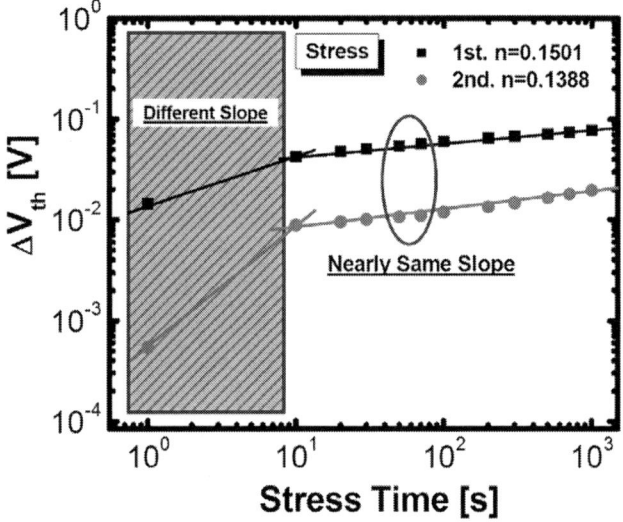

Figure 14. Comparison between V_{th} shifts induced by first and second CVS stresses.

Fig. 14 illustrates the comparison between V_{th} shifts induced by first and second CVS stresses. The most notable difference is in the first stage, i.e. trap filling process. Although the V_{th} shift of the second CVS shown in Fig. 14 is smaller than that of first CVS, the slope of second CVS is larger than that of first CVS instead. In other words, the trap filling process becomes faster after recovery. As shown in Fig. 14, the exponents of stage 2 are nearly identical. This indicates that the additional trap generation induced by CVS is not affected by the recovery cycle. It is only determined by the stress voltage and temperature.

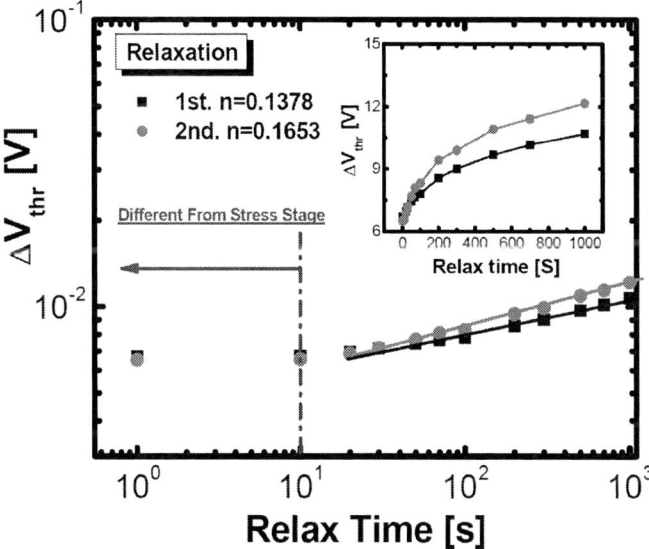

Figure 15. Comparison between V_{th} shifts induced by the first and the second recovery processes.

Fig. 15 shows the comparison between V_{th} shifts induced by the first and the second recovery processes. Being different from the stress stage, the recovery would delay about 10 sec, and then the V_{th} starts recovering. This might be attributed to the 0 V applied as recovery bias condition in this work. The fast recovery characteristics under negative bias were investigated in Ref. 33-35. Once the recovery starts, the recovery rate of the first and the second recovery are similar.

Conclusions

Electrical and reliability characteristics of MOS devices with various interfacial layers (IL) at high-k dielectric/Si substrate were studied in this work. The reduced leakage current and improved reliability for high-k gated MOS devices can be clearly achieved by applying stress-relieved pre-oxide (SRPO) treatment. The improvement can be attributed to the high quality IL formed by SRPO treatment, which suppresses the oxygen diffusion through the IL to form the additional IL during post deposition annealing. Therefore, the process engineering on IL is promising for high-k gated MOS device to further obtain the EOT scaling and reliability improvement.

Furthermore, the trap distribution in high-k dielectric of MOS device was characterized by a promising CP technique. The CVS induced interface trap generation is measured to study the reliability characteristics. The influences of stress and recovery on devices with HfO_2 high-k dielectric are also compared. Results show that the stress induced V_{th} shifts can be separated into two stages, trap filling and generation. The trap generation stage is only determined by the stress voltage and temperature. The CP technique is helpful to facilitate the process engineering of IL for reliability improvement.

Acknowledgments

The authors would like to thank the National Science Council of Taiwan, the Republic of China (R.O.C.) for financially supporting this research. The technical supports from National Nano Device Laboratories (NDL) of the Taiwan, R.O.C. and Nano Facility Center (NFC) of National Chiao Tung University, Taiwan, R.O.C., are also acknowledged.

References

1. J. H. Stathis and D. J. DiMaria, *Int. Electron. Dev. Meet Tech. Dig.*, 167 (1998).
2. G. Bersuker, J. Barnett, N. Moumen, B. Foran, C. D. Young, P. Lysaght, J. Peterson, B. H. Lee, P. M. Zeitzoff and H. R. Huff, *Jpn. J. Appl. Phys.* **43**, 7899, (2004).
3. M. H. Cho, Y. S. Roh, C. N. Whang, K. Jeong, S. W. Nahm, D. H. Ko, J. H. Lee, N. I. Lee, and K. Fujihara, *Appl. Phys. Lett.*, **81**, 472 (2002).
4. K. Choi, H. Jagannathan, C. Choi, L. Edge, T. Ando, M. Frank, P. Jamison, M. Wang, E. Cartier, S. Zafar, J. Bruley, A. Kerber, B. Linder, A. Callegari, Q. Yang, S. Brown, J. Stathis, J. Iacoponi, V. Paruchuri, and V. Narayanan, *in VLSI Symp. Tech. Dig.*, 138 (2009).
5. J. Huang, P. D. Kirsch, J. Oh, S. H. Lee, J. Price, P. Majhi, H.R. Harris, D. C. Gilmer, D. Q. Kelly, P. Sivasubramani, G. Bersuker, D. Heh, C. Young, C.S. Park, Y. N. Tan, N. Goel, C. Park, P.Y. Hung, P. Lysaght, K. J. Choi, B. J. Cho, H.-H. Tseng, B .H. Lee, and R. Jammy, *in VLSI Symp. Tech. Dig.*, 82, (2008).
6. K. Ramani, R.K. Singh, V. Craciun, *Microelectronic Eng.*, 1758 (2008).
7. S. P. Devireddy, B. Min, Z. Çelik-Butler, H. H. Tseng, P. J. Tobin, F. Wang, and A. Zlotnicka, *IEEE Trans. Electron Devices*, **53**, 538 (2006).
8. H. H. Tseng, P. J. Tobin, S. Kalpat,J. K. Schaeffer, M. E. Ramón, L. R. C. Fonseca, Z. X. Jiang, R. I. Hegde, D. H. Triyoso, and S. Semavedam, *IEEE Trans. Electron Devices*, **54**, 3267 (2007).
9. H. Jin, K. J. Weber and A. W. Blakers, *World Conference on Photovoltaic Energy Conversion (WCPEC)*, 1071 (2006).
10. W. Zhu, T. Tamagawa, M.Gibson, T. Furukawa, and T.P. Ma, *IEEE Electron Device Lett.*, **23**, 649 (2002).
11. E. P. Gusev, D.A. Buchanan, E. Cartier, A. Kumar, D. DiMaria, S. Guha, A. Callegari, S. Zafar, P.C. Jamison, D.A. Neumayer, M. Copel, M.A. Gribelyuk, H. Okorn-Schmidt, C. D'Emic, P. Kozlowski, K. Chan, N. Bojarczuk, L.-A. Ragnarsson, P. Ronsheim, K. Rim, R.J. Fleming, A. Mocuta, A. Ajmera, in *IEDM Tech. Dig.*, 20.1.1 (2001).
12. G. Ribes, J.Mitard, M. Denais, S. Bruyere, F. Monsieur, C. Parthasarathy, E. Vincent, and G. Ghibaudo, *IEEE Trans. Device and Materials Reliability*, **5**, 5 (2005).
13. A. Shanware, M.R. Visokay, J.J. Chambers, A.L.P. Rotondaro, J. McPherson, L. Colombo, G.A. Brown, C.H. Lee, Y. Kim, M. Gardner, R.W. Murto, in *IEDM Tech. Dig.*, 939 (2003).
14. Y. Yamamoto, K. Kita, K. Kyuno and A. Toriumi, *Appl. Phys. Lett.*, **89**, 032903 (2006).

15. C. Y. Lu, K. S. Chang-Liao, P. H. Tsai, and T. K. Wang, *IEEE Electron Device Lett.*, **27**, 859 (2006).
16. C. Y. Lu, K. S. Chang-Liao, C. C. Lu, P. H. Tsai, Y. Y. Kyi, T. K. Wang, *Microelectronic Engineering*, **85**, 20 (2008).
17. C. C. Lu, K. S. Chang-Liao, Y. F. Cheng, and T. K. Wang, *Microelectronic Engineering*, **86**, 1703 (2009).
18. X. M. Li, M. J. Deen, in *IEDM Tech. Dig.*, 85 (1990).
19. M. G. Ancona, N. S. Saks, D. McCarthy, *IEEE Trans. Electron Devices*,**35**, 2221 (1988).
20. M. Tsuchiaki, H. Hara, T. Morimoto, H. Iwai, *IEEE Trans. Electron Devices*, **40**, 1768 (1993).
21. P. Mason, Jean-Luc Autran, and Jean Brini, *IEEE Electron Device Lett.*, **20**, 92 (1999).
22. D. Bauza, *IEEE Electron Device Lett.*, **23**, 658 (2002).
23. J. R. Hauser, K. Ahmed. *Characterization Metrology ULSI Technology*, 230, (1998).
24. X. Guo, and T. P. Ma, *IEEE Electron Device Lett*, **19**, 207 (1998).
25. J. C. Wang, S. H. Chiao, C. L. Lee, T. F. Lei, Y. M. Lin, M. F. Wang, S. C. Chen, C. H. Yu, and M. S. Liang, *J. Appl. Phys.*, 3936 (2002).
26. G. D. Wilk, M.L. Green, M.-Y. Ho, B.W. Busch, T.W. Sorsch, F.P. Klemens, B. Brijs, R.B. van Dover, A. Kornblit, T. Gustafsson, E. Garfunkel, S. Hillenius, D. Monroe, P. Kalavade, and J.M. Hergenrother, *in VLSI Symp. Tech. Dig.*, 88 (2002).
27. W. L. Warren, D. M. Fleetwood, M. R. Shaneyfelt, J. R. Schwank, P. S. Winokur, R. A. B. Devine, and D. Mathiot, *Appl. Phys. Lett.*, 3452 (1994).
28. Q. Lu, K.P Cheung, N.A.Ciampa, C.T. Liu, C-P.Chang, J.I Colonell, W-Y-C.Lai, Liu, R. J. F. Miner, H.Vaidya, C-S Pai, and J.T. Clemens, *Int. Rel. Phys. Symp*, (IRPS), 369 (1999).
29. G. Bersuker, C. S. Park, J. Barnett, P. S. Lysaght, P. D. Kirsch, C. D. Young, R. Choi, and B. H. Lee, *J. Appl. Phys.* 094108 (2006).
30. P. F. Hsu, Y. T. Hou, F. Y. Yen, V. S. Chang, P. S. Lim, C. L. Hung, L. G. Yao, J. C. Jiang, H. J. Lin, J. M. Chiou, K. M. Yin, J. J. Lee, R. L. Hwang, Y. Jin, S. M. Chang, H. J. Tao, S. C. Chen, M. S. Liang, and T. P. Ma, in *VLSI Symp. Tech. Dig.*, 14 (2006).
31. E. Y. Wu, A. Vayshenker, E. Nowak, J. Sune, R. P. Vollerston, W. Lai, D. Harmon, *IEEE Trans. Electron Devices.*, **49**, 2244 (2002).
32. R. Degraeve, T. Kauerauf, M. Cho, M. Zahid, L-Å. Ragnarsson, D. P. Brunco, B. Kaczer, Ph. Roussel, S De Gendt, G. Groeseneken, in *IEDM Tech. Dig.*, 408 (2005).
33. D. Heh, C. D. Young, G. Bersuker, *IEEE Electron Device Lett.*, **29**, 180 (2008).
34. C. D. Young, S. Nadkarni, D. Heh, H.R. Harris, R. Choi, J.J. Peterson, J.H. Sim, S.A. Krishnan, J. Barnett, E. Vogel, B.H. Lee, P. Zeitzoff, G.A. Brown, and G. Bersuker, in *Int. Rel. Phys. Symp.*, 169 (2006).
35. K. T. Lee, C. Y. Kang, O. S. Yoo, R. Choi, B. H. Lee, Jack C. Lee, Hi-Deok Lee, Yoon-Ha Jeong, *IEEE Electron Device Lett.*,**29**, 389 (2008).

54

Current Understanding of the Transport Behavior of Hydrogen Species in MOS
Stacks and Their Relation to Reliability Degradation

Ziyuan Liu[a], Shinji Fujieda[b], Hirokazu Ishigaki[a],
Markus Wilde[c] and Katsuyuki Fukutani[c]

[a]Device & Analysis Technology Division, Renesas Electronics Corporation,
1753 Shimonumabe, Nakahara-ku, Kanagawa 211-8668, Japan
Email: ziyuan.liu.jc@renesas.com
[b]Green Innovation Research Laboratories, NEC Corporation,
34, Miyukigaoka, Tsukuba, Ibaraki 305-8501, Japan
[c]Institute of Industrial Science, University of Tokyo and CREST-JST
4-6-1 Komaba, Meguro-ku, Tokyo 153-8505, Japan

We review recent experiments that suggest a comprehensive model
for the reversible hydrogen (H) transport between the poly-Si
interface and the oxide/Si interface of MOS stacks. The H
diffusion in intact model MOS structures is probed by H depth
profiling via resonant ^{15}N-H nuclear reaction analysis (NRA). It is
demonstrated that MOS device degradation correlates with H
accumulation in the oxide/Si interface region. A specific ultra-thin
oxynitride is discovered in the poly-Si/oxynitride interface as well
as in the near-surface region of N_2-annealed nitride films and
shown to function as a potential H-storage layer. The interfacial
storage layer between the poly-Si gate and the oxynitride dielectric
of MOS transistors is found to contain two kinds of H species,
which are mobile and stable, respectively, versus irradiation-
induced relocation. The mobile H, if stimulated by energetic
carriers, migrates across the gate films and relocates to the SiO_2/Si
interface, causing device instabilities. Understanding this basic
hydrogen transport behavior allows conceiving fabrication
countermeasures to improve the device reliability.

Introduction

It is well recognized that hydrogen-related species (H) play an important but intricate role
in reliability issues of gate dielectrics. On one hand, H species present in the oxide
contribute to the superior charge-to-breakdown (Q_{bd}) quality of 'wet' oxides compared to
'dry' oxides (1); on the other hand, H has long been suspected to be a trigger for
dielectric breakdown and a detrimental cause for degradation processes such as negative-
bias-temperature instability (NBTI) and hot-carrier instability (HCI) (2-11). Presumably
these effects relate to the diffusivity of H and to its ability to passivate and to depassivate
dangling bonds in interfacial Si (12-16). Understanding the diffusion of H between the
layers in MOS stacks and its behavior as it approaches the oxide/Si interface is therefore
exceedingly important.

In practice, the inclusion of H impurities is inevitable in current manufacturing
processes, e.g., H_2 forming gas anneal and formation of poly-Si from SiH_4 precursors, as

well as repeated air-contact between different process steps. Beside this 'native' H resident in MOS structures, H has also been shown to penetrate into MOS stacks during back-end processing, which may result in significant degradation of the device reliability (17, 18).

In order to meet the requirements for downscaling, boron penetration into the gate dielectrics from p-type poly-Si electrodes has to be suppressed and the gate leakage current needs to be improved while maintaining the gate capacitance. To this end, nitrogen (N) was introduced into the oxide using various nitridation processes such as NO and N$_2$O gas annealing as well as plasma nitridation. N-incorporation, however, was found to severely augment the H-related degradation (19). This degradation behavior is extremely sensitive to the nitridation conditions, essentially depending on the resulting N concentration distribution near the poly-Si and the SiO$_2$/Si interfaces (20-23). Based on extensive investigations of oxynitride (SiON), we expected that the H diffusion behavior might correlate to the N configuration in oxynitrides. We focused on this concept and consequently investigated the H impurity diffusivity in relation to the N distribution.

A number of previous investigations have used H depth profiling by either familiar secondary ion mass spectroscopy (SIMS) (24) or nuclear reaction analysis (NRA) (15, 25-27) to study H in MOS devices. Among these techniques, only NRA has the principal capability to detect H transport phenomena in *intact* MOS stacks. NRA leaves the target material structure undestroyed and only partially perturbs the original H-distribution due to ion irradiation-induced relocation (27, 28). Rather than being merely detrimental, this effect in fact allows monitoring the H-diffusion as it would likely occur in actual devices under influence of operating stress (energetic carrier flow). By utilizing NRA H depth profiling in this way, we have recently visualized the H transport across MOS stacks successfully and could directly observe the (often postulated but rarely proved) H-accumulation near the SiO$_2$/Si interface (18, 29, 30).

In this paper, we review recent experimental investigations of the H transport in MOS stacks including the basic SiO$_2$/Si structure, N$_2$-annealed nitride/oxide stack (SiN/SiO$_2$/Si), as well as poly-Si/SiON stack. The picture emerging from the results allows constructing a comprehensive model of the reversible H transport between the top poly-Si and the bottom SiO$_2$/Si interface of MOS structures. The concomitant electrical performance characterization of corresponding MOS devices then allows drawing conclusions regarding the consequences of the internal H distribution behavior for the device reliability. On the basis of our findings, novel fabrication schemes may be devised that comprise countermeasures against the latent H redistribution trends and thus may ultimately enable the production of more reliable and durable devices.

Probing Technique for Hydrogen Redistribution in MOS-like Stacked Layers

H depth profiling by nuclear reaction analysis (NRA) was applied to probe the H-redistribution in stacked layer targets modeling decisive parts of the MOS structure. Basic details of the NRA method for hydrogen depth profiling are published elsewhere (27, 31). Briefly, the investigated material is irradiated by a ^{15}N ion beam to induce the resonant ^{1}H(^{15}N,$\alpha\gamma$)^{12}C nuclear reaction with H at 6.385 MeV (E_{res}). At a given incident ^{15}N ion energy (E_i), the emitted γ-radiation signal is proportional to the H concentration at a probing depth z, which is defined by the stopping power (dE/dz) of the material: $z = (E_i - E_{res})/(dE/dz)$. Due to vibration Doppler broadening and energy straggling the NRA 'depth profiles' (obtained by scanning E_i) presented in the following do not directly represent actual H concentration distributions, but rather their convolution with a Gaussian shaped ^{15}N ion energy distribution. The Gaussian energy width limits the NRA depth resolution, which hence decreases at larger probing depths due to straggling. In the near-surface region the resolution can be optimized to below 1.5 nm by glancing ion incidence.

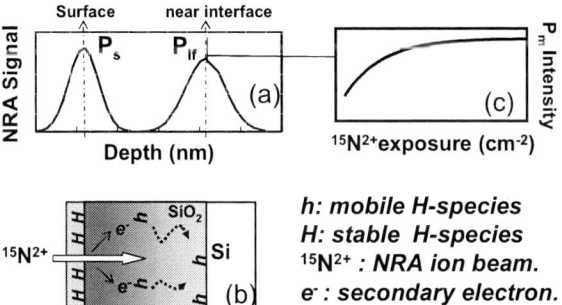

Figure 1. Schematic of NRA H profiling in SiO₂/Si; (a) NRA H depth profile, (b) H-accumulation near SiO₂/Si interface; (c) the secondary electron-induced H redistribution in the stack target caused by the ^{15}N ion irradiations.

Figure 1 (a) schematically illustrates a NRA H depth profile of a 25 nm thick SiO₂ layer on Si. Two well-separated peaked signals denoted P_s and P_{if} indicate thin H-rich layers near the surface and in the SiO₂/Si interfacial region (27). It is important to note that the H-distribution in the target under NRA investigation is not static. The irradiation with MeV ^{15}N ions induces H-redistribution in SiO₂ and SiN films, resulting in H-accumulation near the SiO₂/Si interfaces (18, 27, 28). This phenomenon is understood as follows: the cross section of the nuclear reaction is so small (order of 10^{-24} cm²) that only few ^{15}N ions react with H to produce γ-rays. The dominant ion beam interaction with the sample material is electronic stopping, which generates large numbers of secondary electrons. These energetic electrons cause detachment of H from various binding sites, thereby creating mobile H species (**Fig. 1** (b)). The mobilized H species may diffuse and partially accumulate in a transition layer near the SiO₂/Si interface, where a large number of preferable H interaction sites exists (32, 33). By probing H in the interfacial region on fresh sample area, the H uptake can thus be monitored in real time (**Fig. 1** (c), **Fig. 2**).

To illustrate the ion irradiation effect, **Figure 2** (a) compares NRA H depth profiles of as-grown SiN(10nm)/SiO₂(25nm) stacks recorded in the beginning and after saturation of the ^{15}N ion-induced H-redistribution. The data collected in the initial state (beginning, i.e. small ^{15}N ion dose) represent the near-original H-distribution of the target, while those recorded after the H-diffusion reached saturation reflects the redistributed depth profiles. It can be seen that the signal corresponding to H in the SiN film decreases considerably during NRA profiling, while the H peak near the SiO₂/Si interface increases and finally reaches a saturated level (**Fig. 2** (b)). The H-redistribution in the SiN/SiO₂ stack thus proceeds by depletion of H from the SiN with a simultaneous H pile-up near the Si/SiO_2 interface. This observation is direct evidence for energetic electron-induced H diffusion and for transport of mobile H species across both the SiN and the SiO_2 films, which finally results in H accumulation near the SiO_2/Si interface. It is believed that this H-redistribution reflects the situation that would likely occur in active MOS devices under operating stress.

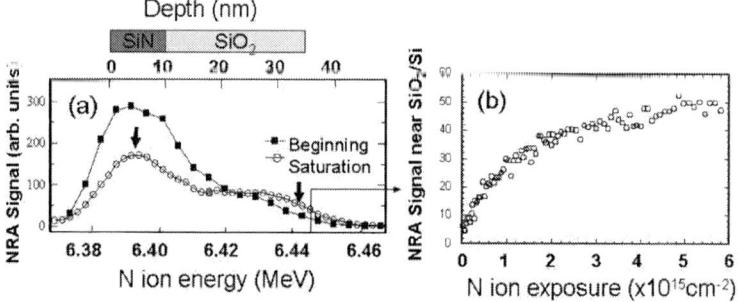

Figure 2. Comparison of NRA H depth profiles in as-grown SiN/SiO_2 stacks before and after H-redistribution (a); H-uptake in the SiO_2 /Si interface (b).

As an analytical method, we may thus apply the effect of the ^{15}N ion irradiation to evaluate the H diffusivity in layered stacks, consisting of a H-rich upper layer on SiO_2 /Si, to generate freely diffusing H species within the upper layer, while simultaneously probing the uptake of the mobilized H near the SiO_2/Si interface. The large quantities of H included in the upper layer act as H reservoir, and the near-interfacial H traps in the SiO_2 film serve as H-acceptors. Possible applications of this NRA technique are the following:

   a) By determining the initial H-concentration value at the beginning of the H-relocation, probed at a certain depth, the original H-distribution can be assessed.
   b) By observing the H depth profile after saturation of the H diffusion, H species can be distinguished into stable H and mobile H. Unlike mobile H, stable H species do not migrate to the bottom layer even under NRA ion irradiation.
   c) By monitoring the H transport from the upper H-rich layer to the SiO_2/Si interface, the stable H retained in the H-storage layer can be determined.
   d) By observing how closely H approaches the SiO_2/Si interface, the position of the accumulation region center can be assessed and locations of H diffusion barriers can be identified.

Relationship Between Interfacial H Accumulation, NBTI and POA in SiO_2

Roy et al. demonstrated that a post-oxidation anneal (POA) process (treating a pregrown SiO_2 film at temperatures above the viscoelastic temperature, typically at 1213-1323 K, in N_2 with <0.1% O_2 ambient) delivers significant enhancement in device reliability, especially with respect to NBTI (34). The POA effect was attributed to the formation of a strain-free and planar Si/SiO$_2$ interface. In this section we focus on the relationship between H diffusion, NBTI and POA in oxide films.

Poly-Si/SiO$_2$/Si MOS diodes isolated by local oxidation of silicon were fabricated on two p-type Si(100) wafers. Gate oxides (25nm) on both wafers were prepared by wet oxidation. One of the two wafers was subsequently treated by POA (rapid thermal anneal at 1050°C, 120 sec, in N$_2$ with <0.1% O$_2$). An arsenic-doped poly-Si layer was used as the electrode. Diodes of 0.4×0.4 mm^2 and 5×15 mm^2 areas were arrayed on a chip, the former for capacitance-voltage (CV) and the latter for NRA measurements. Negative-bias-temperature stress (NBTS) and zero-bias temperature stress (ZBTS) were applied on the diodes. The stress conditions were set as −5MV/cm for NBTS and 0 MV/cm for ZBTS at 360°C for 6 hours in vacuum. ZBTS is a control experiment to verify the importance of the bias during NBTS, singling out the influence of temperature. Vacuum annealing below 400 °C has negligible effects on the activation of interface defects (35). After stress application, the poly Si electrode covering the oxide in the 5×15 mm^2 area diode was etched off chemically to enhance the depth resolution of the NRA analysis. Quasi-static CV measurements were performed before and after the stress application.

CV data are shown in **Fig. 3**. The result observed in ZBTS samples is the same as that before stress application, representing ideal CV curves of MOS capacitors. The CV curves observed for the NBTS samples however appear distorted, especially for the sample without POA, obviously due to interface state (D$_{it}$) generation and positive oxide-fixed charge. The mid-gap D$_{it}$ increase was estimated to be 9.7×10^{11}/cm^2 eV and 1.9×10^{11}/cm^{2}eV for samples without and with POA treatment, respectively. Clearly,

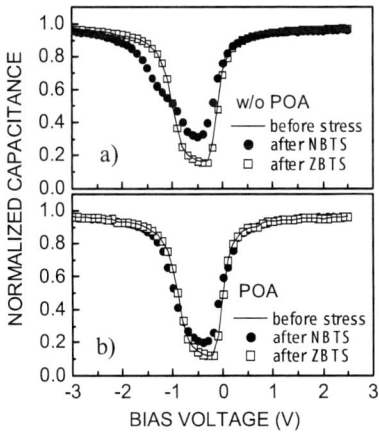

Figure 3. CV curves of poly-Si/SiO$_2$/Si diodes before and after application of ZBTS (0 MV/cm at 360°C for 6 h) and NBTS (-5 MV/cm at 360°C for 6 h). In (a) the oxide layer was not POA (post-oxidation anneal at above 1000°C in N$_2$ with <0.1% O$_2$ ambient) treated, while the samples in (b) received the POA treatment.

devices degraded only due to NBTS (not ZBTS), and the interface state density induced by NBTS was reduced to one fifth after the POA process. The CV result is consistent with previous reports (19, 34).

In order to compare the near-interface H density before and after NBTS, without the effects of NRA ion irradiation, the interfacial H intensities were measured by restricting ion exposures below a threshold of $3 \times 10^{15}/cm^2$, which did not noticeably change the intensity of the near-interface H peak even in native oxides.

Figure 4 shows the comparison of the interfacial H intensities for samples with and without POA treatment. For the sample without POA treatment, the intensity in the ZBTS sample is the same as that in the sample without exposure to stress, while the H concentration of the NBTS sample is about 5 times as large as that before stress application. This result indicates that NBTS induces H redistribution within the gate oxide resulting in H accumulation near the interface. On the other hand, the H concentration in the near-interface region of POA-treated samples is initially slightly lower than in as-grown and ZBTS samples without POA treatment. Like in the latter, the near-interfacial H concentration is increased after NBTS, but only to about 3 times the initial value. Clearly, the NBTS-induced H accumulation in the near-interface region was significantly suppressed by the POA. This trend is consistent with the tendency of the CV curve distortion shown in **Fig. 3**. It therefore can be concluded that H redistribution and accumulation in oxide/Si interface relate to device degradation caused by NBTS, and which is at least partly suppressed by POA.

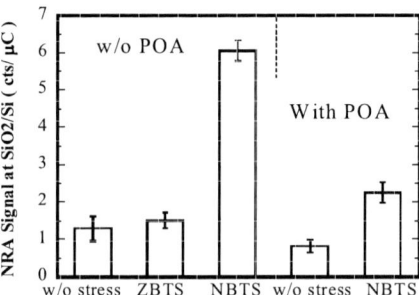

Figure 4. Comparison of the initial hydrogen intensity measured at the near-interface peak maximum of samples with and without POA treatment for different stress conditions.

The NRA data experimentally demonstrate that NBTS causes accumulation of H species in the near-interface region, and that the total amount of accumulated H ($\sim 1.0 \times 10^{14}/cm^2$) is significantly larger than the number of NBTS-induced interface states ($\sim 10^{12}/cm^2$). Clearly, only a very small fraction of the accumulated H can be attributed to the H released from the interfacial Si-H bonds and the majority of the accumulated H species must have a different, not yet clarified origin. The questions related to the NBTS mechanism raised by this study therefore concern the source and nature of the mobile H species and the reason for the improvement by POA on the NBTI.

Hydrogen Transport Between the H-storing SiON Layer and the SiO₂/Si Interface

After repeatedly observing that MOS device degradation is connected to significant H accumulation in the oxide/Si interface region (29, 30, 36), we realized that the H diffusion should trace back to a source of mobile H. In this part we focus on the source of the diffusing H species that approach the oxide/Si interface.

1. Hydrogen Storage Layer on the Surface of N_2-annealed Nitride Films

It was reported that H species prefer to locate in an oxynitride transition layer at the interface between the top oxide and the nitride layer of oxide-nitride-oxide (ONO) stacks (37). This buried H-rich interface layer, can apparently, be created through a 1000°C N_2-anneal step prior the top oxide formation (30). In the region just below the surface of N_2-annealed (furnace annealing around 1000°C, 1h, in N_2 with <1% O_2) SiN/SiO₂/Si stacks, we discovered a specific thin (<1 nm) oxynitride layer of Si_2N_2O-like composition, which was found to have highly efficient H-storage properties (38).

To analyze the surface-near elemental composition, high resolution Rutherford backscattering (HR-RBS) spectra of the as-grown and near 1000°C N_2-annealed SiN/SiO₂ stacks were recorded in channeling-mode. These are shown in **Fig. 5**. The peak at highest energies is due to oxygen in the surface, while the peak at 274-286 keV originates from nitrogen in the Si_3N_4 layer. The peak between the nitrogen and the surface oxygen corresponds to the oxygen in the bottom oxide. A small carbon signal due to surface

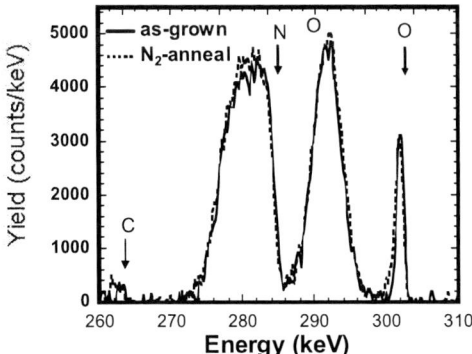

Figure 5. Channeling-mode HR-RBS spectra of SiN (8nm)/SiO₂ (4nm) stacks in as-grown and near 1000°C N_2-annealed conditions. The arrows identify the energy position of surface oxygen and nitrogen, respectively.

contamination was also observed. These data indicate the existence of a 'native' oxide on both SiN films due to air exposure. Note that a small amount of oxygen (low energy shoulder in the signal at 300 keV) was incorporated into a very shallow region (<0.5 nm) beneath the native oxide of the N_2-annealed Si_3N_4, which probably originated from oxygen impurities in the flowing N_2. HR-RBS spectra taken in random mode also confirm this oxygen incorporation. The amount of increased oxygen (6×10^{14} atoms/cm^2) is close to 1 monolayer of Si (100), which implies the possibility that during the N_2-anneal a SiO_xN_y monolayer was formed underneath the native oxide.

Figure 6 shows NRA H-depth profiles of N_2-annealed SiN/SiO$_2$ specimens (No. 1-3) that originated from the same wafer but received different *in situ* treatments in the vacuum chamber before the NRA measurements. No. 1 refers to the sample as it was installed from air. No. 2 is the specimen first heated in UHV at 750°C under a pressure of 10-5 mbar for 5 min, without admitting air to the sample before the NRA measurement. In contrast, No.3 is the No.2 specimen, but exposed to air again for 8 hours prior to NRA. The H depth profiles exhibit two distinctly peaked signals. The peak (Ps) centered at a depth <1.0 nm is attributed to H retaining by a ultra thin SiON layer on the SiN surface, while the peak (Pm) close to the SiO$_2$/Si interface corresponds to accumulated H, which was initially present in the SiN surface but migrated to the SiO$_2$/Si interface under the NRA ion irradiation. **Figure 6** indicates that H is predominantly present near the surface of the N_2-anneal SiN films (not in the SiN film). In case of the specimens that had air contact prior to NRA (No. 1 and 3), this region mainly contains mobile H species that relocate to the SiO$_2$/Si interface as indicated by the pronounced Pm peaks. In contrast, the Ps peak in the H-depth profile of the vacuum-annealed sample without air contact (No.2) is narrow, nearly symmetric, and centers slightly beneath the surface. The Pm peak is much smaller. This demonstrates the formation of a ~0.5 nm thin near-surface layer that contains H species at a concentration of ~10^{20}cm^{-3}, which are stable against ion beam-induced relocation. Before air admission, the penetration of this stable H from the storage

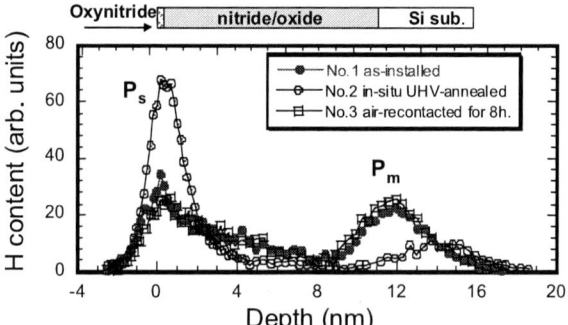

Figure 6. NRA hydrogen depth profiles for near 1000°C N_2-annealed S_3N_4(8nm)/SiO$_2$(4nm) stacks. Sample conditions prior to the NRA measurement (performed in vacuum) were: No.1 (as installed from air); No.2 (*in-situ* vacuum-annealed at 750°C) and No.3 (re-exposed to air after vacuum-annealing).

layer into the bottom oxide is strongly suppressed. These NRA results combined with layer structure and composition analysis (38) demonstrate that a thin, thermally stable, yet air-sensitive Si_2N_2O-like layer (39) is formed just beneath the native surface oxidelayer on N_2-annealed SiN. This layer stores stable H species, which are resistant to energetic electron damage because apparently their diffusion into the bottom oxide is strongly suppressed.

The formation of the surface H-storage layer due to the N_2-anneal becomes evident by comparing the NRA H depth profiles of as-grown and N_2-annealed SiN/SiO_2 stacks shown in **Fig. 7**. The as-grown SiN film exhibits a broad Pn feature at the center of SiN layer, which indicates that the H distribution in the nitride film is nearly homogeneous. The H signal observed near the SiO_2/Si interface corresponds to mobile and hence relocated H. After N_2-annealing, the nitride film as well as the bottom oxide are almost completely swept clear of H. Only a sharp peak Ps remains near the surface, which represents a thin layer holding H at a high concentration. We refer to this near-surface layer as 'H-storage layer' and believe that it is highly desirable for MOS stacks because of its ability to retain stable, i.e., relocation-resistant H species.

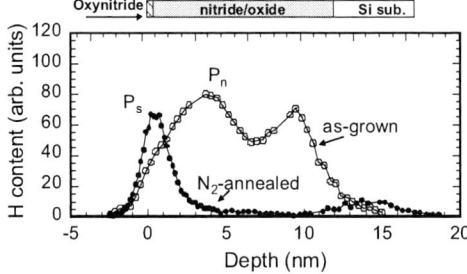

Figure 7. Comparison of NRA H depth profiles between the as-grown and the N_2-annealed SiN/SiO_2 stacks, where letter received *in-situ* UHV anneal prior NRA measurements. Note that the bottom oxide in the N_2-annealed specimen is almost completely swept clear of hydrogen.

2. Mobile and Stable Hydrogen Species

Figure 8 shows the H depth profile and H diffusion kinetics obtained from the N_2-annealed SiN/SiO_2 stack with different in situ treatments in the NRA vacuum system. **Figure 8** (b) provides clear evidence that the mobile H observed near the SiO_2/Si initially resided in the Si_2N_2O-like surface layer but migrated to the SiO_2/Si interface due to the NRA ion irradiation. Note that the proportion of mobile H (Pm) significantly decreased while the stable H (Ps) increased due to in situ UHV anneal. The specific surface oxynitride layer apparently contains two kinds of H species (mobile and stable) of different mobility, which can be converted into each other by thermal stimulation or air contact.

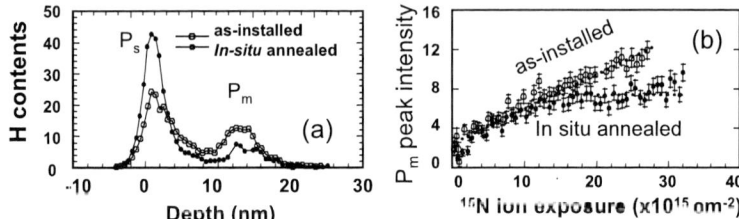

Figure 8. The NRA H-profiles (a), and the intensity variations of mobile H (P_m) with ion exposure (b). The data are observed from the same sample measured as installed (open circles) and after *in-situ* vacuum anneal (full circles), respectively.

Figure 9 shows the variation of the mobile H peak (Pm) intensity while the sample is heated under continuous NRA observation. The measurement starts (t=0) from the Pm peak saturation level at room temperature and monitors the H signal at different temperatures. The mobile H intensity localized in the oxide/Si interface decreases between ~370°C and ~400°C. After a significant reduction of the H intensity above 400°C the heater was stopped. Note that the reduced Pm H intensity almost recovers back to its saturation level after the heater is turned off and the sample cools to room temperature. Since the Pm intensity reflects the balance of the H transport back and forth between the H storage layer and the SiO_2/Si interface, this result means that H is preferably bound in the H storage layer, but migrates into the SiO_2/Si interface when the sample is irradiated by energetic charges. The heating process transfers H back to the H-storage layer.

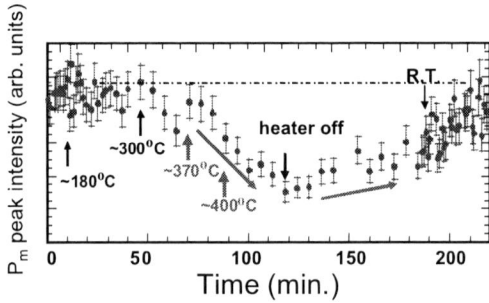

Figure 9. Saturated mobile H intensities variation of the *in-situ* annealed SiN/SiO_2 stack specimen at different temperatures.

Based on above results, we propose a model for the reversible H migration as illustrated in **Fig. 10**. Beside schematic stages of the H-relocation as observed by NRA, (a), (b), and (c), the energy levels of the migrating H species at specific locations within the oxynitride/Si stack are illustrated on the right side. H initially resides in the most stable sites in the H-storing SiON layer near the surface. Energetic electrons (e.g., from the NRA ion irradiation) excite H into a mobile state that allows diffusion within the stack interior to trap sites near the Si interface. Heating thermally activates the H escape from these interfacial traps into the mobile diffusion state and hence promotes the H return back to the near-surface storage layer. This characteristic is important because oxynitride dielectrics usually endure both electrical and temperature stress.

Having established the concept of the H-storage layer and the stable H species it contains, we now recall the POA ($1050°C$, N_2 with <0.1% O_2) effects on the NBTI. As described in the previous section, POA suppresses the H redistribution. The POA process conditions resemble those of the N_2-anneal ($1000°C$, N_2 with <0.1% O_2) in terms of temperature (near $1000°C$) and low oxygen partial pressure (<0.1%), which both limit the film thickness increase to < 0.1 nm. The POA effect may therefore be expected to relate to the H-storage layer formation near the surface of the oxide films and may possibly cause a reduction of mobile H species. Further study and understanding of the H species and their stability in the oxide film is required to clarify this important issue.

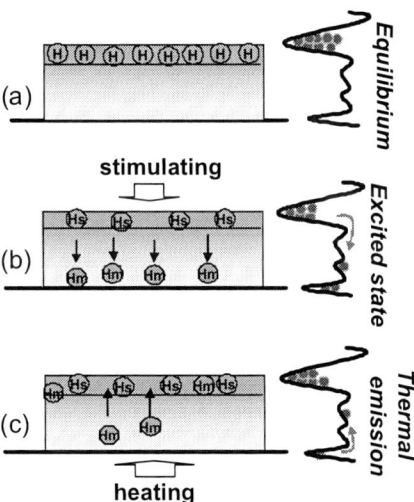

Figure 10. Illustrations of the reversible H transport back and forth between the H-storage layer and the oxide/Si interface depending on the temperature and stimulation by energetic charge irradiation. The right side shows corresponding energy levels of H in the storage layer, the dielectric interior and the Si interface, respectively.

3. Stable H–storage at the Poly-Si/oxynitride Interface

The poly-Si/SiON interface resembles the H-storing surface layer on the N_2-annealed SiN/SiO$_2$ stack in the fact that it apparently contains stable and mobile H species as well (28). **Figure 11** (a) and (b) show thermal desorption spectroscopy (TDS) spectra of H$_2$ (m/e=2) and SiO (m/e=44) species obtained from the as-fabricated SiON films, formed by plasma-nitridation, and from the same material that underwent deposition of an ultra thin poly-Si layer. The deposited Si mimics the poly-Si interface in intense reactive contact with the SiON, where the surface Si is indistinguishable from the underlying oxynitride. These TDS data reveal that H predominantly desorbs below 600°C (the peaks of β_1, β_2) for the as-fabricated oxynitride, but an additional peak at a temperature above 1100°C appears after poly-Si nucleation. This indicates the ability of the poly-Si interface to retain thermally highly stable H species. Note that this stable H species desorbs combined with SiO species above 1100°C. H bonds in ordinary Si-H, N-H, and O-H units cannot explain such high desorption temperatures. We suggest instead that the stability of these H species is attributable to the high thermal stability of the interfacial layer matrix, rather than to the H bond strength itself.

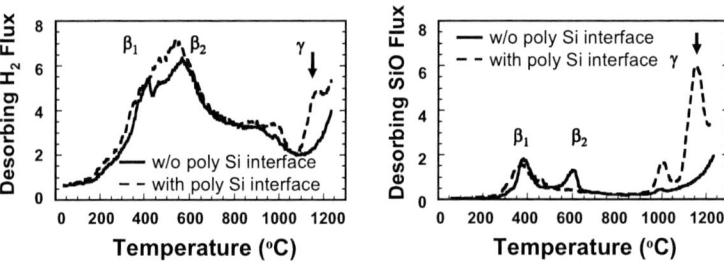

Figure 11. TDS of H$_2$ (a) and SiO (b) species observed from 10 nm SiON films in as-fabricated condition and after nucleation of an ultra thin poly Si layer.

Realizing that the poly-Si/SiON interface contains mobile and stable H species is important. Accordingly, these H species can be released from the layer under device operating stress and subsequently diffuse towards the SiON/Si interface as illustrated in **Fig. 12**. The diffusing H species may react with the SiO$_2$ network, leaving leakage paths across the dielectric along their track, which eventually result in breakdown. This model is consistent with Buchanan et al., who demonstrated that H release and transport induced by hot electrons in Al gate MOS stacks results in H buildup at the substrate interface and generates electrically active defects (25).

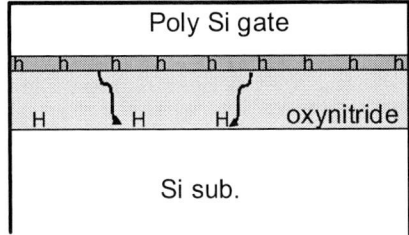

Figure 12. Schematic of H diffusion from the H-storage layer existing in the poly-Si/oxynitride interface towards the bottom Si interface. The diffusion track results in the formation of leakage paths in the oxynitride dielectric.

Since the poly-Si/oxynitride interface apparently acts as a mobile H source, and the properties of the H species depends on the thermal process and air-contact history, we suggest that the processing conditions during the poly Si interface fabrication are of highest importance. Eliminating the mobile H species from the gate interface through advanced processing parameters may be an essential key to improve the reliability of oxynitride dielectrics.

Building a Hydrogen Diffusion Barrier in Front of the Oxide/Si Interface

In this section we compare two test n-MOSFETs that applied the CVD oxide as gate dielectrics. 16 nm gate oxides with the same amount of N were formed by annealing CVD oxide in NO (NO-oxynitride) and N_2O (N_2O-oxynitride) gas, respectively, and were subsequently subjected to NRA H-depth profiling. Q_{bd}, D_{it}, and gate current-voltage (I-V) curves were evaluated to characterize the electrical device properties

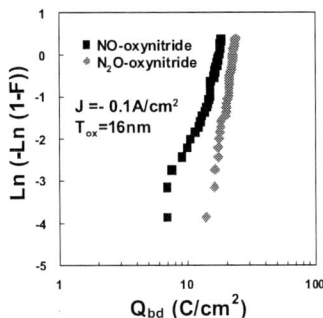

Figure 13. Q_{bd} distributions of NO-oxynitrides and N_2O-oxynitrides under a constant current stress of -0.1 A/cm^2.

Figure 13 shows Q_{bd} distributions of the NO-oxynitride and N_2O-oxynitride under negative gate bias polarity. The N_2O-oxynitride shows a higher intrinsic Q_{bd} than the NO-oxynitride. It has been pointed out previously that the N-depth profile in the SiO_2 films influences the Q_{bd} distribution (40, 41) due to different distribution and chemical configuration of the incorporated N.

The comparison of D_{it} generation measured by CP as a function of injected charge is shown in **Fig. 14**. Although the initial D_{it} values for the two kinds of SiON are comparable (D_{it} of N_2O-oxynitride is slightly higher than that in NO-oxynitride), the D_{it} increase rate induced by charge injection is significantly larger for NO-oxynitride than that in N_2O–oxynitride. This observation can be readily explained by a different number of interface defects, which were initially terminated with H but were reactivated by H release from the interface defects in the course of the charge injection. This result provides us with the important hint that H should be involved in the model to explain the different electrical properties of the NO-oxynitride and the N_2O-oxynitride.

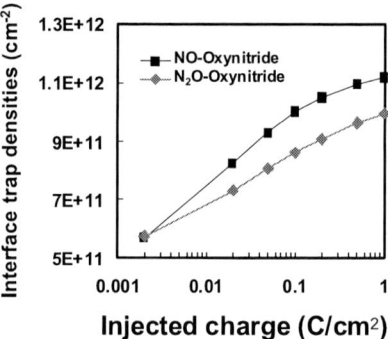

Figure 14. Interface traps density versus injected charge density for NO-oxynitride and N_2O-oxynitride films. Injected charge density is 0.01 A/cm^2.

Figure 15 compares NRA H depth profiles of the two 16 nm SiON films. Rather than the redistributed H depth profile, we focus on the near interface P_m peak position. The center positions of the two broad P_n peaks are significantly different in the two NRA profiles, i.e. they depend on the nitridation process. In the N_2O-oxynitride, the H-distribution is more clearly separated from the optical interface position (dash-dotted line shown as SiO_2/Si). In the fit analysis shown in **Fig. 15**, we decomposed the P_m peaks into a near interface peak (dotted line) and an interface peak (dashed line), respectively. It is seen that in the NO-oxynitride, H mainly accumulates in very close vicinity of the interface, while most of the H appears to have stopped at a larger distance from interface in the N_2O-oxynitride. The resistance against H-diffusion apparently is possibly attributed to an oxynitride H-storage layer.

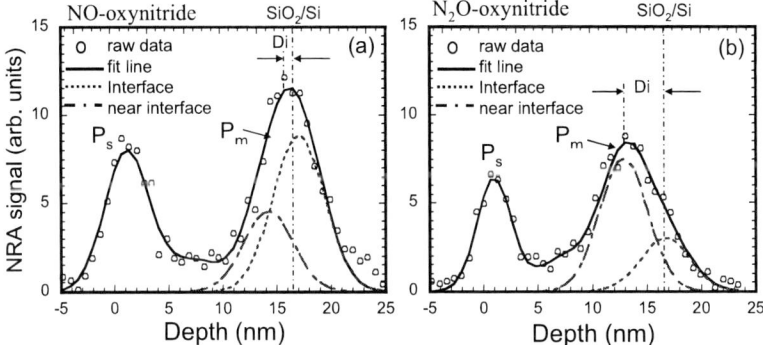

Figure 15. NRA H profiles of NO-oxynitride (a) and N$_2$O-oxynitride (b) films. Surface and near interface peaks and fit components are shown.

Chemical analysis by angle-resolved photoelectron spectroscopy using synchrotron radiation also supports the possibility for the localization of N in the near-interface region of the N$_2$O-oxynitride (42). The above result can be explained by the idea that the high N concentration localized in front of the optical interface constitutes an H-diffusion barrier layer that protects the SiO$_2$/Si interface from the approach by H. Thus the N$_2$O-oxynitride with its interface-protecting H barrier layer is expected to be more robust than the NO-oxynitride.

Novkovski et al. have shown a correlation between the oxidation resistance of the oxynitride interface and its electrical properties (41). Although the nature of the H-species discussed in this study is not yet specified, the oxidation resistance may be related at least partially to the presence of a diffusion barrier layer, which restricts the diffusion of the oxidant as well as that of the H-containing species, which are detected by NRA in this study.

Summary

We surveyed our applications of NRA hydrogen depth profiling to probe the H distribution in MOS stack constituting model layer systems. The main results summarize as follows:

- The interfacial region between the poly-Si gate and the oxynitride dielectric of MOS transistors contains two kinds of H, referred to as mobile and stable H species. This situation is similar to that of H stored in air-exposed oxynitride layers on N$_2$-annealed SiN/SiO$_2$ stacks. The mobile H, if stimulated by energetic carriers, migrates across the gate films and relocates to the SiO$_2$/Si interface.

- The accumulation of H in the oxide/Si interface region correlates with MOS device degradation caused by NBTS. The device stability can be improved by POA, as the latter prevents the H transport towards the SiO_2/Si interface.

- The two H species (mobile and stable) of different mobility in the poly-Si/oxynitride interfacial layer can be converted into each other by thermal stimulation or air contact. This result delivers an important message for the processing optimization, as an optimized anneal process should be capable to stabilize the mobile H by repairing the air-degraded H storage layer. The air-exposure intervals during different manufacturing steps should be taken into account and a repetition of the annealing process may be considered.

- Constructing an H-diffusion barrier layer in front of the SiON/Si interface is an effective way to protect the sensitive dielectric/Si interface from the approach by diffusing H. Such an H-diffusion barrier layer may be realized by optimizing the concentration and distribution of nitrogen near the SiON/Si interface.

In conclusion, the discovery of the specific thin (<1 nm) oxynitride layer (Si_2N_2O-like composition) with the potential to store stable H species that resist energetic electron damage encourages us to reconsider the design of well-known silicon-oxygen-nitrogen materials with the objective to enhance the device reliability by increasing their resistance against internal H redistribution.

Acknowledgments

The author, Z. Liu, wishes to thank N. Nakamura, T. Ishiyama of Renesas Electronics Corporation for the management supports. Thanks are given to K. Ando, T. Matsuda, F. Hayashi, S. Azuma and S. Ito of Renesas Electronics Corporation for wafer preparation, electrical measurements and valuable discussions. M. Wilde and K. Fukutani are grateful for assistance in the MALT tandem accelerator operation by H. Matsuzaki and C. Nakano at the University of Tokyo.

This work was partially supported by New Energy and Industrial Technology Development Organization of Japan (NEDO).

References

1. A. Toriumi and H. Satake, in *Mat. Res. Soc. Sym. Proceedings*, Boston, **592**, p. 323 (1999).
2. G. J. Gerardi, E. H. Poindexter, P. J. Caplan, M. Harmatz, W. R. Buchwald, and N. M. Johnson, *J. Electrochem. Soc.* **136**, 2609 (1989).
3. D. K. Schrodera and J. A. Babcock, *J. Appl. Phys.* **94**, 1 (2003).
4. C. E. Blat, E. H. Nicollian, and E. H. Poindexter, *J. Appl. Phys.* **69**, 1712 (1991).
5. C. R. Helms and E. H. Poindexter, *Rep. Prog. Phys.*, **57**, 791 (1994)
6. Y. Nissan-Cohen and T. Gorczyca,, *IEEE Electron Dev. Letter,* **9**, 287 (1988).
7. L. Dori, J. H. Stathis and J. A. Tornello, *J. Appl. Phys.* **70**, 1510 (1991).
8. P. E. Nicollian, C. N. Berglund, P. F. Schmidt and J. M. Andrews, *J. Appl. Phys.* **42**, 5654 (1971).
9. P. E. Nicollian, A. T. Krishnan, C. A. Chancellor, R. B. Khamankar, S. Chakravarthi, C. Bowen and V. K. Reddy, in *Proc. 46th Int. Reliab. Phys. Symp.*, p. 197 (2007).
10. D. J. DiMaria, *Appl. Phys. Lett.,* **51**, 463 (1987).
11. J. W. Lyding, K. Hess and I. C. Kizilyalli, *Appl. Phys. Lett.* **68**, 2526 (1996).
12. S. Tsujikawa, T. Mine, K. Watanabe, Y. Shimamoto, R. Tsuchiya, K. Ohnishi, T. Onai, J. Yugami and S. Kimura, in *Proc. 41th Int. Reliab. Phys. Symp.*, p.183 (2003).
13. S. Ogawa and N. Shiono, *Phys. Rev. B* **51**, 4218 (1995).
14. Horiuchi, in *Tech. Dig. Symp. VLSI Technol.*, p. 92 (2000).
15. Y. Ohji, Y. Nishioka, K. Yokogawa, K. Mukai, Q. Qiu, E. Arai and T. Sugano, in *Proc. 38th Int. Reliab. Phys. Symp.*, p. 82 (1989).
16. N. S. Saks, D. B. Brown and R. W. Rendell, *IEEE Trans. Nucl. Sci.* **NS-38**, 1130 (1991).
17. S. Shuto, M. Tanaka, M. Sonoda, T. Idaka, K. Sasaki, and S. Mori, in *Proc. 36th Int. Reliab. Phys. Symp.*, p. 17 (1997).
18. Z. Liu, S. Fujieda, F. Hayashi, M. Shimizu, M. Nakata, H. Ishigaki, M. Wilde, and K. Fukutani, in *Proc. 46th Int. Reliab. Phys. Symp.*, p. 190 (2007).
19. N. Kimizuka, K. Yamaguchi, K. Imai, T. Iizuka, C. T. Liu, R. C. Keller, and T. Horiuchi, in *Tech. Dig. Symp. VLSI Technol.*, p. 92 (2000).
20. J. Jee, W. Kwon, W. Lee, J. Park, H. Kim, H. Son, W. Chang, J. Han, Y. Hyung and H. Lee, in *Proc. 46th Int. Reliab. Phys. Symp.*, p. 184 (2007).
21. Y. Okada, P. Tobin, V. Lakhotia, W. Feil, S. Ajuria and R. Hegde, *Appl. Phys. Lett.*, **63**, 194 (1993).
22. C. Chen, Y. Fang, S. Ting, W. Hsieh, C Yang, T. Hsu, M. Yu, T. Lee, S. Chen, C. Yu, and M. Liang, *IEEE Transactions on Electron Dev.*, **49**, 840 (2002).
23. D. Matsushita, K. Muraoka, Y. Nakasaki, K. Kato, S. Kikuchi, K. Sakuma, Y. Mitani, K. Eguchi and M. Takayanagi, in *Tech. Dig. Int. Electron Device Meeting*, p. 847 (2005).
24. R. Gale, F. J. Feigl, C. W. Magee and D. R. Toung, *J. Appl. Phys.*, **54**, 6938 (1983).
25. D. A. Buchanan, A. D. Marwick, and D. J. DiMaria, L. Dori, *J. Appl. Phys.* **76**, 3595 (1994).
26. F. H. P. M. Habraken, R. H. G. Tijhaar, W. F. van der Weg, A. E. T. Kuiper, and M. F. C. Willemsen, *J. Appl. Phys.*, **59**, 447 (1986).

27. M. Wilde, M. Matsumoto, K. Fukutani, Z. Liu, K. Ando, Y. Kawashima and S. Fujieda, *J. Appl. Phys.*, **92**, 4320 (2002).
28. M. A. Briere and D. Braunig, *IEEE Trans. Nuclear Science*, **37**, 1658 (1990).
29. Z. Liu, S. Ito, S. Koyama, M. Makabe, M. Wildeand K. Fukutani, in *Proc. 49th Int. Reliab. Phys. Symp.*, p. 417 (2010).
30. Z. Liu, T. Saito, T. Matsuda, K Ando, S. Ito, M. Wilde and K. Fukutani, in *Proc. 47th Int. Reliab. Phys. Symp.*, 705 (2008).
31. K. Fukutani, H. Iwai, Y. Murata, and H. Yamashita, *Phys. Rev. B,* **59**, 1 (1999).
32. F. J. Himpsel, F. R. McFeely, A. Taleb-Ibrahimi, J. A. Yarmoff and G. Hollinger, *Phys. Rev.*, **B38**, 6084 (1988).
33. V. Afanas'ev and A. Stesmans, *Appl. Phys. Lett.*, **71**, 3844 (1997).
34. P. K. Roy, Y. Chen and S. Chetlur, *IEEE Trans. Electron Devices*, **48**, 2016 (2001).
35. J. Stathis, *J. Appl. Phys.*, **77**, 6205 (1995).
36. Z. Liu, S. Fujieda, K. Terashima, M. Wilde and K. Fukutani, *Appl. Phys. Lett.* **81**, 2397 (2002).
37. G. Rosenman, M. Naich, Ya. Roizin and Rob van Schaijk, *J. Appl. Phys.* **99**, 023702 (2006).
38. Z. Liu, S. Ito, M. Wilde, K. Fukutani, I. Hirozawa, and T. Koganezawa, *Appl. Phys. Lett.*, **92**, 192115 (2008).
39. H. Du, R. E. Tressler, K. E. Spear, and C. G. Pantano, *J. Electrochem. Soc.,* **136**, 1527 (1989).
40. B. Maiti, M. Y. Hao, l. Lee, and J. C. Lee, *Appl. Phys. Lett.,* **61**, 1790 (1992).
41. N. Novkovski, I, Aizenberg, E. Goin, E. Fullin and M. Dutoit, *Appl. Phys. Lett.*, **54**, 2408 (1989).
42. Z. Liu, S. Ito, T. Ide, M. Nakata, H. Ishigaki, M. Makabe, M. Wilde, K. Fukutani, H. Mitoh and Y. Kamigaki, in *Proc. 48th Int. Reliab. Phys. Symp.*, p. 902 (2009).

ECS Transactions, 35 (4) 73-79 (2011)
10.1149/1.3572276 ©The Electrochemical Society

Impact of Silicon Nitride Gate Dielectric Composition on the Stability of Low Temperature Nanocrystalline Silicon Thin Film Transistors

M.R. Esmaeili-Rad[1], G.R. Chaji[1,2], F. Li[3], M. Moradi[2], A. Sazonov[1], and A. Nathan[4]

[1]Electrical and Computer Eng., Univ. of Waterloo, Waterloo, Ontario N2L 3G1, Canada
[2]Ignis Innovation Inc., Kitchener, Ontario N2H 6M6, Canada
[3]Polymer Vision-Wistron, Kastanjelaan 1000, 5616 LZ Eindhoven, The Netherlands
[4]London Centre for Nanotechnology, Univ. College London, London WC1H 0AH, UK

We report on the stability of nanocrystalline silicon (nc-Si) bottom gate (BG) thin film transistors (TFTs) with various compositions ($[N]/[Si]$) of hydrogenated amorphous silicon nitride (a-SiN$_x$:H) gate dielectric, formed at 280°C. The shift in threshold voltage (ΔV_T) is larger for gate dielectrics with lower $[N]/[Si]$ content. For example, after 5 hours of stressing at 15 V, the ΔV_T is 0.3 V, 1 V, and 12.4 V for $[N]/[Si]$ of 1.3, 1.2, and 1, respectively. Relaxation tests on the stressed TFTs show that the charge trapping in the gate dielectric is the primary instability mechanism in nc-Si BG TFTs.

Introduction

It is well known that the threshold voltage in hydrogenated amorphous silicon (a-Si:H) thin film transistors (TFTs) changes under prolonged gate bias. However, this shift has not prevented its application in liquid crystal displays (LCDs) in which the TFT acts as a simple switching element (1,2). But new applications are emerging whereby the TFT is also required to function as an analog circuit element to provide a stable current. This is particularly true in organic light emitting diode (OLED) displays (3,4). Here, the shift in threshold voltage must be minimized for the TFT to source a stable current. It has been shown that hydrogenated nanocrystalline silicon (nc-Si) TFTs can meet this requirement as they are much more stable than their a-Si:H counterpart.

In this paper, we review the progress in nc-Si TFT research with particular focus on the stability of bottom-gate TFTs of various gate dielectric compositions.

Experimental

Nc-Si thin films were deposited by plasma enhanced chemical vapor deposition (PECVD) using silane (SiH$_4$) source gas highly diluted in hydrogen, H$_2$/SiH$_4$ = 100. The a-SiN$_x$:H gate dielectric layers (300 nm thick) were deposited by PECVD on p-type crystalline silicon substrates from a mixture of silane and ammonia (NH$_3$). The NH$_3$ to SiH$_4$ gas flow ratio was varied in the range 5-20 to obtain different compositions. The composition of the nitride was obtained by elastic recoil detection analysis (ERDA) and their chemical bonding was studied by Fourier-transform infrared (FTIR) spectroscopy. The capacitance-voltage characteristics were measured on metal-insulator-semiconductor (MIS) test structures. Bottom-gate nc-Si TFTs were fabricated based on the tri-layer

73

inverted-staggered TFT structure (see Fig. 1). The tri-layer consists of 300 nm a-SiN$_x$:H gate dielectric, 50 nm nc-Si/a-Si:H active bi-layer, and 300 nm a-SiN$_x$:H passivation dielectric. These layers were deposited at 280°C in a multi-chamber 13.56 MHz PECVD system. Further details can be found elsewhere (5). To evaluate device stability, DC gate and drain voltages were applied at room temperature to bias the TFT in the linear operation regime. The threshold voltage shift was subsequently retrieved from the transfer characteristics.

Al			Al
n+ nc-Si	Passivation nitride		n+ nc-Si
a-Si:H			
nc-Si			
Gate dielectric			
Gate			
Substrate			

Figure 1. Cross section of the tri-layer inverted-staggered TFT structure.

Gate Dielectric Properties

The electrical properties of the gate dielectric play a key role in the TFT performance. It is well known that the density of charge traps and leakage current of nitride dielectrics is a function of its composition, typically described in terms of the nitrogen to silicon ratio ([N]/[Si]). In this study, three films of varying composition were evaluated. Fig. 2 shows their FTIR spectra. The main absorption peaks in the spectra are due to Si-N bond stretching, Si-H bond stretching, and N-H bond stretching (6). Increasing the NH$_3$ to SiH$_4$ gas flow ratio appears to strengthen the N-H bond peak and weaken the Si-H bond peak, indicating an increase in [N] content in the film. As the N-content increases, the Si atoms in Si-H bonds are replaced by N atoms, due to the higher electronegativity of nitrogen compared to silicon. The [N]/[Si] ratio increases from 1 to 1.3 by increasing the ammonia gas flow, indicating that the film is becoming nitrogen (N) rich. The composition of the nitride has a strong bearing on TFT stability. This will be discussed in next sections.

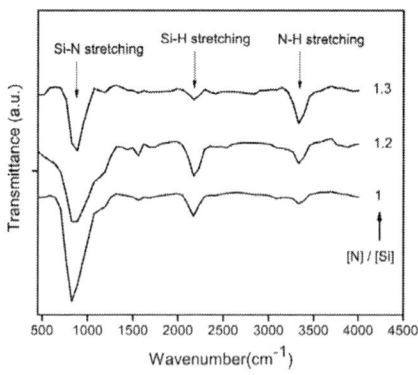

Figure 2. FTIR spectra of various nitride films with [N] / [Si] of 1.3, 1.2, and 1.

The electrical properties of a-SiN$_x$:H films were characterized using MIS structures. Depicted in Fig. 3 are capacitance-voltage (C-V) curves when the voltage applied to the MIS capacitors is swept in forward (negative to positive) and reverse (positive to negative) directions. The forward and reverse C-V curves are shifted relative to each other, exhibiting hysteresis. This shift is due to charge (electron) trapping within the a-SiN$_x$:H bulk. From Fig. 3, it is seen that the hysteresis width is ~ 2 V, 14 V, and 30 V for film compositions [N]/[Si] of 1.3, 1.2, and 1, respectively. Hence, at lower [N]/[Si] ratios, we observe increased charge trapping. Thus from a stability standpoint, N-rich a-SiN$_x$:H is preferred over the silicon(Si)-rich counterpart, which is consistent with previous studies on a-Si:H TFTs (7). Several mechanisms associated with electron injection and trapping have been proposed to explain the hysteresis in C-V characteristics, e.g. Fowler-Nordheim injection, trap-assisted injection, constant-energy tunneling from silicon conduction band, direct tunneling from silicon valence band, and hopping at the Fermi level (8). It has been shown that the Si-rich a-SiN$_x$:H has a higher density of silicon dangling bonds, compared to the N-rich a-SiN$_x$:H, acting as charge trapping centers (7).

Impact on TFT Stability

Nitride films with [N]/[Si] of 1.3, 1.2, and 1 were used as gate dielectric in nc-Si TFTs labeled TFT1, TFT2, and TFT3, respectively. The corresponding transfer characteristics of these TFTs are depicted in Fig. 4. Details of the electrical properties, such as threshold voltage and carrier mobility, can be found elsewhere (5). The effect of gate bias stress on V$_T$ stability was evaluated by subjecting the TFTs to a DC gate voltage stress of 15 and 25 V. The drain-source voltage (V$_{DS}$) was set to 0.1 V to maintain the TFT in the linear operation regime, in which the ΔV$_T$ has been observed to be more profound (9). The total stress time was 5 hours and the stress test was briefly interrupted three times (after 1 hour, 3 hours, and 5 hours) to retrieve the transfer characteristics. As seen in Fig. 5, after 5 hours of stressing at 15 V, the V$_T$ shift is 0.3 V, 1 V, and 12.4 V for TFT1, TFT2, and TFT3, respectively. The same trend is observed for the 25 V stress voltage. One can see that device stability is highly dependent on the nitride composition (7). TFT1 with a nitride composition [N]/[Si] of 1.3 exhibits the best stability, with a 0.3 V shift in V$_T$ at a gate voltage of 15 V. However, TFT3 with a Si-rich dielectric ([N]/[Si] of 1) shows a 12.4 V shift at a gate voltage of 15 V. The shifts in V$_T$ can be explained by charge trapping in the nitride gate dielectric, which can occur even at low applied gate voltages, for example, by tunneling from the silicon conduction band into the nitride Fermi level (8). The availability of charge trapping centers around the Fermi level facilitates electron injection from the channel into empty trap states in the nitride, believed to be silicon dangling bonds (7). TFTs with a Si-rich nitride are expected to possess larger density of trapping centers, and thus show larger shifts in V$_T$, irrespective of the channel material as observed here for nc-Si TFTs and elsewhere for a-Si:H TFTs (10). In contrast, the silicon dangling bond density is minimized in an N-rich a-SiN$_x$:H, thus suppressing the rate of charge trapping and the shift in V$_T$. In Fig. 5 (a), the threshold voltage shift of a circuit-grade a-Si:H TFT is also shown for comparison. As seen, the ΔV$_T$ of TFT1 is around 20-30% that of the a-Si:H counterpart, showing that nc-Si TFT with the proper a-SiN$_x$:H gate dielectric can provide a more stable device than a-Si:H TFTs.

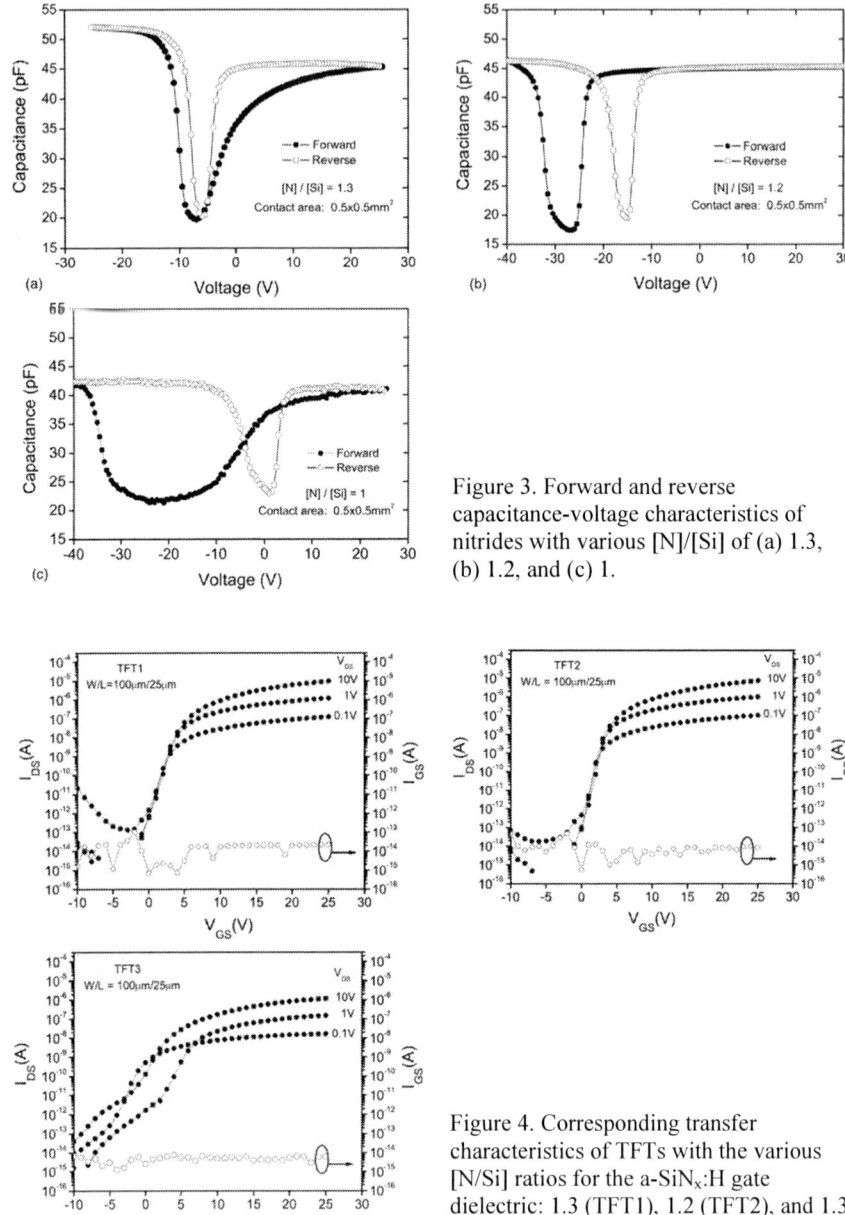

Figure 3. Forward and reverse capacitance-voltage characteristics of nitrides with various [N]/[Si] of (a) 1.3, (b) 1.2, and (c) 1.

Figure 4. Corresponding transfer characteristics of TFTs with the various [N/Si] ratios for the a-SiN$_x$:H gate dielectric: 1.3 (TFT1), 1.2 (TFT2), and 1.3 (TFT3).

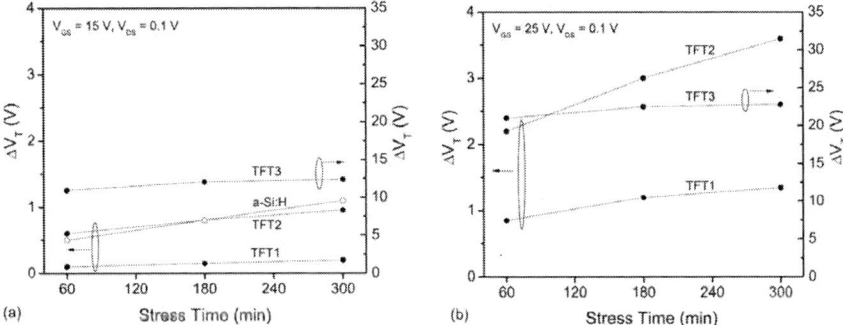

(a) Stress Time (min) (b) Stress Time (min)

Figure 5. Threshold voltage shift (ΔV_T) as a function of stress time for gate voltages of (a) 15 V and (b) 25 V in the linear regime. In part (a), the ΔV_T of a circuit-grade a-Si:H TFT is also shown for comparison (open circles) using data from Ref. 11.

The relaxation of stressed TFTs can provide significant insight on the instability mechanisms. It is known that defect state creation in the channel is indefinitely stable at room temperature (12). But charge trapping in the dielectric is reversible. Relaxation tests were conducted to examine this behavior. Fig. 6 shows transfer characteristics of TFT1 in different states: unstressed, gate bias stressed for 5 hours at 25 V, and relaxed for 5 days at room temperature following gate bias removal. After 5 hours of gate bias stress, ΔV_T is ~1.4 V, and the transfer characteristic returns to its unstressed state after 5 days relaxation at room temperature. This observation is consistent with charge trapping/detrapping into/from the gate dielectric. Further analysis and details can be found in Ref. (5).

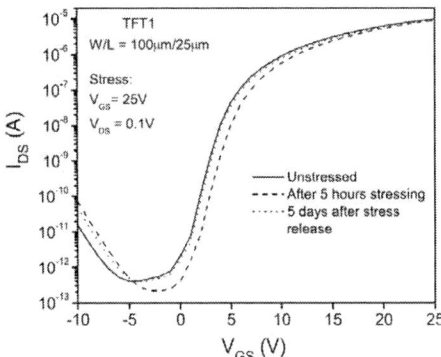

Figure 6. Transfer characteristics of TFT1 in different states: unstressed, stressed for 5 h at 25 V gate stress, and relaxed for 5 days after stress release at room temperature.

The relatively fast relaxation of nitrogen-rich gate dielectrics, consistent with the behavior observed in TFT1 in Fig. 6 above, meets the stability and lifetime requirements in display applications involving current drive, such as that in active matrix organic light emitting diode (AMOLED) displays. Here, even with a-Si:H TFTs, the use of a nitrogen-rich a-SiN$_x$:H gate dielectric can substantially improve the stability behavior and provide the needed recovery in threshold voltage for reasonably high frame rate and low power applications. In particular, for AMOLED displays, the driving scheme forces the TFT to operate under low constant current (of the order of μA) but for long periods of time. Also, it is demonstrated that adding an off cycle (relaxation) to each frame improves the lifetime significantly [13]. Thus, two different TFTs (60 $\mu m/4$ μm) with silicon rich (existing) and nitrogen rich (improved) dielectric are stressed at 500 nA (required for AMOLED displays) with 3-ms relaxation time during each 16-ms frame time. As it is shown in Fig. 7(a), the improved TFT has significantly lower V_T shift. This improvement enables the use of a-Si:H based TFTs in AMOLED displays since the V_T shift becomes manageable by compensation circuits. Fig. 7(b) shows a typical display lifetime under such a current level.

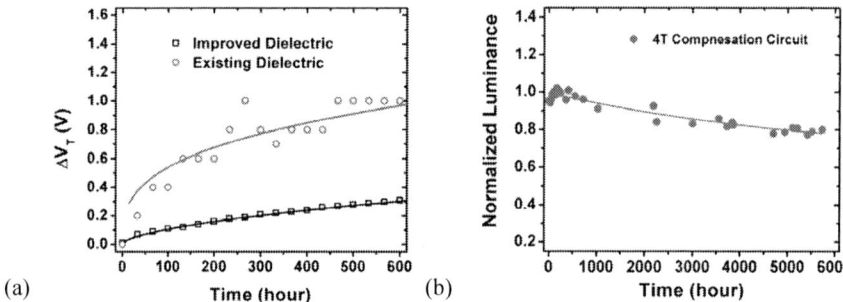

(a)

(b)

Figure 7: (a) Threshold voltage shift for TFTs using silicon and nitride rich dielectric, (b) lifetime of an AMOLED display using 4T compensation circuit [13].

Conclusions

We studied the effect of the a-SiN$_x$:H gate dielectric composition on the stability of nc-Si TFTs. The bond configuration and elemental concentration of the a-SiN$_x$:H thin films have been analyzed using ERDA and FTIR. The stability of nc-Si TFTs was found to be highly dependent on the a-SiN$_x$:H gate dielectric composition. The results show that the gate dielectric with highest [N]/[Si] ratio exhibits the lowest shift in V_T. In addition, the relaxation behavior of the stressed TFTs confirms that the charge trapping in the a-SiN$_x$:H gate dielectric is the underlying mechanism responsible for this shift. The enhanced stability makes nc-Si TFTs very attractive for large-area electronics applications, and in particular, as pixel drivers in active matrix OLED displays.

Acknowledgments

The authors would like to thank the Natural Sciences and Engineering Research Council of Canada (NSERC) for supporting this research and the Royal Society Wolfson Research Merit Award.

References

1. B. Stannowski, J.K. Rath, and R.E.I. Schropp, *Thin Solid Films*, **430**, 220 (2003).
2. S. Sambandan, L. Zhu, D. Striakhilev, P. Servati, and A. Nathan, *IEEE Electron Device Letters*, **26**, 375 (2005).
3. P. R. Cabarrocas, R. Brenot, P. Bulkin, R. Vanderhaghen, B. A. Drevillon, and I. French, *J. Appl. Phys.*, **86**, 7079 (1999).
4. A. Nathan, A. Kumar, K. Sakariya, P. Servati, S. Sambandan, and D. Striakhilev, *IEEE J. Solid State Circuits*, **39**, 1477 (2004).
5. M.R. Esmaeili-Rad, F. Li, A. Sazonov, and A. Nathan, *J. Appl. Phys.*, **102**, 064512-1 (2007).
6. D. Stryahilev, A. Sazonov, and A. Nathan, *J. Vacuum Science and Technology A*, **20**, 1087 (2002).
7. W. S. Lau, S. J. Fonash, and J. Kanicki, *J. Appl. Phys.*, **66**, 2765 (1989).
8. M. J. Powell, *Appl. Phys. Lett.*, **43**, 597 (1983).
9. K. S. Karim, A. Nathan, M. Hack, and W. I. Milne, *IEEE Electron Device Lett.*, **25**, 188 (2004).
10. M. J. Powell, C. van Berkel, and J. R. Hughes, *Appl. Phys. Lett.*, **54**, 1323 (1989).
11. M. R. Esmaeili-Rad, A. Sazonov, and A. Nathan, *IEEE IEDM Proc.*, 303 (2006).
12. R. A. Street, *Hydrogenated Amorphous Silicon*, Cambridge University Press, Cambridge (1991).
13. G.R. Chaji, S. Alexander, A. Nathan, C. Church, and S.J. Tang, *Technical Digest of SID Symposium,* Long Beach, USA, 1580 (2007).

80

ECS Transactions, 35 (4) 81-93 (2011)
10.1149/1.3572277 ©The Electrochemical Society

Development of a Fast Technique for Characterizing Interface States

L. Lin, Z. Ji, J. F. Zhang, and W. D. Zhang

School of Engineering, Liverpool John Moores University,
Byrom Street, Liverpool L3 3AF, UK.

Characterizing interface states is an important task for test
engineers. The existing techniques typically have a measurement
time in the order of seconds. Recent results, however, show that
degradation can recover substantially within seconds and there is a
need for improving the speed for measuring interface states. The
central task of this work is to reduce the time for measuring
interface states to the order of microseconds, so that the recovery
during measurement can be minimized. An analysis of the existing
techniques ruled out the popular techniques, such as conductance,
various capacitance-voltage methods and the subthreshold swing,
since they all require a gate voltage ramp that must be slow enough
to maintain thermal equilibrium with interface states. Although
conventional charge pumping (CCP) does not require such
equilibrium, the DC recombination current used limits the speed.
By replacing the DC current with transient currents, we will show
that interface states can be extracted from a single pulse applied to
the gate, allowing the measurement time being reduced to the order
of microseconds. The results are calibrated against those from the
well-established technique and accuracy in the order of 10^{10} cm^{-2} is
achieved.

Introduction

The near perfect SiO_2/Si interface plays a major role in the success of silicon-based
CMOS technologies. This high quality of interface, however, is achieved only after an
anneal in a hydrogen environment. Without such an anneal, the typical interface states are
in the order of 10^{12} cm^{-2} [1-3]. It is widely believed that the interface states originate from
silicon atoms of a dangling bond [1], although different types of interface states have
been reported [3,4]. Annealing in hydrogen at a temperature around 400 °C forms Si-H
bonds and passivates the interface states. When compared with Si-O bonds, Si-H bonds at
the interface are relatively weak and can be ruptured by a number of physical processes,
such as irradiation [5-7], hot carriers [8-10], positive [11,12] and negative [13-18] biased
temperature stresses. Moreover, the generation of interface states can continue even post
irradiation [6] and electrical stresses [19-21]. It has been proposed that the generation of
interface states controls the hot carrier lifetime of nMOSFETs [8], substantially degrades
the lifetime of pMOSFETs due to the negative bias temperature instability (NBTI) [13-
18], and reduce the transfer efficiency of charge coupled devices (CCDs). Characterizing
interface states is an important task, therefore.

Several techniques have been developed to measure interface states, such as high-frequency capacitance-voltage (HFCV), namely the Terman's method [22], quasi-static capacitance-voltage (QSCV) [23,24], conductance [23,24], subthreshold swing [25-27] and charge pumping [4,28-30]. All of these techniques, however, suffer from a common drawback: slow measurement speed and the typical measurement time is in the order of seconds [22-30]. It has been shown recently that the degradation measured at such speed is only a fraction of the real degradation, because of the rapid recovery of degradation after removing stresses [15-17,31,32]. Ultra-fast measurement techniques have been developed and the results show that measurement time has to be reduced to the order of microseconds to minimize the recovery [15-17,31,32]. However, these fast techniques are based on monitoring the transient transfer characteristics and the measured drain current is affected by both created interface states and charges trapped in the gate dielectrics [26,32]. The interface state density cannot be extracted from these fast measurements. The central task of this work is to develop a fast characterization technique that allows direct evaluation of interface states with a time in the order of microseconds.

The paper will be organized in the followings. We will first briefly analyze the potential of each existing technique for fast measurements and justify the selection of charge pumping method. The experimental setup for overcoming the shortcomings of the conventional charge pumping technique will then be described. After presenting the typical results, the formula needed for extracting interface states will be developed. Finally, we will calibrate the interface states measured by our method against those from the well-accepted charge pumping technique.

Selection of techniques and experimental setup

Selection of techniques

The most sensitive and complete technique for measuring interface states is the conductance method, which can detect interface states as low as the order of 10^9 cm^{-2}. It also gives the capture cross section of interface states [23]. Unfortunately, this is the most time consuming technique. Under each gate bias, a conductance against frequency sweep is typically carried out, that can take tens of seconds. In addition, to find the interface potential under a given gate bias, a low frequency CV has to be measured. As a result, it is impossible to use this technique for fast measurements.

The Terman's method is one of the first techniques used for measuring interface states, developed as early as in 1962 [22]. Although it measures capacitance at high frequency, this does not mean that it can be done at a high speed. During the measurement, the gate bias consists of a quasi-DC ramp and a small high frequency probing signal. The quasi-DC ramp rate must be sufficiently slow that the thermal equilibrium with interface states is maintained. The typical ramp rate used is between 5 and 50 mV/sec [24] and it can take over 20 sec to sweep one volt. The other CV techniques such as quasi-static CV and high-low frequency CV [22] all require the same quasi-DC ramp for the gate bias, so that they cannot be used for fast measurements.

The subthreshold swing (SS) technique [25-27] only requires measuring the transfer characteristics in the low gate bias region, making it suitable for samples with thin gate dielectrics where gate leakage is problematic for other techniques. Unlike charge pumping, the SS does not need a contact to the substrate, so that it can be applied to silicon on insulator devices, where a connection to the substrate is often not available. Unfortunately, this technique is also based on the assumption that the interface states are in thermal equilibrium with the gate bias sweep, so that fast pulse cannot be applied here.

The charge pumping can be used for small MOSFETs and is a popular technique [4,28-30]. It is a technique that actually requires the interface states not being in thermal equilibrium with the measurement signal, offering the potential for fast measurements. Although the base level of gate bias pulse is often swept, such sweep is not essential. In principle, the interface states can be determined by measuring a single charge pumping current, so long that the gate bias pulse covers the range of flatband voltage and threshold voltage. Conventionally, multiple gate pulses are applied during the measurement and the measured charge pumping current is an average DC current. For a typical parameter analyzer, measuring one DC point will take 10-150 ms [17]. Although this is a significant improvement when compared with the seconds needed by other techniques, it is still too slow since recovery was observed in the order of microseconds [15-17,31]. One potential solution to the problem is to replace the DC current measurement by recording transient currents. We will show that this is indeed achievable.

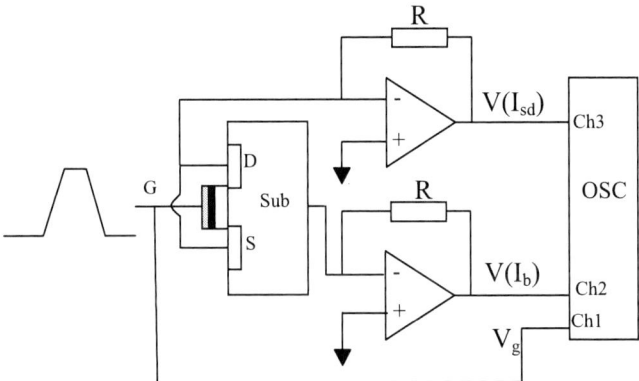

Fig.1 A schematic diagram of the experimental setup for the single pulse charge pumping measurement. When a pulse is applied to the gate, the induced transient channel current, I_{sd}, and the substrate current, I_b, are recorded.

Devices and Experimental setup

The devices used have n-channel, 7 nm SiO_2, and a doping density of 5×10^{17} cm^{-3}. The channel length and width is 10 μm and 200 μm, respectively. The interface state density in a fresh device is below 5×10^{10} cm^{-2}, as measured by the conventional charge pumping technique. Some devices were stressed uniformly under a gate bias of V_g=+7.5V with source and drain grounded. All the stress and measurements were carried out at room temperature in this work.

The experimental setup is schematically illustrated in Fig. 1. During the measurement, a pulse is applied to the gate of the device under test (DUT). At the same time, the oscilloscope is triggered to record this gate pulse together with the corresponding transient voltages at the output of two operational amplifiers, $V(I_{sd})$ and $V(I_b)$, which can be converted into the transient channel current I_{sd} and substrate current I_b by,

$$I_{sd} = \frac{V(I_{sd})}{R}$$

(1)

and

$$I_b = \frac{V(I_b)}{R}.$$

(2)

where the feedback resistor, R, typically has a value of 10 kΩ.

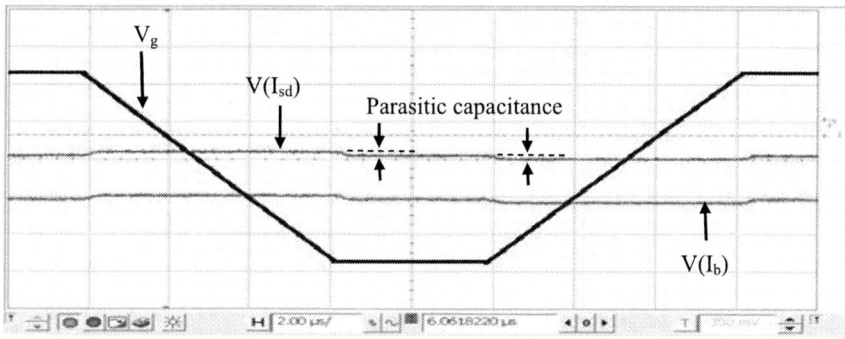

Fig. 2 A screen shot of the signals when a gate pulse, V_g, is applied without connecting the devices. The variations in $V(I_{sd})$ and $V(I_b)$ from the pulse edges to the plateau regions of V_g are caused by parasitic capacitances and are removed in evaluation I_{sd} and I_b.

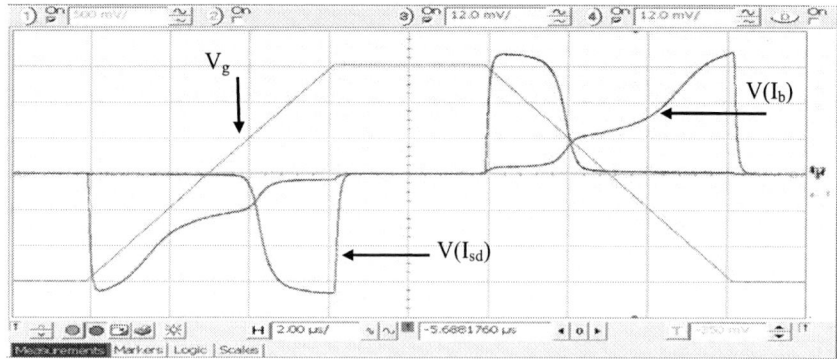

Fig. 3 A screen shot of the signal when a gate pulse, V_g, is applied to a fresh nMOSFET. A change of direction in V_g sweep also changes the polarity of $V(I_{sd})$ and $V(I_b)$.

To find the real zero level for $V(I_{sd})$ and $V(I_b)$, we first carried out a test without connecting the DUT. Fig. 2 shows that $V(I_{sd})$ and $V(I_b)$ have different values for the pulse edges and the plateau regions of V_g. Within each pulse edge, $V(I_{sd})$ and $V(I_b)$ does not change with V_g. When the pulse sweep direction changes, $V(I_{sd})$ and $V(I_b)$ also change the polarity. This indicates that there are parasitic capacitance in the setup and their contribution to $V(I_{sd})$ and $V(I_b)$ will be removed.

Fig. 3 shows the typical waveforms of V_g, $V(I_{sd})$ and $V(I_b)$ when the pulse was applied to the gate of a fresh nMOSFET. It can be seen that there are no overshoots around the corners of the gate pulse for all signals, indicating that the impact of parasitic inductance is suppressed by keeping all connection wires as short as possible. The typical pulse amplitude and edge time used in this work is 3 V and 6 μs, respectively. An pulse edge time in the order of microseconds is typical for the pulsed measurements used in recent works [15-17,31].

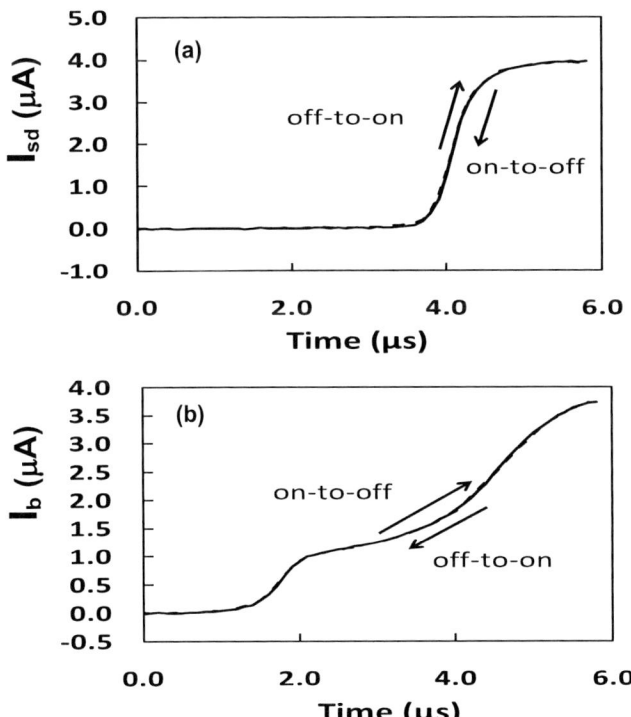

Fig. 4 The transient channel current (a), I_{sd}, and substrate current (b), I_b, converted from the $V(I_{sd})$ and $V(I_b)$ in Fig. 3. The 'Time' for the horizontal axis is the pulse edge time. 'Off-to-on' and 'on-to-off' corresponds to the first and second edges in Fig. 3, respectively. To facilitate the comparison, the negative value of $V(I_{sd})$ and $V(I_b)$ for the first edge is inverted to positive.

Figs. 4a and 4b show the I_{sd} and I_b converted from $V(I_{sd})$ and $V(I_b)$ through equations (1) and (2), respectively. The label 'off-to-on' corresponds to the first pulse edge in Fig. 3, where V_g was swept from negative to positive and the nMOSFET was turned on. Similarly, the label 'on-to-off' represents the second pulse edge. To facilitate the comparison of the currents for the two edges, we inverted the polarity of I_{sd} and I_b for the first edge in Fig. 3. Figs. 4a and 4b show that the transient currents for the two edges agree well for a fresh device.

Waveform analysis

Fig. 5 presents the waveforms of $V(I_{sd})$ and $V(I_b)$ for a stressed nMOSFET. When compared with the result before stress in Fig. 3, peaks can be observed in Fig. 5, especially for $V(I_{sd})$. These peaks cannot originate from the parasitic components and are not artifacts, since they are absent for fresh sample. As a result, they must originate from the stress induced defects. Figs. 6a and 6b compare the I_{sd} and I_b for the two pulse edges. Unlike the results of a fresh sample in Figs. 4a and 4b, the I_{sd} and I_b for the two edges are clearly different here. These differences are analyzed next.

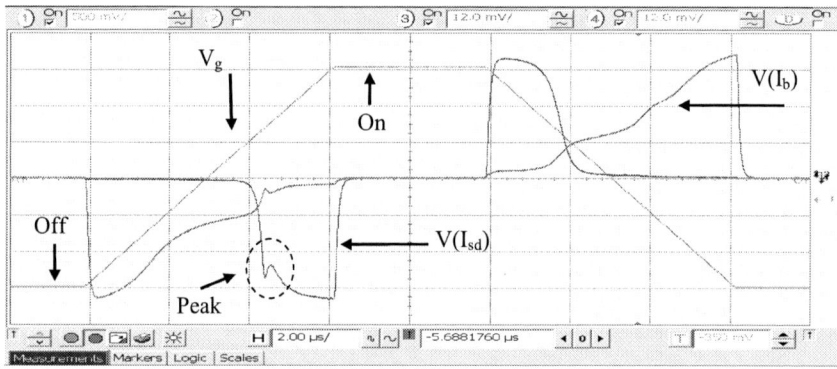

Fig. 5 A screen shot of the signals when a gate pulse, V_g, is applied to a stressed nMOSFET. Apart from a change of polarity, note the differences in $V(I_{sd})$ and $V(I_b)$ for the two pulse edges .

To facilitate the analysis, we re-plot the I_{sd} and I_b of Figs. 6a&b against V_g in Figs. 7a and 7b. Since V_g changes linearly with time during the pulse edge, the shape of I_{sd} and I_b remain the same. It is well accepted that the charge neutrality level for SiO_2/Si interface is close to the midband and the interface states are acceptor-like and donor-like in the upper and lower half of the bandgap, respectively [33,34]. When the nMOSFET is in off mode, $V_g<0$ and Fig. 8a shows that the donor-like states are positively charged, whilst the acceptor-like states are neutral. During the off-to-on transition, electrons flow into the devices. Fig. 8b shows that the donor-like states sufficiently away from the top edge of the valence band will not have enough time to emit its charges and will be neutralized by recombining with the electrons from the source and drain, giving rise to a recombination current. In addition, the neutral acceptor-like states will charged by capturing electrons from the channel, resulting in a charging current. This is to say that I_{sd} here actually contains three components: a displacement current, a recombination current, and a charging current. During the on-to-off switching, however, electrons in the channel will flow out the device and we also have a displacement component in I_{sd}. The electrons used to charge the acceptor-like states will not be able to flow out the devices in time, unless they are close to the edge of the conduction band. Moreover, the electrons used to neutralize the donor-like states will not flow out, either. The loss of the charging and recombination currents is responsible for the missing of the peak during the on-to-off transition in Fig. 7a.

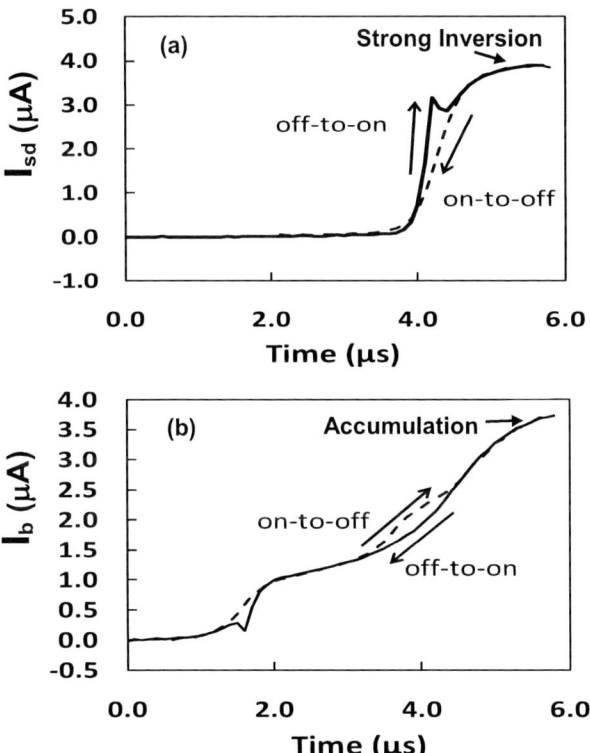

Fig. 6 The transient channel current (a), I_{sd}, and substrate current (b), I_b, converted from the $V(I_{sd})$ and $V(I_b)$ in Fig. 5. After stress, clear differences are observed in both I_{sd} and I_b between the two edges, caused by the stress induced interface states.

Similar analysis can be made for the I_b waveform in Figs. 6b and 7b. Here the holes from the substrate flow towards the interface during the on-to-off switch, neutralizing the acceptor-like states and charging the donor-like states, as illustrated in Figs. 8c and 8d. The recombination and charging currents are absent for the off-to-on transition, leading to the lower I_b in Fig. 7b.

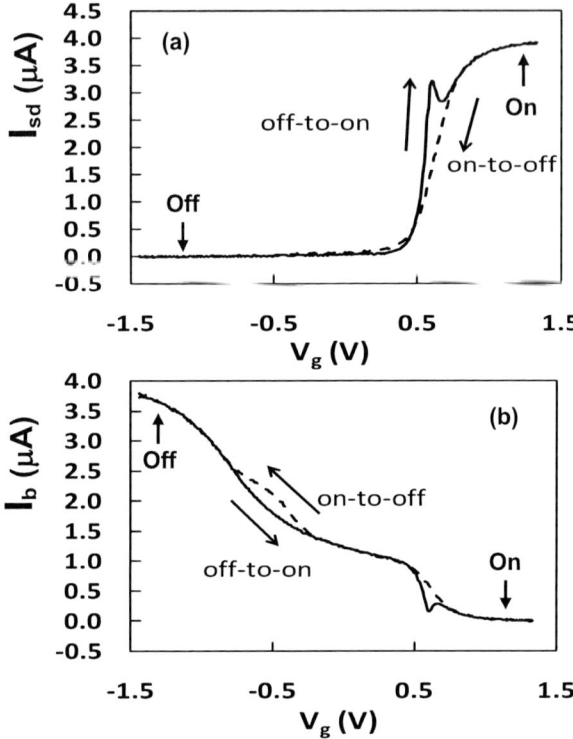

Fig. 7 The transient channel current (a), I_{sd}, and substrate current (b), I_b, against the gate bias, V_g. The I_{sd} and I_b used here are the same as that in Fig. 6.

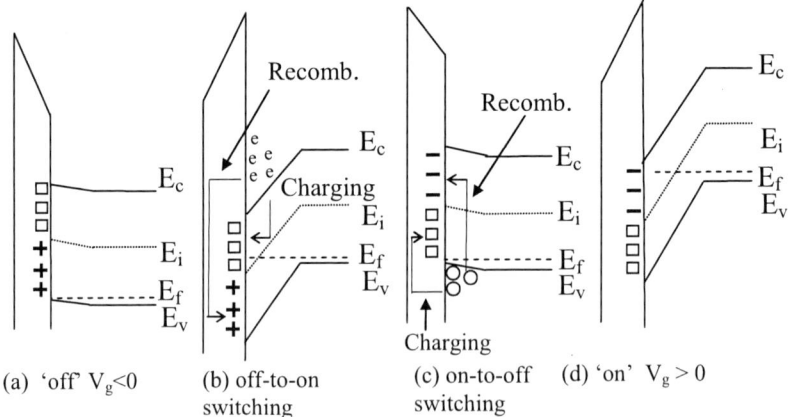

Fig. 8 Energy band diagrams under different operation conditions. '-', '□' and '+' represents negative, neutral and positive states and 'o' and 'e' are holes and electrons.

When compared with the clear peak in I_{sd} in Fig. 6a, there is only a hump in the on-to-off I_b in Fig. 6b, although both of them are caused by the interface states. This is mainly because the transition to strong inversion in Fig. 6a happens over a shorter time than the transition to accumulation given in Fig. 6b. When the number of charge carriers involved for charging and neutralizing interface states is the same, a shorter transition time means a larger current and a clearer peak.

Evaluation of interface states

To extract interface states, we evaluate the charges flowing in or out devices during the two pulse edges. The charges flowing into the device from the source and drain when the device is switched on, Q_{sd}(off-to-on), is calculated from,

$$Q_{sd}(off-to-on) = \int_0^t I_{sd}(off-to-on)dt \qquad (3)$$

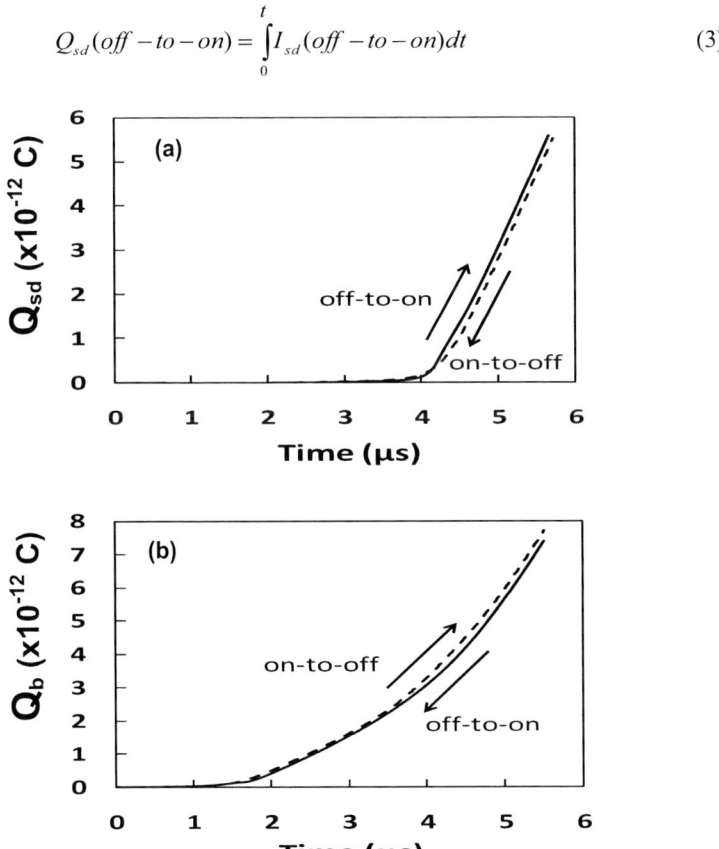

Fig. 9 The charges flowing in and out the devices from the source and drain (a) and the substrate (b) against the edge time.

By replacing I_{sd}(off-to-on) with I_{sd}(on-to-off) in the equation (3), we also can calculate the charges flowing out the device through the source and drain during the on-to-off transition, Q_{sd}(on-to-off). The typical results are given in Fig. 9a. The net charges pumped into the devices from the source and drain for one gate pulse is,

$$\Delta Q_{sd} = Q_{sd}(off-to-on) - Q_{sd}(on-to-off) \qquad (4)$$

Similarly, we can calculate the holes flowing out of the device when the device is turned on, Q_b(off-to-on), and the hole entering the device during the on-to-off transition, Q_b(on-to-off) by replacing the I_{sd} in equation (3) with the corresponding substrate current, I_b. The typical results are given in Fig. 9b. The net charges pumped into the device from the substrate contact during one gate pulse is,

$$\Delta Q_b = Q_b(on-to-off) - Q_b(off-to-on) \qquad (5)$$

The ΔQ_{sd} and ΔQ_b are given in Fig. 10. Although Q_{sd} and Q_b in Figs. 9a and 9b increase continuously during the pulse edge, both ΔQ_{sd} and ΔQ_b saturates. This is what we expect, since the number of interface states are limited. Importantly, we observe that ΔQ_{sd} and ΔQ_b saturate at the same level, despite the different waveform of I_{sd} and I_b in Figs. 6a and 6b. Both the saturation and the common saturation level strongly support that ΔQ_{sd} and ΔQ_b originate from interface states. The interface states per unit area, N_{it}, can be evaluated from this saturation level, ΔQ_{sat}, by,

$$N_{it} = \frac{\Delta Q_{sat}}{qLW}, \qquad (6)$$

where q is one electron charge, L and W are device length and width, respectively. Note that the equation (6) gives the interface states per unit area N_{it}, rather than the interface states "generated" by stress, which is typically represented by ΔN_{it} in literature [19-21].

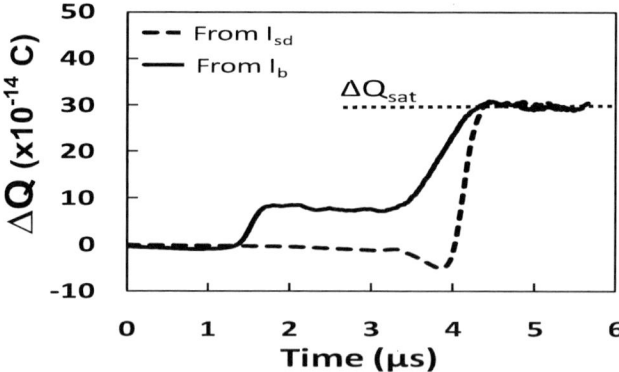

Fig. 10 The net charges pumped into the device from the substrate (the solid line) and the source and drain (the dashed line). Although they have different shapes, both of them saturate at the same level.

Comparison with the conventional charge pumping technique

Fig. 10 demonstrates that the net pumped charges, ΔQ, can be measured with a time in microseconds, improving the measurement speed for interface states by orders of magnitude. To calibrate the interface states evaluated by this single pulse charge pumping (SPCP) technique, we compare them with those evaluated by the conventional charge pumping (CCP) technique. The pulse edge times used for the CCP are the same as those for SPCP, but the pulse was repeatedly applied for the CCP and the DC charge pumping current was measured. Care also was excised to ensure that the same number of interface states were present during both measurements. Fig. 11 shows that the N_{it} extracted from SPCP agrees well with that obtained from CCP. When the stress increases, N_{it} obtained by both techniques increases in step and their ratio remains at one. Moreover, although the accuracy of transient measurements is generally not as good as that of DC measurements, Fig. 11 shows that the SPCP can be used to measure the degradation of N_{it} in the order of 10^{10} cm^{-2}, which is adequate for typical stress tests [26]. This gives us the confidence that the SPCP can be used to monitor the transient behavior of interface states with a time resolution in the order of microseconds in future.

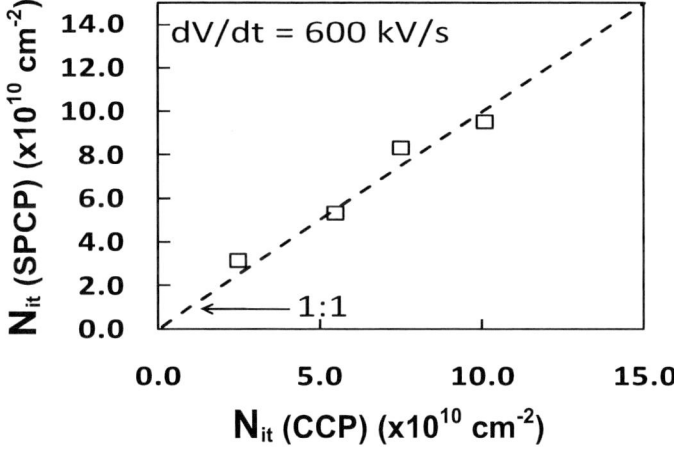

Fig. 11 Calibration of the single pulse CP technique (SPCP) against the conventional CP (CCP) method. The interface states were first measured by CCP and then by SPCP. The interface states measured by SPCP agree well with those by CCP.

Conclusion

The existing techniques for characterizing interface states typically take several seconds and the degradation can recover substantially during the measurement. In this work, we analyzed the potential of the established techniques for fast measurements. The conductance, high-frequency CV, quasi-static CV, and subthreshold swing all requires a gate voltage sweep that must be sufficiently slow to maintain the thermal equilibrium of interface states with the gate bias. This makes them fundamentally unsuitable for fast measurements.

In contrast, charge pumping does not need this thermal equilibrium and can be used for fast measurement in principle. The speed of conventional charge pumping method is limited by measuring a DC charge pumping current. In this work, we replaced the DC current by recording the transient channel and substrate currents when a pulse was applied to the gate. A detailed analysis of the current waveform reveals that the differences between the currents for the two pulse edges originate from the interface states. The net charges flowing into the devices during one pulse can be evaluated from these transient currents and it is found that the net charge supplied from the source and drain saturates at the same level as that from the substrate. This saturation level is used to extract interface state density. The results from the fast single pulse CP measurement agree well with that from the conventional CP and a measurement accuracy of 10^{10} cm^{-2} is achieved. The use of this technique for monitoring the transient behavior of interface states is awaiting future work.

Acknowledgements

This work is supported by the Engineering and Physical Science Research Council of UK under the grant no. EP/I012966/1. L. Lin and Z. Ji would like to express their appreciation for the partial studentships funded by the Higher Education Funding Council of England. The test samples used in this work were provided by IMEC.

References

1. A. Stesmans and V. V. Afanas'ev, *Phys. Rev. B.,* **57**, 10030 (1998).
2. D. J. DiMaria, E. Cartier, and D. Arnold, *J. Appl. Phys.,* **73**, 3367 (1993).
3. J. F. Zhang, H. K. Sii, R. Degraeve, and G. Groeseneken, *J. Appl. Phys.,* **87**, 2967 (2000).
4. W. D. Zhang, J. F. Zhang, M. J. Uren, G. Groeseneken, R. Degraeve, M. Lalor, and, D. Burton, *Appl. Phys. Lett.,* **79**, 3092 (2001).
5. T. P. Ma, *Semicond. Sci. Technol.,* **4**, 1061 (1989).
6. R. E. Stahlbush, A. H. Edwards, D. L. Griscom, and B. J. Mrstik, *J. Appl. Phys.,* **73**, 658 (1993).
7. A. R. Stivers and C. T. Sah, *J. Appl. Phys.,* **51**, 6292 (1980).
8. C. Hu, S. C. Tam, F. C. Hsu, P. K. Ko, T. Y. Chan, and K. W. Terrill, *IEEE Trans. Electron Dev.,* **48**, 1127 (2001).
9. J. F. Zhang and W. Eccleston, *IEEE Trans. Electron Devices,* **42**, 1269 (1995).
10. I. S. Al-kofahi, J. F. Zhang, and G. Groeseneken, *J. Appl. Phys.,* **81**, 2686 (1997).
11. J. F. Zhang and W. Eccleston, *IEEE Trans. Electron Devices,* **41**, 740 (1994).
12. J. F. Zhang and W. Eccleston, *IEEE Trans. Electron Devices,* **45**, 116 (1998).
13. V. Huard, in *Proc. of Int. Reliability Phys. Symp. (IRPS),* p. 33 (2010).
14. T. Grasser, B. Kaczer, W. Goes, Th. Aichinger, Ph. Hehenberger, and M. Nelhiebel, *Microelectronic Engineering,* **86**, 1876 (2009).
15. M. F. Li, D. Huang, W. J. Liu, Z. Y. Liu, and X. Y. Huang, *ECS Trans.,* **19**, 301 (2009).
16. A. E. Islam, S. Mahapatra, S. Deora, V. D. Maheta, and M. A. Alam, in *IEDM Tech. Dig.,* p. 733 (2009).

17. J. F. Zhang, Z. Ji, M. H. Chang, B. Kaczer, and G. Groeseneken, in *IEDM Tech. Dig.*, p. 817 (2007).
18. S. S. Tan, T. P. Chen, J. M. Soon, K. P. Loh, C. H. Ang, and L. Chen, *Appl. Phys. Lett.*, **82**, 1881 (2003).
19. C. Z. Zhao, J. F. Zhang, G. Groeseneken, R. Degraeve, J. N. Ellis, and C. D. Beech, *J. Appl. Phys.*, **90**, 328 (2001).
20. J. F. Zhang, C. Z. Zhao, G. Groeseneken, and R. Degraeve, *J. Appl. Phys.*, **93**, 6107 (2003).
21. J. F. Zhang, H. K. Sii, G. Groeseneken, and R. Degraeve, *IEEE Trans. Electron Dev.*, **47**, 378 (2000).
22. L. M. Terman, *Solid-State Electron.*, **5**, 285 (1962).
23. E. H. Nicollian and J. R. Brews, *MOS (Metal Oxide Semiconductor) Physics and Technology* (New York: Wiley, 1982).
24. M. J. Uren and K. M. Brunson, *Semicond. Sci. Technol.*, **9**, 1504 (1994).
25. C. Tan, M. Xu, and Y. Wang, *IEEE Electron Dev. Lett.*, **15**, 257 (1994).
26. M. H. Chang and J. F. Zhang, *J. Appl. Phys.*, **101**, 024516 (2007).
27. J. F. Zhang, M. H. Chang, Z. Ji, L. Lin, I. Ferain, G. Groeseneken, L. Pantisano, S. De Gendt, and M. M. Heyns, *IEEE Electron Dev. Lett.*, **29**, 1360 (2008).
28. G. Groeseneken, H. E. Maes, N. Beltran, and R. F. De Keersmaecker, *IEEE Trans. Electron Devices*, **ED-31**, 42 (1984).
29. D. Bauza and G. Ghibaudo, *Solid-State Electron.*, **39**, 563 (1996).
30. M. B. Zahid , R. Degraeve, J. F. Zhang, and G. Groeseneken, in *Proc. of Int. Reliability Phys. Symp. (IRPS)*, p.55 (2007).
31. Z. Ji, J. F. Zhang, M. H. Chang, B. Kaczer, and G. Groeseneken, *IEEE Trans. Electron Devices*, **56**, 1086 (2009).
32. Z. Ji, L. Lin, J. F. Zhang, B. Kaczer, and G. Groeseneken, *IEEE Trans. Electron Devices*, **57**, 288 (2010).
33. G. A. Scoggan and T. P. Ma, *J. Appl. Phys.*, **48**, 294 (1977).
34. J. F. Zhang, S. Taylor, and W. Eccleston, *J. Appl. Phys.*, **71**, 725 (1992).

94

ECS Transactions, 35 (4) 95-113 (2011)
10.1149/1.3572278 ©The Electrochemical Society

Detailed Analysis of Si-SiO₂ Interface Traps in MOSFETs Using Charge Pumping

D. Bauza

CNRS, IMEP-LAHC, Minatec, Grenoble INP,
3, Parvis Louis Neel, BP 257,
38016, Grenoble, Cedx 1, France,
bauza@minatec.grenoble-inp.fr

Based on charge pumping (CP) mechanisms pointed out recently, the way Si-SiO₂ interface traps interact with the gate CP signal in MOSFETs with thick SiO₂ gate dielectric is detailed. The consequences that can be deduced or studied from such an analysis are presented. This includes the way the interface is probed, the slope of the CP curves and the resulting interface trap density extraction from them. The dependence of CP curves on gate signal frequency along with the CP signal high frequency limit case is dealt with using simulation. Both are very well accounted for by the general CP model used. This allows the capture cross-section for holes and for electrons of the two kinds of interface traps that dominate at the Si-SiO₂ interface in such devices to be evaluated. This supports that each trap type is Coulomb attractive for capturing one carrier type and neutral for capturing the other carrier type which agrees with acceptor-like and donor-like traps.

Introduction

For more than five decades now, the traps that exist between the silicon substrate and its oxide have attracted much attention due to their crucial role in MOS device technology: charge storage, noise, current-voltage characteristics stretching [1-3]. The poor properties of this interface have been a problem with regard to MOSFET scaling and to parameters stabilization but both the continuous progress in technology reliability and reproducibility [4, 5], along with the use of hydrogen for dangling bonds passivation, lead to interface trap densities, D_{it}, generally between $10^9 \, eV^{-1}.cm^{-2}$ and $2\text{-}4x10^{10} \, eV^{-1}.cm^{-2}$. Nevertheless, to date, decisive points are not clarified yet. D_{it} can be easily evaluated in MOS devices provided the SiO₂ thickness is large enough [6]. But in spite of numerous studies and of the wealth of electrical characterization tools proposed during these decades [1-3, 7], it is not possible yet for a given MOS device with thick SiO₂ gate dielectric to determine the Si-SiO₂ interface trap density as a function of energy and the corresponding trap capture cross-sections for holes and for electrons in a region significantly greater than half of the silicon bandgap [8, 9]. An easy comparison between such properties measured on device from different batches, from different generations and/or manufacturers is therefore difficult and generally reduces to D_{it}. In addition, it is known that different methods used on the same device often provide significantly different results [8, 9]. Finally, and from the MOS technology point of view, significant differences in D_{it} values can be measured on devices having different origins [10, 11].

95

As MOS device dimensions continuously shrunk, some old problems gained importance while new ones emerged: among others, for MOSFETs with channel submicron width and length, increase in electrical characteristics variability due to statistical fluctuations in the number and properties of the few interface or oxide traps present [12]; degradation of the Si-SiO$_2$ interface due to channel hot carriers [13] or to negative bias temperature stress [14] in which both the interface and the oxide traps are involved [14, 15]; exponential increase of the tunneling or leakage current with decreasing SiO$_2$ thickness. This leakage current rendered the conventional Si-SiO$_2$ interface electrical characterization techniques unusable anymore. It also accelerated the replacement of SiO$_2$ by high-κ gate dielectrics, the prerequisite for further scaling while limiting to acceptable values the gate leakage. As a SiO$_2$ interfacial layer (IL) tends to grow between silicon substrate and high-κ dielectrics [16], a Si-SiO$_2$ interface still exists in such devices but is not as easy to study as in conventional MOSFETs with thick SiO$_2$.

A wealth of techniques has been proposed for Si-SiO$_2$ interface electrical characterization. Following methods proposed initially [1, 6], a few of them emerged because they were easy to use in spite of a poor sensitivity (capacitance-voltage (C-V) methods [9]), or because of their good reliability and sensitivity, in spite of a limited energy excursion (conductance method [9]). Such techniques justified thorough analysis and extensions in order to extract D$_{it}$ closer to band edges in case of high frequency C-V [17], or for probing slow oxide traps in that of conductance method [18, 19]. Charge Pumping (CP), proposed primarily in the late 60's [20] and significantly improved fifteen years later [21], appears poorly understood in retrospect when compared with the techniques mentioned above, especially with regard to its wide use in numerous situations [22].

At present, new methods are needed again for characterizing silicon-high-κ interfaces, not for accounting for the slow traps as two decades ago in case of the Si-SiO$_2$ system, but for probing only the fastest ones or the stress induced ones that recover before being measured [23, 24]. Also, alternative high mobility substrates are studied along with the interface of poor quality they form with insulators [25, 26]. There is no doubt in that context that CP will remain a technique of choice for studying interface traps .

In the first section of the paper, the main recent insights into CP mechanism that allowed reliable CP curves simulation to be carried out will be recalled. The way a trapezoidal signal applied to the gate interacts with Si-SiO$_2$ interface traps will be discussed using simulation. Simulation will the extended to the impact of gate signal frequency on the slope of CP curves and to the high frequency limit case which renders the shape of the CP curves independent on gate signal frequency. Then it will be shown that the slope of the CP curves directly account for D$_{it}$ through the CP mechanisms. The interface trap density usually extracted from the CP current maximum can also be obtained accurately from the slope of all well-known basic CP curves. The two traps types entering the CP curves the cross-sections of which agree with Coulomb attractive and/or neutral states for the capture of electrons and/or holes (i.e. acceptor- and donor-like traps) will also be depicted in a very similar manner in experimental and simulated CP curves, allowing for the first time to detail the contribution to the CP curves of the two kinds of traps in their two different states, i.e. Coulomb attractive and neutral. Finally, the slope of the CP curves which always show two components will be discussed with regard to the CP curves recorded in the high frequency limit case which exhibit defect contributions at the same gate biases.

Simulation of MOSFET CP Characteristics

Devices studied and trap characteristics used for simulation

In our studies dealing with Si-SiO$_2$ interface trap properties in MOSFETs with thick SiO$_2$ using CP, two different n-channel devices were employed [11, 27-33]. Those used in this study have an oxide thickness, d_{ox} = 27 nm, a doping concentration $N_a \approx 2 \times 10^{16}$ cm^{-3} and an interface trap density, $D_{it} \approx 4$-6×10^{10} eV^{-1} cm^{-2}. In spite of a tenfold difference in their D_{it} values ($D_{it} \approx 2$-3×10^9 eV^{-1} cm^{-2} vs. $D_{it} \approx 4$-6×10^{10} eV^{-1} cm^{-2}), the two device types always provided identical results in terms of trap properties [10]. They have therefore been regarded as good examples of fresh and fully processed MOSFETs with thick oxides, well suited for Si-SiO$_2$ interface CP studies [10, 27-33].

The simulation of MOSFET experimental CP curves requires long enough devices with regard to the technology node in order recombination in the source and drain regions (SDR) to remain small with regard to the overall CP signal measured. The doping concentration can then be assumed constant in the channel region and the free carrier concentration uniform regardless of gate bias [31, 34]. The devices used here which have been shown to fulfill these criteria have length L and width W equal to 4x800 μm^2 [31].

The three basic CP curves can be recorded by [20-22, 35]:
1) keeping the gate voltage swing, V_{sw} = (V_h – V_l) constant and recording at a gate signal frequency f, Q_{cp} = I_{cp}/f, as a function of V_l or of V_h, where V_h and V_l, are the high and low level of the trapezoidal gate signal, the transition times of which are t_r = t_f = t_{rf}. These plots will be referred to as "Elliot curves" [35]. The time at V_h and at V_l is t_h = t_l = t_{hl} = (1/2f – t_{rf}). E_{fh} and E_{fl} are Fermi level energy position at the Si-SiO$_2$ interface at V_h and V_l, respectively. I_{cp} is the current measured and Q_{cp}, the charge that recombines during one period of the gate signal;
2) keeping V_l constant and increasing V_h, at a given f, and
3) keeping V_h constant and decreasing V_l, at f.

In all cases, the CP curve has been simulated using the CP model based on Shockley-Reed-Hall (SRH) kinetics [32, 33, 36], proposed primarily with no trap time constant distribution (TCD) [36]. The model has been used assuming:
1) two different trap types enter the CP curves at the interface in the energy region probed (~ 0.6 eV) along with one trap type in the near oxide [28];
2) the different trap types are uniformly distributed in energy in the region probed, i.e. either at the Si-SiO$_2$ interface or in the near oxide. The oxide trap has also a constant concentration, N_{to}, towards oxide depth;
3) the two kinds of interface traps have their own capture cross section for electrons and for holes. The values required agree well with Coulomb attractive ($\sigma_{attr} \approx 10^{-14}$ cm^2) and neutral ($\sigma_{neut} \approx 10^{-16}$ cm^2) states of acceptor- and donor-like traps [32, 33]. This is especially true for Elliot curves that cannot be simulated assuming only one trap type at Si-SiO$_2$ interface [32];

4) the three trap types were supposed to fill independently ones from the others, with a capture rate independent of energy, and from the interface throughout the Si-SiO₂ interface traps time constant distribution (TCD) measured [11, 27]. The oxide traps were supposed exchange carriers with the substrate by direct tunneling.

The two interface trap types seem to belong to the Si-SiO₂ interface in such fresh and fully processed MOSFETs and could therefore be observed in such devices provided that small enough V_{sw} are used [28, 29, 32]. When the TCD is supposed to result from carrier tunneling through the Si-SiO₂ barrier [27, 32], the tunneling attenuation length for electrons and for holes, λ_e and λ_h, were taken equal.

Simulation of Elliot curves

The simulation of this first kind of curves has been published primarily in [29, 32] after introducing trap profiles as close as possible to those extracted from the Si-SiO₂ interface trap TCD and measured as in [11, 27, 31]. These profiles, which have been obtained systematically after assuming that the trap cross section decreases exponentially with the parameter that governs capture (the distance between a trap and the Si-SiO₂ interface in case of pure tunneling normalized with regard to the tunnel attenuation length, λ_e or λ_h), are composed of an exponential decrease in trap concentration, N_t, from the faster to the slower interface traps followed by a plateau for even slower or oxide traps, i.e. $N_t(\alpha) = N_{ts}.\exp(-x/\alpha) + N_{to}$, where $\alpha = \lambda_e$ or λ_h in case of tunneling [11, 31].

The result of such simulations obtained after accounting for Si-SiO₂ interface trap profile extracted from the trap TCD and measured as in [11, 27, 31] is shown in Fig. 1a when plotted with regard V_h and in Fig. 1b when plotted with regard to V_l.

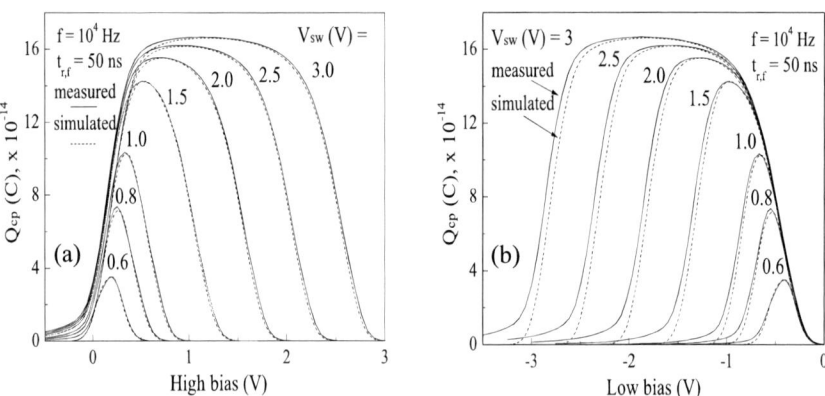

Figure 1. Set of Elliot curves recorded from one of the MOSFETs studied using a large range of V_{sw} values, and corresponding simulation results accounting for the trap profiles extracted from the Si-SiO₂ interface trap TCD [11, 27, 31]. The same measured and simulated curve sets are plotted with regard to V_h in a) and V_l in b) (surface potential fluctuations, $\sigma_s = 1.5$ kT) [31].

Simulation of $Q_{cp}(V_h)_{Vl}$ and $Q_{cp}(V_l)_{Vh}$ curves

The simulation of these two other curve types has been primarily presented in [30]. It is shown in Figs. 2a and 2b. For the ten curves simulated here, the exact experimental bias conditions noted in the figures were used. With regard to the curves in Fig. 1, only one trap type was required, having same capture cross section for electrons and holes ($\sigma_{e,h} = 10^{-15}$ cm^2). The trap concentration profile introduced in the simulations for accounting for the Si-SiO$_2$ interface trap TCD was close to that extracted as in [11, 27, 31].

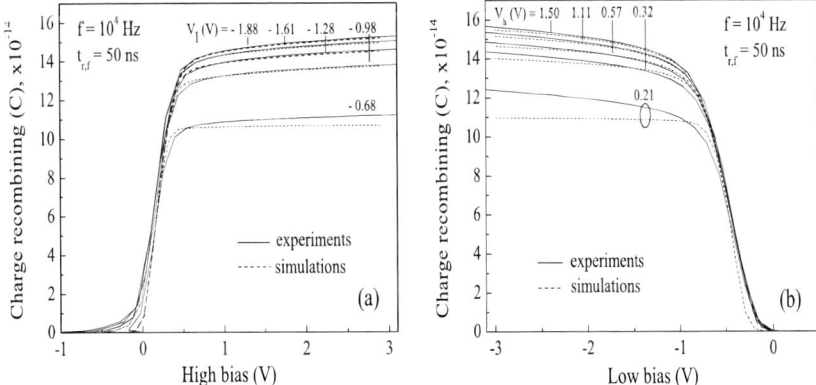

Figure 2. Sets of a) $Q_{cp}(V_h)_{Vl}$ and b) $Q_{cp}(V_l)_{Vh}$ curves recorded from one of the MOSFETs studied along with the corresponding simulations for which all experimental conditions were strictly used. Only one interface trap with a capture cross section identical for electrons and holes ($\sigma_{e,h} = 10^{-15}$ cm^2) was required ($\sigma_s = 1.2$ kT) (After [30]).

Simulation of $Q_{cp}(f)_{Vsw}$ curves and high frequency limit case

Before addressing the next point, one may note that for the simulations reported up to now, the frequency used was always the same and was rather low with regard to CP, i.e. $f = 10^4$ Hz. It results that the impact of f on the shape of CP curves, a key parameter not only in the extraction of trap profiles from the Si-SiO$_2$ interface trap TCD, but also for both the reliability of the model and the trap properties introduced, was not evaluated yet.

The result is shown in Fig. 3a in case of Elliot curves recorded from the same device as that used in Fig. 1. Then $V_{sw} = 2$ V and f was varied from ~ 10^3 Hz to 10^6 Hz. The small contribution of the source and drain regions (SDR) of the device leads to the tail at low Q_{cp} towards negative V_l values, and not accounted for in the simulations [31].

For obtaining the results in Fig. 3a, the same trap profile close to that extracted from the Si-SiO$_2$ interface trap TCD was used for the two interface trap types. As expected, the variation of the curve maximum with f requires an oxide trap concentration N_{to} equal to that measured when recording trap depth profiles from the trap TCD [11, 31, 32], i.e.

around 1/100 to 1/200 that at the interface [31] and therefore several times smaller than that required in [32] and in Fig. 1.

A crucial point with regard to the reliability of the CP model and to the interface trap characteristics introduced is the way the left and right hand edges of the curves in Fig. 3a vary with f. Fig. 3a shows that the curves simulated very well follow those measured in these regions: for both curves types, the width of the bell-shaped curves decreases when increasing f (arrows), increasing the steepness of their edges, as detailed in next section.

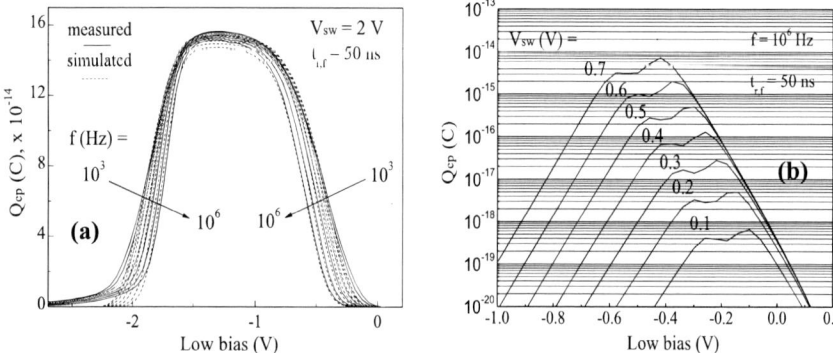

Figure 3. a) Elliot curves recorded at $V_{sw} = 2$ V and frequencies between ~ 10^3 and 10^6 Hz, and corresponding simulations ($\sigma_s = 1.5$ k.T); b) CP high frequency limit in which the traps only partly fill ($f = 10^6$ Hz , 0.1 V $\leq V_{sw} \leq 0.7$ V, and $\sigma_s = 0.0$ kT) [28, 37].

For completeness, the case of small V_{sw} values and high gate signal frequency, i.e. the high frequency limit case discussed in [37] is simulated in Fig. 3b. Then, the traps around midgap, the first ones to be able to be filled successively by minority and majority carriers in inversion at V_h and in accumulations at V_l, only partially fill because the traps filling function variation, ΔF [31, 37], is such that either ($c_n.t_h$) at V_h or ($c_p.t_l$) at V_l, remain much smaller than 1 regardless of V_l and V_{sw} (c_n and c_p are the capture rates at V_h and V_l, respectively) [37]. For a trap to be fully accounted for in Q_{cp}, ΔF requires the filling by an electron in inversion and a hole in accumulation. As ΔF, which is not a step function, varies from 1 to 0, from faster to slower traps, it results that in this peculiar situation, the fastest traps in the center of the silicon bandgap only partially contribute to recombination through the tail of ΔF at the interface [31]. Thus, for a given pulse position, the CP current I_{cp} (and not Q_{cp} here) becomes frequency independent. Increasing f simply compensates for the decrease in I_{cp} which results from the reduced fraction of traps entering ΔF through t_h and t_l [37]. As the same traps are probed regardless of the experimental conditions, the same signal is obtained but shifted to the left when increasing V_{sw}, as expected from CP mechanisms and from Fig. 1b, as dealt with in [37].

From the simulations in Figs. 1 to 3, it seems established that the different MOSFET basic CP curves can now be very well simulated. The impact of gate signal frequency on CP curves and the high frequency limit case dealt with here have also been very well accounted for by the model. This supports further that the essential of the CP mechanisms has been now captured concerning the traps at the Si-SiO$_2$ interface and that the trap characteristics introduced are close enough to the actual ones to allow such simulations. The results presented in the next sections will further support these conclusions.

Nevertheless, the simulations of the different curve types carried out in these figures to account for the CP response of the same (Fig. 1 and 3a) or quite identical devices have been achieved separately and led to somewhat different parameters (surface potential fluctuations, number of interface trap types required and corresponding trap profiles with regard to those extracted from trap TCD, oxide trap density with regard to that measured). The next step in that field will aim at reducing this spread to simulate all curve types with minimum variations in trap and interface parameters differences.

Slope of the CP curves

As shown previously in [32, 33], in the largest part of CP curves, Fermi level closest to the intrinsic level, i.e. E_{fh} or E_{fl} that corresponds to V_h, or V_l, respectively, governs and limits the CP current both in terms of energy swept at the interface and in terms of the fraction of the TCD entering recombination through ΔF [32, 33]. Then, assuming here that $(E_{fh} - E_i) < (E_i - E_{fl})$, i.e. that $E_{fh} = E_{fhc}$ controls/limits Q_{cp} and until re-emission of captured carriers towards the band that provided them definitely impacts the CP current and limits the energy excursion also on the $(E_{fh} - E_i)$ side of the gap, ΔE, the overall energy interval over which recombination proceeds is given by [32, 33]:

$$\Delta E_h \approx (E_{fh} - E_i) \approx (E_i - E_{em,h}) = \Delta E_l \qquad \text{so that:} \qquad \Delta E = \Delta E_h + \Delta E_l \approx 2\,\Delta E_h. \quad (1)$$

In Eq. 1, recombination fully enters the $\Delta E_h = (E_{fh} - E_i)$ region during t_h, and is reduced on the other half of the silicon bandgap from $(E_i - E_{fl})$ to $(E_i - E_{em,h})$ by hole re-emission during t_h through SRH kinetics [32, 33].

For ΔE_h values of the order of 0.3 eV for trap cross sections around a few 10^{-16} cm^2 [32, 33], re-emission of carriers also enters the upper part the energy interval at the interface (fastest traps) so that the energy region over which recombination occurs at the interface does not increase much anymore as capture proceeds now during the transition times of the gate signal.

The fact that $\Delta E = \Delta E_h$ has significant consequences for both understanding CP mechanisms and Si-SiO$_2$ interface traps electrical characterization:

1) in strong asymmetrical capture conditions [32, 33, 36], in which carrier re-emission enters one side of the energy swept at the interface only, the CP current depends on a) surface potential on the side of the signal that controls/limits recombination and b) the filling of the traps during the time spent at the corresponding energy, i.e. t_h or t_l.

2) if D_{it} does not strongly depend on energy in the region probed and provided that the gate bias conditions with regard to interface traps capture are such that the traps have enough time to fill and thus to impact the slope of the CP curves, Q_{cph} and Q_{cpl} measured when increasing V_h or V_l, respectively, may be approximated by [30]:

$$Q_{cph} \approx q \ W \ L \ D_{it} \ [2 \ (E_{fhc} - E_i)] \qquad \text{or to} \qquad Q_{cpl} \approx q \ W \ L \ D_{it} \ [2 \ (E_i - E_{flc})]. \qquad (2)$$

Eq. 2 is fully verified in Figs. 4 and 5, regardless of the CP curve type, provided that the conditions for such an observation are fulfilled (see next section). In Fig. 4a, one of the experimental $Q_{cp}(V_h)_{Vl}$ and $Q_{cp}(V_l)_{Vh}$ curves shown in Fig. 2a and 2b is plotted again and compared with its calculated and simple counterpart obtained using Eq. 2, in which $D_{it} \approx 4.1x10^{10} \ eV^{-1}.cm^{-2}$ as measured at CP curves maxima [36].

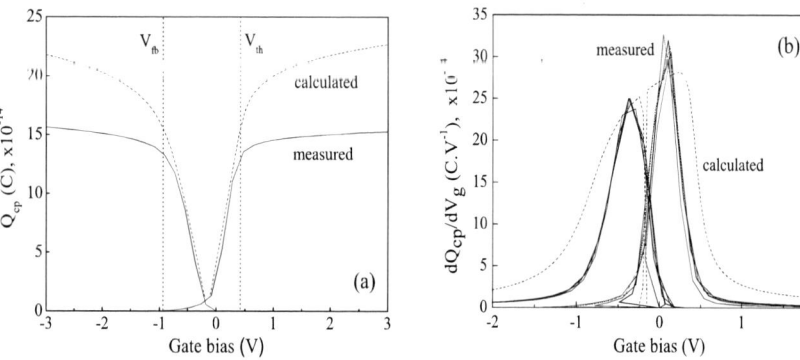

Figure 4. a) One of the experimental curves in Figs. 2a and 2b vs. $Q_{cp}(V_h)$ and $Q_{cp}(V_l)$ ones calculated using Eq. 2; b) slope of all curves in Figs. 2a and 2b vs. that obtained with Eq. 2 in absolute values. The threshold, V_{th}, and flat band, V_{fb}, voltages corresponding to the simulation are marked in Fig. 4a.

It can be seen in Fig. 4a and for the two curve types, that the experimental Q_{cp} variations with $V_g = V_h$ or $V_g = V_l$ are well accounted for by Eq. 2 in most of the region where Q_{cp} strongly varies. When the experimental $Q_{cp}(V_g)$ variations abruptly slow down with V_h or V_l magnitude, due to carrier re-emission on both sides of the CP signal at the interface, Q_{cph} and Q_{cpl}, calculated using Eq. 2 which simply account for [2 ($E_{fh} - E_i$)] or [2 ($E_i - E_{fl}$)] continue to increase with V_h and V_l magnitude. The differences between the calculated and the measured curves at large Q_{cp} therefore account for carrier re-emission.

Besides, in the rising part of the curves in Fig. 4a, it can be seen that at low Q_{cp}, the experimental curves increase primarily more slowly with V_g magnitude than those calculated using Eq. 2 in which D_{it} is constant, before increasing faster.

This point is detailed in Fig. 4b. In this figure, the slope of all the experimental curves in Fig. 2a and Fig. 2b is plotted and compared to that calculated using Eq. 2 as function of the bias that varies, as in Fig. 4a. One can see that the slopes of all the curves of the same type are very close together, but that they are greater when V_h varies than when V_l varies, due to the asymmetry in surface potential, ψ_s, variations with regard to $\psi_s = \Phi_B$, Φ_B being the Fermi potential. For both curve types, the experimental slope maximum (or the inflexion point in $Q_{cp}(V_g)$ curves), is close to that calculated using Eq. 2 in which D_{it} extracted from CP Elliot curves maxima has been determined accurately [36].

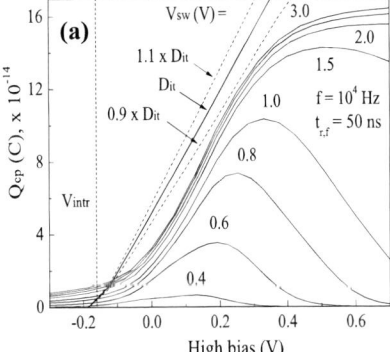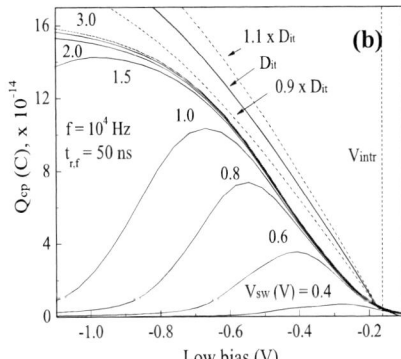

Figure 5. Detail of the a) left hand edge of the curves in Fig. 1a, when plotted as function of V_h and b) right hand edge of the same curves when plotted a function of V_l in Fig. 1b. Curve edges are compared to Q_{cph} and Q_{cpl} plots calculated using Eq. 2 with either $D_{it} \approx 5 \times 10^{10}$ eV^{-1} cm^{-2}, as extracted from curves maxima, or 0.9 and 1.1 times this D_{it} value.

The points discussed above are further confirmed in Figs. 5a and 5b, where the edges of the Elliot curves in Figs. 1a and 1b, i.e. as function of V_h and of V_l, here for V_{sw} values ranging from 0.4 V to 3 V are detailed and compared with the $Q_{cp}(V_h)$ and $Q_{cp}(V_l)$ curves calculated again using Eq. 2. Here, D_{it} extracted from CP curve maxima was found around 4.8×10^{10} eV^{-1} cm^{-2} and for the sake of comparison, 0.9 and 1.1 times this value in Eq. 2 has also been plotted. Fig. 5 further points out that D_{it} given by the slope of the experimental CP curves cannot be far from that obtained at curve maxima and used in Eq. 2.

It is clear that such a slope could not be obtained if most of the traps probed were far from having a single response to capture, as shown by the trap profiles extracted from the TCD in such devices [28]. For instance, this could not be verified in case of a constant trap density from the interface towards oxide depth instead of an oxide trap concentrations 100 to 200 times smaller than that at the interface [28]. Fig. 5 also highlights that coefficient 2 in Eq. 2 is far beyond the uncertainty that may be obtained on the slope of CP curves when overlooking it.

As a result, CP curves recorded from conventional and fresh MOSFET such as those studied here, are supposed to mainly probe Si-SiO$_2$ interface traps so that D_{it} can be extracted from CP curves maximum after accounting for carrier re-emission [29, 36]. The above results demonstrate that D_{it} also straightforwardly enters the slope of the CP curves in a region where capture still governs Q_{cp} magnitude as will be shown.

Extraction of D_{it} from the Slope of CP Curves

In the region discussed above where Q_{cp} is controlled primarily by capture (Fig. 4a) [31, 32], if for a given $E_{fh} = E_{fhc}$ (resp. $E_{fl} = E_{flc}$) position, most of the traps that may enter Q_{cp} already contribute to it, so that the increase in Q_{cp} essentially results from the increase in ΔE, the interface trap density D_{it} may be obtained from the slope of the CP curves using:

$$D_{it,h,l} = \frac{1}{2qWL} \left| \frac{dQ_{cph,l}}{dV_{h,l}} \frac{dV_{h,l}}{dE_{fhc,flc}} \right|_{Max}. \tag{3}$$

Eq. 3, which has been generalized to the cases when the high, E_{fh}, or the low, E_{fl}, Fermi level control/limits recombination also takes into accounts coefficient 2 discussed in the preceding section.

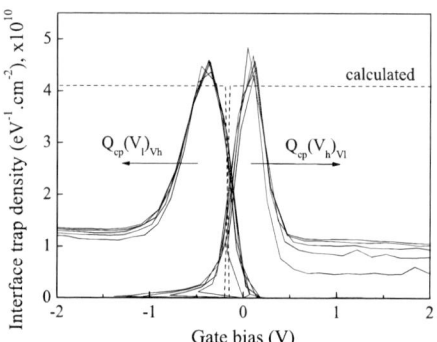

Figure 6. D_{ith} and D_{itl} values extracted from the slope of all $Q_{cp}(V_h)_{Vl}$ and $Q_{cp}(V_l)_{Vh}$ curves plotted in Fig. 2a and Fig. 2b, respectively. $D_{ith} \approx D_{itl} \approx 4.4 \times 10^{10}$ eV^{-1} cm^{-2}.

The result in case of the $Q_{cp}(V_h)_{Vl}$ and $Q_{cp}(V_l)_{Vh}$ curves shown in Figs. 2a and 2b, the slope of which have been compared in Fig. 4a and Fig. 4b to that obtained using Eq. 2 is shown in Fig. 6.

It can be seen that the curves in Figs. 2a and 4a, on the one hand, or in Figs. 2b and 4a, on the other hand, which exhibit significantly different slopes for the same D_{it} measured at Q_{cp} maximum, yield the same $D_{it} \approx D_{ith} \approx D_{itl}$ through Eq. 3: $D_{it} \approx 4.4 \times 10^{10}$ eV^{-1} cm^{-2}. As expected in Fig 6, the "calculated" D_{it} is constant and equal to that introduced in Eq. 2, i.e. $D_{it} = 4.1 \times 10^{10}$ eV^{-1} cm^{-2}.

The case of Elliot curves in Figs. 5a and 5b is dealt with in Figs. 7a and 7b. Again, the significantly different V_h or V_l Q_{cp} dependence observed in Fig. 1 (see Figs. 13a, 13b and [30]), provide nearly the same $D_{it} \approx D_{ith} \approx D_{itl}$. Here, $D_{ith} \approx 5.8 \times 10^{10}$ eV^{-1} cm^{-2} is slightly greater than $D_{itl} \approx 5.5 \times 10^{10}$ eV^{-1} cm^{-2}. This may be attributable at least partly to the source and drain regions in these devices the small contribution of which can be observed, as in Fig. 4a, at low Q_{cp} and negative V_h in the CP curves in Fig. 7a [31]. In the $D_{it}(V_l)_{Vsw}$ curves in Fig. 5b, this contribution is rejected towards the top of the CP curves, i.e. at the extreme left of the curves recorded at large V_{sw} in Fig. 7b, well beyond the bias providing D_{it} [31, 34].

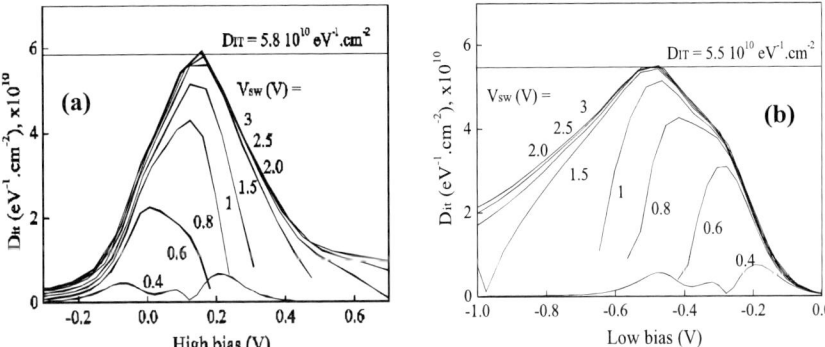

Figure 7. a) $D_{ith,}$ and b) D_{itl} extraction from the slope of the Elliot curves shown in Fig. 1a and 1b, respectively. $D_{ith} \approx 5.8 \times 10^{10}$ eV^{-1} cm^{-2} and $D_{itl} \approx 5.5 \times 10^{10}$ eV^{-1} cm^{-2} [After 30].

Detailed Observation of Si-SiO₂ Interface Traps

Accounting for the CP mechanisms recalled above, the central region of the silicon bandgap and the Si-SiO₂ interface traps located there can be observed and detailed in different conditions. The first ones, pointed out primarily by Wachnik [38] using Elliot curves and small gate pulses have been clarified in [37], studied experimentally in [28] and simulated here in Fig 3b. They are accounted for in the CP model and have been referred to as "the high frequency limit" case.

As pointed out in [28, 32], in strong asymmetrical capture conditions, i.e. when E_{fh} or E_{fl} strongly limit recombination, (($E_{fh} - E_i$) << ($E_i - E_{fl}$) or ($E_i - E_{fl}$) << ($E_{fh} - E_i$)), the center of the silicon bandgap can also be probed once when the CP current is limited by E_{fh} (left hand edge of the CP Elliot curves) and a second time when it is limited E_{fl} (right hand edge of the same curves). The transition from the situation depicted in the preceding paragraph to those ones has been detailed in [28]. They are achieved at large enough V_{sw} values, when recording CP curves at different frequencies and making the difference between them at identical V_g. The result already reported in [28] which simply comes from the difference in the experimental CP curves in Fig. 3a and is recalled in Fig. 8a.

For the peak on the left hand side of the figure, E_{fh} limits the CP current (left hand edge of the experimental curves in Fig. 3a) while for that on the right hand side of Fig. 8a, this is E_{fl} that limits Q_{cp} (right hand edge of the CP curves in Fig. 3a). Between the two peaks, due to the strong increase of the concentration of the carrier type that limits recombination and to the resulting increase in Q_{cp}, the top of the CP curve is reached. The small Q_{cp} variations with f there (plateau region in the trap profiles extracted from the trap TCD) leads through the Q_{cp} difference between successive frequencies to the wide depression between the two lateral peaks [28].

The two different interface trap types entering the CP signal are well visible, especially in the peak on the right hand side of the figure due to the slower Q_{cp} variations with V_l with regard to V_h. A simulation also presented in [28] and accounting for only one trap type at the Si-SiO$_2$ interface is shown for comparison in Fig. 8b.

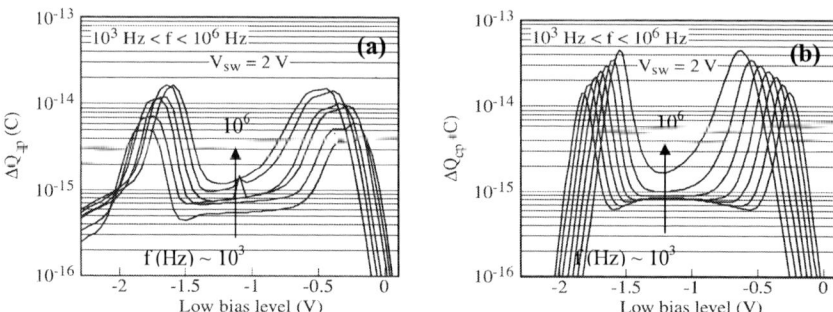

Figure 8. a) Difference between curves recorded at successive frequencies in Fig. 3a, showing in the lateral peaks the contributions to the CP signal of two trap types as also observed each time the silicon bandgap central region is probed; b) Comparison with a simulation accounting for one trap type at the interface [After 28].

This technique has been referred to as Energy and Frequency Resolved CP (EFRCP). Indeed, it accounts for the increase in energy swept at the interface due to the control of recombination by one carrier type only and the consequence on recombination in the other half of the silicon bandgap. It also accounts for the difference between the signal measured at different frequencies which points out the traps entering recombination when allowing them to capture during a longer time at V_h or V_l [28].

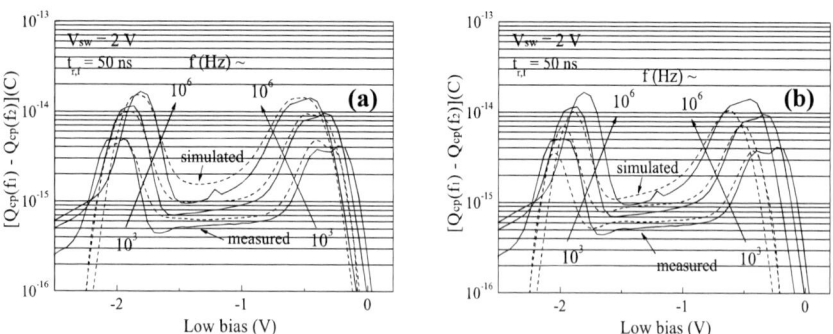

Figure 9. a) Difference between experimental curves recorded using successive gate signal frequencies at $V_{sw} = 2$ V and corresponding simulations, both in Fig. 3a; b) Same as in Fig. 9a, but contribution only to the peaks simulated of the trap type having the larger cross section for electrons and the smaller one for holes.

In Fig. 9a, the curves in Fig. 8a are compared to simulation accounting for now the two Si-SiO$_2$ interface trap types yielding the CP curves simulated in Fig. 3a. In other words, both the experimental and simulated curves account for the Coulomb attractive and neutral states of the two trap types probed at the Si-SiO$_2$ interface. The simulations very well account for what is observed experimentally. As the experimental peaks are very well simulated in Fig. 9a, it becomes possible to separate the contribution of the two kinds of traps to these peaks i.e. to the slope of the CP curves. This is done in Fig. 9b, where only the contribution of the traps having a larger cross section for electron than for holes (donor-like trap type) has been plotted.

As expected but shown here for the first time with regard to actual device characteristics, when shifting the gate pulse from negative to positive biases, and when ($E_{fhc} - E_i$) increases (left hand edge of the CP curves in Fig. 3a and peak on the left hand side in Fig. 9a and 9b), due to their greater cross section for electron capture with regard to that of the acceptor-like traps which are neutral, the positively charges and Coulomb attractive donor-like traps will capture an electron first yielding the contribution simulated in the left hand peak of the experimental and simulated curves in Fig. 9b. At greater ($E_{fhc} - E_i$) values, the neutral acceptor-like traps will also capture an electron allowing the CP current to increase again and yielding the second contribution to the left hand peak in Fig. 9a not simulated in Fig. 9b.

When reaching the left hand edge of the CP curves in Fig. 3a, ($E_i - E_{flc}$) now decreases. Therefore, due to the smaller cross section for holes capture with regard to that of the negatively charged and Coulomb attractive acceptor-like traps, the positively charges donor-like traps disappear first from the CP current (contribution accounted for in the peak on the right hand edge of the curves in Fig. 9b). With additional decrease of ($E_i - E_{flc}$), the contribution of the Coulomb attractive acceptor-like traps will also decrease until the CP current cancels out.

Impact of the Gate Signal Frequency on Interface Traps Filling

For given gate signal frequency and trap cross section, the conditions allowing the interface traps to fill, defined in [28, 36], are given by:

$$E_{fh,l} - E_i = \pm kT \ln \frac{(n_i \sigma_{e,h} V_{th} t_{h,l})}{[-\ln(0.5)]}. \tag{4}$$

The E_{fh} and E_{fl} values extracted from Eq. 4 are shown in Fig. 10 in case of a trap capture cross section of the order of a few 10^{-16} cm^2 as measured on the device studied using the method proposed in [10, 27]. This corresponds to the high and low threshold voltages, V_{thcph} and V_{thcpl}, which allow CP to occur during the high and low bias levels, well before capture proceeds during the transitions times of gate signal [32, 36]. With such cross sections, the thresholds are around 0.15 eV above/below E_i at f = 10^4 Hz, but a ten fold increase (decrease) in these values reduces (augments) the thresholds by 60 meV [36].

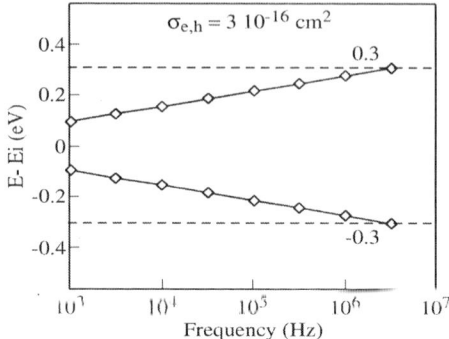

Figure 10. E_{fh} or E_{fl} energy position allowing the filling of the traps with electrons or holes if $\sigma_e \approx \sigma_h \approx 3\times10^{-16}$ cm^2. A ten fold increase (decrease) of σ_e or σ_h reduces (increases) these thresholds by 60 meV [After 36].

The dependence of these thresholds on the time available for capture $t_{h,l} = (1/2f - t_{rf})$, explains why the slope of CP curves varies with f in Fig. 3a as observed experimentally and using simulation. It results that the higher the gate signal frequency in Fig. 3a, the greater the $(E_{fhc} - E_i)$ or $(E_i - E_{flc})$ values required to fill the traps in Fig. 10 and the stepper the edges of CP curves. This also reduces distance between lateral peaks in Figs. 9a and 9b.

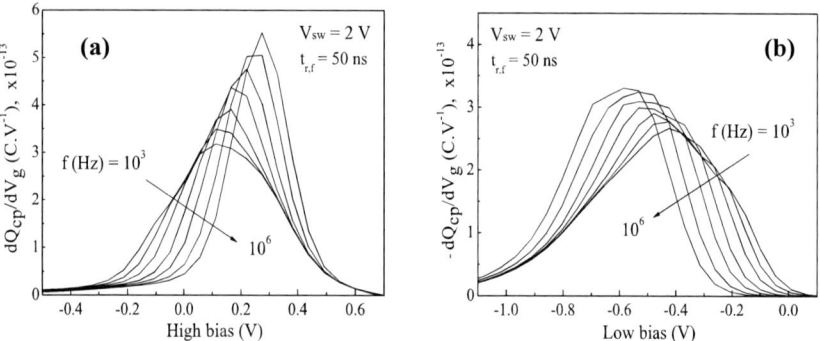

Figure 11. Slope of the a) left and b) right hand side of the experimental curves in Fig. 3a.

As the slope of the CP curves significantly varies with f (Fig. 3a), one may wonder in which conditions D_{it} can be extracted from Eq. 2 through surface potential variations as shown in Figs. 4a or 5a and 5b.

To answer that question, slope of the experimental curves recorded as function of f in Fig. 3a are shown in Figs. 11a and 11b. These figures show, in agreement with Fig. 10 that the shape of the slope changes when varying f. The more abrupt filling of the Si-SiO$_2$ interface traps at high frequencies in Figs. 11 yielding greater slope maxima in

Fig. 11a and 11b should not allow a good correlation between traps filling and D_{it} extracted from CP curves maximum, but should lead to overestimate D_{it}.

Another Approach to Si-SiO$_2$ Interface Traps Filling

Looking in detail at the two peaks in Fig. 6, obtained from the slope of all the curves in Fig. 2a and 2b, and yielding D_{it} in both cases, one can note that two contributions enter the slope of the curves. The first one, yielding the steeper increase in the slopes, appears with increasing V_h or V_l towards more positive or more negative values, respectively. It allows, through the slope, D_{it} to reach $\approx 3/4$ of its final value. The second one, smaller in magnitude yields the greatest slope up to the inflexion point and gives the rest of D_{it}.

The same remarks hold for the curves in Fig. 7a and 7b in which Elliot plots have been dealt with. The steepest part of the slope when E_{fh} governs Q_{cp} in Fig. 7a contributes to around half of the final D_{it} when it seems to provides 3/5 of this value when this is E_{fl} that governs Q_{cp} in Fig. 7b.

Besides, when V_{sw} is increased, for a given V_h in Fig. 7a or a given V_l in Fig. 7b, recombination increases because V_l is greater in the first case while V_h is greater in the second one. In these conditions one observe in both cases that the slope increase and that at $V_{sw} = 0.8$ V, it does not increase anymore and remains therefore constant there in the first or "steeper increase region". Nevertheless, when V_h becomes grater than -0.02 V in Fig. 7a, and V_l more negative than -0.3 V in Fig. 7b, the second component in the slope appears and saturates also at V_{sw} from 1.5 to 2 V, leading to the final D_{it} value.

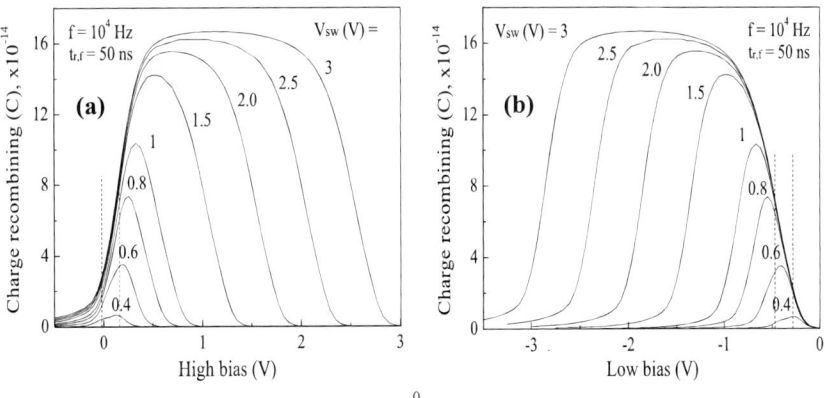

Figure 12. Same set of Elliot curves as in Figs. 1a and 1b, respectively, but in which the bias positions where the two contributions in the slopes, pointed out in Fig. 7a and 7b, fully enter them are marked.

The bias positions in Figs. 1a and 1b where these two contributions fully enter the slope and D_{it} are marked in Fig. 12a and 12b by vertical dashed lines. In these figures the curves reported in Fig. 1a and 1b are shown again. One can note that the first CP curve contribution occurs in the very low CP current region, when the inflexion point is situated at around half the maximum Q_{cp} value obtained at large V_{sw}. One can also note that the biases where these contributions fully enter the slope seem located above the traps entering the CP curve recorded at $V_{sw} = 0.4$ V in Fig. 5 or 12, i. e., in the CP current high frequency limit.

The position of these contributions to the slopes is also reported in Fig. 13a and 13b where the slopes of the CP curves in Fig. 5a and 5b not presented until now are displayed. From these figures and from the CP curves in Fig. 5a and 5b, the first contribution to the Q_{cp} increase in Figs. 5a and 12a corresponds to the first maximum in the left of the curves at $V_{sw} = 0.4$ V when the actual slope maximum in these curves is slightly beyond the second maximum of the curves at $V_{sw} = 0.4$ V and situated at $V_h \approx 0.15$ V.

The reverse applies for the curves in Fig. 5b and 12b for which the first part of the slope increase when increasing V_h is situated above the first maximum from the right of the curve at $V_{sw} = 0.4$ V situated at $V_l \approx - 0.28$ V, when the actual slope maximum, at $V_l \approx - 0.48$ V, is slightly beyond the second maximum in the curves recorded at $V_{sw} = 0.4$ V.

By plotting the curves in Fig. 13a and 13b as function of the E_{fh} and E_{fl} Fermi level positions (not shown), it is found that the strongest increase in the slopes yield $(E_{fh} - E_i) \approx (E_i - E_{fl}) \approx 0.1$ eV while the inflexion points occurs at $(E_{fh} - E_i) \approx (E_i - E_{fl}) \approx 0.2$ eV, i.e. well before $(E_{fh} - E_i) \approx (E_i - E_{fl}) \approx 0.3$ eV where carrier re-emission fully enters and limits Q_{cp} [32].

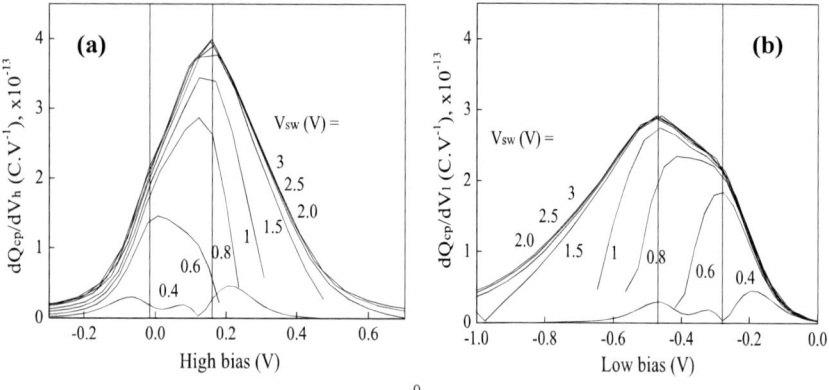

Figure 13. Slope of the CP curves in Figs. 5a and 5b yielding D_{it} in Figs. 7a and 7b, respectively. The biases where the contributions fully enter the slope of the curves are marked with vertical dashed lines.

These two regions decompose differently in Fig. 11a and 11b when increasing gate signal frequency. This may suggest that the strongest increase results from the filling through ΔF of the fastest traps the concentration of which is the greatest in the trap TDC [30, 31]. It may be followed by that of the slower part of the interface traps the concentration of which rapidly decreases with increasing time constant [28, 31]. The conditions allowing D_{it} to be measured may be given by the conditions in which ΔF, which is not a step function [31], a) accounts for the largest part of the trap TCD so that b) further Q_{cp} increase is essentially due to the increase of surface potential in agreement with Eq. 2. Such conditions should not be fulfilled at higher frequencies (Fig. 11), where D_{it} extracted from the slopes should be overestimated due to both the contribution to Eq. 3 of ΔE and ΔF. Nevertheless, these conditions along with their relations with the two trap types should be clarified.

Conclusion

The simulation of CP curves of fully processed MOSFET with thick SiO_2 has been extended to the dependence of the CP curves on gate signal frequency and to the high frequency limit of the CP signal with very good results again. In particular, is has been shown that the variations of the slope of CP curves with gate signal frequency are very well accounted for by the model. This further confirms that at least at the $Si-SiO_2$ interface, the CP mechanisms along with trap properties are now well accounted for.

It has also been confirmed using all basic CP curve types and several devices that provided that the source and drain regions of the transistors do not significantly impact the recombination current, the slope of the CP curves can be well approximated by Eq. 2, that is, by accounting for the CP mechanism in which capture on one half of the silicon forbidden bandgap governs Q_{cp} through surface potential variations. The contribution to the CP current of the second half of the silicon bandgap is obtained through the SRH kinetics and give a coefficient 2 (symmetrical variations with regard to E_i) that very well accounts for what is observed experimentally [32, 33]. However, this "image" of what occurs in the other part of the silicon bandgap results from carrier re-emission and not from Fermi level position. It result that the first and intrinsic cause of CP curves asymmetry is surface potential variations with regard $\psi_s = \Phi_B$.

Due to the above situation and using the left and right hand edge of CP curves which exhibit significantly different slopes, it has been shown that the interface trap density extracted conventionally from the maximum CP current can also be obtained from the slope maximum provided that surface potential variations are accounted for. This confirms that the probing of the $Si-SiO_2$ interface traps in a way that clarifies now, constitutes the essential of the CP curves in such MOSFETs, so that these traps along with the resulting overall D_{it} can be probed using different ways depicted in the paper, all leading in proper conditions to the same D_{it} value.

The excellent results obtained when simulating the frequency dependence of the CP Elliot curves provided evidence through the simulations also, of the contribution of the two kinds of traps widely pointed out at the $Si-SiO_2$ interface in our previous studies. This enabled to extract from CP curves a signal having two components in which the traps that may be donor-like and those that may be acceptor-like have been clearly distinguished and their contribution to the CP curves delineated as never done before.

Finally, regardless of the curve type provided that large enough voltage swings are used, the slope of the CP curves has also systematically been shown to include two parts each of which contributing successively to the D_{it} values finally extracted at large enough

V_{sw} from the inflexion point of the CP curves. At first guess, this point is situated at half Q_{cp} maximum provided a low enough frequency is used. These two components well correspond to the peaks pointed out in the CP curves recorded in the high frequency limit conditions. The corresponding energies in the bandgap, well below those where emission limits Q_{cp} magnitude through the amplitude of the energy excursion, have been found at $E_{fh,l} \approx E_i \pm 0.1$ eV for the first one and $E_{fh,l} \approx E_i \pm 0.20$ eV for that yielding D_{it}. The dependence of the slopes on gate signal frequency suggests the conditions for reliable D_{it} extraction there, but this point along with the relations with traps filling and with both trap types need to be clarified.

Acknowledgements

The research leading to these results has been partially supported by the European network of excellence NANOSIL and Catrene Project UTTERMOST.

References

1. D. Bauza, *Thermal Oxidation of Silicon and Si-SiO₂ Interface Morphology, Structure and Localized States*, Chapter 2, p. 115 in *Handbook of Surfaces and Interfaces of Materials*, Vol. 1, H. S. Nalwa Editor, Academic Press, San Diego-CA (2001).
2. L. M. Terman, *Solid-St. Electron.*, **5**, 285 (1962).
3. R. J. Van Overstraeten, G. Declerck, and G. L. Broux, *IEEE Trans. Electron Devices*, **20**, 1150 ((1973).
4. B. E. Deal, M. Sklar, A. S. Grove, and E. H. Snow, *J. Electrochem. Soc.*, **114**, 266 (1967).
5. B. E. Deal, *IEEE Trans. Electron Devices*, **27**, 606 (1980).
6. D. Bauza, *IEEE Electron Device Lett.*, **23**, 658 (2002).
7. P.V. Gray and D.M. Brown, *Appl. Phys. Lett.*, **8**, 31 (1966).
8. J. L. Autran, F. Seigneur, and B. Balland, *J. Appl. Phys.*, **74**, 3932 (1993).
9. D. Bauza, Y. Morfouli, and G. Pananakakis, *Solid-St. Electron.*, **34**, 933 (1991).
10. E. H. Nicollian and J. R. Brews, *MOS Physics and Technology*, Wiley NY (1982).
11. D. Bauza and Y. Manéglia, *IEEE Trans. Electron Devices*, **44**, 2262 (1997).
12. E. Simoen, B. Dierickx, C. L. Clayes, and G. J. Declerk, *IEEE Trans. Electron Devices*, **39**, 422 (1992).
13. C. Hu, S. C. Tam, F. C. Hsu, P. K. Ko, T. Y. Chan, and K. W. Terrill, *IEEE Trans. Electron Devices*, **48**, 1127 (2001).
14. K. Jeppson and C. Svensson, *J. Appl. Phys.*, **48**, 2004 (1977).
15. M. A. Alam and S. Mahapatra, *Microelectronics Reliability*, **47**, 841 (2007).
16. E. Amat, R. Rodriguez, M. Nafria, X. Aymerich, and J. H. Stathis, *Microelectronics Reliability*, **47**, 844 (2007).
17. H. Katto and Y. Itoh, *Phys. Stat. Sol. (a)*, **21**, 627 (1974).
18. S. Collins, M. J. Kirton, and M. J. Uren, *Appl. Phys. Lett.*, **57**, 372 (1990).
19. M. J. Kirton, M. J. Uren, S. Collins, M. Schulz, A. Karman, and K. Scheffer, *Semicond. Sci. Technol.*, **4**, 1116 (1989).
20. J. S. Brugler and P. G. A. Jespers, *IEEE Trans. Electron Devices*, **16**, 297 (1969).

21. G. Groeseneken, H. E. Maes, N. Beltran, and R. F. de Keersmaecker, *IEEE Trans. Electron Devices*, **31**, 42 (1984).
22. D. Bauza in Chap. 15 in *"Nanoscale CMOS,"* F. Balestra Editor, Wiley, 2010.
23. M. Denais, A. Bravaix, V. Huard, C. Parthasarathy, G. Ribes F. Perrier, Y. Rey-Tauriac, and N. Revil, *IEDM Tech. Digest*, p. 109 (2004).
24. C. Shen, M.-F Li, X. P. Wang, Y.-C. Yeo, and D.-L. Kwong, *IEEE Electron Device Lett.*, **27**, 55 (2006).
25. W. Wang, J. Deng, J. C. M. Hwang, Y. Xuan, Y. Wu, and P. D. Ye, *Appl. Phys. Lett.*, **96**, 0721021 (2010).
26. A. Lubov, S. Ismail-Geigi, and T. P. Ma, *Appl. Phys. Lett.*, **96**, 122105 (2010).
27. Y. Manéglia and D. Bauza, *J. Appl. Phys.*, **79**, 4167 (1996).
28. D. Bauza, S. Bayon, and O. Ghobar, *ECS Transactions*, **6**(3), 3 (2007).
29. D. Bauza, O. Ghobar, N. Guénifi, and S. Bayon, *ECS Transactions,* **19**(2), 19 (2009).
30. D. Bauza, *ECS Transactions*, **28**(2), 251 (2010).
31. Y. Manéglia, F. Rahmoune and D. Bauza, *J. Appl. Phys.*, **97**, 014502 (2005).
32. D. Bauza, *IEEE Trans. Electron Devices*, **56**, 70 (2009).
33. D. Bauza, *IEEE Trans. Electron Devices*, **56**, 77 (2009).
34. P. Heremans, J. Witters, G. Groeseneken, and H.E. Maes, *IEEE Trans. Electron Devices*, **36**, 1318 (1989).
35. A. B. M. Elliot, *Solid-St. Electron.*, **19**, 241 (1976).
36. D. Bauza, *J. Appl. Phys.*, **94**, 3239 (2003).
37. D. Bauza and G. Ghibaudo, *Solid-St. Electron.*, **39**, 563 (1996).
38. R. A. Wachnik, *IEEE Trans. Electron Devices*, **33**, 1054 (1986).

Clear Difference between the Chemical Structure of SiO_2/Si Interfaces Formed Using Oxygen Radicals versus Oxygen Molecules

Tomoyuki Suwa[a], Yuki Kumagai[b], Akinobu Teramoto[a], Takayuki Muro[c],
Toyohiko Kinoshita[c], Tadahiro Ohmi[a,d], and Takeo Hattori[a]

[a] New Industry Creation Hatchery Center, Tohoku University, Sendai, 980-8579, Japan
[b] Graduate School of Engineering, Tohoku University, Sendai, 980-8579, Japan
[c] Japan Synchrotron Radiation Research Institute, Sayo-cho, Hyogo 679-5198, Japan
[d] WPI Research Center, Tohoku University, Sendai, 980-8579, Japan

Soft-x-ray-excited angle-resolved photoelectron spectroscopy studies on silicon dioxide films formed using oxygen radicals (OR) versus oxygen molecules (OM) are reported. Most of intermediate oxidation states of Si, so called suboxides, consisting of Si^{3+}, Si^{2+}, and Si^{1+} are localized at and near the SiO_2/Si interfaces. The suboxides formed utilizing OM are located extremely close to the interfaces, while suboxides formed utilizing OR are located not only at the interface but also in the Si substrate. This implies the penetration of a part of OR into the Si substrate to form Si^{1+} and to relax the interfacial stress. Intermediate chemical bonding states of Si are formed in Si substrate closer to the interface as compared with those formed utilizing OM. This implies that the interfacial stress in the former is smaller than that in the latter.

Introduction

It is strongly anticipated that advanced ultra-large-scale-integrated (ULSI) devices, which can operate at ultrahigh speed with low power consumption, will be used in future electronics. This will require the control of the oxidation process on an atomic scale. To meet this requirement, the oxidation utilizing oxygen (O) atoms (hereafter referred to as O radicals (OR) produced in a microwave-excited high-density Kr/O_2 mixture plasma [1] has been extensively studied in recent years [2-10] and has been used to form high quality oxide films on a Si surfaces with any crystallographic orientation [2]. On the other hand, in the case of conventional thermal oxidation utilizing dry oxygen molecules (OM), the relatively high quality oxide film was obtained only on Si(100) surface. The purpose of this study is to clarify the difference between the chemical structures of the SiO_2/Si interfacial transition layer formed on Si(100) utilizing OR and that formed utilizing OM by measuring angle-resolved photoelectron spectra arising from the Si $2p$ core levels.

Experimental Details

To detect the photoemission arising from the Si $2p$ core levels at and near the SiO_2/Si interface through nearly 1-nm-thick silicon oxide films, an inelastic mean free path [10] of more than 2 nm in silicon oxide is necessary. It is also necessary to measure the Si $2p$ photoelectron spectra with high kinetic energy resolution. Such measurements were performed using photon energy (PE) of 1050 eV at Super Photon Ring 8 GeV (SPring-8).

The wafers used in this study were 10 Ωcm p-type Si(100) substrates whose surfaces are atomically flat on a wafer scale [7-9]. The wet oxidation of these substrates was performed at 1100 °C to form 1-μm-thick silicon oxide films. After etching the oxide films in a HCl/HF mixture solution [11], the substrates were cleaned in five steps [23]. The surface microroughness (Ra) of the cleaned substrates measured by atomic force microscopy (AFM) was less than 0.08 nm. A 0.60-nm-thick oxide film was formed on the cleaned substrate at 400 °C utilizing OR produced in a microwave-excited high-density Kr/O$_2$ mixture plasma [1, 2] at a pressure of 133 Pa. Here, the microwave frequency and power were 2.45 GHz and 2.7 W/cm^2, respectively. A 0.91-nm-thick oxide film was also formed on the cleaned substrate at 900 °C utilizing dry OM. Furthermore, to prevent further oxidation, these oxide films were kept in oxygen free isopropyl alcohol solution, until their photoelectron spectra were measured.

Experimental Results and Discussion

We measured the Si $2p_{3/2}$ spectrum arising from the 0.60-nm-thick oxide films formed on Si(100) surface utilizing OR and that arising from 0.91-nm-thick oxide films formed on Si(100) surface utilizing dry OM at the photoelectron take-off angles (TOAs) of 85°, 52°, 30°, 15° and 10° and a PE of 1050 eV. After removing the background signal based on Tougaard's method from the observed spectrum [13], the spectrum is decomposed into the Si $2p_{1/2}$ and Si $2p_{3/2}$ spin-orbit partner lines. In this decomposition, it was assumed that the spin-orbit splitting of the Si $2p$ spectrum is 0.605 eV and the Si $2p_{1/2}$ to Si $2p_{3/2}$ intensity ratio is 0.5 [14]. Figure 1 show the Si $2p_{3/2}$ spectra arising from (a) oxide film formed utilizing OR and (b) that formed utilizing dry OM measured at TOAs of 85° and 15°. These spectra were normalized by the intensity of the Si $2p_{3/2}$ spectra arising from Si substrate.

Figure 1(a) and 1(b) show Si $2p_{3/2}$ spectra arising from oxide films formed utilizing OR and those formed utilizing dry OM measured at TOAs of 85° and 15°, respectively. Figure 2(a) is obtained by subtracting the Si $2p_{3/2}$ spectrum measured at TOA of 85° from the Si $2p_{3/2}$ spectrum measured at TOA of 15° for the oxide film formed utilizing OR. This subtraction was performed, to eliminate spectrum arising from the bulk Si after multiplying the spectrum measured at a TOA of 85° by an appropriate factor a [6, 15]. Figure 3(a) is obtained by subtracting the Si $2p_{3/2}$ spectrum measured at TOA of 15° from Si $2p_{3/2}$ spectrum measured at TOA of 85° for the oxide film formed utilizing OR. This subtraction was performed, to eliminate spectrum arising from the oxide film after multiplying the spectrum measured at a TOA of 15° by an appropriate factor b [6, 15]. Figures 2(b) and 3(b) show the Si $2p_{3/2}$ spectrum arising from the silicon oxide film and the SiO$_2$/Si interfaces formed utilizing dry OM obtained by the same procedure used to obtain Fig. 2(a) and Fig. 3(a), respectively.

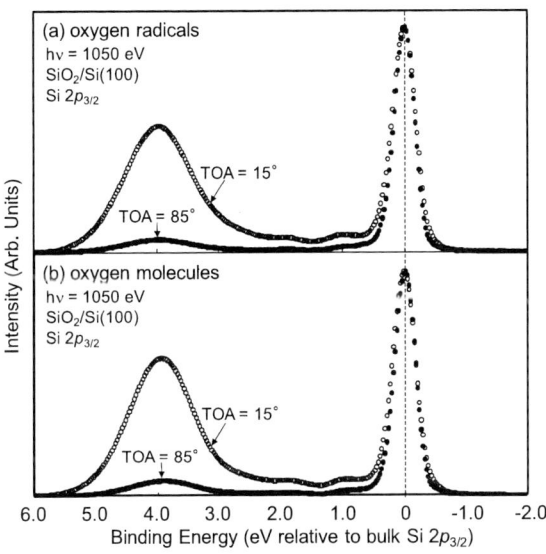

Figure 1. Si $2p_{3/2}$ spectra arising from (a) oxide film formed utilizing OR and (b) that formed utilizing dry OM measured at TOAs of 85° and 15°.

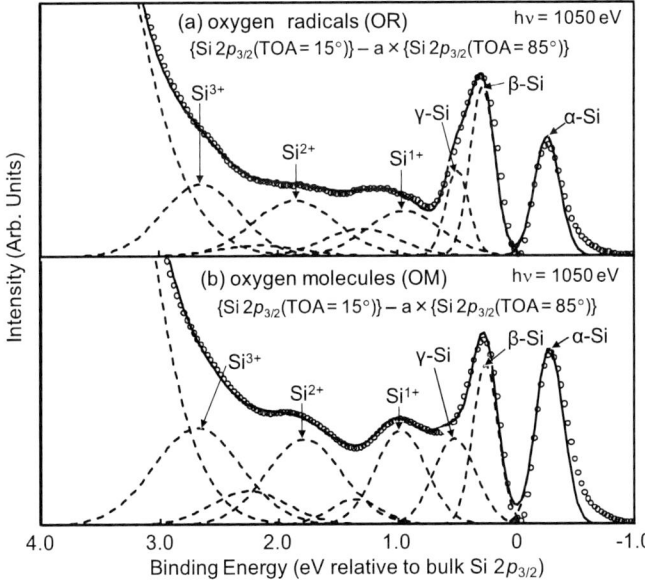

Figure 2. Si $2p_{3/2}$ spectra obtained by taking difference between two spectra measured at TOAs of 15° and 85° to eliminate spectra arising from bulk Si (a) for oxide film formed utilizing OR and (b) for oxide film formed utilizing dry OM.

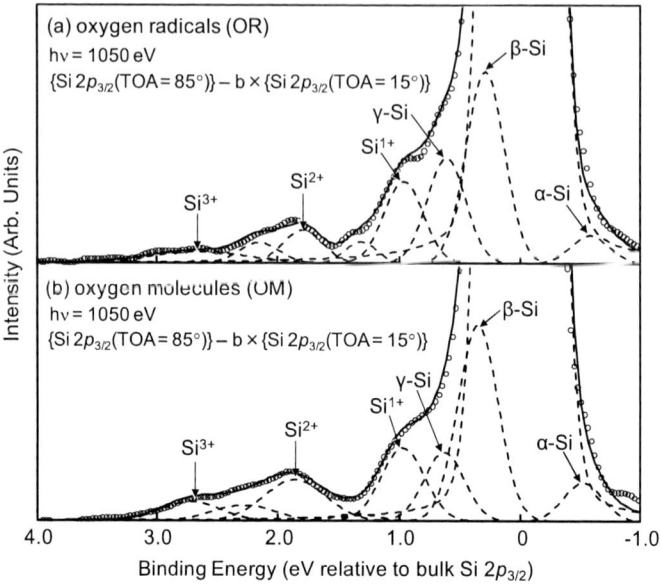

Figure 3. Si $2p_{3/2}$ spectra obtained by taking difference between two spectra measured at TOAs of 85° and 15° to eliminate spectra arising from silicon oxide film (a) for oxide film formed utilizing OR and (b) for oxide film formed utilizing dry OM.

It is found from Figs. 2(a), 2(b), 3(a), and 3(b) that both the Si $2p_{3/2}$ spectrum arising from the SiO_2/Si interface formed utilizing OR and that utilizing dry OM can be decomposed into the spectra having almost the same chemical shifts (CSs) in the binding energies (BEs) of Si $2p$ core levels from that in bulk Si (Si^0) as listed in Table I. The decomposed spectra consist of the spectra arising from $\underline{Si}$-O_4 (Si^{4+}), Si (Si^0), and intermediate chemical bonding states of Si (IMSs) consisting of Si-$\underline{Si}$-O_3 (Si^{3+}), Si_2-$\underline{Si}$-O_2 (Si^{2+}), Si_3-$\underline{Si}$-O (Si^{1+}), α-Si, β-Si, and γ-Si. Using the chemical shifts (CSs) thus determined in Figs. 2 and 3, the Si $2p_{3/2}$ spectra can be decomposed. Here, Si^{3+}, Si^{2+}, and Si^{1+} denote the intermediate oxidation states of Si (IOSs) having a BE between BE(Si^{4+}), that is BE of Si $2p$ core level arising from Si^{4+}, and BE(Si^0) [16]. The spectra arising from α-Si is affected by its second nearest neighbor O atoms [17] and those arising from γ-Si formed utilizing OR is affected strongly by removing H atoms before the oxidation.

Table I Chemical shift in the binding energy of Si $2p$ core level in Si^{4+} and those in intermediate chemical bonding states of Si, from that in bulk Si (Si^0)

	Chemical Shifts (eV)							
	SiO_2	Si^{3+}	Si^{2+}	Si^{1+}	γ-Si	β-Si	bulk-Si	α-Si
oxygen radicals (OR)	3.97	2.65	1.86	0.93	0.50	0.30	0	-0.24
oxygen molecules (OM)	3.90	2.67	1.83	0.97	0.55	0.27	0	-0.25

The quantitative spectral analyses are performed based on the following assumptions: 1) the number of Si atoms in a unit volume of SiO_2 is 2.28×10^{28} m^{-3} [18], 2) inelastic mean free path (IMFP) in the Si denoted by Λ_{bs}, and that in the SiO_2 denoted by Λ_{bo} are 1.59 nm and 2.86 nm, respectively [19], 3) σ_{ct2} and σ_{ct1} are given by following equations (1) and (2).

$$n_{sct2}\sigma_{ct2} = \frac{n_{sbo}\sigma_{bo} + n_{sct1}\sigma_{ct1}}{2} \tag{1}$$

$$n_{sct1}\sigma_{ct1} = \frac{n_{sct2}\sigma_{ct2} + n_{sbs}\sigma_{bs}}{2} \tag{2}$$

Here, n_{sbo}, n_{sct2}, n_{sct1}, and n_{sbs} denote the density of SiO_2 molecules, the total density of SiO_2 and Si_2O_3 molecules in the layer containing SiO_2 and Si_2O_3, the total density of SiO and Si_2O molecules in the layer containing SiO and Si_2O, and the density of Si atoms in the Si substrate, respectively, σ_{bo}, σ_{ct2}, σ_{ct1}, and σ_{bs} denote the inelastic cross section of electron with SiO_2 molecule, that with oxidized Si molecule in the layer containing SiO_2 and Si_2O_3, that with oxidized Si molecule in the layer containing SiO and Si_2O, and that with Si atoms in the Si substrate, respectively. Because IMFP denoted by Λ_i is equal to $1/(n_i\sigma_i)$, IMFP denoted by Λ_{ct2} and IMFP denoted by Λ_{ct1} can be expressed by following eqs. (3) and (4) using n_i and σ_i defined by eqs. (1) and (2).

$$\Lambda_{ct2} = \frac{3\Lambda_{bo}\Lambda_{bs}}{2\Lambda_{bo} + \Lambda_{bs}} \tag{3}$$

$$\Lambda_{ct1} = \frac{3\Lambda_{bo}\Lambda_{bs}}{\Lambda_{bo} + 2\Lambda_{bs}} \tag{4}$$

4) Due to the inner potential of 15.3 eV, the internal emission angle ϕ and the external emission angle θ are not the same compared with ϕ even for extremely large kinetic energy of Si $2p$ photoelectrons.

Figure 4 shows the ratios $I(Si^{3+})/I(Si^{4+})$, $I(Si^{2+})/I(Si^{4+})$, $I(Si^{1+})/I(Si^{4+})$, $I(\alpha)/I(Si^{4+})$, $I(\beta)/I(Si^{4+})$, and $I(\gamma)/I(Si^{4+})$ as a function of TOA for the SiO_2/Si interface formed utilizing OR and that formed utilizing dry OM. Here, Here, $I(Si^{4+})$, $I(Si^{3+})$, $I(Si^{2+})$, $I(Si^{1+})$, $I(\alpha)$, $I(\beta)$, and $I(\gamma)$ denote the integrated spectral intensity of the Si $2p_{3/2}$ spectrum arising from the Si^{4+}, Si^{3+}, Si^{2+}, Si^{1+}, α-Si, β-Si, and γ-Si, respectively.

Figure 5 shows the depth profiles of the IMSs determined from the analyses of the angle-resolved Si $2p_{3/2}$ photoelectron spectra. Solid lines in Fig. 4 were calculated using the areal densities (ADs) of IMS and the distance of IMS from the Si surface shown in Fig. 5. The following results are obtained: 1) in two kinds of oxide films the most of suboxides consisting of Si^{3+}, Si^{2+}, and Si^{1+} are localized extremely near the SiO_2/Si interfaces, 2) suboxides formed utilizing dry OM are located extremely close to the interfaces, while suboxides formed utilizing OR are located not only at the interface but also in the Si substrate. This implies the penetration of a part of OR into the Si substrate to form Si^{1+} and to relax the interfacial stress at the interface, 3) when the oxidation is formed using OR, the intermediate chemical bonding states of Si consisting of α-Si, β-Si and γ-Si are formed in the Si substrate closer to the interface as compared with those formed utilizing dry OM. This implies that the interfacial stress in the former is smaller than that in the latter.

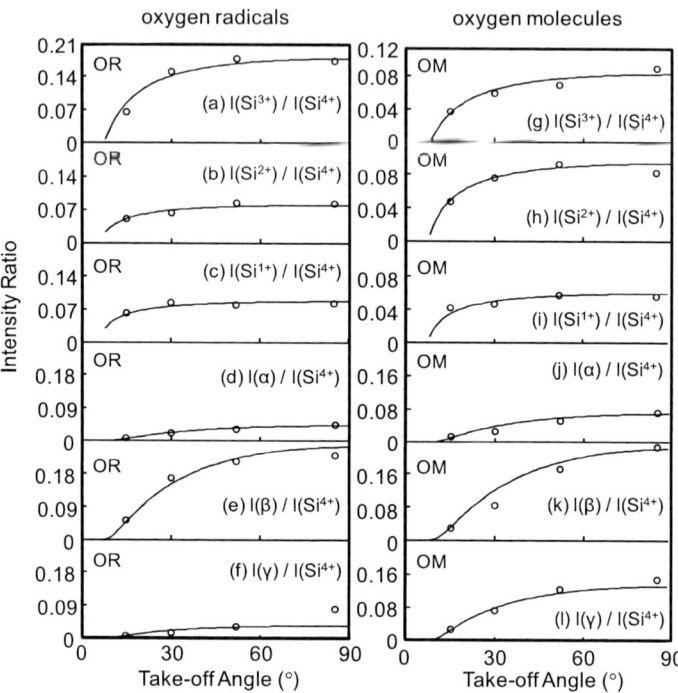

Figure 4. Integrated intensity ratio $I(Si^{3+})/I(Si^{4+})$, $I(Si^{2+})/I(Si^{4+})$, $I(Si^{1+})/I(Si^{4+})$, $I(\alpha)/I(Si^{4+})$, $I(\beta)/I(Si^{4+})$, and $I(\gamma)/I(Si^{4+})$ as a function of TOA for the SiO^2/Si interface formed utilizing OR and that formed utilizing dry OM. Here, Here, $I(Si^{4+})$, $I(Si^{3+})$, $I(Si^{2+})$, $I(Si^{1+})$, $I(A)$, $I(B)$, $I(\alpha)$, $I(\beta)$, and $I(\gamma)$ denote the integrated spectral intensity of the Si $2p_{3/2}$ spectrum arising from the Si^{4+}, Si^{3+}, Si^{2+}, Si^{1+}, α-Si, β-Si, and γ-Si, respectively. solid lines were calculated using the areal densities (ADs) of IMS and the distance of IMS from the Si surface as shown in Fig. 5.

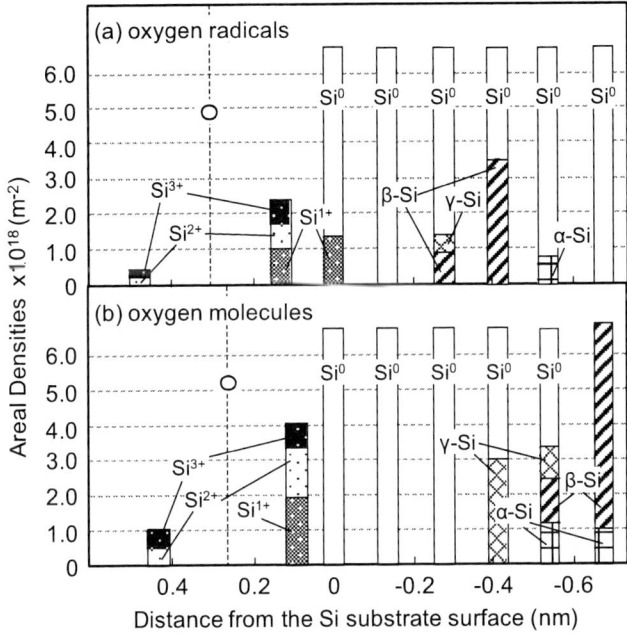

Figure 5. (a) Depth profiles of the areal densities of IMSs for the oxide films formed utilizing OR and (b) those for the oxide film formed utilizing dry OM as a function of the distance from the surface of Si substrate.

Conclusion

The chemical structure of the compositional and structural transition layer at and near the $SiO_2/Si(100)$ interface formed utilizing OR versus dry OM were investigated by measuring soft x-ray-excited angle-resolved Si $2p$ photoelectron spectra. A clear difference between the chemical structure of the SiO_2/Si interface formed using oxygen radicals and that formed using dry oxygen molecules was found and can be summarized as follows: 1) in two kinds of silicon oxide films, most of the suboxides consisting of Si^{3+}, Si^{2+}, and Si^{1+} are localized extremely near the SiO_2/Si interfaces, 2) suboxides formed utilizing dry OM are located extremely close to the interfaces, while suboxides formed utilizing OR are located not only at the interface but also in the Si substrate. This implies the penetration of a part of OR into the Si substrate to form Si^{1+} and to relax the interfacial stress at the interface, 3) when the oxidation is formed using OR, intermediate chemical bonding states of Si consisting of α-Si, β-Si and γ-Si are formed in Si substrate closer to the interface as compared with those formed utilizing dry OM. This implies that the interfacial stress in the former is smaller than that in the latter.

Acknowledgments

This work was supported by the Ministry of Education, Culture, Sports, Science and Technology of Japan (MEXT) under a Grant-in-Aid for Specially Promoted Research (Project No. 18002004 and Project No. 22000010) and a Grant-in-Aid for Scientific Research (B) (Project No. 19360014). The synchrotron radiation experiments were performed at SPring-8 with the approval of JASRI as a part of the projects of Nanotechnology Support commissioned by the MEXT.

References

1. T. Ueno, A. Morioka, S. Chikamura, and Y. Iwasaki, *Jpn. J. Appl. Phys.*, **39**, L327 (2000).
2. M. Hirayama, K. Sekine, Y. Saito, and T. Ohmi, in *IEDM Tech. Dig.*, 1999, p.249.
3. Y. Saito, K. Sekine, N. Ueda, M. Hirayama, S. Sugawa, and T. Ohmi, in *Tech. Dig. 2000 Symp.VLSI Tech.*, Hawaii, p. 176.
4. K. Sekine, Y. Saito, M. Hirayama, and T. Ohmi, *IEEE Trans. Electron Devices* ,**48**, 1550 (2001).
5. T. Hamada, Y. Saito, M. Hirayama, H. Aharoni, and T. Ohmi, *IEEE Elec. Device Lett.*, **22**, 423 (2001).
6. T. Suwa, A. Teramoto, Y. Kumagai, K. Abe, X. Li, Y. Nakao, M. Yamamoto, Y. Kato, T. Muto, T. Kinoshita, T. Ohmi, and T. Hattori, *Appl. Phys. Lett.*, **96**, 173103 (2010).
7. R. Kuroda, A. Teramoto, Y. Nakao, T. Suwa, M. Konda, R. Hasebe, X. Li, T. Isogai, H. Tanaka, S. Sugawa, and T. Ohmi, *Jpn. J. Appl. Phys.*, **48**, 04C048 (2009).
8. R. Kuroda, T. Suwa, A. Teramoto, R. Hasebe, S. Sugawa, and T. Ohmi, *IEEE Trans. Electron Devices,* **56**, 291 (2009).
9. X. Li, T. Suwa, A. Teramoto, R. Kuroda, S. Sugawa, and T. Ohmi, *Electrochem. Soc. Trans.*, **28**, 299 (2010).
10. S. Tanuma, C. J. Powell, and D. R. Penn, *Surf. Interface Anal.*, **20**, 77 (1993).
11. Y. Morita and H. Tokumoto, *Appl. Surf. Sci.*, **100**, 440 (1996).
12. T. Ohmi, *J. Electrochem. Soc.*, **143**, 2957 (1996).
13. K. Ohishi and T. Hattori, *Jpn. J. Appl. Phys.*, **33**, L675 (1994).
14. F. J. Himpsel, F. R. McFeely, A. Talev- Ibrashimi, J. A. Yarmoff, and G. Hollinger, *Phys. Rev. B*, **38**, 6084 (1988).
15. T. Aratani, , M. Higuchi, S. Sugawa, E. Ikenaga, J. Ushio, H. Nohira, T. Suwa, A. Teramoto, T. Ohmi, and T. Hattori, *J. Appl. Phys.*, **104**, 114112 (2008).
16. G. Hollinger and F. J. Himpsel, *Appl. Phys. Lett.*, **44**, 93 (1984).
17. O. V. Yazyev and A. Pasquarello, *Phys. Rev. Lett.*, **96**, 157601 (2006).
18. M. F. Hochella, Jr. and A. H. Carim, *Surf. Sci.*, **197**, L260 (1988).
19. M. Shioji, T. Shiraishi, K. Takahashi, H. Nohira, K. Azuma, Y. Nakata, Y. Takata, S. Shin, K. Kobayashi, and T. Hattori, *Appl. Phys. Lett.*, **84**, 3756 (2004).

CHAPTER 2

ULTRA-THIN FILM/RELIABILITY

124

Impact of Twofold Coordinated Nitrogen on the Generation of Deep-Level Hole Traps under Negative-Bias Temperature Stressing

C. J. Gu, D. S. Ang and Z. Q. Teo

Nanyang Technological University, School of Electrical and Electronic Engineering, Nanyang Avenue, Singapore 639798

In this paper, we discuss the generation of deep-level hole traps (DLHTs) during negative-bias temperature stressing. Because of their ability to trap positive charges for a long time even in the case of the ultra-thin gate oxide, it is imperative to gain a sound understanding on the nature of the DLHT, especially in relation to nitrogen, which has been incorporated in increasing dose into the SiO_2 gate oxide to solve boron penetration and gate tunneling leakage issues. Experimental evidence showing a correlation between the generation of DLHTs and nitrogen is first presented. This is followed by a discussion on the role nitrogen plays in influencing the generation of DLHT via first-principles simulation based on the density functional theory. We focus on the impact of a neighboring nitrogen atom on the structural and electronic properties of the oxygen-vacancy defect (V_O), which is a major precursor for hole traps in the SiO_2. A neighboring twofold coordinated nitrogen atom (i.e. comprising an unpaired electron) is found to induce significant structural relaxation of V_O following the capture of a hole. The charge transition level (CTL) associated with this defect cluster is located at 5.32 eV from the SiO_2 valence band edge (i.e. near the Si conduction band edge), making it a very deep level hole trap. On the other hand, the CTL for V_O in the absence of nitrogen or in the presence of a neighboring threefold coordinated nitrogen atom is found to be much lower, at ~2.5-3 eV from the SiO_2 valence band edge (i.e. below the Si valence band edge). These defect clusters exhibit much smaller structural relaxation upon the capture of a hole and they correspond to shallow hole traps responsible for the fast charging/discharging transient-like behavior of negative-bias temperature instability.

Introduction

Negative-bias temperature instability, or more commonly known as NBTI, is widely recognized as a critical reliability issue of state-of-the-art complementary metal-oxide-semiconductor (CMOS) technology. NBTI is the consequence of a progressive generation of interface states and near-interface positive oxide trapped charges under the combined effect of a negative gate voltage and an elevated temperature. The positively charged defects shift the threshold voltage (V_t) of the p-channel MOS field-effect transistor (p-MOSFET) negatively and degrade the inversion charge mobility, thereby reducing its drive current. When the shift exceeds a certain design tolerance, functional failure of integrated circuits may result. NBTI was observed as early as the beginning of the MOS technology in the mid 1960's (1) but it received very little attention at that point in time because of the buried-channel structure of the p-MOSFET. In the buried-channel structure, the holes (which play an important role in the generation of the oxide/interface defects), are separated from the silicon/silicon dioxide (Si/SiO$_2$) interface. This had helped mitigate the instability problem to a large extent in older CMOS technologies.

The aggressive scaling of the CMOS technology into the deep sub-micrometer regime at the dawn of this century changed the situation drastically. A switch from a buried- to a surface-channel structure had to be made to keep short-channel effects in check. This move places the holes exactly at the interface, exacerbating the NBTI issue. The problem is compounded by the increase in oxide field as a result of the reduction in the gate oxide thickness. Another main reason for NBTI becoming so important in recent years lies in the presence of nitrogen in the gate oxide (2), (3). Nitrogen has been incorporated into the ultra-thin gate oxide because it helps block boron from the heavily doped p-type polysilicon gate (required for a surface-channel structure), thus improving the V_t uniformity of as-processed devices. Nitrogen also increases the dielectric constant of the gate oxide, which in turn allows a thicker SiO$_2$ to be used for suppressing gate leakage without compromising the effective capacitance. But the presence of nitrogen in the gate oxide has substantially worsened the NBTI problem (2), (3).

In the past decade, there were many electrical characterization studies on the NBTI of p-MOSFETs employing the ultra-thin oxynitride (SiON) gate oxide. The degradation of the p-MOSFET is typically characterized by a much weaker dependence on stress time and temperature as compared to that of the SiO$_2$ counterpart (3)-(5). This behavior was ascribed to the increased trapping of holes within the ultra-thin gate oxide by defect precursors associated with nitrogen. It was also observed that more trapped holes could be sensed electrically at a given delay after the termination of stress, implying the existence of a larger density of hole traps having deeper energy levels. Ang *et al.* (6)-(9) were among the first to infer the presence of deep-level hole traps (DLHTs) after negative-bias temperature stressing. These hole traps have energy states above the Si Fermi potential, which result in them having the ability to retain their positive trapped charges for a very long time under negative gate biasing, even in a direct-tunneling gate oxide (cf. Fig. 2(c)).

Although the general consensus is that nitrogen increases hole trapping in the gate oxide, the nature of the defects is seldom explored. A recent study (10) on electron paramagnetic resonance has shown that hole trapping sites in the SiON have a crossover g tensor that is different from that of the E' centers (O$_3\equiv$Si• +Si$\equiv$O$_3$) in the SiO$_2$. The difference was ascribed to the neighboring O atoms being replaced by nitrogen atoms. A defect configuration of the form N$_3\equiv$Si• +Si$\equiv$O$_3$ (referred to as a K center) was shown to have a crossover g tensor similar to that observed experimentally. However, knowledge on how neighboring nitrogen atoms impact the properties of the oxygen vacancy defect (V_O) remains very limited. In particular, the mechanism behind the generation of DLHTs is not known at this stage.

In this paper, we show via first-principles simulation based on the density functional theory that the coordination of the neighboring nitrogen atom plays a paramount role on the structural relaxation of V_O following the capture of a hole. In particular, a twofold coordinated nitrogen atom is shown to induce very significant structural relaxation of V_O. The resultant trapped-hole site has a very deep charge transition level in the SiO$_2$

bandgap. On the other hand, in the absence of the nitrogen atom or if the neighboring nitrogen atom is threefold coordinated (e.g. the dangling nitrogen bond is terminated by a hydrogen atom), the resultant trapped-hole site would have a much shallower charge transition energy level. We first present experimental results showing a clear correlation between nitrogen concentration in the gate oxide and the density of DLHT generation by negative-bias temperature stress. This is then followed by a detailed discussion of the simulation results. Based on the new insights gained, a unified explanation for hole-trap generation and relaxation is given.

Experimental Details

The test devices (DUTs) were p^+ polysilicon gate p-MOSFETs. The SiON gate oxide was formed via decoupled plasma nitridation of an *in-situ* steam generated base oxide. The concentration of nitrogen at the polysilicon/SiO_2 interface varies from ~6 to 13 atomic percent. The equivalent oxide thickness is approximately 1.5 nm, extracted from capacitance-voltage measurement. Negative-bias temperature stressing was carried out at a temperature of 100 °C and at an oxide field of about 10 MV/cm. Device degradation was monitored in terms of threshold voltage shift (ΔV_t), obtained from linear drain current versus gate voltage (I_d-V_g) curves using the constant subthreshold drain current method. An ultra-fast switching method (11), which is essentially a modified form of the pulsed current-voltage (PIV) method, was employed for I_d-V_g measurement. An advantage of this method is that the very narrow gate pulses (100 ns) enhance probing of shallow hole traps (SHTs) with short detrapping time constants.

Deep-Level Hole Trap Generation and Nitrogen

In this section, we present experimental result showing a correlation between DLHT generation and the concentration of nitrogen in the gate oxide. We first outline the experimental procedure for revealing and determining the amount of DLHTs generated by negative-bias temperature stressing before discussing the impact of nitrogen on the generation of DLHTs.

Fig. 1(a) depicts the evolution of ΔV_t during the stressing and relaxation phase of a dynamic NBTI cycle. As expected, ΔV_t increases *negatively* during stressing, indicating the progressive build-up of *positive* oxide and interface trapped charge. Trapping of holes by oxide/interface defects is believed to be one of the major mechanisms. By virtue of their proximity to the Si/SiO_2 interface, one would expect hole trapping at near-interface defect states to occur spontaneously the moment the gate stress voltage is applied (Fig. 2(a)). Hole trapping at defect states further away from the Si/SiO_2 interface would take longer time, but in the case of the direct tunneling gate oxide, these defect states are not likely to be positively charged because of the correspondingly large probability of hole detrapping to the gate (Fig. 2(b)). Based on this viewpoint, the net hole trapping is expected to reach a saturation level very quickly (< 1 s) after the application of the gate stress voltage, with the saturation level limited by the available density of defect states. Some authors attributed the subsequent progressive shift of ΔV_t over prolonged stress period to the generation of interface states, under the framework of the well known reaction-diffusion (R-D) model (12). We will return to this point at a later stage citing

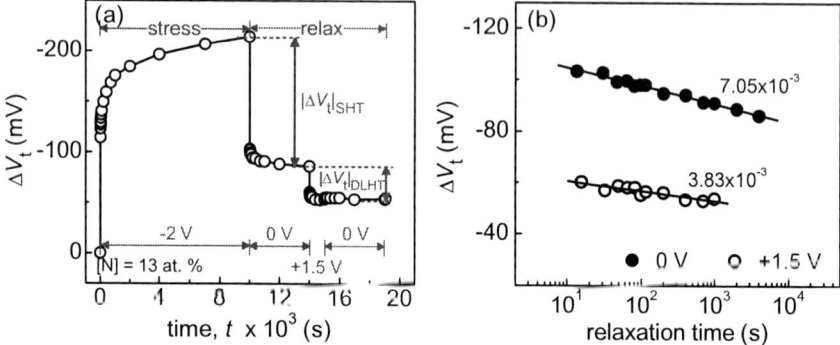

Figure 1. (a) Evolution of threshold voltage shift ΔV_t during negative-bias temperature stressing and relaxation. Gate voltages applied in the respective phases are as indicated. During relaxation, the gate voltage was first switched to 0 V and maintained at that level for 4×10^3 s. It was then switched to +1.5 V for 1×10^3 s (but PIV measurement was carried out with a 0-V offset) after which it was returned to 0 V for another 4×10^3 s. (b) ΔV_t evolution during the first 0 V and the subsequent +1.5 V relaxation phases plotted on a linear-log scale. The slopes of the regression line are as shown.

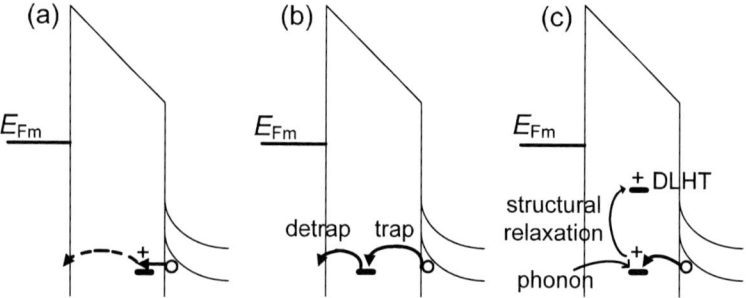

Figure 2. Schematic energy band diagrams of the MOS structure under negative gate biasing. (a) Hole capture by a near-interface defect state where the trapping probability is greater than the detrapping (dashed arrow) probability. (b) In an ultra-thin gate oxide, hole capture by a defect state located further from the Si/SiO_2 would be less likely because of a correspondingly large probability of detrapping to the gate, i.e. the defect is neutral and does not contribute to NBTI. In both (a) and (b), structural relaxation of the defect is ignored. (c) Significant lattice relaxation (via interaction with phonon) upon hole capture results in an "upward" shift of the positively charged defect state (i.e. negative U behavior) and the creation of a deep-level hole trap.

recent experimental evidence which does not support the notion that the long term evolution of NBTI is driven by hydrogen transport.

Just as one expects spontaneous hole trapping to take place at near-interface defect states when the negative gate stress voltage is applied, one also expects spontaneous detrapping of holes from near-interface defect states when the gate stress voltage is removed, triggering a decrease in the magnitude of ΔV_t. This is evident from the abrupt decrease in $|\Delta V_t|$ at the onset of the relaxation cycle. The decrease is also observed to approach quasi-saturation fairly quickly; after the initial abrupt decrease, the subsequent decrease of $|\Delta V_t|$, over an extended recovery period of 4×10^3 s, is less than 10% of its value at the end of the stress cycle. This quasi-saturation behavior tends to signify the near completion of the hole detrapping process.

At this juncture, it is essential to point out that some authors (13), (14) ascribed the recovery of ΔV_t to the repassivation of dangling Si bonds by back-diffusing hydrogen, under the framework of the R-D model (12). This school of thought presumes that hole trapping saturates very quickly and the subsequent evolution of $|\Delta V_t|$ results from the generation of Si dangling bonds via the dissociation of Si-H bonds, with the rate of generation limited by the transport of hydrogen away from the Si/SiO$_2$ interface. Our recent experiments (15)-(17) on repetitive stressing and relaxation did not, however, support this proposition. We found that the amount of ΔV_t recovered per relaxation cycle remain the same regardless of the number of stress and relaxation cycles (see Fig. 3(a) of ref. (16); reproduced here as Fig. 3(a)). This characteristic does not conform to the hydrogen transport model, which stipulates that the amount of interface state repassivation per cycle must decrease with increase in the number of stress and relaxation cycles. This is because repeatedly cycling the DUT between stress and relaxation would result in more hydrogen being transported further away from the Si/SiO$_2$ interface, thus gradually limiting the amount of interface states which could be repassivated per relaxation cycle. Detailed investigation over many stress and relaxation cycles has instead shown a cyclic behavior of ΔV_t, i.e. the increase in the magnitude of ΔV_t in a stress cycle would become

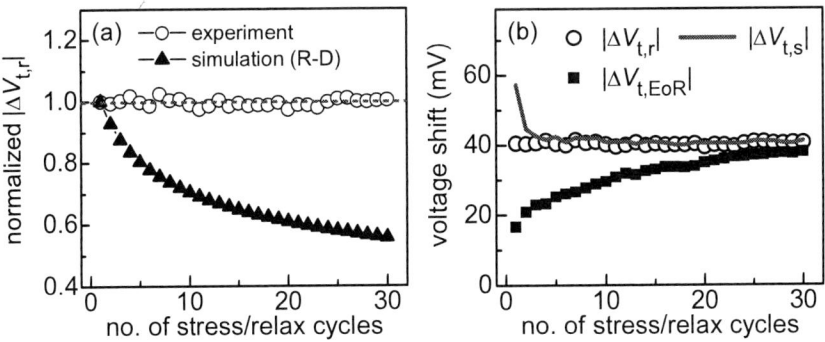

Figure 3. (a) Comparison of $|\Delta V_{t,r}|$, the decrease in the magnitude of ΔV_t per relaxation cycle as a function of the number of stress/relax cycles between experiment and simulation based on the R-D model. For the ease of comparison with the simulation data, $|\Delta V_{t,r}|$ is normalized by the value for the first relaxation cycle. (b) Comparison of $|\Delta V_{t,r}|$ and $|\Delta V_{t,s}|$, the increase in the magnitude of ΔV_t per stress cycle. $|\Delta V_{t,EoR}|$ refers to the magnitude of ΔV_t at the end of each relaxation cycle (16).

almost equal to the decrease in the ensuing relaxation cycle (see Fig. 3(b) of ref. (16); reproduced here as Fig. 3(b)). Our recent results strongly indicate that the ΔV_t fluctuations one typically observed during dynamic NBTI were the consequence of an ensemble of switching hole-trap precursors that were being repeatedly charged (during stressing) and discharged (during relaxation) (18). Oxygen vacancy defects are a major source of switching oxide traps in the SiO_2 (18), (19). These defects are typically found within the intrinsic substoichiometric oxide region near the Si/SiO_2 interface.

It should be pointed out that the above first-order treatment on hole trapping/ detrapping ignores structural relaxation that occurs upon hole capture. This can lead to an underestimation of the extent of hole trapping in an ultra-thin gate oxide (20). As depicted in Fig. 2(c), if there is significant lattice relaxation at the defect site (via interaction with phonons) following the capture of a hole, it could induce an upward shift of the defect level (i.e. negative U behavior). This upward shift renders a significant increase of the hole emission time. If the defect level remains above the Si Fermi potential after the removal of the negative gate stress voltage, hole emission could not readily take place via the capture of electrons from the Si substrate, yielding a DLHT. This is unlike shallow hole traps which are spontaneously discharged when their energy levels are lowered below the Si Fermi potential upon the termination of the stress.

The existence of these DLHTs can be shown by switching the gate voltage to a moderately positive value during the course of the first 0 V relaxation phase, after quasi-saturation is reached. The positive gate voltage lowers DLHTs below the conduction band edge of Si, promoting the emission of trapped holes via the capture of electrons from the n-Si well/substrate. If the hole detrapping process has indeed completed (as suggested by the quasi-saturation level), one would not expect the recovery of ΔV_t to be affected by a positive gate voltage. However, this inference is not supported by Fig. 1(a), which clearly shows a non-negligible, abrupt decrease in $|\Delta V_t|$ immediately following a switch of the gate voltage from 0 to +1.5 V. This implies that a substantial fraction of the subsequent evolution of $|\Delta V_t|$ following the initial transient-like increase during the stressing phase comprises further trapping of holes but at much deeper energy levels.

In Fig. 1(b), we replotted the ΔV_t recovery data of the first 0-V and the subsequent +1.5-V relaxation phases against the relaxation time (defined with respect to the instant when the gate voltage was changed) on a linear-log plot. A logarithmic time dependence spanning many orders of time magnitude is evident, particularly for the 0-V case, as has also been observed in other studies (20), (21). A caution about the tendency for one to conclude that recovery has reached saturation based on the "plateau" observed in Fig. 1(a). As shown in Fig. 1(b), the recovery has in fact not saturated although its rate is decreasing continuously with time. Similar logarithmic time dependence has been observed in the past for the thick gate oxide and has been attributed to the spatial distribution of oxide traps (22). In our case, the gate oxide thickness is less than 2 nm and the electron tunneling time constant is on the order of microseconds or less. Clearly, the spatial distribution of oxide traps could not explain the wide spread of recovery time constants. In order to account for the observation, it is necessary for one to consider a wide spread in the energy levels of the hole traps, whereby relatively deep hole traps are responsible for long recovery time constants.

The decrease in $|\Delta V_t|$ following a switch of the gate voltage to a positive value is permanent, i.e. a subsequent switch of the gate voltage back to 0 V does not return $|\Delta V_t|$ to the level prior to the gate polarity change. A similar statement can be made in regards to the rapid decrease in $|\Delta V_t|$ at the onset of the 0 V relaxation phase, i.e. the level of ΔV_t shift observed in the stressing phase is not seen again unless the DUT is re-stressed at the same negative gate stress voltage for the same period (cf. Fig. 3). The decrease in $|\Delta V_t|$ during the transition from 0 to +1.5 V gate recovery voltage is also found to occur only once at the first instant when the gate voltage is made positive; a subsequent switch to a more positive gate voltage yields no further decrease in $|\Delta V_t|$ (not shown). This implies that a significant part of the hole detrapping process is irreversible, and could possibly involve the reconstruction of broken bonds, which totally eliminate the trapped-hole site. This inference is supported by our simulation work (see later discussion). The remnant ΔV_t is rather permanent, as is evident from the substantially reduced recovery rate under +1.5 V (Fig. 1(b)). This permanent ΔV_t may be attributed to trapped-hole sites which have undergone so substantial structural relaxation that subsequent bond reformation requires extremely long time.

For the purpose of subsequent discussion, we define the difference between $|\Delta V_t|$ at the end of the first 0 V relaxation period and that after the switch to a positive gate voltage was made as the component of $|\Delta V_t|$ due to DLHT ($|\Delta V_t|_{DLHT}$). The difference between $|\Delta V_t|$ at the end of the stress cycle and that at the end of the first 0 V relaxation period is defined as $|\Delta V_t|$ arising from relatively shallow hole traps ($|\Delta V_t|_{SHT}$). Since the recovery of ΔV_t in the first 0 V relaxation phase reaches quasi-saturation fairly quickly, varying its period would not significantly affect the values of $|\Delta V_t|_{DLHT}$ and $|\Delta V_t|_{SHT}$. Our classification in fact underestimates $|\Delta V_t|_{DLHT}$ since the permanent fraction is excluded.

Fig. 4 shows the dependence of $|\Delta V_t|_{SHT}$ and $|\Delta V_t|_{DLHT}$ on the nitrogen concentration in the gate oxide (which corresponds to that at the polysilicon/SiO_2 interface). As expected, hole trapping increases with nitrogen concentration. More importantly, there is also a distinct increase in the amount of DLHTs, indicating that a higher nitrogen concentration tend to promote greater DLHT generation for a given stress period. The time dependence of $|\Delta V_t|_{SHT}$ and $|\Delta V_t|_{DLHT}$ are examined in Fig. 5. The former exhibits very weak dependence on the stress time while the stress time dependence of the latter is clearly much stronger. This observation supports the notion that generation of the relatively permanent DLHTs is generally more sensitive to time since it involves more significant structural relaxation. Nevertheless, an increase in the nitrogen concentration tends to reduce the time dependence (23), implying that the generation of DLHTs is accelerated by more nitrogen. As will be shown later via simulation, a nitrogen atom in a twofold coordinated form facilitates the structural relaxation of an adjacent oxygen vacancy defect.

Simulation Details

We have chosen the cristobalite SiO_2 for our simulation study, in view that its Kohn-Sham density of states is highly similar to that of the amorphous SiO_2 (24), (25). A supercell consisting of eight α-cristobalite unit cells was employed to provide sufficient isolation of defects. Fig. 6 depicts the supercell used in our study. The dimensions of

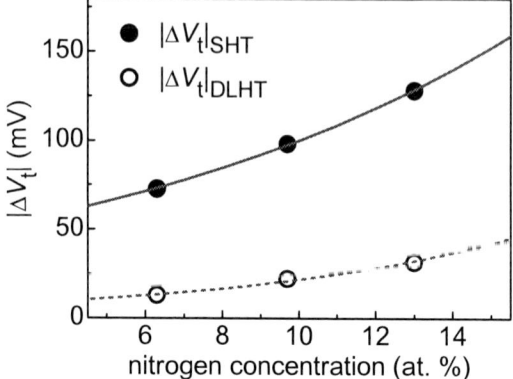

Figure 4. Dependence of threshold voltage shift (extracted after 1×10^4 s stress) contributed by shallow and deep hole traps on the nitrogen concentration at the polysilicon/SiO$_2$ interface in atomic percent.

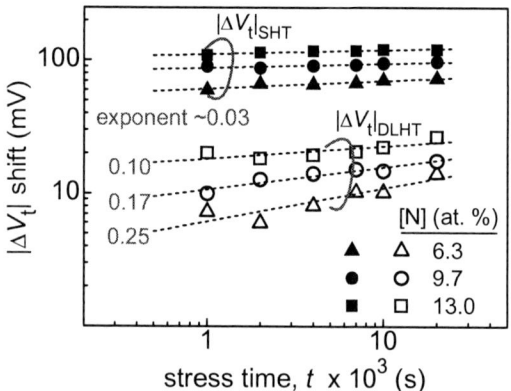

Figure 5. Evolution of threshold voltage shift $|\Delta V_t|$ arising from shallow hole traps (SHTs) and deep-level hole traps (DLHTs) for DUTs having varying nitrogen concentration at the polysilicon/SiO$_2$ interface.

the supercell are 9.96 Å × 9.96 Å × 13.90 Å. The total number of atoms in the supercell is 96 and the density is 2.32 g/cm^3. An oxygen vacancy defect was manually created by removing an oxygen atom to induce the formation of the O$_3$≡Si–Si≡O$_3$ dimmer configuration. In the case of the nitrided SiO$_2$, a neighboring oxygen atom of the Si–Si dimmer was replaced with a nitrogen atom. Since the valency of nitrogen is one less than that of oxygen, a direct replacement would result in the nitrogen atom having an unpaired spin, i.e. a dangling bond. Leaving the dangling bond unpassivated forms our twofold coordinated nitrogen defect model. Both experimental and simulation data supporting the

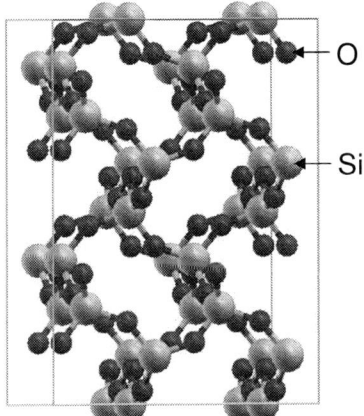

Figure 6. Supercell comprising 8 α-cristobalite unit cells used in our simulation study. The bigger blue spheres denote Si atoms while the red spheres denote oxygen atoms.

formation of this twofold coordinated nitrogen defect in the nitride SiO_2 have been reported (25), (26). Passivating the dangling bond with a hydrogen atom leads to the threefold coordinated nitrogen defect. As will be shown later, the coordination of the nitrogen atom has a significant impact on the structural relaxation and resultant stability of V_O following the capture of a hole.

The calculation was carried out using the PWSCF code of the QUANTUM ESSPRESSO package (27). Based on the density-functional theory, the electronic structure in the calculation was described by generalized gradient approximation developed by Perdew, Burke, and Ernzerhof (PBE) (28). The eigenfunctions were expanded on a plane wave basis sets, and ultra-soft pseudo-potentials were applied to each kind of atoms. For all cases, full relaxation of the system was carried out. We tested and found that a cut-off energy of 50 Ry for the wave-function expansion, and a value of 150 Ry for the charge density yielded well converged results. The Γ point was used during the whole relaxation procedure. For the charged system, a uniform background charge with an opposite sign was adopted to maintain neutrality of the whole system. This avoided divergence during the calculation of the total energy.

A well known issue with the semilocal density-functional approach is the underestimation of the band-gap. This problem prevents an accurate determination of the position of the defect level with respect to the Si band-gap. Fortunately, a recent study has shown that a hybrid functional approach could yield an accurate SiO_2 band-gap (29). Both approaches, however, give similar result for the charge transition level (< 0.1 eV variation). This finding allows us to make use of the semilocal approach to first determine the charge transition level and then position it accurately with respect to the Si band edges by making corrections to the Si/SiO_2 band-edge offsets based on the results of the hybrid functional approach. Fig. 11 illustrates the details.

Discussion on Simulation Results

Structural Relaxation

As expected, the removal of an oxygen atom to form a neutral oxygen vacancy results in the two affected Si atoms moving closer to one another to form a Si-Si bond as this lowers the overall system energy. The spacing between the two Si atoms (labeled as Si(1) and Si(2)) is similar in all three cases (i.e. pure SiO_2, SiON with a threefold coordinated nitrogen atom next to Si(1), and SiON with a twofold coordinated nitrogen atom next to Si(1)), as shown in Fig. 7. From electron density analysis, we confirm that there is a bond formed between the two Si atoms. Moreover, the charge density distribution (dashed circle) between the two Si atoms is similar for all three cases.

However, the neutral oxygen vacancy responds very differently upon the capture of a hole, depending on the coordination of the neighboring nitrogen atom. In the case where the nitrogen atom is absent (i.e. pure SiO_2), the distance between the two Si atoms is increased by ~0.13 Å (Fig. 8(a)(i)). The charge density between Si(1) and Si(2) has thinned down (Fig. 8(a)(ii)) compared to the neutral case (cf. Fig. 7(a)(ii)), implying that the electronic charge is lost (or a hole is trapped) at this location. Nevertheless, the remaining unpaired spin is still almost equally shared between the two atoms. This is consistent with the marginal structural relaxation (Fig. 8(a)(i)), implying that the original bond between the two Si atoms could readily reform upon the recapture of an electron. In the case of the SiON with a threefold coordinated neighboring nitrogen atom, the separation between Si(1) and Si(2) increases to ~0.23 Å. Due to the increased spacing, the distribution of the remaining charge between the two atoms is different compared to the

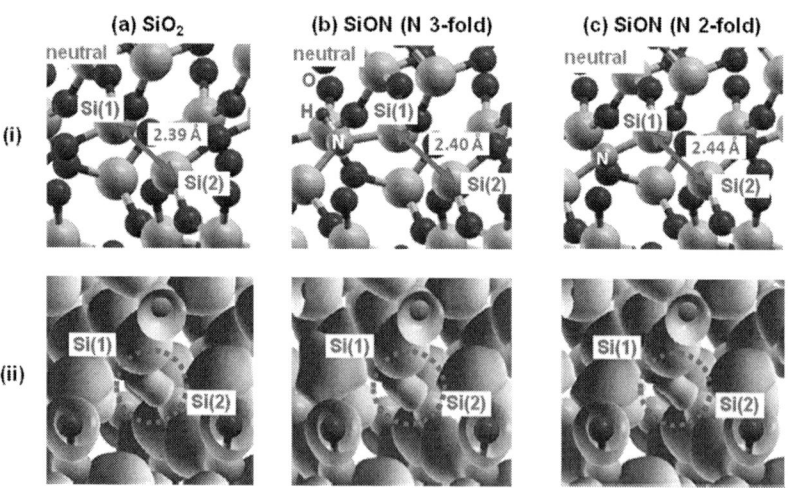

Figure 7. (i) Oxide bonding network in the vicinity of the neutral oxygen vacancy defect, depicting the distance between Si(1) and Si(2). (ii) Electron density analysis showing the distribution of electronic charge (yellow) in the same vicinity of the oxide network. The dotted circle highlights the bonding between Si(1) and Si(2).

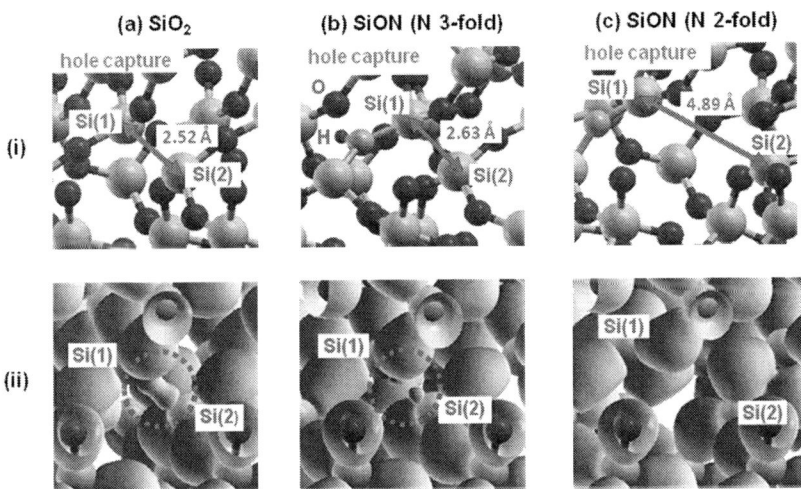

Figure 8. (i) Oxide bonding network in the vicinity of the oxygen vacancy defect following the capture of a hole. (ii) Electron density analysis showing the distribution of electronic charge (yellow) in the same vicinity of the oxide network. The dotted circle highlights the electronic charge density between Si(1) and Si(2). The dotted circle is not shown in part (c)(ii) as there is no electronic charge between Si(1) and Si(2) in this case.

pure SiO_2 case. Instead of a continuous "dumbbell-like" shape in the latter (Fig. 8(a)(ii)), distinct lobes are observed on each of the two Si atoms (Fig. 8(b)(ii)). The similar size of the lobes implies that the remaining unpaired spin is still equally shared between the two Si atoms. But the distinct lobes on each of the Si atoms suggest that the unpaired spin is "hopping" back and forth the two Si atoms continuously. An interesting point is that the N-H bond remains pretty much intact following the hole capture, indicating that the hole is not trapped at this location but is preferentially located in-between the two Si atoms.

The situation for the case of the SiON with a twofold coordinated neighboring nitrogen atom is drastically different compared to the rest. A very significant increase in the spacing between Si(1) and Si(2), by ~2.45 Å, is evident (Fig. 8(c)(i)). The original tetrahedral structure of Si(1)O_2N has changed into an almost planar form (Fig. 9). Compared to the threefold coordinated nitrogen case, the bond length between the Si and N atoms has decreased from the original value of 1.64 Å to 1.58 Å after the hole capture. The former agrees well with that reported for a *single* Si-N bond while the latter is closer to that of a *double* Si=N bond (Table I). We reckon that the significant structural relaxation is made possible by a lowering of the total system energy via the interaction between the Si(1) and nitrogen dangling bonds. This is consistent with electron localization function analysis, revealing a shell-like iso-surface (dotted circle, Fig. 9(b)) between the Si(1) and N atoms after the hole capture. This implies greater localization of electronic charge between the two atoms. The almost planar structure of Si(1)O_2N suggests that the interaction is facilitated by sp^2 re-hybridization of the now threefold coordinated Si(1) atom. The significant structural relaxation has an important implication

(a) 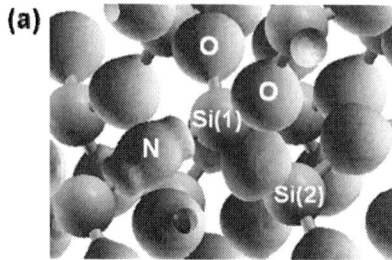 **(b)**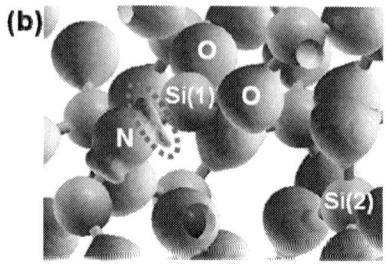

Figure 9. Electron localization function analysis in the vicinity of the oxygen vacancy defect for the case of the twofold coordinated nitrogen atom, before (a) and after (b) the capture of a hole. The yellow regions denote iso-surfaces having a value of 0.85 (a value of 1 means the electronic charge is completely localized). An enlargement of the iso-surface between N and Si(1) following the capture of a hole signifies increased electronic charge localization between the two atoms. There is also a change in the structure of the Si(1)O$_2$N cluster from tetrahedral to planar after the hole capture.

TABLE I. A comparison of Si-N bond length before and after the capture of a hole by the oxygen vacancy defect.

Bond Type (in bold)	Vacancy Charge State		Remarks
	0	+1	
O–Si–Si	1.64 Å	1.64 Å	Si–N (1.65 Å)
N–Si–Si (N is twofold)	1.64 Å	1.58 Å	Si=N (1.55 Å)
N–Si–Si (N is threefold)	1.73 Å	1.71 Å	(Ref. (30))

in that it makes the resultant trapped-hole site more permanent. The corresponding charge transition level for the defect structure would also be raised as elaborated in the subsequent section.

<u>Charge Transition Level</u>

To further probe the defect generation process, we examine the charge transition level (CTL) that corresponds to a change in the charge state of V_O from neutral to positive. The CTL is defined as the value of the electron chemical potential for which two different charge states (neutral and positive in our case) have the same formation energy, E^f (25), (31). Using the neutral state as the reference, the E^f for V_O^+ may be expressed as

$$E^f\left(+q\right) = E_{tot}\left(+q\right) - E_{tot}\left(0\right) + q\left[E_F + E_V + \Delta V\right] + E_{corr} \qquad (1)$$

where $E_{tot}(+q)$ is total system energy after the addition of a positive charge $+q$ (q is the electronic charge (1.6×10^{-19} C)), $E_{tot}(0)$ is total system energy for the neutral state, E_F is Fermi level measured with respect to the valence band maximum of a defect free oxide,

ΔV is a correction term describing the potential shift which occurs when the oxide is changed from a defect-free state to one having a defect structure, and E_{corr} is a correction term which accounts for the spurious electrostatic interactions introduced by the finite boundary condition.

A plot of E^f versus E_F is depicted in Fig. 10. The intersection of the E^f versus E_F characteristic with the $E^f = 0$ line gives the CTL that corresponds to the transition of V_O from a neutral to a positively charged state. For $E_F < $ CTL, formation of the positively charged state is favorable, i.e. V_O^+ is stable since E^f is negative. On the other hand, V_O^+ is unstable (and would have a strong tendency to revert to its neutral state) when $E_F > $ CTL, since E^f is positive in this case. For both the pure SiO_2 and SiON with a threefold coordinated neighboring nitrogen atom, the CTL for V_O^+ are comparable (~2.5-3 eV). A significant increase of the CTL to ~5.32 eV is, however, apparent in the case of the SiON with a twofold coordinated neighboring nitrogen atom. This substantial increase in the CTL translates to a wider range of E_F for which the V_O^+ defect state is stable. We reckon that this significant rise in the CTL is made possible by the significant structural relaxation which helps achieve an even lower total system energy after the hole capture. The positions of the various CTLs in the SiO_2 band-gap are shown in Fig. 11.

A Unified Model for Hole-Trap Generation and Relaxation

The simulation results clearly show that the extent of structural relaxation of V_O, as well as the influence exerted by a neighboring nitrogen atom, play a significant role on determining the CTL of the resultant hole trap. Based on this insight, we attempt to provide a unified explanation for some of the seemingly different positive oxide charges reported in the literature. We focus on the recent work of Zhang et al. (32), which provides

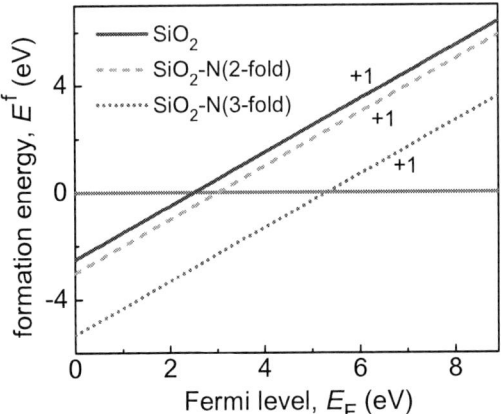

Figure 10. Formation energy versus Fermi level ($E_F = 0$ corresponds to the oxide valence band edge) characteristic for a positively charged oxygen vacancy defect. Intersection with the $E^f = 0$ line gives the charge transition level (CTL), which denotes the electron chemical potential energy for transition from neutral to a positively charged state. For $E_F < $ CTL, the positively charged state is stable.

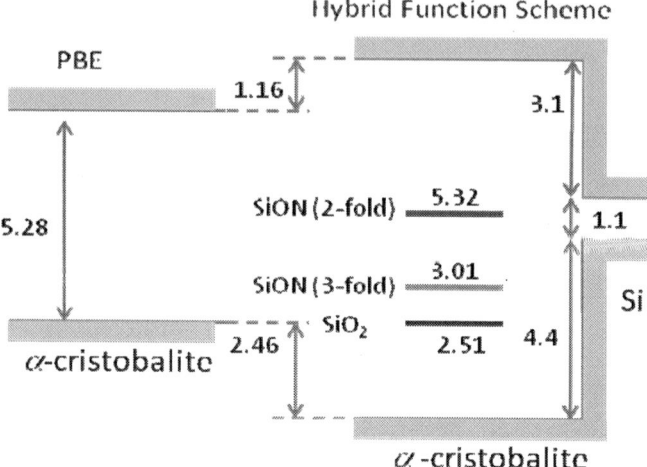

Figure 11. Schematic diagram illustrating the underestimation of the oxide band gap using the semilocal density functional scheme (left), which can be corrected by the hybrid functional scheme (right). After correction, the charge transition levels (CTLs) are more accurately located with respect to the oxide and Si band edges. The oxygen vacancy defect with a twofold neighboring N gives rise to the deepest CTL (i.e. furthest above the oxide valence band edge).

a comprehensive summary on the different electrical characteristics of positive charges in the SiO_2. Based on the different electrical responses observed, the authors proposed three types of positive oxide charge: 1) as-grown hole traps; 2) cyclic positive charge (CPC); 3) anti-neutralization positive charge (ANPC). In our discussion, we assume V_O as the origin of the positive oxide charge although a similar explanation may also be applied on other types of positive oxide charge. The main essence is that the apparent differences in the electrical behavior (32) may actually be different manifestations of the same defect (V_O) which have undergone varying degrees of structural relaxation after hole capture, depending on the surrounding atomic configurations. This in turn results in a considerable spread in the hole-trap energy distribution.

Through changing the gate voltage, one modulates the Fermi level E_F at the Si/SiO_2 interface, thereby changing the occupancy of near-interface defects. For V_O's having shallow CTLs (i.e. close to the SiO_2 valence band edge), they would only charge up positively provided the gate bias is negative enough to "lift them up" to the level of the E_F. Once the negative gate bias is removed, the CTLs would fall below E_F, and the corresponding V_O's would revert to the neutral state (via the capture of an electronic charge). As these V_O's also exhibit very little structural relaxation upon the capture of holes (Fig. 8), their return to a neutral state is likely to be accompanied by a spontaneous reformation of the Si-Si bond. But the overall speed at which this could happen (recovery speed) depends on the distance of a given V_O from the Si/SiO_2 interface, which determines how soon the trapped-hole site captures an electronic charge. In a direct tunneling gate oxide (< 2 nm in thickness), the capture of electronic charge happens almost immediately, since the tunneling time constant is on the order of microseconds or less. Therefore, V_O's with shallow CTLs function as "fast" hole traps which spontaneously discharge when the negative gate bias is removed and one would need a fast electrical measurement method in order to detect them. In a thicker gate oxide, however, the recovery speed involving V_O's situated further away from the interface would be much slower although near-interface V_O's would also recover almost spontaneously. Electron injection under a relatively high oxide field (e.g. Fowler-Nordheim tunneling) would be needed to eliminate the former (32). By virtue of their shallow CTLs, these V_O's are not likely to affect device operation under nominal voltage condition (since the voltage would not be large enough to charge them up positively) but are triggered under more severe gate stress voltage or oxide field conditions. But once they are annihilated (via bond reformation), they would no longer be observed under nominal measurement condition since the voltage involved would not be large enough to lift them up again to the E_F. This group of V_O's fits very well the description of the as-grown hole traps given in ref. (32).

On the other hand, V_O's with deeper CTLs would charge up positively even under a moderate negative gate voltage. However, their transformation to a stable V_O^+ state may take time, since interaction with phonons is needed to create the substantial structural relaxation that stabilizes the V_O^+ state. However, when this happens, the V_O^+ state is very stable owing to the relatively high CTLs. Removing the negative bias may not return these defects to the neutral state if the CTLs remain above E_F, i.e. they behave as DLHTs. Thus, even in a direct tunneling gate oxide, the recovery of these V_O's could take substantial time, explaining the logarithmic dependence spanning many orders of time under a 0-V gate voltage (Fig. 1(b)). In a thick gate oxide, the resultant time dependence may consist of another component arising from the discharging of shallow hole traps located at a distance from the interface. The application of a positive gate voltage lowers some of the CTLs below E_F, thereby neutralizing the positively charged defects, promoting the annihilation of a fraction of the DLHTs (see the abrupt decrease in $|\Delta V_t|$ upon the application of +1.5 V; Fig. 1(a)). At this point, it is essential to highlight that DLHTs situated very close to the interface may be "pinned" by the SiO_2/Si conduction band discontinuity and therefore remain permanently positive even under a positive gate bias (7), (9). In Fig. 1(a), it can be observed that switching the gate voltage from +1.5 V back to 0 V does not restore $|\Delta V_t|$ to the level prior to the application of the +1.5 V gate voltage. This implies that a fraction of the DLHTs is indeed annihilated by the positive gate voltage. It should be mentioned that although a substantial part of $|\Delta V_t|$ is eliminated almost immediately (implying that most of the bond reformation processes occur rather quickly), a subsequent gradual decrease in $|\Delta V_t|$ spanning many orders of time remains apparent (Fig. 1(b)). For DLHTs which are neutralized, whether reformation of the Si-Si bond would occur clearly depends on the extent of structural relaxation. The rate of Si-Si bond reformation obviously decreases with increase in the structural relaxation. The remaining part of V_O's could explain the ANPC of ref. (32). It should be noted that ANPC may also be formed via interaction between trapped holes and H^+ (33), and this mechanism may co-exist under more severe stress condition (e.g. high charge-injection fluence) which triggers the substantial release of hydrogenous species.

For V_O's which have undergone significant structural relaxation, they may be momentarily charge-compensated under a positive gate voltage and would revert to the more stable V_O^+ state upon removal of the gate bias (18). At this point, it is necessary for us to clarify the reason for the negligible change in $|\Delta V_t|$ when the gate voltage is switched from +1.5 V to 0 V (Fig. 1(a)). This is because even though the device was subjected to a +1.5-V gate voltage during relaxation, the gate bias was interrupted when measurement was made (i.e. PIV measurement was made with a 0-V dc offset (11)). To check this point, we repeated the experiment whereby PIV measurements during relaxation were made by maintaining the gate bias at +1.5 V. The results are depicted in Fig. 12. In this case, an increase in $|\Delta V_t|$ is evident following the switch of the gate bias back to 0 V. It is also essential to clarify that the resultant increase of $|\Delta V_t|$ after the gate was switched back to 0 V depends on the relaxation time of the trapped electron in relation to the measurement delay. As measurement of the p-MOSFET I_d-V_g curve requires the gate to be pulsed to the negative polarity, there would be an inevitable loss of trapped electrons depending on the measurement delay. A reduced delay, crucial for the direct tunneling gate oxide, would help suppress the loss of the trapped electrons. Based on this argument, the quasi-saturation level of $|\Delta V_t|$ at +1.5 V observed experimentally is expected to be always higher than the ideal case for which no trapped electrons is lost during measurement.

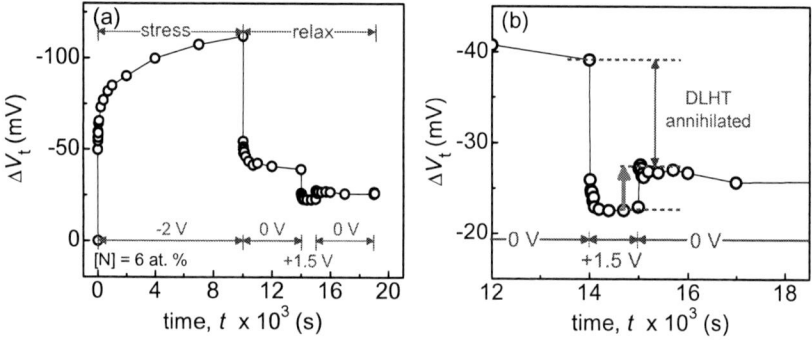

Figure 12. Evolution of threshold voltage shift $|\Delta V_t|$ during negative-bias temperature stressing and relaxation. Unlike in Fig. 1 where the +1.5 V gate voltage was interrupted and set to 0 V during PIV measurement, the gate voltage in this case was maintained at +1.5 V throughout the relaxation (i.e. PIV measurements were made at a dc offset of +1.5 V). A step-like increase in $|\Delta V_t|$ is evident when the gate voltage was subsequently switched to 0 V (big upward arrow). (b) A magnitude view of the $|\Delta V_t|$ increase at the +1.5 V to 0 V transition. Such a step-like change is not observed in a pristine device.

Now, we turn our attention to the CPC mentioned in ref. (32). Prior to the alternating positive and negative gate biasing that gave rise to the cyclic behavior, the device was subjected to Fowler-Nordheim injection (FNI) to neutralize the trapped holes. During FNI under a relatively large oxide field, electrons were captured by almost all the hole traps (including those that are relatively deep or ANPC). V_O's which have undergone minimal structural relaxation would have been annihilated. Thus, it is reasonable for one to assume that only V_O's which had undergone substantial structural relaxation, i.e. permanent and relatively deep-level hole traps remained. At the first negative gate bias, electrons trapped at these hole traps would be detrapped and this might take some time especially if they were situated at some distance from the interface. However, once these electrons were detrapped, some of the relatively deep hole traps might not be re-neutralized again during the following positive gate biasing phase. This was because the moderate positive gate bias only allowed hole traps within a limited energy range to be re-occupied by electrons. This gave rise to an asymmetric first cycle. The subsequent switching of the gate alternately between the same negative and positive polarities resulted in the repetitive detrapping and trapping, respectively, of electrons at these relatively permanent hole traps, resulting in a cyclic behavior. Since the gate oxide used was thick (5.5-7 nm), the spatial distribution of the hole traps from the interface also influenced the logarithmic time dependence of electron trapping/detrapping (Fig. 1(b) of ref. (32)) observed, apart from the energy distribution of the hole traps.

The cyclic behavior (Fig. 1(a) of ref. (32)) implied that it was the same group of V_O's that consistently responded to the given alternating negative/positive electrical stimuli. A cyclic behavior of the $|\Delta V_t|$ fluctuations under dynamic NBTI was also observed in our recent studies (15)-(17), (34). The important point here is that the total $|\Delta V_t|$ fluctuation included the fast recovery component that occurs during stress-to-relaxation transition. Moreover, such as a cyclic behavior is observed regardless whether the gate voltage is at 0 V or a positive voltage during relaxation. In the former, only relatively shallow hole traps are annihilated whereas both shallow and deep hole traps are removed in the latter (Fig. 1). These observations suggest that both the shallow and deep hole traps, corresponding to as-grown hole traps and ANPC in the context of ref. (32), have similar origins. This inference agrees with the result of ref. (33), in which the interaction between trapped holes and hydrogen was studied. The increase in ANPC after exposure to hydrogen corresponded to the decrease in the amount of as-grown hole traps.

Our simulation clearly shows that V_O's with a neighboring twofold coordinated nitrogen atom is one type of defect precursors to avoid since they exhibit very deep CTLs. Therefore, these V_O's would charge up positively even under moderate negative gate voltage and the resultant trapped-hole site would exhibit more significant structural relaxation in a given time compared to the case when the neighboring nitrogen atom is threefold coordinated or is absent. As a consequence, the trapped-hole sites would be permanent, severely impacting the NBTI reliability. Passivating the nitrogen dangling bond with hydrogen (resulting in a threefold coordinated nitrogen) is shown to significantly lower the CTL, and it appears to be a viable means for suppressing DLHT generation. Zhao and Zhang (33) showed that hydrogen could also result in relatively deep hole traps. It should, however, be emphasized that in ref. (33), the increase in the ANPC only occurred when a device with a substantial amount of trapped holes was exposed to hydrogen. Hydrogen was found to have no observable effect on a pristine device or in a stressed device where the trapped holes were first neutralized by electron injection. This observation suggests that the cracking of hydrogen by trapped hole sites, which leads to the formation of H^+, is responsible for the increased ANPC. *Ab-initio* simulation has also shown that the positive trap states related to H^+ are relatively deep (35). With these considerations, hydrogen annealing of as-processed devices, which are expected to contain a low density trapped holes, may remain a viable means for suppressing DLHT generation related to nitrogen.

Conclusion

The issue of DLHT generation under negative-bias temperature stressing is examined in detail both via experiment and simulation. A correlation between DLHT generation and the nitrogen concentration in the gate oxide is demonstrated experimentally. Via first-principles simulation, we show that an oxygen vacancy defect with a neighboring twofold coordinated nitrogen atom is a likely candidate for the DLHT. This defect is found to exhibit very significant structural relaxation upon the capture of a hole, and therefore the resultant positively charged state is relatively stable (corresponding to a deep CTL) On the other hand, the properties of an oxygen vacancy defect with a neighboring threefold coordinated nitrogen atom are shown to be similar to that for the pure SiO_2 (i.e. much shallower CTLs (~2.5 – 3 eV) and reduced structural relaxation). These defects are mostly responsible for the fast transients typically observed in dynamic NBTI and they are normally triggered under more negative gate biasing and may not play a significant role under nominal gate voltage condition. Based on the new insights, a unified explanation for hole-trap generation and relaxation is proposed and is shown to be able to account for the different electrical responses of positive oxide charges reported in recent literature.

Acknowledgements

This work is funded in part by a Singapore Ministry of Education research grant MOE2009-T2-1-050. Z. Q. Teo is grateful to the Singapore Economic Development Board and GLOBALFOUNDRIES Singapore for a joint Ph.D. scholarship grant.

References

1. B. E. Deal, M. Sklar, A. S. Grove, and E. H. Snow, *J. Electrochem. Soc.*, **114**, 266 (1967).
2. N. Kimizuka, Y. Yamaguchi, K. Imai, T. Iizuka, C. T. Liu, R. C. Keller, and T. Horiuchi, *Proc. Symp. VLSI Technol.*, 92 (2000).
3. Y. Mitani, M. Nagamine, H. Satake, A. Toriumi, *Int. Electron Dev. Meet. Tech. Dig.*, 509 (2002).
4. D. S. Ang, S. Wang, and C. H. Ling, *IEEE Electron Dev. Lett.*, **26**, 906 (2005).
5. D. S. Ang and S. Wang, *Appl. Phys. Lett.*, **88**, 093506 (2006).
6. D. S. Ang and K. L. Pey, *IEEE Electron Dev. Lett.*, **25**, 637 (2004).
7. D. S. Ang and S. Wang, *IEEE Electron Dev. Lett.*, **27**, 914 (2006).
8. D. S. Ang and S. Wang, *IEEE Electron Dev. Lett.*, **27**, 755 (2006).
9. D. S. Ang, S. Wang, G. A. Du, and Y. Z. Hu, *IEEE Trans. Mat. Dev. Reliab.*, **8**, 22 (2008).
10. J. P. Campbell, P. M. Lenahan, C. J. Cochrane, A. T. Krishnan, and S. Krishnan, *IEEE Trans. Mat. Dev. Reliab.*, **7**, 540 (2007).
11. G. A. Du, D. S. Ang, Z. Q. Teo, and Y. Z. Hu, *IEEE Electron Dev. Lett.*, **30**, 275 (2009).
12. K. O. Jeppson and C. M. Svensson, *J. Appl. Phys.*, **48**, 2004 (1977).
13. M. A. Alam, in *Int. Electron Dev. Meet. Tech. Dig.*, p. 345 (2003).
14. J. H. Lee, W. H. Wu, A. E. Islam, M. A. Alam, and A. S. Oates, *Proc. Int. Reliab. Phys. Symp.*, 745 (2008).

15. Z. Q. Teo, D. S. Ang, and K. S. See, *Int. Electron Dev. Meet. Tech. Dig.*, 737 (2009).
16. Z. Q. Teo, D. S. Ang, and C. M. Ng, *IEEE Electron Dev. Lett.*, **31**, 269 (2010).
17. D. S. Ang, Z. Q. Teo, T. J. J. Ho, and C. M. Ng, *IEEE Trans. Dev. & Mat. Reliab.*, in press.
18. A. J. Lelis and T. R. Oldham, *IEEE Trans. Nul. Sci.*, **41**, 1835 (1994).
19. C. J. Nicklaw, Z.-Y. Lu, D. M. Fleetwood, R. D. Schrimpf, S. T. Pantelides, *IEEE Trans. Nul. Sci.*, **49**, 2667 (2002).
20. T. Grasser, B. Kaczer, W. Goes, Th. Aichinger, Ph. Hehenberger, and M. Nelheibel, *Proc. Int Reliab. Phys. Symp.*, 33 (2009).
21. H. Reisinger, O. Blank, W. Heinrigs, A. Mühlhoff, W. Gustin, and C. Schlünder, *Proc. Int. Reliab. Phys. Symp.*, 448 (2006).
22. T. R. Oldham, A. J. Lelis, and F. B. McLean, *IEEE Trans. Nucl. Sci.*, **NS-33**, 1203 (1986)
23. D. S. Ang, S. C. S. Lai, G. A. Du, Z. Q. Teo, T. J. J. Ho, and Y. Z. Hu, *IEEE Electron Dev. Lett.*, **30**, 751 (2009).
24. A. Yokozawa and Y. Miyamoto, *Phys. Rev. B*, **55**, 13783 (1997).
25. P. Dahinden, P. Broqvist, and A. Pasquarello, *Phys. Rev. B*, **81**, 085331 (2010).
26. I. A. Chaiyasena, P. M. Lenahan, and G. J. Dunn, *Appl. Phys. Lett.*, **58**, 2142 (1991).
27. P. Giannozzi, S. Baroni, N. Bonini, M. Calandra, R. Car, C. Cavazzoni, D. Ceresoli, G. L. Chiarotti, M. Cococcioni, I. Dabo, A. D. Corso, S. de Gironcoli, S. Fabris, G. Fratesi, R. Gebauer, U. Gerstmann, C. Gougoussis, A. Kokalj, M. Lazzeri, L. Martin-Samos, N. Marzari, F. Mauri, R. Mazzarello, S. Paolini, A. Pasquarello, L. Paulatto, C. Sbraccia, S. Scandolo, G. Sclauzero, A. P. Seitsonen, A. Smogunov, P. Umari, and R. M. Wentzcovitch, *J. Phys.: Cond. Matt.*, **21**, 395502 (2009).
28. J. P. Perdew, K. Burke, and M. Ernzerhof, *Phys. Rev. Lett.*, **77**, 3865 (1996).
29. A. Alkauskas, P. Broqvist, F. Devynck, and A. Pasquarello, *Phys. Rev. Lett.*, **101**, 106802 (2008).
30. H.-T. Yu, H.-G. Fu, Y.-J. Chi, X.-R. Huang, Z.-S. Li, and C.-C. Sun, *Chem. Phys. Lett.*, **359**, 373 (2002).
31. C. G. Van de Walle and J. Neugebauser, *J. Appl. Phys.*, **95**, 3851 (2004).
32. J. F. Zhang, C. Z. Zhao, A. H. Chen, G. Groeseneken, and R. Degraeve, *IEEE Trans. Electron Dev.*, 51, 1267 (2004).
33. C. Z. Zhao and J. F. Zhang, *J. Appl. Phys.*, 97, 073703 (2005).
34. Z. Q. Teo, D. S. Ang, and C. M. Ng, *IEEE Electron Dev. Lett.*, **31**, 656 (2010).
35. J. Godet, F. Giustino, and A. Pasquarello, *Phys. Rev. Lett.*, **99**, 126102 (2007).

144

ECS Transactions, 35 (4) 145-174 (2011)
10.1149/1.3572281 ©The Electrochemical Society

Essential Aspects of Negative Bias Temperature Instability (NBTI)

Ahmad Ehteshamul Islam[a], Souvik Mahapatra[b], Shweta Deora[b], Vrajesh D. Maheta[b], and Muhammad Ashraful Alam[c]

[a]Department of Materials Science and Engineering, University of Illinois at Urbana-Champaign, IL 61801, USA
[b]Department of Electrical Engineering, Indian Institute of Technology Bombay, Mumbai 400076, India
[c]Department of Electrical and Computer Engineering, Purdue University, West Lafayette, Indiana 47907. USA

Email: [a] aeislam@ieee.org, [b] souvik@ee.iitb.ac.in, [c] alam@purdue.edu

(Invited Paper)

We develop a comprehensive theoretical framework for explaining the key and characteristic experimental signatures of NBTI. The framework is based on an uncorrelated dynamics of interface-defect creation/annihilation described by Reaction-Diffusion (R-D) theory and hole trapping/detrapping into/out-of oxide defects based on a generalized Shockley-Read-Hall model. The proposed theory can consistently explain the long-term stress-phase power-law time exponent, stress/relaxation-phase temperature dependence, characteristic feature of duty-cycle dependence, and universal feature of frequency independence - measured in DC and AC stress conditions over a wide variety of transistors. Thus, we confirm the general validity of R-D theory in explaining the universal features (irrespective of dielectric material) of both DC and AC NBTI. The non-universal features of NBTI have correlation with the amount of oxide defects within the dielectric and do not affect AC NBTI measurements at lower duty cycle. Decomposition of these (uncorrelated) universal and non-universal components is, therefore, essential before comparison with any theory.

1. Background

Negative Bias Temperature Instability (NBTI) indicates a temperature accelerated degradation in MOS transistors when it is stressed with a negative gate voltage, *i.e.*, with an oxide electric field that is directed from the channel towards the gate of a MOS transistor. In normal CMOS (which is the basis for today's microprocessor) operation, only the PMOS transistors are subjected to such negative oxide electric field. Therefore, since its introduction in microprocessor in early 1970s [1, 2], NBTI-induced performance degradation in PMOS transistors (*e.g.*, increase in threshold voltage, reduction in current drivability) has always been a concern in CMOS technology. Recent use of thin gate dielectric (a consequent increase in oxide electric field) and use of high-κ (oxynitride and Hf-based) dielectric materials have further enhanced NBTI degradation. As such, NBTI is considered as one of the major reliability concerns in current CMOS technology [3-6]. This industry-wide reliability concern has encouraged numerous efforts (see reference [7] for a chronology of the number of papers being published in this area) to understand the essential aspects of NBTI degradation by using different types of transistor-based measurements and then develop appropriate theory to explain the experiments. However, there is a perception that a definitive interpretation of this PMOS-specific phenomenon remains elusive.

Most of the literature on NBTI is focused on explaining isolated experimental signatures of NBTI degradation and recovery and, in the process, may sometimes overlook the fact that the proffered explanations have not been consistent with other characteristic features of NBTI. Such fragmented modeling efforts have made NBTI literature extremely confusing to the general audience. For capturing the experimental features, NBTI dynamics has sometimes been attributed to interface defect N_{IT} [2, 8, 9], or to hole trapping into pre-existing oxide defects N_{HT} [10-14], or to both N_{IT} and N_{HT} [15-22]. Recent introduction of Spin Dependent Recombination (SDR) experiment [23] have helped shed light on the nature of NBTI-specific defects, but the work is still in flux and have not been conclusive enough to settle the discussion. For example, even though many groups have historically attributed NBTI-related N_{IT} generation to P_b-centers (by ESR and SDR experiments), recent SDR experiment [23] indicates that P_b-centers are apparently absent in transistors with high-κ oxynitride gate dielectric. These experiments have motivated several articles to explain NBTI from the point-of-view of bulk oxide defects, where N_{IT} is either indirectly created via hole trapping into generated oxide defects N_{OT} [22], or N_{IT} is absent altogether [12, 13]. Unfortunately, however, the new theories have not always attempted to explain other well known features of NBTI experiments, i.e., (i) stress-phase power-law time exponent, (ii) stress and relaxation-phase temperature dependence, (iii) duty cycle dependence, and (iv) universal frequency independence, and therefore the generic validity of such bulk-oxide defect oriented approaches remain questionable.

In this manuscript, we develop a robust theoretical framework for explaining these key characteristic experimental features of NBTI, measured in DC and AC stress conditions for a wide variety of transistor technologies. The framework is based on an uncorrelated dynamics for Reaction-Diffusion (R-D) theory based interface defect creation/annihilation and Shockley-Read-Hall theory based hole trapping/detrapping into/out-of oxide defects. Our analysis suggests the necessity of decomposing NBTI measurements into a universal slow component associated with interface defect generation and a non-universal fast component associated with charge trapping/detrapping into/out-of oxide defects. The universal component can be consistently explained using R-D theory based interface defect dynamics and is directly relevant for product qualification. On the other hand, the non-universal component depends on the amount of oxide defect within the dielectric (and hence associated with the gate stack fabrication technology). Hence, this non-universal component is sample dependent and must be accounted only for high-fidelity experimental fitting in short-time scales. Therefore, one should decompose these two uncorrelated (universal and non-universal) components and then explain their respective features using respective theories. Any attempt to violate such decomposition will lead to proliferation of unphysical model parameters to capture the experimental trends.

1.1 Interface Defect in NBTI and Relevance of R-D Theory

Historically, interface defects have always been a technology challenge for semiconductor industry. Unlike oxide defects, interface defects cannot be removed or reduced through purification (or gettering). At the silicon-dielectric interface of a MOS transistor, interface defects (dangling Si- bonds) are terminated or passivated (Si- + H → Si-H) by using hydrogen compounds like silane (SiH_4). Efficacy of such passivation techniques using atomic hydrogen (H) has been extensively studied since 1970s [1, 24-29]. However, H-passivation of dangling bonds only provides a 'time-zero' (pre-use) solution from interface defects. After a period of transistor operation, mostly in the PMOS configuration, the interfacial Si-H bond starts to dissociate or depassivate (Si-H → Si- + H) at normal operating condition in the presence of cold holes near the interface and reforms the dangling Si- bonds. Formation of these interface defects due to Si-H depassivation has been observed by using ESR and SDR experiments, especially on the Si/SiO_2 interface [30, 31]. The same has also been confirmed using capacitance-voltage measurement and mobility degradation experiments [17, 32, 33].

Theory of interface defect formation has evolved around the (1) modeling of Si-H bond dissociation and (2) subsequent handling of the resultant hydrogen species. Si-H bond dissociation is either considered to be – (1a) cold-hole assisted [8, 9, 15, 34-36], (1b) dopant-activated [37, 38], or (1c) oxide-defect induced [22]. Resultant hydrogen species after Si-H bond dissociation is either considered to be – (2a) trapped within the oxide [10, 22, 33], or (2b) diffuse within the oxide/gate region [8, 15]. Mechanisms (1a) and (2b), or (1b) and (2b), constitute the basis of the so-called Reaction-Diffusion (R-D) framework. Note that the environment of Si-H bond (*i.e.*, whether the resultant Si- bond is a P_b-center or a K-center [23]) or details of the dissociation process does not affect the characteristic predictions (*i.e.*, power-law time exponent, frequency/duty cycle dependence, temperature dependence) of the R-D theory.

Originally proposed in 1977 by Jeppson *et al.* [2] (and later refined by Alam *et al.* in early 2000s [8, 15, 34, 39, 40]) to interpret fractional kinetics of NBTI degradation, the implications of the R-D model has been explored in hundreds of papers through various generations of CMOS technology. R-D model has a major advantage over its counterparts in explaining interface defect formation, *i.e.*, it provides a parameter-free interpretation of all the *four* broad universal features of NBTI degradation after non-universal and technology-specific hole trapping into oxide defect component is subtracted out [16, 41]. It can predict a long term time dependence of $t^{1/6}$, long-term Arrhenius-activated temperature dependence, the frequency independence of degradation, and the characteristic shape of duty cycle dependence – regardless of the technology or operating condition. All these have made R-D theory (as discussed in section 3) the starting point for all the discussions of NBTI degradation.

1.2 Hole Trapping into Oxide Defects in NBTI

Dynamics of trapping into oxide defects have been studied since 1970s' [42-44], when Metal-Nitride-Oxide-Semiconductor (MNOS) structure became a popular memory element, because of its simple structure and nonvolatile nature. Detrapping of carriers from pre-existing charged defects in the nitride layers was identified as the main mechanism for limiting the retention time in these memories. Moreover, discharge time was observed to increase with the increase in oxide thickness of the oxide layer, which proves that tunneling of carriers from charged defects within the nitride layer is gradually reduced with the increase in oxide thickness [42]. Similarly, trapping/detrapping of holes and electrons in/out-of pre-existing oxide defects has also been considered as a source of threshold voltage instabilities in MOS transistors during the same period [45-47]. Now-a-days, the incorporation of high-κ materials within the dielectric of MOS structures have caused significant concerns from these hole trapping/detrapping issues during NBTI stress [10, 20, 48-56]. In addition, hole trapping/detrapping phenomena can also give rise to the observation of Random Telegraph Noise (RTN) and 1/f noise in modern transistors having high-κ gate dielectric [57-61].

Hole trapping is generally modeled by considering tunneling of channel carriers into oxide defects. These oxide defects can either be pre-existing [15, 18] or generated by electrical stress [20, 22, 62]. Pre-existing oxide defect is mainly an issue in over-coordinated high-κ (oxynitride or Hf-based) materials [63, 64]. Higher coordination number (i.e., number of nearest neighbors for a particular atom within the material) in high-κ materials makes it difficult to satisfy all the chemical bonding within the amorphous network. Thus, transistors with high-κ dielectric are prone to high density of pre-existing defects and generally suffer from BTI effects due to hole trapping (will be designated using N_{HT} from now on). On the other hand, oxide defect generation (will be designated using N_{OT} from now on) has always been a problem within the reliability community (irrespective of dielectric material) [62, 65-67] that leads to the well-known phenomena, called Time-Dependent Dielectric Breakdown or TDDB. These oxide defect formation has an empirically extracted to have universal nature [68] and is often characterized with a time dependence of $t^{1/3}$ [69]. We will explain the theory of N_{HT} and N_{OT} in sections 4 and 5, respectively and therefore, show its relevance in explaining experimental features of NBTI that is presented in section 2.

2. Broad Empirical Features of NBTI

2.1 Summary of NBTI Measurements

Being one of the major reliability concerns, NBTI has been studied by a large group of researchers all over the world. This breadth of experimental data provides us an opportunity to collect measurements from published reports across industry and academia (see Figure 1) and compare with our own measurements. The comparison is summarized as follows:

➢ Figure 1a shows the power-law time exponent (when ΔV_T is expressed as At^n) measured at long stress time (t_{STS}) in the industrial grade devices. Here, A is a voltage and temperature-dependent constant (see equation (13) for details on this voltage and temperature dependence). In spite of disparate sources, all devices unequivocally show an exponent of n ~1/6, independent of voltage and temperature.

➢ Figure 1b shows the stress-phase ultra-fast on-the-fly (UF-OTF) measurements (where time-zero delay [70] t_0 is 1μs and ΔV_T is estimated ignoring the mobility correction [71], i.e., $\Delta V_T \sim \Delta I_D/I_{D0}(V_G-V_{T0})$) at different temperature for different oxynitride process splits. The measurements reveal the existence of a short-time, temperature independent fast component (that saturates within ~ms) in transistors having high %N for optimized (Type-A) plasma oxynitride dielectric. Decrease in %N within the dielectric reduces the contribution from this fast component and shows the existence of temperature activated slow component, even at short t_{STS}. Moreover, at relatively long $t_{STS} > 1$ s, all devices show some degree of temperature activation (less for high %N). Note that in un-optimized oxynitride process split (Type-B), the temperature-independence at short t_{STS} is even observed for low %N. The readers may wish to review reference [72] for an analysis of nitridation process flow on NBTI characteristics.

➤ Figure 1c shows NBTI relaxation measurements across different type-A oxynitride process splits (ultra-fast V_T or UFV measurement of reference [50] is also shown for comparison). The important point to note here is the initiation of ~5% NBTI relaxation ($t_{REC,start}$; where $\Delta V_T(t_{REC}) / \Delta V_T(t_{STS}) \sim 95\%$) and the time-dependence of NBTI relaxation. Though several studies [22, 50, 73, 74] on NBTI have reported the universality of *log-t relaxation* with $t_{REC,start}$ of ~ µs, our UF-OTF measurements [16, 75, 76] demonstrate that $t_{REC,start}$ and the time-dependence of NBTI relaxation depend on %N of the oxynitride high-κ gate dielectric, as well as on the difference between stress and recovery voltages ($V_{STS}-V_{REC}$). In general, $t_{REC,start}$ is larger (~ ms) for low %N and smaller ($V_{STS} - V_{REC}$), very clearly indicating the *non-universal* nature of NBTI recovery. In addition, NBTI relaxation data also shows temperature independence at short t_{REC} for high %N oxynitride transistors (Figure 2a,b). And similar to the stress-phase measurements of Figure 1b, the temperature independence at short t_{REC} disappears with the reduction in %N.

➤ Figure 1d,e shows a comprehensive summary of duty cycle and frequency dependent NBTI measurement, obtained from a broad range of published reports across industry and academia. When normalized to DC, the duty cycle dependent measurement shows large spread in AC/DC ratio (= $\Delta V_T(AC) / \Delta V_T(DC)$), although the frequency independence in AC/DC ratio at 50% duty cycle is generally observed (some older datasets show slight drop at higher frequency and most researchers now consider this droop to be a measurement artifact [77]).

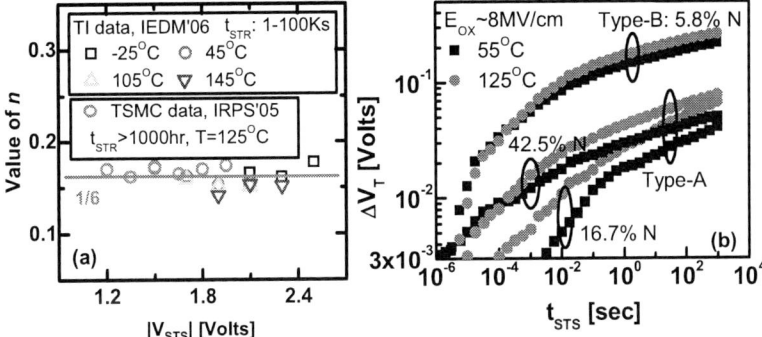

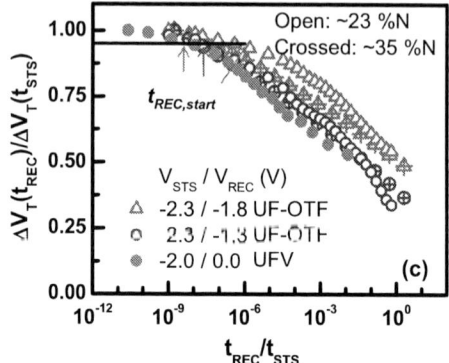

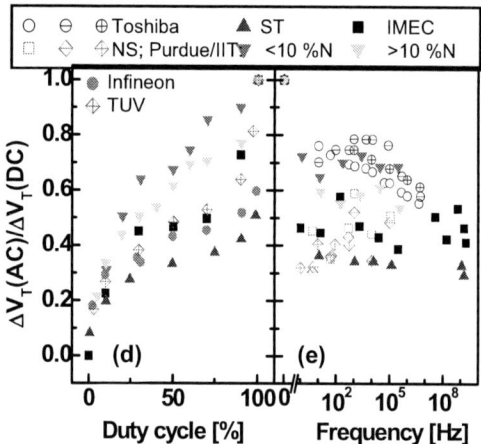

Figure 1. **(a)** Long-term power-law time exponent n (where, $\Delta V_T \sim t^n$) for NBTI, collected from TSMC and TI measurements (taken from reference [78] and [40], respectively), indicates $n \sim 1/6$. Note that Freescale [79] and Infineon [80] data also shows similar time exponent. **(b)** Temperature dependent stress-phase NBTI measurement across different process split of oxynitride gate dielectric (type-A has optimum nitridation; and type-B has non-optimized nitridation) **(c)** Initiation of NBTI relaxation ($t_{REC,start}$) varies with V_{REC} and %N of the oxynitride high-κ dielectric (here, the UF-OTF are our measurements and ultra-fast V_T or UFV measurement is taken from reference [50]). **(d)** AC/DC ratio (when ΔV_T is measured at the end of AC cycles) vs. duty cycle, and **(e)** AC/DC ratio vs. frequency (at 50% duty cycle) plots for different technologies indicate wide spread in measured data. Measurements of (d-e) are taken from the following references: Toshiba [81], ST [21], IMEC [77], NUS [82], Infineon [80], TUV [11].

2.2 Modeling Challenge

The broad scatter in NBTI measurements (as presented in Figure 1) makes NBTI modeling a considerable challenge. In an effort to address this challenge, NBTI researchers have taken widely different strategies: Some articles have used a wide distribution of capture and emission time constants (from 10^{-9}-10^{14} s) for fitting each of the stress and relaxation phase measurements with independent distribution [80, 83]. Unfortunately, this approach results in proliferation of fitting parameters that cannot always be physically justified. Other articles have considered wide distribution of defects within the dielectric [12, 13] to capture the long-term nature of the time exponent. In addition, interaction and transfer of chemical species among finite numbers of energy wells [11, 22] is recently used to capture part of the experimental features of Figure 1. However, none of these approaches are comprehensive enough to consistently interpret the four characteristic features of NBTI degradation without using unphysical parameters, as summarized in Figure 1.

In an attempt to explain the broad experimental features of NBTI, we ask the following questions:

1) Why is the long-term time exponent always ~1/6, regardless of the stress voltage or stress temperature, as shown in Figure 1a? Note that these results are usually obtained with small measurement delay (order of seconds), which however has insignificant effect when the stress time is very long. Obviously, the use of excessive measurement delay may lead to higher time exponent even at long stress time [84].
2) Why does the early part of stress and relaxation experiments so sensitive to process details of dielectric material, as shown in Figure 1b,c? Why is process dependent part insensitive to temperature, especially at higher %N, as shown in Figure 1b and Figure 2a,b? And, in the same context, why $t_{REC,start}$ in Figure 1c has process dependence?
3) Why is there a sudden drop in $\Delta V_T(AC)/\Delta V_T(DC)$ in the range of 80-100% duty cycle (Figure 1d)? Why is the shape of duty cycle vs. $\Delta V_T(AC)/\Delta V_T(DC)$ universal (this is more evident, when Figure 1d is normalized with respect to 50% duty cycle value, as shown in Figure 2c) in the lower duty cycle regime?
4) Why is the measured NBTI at 50% duty cycle always frequency independent, irrespective of transistor technology (as shown in Figure 1e)?

In this manuscript, we explain NBTI in a broader context by answering the aforementioned four questions, rather than focusing on a smaller subset. Therefore, we decompose measured ΔV_T into three uncorrelated components:

A. The first and major part of NBTI-induced ΔV_T comes from the Reaction-Diffusion theory based interface defect (N_{IT}) generation and relaxation. We will show how this component can explain: i) the universal observation of 1/6 power-law time exponent, ii) the slow and temperature dependent part of NBTI stress and relaxation, iii) the universal part of duty cycle dependence upto ~50% duty cycle, and iv) the universal observation of frequency independence.

B. The second component of NBTI comes from hole trapping into pre-existing oxide defects (N_{HT}). This component can explain: i) the fast and temperature independent part of NBTI stress and relaxation, ii) the sharply decreasing part of duty cycle dependence above ~50% duty cycle, and iii) the magnitude of $\Delta V_T(AC)/\Delta V_T(DC)$ at a particular duty cycle.

C. The last component of NBTI comes from hole trapping into generated oxide defects (N_{OT}). This, in addition to N_{HT}, can explain the disparity of $t_{REC,start}$ in different nitrided transistors.

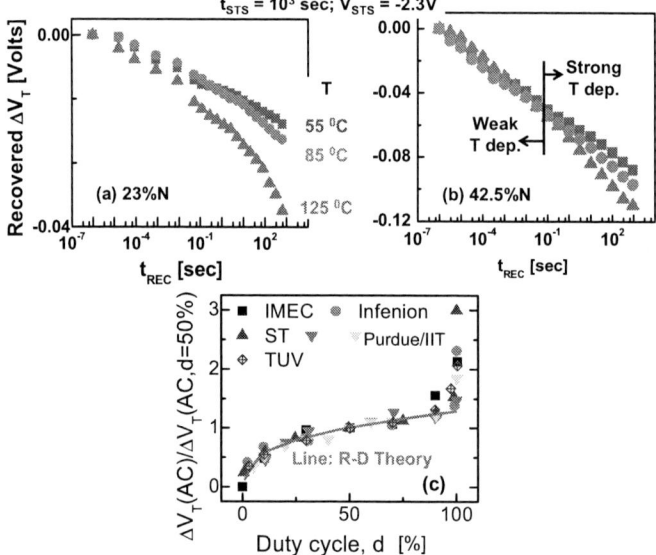

Figure 2: (a-b) Similar to NBTI stress phase (Figure 1b), early relaxation phase is temperature independent. Such temperature independent early relaxation phase is cleanly observed for high %N transistors. (c) The universal shape of duty cycle dependence (upto $d \sim 80\%$) can be nicely captured by R-D theory. Here, the universal shape of duty cycle is obtained by scaling the duty cycle dependent NBTI measurement with respect to the 50% duty cycle value (see section 7 for the justification).

In other words, we express ΔV_T using –

$$\Delta V_T(t) = \Delta V_{IT}(t) + \Delta V_{HT}(t) + \Delta V_{OT}(t)$$

$$= \alpha \frac{q \Delta N_{IT}(t)}{C_{ox}} + \frac{\int_0^{T_{ox}} \int_E x \rho_{HT}(x,E,t)\, dE\, dx}{C_{ox} T_{ox}} + \frac{\int_0^{T_{ox}} \int_E x \rho_{OT}(x,E,t)\, dE\, dx}{C_{ox} T_{ox}}.$$

(1)

Here, ΔV_{IT}, ΔV_{HT}, and ΔV_{OT} refer to the contributions to ΔV_T from N_{IT}, N_{HT}, and N_{OT} components, respectively; C_{ox} is the oxide capacitance; T_{ox} is the oxide thickness, α accounts for the fraction of donor type [85] N_{IT} above the substrate Fermi-level that is contributing to NBTI; $\rho_{HT}(x,E,t)$ represents trapped holes into the pre-existing oxide defects at location x (measured into the oxide from the poly/oxide interface) and at energy E at time t; and $\rho_{OT}(x,E,t)$ represents trapped holes at generated oxide defects.

3. Theory of Interface Defect (N_{IT})

In this section, we summarize the main features of R-D theory, considering both atomic (H) and molecular (H$_2$) diffusion. Our goal is to show how the theory provides a *parameter-free* prediction of most of the experimental features of NBTI. R-D theory with H-H$_2$ diffusion considers dissociation of Si-H bond and subsequent diffusion of hydrogen species, as governed by the following equations [15, 86]:

$$\frac{dN_{IT}}{dt} = k_F \left(N_0 - N_{IT} \right) - k_R N_{IT} N_H^{(0)}, \tag{2}$$

$$\frac{\delta}{2} \frac{dN_H^{(0)}}{dt} = D_H \frac{dN_H^{(0)}}{dx} + \frac{dN_{IT}}{dt} - \delta k_H \left[N_H^{(0)} \right]^2 + \delta k_{H2} N_{H2}^{(0)}, \tag{3}$$

$$\frac{\delta}{2} \frac{dN_{H2}^{(0)}}{dt} = D_{H2} \frac{dN_{H2}^{(0)}}{dx} + \frac{\delta}{2} k_H \left[N_H^{(0)} \right]^2 - \frac{\delta}{2} k_{H2} N_{H2}^{(0)}, \tag{4}$$

$$\frac{dN_H}{dt} = D_H \frac{d^2 N_H}{dx^2} - k_H N_H^{\,2} + k_{H2} N_{H2}, \tag{5}$$

$$\frac{dN_{H2}}{dt} = D_{H2} \frac{d^2 N_{H2}}{dx^2} + \frac{1}{2} k_H N_H^{\,2} - \frac{1}{2} k_{H2} N_{H2}. \tag{6}$$

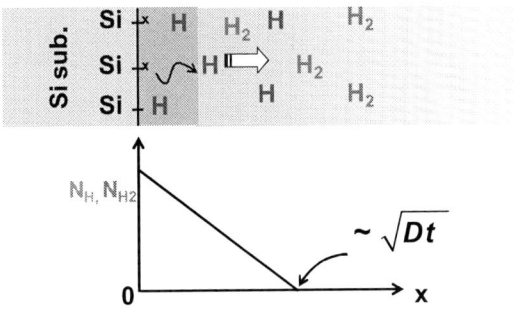

Figure 3: Schematic of Si-H bond dissociation and consequent hydrogen diffusion, as considered in R-D theory based N_{IT} dynamics.

Equation (2) represents passivation/de-passivation effects of Si-H bond, where k_F, k_R, N_0, N_{IT}, $N_H^{(0)}$ are defined as Si-H bond-breaking rate, Si-H bond-annealing rate, initial bond density available before stress, interface defect density, and hydrogen density at the Si/dielectric interface, respectively. Equations (3) and (4) correspond to the conservation of fluxes of diffusing hydrogen species (H and H_2) near the interface (along the x axis), whereas equations (5) and (6) describe diffusion (along the x axis) of H and H_2. $k_H N_H^2$ and $k_{H2} N_{H2}$ terms in equations (3)-(6) incorporate the H-H_2 conversion within the generalized R-D framework. Among the symbols used in equations (3)-(6), k_H and k_{H2} represent generation and dissociation rates of H_2; D_H and D_{H2} represent diffusion coefficients for H and H_2; N_H and N_{H2} represent the concentration of atomic and molecular hydrogen; δ represents the interfacial thickness (~1-2 Å). All parameters are greater than zero for the stress phase. In particular, k_F has the following dependence of equation (7) that serves a physical way of explaining the oxide electric field E_{ox} dependence of interface defect generation [15, 40, 87, 88].

$$k_F \sim p_h * N_0 P_T * \exp(\gamma_T E_{ox}) * \exp(aE_{ox}/kT), \qquad (7)$$

where p_h is the hole concentration within the inversion layer, $P_T \sim exp(-\sqrt{m_{ox}\varphi_{bh}})$ is the field-independent pre-factor for hole tunneling probability (m_{ox}: oxide effective mass and φ_{bh}: barrier height for hole tunneling), $exp(\gamma_T E_{ox})$ is the field-dependent factor for hole tunneling with field acceleration γ_T, and $exp(aE_{ox}/k_B T)$ is field-assisted Si-H bond dissociation enhancement factor (a: effective dipole moment and $k_B T$: thermal voltage). Thus, field acceleration of N_{IT} can be expressed as –

$$\gamma_{IT} = \gamma_T + a/k_B T. \qquad (8)$$

As shown in [15, 40], equation (8) explains the temperature dependence of N_{IT}'s field acceleration (hence provides an experimental way of extracting a and γ_T). In addition, equation (7) can be used to explain the nitridation process dependence [87] and strain dependence [88] of N_{IT} in NBTI measurements.

Now, let us derive the key results of R-D model and show how it anticipates the key experimental features of NBTI.

3.1 Power-law time exponent of $n \sim 1/6$:

Assuming $N_0 >> N_{IT}$ and $dN_{IT}/dt \sim N_{IT}/t$, equation (2) simplifies to,

$$N_H^{(0)} = \frac{k_F N_0 - N_{IT}/t}{k_R N_{IT}}. \qquad (9)$$

Moreover, the numerical solutions indicate that for continuous NBTI stress, $dN_H^{(0)}/dt$ and diffusion of H is negligible at all stress time [15], so that H_2 diffusion part in equation (3) reduces to,

$$\frac{N_{IT}}{t} = \delta k_H N_H^{(0)2} - \delta k_{H2} N_{H2}^{(0)}, \qquad (10)$$

and the conservation of hydrogen species within the system suggests –

$$N_{IT} \approx N_{H2}^{(0)} \sqrt{6D_{H2}t}. \tag{11}$$

Equation (11) requires that the extent of diffusion profile is larger than $\sqrt{D_H t}$, commonly used for approximating the complementary error function solution of diffusion equation [8, 34, 89]; which is more consistent with the use of $\sqrt{16D_H t/\pi}$ in [90, 91]. Now, by eliminating $N_H^{(0)}$ and $N_{H2}^{(0)}$ from equations (9)-(11), we have –

$$\frac{N_{IT}}{t} - \frac{\delta k_H \left(k_F N_0 - N_{IT}/t \right)^2}{k_R^2 N_{IT}^2} + \frac{\delta k_{H2} N_{IT}}{\sqrt{6D_{H2}t}} = 0. \tag{12}$$

Equation (12) is the (implicit) analytical solution of H-H_2 R-D model, presented in equations (2)-(6), which compares very favorably with the detailed numerical simulation [15]. At long stress time (i.e., in the H_2 diffusion limited regime), equation (12) simplifies to (when N_{IT}/t becomes negligible) –

$$N_{IT} = \left(\frac{k_H}{k_{H2}} \right)^{1/3} \left(\frac{k_F N_0}{k_R} \right)^{2/3} \left(6D_{H2}t \right)^{1/6} \equiv At^{n=1/6}. \tag{13}$$

Therefore, R-D theory anticipates the long-term time exponent of $n \sim 1/6$. In addition, R-D theory also establishes the fact that n should not depend on stress voltage and temperature. Both of these observations are consistent with the broad range of NBTI measurements (Figure 1a).

3.2 Long-term Temperature Dependence:

Using appropriate activation energies for k_F, k_R and D_{H2} in equation (13) and assuming H-H_2 conversion process has similar activation for forward and reverse reactions [92], the overall activation energy for $\Delta V_{IT} = qN_{IT}/C_{ox}$ can be written as [15],

$$E_{A,IT} \equiv nE_{A,H_2} + \frac{2}{3}\left(E_{A,F} - E_{A,R} - aE_{ox} \right). \tag{14}$$

where $E_{A,F}$, $E_{A,R}$, $E_{A,H2}$ are activation energies for k_F, k_R and D_{H2}, respectively. Experimentally, one can estimate $E_{A,IT} \sim 0.1$ eV by measuring NBTI in transistors having dominant interface defect generation (i.e., $\Delta V_T \sim \Delta V_{IT}$) [18, 32, 40, 50, 85, 88, 93, 94]. We have also shown that the variation of $E_{A,IT}$ due to aE_{ox} term comes within the error margin of activation energy estimation procedure [87]. This measured value of $E_{A,IT}$ and reported magnitude of $E_{A,H2} \sim 0.6$ eV [95] suggests that $E_{A,F} \sim E_{A,R}$. Thus, activation energy of molecular hydrogen diffusion governs the long-term temperature dependence of NBTI measurements.

3.3 Duty-cycle and Frequency Dependence:

In order to obtain the duty cycle and frequency dependence of N_{IT}, we derive the amount of N_{IT} under AC NBTI stress condition. Here, we assume that the NBTI stress is applied in a transistor for k $(>> 1)$ cycles with $2k$ stress/relaxation steps, defined by the duty cycle $d = t_{STS}/T$; where $T = t_{STS} + t_{REC}$ is the period of the signal. We also define that the degradation at the end of k-cycles (with respect to degradation over a single stress cycle) using $R_{2k} \equiv N_{IT}(kT)/N_{IT}(t_{STS})$ and $R_{2k-1} \equiv N_{IT}((k-1)T + t_{STS})/N_{IT}(t_{STS})$ with $R_1 = 1$, by definition. Following the analysis in Refs. [89] and [96], one can show that –

$$\left(R_{2k-1}\right)^{1/n} \approx \frac{k-1}{1+\sqrt{\xi(1-d)}} + \left(R_1\right)^{1/n} \approx \frac{k}{1+\sqrt{\xi(1-d)}} \equiv pk, \qquad (15)$$

or equivalently, $R_{2k-1} \approx R_{2k} \approx (pk)^n$, where $p = 1/\left(1+\sqrt{\xi(1-d)}\right)$.

If two transistors are stressed at two different frequencies ($f_1 = 1/T_1$ and $f_2 = 1/T_2$) at same duty cycle, then for long (but same) integrated stress times (*i.e.*, $k_1 T_1 = k_2 T_2$ or, $k_1 t_{STS,1} = k_2 t_{STS,2}$), the ratio of net degradation for the two transistors is given by:

$$\frac{N_{IT,f_1}(k_1 T_1)}{N_{IT,f_1}(k_2 T_2)} = \frac{R_{k1} N_{IT}(t_{STS,1})}{R_{k2} N_{IT}(t_{STS,2})} \approx \left(\frac{k_1}{k_2}\right)^n \left(\frac{t_{STS,1}}{t_{STS,2}}\right)^n = 1; \qquad (16)$$

which is frequency-independent (as observed in Figure 1e). And similarly, the ratio of AC NBTI degradation at a given frequency for total duration of $T_T = k_1 T_1$ compared to the DC NBTI degradation for the same period of time is given by –

$$\frac{N_{IT,f_1}(k_1 T_1)}{N_{IT,DC}(T_T = k_1 T_1)} = \frac{(pk_1)^n A(t_{STS,1})^n}{A(k_1 T_1)^n} = \left[\frac{d}{1+\sqrt{\xi(1-d)}}\right]^n; \qquad (17)$$

which gives the characteristic duty cycle dependence of NBTI degradation (line in Figure 2c). We defer the discussion of the scaling algorithm till section 6.3, which is used in Figure 2c for the experimental data scaling. Thus, R-D theory can explain the universal observation of frequency independence (Figure 1d) and the characteristics trend of duty cycle dependence upto $d \sim 80\%$ (Figure 2c) for all NBTI experiments, measured in wide variety of process splits.

4. Theory of Hole Trapping into Pre-existing Oxide Defects (N_{HT})

Let us now return to the second component of NBTI in equation (1) related to the charging of pre-existing oxide defects. Pre-existing defects are commonly observed in almost all high-κ dielectrics (like oxynitrides and Hf-based compounds), which is an inherent feature of such over-coordinated high-κ materials. Trapping into these pre-existing defects leads to ΔV_{HT}, independent of the generation of N_{IT} or N_{OT}. Here, the thinness of modern dielectric film ($\sim$ few nm) requires us to consider the effect of detrapping towards the transistor gate [15]. In addition, typical defect density ($\sim 10^{18}$-10^{19} cm^{-3}, *i.e.*, 1 trap in 100-1000 nm^3) ensures that trap-to-trap transport is implausible.

Therefore, we only need to consider the processes involving single hop to and from the traps, i.e.,

$$\frac{df_T(x,t)}{dt} = \int_{-\infty}^{\infty} dE_T \sigma_{HT} v_{th} \left[p_h T_{S \to T} \left(1 - f_T\right) - n_S T_{T \to S} f_T - n_G T_{T \to G} f_T \right], \qquad (18)$$

where p_h is the inversion layer hole density; v_{th} is the thermal velocity; n_S, n_G are the concentration of detrapping states at substrate and poly-Si respectively; T's are tunneling probabilities between substrate (S), defect/trap (T), and gate (G) of the MOS transistor; and σ_{HT} is the capture cross-section for the hole trapping/detrapping process. The first term in equation (18) represents hole trapping into pre-existing oxide defects from the substrate, whereas the second and third terms represent hole detrapping out of oxide defects towards substrate and gate, respectively. These hole capture (during trapping process) and release (during detrapping process) events can be elastic or inelastic, as discussed in detail in reference [97]. Stochastic nature of such hole capture and release events gives rise to random telegraph noise, as studied in references [14, 57-61]. In the following discussion, we focus on the implication of elastic/inelastic hole trapping into thin dielectric and study its implication in long-term NBTI degradation. The readers may wish to review reference [14] for understanding the short-stress features of hole trapping.

Solution of equation (18) suggests (see reference [97] for details) the followings:

❏ **Time dependence:** The role of detrapping towards the poly-gate – while negligible in thick films (as for the case in [42, 98]) – are fundamentally important in the trapping dynamics of thin films. In ultra-thin dielectrics of current CMOS technology, detrapping process limits the possibility of hole trapping in sites located near the gate. As a result, threshold voltage shift due to hole trapping into pre-existing defects should saturate within orders of milliseconds, where the saturation time depends on the dielectric quality (i.e., nitridation) and thickness. As such, the corresponding threshold voltage shift can be approximated as –

$$\Delta V_{HT} = A_{HT} \left[1 - \exp\left(-t/\tau\right)^{\beta} \right], \qquad (19)$$

where A_{HT}, τ and β are trapping parameters that depends on the dielectric quality and dielectric thickness. In general, τ and β are larger for larger dielectric thickness and A_{HT} is larger for higher %N within the dielectric (i.e., for the dielectric having more N_{HT}).

❏ **Voltage dependence:** Voltage or field dependency of hole trapping arises from the relative position of the quasi-Fermi level E_{FS} within the dielectric as a function of oxide electric field E_{ox}. When E_{ox} is reduced, more trapping sites below E_{FS} are filled with electrons (see the change of trapping sites f_T values from Figure 4b to Figure 4c), with corresponding reduction ΔV_{HT}. As a result, hole trapping process shows significant voltage dependency.

❏ **Temperature dependence:** Since tunneling is a temperature independent process, temperature dependency of hole trapping mainly depends on the temperature dependence of capture cross-section (σ_{HT}) and thermal velocity (v_{th}). Recently, many hole trapping models [22, 59] consider structural relaxation as a part of the hole trapping process. Structural relaxation causes temperature activation (which depends on the energy barrier that is required to overcome for achieving structural relaxation) in the trapping process and can effectively be incorporated within the σ_{HT} parameter of equation (18).

As the hole trapping process is fast for modern transistors having thin gate dielectric, we can attribute the ultra-fast component of NBTI stress phase (in Figure 1b) to ΔV_{HT}. The increase of this ultra-fast component with the increase in nitrogen within the silicon oxynitride (SiO_xN_y with $2x+3y=4$ [99]) dielectric further confirms that this NBTI component is indeed related to pre existing oxide defect. Such increase of oxide defect with nitrogen within the SiO_xN_y dielectric is routinely reported in literature [17, 18, 21, 100]. Moreover, our measurements suggest relatively weak temperature dependence for this ultra-fast ΔV_{HT} component (Figure 1b), which indicates that hole trapping in our transistors have temperature independent σ_{HT}. Finally, as ultra-fast NBTI relaxation (Figure 1c) has similar features like the one for ultra-fast NBTI stress measurements, we expect initial part of NBTI relaxation to be related to an equivalent temperature independent, but nitrogen-dependent hole detrapping process. Therefore, like the stress phase, hole detrapping process in the relaxation phase will be complete within $t_{REC} =$ orders of ~ ms, depending on the nitrogen content within the oxynitride dielectric.

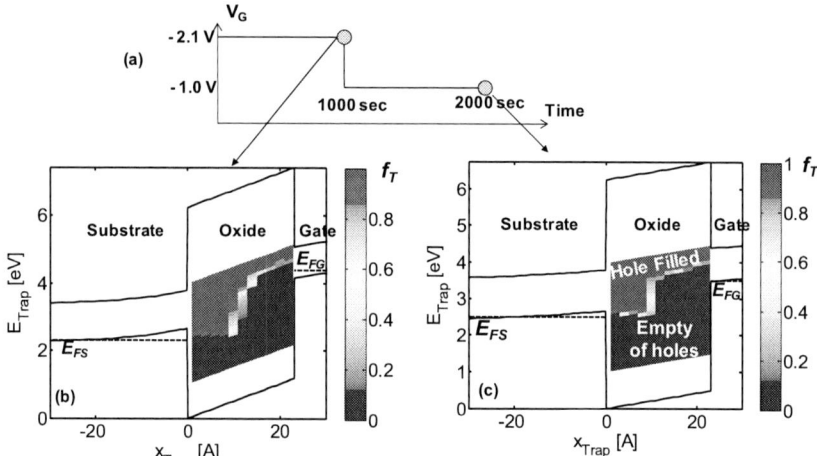

Figure 4: (a) Timing diagram used in the hole trapping simulation for a particular transistor having physical oxide thickness $T_{PHY} \sim 2.3$ nm, effective oxide thickness $EOT \sim 1.35$ nm, and 39.22 %N_2 dose (measured using XPS). The dielectric parameter (hole effective mass, oxide bandgap, and barrier height) for the simulation is calculated following the procedure stated in reference [87]. (b) Occupancy of defects or, hole trapping sites (assumed to have existence within the colored region only), after the transistor is kept at $V_{STS} = -2.1$ V for 1000 s. (c) Occupancy of hole trapping sites after the transistor is switched to $V_{STS} = -1.0$ V at 1000 s and kept at that voltage up to 2000 s.

5. Theory of Hole Trapping into Generated Oxide Defects (N_{OT})

NBTI stress is also a TDDB stress in p-MOSFETs. Therefore, we also need to consider hole trapping/detrapping from newly created bulk oxide defects N_{OT}, and corresponding ΔV_{OT}, at NBTI stress condition. In stress phase, the trapping process follows the same equation (18). However, unlike N_{HT}, in this case (slower) oxide defect generation is the rate-limiting process, rather than the hole trapping. Therefore, contrary to ΔV_{HT}, ΔV_{OT} is expected to show non-saturating behavior, proportional to the time-dependent increase of bulk defect density. Indeed, Figure 5 confirms the existence of universal ΔV_{OT}, even at NBTI stress conditions, as follows: Figure 5a indicates an increase in NBTI time exponent at higher V_{STS}, thus indicating a signature of oxide defect generation [62] or ΔV_{OT} at higher V_{STS} and negligible ΔV_{OT} at lower V_{STS}. Moreover, as the measurement scheme for [78] is slow, we expect the ΔV_{HT} component in ΔV_T to be negligible. Therefore, at lower V_{STS}, we can presume $\Delta V_T \sim \Delta V_{IT}$ and calculate field acceleration for ΔV_{IT}. Later, using the field acceleration for ΔV_{IT}, we can calculate ΔV_{IT} and thereby find $\Delta V_{OT} = \Delta V_T - \Delta V_{IT}$ at higher V_{STS} (Figure 5b). The calculated ΔV_{OT} for the measurements in [78] shows a decrease in time exponent for ΔV_{OT} at higher V_{STS}, as also observed in [68]. Interestingly, the estimated ΔV_{OT} at different V_{STS} scales universally and can be fitted (see Figure 5c) using a dispersive bulk defect generation model [68, 101] –

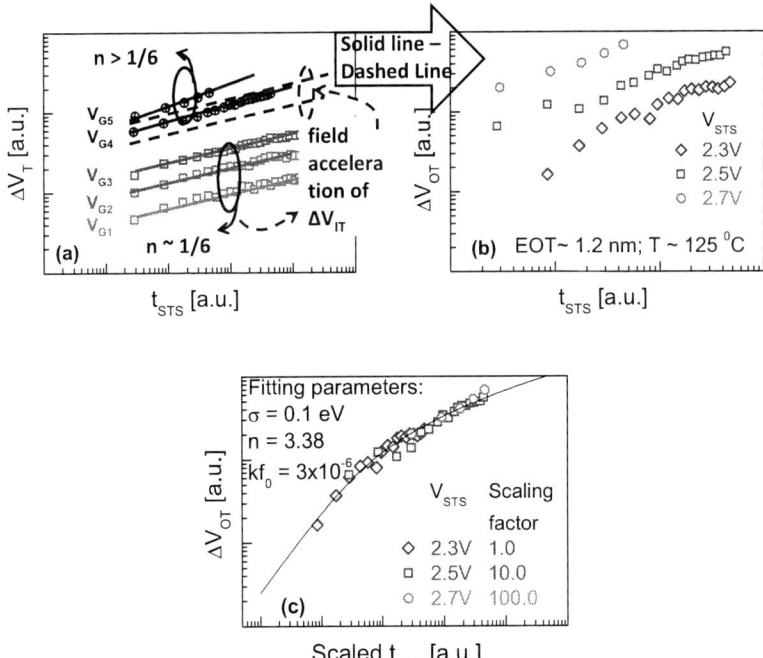

Figure 5: (a) Time exponent of NBTI degradation (fitted solid lines) increases at higher NBTI stress voltage of $V_{STS} \geq V_{G4}$ [78]. (b) Extracted ΔV_{OT} for the

measurements in [78] shows a decrease in time exponent for ΔV_{OT} at higher V_{STS}. (c) Estimated ΔV_{OT} at different V_{STS} scales universally (similar to the observation in [68]) and can be fitted using equation (20).

$$N_{OT} = \int n_{OT}(E)\, dE,$$

$$n_{OT}(E) = g(E)\left[1-\exp\left(-k_F(E)*t\right)\right],$$

$$k_F(E) = k_{F0}\exp\left[-(E-E_0)/kT\right],$$

$$g(E) \sim \frac{1}{\sigma_{OT}}\frac{\exp\left[(E-E_0)/\sigma_{OT}\right]}{\left[1+\exp\left((E-E_0)/\sigma_{OT}\right)\right]^2},$$

(20)

where E_0 is the average energy of bond dissociation that is leading to oxide defects and σ_{OT} is its standard deviation.

Therefore, ΔV_{OT} is another component that we need to consider for NBTI stress phase. In the NBTI relaxation phase, trapped holes from the oxide defects will detrap in a manner similar to the detrapping from pre-existing oxide defects. So, hole detrapping in the NBTI relaxation phase both from the pre-existing and generated oxide defects can be handled by using a single detrapping formula, as done in reference [41].

6. Interpreting NBTI Measurements

Our theoretical understanding of sections 3, 4 and 5 enables us to isolate the ΔV_{IT}, ΔV_{HT} and ΔV_{OT} components of ΔV_T. Note that without such decomposition, it is impossible to interpret and model the dynamics of ΔV_T (and its components), measured during NBTI stress. Without performing such decomposition of interface and oxide defect components, recently several efforts [13, 14, 22, 51, 73, 102] have arraigned the R-D theory to be inconsistent with NBTI relaxation (for an example of the purported inconsistency, see Figure 6a). Such alleged inconsistency has raised questions regarding the general validity of the R-D theory. Many alternative theories [12, 13, 21, 22] have also been proposed to explain ultra-fast NBTI relaxation, although their ability in predicting the broad features of NBTI remains questionable (see section 7 for further details).

6.1 Interpreting NBTI Stress Phase Measurements

If we decompose ΔV_T into its ΔV_{IT}, ΔV_{HT} and ΔV_{OT} components [41] by calculating ΔV_{IT} using equation (12), ΔV_{HT} using equation (19), ΔV_{OT} using $\sim t^{0.3}$ [69], the voltage and temperature dependency of extracted components show remarkable consistency for transistors having wide variation in dielectric material. The analysis shows high-fidelity matching for stress and relaxation phase NBTI measurements and suggests that in the range of $t_{STS} \sim 1\text{-}1000$ s, $\Delta V_{IT} \sim t^{1/6}$, $\Delta V_{HT} \sim$ constant, and ΔV_{OT} is quite small (Figure 6b). Based on these, we can approximate equation (1) in the range of $t_{STS} \sim 1\text{-}1000$ s with –

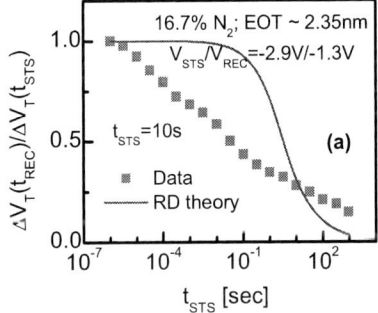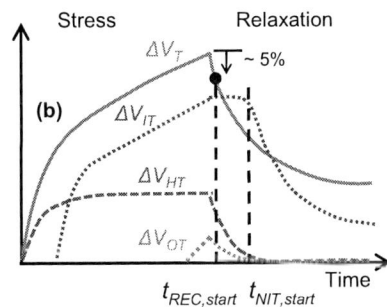

Figure 6: (a) Naive (and unphysical) comparison of the time evolution of measured fractional NBTI relaxation at short t_{REC} shows significant inconsistency with R-D solution. (b) Resolution between the measured ultra-fast ΔV_T relaxation and comparatively slower N_{IT} relaxation becomes possible, after decomposing ΔV_T (solid line) into ΔV_{IT} (dotted line), ΔV_{HT} (dashed line) components; each having separate time-dynamics.

$$\Delta V_T \approx \Delta V_{IT} + \Delta V_{HT} = A_{IT} t^{1/6} + B_{HT}. \tag{21}$$

We use equation (21) and estimate ΔV_{IT} by subtracting a constant (saturated) $\Delta V_{HT} \sim B_{HT}$ from ΔV_T for $t_{STS} > 1$ s in such a way that it provides time exponent $n \sim 1/6$ for ΔV_{IT} at t_{STS} of 1-1000 s (Figure 7a). Here, the effect of mobility [71, 103] and electric field-reduction [15] is taken into account in estimating ΔV_T. Next, we repeat the decomposition at different temperature for same E_{ox} and t_{STS}, so that we can extract the activation energy of the ΔV_{IT} and ΔV_{HT} components. Figure 7b shows the extracted ΔV_{IT} and ΔV_{HT} component at different temperature, which suggests E_A for ΔV_{IT} ($E_{A,IT}$) is ~ 0.094 eV; which is expected for R-D model based interface defect generation with H_2 diffusion (see section 3.2); on the other hand, E_A for ΔV_{HT} or $E_{A,HT}$ is ~ 0.04 eV, which is typically expected in any hole trapping process, involving tunneling and no structural relaxation. The signature of $E_{A,IT} > E_{A,HT}$ is also evident from Figure 7c, which indicates an increase in the extracted $\Delta V_{IT}/\Delta V_T$ (i.e. decrease in $\Delta V_{HT}/\Delta V_T$) with increase in temperature, at fixed t_{STS}.

Identical procedure is followed to isolate ΔV_{IT} and ΔV_{HT} for the transistors of Figure 8, at different voltages and temperatures. The extracted $E_{A,IT}$ (supported by R-D theory of section 3) and $E_{A,HT}$ (supported by hole trapping theory of section 4) is similar for these transistors (see Figure 8a). Moreover, extracted ΔV_{HT} for these nitrided transistors at a particular E_{ox}, T_{STS} and t_{STS} (Figure 8b) is observed to increase significantly with the increase in %N (with a very rapid increase seen for %N > 30), which is indeed a signature of higher hole trapping for higher %N. On the other hand, ΔV_{IT} only increases slightly with %N, which indicates that there is negligible change in the Si-H bond dissociation mechanism and hydrogen diffusion dynamics with the change in %N. Therefore, the extracted parameters for both ΔV_{IT} and ΔV_{HT} are consistent with the theoretical predictions.

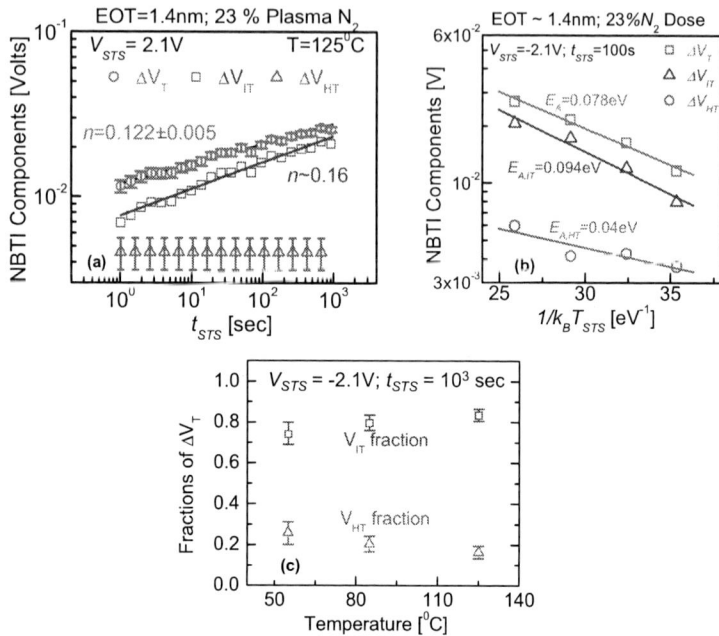

Figure 7: (a) Measured ΔV_T and extracted ΔV_{IT} and ΔV_{HT} components for a transistor having optimized nitrided dielectric. (b) Temperature dependence and corresponding activation energies (E_A) of ΔV_T, ΔV_{IT} and ΔV_{HT} components for the same transistor. (c) Estimated $\Delta V_{HT}/\Delta V_T$ and $\Delta V_{IT}/\Delta V_T$ at different temperature (T_{STS}) indicate an increase (decrease) in $\Delta V_{IT}/\Delta V_T$ ($\Delta V_{HT}/\Delta V_T$) at higher temperature. Here, the error bars represent the noise in $I_{D,lin0}$ measurement for OTF-$I_{D,lin}$ [104], which causes a ± 0.005 error in n for ΔV_T and a ± 1mV error in estimated ΔV_{HT}.

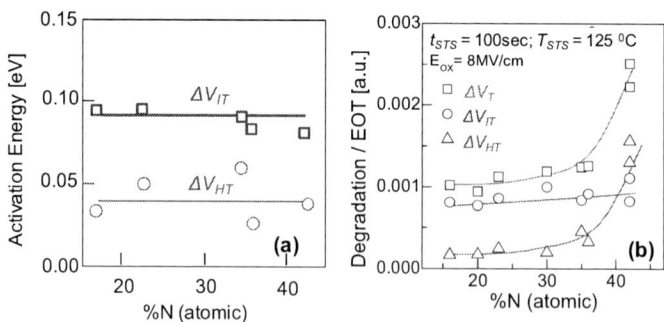

Figure 8: (a) Activation energy for extracted ΔV_{IT} and ΔV_{HT} components indicates negligible $\%N$ dependence. (b) $\%N$ Dependence of measured ΔV_T and extracted

ΔV_{IT} and ΔV_{HT} components for a particular t_{STS}, T_{STS} and E_{ox}. Lines are guide to the eye only.

As shown in reference [41], consideration of ΔV_{OT} in estimating ΔV_{IT} and ΔV_{HT} from ΔV_T merely changes the signatures of ΔV_{IT} and ΔV_{HT}, presented in Figure 7 and Figure 8. In that case, estimated ΔV_{IT}, ΔV_{HT}, and ΔV_{OT} shows good consistently with the theoretical expectations and suggests that ΔV_{OT} is a component that one should consider at higher stress bias.

6.2 Interpreting NBTI Relaxation Phase Measurements

Let us now apply the same $N_{IT}/N_{HT}/N_{OT}$ decomposition for NBTI relaxation measurements on the same transistor that we have previously analyzed through stress-phase decomposition in Figure 7a, which also has the temperature independent NBTI relaxation for $t_{REC} <$ ms (see Figure 2a). Here, we further use the observation of Figure 9a, which suggests that for $V_{STS} = -2.3$ V, hole detrapping occurs (*i.e.*, $t_{REC,start}$ shows sudden decrease) predominantly at $V_{REC} \geq -1.8$ V. Figure 9a also suggests that the amount of hole detrapping is similar from $V_{REC} = -1.3$ V to -1.6 V for the minimum $t_{REC} \sim \mu$s measured in this experiment.

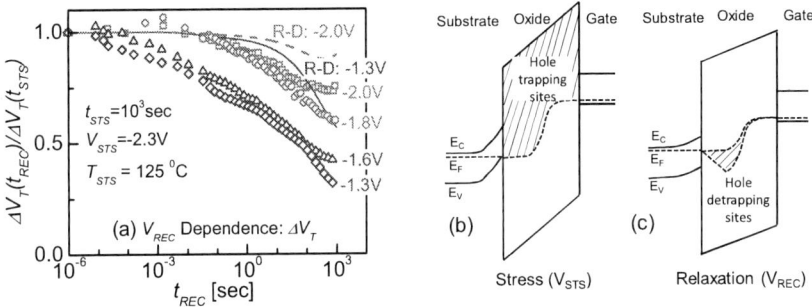

Figure 9: (a) NBTI relaxation experiments at different V_{REC} show significantly different $t_{REC,start}$. Since relaxation for $V_{REC} \leq -1.8$ V is very close the R-D theory, there is an additional relaxation mechanism for $V_{REC} > -1.8$ V. Our observation suggests hole detrapping to be the additional mechanism. (b) Schematic (based on simulation within a Shockley-Read-Hall trapping-detrapping framework; section 4) for expected hole trapping sites (hatched region) at V_{STS}. (c) When gate bias is switched from V_{STS} to V_{REC}, the hatched region will detrap the captured hole in a temperature independent manner.

Thus, trapping sites within the quasi-Fermi levels at $V_{REC} = -1.8$ V and -1.6 V (shown schematically by the hatched region in Figure 9c) will detrap all the holes that were captured by the oxide defects during stress within a timescale of ~ms. Considering such total hole detrapping at $V_{REC} = -1.3$ V both from the pre-existing and generated oxide defects (here, ΔV_{OT} is considered to be ~ 3 mV at $t_{STS} = 10^3$ s, having the time dependence similar to Figure 5c), the resultant N_{IT} component of the NBTI relaxation experiment shows excellent consistency with R-D theory in terms of $t_{NIT,start}$ at all V_{REC} (see Figure 10a). To understand the contribution of ΔV_{OT} on $t_{NIT,start}$, we redo the experiment-theory comparison by assuming $\Delta V_{OT} \sim 0$. Figure 10b suggests that consideration of finite ΔV_{OT} increases $t_{NIT,start}$ by an order of magnitude.

The remaining theory-experiment gap in terms of time dependence for $t_{REC} > t_{NIT,start}$ reflects the inability of 1D diffusion formulation (considered in classical R-D model of section 3) to capture the details of an essentially 3D diffusion problem. Consideration of 3D diffusion ensures high-fidelity matching for the time dependence for $t_{REC} > t_{NIT,start}$ [41]. Therefore, decomposing the contributions from interface and oxide defects enables us to explain the NBTI relaxation features in a theoretically consistent way.

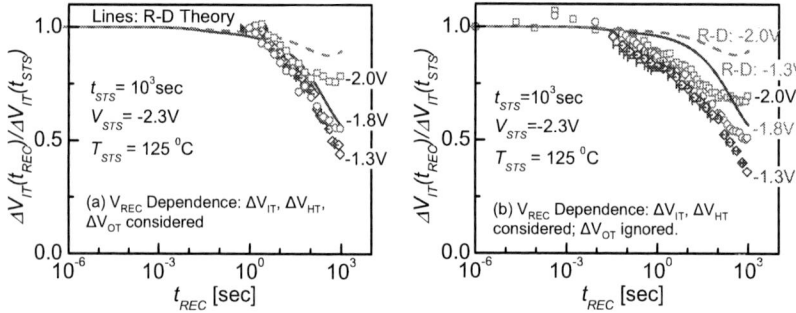

Figure 10: (a) (b) Consideration of ΔV_{IT}, ΔV_{HT}, and ΔV_{OT} and their decomposition indicates excellent consistency of $t_{NIT,start}$ with R-D theory. (b) Neglecting ΔV_{OT} reduces $t_{NIT,start}$ by an order of magnitude. Here, the error bar along the x-axis for $V_{REC} = -1.3$ V is due to the $I_{D,lin0}$ error in calculating $\Delta V_{IT}/\Delta V_T$ (see Figure 7).

6.3 Interpreting AC NBTI Measurements

So far we have applied the decomposition procedure to identify the interface and oxide defect components of NBTI-induced ΔV_T. Our analysis demonstrates that hole trapping and detrapping occur at similar time-scales (for example, compare the time-scale of the temperature independent hole trapping component in Figure 1b, and hole detrapping component in Figure 2b). Therefore, we expect total hole detrapping for a AC NBTI stress (measured at the end of OFF-state) with $\leq 50\%$ duty cycle (Figure 11a). In other words (see Figure 11b), AC/DC ratio for $\leq 50\%$ duty cycle in high $\%N$ transistors will measure $\Delta V_{T,AC}/\Delta V_{T,DC} \sim \Delta V_{IT,AC}/[\Delta V_{HT,DC} + \Delta V_{IT,DC}]$ (considering ΔV_{OT} component is quite small) and hence will always be less than the contribution from N_{IT}'s component, $AC/DC(N_{IT}) = \Delta V_{IT,AC}/\Delta V_{IT,DC}$, predicted by R-D theory. Moreover, as ΔV_{HT} decreases for smaller $\%N$, low $\%N$ transistors will have $\Delta V_{T,AC}/\Delta V_{T,DC} \sim \Delta V_{IT,AC}/\Delta V_{IT,DC}$, and thus the measured AC/DC ratio for low *%N* transistors should be consistent with the one obtained from the R-D theory.

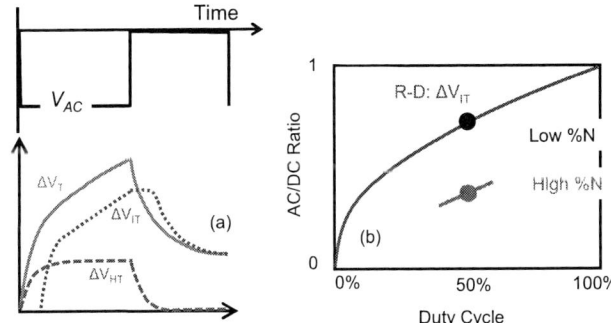

Figure 11: (a) Since hole trapping and detrapping happens at similar time-scale, we expect total hole detrapping at the end of OFF-state during AC NBTI stress. Therefore, (b) AC/DC ratio (with ΔV_T for AC NBTI measured at the end of AC cycles) will be consistent with the R-D theory only for transistors having lower ΔV_{HT} or low *%N*.

Indeed, our $\Delta V_{T,AC}/\Delta V_{T,DC}$ measurement on nitrided transistors show that for the lowest *%N* oxynitride dielectric (where $\Delta V_T \sim \Delta V_{IT}$), the predictions from R-D theory matches very well with the experimental data (Figure 12a). On the other hand, transistors with higher *%N* oxynitride dielectric have significant contribution from ΔV_{HT} in DC NBTI stress, *i.e.*, $\Delta V_{T,DC} > \Delta V_{IT,DC}$. As a result, $\Delta V_{T,AC}/\Delta V_{T,DC}$ is lower than $\Delta V_{IT,AC}/\Delta V_{IT,DC}$ (or the line predicted by the R-D theory) in such higher *%N* transistors. However, because AC NBTI stress at lower duty cycle predominantly reflects ΔV_{IT}, the shape of $\Delta V_{T,AC}/\Delta V_{T,DC}$ vs. duty cycle plot is similar to the R-D's prediction at lower duty cycle for all transistors. In other words, equation (17) is universal at lower duty cycle for any transistors, irrespective of *%N* within the dielectric. Such universal shape of $\Delta V_{T,AC}/\Delta V_{T,DC}$ vs. duty cycle dependence is evident in Figure 2c, where AC data at all duty cycle is scaled to $d \sim 50\%$ data to capture the intrinsic ΔV_{IT} behavior (as total hole detrapping occurs at $d \sim 50\%$). The scaling confirms a robust *universality* of the duty-cycle data from diverse sources, even up to ~80% duty cycle. This indicates that, for the transistors of Figure 2c, time scale for hole-detrapping is much less than the time scale for hole trapping. At higher duty cycle, detrapping of holes is incomplete, and therefore the experimental data deviates from the prediction of R-D theory.

Consideration of total hole detrapping for ~50% duty cycle also explains how R-D theory interprets the widely observed frequency independence in all NBTI measurements, irrespective of the type of dielectric, while overestimating the measured $\Delta V_{T,AC}/\Delta V_{T,DC}$ at different frequencies (Figure 12b and Figure 1e). Since ΔV_{HT} will be absent for AC NBTI stress with ~50% duty cycle, the frequency independence of $\Delta V_{IT,AC}/\Delta V_{IT,DC}$ (following R-D theory; equation (16)) will also result in frequency independence of $\Delta V_{T,AC}/\Delta V_{T,DC}$, irrespective of %N. However, for transistors with larger %N, the magnitude of $\Delta V_{T,AC}/\Delta V_{T,DC}$ will be lower than $\Delta V_{IT,AC}/\Delta V_{IT,DC}$ due to the presence of ΔV_{HT} in $\Delta V_{T,DC}$. Therefore, $\Delta V_{T,AC}/\Delta V_{T,DC}$ measurement on nitrided transistor show frequency independence, irrespective of %N (Figure 12b); while the magnitude of $\Delta V_{T,AC}/\Delta V_{T,DC}$ is only consistent with R-D theory for low %N transistors, when the hole trapping contribution is negligible. Thus, the co-existence of interface and oxide defects and their decomposition can explain both the duty cycle and frequency dependent NBTI experiments on nitrided transistors.

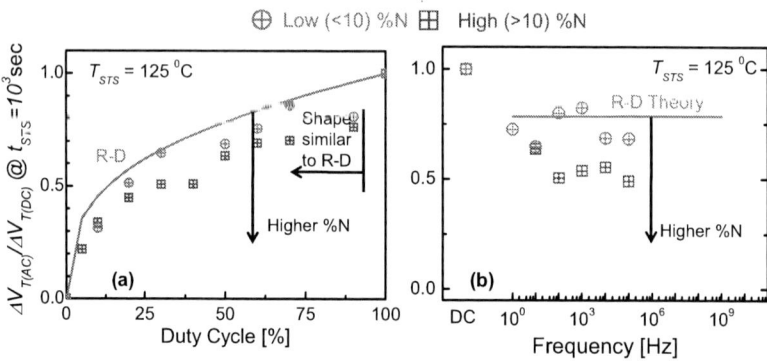

Figure 12: (a) AC/DC ratio (when ΔV_T is measured at the end of AC cycles) vs. duty cycle plot for different nitrided transistors. The experiments show remarkable consistency with the prediction of R-D theory (solid line) for low %N. (b) Though AC/DC ratio for any %N is always frequency independent, there is significant %N dependency due to the presence of ΔV_{HT} in DC NBTI stress.

7. A Critique of Alternate NBTI Theories

Contrary to the uncorrelated dynamics for interface and oxide defects that is used in this paper to explain the experimental features of NBTI, correlated model of these defects has also been recently used [22, 105] to explain the so-called universally observed temperature-dependence at arbitrary time scale. Actually, we observe such universal temperature dependence only in transistors having low %N [15, 16, 18, 19] (see Figure 1b), but not in all transistors as claimed in references [22, 105]. Moreover, the energy-well theory [11] behind such correlated mechanism of N_{IT} and N_{HT} fails to explain the basic features of NBTI degradation, such as the existence of stress-phase power-law time exponent, frequency independence [106, 107] - features that has been universally observed in all SiON transistors (see Figure 1). Indeed, as shown in Figure 13, the power-exponent predicted by energy-well theory (or any of its recent versions [22]) is fragile and its range changes by orders of magnitude with small changes in the parameters like broadening of well-barrier σ_{WB}.

In addition to this energy-well theory, some references consider NBTI (merely) as a trapping-detrapping problem into oxide defects [12, 13]. However, such theoretical analysis requires one to use a wide distribution of hole trapping sites (either spatially [12, 13] or energetically) for capturing the stress-phase power exponent and relaxation dynamics. Experimental support for such wide distribution of hole trapping events (for explaining NBTI time dynamics over more than 15 decades) is still lacking in literature.

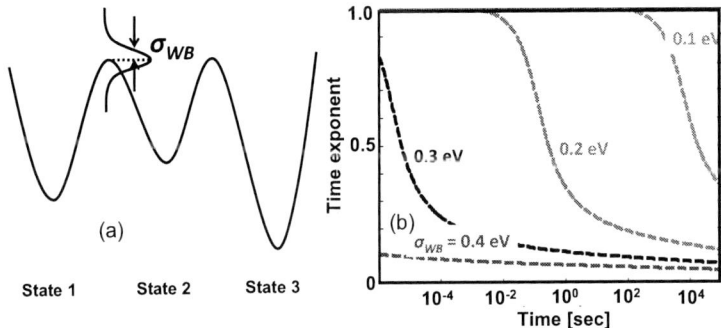

Figure 13: (a) Energy-well model (like the one presented in reference [11]) for explaining the dynamics of NBTI. Such energy-well model or its recent derivatives [22], in general, requires one to use a spread in barrier height (having a standard deviation of σ_{WB}) for capturing the NBTI dynamics. (b) Unless we use large σ_{WB}, it becomes difficult to obtain a robust power-law time exponent using such energy-well model.

8. Conclusion

In this paper, we have highlighted the importance of careful analysis of NBTI experiments, before comparing it with any theory. Our analysis demonstrates the uncorrelated dynamics of threshold voltage shift associated with interface and oxide defects can adequately explain the questions stated in section 2.2 of the manuscript. *The universal features of DC and AC NBTI* (*i.e.*, long-term stress-phase power-law time exponent, long-term stress/relaxation-phase temperature dependence, characteristic feature of duty-cycle dependence, and universal feature of frequency independence) can be consistently explained using R-D theory based interface defect dynamics. The *remaining non-universal features* (*i.e.*, short-term stress/relaxation-phase time dependence, short-term stress/relaxation-phase temperature independence, and the magnitude of AC/DC ratio at a particular duty cycle and frequency) are shown to be directly correlated to the amount of oxide defects within the dielectric. However, these non-universal features disappear for less than ~80% duty cycle and hence have no relevance for practical AC NBTI analysis. Therefore, a decomposition of these (uncorrelated) universal and non-universal components is established as a necessity before comparison with any theory. And without such decomposition, one will only be able to capture part (not all) of the experimental features of NBTI.

Acknowledgments

The work is an accumulation of our last few years' research on NBTI. It has been immensely benefited from the discussion with/contribution from D. Varghese, N. Kumar,

H. Kufluoglu, A. Krishnan, J. H. Lee, A. Oates, A. Jain, K. Ahmed, C. Olsen, E. Murakami, H. Aono, M. Masuduzzaman, G. Gupta, and H. Das. The financial support was obtained from Applied Materials, Taiwan Semiconductor Manufacturing Corporation, SRC-GRC, 2008 IEEE EDS PhD Fellowship, and 2009-2010 Intel Foundation PhD Fellowship.

References

1. B. E. Deal, M. Sklar, A. S. Grove and E. H. Snow, *Journal of the Electrochemical Society*, **114**, 266 (1967).

2. K. O. Jeppson and C. M. Svensson, *Journal of Applied Physics*, **48**, 2004 (1977).

3. K. Bernstein, D. J. Frank, A. E. Gattiker, W. Haensch, B. L. Ji, S. R. Nassif, E. J. Nowak, D. J. Pearson and N. J. Rohrer, *IBM Journal of Research and Development*, **50**, 433 (2006).

4. D. K. Schroder, *Microelectronics Reliability*, **47**, 841 (2007).

5. D. K. Schroder and J. A. Babcock, *Journal of Applied Physics*, **94**, 1 (2003).

6. J. H. Stathis and S. Zafar, *Microelectronics Reliability*, **46**, 270 (2006).

7. J. Campbell, in *Tutorial of IEEE International Integrated Reliability Workshop*, p. T4 (2009).

8. M. A. Alam and S. Mahapatra, *Microelectronics Reliability*, **45**, 71 (2005).

9. S. Chakravarthi, A. Krishnan, V. Reddy, C. F. Machala and S. Krishnan, in *IEEE International Reliability Physics Symposium*, p. 273 (2004).

10. V. Huard, M. Denais and C. Parthasarathy, *Microelectronics Reliability*, **46**, 1 (2006).

11. T. Grasser, B. Kaczer and W. Goes, in *IEEE International Reliability Physics Symposium*, p. 28 (2008).

12. D. Ielmini, M. Manigrasso, F. Gattel and M. G. Valentini, *IEEE Transactions on Electron Devices*, **56**, 1943 (2009).

13. P. M. Lenahan, in *IEEE International Reliability Physics Symposium*, p. 1086 (2010).

14. T. Grasser, H. Reisinger, P. Wagner, F. Schanovsky, W. Goes and B. Kaczer, in *IEEE International Reliability Physics Symposium*, p. 16 (2010).

15. A. E. Islam, H. Kufluoglu, D. Varghese, S. Mahapatra and M. A. Alam, *IEEE Transactions on Electron Devices*, **54**, 2143 (2007).

16. A. E. Islam, S. Mahapatra, S. Deora, V. D. Maheta and M. A. Alam, in *International Electron Devices Meeting (IEDM) Technical Digest*, p. 733 (2009).

17. J. H. Lee and A. S. Oates, *IEEE Transactions on Device and Materials Reliability*, **10**, 174 (2010).

18. S. Mahapatra, K. Ahmed, D. Varghese, A. E. Islam, G. Gupta, L. Madhav, D. Saha and M. A. Alam, in *IEEE International Reliability Physics Symposium*, p. 1 (2007).

19. S. Mahapatra, V. D. Maheta, A. E. Islam and M. A. Alam, *IEEE Transactions on Electron Devices*, **56**, 236 (2009).

20. D. S. Ang, S. Wang, G. A. Du and Y. Z. Hu, *IEEE Transactions on Device and Materials Reliability*, **8**, 22 (2008).

21. V. Huard, in *IEEE International Reliability Physics Symposium*, p. 33 (2010).

22. T. Grasser, B. Kaczer, W. Goes, T. Aichinger, P. Hehenberger and M. Nelhiebel, in *IEEE International Reliability Physics Symposium*, p. 33 (2009).

23. J. P. Campbell, P. M. Lenahan, A. T. Krishnan and S. Krishnan, in *IEEE International Reliability Physics Symposium*, p. 503 (2007).

24. A. S. Grove, *Physics and Technology of Semiconductor Devices*, John Wiley and Sons (1971).

25. P. V. Gray and D. M. Brown, *Applied Physics Letters*, **8**, 31 (1966).

26. Y. Nishi, *Jpn J Appl Phys*, **10**, 52 (1971).

27. P. J. Caplan, E. H. Poindexter, B. E. Deal and R. R. Razouk, *Journal of Applied Physics*, **50**, 5847 (1979).

28. K. L. Brower, *Physical Review B*, **38**, 9657 (1988).

29. E. Cartier, J. H. Stathis and D. A. Buchanan, *Applied Physics Letters*, **63**, 1510 (1993).

30. J. P. Campbell, P. M. Lenahan, A. T. Krishnan and S. Krishnan, in *IEEE International Reliability Physics Symposium*, p. 442 (2006).

31. P. M. Lenahan, in *Defects in Microelectronic Materials and Devices*, D. Fleetwood, S. Pantolides and R. D. Schrimpf, Editors, p. 163, CRC Press (2008).

32. A. T. Krishnan, C. Chancellor, S. Chakravarthi, P. E. Nicollian, V. Reddy, A. Varghese, R. B. Khamankar, S. Krishnan and L. Levitov, in *International Electron Devices Meeting (IEDM) Technical Digest*, p. 688 (2005).

33. S. Tsujikawa, T. Mine, K. Watanabe, Y. Shimamoto, R. Tsuchiya, K. Ohnishi, T. Onai, J. Yugami and S. Kimura, in *IEEE International Reliability Physics Symposium*, p. 183 (2003).

34. M. A. Alam, H. Kufluoglu, D. Varghese and S. Mahapatra, *Microelectronics Reliability*, **47**, 853 (2007).

35. S. Chakravarthi, A. T. Krishnan, V. Reddy and S. Krishnan, *Microelectronics Reliability*, **47**, 863 (2007).

36. S. Mahapatra and M. A. Alam, *IEEE Transactions on Device and Materials Reliability*, **8**, 35 (2008).

37. L. Tsetseris, X. J. Zhou, D. M. Fleetwood, R. D. Schrimpf and S. T. Pantelides, *Applied Physics Letters*, **86**, 142103 (2005).

38. L. Tsetseris, X. J. Zhou, D. M. Fleetwood, R. D. Schrimpf and S. T. Pantelides, *IEEE Transactions on Device and Materials Reliability*, **7**, 502 (2007).

39. M. A. Alam, in *International Electron Devices Meeting (IEDM) Technical Digest*, p. 345 (2003).

40. A. E. Islam, G. Gupta, S. Mahapatra, A. Krishnan, K. Z. Ahmed, F. Nouri, A. Oates and M. A. Alam, in *International Electron Devices Meeting (IEDM) Technical Digest*, p. 329 (2006).

41. S. Mahapatra, A. E. Islam, S. Deora, V. D. Maheta and M. A. Alam, *accepted in IEEE International Reliability Physics Symposium (2011)*.

42. L. Lundkvist, I. Lundstrom and C. Svensson, *Solid-State Electronics*, **16**, 811 (1973).

43. R. A. Williams and M. M. E. Beguwala, *IEEE Transactions on Electron Devices*, **25**, 1019 (1978).

44. G. L. Heyns and H. E. Maes, *Applied Surface Science*, **30**, 153 (1987).

45. F. P. Heiman and G. Warfield, *IEEE Transactions on Electron Devices*, **Ed12**, 167 (1965).

46. P. Rossel, H. Martinot and D. Esteve, *Solid-State Electronics*, **13**, 425 (1970).

47. I. Lundstrom, S. Christensson and C. Svensson, *Phys. Status Solid (A)*, **1**, 395 (1970).

48. V. Huard and M. Denais, in *IEEE International Reliability Physics Symposium*, p. 40 (2004).

49. C. R. Parthasarathy, M. Denais, V. Huard, G. Ribes, E. Vincent and A. Bravaix, in *IEEE International Reliability Physics Symposium*, p. 471 (2006).

50. H. Reisinger, O. Blank, W. Heinrigs, A. Muhlhoff, W. Gustin and C. Schlunder, in *IEEE International Reliability Physics Symposium*, p. 448 (2006).

51. C. Shen, M.-F. Li, C. E. Foo, T. Yang, D. M. Huang, A. Yap, G. S. Samudra and Y.-C. Yeo, in *International Electron Devices Meeting (IEDM) Technical Digest*, p. 333 (2006).

52. T. Yang, C. Shen, M. F. Li, C. H. Ang, C. X. Zhu, Y. C. Yeo, G. Samudra, S. C. Rustagi, M. B. Yu and D. L. Kwong, *IEEE Electron Device Letters*, **26**, 826 (2005).

53. M. F. Li, G. Chen, C. Shen, X. P. Wang, H. Y. Yu, Y. C. Yeo and D. L. Kwong, *Jpn J Appl Phys*, **43**, 7807 (2004).

54. G. Bersuker, J. H. Sim, C. S. Park, C. D. Young, S. V. Nadkarni, R. Choi and B. H. Lee, *IEEE Transactions on Device and Materials Reliability*, **7**, 138 (2007).

55. C. Shen, M. F. Li, X. P. Wang, H. Y. Yu, Y. P. Feng, A. T. L. Lim, Y. C. Yeo, D. S. H. Chan and D. L. Kwong, *International Electron Devices Meeting (IEDM) Technical Digest*, 733 (2004).

56. W. Goes, M. Karner, V. Sverdlov and T. Grasser, in *Proceedings of International Symposium on the Physical & Failure Analysis of Integrated Circuits (IPFA)*, p. 249 (2008).

57. A. Ghetti, C. M. Compagnoni, A. S. Spinelli and A. Visconti, *IEEE Transactions on Electron Devices*, **56**, 1746 (2009).

58. T. H. Morshed, S. P. Devireddy, Z. Celik-Butler, A. Shanware, K. Green, J. J. ChamberS, M. R. Visokay and L. Colombo, *Solid-State Electronics*, **52**, 711 (2008).

59. D. Veksler, H. Park, C. Young, B. Taylor, G. Bersuker and R. Jammy, *Solid State Technology*, **53**, 21 (2010).

60. H. D. Xiong, D. Heh, M. Gurfinkel, Q. Li, Y. Shapira, C. Richter, G. Bersuker, R. Choi and J. S. Suehle, *Microelectronic Engineering*, **84**, 2230 (2007).

61. J. P. Campbell, J. Qin, K. P. Cheung, L. C. Yu, J. S. Suehle, A. Oates and K. Sheng, in *IEEE International Reliability Physics Symposium*, p. 382 (2009).

62. S. Mahapatra, P. B. Kumar and M. A. Alam, *IEEE Transactions on Electron Devices*, **51**, 1371 (2004).

63. J. Robertson, *Solid-State Electronics*, **49**, 283 (2005).

64. J. Robertson, *Reports on Progress in Physics*, **69**, 327 (2006).

65. J. Sune, E. Farres, I. Placencia, N. Barniol, F. Martin and X. Aymerich, *Applied Physics Letters*, **55**, 128 (1989).

66. M. Alam, B. Weir and P. Silverman, *IEEE Circuits & Devices*, **18**, 42 (2002).

67. F. Crupi, R. Degraeve, G. Groeseneken, T. Nigam and H. E. Maes, *IEEE Transactions on Electron Devices*, **45**, 2329 (1998).

68. D. Varghese, H. Kufluoglu, V. Reddy, H. Shichijo, D. Mosher, S. Krishnan and M. A. Alam, *IEEE Transactions on Electron Devices*, **54**, 2669 (2007).

69. P. E. Nicollian, *Microelectronics Reliability*, **48**, 1171 (2008).

70. A. E. Islam, H. Kufluoglu, D. Varghese and M. A. Alam, *Applied Physics Letters*, **90**, 083505 (2007).

71. A. E. Islam, V. D. Maheta, H. Das, S. Mahapatra and M. A. Alam, in *IEEE International Reliability Physics Symposium*, p. 87 (2008).

72. V. D. Maheta, C. Olsen, K. Ahmed and S. Mahapatra, *IEEE Transactions on Electron Devices*, **55**, 1630 (2008).

73. T. Grasser, B. Kaczer, R. Hehenberger, W. Gos, R. O'Connor, H. Reisinger, W. Gustin and C. Schlunder, *International Electron Devices Meeting (IEDM) Technical Digest*, 801 (2007).

74. B. Kaczer, T. Grasser, P. J. Roussel, J. Martin-Martinez, R. O'Connor, B. J. O'Sullivan and G. Groeseneken, in *IEEE International Reliability Physics Symposium*, p. 20 (2008).

75. S. Deora, V. D. Maheta, A. E. Islam, M. A. Alam and S. Mahapatra, *IEEE Electron Device Letters*, **30**, 978 (2009).

76. S. Mahapatra, V. D. Maheta, S. Deora, E. N. Kumar, S. Purawat, C. Olsen, K. Ahmed, A. E. Islam and M. A. Alam, in *Silicon Nitride, Silicon Dioxide, and Emerging Dielectrics*, R. E. Sah, J. Zhang, J. Deen, J. Yota and A. Toriumi, Editors, p. 243, The Electrochemical Society Proceedings Series (2009).

77. R. Fernández, B. Kaczer, A. Nackaerts, S. Demuynck, R. Rodríguez, M. Nafría and G. Groeseneken, in *International Electron Devices Meeting (IEDM) Technical Digest*, p. 337 (2006).

78. C. L. Chen, Y. M. Lin, C. J. Wang and K. Wu, in *IEEE International Reliability Physics Symposium*, p. 704 (2005).

79. A. Haggag, G. Anderson, S. Parihar, D. Burnett, G. Abeln, J. Higman and M. Moosa, in *IEEE International Reliability Physics Symposium*, p. 452 (2007).

80. H. Reisinger, T. Grasser, W. Gustin and C. Schlunder, in *IEEE International Reliability Physics Symposium*, p. 7 (2010).

81. Y. Mitani, H. Satake and A. Toriumi, *IEEE Transactions on Device and Materials Reliability*, **8**, 6 (2008).

82. S. Wang, D. S. Ang and G. A. Du, *IEEE Electron Device Letters*, **29**, 483 (2008).

83. H. Reisinger, in *NBTI Tutorial, IEEE International Integrated Reliability Workshop*, p. T2 (2010).

84. V. Huard, C. Parthasarathy, C. Guerin, T. Valentin, E. Pion, M. Mammasse, N. Planes and L. Camus, in *IEEE International Reliability Physics Symposium*, p. 289 (2008).

85. A. T. Krishnan, S. Chakravarthi, P. Nicollian, V. Reddy and S. Krishnan, *Applied Physics Letters*, **88**, 153518 (2006).

86. H. Kufluoglu and M. A. Alam, *IEEE Transactions on Electron Devices*, **54**, 1101 (2007).

87. A. E. Islam, G. Gupta, K. Z. Ahmed, S. Mahapatra and M. A. Alam, *IEEE Transactions on Electron Devices*, **55**, 1143 (2008).

88. A. E. Islam, J. H. Lee, W. H. Wu, A. Oates and M. A. Alam, in *International Electron Devices Meeting (IEDM) Technical Digest*, p. 107 (2008).

89. S. Kumar, C. H. Kim and S. S. Sapatnekar, in *International Conference on Computer-Aided Design*, p. 6D.1 (2006).

90. J. Crank, *The Mathematics of Diffusion*, Oxford University Press (1980).

91. T. Grasser, W. Gos and B. Kaczer, *IEEE Transactions on Device and Materials Reliability*, **8**, 79 (2008).

92. C. G. Van de Walle and B. R. Tuttle, *IEEE Transactions on Electron Devices*, **47**, 1779 (2000).

93. D. Varghese, D. Saha, S. Mahapatra, K. Ahmed, F. Nouri and M. A. Alam, in *International Electron Devices Meeting (IEDM) Technical Digest*, p. 684 (2005).

94. G. Gupta, S. Mahapatra, L. Madhav, D. Varghese, K. Z. Ahmed and F. Nouri, in *IEEE International Reliability Physics Symposium*, p. 731 (2006).

95. M. L. Reed and J. D. Plummer, *Journal of Applied Physics*, **63**, 5776 (1988).

96. M. A. Alam, Reliability Physics of Nanoelectronic Transistors, in *EE 695-A lecture series, uploaded in* http://cobweb.ecn.purdue.edu/~ee650/handouts.htm.

97. A. E. Islam, *Ph. D. Dissertation*, Theory and Characterization of Defect Formation and Its Implication in Variability of Nanoscale Transistors, Electrical and Computer Engineering, Purdue University, 2010.

98. T. L. Tewksbury and H. S. Lee, *IEEE Journal of Solid-State Circuits*, **29**, 239 (1994).

99. S. V. Hattangady, H. Niimi and G. Lucovsky, *Journal of Vacuum Science & Technology-A*, **14**, 3017 (1996).

100. K. Sakuma, D. Matsushita, K. Muraoka and Y. Mitani, in *IEEE International Reliability Physics Symposium*, p. 454 (2006).

101. K. Hess, A. Haggag, W. McMahon, B. Fischer, K. Cheng, J. Lee and J. Lyding, in *International Electron Devices Meeting (IEDM) Technical Digest*, p. 93 (2000).

102. H. Reisinger, O. Blank, W. Heinrigs, W. Gustin and C. Schlunder, *IEEE Transactions on Device and Materials Reliability*, **7**, 119 (2007).

103. A. E. Islam, E. N. Kumar, H. Das, S. Purawat, V. D. Maheta, H. Aono, E. Murakami, S. Mahapatra and M. A. Alam, in *International Electron Devices Meeting (IEDM) Technical Digest*, p. 805 (2007).

104. V. D. Maheta, E. N. Kumar, S. Purawat, C. Olsen, K. Ahmed and S. Mahapatra, *IEEE Transactions on Electron Devices*, **55**, 2614 (2008).

105. T. Grasser and B. Kaczer, *IEEE Transactions on Electron Devices*, **56**, 1056 (2009).

106. A. E. Islam and M. A. Alam, *unpublished results*

107. M. A. Alam, in *IEEE International Integrated Reliability Workshop*, p. 1 (2010).

ECS Transactions, 35 (4) 175-189 (2011)
10.1149/1.3572282 ©The Electrochemical Society

Atomic Imaging of Atomic H Cleaning of InGaAs and InP for ALD

Wilhelm Melitz[1,2], Jian Shen[1,2], Tyler Kent[1,2], Ravi Droopad[3]

Paul Hurley[4] and Andrew C. Kummel[2]*

[1]Materials Science & Engineering Program, [2]Department of Chemistry & Biochemistry,
University of California, San Diego, La Jolla, CA, 92093 USA
[3]Department of Physics, Texas State University, San Marcos, TX 78666 USA
[4]Tyndall National Institute, University College Cork, Lee Malting, Cork, Ireland

*Corresponding author: akummel@ucsd.edu

Air exposed III-V surfaces nearly always have electronic defects which prevent full modulation of the Fermi level thereby impeding their use in practical semiconductor devices such as metal oxide field effect transistors (MOSFETs). For a high speed device, the air induced defects and contaminants need to be removed to reduce trap states while maintaining an atomically flat surface to minimize interface scattering thereby maintaining a high carrier mobility. Using in-situ atomic scaling imaging with scanning tunneling microscopy, a combination of atomic H dosing, annealing and trimethyl aluminum dosing is observed to produce an ordered passivation layer on air exposed InGaAs(001)-(4×2) surface with only monatomic steps. A similar atomic H cleaning procedure has been demonstrated to produced an ordered passivation layer on air exposed InP(100).

Introduction

Several hundred research papers are published each year upon development of MOSFETs using III-V semiconductors, especially InGaAs [1]. To enable low power MOSFET operation, a low supply voltage is required, the semiconductor must have high mobility and high saturation velocity, the oxide-semiconductor interface must have a low density of trap states (D_{it}), and the oxide-semiconductor interface must be nearly atomically flat. Surface channel III-V MOS devices can be fabricated with atomic layer deposition (ALD) high-K gate-first processes [2-5] which are similar to silicon processes for SiO_2 growth on silicon or ALD of high-K on silicon [6-10]. For silicon, the only commercial ALD high-k fabrication process is a replacement gate process (a type of gate last process) to avoid processing induced damage [11]. By using a gate-last process, the dielectric and oxide/semiconductor interface can avoid major damage from processing; however, the key to a gate-last process is the condition of the III-V channel prior to dielectric deposition. It has been shown that ALD of trimethyl aluminum (TMA) [12, 13] or tetrakis(ethylmethylamino)hafnium (TEMAH) [14] on III-V has self-cleaning properties by reducing the presence of As-O and Ga-O bonds. However, for high quality dielectric semiconductor interfaces, further reduction or cleaning may be required and the interface

175

must be atomically flat. Furthermore, aggressive oxide thickness reduction (EOT scaling) is needed to fabricate small gate length devices with small subthreshold swings, and aggressive EOT scaling requires a very high uniform ALD nucleation density with no pinholes due to surface contaminants [15]. The key barrier to a very practical problem is a simple surface chemistry challenge: development of a chemical process, which removes all air induced defects and contaminants and leaves the III-V surface flat and electrically active for high nucleation density ALD gate oxide deposition which unpins the Fermi level.

While InGaAs(100) is the most common channel material for high mobility channels, InP(100) has been used successfully as a capping layer on the narrower bandgap InGaAs(100) channel. By using the InP layer, the energy level of the defects can be controlled. However, minimal trap and fixed charge density at the oxide/InP(100) interface are still critical issues. The purpose of comparing InP to InGaAs is motivated by recent results using an InP layer on scaled MOSFETs [16, 17]. The wide bandgap layer might influence the energy levels of the density of interface trap states (D_{it}) loacted at the oxide/semiconductor interface and their impact on the on and off currents. If the D_{it} of the dielectric/InP interface moves higher towards the conduction band edge of the InP capping layer and away from the energy levels in the InGaAs bandgap over which the Fermi level is modulated, an improvement in device performance may occur. For InP, the Fermi level pinning position is usually 0.12eV below the conduction band edge [18]. The InP layer may also act as an electrostatic control layer which would confine carriers in the channel, and the mobility of the carriers might improve by decreasing surface scattering. The surface scattering in the channel with an oxide/InP/InGaAs stack should be better than for a simple surface channel oxide/InGaAs stack because the latticed matched $In_{0.53}Ga_{0.47}As$ and InP should produce a well ordered interface with low surface roughness in comparison to an InGaAs/oxide interface.

Atomic H cleaning of GaAs, InGaAs, and related semiconductors has been investigated [19-26]. Traditional atomic H cleaning is performed at elevated temperatures as surface preparation for molecular beam epitaxial growth. Atomic H cleaning of InP has also been studied by many group using RHEED [27, 28], and XPS [26, 29, 30]. This study investigates the atomic H cleaning at 380°C with post cleaning annealing to determine the influences of cleaning and annealing temperature on electronic structure as well as surface defects, roughness, and step density. This is in contrast to previous works which did not study the influence of post cleaning annealing. These post cleaning annealing studies are enabled use of in-situ atomic structure characterization using scanning tunneling microscopy (STM). Atomic H has been shown to unpin GaAs [24]; however, it induces surface etching [23, 31, 32]. Previous STM studies of atomic H cleaning of GaAs(100) showed there was a high step density which could be reduced by annealing in an As_2 flux [25]; while an excellent solution for MBE growth of quantum wells, this technique would be difficult to implement in an ALD tool. For GaAs, the atomic H etch rate increases with temperature as well as pressure of atomic H [33]. Etching must be minimized since it can induce surface roughness, which is incompatible with the thin channel structures required in low power highly scaled devices [10, 34, 35]. By employing STM of atomic H cleaned surfaces, the surface features can be studied at the atomic level.

In the present study, the InGaAs samples are 0.2μm thick $In_{0.53}Ga_{0.47}As$ layer grown by MBE on commercially available InP wafers. The MBE-grown InGaAs layers are doped n-type and p-type with a doping concentration of 2×10^{18} cm^{-3} of Si and Be dopants. Following MBE growth, all samples are capped with a 50nm As_2 layer and shipped/stored under vacuum before being loaded into the UHV chamber. The As_2 capped samples allow for comparison of pristine samples to air exposed/H cleaned samples. A diagram of the system chamber can be seen in Figure 1.

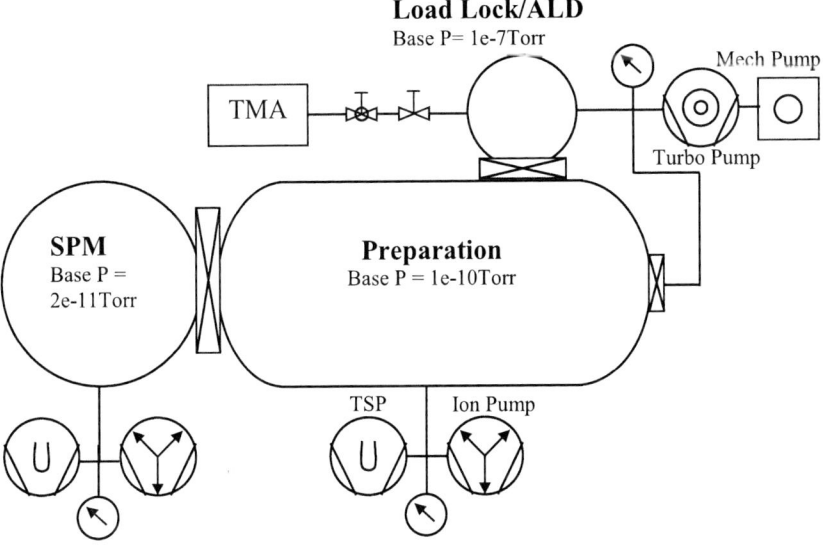

Figure 1: Diagram of ultra high-vacuum (UHV) system chamber. The scanning probe microscopy (SPM) chamber has base pressure $2x10^{-11}$ Torr. The SPM chamber contains an Omicron STM/AFM VT capable of performing STM, STS, atomic force microscopy (AFM), Kelvin probe force microscopy (KPFM). The preparation chamber has base of $1x10^{-10}$ Torr, and the atomic H cracker is located in the preparation chamber. The preparation chamber also contains a dual anode x-ray photoelectron source and analyzer, mass spectrometer, tip cracker, and low energy electron diffraction (LEED). The load lock is pumped by a turbomolecular pump to achieve a base pressure of $1 x 10^{-7}$ torr; the turbomolecular pump is also used to remove any residual water in the load lock that might react with the TMA.

The InGaAs samples are loaded into an Omicron UHV chamber with base pressure below $1x10^{-10}$ Torr. Samples are decapped in UHV and annealed at 450-470 °C to form the InGaAs(001)-(4×2) surface reconstruction in the preparation chamber. Further details concerning the samples and preparation methods are published in reference [36]. Once the decapped surface structure is confirmed with STM in the scanning probe microscopy (SPM) chamber with a base pressure of $2x10^{-11}$ Torr, the decapped sample is transferred to the load lock and exposed to air for 0.5-30 minutes. Using an Oxford Applied Research TC-50 thermal gas cracker in the preparation chamber, the sample is cleaned with atomic H at various sample temperatures and dose times with a H_2 pressure of $1-2x10^{-6}$ Torr. The percent of atomic H from the thermal cracker at 65 watts is ~50%, and the thermal cracker was operated at 60 watts. The flux of all gas at the sample surface is a mixture of the recombined background H_2 and the stream of atomic H from the thermal cracker. The background H_2 should be inert; therefore, the reduction in surface oxides is from the atomic H produced with the thermal cracker; however, the exact dose of atomic H is unknown. Therefore, each hydrogen source and dosing system will have different fluxes and require calibration to the dosing times reported in this manuscript. Following the atomic H cleaning, the sample is annealed to 460-480 °C to regain the InGaAs(001)-(4×2) surface reconstruction, and STM is employed to determine the reconstruction, defect density, and surface roughness. The TC-50 thermal cracker employs a hot iridium tube to crack the H_2 molecules thereby avoiding contamination that can occur with hot tungsten filaments and or formation of ions that can occur with a plasma source; both metal contamination and ion bombardment are likely to negatively influence surface cleaning, ordering, and electronic passivation. The deposition of TMA is performed in the load lock. Prior to ALD dosing, the load lock is baked over night until it reaches a base pressure below 1×10^{-7} Torr to avoid water contamination (commercial ALD tools employ hot walls, which is a similar but faster technique). The sample is exposed to 1×10^{-3} -1×10^{-2} Torr of TMA vapor for 5 seconds at room temperature (RT) followed by a 250 °C anneal in the preparation chamber

A similar sample procedure was used for the InP samples. The InP samples employed in this study are from an InP wafer with 1μm of InGaAs channel layer doped with 4×10^{18} cm^{-3} of Si channel layer and a 2nm InP undoped surface top layer. The samples were first degassed for several hours at 150 °C, followed with an exposure of atomic H to remove the oxide. The surface was inspected with STM to determine the quality of the cleaned surface. A comparison to a decapped surface was not performed because InP does not have an equivalent capping and decapping method used for the InGaAs. The goal is to use the same surface preparation from InGaAs on InP to prepare the surface for an ALD gate oxide deposition. After atomic H cleaning, the InP surface was exposed to 1×10^{-3} -1×10^{-2} Torr of TMA vapor for 5 seconds at room temperature (RT) followed by a 250 °C anneal in the preparation chamber.

Results and Discussion

InGaAs(100):   Figure 2(a) shows the STM image of the decapped InGaAs(001)-(4×2) surface. The decapped surface includes two distinct defects: dark horizontal features (black rectangle) and bright vertical features (white rectangle). After the InGaAs(001)-(4×2) surface is exposed to air for 0.5 minutes, an amorphous film is observed by STM, Figure 2(b). Note the surface was annealed to 200 °C in order to achieve stable STM images. This oxide film should consist of primarily As_2O_3, Ga_2O_3, and In_2O_3 [20]. Figure 2(c) shows a surface exposed to air for 30 minutes followed immediately by atomic H cleaning at 380 °C for 30 minutes. The atomic H cleaned surface shows dark features (red rectangle) consistent with monolayer etch pits; however, there is a reduction in the other defects observed on the decapped surface. Figure 2(d) shows the atomic H cleaned surface after an anneal of 460-480 °C for 10 minutes showing an increase in terrace size and uniformity. The densities of horizontal dark defects (black rectangle) appear to be similar to the decapped surface while the vertical defects are drastically reduced. There are some bright vertical features, but the structures appear different from the features on the initial decapped surfaces.

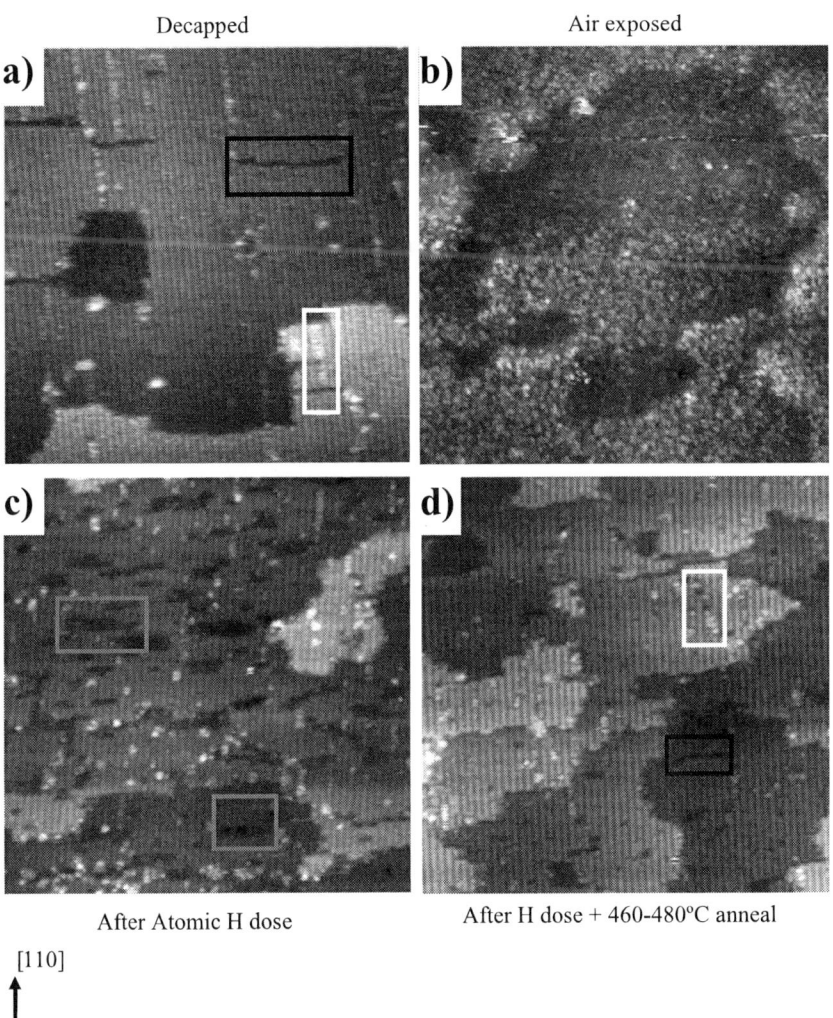

Decapped — Air exposed

a) b)

c) d)

After Atomic H dose — After H dose + 460-480°C anneal

[110]

[1 1̄ 0]

Figure 2: 100×100nm^2 filled state STM image of InGaAs a) decapped surface, b) air exposed and annealed to 200 °C surface, c) surface after 30 minutes air exposure followed by 30 minutes dose of atomic H at 380 °C, and d) surface after 30 minutes air exposure, 30 minutes dose of atomic H at 380 °C, and a high temperature anneal to 460-480 °C. The black rectangles show dark horizontal defect features, white rectangles indicate vertical bright defect features and red rectangles show dark surface features seen after atomic H dosing which might be from surface etching. After annealing to 460-480 °C, the surface contains fewer dark horizontal defects and bright vertical defects.

Figure 3(a) is an STM image of the decapped surface showing ~242 (manually counted) of the bright vertical defect features with large terraces. After exposure to air for 30 minutes and a 30 minute dose of atomic H at 400 °C, the number of bright vertical defects is reduced to ~70 in Figure 3(b); however, the STM image shows reduction in terrace size because of the dark monolayer deep etch features. Finally, if the sample is annealed to 460-480 °C the bright features are further reduced to ~50 in Figure 3(c), and the terrace size is restored to almost the same size as the decapped surface.

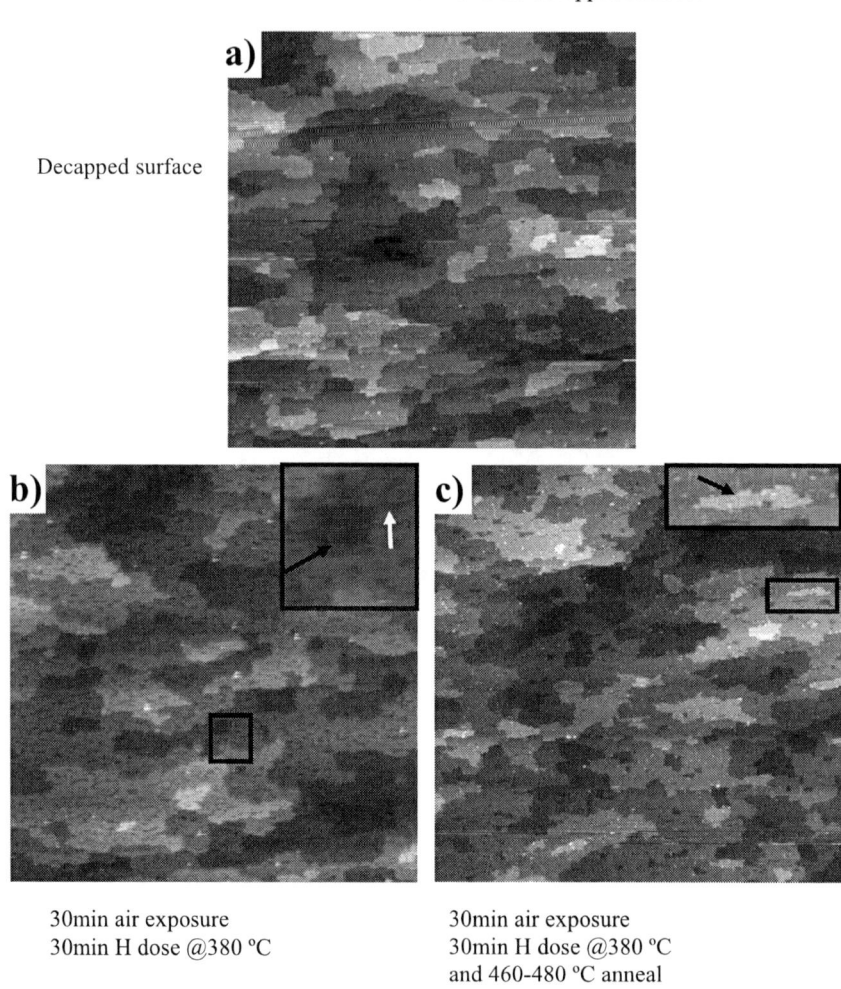

a)

Decapped surface

b)

c)

30min air exposure
30min H dose @380 °C

30min air exposure
30min H dose @380 °C
and 460-480 °C anneal

Figure 3: 500×500nm^2 filled state STM image of InGaAs a) decapped surface, b) 30 minutes air exposure followed by 30 minutes dose of atomic H at 380 °C, and c) after high temperature anneal to 460-480 °C. The terrace sizes improve with high temperature annealing. The atomic H dosed surface shown in (b) has a large density of dark surface features which are no longer present in (c). The inset in (b) includes large (black arrow) and small (white arrow) etch features caused by atomic H exposure. The inset in (c) includes a small terrace (black arrow) consistent with an incomplete terrace.

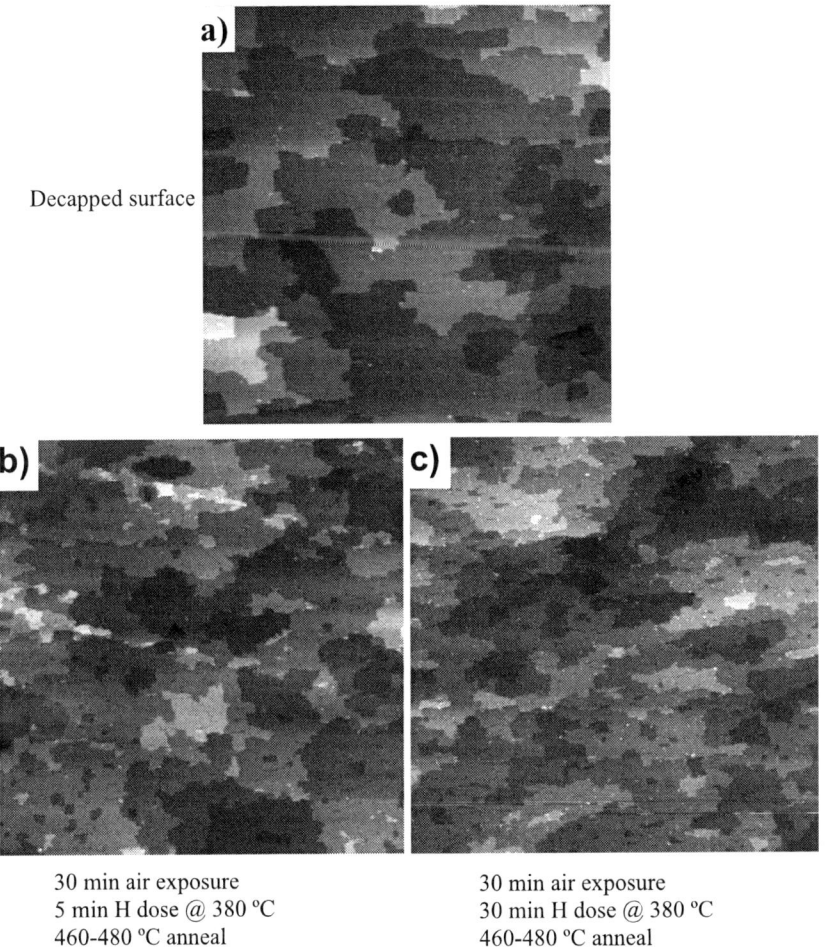

Decapped surface

30 min air exposure
5 min H dose @ 380 °C
460-480 °C anneal

30 min air exposure
30 min H dose @ 380 °C
460-480 °C anneal

Figure 4: 500×500nm^2 filled state STM images of InGaAs a) a typical decapped InGaAs(001) surface, with a surface step coverage (SSC) of 5.66%. In the decapped surface, there are both dark etch like features and incomplete terraces. For STM images b)-c) all samples were decapping, exposed to air for 30 minutes, and atomic H cleaned at 380°C and annealed at 460-480 °C. b) A sample atomic H cleaned at 380 °C for 5 minutes with SSC= 7.65%. Large increases in the dark etch features and incomplete terraces can be seen in comparison to the decapped surface. c) A sample atomic H cleaned at 380 °C for 30 min with SSC=12.8%. The images are corrected for global tilt.

After atomic H exposure, the presence of etch features reduces the terrace sizes. There are two types of etch features, dark pits, inset Figure 3(b), which shows removal of surface atoms in the plane of the terrace and incomplete terraces, inset Figure 3(c), which illustrates a terrace that has been almost completely etched with only a residual amount remaining. Comparing the decapped sample to the 30 minute high temperature (HT) dose without anneal sample, it is clear that there is a sharp increase in the etch pits. After a HT anneal, distinct improvement is observed in the surface morphology as shown in Figure 3(c). The atomic H cleaned and annealed surface has etch feature densities similar to that of the clean surface.

Besides etch pits, step edges are a major defect that can reduce carrier mobility in the channel, because steps usually contain dangling bonds due to the under-coordinated bonding configuration. Figure 4 compares air exposed samples cleaned with atomic H for 5 minutes vs. 30 minutes; both samples are annealed to 460 °C after atomic H dosing. Using Scanning Probe Image Processor's (SPIP) grain analysis tool, the amount of the surface covered in edge features was quantified (step edges, horizontal and vertical defects). Typical decapped surfaces (Figure 4(a)) have a 5.6% surface coverage of step edges. A typical surfaces after a 30 minute air expose and 5 minute atomic H cleaning at 380 °C followed by 460 °C anneal have 7.6% surface coverage of step edges (Figure 4(b)), close to that of the decapped surface. Typical surfaces after a 30 minute air expose and a 30 minute atomic H cleaning at 380 °C followed by 460 °C have 12.8% surface coverage of step edges (Figure 4(c)), consistent with longer exposures generating more etch features.

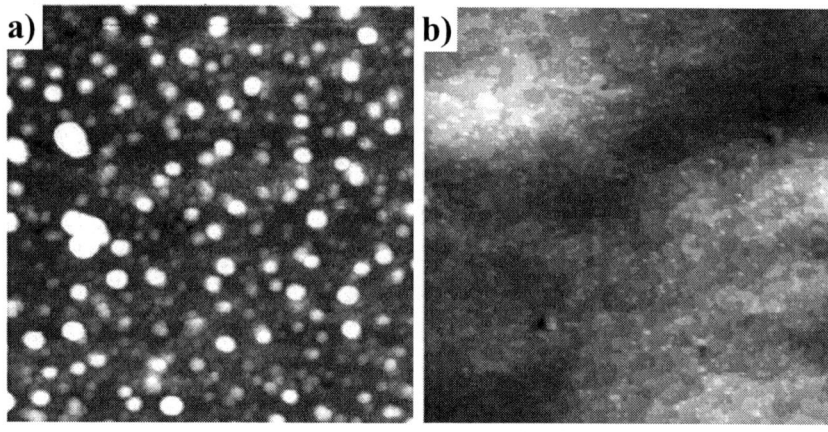

30 min air exposure
30 min H dose @ -40 °C
460-480 °C anneal

30 min air exposure
200 °C anneal
30 min H dose @ -40 °C
460-480 °C anneal

Figure 5: 500×500nm^2 filled state STM image of InGaAs a) dose with atomic H for 30 minutes at -40 °C directly after exposure to air for 30 minutes. The images shows a high density of islands most likely unreduced oxide that was not full desorbed at low temperatures. b) STM image of a surface dose with atomic H at -40 °C for 30 minutes with a 200 °C anneal between air exposure and atomic H cleaning. The initial anneal before cleaning indicates the removal of water and other lower adsorbates can assist in the cleaning procedure.

Figure 5 shows STM images of atomic H cleaning at -40 °C. Directly after exposure to air for 30 minutes, the sample was dosed with atomic H for 30 minutes at -40 °C with a final anneal of 460-480 °C, Figure 5(a). The surface shows large bright islands probably from residual oxide left after the atomic H cleaning. At sub 0°C temperatures, the surface will condense water which might not be able to desorb during atomic H cleaning at -40 °C. However, if the sample is first annealed to drive off any residual water before cleaning, a -40 °C atomic H cleaning results in a surface morphology similar to the RT atomic H cleaned surface. Figure 5(b) shows a surface exposed to air for 30 minutes then annealed to 200 °C in UHV for 1.5 hours prior to atomic H cleaning for 30 minutes at -40 °C with a final anneal of 460-480 °C. The only major difference observed between the RT and the -40 °C atomic H dose surfaces is the number of steps on a 500×500nm² STM image. The surface shown in Figure 5(b) is consistent with inhomogeneous etching during atomic H cleaning at -40 °C.

The atomic H induced removal of the oxide layer involves multiple reactions. Atomic H can reduce Ga_2O_3 to Ga_2O and H_2O [20, 21, 32], and desorption of Ga_2O occurs at ~400°C [21, 37]. The influence of atomic H on In oxides has been assumed to be similar to that of Ga oxides [20]. Atomic H has been show to convert atomic As and As_2O_3 into AsH_3 and/or H_2O[38]. Atomic H at elevated temperature around 400 °C is known to remove As on GaAs[20, 21]. Furthermore, the desorption of As oxides and As_2/As_4 occurs around 400 °C [20, 21]. This atomic H induced volatilization of As is consistent with the reduction of any excess As defects on the atomic H clean surface. The absence of any oxides is likely due to the >300 Langmuir exposure which reduces any Ga_2O_3 or In_2O_3 to suboxides which are volatilized during annealing along with any As oxides.

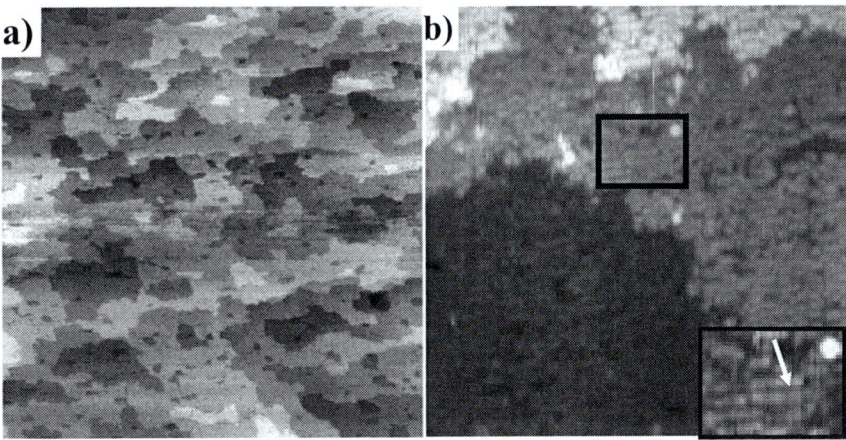

Figure 6: 40×40nm² filled state STM images of InGaAs surface after TMA dose at room temperature and annealed to 250 °C. The inset shows an expanded view of 7×7nm² indicated by black square. The surface shows highly ordered horizontal row features (indicated by white arrow), consistent with a TMA induced surface reconstruction.

Following atomic H cleaning, the surface was exposed to TMA at room temperature and annealed to 250 °C, an STM of the TMA dosed surface can be seen in Figure 6. Figure 6(a) shows a 500×500nm² STM image of the TMA dose on the atomic

H cleaned surface, showing a high nucleation density single monolayer film. The TMA induces a surface reconstruction creating a bulk like bonding configuration between the Al atoms and the surface As atoms[39]. A more detailing image is shown in Figure 6(b) The TMA passivation layer has horizontal rows of dimethyl aluminum. The surface is a highly ordered self-limiting layer that has high nucleation density. The self-limiting and high nucleation density are necessary for EOT scaling. The STM image of the TMA dosed surface is shown to satisfy key processing conditions: an atomically flat, high nucleation density, passivation layer; other scanning tunneling spectroscopy studies have shown this passivation layer unpins the Fermi level on InGaAs(001)-4×2 so the surface is electrically active [39].

InP: Figure 7(a) shows a STM image of InP after degassing an a 10 minute dose of atomic H at 380 °C. The surface shows removal of the native oxide by atomic H exposure. The native oxide reduction by atomic H of InP, $InPO_4$, to H_2O, PH_3, PH_2, InH, and In has been reported in literature [40]. Others report the reductions of $InPO_4$, $In(PO_3)_3$, and In_2O_3 [26, 41]. Most reports are consistent with atomic H lowering the cleaning temperature for InP, and the exposure time for InP needed to clean the native oxide being longer than for GaAs. Five minute exposures to atomic H were also performed, not shown, which also had large bright features and partial coverage of the clean ordered surface. As shown in Fig 7, 10 minute exposures of InP to atomic H showed residual bright features that are most likely uncleaned oxide.

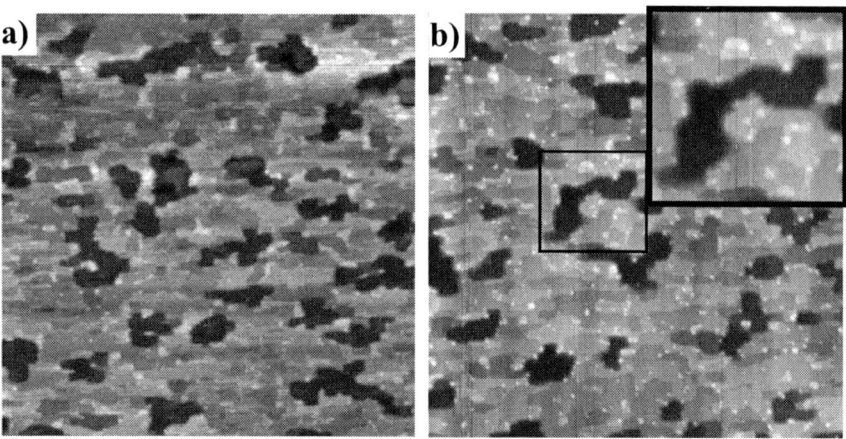

10 min H dose @ 380 °C   10 min H dose@380 °C + 470 °C anneal

Figure 7: 500×500nm² filled state STM image of InP a) after 10 minutes dose of atomic H at 380 °C, and b) after 10 minutes dose of atomic H at 380 °C plus a high temperature anneal to 470°C. The STM images show an InP after atomic H exposure generates a flat surface with the large etch features. The density of the etch features does not significantly decrease with high temperature annealing. The inset shows an example of one etch features present on an InP sample.

After 10 minute atomic H dose and high temperature annealing the InP surface shows some etch features indicated by the inset in Figure 7(b). These etch features are different from that of InGaAs: they are much larger and deeper. Unlike InGaAs etch features, these InP etch features do not decrease in density with a high temperature anneal to 470 °C, Figure 7(b). The InP etch features could either be an etch pit to the InGaAs channel layer or impurities in the InP layer. The origin of these dark etch features is difficult to determine because a comparison to a decapped InP sample can not be performed. Fortunately, the density of the InP etch features can be controlled by the temperature during atomic H cleaning.

Other dosing parameters produced different surface reconstructions. Figure 8(a) shows a mixed surface reconstruction obtained after a 10 minute atomic H dose at 440 °C. Conversely, after atomic H dosing at higher temperatures, 460-480 °C, the results obtained show a single surface reconstruction, Figure 8(b). The surface reconstruction highly resembles that of the InGaAs(001)-(4×2) which is consistent STM of InP reported in literature [42]. Further studies on this surface are needed to ensure that at higher temperatures the InP layer is not fully etched and the remaining surface is the InGaAs channel.

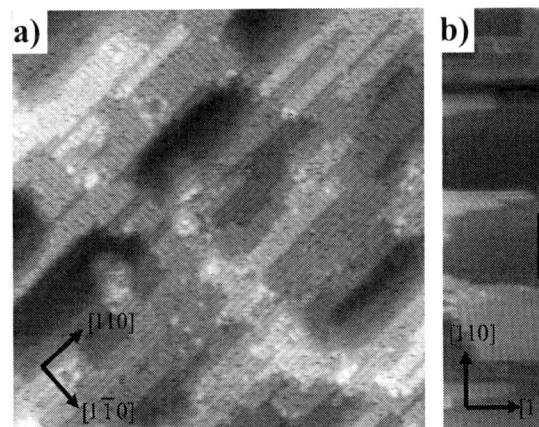 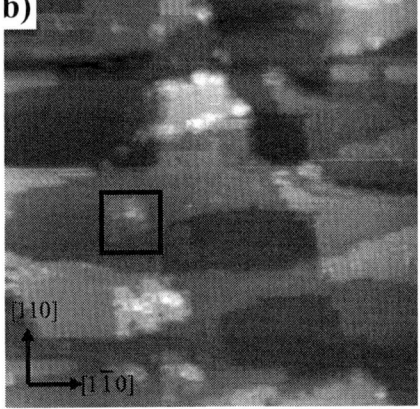

10 min H dose @ 440 °C   5 min H dose @ 460-480 °C

Figure 8: $100 \times 100 \text{nm}^2$ filled state STM image of InP a) 10 minutes dose of atomic H at 440 °C, and b) 5 minute dose of atomic H at 460-480 °C. The STM images shows, at lower dosing temperatures, a mixed surface reconstruction is obtained as shown by the mixture of rows along the [110] and [1 $\overline{1}$ 0] directions. Conversely, for higher temperature atomic H cleaning, a single surface reconstruction is observed with rows directed only along the [110] direction. The surface reconstruction highly resembles that of the InGaAs(001)-(4×2) which is consistent STM of InP reported in literature [42]. The black box in (b) indicates bright defect features. These bright defect features resemble incomplete terraces or material that did not properly arrange into the surface reconstruction.

Figure 9 shows STM images of an InP surface cleaned with atomic H for 5 minutes at 460-480 °C and dosed with TMA at room temperature followed by a 250 °C anneal. The InP surface after TMA exposure resembles that of the InGaAs surface after TMA exposure. The new ordered monolayer is perpendicular to that of the original InP rows, and has 8Å spacing between the horizontal rows. An inset in Figure 9 shows the horizontal ordered monolayer more clearly.

The primary difference in surface quality between InP and InGaAs is the terrace sizes. With the present processing condition, the InP has slightly smaller terrace sizes. Furthermore, the atomic H cleaned InP(100) surface lacks the dark horizontal defects and the bright vertical defects observed on the decapped InGaAs surface. Conversely, the InP surface has more bright island features indicated by the rectangle in Figure 8.

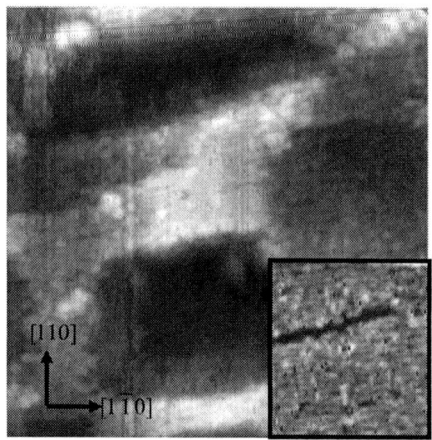

TMA dose at RT
250C anneal

Figure 9: 40×40nm^2 filled state STM images of InP surface cleaned with atomic H for 5 minutes at 460-480 °C after TMA dosing at room temperature and annealing to 250 °C. The inset shows an expanded view of 25×25nm^2 image. The ordered surface reconstruction appears the same as the TMA dose on InGaAs.

Conclusion

In summary, atomic H cleaning is able to restore InGaAs(001)-(4×2) surface after air exposure. With STM, the removal of the oxide layer and restoration of the clean InGaAs surface reconstruction is observed, allowing for a gate-last or replacement-gate process. The difference between room temperature and high temperature is minimal, allowing for a wide process window that must be optimized for etch rates compatible with the device structure. The key process, which is common to both room temperature and high temperature atomic H cleaning, is the high temperature anneal which reduces the step edge density and increases the terrace size. It is likely, the process can easily be implemented with other atomic H sources as long as the atomic H is free from high-energy ions and chemical contaminants.

Atomic H was also performed on a 2nm InP layer on top of the InGaAs channel. STM showed that variation of the temperature of the sample during atomic H cleaning had a strong effect on the surface reconstruction. After cleaning at high temperature, the surface behaved very similar to that of InGaAs for TMA exposure, creating a highly ordered monolayer

Acknowledgments

This work was supported by NSF under Grant Nos. NSF-DMR-0706243, SRC-NCRC-1437.003, and an Applied Materials GRC fellowship. One of the authors (PH) acknowledges the financial support of Science Foundation Ireland under Grant Number: SFI/09/IN.1/I2633.

References

1. J. Robertson and B. Falabretti, *Journal of Applied Physics*, **100**, 014111 (2006).
2. M. Passlack, J. K. Abrokwah, R. Droopad, Z. Y. Yu, C. Overgaard, S. I. Yi, M. Hale, J. Sexton and A. C. Kummel, *Ieee Electron Device Letters*, **23**, 508 (2002).
3. E. J. Kim, L. Q. Wang, P. M. Asbeck, K. C. Saraswat and P. C. McIntyre, *Applied Physics Letters*, **96**, 012906 (2010).
4. Y. Q. Wu, M. Xu, R. S. Wang, O. Koybasi and P. D. Ye, in *Electron Devices Meeting (IEDM), 2009 IEEE International*, p. 1 (2009).
5. E.-H. Roman, H. Yoontae and S. Susanne, *Journal of Applied Physics*, **108**, 124101 (2010).
6. C. H. Diaz, K. Goto, H. T. Huang, Y. Yasuda, C. P. Tsao, T. T. Chu, W. T. Lu, V. Chang, Y. T. Hou, Y. S. Chao, P. F. Hsu, C. L. Chen, K. C. Lin, J. A. Ng, W. C. Yang, C. H. Chen, Y. H. Peng, C. J. Chen, C. C. Chen, M. H. Yu, L. Y. Yeh, K. S. You, K. S. Chen, K. B. Thei, C. H. Lee, S. H. Yang, J. Y. Cheng, K. T. Huang, J. J. Liaw, Y. Ku, S. M. Jang, H. Chuang and M. S. Liang, in *Electron Devices Meeting, 2008. IEDM 2008. IEEE International*, p. 1 (2008).
7. T. Tomimatsu, Y. Goto, H. Kato, M. Amma, M. Igarashi, Y. Kusakabe, M. Takeuchi, S. Ohbayashi, S. Sakashita, T. Kawahara, M. Mizutani, M. Inoue, M. Sawada, Y. Kawasaki, S. Yamanari, Y. Miyagawa, Y. Takeshima, Y. Yamamoto, S. Endo, T. Hayashi, Y. Nishida, K. Horita, T. Yamashita, H. Oda, K. Tsukamoto, Y. Inoue, H. Fujimoto, Y. Sato, K. Yamashita, R. Mitsuhashi, S. Matsuyama, Y. Moriyama, K. Nakanishi, T. Noda, Y. Sahara, N. Koike, J. Hirase, T. Yamada, H. Ogawa and M. Ogura, in *VLSI Technology, 2009 Symposium on*, p. 36 (2009).
8. F. Arnaud, J. Liu, Y. M. Lee, K. Y. Lim, S. Kohler, J. Chen, B. K. Moon, C. W. Lai, M. Lipinski, L. Sang, F. Guarin, C. Hobbs, P. Ferreira, K. Ohuchi, J. Li, H. Zhuang, P. Mora, Q. Zhang, D. R. Nair, D. H. Lee, K. K. Chan, S. Satadru, S. Yang, J. Koshy, W. Hayter, M. Zaleski, D. V. Coolbaugh, H. W. Kim, Y. C. Ee, J. Sudijono, A. Thean, M. Sherony, S. Samavedam, M. Khare, C. Goldberg and A. Steegen, in *Electron Devices Meeting, 2008. IEDM 2008. IEEE International*, p. 1 (2008).
9. K. Choi, H. Jagannathan, C. Choi, L. Edge, T. Ando, M. Frank, P. Jamison, M. Wang, E. Cartier, S. Zafar, J. Bruley, A. Kerber, B. Linder, A. Callegari, Q. Yang, S. Brown, J. Stathis, J. Iacoponi, V. Paruchuri and V. Narayanan, in *VLSI Technology, 2009 Symposium on*, p. 138 (2009).
10. Y. Taur and T. H. Ning, *Fundamentals of modern VLSI devices*, p. p., Cambridge University Press, Cambridge ; New York (2009).
11. P. Packan, S. Akbar, M. Armstrong, D. Bergstrom, M. Brazier, H. Deshpande, K. Dev, G. Ding, T. Ghani, O. Golonzka, W. Han, J. He, R. Heussner, R. James, J.

Jopling, C. Kenyon, S. H. Lee, M. Liu, S. Lodha, B. Mattis, A. Murthy, L. Neiberg, J. Neirynck, S. Pae, C. Parker, L. Pipes, J. Sebastian, J. Seiple, B. Sell, A. Sharma, S. Sivakumar, B. Song, A. St. Amour, K. Tone, T. Troeger, C. Weber, K. Zhang, Y. Luo and S. Natarajan, in *Electron Devices Meeting (IEDM), 2009 IEEE International*, p. 1 (2009).

12. M. Milojevic, F. S. Aguirre-Tostado, C. L. Hinkle, H. C. Kim, E. M. Vogel, J. Kim and R. M. Wallace, *Applied Physics Letters*, **93**, 202902 (2008).

13. C. L. Hinkle, A. M. Sonnet, E. M. Vogel, S. McDonnell, G. J. Hughes, M. Milojevic, B. Lee, F. S. Aguirre-Tostado, K. J. Choi, H. C. Kim, J. Kim and R. M. Wallace, *Applied Physics Letters*, **92**, 071901 (2008).

14. C. H. Chang, Y. K. Chiou, Y. C. Chang, K. Y. Lee, T. D. Lin, T. B. Wu, M. Hong and J. Kwo, *Applied Physics Letters*, **89**, 242911 (2006).

15. M. Radosavljevic, G. Dewey, J. M. Fastenau*, J. Kavalieros, R. Kotlyar, B. Chu-Kung, W. K. Liu*, D. Lubyshev*, M. Metz, K. Millard, N. Mukherjee, L. Pan, R. Pillarisetty, W. Rachmady, U. Shah, and and R. Chau, Non-Planar, Multi-Gate InGaAs Quantum Well Field Effect Transistors with High-K
Gate Dielectric and Ultra-Scaled Gate-to-Drain/Gate-to-Source Separation for Low
Power Logic Applications, in *Electron Devices Meeting, 2010. IEDM 2010. IEEE International* (2010).

16. M. Radosavljevic, B. Chu-Kung, S. Corcoran, G. Dewey, M. K. Hudait, J. M. Fastenau, J. Kavalieros, W. K. Liu, D. Lubyshev, M. Metz, K. Millard, N. Mukherjee, W. Rachmady, U. Shah and R. Chau, in *Electron Devices Meeting (IEDM), 2009 IEEE International*, p. 1 (2009).

17. M. Radosavljevic, G. Dewey, J. M. Fastenau, J. Kavalieros, R. Kotlyar, B. Chu-Kung, W. K. Liu, D. Lubyshev, M. Metz, K. Millard, N. Mukherjee, L. Pan, R. Pillarisetty, W. Rachmady, U. Shah and R. Chau, in *Electron Devices Meeting (IEDM), 2010 IEEE International*, p. 6.1.1 (2010).

18. W. Weiss, R. Hornstein, D. Schmeisser and W. Gopel, *Journal of Vacuum Science & Technology B: Microelectronics and Nanometer Structures*, **8**, 715 (1990).

19. E. J. Petit and F. Houzay, *Journal of Vacuum Science & Technology B*, **12**, 547 (1994).

20. F. S. Aguirre-Tostado, M. Milojevic, C. L. Hinkle, E. M. Vogel, R. M. Wallace, S. McDonnell and G. J. Hughes, *Applied Physics Letters*, **92**, 171906 (2008).

21. M. Yamada, Y. Ide and K. Tone, *Japanese Journal of Applied Physics Part 2-Letters*, **31**, L1157 (1992).

22. A. Khatiri, J. M. Ripalda, T. J. Krzyzewski, G. R. Bell, C. F. McConville and T. S. Jones, *Surface Science*, **548**, L1 (2004).

23. J. A. Schaefer, T. Allinger, C. Stuhlmann, U. Beckers and H. Ibach, *Surface Science*, **251**, 1000 (1991).

24. J. Szuber, *Vacuum*, **57**, 209 (2000).

25. A. Khatiri, T. J. Krzyzewski, C. F. McConville and T. S. Jones, *Journal of Crystal Growth*, **282**, 1 (2005).

26. T. Kikawa, I. Ochiai and S. Takatani, *Surface Science*, **316**, 238 (1994).

27. K. A. Elamrawi, M. A. Hafez and H. E. Elsayed-Ali, *Journal of Applied Physics*, **84**, 4568 (1998).

28. M. A. Hafez and H. E. Elsayed-Ali, *Journal of Applied Physics*, **91**, 1256 (2002).

29. F. Stietz, T. Allinger, V. Polyakov, J. Woll, A. Goldmann, W. Erfurth, G. J. Lapeyre and J. A. Schaefer, *Applied Surface Science*, **104-105**, 169 (1996).

30. T. Kikawa, I. Ochiai and S. Takatani, *Surface Science*, **316**, 238 (1994).

31. R. P. H. Chang, C. C. Chang and S. Darack, *Journal of Vacuum Science & Technology*, **20**, 490 (1982).
32. P. Tomkiewicz, A. Winkler and J. Szuber, *Applied Surface Science*, **252**, 7647 (2006).
33. J. W. Elzey, P. F. A. Meharg and E. A. Ogryzlo, *Journal of Applied Physics*, **77**, 2155 (1995).
34. M. J. W. Rodwell, U. Singisetti, M. Wistey, G. J. Burek, A. Carter, A. Baraskar, J. Law, B. J. Thibeault, K. Eun Ji, B. Shin, L. Yong-ju, S. Steiger, S. Lee, H. Ryu, Y. Tan, G. Hegde, L. Wang, E. Chagarov, A. C. Gossard, W. Frensley, A. Kummel, C. Palmstrom, P. C. McIntyre, T. Boykin, G. Klimek and P. Asbeck, in *Indium Phosphide & Related Materials (IPRM), 2010 International Conference on*, p. 1 (2010).
35. M. J. W. Rodwell, M. Wistey, U. Singisetti, G. Burek, A. Gossard, S. Stemmer, R. Engel-Herbert, Y. Hwang, Y. Zheng, C. Van de Walle, P. Asbeck, Y. Taur, A. Kummel, B. Yu, D. Wang, Y. Yuan, C. Palmstrom, E. Arkun, P. Simmonds, P. McIntyre, J. Harris, M. V. Fischetti and C. Sachs, in *Indium Phosphide and Related Materials, 2008. IPRM 2008. 20th International Conference on*, p. 1 (2008).
36. W. Melitz, J. Shen, S. Lee, J. S. Lee, A. C. Kummel, R. Droopad and E. T. Yu, *Journal of Applied Physics*, **108**, 023711 (2010).
37. M. Yamada, *Japanese Journal of Applied Physics Part 2-Letters*, **35**, L651 (1996).
38. S. J. Pearton, *Materials Science and Engineering B-Solid State Materials for Advanced Technology*, **10**, 187 (1991).
39. J. B. Clemens, E. A. Chagarov, M. Holland, R. Droopad, J. A. Shen and A. C. Kummel, *Journal of Chemical Physics*, **133**, 154704 (2010).
40. P. G. Hofstra, B. J. Robinson, D. A. Thompson and S. A. McMaster, *Journal of Vacuum Science & Technology A: Vacuum, Surfaces, and Films*, **13**, 2146 (1995).
41. Y. J. Chun, T. Sugaya, Y. Okada and M. Kawabe, *Japanese Journal of Applied Physics*, **32**, L287 (1993).
42. L. Li, Q. Fu, C. H. Li, B. K. Han and R. F. Hicks, *Physical Review B*, **61**, 10223 (2000).

190

ECS Transactions, 35 (4) 191-204 (2011)
10.1149/1.3572283 ©The Electrochemical Society

Plasma-Assisted Atomic Layer Deposition of Low Temperature SiO$_2$

G. Dingemans, C. A. A. van Helvoirt, M. C. M. van de Sanden, and W. M. M. Kessels

Department of Applied Physics, Eindhoven University of Technology, P.O. Box 513,
5600 MB Eindhoven, The Netherlands,

Atomic layer deposition (ALD) was used to deposit SiO$_2$ films in
the temperature range of 50-400 °C. H$_2$Si[N(C$_2$H$_5$)$_2$]$_2$ and an O$_2$
plasma were used as Si precursor and oxidant, respectively. The
growth process was characterized in detail, using various in situ
diagnostics. Ultrashort precursor doses (~50 ms) were found to be
sufficient to reach self-limiting ALD growth with a growth-per-
cycle of ~1 Å. The films exhibited a refractive index of 1.46 ±
0.02, a mass density of 2.0 ± 0.1 g/cm^3, and an O/Si ratio of 2.1 ±
0.1, virtually independent of the substrate temperature. The results
therefore demonstrate an efficient ALD process for the conformal
and uniform deposition of SiO$_2$ at low substrate temperatures. Also
the surface chemistry during the plasma ALD process and surface
passivation performance of the ALD SiO$_2$ films on crystalline
silicon surfaces are briefly addressed.

Introduction

The key importance of silicon dioxide (SiO$_2$) for applications in silicon-based
microelectronics needs no introduction. Also in silicon photovoltaics SiO$_2$ is a key
material, as it has for long been the state-of-the-art passivation material leading to a
substantial reduction of the surface recombination losses enabling high solar cell
efficiencies [1-4]. High quality SiO$_2$ is obtained by thermal oxidation of the Si surface at
temperatures > 800°C. Alternative methods for the synthesis of SiO$_2$ have been
developed to avoid such high temperatures and long processing times. They may also
enable single side deposition and a high level of control of the material properties and
film thickness. These alternative methods include (wet) chemical oxidation, (plasma-
enhanced) chemical vapor deposition, sputtering and electron beam evaporation.

Atomic layer deposition (ALD) is an alternative CVD-like method that recently
gained a lot of attention. ALD allows for precise thickness control, optimal large-area
uniformity, and the conformal coating of demanding substrate topologies [5,6]. In a first
report, ALD SiO$_2$ was synthesized employing SiCl$_4$ and H$_2$O [7,8] which required
relatively high substrate temperatures (> 300 °C) and long precursor exposures. In recent
years, various alternative Si precursors have been tested in combination with O$_3$ or H$_2$O
as the oxidants. These processes include the use of pyridine (C$_5$H$_5$N) [9] and Al as
catalysts [10,11]. The approach employing Al was referred to as rapid ALD as it resulted
in deposition rates above the "theoretical" maximum of one monolayer per ALD cycle. In
addition, more recently a thermal ALD process for low-temperature SiO$_2$ was reported
which was free of catalysts or corrosive by-products [12]. In this respect, the use of
precursors with amino ligands has also shown promising results, in particular when
combined with H$_2$O$_2$, O$_3$, or O$_2$ plasma as the oxidant [13-16]. SiO$_2$ films grown with
ALD have been reported to exhibit low carbon content, and a high electrical breakdown

191

field [9,14]. Nevertheless, to improve properties such as the chemical etch rate or the interface defect density, annealing at a temperature of 1000°C was shown to be beneficial [12].

In this contribution an efficient plasma-assisted ALD process is demonstrated for the low-temperature synthesis of SiO_2 using $H_2Si[N(C_2H_5)_2]_2$ as the Si precursor (Figure 1a). This precursor is commercially supplied by Air Liquide under the product name SAM.24 [17,18]. Data are presented for the ALD process within the temperature range of 50-400 °C and the results are compared to the Al_2O_3 ALD processes from $Al(CH_3)_3$ and H_2O/O_2 plasma. These Al_2O_3 processes have been widely studied and are considered rather typical and "ideal" ALD processes [19]. Moreover for ALD Al_2O_3 films excellent surface passivation properties for silicon of arbitrary doping types and different doping levels have been reported [20,21]. In addition, the surface chemistry during the plasma ALD process of SiO_2 is discussed and also the first results on the surface passivation performance of the ALD SiO_2 films on crystalline silicon surfaces are briefly addressed.

Experimental

The SiO_2 films were deposited in the Oxford Instruments OpAL reactor. This is an open-load system, suited for both plasma and thermal ALD and operating at typical pressures of 150 mTorr. A remote O_2 plasma was used during the oxidation step in the ALD cycle. SAM.24 (Air Liquide) was used as the Si precursor (Figure 1a) [17,18]. This is a liquid (melting point < -10 °C) which exhibits a high vapor pressure, i.e. ~100 Torr at 100 °C (Figure 1b). The SAM.24 was held in a stainless steel bubbler heated to 50 °C and the precursor was introduced into the reactor by ultrashort doses (10-120 ms) using fast ALD valves. A flow of Ar as well as the O_2 flow were continuously on during the process. The latter was feasible as no evidence was found for reactions between the Si precursor and O_2 under the experimental conditions used. The substrate temperature during deposition, T_{dep}, was varied between 50 and 400 °C. The reactor wall temperature was 180 °C unless the substrate temperature was lower. Under these conditions the wall and substrate temperature were equal. To allow for direct comparison, Al_2O_3 was synthesized in the same reactor using $Al(CH_3)_3$ as the metal precursor and H_2O or O_2 plasma as the oxidant [19-21]. All films were deposited on Si (100) wafers which received a short treatment in diluted HF (~1% in $DI-H_2O$) to remove the native oxide prior to loading in the ALD reactor.

In situ spectroscopic ellipsometry (SE) measurements were used for optimizing the ALD process. The growth-per-cycle, GPC, and refractive index were determined by using a Cauchy optical model to fit the ellipsometry data. Rutherford backscattering spectroscopy (RBS) and elastic recoil detection (ERD) employing ~2 MeV He^{2+}-ions from the singletron at the Eindhoven University of Technology and transmission Fourier transform infrared absorption (FTIR) measurements were used to analyze the film composition. The surface morphology was investigated by atomic force microscopy (AFM) measurements in semi-contact mode whereas high-resolution transmission electron microscopy (TEM) was used to study the samples in cross-section. The deposition process itself was studied in real time by quadrupole mass spectroscopy (QMS) probing the gas in the exhaust line and by optical emission spectroscopy (OES) through a view port located on top of the ALD reactor.

(a)   (b)

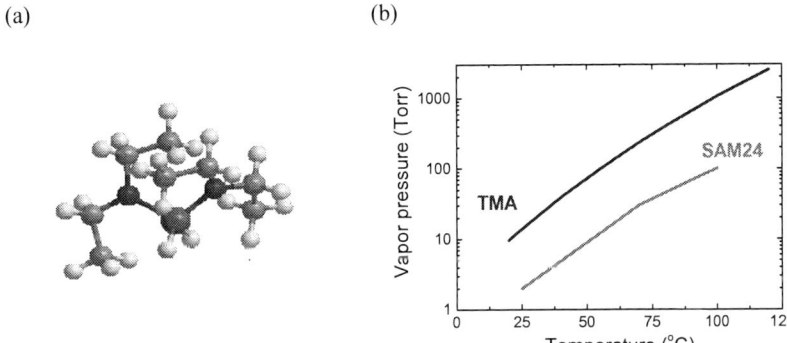

Figure 1. (a) The precursor $H_2Si[N(C_2H_5)_2]_2$ (SAM.24, Air Liquide) used for ALD of SiO_2 and (b) the vapor pressure of the precursor as a function of the temperature. The vapor pressure is compared to the one of $Al(CH_3)_3$ (trimethylaluminum, TMA). This precursor is commonly used for ALD of Al_2O_3.

The passivation performance of the ALD SiO_2 films was evaluated from the effective lifetime τ_{eff} of the minority carriers in double-side coated floatzone n-type Si wafers (~3.5 Ω cm). τ_{eff} was determined with photoconductance decay in the transient mode and quasi-steady-state-mode (for $\tau_{eff} < 100$ µs) using a Sinton lifetime tester (WCT 100). The upper level for the surface recombination velocity $S_{eff,max}$ was extracted at an injection level of 5×10^{14} cm^{-3} by the expression

$$S_{eff,max} = \frac{W}{2 \cdot \tau_{eff}},\qquad(1)$$

with W the thickness of the silicon wafer (~280 µm). In the derivation of this expression it is assumed that all recombination takes place at the surface.

Results and discussion

ALD Growth Process

The ALD process was monitored by in situ spectroscopic ellipsometry by taking data points after a certain number of cycles. Typical thickness data plotted as a function of number of ALD cycles are shown in Figure 2 for the substrate temperatures of 50 °C and 250 °C. For 250 °C the SiO_2 thickness increased linearly with the number of cycles (the data for 50 °C will be discussed below). The slope of the curve yields the growth-per-cycle (GPC) and for a substrate temperature of 250 °C a GPC value of 1.1 Å/cycle is obtained. This GPC value is very comparable to the plasma-assisted ALD process of Al_2O_3 [19,20]. Moreover, no indications for a significant growth delay on the H-terminated Si(100) substrates were observed for the SiO_2 ALD process. This is also similar to what has been observed for plasma ALD of Al_2O_3 [19].

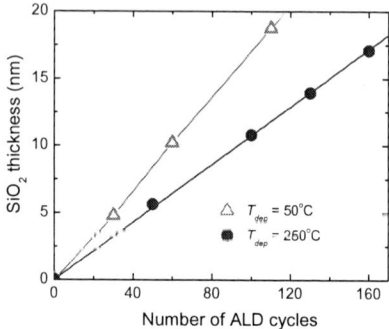

Figure 2. SiO$_2$ film thickness as a function of the number of ALD cycles at substrate temperatures of 50 °C and 250 °C. The film thickness was measured by in situ spectroscopic ellipsometry.

Figure 3 shows the effect of the duration of the successive steps in the ALD recipe, i.e., the precursor dosing, precursor purge step, plasma exposure, and plasma exposure purge step (Figure 1a-1d). In the corresponding experiments one process parameter in the ALD recipe was varied whereas the duration of the other steps was taken sufficiently long to guarantee saturated ALD conditions for the non-varied process parameters. The substrate temperature was set to 250 °C. The growth process of SiO$_2$ is compared to plasma-assisted ALD of Al$_2$O$_3$ at 250 °C (Figure 1e-1h). From the figure it is evident that ultrashort precursor dosing times (~50 ms) were already sufficient to reach a self-limiting growth with a GPC of ~1.1 Å. These short dosing times were only slightly higher compared to those used for the Al(CH$_3$)$_3$ precursor for ALD of Al$_2$O$_3$. The fact that short dosing times are sufficient is in agreement with the expectations based on the relatively high vapor pressure of the SAM.24 precursor. Many other processes from different precursors require much longer dosing times, for instance ALD of TiO$_2$ and Ta$_2$O$_5$ require dosing times > 1s in a similar remote plasma and thermal ALD reactor [22,23]. The duration of the purge step after precursor dosing was required to be > 2 s. For shorter purge times, residual precursor remaining in the reactor volume can react in the plasma, causing parasitic (PE)CVD-like growth and a higher GPC value. Regarding the O$_2$ plasma step, a plasma exposure time > 1 s was found to be sufficient to reach a saturated GPC, indicating the rapid removal of the precursor ligands. This plasma exposure time is slightly shorter than for plasma ALD of Al$_2$O$_3$ which requires plasma times of ~2 s to reach saturated growth. However with in situ spectroscopic ellipsometry only the center of the Si wafer is probed and therefore, to ensure saturation over full wafer surface, a plasma exposure time of 4 s was employed in all subsequent experiments. Interestingly, the purge after the plasma step had a significant impact on the GPC. This is in contrast to the ALD process for Al$_2$O$_3$, where the purge step after O$_2$ plasma exposure was found to have little influence on the GPC and could be reduced well below 0.5 s. We attribute the higher GPC for shorter purges (< 2 s) to reactions between residual H$_2$O, formed during the plasma process, with the Si precursor injected in the subsequent step. Although it is known that the H$_2$Si[N(C$_2$H$_5$)$_2$]$_2$ precursor reacts with H$_2$O, it is relevant to mention here that we were unable to develop a thermal ALD process for SiO$_2$ using

SAM.24 as precursor and H_2O as the oxidant. No film growth was observed. Instead, even with the shortest possible H_2O doses applied, powder formation occurred in the reactor as was noticeable by the naked eye.

Figure 3 also shows the refractive index of the films corresponding to the experiments to verify saturation of the SiO_2 ALD process. At a photon energy of 2 eV a refractive index of 1.46 ± 0.02 was obtained for the SiO_2 films under the saturated ALD conditions. The refractive index was observed to drop for very short dose and purge times, most prominently for a too short plasma exposure time. This can most probably be attributed to a reduced SiO_2 density under these conditions.

The thickness uniformity of the SiO_2 films deposited by plasma-assisted ALD at 250 °C was evaluated by mapping the thickness by spectroscopic ellipsometry. For a 8 inch (200 mm) wafer the nonuniformity, defined by the difference between the maximum and minimum thicknesses divided by the twice the average thickness of all data points measured [19], was < 3.5%. The thickness nonuniformity achieved on 4 inch (100 mm) wafers was ~1%.

Material Properties and Substrate Temperature Dependence

Figure 4 shows the effect of the substrate temperature between 50 and 400 °C on the ALD growth process of SiO_2. The length of the purge steps in the lower temperature regime was extended (up to 10 s at 50 °C) as it is more difficult to remove H_2O at lower temperatures which could impact the saturation behavior of the process. The GPC was observed to decrease with increasing deposition temperature from ~1.7 Å/cycle at 50 °C to 0.8 Å at 400 °C. The refractive index was fairly constant between 100 and 300 °C. Below 100 °C and above 300 °C the refractive index was somewhat lower and it can therefore not be excluded that some non-ideal ALD behavior takes place at the lowest and highest temperatures investigated. At the low substrate temperature of 50 °C, additional CVD reactions may contribute to the higher GPC. This might also explain the slightly non-linear trend between the film thickness and number of cycles observed at this deposition temperature as shown in Figure 2. The slightly increasing GPC with number of ALD cycles points to the accumulation of some residual H_2O in the reactor with which the precursor can react, despite the long purging times (10 s) after the plasma step. For temperatures reaching 400 °C thermal stability issues of the precursor and its ligands can start to play a role.

Table 1 shows RBS and ERD data obtained at substrate temperatures of 100, 200 and 300 °C. The table shows that the number of Si atoms deposited per cycle decreases with increasing substrate temperature. This clearly demonstrates that the decrease of the GPC with increasing substrate temperature can be attributed to reduced precursor adsorption per cycle at higher temperatures. Similar results were obtained for Al_2O_3 synthesized by plasma-assisted ALD [19,20,23]. As shown in Figure 4c, the GPC for this process was also found to decrease significantly when going from 25 to 400 °C. This could be attributed almost fully to the decrease in the number of Al atoms deposited per cycle. Film densification, decreasing the thickness per "monolayer" of Al_2O_3 deposited, was found to play a minor role and only for temperatures well below 100 °C. For Al_2O_3, the decrease in GPC with increasing temperature could be attributed to a loss of –OH surface groups with increasing temperature due to thermally activated dehydroxylation reactions [6,24,25]. Thermal ALD of Al_2O_3 (Figure 4c) exhibited a different trend for substrate temperatures < 200°C. For these temperatures the ALD process was not ruled by the density of –OH surface groups but rather by the reduced oxidation efficiency of H_2O at the lower substrate temperatures [19,24].

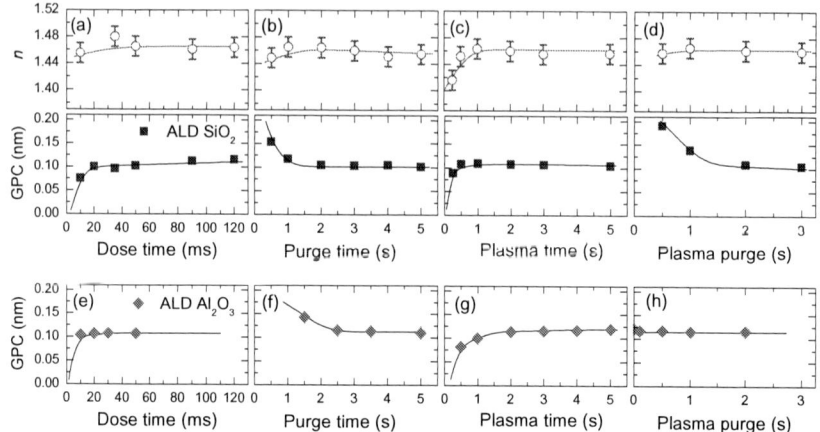

Figure 3. (a)-(d) Saturation curves for the growth-per-cycle GPC and refractive index n of the SiO_2 films as measured by in situ spectroscopic ellipsometry as a function of the 4 process parameters in the ALD recipe. (e)-(h) Saturation curves for the GPC of Al_2O_3 films. The substrate temperature was 250 °C for both the SiO_2 and Al_2O_3 process.

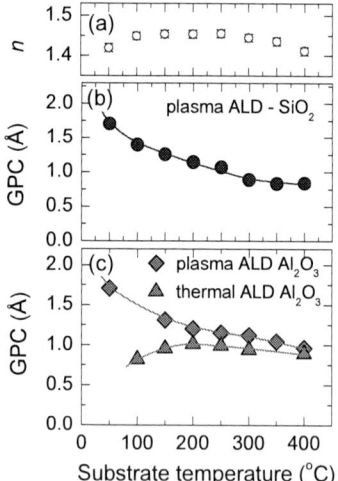

Figure 4. Influence of the substrate temperature during deposition on (a) the refractive index n and (b) the growth-per-cycle GPC of SiO_2 as determined with in situ spectroscopic ellipsometry. In (c) the GPC is given for plasma and thermal ALD of Al_2O_3.

Table 1. Data on ALD SiO_2 as determined from the RBS and ERD measurements. In the calculation of the mass density the film thickness as obtained by SE was used. The thickness of the films was in the range of 35 – 45 nm.

Substrate temperature (°C)	Si atoms per cycle (10^{14} cm^{-2})	[Si] (at.%)	[O] (at.%)	[H] (at.%)	O/Si ratio	mass density (g/cm^3)
100	2.8 ± 0.1	29.1 ± 0.8	61.3 ± 1.5	9.6 ± 0.9	2.1 ± 0.1	2.0 ± 0.1
200	2.3 ± 0.1	29.9 + 0.8	62.9 ± 1.5	7.1 ± 0.7	2.1 ± 0.1	2.0 ± 0.1
300	1.9 ± 0.1	29.6 ± 0.8	62.3 ± 1.5	8.1 ± 0.8	2.1 ⊥ 0.1	2.1 + 0.1

The fact that good SiO_2 material properties were obtained between 100 and 300 °C can be concluded from Table 1. The Si and O content correspond with an O/Si ratio of 2.1 ± 0.1 and the hydrogen content of the films is 7 – 10 at.% depending on the substrate temperature. The carbon and nitrogen content of the films was below the detection limit of ~5 at.%. Apart from the hydrogen content, the materials properties were found virtually independent of the substrate temperature in the range of 100 – 300 °C. This also holds for the mass density which was found to be 2.0 ± 0.1 g/cm^3. The film quality was therefore fairly constant for the plasma ALD SiO_2 process. For the plasma-assisted ALD process of Al_2O_3 the variation in film properties also remained within a narrow range, although the variation with temperature was slightly more pronounced compared to SiO_2 prepared by plasma ALD. For Al_2O_3 deposited at temperatures between 100 and 400 °C, the O/Al ratio varied between 1.5 and 1.7, and the mass density between 2.9 and 3.2 g/cm^3 [19,20]. The hydrogen content of the Al_2O_3 was in the range 1-8 at.% decreasing with increasing substrate temperature whereas carbon could not be detected above the detection limit of the RBS measurements. For temperatures below 100 °C, the Al_2O_3 became more O-rich, less dense, and contained significantly more hydrogen [19,20].

Fourier transform infrared absorption spectroscopy was used to compare the ALD SiO_2 films with thermally-grown SiO_2. Figure 5 shows the FTIR spectra revealing a shift of the Si-O-Si stretching and Si-O-Si rocking modes toward lower wavenumbers for the ALD SiO_2 film. This is in agreement with the slightly non-stoichiometric nature (O/Si ratio = 2.1 ± 0.1) of the films and the fact that the mass density of ALD SiO_2 is slightly lower compared to typical values for wet thermally grown SiO_2 films (~2.2 g/cm^3). The FTIR data also confirm the presence of hydrogen in the ALD SiO_2 films by the observation of SiO-H bending (~920 cm^{-1}) and SiO-H stretching (2500-3600 cm^{-1}) signatures in the spectrum.

The surface morphology of the SiO_2 films was studied by AFM in semi-contact mode. Films deposited at a substrate temperature of 100 °C (film thickness is 51 nm) and 200 °C (film thickness is 48 nm) were compared. The AFM scan of the SiO_2 film deposited at 200 °C is shown in Figure 6 and reveals a root-mean-square surface roughness of 1.6 Å. The film deposited at 100 °C, exhibited only a slightly higher roughness of 1.9 Å. These values were similar to those obtained for uncoated polished Si wafers and this demonstrates that the films show negligible roughness development on the Si(100) substrate. The latter can also be appreciated from the high-resolution TEM image displayed in Fig. 7. From the AFM data it also follows that the aforementioned (PE)CVD growth component at lower substrate temperatures caused by residual H_2O is not pronounced yet at 100 °C.

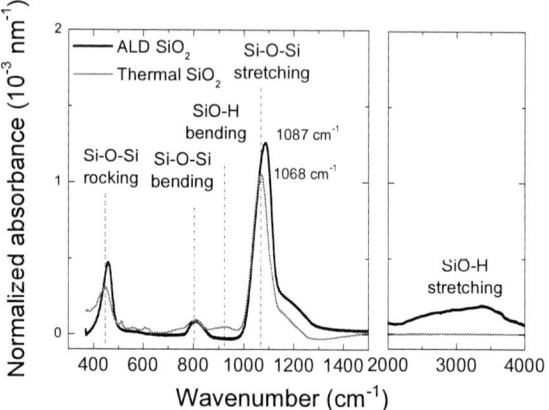

Figure 5. FTIR spectra of ALD SiO_2 prepared at 200 °C (48 nm film thickness) and thermal SiO_2 grown by wet oxidation at ~900°C (295 nm film thickness). The most prominent absorption peaks have been assigned (see Ref. 24 and references therein). The absorbance is normalized by the film thickness. A Si wafer without SiO_2 served as a reference to obtain the absorption spectra.

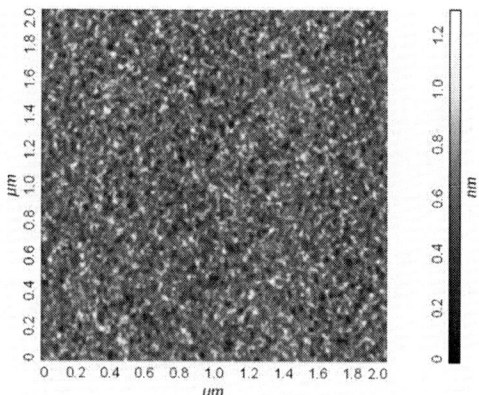

Figure 6. AFM scan of an ALD SiO_2 film deposited at a substrate temperature of 200 °C. The thickness of the film was 48 nm. The scan size was 2 μm × 2 μm.

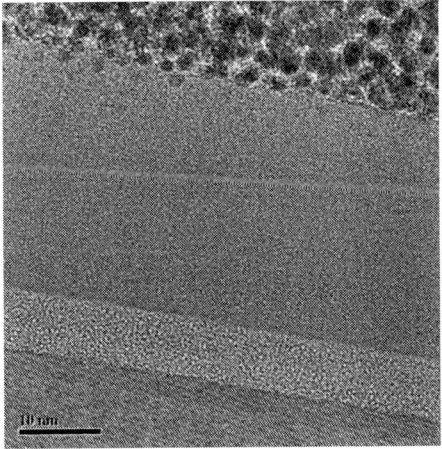

Figure 7. High-resolution TEM image of an ALD SiO$_2$ film of 7.0 ± 0.3 nm thickness deposited on a H-terminated Si(100) wafer. The SiO$_2$ was encapsulated by an Al$_2$O$_3$ film deposited by plasma-assisted ALD to prepare the sample for TEM analysis.

Surface Chemistry

From the data presented and on the basis of literature reports, also a few comments can be made about the surface chemistry. The GPC decreases monotonically with the substrate temperature which is very similar to the case of Al$_2$O$_3$ synthesized by plasma-assisted ALD. It was experimentally verified that the surface chemistry during plasma ALD Al$_2$O$_3$ was ruled by –OH groups, with the number density of these –OH groups decreasing with temperature [24,25]. It can therefore be concluded that also the plasma-assisted ALD SiO$_2$ process is governed by the –OH groups with the decrease of the GPC in Figure 4b ruled by thermally activated dehydroxylation reactions. Moreover, the presence of –OH groups was demonstrated by the FTIR spectrum of the ALD SiO$_2$ film showing a clear signature of SiO-H bonds incorporated in the film (Figure 5). During the precursor step, it is therefore most likely that the –N(C$_2$H$_5$)$_2$ ligands of the precursor react with the surface –OH groups producing volatile HN(C$_2$H$_5$)$_2$. A reaction involving the breaking of the Si-H bond in the precursor is very unlikely [16]. We propose therefore similar surface chemical reactions during the first ALD half cycle as reported by Burton et al. [16] for the SiH(N(CH$_3$)$_2$)$_3$ precursor which is comparable to the present precursor:

$$Si\text{-}OH^* + H_2Si[N(C_2H_5)_2]_2 \rightarrow SiO\text{-}SiH_2[N(C_2H_5)_2]_{2-x}{}^* + x\ HN(C_2H_5)_2,$$

where surface species are indicated by *. In this precursor adsorption reaction, only one ($x = 1$) or both ($x = 2$) of the –N(C$_2$H$_5$)$_2$ ligands may react. In the second half cycle, the surface reactions will be dominated by O radical species delivered by the plasma [27]. From similar cases studied previously (e.g., Al$_2$O$_3$ from Al(CH$_3$)$_3$ and O$_2$ plasma [27] and

Ta_2O_5 from $Ta[N(CH_3)_2]_5$ and O_2 plasma [22]) it can be hypothesized that combustion-like reactions dominate:

$$SiH_2N(C_2H_5)_2* + O \rightarrow Si\text{-}OH* + H_2O + CO_x + other \text{ (N-containing) species.}$$

In the latter expression the species are not balanced as it is unclear what reaction products are actually created. In the second half cycle also N-containing species need to be produced for the case that not all precursor molecules react with the $-OH$ covered surface through the release of both $-N(C_2H_5)_2$ ligands, i.e. when $x \neq 2$ for all precursor molecules adsorbing.

Evidence for the fact that $x \neq 2$ for all precursor molecules was obtained from preliminary quadrupole mass spectrometry (QMS) measurements. Figure 8 shows time-dependent mass spectrometry data for a number of selected mass-over-charge m/z ratios. The enhanced signals at $m/z = 72$ ($N[C_2H_5]_2^+$) and $m/z = 73$ ($HN[C_2H_5]_2^+$) during the first half cycle are consistent with the removal of the precursor ligands during precursor adsorption. However, it should be noted that these signals can also originate from the cracking of the precursor molecule in the mass spectrometer. Slightly enhanced signals at these m/z values were also observed during the second half cycle whereas the signals were absent during steps in which the plasma was ignited without preceding precursor dosing. This suggests that after the first half cycle indeed a fraction of the $-N(C_2H_5)_2$ ligands remain intact on the surface [16]. The latter can also be concluded from the other species observed during the second half cycle. During this plasma step, the prominent m/z ratios that were detected included $m/z = 2$ (H_2^+), $m/z = 18$ (H_2O^+), $m/z = 28$ (CO^+) and $m/z = 44$ (CO_2^+). Figure 8 shows the signals at $m/z = 18$ and $m/z = 44$. The fact that combustion products such as CO_2 are observed during the plasma step clearly indicates that some $-N(C_2H_5)_2$ ligands remain on the surface after precursor adsorption.

The interpretation of mass spectrometry for plasma-assisted ALD processes is more complicated than for thermal ALD as the species released from the surface can react in the plasma leading to the creation of new species. The preliminary mass spectrometry data as shown in Figure 8 should therefore be interpreted with care [27]. The plasma however also allows investigation of the optical emission spectrum during the plasma step [28]. Figure 9 shows two optical emission spectra, one for a plasma step during ALD (recorded immmediately after plasma ignition) and one for a regular O_2 plasma without preceding precursor dosing step. The presence of OH and H emission (i.e., H_α, H_β, H_γ of the Balmer series) is clearly observed for the plasma step during ALD. These excited fragments are formed in the plasma by (electron-induced) dissociation of volatile species (likely mostly from H_2O) originating from the reactor surfaces and substrate. The inset shows the transient H_α emission during the plasma step in the ALD cycle. The increase and subsequent decrease of the signal suggests that the reaction products are formed within the first second after plasma ignition. This interpretation is consistent with the fast saturation behavior as displayed in Figure 3c. The H_α emission disappears within 3-4 s after plasma ignition which is similar to the residence time of the particles in our reactor at the operating pressure used (~150 mTorr). This indicates that the surface reactions take place almost instantly after plasma ignition. Interesting is also that no signal due to CN emission is observed during the plasma step. This emission was prominently present in the emission spectra during plasma ALD from Ta_2O_5 from $Ta[N(CH_3)_2]_5$ and O_2 plasma [22].

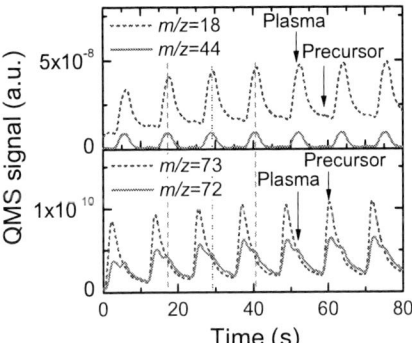

Figure 8. Data from quadrupole mass spectrometry (QMS) for selected mass-over-charge ratios, i.e., m/z = 18 (H_2O^+), 44 (CO_2^+), 72 ($HN[C_2H_5]_2^+$), and 73 ($HN[C_2H_5]_2^+$). During the measurements the substrate temperature was 250 °C and the wall temperature was 180 °C.

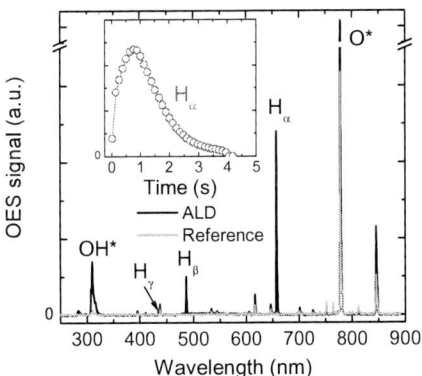

Figure 9. Optical emission spectra (OES) for the plasma step during plasma-assisted ALD and for a regular plasma step without preceding precursor dosing. The latter served as a reference. The most prominent emission lines have been assigned. The inset shows the transient signal due to H_α emission after the ignition of the plasma at 0 s.

Silicon Surface Passivation

Finally, preliminary data on the passivation performance of the ALD SiO_2 films are discussed. As described in the experimental section the passivation properties were evaluated by measuring the effective lifetime τ_{eff} of the minority carriers in floatzone n-

type Si wafers (~3.5 Ω cm) which were deposited by SiO_2 films of 45 nm thickness at both sides after a treatment of the wafer in diluted HF (~1% in DI-H_2O). The substrate temperature was 200 °C. The films afforded no significant surface passivation in the as-deposited state as indicated by a very low effective lifetime of τ_{eff} = ~4 µs. Similar results were reported for as-deposited plasma ALD Al_2O_3 films [20,29]. These Al_2O_3 films were shown to exhibit a very high defect density at mid gap (~10^{13} eV^{-1} cm^{-2}) related to the VUV radiation present in the plasma [30]. Annealing the ALD SiO_2 films in forming gas (10% H_2 in N_2) at 400 °C for 10 min. led to improved surface passivation with τ_{eff} = 450 µs at an injection level of 5×10^{14} cm^{-3}, which corresponds to $S_{eff,max}$ of 31 cm/s. This value is comparable to the $S_{eff,max}$ of 54 cm/s reached on floatzone n-type Si wafers (~1.3 Ω cm) by PECVD SiO_2 films after a 15 min. forming gas anneal at 600 °C [26]. The latter is among the best reported values for films prepared by CVD-like methods. The value of $S_{eff,max}$ is however much higher than the surface recombination velocity achieved by Al_2O_3 films prepared by plasma ALD and thermal ALD after annealing at 400 °C for 10 min. in N_2 [21]. For the same wafers this leads to $S_{eff,max}$ values as low as 0.8 cm/s and 2 cm/s for plasma and thermal ALD, respectively. Moreover, the surface passivation of the ALD SiO_2 films was not stable over time and gradually deteriorated. Issues with the long term stability of the passivation by SiO_2 have been reported before for chemical oxides [31]. Nevertheless, the ALD SiO_2 films are of significant interest for surface passivation of silicon surfaces, for example in combination with Al_2O_3 films as has recently also been demonstrated for PECVD SiO_x [32].

Conclusions

A plasma-assisted ALD process for SiO_2 from the SAM.24 precursor ($H_2Si[N(C_2H_5)_2]_2$) and an O_2 plasma as oxidant has been developed for substrate temperatures between 50 and 400 °C. It is demonstrated that this process is suited for low-temperature synthesis of high-quality SiO_2 by ALD with the SiO_2 properties being relatively insensitive to the substrate temperature for the temperature range of 100 – 300 °C. The process is also relatively fast as it combines a high growth-per-cycle (0.8 – 1.7 Å/cycle) with relatively short dosing and purge times. These results therefore complement earlier work employing the same precursor and O_3 as the oxidant in an ALD process [33]. The ALD SiO_2 processes with this precursor are therefore of interest for high-volume manufacturing applications, for instance using ALD batch processes or inline (plasma) ALD equipment [34]. In a subsequent study, the interface properties of the SiO_2 films on Si will be evaluated in more detail by additional silicon surface passivation studies [35].

Acknowledgments

Dr. S.E. Potts is acknowledged for the fruitful discussions and B. Macco and Dr. M.A Verheijen for carrying out the AFM and TEM analysis, respectively. Ch. Lachaud, N. Blasco, and A. Madec from Air liquide are acknowledged for donating the SAM.24 precursor. This work is supported by Q-Cells, the German Ministry for the Environment, Nature Conservation and Nuclear Safety (BMU) under contract number 0325150 ("ALADIN"), and by the Dutch Technology Foundation STW (Thin Film Nanomanufacturing (TFN) program).

References

1. A. G. Aberle, *Prog. Photovoltaics* **8**, 473 (2000).
2. M. J. Kerr and A. Cuevas, *Semicond. Sci. Technol.* **17**, 35 (2002).
3. J. Zhao, A. Wang, M.A.Green, and F. Ferrazza, *Appl. Phys. Lett.* **73**, 1991 (1998).
4. O. Schultz, A. Mette, M. Hermle, and S. W. Glunz, *Prog. Photovolt: Res. Appl.* **16**, 317 (2008).
5. S.M. George, *Chem. Rev.* **110**, 111 (2010).
6. R. L. Puurunen, *J. Appl. Phys.* **97**, 121301 (2005).
7. J.W. Klaus, O.W. Ott, J.M.Johnson, S.M. George, *Appl. Phys. Lett.* **70**, 1092 (1997).
8. S.M. George, O. Sneh, A.C. Dillon, M.K. Wise, A.W. Ott, L.A. Okada, and J.D. Way, *Appl. Surf. Science* **82**, 460 (1994).
9. J. W. Klaus, O. Sneh, and S. M. George, *Science* **278**, 1934 (1997).
10. D. Hausmann, J. Becker, S. Wang, and R.G. Gordon, *Science* **298**, 402 (2002).
11. B. B. Burton, M. P. Boleslawski, A. T. Desombre, and S. M. George, *Chem. Mater.* **20**, 7031 (2008).
12. D. Hiller, R. Zierold, J. Bachmann, M. Alexe, Y. Yang, J. W. Gerlach, A. Stesmans, M. Jivanescu, U. Müller, J. Vogt, H. Hilmer, P. Löper, M. Künle, F. Munnik, K. Nielsch, and M. Zacharias, *J. Appl. Phys.* **107**, 064314 (2010).
13. S. Kamiyama, T. Miura, and Y. Nara, *Thin Solid Films* **515**, 1517 (2006).
14. S-J. Won, S. Suh, M. Soo Huh, and H. Joon Kim, *IEEE Electron Device Lett.* **31**, 857 (2010).
15. R. Katamreddy, B. Feist, and C. Takoudis, *J. Electrochem. Soc.* **155**, G163 (2008).
16. Burton, B.B., Rang, S.W., Rhee, S.W., and George, S.M., *J. Phys. Chem. C* **113**, 8249 (2009).
17. C. Dussarrat, WO Patent 2006/097525 (2006).
18. Air Liquide, 75 Quai d'Orsay, 75321, Paris Cedex 07, France, http://www.airliquide.com
19. J. L. van Hemmen, S. B. S. Heil, J. H. Klootwijk, F. Roozeboom, C. J. Hodson, M. C. M. van de Sanden, and W. M. M. Kessels, *J. Electrochem. Soc.* **154**, G165 (2007).
20. G. Dingemans, M.C.M. van de Sanden, W.M.M. Kessels, *Electrochem. Solid. State Lett.* **13**, H76 (2010).
21. G. Dingemans, R. Seguin, P. Engelhart, M. C. M. van de Sanden, and W. M. M. Kessels, *Phys. Status Solidi RRL* **4**, 10 (2010).
22. S.B.S. Heil, F. Roozeboom, M.C.M. van de Sanden, and W.M.M. Kessels, *J. Vac. Sci. Technol. A* **26**, 472 (2008).
23. S.E. Potts, W. Keuning, E. Langereis, G. Dingemans, M.C.M. van de Sanden, and W.M.M. Kessels, *J. Electrochem. Soc.* **157**, P66 (2010).
24. C. Dillon, A.W. Ott, J.D. Way, and S.M. George, *Surf. Sci.*, **322**, 230 (1995).
25. E. Langereis, J. Keijmel, M.C.M. van de Sanden, and W.M.M. Kessels, *Appl. Phys. Lett.* **92**, 231904 (2008).
26. B. Hoex, F.J.J. Peeters, M. Creatore, M.A. Blauw, W.M.M. Kessels, and M.C.M. van de Sanden, *J. Vac. Sci. Technol. A* 24, 1823 (2006).
27. S. B. S. Heil, J. L. van Hemmen, M. C. M. van de Sanden, and W. M. M. Kessels, *J. Appl. Phys.* 103, 103302 (2008).
28. A.J.M. Mackus, S.B.S. Heil, E. Langereis, H.C.M. Knoops, M.C.M. van de Sanden, and W.M.M. Kessels, *J. Vac. Sci. Technol. A* **28**, 77 (2010).
29. B. Hoex, S.B.S. Heil, E. Langereis, M.C.M. van de Sanden, and W.M.M. Kessels, *Appl. Phys. Lett.* **89**, 042112 (2006).

30. G. Dingemans, N. M. Terlinden, D. Pierreux, H. B. Profijt, M. C. M. van de Sanden, and W.M.M. Kessels, *Electrochem. Solid. State Lett.* **14**, H1 (2011).
31. N.E. Grant and K.R. McIntosh, *IEEE Electron Device Lett.* **31**, 1002 (2010).
32. G. Dingemans, M.C.M. van de Sanden, and W.M.M. Kessels, *Phys. Status Solidi RRL* (2011). DOI 10.1002/pssr.201004378
33. Air Liquide, Private communication.
34. G. Dingemans, N. M. Terlinden, D. Pierreux, H. B. Profijt, M. C. M. van de Sanden, and W. M. M. Kessels, *Electrochem. Solid-State Lett.* **14**, H1 (2011).
35. G. Dingemans, N.M. Terlinden, M.C.M. van de Sanden, and W.M.M. Kessels, to be published (2011)

Surface Passivation of InGaAs/InP HBTs Using Atomic Layer Deposited Al₂O₃

R. Driad, F. Benkhelifa, L. Kirste, R. Lösch, M. Mikulla, and O. Ambacher,

Fraunhofer Institute for Applied Solid State Physics,
Tullastrasse 72, 79108 Freiburg, Germany

In this contribution, we investigate the Al_2O_3 surface passivation of InGaAs/InP heterostructures using thermal atomic layer deposition (ALD) with water vapor, and plasma ALD with oxygen plasma. The microstructure and optical properties of the Al_2O_3 layers are examined by X-ray reflectivity (XRR) and spectroscopic ellipsometry (SE) on InGaAs/InP epilayers and Si substrates. The dc current gain and breakdown voltage of InGaAs/InP heterostructure bipolar transistors (HBTs) have subsequently been used to evaluate the impact and efficiency of the ALD-Al_2O_3 passivation layers. The thermal-ALD-Al_2O_3 passivated InGaAs/InP HBTs show relatively higher current gains as compared to structures passivated using the plasma-ALD process, suggesting differences in the dielectric-semiconductor interface properties. The common emitter characteristics of both (thermal and plasma) ALD-Al_2O_3 passivated HBTs show, however, fairly comparable device breakdown voltages. These results will be contrasted with results from similar samples passivated with SiO_2 using conventional plasma enhanced chemical vapor deposition (PECVD).

Introduction

Due to the lack of thermodynamically stable insulators, the passivation of III-V semiconductor surfaces has always been a challenge and difficult to control during device and integrated circuit (IC) manufacturing. To improve device performance uniformity, reproducibility and reliability, various wet/dry passivation treatments have been developed to lower the interfacial density of states and reduce the leakage currents originating from surface/interface recombination [1-3].

In InP-based electronic and optoelectronic devices, the surface recombination velocity of InGaAs is low (as compared to GaAs), but it is still large enough to degrade the device performance. Moreover, although the surface recombination can be reduced at bare surfaces by chemical treatments, the device characteristics can be drastically degraded when encapsulating layers such as silicon oxides (SiO_x) or silicon nitrides (SiN_x) are deposited by high-density plasma processes [3-5]. Fukano et al. [4] attributed the increase of the surface recombination current to a generation of surface/interface states during the deposition of the dielectric layers. Ouacha et al. [5] have suggested that low-temperature deposition and the dielectric quality are critical parameters for optimum device performance.

Recently, atomic layer deposition (ALD) has attracted great interests for a wide range of applications [6-17]. Due to its unique features, ALD allows excellent conformal coverage of high-aspect ratio patterns, high quality materials with low defect densities, and growth of various materials with precise composition and thickness control. In particular, ALD-grown high permittivity (high-K) dielectric layers, used as gate dielectrics in metal oxide semiconductor field effect transistors (MOSFETs), were proven to be of higher quality as compared to gate oxides grown by conventional techniques [8,9]. The ALD process has also been reported to enable unpinning of the Fermi level on compound semiconductors [9,10]. However, in contrast to the well-established applications of ALD in Si-based technologies [11,12], there is still a lack of information on the integration and effectiveness of the ALD-based dielectrics in the manufacturing of III-V based electronic and optoelectronic components.

Among major oxide alternatives to silicon dioxide, Al_2O_3 has lately emerged as a robust solution and an important surface passivation material, enabling ultra-low surface recombination velocities on Si. The surface passivation by Al_2O_3 has been found to rely on a combination of chemical passivation and field-effect passivation provided by a high fixed negative charge density located at the semiconductor/Al_2O_3 interface [13,14]. Moreover, the high level of surface passivation based on Al_2O_3 has recently enabled increased efficiencies for silicon solar cells [15,16]. Furthermore, Al_2O_3 layers are also known for their excellent mechanical properties which makes them very useful as wear resistive coatings on micro-electromechanical systems (MEMS) [17].

In this contribution, we investigate the ALD deposition of Al_2O_3 passivation layers, and assess their impact and suitability on MBE (molecular beam epitaxy) grown InGaAs/InP heterostructures. The effects of ALD-based processes (thermal-ALD using H_2O as the oxidant and plasma-ALD using atomic oxygen generated by an O_2 plasma) on the device characteristics of InGaAs/InP heterojunction bipolar transistors (HBTs) have been evaluated. Moreover, results from these films have been contrasted with results using plasma enhanced chemical vapor deposited (PECVD) SiO_2.

Similar to PECVD-SiO_2, the Al_2O_3 passivated structures using a plasma-ALD process show relatively lower gains (β) as compared to uncoated devices. In contrast, InGaAs/InP HBTs passivated using an Al_2O_3 layer deposited by thermal-ALD show more stable gains, suggesting no generation of interface defects. On the other hand, the output characteristics of the PECVD-SiO_2 samples are drastically degraded after PECVD due to an increase in the base-collector leakage current. Both ALD-Al_2O_3 passivated HBTs show no marked degradation in the beakdown voltage. These results will be discussed and compared to unpassivated structures.

Experiment

The InGaAs epilayers (~50 nm, undoped) and InGaAs/InP HBT structures were grown in-house by solid source MBE on 3-inch (100) oriented semi-insulating InP substrates. The single HBT structure included an undoped InGaAs collector, a highly-doped InGaAs base layer, and an InP emitter (Table I). The n- and p-type dopants were Si and C, respectively.

TABLE I. Layer structure of InGaAs/InP HBTs.

Layer	Material	Thickness (nm)	Doping (cm^{-3})	Dopant
Cap	InGaAs	50	1E19	Si
	InP	40	1E19	Si
Emitter	InP	70	3E17	Si
Base	InGaAs	40	5E19	C
Collector	InGaAs	300	nid	-
Subcollector	InGaAs	300	1E19	Si
Substrate	InP		S.I	

To avoid any possible surface damage generally induced by dry etching processes, conventional mesa structure transistors (Figure 1) were fabricated using selective wet chemical etchants ($H_3PO_4:H_2O_2:H_2O$ and $H_3PO_4:HCl$) and lift-off processes. Non-alloyed TiPtAu metallization was used simultaneously for both n- and p-type ohmic contacts. To restrict the effects of the various parameters of the fabrication process, large area devices, which allow direct probing of the transistors, were fabricated. The emitter-base junction areas were typically in the range 50×100 to 100×100 μm^2. The HBT devices were subjected to the ALD or PECVD process only after the device fabrication was completed.

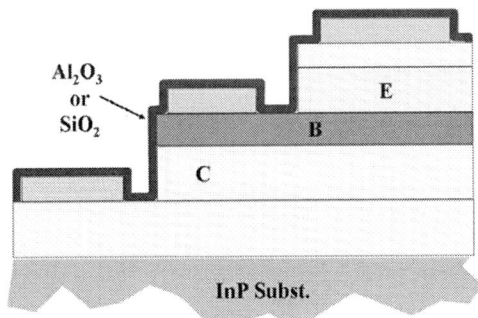

Figure 1. Schematic cross section of Al_2O_3 (or SiO_2) passivated HBTs.

In order to provide direct comparisons between processes on the same wafer, the passivation properties were evaluated on quarter wafers. Before SiO_2 or Al_2O_3 passivation, the InGaAs/InP epilayers and HBT samples, or Si calibration substrates, were dipped in a diluted ammonia solution for 30 seconds at room temperature then rinsed in water. After treatment, the samples were transferred immediately to deposition chambers for ALD or PECVD coating.

The Al_2O_3 thin films were deposited by thermal and plasma ALD in an Oxford Instruments FlexAL system at a substrate temperature of 300°C. The thin layers were deposited by sequential pulsing of trimethylaluminum (TMA) and an oxygen source (O_2 plasma or water vapour), separated by short nitrogen purges. The cycles (precursor-purge-reactant-purge) were repeated until the target film thickness was reached.

After deposition, the microstructural parameters and optical properties of the Al_2O_3 layers are examined by X-ray reflectivity (XRR) and spectroscopic ellipsometry (SE) on the InGaAs/InP epilayers. The PECVD-SiO_2 and ALD-Al_2O_3 (thermal and plasma) film thicknesses and the refractive index, n, were determined at wavelengths from 400 to 750 nm at room temperature using spectroscopic ellipsometry. All ellipsometric data were measured at an incident angle of 70° in air. The measured data were subsequently numerically fitted with a Levenberg-Marquardt regression method. The X-ray reflectivity measurements were carried out using a PANALYTICAL MRD diffraction / reflectometry system with $CuK\alpha_1$-radiation. The XRR specular $2\Theta/\Theta$-profiles were recorded with a parabolically bent graded multilayer mirror for the conditioning of the primary X-ray beam and a flat graphite analyzer before the detector.

Results and Discussion

Characterization of ALD Grown Al_2O_3 Layers

The ALD growth of Al_2O_3 using TMA and H_2O is certainly the most established and most commonly used reaction [6,8,18,19]. Recently, plasma-assisted ALD processes using O_2-plasma have also been developed to allow for deposition at reduced temperatures and increase the choice in chemistry and precursors [19-22].
Figure 2a shows the extracted ellipsometry data recorded from ALD-Al_2O_3 layers deposited on Si substrates using TMA in combination with water vapor (thermal-ALD). It is clear from Figure 2a that the film thickness increases linearly with the number of deposition cycles, resulting in a growth rate of about 0.7 Å/cycle. Hence, by selecting the appropriate number of ALD cycles, the desired film thickness can be precisely controlled.

(a)   (b)

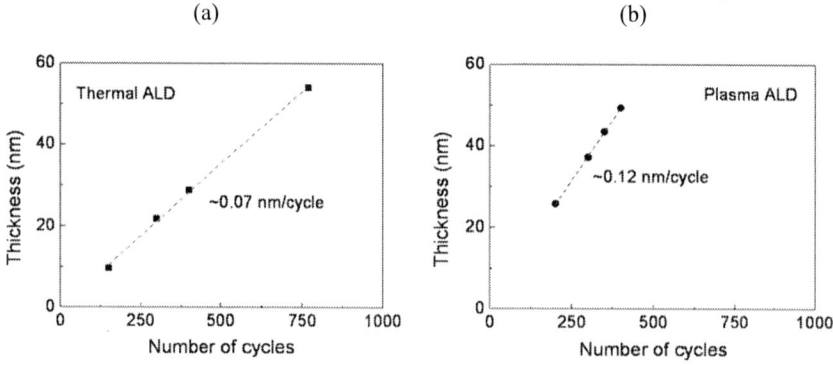

Figure 2. Layer thickness versus number of ALD cycles of Al_2O_3 layers deposited by (a) thermal-ALD and (b) plasma-ALD on Si substrates.

Similarly, the thickness variation as a function of the number of ALD cycles using TMA in combination with O_2 (plasma-ALD) is shown in Figure 2b. The growth rate extracted from the linear fit of these data is about 1.2 Å/cycle, which is in good agreement with literature data [19]. A consequence to the use of the O_2-plasma is the significant increase of the growth rate. The higher growth rate also translates into a reduction of the ALD deposition time, which results in a higher wafer throughput. This is particularly desirable in a production environment.

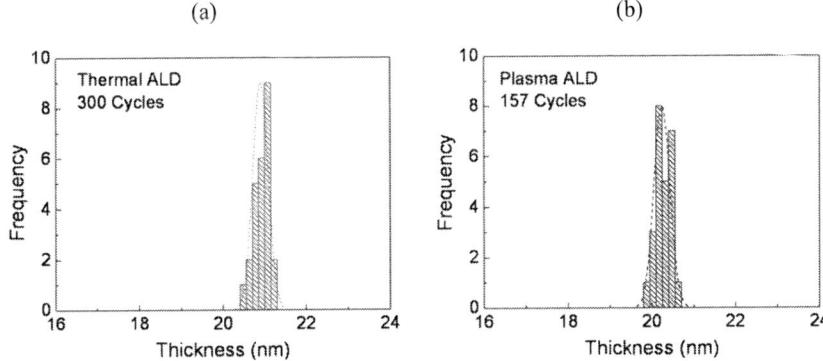

Figure 3. Thickness distribution of ~20 nm Al_2O_3 layers deposited by (a) thermal-ALD and (b) plasma-ALD on 4-inch Si substrates.

Figure 3 shows the Al_2O_3 film thickness distributions across 4-inch Si wafers, measured after thermal-ALD or plasma-ALD deposition, respectively. A mean value of 21 nm and a 3 sigma relative standard deviation of 0.6% were obtained for the Al_2O_3 deposited by thermal-ALD using 300 cycles. Similarly, a mean thickness value of 20 nm with a 3 sigma of 0.6% was measured on the Al_2O_3 deposited by plasma-ALD using only 157 cycles. It is clear from the data extracted from figure 3 that the uniformity is fairly comparable for both processes.

TABLE II. Thickness, refractive index and interlayer thickness from spectroscopic ellipsometry of ALD-Al_2O_3 and PECVD-SiO_2 layers on InGaAs/InP epilayers.

	Thickness (nm)	Refractive index n	InGaAs Thickness (nm)
Thermal-ALD-Al_2O_3	15	1.77	52.6
Plasma-ALD- Al_2O_3	16.5	1.75	53.5
PECVD-SiO_2	10.3	1.71	53.3

The properties of Al_2O_3 deposited by thermal-ALD and plasma-ALD were subsequently investigated on InGaAs/InP epilayers. The resulting Al_2O_3 layer thicknesses on InGaAs/InP structures were found to be similar to those obtained on Si substrates for both thermal-ALD and plasma-ALD, indicating no major substrate effects. Table II summarizes the spectroscopic ellipsometry data for a ~10 nm PECVD-SiO_2 and ~15 nm Al_2O_3 layers obtained using 120 cycles for plasma-ALD and 225 cycles for thermal-ALD.

(a)

(b)

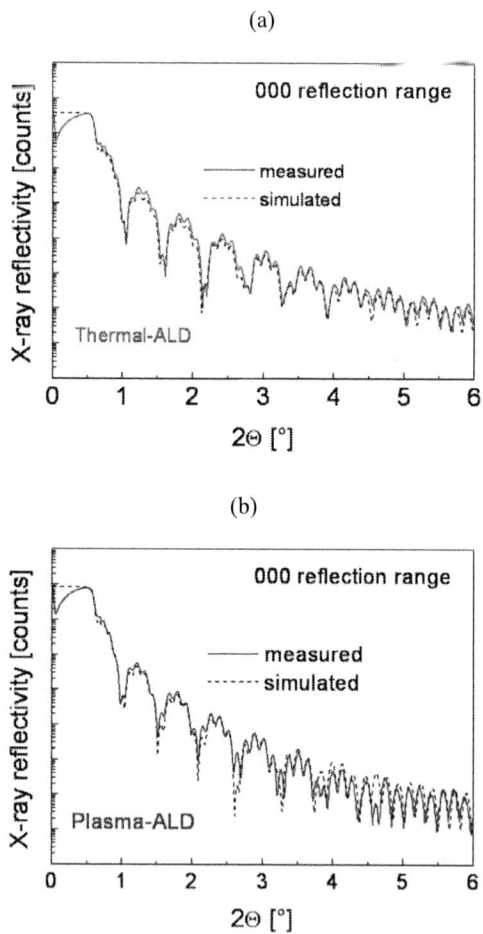

Figure 4. X-ray reflectivity profiles from Al_2O_3 layers deposited using (a) thermal-ALD and (b) plasma-ALD on InGaAs/InP samples.

The corresponding XRR measurements are shown in Figure 4 for the Al_2O_3 layers deposited by thermal-ALD and plasma-ALD, respectively. From the X-ray reflectivity profiles and a multilayer model, we extracted the thicknesses, the surface and interfacial layer roughnesses and the densities of the dielectric and InGaAs epilayers. The thickness of the film is measured from the periodicity of the fringes, the density from the angle at which the intensity begins to drop, and the roughness from the damping of the thickness fringes and the rate of intensity decrease with angle. In order to improve the quality of the fit, the model was refined by adding a 1 nm layer at the dielectric-semiconductor interface.

As can be seen from Figure 4a and 4b, the simulated curves are in good agreement with the experimental ones. Moreover, there are no significant differences in the two XRR profiles. The slight differences in the oscillation forms might be related to a small difference in Al_2O_3 film thickness ($\sim$ 1nm) or to a difference in the interface composition. A summary of the fitted parameters is reported in Table III.

TABLE III. Thickness, density, surface/interface roughness from XRR of thermal-ALD-Al_2O_3 and plasma-ALD-Al_2O_3 deposited on InGaAs/InP epilayers.

	Al_2O_3			InGaAs	
	Thickness	Roughness	Density	Thickness	Roughness
	(nm)	(nm)	(g/cm3)	(nm)	(nm)
Thermal-ALD	13.9	0.6	3.1	55	0.2
Plasma-ALD	15.3	0.7	3.1	55	0.2

According to Table III, both techniques (thermal-ALD and plasma-ALD) result in fairly comparable values. The XRR fitted dielectric thicknesses are fairly similar to the values previously measured by ellipsometry. Moreover, the calculated film densities of 3.1 g/cm^3 for both thermal-ALD-Al_2O_3 and plasma-ALD-Al_2O_3 films are in agreement with values reported in literature [19,20].

Device Results

PECVD-SiO_2 passivation. As mentioned previously, treated III-V semiconductor surfaces are usually encapsulated with dielectric films using CVD-based techniques to improve their stability. Although novel oxides have been developed, SiO_2 and SiN_x, remain the dielectrics of choice for most applications in the microelectronic industry.

The effects of the PECVD process are shown by comparing the electrical characteristics of InGaAs/InP HBTs before and after deposition. Figure 5 shows typical Gummel plots (collector and base currents as a function of the base–emitter voltage) of large area (S_{EB} = 100x100μm^2) InGaAs/InP HBTs before and after PECVD deposition of a $\sim$10 nm SiO_2 passivation layer. While the collector current is almost unaltered by the PECVD deposition, the base current increases drastically. This results in a degradation of the current gain, which reduced after iterative measurements (transient effect) or just with time.

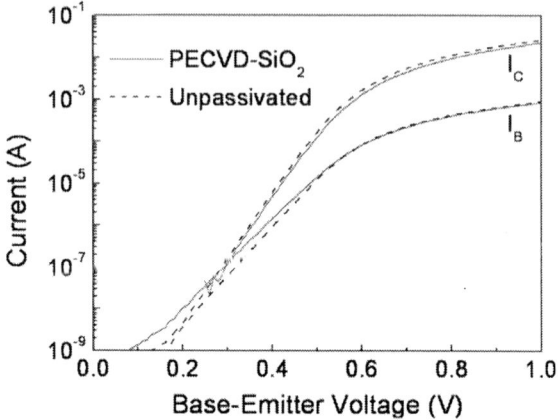

Figure 5. Gummel plots of InGaAs/InP HBTs before and after PECVD-SiO$_2$.

In addition of affecting the current gain, we found that the PECVD process also degrades other parts of the device and consequently transistor characteristics. Figure 6 shows common-emitter characteristics of large area InGaAs/InP HBTs before and after PECVD deposition. The untreated device has a breakdown voltage of about 6 V. After PECVD coating, this decreases to less than 2 V, which is not suitable for major applications. This reduction was associated with an increase in the base-collector leakage current, which might be attributed to an increase of the generation rate. The origin of the degradation could be related to an incorporation of fixed charges in the SiO$_2$ layer or at the dielectric/semiconductor interface [23,24]. Some of the charges can, however, be dissipated in a burn-in process or just with time.

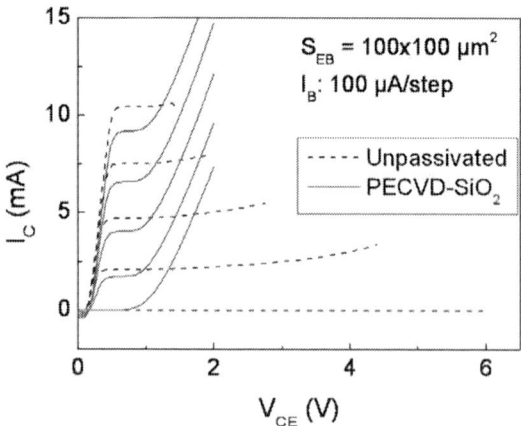

Figure 6. Output characteristics of InGaAs/InP HBTs before and after PECVD-SiO$_2$.

ALD-Al$_2$O$_3$ passivation. Figure 7 shows the Gummel plots for the InGaAs/InP HBTs passivated with thermal-ALD-Al$_2$O$_3$ and plasma-ALD-Al$_2$O$_3$. The characteristics of the reference device (before passivation) are also included for comparison. As shown in Figure 7, no marked differences are observed in the collector currents. In contrast, slight differences in the base current can clearly be noticed. The base current of the thermal-ALD-Al$_2$O$_3$ passivated device is lower than the base current of the plasma-ALD-Al$_2$O$_3$ and fairly comparable to that of the uncoated device.

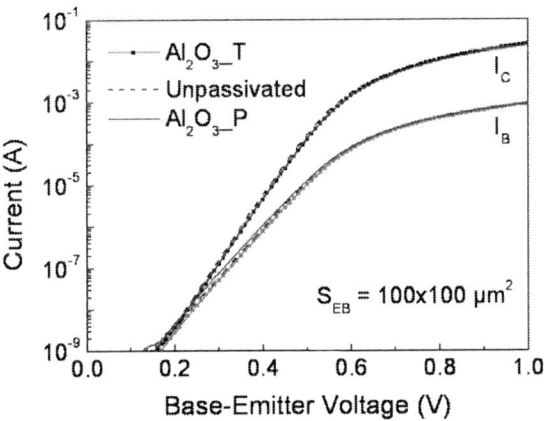

Figure 7. Typical Gummel plots of InGaAs/InP HBTs before and after plasma-ALD-Al$_2$O$_3$ or thermal-ALD-Al$_2$O$_3$, respectively.

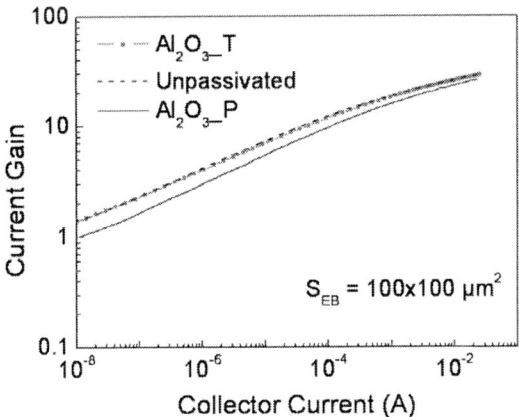

Figure 8. Current gain as a function of collector current of InGaAs/InP HBTs before and after plasma-ALD-Al$_2$O$_3$ or thermal-ALD-Al$_2$O$_3$, respectively.

Figure 8 shows the corresponding variation of the current gain as a function of collector current (Ic) for the same devices before and after plasma-ALD-Al_2O_3 and thermal-ALD-Al_2O_3 passivation, respectively. Obviously, the thermal-ALD process does not affect the current gain over the full range of collector currents. In contrast, the plasma-ALD reduces the current gain, especially at low collector current levels. This is attributed mainly to the increase of the base current, as shown in figure 7. A possible reason for the increase in base leakage current is the activation of defects in the extrinsic base region or at the emitter mesa intersection with the extrinsic base surface. The defects could have also been created during plasma-ALD-Al_2O_3 passivation.

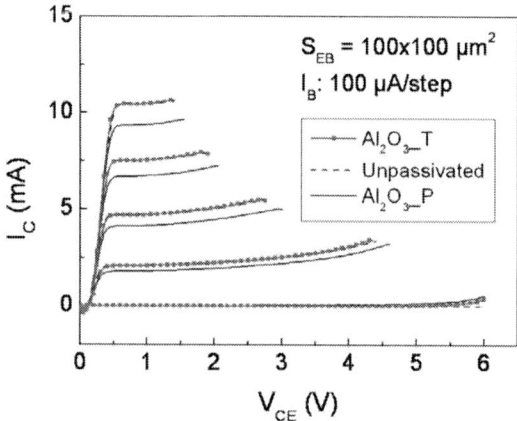

Figure 9. Output characteristics of InGaAs/InP HBTs before and after plasma-ALD-Al_2O_3 or thermal-ALD-Al_2O_3, respectively.

The output characteristics for the same type of transistors are shown in Figure 9. The figure shows again that the thermal-ALD-Al_2O_3 process produces higher collector currents than the plasma-ALD-Al_2O_3 process. Nevertheless, compared to the reference device, fairly similar breakdown and saturation voltages are obtained for both thermal-ALD-Al_2O_3 and plasma-ALD-Al_2O_3 processes. A high and stable breakdown voltage is particularly desirable for analog and power applications.

These results indicate that the conventional PECVD-SiO_2 induces unstable leakage currents in both emitter-base and base-collector junctions. The plasma-ALD-Al_2O_3 process also increases the base leakage current while maintaining a fairly comparable breakdown voltage to untreated devices. In contrast, the thermal-ALD-Al_2O_3 process combines the advantage of maintaining a relatively high breakdown voltage with a stable current gain.

The contrast between the changes of gain and breakdown voltage in the ALD processes might be due to different nucleation mechanisms between the plasma and the thermal methods, which might result in different qualities of oxides at the interface. Further investigation is required to confirm this assumption.

Conclusion

In summary, we have investigated and compared the properties of thermal-ALD-Al_2O_3 and plasma-ALD-Al_2O_3 passivation layers on InGaAs/InP heterostructures using X-ray reflectivity (XRR) and spectroscopic ellipsometry (SE) techniques. Both deposition methods have been shown to be suitable for implementation of high-K dielectrics in III-V semiconductor minority carrier devices. As compared to thermal-ALD, the remote plasma-ALD-Al_2O_3 films exhibited higher growth rates (in excess of 1 Å/cycle) with fairly comparable film properties (surface roughness, mass density and dielectric constants).

To evaluate the potential of these layers for III-V-based applications, leakage current and breakdown voltage variations of ALD-Al_2O_3 passivated InGaAs/InP HBTs were investigated. For comparison, similar structures have also been coated with a thin SiO_2 layer deposited by conventional PECVD. As expected, the PECVD process caused a significant degradation of the breakdown voltage due to an increase in the leakage current of the base-collector junction. In contrast, the breakdown voltage is almost unchanged in ALD-Al_2O_3 passivated structures. The breakdown voltage stability is particularly critical for analog and power devices. As compared to the plasma-ALD-Al_2O_3 process, the thermal-ALD-Al_2O_3 process yields lower leakage currents indicating no additional generation or activation of interface defects. The differences in the electrical behavior of thermal-ALD and plasma-ALD-Al_2O_3 passivated structures might be attributed to differences in the nucleation mechanisms. Further investigation is, however, required to sort out possible effects of the deposition parameters. Nevertheless, compared to conventional CVD-based techniques, these results confirm the potential of combining the ALD process and Al_2O_3 layers as a robust solution for the surface passivation of III-V semiconductor based heterostructure devices.

Acknowledgements

The authors would like to acknowledge the staff of the technology department for their help in wafer processing and M. Prescher for his help in XRR measurements.

References

1. M. Hong, C. T. Liu, H. Reese, and J. Kwo, in *"Encyclopedia of Electrical and Electronics Engineering"*, J. G. Webster, Ed., **19**, p.87, John Wiley & Sons, New York (1999).
2. R. Driad, Z. H. Lu, S. Charbonneau, W. R. McKinnon, S. Laframboise, P. J. Poole, and S. P. McAlister, *Appl. Phys. Lett.*, **73**, 665 (1998).
3. R. Driad, S. R. Laframboise, Z. H. Lu, S. P. McAlister, and W. R. McKinnon, *Solid-State Electronics*, **43**, 1445 (1999).
4. H. Fukano, Y. Takanashi, and M. Fujimoto, *Jpn. J. Appl. Phys.*, **32**, L1788 (1993).
5. A. Ouacha, M. Willander, B. Hammarlund, and R. A. Logan, *J. Appl. Phys.*, **74**, 5602 (1993).

6. R. L. Puurunen, *J. Appl. Phys.*, **97**, 121301 (2005).
7. T. Suntola, *Appl. Surf. Science*, **100**, 391 (1996).
8. P. D. Ye, G. D. Wilk, J. Kwo, B. Yang, H.-J. L. Gossmann, M. Frei, S. N. G. Chu, J. P. Mannaerts, M. Sergent, M. Hong, K. K. Ng, and J. Bude, *IEEE Electron Dev. Lett.*, **24**, 209 (2003).
9. Y. Xuan Y. Q. Wu, H. C. Lin, T. Shen, and P. D. Ye, *IEEE Electron Dev. Lett.*, **28**, 935 (2007).
10. M. L. Huang, Y. C. Chang, C. H. Chang, Y. J. Lee, P. Chang, J. Kwo, T. B. Wu, and M. Hong, *Appl. Phys. Lett.*, **87**, 252104 (2005).
11. K. H. Lee, S. J. Lee, S. H Kim, I. S. Dae., B. C. Cho. J. W. Lee, S. J. Park, S. K. Lee, and T. W. Seo, *ECS Trans.*, **3**, 3 (2007).
12. E. P. Gusev, C. Cabral Jr., M. Copel, C. D'Emic, and M. Gribelyuk, *Microelectronic Eng.*, **69**, 145 (2003)
13. B. Hoex, J. Schmidt, R. Bock, P. P. Altermatt, M. C. M. van de Sanden, and W. M. M. Kessels, *Appl. Phys. Lett.*, **91**, 112107 (2007).
14. B. Hoex, J. Schmidt, P. Pohl, M. C. M. van de Sanden, and W. M. M. Kessels, *J. Appl. Phys.*, **104**, 044903 (2008).
15. J. Benick, B. Hoex, M. C. M. van de Sanden, W. M. M. Kessels, O. Schultz, and S. W. Glunz, *Appl. Phys. Lett.*, **92**, 253504 (2008).
16. P. Saint-Cast, J. Benick, D. Kania, L. Weiss, M. Hofmann, J. Rentsch, R. Preu, and S. W. Glunz, *IEEE Electron Dev. Lett.*, **31**, 695 (2010).
17. T. M. Mayer, J. W. Elam, S. M. George, and P. G. Kotula, *Appl. Phys. Lett.*, **82**, 2883 (2003).
18. M. D. Groner, J. W. Elam, F. H. Fabreguette, and S. M. George, *Thin Solid Films*, **413**, 186 (2002).
19. J. L. van Hemmen, S. B. S. Heil, J. H. Klootwijk, F. Roozeboom, C. J. Hodson, M. C. M. van de Sanden, and W. M. M. Kesselsa, *J. Electrochemical Society.*, **154**, G165 (2007).
20. C. Detavernier, J. Dendooven, D. Deduytsche, and J. Musschoot, *ECS Trans.*, **16**, 239 (2008).
21. J-W. Lim, S-J. Yun, Y-H. Kim, C-Y. Sohn, and J-H. Lee, *Electrochem. Solid-State Lett.*, **7**, G185 (2004).
22. A. Niskanen, K. Arstila, M. Ritala, and M: Leskelä, *J. Electrochem. Soc.*, **152**, F90 (2005).
23. Z. Jin, W. Prost, S. Neumann, and F. J. Tegude, *J. Appl. Phys.*, **96**, 777 (2004).
24. H. Wang, G. I. Ng, H. Yang, and K. Radhakrishnan, *Jpn. J. Appl. Phys.*, **41**, 1059 (2002).

Nitric Acid Oxidation to Form a Gate Oxide Layer in Sub-Micrometer TFT

T. Matsumoto[a,b], Y. Kubota[c], S. Imai[b,d] and H. Kobayashi[a,b]

[a] ISIR, Osaka University, Osaka 567-0047, Japan
[b] CREST, Japan Scientific and Technology Agency, Japan
[c] Liquid Crystal Display Group, Sharp Corporation and CREST, Japan Science and Technology Agency, Taki, Mie 519-2192, Japan.
[d] Display Technology Development Group, Sharp Corporation, Nara 632-8567, Japan

We have succeeded in fabrication of sub-micrometer TFT which can be operated at 1.5 V, having stack gate oxide structure formed by the nitric acid oxidation of Si (NAOS) method. The ultrathin NAOS SiO_2 layer possesses a low leakage current density, and consequently, the gate oxide layer afterward deposited by the CVD method can be made as thin as 20 nm. Because of the thin gate oxide, miniaturization of TFT with the 0.6 μm gate length is achieved. The operation voltage of the TFTs can be set at as low as 1.5 V because of the low threshold voltages (i.e., −0.5 V for P-ch TFT and 0.5 V for N-ch TFT). The sub-threshold swing value is ~80 mV/decade and the on/off ratio is ~10^9 for both the P-ch and N-ch TFTs. The channel mobility is ~100 $cm^2/V \cdot s$ for the P-ch TFT and ~200 $cm^2/V \cdot s$ for the N-ch TFT.

Introduction

Low power consumption is the most important issue for the thin film transistors (TFTs) in the mobile electronic devices, which requires extremely a low leakage current. Gate insulators in these TFTs are usually produced by deposition methods such as chemical vapor deposition (CVD) [1-4] because of use of glass substrates which do not allow high temperature processes. However, in contrast to thermal oxidation which requires high temperature heating above 800 °C, the deposition methods cannot form a SiO_2 layer with good electrical characteristics. The poor electrical characteristics are partly due to poor interfacial characteristics because of incomplete interfacial bond formation and presence of contaminants, leading to high interface state densities [5,6]. Bulk characteristics of a deposited SiO_2 layer is also poor due to porous structure and inclusion of undesirable species such as carbon and water [7,8]. Moreover, deposition methods cannot form a uniform thickness SiO_2 layer especially on a rough poly-polycrystalline Si (poly-Si) surface with ridge structure arising from laser annealing to crystallize amorphous Si (a-Si) [9].

In the case of direct oxidation methods. on the other hand, an SiO_2/Si interface is formed in the initial Si bulk, and therefore initial contamination on Si surfaces does not seriously degrade the SiO_2 characteristics. Moreover, the density of interface states for direct oxidation is much lower than that for deposition methods because of nearly complete interfacial bond formation. The density of a SiO_2 layer formed by direct oxidation methods is higher than that for deposition methods [10], leading to better electrical characteristics. Direct oxidation methods can form a uniform thickness SiO_2 layer even on rough Si surfaces such as poly-Si thin films [11] if the oxidation rate is independent of surface orientations.

Low temperature direct oxidation methods developed so far include plasma oxidation [12,13], photo-oxidation [14,15], ozone oxidation [16,17], metal-promoted oxidation [18,19], etc. For a direct chemical oxidation method, we have developed the nitric acid oxidation of Si (NAOS) method. This method utilizes the strong oxidizing activity of nitric acid (HNO_3) which generates oxygen atoms and/or dissociated oxygen ions (O^-) by its decomposition. Simply by immersion of Si in high concentration (i.e., 68~98wt%) HNO_3 aqueous solutions, an ultrathin (i.e., 1.2~1.4 nm) SiO_2 layer with a leakage current density much lower than that of a thermal SiO_2 layer with the same thickness can be formed [20-24].

In the present study, stack gate oxide structure, i.e., the ultrathin NAOS SiO_2 layer/CVD SiO_2 layer, has been fabricated. The NAOS SiO_2 layer effectively blocks a leakage current flowing through the gate oxide layer, and hence the total thickness of the gate oxide layer can be decreased to 20 nm. The thin gate oxide layer makes it possible to fabricate sub-micrometer TFTs with the operation voltage of 1.5 V.

Nitric Acid Oxidation of Si (NAOS) Method on Si surfaces

NAOS SiO_2 on n-Si(100) in Azeotropic Nitric Acid Aqueous Solution at 120°C

Figure 1 shows the I-V curves for the <Al/~1.4 nm NAOS SiO_2/n-Si(100)> MOS diodes with the SiO_2 layer formed in azeotropic (i.e., 68wt%) HNO_3 aqueous solutions at 120°C for 10 min. The thickness of the oxide layer was determined by X-ray photoelectron spectroscopy (XPS) and transmission electron microscopy (TEM)

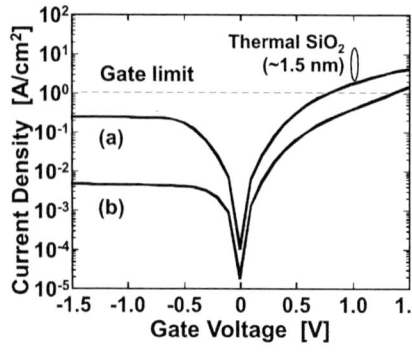 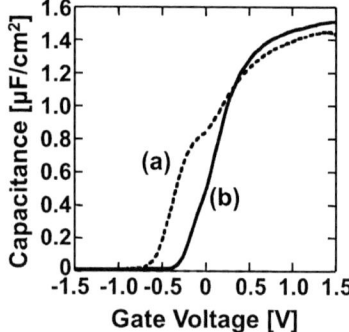

Figure 1. I-V curves for the <Al/~1.4 nm NAOS SiO_2/n-Si(100)> MOS diodes with the SiO_2 layer formed in 68wt% HNO_3 aqueous solutions at 120°C for 10 min: a) with no PMA; b) with PMA at 200°C.

Figure 2. C-V curves for the <Al/~1.4 nm NAOS SiO_2/n-Si(100)> MOS diodes with the SiO_2 layer formed in 68wt% HNO_3 at 120°C for 10 min: a) with no PMA; b) with PMA at 200°C.

measurements. Even with no treatment (curve a), the leakage current density flowing through the NAOS SiO_2 layer was lower than that of SiO_2 thermally grown at 900°C with the same thickness [25-28]. The leakage current density further decreased approximately by one order of magnitude after post-metallization annealing (PMA) in H_2 at 200°C for 20 min (curve b).

Figure 2 shows the C-V curves for the <Al/~1.4 nm NAOS SiO_2/p-Si(100)> MOS diodes with the SiO_2 layer formed in 68wt% HNO_3 at 120°C for 10 min. Even with no PMA (curve a), a C–V curve could be measured because of the low leakage current density (cf. Figure 1). Although the C–V curve had a typical high-frequency structure with a saturation capacitance in the accumulation condition, a hump was present in the C–V curve, indicating the presence of high-density interface states. This hump disappeared after PMA at 200 °C (curve b), showing elimination of the interface states. We think that passivation of interface states is one of the important reasons for the decrease in the leakage current density because a leakage current can flow via interface states [28].

<u>NAOS SiO_2 on Si(100) in Azeotropic Nitric Acid Aqueous Solution at Room Temperature</u>

The room temperature NAOS method is very simple, i.e., immersion of Si in HNO_3 aqueous solutions at room temperature, and it is found that it causes no contamination from the glass substrates. We have fabricated TFTs with the room temperature NAOS SiO_2/CVD SiO_2 stack gate oxide structure in the present study.

Figure 3 shows an XPS spectrum in the Si 2p region for the SiO_2/Si structure formed by immersion of p-Si(100) wafers in 68wt% HNO_3 aqueous solutions at room temperature for 10 min followed by PMA in 5% H_2 at 200°C for 10 min. The doublet

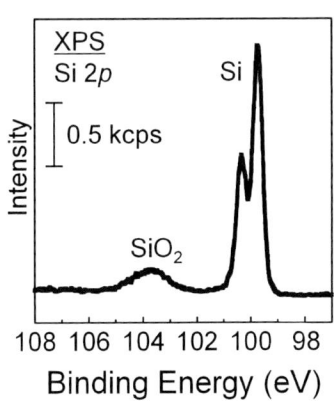

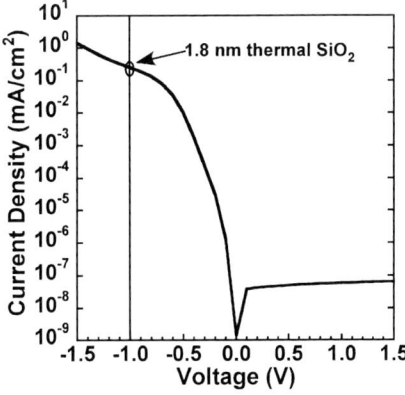

Figure 3. XPS spectrum in the Si 2p region for the SiO_2/poly-Si thin film structure formed in 68wt% HNO_3 aqueous solutions at room temperature.

Figure 4. I-V curves for the <Al/1.8 nm SiO_2/p-Si(100)> MOS diodes with the SiO_2 layer formed in 68wt% HNO_3 aqueous solutions at room temperature followed by with PMA at 200°C.

peaks were attributable to Si $2p_{3/2}$ and $2p_{1/2}$ levels of the substrate and the broad peak in the higher energy region was due to an SiO_2 layer. The thickness of the SiO_2 layer was estimated to be 1.8 nm from the ratio in the area intensity between these peaks [29,30]. In the estimation, the values of 2.7 and 3.2 nm were adopted for the electron mean free paths in SiO_2 and Si, respectively [28].

Figure 4 shows the I-V curves for the <Al/1.8 nm SiO_2/Si(100)> MOS diodes in which the SiO_2 layer was formed in 68wt% HNO_3 at room temperature. The leakage current density was as low as that of SiO_2 thermally grown at 900°C with the same thickness [27-29].

TFTs with Stacked Gate Oxide Fabricated by the NAOS Method

TFT Fabrication

TFTs were fabricated on non-alkali metal glass substrates. A 100 nm thick silicon nitride layer was deposited on the glass substrates to prevent diffusion of undesirable species from the glass substrates. Then, 50 nm thick a-Si films were deposited by use of the plasma-enhanced chemical vapor deposition (PECVD) method. The a-Si films were irradiated using an excimer laser in order to crystallize. On the poly-crystalline Si (poly-Si) film, an ultrathin SiO_2 layer was formed by immersion in 68wt% HNO_3 aqueous solutions (i.e., azeotropic mixture of HNO_3 and water) at room temperature. On the NAOS SiO_2 layer, an SiO_2 layer of ~40 and ~20 nm [31] thickness was deposited using the PECVD method. Tungsten gate electrodes were fabricated, and source and drain regions were produced by implanting boron ions for P-ch TFT and phosphorus ions for N-ch TFT. The TFTs without the NAOS interfacial layer could not be operated mainly because of a high leakage current density due to the thin gate oxide layer.

TFTs with 40 nm Gate Oxide

Figure 5 shows the threshold voltage, Vth, of the P-ch and N-ch TFTs vs. the channel length, L. The threshold voltage was approximately -1 V for the P-ch TFT (Fig. 5a) and 0.7 V for the N-ch TFT (Fig. 5b). The low Vth values made it possible for the TFTs to be operated at 2.5 V, as described below. In the sub-micron channel length region, the threshold voltage decreased most probably due to the short channel effect [32]. For the TFT with an 80 nm CVD gate oxide layer, on the other hand, the threshold voltage was as high as 3 V, and the short channel effect was observed in the channel length region shorter than 3 μm [33]. This result clearly shows that by use of the thinner gate oxide layer, miniaturization of TFT is possible because of avoidance of the short-channel effect [34].

Figure 6 shows the drain current vs. the source-drain voltage (I_d-V_{ds}) characteristics for the P-ch and N-ch TFTs with the room temperature NAOS SiO_2/40 nm CVD SiO_2 stack gate oxide structure. The curves possessed ideal features with high saturation current densities. The I_d-V_{ds} curves showed saturation current behavior with a sufficiently high density even at 2.5 V, indicating that the TFTs could be operated at 2.5 V. Since the TFT power consumption, P, is given by [35]

$$P = fCV^2 \tag{1}$$

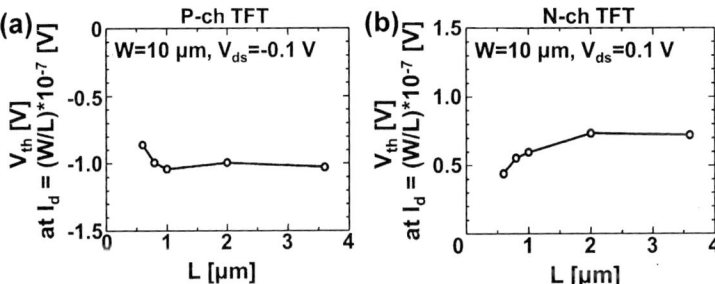

Figure 5. Threshold voltage vs. the channel length for the TFTs with the 1.8 nm NAOS SiO$_2$/40 nm CVD SiO$_2$ stack gate oxide structure: a) P-ch TFT; b) N-ch. TFT.

where f is the signal frequency, C is the charging and discharging equivalent capacitance at the gate of the load TFTs, and V is the operation voltage, P is decreased to $(3/12)^2=1/16$ of that for currently commercial TFTs with the operation voltage of 12 V.

Figure 7 shows the sub-threshold swing (S) values of the P-ch and N-ch TFTs with the room temperature NAOS SiO$_2$/40 nm CVD SiO$_2$ stack gate oxide structure. For the TFTs with no NAOS SiO$_2$ layer and with an 80 nm CVD oxide layer, the S value was ~200 mV [33]. For the P-ch (plot a) and N-ch (plot b) TFTs with the stack gate oxide structure, on the other hand, the S value was ~100 mV/dec. The decrease in the S value is attributable to the reduced gate oxide thickness and the high quality Si/SiO$_2$ interface.

Figure 8 shows the channel mobility of the fabricated TFTs. The mobility was ~100 cm^2/V·s for the P-ch TFTs and ~200 cm^2/V·s for the N-ch TFTs, which could be considered to be sufficiently high for poly-Si-based TFTs. The difference between the P-ch and N-ch mobilities resulted from the higher electron mobility (i.e., 1500 cm^2/V·s) than the hole mobility (i.e., 450 cm^2/V·s) [36].

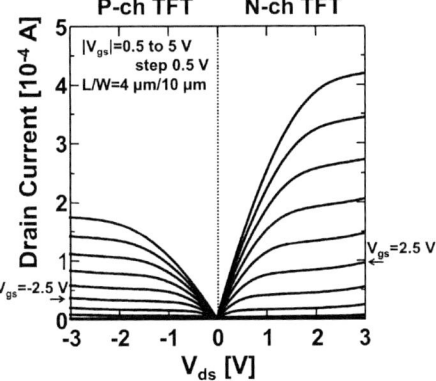

Figure 6. I$_d$-V$_{ds}$ curves for the P-ch (left) and N-ch (right) TFTs with the 1.8 nm room temperature NAOS SiO$_2$/40 nm CVD SiO$_2$ stack gate oxide structure having the gate length of 4 μm.

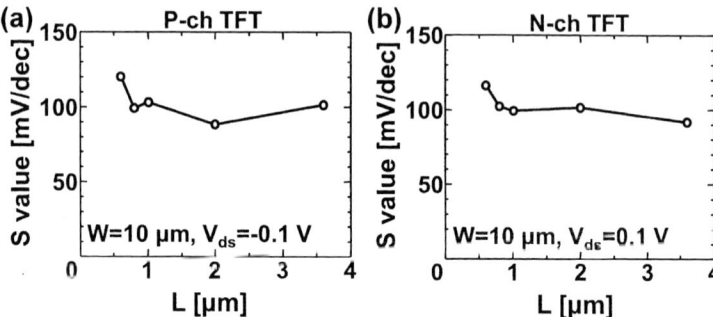

Figure 7. Sub-threshold swing value vs. the gate length for the TFTs with the 1.8 nm room temperature NAOS SiO$_2$/40 nm CVD SiO$_2$ stack gate oxide structure: a) P-ch TFT; b) N-ch TFT.

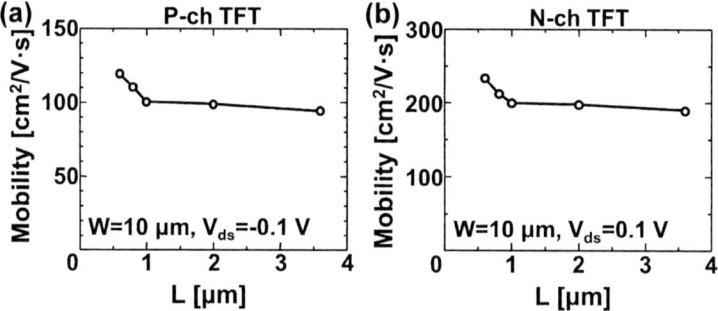

Figure 8. Channel mobility vs. the gate length for the TFTs with the 1.8 nm room temperature NAOS SiO$_2$/40 nm CVD SiO$_2$ stack gate oxide structure: a) P-ch TFT; b) N-ch TFT.

TFTs with 20 nm Gate Oxide

Figure 9 shows the threshold voltage, Vth, of the TFTs with the room temperature NAOS SiO$_2$/20 nm CVD SiO$_2$ stack gate oxide structure vs. the gate length. For both the P-ch and N-ch TFTs, Vth did not strongly depend on the gate length, and it was it was approximately −0.6 for the P-ch TFT (Fig. 9a) and 0.6 V for the N-ch TFTs (Fig. 9b). Because of the thin gate oxide layer, no short channel effect [32] was observed even in the sub-micrometer gate length region [36]. The low threshold voltage enables the low voltage operation of TFT, as described below.

Figure 10 shows the drain current vs. the source-drain voltage (I$_d$-V$_{ds}$) curves for the TFTs with the 1.8 nm room temperature NAOS SiO$_2$/20 nm CVD SiO$_2$ stack gate oxide structure. The I$_d$-V$_{ds}$ curves possessed saturation behavior. The saturation current density increased with a decrease in the channel length, i.e., miniaturization increased the saturation current. In the case of the 0.9 (Fig. 10c) and 0.6 μm (Fig. 10 d) channel length, the saturation current densities were sufficiently high even at the operation voltage of 1.5

V, indicating that the TFTs could be driven at this voltage. Because power consumption, P, is proportional to the square of the operation voltage (cf. Eq. (1)), it can be concluded that P for the fabricated sub-micron TFTs is only $(1.5/12)^2 = 1/64$ of the commercial TFTs driven at 12 V [35]. It should be noted that the ultra-low power TFTs could be achieved by the low leakage current characteristic of the NAOS SiO_2 layer.

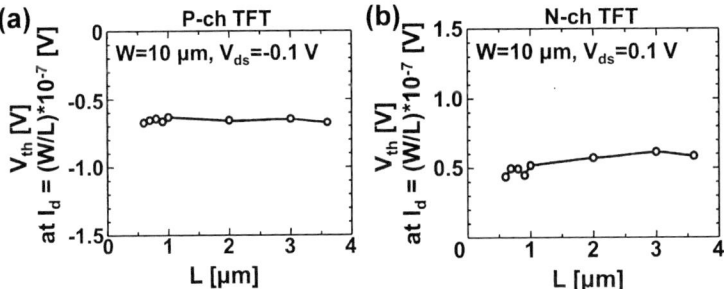

Figure 9. Threshold voltage vs. the gate length for the TFTs with the 1.8 nm room temperature NAOS SiO_2/20 nm CVD SiO_2 stack gate oxide structure: a) P-ch TFT; b) N-ch TFT.

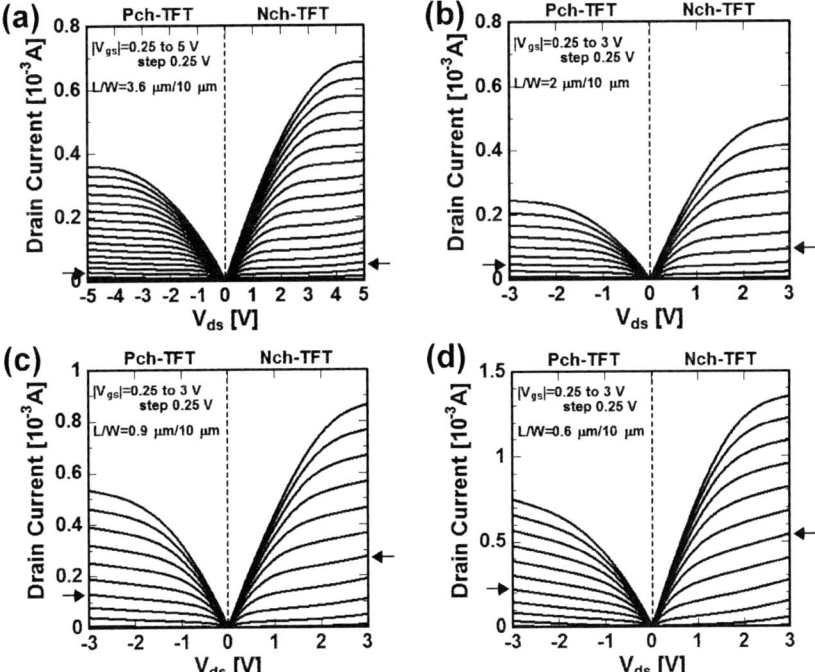

Figure 10. I_d-V_{ds} curves for the P-ch (left) and N-ch (right) TFTs with the 1.8 nm room temperature NAOS SiO_2/20 nm CVD SiO_2 stack gate oxide structure having following gate length: a) 4 μm; b) 2 μm; c) 0.9 μm, d) 0.6μm.

Figure 11 shows the drain current vs. the gate voltage (I_d-V_g) curves for the TFTs with the room temperature NAOS/20 nm CVD SiO_2 stack gate oxide structure. In spite of the thin gate oxide layer, the off-current was sufficiently low, i.e., in the range between 10^{-13} and 10^{-14} A. The low off-current resulted from effective block of the leakage current by the NAOS SiO_2 layer. Due to the low off-current, the on/off ratio of thefabricated TFTs was 10^9 which was higher by about two orders of magnitude than that of conventional TFTs. The drain current increased steeply with the gate voltage, indicating the low S values as described below.

Figure 12 shows the S values for the room temperature NAOS/20 nm CVD SiO_2 stack gate oxide structure. For both the P-ch and N-ch TFTs, the S values were in the range between 70 and 80 mV/dec, which was close to the theoretical value of 60 mV/dec [33,37,38]. The nearly ideal S values resulted from the thin gate oxide layer and low interface state density, leading to a low net bias voltage across the gate oxide layer.

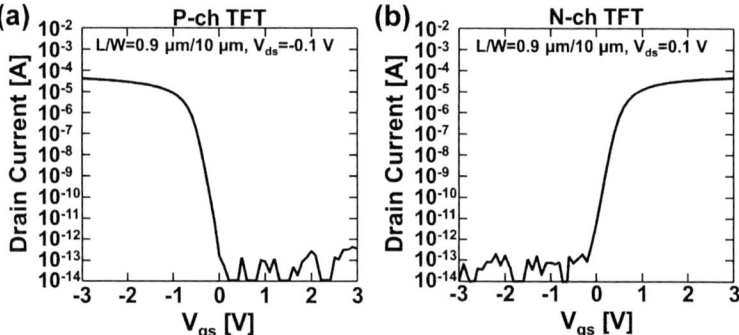

Figure 11. I_d-V_g curves for the TFTs with the 1.8 nm NAOS SiO_2/20 nm CVD SiO_2 stack gate oxide structure: a) P-ch TFT; b) N-ch TFT.

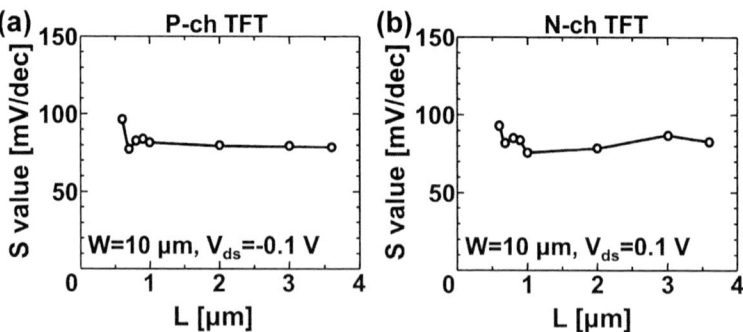

Figure 12. Sub-threshold swing value vs. the gate length for the TFTs with the 1.8 nm room temperature NAOS SiO_2/20 nm CVD SiO_2 stack gate oxide structure: a) P-ch TFT; b) N-ch. TFT.

Figure 13 shows the channel mobility of the fabricated TFTs with the room temperature NAOS SiO_2/20 nm CVD SiO_2 stack gate oxide structure. The mobility did not strongly depend on the channel length, and that for the P-ch (Fig. 12 a) and N-ch (Fig. 12b) was approximately 100 and 200 cm^2/V·s, respectively, which was nearly the same as that for the TFTs with the 40 nm gate oxide thickness (Fig. 8). Therefore, it is highly probable that the mobility is mainly determined by the quality of poly-Si thin films because of the sufficiently good gate oxide and interface characteristics which do not decrease the mobility.

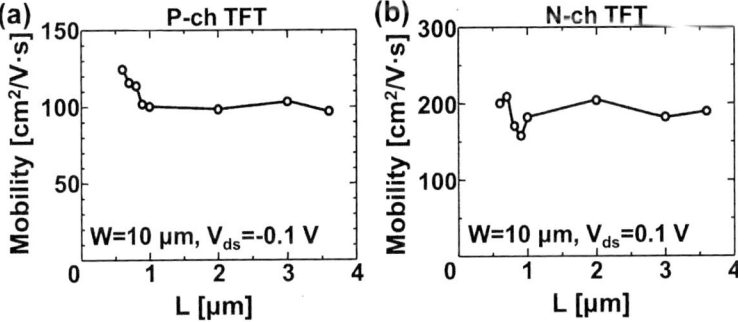

Figure 13. Channel mobility vs. the gate length for the TFTs with the 1.8 nm room temperature NAOS SiO_2/20 nm CVD SiO_2 stack gate oxide structure: a) P-ch TFT; b) N-ch TFT.

Conclusion

Immersion of Si in 68 wt% HNO_3 aqueous solutions at room temperature, i.e., the room temperature NAOS method, forms a 1.8 nm SiO_2 layer with the leakage current density as low as that for a thermal SiO_2 layer formed at 900°C with the same thickness. Using this low leakage current characteristic, the TFTs with the 40 nm and 20 nm gate oxide thickness could be fabricated without increasing the off-current. The threshold voltages for the P-ch and N-ch TFTs were both sufficiently low, i.e., −0.5 V for the P-ch TFT and 0.5 V for the N-ch TFT. Due to the low threshold voltages, both the P-ch and N-ch TFTs could be operated at 1.5 V with sufficiently high saturation currents (i.e., 2.0×10^{-5} A/μm at V_{ds}=−0.1 V for the P-ch TFT and 4.4×10^{-5} A/μm for the N-ch TFT), resulting in the power consumption of 1/64 of that for commercial TFTs driven at 12 V. The low off-current of $10^{-13} \sim 10^{-14}$ A was achieved, leading to the high on/off ratio of $\sim 10^{9}$. The S-values for both the P-ch and N-ch TFTs were in the range between 70 and 80 mV/dec, due to the thin gate oxide layer and excellent interfacial characteristics. The channel mobility was ~ 100 cm^2/V·s for the P-ch TFT and ~ 200 cm^2/V·s for the N-ch TFT. A short channel effect did not appear for both the P-ch and N-ch TFTs even with sub-micrometer gate length.

Acknowledgments

The authors would like to thank SEL (Semiconductor Energy Laboratory) Co., Ltd. for TFT fabrication.

References

1. Y. Z. Wang, O. O. Awadelkarim, J. G. Couillard and D. G. Ast, *Solid-State Electron.*, **42**, 1689 (1998).
2. B. Stannowski, J. K. Rath and R. E. I. Schropp, *Thin Solid Films*, **430**, 220 (2003).
3. Y. Chen, K. Pangal, J. C. Sturm and S. Wagner, *J. Non-Crystal. Solids*, **266-269**, 1274 (2000).
4. A. Saboundji, N. Coulon, A. Gorin, H. Lhermite, T. Mohammed-Brahim, M. Fonrodona, J. Bertomeu and J. Andreu, *Thin Solid Films*, **487**, 227 (2005).
5. S. V. Nguyen, D. Dobuzinsky, D. Dopp, R. Gleason, M. Gibson, and S. Fridmann, *Thin Solid Films*, **193/194**, 595 (1990).
6. O. Maida, H. Yamamoto, N. Okada, T. Kanashima and M. Okuyama, *Appl. Surf. Sci.* **130-132**, 214 (1998).
7. H. Kinoshita, T. Murakami and F. Fukushima, *Vacuum*, **76**, 19 (2004).
8. V. E. Vamvakas and D. Davazoglou, Microelectron. Reliab., **38**, 265 (1998).
9. S. Uchikoga and N. Ibaraki, *Thin Solid Films*, **383**, 19 (2001).
10. M. Creatore, S. M. Rieter, Y. Barrell, M. C. M. van de Sanden, R. Vernhes and L. Martinu, *Thin Solid Films*, **516**, 8547, (2008).
11. S. Mizushima, S. Imai, Asuha, M. Tanaka and H. Kobayashi, *Appl. Surf. Sci.*, **254**, 3685 (2008).
12. H. Niimi, K. Koh and G. Lucovsky, *Nucl. Instrum. Methods Phys. Res. B*, **127/128**, 364 (1997).
13. H. Kakiuchi, H. Ohmi, M Harada, H. Watanabe and K. Yasutake, *Sci. Technol. Adv. Mater.*, **8**, 137 (2007).
14. J.-Y. Zhang and I. W. Boyd, *Appl. Surf. Sci.*, **186**, 64 (2002).
15. N. Kaliwoh, J.-Y. Zhang and I. W. Boyd, *Appl. Surf. Sci.*, **168**, 288 (2000).
16. S. Ichimura, A. Kurokawa, K. Nakamura, H. Itoh, H. Nonaka and K. Koike, *Thin Solid Films*, **377-378**, 518 (2000).
17. K. Koike, K. Izumi, S. Nakamura, G. Inoue, A. Kurokawa and S. Ichimura, *J. Electron. Mater.*, **34**, 240 (2005).
18. H. Kobayashi, T. Yuasa, Y. Nakato, K. Yoneda and Y. Todokoro, *J. Appl. Phys.*, . **80**, 4124 (1996).
19. H. Kobayashi, T. Yuasa, K. Yamanaka, K. Yoneda and Y. Todokoro, *J. Chem. Phys.*, **109**, 4997 (1998).
20. Asuha, Y. Yuasa, O. Maida and H. Kobayashi, *Appl. Phys. Lett.*, **80**, 4175 (2002).
21. H. Kobayashi, Asuha, O. Maida, M. Takahashi and H. Iwasa, *J. Appl. Phys.*, **94**, 7328 (2003).
22. Asuha, T. Kobayashi, M. Takahashi and H. Kobayashi, *Surf. Sci.*, **547**, 275 (2003).
23. Asuha, Y.-L. Liu, O. Maida, M. Takahashi and H. Kobayashi, *J. Electrochem. Soc.*, **151**, G824 (2004).
24. W.-B. Kim, T. Matsumoto and H. Kobayashi, *J. Appl. Phys.*, **105**, 103709 (2009).
25. H. S. Momose, M. Ono, Y. Yoshitomi, T. Ohguro, S. Nakamura, M. Saito and H. Iwai, IEEE Trans. Electron. Device, **43**, 1233 (1996).

26. S.-H. Lo, D. A. Buchanan, Y. Taur, and W. Wang, *IEEE Electron Device Lett.*, **18**, 209 (1997).
27. J. Joseph, Y. Z. Hu, and E. A. Irene, *J. Vac. Sci. Technol. B*, **10**, 611 (1992).
28. Asuha, T. Kobayashi, O. Mauda, M. Inoue, M. Takahashi, Y. Todokoro and H. Kobayashi, *Appl. Phys. Lett.*, **81**, 3410 (1996).
29. F. J. Himpsel., F. R. McFeely, A. Taleb-Ibahimi, J. A. Yarmoff and G. Hollinger, *Phys. Rev. B*, **38**, 6084 (1988).
30. H. Kobayashi, T. Ishida, Y. Nakato and H. Tsubomura, *J. Appl. Phys.*, **69**, 1736 (1991).
31. Y. Kubota, T. Matsumoto, M. Yamada, H. Tsuji, K. Taniguchi, S. Imai, S. Terakawa and H. Kobayashi, IEEE Trans. Electron Device, accepted.
32. K.-M. Chang, W.-C. Yang and C.-P. Tsai, *IEEE Electron Device Lett.*, **24**, 512 (2003).
33. G. Fortunato, A. Valletta, P. Gaucci, L. Mariucci and S. D. Brotherton, *Thin solid films*, **487**, 221 (2005).
34. T. Matsumoto, Y. Kubota, M. Yamada, H. Tsuji, T. Shimatani, Y. Hirayama, S. Terakawa, S. Imai and H. Kobayashi, *IEEE Electron Device Lett.*, **31**, 821 (2010).
35. G.W. Taylor, *Solid-State Electronics*, **22**, 701 (1979).
36. S. M. Sze, *Physics of Semiconductor Devices*, 2nd ed., Wiley, New York, 1981.
37. H.-C. Lin, C.-H. Kuo, G.-J. Li, C.-J. Su, and T.-Y. Huang, *IEEE Electron Device Lett.*, **31**, 384 (2010).
38. D.K. Shroder, *Semiconductor Material and Device Characterization*, 3rd ed., Wiley, New Jersey, 2006.

228

Effects of Deposition Method of PECVD Silicon Nitride as MIM Capacitor Dielectric for GaAs HBT Technology

Jiro Yota

GaAs Technology, Skyworks Solutions, Inc.
2427 W. Hillcrest Drive, Newbury Park, CA 91320, USA
jiro.yota@skyworksinc.com

Thin silicon nitride (Si_3N_4) films deposited using plasma-enhanced chemical deposition (PECVD) method have been used as metal-insulator-metal (MIM) capacitor dielectric for GaAs hetero-junction bipolar transistor (HBT) technology. The characteristics of the films, which were deposited at 300°C, were found to be dependent on how the PECVD film was deposited. A silicon nitride film deposited as a multi-layer-layer film has different properties compared to a film deposited under the same processing conditions as a single layer film. When used as MIM capacitor dielectric, the multi-layer Si_3N_4 film is shown to have significantly superior and higher dielectric breakdown voltage and lower leakage current characteristics, as compared to the single layer film, while the capacitance density is found to be similar. Additionally, the multi-layer Si_3N_4 film is shown to have lower compressive stress and lower refractive index, indicating that the characteristics of the film are influenced by the additional interfaces present in a multi-layer film.

Introduction

Due to the increasing demand for capacity and revenue, the die size in semiconductor wafer manufacturing must be reduced. One method to reduce the die size of circuit designs is to increase the capacitance density of metal-insulator-metal (MIM) capacitor device, which is a key passive component in GaAs-based technologies, including hetero-junction bipolar transistor (HBT) technology [1-10]. The capacitance density of capacitors can be increased by simply reducing the thickness of the capacitor dielectric or insulator. However, reducing this dielectric thickness will typically also reduce the breakdown voltage and increase the leakage current of the capacitor, and thereby, degrading the performance of the device. Therefore, it is imperative that the dielectric film is well optimized, resulting in a film with desired thickness, and electrical, physical, and chemical characteristics. The electrical characteristics of a MIM capacitor dielectric in GaAs HBT technology typically should include high capacitance density, high dielectric breakdown voltage, and low leakage current, and the capacitor needs to meet and satisfy the performance and reliability requirements of the devices and circuits [2,4,8,11].

In semiconductor technology, there are multiple materials that can be used as MIM capacitor dielectric. They include silicon nitride (Si_3N_4), silicon oxynitride, and various higher dielectric constant materials, such as tantalum oxide, hafnium oxide, aluminum oxide, niobium oxide, strontium oxide, and many others, including composite oxides [8,12-22]. The silicon nitride and silicon oxynitride as capacitor dielectric is typically deposited using plasma-enhanced chemical vapor deposition (PECVD) [3,4,7,8,23], while the higher dielectric constant materials are typically deposited using sputtering, atomic layer deposition (ALD), or other deposition methods [17,19,20]. The most common dielectric material used as the insulator or dielectric for MIM capacitors in GaAs HBT technology is PECVD silicon nitride, due to its good electrical characteristics, including relatively high dielectric constant, high dielectric breakdown voltage, and low leakage current, and due to its compatibility with GaAs processing [1-4,7,8] The application of PECVD Si₃N₄ film, which can be deposited at temperature of 300°C or lower, will minimize device degradation that may occur when GaAs devices are exposed to higher processing temperatures [4,8,10,24-26].

A PECVD silicon nitride film can be deposited as a single layer film or as a multi-layer film, depending on the tool and process configuration. A single layer film will only have two surfaces, one on each side of the film, while a multi-layer film, in addition to the two surfaces, will have multiple interfaces separating the layers, and which is determined by how many layers are deposited. Previous studies on PECVD multi-layer films, including silicon oxynitride, carbon-doped silicon dioxide, and silicon nitride, show that the elemental concentration at the surfaces and interfaces are different compared to the concentration within the bulk of the film or within each layer [27-32]. The different elemental concentration at surfaces and the interfaces of the film is most likely caused by the deposition method [27-32]. During a silicon nitride deposition process, typically, reactant gases are flowed into the PECVD deposition chamber, before the RF power is turned on and any of the main reactions occur. In this initial stage, even though the RF power is still not on, some reactions or deposition can already occur on the wafer substrate surface. These reactions will result in a film with different elemental concentration at the surfaces and interfaces within the film that is different compared to the concentration of the film when the RF power is on during the deposition and the process condition is stable. It has been shown in previous studies that multi-layer PECVD films have different glass transition temperature [28], optical properties [29], and dielectric breakdown strength [30,31], compared to single layer films. Additionally, a multi-layer PECVD film is known to have no or minimal pinholes within the film, as compared to single layer film. This is due to the fact that during a multi-layer deposition, any pinholes that may be present or form in a particular layer of the film, will be covered by the overlying and subsequent layers. This lack of pinholes has been shown to improve the reliability of devices utilizing the multi-layer film and make the film to be more suitable for various applications, such as capacitor dielectric, pre-metal dielectric, surface passivation, final passivation, and interlevel dielectric [4,8,24,32-35].

In this study, we have deposited and characterized various PECVD silicon nitride films for MIM capacitor dielectric applications in GaAs HBT technology. We have compared the properties of single layer and multi-layer films, deposited under the same processing conditions. Physical and optical characterization was performed by studying the stress and refractive index of these silicon nitride films. Furthermore, the conformality and surface roughness of the PECVD Si₃N₄ film were also evaluated, in addition to the chemical bonds present in the film. Electrical characterization was performed by studying the capacitance density, breakdown voltage, and current-voltage

(I-V) characteristics of the silicon nitride film, when used as a capacitor dielectric of a metal-insulator-metal capacitor.

Experimental

The deposition of silicon nitride (Si_3N_4) films used as the capacitor insulator or dielectric in a metal-insulator-metal (MIM) capacitor was performed using plasma-enhanced chemical vapor deposition (PECVD) method. The films, ranging in thickness from 500 Å to 700 Å, were deposited on both 4-inch GaAs bare and device wafers, and were processed using GaAs hetero-junction bipolar transistor (HBT) technology. The devices on the wafers include HBT devices and metal-insulator-metal (MIM) capacitors. The bottom conductor (Metal 1) of the MIM capacitor consists of 1 μm Ti/Au/Ti metal stack, while the top conductor (Metal 2) consists of 2 μm Ti/Au/Ti stack. Two different areas of MIM capacitor structure were investigated in this study, which are 250 μm^2 and 23,000 μm^2.

The chemical vapor deposition system used to deposit the PECVD silicon nitride films is a Novellus Concept One (C-1) multi-station sequential deposition system [4,8]. This system allows both single layer and multi-layer films to be deposited on the wafer substrates. There are seven deposition stations in the process chamber. Each deposition station consists of a single wafer pedestal and a gas shower head. All wafers in the chamber are placed on a resistively heated pedestal or heater block which results in quick and uniform wafer heating. Process gases are introduced into and distributed to each of the seven showerheads for the deposition. All the process gases and process by-products for the entire reactor are evacuated through a single vacuum line at the center of this PECVD process chamber.

Single layer deposition is performed by placing one wafer on each of the seven stations, and then by depositing the total desired thickness at the same time on all seven wafers. For a multi-layer deposition, wafers are sequentially processed through each deposition station where they receive one-seventh of the total deposition thickness. Therefore, the resulting thickness is the same for both the single layer film and the multi-layer film, when the total deposition time is the same. Figure 1 shows a diagram comparing the silicon nitride film deposited using a single layer or multi-layer method. As can be seen in the figure, the single layer film has one surface on each side, while the multi-layer film, deposited on seven stations and resulting in seven layers, has six interfaces, in addition to the two surfaces.

In this study, the PECVD process condition of the Si_3N_4 is kept the same for both single and multi-layer films, including the same gas flow rates, deposition pressure, temperature, and the RF power. The deposition temperature is 300°C. The RF power used is a combination of both high frequency (HF) and low frequency (LF) power. The gases used are silane (SiH_4), ammonia (NH_3), and nitrogen (N_2). Both silane and ammonia are the reactant gases, while the nitrogen is used as a diluents gas. The NH_3 gas is flown continuously during the process to maintain chamber pressure, while the SiH_4 and N_2 gases are flown during the deposition process only. The sequence of events, including the time when the gases are introduced into the deposition chamber and the time when the RF power is turned on in this study, is shown in Figure 2. As can be seen, in the initial stages of the process, the SiH$_4$ and N$_2$ are introduced into the process chamber, before the RF power is turned on and the deposition occurs. At the end of the deposition process, the SiH$_4$ and N$_2$ stop flowing, followed by the RF power being turned off. Since the process conditions are the same for both single layer and multi-layer films, the deposition rate of both films deposited is also the same. The only difference between these two deposition methods is the deposition time in each station. As discussed before, the deposition time of the multi-layer film per station is one-seventh of that of a single layer film. However, the total deposition time and the resulting thickness for both films are the same.

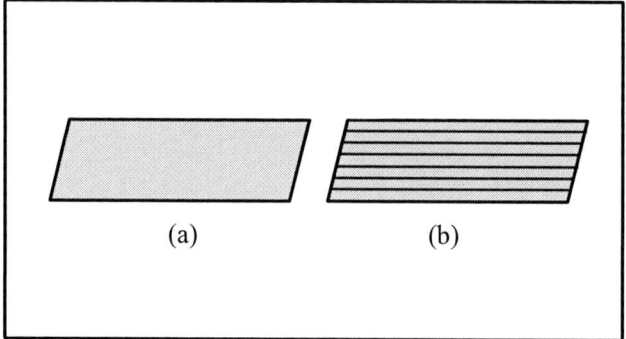

Figure 1. Comparison between (a) single layer and (b) multi-layer PECVD film deposited in this study.

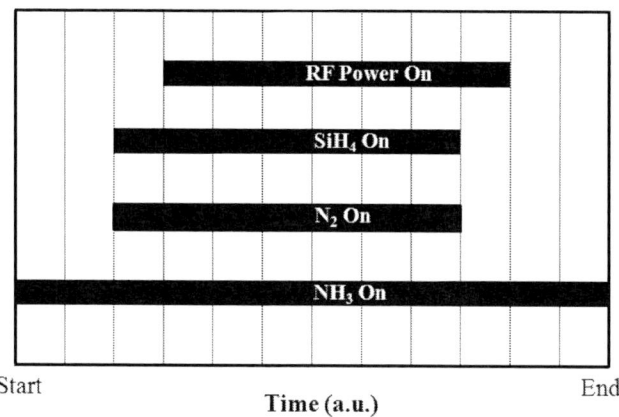

Figure 2. The sequence of events during the deposition of PECVD silicon nitride film in this study.

Both the thickness and refractive index of the silicon nitride films were measured using a Rudolph FE-VII ellipsometer, while the stress was obtained using a Frontier FSM 8800 system. Fourier Transform Infrared (FTIR) analysis was performed using a Nicolet Magan 550 at 4 cm^{-1} resolution. The FTIR analysis was performed on 600 Å PECVD Si$_3$N$_4$ films deposited on bare GaAs test wafers. Current-Voltage (I-V) and capacitance measurements were obtained using both a manual probe station (Agilent B1500A semiconductor device analyzer) and an automatic parametric tester (Agilent 4284A Precision LCR meter). Ramped voltage measurements were performed with a ramp rate of 1 V/sec, while monitoring the current until breakdown occurs. Focused-Ion Beam/Scanning Electron Microscopy (FIB/SEM) analysis was performed using a FEI 820 dual beam system.

Results and Discussion

Stress and Refractive Index

Figure 3 shows the stress of 600 Å silicon nitride films deposited using two different methods of single layer and multi-layer deposition processes. As can be seen, the single layer film has a compressive stress of -118 MPa, while the multi-layer film has a lower compressive stress of -72 MPa. Figure 4 shows the refractive index of a single layer 600 Å Si$_3$N$_4$ film and multi-layer films with different thicknesses, ranging from 500 Å to 700 Å. It can be observed that for this range of thicknesses investigated, there is no significant difference in refractive index for the multi-layer films, with refractive index ranging from 1.884 to 1.889, while the 600 Å single layer silicon nitride film has a much higher refractive index of 1.918. The lower refractive index and compressive stress of the multi-layer film most likely are caused by the presence of interfaces within the film, which results in a silicon nitride film containing lower concentration of Si and higher concentration of N.

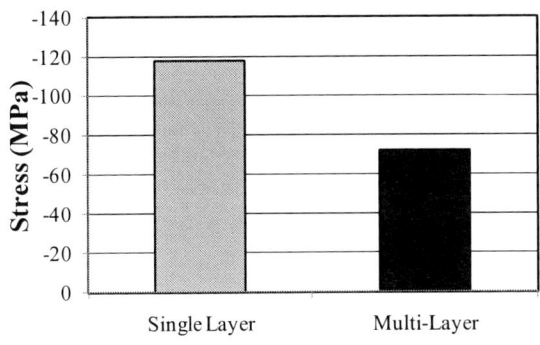

Silicon Nitride Deposition Method

Figure 3.   The compressive stress of 600 Å PECVD silicon nitride deposited using single layer and multi-layer deposition methods.

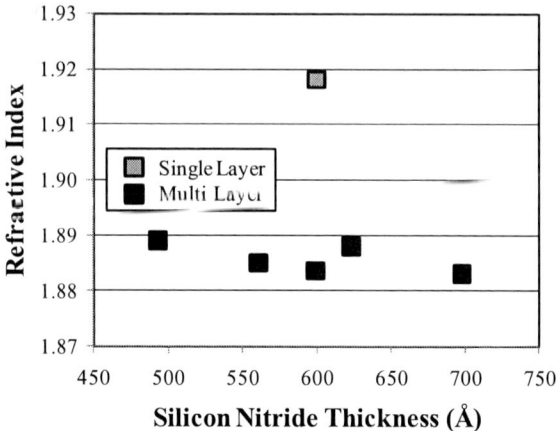

Figure 4. The refractive index of the PECVD silicon nitride deposited using single layer and multi-layer deposition methods.

FTIR Analysis

Fourier Transform Infrared (FTIR) analysis was performed to investigate the nature of chemical bonds present in the PECVD silicon nitride films, and to evaluate whether there are any significant differences that can be detected or observed between the deposited single layer and multi-layer films. Figure 5 shows the infrared spectra of these two types of Si_3N_4 films. It can be seen that the infrared absorbance peaks that are present in both spectra include those of the Si-N stretching mode (860 cm^{-1}), the N-H bending mode (1160 cm^{-1}), the N-H stretching mode (3340 cm^{-1}) and the Si-H stretching bonds (2200 cm^{-1}) [4,32,33]. These spectra show that the two silicon nitride films contain a significant amount of hydrogen, as indicated by the presence of the absorbance peaks of Si-H and N-H bonds. These results are typical for PECVD Si_3N_4 films and are consistent with previous studies [4,8,32,33]. There may be some slight difference in some of these infrared peaks present in the spectra of these two films. However, the differences, if any, cannot be detected and observed using this FTIR analysis.

Electrical Characterization

Electrical characterization of the two types of Si_3N_4 film, when used as capacitor dielectric of an MIM capacitor, was performed by evaluating the capacitance, breakdown voltage, and current-voltage (I-V) characteristics. Figure 6 shows the capacitance density data of a capacitor using a dielectric of a single layer 600 Å silicon nitride film, and of multi-layer silicon nitride films with different thicknesses, ranging from 500 Å to 700 Å. As can be seen, the capacitance density obtained is about 0.942 $fF/\mu m^2$, and there is no significant difference observed between the 600 Å single layer and the 600 Å thick multi-layer Si_3N_4 capacitor dielectric films. Furthermore, as expected, the data also shows that the capacitance density is reduced as the silicon nitride film thickness is increased. Figure 7 shows the capacitance of MIM capacitors with a small area of 250 μm^2 and a much larger area of 23,000 μm^2, using these two types of 600 Å Si_3N_4 films. As shown, Figure 7 shows that there is no significant difference in capacitance between these two films for both capacitor areas. All the above data indicate that the capacitance characteristics of these films are very similar.

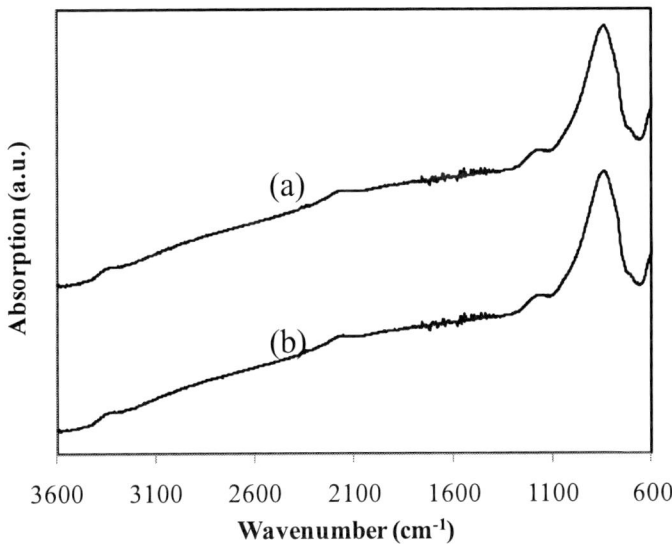

Figure 5. The infrared spectra of the PECVD silicon nitride deposited using (a) single layer and (b) multi-layer deposition methods.

Figure 8 shows the current voltage (I-V) curves obtained from MIM capacitors using a capacitor dielectric of a single layer 600 Å PECVD silicon nitride film, and of multi-layer silicon nitride films with different thicknesses. As expected, the leakage current increases when the capacitor dielectric film thickness is decreased for the Si_3N_4 films deposited using the same method. However, as can be seen, the capacitor with the single layer 600 Å Si_3N_4 film has significantly higher leakage current, as compared to the capacitor with multi-layer films, irrespective of the thickness in the range of 500 Å to 700 Å. The observed results indicate that the multi-layer film has superior and lower leakage current characteristics than the single layer. The multi-layer silicon nitride films also show superior and higher breakdown voltage characteristics. Figure 9 shows the dielectric breakdown voltage of the 600 Å Si_3N_4 deposited as a multi-layer film is 63 V, which is 7.23% higher than that of the 600 Å deposited as a single layer film, which breakdown voltage is 58.75 V. Furthermore, Figure 9 also shows that the breakdown voltage of multi-layer silicon nitride films increases with increasing film thickness.

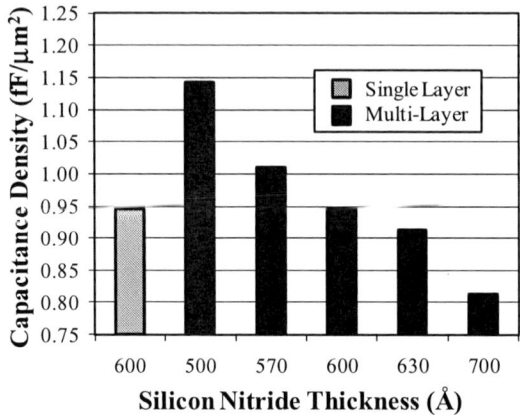

Figure 6. The capacitance density of MIM capacitor using the PECVD silicon nitride as capacitor dielectric with different thicknesses, deposited using single layer and multi-layer deposition methods.

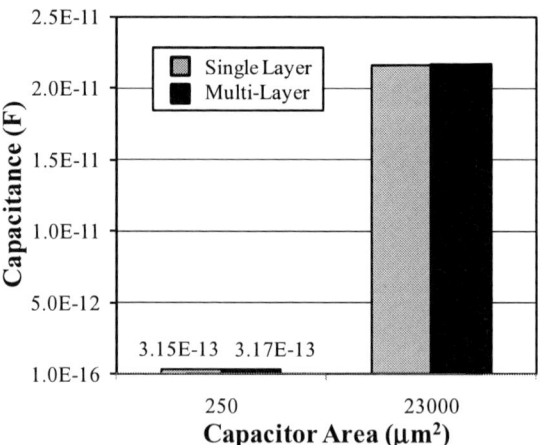

Figure 7. The capacitance density of MIM capacitor of different areas, using 600 Å PECVD silicon nitride film as capacitor dielectric, deposited using single layer and multi-layer deposition methods.

The MIM capacitor electrical characterization data show that the Si_3N_4 deposited as a multi-layer film is significantly superior in electrical characteristics than the single layer film. For the same thickness, both types of film yield similar capacitances, while at the same time, the multi-layer Si_3N_4 film shows much lower leakage current and higher dielectric breakdown voltage. These superior characteristics may be attributed to the presence of multiple interfaces with differing elemental concentrations and the presence of minimal pinholes, if any, in a multi-layer film. This is especially important, if the underlying metal or electrode used in the MIM capacitor has a rough surface, which is typical for evaporated metal usually used in the GaAs processing [4,8]. During this multi-layer PECVD silicon nitride film deposition, each layer that is deposited will cover conformally and eliminate or reduce any surface roughness and sharp features of the underlying metal. All these make this multi-layer dielectric film to be very much suitable for MIM capacitor application in GaAs HBT technology [4,8].

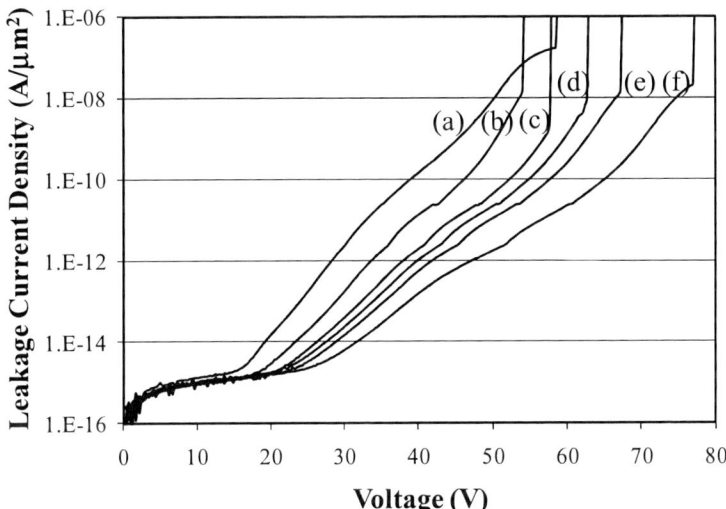

Figure 8. The current-voltage (I-V) curves of MIM capacitor, with (a) 600 Å single layer, and (b) 500 Å, (c) 570 Å, (d) 600 Å, (e) 630 Å, and (f) 700 Å multi-layer PECVD silicon nitride capacitor dielectric.

FIB/SEM Analysis

Figure 10 shows the Focused-Ion Beam/Scanning Electron Microscopy (FIB/SEM) images of an MIM capacitor with 600 Å multi-layer PECVD Si_3N_4 as capacitor dielectric, and the hetero-junction bipolar transistor (HBT) with the emitter, base, and collector, on a GaAs wafer manufactured using GaAs HBT technology. As shown, the PECVD silicon nitride film is conformal and has excellent step coverage over the rough underlying Metal 1 surface, and in fact, reduces this metal surface roughness. This good conformality is critical in order to reduce increased electric field and fringe capacitance due to the sharp features and non-uniformity of the underlying metal conductor surface [4,8]. The images in Figure 10 show that the multi-layer Si_3N_4 is very suitable and excellent for MIM capacitor dielectric application in GaAs HBT technology.

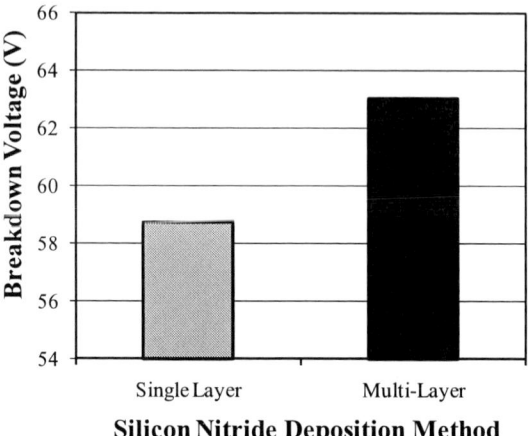

Silicon Nitride Deposition Method

Figure 9.   The dielectric breakdown voltage of MIM capacitor using 600 Å PECVD silicon nitride film as capacitor dielectric, deposited using single layer and multi-layer deposition methods.

Conclusions

We have evaluated the use of thin silicon nitride (Si_3N_4) films deposited using plasma-enhanced chemical deposition (PECVD) method as metal-insulator-metal (MIM) capacitor dielectric.   The film characteristics were found to be dependent on the deposition method, including whether the film is deposited as a single layer or multi-layer film.   The multi-layer Si_3N_4 film is shown to have lower compressive stress and lower refractive index, compared to the single layer film.   FTIR data show that there is no significant difference in chemical bonds present in the film between the two types of silicon nitride.   Electrical characterization of MIM capacitor shows that there is no significant difference in the capacitance density obtained when a single layer silicon nitride or a multi-layer silicon nitride is used as the capacitor dielectric.   However, the multi-layer silicon nitride film is shown to have significantly superior and higher dielectric breakdown voltage and lower leakage current characteristics, as compared to the single layer film, making it very suitable for capacitor dielectric application in GaAs HBT technology.

Acknowledgments

The author would like to acknowledge Ravi Ramanathan, Jose Arreaga, Bruce Darley, Daniel Weaver, David Tuunanen, and Mike Sun of Skyworks Solutions, Inc. for their help in this study.

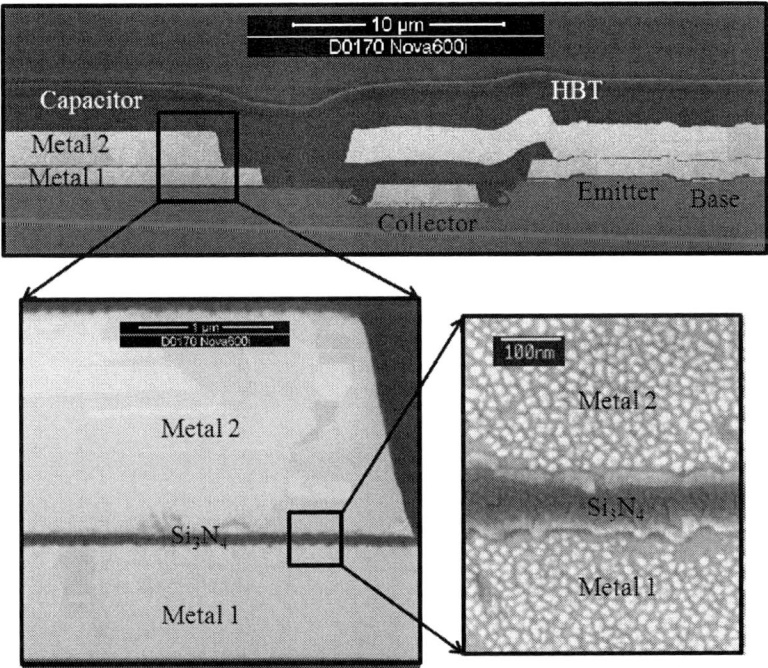

Figure 10. X-Section FIB images of a metal-insulator-metal (MIM) capacitor with 600 Å multi-layer PECVD silicon nitride as capacitor dielectric and hetero-junction bipolar transistor (HBT) on wafers manufactured using GaAs HBT technology.

References

1. C. Whitman and M. Meeder, in *Reliability of Compound Semiconductors Workshop*, pp. 91-102 (2004).
2. J. Scarpulla, E.D. Ahlers, D.C. Eng, D.L. Leung, S.R. Olson, and C.S. Wu, in *GaAs Reliability Workshop*, pp. 92-105 (1998).
3. G. Dandrova, J.M. Beall, and K.D. Decker, *ECS Trans.*, **6** (3), 397 (2007).
4. J. Yota, R. Ramanathan, J. Arreaga, P. Dai, C. Cismaru, R. Burton, P. Bal, and L. Rushing, in *2003 GaAs MANTECH Tech. Digest*, pp. 65-68 (2003).
5. A. Muller, S. Simion, M. Dragoman, S. Iordanescu, I. Petrini, C. Anton, D. Vasilache, V. Avramescu, A. Coraci, and F. Craciunoiu, in *Proc. International Semiconductor Conference*, pp. 185-188 (1996).
6. C. Nevers, A.T. Ping, T. Rivers, S. Varma, F. Pool, M. Minkoff, E. Etzkorn, and O. Berger, in *2009 CS MANTECH Tech. Digest*, pp. 117-120 (2009).
7. B.N. De and M. Shokrani, in *2005 CS MANTECH Tech. Digest*, pp. 145-148 (2005).
8. J. Yota, R. Ramanathan, K. Kwok, J. Arreaga, T. Ko, and H. Shao, in *Proc. 2005 Electrochemical Society Meeting: State-of-the-Art-Program on Compound*

Semiconductors (SOTAPOCS), PV 2005-04, pp. 315, The Electrochemical Society Proceeding Series, Pennington, NJ (2005).

9. T. Kagiyama, Y. Tosaka, R, Yamabi, and H. Yano, in *2007 CS MANTECH Tech. Digest*, pp. 51-54 (2007).

10. R. Williams, *Modern GaAs Processing Methods*, Artech House, Boston (1990).

11. B. Yeats, *IEEE Trans. Electron. Dev.*, **45** (4), 939 (1998).

12. A.S. Zoolfakar and H. Hasim, in *Proc. IEEE International Conference on Semiconductor Electronics (ICSE)*, pp. 445-449 (2008).

13. H. Sanchez, *et al.*, in *Proc. International Interconnect Technology Conference (IITC)*, pp. 84-89 (2007).

14. N. Inouc, H. Ohtake, I. Kume, N. Furutake, T. Onodera, S. Saito, A. Tanabe, M. Tagami, M. Tada, and Y. Hayashi, in *Proc. International Interconnect Technology Conference (IETC)*, pp. 63-65 (2006).

15. S.J. Kim, B.J. Cho, M.B. Yu, M.F. Li; Y.Z. Xiong, C.X. Zhu, A. Chin, and D.L. Kwong, *IEEE Electron. Dev. Lett.*, **26**, 625 (2005).

16. K.C. Chiang, C.H. Lai, A. Chin, H.L. Kao, S.P. McAlister, and C.C. Chi, in *Microwave Symposium Digest*, pp. 287-290 (2005).

17. H. Hang, *et al.*, in *Proc. IEEE International Electron Devices Meeting (IEDM)*, pp. 379-382 (2003).

18. T.T. Vo, T. Lacrevaz, B. Flechet, A. Farcy, Y. Morand, S. Blonkowski, J. Torres, and E. Defay, in *Proc. Asia-Pacific Microwave Conference*, pp. 1-4 (2007).

19. N. Menou, *et al.*, in *Proc. IEEE International Electron Devices Meeting (IEDM)*, pp. 1-4 (2008).

20. N.A. Safford, R. Katamreddy, L. Guerin, B. Feist, C. Dussarat, V. Pallem, C. Weiland, and R. Opila, *ECS Trans.*, **19** (2), 525 (2009).

21. C. Dubourdieu, O. Salicio, S. Lhostis, L. Auvray, Y. Rozier, F. Ducroquet, and S. Daniele, *ECS Trans.*, **19** (2), 669 (2009).

22. J. Robertson, *ECS Trans.*, **19** (2), 579 (2009).

23. M. Maeda, E. Yamamoto, S. Ohfuji, and M. Itsumi, *J. Vac. Sci. & Technol. B*, **17**, 201 (2009).

24. J. Yota, *J. Electrochem. Soc.*, **156**, G173 (2009).

25. W. Liu, *Handbook of III-V Heterojunction Bipolar Transistors*, New York, John Wiley (1998).

26. J. Yota, H. Ly, R. Ramanathan, M. Sun, D. Barone, T. Nguyen, K. Katoh, M. Ohe, R. L. Hubbard and K. Hicks, *IEEE Trans. Semicond. Manuf.*, **20**, 323 (2007).

27. A.A. Saleh, J.B. Rothman, J.F. Kirchoff, J. Yota, and C. Nguyen, *Thin Solid Films*, **355/356**, 363 (1999).

28. H. Zhou, H.K. Kim, F.G. Shi, B. Zhao, and J. Yota, *Microelectron. J.*, **33**, 221 (2002).

29. H. Zhou, H.K. Kim, F.G. Shi, B. Zhao, and J. Yota, *Microelectron. J.*, **33**, 999 (2002).

30. H. Zhou, F.G. Shi, B. Zhao, and J. Yota, *Microelectron. J.*, **35**, 571 (2004).

31. H. Zhou, F.G. Shi, B. Zhao, and J. Yota, *Appl. Phys. A*, **81**, 767 (2005).

32. J. Yota, J. Hander, and A.A. Saleh, *J. Vac. Sci. Technol. A*, **18** (2), 372 (2000).

33. S. Wolf and R.N. Tauber, *Silicon Processing for the VLSI Era Volume 1 - Process Technology*, Lattice Press, Sunset Beach (1986).

34. S.K. Gandhi, *VLSI Fabrication Principles, Silicon and Gallium Arsenide*, New York, John Wiley (1983).

35. J. Yota, *et al.*, in *Proc. 4th Inter. Dielectrics ULSI Multilevel Interconnect. Conf. (DUMIC)*, pp. 185-192 (1998).

Low Temperature Processing of Si-based Dielectric Thin Films

P. C. Joshi, A. T. Voutsas, and J. W. Hartzell

Material and Device Applications Laboratory,
SHARP Laboratories of America, Inc.
5700 NW Pacific Rim Blvd., Camas, Washington 98607, USA

Low temperature processing of dielectric thin films is critical to realize high performance TFTs, solar cells, and sensors integrated on glass or plastic substrates. The plasma based deposition techniques have successfully met the dielectric performance requirements at process temperatures above 300 °C. However, the dielectric quality and reliability degrade severely at lower process temperatures indicating that the substrate surface reaction kinetics strongly dictate the thin film growth and properties. In the present work, we report on the low temperature processing (100-300 °C) of Si-based dielectric films by high-density PECVD technique which offers significant process and thin film property control. The microstructure and optical properties of the thin films, and the bulk and interfacial electrical quality and reliability of the MIS capacitors and TFTs as a function of process temperature are discussed in this report. The observed thin film and device performances at a low processing temperature of 100 °C show promise for low temperature electronic applications.

Introduction

Recently there has been growing interest in the development of low temperature thin film and device fabrication methods for novel displays, sensors, and photovoltaic applications. Low temperature processing of high performance dielectric and semiconductor films is critical to realize flexible displays integrated on plastic or other low temperature substrates. The development of high quality gate dielectric films at low temperatures is a key challenge for various R&D efforts focused on flexible electronic device development. Silicon CMOS technology has successfully exploited the high temperature oxidation, nitridation, and CVD processes for the fabrication of high quality SiO_2 and Si_3N_4 thin films. However, these high temperature processes are not suitable for device integration on glass or plastic substrates. The developments in PECVD technology have enabled the successful integration of Si based dielectric thin films on glass substrates at process temperatures as low as 400°C. A further reduction in the process temperature (<200°C) is desired to realize flexible electronic devices integrated on plastic substrates. However, the dielectric thin film performance degrades rapidly with a decrease in process temperature below 400°C. Additionally, the issue of high film reliability at smaller film thicknesses required for future sub-micron devices is driving the R&D efforts on the evaluation of a variety of deposition processes and plasma techniques. Various thin film deposition techniques such as e-beam evaporation, RTCVD, PECVD, ECR-PECVD, chemical solution deposition, hot-wire CVD, etc., have been investigated for the low temperature processing of thin films [1-10]. However, there is no unique solution addressing the issues of low temperature processing, high throughput, and large area uniform deposition. In the present paper, we report on the high density plasma processing of Si-based thin films. The high density plasma process offers superior thin film process and property control due to the following key characteristics: high plasma concentration, high ion/neutral ratio, low plasma potential, and independent control of plasma energy and density. The high density plasma based processes offer to overcome the major limitations of the current PECVD techniques; mainly, the plasma induced bulk/interface damage, impurities, and device reliability. The low pressure operation (<100 mTorr) of the high density plasma provides extended process space for the fabrication of high performance thin films. The high density plasma is suitable for the generation of active radicals in the plasma by efficient energy transfer form the inert metastables to the reactive gases. The active plasma species offer the possibility of efficient oxidation, nitridation, and defect passivation at low process temperatures. In the present paper, we report on the high density plasma growth of SiO_2 and SiN_x thin films on Si substrates at process temperatures lower than 300°C. The properties of high density plasma deposited SiO_2 and SiN_x thin films have also been discussed. The effects of low process temperatures (90-300°C) on the HDP processed thin film properties have been discussed in close correlation with the process parameters.

Experimental

In the present work, we report on the growth and deposition of Si based oxide and nitride dielectric thin films by high density plasma technique employing an inductively coupled plasma source. The 13.56 MHz rf power was coupled to the top electrode through a matching network while no bias was applied to the bottom electrode. The various thin films were processed in the pressure range of 5-100 mTorr while the substrate temperature was maintained in the range of 90-300°C. The dielectric thin films were grown/deposited on p-type prime grade Si wafers. The wafers were cleaned by the standard RCA cleaning process and dipped in BOE (50:1) solution prior to loading into the chamber. The dielectric thin film processes were optimized in terms of the effects of the system pressure, substrate temperature, rf power, and gas flow rates and ratios on the microstructure, optical, electrical, and reliability characteristics of thin films and devices. The density and optical properties of the films were analyzed in terms of the etch rate and n-k dispersion characteristics. The microstructure of the films was analyzed optically by Fourier transform infrared (FTIR) spectroscopy and variable angle spectroscopic ellipsometry (VASE) techniques. The electrical performance of the films was evaluated in terms of the MOS capacitor and TFT device characteristics. The leakage current, dominant conduction mechanism, and the breakdown field strength of the films were analyzed from the current-voltage (I-V) characteristics measured by HP4156A semiconductor parameter analyzer. The interfacial characteristics of the dielectric thin films were evaluated from the small signal capacitance-voltage (C-V) measurements. The midgap interface trap concentration (D_{it}) was determined from the high frequency (1MHz) and quasi-static C-V measurements. The reliability of the thin film and devices was established in terms of the bias-temperature stress (BTS) induced shifts in the MOS capacitor and TFT device characteristics. The BTS measurements were conducted at a temperature of 150°C for 30 minutes under different bias conditions depending on the device configuration.

Results and Discussion

<u>HDP Techniques for High Performance Thin Films</u>

In recent years, extensive R&D work has been done on exploring the effects of enhanced plasma characteristics on the thin film growth and properties. The capacitively coupled plasma (CCP) source has been extensively used in various PECVD techniques. The reaction kinetics in CCP plasma can be enhanced by increasing the frequency of the applied rf power, changes in electrode configuration, decreasing the process pressure, or increasing the rf power coupled to the top electrode. However, any change in these variables from the optimum value leads to major design issues, and significantly lower bulk and interfacial film quality due to increased ion bombardment. The drive towards high density plasma based system and processes is aimed at overcoming the limitations of the existing PECVD technology. Figure 1 shows the high density plasma process space for the fabrication of high performance thin films. The CCP plasma based techniques have resulted in superior thin film performance at higher operating frequencies (>13.56MHz). However, the process flexibility and large area thin film processing capabilities are compromised drastically at higher operating frequencies. The HDP techniques are gaining interest in the area of thin film processing due to the following key characteristics: high plasma density, low plasma potential, and independent control of ion energy and density. The HDP characteristics are suitable for the low temperature processing of high performance thin films due to plasma controlled reaction kinetics dominating the film growth, minimal plasma induced bulk or interfacial damage, and decoupled plasma energy and density. The low pressure operation of the high density plasma provides extended process space for the fabrication of high performance thin films due to enhanced reaction kinetics in the bulk of the plasma and on the growing film surface. In the present work, we have attempted to exploit the HDP characteristics for the growth and deposition of Si-based dielectric thin films at low process temperatures suitable for device integration on flexible plastic substrates.

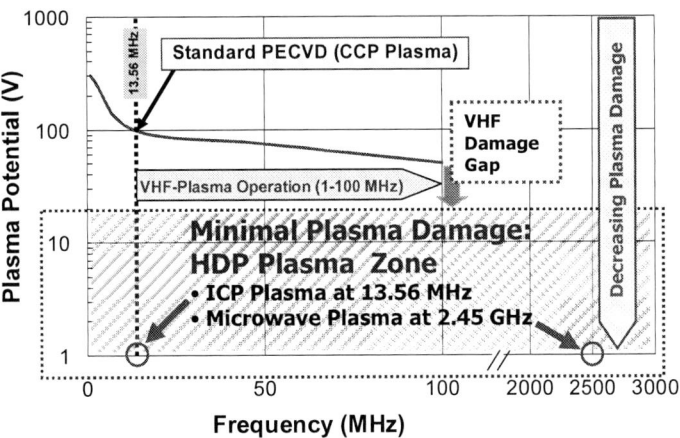

Figure 1. The HDP process space for the low temperature processing of high performance thin films.

Enhancing Gate Dielectric Performance

The silane based chemistries have been extensively used for the fabrication of SiO_2 thin films. However, the TEOS (tetraethyl orthosilicate) oxide films have been successfully exploited for display applications because of high step coverage and uniform deposition on large area substrates by standard PECVD technique employing CCP source. In this paper, the properties of the TEOS oxide films deposited by standard PECVD technique have been compared to high density plasma processed oxide films in an attempt to understand and establish the performance as a function of process temperature. Figure 2 shows the step coverage of TEOS oxide films as a function of deposition temperature in the range of 300-400°C. It is clear from Figure 2 that the step coverage of TEOS oxide films improves significantly with decreasing process temperature. However; the microstructure, and the electrical and reliability characteristics of the TEOS oxide films degrade severely with decreasing process temperature below 400°C, making it unsuitable for device applications.

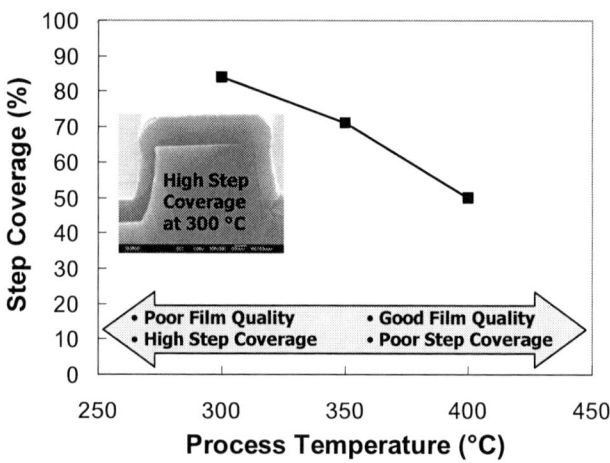

Figure 2. The effect of the process temperature on the step coverage of TEOS oxide thin films deposited by standard PECVD technique employing capacitively coupled plasma source.

Figure 3 shows the effects of the process temperature on the fixed oxide charge concentration (N_f) in TEOS oxide thin films. It is clear from Figure 3 that the fixed oxide charge concentration increases rapidly with a decrease in process temperature below 400°C. The efficient TEOS precursor ionization in the plasma and subsequent conversion to oxide on the substrate surface are critical to form high quality stoichiometric thin films. An inefficient breakdown of the TEOS precursor can lead to large carbon and -OH impurity bonds in the films as verified by SIMS and FTIR analysis. The low temperature (~90°C) HD-PECVD processed silane oxide films showed N_f value comparable to TEOS oxide films processed at a much higher temperature of 400 °C. As the TEOS oxide deposition temperature can not be reduced below 300°C; we made an attempt to improve the TEOS oxide performance by high density oxygen plasma treatment at a low process temperature of 200°C. The high density plasma oxidation was carried out in a He/O$_2$ plasma which is effective in oxidizing the Si substrate at growth rates significantly higher than the thermal growth rates (at 1000°C) as discussed later in this paper. As shown in Figure 3, the high density plasma generated oxygen radicals were effective in significantly reducing the N_f value for 300°C deposited TEOS oxide films. The observed results clearly suggest that the microstructure of the TEOS oxide films strongly depends on the substrate temperature apart from the plasma process parameters. It is not possible to efficiently control the TEOS oxide thin film properties at lower temperatures in CCP plasma as the thermal state of the substrate has a strong influence on the thin film growth and properties. Additionally, any attempt to further tune the plasma from the optimum values at lower process temperatures results in lower quality thin films due to increased plasma induced damage.

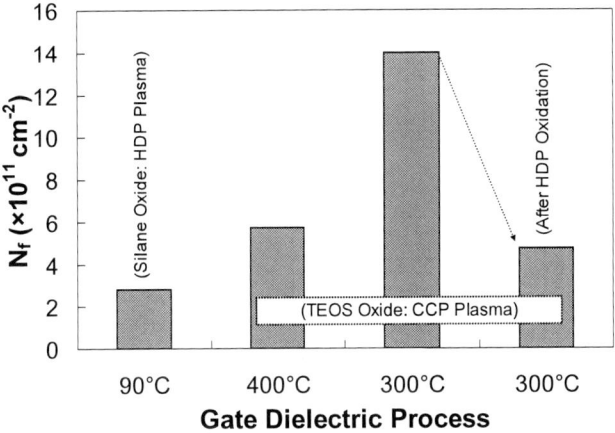

Figure 3. The fixed oxide charge density of TEOS oxide thin films as a function of process temperature and high density plasma oxidation treatment.

We made an attempt to further understand the effects of the high density plasma oxidation on the TEOS oxide thin film microstructure. The SIMS and FTIR analyses were conducted, as shown in Figure 4(a and b), to understand the effects of HDP oxidation on impurity related bonds. The C-content in TEOS oxide films increases rapidly with decreasing process temperatures below 400°C leading to poor electrical characteristics. As shown in Figure 4(a), the high density plasma oxidation treatment was effective in reducing the C-content in TEOS oxide films by about an order of magnitude. The HDP oxidation treatment leads to an efficient reduction of the C-impurities by oxidation to CO. The FTIR analysis also substantiated the SIMS data as shown in Figure 4(b). The carbonyl signal in TEOS oxide thin films was found to decrease as a result of HDP oxidation treatment. A similar approach is employed to reduce C-content in bulk SiC devices by thermal oxidation of the residual C-species at the SiC/dielectric interface. The HDP generated oxygen species are effective in efficient ionization of C-species even at a low temperature of 200°C where thermal processes can not perform. The observed results are promising for the development of high performance thin film and devices integrated on flexible substrates.

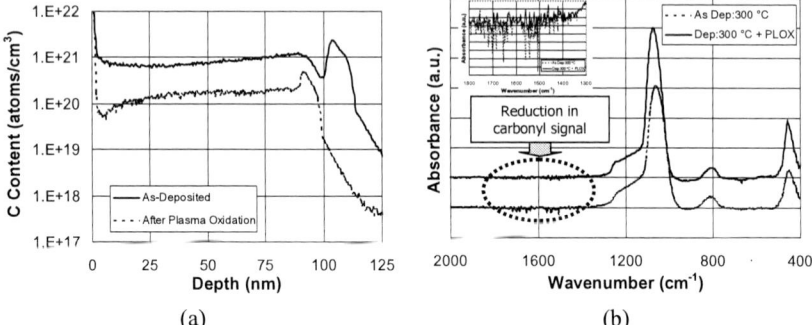

(a)   (b)

Figure 4. The effects of high density plasma oxidation treatment on the C-related impurities in the TEOS oxide films processed at 300°C: (a) SIMS data, (b) FTIR analysis.

Plasma Growth of Dielectric Thin Films

The high density plasma process space was explored for the growth of SiO_2 and SiN_x thin films on Si substrates at process temperatures below 400 °C where thermal growth rates are impractical. The details of our process development, and the thin film growth and properties are described in the following sections.

SiO_2 Growth: A combination of the inert and oxygen gases was used to generate active oxygen radicals for the low temperature oxidation of Si. For the present study; He, Ar, and Kr were used as the inert gases to create metastable states in the high density plasma to generate neutral excited oxygen radicals. The SiO_2 thin films were grown on Si substrates in the temperature range of 100-300°C while the system pressure was maintained in the range of 10-1000 mTorr. Figure 5 shows the SiO_2 film thickness as a function of growth time in various atmospheres containing 3% oxygen gas. The SiO_2 growth rate was found to decrease with an increase in the atomic weight of the inert gas. indicating a difference in the nature of the excited oxygen species in different ambients dictating the SiO_2 growth. The SiO_2 film thickness was found to increase nonlinearly with time in the investigated range of 10s-60min. For example; in He/O_2 plasma, a rapid growth of SiO_2 was observed during the initial 60s and a film thickness of 44Å was obtained after a deposition time of 1 minute. The observed nonlinear growth rate clearly shows that the initial SiO_2 growth is dictated by the silicon surface reaction rate while the diffusion effects lead to lower growth rates at longer growth times. However, the SiO_2 growth rates in the high density plasma are significantly higher than the thermal growth rates at a temperature of 1000 °C suggesting that the oxidation kinetics are quite different for the atomic oxygen species as compared to molecular oxygen.

The SiO_2 growth rate (100Å/10min) was found to be the highest in He/O_2 atmosphere. The high SiO_2 growth rate in He/O_2 plasma is possibly the result of high He metastable energy level. The first metastable state of He is at 19.8 eV while Ar and Kr metastable states lie at lower energies of 11.6 and 9.9 eV, respectively. The high density plasma excited He metastables would be more effective in generating higher energy O^1D (first excited state) radicals as compared to O^3P (ground state); while the Ar and Kr metastables would provide a higher concentration of O^3P species. The observed high growth rate in He/O_2 plasma is possibly due to higher concentration and greater reactivity of O^1D radicals as compared to O^3P radicals. However, the clear role of a particular atomic oxygen radical (ground state/excited) is not clearly established as yet. The observed high growth rate in the He/O_2 atmosphere is desirable as He is the lightest among the investigated gases and would cause minimal SiO_2/Si interface damage; which is critical for the fabrication of stable and reliable devices and circuits.

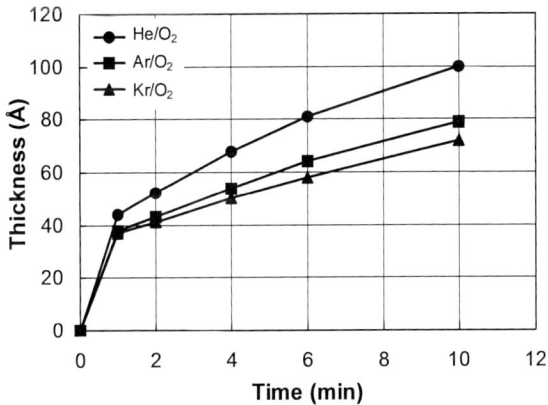

Figure 5. The high density plasma growth of SiO_2 thin films on Si substrates in various atmospheres containing 3% oxygen.

The SiO_2 growth on Si was also found to depend strongly on O_2 concentration, and system pressure. The growth rate was found to be the highest for an oxygen concentration of 3% in the investigated oxygen concentration range of 1-6%. At lower oxygen concentrations (<3%); the growth rate increased with increasing oxygen partial pressure suggesting an increase in active oxygen radicals. However, the growth rate was found to decrease with further increase in oxygen concentration above 3% indicating a quenching of the active oxygen species; possibly though various recombination pathways. The effects of the system pressure on the SiO_2 growth rate were investigated in the range of 10-1000 mTorr for a constant SiO_2 growth time of 10 min. The SiO_2 growth rate did not show a strong dependence on the system pressure at levels below 150 mTorr while it was found to decrease appreciably with pressure in the investigated range up to 1000 mTorr. A thickness of about 42Å was measured at a pressure of 1000 mTorr. The observed growth behavior suggested that lower pressures were necessary to achieve higher growth rates by minimizing the plasma interactions among the oxygen species.

The effects of the Si surface orientation on the growth of SiO_2 thin films were investigated on <100> and <111> Si wafers to gain a further insight into the high density plasma growth kinetics. There was no appreciable difference in the SiO_2 film thickness measured after a growth time of 30s and 10min on <100> and <111> Si surfaces indicating that the high-density plasma growth was not limited by surface concentration of silicon atoms. The observed high SiO_2 growth rate in the rf driven inductively coupled plasma independent of the Si surface orientation is attractive for achieving high step coverage in STI structures and gate dielectric applications in low temperature poly-Si TFT devices integrated on glass or other low temperature substrates.

Figure 6 shows a typical room temperature J-E curve for a 100 Å thick SiO_2 film measured by applying dc voltages with a step height of 0.1V. The J-E curve exhibited good low and high field insulating characteristics with current density lower than 10^{-8} A/cm^2 at an applied electric field of 2 MV/cm. The SiO_2 films showed high breakdown field strength of about 14.2 MV/cm while the leakage current density remained lower than 10^{-6} A/cm^2 even at an applied electric field of 10 MV/cm. The leakage current characteristics of a 3-nm-thick SiO_2 film were measured as a function of applied electric field and substrate temperature to further evaluate the interface quality. The J-E characteristics of a 3-nm-thick film exhibited a sharp increase in current at an applied electric field greater than 4 MV/cm indicating that the current was dominated by Fowler-Nordheim (F-N) tunneling. The J-E relation for F-N tunneling is given by the equation:

$$J = AE^2 \exp\left(-\frac{B}{E}\right)$$

where J is the current density, E is the applied electric field, and A and B are constants. The constant B is given by the relation:

$$B = \frac{8\pi\left(2qm^*\right)^{1/2}\left(\phi_o\right)^{3/2}}{3h}$$

where m^* is the effective mass of the charge carrier in oxide, q is the electron charge, h is Planck's constant, and ϕ_o is the barrier height at the Si/SiO_2 interface. The value of the barrier height can be calculated from the slope of the $\ln(J/E^2)$ vs. $1/E$ plot as a function of substrate temperature, as shown in Figure 6. For the HDP grown SiO_2 thin films, a potential barrier height (ϕ_o) of 3.2 eV was calculated which is similar to the values reported for high quality SiO_2 thin films prepared by various methods [19-21]. The observed high value of the potential barrier at the Si/SiO_2 interface shows the effectiveness of the high-density plasma process in minimizing any plasma induced interface damage and providing a high quality SiO_x transition layer at the interface.

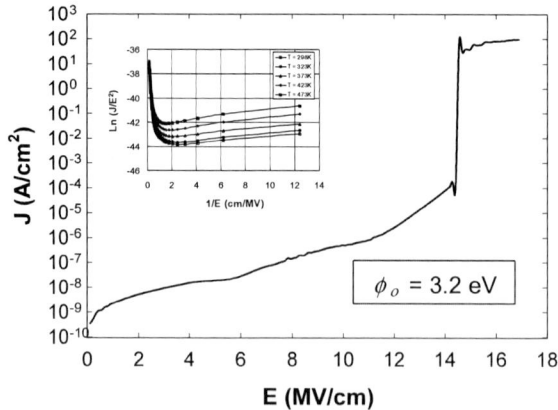

Figure 6. The typical J-E characteristic of a 100 Å thick SiO_2 film grown in high density He/O_2 plasma. The inset shows the F-N plots as a function of measurement temperature.

The SiO$_2$/Si interface quality was analyzed from the high frequency (1 MHz) C-V characteristics of the MOS capacitors as shown in Figure 7. The MOS capacitors did not exhibit any appreciable frequency dispersion in the frequency range of 1-1000 KHz indicating good film/electrode interfacial characteristics. The SiO$_2$ thin films exhibited a low flat band voltage of about -0.4 V indicating good SiO$_2$/Si interfacial characteristics. The corresponding fixed oxide charge density was about 2.7×10^{11} cm^{-2}. The C-V characteristics of the HDP grown SiO$_2$ thin films exhibited a small hysteresis with a voltage width of about 25 mV indicating a low density of the interface traps. The flat band voltage and the hysteresis voltage width were comparable to thermal oxide films; indicating high interface quality of the high-density plasma grown SiO$_2$ thin films.

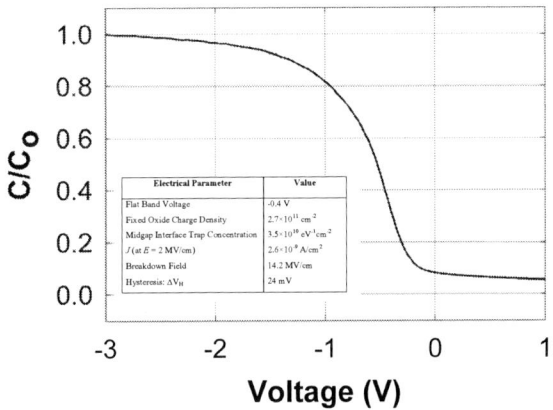

Figure 7. The typical C-V characteristics of a 100Å thick high density plasma grown SiO$_2$ film. The films were grown on Si substrate in He/O$_2$ atmosphere.

SiN$_x$ Growth: Silicon nitride films are widely used for diverse electronic applications exploiting its excellent insulating, dielectric, and diffusion resistance characteristics. The high dielectric constant, effective diffusion barrier resistance for dopant species, radiation hardness, and high breakdown field characteristics of SiN$_x$ are attractive for gate dielectric applications. The thermal growth of nitride is impractical at process temperatures lower than 1100 °C. The high SiN$_x$ thermal growth temperatures are unsuitable for device integration on glass or plastic substrates. The HDP process was further explored for the growth of SiN$_x$ thin films on Si substrates. A combination of He and N$_2$ gases was used for the generation of active nitrogen species for SiN$_x$ growth at a low temperature of 100°C. Figure 8 shows the high-density plasma growth of SiN$_x$ in He/N$_2$ (3%) atmosphere at a substrate temperature of 100 °C. The HDP growth rate was found to be significantly higher than the thermal growth rate at a temperature of 1150 °C. The SiN$_x$ growth rate was well maintained in the temperature range of 100-300°C suggesting that the growth kinetics were controlled by the high-density plasma rather than the thermal state of the substrate. The most significant aspect of the HDP growth of SiN$_x$ is the initial rapid growth of the thin film. It was possible to grow a thickness of 25Å after 1 minute. This initial high growth rate can be exploited for low thermal budget processing of thicker films on novel device structures.

The quality of the SiN$_x$ thin films was further evaluated in terms of the oxygen diffusion resistance at a temperature of 1000 °C. The 27-50Å thick HDP grown SiN$_x$ films were subjected to dry O$_2$ atmosphere to investigate the diffusion resistance. A bare Si wafer was also included in the study to establish the oxide growth on Si. As shown in Figure 8, the thermal anneal in dry O$_2$ at a temperature of 1000 °C for 14 min resulted in an oxide growth of 213 Å on bare Si wafer while no appreciable oxide growth was observed on Si wafers with 27-50Å SiN$_x$ overlayer. The observed results establish high diffusion resistance characteristics of HDP grown SiN$_x$ thin films. The observed SiN$_x$ growth independent of the thermal state of the substrate and high diffusion resistance characteristics suggest the suitability of the HDP growth process for novel device development exploiting the unique properties of silicon nitride thin films.

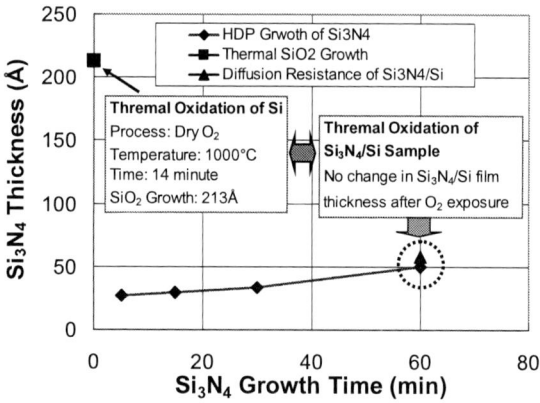

Figure 8. The high density plasma growth of SiN$_x$ thin film as a function of time, and oxygen diffusion resistance characteristics at a temperature 1000°C.

<u>High Density Plasma Deposition of Dielectric Thin Films</u>

The high density plasma technique was further explored for the deposition of SiO$_2$ and SiN$_x$ thin films at process temperatures lower than 400°C. The SiO$_2$ and SiN$_x$ thin films were deposited using various combinations of SiH$_4$, N$_2$O, N$_2$, and He gases as discussed in the following sections.

<u>SiO$_2$ Thin Films:</u> The SiO$_2$ deposition process was optimized in terms of the influence of the processing parameters on the thin film properties. At the first step, the effects of the HDP process on the film microstructure were analyzed by evaluating the breakdown field strength as a function of film thickness. As shown in Figure 9, the physical breakdown field strength of the MOS capacitors was maintained in the film thickness range of 10-100 nm. The observed results are promising as; in general, the film density and the electrical performance usually degrade with decreasing film thickness. The high breakdown field strength of the HDP processed SiO$_2$ thin films clearly suggests that the film density is well maintained even at a film thickness of about 10 nm which is

critical for the realization of sub-micron TFT devices and circuits integrated on low temperature substrates.

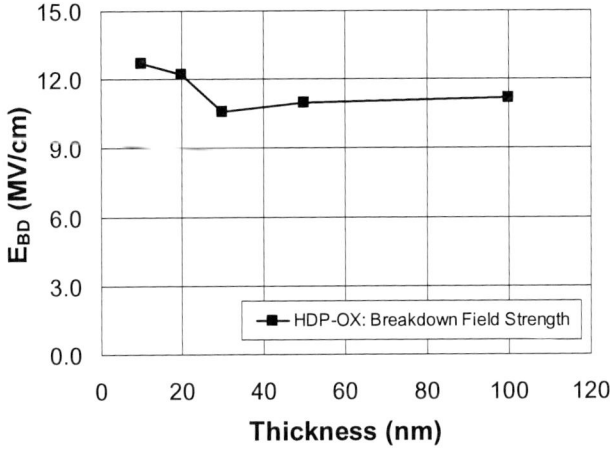

Figure 9. Effect of film thickness on the breakdown field strength of SiO_2 thin films.

One of the key advantages of the high density plasma technique is the process development at low pressures (<100 mTorr). The high plasma density and low plasma potential enable the deposition of high quality thin films with minimal plasma induced bulk or interface damage. The effects of the high density plasma on the SiO_2 thin film properties were further evaluated at a system pressure of 10 mTorr. The substrate temperature effects were analyzed in the range of 90-270 °C. A comparison with the TEOS oxide films, deposited at a temperature of 400 °C, was made to establish the thin film quality. The effects of the substrate temperature on the film density were analyzed in terms of the wet etch rate characteristics. The wet etch rate of the HDP processed SiO_2 thin films deposited in the temperature range of 90-270 °C was comparable to TEOS oxide films deposited at a temperature of 400 °C by a standard PECVD system employing CCP source. The observed results indicate that the film density was well maintained even at a substrate temperature of 90 °C.

The effects of the lower system pressure of 10 mTorr on the electrical properties of the MOS capacitors were also analyzed. Figure 10(a) shows the flat band voltage of the HDP processed SiO_2 thin films as a function of substrate temperature. The observed results show the key impact of the low pressure operation in high density plasma which promotes enhanced surface reactivity of the impinging plasma species. The leakage current characteristics of the high density plasma deposited SiO_2 thin films were far superior, as shown in Figure 10(b), to the TEOS oxide films deposited at a higher temperature of 400 °C. It was possible to further improve the electrical performance of HDP processed SiO_2 thin films by oxidation treatment as shown in Figure 10(a and b). The observed leakage current and flat band voltage of the SiO_2 thin films even at a low deposition temperature of 90 °C clearly indicate high bulk and SiO_2/Si interface quality.

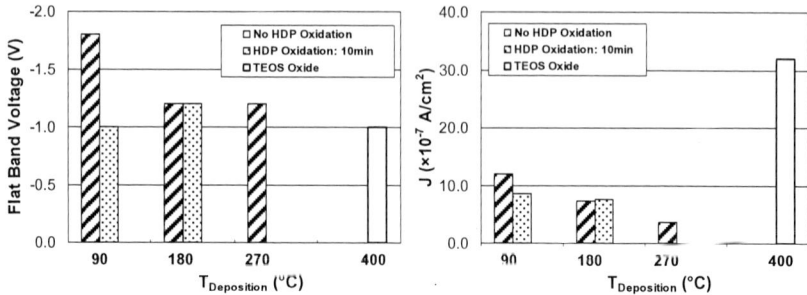

Figure 10. The effects of substrate temperature on the (a) flat-band voltage and (b) leakage current of high density plasma processed SiO_2 thin films.

The performance of low temperature TFTs strongly depends on effective defect passivation in the bulk of the dielectric and at the Si/dielectric interface. The forming gas anneal at high temperatures, exceeding 400°C, is effective in passivating the defects and providing stable device characteristics. However, flexible electronics demands low temperature and low thermal budget passivation of the dangling bonds and defects. The high density plasma generated reactive species offer efficient oxidation, nitridation, and hydrogenation possibilities at low thermal budgets. We were successful in significantly enhancing the leakage current characteristics at low temperatures; as shown in Figure 11, by efficient passivation of the defects by reactive hydrogen species generated in the high density plasma. A 30s exposure to high density H_2 plasma at a process temperature of 150°C was effective in reducing the leakage current of the oxide thin films by more than an order of magnitude which is significant for low temperature device development. Overall, the HDP hydrogenation process exhibits significantly higher efficiency at a much lower thermal budget as compared to forming gas annealing process in improving the electrical characteristics of MOS capacitors and TFTs.

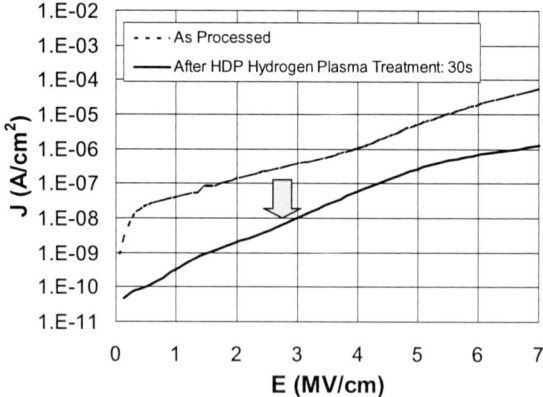

Figure 11. The active hydrogen species in high density plasma significantly improve the leakage current characteristics at a low processing temperature of 150°C.

SiN$_x$ Thin Films: The high density plasma deposition of SiN$_x$ thin films was carried out at 150°C to understand the process and thin film property control. Figure 12 shows the stress, wet etch rate, and optical properties of HD-PECVD SiN$_x$ thin films. It is clear from Figure 12. that the growth and properties of SiN$_x$ thin films can be effectively controlled by HDP technique at a low temperature of 150°C. It was possible to tune the stress in the films over a wide range by controlling the He gas flow rate while the optical refractive index of the films was maintained in the range of 1.99-2.01. The SiN$_x$ thin films with a refractive index value of 2.01 exhibited a low wet etch rate of 7.2Å/min in BOE(10:1) solution which is comparable to the value typically reported for thermal quality SiN$_x$ thin films. The N/Si ratio of 1.39, as established by RBS analysis, also indicated an efficient ionization of the reactive species in the plasma for the formation of stoichiometric thin films. The observed results on the SiN$_x$ thin films are promising and show the possibility of fabricating single or multilayer stacks, involving SiN$_x$ and SiO$_2$ thin films, on low temperature substrates to realize various electronic devices such as memories, TFTs, and MOS capacitors.

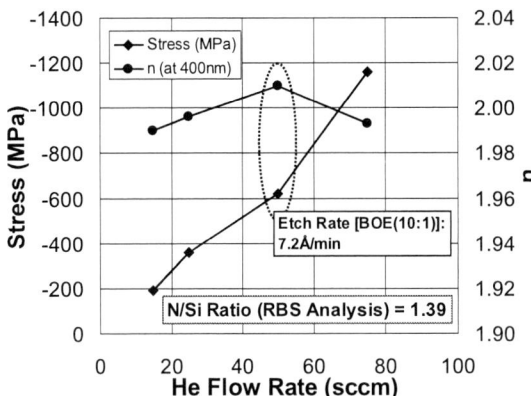

Figure 12. The effects of He flow rate on the stress, wet etch rate, and refractive index of SiN$_x$ thin films deposited at a low processing temperature of 150°C.

Device Reliability

MOS Capacitor: The electrical reliability of the HDP processed SiO$_2$ thin films was evaluated in terms of the bias temperature stress (BTS) effects on the high frequency C-V characteristics. The C-V curves were measured before and after BTS to analyze the impact on the flat band voltage. The BTS measurements were conducted at an applied electric field of -2MV/cm and a substrate temperature of 150 °C. The C-V curves were measured after a stress time of 30 minutes. As shown in Figure 13, the C-V curves showed a flat band voltage shift of about -4V for films deposited at a process temperature of 90 °C. The effects of the high density plasma oxidation on the BTS reliability of the films were also analyzed. After the high density plasma oxidation treatment, as shown in Figure 13, the flat band voltage of the films reduced to the levels measured for films deposited at a higher temperature of 270 °C. Additionally, the BTS induced shifts in the flat band voltage also decreased significantly to a value of -2.5V indicating significant improvement in the film microstructure. The observed results show the effectiveness of the high density plasma characteristics in processing high quality thin films at low temperatures suitable for integration on plastic.

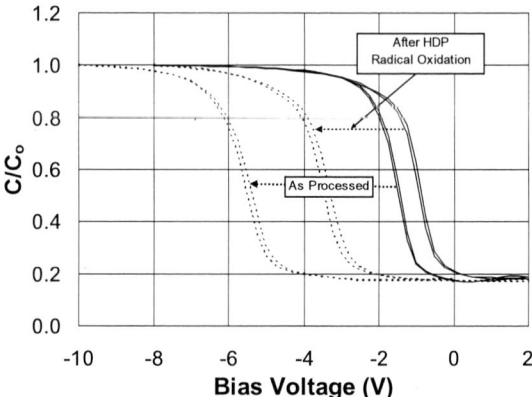

Figure 13. The bias-temperature stress (BTS) induced shift in the flat band voltage of MOS capacitors employing 90°C deposited SiO_2 thin films: (a) as-processed film, (b) after high density plasma oxidation treatment.

TFT Devices: Future high-performance TFT devices require thinner gate dielectric with enhanced reliability and stability. One of the key electrical parameters to fabricate reliable and stable TFT devices at smaller gate dielectric thicknesses is the leakage current. The gate leakage current is strongly influenced by the fabrication method, film thickness, and surface and interface quality. The reliability of the HD-PECVD SiO_2 thin films was evaluated in terms of the bias-temperature stress (BTS) effects on the TFT gate leakage characteristics. The BTS measurements were conducted on TFTs employing 300Å HD-PECVD SiO_2 thin films as gate dielectrics. Figure 14 shows the gate leakage characteristics of TFTs employing 30-nm-thick HD-PECVD and standard TEOS oxide films. The leakage current density of the HD-PECVD gate oxide TFTs did not show any appreciable dependence on the applied electric field up to 4 MV/cm and was limited by the measurement range of the instrument. It was not possible to analyze any film thickness or deposition technique dependence in this low field region up to 4 MV/cm. The field independence of the leakage current indicates that the current was not dictated by the bulk of the gate insulator but the semiconductor. The onset field for the HD-PECVD gate oxide TFTs was found to be higher as compared to the standard TEOS oxide TFTs indicating superior leakage current and interfacial quality of the HD-PECVD gate dielectric film. The HD-PECVD gate oxide TFTs exhibited at least two orders of magnitude lower leakage current density in the applied electric field range of 4-10 MV/cm. The observed high-field leakage characteristics suggest that the high-density plasma process is effective in minimizing the interface damage and maintaining film density at smaller thicknesses.

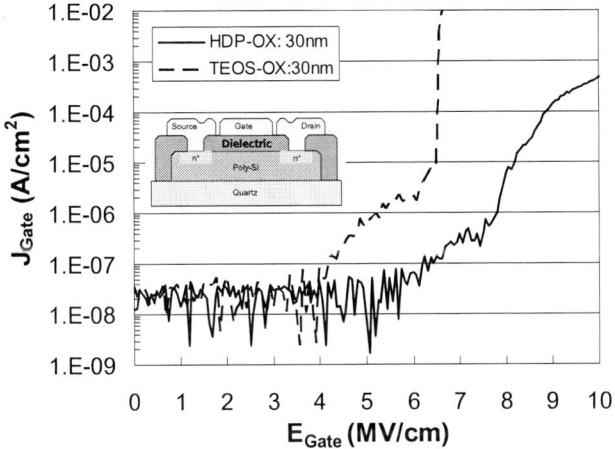

Figure 14. The gate leakage characteristics of TFTs employing HD-PECVD and standard TEOS SiO$_2$ thin films.

The interface and reliability characteristics of the TFTs become increasingly important with decreasing feature size, lower drive voltage, and low power consumption restrictions. The thermal stability measurements were conducted on 2×2 μm TFTs with 30 and 50 nm thick gate dielectric films deposited by HD-PECVD silane oxide and standard TEOS oxide processes. As highlighted in previous section, the TFTs employing TEOS gate oxide exhibited higher leakage current, as shown in Figure 14, as compared to those using HD-PECVD gate oxide. The gate leakage characteristics suggest that it is important to establish the TFT gate and transfer characteristics in close correlation with their thermal and bias stress reliability when evaluating device feasibility. The thermal stress measurements were conducted at a temperature of 150 °C in the ON and OFF TFT states for a time of 30min. Figure 15(a and b) show the effects of ON and OFF state thermal stress on the threshold voltage of the TFTs. The high-density PECVD deposited films showed a lower variation in the threshold voltage as compared to TEOS oxide films.

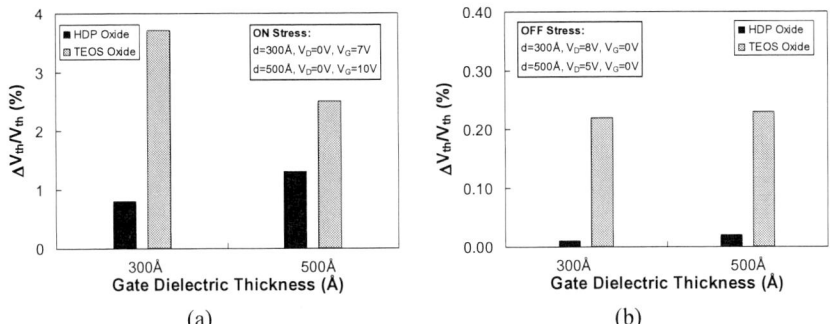

Figure 15. The (a) ON and (b) OFF state thermal stress effects on the threshold voltage of the TFTs employing 30 and 50nm thick HD-PECVD and TEOS oxides.

The HD-PECVD deposited 30 nm thick oxide films exhibited far superior threshold voltage stability characteristics as compared to TEOS oxide films. The observed thermal and bias stress measurements are consistent with the gate leakage characteristics of the TFTs. The TFTs employing HD-PECVD gate oxide films showed superior electrical and reliability characteristics. The observed results suggest that the high-density plasma process is effective in minimizing any plasma induced interface damage, and maintaining the dielectric quality and reliability even at smaller thicknesses.

Conclusions

The high density plasma technique shows promise for the low temperature processing of dielectric thin films for flexible electronic applications. We have successfully demonstrated the growth of high quality SiO_2 and SiN_x thin films on Si substrates at process temperatures lower than 200°C. The thin film growth rates in the high density plasma were found to be higher than the thermal growth rates at temperatures greater than 1000°C indicating efficient oxidation and nitridation of the Si substrate by reactive plasma species. The high density plasma process was also effective in the deposition of high performance oxide and nitride dielectric films at process temperatures as low as 90°C. The electrical characteristics of the high density plasma grown and deposited oxide films was found to be superior to TEOS oxide films deposited at a temperature of 400°C by standard PECVD technique employing capacitively coupled plasma source. The reliability of the MOS capacitors and TFT devices also reflected the superior bulk and interfacial characteristics of the HDP processed dielectric thin films. The overall performance of the SiO_2 and SiN_x dielectric thin films suggests that the low plasma potential and high plasma density are effective in minimizing the interface damage while enabling enhanced reaction kinetics in the plasma and on the substrate surface. The low processing temperatures suggest that the thin film properties are dominantly controlled by the high density plasma characteristics rather than the thermal state of the substrate; which is key to low temperature integration of flexible electronic devices.

References

1. T. Kaspar, A. Tuan, R. Tonkyn, W. P. Hess, J. W. Rogers, Jr., and Y. Ono, *J. Vac. Sci. Technol.* **B21**, 895 (2003).
2. T. Ueno, A. Morioka, S. Chikamura, and Y. Iwasaki, *Jpn. J. Appl. Phys.*, **39**, L327 (2000).
3. R. E. I. Schropp, *Thin Solid Films*, **517**, 3415 (2009).
4. J. Z. Chen and I.-C. Cheng, *J. Appl. Phys.*, **104**, 044508 (2008).
5. S-M Han, J-H Park, S-G Park, S-J Kim, and M-K Han, *Thin Solid Films*, **515**, 7442 (2007).
6. K. Diallo, M. Lemiti, J. Tardy, F. Bessueille, and N. Jaffrezic-Renault, *Appl. Phys. Lett.*, **93**, 183305 (2008).
7. A. Sazonov, M. Meitine, D. Stryakhilev, and A. Nathan, *Semiconductors*, **40**, 959 (2006).

8. T. Hattori, K. Azuma, Y. Nakata, M. Shioji, T. Shiraishi, T. Yoshida, K. Takahashi, H. Nohira, Y. Takata, S. Shin, and K. Kobayashi, *Appl. Surf. Sci.*, **234**, 197 (2004).
9. J. Moon, Y-H Kim, C-H Chung, S-J Lee, D-J Park, and Y-H Song, *Appl. Surf. Sci.*, **254**, 6422 (2008).
10. H. Kim, Y. Lee, Y. Ra, G. P. Li, and J. Yota, *ECS Trans.*, **6**, 531 (2007).

258

ECS Transactions, 35 (4) 259-272 (2011)
10.1149/1.3572288 ©The Electrochemical Society

Negative Charge in Plasma Oxidized SiO₂ Layers

A. Boogaard, A.Y. Kovalgin and R.A.M. Wolters

MESA+ Institute for Nanotechnology, Chair of Semiconductor Components,
University of Twente, P.O. Box 217, 7500 AE Enschede, The Netherlands.
Email: a.y.kovalgin@utwente.nl

Silicon dioxide (SiO_2) gate dielectric layers (4-60 nm thick) were
deposited (0.6 nm/min) on n-type Si by inductively-coupled plasma-
enhanced chemical vapor deposition (ICPECVD) in strongly diluted
silane plasmas at 150°C . In contrast to the well-accepted positive
charge for thermally grown SiO_2, the net oxide charge was negative
and a function of the layer thickness. Our experiments suggested
that the negative charge was created due to unavoidable oxidation of
the silicon surface by plasma species, and the CVD component
adding a positive space charge to the deposited oxide. The net
charge was negative under process conditions where plasma
oxidation played a major role. Such conditions included low
deposition rates and relatively thin grown layers. Additional
measurements showed that the negative charge in SiO_2 also
persisted on p-type substrates. We suggest that plasma oxidation of
the silicon surface results in SiO_2 layers with a surplus of oxygen.
This surplus of oxygen is able to accumulate a negative charge. This
assumption is addressed in this paper by a review of earlier work on
silicon oxidation, and by a first series of experiments wherein
oxygen is implanted into thermal SiO_2. It is shown that the
implantation can result in a negative charge to the bulk oxide layer.
The effect of the negative charge on the flatband voltage can be
described by the implantation profile.

Introduction

Present-day semiconductor device manufacturing involves hundreds of process steps.
Many of these steps are carried out at high temperatures, between 400 and 1000 °C.
There is however a great demand for lower temperature processing. An expanded number
of electronic devices (e.g., thin film transistors (TFT), non-volatile memory cells (NVM),
MEMS/MOEMS, etc.) can be realized when lower temperatures (20-400 °C) are applied
to the manufacturing process. One of the needs is to produce gate dielectrics (e.g. silicon
dioxide – SiO_2) at low substrate temperatures. The temperature reduction normally leads
to deterioration of the electrical and physical characteristics of SiO_2 layers such as
leakage currents, dielectric strength, fixed and mobile oxide charge, defect density at the
interface with silicon, etc.

In our previous study, we deposited SiO_2 films by Ar-N_2O-SiH_4 ICPECVD at
150°C, and a total pressure of 1-6 Pa. For the best-quality films, the gas-phase contained
0.08% of SiH_4 and 18% of N_2O. The films exhibited excellent I-V characteristics and low

259

interface state densities (D_{it}) [1-2]. These oxides were applied as gate dielectrics in low-temperature TFTs and demonstrated competitive mobility values and low off-currents.

In contrast to thermally grown SiO_2, in our ICPECVD-oxide the net charge appeared to be negative and a function of layer thickness [2]. We demonstrated that two mechanisms contributed to the film growth and charge formation, namely plasma oxidation of the silicon substrate and chemical vapor deposition. We suggested that the first nm-range of oxide thickness is formed by plasma oxidation that resulted in a negatively-charged interfacial oxide layer, while the CVD component added a positive charge to the bulk oxide.

In the current work, we extended the earlier results presented in [2] with a series of new experiments to better understand the physical nature of the negative charge observed in the ICPECVD-SiO_2 films. The negative charge might be of interest because of its novelty and the possible utilization in particular applications, e.g., for the reduction of surface recombination losses in photovoltaic devices by electrostatically shielding the minority charge carriers using internal electric fields [3].

Experimental

For this study, SiO_2 films were grown by means of ICPECVD in $Ar-N_2O-SiH_4$ plasma at 150°C and at a total pressure of 1 Pa, as described in [2, 4]. The gas phase contained 0.08% of SiH_4 and 18% of N_2O.

The films were deposited on H-terminated Si-wafers (n- or p-type <100>) that received standard cleaning. First, fuming nitric acid (HNO_3 100%) and boiling nitric acid (NHO_3 69%) were used in order to remove organic and metallic contaminants. The cleaning process was concluded by a 1% HF dip in order to remove the native silicon oxide. The substrates were rinsed with de-ionized water after all cleaning steps. The SiO_2 deposition process was monitored *in situ* using a J.A. Woollam M2000 spectroscopic ellipsometer (SE) with near-infrared (NIR) extension, to determine evolution of the layer thickness in time.

To electrically characterize the films, metal-oxide-semiconductor (MOS) capacitors were implemented by sputtering 1−μm Al over the oxide, followed by lithography and etching processes to define 0.06, 0.1, and 0.2 mm^2 square capacitors. An Al layer was also sputtered on the backside of the Si wafer. Some wafers were subjected to post-metallization annealing (PMA) for 10 min at 400°C in humid, ambient N_2 (N_2 bubbled through de-ionized water at room temperature). Selected wafers received a post-oxidation anneal (POA) in N_2 at 900°C for 30 min prior to the Al metallization.

The charges in thin SiO_2 films were detected by measuring their capacitance-voltage (C-V) characteristics. The high-frequency C-V measurements of the MOS structures were carried out by superimposing a small ac signal (10 kHz – 1 MHz) on a ramped dc bias between the Al gate and the substrate, by using a Hewlett-Packard 4275A multi-frequency meter. The quasi-static C-V curves were measured with a Hewlett-Packard 4140B pA meter, by applying only a dc bias with a sweep rate of 0.1 V/s. The bias was applied to the metal gate. The measurements started in inversion through depletion to accumulation, then back through depletion to inversion.

Results and Discussion

In Figure 1, *in situ* SE measurements (i.e., real-time observation of the SiO_2 growth) are presented from two experiments. The first experiment (open circles) represents the first few minutes of a typical deposition process in $Ar-N_2O-SiH_4$ plasma. Whereas the second experiment (solid triangles) is carried out without SiH_4 in the gas phase, i.e., in $Ar-N_2O$ plasma. Clearly, the oxide growth can also be observed without SiH_4, i.e., by plasma oxidation (solid triangles). This leads to the conclusion that (at these process conditions) the formation of silicon dioxide films is due to two mechanisms: oxidation and CVD. The inset of Figure 1 shows the calculated growth rates of the oxidation- and CVD-components. One can observe a non-linear fast-initial oxide-growth regime that gradually transfers into a linear regime with a constant deposition rate of 0.6 nm/min. We conclude that initial oxide formation is due to the oxidation of Si. Plasma oxidation dominates for the first 2.5 nanometers of the oxide growth, followed by mainly CVD for the thicker layers.

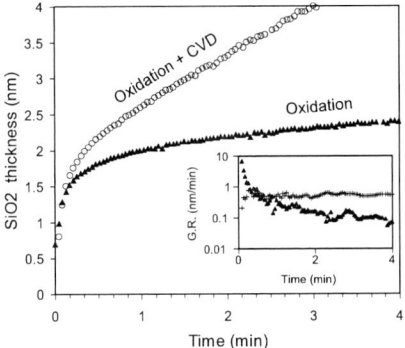

Fig. 1. *In situ* SE thickness measurements without (▲) and with (o) silane (0.08%). In the inset, the growth rates are shown for the CVD (+) and oxidation (▲) components.

One can expect the electrical properties to vary for the layers formed by the two different mechanisms. This variation can be observed particularly for very thin layers, when thicknesses of the two differently formed sub-layers are comparable. Instead of a positive charge, as normally is observed in SiO_2 films, a *negative* oxide charge of $5 \cdot 10^{11}$ cm^{-2} can be calculated from the high-frequency $C-V$ curves [2].

The oxide charge is studied in more detail as a function of the layer thickness. The extracted value of the oxide charge is largely dependent on the absolute value of the metal-semiconductor workfunction difference, φ_{ms}. There is however variation in published φ_{ms} values. As an example, for Al on n-type Si with a doping level of $1.5 \cdot 10^{15}$

cm^{-3}, Sze [5] published a value of -0.2 V while Pierret [6] reported -0.3 V. To determine φ_{ms} for our system, φ_{ms} was measured on thermally grown oxide (dry oxidation at 950°C followed by 20-min POA). For thermally grown oxides, the flatband voltage can be calculated using the standard expression:

$$V_{FB} = \varphi_{ms} - \frac{Q_f T_{ox}}{K_{ox}\varepsilon_0},$$ (1)

where Q_f is the fixed charge near the Si-SiO$_2$ interface, T_{ox} is the oxide thickness, K_{ox} is the oxide dielectric constant and ε_0 is the permittivity of vacuum.

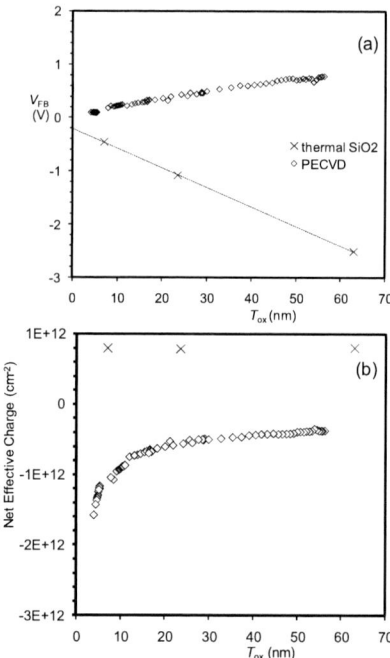

Fig. 2. Flatband voltage (a) and Net Oxide Charge (b) plotted versus oxide thickness; n-type substrates and Al-gates; after PMA. (X) Reference oxides grown by dry oxidation at 950°C followed by a POA; (◊) the ICPECVD oxides. The slope of the dotted line is proportional to the fixed charge while the intercept equals φ_{ms}.

Figure 2 shows a plot of V_{FB} versus oxide thickness for thermally grown oxides, and for our ICPECVD layers. If we compare the curves of thermal oxides and the ICPECVD oxides in Figure 2a, we observe different signs of the slopes, indicating positive and negative net-charges. The data set of thermally grown oxides has a slope of $-Q_f T_{ox}/K_{ox}\varepsilon_0$ and an intercept on the V_{FB} axis of $\varphi_{ms} = -0.214$ V (Q_f is assumed to be the same and near the Si interface for all data points) [7]. The φ_{ms} thus obtained can be used to calculate Q_f. These values are shown in Figure 2b. A *positive* charge of $8 \cdot 10^{11}$ cm^{-2} was calculated for the thermally grown oxide (see crosses in Figure 2b). If Q_{ox} is calculated for the given ICPECVD oxides using the expression above, it appears that the thinner layers contain a higher amount of negative charge compared to the thicker films (see open diamonds in Figure 2b). We attribute this important and novel result to the initial plasma oxidation step, which cannot be avoided (see Fig. 1). Thus, the plasma-oxidized region near the Si-interface becomes dominant for thinner layers. This plasma oxide may therefore be responsible for the negative charge formation.

The effect of the negatively charged plasma oxide on flatband voltage can be described by general equation:

$$V_{FB} = \varphi_{ms} - \frac{1}{C_{ox}} \int_0^{T_{ox}} \frac{x}{T_{ox}} \rho_{ox}(x) dx, \tag{2}$$

where x is the distance from the gate, T_{ox} is the thickness of the oxide, and $\rho_{ox}(x)$ is the oxide charge density in a volume [8]. A determination of the charge distribution and its location in the oxide is needed to solve the integral. For that, we assume a homogeneous distribution of the negative charge density (ρ_{PO}) in oxide volume between the silicon interface, $x = T_{ox}$, and $x = T_{ox} - T_{PO}$, where T_{PO} is the thickness of the plasma-oxidized region (see Figure 3). Integrating reveals

$$V_{FB}^{PO} = \frac{\rho_{PO} T_{PO} (2T_{ox} - T_{PO})}{2 C_{ox} T_{ox}}, \tag{3}$$

where V_{FB}^{PO} is the flatband voltage change due to the plasma oxide. Equation (3) simplifies to Q_{PO}/C_{ox} when T_{ox} is much larger than T_{PO} ($Q_{PO} = \rho_{PO} \cdot T_{PO}$).

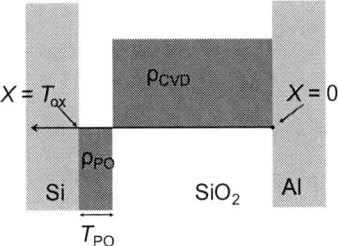

Fig. 3. Graphical representation of the suggested double-layer model. To clarify the actual (inhomogeneous) charge distribution, an additional study is required.

The pure CVD component adds the expected positive charge, thus compensating for the interface-located negative charge (the thicker the film, the more negative charge is compensated). To describe the effect of the positively charged CVD oxide on flatband voltage, we assume a homogeneous distribution of the positive charge density (ρ_{CVD}) in the volume between the gate, $x = 0$, and $x = (T_{ox} - T_{PO})$, see Figure 3.

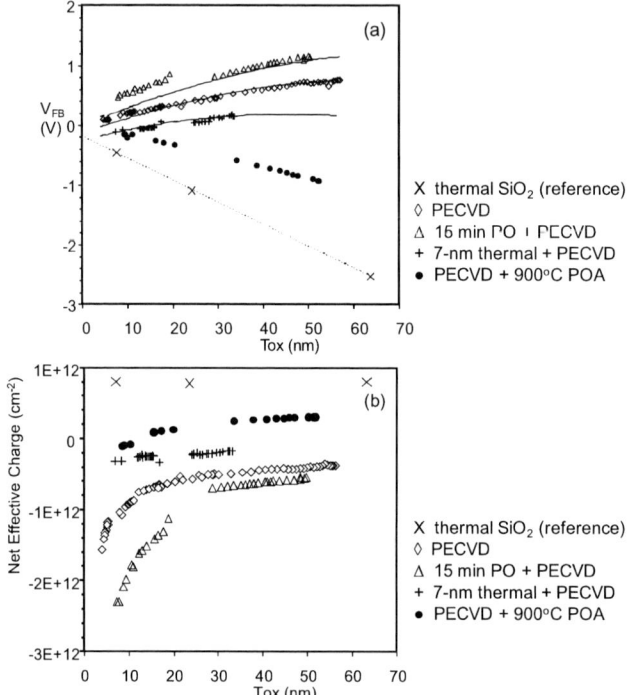

Fig. 4. Flatband voltage (a) and Net Oxide Charge (b) plotted versus oxide thickness; n-type substrates and Al-gates; after PMA. (X) Reference oxides grown by dry oxidation at 950°C followed by a POA; (◊) ICPECVD oxides; (Δ) 15-min plasma oxidation followed by ICPECVD; (+) ICPECVD oxides deposited on 7-nm thick thermal SiO_2; (●) ICPECVD oxides with POA and PMA. The slope of the dotted line is proportional to the fixed charge while the intercept equals φ_{ms}. Solid lines represent double-layer model fitting (see text).

Integrating equation (2) reveals

$$V_{FB}^{CVD} = \frac{\rho_{CVD}(T_{ox} - T_{PO})^2}{2C_{ox}T_{ox}}, \qquad (4)$$

where V_{FB}^{CVD} is the flatband voltage change due to the CVD oxide. Equation (4) simplifies to $Q_{CVD}/2C_{ox}$ for $T_{ox} >> T_{PO}$ ($Q_{CVD} = \rho_{CVD} \cdot T_{ox}$). A uniform charge distribution can occur if defects, such as silicon dangling bonds or plasma damage, are continuously added to the oxide during its deposition.

The effect of the two layers on the flatband voltage can now be calculated. The double layer model fits our data (open diamonds in Figure 4a) given a *negative* oxide charge density ρ_{PO} of $2.3 \cdot 10^{18}$ cm^{-3} within T_{PO}=2.5 nm, and a *positive* space charge density ρ_{CVD} of $9 \cdot 10^{16}$ cm^{-3}.

We further performed additional experiments to confirm the suggested double-layer model. These are presented in Figure 4. The results of thermally grown oxide and our standard ICPECVD process are included in Figure 4 for reasons of clarity. In a first experiment, initial oxide was grown by plasma oxidation only, i.e., in Ar-N$_2$O plasma without SiH$_4$. After 15 minutes of oxidation, SiH$_4$ was introduced into the system and the ICPECVD mode was thus activated. The prolonged plasma oxidation increased the flatband voltage (see open triangles in Figure 4a). Our model described this effect by an increase in T_{PO} from 2.5 to 2.9 nm with the same negative ρ_{PO} of $2.3 \cdot 10^{18}$ cm^{-3}, leaving the ρ_{CVD} unchanged (i.e. $9 \cdot 10^{16}$ cm^{-3}). The latter indicated that the pure CVD mode was not influenced.

In Fig. 4, one can notice a mismatch between the open triangles for T_{ox} in the range between 20 and 30 nm. This is due to the fact that two separate-in-time series of experiments were used to plot the entire curve. As the amount of negative charge is very sensitive to the initial PO step, a small deviation in the PO conditions between the series could cause the observed mismatch. Only the triangles for $T_{ox} > 30$ nm were used for fitting.

In a second experiment, we performed ICPECVD on a 7-nm thick thermally grown SiO$_2$, to minimize the influence of the initial plasma oxidation step (see plus signs in Figure 4a and Figure 4b). We obtained much less negative-charge (ρ_{PO} of $5 \cdot 10^{17}$ cm^{-3}) again without changing ρ_{CVD}, which reflected a similar trend of the flatband voltage in relation to oxide thickness. However, the net effective charge was still negative, indicating that plasma oxidation could not be ruled out completely.

The negative charge can be reduced during POA (see solid circles in Figure 4b), resulting in a net effective positive charge for the thicker layers (15-50 nm). However, the charge for thinner layers remains negative.

Negative effective charges were reported occasionally in PECVD silicon oxides when relatively thin layers (10-50 nm) were deposited in highly-diluted plasmas at low deposition rates [9-11]. Negative charges were also reported for silicon dioxide layers grown solely by plasma oxidation [12-13]. These publications support our conclusion on the influence of plasma oxidation on the oxide charge that always occurs parallel to deposition. With this in mind, a detailed study and model of oxidation mechanisms is needed. The majority of the PECVD oxide layers are deposited at much higher rates, and films are usually (considerably) thicker. These conditions are expected to minimize the influence of plasma oxidation, and a positive oxide charge is likely to be measured.

The electronic nature of the negative charge

Additional measurements showed the impossibility to de-trap the negative charge by applying a negative voltage to the Al gate. This indicates that charge traps, that accumulate the negative charge in our material, have energy levels in the SiO$_2$ band gap situated below the Fermi level of a substrate, see Figure 5. For n-type Si (the experiments described above), the electron traps may lie i) within Si band gap (blue levels), or ii) below the Si valence band offset (red levels). The next step would be to narrow down this energy interval. For case i), one should observe no negative charge accumulation on p-type substrates. Figure 6 however indicates that the negative charge in SiO_2 persists on p-type substrates. Therefore, we conclude that the charge trap levels are located below the Si valence band edge (red levels).

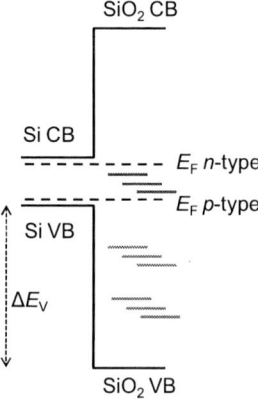

Fig. 5. Energy band diagram of the Si-SiO_2 structure under flatband conditions; $\Delta E_v = 4.4$ eV [14].

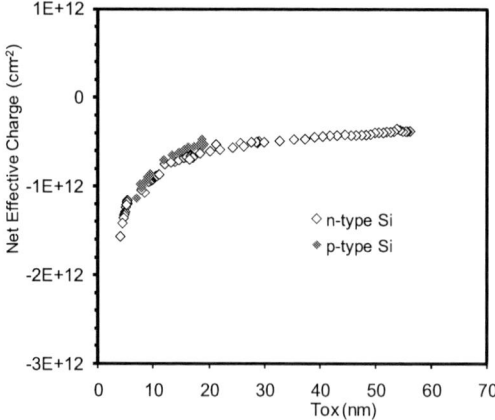

Fig. 6. Net Oxide Charge plotted versus oxide thickness; n-type and p-type substrates and Al-gates; after PMA.

The physical nature of the negative charge

During the last two decades, a number of theoretical models of silicon oxidation in the ultra-thin regime were constructed in order to surpass limitations of the commonly used Deal-Grove model. The Beck-Majkusiak model gives precise predictions even for ultrathin oxide thickness regime, for both classical oxidation in a furnace and processing in a rapid thermal oxidation (RTO) reactor, and is consistent with description of plasma oxidation processes [15]. The Beck-Majkusiak model assumes that the oxidation rate in the first phases is limited by equilibrium between the forward flux of tunneled or thermo-emitted electrons, ionizing oxygen atoms at the outer SiO_2 surface, and the return flux of the ionized species that diffuses back to Si through the already grown oxide. According to the model, the volume of SiO_2 should be full of negatively ionized oxygen atoms. If one would suddenly freeze this distribution, the total effective charge density should then be negative. This cannot be confirmed experimentally for thermal oxidation since the high temperature will also anneal and reduce the negative charge. This is also true for our ICPECVD layers, which exhibit a positive charge after POA; see solid circles in Figure 4b. For the experiments without POA, however, the temperature is much lower, which may lead to the 'freezing' effect.

Therefore, the existence of the negative charge can be related to a surplus of oxygen in the PO layer. The extra (over-stoichiometric) oxygen can be incorporated e.g. in the form of a peroxy bridge ($O_3\equiv Si–O–O–Si\equiv O_3$) [16], or non-bridging O atom (i.e., formation of two $O_3\equiv Si-O\cdot$ $\cdot O-Si\equiv O_3$ groups with oxygen dangling bonds instead of one $O_3\equiv Si-O-Si\equiv O_3$ group). It is well-known that such over-stoichiometric SiO_x does not exist in thermally-oxidized layers [16], besides some peroxy bridges in a very low concentration. Over-stoichiometric SiO_x is not stable at temperatures typical for thermal oxidation. It can possibly exist in the lower-temperature materials. This is supported by our earlier study on plasma deposition of SiO_2 films at 100°C, where we measured (by XPS) a correlation between the negative charge and the O/Si ratio slightly higher than 2 for thin films [10].This is also in agreement with the work of Afanas'ev and Stesmans, who mentioned the creation of $O_3\equiv Si-OH$ centers (by irradiation with 10 eV photons) as neutral traps for electrons [17].

The extra oxygen can form negatively-charged O_2^- ions after trapping electrons. The computations of Ewig and Tellinghuizen showed that O_2^- is stable against electron auto-detachment in ionic crystals [18]. Although the ICPECVD oxide is amorphous, we speculate that the O_2^- ions can also be stable within the silica network. Salh, Von Czarnowski, and Fitting have studied over-stoichiometric SiO_x by cathode-luminescence (CL) [19-21]. For that, they implanted O^+ ions into dry-oxidized SiO_2 layers. An interesting result was the multiple regular-shaped spectra in the green-near IR (500–820 nm) region. The sub-band positions corresponded to almost equidistant energy steps of about 120 meV. Based on [18], they associated their findings to the absorption and emission spectra of O_2^- ions [20].

C-V measurements of SiO_2:O layers

The presence of O_2^- (or any other form of negative ions) is expected to induce a shift of flatband voltage as measured by *C-V* measurements. Our *C-V* measurement results on the oxygen-implanted thermal-SiO_2 samples (kindly provided by Dr. R. Sahl of Rostock University) are presented in this section. Briefly, the SiO_2-layers prepared by dry oxidation (214-nm thick) where implanted at 45 keV with O^+ at doses of 1E+16 cm^{-2}, 5E+16 cm^{-2} and 1E+17 cm^{-2}. The samples were then annealed in N_2 at 1000°C for 30 min. Further processing at MESA$^+$ involved deposition of Al through shadow masks to define capacitors, and PMA at 400°C for 10min in humid N_2.

We will start our discussion with describing the distribution of the implanted ions. The next step will be to formulate an expression that can be used to calculate the influence of the oxide charges on the measured flatband voltage, V_{FB}. It should account for Q_f at Si-SiO_2 interface, and for Q_{bulk} according to the implantation profile. Finally, the measurement results will be presented.

According to the Lindhard, Scharff, and Schiott (LSS) model [22], the ion concentration as a function of depth, $\rho_{ion}(x)$, in amorphous materials can be described by a Gaussian curve:

$$\rho_{ion}(x) = \frac{\theta}{\sqrt{2\pi}\Delta R_p} e\left[-\frac{(x - R_p)^2}{2\Delta R_p^2}\right] \tag{5}$$

where θ is the implantation dose (ions/cm^2), ΔR_p is the standard deviation of the Gaussian distribution (also known as projectred straggle), and R_p is the projected range (see Figure 7). R_p and ΔR_p can be calculated from the mass of the implanted ions and the target atoms, and the mass density of the target material [22]. The results of these calculations (using Silvaco TCAD tools) are summarized in Table 1 and plotted in Figure 7. The table and figure show that the maximum of the distribution can be found at approximately half the oxide thickness. Please note that the SiO_2-Si interface, located at 0.214 µm, is not included into the simulation. Oxygen atoms will be implanted at substantial concentrations into the Si, but their exact distribution in the silicon is not relevant for the considerations we make. It is important however to be aware of the effect, because such concentrations (around $5 \cdot 10^{19}$ cm^{-3} at the SiO_2-Si interface) are likely to deteriorate the interface and quality of Si, and might thus affect the C-V measurements. On the other hand, the damage of Si and the interface can be restored by the following annealing step at 1000°C for 30 min [23]. Choosing a lower implantation energy (e.g., 20 keV) would also minimize the effect.

Table 1. Calculated parameters used to describe the distribution of the implanted O^+ ions (with dose θ) into amorphous SiO_2.

Implantation energy		45 keV	
θ (ions/cm^2)	1E+16	5E+16	1E+17
R_p (nm)		100	
ΔR_p (nm)		36	
$\rho_{ion}(x = R_p)$ (cm^{-3})	1.1E+21	5.6E+21	1.1E+22

The expression used to calculate the influence of the oxide charges on V_{FB} should account for ρ_f ($Q_f = \rho_f T_f$) at Si-SiO$_2$ interface, and for ρ_{bulk} ($Q_{bulk} = \rho_{bulk} \cdot T_{ox}$) according to the implantation profile. The effect of the fixed charge, ρ_f, on the flatband voltage can be calculated using equation (3). To describe the effect of the negatively charged implanted oxide, having the charge density ρ_{bulk}, on flatband voltage, we insert the simulated implantation profile (equation (5)) of the oxygen atoms (ρ_{bulk}) into equation (2) (see Figure 8). Integrating reveals

$$V_{FB}^{bulk} = -\frac{1}{2} \frac{\theta\left[-\sqrt{2}\Delta R_p e^{-\frac{R_p^2}{\Delta R_p^2}} - \sqrt{\pi}R_p erf\left(\frac{\sqrt{2}R_p}{2\Delta R_p}\right) + \sqrt{2}\Delta R_p e^{-\frac{(R_p-t_{ox})^2}{\Delta R_p^2}} + \sqrt{\pi}R_p erf\left(\frac{\sqrt{2}(R_p - T_{ox})}{2\Delta R_p}\right)\right]}{\sqrt{\pi}C_{ox}T_{ox}}, \quad (6)$$

However, it is assumed that only a fraction, $f(\theta)$, of the implanted ions is electrically active as O$_2^-$. So, a small change in the expression is necessary: θ is to be replaced by $f(\theta) \times \theta$.

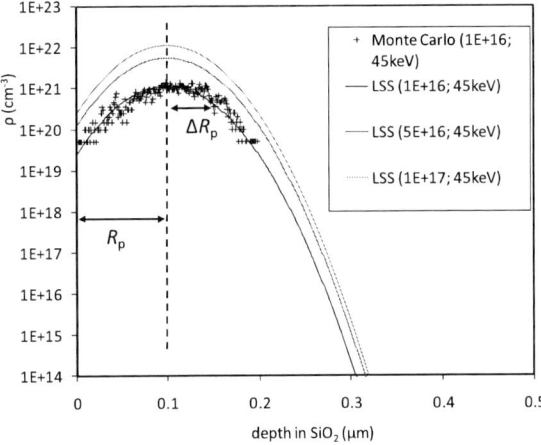

Fig. 7. The implantation profiles of O$^+$ implanted into amorphous SiO$_2$. The profiles are shown for both the LSS model and Monte Carlo simulations.

The measured C-V curves are shown in Figure 9. The figure reveals that V_{FB} of the SiO$_2$:O films is more negative than φ_{ms}. (The V_{FB} of the reference SiO$_2$ thermally grown on n-type Si and without implantation of oxygen was measured to be -0.214 V, see text under Fig. 2.) On one hand, the O$^+$-implanted films clearly have an increased positive effective charge (cf. equation (3) and Figure 9) compared to the reference SiO$_2$. This could be due to the mentioned deterioration of the interface after implantation. On the other hand, V_{FB} clearly shifts towards φ_{ms} for the higher implantation doses. This means that the effective positive charge is reduced when the implantation dose is increased. To explain this observation, one should bear in mind that thermal oxidation is known to create a positive charge (Q_f) near the Si-SiO$_2$ interface. This Q_f is partly compensated by a negative charge of the over-stoichiometric oxygen ions additionally implanted into the bulk oxide. A higher implantation dose will thus reduce the effective positive charge to lower values.

The effect of the implantations on the flatband voltage can also be calculated by combining equations (3) and (6). The results are shown in Table 2. Our model can describe the measured flatband voltages by a constant Q_f of $3.2 \cdot 10^{+11}$ cm^{-2} for all the samples. This indicates that annealing can equally restored the Si-SiO$_2$ interface for all the films. The Q_f is partly compensated by the negatively charged oxygen atoms.

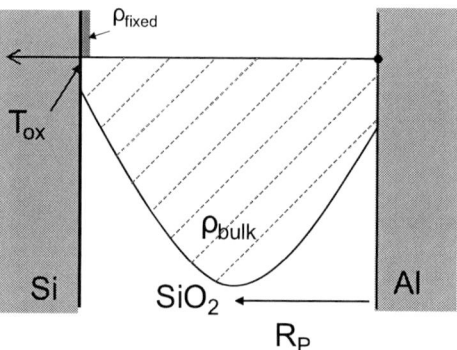

Fig. 8. Graphical representation of the charge distribution in the Si$_2$O:O (i.e., O$^+$-implanted) layer.

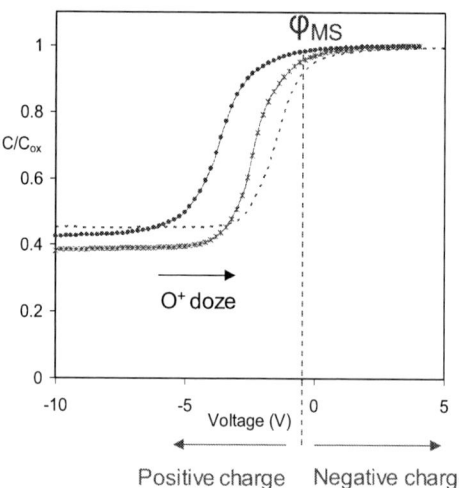

Fig. 9. Measured C-V curves of the SiO$_2$:O layers. The O$^+$ implantation dose (see Tables 1 and 2 for details) increases from left to right.

According to our model, only a small fraction (about $5 \cdot 10^{-6}$) of the implanted ions is electrically activated in the form of negatively charged ions. The negative-charge distribution can easily be found from Figure 7 by multiplying $\frac{1}{2} \cdot \rho_{bulk}$ by the activation fraction from Table 2 ($\frac{1}{2}$ is added because 2 implanted oxygen atoms are needed to form one O_2^-). The highest concentration (according to our model) of O_2^- is $2 \cdot 10^{+16}$ cm^{-3} (located at R_p and for $\theta = 1 \cdot 10^{+17}$ ions/cm^2). This is just below the maximum solubility of O_2 in SiO_2, which is $5 \cdot 10^{+16}$ cm^{-3} [24], and is therefore a physically feasible and relevant number.

Table 2. Calculated (based on the *C-V* data of Fig. 9) V_{FB} and Q_f as a function of the oxygen implantation dose into amorphous SiO_2.

θ (ions/cm^2)	1E+16	5E+16	1E+17
measured V_{FB} (V)	-3.09	-1.93	-0.94
modelled V_{FB} (V)	-3.08	-1.92	-0.93
$f(\theta)$ (-)	6.3E-06	5.3E-06	4.4E-6
Q_f (cm^{-2})		3.2E+11	

Conclusions

The net charge of the studied ICPECVD-SiO_2 films was found to be negative, and was a function of the film thickness. This was explained by the plasma oxidation of silicon, which added a negative charge ($2.3 \cdot 10^{18}$ cm^{-3}) to the interfacial oxide layer (2.5-nm thick), while the CVD-component added a nearly homogeneous distribution of a positive charge ($9 \cdot 10^{16}$ cm^{-3}) to the bulk oxide. The charge traps of the negatively-charged PO layer are located below the Si valence band edge.

The thermally-grown O^+-implanted SiO_2 films clarified the physical nature of the negative charge. Although the net-oxide charge was positive, it was a function of the implantation dose. Our model assumed a positive fixed charge at the interface ($3.2 \cdot 10^{+11}$ cm^{-2}) and a negative charge, distributed similarly to the implantation profile. We therefore attributed the negative charge to a surplus of oxygen. This over-stoichiometric oxygen was possibly able to accumulate a negative charge.

Acknowledgements

The authors are very grateful to Dr. R. Salh of Rostock University for providing the thermally-grown O^+-implanted SiO_2 samples.

References

1. I. Brunets, J. Holleman, A. Y. Kovalgin, A. Boogaard and J. Schmitz, *IEEE Transactions on Electron Devices*, **56**, 1637 (2009).

2.	A. Boogaard, A. Y. Kovalgin and R. A. M. Wolters, *Microelectronic Engineering*, **86**, 1707 (2009).
3.	B. Hoex, J. J. H. Gielis, M. C. M. v. d. Sanden and W. M. M. Kessels, *Journal of Applied Physics*, **104**, 113703 (2008).
4.	A. Boogaard, A. Y. Kovalgin, I. Brunets, A. A. I. Aarnink, J. Holleman, R. A. M. Wolters and J. Schmitz, *Surface and Coatings Technology*, **201**, 8976 (2007).
5.	S. M. Sze and K. K. Ng, in *Physiscs of Semiconductor Devices*, third edition ed., p. 197, Wiley-Interscience, New York (2007).
6.	R. F. Pierret, in *Semiconductor Device Fundamentals*, p. 649, Addison Wesley Longman, Amsterdam (1996).
7.	A. S. Grove, in *Physics and technology of semiconductor devices*, p. 345, John Wiley and Sons, Inc., New York (1967).
8.	D. K. Schroder, in *Semiconductor material and device characterization*, 3rd ed., p. 321, John Wiley and Sons, Inc., Hoboken (2006).
9.	S. S. Iyer, P. M. Solomon, V. P. Kesan, A. A. Bright, J. L. Freeouf, T. N. Nguyen and A. C. Warren, *Electron Device Letters, IEEE*, **12**, 246 (1991).
10.	A. Y. Kovalgin, G. Isai, J. Holleman and J. Schmitz, *Journal of The Electrochemical Society*, **155**, G21 (2008).
11.	M. J. Hernandez, J. Garrido, J. Martinez and J. Piqueras, *Semiconductor Science and Technology*, **11**, 422 (1996).
12.	T. Yasuda, Y. Ma, S. Habermehl and G. Lucovsky, *Applied Physics Letters*, **60**, 434 (1992).
13.	M. Tabakomori and H. Ikoma, *Japanese Journal of Applied Physics Part 1-Regular Papers Short Notes & Review Papers*, **36**, 5409 (1997).
14.	J. Robertson, *Journal of Non-Crystalline Solids*, **303**, 94 (2002).
15.	R. B. Beck, *Materials Science in Semiconductor Processing*, **6**, 49 (2003).
16.	C. R. Helm and B. E. Deal, *The Physics and Chemistry of SiO_2 and the $Si–SiO_2$ Interface*, Plenum, New York (1993).
17.	V. V. Afanas'ev and A. Stesmans, *Applied Physics Letters*, **71**, 3844 (1997).
18.	C. S. Ewig and J. Tellinghuisen, *The Journal of Chemical Physics*, **95**, 1097 (1991).
19.	H. J. Fitting, T. Barfels, A. N. Trukhin, B. Schmidt, A. Gulans and A. von Czarnowski, *Journal of Non-Crystalline Solids*, **303**, 218 (2002).
20.	H. J. Fitting, R. Salh, T. Barfels and B. Schmidt, *physica status solidi (a)*, **202**, R142 (2005).
21.	R. Salh, A. von Czarnowski and H. J. Fitting, *Journal of Non-Crystalline Solids*, **353**, 546 (2007).
22.	J. Lindhard and M. Scharff, *Physical Review*, **124**, 128 (1961).
23.	S. Wolf and R. N. Tauber, in *Silicon processing for the VLSI Era. Vol. 1. Process technology*, 1st ed., p. 198, Lattice Press, Sunset Beach (1986).
24.	N. Mott, *Proceedings of the Royal Society of London. Series A, Mathematical and Physical Sciences*, **376**, 207 (1981).

Optical And Electrical Properties Of Si-based Multilayer Structures For Solar Cell Applications

R.Pratibha Nalini, J.Cardin, K.R Dey, X. Portier, C.Dufour and F.Gourbilleau

CIMAP UMR CNRS/CEA/ENSICAEN/UCBN, 6 Boulevard Maréchal Juin 14050 CAEN, FRANCE

> Among the materials of interest in the field of solar cells, Silicon Rich Silicon Oxide (SRSO) has proved convincing results. In order to benefit from the carrier confinement within the active material, a multilayered arrangement of SRSO alternated with an insulating matrix is preferred. SiO_2 matrix has been used as a host matrix, but its limitations appear due to the poor electrical conductivity. Hence, SiN_x has become an alternative choice for such matrices in the recent past. This work compares the structure and luminescence properties of $SRSO/SiO_2$ and $SRSO/SiN_x$ multilayers and demonstrates the interest in replacing SiO_2 sublayer by SiN_x. Optical and electrical characterizations have been made on these materials.We eventually propose a promising type of multilayer structure with enhanced photoluminescence and electrical conductivity with a control over the thermal budget.

Introduction

Solar cell research is a hot topic in the field of energy production since ways must be found in order to decrease the production cost and increase the efficiency of device performances. The compatibility with the existing technologies also becomes a challenge faced by this research sector. Research on solar cells can be categorized into three directions: i) improving the already existing technologies, ii) developing new technologies based on new architectural designs and iii) developing new materials to serve as light absorbers, creators of charge carriers and transporters.

Silicon has dominated the semiconductor industry and microelectronic industry. But, the indirect band gap of silicon prevents the efficient conversion of electronic energy into photons and thus offers limitations to the photonic needs. When reducing the size down to nanometer scale, silicon based materials became attractive in the field of photonics. For sizes lower than the exciton Bohr radius, there is a widening of the bandgap and also a possibility of direct transition from conduction to valence band. This makes the photoluminescence in the visible range more efficient. Among the various models, the "quantum confinement theory" is widely accepted as the most successful model (1,2,3) With the discovery of visible luminescence from porous Si in 1990 (1), silicon nanostructures started to dominate the nanophotonic research. Porous silicon and nanocrystalline Si structures became the materials of interest and a wide range of luminescent wavelengths has been observed. The visible luminescence was attributed to the quantum confinement effect (4,5). But, the most commonly observed blue or green luminescence from porous silicon and nanocrystalline silicon was attributed to defects (6,7).

The solar cells in general require the use of a material that is able to absorb the incident photons over the whole solar spectrum and convert them into electric current. The thin film approach helps in reducing the cost of the light absorbing material by reducing the amount of material in comparison to the bulk approach. One of the most promising materials for the third generation of solar cell is based on nanostructured Si-based thin films. These nanostructured Si based thin films take advantage of the size dependent quantum confinement effect for managing the Si light absorption band. Since numerous studies have been made already on silicon and due its compatibility with the existing technologies, research on silicon based solar cells has gained much importance. Though porous Si exhibited intense luminescence in the visible region, its lack of mechanical strength triggered research on alternate Si based nanostructures. Quantum dot and multilayer approaches became the two major means of obtaining Si based nanostructures. The former approach involves nanograins of silicon embedded in an insulating matrix. Common fabrication techniques of quantum dots include Si^+ ion implantation in thermally grown silica (8), PECVD (9) and Magnetron Sputtering (10,11). The multilayer approach of Si has been adopted since the recent past. It involves a sandwich type of structure where the Si sublayer is sandwiched between two insulating layers, the most comon insulating layer being SiO_2. This kind of multilayer structures can be obtained by various methods that include Molecular Beam Epitaxy (12), PECVD (13,14) and magnetron sputtering (15,16).

Systems composed of silicon nanoclusters (Si-nc) embedded in an amorphous matrix constitute an important part of the ongoing research. The matrix could be silicon oxide or amorphous silicon. Though SiO_2 is the most common matrix, the problem of carrier injection comes as a major drawback owing to the large bandgap of SiO_2. Hence, the replacement of SiO_2 by other dielectric matrices with smaller bandgap than SiO_2 turns out to be the solution. SiN_x matrix meets up these requirements. Visible photoluminescence from Si-nc embedded in SiN_x matrix has been recently demonstrated (15,17-19). However, the reason for luminescence and the underlying mechanism is still under hot debate.

This work is a comparative study on the photoluminescence and electrical conductivity of the two different host matrices (SiO_2 and SiN_x) within multilayered structures such as silicon rich silicon Oxide (SRSO)/SiO_2 and SRSO/SiN_x films.

Experimental

The multilayers of SRSO/SiO_2 were grown using the 'reactive magnetron sputtering' technique elaborated by our team using a pure silica target (20). The term "reactive" is used since the process takes advantage of the capability of hydrogen to reduce the oxygen in the plasma from the sputtered target. In order to facilitate a higher incorporation of Si, both SiO_2 and Si cathodes were used for the fabrication of SRSO sublayer, the power density of which were 7.4 W/cm^2 and 2.2 W/cm^2 , respectively. The gas flow rate is defined as $r_g = f_g/(f_g+f_{Ar})$ where f_g represents the N_2 or H_2 flow and f_{Ar} represents the Argon flow . The films were deposited on (1 0 0) silicon substrates at $T_{substrate}= 500°C$. In addition, few films were grown on quartz substrates. The substrates were placed at a distance of 38cm from the target and were kept rotating throughout the process of deposition in order to ensure homogeneous deposition. Before the fabrications of multilayers, several single layers of SRSO were deposited at substrate temperatures (T_s) ranging from 100°C to 500°C and also at varying hydrogen partial pressure (r_H) ranging from 10% to 80%. In order to tune the rate of deposition, from the analysis of monolayers it was decided to fabricate multilayers at T_s = 500°C and r_H = 50%. The SiO_2 sublayer was fabricated by sputtering the SiO_2 cathode under pure Ar plasma. The SiN_x sublayer was fabricated by sputtering the Si cathode while simultaneously introducing nitrogen into the Ar plasma. The nitrogen rate r_N was kept at 10% while the total flow rate was fixed at 10sccm. The thickness of each sublayer was chosen to be within the quantum confinement regime by adjusting the deposition time obtained from each of its constituent monolayers. In all the cases the thickness of the SRSO sublayer and SiO_2 sublayer were fixed to be 3.5nm. In order to understand the influence of SiN_x matrix, two different thickness of the SiN_x sublayer (3.5nm and 5nm) were fabricated and analysed.

In order to determine the refractive index and thickness of the films, the ellipsometric spectroscopy analysis was carried out between 1.5 and 4.5 eV, using a Jobin Yvon ellipsometer (UVISEL) at an incidence angle of 67° on the SRSO monolayers. The FTIR transmission spectra of these samples were recorded using Nicolet Nexus spectrometer at Brewsters angle (65°). The Photoluminescence (PL) spectra of the annealed samples were obtained using TRIAX 180 Jobin Yvon monochromator in the wavelength range 550-1100nm. Two different annealing treatments were investigated-1100°C for 1 hour which is the classical annealing treatment used for recovering defect in SiO_2 matrix to favour luminescence from Si-nc in the SiO_2 matrix (21) and a Rapid Thermal Annealing of 1000°C for 1min, abbreviated RTA throughout the paper. The excitation wavelength of 488 nm (Ar laser) was used for measurements. The XRD and XRR measurements on these samples were made using Philips XPERT MPD PRO diffractometer. TEM and HRTEM observations were performed in Jeol 2010 and ABT EM002B apparatus, respectively.

Results and Discussions

SRSO monolayers

The influence of substrate temperature and hydrogen partial pressure has been analysed on the SRSO monolayers.

Effect of Hydrogen Partial Pressure. The deposition rate and the refractive index of the deposited monolayers were estimated from the ellipsometry simulations. The left part of Fig.1. depicts the influence of hydrogen partial pressure (r_H) on the properties of SRSO layers. The deposition rate (nm/s) decreased steadily with increase in the partial pressure of hydrogen. This is consistent with our earlier observations (16,22) and is indicative of the etching mechanism that competes with the deposition process. The refractive index increases with increasing partial pressure of hydrogen. However, the refractive index value does not significantly change for r_H=50% and 80%. Though the deposition rate is comparatively higher at r_H= 10% and 30%, the refractive index from these samples are lower. Regarding the compromise between a high depostion rate and a necessary high refractive index, we now consider the two interesting values of r_H between 50% and 80% for the fabrication of multilayers. Meanwhile, the percentage concentration x of Si excess was calculated on the as-deposited samples from the FTIR spectra recorded at normal incidence using the formula $x = (\upsilon_{TO3}/60)-16$. It can be seen from the figure that the percentage of silicon excess is high at $r_H=50\%$. On the other hand it can be noticed that the value goes very low for $r_H = 80\%$. This is contradictory to the ellipsometry results that show $r_H = 50\%$ and 80% have similar values of refractive index. This contradiction may be due to the fact that the FTIR spectra could not either detect all the phonons or Si nanoparticles have been already formed and hence the under estimation of silicon excess. The value $r_H = 50\%$ seemed satisfactory regarding the parameters estimated above, it has been chosen for the multilayer depositions.

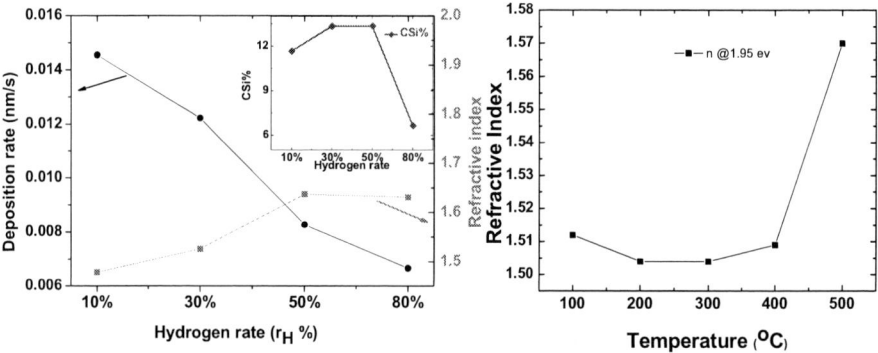

Figure 1. Influence of hydrogen rate (left) and substrate temperature (right) on the properties of SRSO layers.

Effect of Temperature. The effect of deposition temperature on refractive index is shown in the right side part of Fig.1. It can be seen from the figure that with increasing temperature (above 200°C), the refractive index is increasing which is a proof of higher incorporation of silicon. We note the unusual behavior from 100°C temperature deposition. These set of monolayers were made on a trial basis without using two cathodes for higher silicon incorporation and hence the lesser values of refractive index. However, it was observed that the deposition with Ts = 500°C incorporated more Si excess and this substrate temperature was chosen for the fabrication of multilayer structures.

Two cathode depositions. After having chosen r_H =50% and Ts = 500°C, the next objective was to incorporate a higher excess of Si in the SRSO layers. Hence in addition to the reactive magnetron approach with an optimal amount of hydrogen introduced into the plasma, a Si cathode was also sputtered simultaneously with the SiO_2 cathode. Two different powers (P_{Si}) have been applied on the silicon cathode: a) 1.628 W/cm^2 and b) 2.22 W/cm^2. These monolayers were subjected to ellipsometry measurements. Simulations from the ellipsometry data revealed that the SRSO with P_{Si} = 1.63 W/cm^2 and P_{Si} = 2.22 W/cm^2 have refractive indices 1.893 and 2.109 respectively. These values correspond to silicon volumic fraction of 23.5% and 32.5%. Hence The SRSO multilayers have been fabricated using P_{Si} = 2.22 W/cm^2.

SRSO/SiO₂ multilayers

Multilayer structure composed of 50 SRSO/SiO₂ patterns was made, each sublayer measuring 3.5 nm, using the reactive magnetron co-sputtering approach at Ts = 500°C. The multilayer was subjected to two kinds of annealing: 1100°C for 1 hour and 1000°C for 1min. Micro structural investigations were carried out on the above mentioned sample after annealing treatments, following which optical and electrical properties were studied.

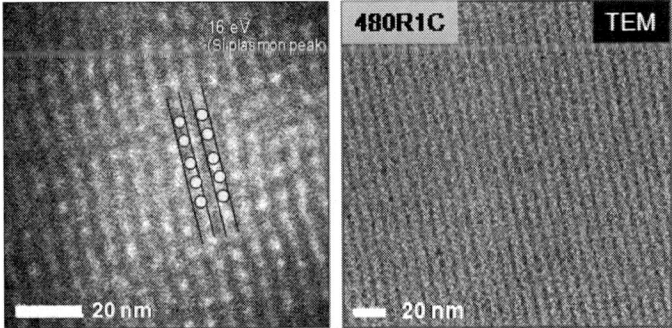

Figure 2. TEM micrographs of SRSO/SiO₂ multilayer structure (21).

Microstructure. The EFTEM image (Fig.2) shows the multilayer structure comprising alternating layers of SRSO/SiO₂ with estimated thickness of 3.5nm for each sublayer. A clear view of the formation of Si-nc can be seen from the image. The Si-nc particle density was estimated to be 10^{19} np/cm³.The micrograph confirms that we are able to achieve significant concentration of Si-nc with the reactive magnetron sputtering approach with good control over the size of nanoclusters .

FTIR spectroscopic study. The transmission FTIR spectroscopy was made to see the characteristic peaks arising from the Si-O bonds within the silica matrix and at the Si-nc / SiO₂ interface. Fig.3 shows the FTIR spectra of SRSO/SiO₂ recorded at the Brewster angle of 65° that enables the detection of the LO_3 mode of silica at about 1250 cm⁻¹ in addition to the TO_3 mode located near 1080 cm⁻¹. In SRSO/SiO₂ around 1225 and 1080 cm⁻¹ we notice the LO_3 and TO_3 peak from the Si-O stretching, the TO_4-LO_4 doublet between the 1100-1200 cm⁻¹ and the TO_2 - LO_2 asymmetric stretching of Si-O from SiO₂ at 810 and 820 cm⁻¹ respectively (16). The presence of Si-nc is attested by the intensity of the LO_3 peak which is representative of the Si-O bond at the interface (23) between silicon and silica while the TO_3 vibration mode at about 1080 cm⁻¹ is the signature of the volumic silica.

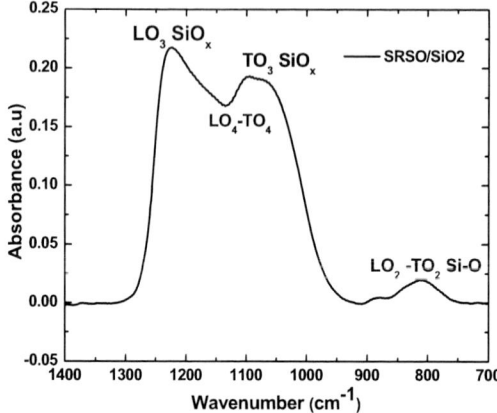

Figure 3.FTIR spectrum of SRSO/SiO₂ multilayer structure recorded at Brewsters incidence.

Photoluminescence. The PL spectra were recorded on this sample after being subjected to two different aforesaid annealing treatments (Fig.4).

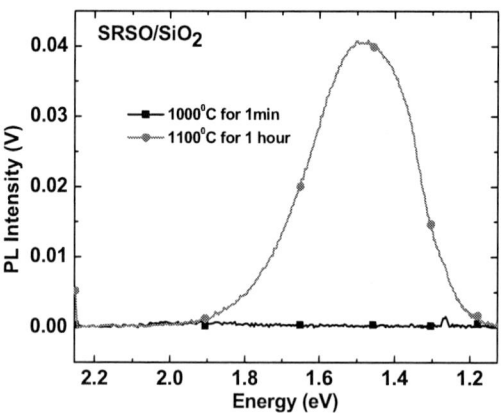

Figure 4. PL spectra of SRSO/SiO₂ multilayer structure with two different annealing treatments

The PL spectra of SRSO/SiO₂ annealed 1100°C for 1 hour presents a luminescence peak at about 1.5eV consistent with earlier reports. On the other hand there is no observable luminescence from the sample annealed at 1000°C for 1min.

Electrical properties. A preliminary study on the electrical properties of non annealed SRSO/SiO$_2$ and on SRSO/SiO$_2$ annealed at 1100°C for 1 hour, was made using a two probe apparatus. Fig.5. shows the I-V characteristic curve of the aforesaid samples.
It can be seen from the figure that in this material for a given voltage there is a decrease in the current when the sample is annealed. The SRSO/SiO$_2$ structure annealed at 1100°C for 1 hour favored a higher incorporation of Si-nc and consequently improved the PL properties. But the electrical properties are not favorable enough to meet the requirements for device application. From this curve the material's resistance and resistivity were calculated to be 14500 Ω and 0.73x 10^8 Ω.cm respectively.

Concerning this study on the SRSO/SiO$_2$ multilayers, we conclude that our fabrication technique was successful in producing a significant silicon nanocluster density. It can also be noted from the micrograph image that the grain size can be almost accurately controlled via the sublayers thickness. The increased PL intensity from the classically annealed sample supports the presence of Si-nc formed after an annealing treatment of 1100°C for 1 hour. It can be seen that this structure has high electrical resistivity and hence offers limitations for device applications. This study was made on two basis: i) to observe the properties on this peculiar fabrication technique of SRSO using two cathodes in addition to the reactive magnetron sputtering approach and ii) to make a study on a material of 50 periods with 3.5nm sublayer thickness in order to make a comparative study with its SiN$_x$ counterpart.

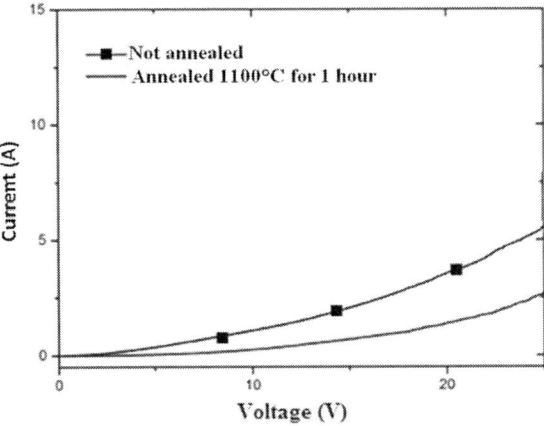

Figure 5. I-V characteristic curve of SRSO/SiO$_2$ multilayered structure

SRSO/SiN$_x$ multilayers

Multilayer structures composed of 50 SRSO/SiN$_x$ patterns have been fabricated keeping the same deposition conditions of SRSO as used in SRSO/SiO$_2$ multilayer structure. In order to have a uniform basis of comparison, each layer was fabricated to have a thickness of 3.5nm.

Photoluminescence. The SRSO/SiN$_x$ multilayer structure was first subjected to an annealing treatment of 1100°C for 1 hour leading to a weak intensity of PL emission. Hence, among the possible annealing treatments, Rapid Thermal Annealing (1min, 1000°C) yielded the highest emission observed so far. Hence this analytical study of two kinds of multilayer structures focuses only on these two annealing treatments (classical and RTA). It was noticed that the PL emission from SRSO/SiN$_x$ showed a reverse trend of what was observed with SRSO/SiO$_2$. The emission behavior from this multilayer structure on RTA has to be understood better. Assuming that the SiN$_x$ sublayer plays a role in the enhancement of emission, another multilayer structure of 50 patterns with SiN$_x$ sublayer thickness increased to 5nm was fabricated. Fig.6 shows the PL spectra of SRSO/SiN$_x$ with (left) two annealing treatments and (right) with two sublayer thicknesses of SiN$_x$.

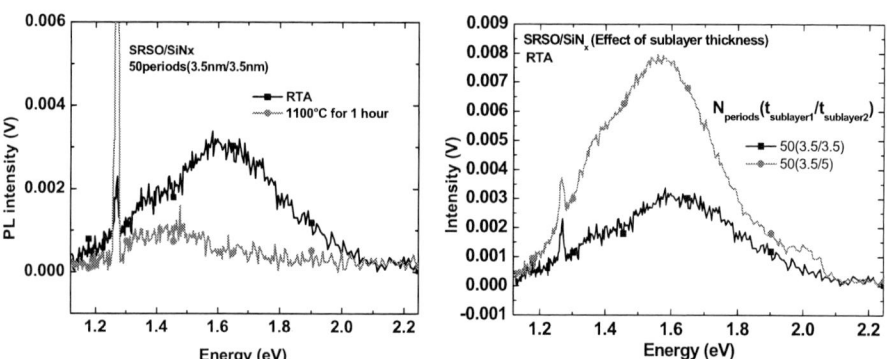

Figure 6. PL spectra of SRSO/SiN$_x$ (left) with two annealing treatments and (right) with two different thicknesses of SiN$_x$ sublayer

The left side of Fig.6 shows that classical annealing treatment is not favorable with regard to PL emission on this kind of multilayer structure. Contrary to the RTA SRSO/SiO$_2$ which did not show any considerable PL emission, SRSO/SiN$_x$ multilayer structure recorded the highest PL emission after RTA. Despite RTA enhancing the PL emission in comparison to the 1h1100°C annealed sample, the intensity is still lesser in comparison to the emission from classically annealed SRSO/SiO$_2$ structure. This could be due to the larger silicon grains formed by the merging of excess Si from SRSO and SiN$_x$ sublayers. Hence in order to inhibit silicon diffusion, the thickness of SiN$_x$ was increased to 5nm and the PL intensity was measured. The right side of Fig.6 displays the effect of SiN$_x$ sublayer thickness on the PL intensity. The intensity increases by a factor of 2.4 with an increase in the SiN$_x$ barrier thickness to 5nm.

Thus, it can be seen that SRSO(3.5nm)/SiN$_x$(5nm) multilayers exhibit enhanced PL emission. But it cannot be confirmed that there is a change of the Si grain size as the PL peaks in both cases are around the same energy range. The sharp peak that is noticed between 1.2 -1.3 eV is due to the second order reflection of the laser beam from the grating. The next trial in understanding the trend of the PL was to fabricate a multilayer structure with the better of the two sublayer thicknesses while doubling the number of patterns. Hence, 100 periods of SRSO(3.5nm)/SiN$_x$(5nm) multilayered structure was made. Fig.7. shows comparative PL spectra of the three kinds of SRSO/SiN$_x$ described above and the inset of Fig.7 compares the PL spectrum of RTA SRSO/SiN$_x$ with the SRSO/SiO$_2$ matrix annealed at 1100°C for 1 hour.

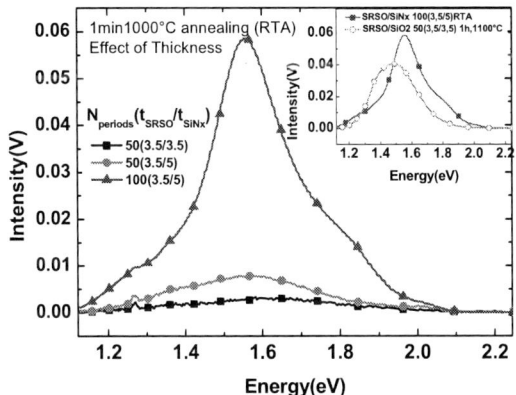

Figure 7. Effect of three types of SRSO/SiN$_x$ multilayers on photoluminescence emission. (Inset) A comparison between the classically annealed SRSO/SiO$_2$ with the RTA SRSO/SiN$_x$

It can be seen from Fig.7, that doubling the number of patterns has resulted in a PL enhancement by a factor of 7.5. This observation was quite interesting as this PL intensity from RTA SRSO/SiN$_x$ was 1.43 times more intense than the 1100°C for 1 hour annealed SRSO/SiO$_2$ emission (Inset). The reason behind this huge increase in the PL emission is still under investigations. However, it gives a positive idea that this material's optical properties can be exploited for device applications, if the origin and mechanism of PL is better understood. If the emission results from the Si-nc and if it can be confirmed that this PL emission is not from the defects or stresses, this "RTA SRSO/SiN$_x$" would prove to be a better material than SRSO/SiO$_2$ due to its lower thermal budget.

FTIR spectra. Fig.8 shows the FTIR spectra of the SRSO/SiN$_x$ multilayer structure recorded at Brewster incidence. In the FTIR spectra of SRSO/SiN$_x$ structure, the peak found between 1050 – 1070 cm^{-1} is attributed to a combined contribution of TO mode from Si-O and LO mode of a-Si$_x$N$_y$H$_z$ from Si-N. The shoulder around 1100 cm^{-1} may be either due to N-H bond (24,25) or due to a contribution of the LO3 mode of Si-O-Si bonds at 180° (16). Such a shoulder is the signature of the Si nanoparticles formation within either the SiN$_x$ (26) and/or the SRSO (16). The absorption band located around 860 cm^{-1} could be attributed to the Si-N asymmetric stretching mode. It can also be noted from the figure that we evidence a better phase separation and a higher incorporation of Si-nc with an increase in the SiN$_x$ sublayer thickness. The same is observed with an increase in the total number of patterns also.

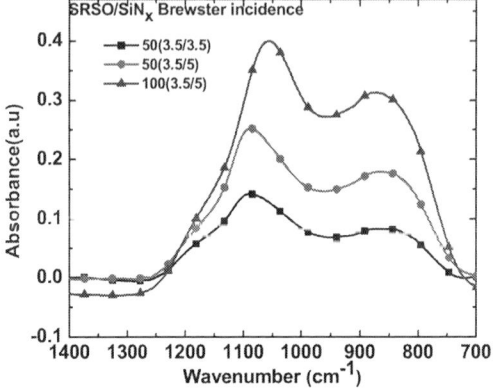

Figure 8. FTIR spectra of three types of SRSO/SiN$_x$ multilayer structures recorded at Brewster's incidence.

TEM observation. HRTEM observations were made on the SRSO(3.5nm)/SiN$_x$(5nm) sample . Fig. 9 shows the micrograph image of SRSO/SiN$_x$ multilayers. It could be seen from the micrograph image that the sublayers are perfectly alternated. The thickness of each sublayer is slightly higher than what was expected. However, it is almost equal to the estimated thickness of 3.5nm and 5nm for SRSO and SiN$_x$, respectively.

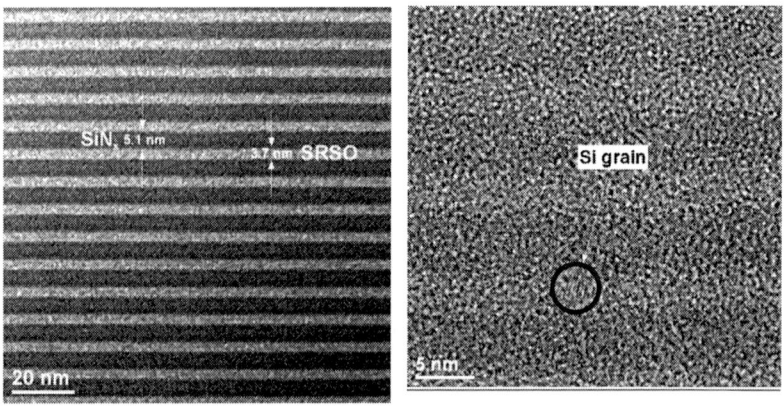

Figure 9. Micrograph of SRSO/SiNx multilayer structure

Electrical properties. Electrical measurements were made on the non annealed and RTA SRSO/SiNx multilayers. The I-V characteristics obtained is shown in Fig. 10.

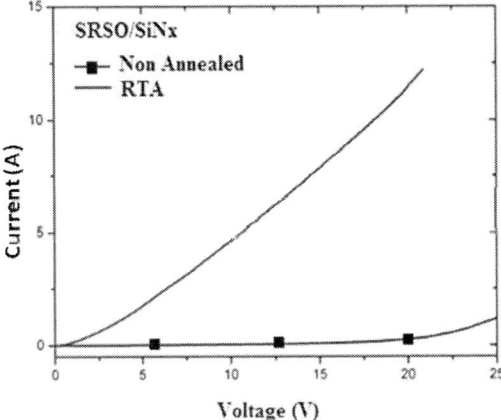

Figure 10. I-V characteristics of SRSO/SiN$_x$

The main objective of comparing SRSO in SiO$_2$ matrix and in SiN$_x$ matrix is the electrical properties. Contrary to what was observed with SRSO/SiO$_2$ multilayers, the annealing treatment increases the current for a given voltage in comparison to the non annealed sample thereby indicating better conduction. The values of resistance and resistivity on this RTA sample were calculated to be 1753 Ω and 0.14 x 10^7 Ω.cm respectively. This value is almost two orders of magnitude lesser in comparison to the resistivity of SRSO/SiO$_2$.

To summarize this study on SRSO/SiN$_x$ multilayer structure, it was seen that a thickness of 5nm is better than 3.5nm for the SiN$_x$ sublayer with regard to the photoluminescence intensity. Perfect alternations of each of these sublayers were achieved by our fabrication technique and silicon grains were observed from the HRTEM measurements. The PL intensity was high for RTA sample and negligible after 1100°C for 1 hour annealing treatments. The intensity after Rapid Thermal Annealing could be enhanced by playing with different parameters of the multilayer structure like the sublayer thickness and the total number of patterns. Increasing the SiN$_x$ sublayer thickness enhanced the PL by a factor of 2.4 and doubling the number of patterns enhanced the PL emission by a factor of 7.4. Electrical properties are enhanced in comparison to those obtained from SRSO/SiO$_2$ multilayers by almost two orders of magnitude.

Conclusion

A systematic optimization of the deposition condition was made for fabricating Silicon Rich Silicon Oxide layers. This study was based upon two major parameters; the substrate temperature and the hydrogen partial pressure in the plasma. Basic analysis such as the change in refractive index, deposition rate (nm/s) and concentration of silicon excess were made using ellipsometry and FTIR spectroscopy. It was found that SRSO monolayers fabricated at $500^{\circ}C$ and $r_H = 50\%$ are better in terms of the parameters analysed so far. However the incorporation of Si excess was still too low and hence the SRSO monolayers were made by sputtering simultaneously SiO_2 and Si cathode in addition to the hydrogen introduced in the plasma to make SiO_x (x<2) layers. The aforesaid parameters were calculated on SRSO monolayers made with this modified technique fabricated with two different powers applied on the Si cathode ($P_{Si} = 11\%$ and $P_{Si} = 15\%$). Finally, it was seen that SRSO layers fabricated with $r_H= 50\%$, $P_{Si} = 15\%$ at $500^{0}C$ were better and these conditions were adapted to the SRSO sublayers in all the multilayers fabricated. Multilayers of $SRSO/SiO_2$ and $SRSO/SiN_x$ were made and compared. The micrographs revealed that our technique is successful to achieve the multilayered configuration with a significant silicon nanoclusters density of 10^{19} cm^{-3}.

Though the $SRSO/SiO_2$ samples can be made to exhibit a strong PL emission using suitable annealing treatment, their electrical resistivity as demonstrated by the I-V curve is quite high. This becomes a major barrier for electrical transport and consequently for device applications. On the other hand we could discover from the above set of investigations that the $SRSO/SiNx$ subjected to an optimized annealing treatment also gives rise to promising optical electrical properties. It not only enhances the luminescence but also helps to achieve a control over the thermal budget. The peak of photoluminescence emission from this material coincides well with the literature and has been attributed to the presence of Si-nc. However the exact mechanism and origin for luminescence is still under systematic investigations. With the RTA treatment we could achieve decrease in resistivity of $SRSO/SiN_x$ matrix by almost two orders of magnitude in comparison to the $SRS0/SiO_2$ structure. This enhanced electrical property from this material becomes an added advantage of exploiting the material further for device applications.

Acknowledgments

This work is supported by the DGA (Defence Procurement Agency) through the research program n°2008.34.0031.

References

1. L.T.Canham, *Appl. Phys. Lett.* **57**, p. 1046 (1990).
2. V. Lehman and U. Gösele, Appl. Phys. Lett. **58**, p. 865 (1991).
3. J. P. Proot, C. Delerue, and G. Allan, Appl. Phys. Lett. **61**, p. 1948 (1992).
4. A. G. Cullis, L. T. Canham, P. D. J. Calcott, *J. Appl. Phys.,* **82**, p. 909 (1997).
5. S. Charvet, R. Madelon, F. Gourbilleau, R. Rizk, *J. Appl. Phys.* **85**, p. 4032 (1999).
6. F. Koch, V. Petrova-Koch, and T. Muschik, J. Lumin. **57**,271 (1993).
7. S. M. Prokes, Appl. Phys. Lett. **62**, 3244 (1993).
8. L. S. Liao, X. M. Bao, X. Q. Zheng, N. S. Li,N. B. Min, *Appl. Phys. Lett.*, **68**, 850 (1996).

9. L. S. Liao, X. M. Bao, X. Q. Zheng, N. S. Li,N. B. Min, *Appl. Phys. Lett.*, **68**, p. 850 (1996).
10. T. Inokuma, Y. Wakayama, T. Muramoto, R. Aoki, Y. Kurata, S. Hasegwa, *J. Appl. Phys.*, **83**, p. 2228 (1998).
11. Y. Kansagawa, T. Kageyama, S. Takeoka, M. Fugiti, S. Hayashi, K. Yammoto, *Solid State Commun.*, **102**, p. 533 (1997).
12. S. Charvet, R. Madelon, F. Gourbilleau, R. Rizk, *J. Lumin.*, **80**, p. 257 (1999).
13. D. J. Lockwood, Z. H. Lu, J. M. Baribeau, *Phys. Rev. Lett.*, **76**, p. 539 (1996).
14. X. L. Wu, G. G. Siu, S. Tong, X. N. Liu, F. Yan, S. S. Jiang, X. K. Zhang, D. Feng, *Appl. Phys. Lett.*, **69**, p. 523 (1996).
15. L. Tsybeskov, K. D. Hirschman, S. P. Dasgupta, M. Zacharias, P. M. Fauchet, J. P. McCaffrey and D. J. Lockwood, *Appl. Phys. Lett.*, **72**, p. 43 (1998).
16. N. M. Park, C. J. Choi, T. Y. Seong and S. J. Park, *Phys. Rev. Lett.*, **86**, p. 1355 (2001).
17. C. Ternon, F. Gourbilleau, X. Portier, P. Voivenel and C. Dufour, *Thin Solid Films.*, **419**, p. 5 (2002).
18. B. H. Kim, C. H. Cho, T. W. Kim, N. M. Park, G. Y. Sung and S. J. Park, *Appl. Phys. Lett.*, **86**, p. 091908 (2005).
19. L. Dal Negro, J. H. Yi, V. Nguyen, Y. Yi, J. Michel and L.C. Kimerling, *Appl. Phys. Lett.*, **86**, p. 261905 (2005).
20. K. S. Cho, N. M. Park, T. Y. Kim, K. H. Kim, G. Y. Sung and J. H. Shin, *Appl. Phys. Lett.*, **86**, 071909 (2005).
21. F. Gourbilleau, X. Portier, C. Ternon, P. Voivenel, R. Madelon and R. Rizk, *Appl. Phys. Lett.*, **78**, p. 3058 (2001).
22. F. Gourbilleau, C. Ternon, D. Maestre, O. Palais and C. Dufour, *J. Appl. Phys.*, **106**, p. 013501 (2009).
23. A. Achiq, R. Rizk, F. Gourbilleau and P. Voivenel, *Thin Solid Films.*, **348**, p. 74 (1999).
24. J. E. Olsen and F. Shimura, *J. Appl. Phys.*, **66**, p. 1353 (1989).
25. G. Scardera, T. Puzzer, G. Conibeer and M.A Green, *J. Appl. Phys.*, **104**, p. 104310 (2008).
26. Sanghoon Bae, David G. Farber and Stephen J. Fonash, *Solid-State Electronics.*, **44**, p. 1355 (2000).
27. F. Delachat, M. Carrada, G. Ferblantier, J-J Grob, A. Slaoui and H. Rinnert, *Nanotechnology.*, **20**, p. 275608 (2009).

Context Dependence Effects in Si/SiON Based Advanced CMOS Devices

O. O. Olubuyide[a]

[a]Texas Instruments, Inc.; MS 366, 13121 TI Blvd., Dallas, TX 75243

Significant sources of mechanical stress have been deliberately added into the active silicon of CMOS devices to improve carrier mobility. These include Dual Stress nitride Liners (DSL) and embedded Silicon Germanium (eSiGe) in the source and drain regions of pMOS transistors. Moreover, as transistor dimensions have shrunk, unintentional sources of variation such as Shallow Trench Isolation (STI) mechanical stress and dopant redistribution due to proximity to STI boundary or the well implant boundary (WPE) have also increased the systematic transistor variability. This increased systematic variability has impacted the product yield and provided the impetus to include context dependence in SPICE models of transistor electrical performance. This paper outlines a rigorous approach to not only quantify the physical mechanisms and impact of context effects, but also how to model these effects on the electrical transistor performance and various strategies to mitigate the impact of these context dependencies on digital circuit performance.

Introduction

Background

Since the 90 nm node it has become widely recognized that in order to maintain the historical performance improvements that was achieved by scaling in prior nodes, sources of mechanical stress has to be deliberately added into the active silicon of CMOS devices. These stress sources improve the carrier mobility, and thus have become critical to maintaining the device scaling performance trends in the microelectronics industry. These stress sources include Dual Stress nitride Liners (DSL) and embedded Silicon Germanium (eSiGe) in the source and drain regions of pMOS transistors. These stress sources also impact the electrical performance of adjacent transistors and therefore increase the localized, systematic variability in transistor layouts. Moreover, as transistor dimensions have shrunk, unintentional sources of variation such as Shallow Trench Isolation (STI) mechanical stress and dopant redistribution due to proximity to STI boundary or the well implant boundary (WPE) not only have a noticeable impact on transistor performance but have also increased the systematic transistor variability . All these sources of increased systematic variability have impacted the product yield and provided the impetus to include the impact of the surrounding environment, or context dependence, in SPICE models of transistor electrical performance as shown in Figure 1. Ensuring minimal context dependence is now quickly becoming one of the major factors that designers are utilizing to qualify their circuits layouts [1].

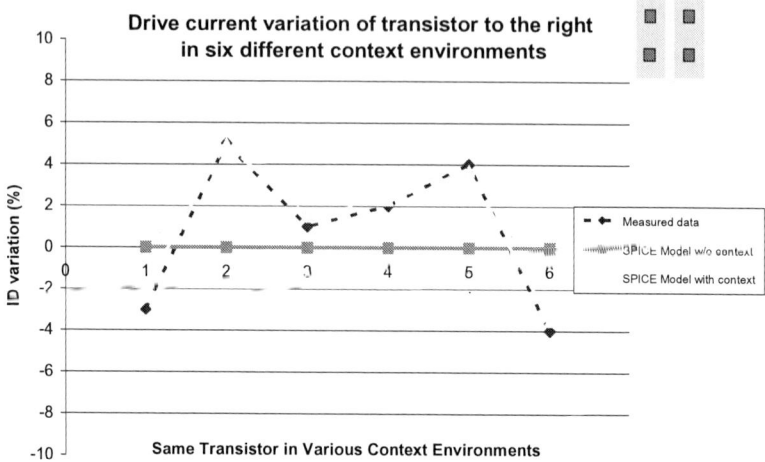

Figure 1. The same transistor is placed in six different environments that have varying values for Active Proximity Effect (APE), DSL, WPE. Length of gate Oxide Definition (LOD), Active Jog Effect (AJE), Contact Pitch Effect (CPE) and Poly Active Space (PAS) effect stay the same as it is the same transistor and active shape in this study. As can be seen, there is a ± 5% variation in performance due to context for this specific transistor that is accurately captured by a SPICE model that includes context variables.

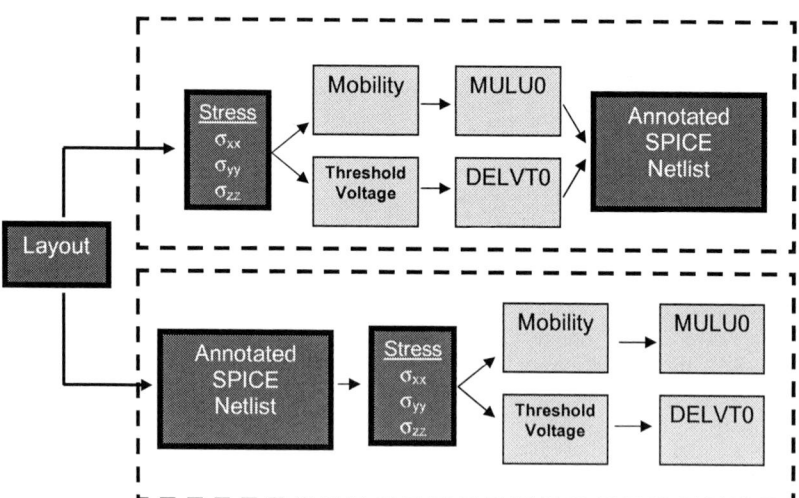

Figure 2. Two different methods of translating extracted layout parameters into BSIM variables that are used to capture context dependent effects (CDE).

Modeling methodology for Context Dependence

Approaches to model context effects in SPICE models is based on an approximation of the final stress and dopant distribution after the complete silicon process flow. This approach requires calibration to actual silicon test structures that capture the design rule space of interest and/or possibly utilizing three dimensional numerical simulations. The final stress and dopant distributions for each target transistor are then translated into value offsets to nominal simulator parameters such as for mobility, velocity saturation, channel resistance and threshold voltage. The final step of translating to offsets to nominal simulator parameters can be performed either in the layout parameter extraction software or in the SPICE modelcard as shown in Figure 2. With either approach, the SPICE model implementation, although it is not strictly a predictive approach, can simulate large circuits and is the preferred method for design verification. This paper will touch on what physical parameters need to be extracted from the layout in order to accurately model these context effects, how to optimize transistor performance for each context effect and how to mitigate the variability of these context effects.

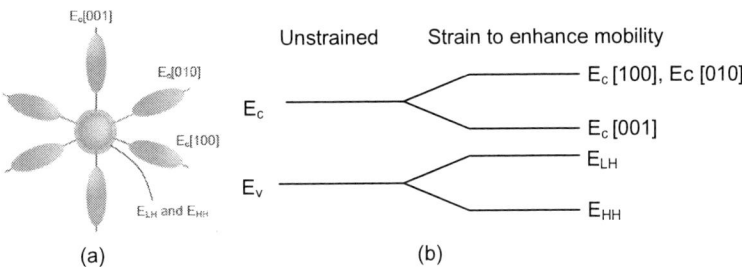

Figure 3. (a) Constant-energy surfaces of ellipsoidal minima that form the conduction band in silicon and also the spherical maxima of light and heavy hole bands. (b) The splitting of the equivalent energy bands after the application of tensile uniaxial stress for the conduction band and compressive uniaxial stress for the valence band.

Stress impact on Transistor Parameters

The inclusion of mechanical stress improves device performance via its impact on the band structure of silicon. A simple explanation can be achieved by examining the silicon conduction band. In silicon, the conduction band is made up of states from six ellipsoidal shaped minima that lie in the <100> direction, as shown in Figure 3a. Each minima has a large longitudinal effective mass and a small transverse effective mass. In the absence of strain, all six minima are at the same energy level and are equivalent as far as electron occupation. The introduction of <100> uniaxial tensile stress splits these six equivalent minima by lowering the energies of the minima in the <100> tensile stress direction (Δ_2 valley) and lifting the energies of the minima in the other two directions (Δ_4 valley) as shown in Figure 3b. This shift in the band structure leads a higher electron population in the Δ_2 valley which also has a lower in-plane effective mass since it is determined by the small transverse effective mass. Moreover the out-of-plane effective mass—determined by the large longitudinal effective mass—creates strong confinement for the Δ_2 valley, ensuring that a greater number of electrons experience the reduced effective mass.

Furthermore, the separation between the Δ_2 valley and Δ_4 valley leads to a suppression of the intervalley scattering rates that further boosts the mobility. These three effects, lowering of the bandgap, changing of the effective mass, and the scattering rate captures the major impacts of stress on the band structure of silicon. Similarly to the example given for electrons above, the same treatment holds for holes, where if stress is applied a two fold energy degeneracy is broken and a heavy hole band and light hole band are created. If the stress is compressive, the light hole band is lifted above the heavy hole band and this leads to an increase in the hole mobility in the valence band.

The relationship between the mobility and the velocity in a material is captured by the equation shown below:

$$v_d = \mu \frac{E}{\left[1 + \left(E/E_c\right)^\beta\right]^{1/\beta}} \qquad [1]$$

where v_d is the velocity, μ is the low-field mobility, E is the electric field, E_c is the saturation electric field and is a fitting parameter, β is a fitting parameter and vsat is equivalent to μE_c. This equation shows that an improvement in the electron mobility for applied tensile stress in the conduction band or similarly for the hole mobility with applied compressive stress in the valence band also leads to corresponding increase in the number of electrons or holes that reach a higher transit velocity at a given electric field. This improvement comes from a lower in-plane effective mass that improves the carrier mobility (as shown in equation (1)) and also the reduction in the scattering rates that increases the maximum transit velocity achieved. These two effects together effectively increase the saturation velocity value with the application of the right type of stress to the conduction and valence bands respectively.

Moreover mechanical stress can also influence the dopant distribution in the channel of the transistor as shown in equation (2):

$$D = D_i \exp\left(-\frac{Q_i \varepsilon}{kT}\right) + D_v \exp\left(-\frac{Q_v \varepsilon}{kT}\right) \qquad [2]$$

where D_i and D_v are the diffusivity due to interstitials and vacancies, ε is the in-plane strain, Q_i and Q_v are the strain coefficients for the interstitial and vacancy fluxes, k is Boltzmann's constant and T is the temperature.

For example, compressive stress reduces the amount of interstitials that are available inside the silicon lattice [2]. Hence atoms such as boron that have a noticeable interstitial contribution to their overall diffusivity will exhibit reduced diffusivity in the presence of compressive stress, which will impact the threshold voltage of the transistor. Similarly, in a region with high tensile stress the overall diffusivity of boron will be increased which will also impact the threshold voltage of the transistor.

Furthermore, for the case of applied biaxial mechanical stress, the threshold voltage is also impacted by a few additional factors. The most common attributed source of threshold voltage shift from applied mechanical stress is due to bandgap narrowing, but biaxial mechanical stress also causes changes to the electron affinity and also the density of states, all factors that impact the threshold voltage [3]. This implies that context dependent effects (CDE) from biaxial stress sources such as DSL also cause a noticeable threshold voltage shift in addition to the expected mobility shifts.

As shown in Figure 4, generally for transistors in <100> silicon with the channel oriented in the <110> direction, nMOS transistors prefer tensile stress in the lateral and transverse directions relative to current flow, while pMOS transistors prefer compressive stress in the lateral direction and tensile stress in the transverse direction. Therefore variations in the channel stress of silicon transistors can cause shifts in carrier mobility, saturation velocity, threshold voltage and the transistor leakage due to dopant redistribution, bandgap narrowing, electron affinity shifts, and density of states [4].

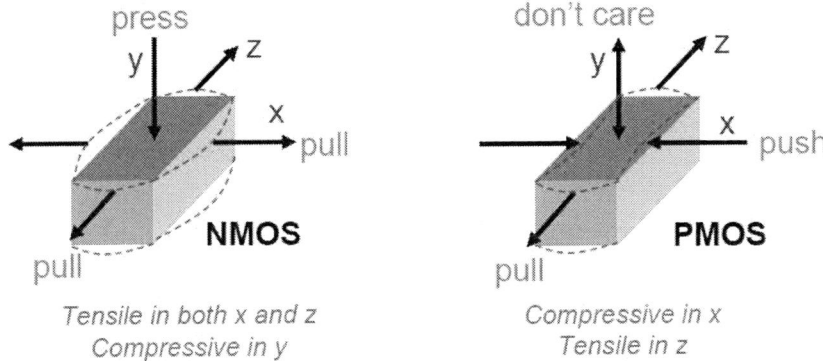

Figure 4. Preferred stress orientation of nMOS and pMOS transistors where press/push refers to compressive stress and pull refers to tensile stress. Image courtesy of D. Pramanik *et al*, Synopsys, Inc 2006.

Intra-Active Context Dependent Effects

Length of gate Oxide Definition Effect

Intra-active CDE derive their impact from the geometry of a given silicon active region. One example is the Length of gate Oxide Definition (LOD) effect. This effect originates from the mechanical stress sources within an active and the discontinuity that occurs at the active boundary with the Shallow Trench Isolation (STI). Thus LOD is measured as a function of the distance from the edge of the gate to the edge of the active, with one measurement on the left side, named SA, and another measurement on the right named SB as shown in Figure 5. The mechanical stress sources for LOD are basically the STI stress that is generally compressive and the eSiGe in the source and drain. For nMOS transistors, only the compressive STI is a valid stress source. This stress builds up during processing due to the difference in the thermal expansion coefficient of silicon and silicon dioxide and is usually in the range of a few hundred megapascals. Since the nMOS transistors prefer tensile stress in both the horizontal and vertical direction (as shown in Figure 4), the stress from the STI degrades the performance of the nMOS

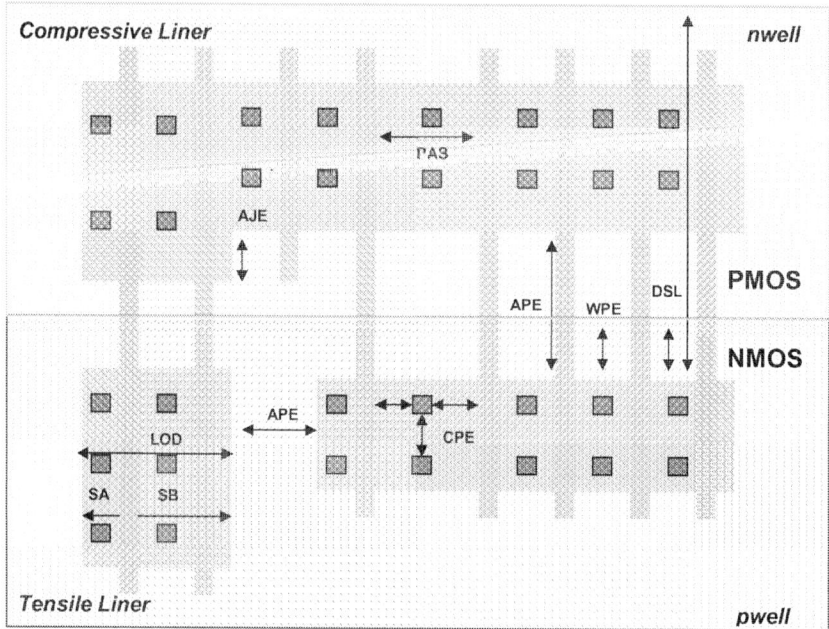

Figure 5. A representation of some of the major context effects in sub-90 nm technology transistors.

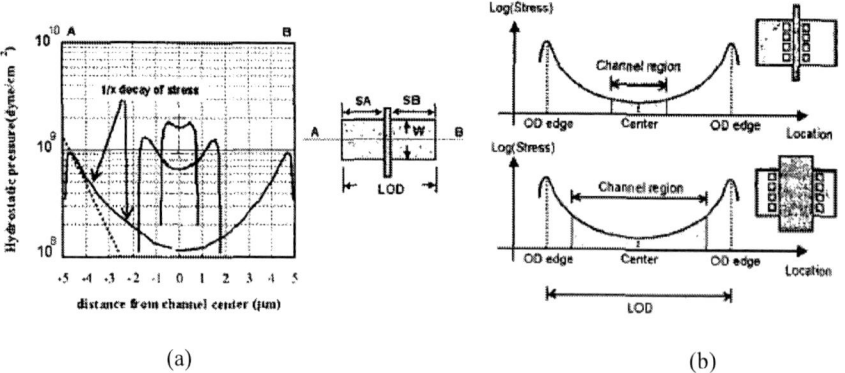

(a)   (b)

Figure 6. (a) STI stress curve as function of horizontal in a Si active. (b) The effect of longer gate lengths on STI stress, the longer the active, the higher the average STI stress the channel region experiences. Image courtesy of K.-W. Su et al, IEEE CICC 2003.

transistor. The degradation from the vertical STI is effectively a constant that is determined by the transistor width, hence it is only the horizontal decrease in drive current that is modeled. The closer the proximity of the gate to the STI edge the lower the drive current. This degradation is primarily mobility parameter dominated as the magnitude of mechanical stress is not large enough to have an appreciable impact on threshold voltage parameters. It is worth noting though that for the transistor closest to STI edge, there can also be dopant redistribution effects that will lead to threshold voltage shifts. Moreover, given the same active size, the LOD effect is stronger for longer gate lengths, since they not only will have a greater average stress from the STI (as shown in Figure 6b), but since LOD also primarily impacts the mobility parameters that have a greater impact at longer gate lengths, the impact of LOD is also more apparent at these longer gate lengths. Finally, given the same channel length a larger gate width will generally have a lower average stress than a shorter gate width and will thus have a lower LOD effect and improve the nMOS transistor performance.

The physical behavior captured above for nMOS transistors applies directly to pMOS transistors except that in the horizontal <110> direction, as the gate gets closer to the compressive STI edge, the pMOS transistor drive current increases. So the pMOS LOD effect from STI improves the transistor performance as gate length increases. Similarly as the width increases, the pMOS LOD effect from STI also improves the transistor performance. This is not surprising as the pMOS and nMOS transistors improve in performance with tensile stress in the vertical directions.

The pMOS transistor also has eSiGe that has an effect on the LOD behavior. The eSiGe compressive stress is a cumulative stress value that includes the effects of not just the eSiGe adjacent to the channel but also the eSiGe from the adjacent transistors. Thus for pMOS transistors as the value of SA or SB decreases, the cumulative volume of eSiGe available also drops and the drive current decreases. This effect is particularly evident for the final 2 transistors adjacent to the STI edge as shown in Figure 7. Typically the decrease in performance from eSiGe will outweigh the improvement from coming closer to the STI, although this depends on the volume of the eSiGe, composition of germanium in the silicon, and stress of the STI. Furthermore, for a fixed volume of eSiGe, as the gate length increases, the average compressive stress in the channel from the eSiGe decreases and the impact of the LOD effect decreases. Finally, for a fixed gate length, as the width increases the performance correspondingly increases since the volume of eSiGe increases as a function of the transistor width. In general the effect of the DSL is not a significant factor in the determination of the impact of the LOD effect.

In order to mitigate the impact of the LOD effect, on the process front the STI layer can be grown and/or deposited to be as close to zero stress as possible. On the design rule front, in order to mitigate the LOD effect, a restriction can be created so that there should be a dummy transistor gate on the silicon active at the STI edge. This dummy gate will increase the area of circuits, and if this is a concern, another option is to ensure that the volume of silicon or eSiGe for the STI edge should be as large as possible without violating the area constraints.

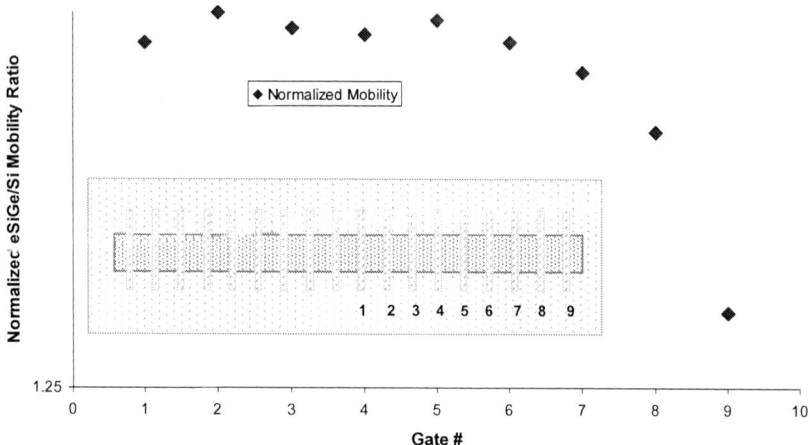

Figure 7. Plot of the effective mobility ratio of a pMOS transistor with embedded eSiGe and Si as a function of gate position in an active region.

Active Jog Effect

Another intra-active effect is the Active Jog Effect (AJE). This effect deals with the impact of changes to the active width in the source and/or drain regions. A significant fraction of the context impact of AJE comes from lithographic printing and etching errors, but due to the variability of these effects as a function of process conditions they will not be covered in this discussion.

Disregarding lithographic effects, the nMOS AJE comes from proximity to the STI edge. It is only in the case that the AJE comes from a jog to reduce the width in the source and/or drain regions (inner jog) that the compressive stress from the STI edge comes into play. The impact of this effect can be captured as a weighted average of the LOD degradation from the portion of the transistor width that is exposed to the STI edge and the portion that is still shielded from the compressive STI by additional active silicon. In the case of the jog that increases the width in the source and/or drain regions (outer jog), there appears to be no strong impact of stress on the transistor parameters. The width and length dependence of the AJE for the inner jog conditions follows that of the nMOS LOD effect.

When the lithographic effects are removed, the pMOS AJE comes from proximity to the STI edge and the change in the volume of eSiGe. For an inner jog, the proximity to the compressive stress from the STI improves the transistor drive current, but this is typically more than compensated for by the loss of the compressive stress from the missing eSiGe that decreases the drive current. In the case of an outer jog, the increase in volume of eSiGe leads to an increase in the drive current. As captured in equation (3) below, the change in the drive current is directly proportional to the change in the area of the source and drain region:

$$\Delta Id \alpha \frac{Jogged\_Area}{Transistor\,Width \times Gate\,Space} \qquad [3]$$

The width and length dependence of AJE also closely follows the width and length dependence of pMOS LOD effects. Finally, in the case of pMOS transistors, due to the presence of eSiGe, the effect of a jog affects not only the nearest gate but also the next nearest neighbor due to the cumulative stress effects of eSiGe. Similarly as the LOD effect, the impact of the DSL does not factor significantly into the nMOS and/or pMOS AJE.

In order to mitigate the impact of AJE on the process front the STI layer can be created to have as close to zero stress as possible. On the design rule front, the distance from a gate to an inner jog can be optimized to mitigate the negative impacts of inner jogs. Moreover the magnitudes of allowed jogs can be minimized or restricted to zero to eliminate the AJE.

Poly Active Space Effect

The next intra-active effect is the Poly Active Space (PAS) effect. As shown in Figure 5, this effect deals with the variation in the space between the adjacent gates. For a fixed SA and SB, the nMOS PAS effect from stress comes from the DSL effect. As the space between adjacent gate decreases, the amount of stress transferred to the channel from the DSL layer decreases proportionally. Although there are also threshold voltage impacts for the nMOS PAS effect, these typically stem from process impacts due to doping and diffusion variations and thus cannot be simply generalized. The width and length dependence of this CDE also follows similar width and length dependence described for the nMOS LOD effect.

For the pMOS transistor the PAS effect leads to changes in eSiGe volume and also the DSL volume for each individual transistor. For a fixed SA and SB, changes to PAS is directly proportional to the change in the drive current. Similar to the nMOS transistor, the threshold voltage shift from process impacts cannot be simply generalized and so will not be included in this discussion. The width and length dependence of this CDE also follows similar width and length dependence described for the pMOS LOD effect. In order to mitigate the PAS effect, design rules should be created so that the gate space between transistors varies by a minimal amount or is even fixed at a constant value.

Contact Pitch Effect

The final intra-active effect is the Contact Pitch effect (CPE). This effect deals with the impact of the space between the gate and the contact spacing. For a given gate space, the placement and number of contacts impacts the stress transferred from the DSL layer to the channel and the channel resistance respectively. As the number of contacts increase, the amount of stress transferred to the channel from the DSL layer decreases proportionally, due to the fact that the holes etched in the nitride liner for the contact reduces the liner stress in the contact region to zero. Similarly, as the distance between the contact and gate decreases, the proximity of the zero liner stress region to the channel decreases, making the channel stress decreases proportionally. In order to mitigate the CPE, design rules should be created so that the contact distance between gates varies by a minimal amount or is fixed to a constant value.

Inter-Active Context Dependent Effects

Active Proximity Effect

The next set of CDEs is inter-active effects or effects that take place between actives. The first of these is the Active Proximity Effect (APE). This effect refers to the impact of the space of one active to another active. The first subset of this will be for the Same APE (SAPE), which refers to the case when similar active types (nMOS to nMOS or pMOS to pMOS) are interacting. In the case of nMOS SAPE, the only variable of interest is the volume of STI between the actives. Since the STI is generally compressive, it is preferable to have a minimal distance between the adjacent actives in either the horizontal, vertical and/or radial directions to minimize the compressive impact of the STI. This leads to a dense layout being preferred for the nMOS SAPE. As the transistor width increases the impact of the STI on transistor performance drops as expected. On the other hand, as the transistor gate length increases, the impact of the SAPE increases due to the channel seeing a greater average stress. In order to mitigate the nMOS SAPE, from a process perspective essentially the STI layer has to be deposited and/or grown so that it transfers minimal stress to the adjacent active layers. Alternatively from a design rule perspective to minimize nMOS SAPE spaces between adjacent actives has to be allowed to vary minimally or set to a constant value.

For the pMOS SAPE, there are two stress sources, the compressive STI and the compressive eSiGe. Initially, only the STI stress source will be considered. In the horizontal direction for pMOS transistors, it is preferable to have compressive stress hence a large volume of compressive STI is preferred. In the vertical direction, pMOS transistors prefer tensile stress hence it is preferred to minimize the compressive stress by minimizing the distance between the adjacent transistors as much as possible. The compressive eSiGe stress between adjacent transistors will generally be a shorter range effect relative to the STI stress, due to volume considerations and boundary effects. The range of the eSiGe depends on the germanium concentration and the volume of the eSiGe that is contained in the adjacent transistor. Thus in the horizontal direction there will be an optimal distance where the combined impact of the STI and eSiGe compressive stress is maximized without exceeding area constraints. On the other hand, this also implies that in the vertical direction, there will be an optimal minimum distance where the compressive stress of the combined STI and eSiGe will be minimized. Since the piezoresistance coefficients for the PMOS are more sensitive to stress in the horizontal direction, when looking for the optimum spacing radially, it is preferred to start out by maximizing the compressive STI and utilizing the point where the contribution from the STI and eSiGe is maximized.

As the transistor width increases the impact of the STI on transistor performance drops as expected. On the other hand, as the transistor gate length increases, the impact of the SAPE increases due to the channel seeing a greater average stress. Finally, in order to mitigate the pMOS SAPE, essentially the STI layer has to be deposited and/or grown so that it transfers minimal stress to the adjacent active layers or from a design rule perspective spaces between adjacent actives has to be allowed to vary minimally or set to a constant value.

The final subset of the APE inter-active effects is the Opposite APE (OAPE). This case addresses the situation when there is an nMOS active adjacent to a pMOS active. Since both nMOS and pMOS transistors would prefer tensile stress in the vertical direction, it is preferable to increase the drive current for both transistors by having them close to each other vertically. This proximity will still have to be bounded by the optimal minimum distance where the impact of the pMOS eSiGe has not compensated for the loss of the compressive STI. In the horizontal direction since the nMOS and pMOS have opposite stress preferences, it depends on the circuit application as to the spacing that will optimize performance. Generally, pMOS transistors are slower devices and thus it is preferable to optimize their performance by maximizing the horizontal space between the nMOS and pMOS transistors. This approach will also apply to the radial direction.

As the transistor width increases the impact of the STI on transistor performance drops as expected. On the other hand, as the transistor gate length increases, the impact of the OAPE increases due to the channel seeing a greater average stress. Finally, in order to mitigate the OAPE, essentially the STI layer has to be deposited and/or grown so that it transfers minimal stress to the adjacent active layers or from a design rule perspective spaces between adjacent actives has to be allowed to vary minimally or set to a constant value.

Well Proximity Effect

The next inter-active effect is the Well Proximity Effect (WPE). WPE was first published by Hook and was initially described as coming from the horizontal scattering of high energy ions from the photoresist edge that is deposited when creating threshold voltage wells [5]. The scattering of these ions (creating extra dopant atoms in the well) is what causes changes in the threshold voltage as a function of the transistor distance to the well boundary. In general as the distance between the gate and the well boundary grows smaller, the well region beneath the gate gains a greater amount of dopant atoms and hence has a higher threshold voltage. Another, more recently recognized component, of WPE is pocket shadowing which means the full pocket dose does not enter the silicon but a portion of the pocket dose is implanted into the photoresist or is "shadowed" [6]. This effect will lower the dose of dopant atoms available to increase the threshold voltage and will thus reduce the threshold voltage.

In general to model this effect, all the well edges within a 1 – 2 micron region (process dependent) need to be taken into account and the overall voltage shift from these edges summed together to reach a total voltage shift [6]. As shown in Figure 8, the general threshold voltage shift as a function of distance from the well edge is captured. For a fixed distance from the well edge, since WPE does not depend on transistor dimensions the width and length dependence of this effect tracks those of the nominal transistor width and length dependence once the threshold voltage shift from the implant dose is taken into account. In order to mitigate WPE, from a process perspective the implant energy can be modified to lower the scattering probability or depth of scattered

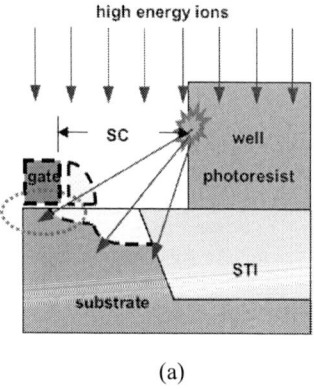 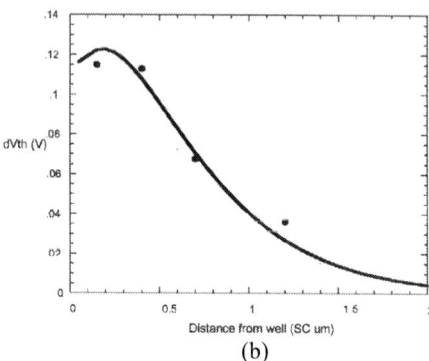

(a)   (b)

Figure 8. (a) The origin of the Well Proximity Effect (WPE) (b) As the gate distance from the well boundary decreases, the magnitude of the threshold voltage increases but at distances very close to the well boundary effects such as pocket shadowing and shallow angle implants take over and the magnitude of the threshold voltage begins to decrease. Image courtesy of Y. Sheu et al., IEEE CICC 2005.

atoms and/or the photoresist height can also be lowered to reduce the effective range of the scattered atoms and also reduce the amount of dopants lost to pocket shadowing.

Dual Stress Liner Effect

The final inter-active effect is the impact of the Dual Stress Liner (DSL) effect. nMOS transistors have a biaxial tensile nitride stress liner (stress values up to 2 GPa) covering the n-active silicon region while pMOS transistors have a biaxial compressive nitride stress liner (stress values up to -3 GPa) covering the p-active silicon region. In general for nMOS transistors in order to increase the drive current, it is preferred that the vertical and horizontal extension of the tensile liner is as far as possible from the transistor gate edges. Moreover, since at the boundary of the tensile liner will typically be a compressive liner, this boundary region effectively has significantly reduced liner stress. The spatial extent of this reduced liner stress region depends on the relative volume and stress level of each liner. For example a 2 μm long compressive (-2 GPa) stress liner next to a 0.2 μm long tensile (1.6 GPa) stress liner can significantly reduce tensile stress in shorter tensile liner. Therefore, it is preferred that for the nMOS transistor to gain the full benefits of the tensile liner to have at least a 1 μm overlap of the gate edge in all directions and equal or greater volume as the adjacent compressive liner.

For a pMOS transistor the compressive liner is the preferred orientation in the horizontal <110> direction but is not the preferred stress orientation in the vertical direction. Hence the pMOS transistor would prefer to have close proximity to a tensile liner in the vertical direction. On the other hand, in the horizontal direction, the pMOS transistor would prefer at least a 1 μm overlap of the gate edge and also equal or greater volume as the adjacent tensile liner.

For both the nMOS and pMOS transistors the general width and length dependence of the DSL effect follow those of the nominal transistor width and length dependence without stress. Finally, in order to mitigate the impact of the DSL effect, from a design rule perspective it is preferred that there is significant overlap of the gate by the respective liners and generally equal volume of the liners at the boundaries.

Conclusion

General strategies to mitigate CDEs have been presented for each effect discussed above. It is worth mentioning that designers should in general put as much circuits on the same silicon active block as possible as this will shield more transistors from both the intra and inter-active effects. So in order to reduce the impact of context dependence, it is preferred that more complex layouts are utilized. Moreover in the design methodology, outside of extracting parameters, a context variability index should be provided, so that for a given circuit topology the designer will be aware of the range of different results that can be expected for varying environments.

As shown in Table I below, this paper has discussed the basic underlying physical causes of context dependent effects, touched on the physical parameters that need to be extracted from the layout in order to accurately model these context effects, suggested how to optimize transistor performance for each context effect and finally shown how to mitigate the variability of these context effects using process optimizations and restrictive design rule choices.

TABLE I. Summary of CDE Modeling and Mitigation Strategies.

Context Dependent Effect	Layout Variables Extracted	Optimize Transistor Performance	Minimize Variability
AJE	Jog Height, Jog Space	NMOS: Minimize Jog PMOS: Maximize Jog	Process: Minimize STI stress Design Rule: Minimize Jog
APE	Space to adjacent active, Size of adjacent active	NMOS: Minimize Space PMOS: Maximize Space (Horiz.), Minimize Space (Vert.)	Process: Minimize STI stress Design Rule: Keep constant spacings
CPE	Left and Right Contact to Gate spacing	Maximize distance to Gate	Design Rule: Keep a constant spacing
DSL	Space to first and second DSL boundary	NMOS: Maximize Overlap of Gate PMOS: Maximize Overlap of Gate (Horz.), Minimize Overlap of Gate (Vert.)	Design Rule: Maximize extension of DSL overlap of the gate
LOD	Left and Right Gate to STI edge spacing	Maximize distance to STI edge	Process: Minimize STI stress Design Rule: Add dummy transistor at STI edge
PAS	Left and Right Gate to Adjacent Gate/STI edge spacing	Increase Gate Space	Design Rule: Keep Gate Space constant
WPE	Space to first Well boundary	Increase Well Spacing	Process: Vary implant dose and lower photoresist height Design Rule: Maximize space to Well boundary

Acknowledgments

The author would like to thank Xin Zhang, Donald Kolarik, Youn Sung Choi, Greg Baldwin, Kayvan Sadra, Thomas Aton, Sihan Lin, Sagnik Dey, Steve Prins, Tamer Cakici, and Claude Cirba for their useful discussions and suggestions.

References

1. R. A. Bianchi, G. Bouche, O. Roux-dit-Buisson, *Proc. IEDM*, **117**, (2002).
2. N. R. Zangenberg, J. Fage-Pedersen, J. L. Hansen, and A. N. Larsen, *J. Appl. Phys.*, **3883**, 94 (2003)
3. W. Zhang and J.G. Fossum, *IEEE TED*, **263**, 52 (2005).
4. V. Moroz, G. Eneman, P. Verheyen, F. Nouri, L. Washington, L. Smith, M. Jurczak, D. Pramanik and X. Xu, *Proc. SISPAD*, **143**, (2005).
5. T. Hook, J. Brown, P. Cottrell, E. Adler, D. Hoyniak, J. Johnson and R. Mann, *IEEE TED*, **1946**, 50 (2003).

6. J. Watts, K.-W. Su and M. Basel, *IEEE TED*, **2179**, 53 (2006).

302

ECS Transactions, 35 (4) 303-320 (2011)
10.1149/1.3572291 ©The Electrochemical Society

Quantitative Discussion on Electron-hole Universal Tunnel Mass in Ultrathin Dielectric of Oxide and Oxide-Nitride

Hiroshi Watanabe [a]

[a] Department of Electrical Engineering, National Chiao Tung University, Hsinch, Taiwan

MOS and MIS capacitor has been extensively studied in past several decades by many authors. It has been expected to reveal how basic physics relate electron device operation. In this structure, several physical phenomena co-work and then exhibit the electrical properties measured in IV- and CV-characteristics. On the other hand, the conventional models were established separately for the inversion layer (positive gate voltage), the depletion region (negative-low gate voltage), and the accumulation region (negative-high gate voltage). In addition, actual MOS/MIS samples have interfacial transition layer where physical properties are gradually changed from Si to oxide or other dielectric, the varying composition ratio of molecules and local traps owing to atomistic dangling bonds through dielectric layer. *What will happen if we self-consistently unify all the physical models that are separately developed?*

Introduction

In electron devices engineering, the device dimension is shrunk according to Moore's law [1] and the scaling rule [2], and then the thickness of gate dielectric layers fabricated on the silicon surface is thinned annually. Accordingly, the device modeling is made complicated by the interface factors that become notable with the scaling. The interface factors are composed of trap-induced issues or some inconsistency in lattice and contaminations at the interface, and the intrinsic issues that are independent of the traps and the lattice inconsistency. However, it can be regarded that the extrinsic issues, which is from traps, lattice-inconsistency and contaminations, can also exist inside the gate dielectric layers, apart from the interface. It is thereby indispensable to distinguish the intrinsic and extrinsic factors firstly. Here note that the intrinsic interface factor must become notable as the gate dielectric thickness is decreased. This, as a result, relatively enhances the influence of interface physics between the dielectrics and the silicon surface on the property of gate dielectrics [3]-[16]. It is accordingly convenient to carefully study the thickness dependence of physical and electrical properties of gate capacitor samples from which the extrinsic factors have been removed appropriately. MOS capacitor is a first preferable sample. The interface physics made notable by the scaling must be self-consistently involved in the device modeling. Provided that the unified modeling of interface physics and device modeling is succeeded, we must be able to reproduce the measured curves of both CV- and JV-characteristics in both polarity of gate voltage, irrespectively of oxide thickness.

303

In this paper, firstly, we will validate the intrinsic interface modeling with five samples of MOS capacitor with different gate oxide thicknesses. Subsequently, for the modeling of the extrinsic factor inside the gate dielectric, experimental and analyzing technique of molecular composition of dielectric film will be carefully discussed.

Intrinsic factor at the interface

Let us regard that MOS capacitor is an appropriate sample for investigating the intrinsic factor for interface physics, because the extrinsic factors have been extensively studied and removed as possible. Fig. 1 is the first evidence of our successful unification [15] in MOS capacitor. In this comparison, all fitting parameter we used is the tunnel mass whose oxide thickness dependence is shown in Fig. 2. It sounds that the tunnel mass of electron is 0.85 m_0, where m_0 is the rest electron mass, and independent of oxide thickness within a reasonable error. If we ignore anyone of the co-working models, a discrepancy occurs between theory and measurements. Trying to compensate this discrepancy, the tunnel mass becomes dependent of oxide thickness, and is distorted with respect of the polarity of gate voltage.

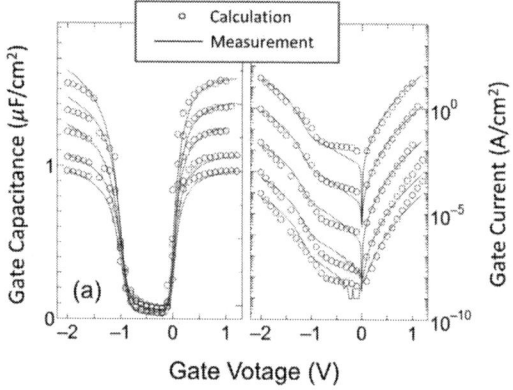

Fig. 1. Comparison of CV and JV curves using the unified model.

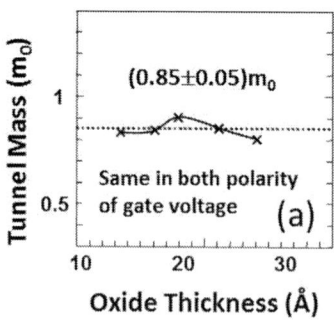

Fig. 2. Obtained tunnel mass using the unified model.

In Fig. 3, the options considered in the implementation of models for interfacial transition (IFT) layers and their corresponding equivalent circuits: (b) and (c) show the cases of removing the IFT layers from the poly-Si side and from both sides, respectively. In the upper line, are shown the corresponding equivalent circuits composed of capacitance of Si surface, of IFT layers, of pure oxide (SiO_2), and of poly-Si bottom. The IFT layers are considered in both of interfaces with Si substrate and with gate poly-Si, whose widths are four angstroms. The band gap (EG) of SiO_2 is assumed to be 8.95 eV from [17], while that of Si is 1.12 eV. Since the valence band affinity (VBA) is 4.49 eV from [18], the tunnel barrier (discontinuity of conduction band edges) is 3.34 eV. The oxide thickness (T_{OX}) is defined as the distance between the centers of the IFT layers. In other words, the interfaces are the centers of IFT layers, provided that the IFT layers have finite width. The dielectric constant (K) is 11.7 and 3.9 in Si and SiO_2, respectively. Here note that the conduction band edge, the valence band edge, and K are linearly changed within IFT layers from those values of pure Si to SiO_2 (**linear approximation**).

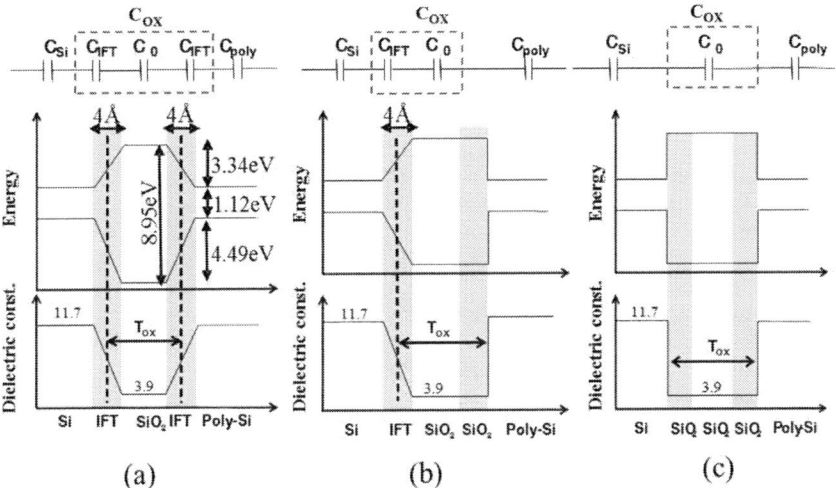

Fig. 3. Options considered in the implementation of models for IFT layers and their corresponding equivalent circuits.

<u>CV-JV fitting</u>

In option (a), we have adjusted T_{OX} to fit the calculated CV-curves with the measured ones in a set of five samples (Sample No. 1, 2, ..., 5) of MOS capacitar. Subsequently, we have adjusted the tunnel mass to fit the calculated JV-curves with the measured ones using the T_{OX} calibrated in the CV-fitting in a set of five samples of MOS capacitor. As shown in Fig. 1, we have successfully reproduced the measured curves of both CV- and JV-characteristics in both polarity of gate voltage, irrespectively of oxide thickness. As shown in Fig. 2, the obtained tunnel mass is around 0.85 m_0 without T_{OX}-dependency. The T_{OX} calibrated in the CV-fitting and used in the JV-fitting is plotted by diamond in Fig. 4. In ref. [15], it is reported that these T_{OX} are consistent with measured by the elipsometry method. It might appear that the elipsometry measures the thickness of pure oxide without the IFT layers.

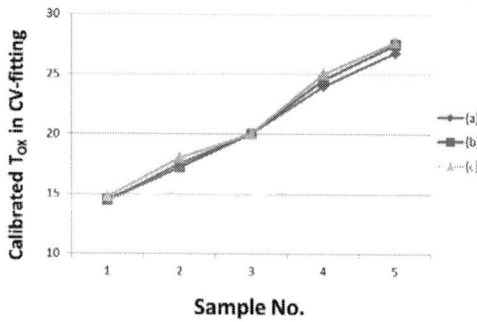

Fig. 4. T_{OX} calibrated via CV-fitting and used in JV-fitting.

In option (b), the IFT layer is removed from the gate poly-Si interface, and the T_{OX} is defined as the distance between the center of the Si IFT layer and the interface with poly-Si. We perform the subsequent fitting procedures of CV and JV characteristics in a similar manner with the option (a). The calibrated T_{OX} (plotted by squares) is consistent with the option (a), as shown in Figure 4, whereas a notable discrepancy in JV-curve appears in a negative gate bias region, as shown in Fig. 5. Here we have fitted the calculated JV-curves with the measured ones in the positive gate bias region. If we fit the calcuated JV-curves with the measured ones in the negative gate bias region, the discrepancy appears in the positive gate bias region. The option (b), thereby, cannot fit the JV-curves, while the CV-curve can be fitted with T_{OX} consistent with the option (a). The tunnel mass obtained in a case that the JV-curves are fitted in the positive gate bias region is dipicted by crosses in Fig. 6. It appears that this shows no T_{OX}-dependent tunnel mass in the positive gate bias case, around 0.59 m_0, whereas the negative gate bias case shows the T_{OX}-dependent tunnel mass appears, as depicted by squares in Fig. 6.

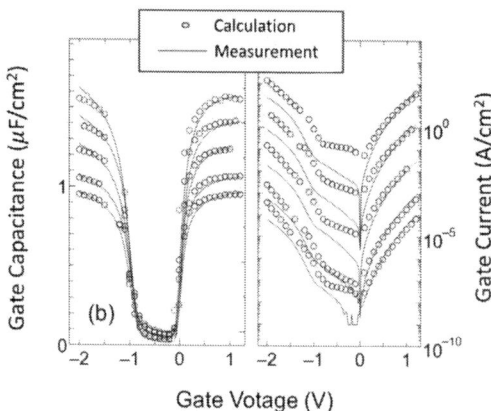

Fig. 5. Comparison of CV and JV curves in MOS capacitor in option (b).

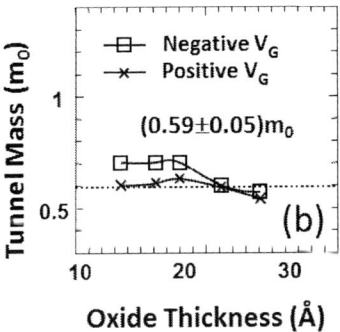

Fig. 6. Obtained tunnel mass in option (b).

In option (c), both IFT layers are removed, and the T_{OX} is defined as the distance between the abrupt interfaces. We perform the subsequent fitting procedures of CV and JV characteristics in a similar manner with the option (a). The calibrated T_{OX} (depicted by triangles) is consistent with the option (a), as shown in Fig. 4, whereas a notable discrepancy in JV-curve appears in a negative gate bias region, as shown in Fig. 7. Here we have fitted the calculated JV-curves with the measured ones in the positive gate bias region. If we fit the calcuated JV-curves with the measured ones in the negative gate bias region, the discrepancy appears in the positive gate bias region. The option (c), thereby, cannot fit the JV-curves, while the CV-curve can be fitted with T_{OX} consistent with the option (a). The tunnel mass obtained in a case that the JV-curves are fitted in the positive gate bias region is dipicted by crosses in Fig. 8. It appears that this shows no T_{OX}-dependent tunnel mass in the positive gate bias case, around 0.49 m_0, whereas the negative gate bias case shows that T_{OX}-dependent tunnel mass appears, as dipicted by squares inFig. 8.

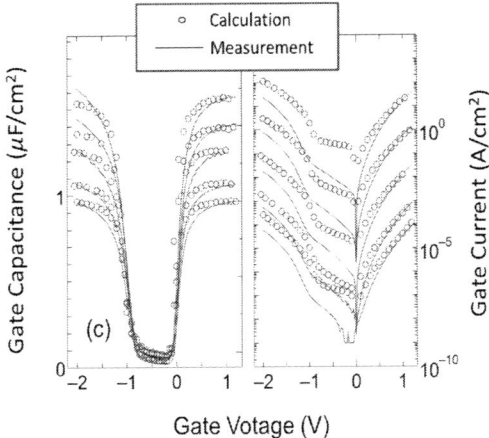

Fig. 7. Comparison of CV and JV curves in MOS capacitor in option (c).

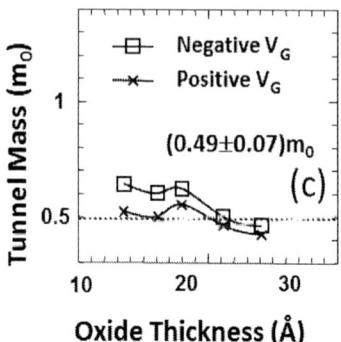

Fig. 8. Obtained tunnel mass in option (c).

The T_{OX}-dependent tunnel mass is not only unphysical, but also inconvenient to the device modeling. Among these options, it can be regarded the option (a) as closest to the actual profile of K and EG within a zero-th order approximation. We will assume the option (a) below in this paper, that is, the linear approximation.

An extrinsic factor inside gate dielectric

As mentioned above, we have appropriately modeled the IFT layers, i.e., the intrinsic factor, which removes the T_{OX}-dependency from the calibrated tunnel mass. To carefully study the impact of extrinsic factors inside the gate dielectric, we need to control the issue of the inconsistency in lattice and traps. Firstly, we should remove the extrinsic issues from the interfaces. Second, we should measure the profile of extrinsic issues of sample capacitors that we will measure. Third, we should remove the issues related to contamination as possible as we can, because it is hard to control it in the fabrication process. If not, we need to know which contamination will enter into the dielectric film and its amount. Fourth, we should determine the issue of extrinsic factors. Fortunately, it is possible to fabricate an appropriate sample capacitor for this aim, that is, ultra-thin gate SiON [19]. It is reported in [20] that the hysteresis of the CV measurements shows less than 5mV flat-band potential shift. This means no evidence for slow state. Let us regard this as satisfying the first requirement, i.e., to remove the extrinsic issues from the interfaces.

Next, as shown in Figure 9, the profiles of the nitride atom concentration, [N], and the oxide atom concentration, [O], are measured by the angle-resolved X-ray photoelectron spectroscopy (AR-XPS) method [21]. The pure oxide (SiO_2) was made at the dielectric surface and may then absorb the boron atoms if the film is covered by the gate polysilicon where the boron atoms are doped. This means that we can regard no contamination in n$^+$ poly-Si gate, whereas we need to consider the influence of boron contamination from p$^+$ poly-Si gate. The shaded layers at the surface and the Si/SiO$_2$ interface depict the IFT layers, where EG and K are assumed to be linearly changed from the values of Si to those of SiO$_2$ [15]. Finally, we assume that the major extrinsic issue inside this film is dangling-bond of atomistic network composed of Si, N, and O atoms.

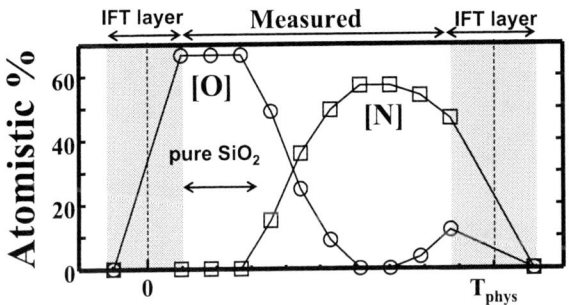

Fig. 9. Measured profile of [O] and [N].

Alloy Model for ultra-thin gate SiON film with IFT layers

The SiON film considered here basically stays on the tie-line of $(SiO_2)_{1-x}$ $(Si_3N_4)_x$. The K on the tie-line is dependent of Si-N bond rate (R) but not of the molecular compound ratio (x) [22], [23]. An excellent agreement is thereby achieved between the calculated and measured values of K in the literature [23]-[25]. We, a priori, regard this property as valid in the following quantities, e.g., EG, K, VBA, the tunnel masses of electrons and holes (m_e and m_h, respectively), and the IFT-layer width (W_{IFT}). They are accordingly assumed to satisfy the tie-line law:

$$Q = (1 - R)Q_{OX} + RQ_{SiN} \qquad (1)$$

$$R = \frac{3x}{1+2x}, \quad x = \frac{0.75[N]}{1-[N]} \qquad (2)$$

with Q being K, EG, VBA, m_e, m_h, and W_{IFT} on the tie-line, Q_{OX} being those at x=0, and Q_{SiN} being those at x=1. These quantities can be estimated on the tie-line according to Table I with [N] shown inFig. 9.

TABLE I. Material Variable: (*) Present results.

Variables	SiO₂	Si₃N₄	Si
K	3.9	7.5	11.7
EG (eV)	8.95	5.4	1.12
VBA (eV)	4.49	1.9	-
m_e (m0)	0.85	0.85*	-
m_h (m0)	0.85*	0.85*	-
W_{IFT} (nm)	0.4	0.5	-

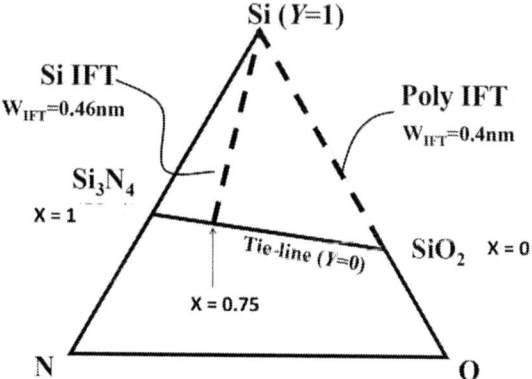

Fig. 10. Tie-line of SiON with IFT layers.

The IFT layers become notable as the gate dielectric film is thinned. Accordingly, we should take into account IFT layers assuming the extension of alloy-model, i.e., $[(SiO_2)_{1-x} (Si_3N_4)_x]_{1-y} Si_y$. As shown in Fig. 10, we have the tie-line at $y = 0$ and Si irrespective of x at $y = 1$. The IFT layers belong to the broken lines with $0 < x < 1$. The physical thickness of this film (T_{phys}) is regarded as the distance between the centers of the IFT layers in a similar manner with gate oxide. In this work, we prepared four samples of $T_{phys} = 1.1$ nm and 1.3 nm each having n^+ poly gate and p^+ poly gate. The W_{IFT} at the surface is therefore regarded as 0.4 nm (pure oxide value) from Table I, whereas that at the interface is found to be 0.485 nm using the tie-line law mentioned above since the measured profile of [N] is 46% (R = 0.85) at the interface. In Fig. 11, we show the profiles of x and R that are calculated by substituting the profile of [N] shown in Fig. 9 to (2). The calculated profile of R is more rounded than that of x and no Si-N atomistic bonds exist at the surface, i.e., R = 0, whereas they exist at the interface (R = 0.85). In Fig. 12, we show the profiles of K and EG that are calculated by using the profile of R shown in Figure 11 according to (1). The K is decreased as EG is increased. The profile of VBA is also calculated in a similar way and shown in Fig. 13. The VBA is changed with Eg, but not with K, as expected.

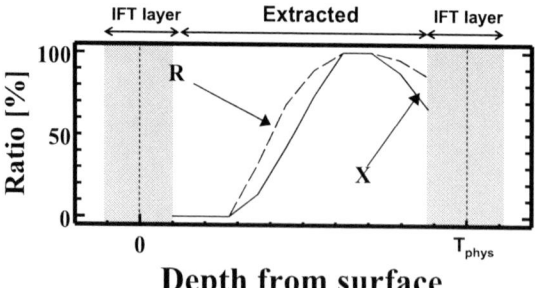

Fig. 11. Extracted profiles of x and R.

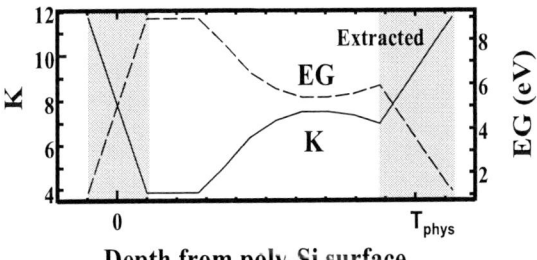

Depth from poly–Si surface

Fig. 12. Extracted profiles of EG and K in SiON.

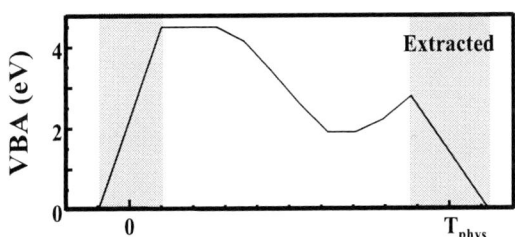

Depth from poly–Si surface

Fig. 13. Extracted profile of VBA.

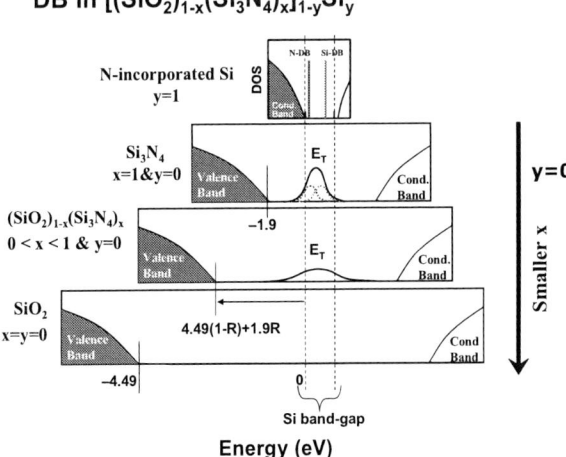

Fig. 14. Illustration for traps owing to dangling bond in SiON.

CV-fitting with Charge Trapping Model

Let us consider the breaking of Si-N atomistic bond in N-incorporated Si crystal. It is considered that we have two types of dangling bond, i.e., N-DB and Si-DB, as shown at the top line in Fig. 14 [26]. In the first step, the band structure is extended from N-incorporated Si-crystal (y=1) to Si_3N_4 (x=1 and y=0), in which EG is increased by nearly 5 times (according to the ratio of EG, 5.4/1.12) and the two sharp peaks due to N-DB and Si-DB become broader in proportion to the spread of EG, as shown at the second line. We can thereby consider that these peaks make a combined broader peak, the center of which is the trap level (E_T). According to [27], E_T is the level 0.4eV lower than the Si mid-gap with the half-value width being 0.1eV. On the other hand, let us regard the peak height in the distribution as proportional to the profile of R, that is, the number of Si-N atomistic bonds. Since the half-value width of the distribution is also extended as EG is increased according to $\Delta E_T = 0.1eV \times EG /1.12eV$, that is, 0.5eV in Si_3N_4 (x = 1 and y = 0). Next, the film is changed from Si_3N_4 (x = 1) to SiON (x < 1) at the third line. As EG and VBA are increased according to the tie-line law mentioned above, the combined peak is further broadened and lowered. The height of the peak is proportional to the trap density. At the last line, the film merges to pure oxide with no traps as the limit of the trap peak height disappears.

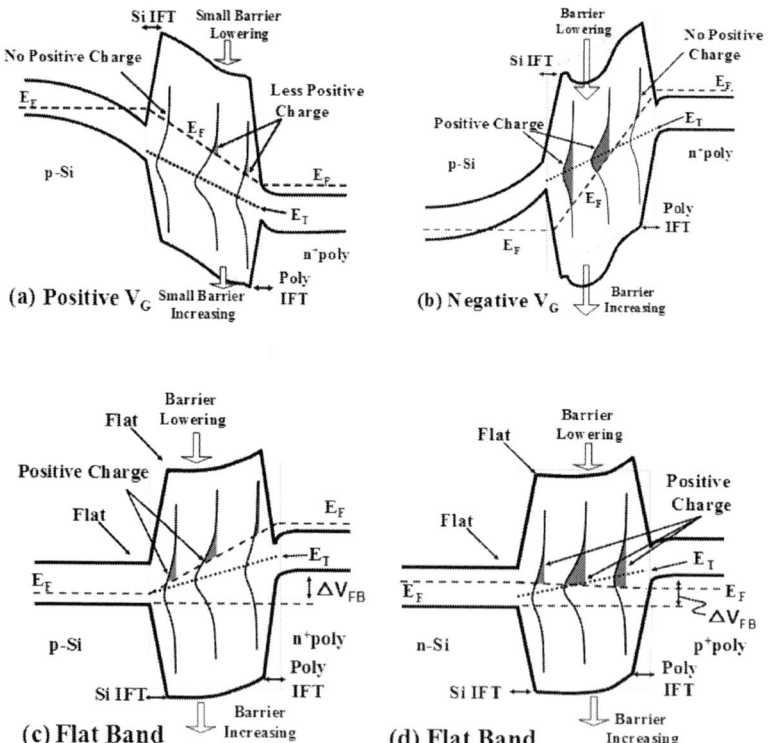

Fig. 15. Illustrations for trapped positive charge. The nMOS case is dealt with in (a) to (c), and the PMOS case corresponds to (d).

We can thus consider that the upper tail of the combined peak above the Fermi level (E_F) contributes to the positive charge, as shown in Fig. 15. The lines of E_T and E_F inside SiON film are represented by straight lines. The relation between E_T and E_F are dependent of bias condition, which affects the profile of charge and barrier modulation. From (a)-(c), we illustrate the band structure of an nMOS at a positive gate voltage (V_G), at a negative V_G, where the absolute value of V_G is larger than that of flat-band potential (V_{FB}), and at the flat-band, respectively. Here note that the flat-band means no electric field at the Si interface, while the electric field at the gate polysilicon (n^+ poly) interface and the amount of trapped charges are cancelled, as illustrated in (c). In these figures, we assumed the Si-N atomistic bond profile has a peak inside the film. In (a), since E_F is made higher than E_T by the band bending of inversion layer at the substrate surface, we have less amount of the positive charges, which is in proportion to that of the dangling bonds with $E_T > E_F$. This positive charge lowers the barrier height for tunneling electron and increases that for tunneling holes slightly at the n^+ poly side. In (b), since E_F is made much lower than E_T by the depletion layer at the substrate surface, we have a much greater amount of positive charge inside the film. This lowers the tunnel barrier for electrons and increases that for holes much more. In (c), E_F is still so high at the n^+ poly side that the positive charge lowers the tunnel barrier for electrons and increases that for holes, and negates the electric field at the interface with n^+ poly to make the flat-band at the Si interface. In (d), we illustrate the band structure of pMOS at the flat-band, where we have a greater amount of positive charge compared with nMOS at the side of gate polysilicon (p^+ poly), since E_F there is much lower in pMOS than in nMOS. The positive charge lowers the tunnel barrier for electrons and increases that for holes much more than in nMOS.

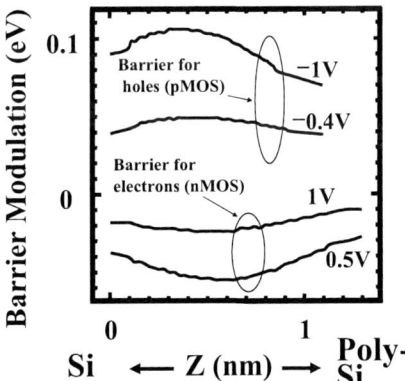

Fig. 16. Calculated barrier modulation due to the trapped positive charge.

The calculated barrier modulation is shown in Fig. 16. This clearly shows that the barrier modulation is more notable in negatively-biased pMOS than in positively-biased nMOS. The positive charge stored by dangling bonds is, in this way, dependent of the electric field that determines $E_T - E_F$ inside the SiON film, which has a notable influence on the gate current through the SiON film, as discussed below.

It is noteworthy that the precise calculation of E_F is indispensable for estimating the amount of the positive charge, which is described in detail in [28], [29]. In addition, E_F has a significant influence on the estimation of the incomplete depletion layer [30] and the weak accumulation layer [31] at the interface with the gate polysilicon. The subband at the Si interface is calculated by the real-space transfer matrix method [32]. The IFT layers are self-consistently implemented with these detailed-advanced models [15].

Influence of boron-contamination on flat band potential

Dividing the film into the grids (i=1, 2…M), the Poisson equation is found to be

$$K_i \varepsilon_0 \frac{\Delta V_i}{\Delta T_i} = -q\Delta N_i , \qquad (3)$$

where ε_0 is the vacuum permittivity, q is the elementary charge, and ΔT_i is the width of i-th grid layer. The ΔN_i and ΔV_i are the local amount of charges divided by q in the i-th grid layer and the potential drop across the i-th grid layer related to ΔN_i, respectively. The K_i is the dielectric constant of i-th grid layer. Integrating these grid layers, we have the flat-band shift,

$$\Delta V_{FB} = \sum \Delta V_i = -\frac{q}{\varepsilon_0 T_{phys}} , \qquad (4)$$

where

$$\gamma = \frac{\sum \Delta N_i}{K_i M} . \qquad (5)$$

In Fig. 17, it is found that the measured flat-band shift is clearly proportional to γT_{phys}. The proportional coefficients are 0.7 MV/cm with the fit-variance being 3.1×10^{-3} in nMOS and 1.4 MV/cm with the fit-variance being 7.8×10^{-3} in pMOS. This means that ΔN_i in pMOS differs from that in nMOS. If the amount of positive charge was equivalent to the number of Si-N bonds, these amounts in nMOS and pMOS would be almost the same. However, the slope in pMOS is twice that in nMOS, as shown in Fig. 17. This may come from the boron-contamination, because p+ poly is fabricated on the oxide surface of the film in pMOS capacitor samples.

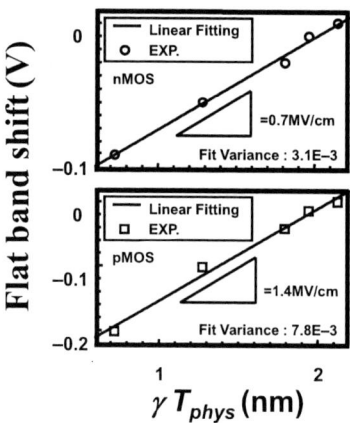

Fig. 17. Analysis of flat-band shift by linear fitting.

On the other hand, since all of the Si-N bonds cannot give the dangling bonds, we cannot directly extract ΔN_i from the measured profile of R. We should therefore carry out a careful study of the trap density (N_T) that is equal to $Y_{DB} \times R$, where Y_{DB} is the dangling-bond (trap site) yield from Si-N atomistic bonds. In such a way, we self-consistently solve (3) considering Y_{DB} as well as all the above-mentioned physical models [28]-[32], and can then obtain CV-characteristics. Fig. 18 is the test calculation of CV-characteristics of an nMOS capacitor. The flat-band potential is decreased as Y_{DB} is increased, since the positive charge inside the gate dielectric SiON film is increased. The influences in inversion and accumulation layers are opposite.

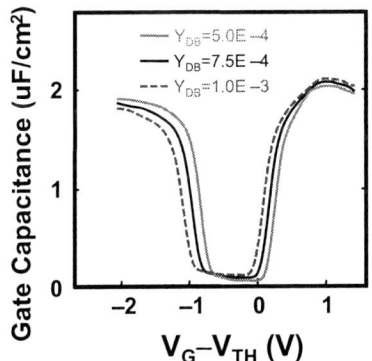

Fig. 18. Influence of Y_{DB} on CV-characteristics.

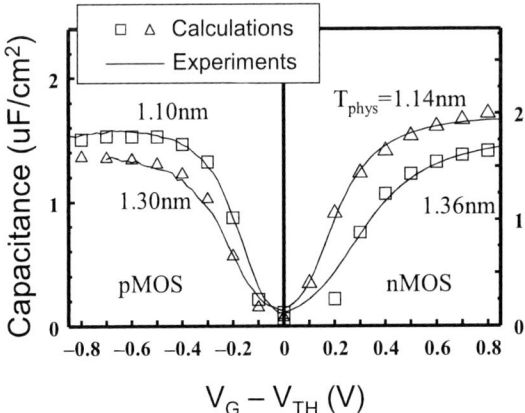

Fig. 19. Comparison of calculated and measured CV-characteristic.

As mentioned above, experimental and analyzing technique of molecular composition of dielectric film has been notably improved. Nowadays, we can measure the profile of Si-N atomistic bonding distribution (R) in ultrathin SiON dielectric film, as shown in Fig.

11. The energy band structure and the dielectric constant (K) are extracted according to the Si-N bonding profile, as shown in Fig. 12. The shaded regions are inserted as models for the IFT layers at polysilicon/dielectric and dielectric/Si in the unified modeling [16]. Provided that the SiON dielectric partially differs from the stoichiometric of $(SiO_2)_{1-x}$ $(Si_3N_4)_x$, some of Si-N bonds are broken; then leaving the dangling bonds in dielectric. Thus, Y_{DB} (dangling bond number / Si-N bond number) can be regarded as the off-stoichiometric ratio. Using this Y_{DB} as fitting parameter in the CV-fitting, the unified modeling can predict the CV-characteristics. As shown in Fig. 19, we have an excellent agreement between the measured CV characteristics and those calculated with the parameters summarized in Table II and the nitride profile measured by AR-XPS. It is found from the present analysis that 0.075% and 0.17±0.02% of Si-N bonds are broken to be dangling bonds in nMOS and pMOS, respectively. The extracted Y_{DB} in pMOS is much larger than in nMOS, which corresponds to the slopes shown in Fig. 17. It appears that the boron atoms diffused from p^+ poly cause the extrinsic dangling bonds inside the SiON dielectric film, which may be associated with B-N-B bonds [33]. As shown in Fig. 9, we have pure SiO_2 at the poly-Si side, which degrades the barrier for boron penetration. The boron contamination can therefore penetrate from p^+ poly. In addition, the dangling bonds at the p^+ poly side contribute to more positive charges at the flat-band in pMOS, as illustrated in Fig. 15 (d). This extrinsic dangling bond causes the excess charge at the p^+ poly side, which results in an apparent increase of Y_{DB}. In other words, *Y_{DB} can compensate the influence of boron contamination on the calibration of universal tunnel mass.*

TABLE II. Parameters used in the present analysis

Y_{DB} (%)	T_{Phys} (nm)	N_{poly} (cm^{-3})	N_{sub} (cm^{-3})
0.19	1.10	6×10^{19}	5×10^{16}
0.15	1.30	4×10^{19}	7×10^{16}
0.075	1.14	1×10^{20}	2.6×10^{16}
0.075	1.36	3×10^{20}	2.6×10^{19}

JV-fitting composed of trap-assisted tunneling and direct tunneling

The comparison of the calculated CV-curves with the measured ones has determined Y_{DB}, as mentioned above. *Regarding the product of Y_{DB} and R as the profile of local traps caused*, we can take into consideration the location effect of local traps causing the trap-assisted tunneling (TAT) through the SiON dielectric film. Before demonstrating this, let us extract the tunnel masses for electrons and holes from an individual experiment. We compare the calculated and measured gate currents of samples in which the profiles of nitride are uniform [21], taking into consideration the IFT layers. In Fig. 20, we plot the four graphs to depict the electron currents and hole ones at 5 MV/cm, which are normalized by the electron current at [N] = 0 %. In the top graph (m_h / m_e = 0.75/0.85), we have the off-set between the electron currents and the hole ones at [N] = 23%. Since the tunnel barrier for holes, i.e., VBA, is decreased with the increase of [N], this off-set is caused by the turnover of the hole current and the electron one. In the second graph (m_h / m_e = 0.85/0.85), we have the off-set at [N] = 31%, which is consistent with the experiment shown at the bottom graph. In the third graph (m_h/m_e = 0.95/0.85), we have the off-set at [N] = 40%. It is therefore found that the tunnel masses for electrons and holes are the same, $0.85m_0$, in the SiON gate dielectric film.

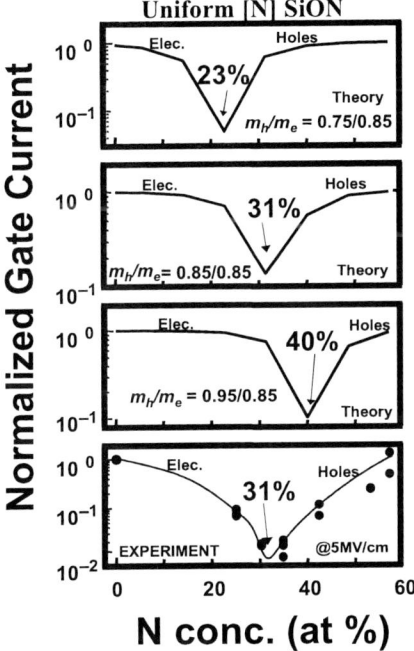

Fig. 20. Extraction of holes tunnel mass.

Since the TAT occurs when an electron scatters with a trap (dangling bond) having the scattering cross-section (σ), the total tunneling probability (D_{TOT}) is written by:

$$D_{TOT}(z, E) = \sigma Y_{DB} R(z) D_{TAT}(E) + (1 - \sigma) D_{DT}(E), \qquad (6)$$

where E is the energy of tunneling electron, z is a spatial location of dangling bond, and D_{TAT} and D_{DT} depict the tunneling probabilities of trap-assisted tunneling and direct tunneling, respectively. While D_{DT} is calculated using the Wenzel-Kramers-Brillouin (WKB) approximation, D_{TAT} is calculated by $D_{TAT} = 1/(D_1^{-1} + D_2^{-2})$, where D_1 and D_2 are the tunneling probability from the beginning of tunneling to the considered dangling bond and that from the dangling bond to the end of tunneling, respectitively. Both D_1 and D_2 are calculated in the same manner as D_{DT}. The formula for leakage current (J_G) is, therefore,

$$J_G = \sigma Y_{DB} \int dE \int dz R(z) D_{TAT}(E) + (1 - \sigma) \int dE \, D_{DT}(E). \qquad (7)$$

This formula is also used in the analysis of tunnel mass shown in Fig. 2, and thereby, we have $m_e = m_h = 0.85 m_0$. Here we assume that σ is a constant in the integrals of E and z. In Fig. 21, we show the influence of σ on the current. It is found that the TAT vanishes, while σ is less than 10^{-15} cm^2, i.e., the capture radius is a few Å. If we can reproduce the leakage current with σ being less than this value, the leakage mechanism is not TAT but direct tunneling.

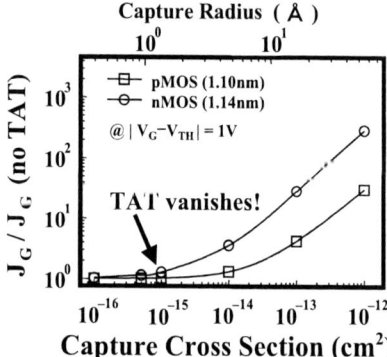

Fig. 21. Impact of capture-cross section on current.

In Fig. 22, we compare the calculated currents with the measured ones, using the measured profiles of R, the parameters shown in Table II, the universal tunnel mass $(0.85m_0)$ and $\sigma = 0$. The excellent agreement is obtained, irrespective of film thickness. This decisively suggests that the leakage mechanism of ultra-thin SiON film is the direct tunneling enhanced by barrier modulation due to the positive charges trapped by dangling bonds whose profile is proportional to that of Si-N atomistic bonds with the proportional coefficient being Y_{DB}. This is quite similar to the degradation mechanism of data retention of floating gate memory cell [34].

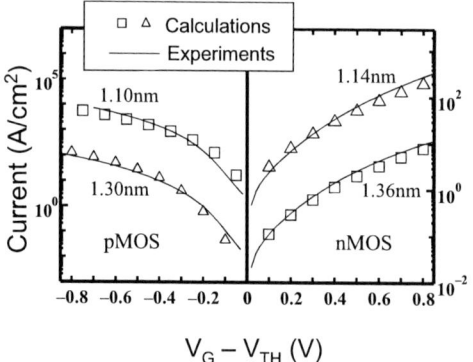

Fig. 22. Comparison of calculated and measured gate currents

Conclusion

From the literature where the IFT layers are ignored, the tunnel masses for holes and electrons in Si_3N_4 have been regarded as 70% and 54.3% of the electron tunnel mass in

SiO_2, respectively [35]. If the electron tunnel mass in SiO_2 was $0.42m_0$ [36], then the hole tunnel mass in Si_3N_4 would be $0.3m_0$ [37]. The unified model with the IFT layers makes the calibrated values of tunnel masses for electrons and holes are the same, $0.85m_0$, irrespective of film thickness as long as the film component exists around the tie-line of $(SiO_2)_{1-x}(Si_3N_4)_x$ with the IFT layers.

In conclusion, we have unified the physical models co-working in MOS/MIS capacitor, to reproduce CV- and JV-characteristics. The parameters calibrated individually are the tunnel masses for electrons and holes, and the off-stoichiometric ratio of dielectric material (abbreviated Y_{DB}). Expanding the present achievement to general dielectric film, the requirement for "universal" tunnel mass would be:

1) Same between electrons and holes.
2) Independent of dielectric thickness.
3) Irrespective of dielectric material.

Further study of metal gate with high-K dielectrics is indispensable for validating the universal tunnel mass. The Y_{DB} assumed as the discrepancy from the stoichiometry is also important to distinguish the intrinsic factors from the extrinsic factors. If we can establish this universality, the device modeling would be much advanced.

Acknowledgments

The author would like to thank D. Matsushita, K. Muraoka, and K. Kato for their collaboration in original work of this topic, when the authors worked for Toshiba Corporation. He also would like to thank K. S. Pua from Phison Electric for totally encouraging his works in National Chiao Tung University. The work is also supported by the National Science Council of Taiwan under Grant NSC99-2218-E-009-021.

References

1. G. Moore, in *IEDM/1975*, Tech. Dig. p. 11.
2. R. H. Dennard, F. H. Gaensslen, H-N Yu, V. L. Rideout, E. Bassous, and A. R. LeBlanc, *IEEE J. of Solid State Circuit*, **sc-9**, 256 (1974).
3. D. A. Muller, T. Sorsch, S. Moccio, F. H. Baumann, K. Evans-Lutterodt, and G. Timp, *Nature*, **399**, 758 (1999).
4. C. Kaneta, T. Yamasaki, T. Uchiyama, T. Uda, and K. Terakura, *Microelectron. Eng.*, **48**, 117 (1999).
5. A. A. Demkov and O. F. Sankey, *Phys. Rev. Lett.*, **83**, 2038 (1999).
6. J. B. Neaton, D. A. Muller, and N. W. Ashcroft, *Phys. Rev. Lett.*, **85**, 1298 (2000).
7. S. T. Pantelides, S. N. Rashkeev, R. Buczko, D. M. Fleetwood, and R. D. Schrimpf, *IEEE Trans. Nucl. Sci.*, **47**, 2262, (2000).
8. H. Niimi, H. Yang, G. Lucovsky, J. W. Keister, and J. E. Rowe, *Appl. Surf. Sci.*, **166**, 485 (2000).
9. T. Yamasaki, C. Kaneta, T. Uchiyama, and T. Uda, *Phys. Rev. B. Condens. Matter*, **B63**, 115314 (2001).

10. K. Takahashi, M. B. Seman, K. Hirose, and T. Hattori, *Jpn. J. Appl. Phys.*, **41**, 223 (2002).
11. T. Hattori, K. Takahashi, M. B. Seman, H. Nohira, K. Hirose, N. Kamakura, Y. Takata, S. Shin, and K. Kobayashi, *Appl. Surf. Sci.*, **212/213**, 547 (2003).
12. F. Giustino, P. Umari, and A. Pasquarello, *Phys. Rev. Lett.*, **91**, 267601 (2003).
13. F. Giustino, A. Bongiorno, and A. Pasquarello, in *IWDTF/2004*, Proc. p. 45.
14. M. Watari, J. Nakamura, and A. Natori, *Phys. Rev. B. Condens. Matter*, **69**, 035312-1–035312-6 (2004).
15. H. Watanabe, D. Matsushita, and K. Muraoka, IEEE TED, **53**, 1323 (2006).
16. H. Watanabe, D. Matsushita, K. Muraoka, and K. Kato, IEEE TED, **57**, 1129 (2010).
17. S. Miyazaki, H. Nishimura, M. Fukuda, L. Ley, and J. Ristein, *Appl. Surf. Sci.*, **113/114**, 585 (1997).
18. J. L. Alay and M. Hirose, *J. Appl. Phys.*, **81**, 1606 (1997).
19. D. Matsushita, K. Muraoka, Y. Nakasaki, K. Kato, S. Inumiya, K. Eguchi, and M. Takayanagi in *VLSI Symp/2004*, Tech. Dig. p. 172.
20. D. Matsushita, K. Muraoka, K. Kato, Y. Nakasaki, S. Inumiya, K. Eguchi, and M. Takayanagi, in *INFOS/ 2005*, Proc. p. 424.
21. K. Muraoka, K. Kurihara, N. Yasuda, and H. Satake, *J. Appl. Phys.*, **94**, 2038 (2003).
22. G. Lucovsky and H.-Y. Yang, *Ext. Abs. SSDM/1996*, p. 356.
23. N. Yasuda, K. Muraoka, M. Koike, and H. Satake, in *SSDM/2001*, Ext. Abs. p. 486.
24. D. M. Brown, P. V. Gray, F. K. Heumann, H. R. Philipp, and E. A. Taft, *J. Electrochem. Soc.* **115**, 311 (1968).
25. T. S. Eriksson and C. G. Granqvist, *J. Appl. Phys.* **60**, 2081 (1986).
26. K. Kato, private communication; K. Kato, Y. Nakasaki, D. Matsushita, and K. Muraoka, in *ICPS-27/2005*, Proc., p. 395.
27. N. Yasuda and H. Satake, *SSDM/2001*, Ext. Abs., p. 202.
28. H. Watanabe and S. Takagi, *J. Appl. Phys.* **90**, 1600 (2001).
29. H. Watanabe, K. Matsuzawa, S. Takagi, IEEE TED, **50**, 1779 (2003).
30. H. Watanabe, IEEE Trans. Electron Devices, **52**, 2265 (2005).
31. H. Watanabe, K. Nakajima, K. Matsuo, T. Saito, and T. Kobayashi, in *SSDM/2005*, Ext. Abs. p. 504.
32. H. Watanabe, K. Uchida, and A. Kinoshita, IEEE TED, **52**, 52 (2007).
33. K. Kato, D. Matsushita, K. Muraoka, and Y. Nakasaki, in *ICPS-28/2007*, Proc. 1403.
34. H. Watanabe, IEEE TED, vol. **57**, 1873 (2010).
35. F. R. Libsch and M. H. White, *Solid-State Elec.* **33**, 105 (1990).
36. B. Brar, G. D. Wilk, and A. C. Seabaugh, *Appl. Phys. Lett.*, **69**, 2728 (1996).
37. T. Maruyama and R. Shirota, *J. Appl. Phys.* **78**, 3912 (1995).

Physics-Based Hot-Carrier Degradation Models

S.E. Tyaginov[*], I.A. Starkov[°], H. Enichlmair[†], J.M. Park[†], Ch. Jungemann[‡], and T. Grasser[*]

[*] Institute for Microelectronics, TU Wien, Wien, Austria

[°] Christian Doppler Laboratory for Reliability Issues in Microelectronics at the Institute for Microelectronics, TU Wien, Wien, Austria

[†] Process Development and Implementation Department, Austriamicrosystems AG, Unterpremstätten, Austia

[‡] Institute for Microelectronics and Circuit Theory, Bundeswehr University, München, Germany

Abstract

We present a thorough analysis of physics-based hot-carrier degradation (HCD) models. We discuss the main features of HCD such as its strong localization at the drain side of the device, the weakening of the degradation at higher temperatures, and the change of the worst-case condition in small devices. The first feature is related to "hot" carriers, while the second is controlled by the fraction of "colder" particles. The latter feature is related to the change of the silicon-hydrogen bond-breakage mechanism from the single- to multiple-carrier process. All these findings suggest that the interface state creation process is controlled by the manner how the carriers are distributed over energy, that is, by the carrier energy distribution function. We distinguish between three main aspects of the physical picture behind hot-carrier degradation: carrier transport, microscopic mechanisms of defect creation and simulation of degraded devices. Therefore, we analyze and classify the existing HCD models in this context. Finally we present our hot-carrier degradation model based on a thorough evaluation of this distribution function by means of a full-band Monte-Carlo device simulator. Our approach tries to address the whole hierarchy of physical phenomena in order to capture all the essential aspects of hot-carrier degradation.

Introduction

Hot-carrier degradation (HCD) is associated with the build-up of defects at or near the silicon/silicon dioxide interface of an MOS transistor. The degradation is due to the bombardment of the interface by carriers which have gained sufficiently high energy and are thus called "hot" carriers, see [1–3] and references therein. The interface states that are created by this process are characterized by a density N_{it}. They are able to capture electrons/holes and, hence, become charged. The density N_{it} is a distributed quantity, that is, varies with the coordinate along the Si/SiO_2 interface as well as in energy. These additional charges introduced into the system are distributed along the channel and perturb the electrostatics of the device resulting for instance in a shift of the threshold voltage. Furthermore, they act as additional scattering centers, thereby degrading the mobility and, as a result, the transconductance G_m and linear drain current I_{dlin}.

Several hot-carrier degradation mechanisms have been suggested in the literature: channel hot-carrier, drain avalanche hot-carrier, secondary generated hot-carrier, substrate hot-carrier, and Fowler-Nordheim and direct tunneling injection, e.g.[1, 4, 5]. The first mechanism is directly linked to the electric field in the channel of Field-Effect-Transistor (MOSFET) which accelerates carriers. Particles with sufficiently high energy, whereby "sufficiently" depends on a concrete HCD model and may mean energy required to overcome the potential barrier at the interface or trigger the Si-H bond rupture, are called "hot". The drain avalanche and secondary generated hot-carrier mechanisms assume a cascade of impact ionization events caused by a solitary hot carrier, which leads to avalanche generation of electron-hole pairs. Under the substrate hot-carrier mode a uniform injection from the channel-substrate p-n junction occurs; this mode has been actively used in pioneering works devoted to hot-carrier reliability [5]. As for the Fowler-Nordheim and direct tunneling mechanisms, carriers are injected either into the SiO_2 conduction band through the triangular potential barrier (Fowler-Nordheim regime) or to the channel overcoming a trapezoidal barrier (direct tunneling). In modern ultra-scaled devices and/or in high-voltage transistors only the channel hot-carrier regime is relevant and drain avalanche and secondary generated hot carriers are considered as a part of the channel hot-carrier degradation phenomenon. Therefore, further in the text we mean just the channel hot-carrier mode when referring to "hot-carrier degradation".

The first successful attempt to HCD modeling was the so-called "lucky-electron" model proposed by Hu [6, 7]. This concept is based on the following assumptions: (i) an electron characterized by an energy high enough to overcome the potential barrier at the interface (ii) impinges onto the interface without collision, that is, without energy loss and (iii) without being scattered back into the channel and being emitted into the SiO_2 conduction band thereby producing a defect. The "lucky electron" model claims that the threshold of HCD is 3.7 eV, however, hot-carrier stresses performed at $V_{ds} < 3$ V demonstrated that device aging can also occur at lower voltages [8]. As a consequence, this approach fails for short-channel devices. However, due to its simplicity, the model still remains one of the most popular approaches.

An empirical extension of the "lucky electron" model was proposed by Takeda and Suzuki [9, 10]. This simple time dependent model expresses the transconductance degradation ΔG_m and/or threshold voltage shift ΔV_{th} by a time power law t^n. The exponent and proportionality coefficients are fitting parameters adjusted independently for a particular device architecture. The advantage of such an approach is that it allows easy extrapolation the device life-time from accelerated hot-carrier stress conditions to real operation biases. However, the applicability of the model is rather limited as demonstrated by investigations employing lightly doped drain structures where the saturation of degradation after a certain value has been observed ([11] and references therein). Although inaccurate for describing hot-carrier degradation, the Takeda model inspired a number of fitting models. These models try to represent device parameter degradation employing some combinations of time exponents. Among them are the Goo model based on the "lucky electron" concept, which can capture saturation of degradation [11], the Dreesen model [12, 13], which follows the same strategy but was adapted for lightly doped drain MOSFETs and is able to successfully represent the I_{dlin} degradation in the range of $\Delta I_{\mathrm{dlin}}{=}0.02\%...10\%$.

Other extensions of the Hu concept have been proposed by Woltjer [14, 15] and by Mistry et al. [16, 17]. In contrast to the "lucky electron" model, which deals with interface trap generation under maximum substrate current conditions but fails at other stress conditions, the Woltjer model considers the oxide field as crucial for the creation of interface states. As a result, a field-driven correction is incorporated into the "lucky electron" model. This extension allows description of the degradation behavior of devices with various dimensions and oxide thicknesses. Mistry and co-workers reported that a single degradation mechanism is not sufficient for proper degradation modeling and three different modes of damage were proposed: at low V_{gs} creation of interface states and oxide neutral electron traps occurs while for mid and high V_{gs} only interface state build-up and oxide electron traps, respectively, contribute. All of them are present during DC-stress and each of them can dominate the AC-stress life-time [17]. However, the life-times predicted by this model were rather inaccurate and thus only of limited applicability. Moreover, the general shortcoming of these approaches is that starting from a certain node and beyond, the field-driven paradigm and related modeling approaches, such as extensions of the "lucky electron" model should be substituted by energy-driven concepts [8, 18, 19].

The idea that two (or several) competing degradation mechanism are required to describe the overall degradation has been further extended by Moens et al. in order to capture degradation in LDMOS transistors [20–23]. In a series of papers Moens demonstrated that for high-voltage devices one should consider defect build-up in different transistor sections, namely in the channel, accumulation, and bird's beak regions. As a result, different components of the damage are characterized by different time exponents, which explains the different slopes of parameter degradation. The dynamic behavior was reported to be determined by hole trapping/detrapping processes [20, 22].

All models described above have been developed for the description of HCD observed in a particular class of devices. As such, they are empirical or at the best phenomenological. But a proper description of HCD may only be possible when the physical picture is accurately understood and captured by the model. There are five main physics-based concepts for hot-carrier degradation modeling elaborated so far:

- the approach presented by Hess and co-authors [24, 25];

- the empirical extension of the Hess model to make it suitable for TCAD device simulators by development in the work of Penzin et al. [26];

- the extension of the reaction-diffusion framework proposed by Alam [27, 28];

- the energy-driven paradigm by Rauch and LaRosa [2, 29];

- the Bravaix model based on the Hess approach [30, 31].

The most important breakthrough in HCD modeling is due to Hess who introduced the interplay between a single- and a multiple-carrier mechanism for Si-H bond-breakage. Since these mechanisms are related to the fractions of "hot" and "colder" carriers, the idea that the matter is controlled by the carrier energy distribution function (DF) was first acknowledged [32]. Notwithstanding the fact that the model is able to explain such a crucial feature of HCD such as the hydrogen/deuterium isotope effect [33], the link between the device microscopic picture of the defect build-up and degradation of device characteristics is missing. An attempt of linking these levels has been undertaken in the successor of the Hess approach, in the Penzin model [26] presenting, in fact, a phenomenological approach for HCD modeling. Another approach is the extension of the reaction-diffusion framework of the negative bias temperature instability (NBTI) in order to capture HCD [27, 28]. This implies, however, that once stress is removed, full recovery should be observable within reasonable times. In reality, the recovery of HCD is very slow, thus suggesting that HCD is a reaction-limited process [34]. One more strategy for HCD modeling proposed by Rauch and LaRosa is called "energy-driven paradigm" [2, 29]. For channel lengths less than 180 nm, HCD was shown to be controlled by the single "knee" energy. This energy is related to the stress bias. Therefore, instead of operating with coordinate-dependent quantities (electric field, dynamic temperature, DF, etc) only a single bias-dependent parameter is considered. A combination of the Hess and Rauch approaches was proposed by Bravaix et al. [30, 31]. In this concept the interaction between the single- and multiple-carrier mechanisms for Si-H bond-breakage has been considered. However, crucial point is that the information about the carrier DF is substituted by some empirical factors. In spite of a certain success of all these approaches the main problem is that they capture just a fragment of the whole HCD mosaic. Therefore, the whole hierarchical ladder connecting the microscopic level of defect creation and the device simulation level is still not fully understood.

To summarize, over the last decades hot-carrier degradation modeling has evolved from simple empirical models to a more detailed understanding of the microscopic physics involving single- and multiple-particle processes (SP- and MP-mechanisms). A detailed description of the physics requires knowledge of the carrier energy distribution function (DF) which can only be obtained from a solution of the Boltzmann transport equation. Most models in use today employ simplified solutions based on the average energy or, even more dramatic, the electric field, while in the ultimate simplification it is tried to capture the physics using closed analytic expressions. Although computationally more efficient, these approaches are inevitably inaccurate, even though their limitations might not be that obvious when a limited range of bias conditions, temperatures, and channel-lengths is investigated. Therefore, after describing the main features of hot-carrier degradation, we proceed to the detailed analysis of the existing physics-based HCD models finishing with the presentation and validation of a detailed model.

Characteristic Features of Hot-Carrier Degradation

Although the detrimental phenomenon of hot-carrier degradation has been known for more than four decades, it remains one of the most crucial concerns in transistor reliability. Since during this period of time several generations of Metal-Oxide-Semiconductor

MOSFETs have been in production, the characteristic features of HCD, their understanding, and the modeling approaches also reflect these trends. For instance, in the eighties, the device dimensions have been reduced rather quickly, accompanied by a slower scaling of the transistor power supply. This tendency led to high electric fields in the MOSFET channel, which accelerated carriers up to energies high enough to directly trigger a Si-H bond-breakage process by a solitary carrier, which then was considered "hot" [1, 7, 9]. Such a situation required specific measures in order to suppress carrier heating. Among them was the demand that the supply voltage should scale faster than device dimensions [35–38] in addition to requirements for doping profiles and device geometry, which for instance resulted in lightly doped drain structures [30, 39].

In particular, even though in the 0.25 μm node hot-carrier degradation could be rather dramatic, its importance was expected to reduce drastically for coming nodes [1]. The physical reason behind this expectation was that the source-drain voltage V_{ds} had already been scaled down to 1 – 1.5 V while the threshold energy required for triggering the Si-H bond dissociation process is about ∼3.0 – 3.5 eV. Therefore, it was expected that the carrier would not be heated up to energies sufficient for the Si-H bond-breakage, resulting in a suppression of HCD. Overall, a complete absence of HCD was expected for extremely-scaled devices [3, 25, 30, 40].

In reality, however, even ultra-scaled modern MOSFETs can show severe HCD [3, 30, 31]. This was first demonstrated for gate lengths less than than 0.2 μm and supply voltages below 1.0 V by Mizuno et al. [41]. The authors related this finding to an energy exchange mechanism populating the "hot" fraction of carrier ensemble. energies substantially higher than the lattice temperature. Possible mechanisms responsible for such an energy gain include impact ionization [42], Auger recombination [43], electron-phonon [44], and electron-electron scattering [18, 19, 45].

Note that electron-electron scattering is of particular importance for nano-scale devices [3, 30]. Particularly for these devices the situation is even more complicated because the dominant mechanism for Si-H bond-breakage changes from a single-carrier to a multiple-carrier mechanism [3, 25, 30, 40]. For example, in a long-channel or high-voltage device carriers striking the interface are already rather hot and are able to trigger silicon-hydrogen bond rupture by a single collision, which is referred to as the *single-carrier mechanism*. In contrast, such extremely hot carriers do not exist in sufficient quantity in scaled devices. Rather, several particles subsequently bombard a bond, thereby exciting and eventually rupture it, which is referred to as the *multiple-carrier process*. However, these two scenarios are just limiting cases and in a particular device geometry under certain operating/stress conditions a superposition of these two mechanisms has to be expected [46, 47].

The most important consequence of the interplay between single- and multiple-carrier processes is the change of the worst-case condition of hot-carrier degradation: traditionally, the worst-case of HCD occurred at $V_{gs} = (0.4 - 0.5)V_{ds}$, corresponding to the maximal substrate current or – in other words – to the largest impact ionization rate [8, 48–50]. However, this is not always the case even for long-channel devices; for example, in high-voltage p-MOSFETs the worst-case conditions are observed at the maximum gate current and no empirical law exists for this case [51–53]. This regime corresponds to the situation where the average carrier energy is maximal, that is, the carrier ensemble includes a substantial fraction of particles with energies high enough to induce the bond dissociation following a single impact.

In contrast, in scaled devices the operating voltages are such that a single carrier is unlikely to reach energies sufficiently large to trigger an SP-process. The process of energy interchange between carriers is of a stochastic nature and therefore one may expect that a certain fraction of particles — however small — may still obtain a relatively large energy. Still, although particles able to launch the SP-mechanism are in principle present, their relative number is rather small and, hence, the MP-process becomes dominant [40,54]. Contrary to the SP-mechanism, the individual carriers contributing to the MP-mechanism require only a relatively low energy. However, a large number of those carriers is needed. Thus, the carrier flux rather than the single-carrier energy becomes important in this case. The maximum carrier flux is obtained at $V_{ds} = V_{gs}$ for both scaled n- and p-MOSFETs [55–58], which now becomes the region of maximum HCD.

As a final note we remark that even in the case of ultra-short devices a certain fraction of "hot" carriers exists because the high-energy tail of the carrier distribution function is populated for instance by the electron-electron scattering process [2,3]. Therefore, the SP-mechanism will still contribute in these devices. Also, thermalized, that is, "cold", particles still exist even in the case of high-voltage devices, thereby also leading to HCD by the MP-process. To conclude, in a real device under real operating/stress conditions, the interplay between the SP- and MP-modes of bond-breakage has to be considered and is controlled by the way carriers are distributed over energy, that is, by the carrier DF.

Another characteristic feature of HCD is its strong localization near the pinch-off region (or the drain end of the gate), just near the area where the electric field peaks [1,3,59–63]. Such a peculiarity is again related to carriers heating up to energies required to launch the bond-breakage process. Since the driving force of this acceleration is the electric field, for the sake of simplicity it is often assumed that the maximum of the interface state generation rate just corresponds to the electric field peak. However, it has been long understood that the DF can follow changes in the electric field only with a certain delay [64]. Therefore, in order to improve over the electric field approximation, such quantities as the carrier temperature have been used to estimate the location of the maximum damage. However, as it was demonstrated for instance in [65,66], the maxima of different quantities are observed at different positions and therefore the N_{it} peak never directly coincides with that of the electric field. Moreover, Zaka et al. showed that different simplified treatments of carrier transport employing the drift-diffusion, energy-transport and spherical harmonics expansion methods (keeping only the 0th and 1st order polynomials) lead to spurious description of hot-carrier injection [67]. As a result, the spherical harmonics expansion method for Boltzmann transport equation solution with a higher expansion number of the stochastic Monte-Carlo based solver have to be used. This finding is very important because the Si-H bond-breakage process is described by an energy-dependent reaction cross section [30,68,69]. Hence, it is important to know the magnitude of the carrier fraction which corresponds to the given energy.

To make the picture complete one should pay attention to the temperature behavior of HCD. Contrary to NBTI, which is made more severe at higher temperatures (see [70,71]), hot-carrier induced damage usually becomes less pronounced at elevated temperatures [72–77]. Note that this traditional tendency is typical only for (relatively) long-channel devices while for ultra-scaled MOSFETs HCD becomes more significant at higher temperatures due to the dominant role of electron-electron scattering and its impact on the carrier distribution function [3, 78–80].

To summarize, the essential features of hot-carrier induced degradation unequivocally demonstrate that the matter is controlled by the carrier distribution function. The DF allows us to judge how efficiently the carriers interact with the bonds or – in other words – how intensive the bond dissociation reactions are. As a result, a comprehensive physics-based HCD model is expected to rely on consistent consideration of the microscopic mechanisms of defect creation and the carrier DF. For the calculation of the carrier distribution function a carrier transport module has to be incorporated into the model.

Hot-Carrier Degradation Models

Hess Model

The main breakthrough in the area of HCD modeling associated with the Hess concept was the introduction of two competing mechanisms for Si-H bond-breakage, namely the single- and multiple-carrier processes, see Fig. 1 [24, 32, 69]. A single-particle process is due to the interaction of a high-energetic solitary carrier with the bond. During this interaction energy is transfered to the bond followed by its dissociation. Due to the large disparity of the electron mass and the mass of hydrogen nucleus, the most probable way to deliver such an energy is via excitation of one of the bonding electrons to an antibonding state. As a consequence, a repulsive force acting on the H atom is induced followed by the release of hydrogen. The desorption rate of this process is [32]:

$$R_{\mathrm{SP}} \sim \int_{E_{\mathrm{th}}}^{\infty} I(E)P(E)\sigma(E)\mathrm{d}E, \tag{1}$$

where $I(E)$ is the flux of carriers with energies in the range of $[E; E + \mathrm{d}E]$, $\sigma(E)$ energy-dependent Keldysh-like reaction cross section, $P(E)$ the desorption probability, while the integration starts from the threshold energy E_{th}.

The first success of the theory was achieved when hydrogen/deuterium desorption induced by subsequent bombardment by several ("cold") carriers from the tip of a scanning tunneling microscope (STM) was investigated on hydrogen- and deuterium-passivated Si surfaces [68, 81–84]. These experiments showed that the D-passivated surfaces are much more resistant with respect to electron bombardment compared to hydrogenated ones. In other words, substantially higher densities of STM currents are required to release the same amount of D atoms vs. H atoms. The difference in depassivation rates (Fig. 2) may be more than two orders of magnitude at high voltages, which gave rise to the name "giant isotope effect". The similarities between the dangling bonds at surfaces and interfaces lead to the application of the theory to H-passivated Si/SiO$_2$ interfaces subjected to HC stress [54, 85–88].

This giant isotope effect was explained by the concept of multivibrational mode ex-

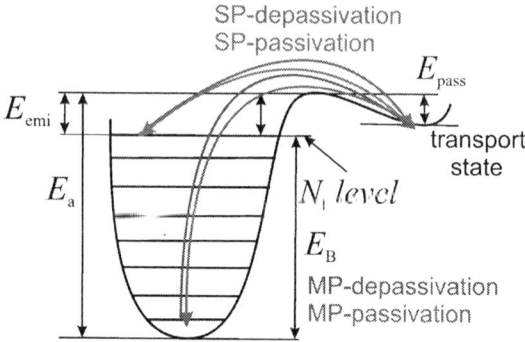

Figure 1: Two competing processes of Si-H bond-breakage: the single- and multiple-carrier mechanisms. The bond is treated as a truncated harmonic oscillator.

citation by linking to the excitation of the phonon modes to a cascade of subsequent bombardments by interfacial carriers. Note that the Si-H bond can relax from an excited state to a lower one and the balance with a reciprocal process has to be considered as well. The bond is treated as a truncated harmonic oscillator (Fig. 1) characterized by a ladder of bonded levels with the last level designated as N_1. The Si-H bond-breakage process is described by the system gradually climbing the ladder of energetic states, a process which is eventually terminated when hydrogen leaves the last bonded level towards the transport state (Fig. 1).

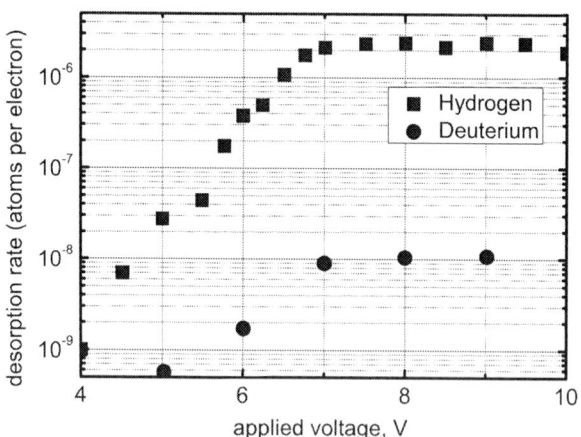

Figure 2: Disparity between H and D desorption rates induced by electrons tunneling from the STM tip on the passivated Si surface (data from [82]).

The reaction rate is defined by the height of the barrier E_{emi} separating the last level N_1 and the transport state (Fig. 1). Similarly, the passivation process is related to the hydrogen jumping into the opposite direction, determined by the barrier height E_{pass}. The corresponding rates (P_{emi}, P_{pass}) are assumed to obey an Arrhenius law.

To obtain an expression for the phonon excitation and decay rates P_{u}, P_{d} (Fig. 1), the formalism described in [32, 69] is applied. The electron flux can induce either phonon absorption (that is, bond heating) or phonon emission (related to the multivibrational mode decay). Therefore, these absorption and emission rates, which are just the product of the electron flux and the process capture cross section divided by the phonon occupation number plus one or by the occupation number, respectively. Summarizing all these considerations one obtains the expression for P_{u}, P_{d} [69]:

$$
P_{\text{d}} \sim \int_{E_{\text{th}}}^{\infty} I(E)\sigma_{\text{ab}}(E)[1 - f_{\text{ph}}(E - \hbar\omega)]\mathrm{d}E,
$$

$$
P_{\text{u}} \sim \int_{E_{\text{th}}}^{\infty} I(E)\sigma_{\text{emi}}(E)[1 - f_{\text{ph}}(E + \hbar\omega)]\mathrm{d}E,
$$

(2)

where $\sigma_{\text{ab}}(E)/\sigma_{\text{emi}}(E)$ are phonon absorption/emission reaction cross sections, $\hbar\omega$ phonon energy and phonon occupation numbers entering the expressions as $f_{\text{ph}}(E)$. Summarizing the findings of (2), one obtains the bond-breakage rate corresponding to the MP-process as

$$
R_{\text{MP}} = \left(\frac{E_{\text{B}}}{\hbar\omega} + 1\right)\left[P_{\text{d}} + \exp\left(\frac{-\hbar\omega}{k_{\text{B}}T_{\text{L}}}\right)\right]\left[\frac{P_{\text{u}} + \omega_{\text{e}}}{P_{\text{d}} + \exp(-\hbar\omega/k_{\text{B}}T_{\text{L}})}\right]^{-E_{\text{B}}/\hbar\omega},
$$

(3)

with E_{B} being the energy of the last bonded level in the quantum well (Fig. 1) and the phonon reciprocal life-time ω_{e}; k_{B} and T_{L} are the Boltzmann constant and the lattice temperature, respectively.

It is worth emphasizing that the particle flux differential $I(E)$ entering formulae (1,2) assumes that the carrier DF implicitly enters the expression. Thus, one of the main conclusions of the works by the group of Hess is the idea that for a *proper description of HCD, the carrier energy distribution function* is required. Another important achievement of this concept the isotope effect is essentially explained because different energetics of Si-H and Si-D lead to different parameters of the corresponding quantum wells (see Fig. 1), that is, to different positions of the last level E_{B} and phonon life-time τ.

Another characteristic feature of the Hess model is the assumption that the *activation energy* E_{a} for the Si-H bond-breakage rate is statistically distributed, see Fig. 1. This assumption is supported by *ab initio* calculations using density functional theory [69, 89]. As a consequence, the dispersion of E_{a} leads to different power-law slopes during degradation, see Fig. 3 [24, 25, 40]. This is essential as simple first-order kinetics of Si-H bond-breakage with a single-valued activation energy lead to an exponential transition between the bonded and broken states within about a single decade in time. However, experimental observations demonstrate a double-power law of degradation:

$$
N_{\text{it}} \sim \frac{p_1}{1 + (t/\tau_1)^{-\alpha_1}} + \frac{p_2}{1 + (t/\tau_2)^{-\alpha_2}},
$$

(4)

where τ_1/τ_2 are characteristic times and α_1/α_2 are two different sublinear time slopes ($\sim 1/2$). This time evolution has been explained assuming that two different types of

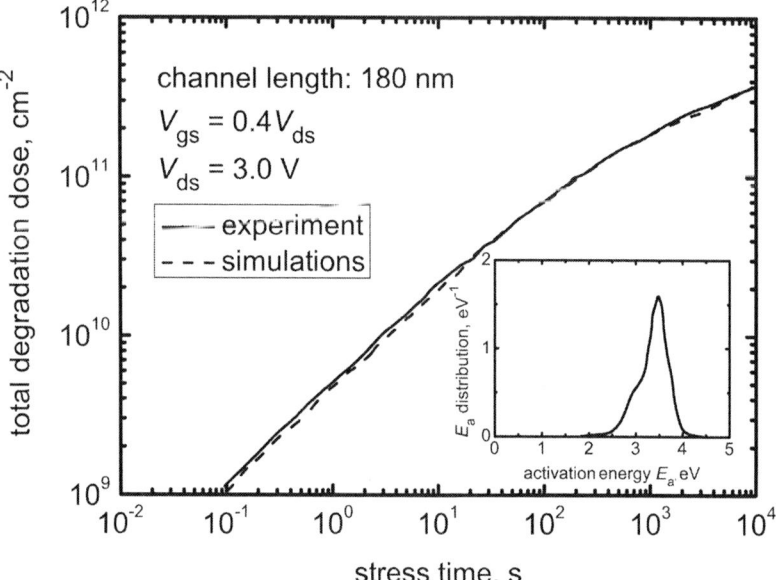

Figure 3: The total degradation dose (cumulative N_{it}) as a function of stress time: experiment vs. theory obtained for a 180 nm device under worst-case stress conditions, i.e. $V_{gs} = 0.4V_{ds}$. Inset: distribution of Si-H bond-breakage activation energy. The data are borrowed from [25].

traps (realized with the probabilities p_1 and p_2) contribute to HCD. These traps are similarly distributed and can be fit by the derivative of the Fermi-Dirac function with different mean values $E_{am,1}/E_{am,2}$ and standard deviations $\sigma_{a,1}/\sigma_{a,2}$ [90], see Fig. 3, inset:

$$E_{a,1/2} \propto \frac{1}{\sigma_{a,1/2}} \frac{\exp\left(\dfrac{E_{am,1/2} - E_{a,1/2}}{\sigma_{a,1/2}}\right)}{\left[1 + \exp\left(\dfrac{E_{am,1/2} - E_{a,1/2}}{\sigma_{a,1/2}}\right)\right]^2} \tag{5}$$

Despite the significant progress due to the work of Hess *et al.*, the interface traps are considered on a microscopic level and remain unconnected to the device level. For instance, the device life-time is estimated as the time when the concentration N_{it} reaches a certain level. Also, the degradation of such parameters as transconductance, linear drain current and so forth, is not really addressed. Furthermore, although the necessity of evaluation of the carrier DF is acknowledged, in practice this information has not been incorporated into the approach. As a result, the model operates with some overall N_{it}, thereby not considering its distributed nature and that the details in the N_{it} distribution follows the features found in the DF.

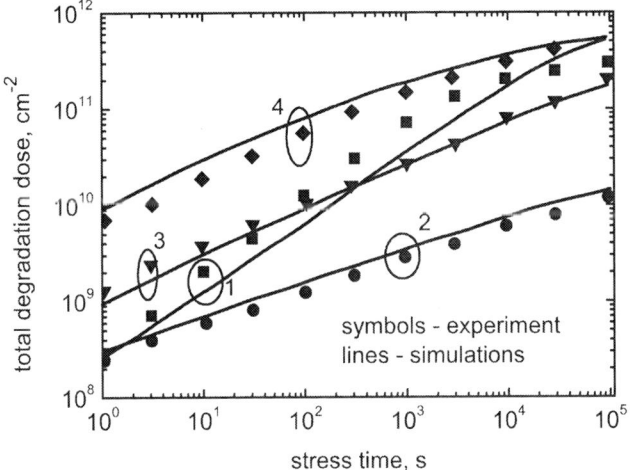

Figure 4: The interface state concentration N_{it}, simulation vs. experiment. An n-MOSFET with a gate length of 0.35 μm and an oxide thickness of 6.5 nm was subjected to hot-carrier stress at (1): $V_{gs} = $ -9V, $V_{ds} = V_b = $ 0V (V_b is the substrate voltage); (2): $V_{gs} = $ 12V, $V_b = $ 0V and floating source and drain; (3): $V_{gs} = $ 1V, $V_{ds} = $ 0V, $V_b = $ -11V; (4): $V_{gs} = $ 2.5V, $V_{ds} = $ 5V, $V_b = $ 0V. Data from [26].

Penzin Model

The Hess model was adapted for TCAD device simulations by Penzin *et al.* [26, 91] by employing phenomenological approximation. The model omits the microscopic level of defect generation (with the interplay of the SP- and MP-mechanisms as the essential attribute) but operates already on the device level. The bond rupture process is described by a kinetic equation for the passivated bond concentration n:

$$\frac{dn}{dt} = -kn + \gamma(N_0 - n), \tag{6}$$

where k is the forward (depassivation) reaction rate while γ is the backward (passivation) rate and N_0 the total concentration of both "virgin" and broken bonds. The forward reaction rate has the following structure: $k = k_0 \exp(-E_a/k_B T_L) k_H$ with the attempt frequency k_0 and k_H being the hot-carrier acceleration factor. This term is controlled by the "local hot carrier current" [26] I_{HC}:

$$k_H = 1 + \delta_{HC} |I_{HC}|^{\rho_{HC}}, \tag{7}$$

where δ_{HC} and ρ_{HC} are fitting parameters.

An important peculiarity of the Penzin approach is that the *activation energy* of bond dissociation *depends on the hydrogen density* and the transversal component of the electric field. The Si/SiO$_2$ interface (and its vicinity) is considered as a capacitor. The released hydrogen is assumed to be charged as well as the remaining dangling bonds. As a results, an additional electric field related to these charges is introduced into the system. This field prevents subsequent hydrogen ions from leaving the system and thus the potential barrier separating bonded and transport states is increased:

$$E_{\mathrm{a}} = E_{\mathrm{a}}^0 + \delta|F|^\rho + \beta k_{\mathrm{B}} T_{\mathrm{L}} \ln \frac{N_0 - n}{N_0 - n^{(0)}}$$
$$\beta = 1 + \beta_\perp F_\perp,$$

(8)

with E_{a}^0 being the activation energy in the absence of mobile H and $n^{(0)}$ the preexisting mobile hydrogen concentration. Since the system is considered as a capacitor, removal of charge from the capacitor is related to an additional energy required to compensate the change of the electric field. This energy is proportional to the capacitor electric field, that is in this case the normal (to the interface) component $F_\perp$ entering the expression. Additionally, the external electric field F can stretch or squeeze the bond, thereby changing the activation energy which is controlled by the term $\delta|F|^\rho$.

Similarly to the approach of Hess, the model employs a distribution of the activation energy and thus is able to represent the sublinear slope of degradation; Fig. 4 demonstrates a reasonable agreement between experiment and theory. Although the model attempts to capture the carrier transport, this issue still remains vague. The formula 7 includes the acceleration factor related to the *"local hot carrier current" which is the equivocal matter*, i.e. it would be reasonable to define a *criterion to distinguish "hot" and "cold" carriers*. This criterion may be based for instance on the *carrier temperature, which is related to the average energy of the distribution function* (compare with the expressions (1, 2)). As an adjacent problem, the information about the N_{it} profile is hardly achievable. Moreover, in spite of the efforts to link the kinetics of the trap generation and the device characteristics, we are not aware of a rigorous comparision against experimental device characteristics. Instead the soundness of the model is only proven by representing the experimental value of some cumulative N_{it} (Fig. 4), but such a representation has already been obtained within the Hess approach, see Fig. 3.

Reaction-Diffusion Framework

Another approach focused on the physical picture behind hot-carrier degradation was developed by the group of Alam [27, 92]. The assumption was that NBTI and HCD are related to the breakage of silicon-hydrogen bonds, differing only in the driving force triggering this dissociation. Therefore, both phenomena are to be coupled within the same modeling framework. The authors claimed that since NBTI is just the breakage of Si-H bonds followed by hydrogen release and diffusion, NBTI and HCD are to be united within the reaction-diffusion concept.

Experimental observations demonstrated that time signatures of NBTI and HCD have different power-law slopes, i.e. the former one can be approximated by a $t^{1/4}$ law while the latter one better obeys a $t^{1/2}$ dependence, see Fig. 5 and [27, 92]. The reaction-diffusion framework includes the following stages [92, 93]:

    1. Creation of interface states via breaking Si-H bonds. This stage is reaction-limited and described by a t^1 dependence.

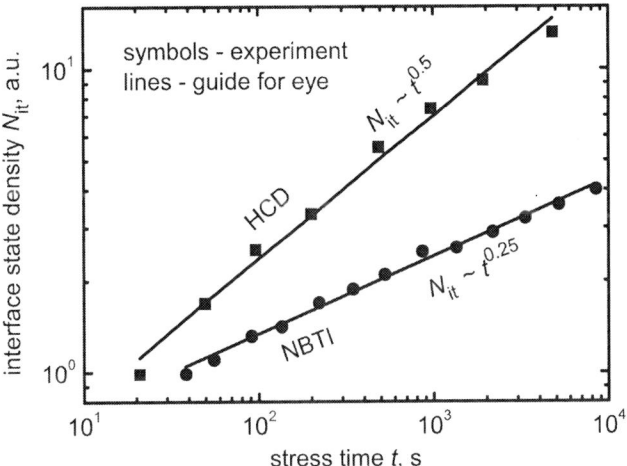

Figure 5: Different time slopes of hot-carrier induced degradation and NBTI. The data are borrowed from [27].

2. Hydrogen diffusion begins to take over with no more interface states created: $N_{it} \sim t^0$.

3. Diffusion-limited phase with $t^{1/4}$ behavior.

4. Hydrogen diffuses away with unlimited diffusion velocity resulting in the $1/2$ degradation time slope, i.e. $N_{it} \sim t^{1/2}$.

5. Finally, saturation occurs when all the "virgin" Si-H bonds are depassivated: $N_{it} \sim t^0$.

Therefore, it was assumed that NBTI is diffusion-limited which describes its $t^{1/4}$ behavior while HCD is controlled by the 4th phase. However, this scenario presumes that in the case of HCD a transition from $t^{1/4}$ to $t^{1/2}$ is to be observed but the authors of [92] claim that in practice no experimental evidence of such a transition is known. Instead, they suggest that the difference in time slopes is related to the circumstance that NBTI is a 1D problem while HCD is a 2D phenomenon due to the non-uniform N_{it} distribution over the lateral coordinate. Since the Si-H bond-breakage event generates one mobile hydrogen and one interface trap one writes $N_{it} = \int N_H(r,t) \mathrm{d}^3 r$ ($N_H(r,t)$ is the coordinate-dependent hydrogen concentration). The diffusion front moves like $(D_H t)^{1/2}$

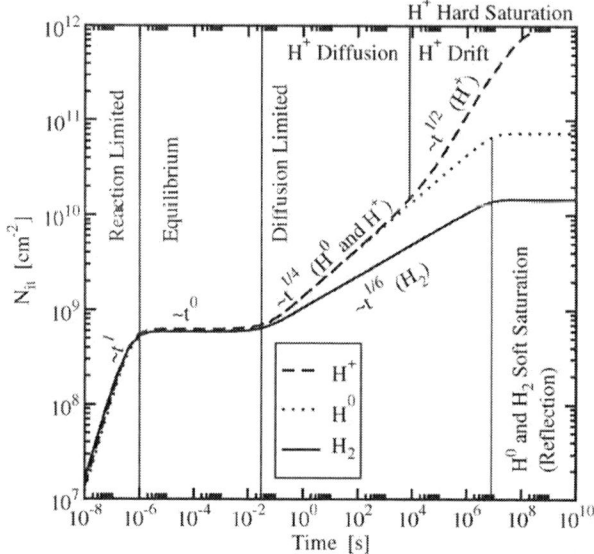

Figure 6: The main phases of the reaction-diffusion model applied to NBTI with different time slopes being marked.

and thus NBTI- and HCD-related N_{it} are:

$$N_{it}^{(NBTI)} = (1/A_d) \int_0^{(D_H t)^{1/2}} N_H^{(0)} \left[1 - r/(D_H t)^{1/2}\right] A_d dr = (1/2) N_H^{(0)} (D_H t)^{1/2}$$

$$N_{it}^{(HCD)} = \frac{\pi}{2A_d} \int_0^{(D_H t)^{1/2}} N_H^{(0)} \left[1 - r/(D_H t)^{1/2}\right] r dr = \frac{\pi}{12 A_d} N_H^{(0)} (D_H t),$$

(9)

where D_H is the hydrogen diffusivity, A_d the area of the degraded spot and $N_H^{(0)}$ is the H density at the interface. Assuming that $N_{it} N_H^{(0)} \sim$ const, one obtains that $N_{it}^{(NBTI)} \sim (D_H t)^{1/4}$ and $N_{it}^{(HCD)} \sim (D_H t)^{1/2}$.

Despite its ability to explain the different time slopes of NBTI and HCD, this reaction-diffusion model suffers from serious shortcomings. First, within this framework it is assumed that both phenomena are diffusion limited. This implies, however, that once the stress is removed recovery should occur rather quickly. Recent NBTI data suggest, however, that *interface state creation is reaction rather than diffusion limited* [71, 94, 95]. Concerning HCD, the *recovery is in general rather weak* if there is any recovery at all. Second, the model does not rely on carrier transport, that is, it does not consider the driving force behind the trap generation. As a consequence, the N_{it} distribution and the localized nature of the damage are not addressed.

The Energy-Driven Paradigm of Rauch

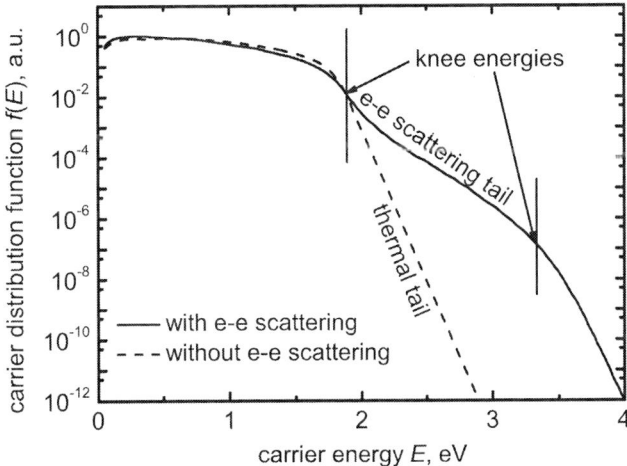

Figure 7: The impact of electron-electron scattering on the shape of the carrier energy distribution function. In the former case an additional hump in the DF high-energy tail appears. Data from [2].

Two main issues associated with the works of Rauch and LaRosa are:

- the increasing impact of electron-electron scattering on HCD at reduced channel lengths [8, 18]

- and the idea that in the case of scaled devices with channel lengths less than 180 nm, the driving force of HCD is the carrier energy rather than the electric field[2, 29, 96].

Electron-electron scattering is of special interest in the case of ultra-scaled MOSFETs because in these devices the supply voltage is rather low and therefore the single-carrier mechanisms of Si-H bond-breakage were expected to be suppressed. This energy exchange mechanism, however, populates the "hot" fraction of the DF and modifies the shape of the DF, that is, results in a pronounced hump in the carrier distribution function, see Fig. 7. Thus, the high-energy tail of the DF can expand deeper into energy than expected from the supply voltage. As a result, the contribution from the SP-mechanism is increased. Additionally, just electron-electron scattering defines the acceleration of HCD at elevated temperatures, which is pronounced in the case of extremely-scaled MOSFETs [3, 78–80].

The energy-driven paradigm presented by Rauch and LaRosa claims that beyond the 180 nm node the driving force of HCD is the energy deposited by carriers, not the maximal electric field in the channel as it was in the "lucky electron model" [7]. Both the impact ionization rate as well as the rate of hot-carrier induced interface state generation is controlled by integrals of the form $\int f(E)S(E)\mathrm{d}E$, where $f(E)$ is the carrier DF and $S(E)$ the reaction cross section; compare this to the formula (1) used previously which has the

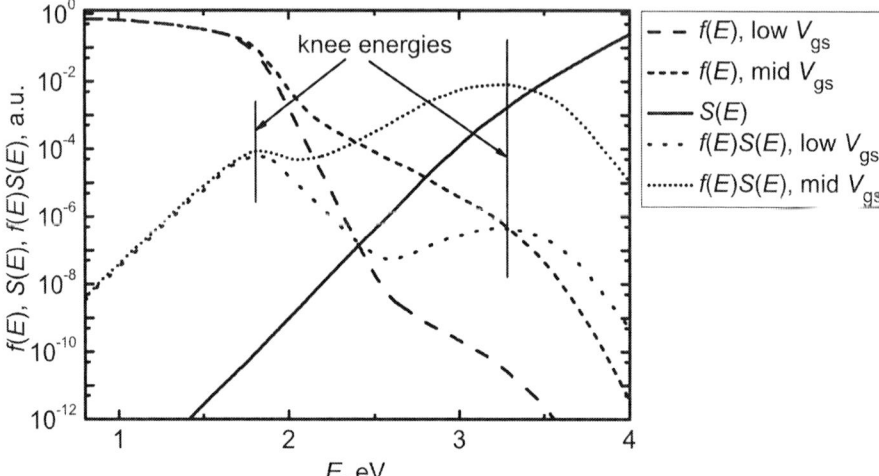

Figure 8: Schematic representation of the energy-driven paradigm. Knee energies shift depending on the applied voltage (the data borrowed from [2]).

same structure. The DF is a strongly decaying function of energy while $S(E)$ grows as a power-law. Hence, this trade-off results in a maximum of the rate pronounced at a certain energy (Fig. 8) determined according to the criterion $\mathrm{d}\ln f/\mathrm{d}E = -\mathrm{d}\ln S/\mathrm{d}E$. This energy E_{knee} is called *"knee" energy* and *is a weak function of the applied bias* V_{ds}. Therefore, if the maximum of the product $f(E)S(E)$ is sufficiently narrow, it can be approximated by a delta-function and instead of integration in the whole energy range one can only calculate the value of the integrand for this energy. To conclude, *the main message of the energy-driven paradigm is that one may avoid time-consuming calculations of the carrier DF* substituting it by the empirical parameter. This parameter is proportional to the reaction rate calculated for $E = E_{\mathrm{knee}}$ which is defined by the bias conditions. This dependence will be discussed in the next subsection devoted to the Bravaix model.

Although this paradigm substantially simplifies the treatment of HCD, it suffers from some shortcomings. Indeed, one can see in Fig. 8 that the maximum is of the integrand $f(E)S(E)$ is not necessarily narrow and in the particular case shown by the authors [2] has a width of 1.5 - 2 eV. Therefore, the concept of a dominant energy sounds doubtful. Furthermore, such a treatment of HCD does not deal with N_{it} as a distributed quantity and thus one of the main features of HCD – its strong localization – is not captured. Finally, as it was in the Hess approach, the device life-time is estimated by the interface state generation rate. However, it would be more reasonable to define it as the time when the degradation of V_{th} or I_{dlin}, etc., has reached a critical value.

Bravaix Model

The model of Bravaix *et al.* inherits the main features of both the Hess and the Rauch/LaRosa approaches: the interplay between single- and multiple-carrier mechanisms as well as the idea that the damage is defined by the carrier DF, which is implemented using Rauch/LaRosa's "fashion", that is, calculations of the DF are substituted by operation/stress condition-related empirical factors.

To describe the MP-process the authors use the formalism of Hess where the Si-H bond is treated as a truncated harmonic oscillator. Following Hess they employ a system of rate equations to describe the kinetics of the oscillator [30, 32]:

$$
\begin{aligned}
\frac{dn_0}{dt} &= P_d n_1 - P_u n_0 \\
\frac{dn_i}{dt} &= P_d(n_{i+1} - n_i) - P_u(n_i - n_{i-1}) \\
\frac{dn_{N_l}}{dt} &= P_u n_{N_l-1} - \lambda_{emi} N_{it}[H^*],
\end{aligned}
\tag{10}
$$

where $[H^*]$ is the concentration of the mobile hydrogen and n_i is the occupancy of the the ith oscillator level. In the last equation corresponding to the last bonded level (labeled as N_l, see Fig. 1) the terms representing the passivation (i.e. from the transport to the last bonded state) and transition from the N_l to $N_l - 1$ state are omitted. The hydrogen released to the transport state is characterized by the rate $\lambda_{emi} = \nu_{emi}\exp(-E_{emi}/k_B T_L)$ with E_{emi} being the height of the barrier separating bonded and transport states (see Fig. 1) and ν_{emi} the attempt frequency.

Similarly to [32], the phonon excitation/decay rates are written in a slightly modified form compared to (2):

$$
\begin{aligned}
P_u &= \int I_d \sigma dE_e + \omega_e \exp(-\hbar\omega/k_B T_L) \\
P_d &= \int I_d \sigma dE_e + \omega_e,
\end{aligned}
\tag{11}
$$

with I_d being the source-drain current. Employing the energy-driven paradigm the hot carrier acceleration factor – which is the first terms in (11) – substituted by the empirical factor S_{MP}:

$$
\begin{aligned}
P_u &= S_{MP}(I_e/e) + \omega_e \exp(-\hbar\omega/k_B T_L) \\
P_d &= S_{MP}(I_e/e) + \omega_e.
\end{aligned}
\tag{12}
$$

The solution of the system (10) for the case of weak bond-breakage rate ($\lambda_{emi} t \ll 1$) leads to a square root time dependence of N_{it} [30]:

$$
N_{it} = (N_0 \lambda_{emi}[P_u/P_d]^{N_l})t^{1/2}.
\tag{13}
$$

In addition it was assumed that the bond is predominately situated in the ground state, i.e. $n_0 \approx \sum n_i \approx N_0$. The MP-related interface state generation rate is:

$$
R_{MP} \sim N_0 \left[\frac{S_{MP}(I_d/e) + \omega_e \exp(-\hbar\omega/k_B T_L)}{S_{MP}(I_d/e) + \omega_e} \right]^{E_B/\hbar\omega} \exp(-E_{emi}/k_B T_L).
\tag{14}
$$

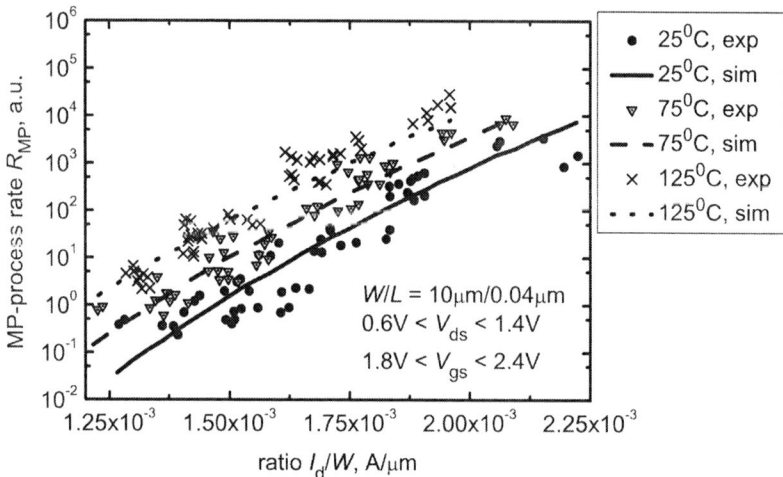

Figure 9: Experimental bond dissociation rate for the MP-process vs. the theoretical one. The information about stress conditions is shown on the canvas. The data are borrowed from [31].

An important question is the choice of quantities (such as E_B, E_{emi}, $\hbar\omega$) defining the energetics of the Si-H bond. In fact, the two main vibrational modes of the Si-H bond are the stretching and bending mode [87] with the main parameters summarized in Table 1 [30]. However, as was previously shown, the experimental data is better fitted by the bending mode and therefore the values corresponding to this mode are employed. The formalism elaborated by Hess and co-authors and refined by Bravaix *et al.* with reasonably chosen simulation parameters allows for perfect representation of the bond dissociation rate by the MP-mechanism, see [31] and the graph from there (Fig. 9).

Table 1: The parameters of the stretching and bending vibrational modes of the Si-H bond.

Parameters	Stretching	Bending
E_b, eV	2.5	1.5
$\hbar\omega$, eV	0.25	0.075
w_e, ps^{-1}	1/295	1/10

Furthermore, the SP- and MP-mechanisms for defect creation are considered within Rauch's energy-driven paradigm, that is, are related to the regimes distinguished by Rauch *et al.* [18, 29]:

The regime with low drain current and high carrier energies corresponds to the "hot-carrier" regime where the SP-mechanism plays the dominant role [31]. In this case the "lucky electron" model is valid and the device life-time is:

$$1/\tau_{\text{SP}} \sim (I_d/W)(I_s/I_d)^m, \tag{15}$$

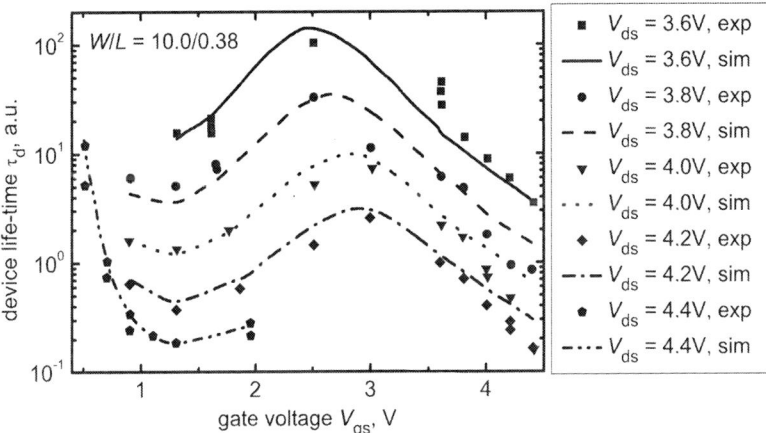

Figure 10: Comparison between the experimental device life-time and that calculated within the Bravaix framework (for devices fabricated in a 65 nm node). The data are taken from [19].

where I_s is the substrate current, W the device width and the factor m is the ratio between the powers in the impact ionization and interface state creation cross sections, i.e. $m \cong 11.0/4.0 \approx 2.7$.

Another limiting case corresponds to the high electron flux with low carrier energies. In this situation the MP-process dominates the bond dissociation and the device life-time is $1/R_{MP}$ (14). According to the knee energy concept, $S_{MP} \sim (V_{ds} - \hbar\omega)^{1/2}$, and we have:

$$1/\tau_{MP} \sim [(V_{ds} - \hbar\omega)^{1/2}(I_s/W)]^{E_B/\hbar\omega}\exp(-E_{emi}/k_BT_L) \approx [V_{ds}^{1/2}(I_d/W)]^{E_B/\hbar\omega}. \qquad (16)$$

The intermediate case with moderate drain current and moderate V_{ds} is governed by electron-electron scattering with the corresponding life-time [31]:

$$1/\tau_{EES} \sim (I_d/W)^2(I_s/I_d)^m. \qquad (17)$$

This quadratic signature is due to impact ionization which generates electron-hole pairs which are still cold in terms of bond-breakage but being further accelerated by electron-electron scattering up to energies ensuring triggering bond dissociation. Since under real device stress/operation conditions all the modes are present, one writes the device life-time considering these competing mechanisms as

$$1/\tau_d = K_{SP}/\tau_{SP} + K_{EES}/\tau_{EES} + K_{MP}/\tau_{MP}, \qquad (18)$$

that is, different contributions are weighted with corresponding probabilities (K_{SP}, K_{EES}, K_{MP}, which are fitting parameters) and summed. Fig. 10 shows a fit of the model to experimental life-times.

Hot-Carrier Degradation Model Based on the Carrier Distribution Function

We have proposed and verified a more detailed approach for hot-carrier degradation modeling which tries to more accurately capture the physical picture behind this phenomenon [46,47,97,98]. This model incorporates the crucial features of the previous approaches for hot-carrier degradation modeling. But contrary to the previous HCD models we aim at covering and linking all the levels related to this effect, starting from microscopic mechanisms of defect generation and ending at the device level. To be concrete, a physics-based model of HCD may be conditionally separated into three main sub-tasks: the carrier transport module, a module describing the defect build-up during the stress and a module responsible for the simulation of the degraded devices. This concept is sketched in Fig. 11, showing the whole chain of simulation tools employed for the model implementation. Carrier transport is treated with the full-band Monte-Carlo device simulator MONJU [99]. Simultaneously, results obtained with MONJU are verified by the device and circuit simulator developed by our Institute [100]. The drift-diffusion and hydrodynamic schemes implemented into MINIMOS-NT are suitable for carrier transport description in long-channel devices; otherwise MONJU or another Boltzmann transport equation solver is used. The carrier transport module allows us to thoroughly evaluate the carrier energy distribution function for a particular device architecture. The distribution function represents populations of "hot" and "colder" carriers and thus controls the interplay between the SP- and MP-mechanisms.

This DF is then used to calculate the carrier acceleration integral (AI) as a function of the coordinate x along the Si/SiO$_2$ interface which controls both SP- and MP-mechanisms (the structure is similar to Eq. 2):

$$
\begin{aligned}
I_{\mathrm{SP}} &= \int_{E_{\mathrm{th,SP}}}^{\infty} f(E)g(E)\sigma_{\mathrm{SP}}(E)v(E)\mathrm{d}E \\
I_{\mathrm{MP}} &= \int_{E_{\mathrm{th,MP}}}^{\infty} f(E)g(E)\sigma_{\mathrm{MP}}(E)v(E)\mathrm{d}E,
\end{aligned}
\tag{19}
$$

where $f(E)$, $g(E)$, $\sigma_{\mathrm{SP/MP}}(E)$, $v(E)$ are the carrier DF obtained for certain device topology and stress conditions, the density-of-states (DOS), Keldysh-like reaction cross section for the SP/MP-processes $\sigma_{\mathrm{SP/MP}} = \sigma_{0,\mathrm{SP/MP}}(E - E_{\mathrm{th,SP/MP}})^{p_{\mathrm{it}}}$ ($\sigma_{0,\mathrm{SP/MP}}$ is the attempt rate and $p_{\mathrm{it}} = 11$) and carrier velocity [25,30,31]. $E_{\mathrm{th}} = 1.5$ eV for both processes. Since the AI defines the interface state generation rates, this factor also defines the evolution of interface state density profiles $N_{\mathrm{it}}(x)$ with time. These profiles with the information about trap density-of-states (DOS) are used as input data for MINIMOS-NT. MINIMOS-NT performs device simulations considering the distortion of the device electrostatics and the additional scattering events induced by charged traps. Furthermore, it calculates the device characteristics (output and transfer characteristics, G_{m}, V_{th}, etc.) of the degraded transistor. The feedback to calibrate the model is given by comparison with the experimental device characteristics, Fig. 11.

For the SP-process, I_{SP} directly enters the interface state generation rate, i.e. $\lambda_{\mathrm{SP}} = \nu_{\mathrm{SP}}I_{\mathrm{SP}}$ with ν_{SP} being the attempt rate. Treating the interface state generation as a

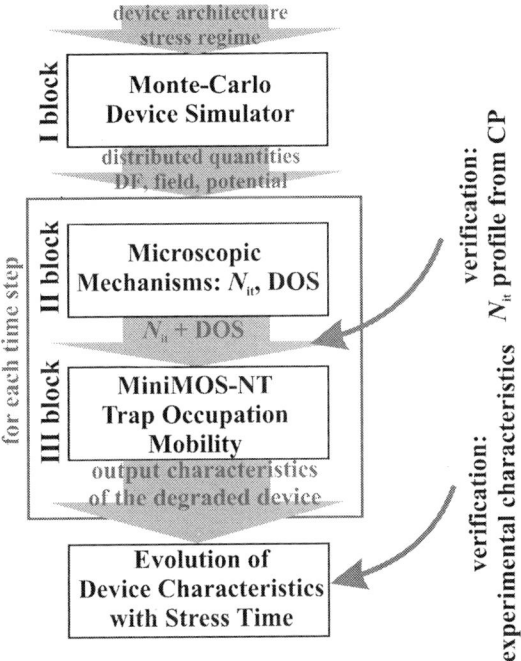

Figure 11: The flowchart of our model for hot-carrier degradation depicting three main modules: carrier transport module, module for microscopic mechanisms for defect creation and module for simulations of degraded devices.

first-order chemical reaction we write:

$$N_{SP} = N_0 \left(1 - e^{-\lambda_{SP}t}\right) \tag{20}$$

As for the MP-process, to describe the kinetics of the oscillator we employ a system of equations similar to those used by Bravaix (cf. with 10). However, in our version of the rate equation system we keep the four terms for the last bonded state N_l, that is consider the bond passivation and exchange with the $N_l - 1$ level:

$$\begin{aligned}
\frac{dn_0}{dt} &= P_d n_1 - P_u n_0 \\
\frac{dn_i}{dt} &= P_d(n_{i+1} - n_i) - P_u(n_i - n_{i-1}) \\
\frac{dn_{N_l}}{dt} &= P_u n_{N_l-1} - P_d n_{N_l} - P_{emi} n_{N_l} + \tilde{P}_{pass} N_{MP}^2,
\end{aligned} \tag{21}$$

where to satisfy the dimensionality we use $\tilde{P}_{pass} = P_{pass}/N_0$. This system of rate equations is solved taking into account the time scale hierarchy. The steady-state of the oscillator is established practically momentary as compared to the hydrogen exchange between the highest bonded and the transport state. Therefore, first the sub-task describing the steady-state of the oscillator is solved recurrently, which results in the following relation between the occupancies of the different levels: $n_i/n_0 = (P_u/P_d)^i$. Then, the passivation/depassivation rates are considered and the solution of the system (21) is written as:

$$N_{MP} = N_0 \left(\frac{\lambda_{emi}}{P_{pass}} \left(\frac{P_u}{P_d} \right)^{N_l} \left(1 - e^{\lambda_{MP}t} \right) \right)^{1/2}. \tag{22}$$

Note that for weak stresses and/or short stress times ($\lambda_{emi}t \ll 1$) this expression transforms to the root time dependence of (14) and in general has a similar structure. While considering the total concentration of the interface states one should take into account the competing nature of SP- and MP-modes and weight their contributions with certain probabilities, i.e. $N_{it} = p_{SP}N_{SP} + p_{MP}N_{MP}$.

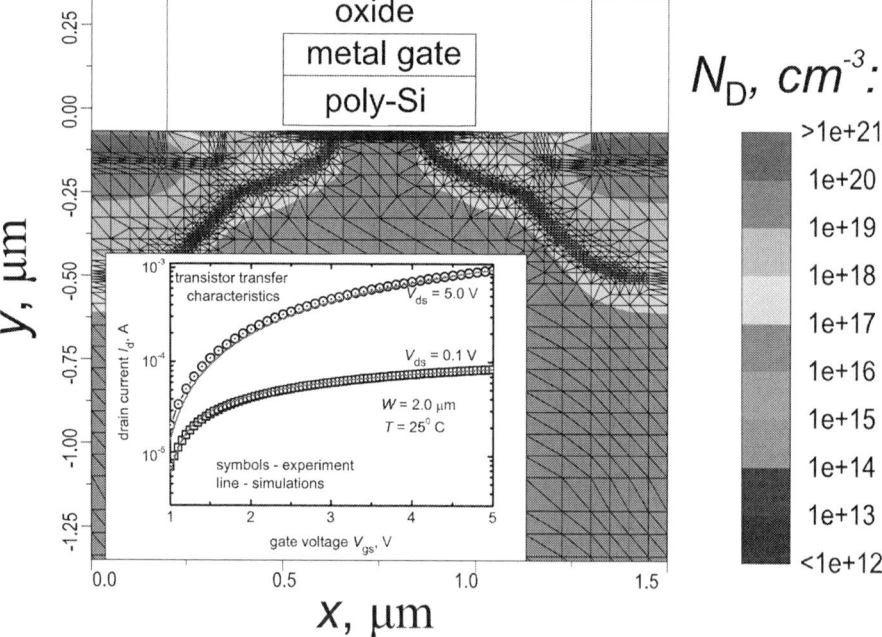

Figure 12: Schematic representation of a 5V n-MOSFETs subjected to hot-carrier stress. Inset: the transfer characteristics of a fresh device represented by our device simulator MINIMOS-NT.

Only charged interface states contribute to the device performance degradation. Therefore, while modeling the transfer characteristic evolution during the hot-carrier stress one should consider effective charges stored in the interface states, not the total concentration

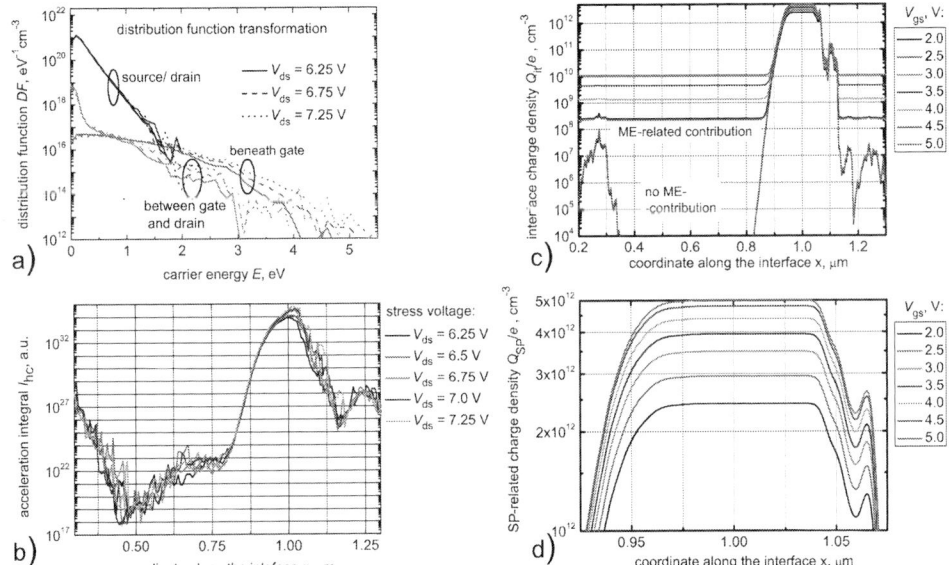

Figure 13: Evolution of crucial characteristics of the degradation with the lateral coordinate: (a) carrier distribution function along the interface; (b) the carrier acceleration integral featuring a peak near the position of most prolonged high-energy tails of the DF; (c) the total interface charge density Q_{it} and (d) stored on the SP-related traps Q_{MP} in the region where the AI peaks.

N_{it}. These effective charges (Q_{SP} and Q_{MP}, the total Q_{it} is their sum) are defined as:

$$Q_{\text{SP/MP}} = \int\limits_{-\infty}^{\infty} g_{\text{SP/MP}}(E) f_{\text{oc}}(E, E_{\text{F}_\text{n}}(x)) dE, \qquad (23)$$

where $g_{\text{SP}}(g_{\text{MP}})$ are the DOS for the SP(ME)-related traps and f_{oc} is the carrier distribution function obtained for device operation conditions. The coordinate-dependent position of the quasi-Fermi level of electrons is designated as E_{F_n}. Note that the functions g_{SP} and g_{MP} are coordinate dependent because of the normalization conditions, i.e. $\int\limits_{-\infty}^{\infty} g_{\text{SP/MP}}(E, x) dE = N_{\text{SP/MP}}(x)$. The lateral coordinate also enters the DF for operation conditions because the quasi-Fermi level for carriers captured by traps is position dependent as well. The model is thus calibrated in order to represent the degradation of the linear drain current I_{dlin} over a wide range of stress and/or operation conditions by proper determination of (Q_{SP} and Q_{MP}).

For the evaluation of the model we used a high voltage 5V n-MOSFETs fabricated on a standard 0.35 μm technology shown in Fig. 12. Fig. 13a demonstrates the evolution of the carrier distribution function along the channel. One can see that near the source and drain the DF behaves like a heated Maxwellian but has deep high energy-tails at

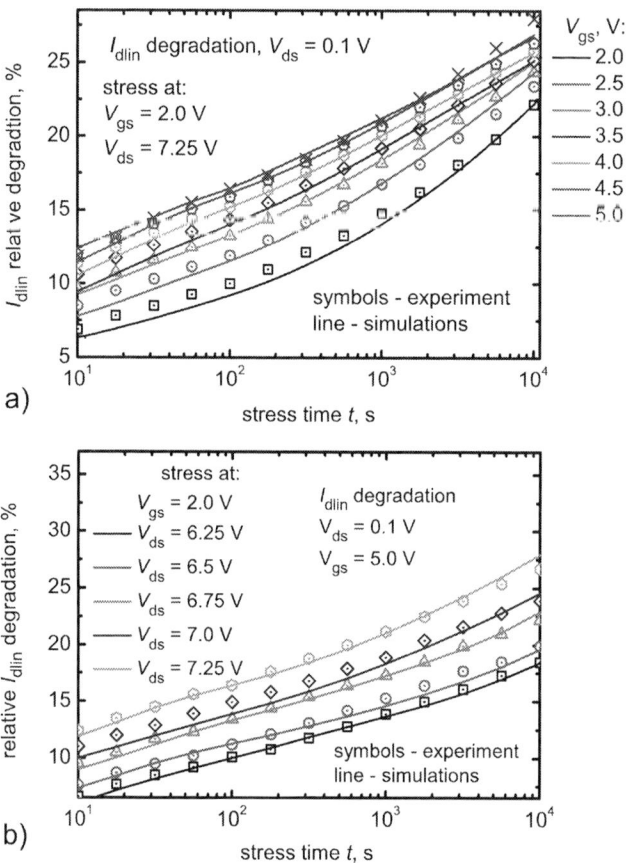

Figure 14: I_{dlin} degradation for different operation V_{gs} and fixed stress conditions $V_{gs} = 2.0$V, $V_{ds} = 7.25$V (a) and for different stress V_{ds} and fixed operation $V_{ds} = 0.1$V, $V_{gs} = 5.0$V (b).

the drain end of the gate. The carrier acceleration integral plotted vs. the coordinate x (Fig. 13b) features its peak near the area with the most extended high-energy tails of the DF. Such a behavior proves that the hot-carrier induced damage is controlled by the carrier AI which is defined by the shape of the DF. The family of the effective $Q_{\text{it}}(x)$ profiles calculated for various operation conditions at a fixed stress time t of 10s is shown in Fig. 13c,d. One can see that MP-induced defects come into play only for $V_{gs} \geq 3.0$V This circumstance means that the SP- and MP-related states are differently distributed over energy with the latter shifted to higher energies. This result agrees with the concept by Hess *et al.* where the double-power law dependence of the degradation was explained by introducing two time slopes for defects created by different processes [25, 40].

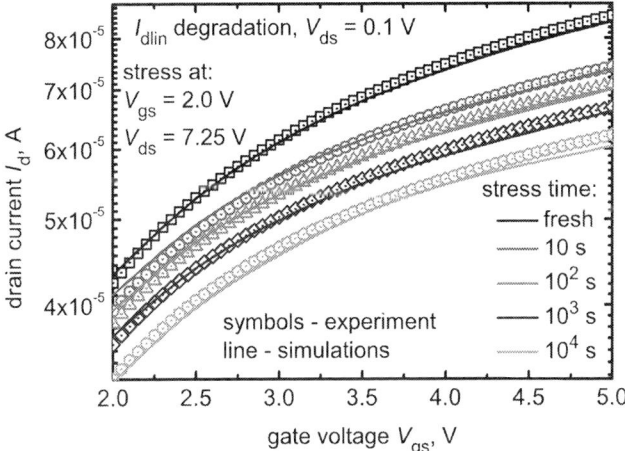

Figure 15: The transformation of the transfer characteristics during the stress: experiment vs. theory.

Fig. 13d resolves the density of particles captured by the SP-traps in the region where the total trap concentration N_{it} is plotted for different V_{gs}. Since out of the N_{it} peak the main contribution to the total density Q_{it} is provided by the MP-process one may compare the behavior of densities related to the different types of traps. For the SP-process the distance between the curves saturates, meaning that interface states of this type are almost fully occupied. In contrast, for the MP-process the increase of charge density continues, indicating that ME-traps are shifted to higher energies.

The model calibrated in the aforementioned manner allows us to represent the I_{dlin} degradation at various V_{gs} (Fig. 14a) and for different stress conditions, i.e. different V_{ds} (Fig. 14b). We do not introduce any additional fitting parameters into the model meaning that N_{it} effectively changes while switching from certain stress conditions to another. Finally, using this approach we are now able to represent the transfer characteristics of the degraded device at each time step, see Fig. 15.

In addition to the interface state generation the bulk oxide charge build-up (with concentration N_{ot}) is another important component often linked in the literature with the hot-carrier stress, which is of special significant in high-voltage devices. In order to check whether this trapped charge considerably contribute to the degradation of device characteristics, we used charge-pumping measurements. In this context, the constant-base-level charge-pumping technique has been employed. Using the approach suggested in [101–103] we were able to resolve the threshold voltage lateral profiles as a function of stress time (Fig. 16.) as well as spatial position of the N_{it} peak which is in good agreement with that obtained from our HCD model (Fig. 17); for the details see [98]. Fig. 16 and its inset representing the threshold voltage shift obtained using the maximal transconductance method show that V_{th} steadily decreases with stress time, however, N_{ot}

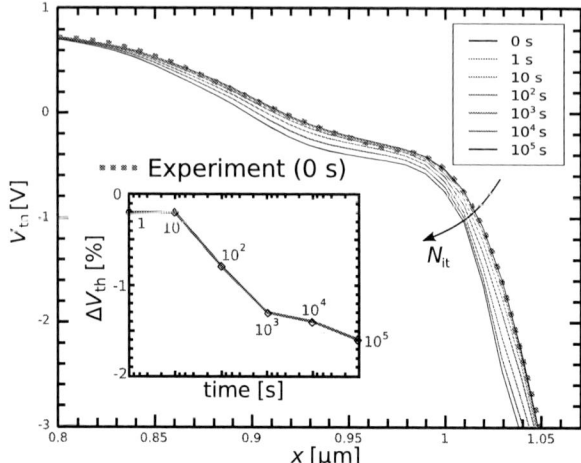

Figure 16: The threshold voltage lateral profile simulated at each time step. Inset: the experimanal change of V_{th} obtained with the maximal transconductance method.

build-up should result in a V_{th} increase or turn-around (when tw o tendencies compensate each other). Another important circumstance confirming that N_{ot} may be neglected is that the charge-pumping current does not shift laterally with the stress time (data not shown).

Some HCD models link the interface state build-up to either the maximum of the electric field, or carrier temperature, or the average electron energy, etc. Fig. 17 provides a short summary showing the spatial positions of the maxima of various quantities which have been used as the driving force of HCD. This information is accompanied by charge-pumping measurements results revealing that the peak of experimental N_{it} coincides with the maximum of the AI (which defines the peak of the simulated N_{it} profile). This tendency confirms once again that the AI is the crucial quantity controlling hot-carrier degradation and just this parameter should primarily been used rather than the electric field or the carrier temperature.

The main advantage of our model is that one does not have to recalibrate it while switching from one device architecture and/or stress conditions to another one(s). In other words, the set of parameters describing the Si-H dissociation kinetics is fitted once and does not depend on process conditions. As for the transport module, the Boltzmann transport equation is to be solved each time we change the device topology and stress conditions to obtain the new set of the DFs corresponding to the current situation. As a result, the model is suitable to predict the device life-time not only for accelerated stress conditions but also under normal operation conditions and thus is useful for development and reliability engineers. Another advantage is that the model provides information about the N_{it} profiles, and in in particular captures the strong localization of HCD. Two additional important peculiarities of hot-carrier degradation captured by the model are the

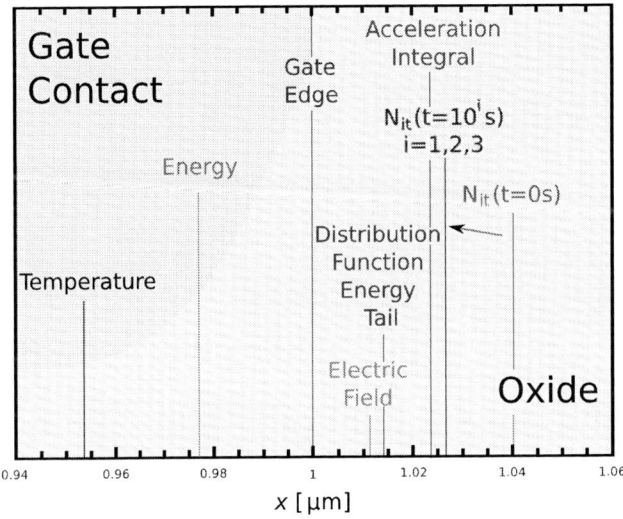

Figure 17: The position of maxima of main crucial quantities: the electric field, average carrier energy, acceleration integral, position of the most extended tails of the, etc.

saturation of the damage as well as the representation of the worst-case conditions [97].

Conclusion

We have carefully analyzed the main approaches to hot-carrier degradation modeling and established a comprehensive framework of a physics-based HCD model. As was demonstrated within the model by Hess, the degradation is controlled by the interplay between single- and multiple-carrier mechanisms of Si-H bond dissociation. This interplay is controlled by the way the carriers are distributed over energy, that is by the carrier energy distribution function. These considerations suggest that carrier transport and microscopic mechanisms of defect creation are two essential sub-tasks of the general problem. While the energy-driven paradigm elaborated by Rauch and LaRosa is focused on the substitution of the DF by some simple approximations, the Bravaix model combines this paradigm with defect generation concepts. After discussion of the advantages and limitations of the models we have introduced our approach based on a thorough evaluation of the carrier DF by solving the Boltzmann transport equation. This concept arranges the whole hierarchical ladder by integrating carrier transport aspects, the microscopic picture of defect creation, and the simulation of degraded devices within the same framework. We have proven that for proper HCD description one should deal with the carrier acceleration integral, not with other factors such as the electric field or average carrier energies.

Acknowledgments

This work has received funding from the EC's FP7 grant agreement n°216436 (ATHENIS) and from the ENIAC MODERN project n°820379.

References

1. A. Acovic, G. L. Rosa, and Y. Sun, *Microel. Reliab.* **36**, 845–869 (1996).
2. S. Rauch, and G. L. Rosa, "CMOS Hot Carrier: From Physics to End Of Life Projections, and Qualification," in *Proc. International Reliability Physics Symposium (IRPS), tutorial*, 2010
3. A. Bravaix, and V. Huard, "Hot-Carrier Degradation Issues in advanced CMOS nodes," in *Proc. European Symposium on Reliability of Electron Devices Failure Physics and Analysis (ESREF), tutorial*, 2010.
4. Y. Liu, *Study of oxide breakdown, hot carrier and NBTI effect on MOS device and circuit reliability*, Ph.D. thesis, University of Central Florida, Orlando, Florida (2005).
5. G. Groeseneken, R. Bellens, and G. V. den Bosch, *Semicond. Sci. Technol.* **10**, 1208–1220 (1995).
6. C. Hu, "Lucky electron model for channel hot electron emission," in *Proc. International Electron Devices Meeting (IEDM)*, 1979, pp. 22–25.
7. C. Hu, S. Tam, F. Hsu, P.-K. Ko, T.-Y. Chan, and K. Terrill, *IEEE Trans. Electron Dev.* **48**, 375–385 (1985).
8. S. Rauch, F. Guarin, and G. LaRosa, *IEEE Electron Dev. Lett.* **19**, 463–465 (1998).
9. E. Takeda, *IEEE Proc.* **131**, 153–162 (1984).
10. E. Takeda, and N. Suzuki, *IEEE Electron Dev. Lett.* **4**, 111–113 (1983).
11. J.-S. Goo, Y.-G. Kim, H. Lee, H.-Y. Kwon, and H. Shin, *Solid-State Electron.* **38**, 1191–1196 (1995).
12. R. Dreesen, K. Croes, J. Manca, W. D. Ceunick, L. D. Schepper, A. Pergoot, and G. Groeseneken, *Microel. Reliab.* **39**, 785–790 (1999).
13. R. Dreesen, K. Croes, J. Manca, W. D. Ceunick, L. D. Schepper, A. Pergoot, and G. Groeseneken, *Microel. Reliab.* **41**, 437–443 (2001).
14. R. Woltjer, and G. Paulzen, "Universal description of hot-carrier-induced interface states in NMOSFETs," in *Proc. International Electron Devices Meeting (IEDM)*, 1992, pp. 535–538.
15. R. Woltjer, G. Paulzen, H. Pomp, H. Lifka, and P. Woerlee, *IEEE Trans. Electron Dev.* **42**, 109–115 (1995).
16. K. Mistry, and B. Doyle, *IEEE Electron Dev. Lett.* **12**, 492–494 (1991).
17. K. Mistry, and B. Doyle, *IEEE Trans. Electron Dev.* **40**, 96–104 (1993).
18. S. Rauch, G. LaRosa, and F. Guarin, *IEEE Trans Dev. Material. Reliab.* **1**, 113–119 (2001).
19. C. Guerin, V. Huard, and A. Bravaix, *IEEE Trans. Dev. Material. Reliab.* **7**, 225–235 (2007).
20. P. Moens, and G. van den Bosch, *IEEE Trans Electron Dev.* **6**, 349–357 (2006).
21. P. Moens, G. van den Bosch, and G. Groeseneken, "Competing hot carrier degradation mechanisms in lateral n-type DMOS transistors," in *Proc. International Reliability Physics Symposium (IRPS)*, 2003, pp. 214–221.
22. P. Moens, and M. Tack, "Hole trapping and de-trapping effects in LDMOS devices under dynamic stress," in *Proc. International Electron Devices Meeting (IEDM)*, 2006.

23. P. Moens, F. Bauwens, M. Nelson, and M. Tack, "Electron trapping and interface trap generation in drain extended pMOS transistors," in *Proc. International Reliability Physics Symposium (IRPS)*, 2005, pp. 93–96.
24. W. McMahon, A. Haggaag, and K. Hess, *IEEE Trans. Nanotech.* **2**, 33–38 (2003).
25. K. Hess, A. Haggag, W. McMahon, K. Cheng, J. Lee, and J. Lyding, *Circuits and Devices Mag.* pp. 33–38 (2001).
26. O. Penzin, A. Haggag, W. McMahon, E. Lyumkis, and K. Hess, *IEEE Trans. Electron Dev.* **50**, 1445–1450 (2003).
27. H. Kufluoglu, and M. Alam, *Journ. Comput. Electron.* **3**, 165–169 (2004).
28. H. Kufluoglu, *MOSFET degradation due to negative bias temperature instability (NBTI) and hot carrier degradation (HCI) and its applications for reliability-aware VLSI design*, Ph.D. thesis, Purdue University, West Lafayette, Indiana, USA (2007).
29. S. Rauch, and G. L. Rosa, "The Energy Driven Paradigm of NMOSFET Hot Carrier Effects," in *Proc. International Reliability Physics Symposium (IRPS)*, 2005.
30. A. Bravaix, C. Guerin, V. Huard, D. Roy, J. Roux, and E. Vincent, *Proc. IRPS* pp. 531–546 (2009).
31. C. Guerin, V. Huard, and A. Bravaix, *Journ. Appl. Phys.* **105** (2009).
32. W. McMahon, K. Matsuda, J. Lee, K. Hess, and J. Lyding, "The Effects of a multiple carrier model of interface states generation of lifetime extraction for MOSFETs," in *Proc. Int. Conf. Mod. Sim. Micro*, 2002, vol. 1, pp. 576–579.
33. K. Hess, I. C. Kizilyalli, and J. W. Lyding, *IEEE Trans Electron Dev.* **45**, 406–416 (1998).
34. T. Grasser, W. Gös, and B. Kaczer, *ECS Transactions* **19**, 265–287 (2009).
35. W. Chang, B. Davari, M. Wordeman, Y. Taur, C. Hsu, and M. Rodriguez, *IEEE Trans. Electron Dev.* **39**, 959 (1992).
36. D. Bursky, *Electronic Design* **41**, 111–116 (1993).
37. D. Frank, R. Dennard, E. Nowak, P. Solomon, M. Stettler, S. Tyagi, and M. Bohr, "Scaling challenges and device design requirements for high performance sub-50 nm gate length planar CMOS transistors," in *Proc. VLSI Symposium Tech. Digest*, 2000, pp. 174–175.
38. L. Hong, *Characterization of hot carrier reliability in deep submicrometer MOSFETsL. Hong*, Ph.D. thesis, National University of Singapore (2005).
39. F.-C. Hsu, and K.-Y. Chu, *IEEE Electron Dev. Lett.* **5**, 162–165 (1984).
40. A. Haggag, W. McMahon, K. Hess, K. Cheng, J. Lee, and J. Lyding, "High-performance chip reliability from short-time-tests. Statistical models for optical interconnect and HCI/TDDB/NBTI deep-submicron transistor failures," in *Proc. International Reliability Physics Symposium (IRPS)*, 2001, pp. 271–279.
41. T. Mizuno, A. Toriumi, M. Iwase, M. Takanashi, H. Niiyama, M. Fukmoto, and M. Yoshimi, "Hot-carrier effects in 0.1 μm gate length CMOS devices," in *Proc. International Electron Devices Meeting (IEDM)*, 1992, pp. 695–698.
42. J. Bude, "Gate-Current by Impact Ionization Feedback in submicron MOSFET Technologies," in *Proc. VLSI Symposium Tech. Digest*, 1995, pp. 101–102.
43. F. Venturi, E. Sangiorgi, and B. Ricco, *IEEE Trans. Electron Dev.* **38**, 1895–1904 (1991).
44. J. Chung, M. Jeng, J. Moon, P. Ko, and C. Hu, *IEEE Trans. Electron Dev.* **37**, 1651–1657 (1990).
45. M. Fischetti, and S. Laux, "Monte-Carlo study of sub-band-gap impact ionization in

small silicon field-effect transistors," in *Proc. International Electron Devices Meeting (IEDM)*, 1995, pp. 305–308.

46. S. Tyaginov, I. Starkov, O. Triebl, J. Cervenka, C. Jungemann, S. Carniello, J. Park, H. Enichlmair, M. Karner, C. Kernstock, E. Seebacher, R. Minixhofer, H. Ceric, and T. Grasser, "Hot-Carrier Degradation Modeling Using Full-Band Monte-Carlo Simulations," in *Proc. International Symposium on the Physical & Failure Analysis of Integrated Circuits (IPFA)*, 2010.

47. S. Tyaginov, I. Starkov, O. Triebl, J. Cervenka, C. Jungemann, S. Carniello, J. Park, H. Enichlmair, C. Kernstock, E. Seebacher, R. Minixhofer, H. Ceric, and T. Grasser, *Microel. Reliab.* **50**, 1267–1272 (2010).

48. D. Brisbin, P. Lindorfer, and P. Chaparala, "Substrate current independent hot carrier degradation in NLDMOS devices," in *Proc. International Reliability Physics Symposium (IRPS)*, 2006, pp. 329–333.

49. M. Annese, S. Carniello, and S.Manzini, *IEEE Trans. Electron Dev.* **52**, 1634–1639 (2005).

50. P. Santos, H. Quaresma, A. Silva, and M. Lanca, *Microel. Jour.* **35**, 723–730 (2004).

51. W. Qin, W. Chim, D. H. Chan, and C. Lou, *Semicond. Sci. Technol.* **13**, 453–459 (1998).

52. S. Manzini, and A. Gallerano, *Solid-State Electron.* **44**, 1325–1330 (2000).

53. V. Reddy, "An introduction to CMOS semiconductor reliability," in *Proc. International Reliability Physics Symposium (IRPS), tutorial*, 2003.

54. Z. Chen, P. Ong, A. Mylin, V. Singh, and S. Cheltur, *Appl. Phys. Lett.* **81**, 3278–3280 (2002).

55. E. Li, E. Rosenbaum, J. Tao, G.-F. Yeap, M. Lin, and P. Fang, "Hot-carrier effects in nMOSFETs in 0.1 μm CMOS technology," in *Proc. International Reliability Physics Symposium (IRPS)*, 1999, pp. 253–258.

56. C. Lin, S. Biesemans, L. Han, K. Houlihan, T. Schiml, K. Schruefer, C. Wann, and R. Markhopf, "Hot carrier reliability for 0.13 μm CMOS technology with dual gate oxide thickness," in *Proc. International Electron Devices Meeting (IEDM)*, 2000, pp. 135–138.

57. R. Woltjer, A. Hamada, and E. Takeda, *Semicond Sci. Technol.* **7**, pp. B581–B584 (1992).

58. A. Bravaix, D. Goguenheim, N. Revil, and E. Vincent, *Microel. Reliab.* **44**, 65–77 (2004).

59. M. Ancona, N. Saks, and D. McCarthy, *IEEE Trans Electron Dev.* **35**, 221–2228 (1988).

60. M. Pagey, *Characterization and modeling of hot-carrier degradation in sub-micron NMOSFETs*, Master's thesis, Vanderbilt University (2002).

61. Q. Wang, L. Sun, and A. Yap, *Microel. Reliab.* pp. 508–513 (2008).

62. Q. Wang, L. Sun, Z. Zhang, A. Yap, H. Li, and S. Liu, *Journ. Non-Crystalline Solids* **354**, 1871–1875 (2008).

63. K.-M. Wu, J. Chen, Y. Su, J. Lee, K. Lin, J. Shih, and S. Hsu, *Appl. Phys. Lett.* **89** (2006).

64. T. Grasser, H. Kosina, and S. Selberherr, *Journ. Appl. Phys.* **90**, 6165–6171 (2001).

65. A. Gehring, T. Grasser, H. Kosina, and S. Selberherr, *Journal of Applied Physics* **92**, 6019–6027 (2002).

66. T. Grasser, H. Kosina, and S. Selberherr, *International Journal of High Speed Elec-*

tronics and Systems **13**, 873–901 (2003).

67. A. Zaka, Q. Rafhay, M. Iellina, P. Palestri, R. Clerc, D. Rideau, D. Garetto, J. Singer, G. Pananakakis, C. Tavernier, and H. Jaouen, *Solid-State Electron.* **in press** (2010).

68. B. Persson, and P. Avouris, *Surface Science* **390**, 45–54 (1997).

69. K. Hess, L. Register, B. Tuttle, J. Lyding, and I. Kizilyalli, *Physica E* **3**, 1–7 (1998).

70. T. Aichinger, M. Nelhiebel, and T. Grasser, *Microel. Reliab.* pp. 1178–1184 (2008).

71. T. Grasser, B. Kaczer, W. Goes, H. Reisinger, T. Aichinger, P. Hehenberger, P.-J. Wagner, F. Schanowsky, J. Franco, P. Roussel, and M. Nelhiebel, "Recent Advances in Understanding the Bias Temperature Instability," in *Proc. International Electron Devices Meeting (IEDM)*, 2010, pp. 82–85.

72. F.-C. Hsu, and K.-Y. Chu, *IEEE Electron Dev. Lett.* **5**, 148–150 (1984).

73. P. Heremans, G. V. den Bosch, R. Bellens, G. Groseneken, and H. Maes, *IEEE Trans. Electron Dev.* **37**, 980–992 (1990).

74. M. Song, K. MacWilliams, and C. Woo, *IEEE Trans Electron Dev.* **44**, 268–276 (1997).

75. A. Bravaix, D. Goguenheim, N. Revil, E. Vincent, M. Varrot, and P. Mortini, *Microel. Reliab.* **39**, 35–44 (1999).

76. P. Moens, J. Mertens, F. bauwens, P. Joris, W. D. Ceuninck, and M. Tack, "A comprehensive model for hot carrier degradation in LDMOS transistors," in *Proc. International Reliability Physics Symposium (IRPS)*, 2007, pp. 492–497.

77. H. Enichlmair, S. Carniello, J. Park, and R. Minixhofer, *Microel. Reliab.* **47**, 1439–1443 (2007).

78. K. Lee, C. Kang, O. yoo, R. Choi, B. Lee, J. Lee, H.-D. Lee, and Y.-H. Jeong, *IEEE Electron Dev. Lett.* **29**, 389–391 (2008).

79. M. Jo, S. Kim, C. Cho, M. Chang, and H. Hwang, *Appl. Phys. Lett.* **94**, 053505–1–053505–3 (2009).

80. E. Amat, T. Kauerauf, R. Degraeve, R. Rodriguez, M. Nafria, X. Aymerich, and G. Groeseneken, *Microel. Engineering* **87**, 47–50 (2010).

81. R. Walkup, D. Newns, and P. Avouris, *Phys. Rev. B* **48**, 1858–1861 (1993).

82. J. Lyding, K. Hess, G. Abeln, D. Thompson, J. Moore, M. Hersam, E. Foley, J. Lee, S. Hwang, H. Choi, P. Avouris, and I. Kizialli, *Appl. Surf. Sci.* **13-132**, 221–230 (1998).

83. K. Stokbro, C. Thirstrup, M. Sakurai, U. Quaade, B. Y.-K. Hu, F. Perez-Murano, and F. Grey, *Phys. Rev. Lett.* **80**, 2618–2621 (1998).

84. M. Budde, G. Lüpke, E. Chen, X. Zhang, N. H. Tolk, L. C. Feldman, E. Tarhan, A. K. Ramdas, and M. Stavola, *Phys. Rev. Lett.* **87**, 1455–1461 (2001).

85. J. Sune, and Y. Wu, *Phys. Rev. Lett.* **92**, 087601 (1–4) (2004).

86. J. Sune, and Y. Wu, "Mechanisms of hydrogen release in the breakdown of SiO_2-based oxides," in *Proc. International Electron Devices Meeting (IEDM)*, 2005, pp. 388–391.

87. R. Biswas, Y.-P. Li, and B. C. Pan, *Appl. Phys. Lett.* **72**, 3500–3503 (1998).

88. G. Ribes, S. Bruyere, M. Denais, F. Monsieur, V. Huard, D. Roy, and G. Ghibaudo, *Microel. Reliab.* **45**, 1842–1854 (2005).

89. B. Tuttle, and C. V. de Walle, *Phys. Rev. B* **59**, 12884–12889 (1999).

90. K. Hess, A. Haggag, W. McMahon, B. Fischer, K. Cheng, J. Lee, and L. Lyding, "Simulation of Si-SiO2 Defect Generation in CMOS Chips: From Atomistic Structure to Chip Failure Rates," in *Proc. International Electron Devices Meeting*

(IEDM), 2000, pp. 93–96.

91. *DESSIS manual.*

92. H. Kufluoglu, and M. Alam, "A geometrical unification of the theories of NBTI and HCI time exponents and its implications for ultra-scaled planar and surround-gate MOSFETs," in *Proc. International Electron Devices Meeting (IEDM)*, 2004, pp. 113–116.

93. T.Grasser, W. Gös, and B. Kaczer, *IEEE Trans Dev. Material. Reliab.* 8, 79–97 (2008).

94. T. Grasser, H. Reisinger, W. Goes, T. Aichinger, P. Hehenberger, P.-J. Wagner, M. Nelhiebel, J. Franco, and B. Kaczer, "Switching Oxide Traps as the Missing Link Between Negative Bias Temperature Instability and Random Telegraph Noise," in *Proc. International Electron Devices Meeting (IEDM)*, 2009.

95. T. Grasser, H. Reisinger, P.-J. Wagner, D. Kaczer, F. Schanowsky, and W. Gös, "The time dependent defect spectroscopy (TDDS) for the characterization of the bias temperature instability," in *Proc. International Reliability Physics Symposium (IRPS)*, 2010, pp. 16–25.

96. S. Rauch, and G. L. R. and, *IEEE Trans Dev. Material. Reliab.* 5, 701–705 (2005).

97. I. Starkov, S. Tyaginov, O. Triebl, J. Cervenka, C. Jungemann, J. Park, H. Enichlmair, M. Karner, C. Kernstock, E. Seebacher, R. Minixhofer, H. Ceric, and T. Grasser, "Analysis of Worst-Case Hot-Carrier Conditions for High Voltage Transistors Based on Monte-Carlo Simulations of Distribution Function," in *Proc. International Symposium on the Physical & Failure Analysis of Integrated Circuits (IPFA)*, 2010.

98. I. Starkov, S. Tyaginov, H. Enichlmair, J. Cervenka, C. Jungemann, S. Carniello, J. Park, H. Ceric, and T. Grasser, *Journal of Vacuum Science and Technology - B* in press (2010).

99. C. Jungemann, and B. Meinerzhagen, *Hierarchical Device Simulation*, Springer Verlag Wien/New York, 2003.

100. Institute for Microelectronic, TU Wien, *MiniMOS-NT Device and Circuit Simulator.*

101. C. Chen, and T. Ma, *IEEE Trans Electron Dev.* 45, 512–520 (1998).

102. S. Chung, and J.-J. Yang, *IEEE Trans Electron Dev.* 46, 1371–1377 (1999).

103. Y.-L. Chu, D.-W. Lin, and C.-Y. Wu, *IEEE Trans. Electron Dev.* 47, 348–353 (2000).

Intrinsic Variability and Reliability in Nano-CMOS

Jyothi Velamala, Chi-Chao Wang, Rui Zheng, Yun Ye, Yu Cao

School of ECEE, Arizona State University, Tempe, Arizona 85287, USA

Random variations have been regarded as one of the major barriers of CMOS technology scaling. Besides profound physical effects that result from the vastly increased parameter variations due to manufacturing, performance is also affected with temporal conditions due to reliability degradation. Compact models that physically capture these effects are crucial to bridge variability and reliability effects with design solutions. By understanding the underlying physics and analyzing the results from atomistic simulations, intrinsic variations from random dopant fluctuation (RDF), line-edge roughness (LER), and oxide thickness fluctuation (OTF) are presented in this paper. Temporal parameter shift from aging mechanisms like negative bias temperature instability (NBTI) effect along with their models are also discussed. The statistical interaction of aging effects with static variability is further discussed. Finally, circuit performance variability impacted by random threshold voltage variation is benchmarked.

Introduction

CMOS technology is expected to enter the 10nm regime for further integrated circuits (IC) [1]. Such aggressive scaling leads to vastly increased variability, posing a grand challenge to robust IC design. Depending on their sources, variations are often categorized into two types: intrinsic fluctuations and process-induced change [2-5]. Process-induced variations are caused by the imperfection in silicon fabrication, varying from foundries to foundries. On the other side, intrinsic variability and reliability, induced by atom-level charge and geometry fluctuations, are inherent to the device structure. They are limited by fundamental physics, posing one of the ultimate barriers to continual technology scaling. Their importance is rapidly increasing as device feature size approaches the atom dimension.

The primary intrinsic variations include random dopant fluctuation (RDF), line-edge roughness (LER) and oxide thickness fluctuation (OTF), as illustrated in Fig. 1.

- RDF: This well known effect is caused by the uncertainty in charge location and numbers, such as the discrete placement of dopant atoms in the channel region that follow a Poisson distribution [2]. As the device size scales down, the total number of channel dopants decreases, resulting in a larger variation of dopant numbers, and significantly impacting threshold voltage (V_{th}).

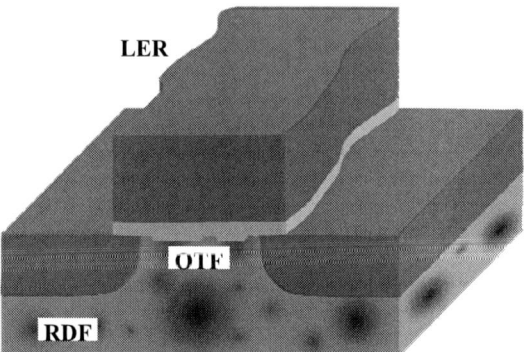

Figure 1. Fundamental variations in a CMOS

- LER: Related to gate material, LER is the distortion of the gate edge, which is induced by gate etching and the lithography process [3]. Although the etching technology has been improved, the trend of LER induced V_{th} variation does not scale accordingly [4]: due to increasingly severe short-channel effect, such as DIBL, LER contributes to a significant amount of V_{th} variation.

- OTF: It is induced by the atom-level interface roughness between silicon and gate dielectric [5]. Such a surface roughness causes the fluctuation of the voltage drop across the oxide layer, further changing V_{th}. OTF becomes more pronounced as gate dielectric thickness (t_{ox}) is approaching the height of the atoms.

Transistor performance not only depends on static process variations as mentioned above, but also changes over the period of dynamic operation because of the effect of reliability degradation [6-7]. As CMOS technology is scaling to the 10nm regime, equivalent oxide thickness will be as thin as 5Å [8]. Such an aggressive pace inevitably leads to multiple reliability concerns, including negative-bias-temperature-instability (NBTI), channel-hot-carrier (CHC), and time-dependent-dielectric-breakdown (TDDB). In particular, there has been a recent increase in interest on the reliability impact of PMOS NBTI. NBTI occurs under negative gate voltage (e.g., V_{gs}= -V_{DD}) and is measured as an increase in the magnitude of threshold voltage [9]. It mostly affects the PMOS transistor and degrades the device driving current, circuit speed, noise margin, the matching property, as well as the device life time. Indeed, as gate oxide gets thinner than 4nm, the threshold voltage change caused by NBTI for the PMOS transistor has become the dominant factor to limit the life time, which is much shorter than that defined by hot-carrier induced degradation of the NMOS transistor [8].

Furthermore, different from CHC that occurs only during dynamic switching, NBTI is caused during static stress on the oxide even without current flow. Consequently, the situation of the NBTI degradation is exacerbated in the nano-scale design as advanced digital systems tend to have longer standby time for lower power consumption. Figure 2 shows the threshold voltage (V_{th}) degradation at different temperature and stress voltages in 65nm PMOS device. As the NBTI effect becomes more severe with continuous scaling,

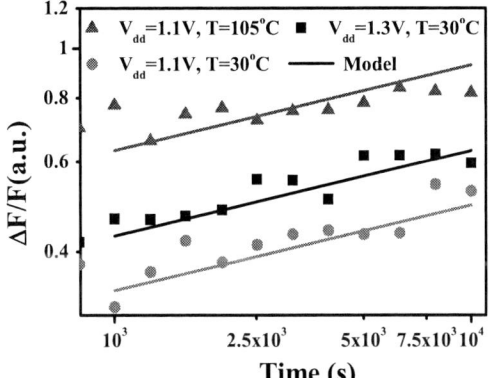

Time (s)

Figure 2. V_{th} degradation under static NBTI for different T and V_{gs} for a 65nm technology device.

it is critical to understand, simulate, and minimize the impact of NBTI in the early design stage to ensure the reliable operation of circuits for a desired period of time.

In this paper, predictive models for intrinsic variations and reliability effects are presented. These models are derived from first principles and are calibrated with atomistic simulations and available data. Leveraging long-range potential based equivalent charge density model [10], RDF effect is able to be simulated in a commercial TCAD tool [11]. Moreover, the geometric roughness due to LER and OTF is generated by Inverse Fourier Transform (IFT) from the power spectrum [5, 12], which is further integrated into the TCAD simulation. NBTI and CHC models are based on the classical reaction-diffusion (RD) model. Further, the statistical interactions of aging effects on intrinsic variations are also investigated. These predictive models indicate the scaling trend of random V_{th} variations and aging, and help benchmark the impact on circuit performance.

Predictive Modeling

Managing variability and reliability in the design process requires accurate and efficient models, especially the dependence on material, device and design parameters. The prediction of nominal device characteristics is based on the Predictive Technology Model (PTM). On top of the nominal model, the effects of intrinsic fluctuations are incorporated by identifying key device parameters affected and modeling their statistics. Moreover, temporal shifts, such as aging effect due to BTI and CHC, are modeled as functions of process parameters and design conditions. These effects systematically shift device parameters, especially the threshold voltage (V_{th}) and mobility. Due to the unique recovery property of BTI, the amount of temporal shift depends on dynamic circuit operation patterns in reality [13].

Intrinsic Parametric Variability

Based on the customized 3-D atomistic simulation result, a suite of scalable models is derived in this section. From first principles, the variance of V_{th} is modeled as functions of key device parameters, such as N_{ch} and t_{ox}.

RDF: In our 22nm simulation, σV_{th} due to body RDF is 35.2 mV, which is indeed the dominant one among all variations. V_{th} variation due to RDF is expressed as [14]:

$$\sigma V_{th} = \frac{q}{C_{INV}} \sqrt{\frac{N_{ch}W_{dep}}{3WL}} \times 1.2$$

(1)

where W, L, N_{ch}, W_{dep} are the channel width, channel length, effective channel doping (N_{ch}) and depletion width respectively. In this model, the non-uniformity along lateral directions and the fluctuation of W_{dep} are ignored and thus, a factor of 1.2 is used to correct the result. By expanding the W_{dep} term and ignoring other second order terms, a more explicit expression is obtained:

$$\sigma V_{th(RDF)} = C_1 \frac{q}{\sqrt{3WL}} \frac{t_{oxe}}{\varepsilon_{ox}} \left(\frac{2\varepsilon_{Si}N_{ch}}{q} \right)^{\frac{1}{4}}$$

(2)

where C_1 is a fitting parameter accounting for surface potential and the correction term; t_{ox}, ε_{Si}, ε_{ox}, and q are the equivalent oxide thickness, permittivity of silicon, permittivity of the oxide layer, and elementary charge, respectively. Eq. (2) suggests that RDF induced V_{th} variation is proportional to t_{ox} and $N_{ch}^{0.25}$. Note that in this work other potential RDF related variation sources, such as RDF induced mobility variation [15], have not been included. Upon the availability of atomistic simulation tools or experimental data, our methodology is able to cover those additional factors.

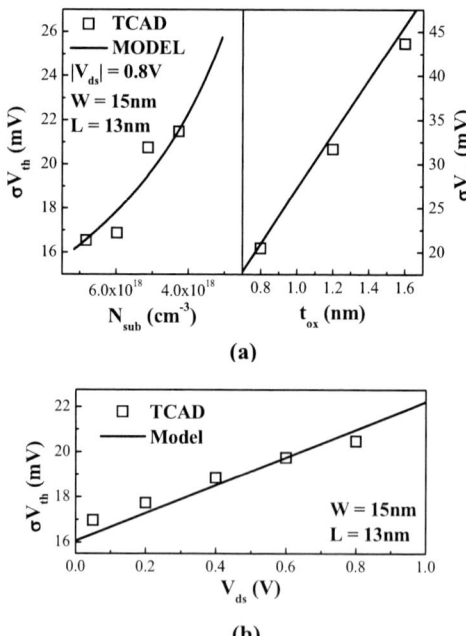

Figure 3. (a) The dependence of LER induced σV_{th} on N_{ch} and t_{ox}; (b) V_{ds}

LER: To the first order, the shift in V_{th} due to short-channel effect can be expressed as Eqs. (3) and (4) [16]:

$$\Delta V_{th} = -\frac{1}{2}\frac{(2V_{bi} - \phi_s) + V_{ds}}{\cosh(L/l') - 1} \tag{3}$$

$$l' = \sqrt{\frac{\varepsilon_{Si} t_{oxe}}{\varepsilon_{ox} \eta}} \cdot \left(\frac{2\varepsilon_{Si}\phi_s}{qN_{ch}}\right)^{\frac{1}{4}} \tag{4}$$

where V_{bi} is the built-in voltage of the source/drain junction, and η is a parameter to model the average depletion width along channel. Assuming the fluctuations of two gate edges are uncorrelated, random variation of channel length due to LER is calculated by using the following Eq. (5):

$$\sigma L = \sqrt{\frac{2}{1 + W/W_c}} \cdot \sigma LER \tag{5}$$

where σLER and W_c is the standard deviation and auto correlation length of the gate edge, respectively. By differentiating Eq. (3), and substituting Eq. (5), the following expression is obtained:

$$\sigma V_{th(LER)} = \frac{(C_2 + V_{ds})\sinh(L/l')}{2l'(\cosh(L/l') - 1)^2}\sqrt{\frac{2}{1 + W/W_c}} \cdot \sigma LER \tag{6}$$

$$l' = C_3\sqrt{\frac{\varepsilon_{Si} t_{oxe}}{\varepsilon_{ox}}} \cdot \left(\frac{2\varepsilon_{Si}}{qN_{ch}}\right)^{\frac{1}{4}} \tag{7}$$

where C_2 is a fitting parameter that is associated with junction built-in voltage induced short-channel effect [16]. C_3 is a fitting parameter associated with surface potential. Figure 3 shows the comparison of model scalability with TCAD simulations.

OTF: Similar to LER, OTF leads to the geometric fluctuation of averaged oxide thickness, and further affects the voltage drop across the oxide layer. For a bulk device, V_{th} is expressed as the following [17]:

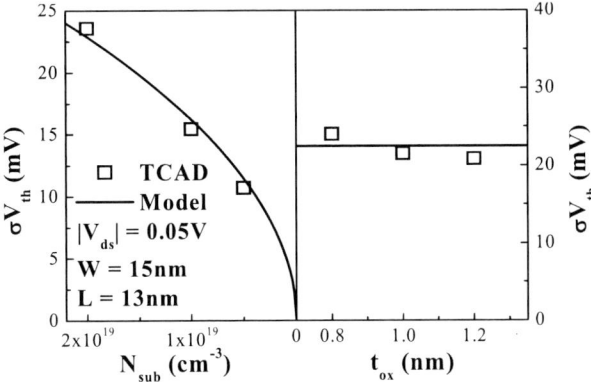

Figure 4. N_{ch} and t_{ox} dependence of OTF induced σV_{th}.

$$V_{th} = V_{FB} + \phi_s + \frac{t_{oxe}}{\varepsilon_{ox}} \sqrt{2qN_{ch}\varepsilon_{Si}\phi_s}$$

(8)

The oxide thickness changes with surface roughness in the interfaces of gate-SiO$_2$ and SiO$_2$-substrate. The minimum magnitude of OTF is the height of one silicon atom layer (ΔH = 2.71Å). The correlation length (λ) of OTF is typically from 1-3nm [14], which is still much smaller than gate length. Assuming that the two interfaces are uncorrelated, the standard deviation of oxide thickness fluctuation is expressed as Eq. (9):

$$\sigma t_{ox} = \Delta H \frac{B\lambda}{\sqrt{2WL}}$$

(9)

where λ denotes the correlation length of oxide surface roughness, and B is a fitting parameter. Moreover, from Eqs. (8) and (9), the standard deviation of OTF induced V_{th} variation is derived as:

$$\sigma V_{th} = C_4 \frac{\sqrt{qN_{ch}\varepsilon_{Si}}}{\varepsilon_{ox}} \frac{\lambda}{\sqrt{2WL}} \Delta H$$

(10)

where C_4 is a fitting parameter. Figure 4 validates this equation with TCAD simulations. From Fig. 4, OTF induced V_{th} change is independent on oxide thickness. Moreover, V_{th} variation due to OTF is much more sensitive to N_{ch}, as compared to that due to RDF (Eq. (2)). This fact suggests that as channel doping concentration increases in future CMOS devices, OTF induced V_{th} variation will become increasingly important over that induced by traditional RDF effect. Note that the V$_{th}$ variation due to OTF has an opposite trend on N$_{sub}$, as compared to that from LER.

Total V$_{th}$ Variation: Assuming that σV_{th} due to these three sources is independent on each other, the total σV_{th} is obtained as Eq. (11):

$$\sigma V_{th,(total)}^2 = \sigma V_{th,(RDF)}^2 + \sigma V_{th,(LER)}^2 + \sigma V_{th,(OTF)}^2$$

(11)

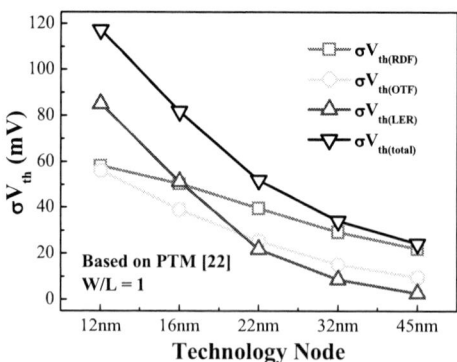

Figure 5. The trend of σV_{th} in device scaling.

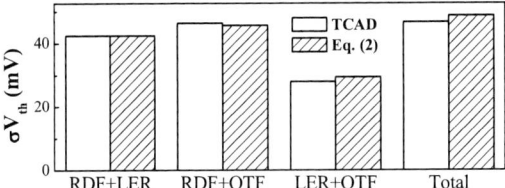

Figure 6. Simulated V_{th} variation due to combined sources.

Figure 6 validates this equation. From TCAD simulation results, RDF is still the major source at the 22nm node, while OTF is the second contributor to V_{th} variability. LER induced variability is relatively small, due to better control in advanced etching process, as well as the retrograde doping with high peak concentration.

Based on the PTM model, Fig. 5 illustrates the projection of V_{th} variation toward the 12nm node. The amplitude of LER is strongly dependent on the lithography and etching process. In this study, the standard deviation of LER is assumed to be fixed at 0.5nm [1, 18]. Under this assumption, it is observed that LER induced V_{th} variation may dominate total V_{th} variability in future technology nodes, due to the ever-increasing short-channel effect. On the other hand, OTF induced V_{th} variation exhibits a faster increasing rate with technology scaling, because of its square-root dependence on N_{ch}.

Temporal Parameter Shift

NBTI and CHC aging mechanisms become severe with continuous technology scaling. It is critical to understand and estimate the impact of these effects to predict the device and circuit lifetime. Both these effects can be physically described as the generation of charges in the region close to the Si-SiO$_2$ interface. A common theoretical framework, reaction-diffusion (RD) model [9], explains the power-law dependence of these effects with time.

Based on the equations that govern the reaction diffusion mechanism, the number of interface traps generated (N_{IT}) can be determined. The generated interface traps manifest as an increase in the threshold voltage (ΔV_{th}) of the PMOS transistor.

$$\Delta V_{th} = \frac{qN_{IT}}{C_{ox}}$$

(12)

The shift in the threshold voltage due to NBTI as a function of operation conditions and device parameters is given by

$$\Delta V_{th} = Kt^n \text{ where } K \propto C_{ox}(V_{gs} - V_{th})\exp(E_{ox} / E_0)\exp(-E_a / kT)$$

(13)

where $E_{ox} = V_{gs}/t_{ox}$ (considering both depletion and inversion charges), k is the Boltzmann constant and n is the time exponent defined by the diffusion species. The value of n is 0.16 if H_2 is the diffusion species and n equals 0.25 for H diffusion. Figure 7 shows the dependence of ΔV_{th} on the stress voltage applied at the gate, validating our model in Eq. (13) with the 65nm technology data.

The shift in the V_{th} due to CHC using RD model is

$$\Delta V_{th} = \frac{q}{C_{ox}} K_2 \sqrt{Q_i} \exp(\frac{E_{ox}}{E_{O2}}) \exp(\frac{-\varphi_{it}}{q\lambda E_m}) t^{n'}$$

(14)

where E_m is given by

$$E_m = (V_{ds} - V_{dsat})/l \text{ and } V_{dsat} = \frac{(V_{gs} - V_{theff} + 2V_t)L_{eff}E_{sat}}{V_{gs} - V_{theff} + 2V_t + A_{bulk}L_{eff}E_{sat}}$$

(15)

NBTI depends only on the gate bias where as CHC depends on the drain voltage as well, as illustrated in Eqs (14) and (15). CHC is permanent effect while NBTI is partially recoverable when the stress is removed. The stress and recovery behavior of NBTI is shown in Fig. 8. In a realistic circuit operation, the gate voltage switches between 0 and V_{DD}. For a PMOS transistor, the condition of $V_g = V_{DD}$ removes NBTI stress and anneals interface traps [13]. Such a process solely relies on the diffusion of neutral H_2 and thus,

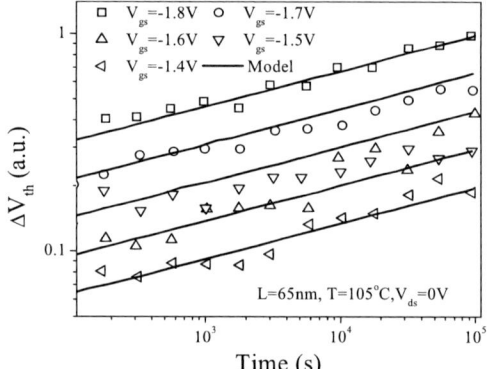

Figure 7. V_{gs} dependence of static NBTI for a 65nm technology.

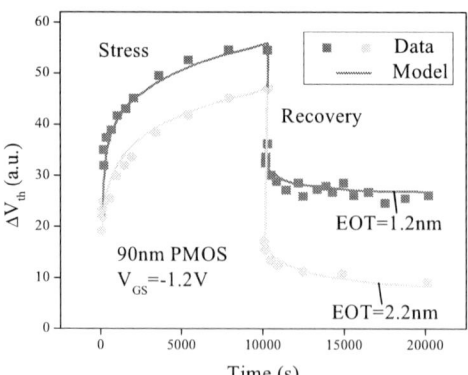

Figure 8. Validation of dynamic NBTI in a 90nm PMOS

has no field dependence. Assuming the recovery happens at t=t0 with $\Delta V_{th}=\Delta V_{th0}$, then the change in V_{th} can be modeled as

$$\Delta V_{th} = \Delta V_{th0}(1-\sqrt{\eta(t-t_0)/t})$$

(16)

In order to predict the long term threshold voltage degradation due to NBTI at a time t, the stress and recovery cycles given in Eqs. (13) and (16) can be simulated for m=t/T$_{clk}$ cycles to obtain the long term degradation. In high performance circuits, m can be very large even for t=1 month and hence is impractical to perform cycle-to-cycle simulation in order to predict the ΔV_{th}. Based on Eqs (13) and (16), it is possible to obtain a closed form for the upper bound on the long term ΔV_{th} as a function of the duty cycle α, T$_{clk}$ and time t:

$$\Delta V_{th} = \left(\frac{\sqrt{K^2 \alpha T_{clk}}}{1-\beta^{1/2n}}\right)^{2n}$$

(17)

where β is a function of oxide thickness, T$_{clk}$, α and t.

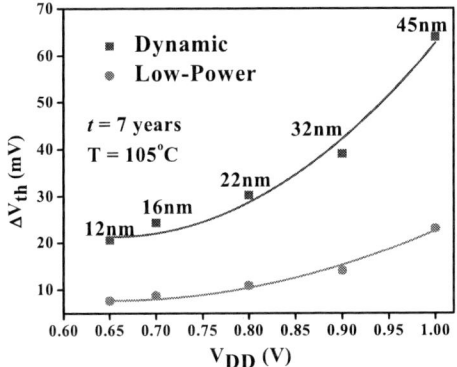

Figure 9. Shift in V$_{th}$ under different technologies in dynamic (50% duty cycle) and low power (1% duty cycle) operations

Further, the NBTI effect on CMOS technology scaling is evaluated. Fig. 9 illustrates the scaling of circuit aging under V$_{DD}$ tuning. Since the amount of the degradation is an exponential function of V$_{DD}$ (Eq. 13), lower V$_{DD}$ helps reduce the degradation rate. On the other side, if V$_{DD}$ is too low, then the sensitivity to V$_{th}$ shift is elevated, which eventually cancels the benefit. Also, the oxide thickness do not scale much with scaling, not impacting electric field across the oxide, resulting in a lower rate in a 12nm process. The degradation in low-power operation where the power supply is switched off is also examined. This trend in both dynamic and low-power operations is shown in Fig. 9, where the reduction rate in ΔV$_{th}$ is much smaller when V$_{DD}$ is lower at modern technology node.

Statistical Interaction

Since NBTI effect is dependent on threshold voltage (Eq. 13), device reliability degradation strongly interacts with process variations, significantly shifting both the mean and the variance of the circuit performance. Both static process variations and dynamic operation affect the performance and its variability [19]. Therefore, accurate prediction of the reliability during the life time should consider the impact of static variations, primary reliability mechanisms, and more importantly, their interactions. This prediction is essential for designers to safely guard-band the circuit for a sufficient life time. Otherwise, we have to either use an overly pessimistic bound, or resort to expensive stress tests in order to collect enough statistical information.

A few works have been published in the literature to estimate the statistical variations in temporal NBTI degradation [19-20]. Their assumption is the number of broken bonds in the interface is a Poisson random variable, and correspondingly V_{th} follows the Poisson distribution. With technology scaling, additional V_{th} variations, such as random dopant fluctuation and short channel effects, need to be considered [21-22]. The measurement data show that the distribution of V_{th} variations follows the Gaussian distribution. In addition, the correlations between process variation and NBTI are ignored in previous work. We begin with the assumption that process variation induced V_{th} change is a Gaussian random variable. We leverage compact models of transistor degradation, such as those presented in precious sub-section, to achieve the reliability prediction.

The model in Eq. (13) assumes nominal degradation without considering the statistical process variations. If there are global and local process variations, V_{th} in Eq. (12-13) should be expressed as

$$V_{th} = V_{th0} + \Delta V_{th-g} + \Delta V_{th-l} \qquad (18)$$

where V_{th0} is the nominal threshold voltage, ΔV_{th-g} and ΔV_{th-l} represent the change of threshold voltage due to global and local variations, respectively. Eq. (18) shows that positive variation results in V_{th} increase, which correspondingly leads to smaller V_{th} degradation (according to Eq.13), while negative variation results in larger V_{th} degradation. Figure 10 shows V_{th} degradation over time for three different transistors at 65nm. Due to static process variations, device 1 starts with a larger V_{th} and device 3 starts with a smaller V_{th}. Under the same stress conditions, the degradation of V_{th} for these three devices is shown in Fig. 10. At the beginning, the difference in V_{th} between device

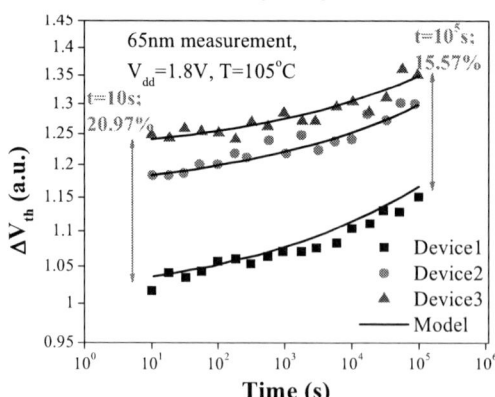

Figure 10. Threshold voltage degradation for different 65nm devices

1 and device 3 is 20.97%. With the increase of stress time, the difference becomes smaller and smaller. After 10^5s stress, it decreases to 15.57%. Such compensation between process variations and reliability degradation is well captured by our models.

In summary, it is widely recognized that process variation and reliability degradation are emerging as fundamental challenges to IC design with scaled CMOS technology; and they will have profound impact on nearly all aspects of circuit performance. Although traditionally their negative effects are mostly dealt with during data preparation and/or manufacturing process, the industry is starting to accept the fact that some of the effects can be better mitigated during the design phase. To facilitate robust design, compact models of variability and reliability are essential to provide coherent and consistent abstraction of the underlying technology, so that designers can take them into consideration.

Impact on Scaled CMOS Design

It is observed that LER induced V_{th} variation may dominate total V_{th} variability in future technology nodes, due to the ever-increasing short-channel effect (Fig. 5). On the other hand, OTF induced V_{th} variation exhibits a faster increasing rate with technology scaling, because of its square-root dependence on N_{ch}. Representative digital circuits, including an inverter chain and SRAM cell, are used to study the impact of random of V_{th} variation on circuit performance variability. During statistical circuit simulation, V_{th} is treated as a random variable with its variance from the model, and other model parameters are fixed at the nominal values.

Inverter Chain

A 7-stage inverter chain is adopted, with both NMOS and PMOS device at the minimum gate length. The width of the NMOS device is assumed to be 8 times the length, while the ratio of PMOS to NMOS width in the inverter is optimized by equating the rising and falling times. Table 1 lists the P to N ratios obtained in this way for the different technology nodes. To study the effect of variation of V_{th} we consider the delay metric. This is measured across fourth inverter because it is well isolated from effects of input waveform and output loading.

Table I. Minimum L, Vdd and P to N ratios
for different technologies

Technology	L(nm)	Vdd(V)	P to N ratio
45nm	45	1.0	1.02
32nm	32	0.9	0.96
22nm	22	0.8	0.91
16nm	16	0.7	0.80
12nm	12	0.65	0.84

Random V_{th} variations, as described in previous sections, are considered in the simulation, assuming they are uncorrelated in all 14 transistors in the 7-stage inverter chain. Figure 11 illustrates the mean and standard deviation of inverter delay during technology scaling. While the scaling successfully speeds up the nominal circuit

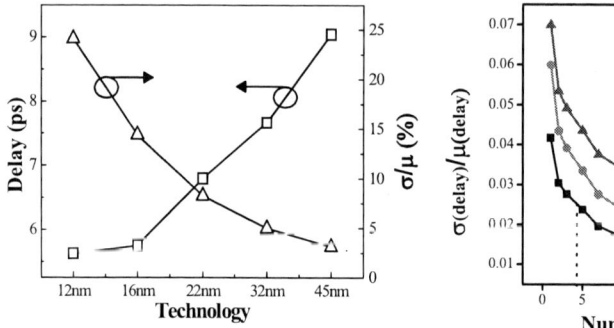

Figure 11. The scaling trend of the mean and standard deviation of inverter delay under random variations.

Figure 12: σ/μ of delay with number of inverters and sizing

performance, the variability of inverter delay keeps increasing, as the result of the rapidly exacerbated random variations.

Figure 12 shows the σ/μ ratio of the inverter chain with change in number of inverters (N) and sizing. The ratio depends on random V_{th} variations, along with a systematic variation due to NBTI on top of it. From the figure, it is evident that σ/μ ratio decreases with N and ratio from NBTI is comparable to that from random variations when N~20 for the inverter sizing 8X. If the sizing is decreased to 4X, σ/μ ratio decreases, but the ratio from NBTI is significant even for a lower N~4.

SRAM cell

A SRAM cell represents the most sensitive circuit unit to process variations. A typical 6-T SRAM is used in this analysis. All six transistors have the minimum gate

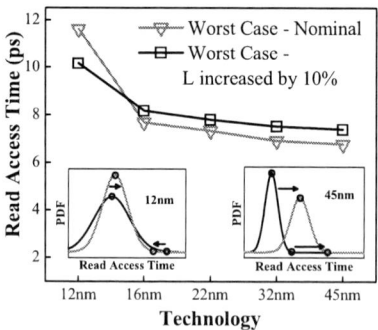

Figure 13. The 3σ corner of SRAM RAT can be effectively reduced by suppressing random V_{th} variations.

length. The pull up PMOS is assumed to be at the minimum width. The widths of access transistors and pull down NMOS are tuned to achieve the same read and write noise margins. To evaluate the operation speed of a SRAM cell, Read Access Time (RAT) is examined. Assuming that sense amplifiers are able to measure 10% of V_{dd} drop on either BL or $\overline{BL}$, read access time is calculated as the time when BL or $\overline{BL}$ reaches 90% of V_{dd}. Monte Carlo simulations are performed to extract the statistics, considering V_{th} variations in all transistors are uncorrelated. Figure 13 illustrates the scaling trend of the 3σ corner of RAT.

To suppress the variation, one technique is L biasing: increasing gate length by 10% is able to reduce V_{th} variation by >28% at the 12nm node (previous section). Although this may not be practical today due to the overhead in the nominal RAT (Fig. 13), such a technique benefits future technology nodes: with 10% L biasing at the 16nm node and below, the reduction in the excessive variability overwhelms the change of the nominal value (Fig. 13) and thus, the corner value of RAT decreases. This tradeoff highlights the importance of variability control for future IC design. Finally, the variability in SRAM Read Noise Margin (RNM) is decomposed into different variation sources, as shown in Fig. 15. For a first order analysis Read Noise Margin of SRAM is considered to be linear function of mismatches between V_{th} of transistors. The following six mismatches are considered and all are taken to be independent (SRAM transistor labels shown in Fig. 14).

1. Mismatch between M1, M2 and between M3, M4

2. Mismatch between M1, M3 and between M2, M4

3. Mismatch between M2, M5 and between M4, M6

The variations in V_{th} of each transistor are directly mapped to mismatch between pairs as listed above. The variation of mismatch is considered to be summation of variation of both transistors as given in Eq. (11)

$$\sigma V_{th,(M1M2)}^{2} = \sigma V_{th,(M1)}^{2} + \sigma V_{th,(M2)}^{2} \tag{22}$$

The variation in RNM is calculated from variation of mismatches and β coefficients as given in:

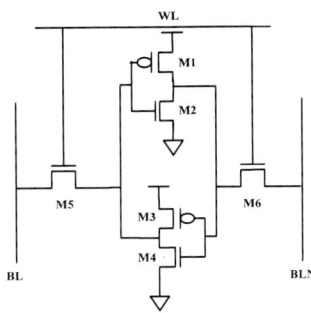

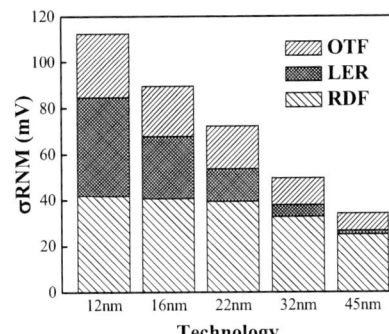

Figure 14. Schematic of a SRAM cell   Figure 15. The decomposition of RNM variability.

$$\sigma RNM^2 = \beta_1^2 \sigma V_{th,(M1M2)}^2 + \beta_2^2 \sigma V_{th,(M4M4)}^2$$
$$+ \beta_3^2 \sigma V_{th,(M1M3)}^2 + \beta_4^2 \sigma V_{th,(M2M4)}^2$$
$$+ \beta_5^2 \sigma V_{th,(M2M5)}^2 + \beta_6^2 \sigma V_{th,(M4M6)}^2 \tag{23}$$

Similar as that in inverter delay, LER and OTF rapidly increase as major contributors to RNM variability, with RDF being relatively constant along with the technology scaling. This behavior is mainly because the device dimension is approaching fundamental atomistic limits, which are not scalable.

Conclusion

In this paper, random V_{th} variation under RDF, LER, and OTF is studied through 3-D atomistic simulation with commercial TCAD device simulator. With the simulated result, a suite of scalable and predictive compact models are proposed. Reliability effects like NBTI and CHC that limit the device and circuit life time are also studied and modeled. Further, statistical interaction of aging effects on intrinsic variations is presented, indicating the self compensation between aging and static variations. Random V_{th} variation is projected to advanced technology nodes, illustrating the trend and importance to future device and circuit performance. The developed predictive models are used to illustrate the scaling trends of variability and reliability in benchmark circuits.

References

1. International Technology Roadmap for Semiconductors, 2008 (available at http://public.itrs.net).

2. K. Bernstein, D. J. Frank, A. E. Gattiker, W. Haensch, B. L. Ji, S. R. Nassif, E. J. Nowak, D. J. Pearson and N. J. Rohrer, "High-performance CMOS variability in the 65-nm regime and beyond," *IBM J. Res. & Dev.*, vol. 50, no. 4/5, pp. 433-449, Jul. 2006.

3. A. T. Putra, A. Nishida, S. Kamohara, and T. Hiramoto, "Random V_{th} variation induced by gate edge fluctuations in nanoscale MOSFETs," *Silicon Nanoelectronics Workshop*, pp. 73-74, 2007.

4. A. Asenov, S. Kaya, and A. Brown, "Intrinsic parameter fluctuation s in decananometer mosfets introduced by gate line edge roughness," *IEEE Transactions on Electron Devices,* 50(5):1254-1260, 2003.

5. S. M. Goodnick, D.K. Ferry, and C.W. Wilmsen, "Surface roughness at the Si(100)-SiO$_2$ interface," *Physical Review B*, vol. 32, no. 12, pp. 8171-8182, Dec. 1985.

6. N. Kimizuka, T. Yamamoto, T. Mogami, K. Yamaguchi, K. Imai, and T. Horiuchi, "The impact of bias temperature instability for direct-tunneling ultra-thin gate oxide on MOSFET scaling," *VLSI Symp. on Tech.*, pp. 73-74, 1999.

7. A. T. Krishnan, C. Chancellor, S. Chakravarthi, P. E. Nicollian, V. Reddy, and A. Varghese, "Material dependence of hydrogen diffusion: Implication for NBTI degradation," *IEDM*, 2005.

8. W. Wang, V. Reddy, A. T. Krishnan, R. Vattikonda, S. Krishnan, Y. Cao, "Compact modeling and simulation of circuit reliability for 65nm CMOS technology," *IEEE Trans. on Device and Materials Reliability*, vol. 7, no. 4, pp. 509-517, Dec. 2007.

9. K. Kang, S. P. Park, K. Roy, and M. A. Alam, "Estimation of statistical variation in temporal NBTI degradation and its impact on lifetime circuit performance," *IEEE/ACM Intnl. Conference on Computer-Aided Design*, pp. 730–734, Nov. 2007.

10. T. Ezaki, T. Ikezawa, A. Notsu, K. Tanaka, and M. Hane, "3D MOSFET simulation considering long-range coulomb potential effects for analyzing statistical dopant-induced fluctuations associated with atomistic process simulator," *Proc. SISPAD*, pp. 91-94, 2002.

11. *Sentaurus User's Manual*, Synopsys, Inc., Mountain View, CA, v. 2009.6.

12. S. Xiong, and J. Bokor, "Study of gate line edge roughness Effect in 50nm bulk MOSFET devices," *Proc. SPIE*, vol. 4689, pp. 733, 2002.

13. S. Rangan, N. Mielke, E. C. C. Yeh, "Universal recovery behavior of negative bias temperature instability," *IEDM*, pp. 341-344, 2003.

14. K. Takeuchi, et al., "Understanding random threshold voltage fluctuation by comparing multiple fabs and technologies," *IEEE Transactions on Electron Devices*, vol., no., pp.467-470, Dec. 2007.

15. C. Alexander, G. Roy, A. Asenov, "Random-dopant-induced drain current variation in nano-MOSFETs: A three-dimensional self-consistent Monte Carlo simulation study using "ab initio" ionized impurity scattering," *IEEE Trans. Electron Devices*, vol. 55, no. 11, pp. 3251–3258, Nov. 2008.

16. Z. H. Liu, et al., "Threshold voltage model for deep-submicrometer MOSFETs," *IEEE Transactions on Electron Devices*, vol.40, no.1, pp.86-95, Jan. 1993.

17. X. Xi, M. Dunga, J. He, W. Liu, K. M. Cao, X. Jin, J. J. Ou, M. Chan, A. M. Niknejad, and C. Hu, BSIM4 Manual, UC Berkeley Device Group.

18. S. Sardo, F. Gicometti, S. Doneda, U. Colombo, M. D. Muri, A. Donghi, R. Morson, G. Mutinati, A. Nottola, M. Gentili and M. C. Ubaldi, "Line edge roughness (LER) reduction strategy for SOI waveguides fabrication," *Microelectronic Engineering*, vol. 85, iss. 5-6, pp. 1210-1213, 2008.

19. G. L. Rosa, W. L. Ng, S. Rauch, R. Wong, and J. Sudijono, "Impact of NBTI induced statistical variation to SRAM cell stability," *IEEE International Reliability Physics Symposium*, pp. 274–282, Mar. 2006.

20. S. E. Rauch, , "The statistics of NBTI induced vt and β mismatch shifts in pmosfets," *IEEE Trans. on Device Material Reliability*, pp. 89–93, 2002.

21. W. Zhao, Y. Cao, "New generation of predictive technology model for sub-45nm early design exploration," *IEEE Transactions on Electron Devices*, vol. 53, no. 11, pp. 2816-2823, Nov. 2006. (Available at http://www.eas.asu.edu/~ptm)

22. N. Sano, K. Matsuzawa, M. Mukai, and N. Nakayama, "On discrete random dopant modeling in drift-diffusion simulations: Physical meaning of 'atomistic' dopants," *Microelectronics Reliability*, vol. 42, no. 2, pp. 189–199, Feb. 2002.

ECS Transactions, 35 (4) 369-380 (2011)
10.1149/1.3572294 ©The Electrochemical Society

Bias-Temperature Instabilities and Radiation Effects on SiC MOSFETs

E. X. Zhang[a], C. X. Zhang[a], D. M. Fleetwood[a,b], R. D. Schrimpf[b],
S. Dhar[c], S.-H. Ryu[c], X. Shen[b], and Sokrates T. Pantelides[b,a]

[a]Electrical Engineering and Computer Science Department,
Vanderbilt University, Nashville, TN 37235, USA
[b]Department of Physics and Astronomy, Vanderbilt University, Nashville, TN 37235, USA
[c]Cree Inc., 4600 Silicon Drive, Durham, NC 27703, USA

> Bias-temperature-instabilities (BTIs) are investigated for 4H-SiC
> based nMOSFETs before and after total ionizing dose irradiation.
> We find that the threshold voltage shifts of unirradiated devices
> decrease significantly with elevated-temperature stress under
> negative bias (accumulation); in contrast, devices stressed under
> positive bias (inversion) do not exhibit significant threshold
> voltage shifts. Threshold voltage shifts due to BTI for unirradiated
> devices stressed under negative bias correlate strongly with the
> additional ionization of deep dopants in SiC at elevated
> temperatures. The charge that leads to BTI lies in deep interface
> traps (more than 0.6 eV away from the SiC conduction or valence
> bands) and O vacancies in the SiO_2. Hole trapping at O vacancies
> dominates the ionizing radiation response. The magnitudes of the
> changes in threshold voltage shifts increase with switched bias-
> temperature stress after irradiation, relative to those in unirradiated
> devices.

INTRODUCTION

Silicon carbide is attractive for high power and high temperature applications because of
its wide band gap (~3.26 eV for 4H-SiC), high breakdown field strength, high saturation
electron drift velocity, high thermal conductivity, and compatibility with Si processing
(1-4). Although bias-temperature instability (BTI) is a critical reliability issue for SiC
devices and circuits (5-11), only limited information on its underlying mechanisms is
available. In Si based MOS devices, the release of hydrogen from passivated interfacial Si
dangling bonds under BTI stress leads to interface trap buildup and oxide-trap charge
(12). But hydrogen processing is much less effective in reducing SiC-SiO_2 interface trap
density (12,13). The quality of the SiC-SiO_2 interface has been improved dramatically by
NO processing, which can lead to a significant reduction in the density of interface traps
(13-18). Because of its wide band gap and differences in processing as compared with Si
MOS devices, the mechanisms of BTI may be expected to differ significantly in SiC
MOS devices from those observed in Si MOS devices.

In this work, we investigate bias-temperature-instabilities (BTIs) for 4H-SiC based
nMOSFETs before and after total ionizing dose (TID) irradiation. Both steady-state and

369

switched-bias stresses were performed under positive and negative bias. Significant shifts are observed for steady-state stresses at elevated temperatures under negative bias (accumulation); negligible shifts are observed for steady-state stresses at elevated temperatures under positive bias (inversion). The BTI response after switched-stress bias is significantly different before and after total dose radiation. We describe the dynamics of the resulting charge trapping and recovery in detail.

EXPERIMENTAL DETAILS

nMOSFETs were fabricated on an aluminum doped (p type) 4H-SiC epitaxial layer with a 55 nm, NO-nitrided gate oxide. All devices received a post-oxidation NO anneal at 1175 °C for 2 h to reduce their interface trap densities (12,16). Steady state and switched BTI experiments were performed at gate bias of ±17.5 V with the other terminals grounded at 150 °C and 250 °C. These electric fields are well below the thresholds for Fowler-Nordheim charge injection (19,20). TID irradiation was performed with 10-keV x-rays at room temperature with an applied gate bias of 8.5 V, with the other terminals grounded. Drain current vs. gate voltage (I_D-V_G) measurements were performed at room temperature with a HP4156A parameter analyzer. At least three devices were measured for each case shown; the results represent the average response of the tested MOSFETs.

RESULTS AND DISCUSSION

A. Unirradiated devices.

Figure 1 shows I_D-V_G curves measured at room temperature as a function of time for nMOSFETs stressed in inversion at an applied gate bias of 17.5 V at 150 °C. The curves do not exhibit any significant shift with stressing time; the threshold voltage shift (ΔV_{TH}) is less than 1 mV. Even if we increase the stress temperature up to 250 °C, the magnitude of ΔV_{TH} is only ~ 4 mV after 20 hour stress, as shown in Fig. 2. This shows the relative stability of these 4H SiC devices under positive bias (inversion) at elevated temperatures.

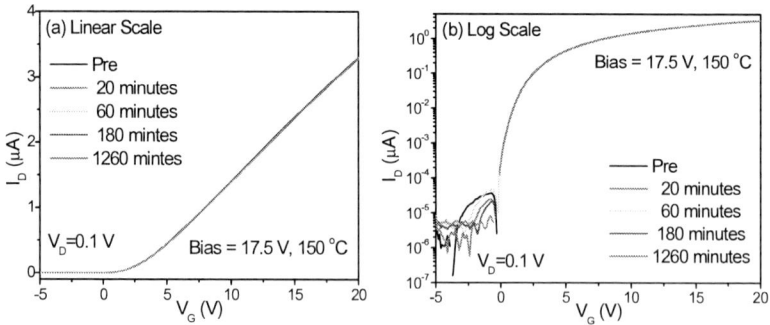

Figure 1. Drain current as a function of gate voltage (I_D-V_G) and stress time for 4H-SiC nMOSFETs, for (a) linear scale and (b) logarithmic scale. The applied gate bias is 17.5 V with the other terminals grounded, and the temperature is 150 °C.

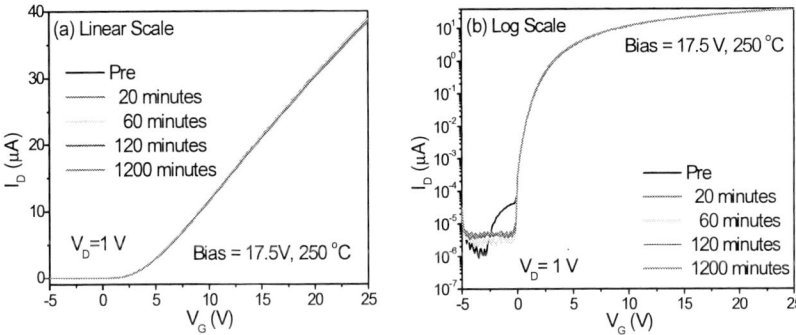

Figure 2. Drain current as a function of gate voltage (I_D-V_G) and stress time for 4H-SiC nMOSFETs, for (a) linear scale and (b) log scale. The applied gate bias is 17.5 V with the other terminals grounded, and the temperature is 250 °C.

Figure 3 shows I_D-V_G curves measured at room temperature as a function of time for nMOSFETs stressed at 150 °C under negative bias at an applied gate bias of -17.5 V with the other terminals grounded. For these stress conditions, the curves shift monotonically negatively with stress time due to hole trapping. The corresponding threshold voltage shift is shown as a function of stressing time in Fig. 4. The magnitude of the midgap voltage increases rapidly at early stress times, with a significant decrease in degradation rate at longer times. These negative shifts are strongly correlated with the ionization of deeper acceptor levels (21) in SiC MOSFETs, as we discuss in detail elsewhere (22). We now provide a brief summary of several converging lines of evidence from capacitor and transistor experiments that reinforce the link between dopant ionization (or carrier availability) and BTIs.

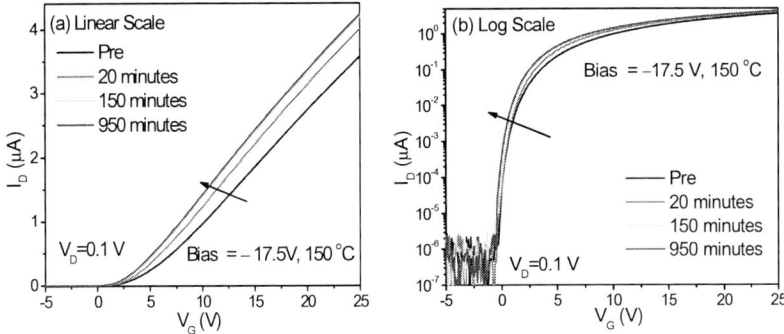

Figure 3. Drain current as a function of gate voltage (I_D-V_G) and stress time for 4H-SiC nMOSFETs, for (a) linear scale and (b) log scale. The applied gate bias is - 17.5 V with the other terminals grounded, and the temperature is 150 °C.

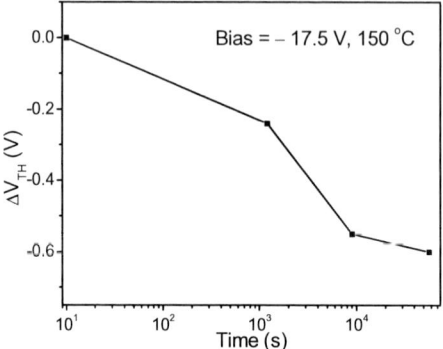

Figure 4. Threshold voltage shifts (ΔV_{TH}) as a function of stress time for the 4H-SiC nMOSFETs and stress conditions of Fig. 3.

Figure 5 shows bias-temperature stress experiments on p-substrate capacitors processed in the same lot as the nMOSFETs. These results show that the effective activation energy for these devices ($\sim 0.23 \pm 0.02$ eV) coincides closely with the known ionization energies of the Al dopants in 4H-SiC, 0.20 to 0.22 eV (21-24). These ionization energies are larger than the ionization energies of typical dopants in Si. Whereas dopants in Si are fully ionized at room temperature, the deeper dopants in SiC are only partially ionized (22,25). The strong correlation of NBTI activation energy and dopant ionization energy strongly suggests that the dopants play an important role in the carrier dynamics and trapping.

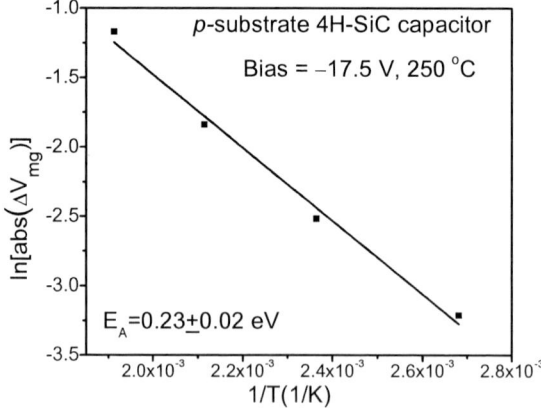

Figure 5. Logarithm of the absolute value of the midgap voltage shift as a function of the reciprocal of the temperature for p-substrate/4H-SiC MOS capacitors stressed for 20 minutes at an applied gate bias of -17.5 V.

Simple calculations show that, for Al acceptors, assuming an ionization energy of ~ 0.22 (0.20) eV, only ~ 37 (49) % are ionized at room temperature (22). At the elevated temperatures of the BT stress, additional carriers are released from the dopants under these accumulation stress-bias conditions. The excess carriers are available to tunnel into border traps (26) and increase the midgap voltage shift. This mechanism appears to be more significant here than in Si because SiC has more defects and fewer carriers than Si. We also note than no significant PBTI is observed for transistors biased in inversion in Fig. 2, under conditions that are higher in temperature and longer in time than those shown for devices in accumulation in Figs. 3-5. This is because carrier densities are much lower in inversion than accumulation for these devices (21,27,28). Moreover, interface trap levels that are away from the band edges are quite slow to populate in SiC MOS devices (27,28). Hence, carrier availability can affect BTI due to both interface and oxide trap charge effects in SiC MOS devices much more significantly than in Si MOS devices.

There are other differences between Si MOS and SiC MOS devices that must be considered. The gate oxides on SiC are thicker and grown at higher temperatures than the ultrathin nitrided SiO_2 and/or high-K gate dielectrics used in Si-based integrated circuit technologies (29-36). For Si MOS technologies, hydrogen is known to play an important role in both the passivation of process-induced interface traps and the creation of defects during bias-temperature stress (11,37). Hydrogen reactions at the SiC interface are typically not as effective in passivating interface traps as at the Si/SiO_2 interface (12-16,38-40), so the role of hydrogen in BTI on SiC is less clear. Excess nitrogen associated with the NO nitridation treatments used to reduce interface-trap densities in these SiC devices (1,15,16,40) may also introduce a relatively small but finite density of N-related trap levels in the near-interfacial SiO_2. N-related centers are observed to function as both electron traps and hole traps (41-43). Moreover, nitrogen has been demonstrated to increase NBTI in Si/SiO_2 structures (41,44,45). On the other hand, NO processing has been found to reduce the density of interface defects (15,16,46,47). Thus, if N-related defects play a significant role in the observed response, it is the excess nitrogen concentration above and beyond the levels required to passivate the process-induced defects that is likely contributing to the BTI. Further, holes trapped by N-related defects typically are re-emitted when the temperature is raised past 125 °C (15). So it seems more likely that O vacancies play a key role in the hole trapping in these devices (22), similar to what is found for Si MOS devices stressed at similar temperatures but higher electric fields (e.g., > 5 MV/cm) (33-36).

Figure 6 shows the values of ΔV_{TH} as a function of switched stressing biases for 4H-SiC based nMOSFETs. The characterization was performed at room temperature before and after BT stressing at 150 °C. The V_{TH} shifts negatively after the first 20 minute negative-bias stress, and then recovers somewhat during the next 20 minute positive-bias stress. In contrast, an additional increase in V_{TH} shift is observed in the second period (60 minutes) of positive-bias stress that follows the second negative-bias stress (60 minutes). Smaller shifts but significantly more reversibility is typically observed for Si MOS capacitors subjected to similar switched-bias stressing conditions (31,32). For the SiC MOS devices of Fig. 6, the majority of trapped holes remain in interface and/or oxide traps despite the reverse bias stress, showing that hole traps are more stable than electron traps in these devices (22). The significant increase in the magnitude of the midgap voltage shift during the second period of PBT stress may result from the bias-induced motion of trapped positive charge in the SiO_2 layer from trapping sites within the oxide to sites at or closer to the SiC/SiO_2 interface. The trapped charge is likely to comprise mostly trapped holes. In oxides in Si, many of these holes would be neutralized via electron compensation near the interface (48,49), but charge generation rates in SiC are significantly less than in Si (27,28), so fewer electrons are available to compensate the trapped holes in SiC MOS capacitors than Si MOS capacitors. It is also possible that some of the trapped positive charge is in the form of H^+ (50,51). If that is the case in these oxides on SiC, then the results of Fig. 6 suggest that the H^+ in the SiO_2 can move closer to the SiO_2/SiC interface during the application of PBT stress. However, there must not be a significant amount of interface trap formation, or the midgap voltage shift would be positive instead of negative, under positive bias. Trapping of H^+ at defects in the near-interfacial SiO_2 regions of oxides with high O vacancy densities has been observed in Si MOS devices (52), so this interpretation of the results is also consistent with Fig. 6.

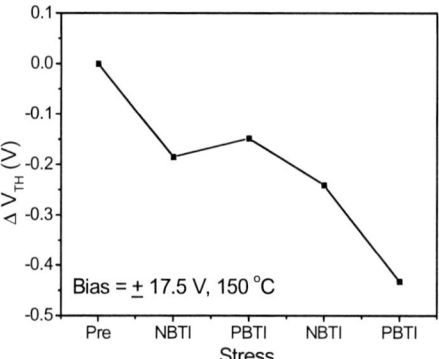

Figure 6. ΔV_{TH} as a function of switched-bias stress for 4H-SiC nMOSFETs. The applied gate bias during the stress was ±17.5 V; the temperature was 150 °C; the stress times were 20 min for the first pair of stresses, and 60 min for the second pair.

B. Irradiated Devices.

Figure 7 shows I_D-V_G curves for 4H-SiC nMOSFETs irradiated with 10-keV x-rays at a dose rate of ~ 31 krad(SiO_2)/min at a gate oxide electric field of ~1.5 MV/cm. The drain leakage increases significantly and the I_D-V_G curves shift negatively with increasing TID. The increase in drain leakage current is likely due to edge leakage (53). The threshold voltage shifts negatively with increasing TID, due primarily to radiation-induced-hole trapping in the SiO_2, as shown in Fig. 8. A threshold voltage shift of ~ -1.5 V is observed after 100 krad(SiO$_2$) irradiation for this 4H-SiC transistor with a gate oxide thickness of 55 nm. Similar or even somewhat larger shifts were observed for 34 nm nitrided oxides by Dixit et al. (54). At these electric fields, the TID response of SiO$_2$ varies as ~ t_{ox}^2, where t_{ox} is the oxide thickness (55,56). Hence, the density of radiation-induced oxide-trap charge is ~ 2-3 times lower in these devices than in those evaluated by Dixit et al. (54). After adjusting for differences in t_{ox}, these devices also show ~ 1.5-2 times lower radiation-induced oxide-trap charge densities than the ~25 nm nitrided oxides on 4H-SiC evaluated by Arora et al. (7).

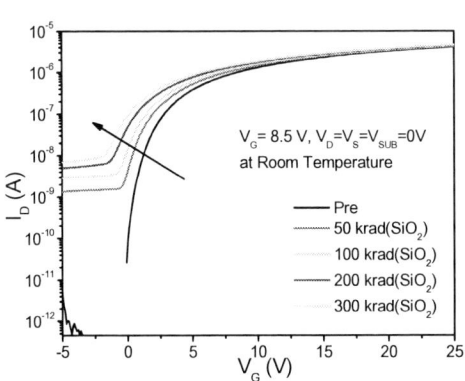

Figure 7. Drain current as a function of gate voltage (I_D-V_G) and total ionizing radiation dose for 4H-SiC nMOSFETs. The irradiation was performed at room temperature, and the applied gate bias was 8.5 V.

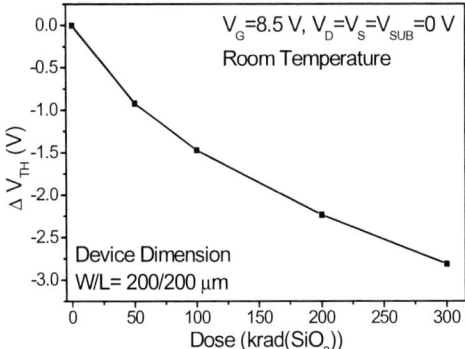

Figure 8. Threshold voltage shift (ΔV_{TH}) as a function of total ionizing dose for the 4H-SiC nMOSFETs for the irradiation and bias conditions of Fig. 7.

Enhanced degradation relative to either individual type of stress has been reported for some SiO$_2$ and/or high-K oxides on Si exposed to bias-temperature (BT) stress after irradiation (31,32). Thus, it is important (e.g., for potential space applications) to evaluate the combined effects of irradiation and BT stress on 4H-SiC based MOSFETs. Figure 9 shows ΔV_{TH} as a function of BTI stress time at 150 °C for an applied oxide electric field of approximately ±3 MV/cm after the TID. The value of ΔV_{TH} shifts positively with time in all cases. The recovery of the threshold voltage shift is faster under positive bias temperature (PBT) stress than under NBT stress. PBT stress accelerates the annealing of positive-oxide-trap charge and the buildup of radiation-induced interface traps (57,58). Each of these reduces the magnitude of the threshold voltage shift, leading to an enhanced rate of recovery for nMOSFETs under these irradiation and stress conditions. No evidence of enhanced degradation is observed for negative BT stress following irradiation, although the recovery rate is substantially reduced.

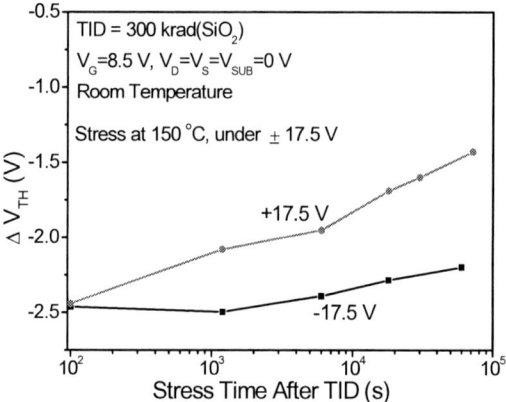

Figure 9. ΔV_{TH} as a function of post-irradiation bias-temperature stress time at 150 °C, for the devices and irradiation conditions of Fig. 8.

Figure 10 shows the results of switched BT stress for nMOSFETs after TID irradiation. The radiation exposure leads to hole trapping in the gate oxide. After irradiation, the value of ΔV_{TH} increases in magnitude relative to its post-irradiation value (negative threshold voltage shifts) under negative-bias stress and decreases in magnitude (positive threshold voltage shifts) for positive-bias stress (48,49). During negative-bias stress, electrons in border traps are pushed out of the oxide and into the SiC, leading to an increase in net trapped positive charge. During positive-bias stress, the annealing of radiation-induced positive oxide-trap charge dominates the device response. Qualitatively similar effects are observed for oxides on Si, which typically exhibit smaller radiation-induced hole trapping and a greater degree of reversibility in midgap voltage shifts (31,32,48) than these oxides on SiC. It is likely that the compensating electron traps that cause this reversibility are associated primarily with O vacancy-related defects in the near-interfacial SiO_2 (46). These electron traps can be stabilized by radiation-induced holes (59), accounting for the additional reversibility that is observed for irradiated devices (Fig. 8), as compared to that observed for unirradiated devices (Fig. 6).

Interestingly, we do not see any increase in the magnitudes of the threshold voltage shifts during the positive-bias annealing periods for these devices, in contrast to the unirradiated devices in Fig. 6. This is because these devices are irradiated under positive bias, so holes are already driven towards the SiC-SiO$_2$ interface. For the unirradiated devices, the initial hole trapping occurs at negative bias, when motion into the bulk of the oxide can occur more easily.

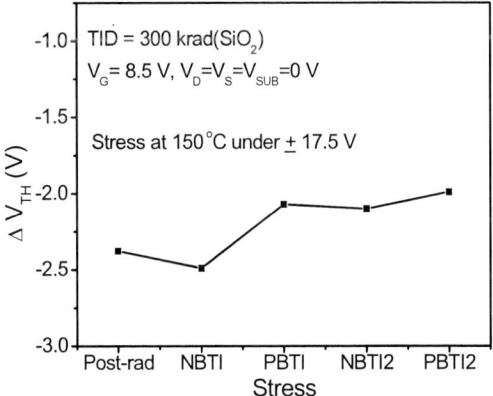

Figure 10. ΔV_{TH} as a function of switched BT stress after TID for nMOSFET transistors irradiated to 300 krad(SiO$_2$) and annealed at 150 °C for 20 minutes, and then for 60 minutes.

CONCLUSION

We have performed a detailed experimental study of bias-temperature-instabilities in 4H-SiC based metal-oxide-semiconductor (MOS) transistors before and after total ionizing dose irradiation. For unirradiated devices, the threshold voltage shifts negatively under negative bias due primarily to hole trapping at or near the SiC-SiO$_2$ interface. The measured activation energy for BTI in p-type 4H-SiC metal-oxide-semiconductor capacitors is 0.23 ± 0.02 eV, which correlates strongly with the ionization energy of the dopant. Coupled with the lack of significant BTI for stress under positive bias, these results suggest a key role for dopant ionization in SiC BTI. O vacancies are found to play a significant role in this degradation, but a supporting role for deep interface traps and N-related defects is also possible. No significant shifts are observed for similar bias-temperature stress performed under positive bias. After irradiation, the midgap voltage shifts in these nitrided oxides are significantly less than those observed in previous work on nitrided oxides on SiC. BT stress after radiation exposure leads primarily to the annealing of positive oxide-trap charge, with more annealing under positive BT stress than negative BT stress. A more significant role is found for compensating electron trapping in switched-bias stress after irradiation than for switched-bias stress for unirradiated devices.

ACKNOWLEDGMENTS

We would like to thank C. Scozzie, B. Geil, and A. J. Lelis at the US Army Research Laboratory, Adelphi, MD, for supporting the fabrication of the SiC MOS capacitors at Cree, Inc. This work was supported in part by the Air Force Office of Scientific Research through a MURI program, by the Defense Threat Reduction Agency's Basic and Applied Sciences Program, by National Science Foundation grant DMR-09-07385, and the McMinn Endowment at Vanderbilt University.

REFERENCES

1. A. Agarwal and S.H. Ryu, *Proc. CS MANTECH Conference*, Vancouver, British Columbia, Canada (2006), pp. 215-218.
2. M. Bhatnagar and B. J. Baliga, *IEEE Trans. Electron Dev.* **40**, 645 (1993).
3. A. A. Orouji and H. Elahipanah, *IEEE Trans. Dev. Mater. Reliab.* **10**, 92 (2010).
4. T. Okayama, S. D. Arthur, J. L. Garrett, and M. V. Rao, *Solid-State Electron.* **52**, 164 (2008).
5. J. A. Cooper, Jr., M. R. Melloch, R. Singh, A. Agarwal, and J. W. Palmour, *IEEE Trans. Electron Dev.* **49**, 658 (2002).
6. A. V. Kuchuk, M. Guziewicz, R. Ratajczak, M. Wzorek, V.P. Kladko, and A. Piotrowska, *Microelectron. Engineering* **85**, 2142 (2008).
7. R. Arora, J. Rozen, D. M. Fleetwood, K. F. Galloway, C. X. Zhang, J. Han, S. Dimitrijev, F. Kong, L. C. Feldman, S. T. Pantelides, and R. D. Schrimpf, *IEEE Trans. Nucl. Sci.* **56**, 3185 (2009).
8. K.Y. Cheong, J. H. Moon, H. J. Kim, W. Bahng, and N. K. Kim, *Thin Solid Films* **518**, 3255 (2010) .
9. K. Kawahara, M. Krieger, J. Suda, and T. Kimoto, *J. Appl. Phys.* **108**, 023706 (2010)
10. M. Gurfinkel, H. D. Xiong, K. P. Cheung, J. S. Suehle, J. B. Bernstein, Y. Shapira, A. J. Lelis, D. Habersat, and N. Goldsman, *IEEE Trans. Electron Dev.* **55**, 2004 (2008).
11. L. Tsetseris, X. J. Zhou, D. M. Fleetwood, R. D. Schrimpf, L. Tsetseris, and S. T. Pantelides, *Appl. Phys. Lett.* **86**, 142103 (2005).
12. J. Rozen, S. Dhar, S. T. Pantelides, L. C. Feldman, S. Wang, J. R. Williams, and V. V. Afanas'ev, *Appl. Phys. Lett.* **91**, 153503 (2007).
13. S. T. Pantelides, S. Wang, A. Franceschetti, R. Buczko, M. Di Ventra, S. N. Rashkeev, L. Tsetseris, M. H. Evans, I. G. Batyrev, L. C. Feldman, S. Dhar, K. McDonald, R. A. Weller, R. D. Schrimpf, D. M. Fleetwood, X. J. Zhou, J. R. Williams, C. C. Tin, G. Y. Chung, T. Isaacs-Smith, S. R. Wang, S. J. Pennycook, G. Duscher, K. Van Benthem, and L. M. Porter, *Mater. Sci. Forum* **527**, 935 (2006).
14. S. Wang, S. Dhar, S.-R. Wang, A. C. Ahyi, A. Franceschetti, J. R. Williams, L. C. Feldman, and S. T. Pantelides, *Phys. Rev. Lett.* **98**, 026101 (2007).
15. J. Rozen, S. Dhar, S. K. Dixit, V. V. Afanas'ev, F. O. Roberts, H. L. Dang, S. Wang, S. T. Pantelides, J. R. Williams, and L. C. Feldman, *J. Appl. Phys.* **103**, 124513 (2008).

16. J. Rozen, S. Dhar, M. E. Zvanut, J. R. Williams, and L. C. Feldman, *J. Appl. Phys.* **105**, 124506 (2009).

17. G. Y. Chung, C. C. Tin, J. R. Williams, K. McDonald, M. D. Ventra, S. T. Pantelides, L. C. Feldman, and R. A. Weller, *Appl. Phys. Lett.* **76**, 1713 (2000).

18. H. Li, S. Dimitrijev, H. B. Harrison, and D. Sweatman, *Appl. Phys. Lett.* 70, 2028 (1997).

19. H.-F. Li, S. Dimitrijev, D. Sweatman, and H. B. Harrison, *Microelectron. Reliab.* 40, 283 (2000).

20. A. K. Agarwal, S. Seshadri, and L. B. Rowland, *IEEE Electron Dev. Lett.* **18**, 592 (1997).

21. I. G. Ivanov, A. Magnusson, and E. Janzen, *Phys. Rev. B* **67**, 165212 (2003).

22. X. Shen, E. X. Zhang, C. X. Zhang, D. M. Fleetwood, R. D. Schrimpf, S. Dhar, S. H. Ryu, and S. T. Pantelides, *Appl. Phys. Lett.*, submitted (Jan. 2011).

23. I. G. Ivanov, A. Henry, and E. Janzen, *Phys. Rev. B* **98**, 241201(R) (2005).

24. Z. Q. Fang, B. Claflin, D. C. Look, L. Polenta, and W. C. Mitchel, *J. Electron. Mater.* **34**, 336 (2005).

25. A. V. Los and M. S. Mazzola, J. Electron. Mater. **30**, 235 (2001).

26. D. M. Fleetwood and N. S. Saks, *J. Appl. Phys.* **79**, 1583 (1996).

27. P. Neudeck, S. Kang, J. Petit, and M. Tabibazar, J. Appl. Phys. **75**, 7949 (1994).

28. J. N. Shenoy, G. L. Chindalore, M. R. Melloch, J. A. Cooper, J. W. Palmour, and K. G. Irvine, *J. Electron. Mater.* **24**, 303 (1995).

29. M. J. Marinella, D. K. Schroder, T. Isaacs-Smith, A. C. Ahyi, J. R. Williams, G. Y. Chung, J. W. Wan, and M. J. Loboda, *Appl. Phys. Lett.* **90**, 253508 (2007).

30. X. J. Zhou, L. Tsetseris, S. N. Rashkeev, D. M. Fleetwood, R. D. Schrimpf, S. T. Pantelides, J. A. Felix, E. P. Gusev, and C. D'Emic, *Appl. Phys. Lett.* **84**, 4394 (2004).

31. X. J. Zhou, D. M. Fleetwood, J. A. Felix, E. P. Gusev, and C. D'Emic, *IEEE Trans. Nucl. Sci.* **52**, 2231 (2005).

32. X. J. Zhou, D. M. Fleetwood, L. Tsetseris, R. D. Schrimpf, and S. T. Pantelides, IEEE Trans. Nucl. Sci. **53**, 3636 (2006).

33. D. K. Schroder and J. A. Babcock, *J. Appl. Phys.* **94**, 1 (2003).

34. J. P. Campbell, P. M. Lenahan, C. J. Cochrane, A. T. Krishnan, and S. Krishnan, *IEEE Trans. Dev. Mater. Reliab.* **7**, 540 (2007).

35. J. H. Stathis and S. Zafar, *Microelectron. Reliab.* **46**, 270 (2006).

36. T. Grasser, W. Gos, and B. Kaczer, *IEEE Trans. Dev. Mater. Reliab.* **8**, 79 (2008).

37. D. M. Fleetwood, *Microelectron. Reliab.* **42**, 523 (2002).

38. S. Wang, M. Di Ventra, S. G. Kim, and S. T. Pantelides, *Phys. Rev. Lett.* **86**, 5946 (2001).

39. K. Fukuda, S. Suzuki, T. Tanaka, and K. Arai, *Appl. Phys. Lett.* **73**, 1585 (2000).

40. S. Dhar, L. C. Feldman, S. Wang, T. Isaacs-Smith, and J. R. Williams, *J. Appl. Phys.* **98**, 014902 (2005).

41. J. P. Campbell, P. M. Lenahan, C. J. Cochrane, A. T. Krishnan, and S. Krishnan, *IEEE Trans. Dev. Mater. Reliab.* **7**, 540 (2007).

42. D. T. Krick, P. M. Lenahan, and J. Kanicki, *J. Appl. Phys.* **64**, 3558 (1988).

43. D. J. DiMaria and J. H. Stathis, *J. Appl. Phys.* **70**, 1500 (1991).

44. S. S. Tan, T. P. Chen, J. M. Soon, K. P. Loh, C. H. Ang, and L. Chan, *Appl. Phys. Lett.* **82**, 1881 (2003).

45. P. M. Lenahan, *Proc. IRPS*, doi:10.1109/IRPS.2010.5488669, p. XT11 (2010)

46. A. J. Lelis, D. Habersat, G. Lopez, J. M. McGarrity, F. B. McLean, and N. Goldsman, *Mater. Sci. Forum* **527-529**, 1317 (2006).

47. A. J. Lelis, D. Haberstat, R. Green, A. Ogunniyi, M. Gurfinkel, J. Suehle, and N. Goldsman, *IEEE Trans. Electron Dev.* **55**, 1835 (2008).

48. A. J. Lelis, T. R. Oldham, H. E. Boesch, Jr., and F. B. McLean, *IEEE Trans. Nucl. Sci.* **36**, 1808 (1989)

49. D. M. Fleetwood, S. L. Kosier, R. N. Nowlin, R. D. Schrimpf, R. A. Reber, Jr., M. DeLau, P. S. Winokur, A. Wei, W. E. Combs, and R. L. Pease, *IEEE Trans. Nucl. Sci.* **41**, 1871 (1994).

50. S. T. Pantelides, S. N. Rashkeev, R. Buczko, D. M. Fleetwood, and R. D. Schrimpf, *IEEE Trans. Nucl. Sci.* **47**, 2262 (2000).

51. D. M. Fleetwood, W.L. Warren, J. R. Schwank, P. S. Winokur, M. R. Shaneyfelt, and L. C. Riewe, *IEEE Trans. Nucl. Sci.* **42**, 1698 (1995).

52. D. M. Fleetwood, M. J. Johnson, T. L. Meisenheimer, P. S. Winokur, W. L. Warren, and S. C. Witczak, *IEEE Trans. Nucl. Sci.* **44**, 1810 (1997).

53. M. R. Shaneyfelt, P. E. Dodd, B. L. Draper, and R. S. Flores, *IEEE Trans. Nucl. Sci.* **45**, 2584 (1998).

54. S. K. Dixit, S. Dhar, J. Rozen, S. Wang, R. D. Schrimpf, D. M. Fleetwood, S. T. Pantelides, J. R. Williams, and L. C. Feldman, IEEE Trans. Nucl. Sci. **53**, 3687 (2006).

55. D. M. Fleetwood and J. H. Scofield, Phys. Rev. Lett. **64**, 579 (1990).

56. T. R. Oldham and F. B. McLean, IEEE Trans. Nucl. Sci. **50**, 483 (2003).

57. J. R. Schwank, P. S. Winokur, P. J. McWhorter, F. W. Sexton, P. V. Dressendorfer, and D. C. Turpin, *IEEE Trans. Nucl. Sci.* **31**, 1434 (1984).

58. D. M. Fleetwood, P. S. Winokur, and J. R. Schwank, *IEEE Trans. Nucl. Sci.* **35**, 1497 (1988).

59. D. M. Fleetwood, P. S. Winokur, O. Flament, and J. L. Leray, *Appl. Phys. Lett.* **74**, 2969 (1999).

CHAPTER 3

EMERGING DIELECTRICS

382

Impact of Gate Dielectric Geometry on the Nanowire MOSFETs

Performance and Scaling

Ming-Fu Li, Wei Cao, D. M. Huang, Chen Shen, S. Q. Cheng, C. J. Yao, H. Y. Yu[#]

State Key Lab ASIC & Syst., Dept. Microelectronics, Fudan University, China, 200433,

mfli@fudan.edu.cn; dmhuang@fudan.edu.cn

[#]School of EEE, Nanyang Technological University, Singapore, 639798

In this paper, we discuss the unique property of electrostatics in the gate dielectric of the cylindrical nanowire (NW) MOSFETs, and its impact on the transistor performance and scaling, particularly on the gate tunneling leakage. A dielectric curvature parameter $\theta = T_{OX}/R_s$ with dielectric physical thickness T_{OX} and NW radius R_s is introduced. (1) A 2D tunneling model is developed and used to assess the tunneling rate reduction in cylindrical gate (CG) in NW transistor comparing with the planar gate (PG) double gate (DG) transistor. This effect can be very significant when θ is large in the practical NW devices. High-k gate dielectric is more effective to suppress the gate leakage in NW than in PG transistors, since High-k dielectric not only increases T_{OX}, but also increases θ, both reduce the tunneling rate. (2) On the other hand, comparing NW with radius R_s and DG-PG transistor with body thickness $T_{semi} = 2R_s$, the carrier quantization induced tunneling barrier reduction is larger in NW than in DG-PG transistors. For n-MOSFETs, the lowest conduction valley transition effect between Γ, L, and Δ (or X) valleys should also be considered due to carrier quantization in the channels. The gate tunneling rate is finally determined by combining two effects of (1) and (2). Different results due to different conduction band structures are demonstrated for Si, Ge, and InGaAs NWs respectively. (3) The overdrive gate voltage in NW transistor can be reduced significantly, comparing with PG transistor with the same T_{OX}. (4) For the device reliability assessment, at the same time evolution of dielectric charge generation, the NW transistor lifetime can be extended by as much as one order or more, comparing with the PG transistor with the same T_{OX}, when θ is large.

Introduction

Nanowire (NW) MOS transistor with gate all around (GAA) structure has attracted great attention in the recent years [1-11]. Due to its superior electric performance, it is possible to replace the conventional planar structure in scaling CMOS transistors to 10 nm gate length and below. In this work, we focus on the discussion of cylindrical NW transistors. The new physics in NW transistor can be classified into the following categories: (1) The better electrostatic control of the channel region and hence more effective suppression of short channel effect (SS, DIBL etc), due to the unique property of Poisson equation solution in the cylindrical channel [9,12,13]. (2) Quantum confinement effect in NW. From the low dimensional semiconductor point of view [14, 15], NW is a 1D semiconductor in different from 2D semiconductor in the double gate (DG) transistor or in the inversion layer in the bulk planar transistor. Therefore the quantum sub-band lift in NW is larger with very different density of states (DOS) in comparing with the 2D layer. It will affect the charge density in the inversion layer and therefore the quantum capacitance [16] and transport current [9, 10, 17, 18] performances. An excellent paper for quantum confinement effect with anisotropic effective mass is given in [9]. On the other hand, as we first raised in 2003 the idea of lowest conduction valley transition in the quantum confined DG transistor [19], the lowest conduction valley may be changed in very thin DG transistor in comparing with the bulk transistor, since the quantum sub-band lifts are different for different conduction valleys with different quantization masses. This effect could be more severe in NW transistors and has never been discussed. (3) The unique property of electrostatic in the gate dielectric, and its affect to the transistor performance and scaling, particularly the gate tunneling leakage. This has rarely been discussed and is the major topic to be investigated in this work.

2D Tunneling Model in the Cylindrical Gate

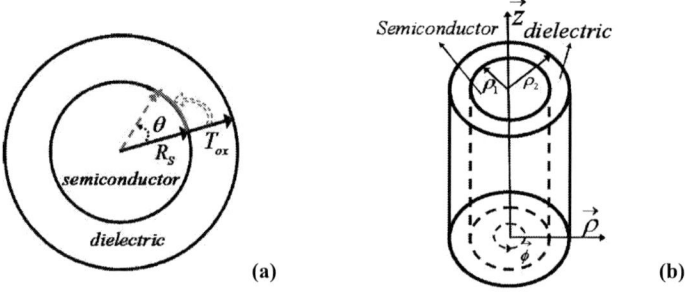

Fig. 1 Schematic diagram of (a): the cross section of a semiconductor nanowire with radius R_S and gate dielectric physical thickness T_{OX}. Projecting T_{OX} on the semiconductor surface circle to define the dielectric film curvature $\theta = T_{OX} / R_S$. (b): cylindrical coordinates and notations used in the calculation.

We first propose a figure of merit parameter $\theta = T_{ox}/R_s$ to describe the curvature of the gate dielectric in a transistor with a cylindrical gate (CG) as shown in Fig. 1. In practical case of existing NW transistor technology, the thinnest reported R_s is less than $1.5nm$

however the T_{OX} cannot be grown as thin as in the planar transistor, probably due to difficulty in growing uniform ultra-thin dielectric film on the NW. Table I lists the T_{OX} and R_S values reported by some leading institutes recently [1-7]. θ can be as large as 0.4-2.7. When $\theta \rightarrow 0$, the gate dielectric performance approaches the case of planar gate (PG) transistor. In this paper, PG transistor is defined as planar DG ultrathin body MOSFET with semiconductor body thickness $T_{semi} = 2R_S$, where R_S is the radius of the semiconductor NW. Both PG and CG transistors have similar low channel doping. Therefore, channel impurity charge can be neglected and the same Si inversion charge density $qN_{inv} = \varepsilon_{dielectric}F_{inv}$ induces the same dielectric field F_{inv} at the semiconductor/dielectric interface in both PG and CG cases. In this section, we neglect the carrier quantization effect in the channel, and will consider the carrier quantization effect in the following section.

TABLE I. θ values for seven existing Si NW transistors.

NW FET	Smallest R_{Si} (nm)	T_{ox} (nm)	Dielectric	θ
Ref. 1	5	2	SiO_2	0.4
Ref. 2	4	3.5	SiO_2	0.88
Ref. 3	1.5-3	4	SiO_2	1.3-2.7
Ref. 4	5	2.5	SiO_2	0.5
Ref. 5	5	3.5	SiO_2	0.7
Ref. 6	$3^{\$\$}$	EOT=1.5	Hf-based	2 (estimated)
Ref.7	**1.3**	**$2^{\$\$}$**	**Hf-based**	**1.5**

$^{\$\$}$ estimated from TEM Figures

Assuming no charge in the gate dielectric, the barrier potential $V(\rho)$ in the gate dielectric in CG transistor can be solved directly from Poisson's equation

$$V(\rho) = E_{offset} - qF_{inv}\frac{T_{ox}}{\theta}\ln\left(\frac{\theta\rho}{T_{ox}}\right), \quad R_s < \rho < R_s + T_{ox}, \tag{1}$$

As shown in Fig. 2, comparing with the case of PG transistor at the same surface field F_{inv}, i.e. the same Si surface inversion charge density qN_{inv}, the electron tunneling barrier is increased in the CG transistor as θ increases. Since the tunneling is extremely sensitive to the change of barrier, we expect a significant reduction of tunneling as θ increases.

To develop a compact model for electron tunneling in CG dielectric, we solve the 2D Schrodinger equation in the cylindrical coordinates and neglect the potential variation along the z axes when $V_D = V_S = 0$. Since the gate dielectric is in the amorphous state, the tunneling effective mass of the dielectric is an isotropic scalar m_T. The electron wave function in the dielectric between ρ_1 and ρ_2 (Fig. 1(b)) is plane wave in the z direction.

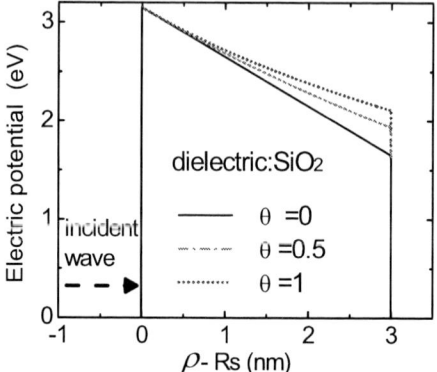

Fig. 2 Oxide barrier potential energy $V(\rho)$ for NW transistors with different θ and PG transistor ($\theta = 0$), all with the same dielectric field $F_{inv} = 5MV/cm$ at the Si/SiO$_2$ interface $\rho = R_S$. $V(\rho)$ is expressed by Eq. (1).

The radial and angular wave function is $R(\rho)\,\Phi(\phi)$ which satisfies the following Schroedinger equation:

$$-\frac{\hbar^2}{2m_T}\left[\Phi\frac{d^2R}{d\rho^2}+\frac{\Phi}{\rho}\frac{dR}{d\rho}+\frac{R}{\rho^2}\frac{d^2\Phi}{d\phi^2}\right]+R\Phi[V(\rho)-E]=0,\tag{2}$$

where E is the energy of electron. By variable separation, the solution of Eq. (2) in angle part is simply $\Phi = e^{jm\phi}$, where m is an integer. Moving on to get the radial solution, considering $V(\rho) = V$ as a constant approximately in ρ to $\rho+\Delta\rho$, and in the dielectric tunneling region $E < V$, we define the decay length λ:

$$\frac{1}{\lambda}=\frac{\sqrt{2m_T(V-E)}}{\hbar},\tag{3}$$

$$x=\rho/\lambda,\tag{4}$$

Changing variable, the radial wave function satisfies:

$$x^2\frac{d^2R_m}{dx^2}+x\frac{dR_m}{dx}-(x^2+m^2)R_m=0\tag{5}$$

Eq. (5) is the well known Bessel equation. The solution can be expressed as linear combination of modified Bessel functions [20]. Noting the boundary condition at $x = \infty$ is $R(\infty) = 0$, only the modified Bessel function of second kind K(x) is possible [20], therefore

$$R(\rho) = K_m\left(\frac{\rho}{\lambda}\right) \approx \sqrt{\frac{\pi\lambda}{2\rho}}\, e^{-\frac{\rho}{\lambda}},$$

(6)

where $K_m(x)$ is the mth order modified Bessel function. The approximation is valid for $(\rho/\lambda) \gg 1$ (Fig. 3). For Si NW with SiO$_2$ gate dielectric, $(V_0\text{-}E) \approx E_{offset} = 3.12\ eV$, $m = m_T = 0.5m_0$ [21], therefore $\lambda = 0.155\ nm$. Assuming the minimum $R_s = 1.5\ nm$, $(\rho/\lambda) \geq 9.7 \gg 1$. Thus, the probability of finding electron in the gate dielectric decays as $|R(\rho)|^2 \propto [\lambda/\rho]\exp(-2\rho/\lambda)$. In the case of arbitrary barrier with finite width, the direct tunneling rate through the dielectric from the circle with radius R_s to the circle with radius $R_s + T_{ox}$ should be

$$T_{WKB} = \frac{(R_s + T_{ox})\left|R(R_s + T_{ox})\right|^2}{R_s\left|R(R_s)\right|^2} \approx \exp\left(-2\int_{R_s}^{R_s + T_{ox}} \sqrt{2m_T[V(\rho) - E]}/\hbar\, d\rho\right),$$

(7)

which has the same formula as 1D tunneling under WKB approximation [22].

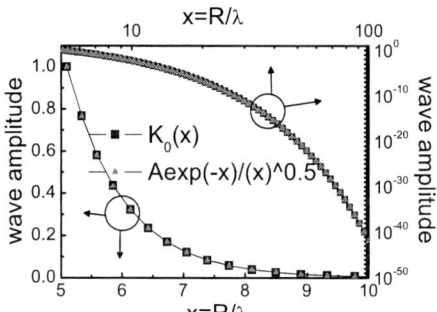

Fig. 3 In the dielectric region, Eq. (6) is a perfect approximation of $K_0(x)$ as shown in the Figure. For the case of SiO$_2$ on Si, the range of $x = 5\text{-}100$ in the figure corresponds to Si NW $R_s = 0.78\ nm$ with SiO$_2$ T_{ox} up to $14\ nm$.

Further, the tunneling rate from channel to gate should include a factor T_R of reflection from both semiconductor/dielectric and dielectric/gate interfaces. In the PG case [22]:

$$T_R = \left(\frac{4k_1\kappa_1}{k_1^2 + \kappa_1^2}\right)\left(\frac{4k_2\kappa_2}{k_2^2 + \kappa_2^2}\right),$$

(8)

where $k_1 = \sqrt{2m_{semi}E}/\hbar$, $\kappa_1 = \sqrt{2m_{ox}(E_{offset} - E)}/\hbar$, $\kappa_2 = \sqrt{2m_{ox}(E_{offset} - E - V_{ox})}/\hbar$, $k_2 = \sqrt{2m_{semi}(V_{ox} + E)}/\hbar$. For the CG, the ground state solution of the Schroedinger Eq. (2)

in Si is the Bessel function $J_0(\rho/\lambda)$ or the combination of the Hankel functions $H_0^{(1)}(\rho/\lambda)$ and $H_0^{(2)}(\rho/\lambda)$ [20] which can be replaced by their asymptotic forms:

$$H_0^{(1)}\left(\frac{\rho}{\lambda}\right) \approx \sqrt{\frac{2\lambda}{\pi\rho}}e^{i[\frac{\rho}{\lambda}-\pi/4]}, \qquad H_0^{(2)}\left(\frac{\rho}{\lambda}\right) \approx \sqrt{\frac{2\lambda}{\pi\rho}}e^{-i[\frac{\rho}{\lambda}-\pi/4]} \tag{9}$$

for $\rho \geq R_s \gg \lambda$ as in our case. Following the derivation for the PG in Ref. [22], using Eq. (9) and (1), the T_R is demonstrated to have the same form of Eq. (8) for both CG and PG.

Compact Model of Tunneling Rate in CG

We use the following approximation for $\sqrt{V(\rho)-E}$ in Eq. (7):

$$\sqrt{V(\rho)-E} \approx \sqrt{E_{offset}-qF_{inv}(\rho-R_s)-E} + \frac{1}{2}qF_{inv}\frac{(\rho-R_s)-R_s\ln(\rho/R_s)}{(E_{offset}-qF_{inv}T_{ox}\chi-E)^{1/2}} \qquad \rho_1<\rho<\rho_2 \tag{10}$$

The second term in the RHS is a correction term approaching zero when θ approaching zero. The parameter $\chi = 0.618$ is used to best fit $\sqrt{V(\rho)-E}$ for all practical cases. Using Eq. (10), the channel to gate tunneling rate can be expressed in a compact form,

$$T_{CG} = T_R T_{PG,WKB}\exp\left\{-\frac{\sqrt{m}qF_{inv}R_s^2\left[\theta(2+\theta)-2(1+\theta)\ln(1+\theta)\right]}{\sqrt{2}\hbar(E_{offset}-qF_{inv}T_{ox}\chi-E)^{1/2}}\right\} \tag{11}$$

Here $T_{PG,WKB}$ is the T_{WKB} in Eq. (7) for PG with $\sqrt{V(\rho)-E} = \sqrt{E_{offset}-qF_{inv}(\rho-R_s)-E}$ as given by [22]. As demonstrated in Fig. 4, the analytical results calculated by (11) are in overall excellent agreement with the numerical calculations using Eqs. (1), (7), and (8).

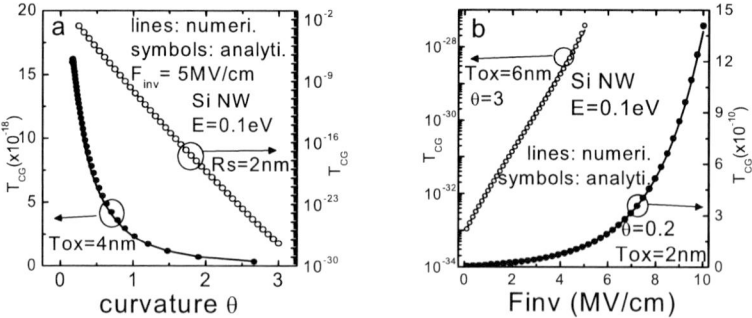

Fig. 4 Comparison of tunneling rate T_{CG} calculated by numerical and compact analytical model respectively. (a) T_{CG} versus θ with fixed R_s or fixed T_{ox}. (b) T_{CG} versus dielectric field at Si/dielectric interface with two very different θ values.

Impact of Dielectric Curvature on CG Tunneling Rate Reduction

Figs. 5-8 show the gate tunneling rate reduction for Si NW n-MOSFETs with SiO_2, Si_3N_4, and HfO_2 gate dielectrics [21, 23, 24] respectively. The interfacial layer between Si and HfO_2 is ignored for simplicity. Figs. 9-12 show the similar results for Ge NW transistors with HfO_2 gate dielectric [25], and $In_{0.7}Ga_{0.3}As$ NW transistors with HfO_2 and Al_2O_3 gate dielectric [27-29] respectively. Table II lists the tunneling parameters. Overall the gate tunneling rate reduction is significant and more reduction is observed with increasing θ, and the reduction can exceed one order of magnitude. HK dielectric has more effective suppression of gate tunneling in CG transistors than in PG transistors, since for fixed nanowire radius and gate dielectric EOT in CG, using HK not only increases the physical dielectric thickness T_{OX}, but also increases θ, both reduce the gate tunneling rate.

Table II. Parameters used for tunneling rate calculations, m_T and dielectric constant ε from [21, 23, 24], E_{offset} from [21, 23, 24, 26, 29].

Semiconductor	Dielectric	E_{offset} (eV)	ε	m_T/m_O
Si	SiO_2	3.12	3.9	0.5
	Si_3N_4	2.1	7.8	0.5
	HfO_2	1.9	22	0.18
Ge	HfO_2	1.68	22	0.18
$In_{0.7}Ga_{0.3}As$	HfO_2	2.19	22	0.18
	Al_2O_3	2.4	9	0.28

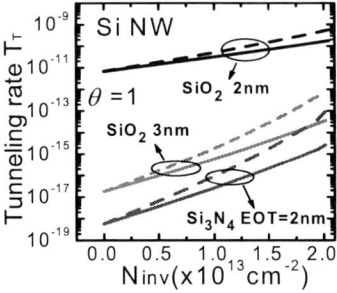

Fig. 5 Electron gate tunneling rate versus Si surface inversion charge density N_{inv} for SiO_2 and Si_3N_4 on Si with planar gate (dash line) and cylindrical gate (solid line) with curvature $\theta=1.0$.

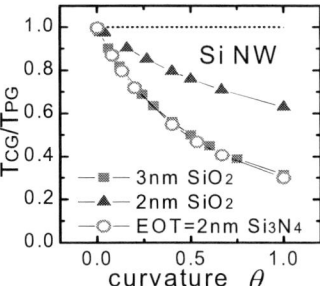

Fig. 6 Gate tunneling rate ratio T_{CG}/T_{PG} between CG and PG transistors (with same T_{OX}) versus curvature θ for SiO_2 and Si_3N_4 on Si transistors, at the surface inversion charge electron density of $1x10^{13}$ cm^{-2}.

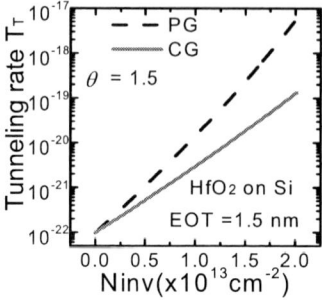

Fig. 7 The same as Fig. 5, however for HfO$_2$ dielectric of PG and CG with $\theta=1.5$.

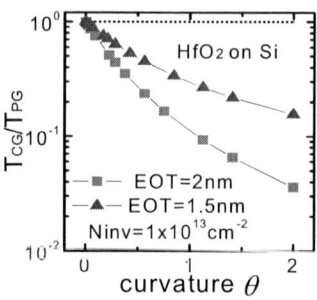

Fig. 8 The same as Fig. 6, however for HfO$_2$ dielectric.

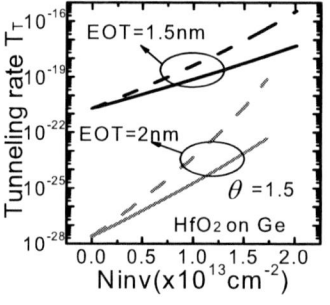

Fig. 9 The same as Fig. 7, however for Ge NW transistors.

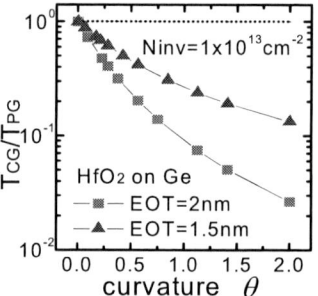

Fig. 10 The same as Fig. 8, however for Ge NW transistors.

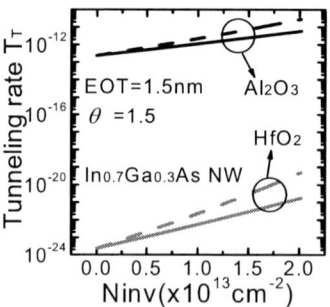

Fig. 11 The same as Fig. 7, however for HfO$_2$ and Al$_2$O$_3$ dielectrics on In$_{0.7}$Ga$_{03}$As NW transistors.

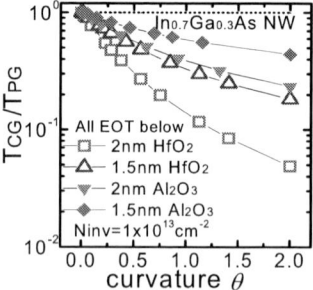

Fig. 12 The same as Fig.8, however for HfO$_2$ and Al$_2$O$_3$ dielectrics on In$_{0.7}$GaAs$_{0.3}$ NW transistors.

Impact of Quantum Confinement on the Gate Tunneling Rate Increment

Electron quantum confinement in NW or DG transistor channel will cause the conduction sub-band lift ΔE_C (energy quantization) [15] and gate tunneling barrier height reduction from E_{offset} to $E_{offset} - \Delta E_C$. Consequently, the gate tunneling rate is increased and the results in the last section should be modified. Comparing the NW transistor with radius R_S and the DG-PG transistor with semiconductor film thickness $T_{semi}=2R_S$, the energy quantization ΔE_C is different. We follow [9] using the approximations of infinitely high barrier and flat bottom in the well, and energy quantization calculation with isotropic or anisotropic effective mass Schrodinger equations. There are two facts different in energy quantization between NW and DG cases: (1) the energy quantization level is determined by sine or cosine wave function node at the boundary in DG case, however by Bessel or Mathieu function node at the boundary in NW case. (2) For anisotropic effective mass, the quantization mass is different in NW and DG cases. These are explained in the following.

We use X and Y coordinates in the NW to replace ρ, θ in Fig. 1(b). For DG- PG case, consider the semiconductor film vertical to Z. For specific, Z is along the <100> direction in the cubic semiconductor in both NW and DG cases. Following [9], for those valleys with isotropic effective mass m_x in the X-Y plane, ΔE_C in the NW can be expressed by

$$\Delta E_C = \Delta E_{mn} = \frac{2\hbar^2}{m_x R_s^2} q_{mn} , \qquad (12)$$

where q_{mn} is determined by the boundary condition

$$J_m(2\sqrt{q_{mn}}) = 0 , \qquad \text{for } m=0, 1, 2 .... \qquad (13)$$

here $J_m(\rho)$ denotes the Bessel function of the first kind. The subscript n indicates that $2\sqrt{q_{mn}}$ is nth zero of $J_m(\rho)$. There is an extra twofold degeneracy for $m \neq 0$.

For valleys with anisotropic effective mass $m_x \gg m_y$ in the X-Y plane, ΔE_C can be expressed by

$$\Delta E_C = \Delta E_{mn}^{(l)} = \frac{2\hbar^2}{(m_x - m_y)R_s^2} q_{mn}^{(l)} , \qquad (14)$$

where $q^{(l)}_{mn}$ ($l = c \text{ or } s$) can be obtained from the following

$$C_m(\xi_b, q^{(c)}_{mn})=0, \qquad \text{for m=0, 1, 2......} \qquad (15)$$
$$\text{Or} \quad S_m(\xi_b, q^{(s)}_{mn})=0, \qquad \text{for m=1, 2, 3 ......} \qquad (16)$$

Here $C_m(\xi_b, q)$ and $S_m(\xi_b, q)$ denote the modified Mathieu functions with order m of cosine and sine types, respectively [30]. $\xi_b = \cosh^{-1}\sqrt{m_x/(m_x - m_y)}$. The subscript n indicates the nth zero of wave function $C_m(\xi_b, q^{(c)}_{mn})$ and $S_m(\xi_b, q^{(s)}_{mn})$ at the NW boundary. The values of q_{mn} and $q^{(l)}_{mn}$ can be obtained from MATHEMATICA.

391

On the other hand, the quantized energy in DG-PG case with $T_{semi}=2R_s$ is determined by the node of sine wave function with the effective mass along the z direction [15]:

$$\Delta E_C = \Delta E_n = \frac{2\hbar^2}{m_z R_s^2}(0.617)n^2, \qquad n=1, 2, 3\ldots \qquad (17)$$

Table III. Electron effective masses and valley energies of bulk Si, Ge, GaAs and InAs used in energy quantization calculation [19,32-34]. For DG transistor, the surface orientation is [100]. The quantization mass is m_z. For NW transistor, the cylinder orientation is <100>. The quantization mass for anisotropic effective mass is m_x - m_y, m_x, m_y, m_z are calculated by the method in [31] table I. For L valley, when rotate x and y axes 45° along the z axes to align to $<1,\bar{1},0>$ and $<1,1,0>$ directions, $m_x=m_t$, $m_y=(m_t+2m_l)/3$, $m_z=(3m_tm_l)/(m_l+2m_t)$.

	ml/mt (m₀)	valley	E - E$_v$ (eV)	m$_X$ (m₀)	m$_Y$ (m₀)	m$_Z$ (m₀)	g
Si	0.916/0.16	Δ_2	1.12	0.19	0.19	0.916	2
		Δ_4	1.12	0.916	0.19	0.19	4
Ge	1.56/0.08	L_4	0.66	1.066	0.08	0.117	4
	0.038	Γ	0.8	0.038	0.038	0.038	1
	0.95/0.2	Δ_2	0.82	0.20	0.20	0.95	2
		Δ_4	0.82	0.95	0.20	0.20	4
GaAs	0.067	Γ	1.42	0.067	0.067	0.067	1
	1.9/0.075	L_4	1.71	1.29	0.075	0.11	4
	1.9/0.19	X_1	1.9	0.19	0.19	1.9	1
		X_2		1.9	0.19	0.19	2
InAs	0.023	Γ	0.36	0.023	0.023	0.023	1
	1.543/0.094	L_4	1.08	1.06	0.094	0.137	4
	1.126/0.175	X_1	1.37	0.175	0.175	1.126	1
		X_2		1.126	0.175	0.175	2

We like to mention that the flat bottom approximation used to derive Eqs. (12-17) is good for small R_s and low field. However in very large R_s and high field limit, the quantum confinement should be determined by two decoupled triangular potential approximation at two sides of DG or NW surface. However, when R_s is large, θ approaches zero and is out of our interests in this work.

In the following, we shall discuss the Si, Ge, and InGaAs NWs separately. The conduction band structure data are listed in Table III. The corresponding q values are listed in Table IV.

	valley	ξ_b	q_{01}	q_{11}	$q_{01}^{(c)}$	$q_{11}^{(c)}$	$q_{21}^{(c)}$	$q_{11}^{(s)}$
Si	Δ_2		1.446	3.671				
	Δ_4	0.4915			3.278	5.517	8.569	11.027
Ge	L_4	0.2811			9.159	12.677	17.049	33.293
	Γ		1.446	3.671				
	Δ_2		1.446	3.671				
	Δ_4	0.4958			3.227	5.449	8.483	10.831
In$_{0.7}$Ga$_{0.3}$As	Γ		1.446	3.671				
	L_4	0.2868			8.823	12.285	16.6	34.389
	X_1		1.446	3.671				
	X_2	0.3392			6.467	9.493	13.373	23.016

Table IV. The characteristic values ξ_b and lowest q_{mn} and $q''_{mn}^{(l)}$ in Eqs. (12, 14) for Si, Ge, and In$_{0.7}$Ga$_{0.3}$As NW.

(1) Si NW with CG and DG with PG :

The bulk Si has six Δ lowest conduction valleys [15, 33]. ΔE_c determined by (12), (14), (17) and Tables III, IV are plotted in Fig. 13. The next lowest L valley is located at $E_L - E_v = 2.0\ eV$ and is too high to be considered. Tunneling rate T_T in the last section should be modified by replacing the barrier E_{offset} in Eqs. (1), (7), (10) and (11) by $E_{offset} - \Delta E_c$. The results are plotted in Figs. 14 and 15. T_T is finally determined by two effects in the opposite direction: the sub-band lift effect tends to reduce the tunneling barrier and increase T_T (denoted by ΔE_c effect), and the dielectric curvature θ effect tends to reduce T_C (denoted by θ effect) as indicated in the last section. θ effect is stronger when N_{inv} is higher as indicated in Figs. 5, 7, 9, 11. ΔE_C effect is stronger when R_s is smaller as shown in Fig. 13.

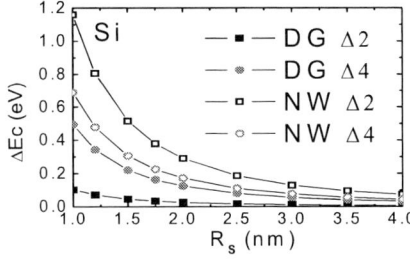

Fig. 13 Sub-band energy lift ΔE_C in Si NW with radius R_s or in DG Si film with thickness $2R_S$. Only lowest sub-bands of Δ valleys are plotted.

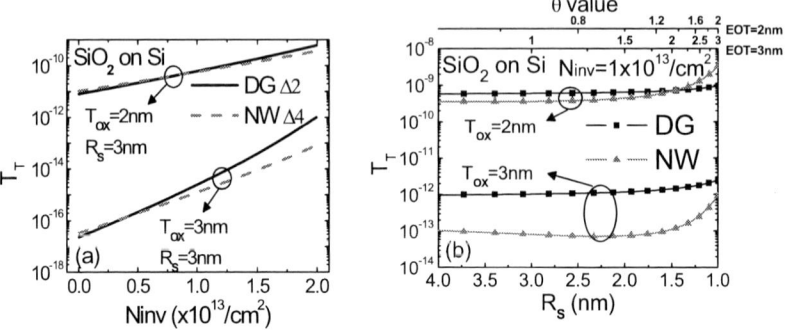

Fig.14 (a) Electron gate tunneling rates T_T versus Si surface inversion charge density N_{inv} for SiO$_2$ on Si with CG-NW (dashed line) with $R_s=3nm$, $\theta=0.67$ and 1 respectively, and with DG-PG (solid line) with same T_{ox} and $T_{semi}=2R_s$. In the low N_{inv} range, T_T is higher in NW than in DG since the larger ΔE_c effect in NW plays the major role in determining T_T, while in the high N_{inv} range, T_T is lower in NW than in DG since the θ effect in NW plays the major role. (b) The same as (a), however T_T versus R_S. In the most R_S range, T_T is lower in NW than in DG. The θ effect overrides the ΔE_c effect, although ΔE_c is larger in NW than in DG as indicated in Fig. 13.

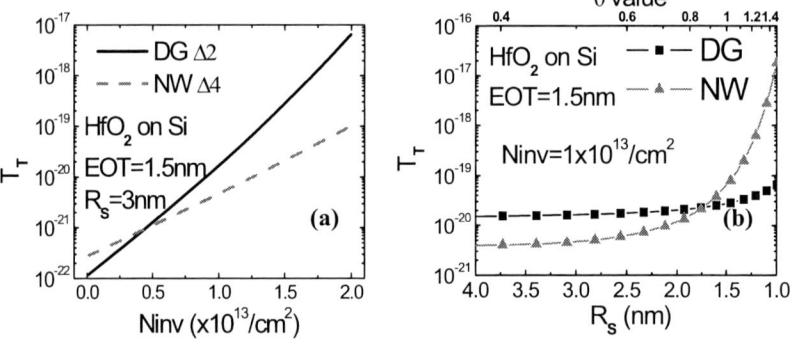

Fig. 15 The same as Fig. 14, however for HfO$_2$ gate dielectric on Si.

(2) Ge NW with CG and DG with PG:

The bulk Ge has four lowest L valleys [15, 33]. Due to the lowest conduction valley transition effect in low dimensional semiconductors [19], we should also consider six Δ valleys with higher energy. Using data in Tables III, IV, the ΔE_c calculation results are plotted in Fig. 16. Note that when R_s is less than 2.2 nm , the lowest conduction valleys are no longer L valleys as in the bulk Ge, but Δ_2 valleys in DG or Δ_4 valleys in NW. The corresponding gate tunneling rates are shown in Fig. 17.

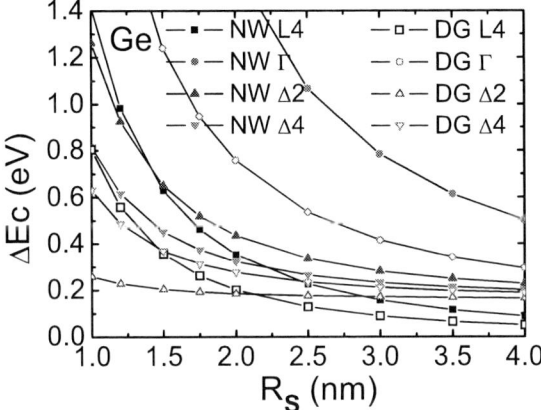

Fig. 16 Sub-band energy lift ΔE_C in Ge NW with radius R_s or in DG Ge film with thickness $T_{Ge} = 2R_S$. Lowest sub-bands of L and Δ valleys are plotted. Note that Δ_4 valley in NW (or Δ_2 valley in DG) crosses the L_4 valley at around *2 nm* of R_s (or *4nm* of T_{Ge} in DG) and become the main occupied lowest valley for smaller R_s.

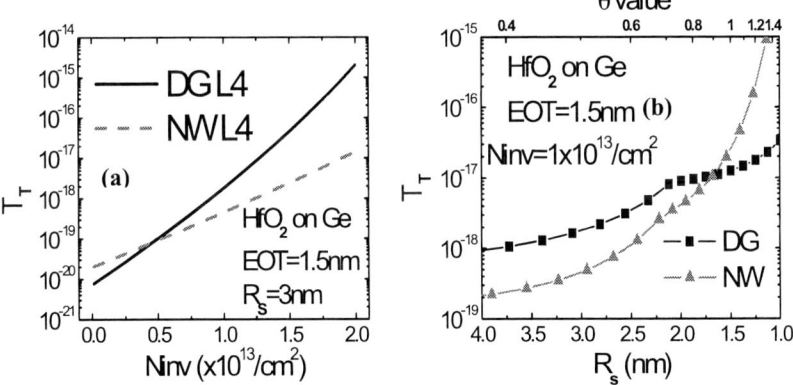

Fig. 17 The same as Fig. 14, however for HfO_2 gate dielectric on Ge.

(3) $In_{0.7}Ga_{0.3}As$ NW with CG and DG with PG :

The effective masses and the valley energies $E\text{-}E_v$ used for $In_{0.7}Ga_{0.3}As$ are obtained by linear interpolation of the data in Table III for GaAs and InAs respectively. The bulk $In_{0.7}Ga_{0.3}As$ has lowest isotropic Γ valley [15, 33]. Due to the lowest conduction valley transition effect, L and X valleys should also be considered [32]. The ΔE_c and T_T calculation results are plotted in Figs. 18-20.

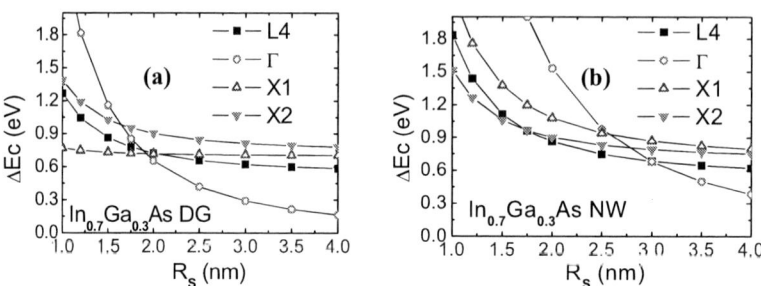

Fig. 18 Sub-band energy lift ΔE_C in In$_{0.7}$Ga$_{0.3}$As DG transistor with thickness T_{InGaAs}= $2R_S$ (plot (a)), or in NW transistor with radius R_s (plot (b)). Lowest sub-bands of Γ, L and X valleys are plotted. Note that L$_4$ valley in NW (or X$_1$ valley in DG) crosses the Γ valley at around *3 nm* of R_s in NW (or *4nm* of T_{Ge} in DG) and become the main occupied lowest valleys for smaller R$_s$.

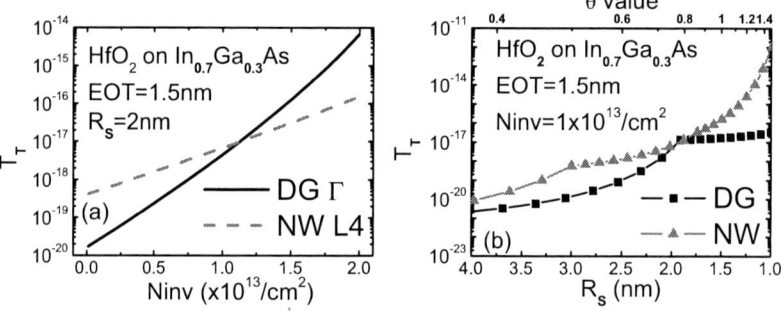

Fig. 19 The same as Fig. 14, however for HfO$_2$ on In$_{0.7}$Ga$_{0.3}$As. In (b), the ΔE_C effect plays a major role in all range of R_S therefore T_T is higher in NW than in DG, since as indicated in Fig. 18, ΔE_C is very large due to the small effective mass of Γ valley in InGaAs, and ΔE_C is larger in NW than in DG.

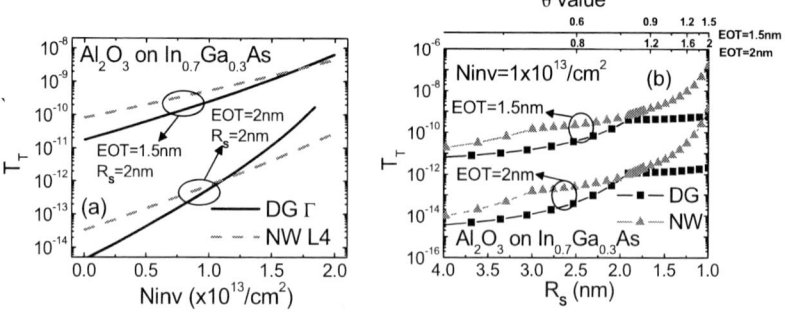

Fig. 20 The same as Fig. 19, however for Al$_2$O$_3$ on In$_{0.7}$Ga$_{0.3}$As.

Reduction of Operation Gate voltage

From Eq. (1), the overdrive gate voltage $V_{god}=V_g-V_{th}$ required to induce the same channel surface inversion charge density Q_{inv} is reduced significantly in the CG compared to the PG transistor. For instance for Si transistors, when $T_{OX}=3nm$ and $\theta=1$, the reduction of V_{god} can be more than $0.4V$ at strong inversion. This is shown in Fig. 21, and is beneficial to low voltage application.

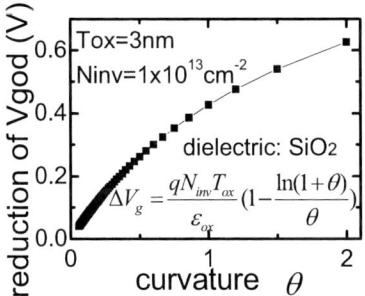

Fig. 21 Comparing with the PG transistor with same T_{OX} and surface charge inversion, the overdrive gate voltage $V_{god}=V_g-V_{th}$ of the NW transistor is significantly reduced when increasing θ.

Impact on the Transistor Reliability Performance

Kufluoglu et al have analyzed the interface trap generation in the NW transistors under the negative bias temperature instability (NBTI) stress [35]. Their analysis based on the H_2 reaction-diffusion (R-D) model indicates that due to the 2D diffusion of H_2 in the NW gate oxide, the curvature of the oxide semiconductor interface effectively increases the diffusion rate while slow down the annealing , causing the increase of NBTI damage and reduction of device lifetime. However, the dielectric electrostatic analysis in this paper indicates that one should also consider another opposite effect. When there is interface trap charge Q_{it} or oxide charge Q_{OX} created under stress, located at the Si-SiO$_2$ interface or at a distance d from the interface, it induces a threshold voltage shift δV_{th}. It is the major reliability concern under stress [36]. In the PG transistor, δV_{th} is proportional to the charge moment to the gate as expressed by Eq. (18P). However in the CG transistor, δV_{th} is expressed by Eq. (18C). δV_{th} is not simply related to the charge moment, but also related to the dielectric film curvature θ.

$$[\delta V_{th}]_{PG} = (\frac{T_{ox}-d}{\varepsilon_{ox}})\delta Q_s , \qquad \text{for Planar Gate transistor} \qquad (18P)$$

$$[\delta V_{th}]_{CG} = \frac{((T_{ox}/\theta)+d)}{\varepsilon_{ox}}\ln(1+\frac{T_{ox}-d}{(\frac{T_{ox}}{\theta})+d})\delta Q_s , \qquad \text{for cylindrical gate transistor} \qquad (18C)$$

Overall this is good news showed in Fig. 22 that for same Q_{it} or Q_{OX} charge density, δV_{th} can be significantly reduced with increasing θ. Assuming the same power law time evolution of Q generation under stress, the corresponding lifetime of the NW device is much longer than the planar device as shown in Fig. 23.

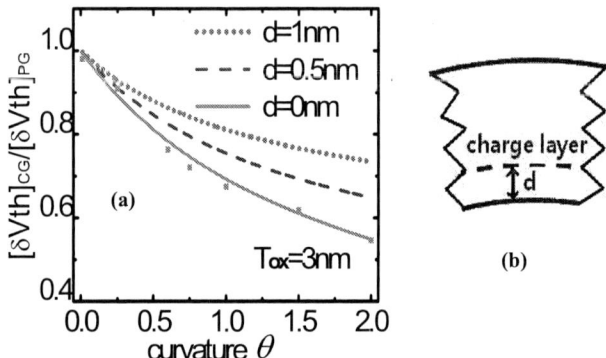

Fig. 22 Comparing with the planar transistor with same T_{OX}, the threshold voltage shift δV_{th} due to same amount of interface trap charge or oxide charge δQ_S generation under stress, is reduced in the CG transistor with increasing θ, as indicated in Eqs. (18P) and (18C), and plotted in (a). d is the distance between the oxide charge layer and the semiconductor surface as shown in (b). For the case of interface traps, $d=0$. The square data in the Figure are numerical simulation results by SENTAURUS program.

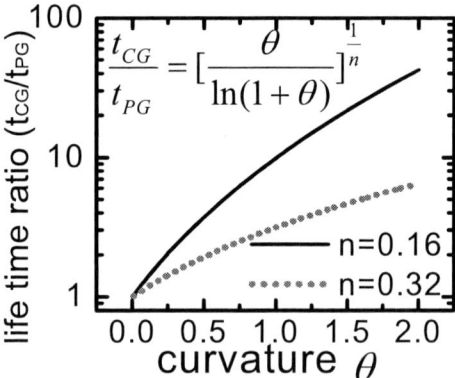

Fig. 23 Assuming the same power law time evolution At^n of interface trap or oxide charge generation under stress, the dielectric curvature effect can prolong the NW transistor lifetime comparing to the planar transistor.

Summary

The impact of gate dielectric curvature on the cylindrical NW transistor performance and scaling is demonstrated. A dielectric curvature parameter $\theta = T_{OX}/R_S$ with dielectric physical thickness T_{OX} and NW radius R_S is proposed to assess the effects quantitatively. A compact analytical 2D tunneling model in the NW cylindrical gate (CG) dielectric is derived. The method is used to assess the performance and scaling of Si, Ge, and InGaAs NW n-MOSFETs. Compared with the planar gate (PG) transistors with the same T_{OX}, the NW transistor with practically achievable θ value has the following characteristics: (1) Overall the gate tunneling rate is significantly reduced when θ is increased. The use of HK dielectric is more effective to reduce the gate tunneling in NW transistors than in the planar transistors, since High-k dielectric not only increases T_{OX}, but also increases θ, both reduce the tunneling rate. The reduction due to dielectric curvature can be more than one order of magnitudes. (2) On the other hand, the carrier quantization induced sub-band lift ΔE_C and corresponding tunneling barrier reduction and tunneling rate increment are investigated in NW with radius R_S, and comparing with the double gate planar gate (DG-PG) transistor with body thickness $T_{SEMI} = 2R_S$. The ΔE_C induced tunneling barrier reduction is larger in NW than in DG-PG transistors, causing more tunneling rate increment in NW. The lowest conduction valley transition effect between Γ, L and Δ (or X) should also be considered due to carrier quantization in the channels. The gate tunneling rate is finally determined by combining two effects of (1) and (2). Different results due to different band structures are demonstrated for Si, Ge, and InGaAs NWs respectively. (3) The overdrive gate voltage can be reduced significantly when θ is increased, beneficial to low voltage application. (4) At the same time evolution of interface trap or oxide charge generation, the device life time can be extended by as much as one order of magnitude or more in NW than in planar transistors.

Acknowledgements

This work was supported by the National Natural Science Foundation of China project # 60936005, the National VLSI project # 2009ZX02035, and the Micro/Nano-electronics Science and Technology Innovation Platform of Fudan University. The work in Singapore was supported by A*STAR Singapore. The authors would like to thank Drs. N. Singh, Partick Lo and D. L. Kwong for useful technical discussion and Dr. Tony Low for valuable theoretical discussion.

References

1. S. D. Suk, S. Y. Lee, S. M. Kim, E. J. Yoon, M. S. Kim, M. Li, C. W. Oh, K. H. Yeo, S. H. Kim, D. S. Shin, K. H. Lee, H. S. Park, J. N. Han, C. J. Park, J. B. Park, D. W. Kim, D. Park, and B. I. Ryu, *IEDM Tech. Dig.*, p. 717 (2005).
2. K. H. Yeo, S. D. Suk, M. Li, Y. Y. Yeoh, K. H. Cho, K. H. Hong, S. K. Yun, M. S. Lee, N. Cho, K. Lee, D. Hwang, B. Park, D. W. Kim, D. Park, and B. I. Ryu, *IEDM Tech. Dig.*, p. 539, (2006).
3. N. Singh, F. Y. Lim, W. W. Fang, S. C. Rustagi, L. K. Bera, A. Agarwal, C. H. Tung, K. M. Hoe, S. R. Omampuliyur, D. Tripathi, A. O. Adeyeye, G. Q. Lo, N. Balasubramanian, and D. L. Kwong, *IEDM Tech. Dig.*, p. 548, (2006).
4. M. Li, K. H. Yeo, S. D. Suk, Y. Y. Yeoh, D. W. Kim, T. Y. Chung, K. S. Oh, and W. S. Lee, *VLSI-TSA.*, p. 94 (2009).
5. J. Zhuge, R. S. Wang, R. Huang, J. B. Zou, X. Huang, D. W. Kim, D. Park, X. Zhang, and Y. Y. Wang, *IEDM Tech. Dig.*, p. 61 (2009).
6. S. Bangsaruntip, G. M. Cohen, A. Majumdar, Y. Zhang, S. U. Engelmann, N. Fuller, L. M. Gignac, S. Mittal, J. S. Newbury, M. Guillorn, T. Barwicz, L. Sekaric, M. M. Frank, and J. W. Sleight, *IEDM Tech. Dig.*, p. 297, (2009).
7. S. Bangsaruntip et al, *Symp. VLSI Tech.*, p. 21 (2010).
8. N. Singh, K. D. Budharaju, S. K. Manhas, A. Agarawal, S. C. Rustagi, G. Q. Lo, N. Balasubramanian, and D. L. Kwong, *IEEE Trans. Electron Devices*, **55**, 3102 (2008).
9. B. Yu, L. Q. Wang, Y. Yuan, P. M. Asbeck, and Y. Taur, *IEEE Trans. Electron Devices*, **55**, 2846 (2008).
10. J. Appenzeller, J. Knoch, M. T. Bjork, H. Riel, H. Schmid and W. Riess, *IEEE Trans. Electron Devices*, **55**, 2827 (2008).
11. Papers in Special issue on nanowire transistors: modeling, device design, and technology, *IEEE Trans. Electron Devices,* **55**, no. 11 (2008).
12. C. P. Auth and J. D. Plummer, *IEEE Electron Device Lett.*, **18**, 74 (1997).
13. D. Jimenez, B. Iniguez, J. Sune, L. F. Marsal, J. Pallares, J. Roig and D. Flores, *IEEE Electron Device Lett.*, **25**, 571 (2004).
14. L. Esaki, in *Electronic properties of multilayers and low dimensional semiconductor structures*, NATO ASI series B., Vol.231, p.3, eds J. M. Chamberlain, L. Eaves, J. C. Portal, Plenum (1990).
15. Ming-Fu Li, *Modern Semiconductor Quantum Physics*, World Scientific (1994).
16. S. Luryi, *Appl. Phys. Lett.*, **52**, 501 (1988).
17. S. Jin, M. V. Fischetti and T. W. Tang, *IEEE Trans. Electron Devices.*, **55**, 2886 (2008).
18. A. Rahman, J. Guo, S. Datta and M. S. Lundstrom, *IEEE Trans. Electron Devices*, **50**, 1653 (2003).
19. Tony Low, Y. T. Hou, M. F. Li, C. X. Zhu, A. Chin, G. Samudra, L. Chan and D. L. Kwong , *IEDM Tech. Digest*, p. 691 (2003).
20. Z. X. Wang and D. R. Guo, *Special Functions,* World Scientific (1989).
21. Y. T. Hou, M. F. Li, Y. Jin, and W. H. Lai, *J. Appl. Phys.*, **91**, 258 (2002).

22. F. Li, S. P. Mudanai, Y. Y. Fan, L. F. Register, and S. K. Banerjee, *IEEE Trans. Electron Devices*, **53**, 1096 (2006).
23. Y. T. Hou, M. F. Li, H. Y. Yu, Y. Jin, and D. L. Kwong, *IEDM Tech. Dig.*, p. 731 (2002).
24. H. Y. Yu, Y. T. Hou, M. F. Li, and D. L. Kwong, *IEEE Trans. Electron Devices*, **49**, 1158 (2002).
25. R. Xie, T. H. Phung, W. He, Z. Q. Sun, M. B. Yu, Z. Y. Cheng and C. X. Zhu, *IEDM Tech. Dig.*, p. 393 (2008).
26. J. Robertson and B. Falabretti, *J. Appl. Phys.*, **100**, 14111(2006).
27. H. Oh, J. Lin, S. Suleiman, G.LO, D. L. Kwomg, D. Chi and S. J. Lee, *IEDM Tech. Dig.*, p. 339 (2009).
28. Y. Q. Wu, M. Xu, R. S. Wang, O. Koybasi and P. D. Ye, *IEDM Tech. Dig*, p. 323(2009).
29. N. V. Nguyen, M. Xu, O. A. Kirillov, P. D. Ye, C. Wang, K. Cheung and J. S. Suehle, *Appl. Phys. Lett.*, **96**, 52107 (2010).
30. N. W. McLachlan, *Theory and application of Mathieu functions*, Dover, N. Y., (1964).
31. F. Stern and W. E. Howard, *Phys. Rev.*, **163**, 816 (1967).
32. A. Pethe, T. Krishnamohan, D. Kim, S. Oh, H.S.P Wong, Y. Nishi and C. Saraswat, *IEDM Tech. Dig.*, p. 619(2005).
33. J. R. Chelikovsky and M. L. Cohen, *Phys. Rev.*, **B14**, 556 (1976).
34. http://www.ioffe.ru/SVA/NSM//Semicond/
35. H. Kufluoglu and M. A. Alam, *IEDM Tech. Dig.*, p. 113 (2004).
36. M. F. Li, D. M. Huang, C. Shen, T. Yang, W. J. Liu and Z. Y. Liu, *IEEE Trans. Device and Materials Reliability,* **8**, 62 (2008).

402

ECS Transactions, 35 (4) 403-416 (2011)
10.1149/1.3572296 ©The Electrochemical Society

Role of oxygen transfer for high-k/SiO$_2$/Si stack structure
on flatband voltage shift

Toshihide Nabatame, Akihiko Ohi and Toyohiro Chikyow

MANA Foundry and Advanced Electronic Materials Center, National Institute for
Materials Science, 1-1 Namiki, Tsukuba, 305-0044, Japan
Email: NABATAME.Toshihide@nims.go.jp

We investigate the difference of flatband voltage (Vfb) behavior in
high-k/SiO$_2$/Si stack structure due to oxygen vacancy (Vo) and
additional oxygen generated by the reduction and oxidation
annealing processes, respectively. The Vfb of Mg and La-
incorporated Hf-based high-k dielectrics is also influenced by Vo
generation in high-k layer. We found that the non-linear
relationships of Vfb behavior appears in HfSiO$_x$, Mg^{2+}-HfSiO$_x$, and
La^{3+}-HfSiO$_x$ dielectrics as a function of the oxidation annealing
temperature, while the HfO$_2$, N^{3+}-HfSiO$_x$, Mg^{2+}-HfO$_2$, and La^{3+}-
HfO$_2$ dielectrics show the linear relationships of Vfb shift by
introducing additional oxygen. Furthermore, it is clear that the Vfb
shift of all high-k materials satisfies the diffusion equation; which
indicates that the oxygen transfer in high-k layer is a dominant
factor in determining Vfb. We found that the oxygen diffusion in
high-k materials can be ordered as follows: Mg^{2+}-HfO$_2$, La^{3+}-HfO$_2$
and N^{3+}-HfSiO$_x$ > HfO$_2$ >> La^{3+}-HfSiO$_x$ > HfSiO$_x$ and Mg^{2+}-HfSiO$_x$.
Note that the oxygen transfer in high-k materials is very important to
recognize mechanism of Vfb shift for high-k/SiO$_2$/Si stack structure.

Introduction

Complementary metal-oxide-semiconductors (CMOSs) with metal gate
electrodes/high-k dielectrics fabricated by the gate-first or gate-last process are being
mass produced. However, the anomalous shift of the flatband voltage (Vfb) remains a big
issue. The Vfb has been reported to be changed by various factors, such as the top
interface dipole at the gate electrode/high-k dielectric interface, the charge and dipole due
to the oxygen vacancy (Vo) in high-k dielectrics, and the bottom interface dipole at the
high-k/interfacial layer (IL)-SiO$_2$ interface [1-7].

To clarify and control of the anomalous Vfb shift, several approaches have been
demonstrated. One is to study the effects of the capping layer that is deposited on Hf-
based high-k dielectrics on the Vfb shift in CMOSs. A capping layer of MgO, SrO, Y$_2$O$_3$,
or La$_2$O$_3$ materials has been used in n-type metal-oxide-semiconductor field-effect
transistors (nMOSFETs), while an Al$_2$O$_3$ capping layer has been used in p-type ones
(pMOSFETs) to control Vfb [8-10]. The Vfb shift in a high-k/SiO$_2$ stack structure has
been attributed to the bottom interface dipole at the high-k/SiO$_2$ interface as shown in Fig.
1. The direction and strength of the dipole moment strongly depend on the high-k

403

materials. In Al_2O_3 and HfO_2 dielectrics, the effective work function ($\phi_{m,eff}$) of the gate electrode become larger value due to the dipole. In contrast, the $\phi_{m,eff}$ value appears smaller in MgO, Y_2O_3, and La_2O_3 high-k materials.

Another approach is to clarify the mechanism of the Vfb shift due to Vo and oxygen transfer in a high-k/SiO_2 stack structure. The impact of the bottom interface dipole on Vfb shift has recently been discussed in terms of structural stability of the high-k/SiO_2 interface [3]. The idea is based on the oxygen transfer at the hetero interface between a high-k ionic oxide and SiO_2 covalent oxide. Large positive and negative Vfb shifts were also reported to occur due to oxidation and reduction annealing, respectively [11-13]. The Vfb shift is thought to be caused by the oxygen transfer in high-k films [2, 3, 4-7, 11, 13, 14].

We also consider that the ionicity of high-k materials contacting SiO_2, which is a covalent oxide, must be related to the dipole formation. The amount of oxygen diffusion in the ionic oxides can be ordered as follows: Y_2O_3- and CaO-stabilized $ZrO_2 > ZrO_2$ and $HfO_2 \gg Al_2O_3$ materials [15, 16]. This means that doped cations with different valences strongly affect oxygen diffusion. Therefore, we focused on the difference in the ionicity of high-k materials, such as pure HfO_2, Mg- and La-incorporated HfO_2 (Mg^{2+}-HfO_2 and La^{3+}-HfO_2), $HfSiO_x$, Mg- and La-incorporated $HfSiO_x$ (Mg^{2+}-$HfSiO_x$ and La^{3+}-$HfSiO_x$), and N-incorporated $HfSiO_x$ (N^{3+}-$HfSiO_x$).

In this paper, we have investigated the Vfb behavior in Pt-gated MOS capacitors having different high-k dielectrics, when the Vo and additional oxygen were generated in high-k layer by the reduction and oxidation annealing processes, respectively. We also discuss the effects of oxygen transfer into the high-k layer on the Vfb shift, based on experiments on the annealing-time dependence of the Vfb shift.

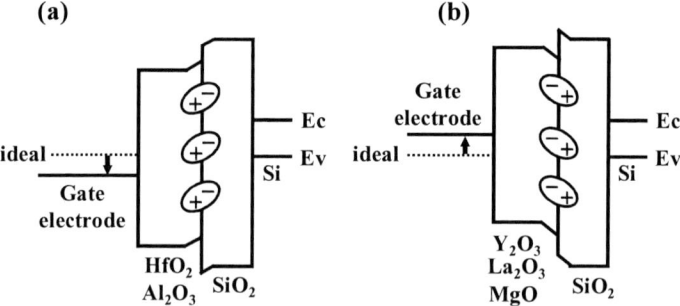

Figure 1 Schematics of two types of bottom interface dipole at high-k/SiO_2 interface for gate electrode/high-k/SiO_2/Si gate stack structure of (a) HfO_2 and Al_2O_3, and (b) Y_2O_3, La_2O_3, and MgO high-k materials, respectively.

Experimental

Typically, a 3-nm-thick SiO_2 film is formed on a p-type Si substrate using a thermal process. In our experiment, 3.5-, 4.2-, and 4.9-nm-thick HfO_2 and $HfSiO_x$ films with 60at.% Hf were deposited on SiO_2 film in a metalorganic chemical vapor deposition process. Post deposition annealing (PDA) was then carried out at 1050°C in N_2 ambient. Mg- or La-incorporated HfO_2 and $HfSiO_x$ films were also prepared as follows: MgO or LaO_x capping layer with a few angstrom thickness was deposited on HfO_2 and $HfSiO_x$ films by sputtering. The annealing process was applied at 900°C to diffuse Mg or La atom into HfO_2 and $HfSiO_x$ films [10]. N^{3+}-$HfSiO_x$ films were also prepared using plasma nitridation after fabricating 4.9-nm-thick $HfSiO_x$ on SiO_2 film. Finally, a 50-nm-thick Pt film was deposited on the HfO_2 and $HfSiO_x$-based films as a gate electrode.

To remove oxygen from the high-k films, forming gas annealing (FGA) was performed at temperatures in the range of 400 - 550°C in $3\%H_2$ ambient by using the effect of the Pt catalyst. On the other hand, to introduce additional oxygen into the high-k films, oxidation annealing (ODA) was performed at temperatures in the range of 100 - 450°C for 30 min in O_2 ambient after FGA at 400°C (FGA400) or 600°C (FGA600). The ODA time was varied between 1 to 100 min to investigate the oxygen diffusion of high-k films, i.e., HfO_2, Mg^{2+}- and La^{3+}-HfO_2, $HfSiO_x$, and Mg^{2+}-, La^{3+}-, and N^{3+}-$HfSiO_x$. The Vfb value was estimated from capacitance-voltage (C–V) characteristics using the program MIRAI-ACCEPT [17].

Results and Discussion

Vfb behavior due to Vo generated by reduction annealing process

We examined the effects of Vo in high-k film, introduced by FGA, on Vfb shift. The C-V characteristics of Pt-gated MOS capacitors with HfO_2 (4.9 nm)/SiO_2 (3 nm) are

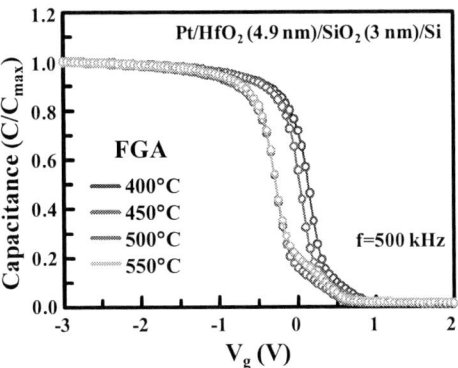

Figure 2 C-V characteristics of Pt/HfO_2 (4.9 nm)/SiO_2 (3 nm)/p-Si MOS capacitors. FGA temperature ranged from 400 to 550°C. FGA time was 30 min.

shown in Fig. 2. FGA was varied from 400 and 550°C in steps of 50°C while the annealing time was kept constant at 30 min. The C–V curves shift in the negative direction as the FGA temperature increaeses.

The C-V characteristics of Pt/HfSiO$_x$ (4.9 nm)/SiO$_2$ (3 nm)/Si MOS capacitors for the same change in the FGA temperature are shown in Fig. 3. A similar negative Vfb shift with increasing FGA temperature was also observed. The temperature dependence of the negative C-V curve shift for HfSiO$_x$ dielectric seems to be large compared to that of the HfO$_2$ dielectric. This behavior has been well explained in terms of the effects of charge transfer due to Vo formation in HfO$_2$ and HfSiO$_x$ dielectrics [2, 6, 11].

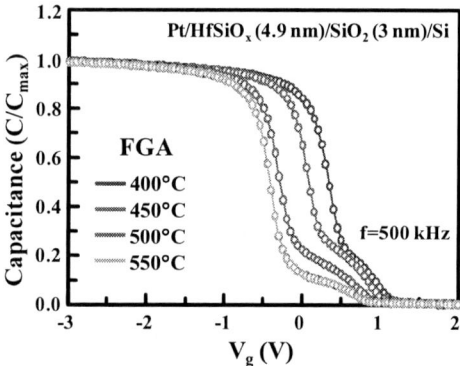

Figure 3 C-V characteristics of Pt/HfSiO$_x$ (4.9 nm)/SiO$_2$ (3 nm)/p-Si MOS capacitors. FGA temperature ranged from 400 to 550°C. FGA time was 30 min.

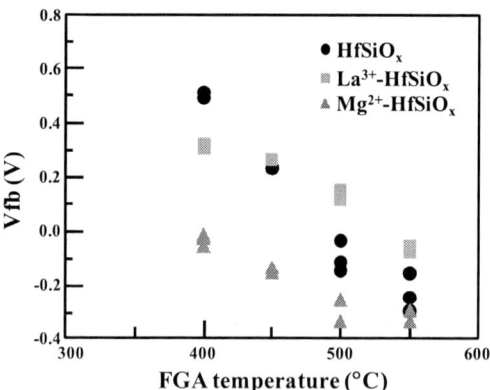

Figure 4 Vfb behavior of Pt-gated MOS capacitors with HfSiO$_x$, Mg^{2+}- HfSiO$_x$, and La^{3+}- HfSiO$_x$ dielectrics as function of FGA temperature. FGA time was 30 min.

The Vfb behaviors of Pt-gateds MOS capacitors with HfSiO$_x$-based high-k dielectric as a function of FGA temperature has been summarized in Fig. 4. Here, 4.9-nm-thick HfSiO$_x$, Mg^{2+}-HfSiO$_x$, and La^{3+}-HfSiO$_x$ films were used as HfSiO$_x$-based high-k dielectrics. Large negative Vfb shifts in the Mg^{2+}- and La^{3+}-HfSiO$_x$ capacitors were clearly observed compared to the pure HfSiO$_x$ capacitor due to the capping effect of the Mg and La elements at FGA400 [8–10]. We found that the Vfb of all the samples shifted in the negative direction with increasing FGA temperature. This suggests that the Vfb values of Mg- and La-incorporated HfSiO$_x$-based high-k dielectrics are also affected by Vo generation in the high-k layer. To control Vfb, we must consider the annealing conditions during CMOS fabrication.

Vfb behavior due to additional oxygen by the oxidation annealing process

To examine the effects of additional oxygen in high-k film on Vfb shift, ODA was performed. Figure 5 shows C–V curves of Pt-gated MOS capacitors with 4.9-nm-thick HfO$_2$ dielectrics. The ODA temperature was varied from 100 to 300°C, and the ODA duration was 30 min. The C–V curves significantly shifted in the positive direction with increasing ODA temperature, as previously reported [14].

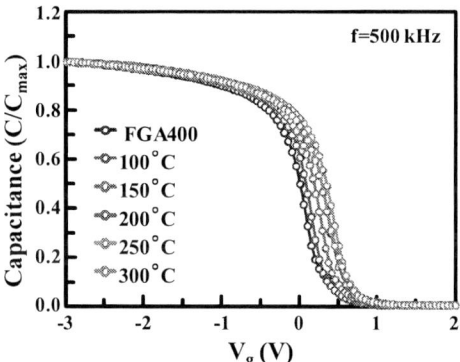

Figure 5 C-V characteristics of Pt/HfO$_2$ (4.9 nm)/SiO$_2$ (3 nm)/p-Si MOS capacitors. ODA temperature ranged from 100 to 300°C after FGA at 400°C. ODA time was 30 min.

The Vfb change as a function of the equivalent oxide thickness (EOT) of the HfO$_2$ film for Pt/HfO$_2$/SiO$_2$/p-Si MOS capacitors is shown in Fig. 6. We also found that the Vfb shifts in the positive direction with increasing ODA temperature irrespective of the HfO$_2$ film thickness. Note that the positive Vfb shift is caused by the dipole rather than the fixed charge in HfO$_2$ and the HfO$_2$/SiO$_2$ interface because of its linear relationship having almost the same slope as a function of the EOT of HfO$_2$. Furthermore, the dipole occurs dominantly at HfO$_2$/SiO$_2$ bottom interface as reported in Refs. 3 and 4. This means that the bottom interface dipole at the high-k/SiO$_2$ interface plays a significant role in Vfb control in high-k MOS devices.

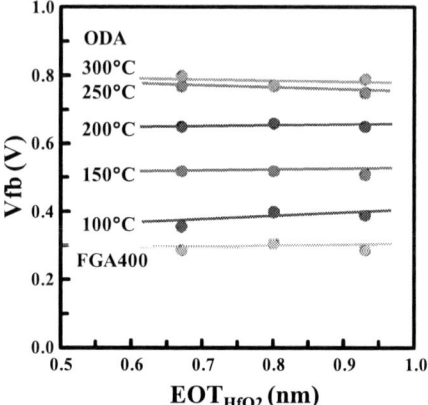

Figure 6 Relationship between Vfb and EOT of HfO_2 films for $Pt/HfO_2/SiO_2/Si$ MOS capacitors with different ODA temperatures. HfO_2 thicknesses were 3.5, 4.2, and 4.9 nm.

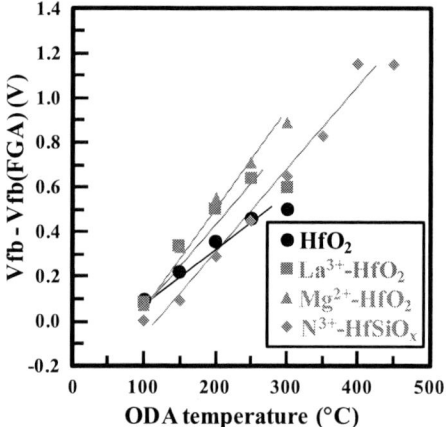

Figure 7 Relationship between normalized Vfb and ODA temperatures for Pt-gated MOS capacitors with HfO_2, N^{3+}-$HfSiO_x$, Mg^{2+}-HfO_2, and La^{3+}-HfO_2 dielectrics. Thickness of all high-k dielectrics was 4.9 nm. FGA was carried out at 400 and 600°C for HfO_2-based high-k and N^{3+}-$HfSiO_x$, respectively, before ODA. ODA time was 30 min. ODA temperatures of HfO_2-based high-k and N^{3+}-$HfSiO_x$ ranged from 100 to 300°C and from 100 to 450°C, respectively.

Figure 7 shows the relationship between normalized Vfb and the ODA temperature for Pt-gated MOS capacitors with HfO_2, N^{3+}-$HfSiO_x$, Mg^{2+}-HfO_2, and La^{3+}-HfO_2

dielectrics. The ODA duration was 30 min. The Vfb of all the samples shifted in the positive direction with increasing ODA temperature. As explained above, the positive Vfb shift was caused by the dipole, as shown by the linear relationships in Fig. 6 [14]. Therefore, these positive Vfb shifts were derived from dipole formation due to the introduced additional oxygen. Clear linear relationships of Vfb shift were also obtained for all the samples.

The different Vfb behaviors of Pt-gated MOS capacitors with HfSiO$_x$, and Mg^{2+}-HfSiO$_x$, and La^{3+}-HfSiO$_x$ dielectrics as a function of ODA temperature are shown in Fig. 8. The ODA duration was 30min. The behaviors can be divided into three cases according to ODA temperature. Vfb is almost unchanged in case I ($\leq$250°C), but a significant Vfb shift appears in case II (250 – 400°C). In case III ($\geq$400°C), it seems that the Vfb shift in all the samples was saturated.

We consider that the difference of Vfb behavior among high-k materials may depend on the difference in oxygen diffusion in high-k films, as shown in Figs. 7 and 8.

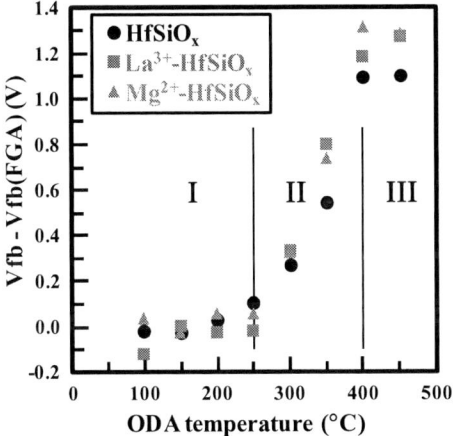

Figure 8 Vfb behavior of Pt-gated MOS capacitors with HfSiO$_x$, Mg^{2+}-HfSiO$_x$, and La^{3+}-HfSiO$_x$ as function of ODA temperature. Thickness of all high-k dielectrics was 4.9 nm. ODA temperature ranged from 100 to 450°C after FGA at 600°C. ODA time was 30 min.

The $\phi_{m,eff}$ behaviors of HfO$_2$ and HfSiO$_x$ dielectrics against the FGA and ODA temperatures are shown in Figs. 9 (a) and (b), respectively. The $\phi_{m,eff}$ values of the HfO$_2$ and HfSiO$_x$ dielectrics as a function of the FGA and ODA temperatures are determined by extrapolating the Vfb versus EOT$_{high-k}$ plots with following the equation:

$$Vfb = (\phi_{m,eff} - \phi_{Si}) - Q_{ox} / \varepsilon_{ox} \times EOT_{high-k}, \qquad (1)$$

where $\phi_{m,eff}$, ϕ_{Si}, Q_{ox}, ε_{ox}, and EOT_{high-k} are the effective work function of the gate electrode on high-k film, the Fermi-level of silicon substrate, the fixed charges, the dielectric constant of SiO_2, and the EOT of the high-k layer, respectively. In the FGA case, the $\phi_{m,eff}$ values of the HfO_2 and $HfSiO_x$ dielectrics clearly became smaller as the FGA temperature increased. In contrast, the $\phi_{m,eff}$ values of the HfO_2 and $HfSiO_x$ dielectrics increased from 5.2 to 5.8 eV and from 4.7 to 5.8 eV, respectively, as the ODA temperature increased. These results indicate that the $\phi_{m,eff}$ values easily shift due to Vo and additional oxygen in the high-k layer generated by changing the ambient and temperature for annealing. This suggests that the $\phi_{m,eff}$ values can be controlled in a range of about 0.8 and 1.1 eV for HfO_2 and $HfSiO_x$ dielectrics, respectively.

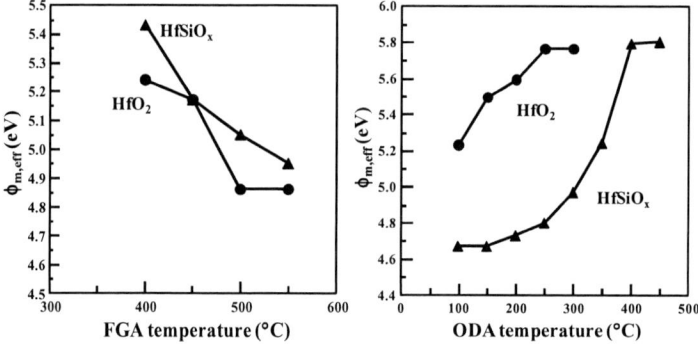

Figure 9 Effective work function behaviors of Pt-gated MOS capacitors with HfO_2 and $HfSiO_x$ dielectrics as function of (a) FGA and (b) ODA temperatures, respectively. FGA temperature ranged from 400 to 550°C. ODA temperature of HfO_2 ranged from 100 to 300°C after FGA at 400°C. ODA temperature of $HfSiO_x$ ranged from 100 to 450°C after FGA at 600°C.

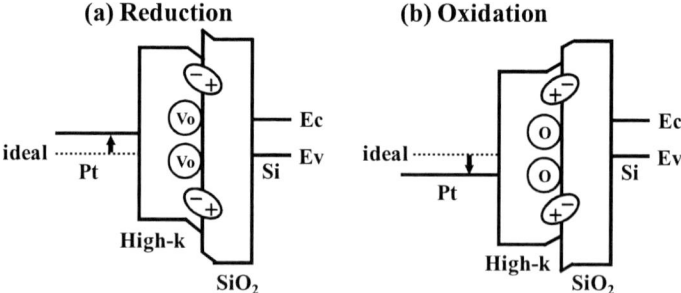

Figure 10 Schematics of bottom interface dipole at high-k/SiO_2 interface due to Vo and additional oxygen generated by annealing in (a) reduction and (b) oxidation ambient, respectively.

Figure 10 is a schematic explaining the difference of the dipole direction for the Pt/high-k/SiO$_2$/Si stack structure after the reduction and oxidation annealing processes. Vo is formed in the high-k layer by FGA because oxygen is removed from that layer. Furthermore, a dipole due to Vo occurs at the high-k/SiO$_2$ bottom interface [3, 4, 7] and is attributed to a small $\phi_{m,eff}$ value. In contrast, a dipole in the opposite direction is formed at the high-k/SiO$_2$ interface through the introduction of additional oxygen during ODA. As a result, a large $\phi_{m,eff}$ value can be obtained.

Role of the oxygen transfer of high-k layer on Vfb

To investigate the effect of the oxygen transfer in high-k materials on Vfb shift, we examined the relationship between Vfb and the annealing time of the ODA process for several high-k dielectrics. Figure 11 shows the annealing-time dependence of the C–V characteristics of Pt/HfSiO$_x$ (4.9 nm)/SiO$_2$/Si MOS capacitors. The ODA temperature was 400°C. The C-V curves clearly shift in the positive direction with increasing ODA time. Furthermore, change in the C–V curve saturates for ODA times above 30 min.

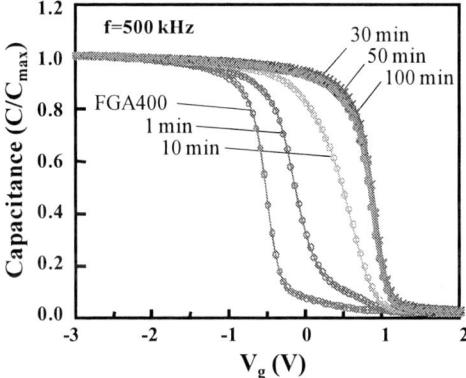

Figure 11 Annealing-time dependence of C-V profiles of Pt/HfSiO$_x$ (4.9 nm)/SiO$_2$/Si MOS capacitor in ODA treatment at 400°C. Annealing time ranged from 1 to 100 min.

Normalized Vfb behavior of Pt/HfSiO$_x$/SiO$_2$/Si MOS capacitors as a function of the square root of the annealing time is shown in Fig. 12. Note that the Vfb shift shows a strong linear relation to the square root of the annealing time until 30 min. This behavior indicates that the Vfb shift must be determined by the oxygen diffusion in the HfSiO$_x$ layer because it satisfies the following equation:

$$l = \sqrt{(D \cdot t)}, \qquad (2)$$

where l, D, and t are the penetration depth, diffusion coefficient of oxygen of the high-k film, and the annealing duration, respectively. However, the Vfb shift was found to saturate above 30 min of annealing. This means that the amount of oxygen that affects the Vfb shift saturates in the $HfSiO_x/SiO_2$ stack structure.

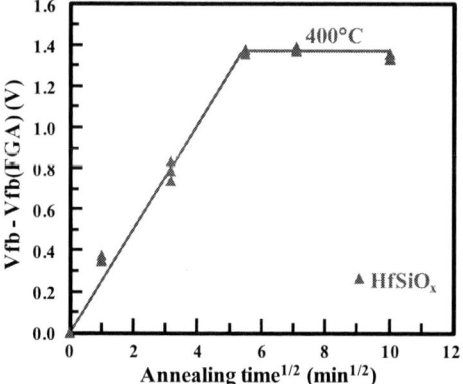

Figure 12 Normalized Vfb behavior of $Pt/HfSiO_x$ (4.9 nm)/SiO_2/Si MOS capacitor as function of square root of annealing time for ODA at 400°C.

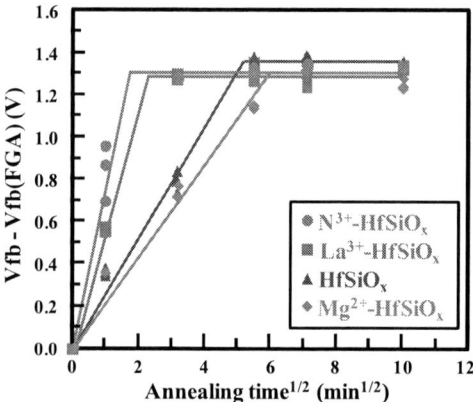

Figure 13 Normalized Vfb behaviors of four $HfSiO_x$-based high-k samples, i.e., $HfSiO_x$, and Mg^{2+}-, La^{3+}-, and N^{3+}-$HfSiO_x$, as function of square root of annealing time for ODA at 400°C.

Figure 13 shows normalized Vfb behaviours of four $HfSiO_x$-based high-k materials, i.e., $HfSiO_x$, and Mg^{2+}-, La^{3+}-, and N^{3+}-$HfSiO_x$, as a function of the square root of the annealing time of ODA at 400°C. A linear dependence on the square root of the annealing time was observed for all samples in a short time range. We found two types of linear dependences: steep slopes for N^{3+}-HfSiO$_x$ and La^{3+}-HfSiO$_x$ dielectrics, and gentler slopes for HfSiO$_x$ and Mg^{2+}-HfSiO$_x$ dielectrics. We consider that the difference in slopes is due to the difference in the oxygen diffusion in the high-k materials. In fact, it has been reported that Y^{3+} and Ca^{2+} cation doped ZrO$_2$ dielectrics have a large oxygen diffusion coefficient compared to a pure ZrO$_2$ dielectric [15, 16]. Furthermore, it is surprising that the Vfb shift saturated at about the same value of 1.3 V in all the high-k dielectrics. We found that additional oxygen which is introduced into a high-k/SiO$_2$ stack structure relates to Vfb shift. Therefore, this suggests that the additional oxygen is limited, regardless of the HfSiO$_x$-based high-k material.

The normalized Vfb behaviors of three HfO$_2$-based high-k dielectrics, i.e., HfO$_2$, Mg^{2+}-HfO$_2$, and La^{3+}-HfO$_2$, as a function of the square root of the annealing time are shown in Fig. 14. The ODA was carried out at 300°C. We also found that the Vfb shift varies linearly with the square root of the annealing time; especially, at shorter annealing times. Furthermore, the slopes for the Mg^{2+}- and La^{3+}-HfO$_2$ dielectrics are significantly larger than that for the HfO$_2$ dielectric. This is possibly due to be the differences in oxygen diffusion among the three high-k materials. We also found that the Vfb shift for the HfO$_2$ dielectric saturated when the ODA was above 200 min.

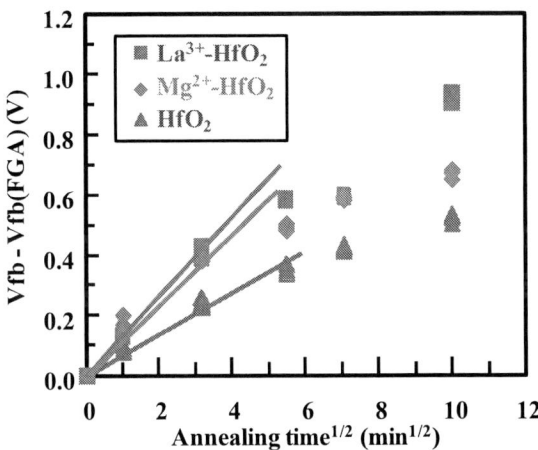

Figure 14 Normalized Vfb behaviors of three HfO$_2$-based high-k dielectrics, i.e., HfO$_2$, Mg^{2+}-HfO$_2$, and La^{3+}-HfO$_2$, as function of square root of annealing time for ODA at 300°C. HfO$_2$ dielectric was only annealed for up to 200 min.

Figure 15 is a schematic explaining oxygen diffusion in the high-k layer of a Pt/high-k/SiO$_2$/Si stack structure during the oxidation annealing process. When oxygen molecules are absorbed on the Pt surface, the molecules are decomposed to activated oxygen atoms by the effect of Pt catalyst. The activated oxygen diffuses into the high-k layer and approaches the high-k/SiO$_2$ interface. Finally, the oxygen accumulates at the high-k/SiO$_2$ interface because oxygen diffusion is slower in SiO$_2$ material than in HfO$_2$ material.

If Vfb shift is assumed to be limited only by the oxygen diffusion in high-k films, the difference in Vfb behavior among several high-k materials can be recorgnized. We have reported that HfO_2-based high-k materials show a linear relationship between the Vfb shift and the oxidation annealing temperature, while $HfSiO_x$-based high-k materials show a non-linear relationship between Vfb shift and temperatures under 250°C as shown in Figs. 7 and 8. There is not enough additional oxygen to affect Vfb shift in the $HfSiO_x$-based high-k materials while the oxygen of the HfO_2-based high-k materials saturates at 30 min of annealing. This suggests that oxygen diffusion in $HfSiO_x$-based high-k materials is lower than in HfO_2-based high-k materials. Therefore, the amount of oxygen diffusion in high-k materials can be ordered as follows: Mg^{2+}-HfO_2, La^{3+}-HfO_2 and N^{3+}-$HfSiO_x$ > HfO_2 >> La^{3+}-$HfSiO_x$ > $HfSiO_x$ and Mg^{2+}-$HfSiO_x$.

These results suggest that oxygen transfer in high-k dielectrics strongly affects Vfb shift in a high-k/SiO_2 stack structure.

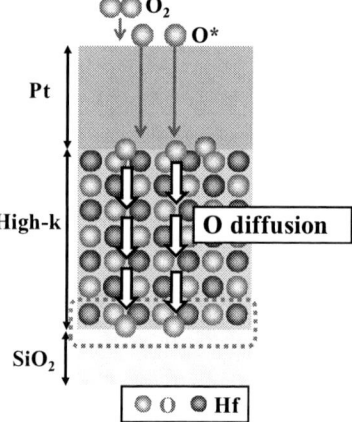

Figure 15 Schematic of oxygen diffusion into high-k layer during ODA process by the effect of Pt catalyst.

Conclusion

We investigated the different Vfb behaviors in Pt-gated MOS capacitors with different high-k dielectrics due to Vo and additional oxygen generated by reduction and oxidation annealing processes, respectively. We found that the Vfb of HfO_2, $HfSiO_x$, and Mg^{2+}- and La^{3+}-$HfSiO_x$ dielectrics shifts in the negative direction with increasing FGA temperature. This suggests that the Vfb of Mg- and La-incorporated Hf-based high-k materials is influenced by Vo generation in the high-k layer. In contrast, we have found that non-linear relationships of Vfb behavior appeared in $HfSiO_x$, Mg^{2+}-$HfSiO_x$, and La^{3+}-$HfSiO_x$ materials as a function of the ODA temperature, while the HfO_2, N^{3+}-$HfSiO_x$, Mg^{2+}-HfO_2, and La^{3+}-HfO_2 materials showed linear Vfb shift relationships when additional oxygen was introduced. Furthermore, the Vfb shifts in all the high-k materials clearly satisfied the diffusion equation. If Vfb shift is assumed to occur only due to the bottom interface dipole, oxygen transfer in the high-k layer is a dominant factor in determining Vfb. We found that the amount of oxygen diffusion in high-k materials can be ordered as follows: Mg^{2+}-HfO_2, La^{3+}-HfO_2 and N^{3+}-$HfSiO_x$ > HfO_2 >> La^{3+}-$HfSiO_x$ > $HfSiO_x$ and Mg^{2+}-$HfSiO_x$. Based on the experimental data, it is clear that oxygen transfer in high-k materials is very important not only for recognizing the mechanism of Vfb shift but also for designing advanced high-k/SiO_2/Si stack structures.

Acknowledgements

We thank Mrs. H. Yamada, and M. Kimura of Shibaura Institute of Technology and the members of MANA Foundry of International Center for Materials Nanoarchitectonics of National Institute for Materials Science.

References

1. C-C. Hobbs, L-R. C. Fonseca, A. Knizhnik, V. Dhandapani, S-B. Samavedam, W-J. Taylor, J-M. Grant, L-G. Dip, D-H. Triyoso, R-I. Hegde, D-C. Gilmer, R. Garcia, D. Roan, M. L. Lovejoy. R. S. Rai, E-A. Herbert, H-H. Tseng, G. H. Anderson, B-E. White and P-J. Tobin, *IEEE Trans. Electron Device*, **51**, 971(2004).
2. K. Shiraishi, K. Yamada, K. Torii, Y. Akasaka, K. Nakajima, M. Konno, T. Chikyow, H. Kitajima, T. Arikado and Y. Nara, *ThinSolid Films*, **508**, 305(2006).
3. K. Kita and A. Totiumi, *Appl. Phys. Lett.*, **94**, 132902(2009).
4. K. Iwamoto, Y. Kamimuta, A. Ogawa, Y. Watanabe, S. Migita, W. Mizubayashi, Y. Morita, M. Takahashi, H. Ota, T. Nabatame and A. Toriumi, *Appl. Phys. Lett.*, **92**, 132907(2008).
5. E. Cartier, M. Hopstaken and M. Cople, *Appl. Phys. Lett.*, **95**, 042901(2009).
6. S. Guha and V. Narayanan, *Phys. Rev. Lett.*, **98**, 196101(2007).
7. T. Nabatame, K. Iwamoto, K. Akiyama, Y. Nunoshige, H. Ota, T. Ohishi and A. Toriumi, *ECS Transactions*, **11**, 543(2007).
8. V. Narayanan, V. K. Paruchuri, N. A. Bojarczuk, B. P. Linder, B. Doris, Y. H. Kim, S. Zafar, J. Stathis, S. Brown, J. Arnold, M. Copel, M. Steen, E. Cartier, A. Callegari, P. Jamison, J.-P. Locquetl, D. L. Lacey, Y. Wang, P. E. Batson, P. Ronsheim, R. Jammy, M. P. Chudzik, M. Leong, S. Guha, G. Shahidi and T. C. Chen, *VLSI Tech. Dig.*, p.224(2006).
9. N. Mise, T. Morooka, T. Eimori, S. Kamiyama, K. Murayama, M. Sato, T. Ono, Y. Nara and Y. Ohji, *IEDM Tech. Dig.*, p.527(2007).
10. T. Morooka, T. Sato, T. Matsuki, T. Suzuki, K. Shiraishi, A. Uedono, S. Miyazaki, K. Yamada, T. Nabatame, T. Chikyow, J. Yugami, K. Ikeda and Y. Ohji, *VLSI Tech. Dig.*, p.33(2010).
11. E. Cartier, F. R. McFeely, V. Narayanan, P. Jamison, B. P. Linder, M. Copel, V. K. Paruchuri, V. S. Basker, R. Haight, D. Lim, R. Carruthers, T. Shaw, M. Steen, J. Sleight, J. Rubino, H. Deligianni, S. Guha, R. Jammy and G. Shahidi, *VLSI Tech. Dig.*, p.230(2005).
12. J. K. Schaeffer, L. R. C. Fonseca, S. B. Samavedam and Y. Liang, *Appl, Phys. Lett.*, **85**, 1826(2004).

13. Y. Kamimuta, K. Iwamoto, Y. Nunoshige, A. Hirano, W. Mizubayashi, Y. Watanabe, S. Migita, A. Ogawa, H. Ota, T. Nabatame and A. Toriumi, *IEDM Tech. Dig.*, p.341(2007).
14. T. Nabatame, A. Ohi and T. Chikyow, *ECS Transaction*, **33**, 59(2010).
15. T. Nabatame, T. Yasuda, M. Nishizawa, M. Ikeda, T. Horikawa and A. Toriumi, *Jpn. J. Appl. Phys.*, **42**, 7205(2003).
16. U. Brossmann, R. Wurschum, U. Sodervall and H. E. Schaefer, *J. Appl. Phys.*, **85**, 7646(1999).
17. N. Yasuda, H. Ota, T. Horikawa, T. Nabatame, H. Satake, A. Toriumi, Y. Tamura, T. Sasaki, and F. Ootsuka. *Ext. Abstr. Solid State Devices and Materials* (2005) p. 250.

Charge Trapping and Reliability Properties of MONOS Memory with High-k Blocking Layer

Naoki Yasuda, Shosuke Fujii, Jun Fujiki, and Haruka Kusai

Advanced LSI Technology Laboratory, Corporate R&D Center, Toshiba Corporation
8, Shinsugita-cho, Isogo-ku, Yokohama 235-8522, Japan

Carrier trapping and reliability characteristics during the cycling operation of metal-oxide-nitride-oxide-semiconductor (MONOS) devices with a high-k blocking layer (Al_2O_3) have been studied by measuring the injected charge during programming and erasing (P/E). The evaluated charge centroid during P/E indicates that the traps near the SiN/Al_2O_3 interface are filled at high electric field, while bulk traps in the SiN layer are also available at low electric field. The degradation of MONOS devices in cycling operation is found to be strongly correlated with injected charge during erasing. It is indispensable to suppress hole injection, as well as lowering the electric field to improve endurance of MONOS devices. The discussion of this paper also extends to the structure of the charge trapping layer to improve both cycling endurance and retention simultaneously. Finally, we report that a high-k blocking layer has significant impact on data readout and programming characteristics through its transient dielectric properties.

Introduction

Metal-oxide-nitride-oxide-semiconductor (MONOS) device with a high-k blocking layer is one of the candidates to replace conventional floating-gate memory devices [1,2]. For the use of MONOS memory cells, it is necessary to have accurate knowledge on carrier trapping properties such as charge centroid [3-8] and trapping efficiency [9]. Since MONOS memory has finite electrical thickness in the charge trapping layer, the evaluation of charge centroid (i.e., the vertical position of trapped charge) gives an important device parameter related to the coupling ratio and the cell-to-cell interference. In this paper we report our evaluation method of charge centroid first, and then apply it to find the dependences of charge centroid on the thickness of a SiN layer and on the electric field in a tunneling oxide layer.

Another primary concern in the use of MONOS memory is reliability from the viewpoints of both endurance [7,10-14] and retention [10,15-18]. In this paper we discuss the degradation of MONOS characteristics during cycling operation by measuring the amount of injected charge during programming and erasing (P/E). It is found that the injected charge density in erase operation is strongly correlated with the cycling degradation in a wide range of pulse voltage and duration. From the dependence of cycling degradation on the atomic composition of a SiN layer, we infer that the injected holes are the main cause of cycling degradation. Although the MONOS devices with a Si-rich SiN layer (showing erase by electron current) can increase the cycling endurance, poor data retention has been a problem [7,15,16]. Recently, it has been reported that

MONOS devices with a laminate SiN layer [19] can improve both endurance and retention simultaneously. Similar improvements have also been reported for a compositionally graded SiN layer [10]. In this paper we look into the physical meaning of the improvement in the laminate SiN layer [19] by introducing the retention experiments where external bias is applied at various temperatures. Thus, we can identify the leakage path and the activation energy for electron detrapping.

Finally, we discuss the influence of a high-k blocking layer on reading and programming characteristics of MONOS devices. It is known that high-k dielectric has slow polarization [20]. The dielectric relaxation of high k materials causes the transient channel current during the data read-out of MONOS devices. The properties of high-k materials also appear in the temperature dependence of programming characteristics. A solution to these issues is presented by using a stacked blocking layer such as $Al_2O_3/SiO_2/Al_2O_3$.

Charge Centroid and Trapping Efficiency in MONOS Memory Devices

The Extraction Method of Charge Centroid

There are several methods of measuring charge distribution in a MONOS gate stack. One of the methods [8] uses several MONOS capacitors with different SiN thickness, and the saturation level of flatband voltage shift during avalanche injection is measured for estimating the bulk and interface trap densities. Another method uses two kinds of samples with a low dopant impurity density in a poly-Si gate and in a channel region, respectively [5], and the two types of flatband-voltage shifts with respect to the gate and the substrate gives the trapped charge density and its vertical location. Yet another method uses the transient J-E characteristics extracted from the measured flatband voltage shift as a function of time, and examines its negligible dependence on the gate voltage [6]. There is also a method to measure the threshold voltage shift as a function of time in a MONOS with a very thin tunneling oxide (typically ~1 nm, allowing direct tunneling) for the spectroscopic extraction of trapped charge location and its energy level [21]. This method, however, is not suitable for a MONOS gate stack used in flash memory application where the tunneling oxide is much thicker.

One of the successful methods of charge centroid evaluation is to measure the injected charge during P/E, in addition to the flatband voltage shift [22,23]. In the case of 100% capture of injected charge into a charge trapping layer, the charge centroid is extracted as

$$z_{eff} = -\frac{\varepsilon_{ox}\Delta V_{fb}}{Q_{trap}} \approx -\frac{\varepsilon_{ox}\Delta V_{fb}}{Q_{inj}} \qquad (1)$$

where z_{eff} is the electric (SiO$_2$ equivalent) distance of the charge centroid from the gate, ε_{ox} is the dielectric constant of SiO$_2$, ΔV_{fb} is the flatband voltage shift, Q_{trap} is the trapped charge density, and Q_{inj} is the injected carrier density. The condition of 100 % carrier capture is typically realized at a low gate voltage and in the early stage of P/E operation just after the application of a P/E pulse. One of the ways to confirm the validity of 100% carrier capture ($Q_{inj} = Q_{trap}$) is to examine whether there is negligible dependence of z_{eff} on the gate voltage. This is because the penetration current across the MONOS dielectrics is a strong function of gate voltage, thereby producing an artificial z_{eff} shift which is dependent on the gate voltage. Alternatively, a more sophisticated method [24] using ISPP (incremental step pulse programming) [25] can also be introduced to ensure 100% carrier capture, as discussed later.

Similar method measuring the injected charge was also proposed by Arreghini *et al.* [4], where a three-level pulse as shown in Fig. 1 (a) was used to cancel out the displacement current during P/E operations. Note that the method in [4] requires the measurement of programming characteristics either twice or in two different samples so as to determine the final voltage of the three-level pulse [4,9]. In contrast, our method [22] is based on a rectangular gate-voltage pulse (Fig. 1(a)) which is compatible with a standard memory device characterization. Besides, we do not need to measure programming characteristics repeatedly. In stead, our method cancels out the displacement current by subtracting the net variation of the Si-substrate surface charge density before and after a P/E operation (ΔQ_{sub}) with the help of the integration of a C-V curve. As shown in Fig. 1(b), the surface condition of a Si-substrate is varied owing to the modulation of electric field caused by the trapped charge in the SiN layer (Q_{trap}). This change in surface charge density, ΔQ_{sub}, can be evaluated with the integration of a C-V curve as shown by the shaded area in Fig. 2. The integration is performed in the portion of a C-V curve crossing $V_g = 0$ (vertical axis) when the C-V curve moves toward the positive direction with the progress of carrier trapping during programming operation.

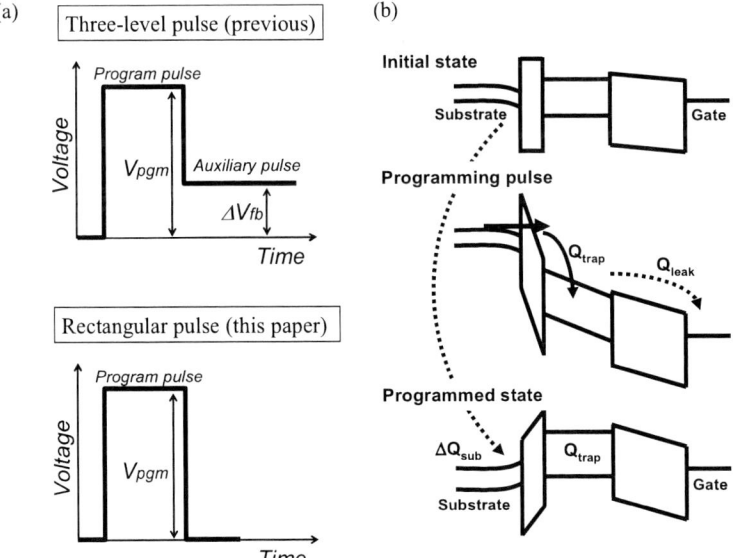

Figure 1. (a): Three-level pulse in used in [4], and rectangular pulse used in this paper. (b): Band diagrams during the program operation of a MONOS device when a rectangular pulse is applied. The measured charge is composed of $Q_{trap} + Q_{leak} + \Delta Q_{sub}$.

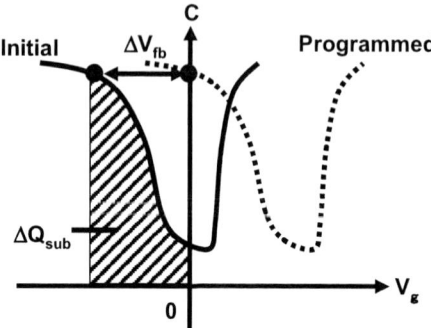

Figure 2. C-V curves before (solid line) and after (dashed line) a program operation. ΔQ_{sub} corresponds to the integration of the C-V curve in the shaded area.

Thus, ΔQ_{sub} is expressed with the integration in the initial C-V curve as

$$\Delta Q_{sub} = -\int_{-\Delta V_{fb}}^{0} C(V_g)dV_g . \qquad (2)$$

By subtracting the displacement charge ΔQ_{sub}, the injected charge across the tunneling oxide is expressed as

$$Q_{inj} = Q_{meas} - \Delta Q_{sub} \qquad (3)$$

where Q_{meas} is the charge measured with a Coulomb meter attached to the Si-substrate of a MONOS capacitor. Then, as mentioned above, the charge centroid z_{eff} is expressed as

$$z_{eff} = -\frac{\varepsilon_{ox}\Delta V_{fb}}{Q_{trap}} \approx -\frac{\varepsilon_{ox}\Delta V_{fb}}{Q_{meas} + \int_{-\Delta V_{fb}}^{0} C(V_g)dV_g} \qquad (4)$$

under the assumption that all the injected charges are trapped in the SiN layer. In other words, the charge centroid in eq. (4) is valid only in the condition where the leakage current across the MONOS gate stack is negligible ($Q_{inj} = Q_{trap}$).

We have confirmed [22] that the charge centroid evaluated with eq. (4) is equivalent to the result of the three-level pulse method. Thus, our method can be used in the charge centroid evaluation described in the next subsections, as well as in the evaluation of injected charge density in a cycling endurance test.

Dependence of Charge Centroid on SiN thickness during Fowler-Nordheim (FN) injection

Using our measurement method, the charge centroid of a MONOS capacitor with 14 nm-thick SiN layer (SiO$_2$(4 nm)/SiN(14 nm)/Al$_2$O$_3$(15 nm)) has been evaluated, and is plotted as a function of flatband voltage shift (ΔV_{fb}) in Fig. 3(a). The figure indicates that the centroid of the trapped charge is located around the middle of the SiN layer for small ΔV_{fb} (< 1V). Note that the downward shift of the charge centroid toward the Al$_2$O$_3$ blocking layer at a large ΔV_{fb} (> 1 V) is an artifact due to the leakage current across the MONOS dielectrics [22], considering that the charge centroid extracted in this region has strong dependence on the programming gate voltage.

In contrast, for a MONOS capacitor with a thin SiN layer of 5 nm (SiO$_2$(4 nm)/SiN(5 nm)/Al$_2$O$_3$(15 nm)), it has been found that the charge centroid is located at around the SiN/Al$_2$O$_3$ interface as shown in Fig. 3(b). Although there is disturbance due to leakage current in the evaluated charge centroid for ΔV_{fb} > 0.5 V (as evidenced by the gate-voltage dependence), the position of the charge centroid for smaller ΔV_{fb} appears to exhibit negligible dependence on the applied gate voltage. Thus, we conclude that the charge centroid for the 5-nm SiN layer is located close to the SiN/Al$_2$O$_3$ interface as indicated by the dotted circle in Fig. 3(b). This result is substantially different from the case of a thicker SiN layer in Fig. 3(a).

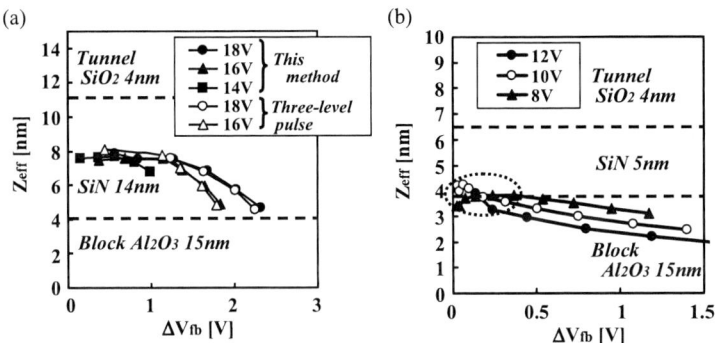

Figure 3. (a): Charge centroid (z_{eff}) of the MONOS with 14-nm SiN layer extracted by our method (filled symbols) and the three-level pulse method (open symbols). The centroid is initially positioned close to the middle of the SiN layer. (b): Charge centroid (z_{eff}) of the MONOS with 5-nm SiN layer. The centroid is located around the SiN/Al$_2$O$_3$ interface.

The location of the charge centroid near the SiN/Al$_2$O$_3$ interface for a thin SiN layer (5 nm) during programming can also be confirmed with the help of charge centroid measurement in the erase operation. As discussed later, the erase operation proceeds either with the detrapping of captured electrons in a SiN layer, or with the injection of holes from a Si-substrate and possible subsequent recombination with trapped electrons. Therefore, the charge centroid measurement in the erase operation can give information on the location of trapped electrons during programming from another aspect. (Note that the detrapping of captured electrons does not contain the issue of extra leakage current, and this is realized as the SiN layer becomes more Si-rich, as we discuss later.) In Fig. 4

we show the charge centroid of two samples with Si-rich SiN layers in both program and erase operations. (Sample structures are later shown in Table I, with the refractive indices (R.I.) in the SiN layer of 2.09 and 2.23, respectively.) We have found that the charge centroids in both P/E operations are located near the SiN/Al_2O_3 interface. This result provides additional evidence for the charge-centroid location near the SiN/Al_2O_3 interface in the MONOS devices with a 5-nm SiN layer.

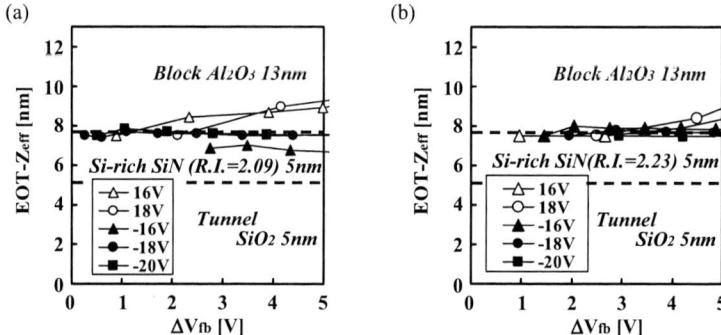

Figure 4. (a): Charge centroid of the MONOS with Si-rich composition (R.I.=2.09) during P/E operations. (b): Charge centroid of the MONOS with Si-rich composition (R.I.=2.23) during P/E operations. In both cases, the thickness of the SiN layer is 5 nm. The charge centroid is located near the SiN/Al_2O_3 interface for both P/E operations. R.I. refers to refractive index, and EOT indicates to the total electrical thickness of the MONOS dielectrics.

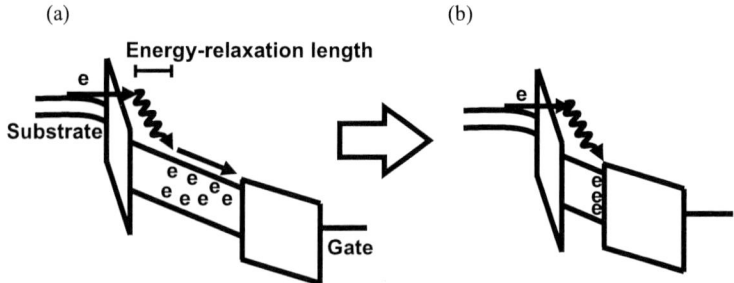

Figure 5. Schematic illustrations of trapping mechanisms for the MONOS with (a): 14 nm-thick and (b): 5 nm-thick SiN layer.

Now, we discuss the mechanism for different location of charge centroid depending on SiN thickness. We infer that the energy relaxation of injected electrons into the SiN layers is the key to this result (Fig. 3 and 4). It was reported that the carrier heating characteristics in SiO_2 and SiN films are nearly the same [26]. It was also experimentally found that the energy of electrons traveling in SiO_2 saturates in the order of ~ 5 nm [27]. Based on these previous reports, it is reasonable to assume that the energy relaxation length in a SiN layer is in the range of several nanometers. This assumption is also consistent with a recent simulation model [28]. Thus, for a thick SiN layer, the energy relaxation of injected electrons is completed in the bulk region of the SiN layer. The electrons after energy relaxation are expected to be captured into the bulk traps of the SiN layer, with negligible detrapping probability. In contrast, for a thin SiN layer, the energy of injected electrons is not fully relaxed and the remaining energy would be lost at the potential barrier of the SiN/Al$_2$O$_3$ interface. This would lead to the predominant carrier trapping near the SiN/Al$_2$O$_3$ interface. The physical interpretation described above is consistent with the dependence of charge centroid on electric field, as discussed in the next subsection. (Note that the electron energy is closely correlated with the applied electric field.)

Finally, it is noted that our result for the thin SiN layer (5 nm) indicating the charge centroid near the SiN/Al$_2$O$_3$ interface is also consistent with the results in reference [7] where MONOS devices with similar dielectric thickness to our samples were evaluated. In contrast, the reference [3] with a high-k (HfO$_2$) blocking layer and a thin SiN layer (5.7 nm) shows significantly different result, i.e. the charge centroid is located in the bulk region of the SiN layer. We believe that the difference originates from a thinner tunneling oxide of 2.2 nm used in [3] (i.e. direct tunneling region) where the electron energy is small in the SiN layer.

Dependence of Charge Centroid and Trapping Efficiency on Electric Field

As discussed in the previous subsection, the existing traps in a SiN layer (both bulk and interface) and the available traps during P/E operations can be substantially different owing to the trapping efficiency which is dependent on carrier energy. Therefore, in this section we discuss the charge centroid and the trapping efficiency of a SiN layer from the viewpoint of electric field dependence, since the electric field is an experimentally available parameter that is closely related to the carrier energy. Avalanche injection is introduced to perform low-field carrier injection, while Fowler-Nordheim (FN) injection is also used for carrier injection at a high electric field. Meanwhile, in some cases we have also performed ISPP (incremental step pulse programming) [25] to confirm 100% carrier capture.

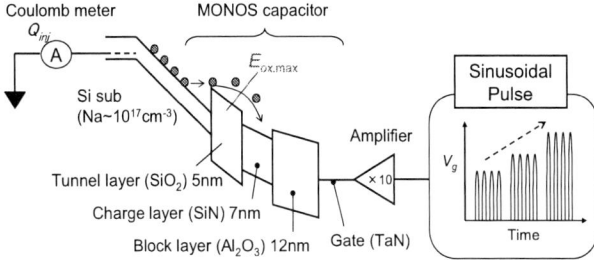

Figure 6. Experimental setup for avalanche-ISPP. High-frequency sinusoidal pulse stream (with a linear increase of its maximum voltage as a function of time) is applied to the gate electrode of a MONOS memory cell. The MONOS device has a p-type substrate without n$^+$ diffusion layers. The injected charge is monitored with a coulomb meter connected to the Si-substrate.

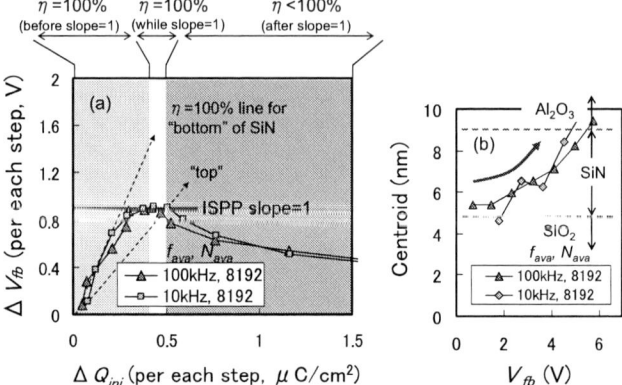

Figure 7. (a) Flatband voltage shift (ΔV_{fb} per each ISPP step) as a function of injected charge (ΔQ_{inj} per each ISPP step). The gate voltage step in ISPP ($\Delta V_{g.step}$) is 0.9 V. While $\Delta V_{fb} = 0.9$V (ISPP slope = 1), $\Delta V_{fb} - \Delta Q_{inj}$ characteristics move from the "bottom" guide line to the "top" guide line of a SiN layer. The notations of η, f_{ava}, and N_{ava} represent trapping efficiency, the frequency of sinusoidal pulse stream, and the number of sinusoidal pulses in each voltage step, respectively. (b) Charge centroid plotted as a function of flatband voltage V_{fb}. The charge centroid shifts from the bottom to the top interface of the SiN layer during the avalanche ISPP.

Figure 6 is a schematic diagram of the experimental setup for the ISPP measurement using avalanche injection. The avalanche-ISPP can be performed by stepping up the maximum voltage of sinusoidal wave forms. The 100% carrier capture is guaranteed in the region where $\Delta V_{fb} / \Delta V_{g.step} = 1$ is satisfied, where $\Delta V_{g.step}$ is the step-up voltage applied to the gate (i.e. increment in the maximum voltage of sinusoidal waves), and ΔV_{fb} is the flatband voltage shift in each step. Figure 7 shows a result of avalanche-ISPP where the electric field of the tunnel oxide was in the range of 3~4 MV/cm. In Fig. 7(a), there is a region indicated by "ISPP slope = 1" (or "while slope = 1") where 100% carrier capture is assured. On the left side of this region, there is a region named "before slope = 1" where we can also reasonably assume 100% carrier capture. (This region is closer to the beginning of ISPP measurement. The smaller electric field at the beginning of ISPP would be favorable for efficient carrier capture. In addition, there are more unfilled traps at the initial stage of ISPP.) Thus, 100 % carrier capture can be assumed in both "before slope = 1" and "while slope = 1" regions. Accordingly, we can exactly extract the charge centroid in these regions without the disturbance from the leakage current, as shown in Fig. 7(b). Note that the vertical axis of Fig. 7(b) represents the charge centroid for *instantaneous* charge injection (i.e. a small amount of charge injection at each V_{fb}), whereas the previous figures (Figs. 3 and 4) show the average charge centroid during P/E operations. The extracted charge centroid in Fig. 7(b) indicates that the injected charges are first trapped near the bottom interface (SiO_2/SiN) of the SiN layer. It also indicates that the charge trapping location gradually shifts towards the top interface (SiN/Al_2O_3) as the carrier capture proceeds during avalanche-ISPP. Thus, this result means that the bulk traps in the SiN layer are available and sequentially filled from the bottom to the top interface during the electron injection at a low tunneling-oxide electric field (E_{ox} = 3 ~ 4 MV/cm).

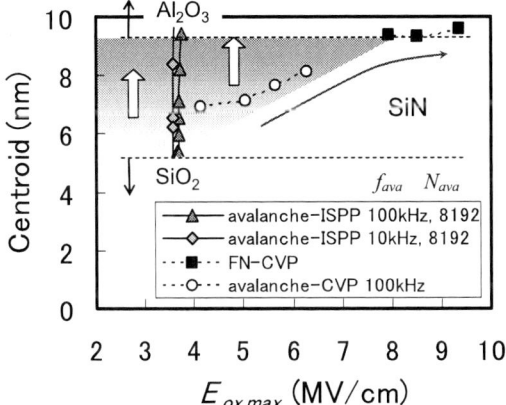

Figure 8. The dependence of charge centroid on the electric field of the tunneling oxide. The charge centroids in avalanche-ISPP, avalanche-CVP (constant voltage programming), and FN-CVP are compared. The accessible trap sites are limited near the charge layer / blocking layer interface when programming is performed at a high electric field.

Now, the electric field dependence of the charge centroid in both avalanche injection and FN injection has been measured, as shown in Fig. 8. The evaluation of the charge centroid has been performed with the methods of ISPP and constant voltage programming (CVP). In the case of CVP, only the first data point in the measurement time sequence have been adopted to ensure 100% carrier capture for all the electric field $E_{ox,max}$. It is clearly seen from the CVP data in Fig. 8 that the charge centroid gradually shifts towards the upper interface (SiN/Al₂O₃) as the electric field of the tunnel oxide increases. The shifting trends of the charge centroid in both avalanche injection and FN injection are continuous, as indicated by the long arrow in Fig. 8. This result suggests that the location of charge centroid is determined only by the electric field of the tunneling oxide, irrespective of the charge injection method. From this result, we can conclude that only the traps at the upper interface (SiN/Al₂O₃) are available in the actual application of MONOS devices at high field (as in NAND flash memory), although bulk traps and/or the traps near the lower interface (SiO₂/SiN) may also contribute in trapping if carrier injection at a lower field is possible.

Finally, the electric field dependence of carrier capture efficiency has been explored. As shown in Fig. 9, the carrier trapping kinetics indicated by the flatband voltage shift ΔV_{fb} vs. injected charge Q_{inj} at a low field (6 MV/cm, avalanche injection) follow 100% capture curve. In contrast, there is significant deviation from the 100% capture at a high field (8 MV/cm, FN injection), resulting in the degradation of carrier capture efficiency. Thus, in accordance with the behavior of charge centroid in Fig. 8, the carrier trapping efficiency is also affected significantly by the electric field. The result in Fig. 9 indicates that the carrier injection at low electric field is necessary to realize the programming operation with high carrier capture efficiency. In this sense, it is required to establish a band engineering of the tunneling oxide so that the electric field in carrier injection can be lowered while maintaining the data retention capability [29, 30].

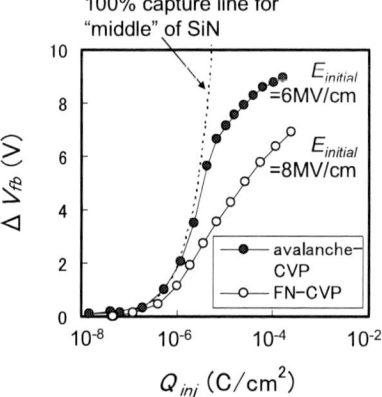

Figure 9. The flatband voltage shift (ΔV_{fb}) as a function of total injected charge (Q_{inj}). Avalanche injection and FN injection are compared. Both data have been measured with CVP. The electric field for programming is set to be 6 MV/cm and 8 MV/cm for avalanche and FN injection, respectively. The avalanche injection follows the 100% capture curve, while the FN injection deviates from it significantly.

Degradation during Program/Erase Cycling

Device degradation during cycling operation (repetitive P/E) is one of the major concerns in the use of MONOS memory [7,10-14]. Although cycling induced endurance was also the issue in the conventional floating-gate (FG) type non-volatile memory, the mechanism of endurance degradation in the FG memory is relatively simple. It is considered to be the trap generation in the tunneling oxide due to electron injection in both program and erase operations [14]. In contrast, in the case of MONOS memory, the erasing is considered to proceed mainly with hole injection from a Si-substrate [31]. Thus, the erase operation is generally slower in MONOS relative to FG memory. Therefore, higher voltage is necessary to perform erasing at a moderate speed, leading to cycling degradation. The detailed origin of the endurance degradation in MONOS memory is, however, believed to be electrons with high energy [14] which cause anode hole injection [11,32]. To suppress cycling degradation, it appears that the improvement of erasing speed is essential. There are actually two directions in the enhancement of erase speed. One of them is to enhance hole injection by introducing a tunneling oxide which can increase the tunneling probability of holes from a Si-substrate. Typical example of this approach is to use ONO ($SiO_2/SiN/SiO_2$) tunneling oxide [30]. The other approach is to use a SiN layer with Si-rich composition, so that the detrapping of electrons during the erasing can be enhanced [10,11,15,16,33].

In this section we evaluate the cycling degradation of a TANOS (MONOS with an Al_2O_3 blocking layer and a TaN gate) while measuring the injected charge density during the programming and erasing. By using a TaN gate, the back tunneling current from the gate electrode is suppressed during the erasing. In our cycling endurance tests, the pulse voltage and duration are varied in a wide range, and we try to correlate the cycling degradation with the total injected charge during programming and during erasing, respectively [34]. Based on the results, we show that the fluence of holes from a Si-substrate is the key factor in the cycling degradation of MONOS devices. Besides, we also discuss the *pros and cons* of the two different directions in erase-speed enhancement as a solution to reduce cycling degradation in MONOS devices.

TABLE I. MONOS devices used for the evaluation of cycling degradation.

Sample name	Tunnel oxide	Charge SiN	Block Al_2O_3	EOT
2.09		R.I.=2.09 5 nm		12.9 nm
2.13		R.I.=2.13 5 nm		13.1 nm
2.23	SiO_2 5 nm	R.I.=2.23 5 nm	13nm	12.9 nm
2.30		R.I.=2.30 5 nm		13.2 nm
ONO+2.09	ONO 5 nm	R.I.=2.09 5 nm		13.0 nm

TABLE II. Program / Erase conditions for cycling endurance test.

Endurance conditions	Program V_{pgm}, T_{pgm}	Erase V_{erase}, T_{erase}
1	20 V, 1 ms	-18 V, 1 or 10 ms
2	18 V, 1 ms	-20 V, 1 or 10 ms
3	20 V, 100 µs	-20 V, 1 or 10 ms
4	20 V, 100 µs	-18 V, 10 or 100 ms
5	18 V, 1 ms	-18 V, 10 or 100 ms
6	18 V, 100 µs	-18 V, 1 or 10 ms
7	20 V, 1 ms	-20 V, 10 or 100 ms
8	18 V, 100 µs	-20 V, 10 or 100 ms

The sample structures used in the measurements are listed in Table I. The samples are TANOS with different SiN compositions which are named after the refractive index (R.I.) of the silicon nitride layer. Note that R.I. = 1.99 corresponds to the stoichiometric Si_3N_4 composition, and the larger R.I. in Table I indicates more Si-rich SiN layer. We have performed cycling operations in the TANOS capacitors by applying various pulse voltage and duration, as indicated in Table II. Since it is known that the interface-state generation is the predominant degradation mechanism in MONOS devices [13,14], the amount of "C-V stretch" defined in Fig. 10 has been evaluated as an indicator of the cycling degradation. The "C-V stretch" is plotted as a function of the total injected charge density during programming Q_{pgm}, and during the erasing Q_{era}, respectively. As shown in Fig. 11, the "C-V stretch" has strong correlation with Q_{era}, while the correlation of the "C-V stretch" with Q_{pgm} is very weak. Considering that these results have been obtained from the endurance conditions with a variety of pulse voltage and duration (Table II), it is reasonable to assume that the cycling degradation is attributed to the charge injection during the erase operation.

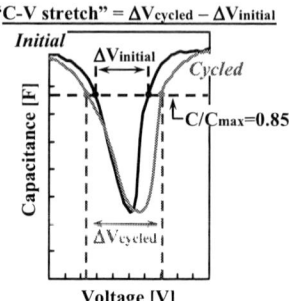

Figure 10. Definition of "C-V stretch" evaluated in this study. The "C-V stretch" is defined as the difference of the gate-voltage widths in the valley of the C-V curves before and after a cycling operation.

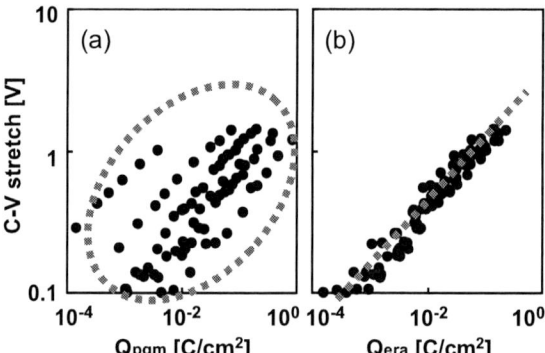

Figure 11. "C-V stretch" as a function of (a): Q_{pgm} and (b): Q_{era}. "C-V stretch" strongly correlates with the total injected charge density during the erase operations: Q_{era}. The measured device is "2.09" (i.e. R.I. = 2.09 in the SiN layer).

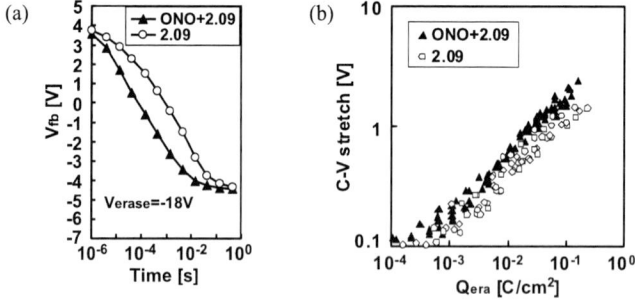

Figure 12. (a): Flatband voltage shift as a function of pulse duration. (b): "C-V stretch" as a function of Q_{era}. The devices are "ONO+2.09" and "2.09".

In Fig. 12, we compare the cycling degradation in different tunneling oxides, i.e. SiO_2 and ONO tunneling layers. As shown in Fig. 12(a), the ONO tunnel oxide has faster erase speed because of the enhancement in hole injection during the erase operation. However, when "C-V stretch" is plotted as a function of Q_{era} as shown in Fig. 12(b), there is not substantial difference between SiO_2 and ONO tunneling oxides. Thus, the cycling degradation characteristics are determined by the charge fluence Q_{era}, irrespective of the tunnel oxide structures. This means that the MONOS devices with ONO tunneling oxide do not have advantage in cycling endurance over those with SiO_2 tunneling oxide, under the assumption that the threshold voltage window (or V_{fb} window) during the P/E cycling is the same. Therefore, care must be taken to suppress the degradation of tunneling oxide and its interface in the use of ONO tunneling oxide [35], as well as in the conventional SiO_2 tunneling oxide.

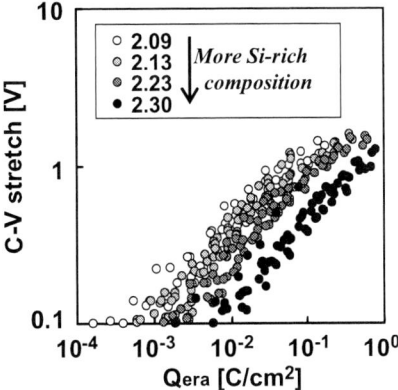

Figure 13. "C-V stretch" vs. Q_{era} for MONOS devices with different SiN compositions: R.I. = 2.09, 2.13, 2.23, and 2.30.

In Fig. 13 we show "C-V stretch" for MONOS capacitors with different SiN compositions as a function of Q_{era}. It is evident from Fig. 13 that MONOS with more Si-rich SiN layer (i.e. larger R.I.) has less cycling degradation as a function of Q_{era}. We infer that this result originates from the predominance of electron detrapping out of a Si-rich SiN layer, relative to hole injection from a Si-substrate, during the erase operation. Since the electron current does not contribute to the cycling degradation as suggested in Fig. 11(a), it turns out that the MONOS with a Si-rich SiN layer requires more Q_{era} to reach an equal amount of "C-V stretch" (i.e. equal amount of hole fluence).

The predominance of electron current during the erase operation in a Si-rich SiN layer has been confirmed as follows. In our method, the instantaneous injection current J is given by the incremental injection charge density ΔQ_{inj} divided by the pulse duration Δt:

$$J = \frac{\Delta Q_{inj}}{\Delta t}. \tag{5}$$

The electric field across the tunneling oxide, corresponding to J, is expressed [31] as

$$E_{ox} = \frac{V_g - V_{fb}}{EOT} \tag{6}$$

where V_g is the pulse gate voltage, and V_{fb} is the flatband voltage during the application of the pulse voltage. From these relations, we can extract J-E characteristics for the carrier injection across the tunneling oxide. As shown in Fig. 14(a), J-E characteristics during programming are independent of the composition of SiN layers. This result means that the injection current during programming is the electron current from the inversion layer of a Si-substrate, which is independent of the SiN composition. In contrast, J-E characteristics in the erase operation have strong dependence on the composition of SiN layers, as shown in Fig. 14(b). In this figure, the theoretical hole FN tunneling current from a Si-substrate is also plotted with a solid curve. It has been observed that the J-E characteristics become closer to the theoretical hole FN tunneling current as the composition of the SiN layer is nearer to the stoichiometric state. Thus, the current flow above the hole FN tunneling is considered to be due to the electron detrapping out of the SiN layer. (Note that electron back tunneling current from the gate electrode is negligible in this figure, since it increases as the progress of erase operation (i.e. as $|E_{ox}|$ becoming smaller).) From this result, it has been found that orders of magnitude larger electron current flows relative to hole current in the erase operation of MONOS devices with Si-rich SiN layers.

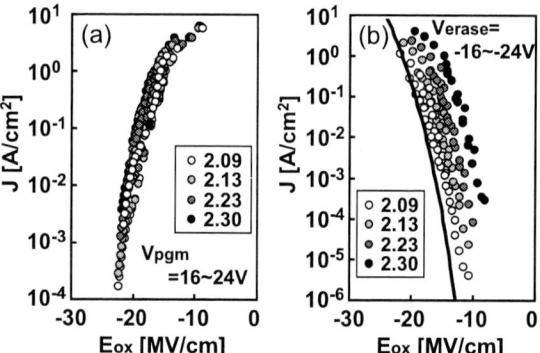

Figure 14. J-E_{ox} characteristics during (a): program and (b): erase operation extracted with eqs. (5) and (6) for MONOS devices having different SiN compositions: R.I. = 2.09, 2.13, 2.23, and 2.30.

The combination of the results in Fig. 13 and Fig. 14(b) indicates the importance of *hole injection from a Si-substrate* in the cycling degradation of MONOS devices, since the near-stoichiometric SiN layer (R.I. = 2.09) with predominant hole current from a Si-substrate shows the largest degradation. This conclusion is quite different from the reports in previous papers. It is true that our results are similar to reference [14] where the interface state generation in the erase operation is reported to have significant role in the endurance degradation of TANOS devices. In reference [14], however, high-energy electrons from the gate electrode and/or the SiN layer were considered to be the origin of endurance degradation. It is also common to believe that anode hole injection caused by high-energy electrons are responsible for endurance degradation [11,32]. In contrast, our results show that high-energy electrons and resultant anode hole injection are not the main cause. The reasons are as follows: On one hand, if the electrons in program operation are responsible, the "C-V stretch" should be a unique function of Q_{pgm}, which is not the case. On the other hand, if the detrapped electrons out of a SiN layer are responsible, the amount of interface states (~ "C-V stretch") should be larger as the composition of the SiN layer becomes more Si-rich (i.e. electron-current dominant), which is not the case, either. Yet another possibility is the electron injection from the TaN gate during the erase. However, this factor is negligible in the usual P/E conditions indicated in Table II, as demonstrated by the J - E_{ox} curve in Fig. 14(b). Therefore, we can conclude that the interface state generation during P/E cycling is mainly due to the holes injected from a Si-substrate. The role of high-energy electrons in endurance degradation is much smaller, at least in the normal operation of TANOS devices.

Finally, as a summary of this subsection, our results indicate the importance of injected carrier type (electron or hole) in the cycling degradation of MONOS devices. Although it was generally believed that lowering the gate voltage during erasing was beneficial to improve cycling endurance, our results have demonstrated that the improvement of erase speed via enhancement of hole injection from a Si-substrate (typically with a ONO tunneling oxide) does not lead to the improvement of endurance characteristics. In contrast, the enhancement of electron current during the erasing by the introduction of a Si-rich SiN layer brings about substantial improvement of endurance characteristics, since the detrapped electrons from a SiN layer has negligible contribution to the degradation of the tunneling oxide and its interface. Therefore, the direction of the solution to improve the erasing performance and the cycling degradation simultaneously is definitely to adopt a MONOS structure with dominant electron current conduction in both programming and erasing. A MONOS with a Si-rich SiN layer is one of the solutions in this direction. However, it has been reported that the MONOS with Si-rich SiN has poor data retention characteristics [10,11,15,16,33]. In the next section, we will discuss how to obtain the simultaneous improvements of P/E, endurance and retention through physical understanding of data retention characteristics.

Understanding Data Retention Characteristics

As discussed in the previous section, it is an important issue to realize MONOS devices where erase performance and endurance properties are improved without sacrificing the data retention characteristics. The problem of the MONOS with a Si-rich SiN layer is that trapped electrons are relatively mobile during the data retention. To improve this situation, it has been reported that laminate SiN layers can achieve simultaneous improvement of endurance and retention [19]. In this paper, we try to understand the underlying physics of the laminate SiN layers through the retention measurements accelerated by external bias and temperature. In the following, we compare the measurement results of MONOS with a Si-rich single SiN layer and those with the laminate SiN layers [36].

The "bias retention" is the retention measurement where an external gate bias is applied [7,37,38]. Through this method it is possible to separate the contributions of the tunneling oxide and the blocking oxide in the retention deterioration characteristics. This is because the application of a positive gate bias accelerates the carrier detrapping across the blocking oxide layer, while the application of a negative bias enhances the carrier detrapping towards the tunneling oxide. In addition, the retention characteristics evaluated with this method at several different temperatures give the activation energies for the charge loss in both of the detrapping directions.

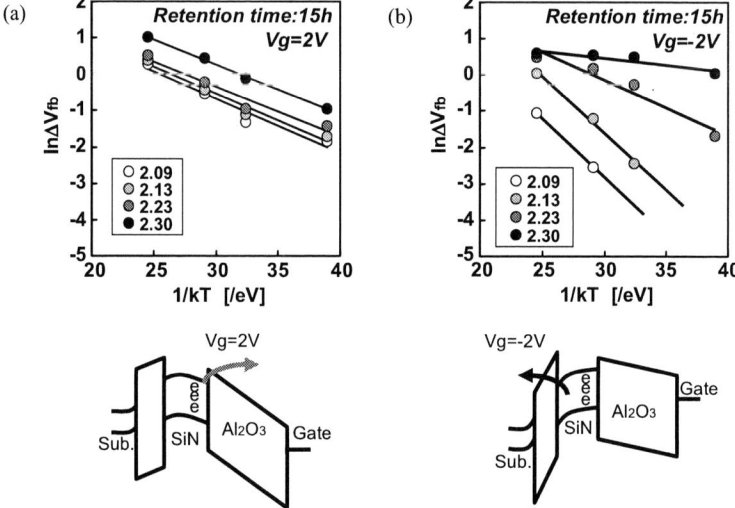

Figure 15. Arrhenius plots for data retention characteristics under (a) $V_g = 2$ V and (b) $V_g = -2$ V for 15 hours. Schematic band diagrams indicating the directions of carrier detrapping are also shown below.

First, we have evaluated the retention characteristics of MONOS devices with a Si-rich single SiN layer. The MONOS devices were programmed up to $V_{fb} = 4$ V at room temperature, and then temperature was risen to measure the flatband voltage shift ΔV_{fb}. Figure 15 shows the Arrhenius plots of data retention characteristics measured at $V_g = 2$ V and -2 V for MONOS devices with different SiN compositions as listed in Table I. The activation energy at $V_g = 2$V is found to be almost constant (~0.15 eV), irrespective of the composition of the SiN layers. In contrast, the temperature dependence of data retention measured at $V_g = -2$V has a strong dependence on the composition of SiN layers, with the activation energy varying in the range of 0.04 ~ 0.32 eV. These results are understandable considering that the trapped charge in a SiN layer is located near the SiN/Al$_2$O$_3$ interface after the programming at a high field (FN tunneling): The trapped electrons do not need to travel across the SiN layer when the detrapping is accelerated towards the Al$_2$O$_3$ blocking layer ($V_g > 0$), leading to the activation energy independent of the SiN composition. Considering the charge loss mechanism through Al$_2$O$_3$, it is possible that the activation energy of ~ 0.15 eV corresponds to the barrier height for Schottky emission at the SiN/Al$_2$O$_3$ interface and/or the thermally activated bulk conduction in the Al$_2$O$_3$ blocking layer. Note that there is only a small potential barrier at the SiN/Al$_2$O$_3$ interface [39,40]. In contrast, when electron detrapping is accelerated towards the tunneling oxide ($V_g < 0$), the trapped electrons near the SiN/Al$_2$O$_3$ interface need to move across the SiN layer to reach the tunneling oxide. Thus, the different composition of SiN layers has significant impact on the retention activation energy.

In the case without the application of an external bias, the activation energy for the data retention is located somewhat in-between. We have found that the activation energy at $V_g = 0$V for the retention after a short time (1 hour) is not dependent on the SiN composition, while the activation energy after a long time (15 hours) has some dependence on the SiN composition, as shown in Fig. 16(a) and (b), respectively. From these results, we infer that the short-term detrapping is caused by the charge emission across the blocking oxide, while the long-term retention is governed by the charge loss through the tunneling oxide. We should point out that the charge leakage path has a very important role during the data retention of MONOS memory devices in actual operations.

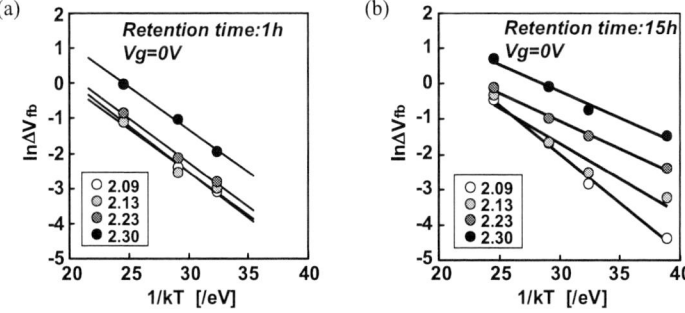

Figure 16. (a) Arrhenius plots for data retention characteristics at $V_g = 0$ V for 1 hour. (b) Arrhenius plots for data retention at $V_g = 0$ V for 15 hours.

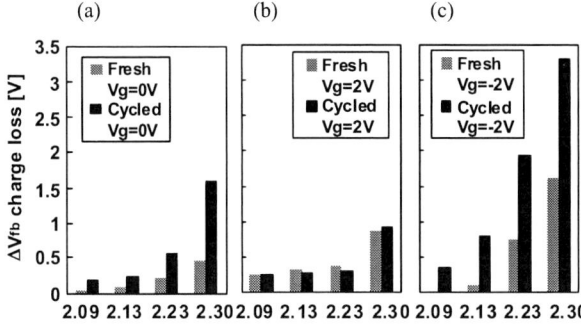

Figure 17. Comparison of data retention characteristics before and after 1.2 k P/E cycling operations. Different gate biases ($V_g =$ (a) 0V, (b) 2V, and (c) -2V) were applied during retention. Cycling operation and retention measurement were performed at room temperature and 85 °C, respectively.

The discussion so far has been focused on the retention characteristics of "fresh" samples with Si-rich SiN composition. Another important issue is the degradation of retention characteristics due to P/E cycling operations, and we can look into those characteristics by using the "bias retention" method, as well. In other words, we can identify whether the tunneling oxide or the blocking oxide is responsible for the device deterioration during cycling. As shown in Fig. 17(a), the data retention after a cycling test at $V_g = 0$ V shows significant degradation, as expected. By performing bias retention experiments, it has been observed that the retention at $V_g = 2$ V (i.e. detrapping into the blocking oxide) does not degrade owing to the cycling operation, as shown in Fig. 17(b). In contrast, the retention at $V_g = -2$ V (i.e. detrapping into the tunneling oxide) severely degrades after the cycling, as indicated in Fig. 17(c). These results mean that the degradation of data retention due to P/E cycling is caused by the damage to the tunneling oxide and/or the SiN layer, while there is no damage to the blocking oxide. The activation energy in the bias retention experiments after cycling has also shown the followings [36]: On one hand, the activation energy at $V_g = 2$ V remains to be ~ 0.15 eV during the cycling. On the other hand, the activation energy at $V_g = -2$ V becomes smaller after the cycling relative to the "fresh" samples. These results have been observed for all the SiN compositions. The results on activation energy are also consistent with the model of damage generation to the tunneling oxide and/or the SiN layer.

TABLE III. MONOS devices used for demonstrating reliability optimization.

Sample name	Tunnel oxide	Charge SiN	Block Al$_2$O$_3$
2.25 + 2.09	SiO$_2$ 5 nm	R.I. = 2.25 2.5 nm / R.I. = 2.09 2.5 nm	15 nm
2.25		R.I. = 2.25 5 nm	

From the retention characteristics of both fresh and cycled samples, it has been found that the composition of SiN layers has significant impact on the data retention properties. Considering that Si-rich SiN layers have better endurance (i.e. less damage to the tunneling oxide) relative to near-stoichiometric SiN as shown in Fig. 13, it is worthwhile to focus on how to improve the initial data-retention characteristics of MONOS with Si-rich SiN layers, so that both better endurance and retention (before and after cycling) are realized simultaneously. From this perspective, we have examined the MONOS with a laminate SiN structure as listed in Table III (sample name: "2.25 + 2.09") which has achieved fast erasing speed and excellent retention characteristics [19]. The evaluation of "bias retention" is now applied to the set of fresh MONOS devices in Table III. Note that the laminate SiN consists of a stack of SiN layers where more Si-rich (R.I. = 2.25) and less Si-rich (R.I. = 2.09) SiN layers are located near the tunneling oxide and the blocking oxide, respectively. The MONOS with a Si-rich single SiN layer (R.I. = 2.25) is evaluated as a reference sample.

We have measured erase characteristics and the data retention of these MONOS devices. In Fig. 18 we show the instantaneous J-E curves during erasing, and it has been found that the injection current across the tunneling oxide is almost the same for both the Si-rich single SiN layer and the laminate SiN layers during an erase operation. The injection current is much larger than that of the theoretical FN hole tunneling from a Si-substrate. Thus, Fig. 18 indicates that the erase operation proceeds mainly by electron detrapping from the SiN layers for both of the samples in quite a similar manner.

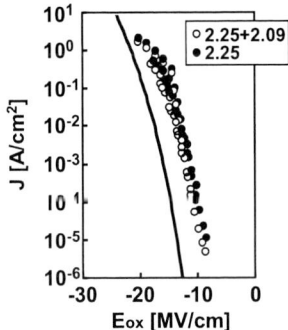

Figure 18. *J-E$_{ox}$* characteristics of "2.25 + 2.09" and "2.25" samples during erase operation. Theoretical hole FN tunneling current from a Si substrate, which is calculated with WKB approximation, is also plotted with a solid curve.

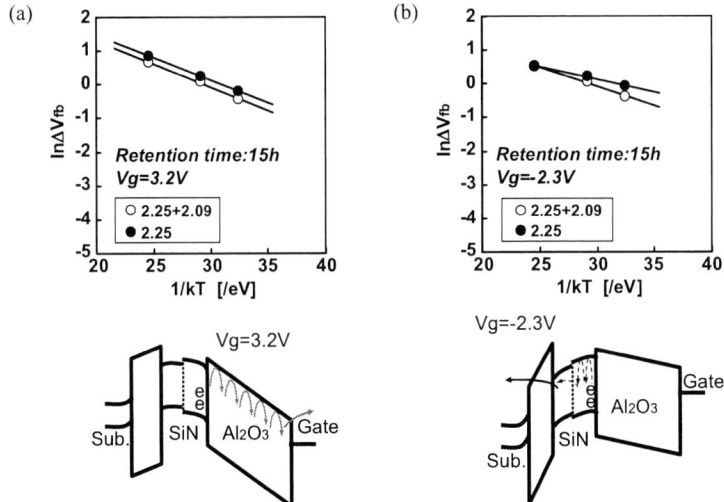

Figure 19. Arrhenius plots for data retention characteristics of MONOS devices listed in Table III under (a) V_g = 3.2 V and (b) V_g = -2.3 V. Schematic band diagrams indicating the directions of carrier detrapping are also shown below.

In Fig. 19 we show Arrhenius plots for the "bias retention" measurements after programming up to V_{fb} = 4 V in the MONOS devices with single-layer SiN ("2.25") and the laminate SiN ("2.25 + 2.09"). The activation energy for the charge loss through the blocking layer (Fig. 19(a) for V_g = 3.2 V) is in the range of 0.13 ~ 0.14 eV without any dependence on the SiN structures. In contrast, the activation energy for the charge loss toward the tunneling oxide layer (Fig. 19(b): V_g = -2.3 V) depends on the SiN structures: The activation energy is 0.11 eV for the laminate SiN layers ("2.25 + 2.09"), and 0.07 eV for the single Si-rich SiN layer ("2.25"). From the larger activation energy in the MONOS with the laminate SiN layer, we infer that the presence of a less Si-rich layer ("2.09") between the SiN/Al_2O_3 and SiO_2/SiN interfaces suppresses the electron movement across the SiN layer. Meanwhile, the fast speed during erasing operation is maintained with the electron current out of the Si-rich SiN layer ("2.25") adjacent to the tunneling oxide layer. Thus, the MONOS with the laminate SiN layer can achieve the simultaneous improvement of cycling endurance and data retention characteristics. The laminate SiN layer [19] is one of the candidates for the charge trapping layer in future MONOS memory devices. Similar effects can also be expected for the compositionally graded SiN layer [10].

The Influence of High-k Blocking Layer on MONOS Device Characteristics

High-k dielectrics have been studied for more than ten years as a gate dielectric of logic CMOS devices. However, the history of introducing high-k dielectrics into the MONOS memory devices, beginning with TANOS [41], is not so long. The influence of high-k dielectrics on memory operation is not yet fully understood. In this section we take up two topics related to high-k materials used in MONOS memory devices.

Transient channel current during data read-out

Recent MONOS memory devices have high-k dielectrics in the blocking oxide. It is known that high-k dielectrics have different properties from the conventional SiO_2. One of them is the presence of slow polarization component [20] as schematically shown in Fig. 20(a), in addition to the instantaneous polarization. This means that the effective oxide thickness (EOT) of the MONOS gate stack with a high-k dielectric layer appears to change during memory operations. As a result, MONOS memory characteristics such as the channel current (See Fig. 20(a)) are influenced through the change in EOT. We have observed transient behavior in the channel current of MONOS transistors with a high-k (Al_2O_3) blocking layer [42]. In Fig. 20 (b) we indicate that the drain current of a MANOS device (SiO_2(4 nm)/SiN(5 nm)/Al_2O_3(15 nm)) shows transient characteristics when a step pulse voltage is applied to the control gate. The transient current component is proportional to the height of a step voltage. This result can be represented as the threshold voltage shift (ΔV_t), as shown in Fig. 20(c), with the use of the expression $\Delta V_t = \Delta I_d / g_m$ where g_m is transconductance. It has been found that the maximum threshold voltage shift can be as large as ~ 0.8 V when the gate voltage of 10 V is applied. This is not a small threshold voltage shift, considering the multi-level operation of NAND flash memory. As this example shows, the slow polarization of high-k dielectric will have strong impact on the data read-out when MONOS memory devices with high-k dielectrics are implemented into a NAND memory string.

In order to understand the origin of the transient channel current, measurements have been performed on a set of transistors with different dielectric layer structures. One is "MAOS" without a SiN layer (Fig. 21(a)), and the other is "SONOS" with a SiO_2 blocking layer in stead of Al_2O_3 (Fig. 21(b)). In addition, a "MOS" transistor with a SiO_2 gate dielectric (Fig. 21(c)) has also been evaluated to confirm that the transient behavior does not originate from the measurement system. As shown in Fig. 21, transient channel current appears in the presence of an Al_2O_3 layer, while the channel current is constant when all the dielectric layers are composed of SiO_2 and SiN. From these results, it is clear that the transient channel current originates from the high-k (Al_2O_3) blocking layer. The linearity of the threshold voltage shift ($\Delta V_{th} = \Delta I_d / g_m$) on the height of a step voltage in Fig. 20(c) suggests that the cause of the transient channel current is the slow polarization of the Al_2O_3 layer rather than the carrier trapping into Al_2O_3 from the gate electrode, because the injection current of carrier trapping is generally a non-linear function of the applied voltage.

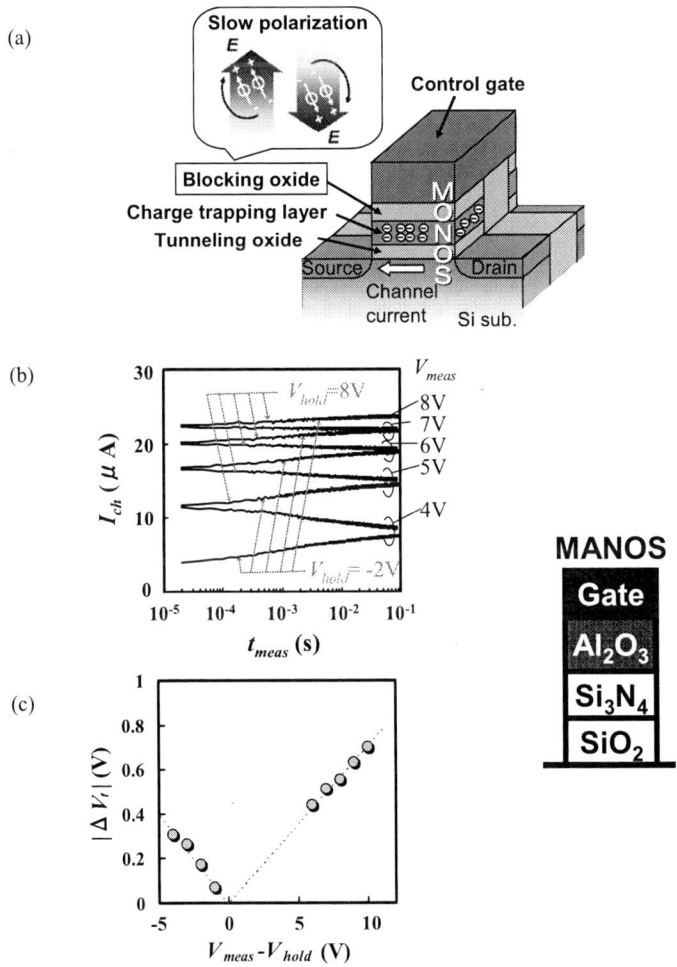

Figure 20. (a) Structure of MONOS with high-k dielectric. (b) The transient channel current as a function of measurement time t_{meas} after a step voltage is applied to the control gate, followed by a hold voltage of 100 ms. (c) The threshold voltage shift (ΔV_t) vs. the step voltage height. Device structure of MANOS is also shown schematically.

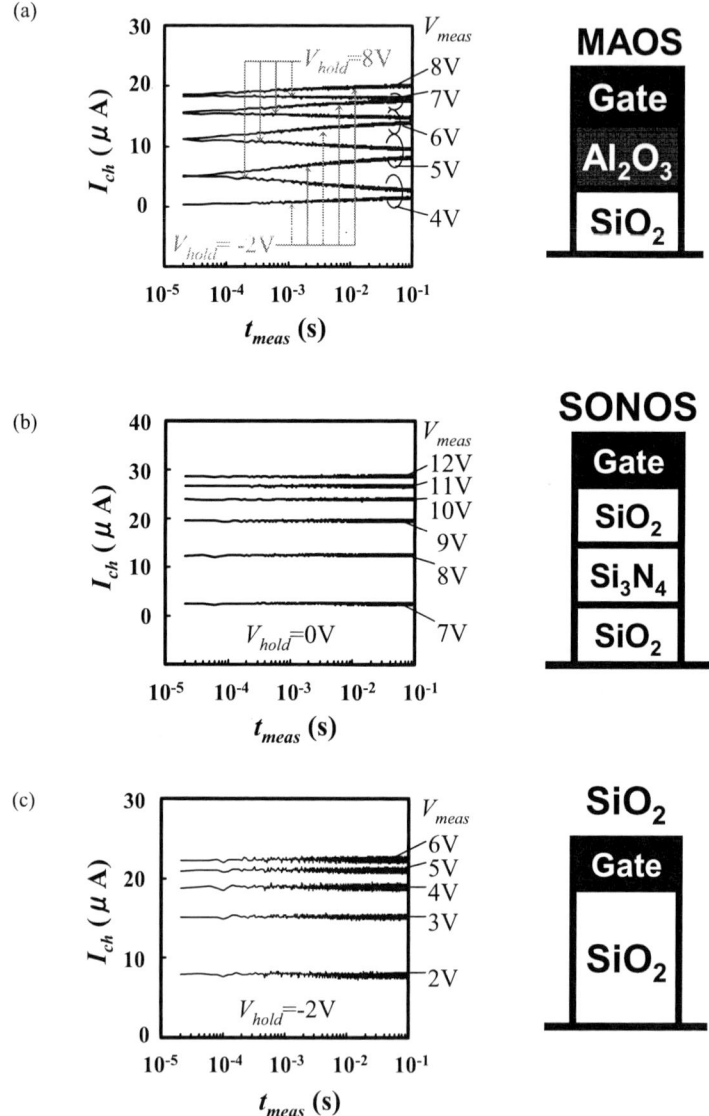

Figure 21. The transient channel current characteristics of (a): MAOS, (b): SONOS and (c): SiO_2 single layer transistors. Gate stack structures are also shown schematically. MAOS has similar transient channel current characteristic as that of MANOS in Fig. 20.

For the suppression of the transient channel current, we have developed a new stacked blocking layer structure composed of $Al_2O_3/SiO_2/Al_2O_3$ (referred to as AOA) [42]. This stacked blocking layer has the advantage of thinner film thickness relative to a single Al_2O_3 layer, while maintaining an equal EOT. The thinner film thickness is suitable for scaled memory devices in the future. The stacked blocking layer also has the advantage of lower leakage current relative to a single Al_2O_3 layer, as shown in Fig. 22. It has been found that the lower leakage current is due to the intermediate SiO_2 layer which works as an additional potential barrier. In Fig. 23 we show that the frequency dispersion of the capacitance in an AOA stacked film is much smaller than that of a single Al_2O_3 layer. This indicates that the slow polarization is smaller in AOA stacked films.

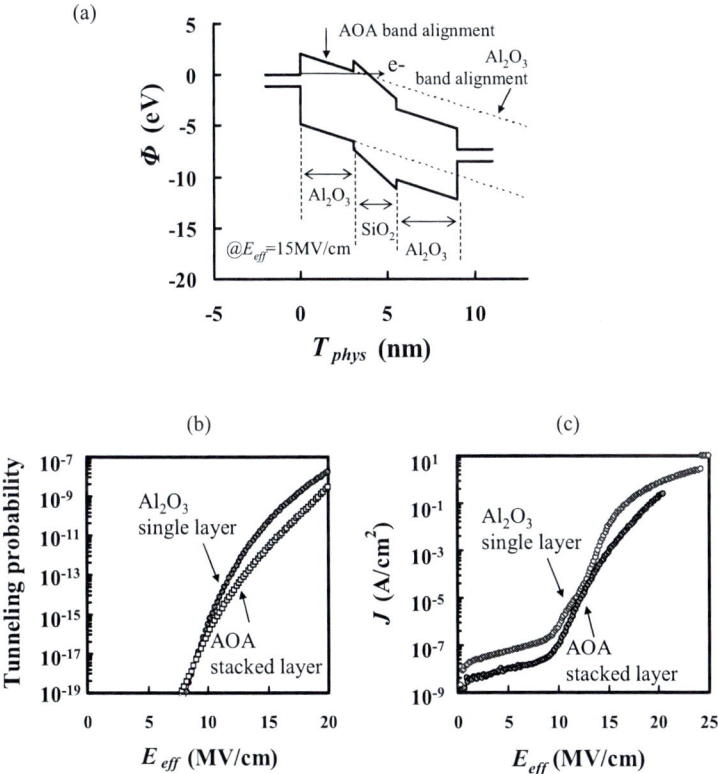

Figure 22. (a): Band diagram of AOA stacked film. SiO_2 middle layer in AOA has a larger band offset relative to Al_2O_3. (b): The tunneling probability calculated with WKB approximation for AOA and Al_2O_3. (c): Measured leakage current in the capacitors of Al_2O_3 (15 nm) and AOA (Al_2O_3(4 nm)/SiO_2(3.5 nm)/Al_2O_3(4 nm)).

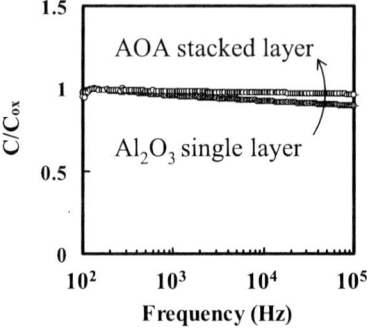

Figure 23. Frequency dispersion of Al_2O_3 (15 nm) single layer and AOA (Al_2O_3(4 nm)/SiO_2(3.5 nm)/Al_2O_3(4 nm)) stacked capacitors. Negligible frequency dispersion in the AOA capacitor indicates substantially less slow polarization relative to the Al_2O_3 single layer.

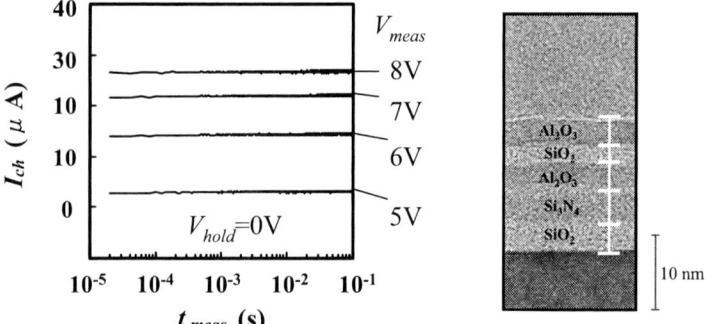

Figure 24. The transient channel current characteristics of a MONOS transistor with an AOA blocking layer. The channel current is stable with the use of an AOA film, instead of an Al_2O_3 blocking layer. Cross sectional TEM image of the MONOS gate stack with AOA blocking layer is also shown.

In Fig. 24 we show the channel current characteristics in a MONOS transistor when an AOA blocking layer is used instead of an Al_2O_3 single layer. It has been found that the channel current is constant during the application of a step voltage. This result is consistent with the negligible frequency dependence of the capacitance of an AOA film in Fig. 23. In summary, the MONOS device with an AOA blocking oxide layer can effectively suppress the transient response of the channel current originating from the slow polarization of the Al_2O_3 blocking layer. Together with the leakage current reduction across the blocking layer, this technology can achieve high P/E performance, stable memory cell current, and reliable data readout in MONOS-type NAND flash memory.

Temperature Dependence in Program Operation

The programming characteristics of MONOS memory is supposed to have no temperature dependence, since the mechanism of injected current during programming is tunneling (typically, FN tunneling). However, we have observed that the actual programming characteristics of SANOS (i.e. MONOS with an Al_2O_3 blocking layer and a poly-Si gate) have temperature dependence, as shown in Fig. 25 (a). The programming speed of SANOS becomes faster as temperature rises. In contrast, for a SONOS (i.e. MONOS with a SiO_2 blocking layer and a poly-Si gate), there is no temperature dependence in the programming characteristics, as shown in Fig. 25 (b).

In order to understand the temperature dependence in the programming characteristics of SANOS devices, we have used our evaluation method to measure the injection charge Q_{inj} and the flatband voltage shift ΔV_{fb} simultaneously. With this method we can separate the temperature dependence of program characteristics into two factors: One is the charge injection and the other is the charge trapping efficiency. The temperature dependence of the charge injection has been analyzed with J-E curves (See eqs. (5) and (6)), while that of the charge trapping efficiency has been estimated from ΔV_{fb} vs. Q_{inj} characteristics. As shown in Fig. 26, it has been found, for the case of SANOS, that the injection current is the main factor to produce temperature dependence, while there is only negligible temperature dependence in the charge trapping efficiency, particularly in the early stage of programming operation ($Q_{inj} < 10^{-5}$ C/cm^2).

From this result, it may seem that the tunneling oxide is primarily responsible for the temperature dependence in the programming characteristics. From the shape of the J-E characteristics of SANOS (Fig. 26(a)), it may also appear that the tunneling oxide has a significant amount of thermally activated leakage current. Nevertheless, the tunneling oxide of our MONOS devices was prepared by thermal oxidation of a Si-substrate. It is unusual that the J-E curve of a thermal oxide is quite different from the theoretical J-E curve of FN tunneling. In order to look into this point, we have also evaluated the J-E characteristics and the trapping efficiency of a SONOS device with a SiO_2 blocking layer where the tunneling oxide and the SiN layer were prepared in exactly the same process as in the SANOS device. As shown in Fig. 27, the J-E characteristics of SONOS have been found to follow the theoretical FN tunneling current, with no temperature dependence. Thus, the tunneling oxide in itself has the feature of FN tunneling current. This observation suggests that the large deviation of the J-E curve from the theoretical FN tunneling current in SANOS devices should be attributed to another factor: the presence of an Al_2O_3 blocking layer.

Recently, Padovani et al. [43] has also reported that there is temperature dependence in the program characteristics of TANOS (with Al_2O_3 layer). They ascribe the temperature-dependent programming characteristics to the increase of Al_2O_3 dielectric constant with temperature rise. It is true that our Al_2O_3 blocking layer also shows temperature dependence of dielectric constant to some extent, which appears as a temperature dependence of EOT in the SANOS devices. However, in our extraction of J-E characteristics in Fig. 26, the change in EOT with temperature is already included into the electric field of the tunneling oxide with eq. (6), where EOT is experimentally evaluated at each temperature. In spite of this analysis method, we still observe anomaly in the J-E curve.

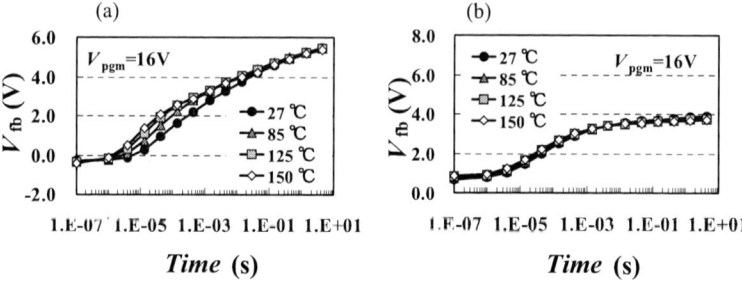

Figure 25. Temperature dependence of program characteristics in (a): SANOS (with Al$_2$O$_3$ blocking layer), and (b): SONOS (with SiO$_2$ blocking layer).

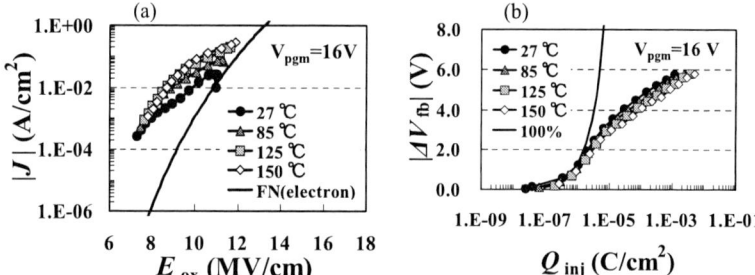

Figure 26. Separation of two factors contributing to the temperature dependence of SANOS devices with 15-nm Al$_2$O$_3$ blocking layer: (a) J-E characteristics, and (b) ΔV_{fb}-Q_{inj} characteristics in the program operation.

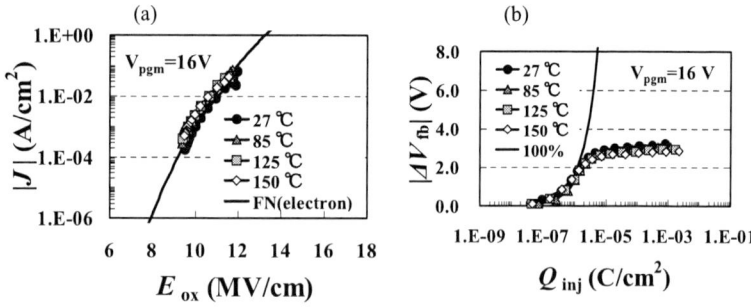

Figure 27. Separation of two factors contributing to the temperature dependence of SONOS devices with 6-nm SiO$_2$ blocking layer: (a) J-E characteristics, and (b) ΔV_{fb}-Q_{inj} characteristics in the program operation.

Therefore, instead of *static* change of dielectric constant, we should consider that the dielectric constant of an Al_2O_3 blocking layer changes in a *transient* manner during the program operation, thereby increasing the electric field across the tunneling oxide and enhancing the injected current from a Si-substrate. However, this physical picture demands transient *increase* of Al_2O_3 dielectric constant, which is not consistent with the emergence of slow polarization as discussed in the previous subsection. The physical and/or chemical mechanisms for the transient nature of Al_2O_3 during the programming operation at high electric field are still under investigation now.

In summary, we have separated the factors causing the temperature dependence in the program characteristics of SANOS devices. It has been found that charge injection is responsible for the temperature dependence of programming characteristics, while the charge trapping efficiency has negligible effect. The temperature dependence of the injection current originates not directly from the tunneling oxide, but from the presence of the Al_2O_3 blocking layer. It looks that Al_2O_3 has temperature-dependent *transient* dielectric response during the programming operation at high field. This conclusion also suggests the possibility that the introduction of an AOA blocking layer, which suppresses transient response, may be beneficial to improve the temperature dependence of programming characteristics.

Summary and Conclusions

We have developed a measurement method to evaluate the charge centroid of MONOS memory during P/E operations. In this method we use a standard rectangular gate pulse and evaluate the injected charge and the flatband voltage shift simultaneously. It has been found through this method that the charge centroid of MANOS devices during programming at high field is located near the SiN/Al_2O_3 interface. In contrast, for MANOS with a thick SiN layer or at a low field, the charge centroid is located in the bulk region of the SiN layer, suggesting the availability of bulk traps. From these results, carrier injection at low field during P/E operations is desired in MONOS memory devices.

The measurement of injected charge has also been applied to the analysis of cycling endurance. The cycling degradation has been found to be strongly correlated with the injected charge density during the erase operation, without regard to the tunneling oxide structure. Meanwhile, the electron current during programming is irrelevant to the cycling degradation. Thus, Si-rich SiN layer, with much electron detrapping current during the erasing, is favorable for the improvement of both the erasing speed and the cycling endurance, although it has poor data retention. The remaining issue of improving retention characteristics can be achieved with a laminate SiN layer by stacking Si-rich and near-stoichiometric SiN layers [19]. The physical reason for the retention improvement has been investigated with the use of bias-retention measurement and its temperature dependence. The presence of a near-stoichiometric SiN layer has been found to prevent trapped electrons near the SiN/Al_2O_3 interface from moving towards the tunneling oxide, while the presence of a Si-rich SiN layer adjacent to the tunneling oxide has the role of enhancing electron current during the erasing operation. As a result, simultaneous improvement of endurance and retention can be attained with the laminate SiN layer.

Finally, the effects of high-k dielectrics on the MONOS memory characteristics have been studied. The high-k dielectrics in a MONOS gate stack have significant impact on the reading and programming characteristics of MONOS memory devices through their transient dielectric properties. It has been found that a composite film such as AOA can be a good solution to suppress the transient response of high-k materials while maintaining a low leakage current and thin film thickness.

In conclusion, MONOS device with high-k dielectrics is one of the advantageous candidates for the future memory cells. The understanding of trapping characteristics and reliability is crucial for its implementation. The accurate measurements described in this paper have brought about much information which is beneficial for in-depth understanding of MONOS device properties. Several important principles to improve MONOS characteristics have been substantiated with the help of the advanced evaluations in this work, providing solid guidelines for future MONOS memory devices. Among them, we suggest that the introduction of the laminate SiN layers and the AOA blocking layer is helpful to enhance the performance and reliability of MONOS devices.

Acknowledgments

The authors are grateful to K. Sekine, R. Fujitsuka, W. Sakamoto, Semiconductor Company, Toshiba Corporation, and K. Sakuma, R&D Center, Toshiba Corporation for sample preparations used in this work. The authors also thank K. Muraoka, R&D Center, Toshiba Corporation for reviewing this paper.

References

1. T. Yaegashi, T. Okamura, W. Sakamoto, Y. Matsunaga, T. Toba, K. Sakuma, K. Gomikawa, K. Komiya, H. Nagashima, H. Akahori, K. Sekine, T. Kai, Y. Ozawa, M. Sugi, S. Watanabe. K. Narita, M. Umemura, H. Kutsukake, M. Sakuma, H. Maekawa, Y. Ishibashi, K. Sugimae, H. Koyama, T. Izumida, M. Kondo, N. Aoki and T. Watanabe, *Symp. VLSI Tech. Dig. Papers*, p. 190 (2009).
2. W. Sakamoto, T. Yaegashi, T. Okamura, T. Toba, K. Komiya, K. Sakuma, Y. Matsunaga, Y. Ishibashi, H. Nagashima, M. Sugi, N. Kawada, M. Umemura, M. Kondo, T. Izumida, N. Aoki and T. Watanabe, *IEDM Tech. Dig.*, p. 831 (2009).
3. A. Arreghini, F. Driussi, E. Vianello, D. Esseni, M. J. van Duuren, D. S. Golubovic, N. Akil, and R. van Schaijk, *IEEE Trans. Electron Devices*, **55**, 1211 (2008).
4. A. Arreghini, F. Driussi, D. Esseni, L. Selmi, M. van Duuren, and R. van Schaijk, *IEDM Tech. Dig.*, p. 499 (2006).
5. P.-Y. Du, H.-T. Lue, S.-Y. Wang, E.-K. Lai, T.-Y. Huang, K.-Y. Hsieh, R. Liu, and C.-Y. Lu, *IEEE Trans. Device Materials Reliability*, **7**, 407 (2007).
6. H.-T. Lue, Y.-H. Shih, K.-Y. Hsieh, R. Liu, and C.-Y. Lu, *IEEE Electron Device Lett.*, **25**, 816 (2004).
7. C. Sandhya, A. B. Oak, N. Chattar, A. S. Joshi, U. Ganguly, C. Olsen, S. M. Seutter, L. Date, R. Hung, J. Vasi, and S. Mahapatra, *IEEE Trans. Electron Devices*, **56**, 3123 (2009).
8. T. Ishida, Y. Okuyama, and R. Yamada, *Proc. IRPS*, p. 516 (2006).
9. A. Suhane, A. Arreghini, G. Van den bosch, L. Breuil, A. Cacciato, A. Rothschild, M. Jurczak, J. Van Houdt and K. De Meyer, *Proc. of ESSDERC* p. 276 (2009).

10. N. Goel, D. C. Gilmer, H. Park, V. Diaz, Y. Sun, J. Price, C. Park, P. Pianetta, P. D. Kirsch, and R. Jammy, *IEEE Electron Device Lett.*, **30**, 216 (2009).
11. C. Sandhya, A. B. Oak, N. Chattar, U. Ganguly, C. Olsen, S. M. Seutter, *IEEE Trans. Electron Devices*, **57**, 1548 (2010).
12. G. Ghidini, C. Scozzari, N. Galbiati, A. Modelli, E. Camerlenghi, M. Alessandri, A. Del Vitto, G. Albini, A. Grossi, T. Ghilardi, and P. Tessariol, *Microelectron. Eng.*, **86**, 1822 (2009).
13. C. H. Lee, W. H. Tu, S. H. Gu, C.W. Wu, S. W. Lin, T. H. Yeh, K.F. Chen, Y. J. Chen, J. Y. Hsieh, I. J. Huang, N. K. Zous, T. T. Han, M. S. Chen, W. P. Lu, K. C. Chen, T. Wang, and C. Y. Lu, *Proc. IRPS*, p. 891 (2009).
14. G. Van den bosch, L. Breuil, A. Cacciato, A. Rothschild, M. Jurczak and J. Van Houdt, *Inernational Memory Workshop*, p. 84 (2009).
15. C. Sandhya, U. Ganguly, N. Chattar, C. Olsen, S. M. Seutter, L. Date, R. Hung, J. M. Vasi, and S. Mahapatra, *IEEE Electron Device Lett.*, **30**, 171 (2009).
16. G. Van den bosch, A. Funemont, M. B. Zahid, R. Degraeve, L. Breuil, A. Cacciato, A. Rothschild, C. Olsen, U. Ganguly, and J. Van Houdt, *Proc. Joint NVSMW and ICMTD* p.128 (2008).
17. T. H. Kim, I. H. Park, J. D. Lee, H. C. Shin, and B.-G. Park, *Appl. Phys. Lett.*, **89**, 063508 (2006).
18. A. Suhane, A. Arreghini, R. Degraeve, G. Van den bosch, L. Breuil, M. B. Zahid, M. Jurczak, K. De Meyer, and J. Van Houdt, *IEEE Electron Device Lett.*, **31**, 77 (2010).
19. R. Fujitsuka, K. Sekine, A. Sekihara, A. Fukumoto, J. Fujita, F. Aiso and Y. Ozawa, *Ext. Abst. of SSDM*, p.861 (2009).
20. B. Lee, T. Moon, T.-G. Kim, D.-K. Choi and B. Park, *Appl. Phys. Lett.*, **87**, 012901 (2005).
21. R. Degraeve, M. Cho, B. Govoreanu, B. Kaczer, M. B. Zahid, J. Van Houdt, M. Jurczak, and G. Groeseneken, *IEDM Tech. Dig.*, p. 775 (2008).
22. S. Fujii, N. Yasuda, J. Fujiki, and K. Muraoka, *Jpn. J. Appl. Phys.*, **49**, 04DD06 (2010).
23. J. Fujiki, S. Fujii, N. Yasuda, and K. Muraoka, *Jpn. J. Appl. Phys.*, **49**, 04DD07 (2010).
24. J. Fujiki, T. Haimoto, N. Yasuda, and M. Koyama, *Ext. Abst. SSDM*, p.756 (2010).
25. K. Suh, B. Suh, Y. Lim, J. Kim, Y. Choi, Y. Koh, S. Lee, S. Kwon, B. Choi, J. Yum, J. Choi, J. Kim, and H. Lim, *ISSCC Tech. Dig.*, p.128 (1995).
26. D. J. DiMaria and J. R. Abernathey, *J. Appl. Phys.*, **60**, 1729 (1986).
27. T. Tomita, Y. Kamakura, and K. Taniguchi, *Phys. Stat. Sol. (b)*, **204**, 129 (1997).
28. A. Mauri, C. M. Compagnoni, S. Amoroso, A. Maconi, F. Cattaneo, A. Benvenuti, A. S. Spinelli, and A. L. Lacaita, *IEDM Tech. Dig.*, p. 555 (2008).
29. R. Ohba, Y. Mitani, N. Sugiyama and S. Fujita, *IEDM Tech. Dig.*, p. 839 (2008).
30. H. T. Lue, S. Y. Wang, Y. H. Hsiao, E. K. Lai, L. W. Yang, T. Yang, K. C. Chen, K. Y. Hsieh, R. Liu, and C. Y. Lu, *IEDM Tech. Dig.*, p.495 (2006).
31. H. Bachhofer, H. Reisinger, E. Bertagnolli, and H. von Philipsborn, *J. Appl. Phys.* **89**, 2791 (2001).
32. J.-H. Yi, H. Shin, Y.-J. Park, and H. S. Min, *IEEE Trans. Device and Materials Reliability*, **6**, 334 (2006).
33. E. Vianello, L. Perniola, P. Blaise, G. Molas, J. P. Colonna, F. Driussi, P. Palestri, D. Esseni, L. Selmi, N. Rochat, C. Licitra, D. Lafond, R. Kies, G. Reimbold, B. De Salvo, and F. Boulanger, *IEDM Tech. Dig.*, p. 83 (2009).

34. S. Fujii, J. Fujiki, N. Yasuda, R. Fujitsuka, and K. Sekine, *Proc. IRPS*, p. 956 (2010).
35. S.-Y. Wang, H.-T. Lue, T.-H. Hsu, P.-Y. Du, S.-C. Lai, Y.-H. Hsiao, S.-P. Hong, M.-T. Wu, F.-H. Hsu, N.-T. Lian, C.-P. Lu, J.-Y. Hsieh, L.-W. Yang, T. Yang, K.-C. Chen, K.-Y. Hsieh, and C.-Y. Lu, *Proc. IRPS*, p. 951 (2010).
36. S. Fujii, R. Fujitsuka, K. Sekine, and N. Yasuda, Accepted for presentation at *IRPS* (2011).
37. M. Chang, H. Hwang, and S. Jeon, *Appl. Phys. Lett.*, **96**, 052106 (2010).
38. J. Postel-Pellerin,R. Laffont, G. Micolau, F. Lalande, A. Regnier, and B. Bouteille, *Microelectronics Reliability*, **50**, 1474 (2010).
39. H. Y. Yu, M. F. Li, B. J. Cho, C. C. Yeo, and M. S. Joo, D.-L. Kwong, J. S. Pan, C. H. Ang, J. Z. Zheng, and S. Ramanathan, *Appl. Phys. Lett.*, **81**, 376 (2002).
40. K. Muraoka, K. Kurihara, N. Yasuda, and H. Satake, *J. Appl. Phys.*, **94**, 2038 (2003).
41. C.-H. Lee, K.-I. Choi, M.-K. Cho, Y.-H. Song, K.-C. Park, and K. Kim, *IEDM Tech. Dig.*, p. 613 (2003).
42. J. Fujiki, N. Yasuda, R. Fujitsuka, W. Sakamoto, and K. Muraoka, *IEDM Tech. Dig.*, p. 952 (2009).
43. A. Padovani, L. Larcher, D. Heh, G. Bersuker, V. Della Marca, and P. Pavan, *Appl. Phys. Lett.*, **96**, 223505 (2010).

Dynamic Negative Bias Stress Instability Effects in Hafnium Silicon Oxynitride and Silicon Dioxide

J. K. Mee[a], R. A. B. Devine[a,b], H. P. Hjalmarson[c] and K. Kambour[c]

[a]AFRL/RSVE, Kirtland AFB, NM 87117
EMail: Jesse.Mee@kirtland.af.mil
[b]EMRTC/NMT, 801 Leroy Place, Socorro, NM 87801
[c]Sandia National Laboratories, Albuquerque, New Mexico 87185, USA

Negative bias temperature instability (NBTI) is an issue of critical importance as the space electronics industry evolves because it may dominate the reliability lifetime of space based assets. Understanding its physical origin is therefore essential in determining how best to search for methods of mitigation. It has been suggested that the magnitude of the effect is strongly dependent on circuit operation conditions (static or dynamic modes). In the present work, we examine the time constants related to the charging and recovery of trapped charged induced by NBTI in HfSiON and SiO_2 gate dielectric devices at room temperature.

Introduction

The issue of reliability in advanced electronics for space based assets is rapidly superseding that which is traditionally assumed to be the most important: radiation effects. The primary reason is that the more advanced technologies have lifetimes which are becoming substantially shorter whereas older technologies had perfectly acceptable reliability lifetimes (> 10 years). Though this may be acceptable in the terrestrial commercial scenario given that the general public accepts the concept of "outmoding", which hides the reliability issue to some extent, this is not an acceptable logic for the military and space avionics.

Negative bias temperature instability (NBTI) is one of a variety of mechanisms [1] leading to failure of microelectronic circuits based upon Si technology. In particular, it is believed to be one of the most important factors in defining reliability lifetime in the more advanced technologies which already/will incorporate newer materials such as the "high-κ" gate dielectrics in field effect transistors. Though the full physics of the bias temperature instability phenomena (NBTI and positive PBTI) remains unclear, it has been recognized and demonstrated that there are dynamic and "static" components which can influence the operation of a micro-circuit in different ways [2]. Accurate modeling of the phenomenon of NBTI must therefore take careful consideration of the dynamic effects related to rapid charge capture and release at the silicon/gate dielectric interface. The dynamic effects, at least in PBTI [3], can occur very rapidly (<1μs) upon the application or removal of a stressing bias at the gate electrode. As a result, such effects have largely gone unnoticed due to the inadequacies of the measurement equipment to detect such rapid changes.

Given the evolving interest in measurement of dynamic effects, we have acquired a rapid data acquisition system (Keithley 4200 SCS), and begun a study of NBTI induced charging and recovery with sub-microsecond stressing/recovery intervals between microsecond measurements giving us maximum resolution in the early time domain. These initial experiments have been performed on metal-oxide-semiconductor field effect transistor (MOSFET) devices with high-κ dielectric and SiO_2 gate insulators. Our initial results already reveal that in the early phase, and at room temperature, we have a fully recoverable mechanism related to NBTI. Preliminary theoretical calculations have been carried out to confirm the origin of this mechanism.

Experiment Methodology

The high-κ MOSFET devices used in the experiments reported here were primarily p-channel with 250, 300, and 350 nm channel lengths. The gate dielectric stacks were formed of chemically oxidized Si (approximately 1 nm of SiO_2) upon which HfSiO was deposited using the technique of atomic layer deposition (ALD). The stack was then nitrided using the process of post-deposition annealing in NH_3. The metal electrodes were formed of physical vapor deposited (PVD) TaN. A conventional CMOS flow completed the fabrications process. Some additional devices were used from IBM's 130nm process which contained a nominal 3.4 nm SiO_2 gate oxide.

Device characteristics [4] were obtained using the Keithley 4200 SCS system equipped with a new state-of-the-art pulse generator. We have developed two types of experiments; the first monitors rapid charging and recovery mechanics, the second monitors charging under AC stress. Schematic pulse trains for the charging/recovery stress and for the AC test are shown in Figs. 1a and 1b respectively. Following a procedure adopted previously [5], in all cases we performed single point source/drain current (I_{ds}) measurements for a chosen gate-source voltage (V_{gs}) and a chosen drain-source voltage (V_{ds}).

Charging/Recovery Stress Procedure

HfSiON and SiO_2 devices were subjected to a continuous stress stressing bias on the gate followed by a recovery time. More specifically, each device was stressed at V_{bias} (V) while I_{ds} was periodically probed using a 1μs pulse at $V_{gs} = -1.0$ V, and $V_{ds} = -0.2$ V. The rise and fall times of the measurement pulse were ~ 20 ns. Although the measurement pulse duration was 1 μs, the source-drain current was actually measured during the time window from 750ns to 880ns. This window ensured that ringing and/or overshoot effects had relaxed and the current actually represented the mean current. With a sampling rate ~ 200 M samples per second, we average 26 individual measurements in the 130 ns measurement window. The frequency of the data acquisitions was logarithmic in time with the first and second points occurring at t = 0 and t = 5 μs. Following the stressing period, which amounted to a total of 100 s, the devices were allowed to recover at a selected voltage V_{rec} (V). As with the charging curve, I_{ds}

measurements were taken starting with μs resolution and going out to 100 s. For the work presented here, the recovery was done at 0 V on the gate/source/drain and body contacts. For the charging curves, we stressed at V_{bias} = -2.0 V.

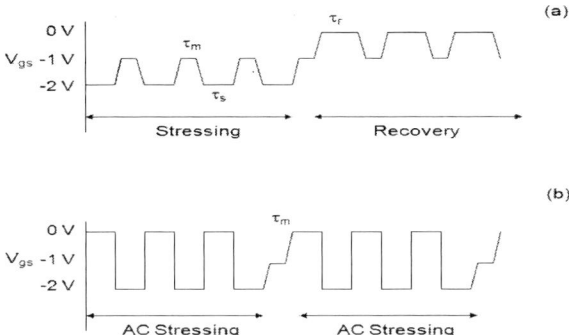

Figure 1. (a) The V_{gs} pulse train for charging/recovery data. (b) V_{gs} pulse train for the AC stress data.

AC Stress Procedure

Additional HfSiON and SiO$_2$ devices were subjected to an AC stressing bias on the gate with a pulse top of 0V and a pulse bottom of -2 V (50% duty cycle). During this stressing the source, drain and body contacts were held at 0 V. Seven frequencies were used in the range 2.5 MHz to 1 Hz. Some additional devices were stressed with a constant bias which provided a reference DC curve. Upon completing a stressing period, which will contain 'n' number of stressing cycles related to frequency and duration of the AC stress, a 1 μs pulse at -1.0 V is applied to the gate and an I_{ds} measurement is extracted. Again, V_{ds} = –0.2 V which ensured the devices operated in the linear regime [4]. Note the length of the stressing periods should be an integral number of single stressing cycles. Otherwise, it is possible to create a scenario whereby a single stressing cycle is longer than the requested stress time.

In all cases, both the HfSiON and SiO$_2$ devices were stressed at 25 °C. Previous attempts to measure these dynamics have been approximate because it is hard to extract the data without significantly perturbing the system. We are optimistic that the 1 μs measurement pulse, τ_m, that we have incorporated here has minimized the impact of charging variation during the data acquisitions; however, it is believed that even at this level we may still be offset by very energetically and or spatially close traps which capture and release faster that we can measure.

Interpretation of Single Point Measurement

We are primarily interested in the device threshold voltage, V_{th}, as the relevant parameter for the MOSFET degradation. Usually such a determination requires acquisition of a full $I_{ds}(V_{gs})$ characteristic [4] for fixed V_{ds}, but such measurement requires a finite time during which the stressing voltage is removed or changed. We have suggested previously [5] that to a good approximation one can use a single point measurement, I_{ds} (V_{gs} fixed, V_{ds} fixed) to obtain an estimate of the V_{th}. In the linear regime ($|V_{ds}| << |V_{gs} - V_{th}^{o}|$),

$$I_{ds} = [W/2L] \, C_{ox} \, \mu [\{V_{gs} - (V_{th}^{o} + \Delta V_{th})\} V_{ds} - 1/2 V_{ds}^{2}] , \qquad (1)$$

where W is the channel width, L the channel length, C_{ox} the gate dielectric stack capacitance, and μ is the mobility of the inversion channel carriers (holes). This can be reduced to the simplified expression for ΔV_{th} as a function of I_{ds}/I_{ds}^{o}:

$$\Delta V_{th} = [V_{gs} - V_{th}^{o}](1 - I_{ds}/I_{ds}^{o}) . \qquad (2)$$

This method requires certain approximations with respect to electric field dependent mobility, all of which have been justified experimentally [5,6]. A greater issue with the approach revolves around the uncertainty in determination of the initial current value, I_{ds}^{o}. With previous equipment (e.g. HP 4156), our first current measurement was not at time zero but rather at approximately 0.5 – 1 second. Our initial intuition was to make a linear fit to the first few $I_{ds}(t)$ points and extrapolate back to t = 0. The introduction of the new Keithley 4200 SCS pulse card has given us the ability to measure I_{ds} at t = 0, and to monitor $I_{ds}(t)$ with microsecond resolution. We have since learned (Fig. 3) that the early time behaves much more like an exponential. This means that previous methods introduced a large underestimation of the I_{ds}^{o} value that translates into an underestimation of the full magnitude of the NBTI effect.

Experimental Results and Discussion

It is suspected that the magnitude and rate of charge trapping would be much greater in HfSiON than in SiO_2. This suspicion is based upon the assumption that there is a high defect density at the Si/dielectric interface in HfSiON as compared to the 'renowned" SiO_2/Si interface. The curves shown in Fig. 2 are the result of room temperature continuous stress experiments out to 300 seconds for both high-k and standard SiO_2. As anticipated, SiO_2 devices degrade significantly less than HfSiON for a nominally equivalent stressing voltage on the gate electrode and equivalent physical thickness of dielectric. We furthermore verified our data interpretation methodology by performing full $I_{ds}(V_{gs})$ measurements and by acquiring a single I_{ds} point during each measurement cycle. For the former, we apply a staircase waveform to the gate and take an I_{ds} measurement at the top of each V_{gs} step. The result is a complete $I_{ds}(V_{gs})$ characteristic in ~ 10 µs. We can then extract $\Delta V_{th}(t)$ using classical "exact methods" [4], or from a single point on this curve by applying Eqn. 2. As seen in Fig. 2, ΔV_{th} determinations using both methods track well within experimental error.

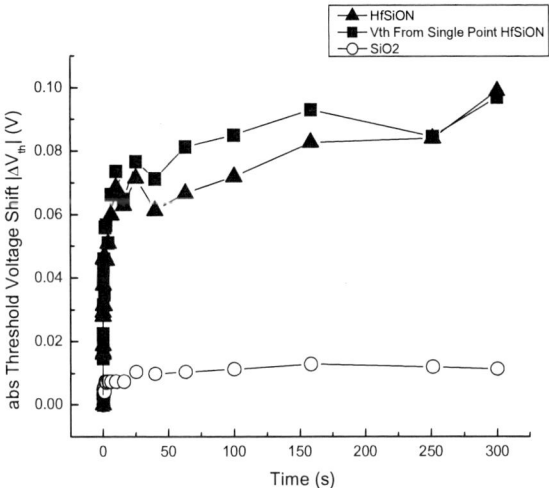

Figure 2. Room temperature continuous stress measurements at V_{gs}=-2.0 V for SiO_2 (circles), and HfSiON (Squares, triangles). Two methods of ΔV_{th} determination were employed for the HfSiON data; (triangles): ΔV_{th} determination from complete $I_{ds}(V_{gs})$ curve, (squares): ΔV_{th} determination from a single point on the $I_{ds}(V_{gs})$ curve using Eqn. 2 with $I_{ds}(t)$ measured at V_{gs} = - 1.0 V.

As observed in Fig.2, charging clearly begins very rapidly upon application of a stressing bias to the gate of a MOSFET. Using very short duration stressing pulses followed by rapid single point measurements, we have examined the magnitude of this effect in multiple devices. Fig. 3 shows the charging and recovery behavior of the threshold voltage as a result of room temperature stressing on HfSiON technology – the effects are emphasized by the logarithmic time plot for the total stress or recovery time.

All three curves in Fig. 3a and 3b are the result of threshold voltage degradation under identical stress conditions on the same device. A 5 minute recovery time was allotted in between stressing experiments (Fig. 3a). Note that the overall stressing times are short enough that apparently these samples do not enter the permanent degradation regime generally associated [7] with NBTI. Clearly a significant amount of V_{th} degradation occurs before 1 second. As mentioned previously, this information would have been lost if this data was collected using previous techniques [5]. Examination of Fig. 3 reveals that the time constants for charging are different than those for recovery. More specifically, it appears that the recovery dynamic is more rapid than the charging dynamic. This asymmetry has very interesting implications for the reliability lifetime characterization because it means that a

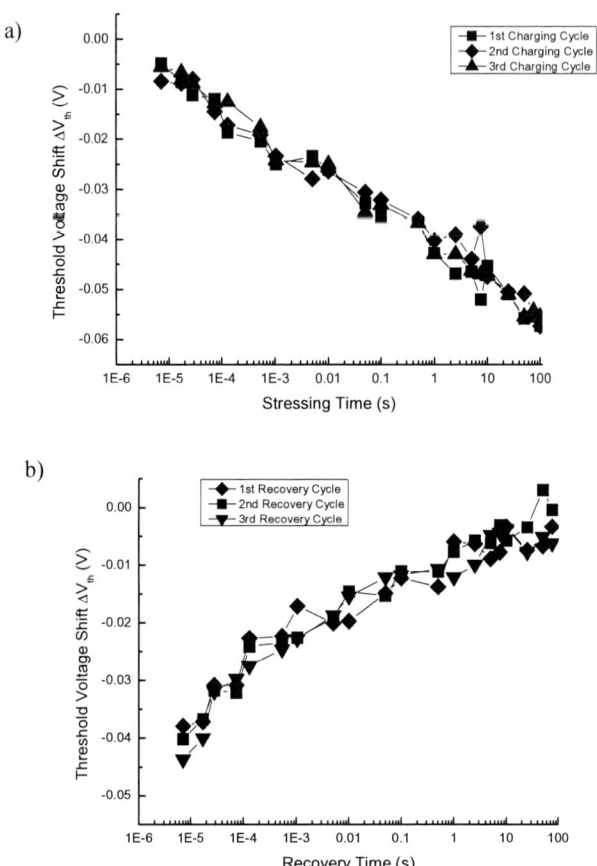

Figure 3. Room temperature continuous stress measurements on HfSiON devices for a) charging at V_{gs}=-2.0 V and b) recovery at V_{gs}=0 V. $\Delta V_{th}(t)$ is calculated by applying equation (2) to single I_{ds} measurements at V_{gs}=-1.0V.

device which is stressed with an AC single pulse at 50% duty cycle should show no net threshold voltage degradation under NBTI stressing. Note that these results are at 25°C; we have not yet determined how curves will behave under elevated temperature stressing. However, one research group [3] has already suggested that the early time behavior is temperature independent – at least for PBTI.

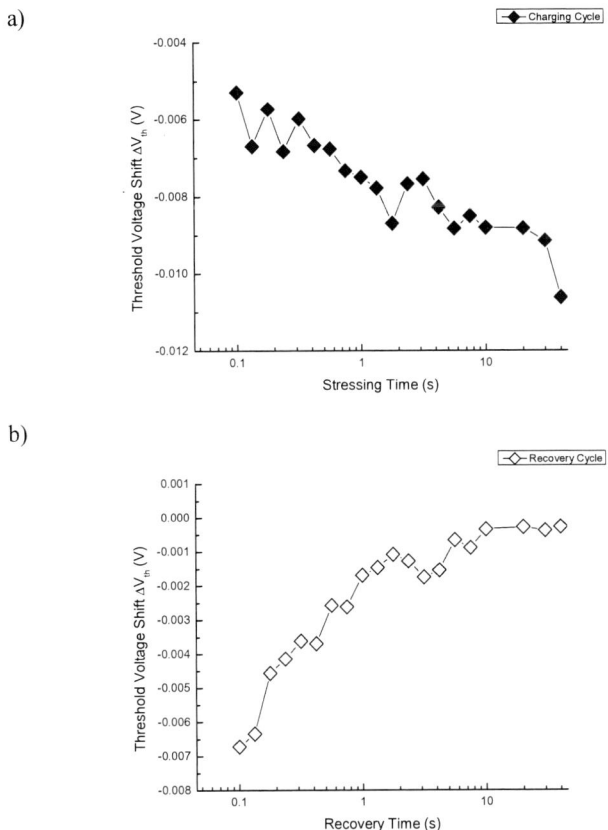

Figure 4. Room temperature continuous stress measurements on SiO_2 devices for a) charging at V_{gs}=-2.0 V and b) recovery at V_{gs}=0V. As with the HfSiON curves, $\Delta V_{th}(t)$ is calculated by applying equation (2) to single I_{ds} measurements at V_{gs}=-1.0V.

Results for SiO_2 are shown in Fig. 4 and are similar to those of the high-κ samples but with obvious variation in the magnitude of the effect. Again, the recovery dynamic appears to be the faster than the charging dynamic. Note that the time base for the SiO_2 experiments is less resolved in the early time domain. The scatter in the sub 1 ms time regime on these devices can be associated with slower charging times for SiO_2.

In light of the observed charging and recovery time constants, we performed a series of NBTI stress experiments in which AC stress was applied at different frequencies for chosen total stress times, and the associated threshold voltage shift then measured (Fig. 1b). These experiments should have simulated the effect of alternating stress which we anticipated would be non-zero and cumulative [2]. In Fig. 5, we show the results of AC stressing at 7 different frequencies ranging from 2.5 MHz to 1 Hz, and include the continuous stress measurement (DC) for HfSiON devices. Similar data for SiO_2 based devices is shown in Fig. 6. Measurement points are equally spaced in linear time to reduce confusion related to the minimum stressing period. The Keithley pulsed system in fact chooses a stress time equal to an integral number of stress pulses so that the actual point in time at which the AC stress stopped and V_{th} measurement began corresponded to the negative maximum of the stressing pulse (see Fig. 1).

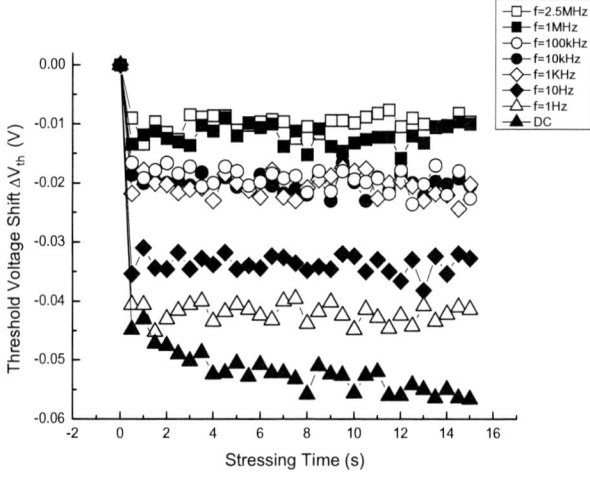

Figure 5. Room temperature AC stress measurements on HfSiON devices. The AC pulse top was 0 V, the pulse bottom was -2.0 V. Frequencies ranged from 2.5 MHz to 1 Hz. The DC curve (Full, up triangles) is shown for reference.

As seen in Fig. 5, the "net" threshold voltage shift varied significantly with the frequency of the AC stressing pulses. However, it would appear that for both HfSiON and SiO_2 based devices, there is no accumulation of threshold voltage shift with total stressing time prior to measurement. This result appeared to us to be rather confusing initially but became clearer. Irrespective of the pulse frequency, for each pulse there was a more or less short relaxation time followed by an equal charging time and then finally followed by rapid measurement of I_{ds}. In fact, it appears that the ΔI_{ds} is induced by the final stressing pulse of the sequence determined by the total stressing time; the other pulses comprising the total pulse chain appear irrelevant. The difference in amplitude of ΔV_{th} shown in Figs. 5 and 6 as a function of frequency is then explained. For a 1 Hz pulse there is an effective stressing time of 0.5 seconds while for a 1 MHz pulse the stressing time is 5 x 10^{-7} seconds. From Figs. 3 and 4, one clearly understands that very short stressing times induce almost negligible threshold voltage shifts while much longer times (e.g. 0.5 seconds) induced threshold voltage shifts almost comparable to those observed for a continuous, DC bias, situation.

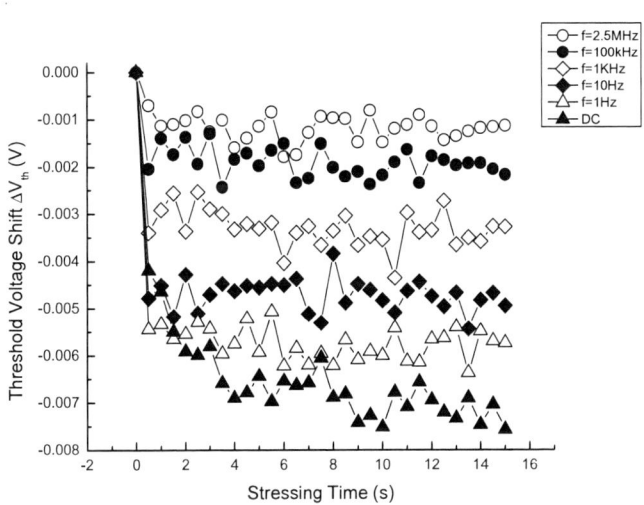

Figure 6. Room temperature AC stress measurements on SiO_2 technology. In this case the frequencies ranged from 2.5 MHz to 10 Hz. Again, the DC curve is shown for reference.

The data presented above reveals the presence of charge trapping and detrapping phenomena in both SiO_2 and HfSiON based devices which can take place at room temperature and can be readily cycled backwards and forwards. Cyclic bias stressing for the frequencies we have used (maximum 2.5 MHz) indicate that there is no net NBTI induced shift in the threshold voltage of the device, at least at room temperature. Preliminary theoretical modeling has been carried out to probe the origin of the observed NBTI effect.

Theoretical Calculations and Discussion

The role of defects in the gate layer was investigated by performing calculations of the threshold voltage shifts for some simple models. The basic physics involves tunneling between the Si substrate and defects in the oxide region. These calculations were performed on a representative one-dimensional structure. In principle, these calculations can mimic the transient gate voltages applied during the measurements of the threshold voltage shifts. Such calculations are used to obtain the density of trapped charge in the HfSiON layer. From the trapped charge density, the threshold voltages shifts are also obtained.

The simplified bandstructure we have assumed is shown in Fig. 7. This figure shows the conduction and valence bands of the constituent materials. Proceeding from the left, this figure shows the heavily doped Si region that serves as the gate electrode in these calculations. The oxide layer is composed of a 3 nm-thick HfSiON layer (bandgap of the order of 4eV) and a 1 nm-thick SiO_2 layer (bandgap of 9eV). To the right is the lightly doped p-type Si substrate In these calculations, the gate contact has an n-type density of 10^{19} cm^{-3} and the substrate has a p-type density of 10^{16} cm^{-3}.

The transient electrical effects are computed using the radiation effects on semiconductors (REOS) [8] program to solve the kinetic equations for the electrons, holes, and the defect densities. The tunneling current is obtained from a reaction of the form

$$T^0 + p \leftrightarrow T^+ \tag{3}$$

in which the holes p in the substrate tunnel to neutral traps T^0 in the HfSiON layer to produce positively-charged traps T^+. These positively-charged traps contribute to a threshold voltage shift.

The tunneling reaction leads to a rate equation for the trapped charge:

$$d[T^+] / dt = k_f [T^0][p] - k_r[T^+] \tag{4}$$

in which

$$k_f = \alpha \exp(-\beta| x_f - x_i|) \tag{5}$$

and

$$k_r = n_{th} k_f. \tag{6}$$

In these equations, the physical parameters have the values $\alpha = 10^{13}$ sec^{-1} and $\beta = 5 \times 10^7$ cm^{-1} [8]. However, for the present calculations, the case of non-resonant tunneling must also be included. In these cases, the energy of the holes is conserved by emission and absorption of phonons. The absorption process causes the trapping rate to become much reduced.

The kinetics explored in the experiments can be understood by using these equations. The filling of the traps is governed by the forward rate constant k_f, and the release is governed by the reverse rate constant k_r. Both of these quantities depend strongly on the tunneling distance $d = | x_f - x_i|$. For these calculations, a uniform distribution of traps in the HSiON is assumed. The varying distance d leads to a distribution of forward and reverse rates defined by the distance of the trap from the Si substrate.

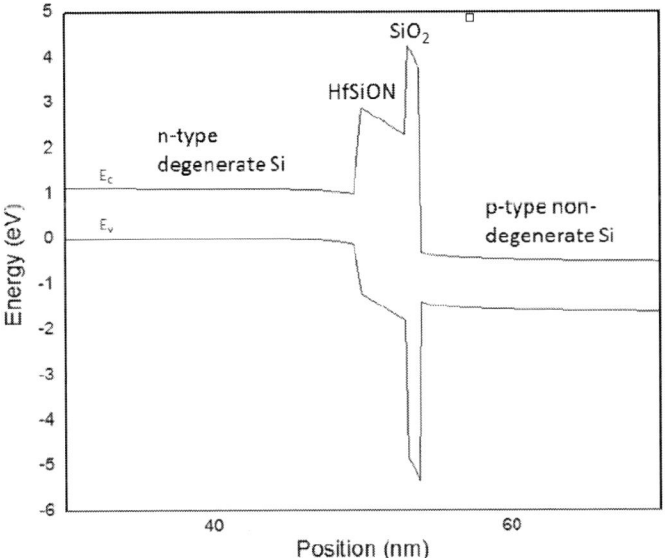

Figure 7: Shows the conduction and valence bands as a function of position for the gate-stack structure.

The computed shift in threshold voltage as a function of time is shown in Fig. 8. This voltage shift is caused by the increase in trapped holes in the HfSiON layer that have tunneled from the Si substrate into the empty traps. The characteristic shape of the threshold voltage is in good agreement with the data shown in the experimental section. The shift in threshold voltage as the holes are released shows similar but faster kinetics.

The kinetics can be understood by doing a simple calculation for traps located at one location a distance d from the substrate. The trapped hole density at a time τ_f for traps at a distance d from the substrate can be written as

$$[T^+(\tau_f)] = T^0(1 - \exp(-k_f\tau_f))$$ (7)

in which

$$k_f = k_0\exp(-\beta d).$$ (8)

In the calculations shown in Fig. 8, each sheet of traps contributes to the total trap density. The shift shown in this figure is the sum of these contributions at the various trap locations. The fact that there is a distribution leads to the log(t) behavior seen in the data and the calculations.

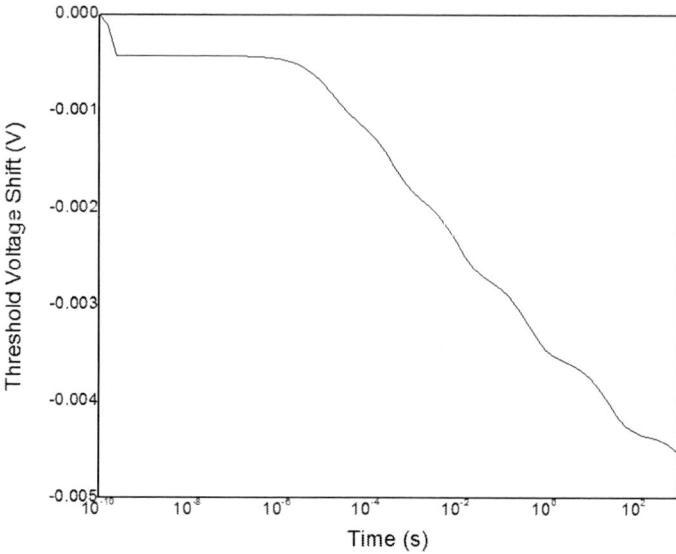

Figure 8: Shows the shift in threshold voltage as the traps are filled in the HfSiON layer.

The results of calculations shown in Fig. 8 are to be compared with the experimental values shown in Fig. 3a. Although the absolute magnitude is not reproduced, the time dependent behavior is relatively respected. To understand the data obtained from the pulse train experiments, one can once again assume that only a sheet of traps is involved. First one can assume that a filling pulse has been applied for a time τ_f to produce filled traps governed by the previous equations. If now the gate pulse is reduced for another period of time τ_r, the trapped hole density becomes:

$$[T^+(\tau_r)] = [T^+(\tau_f)] \exp(-k_r \tau_f) + [T^+(\infty)] \qquad (9)$$

This expression follows because the kinetics are now governed by the reverse rate, but at long times the trap density falls to a value determined by the new gate bias (in this case 0 V). In these formulae, the rates are k_f and k_r, respectively. Combining these expressions, the final form for the filling of these traps with holes and then releasing the holes is obtained:

$$[T^+(\tau_r)] = T^0(1 - \exp(-k_f \tau_f)) \exp(-k_r \tau_r) + [T^+(\infty)] \qquad (10)$$

The factor $[T^+(\infty)]$ is the trap density at long times. To understand this expression, it is useful to assume that the time durations are small. Also, one can assume that the long time density is very small. Then

$$[T^+(\tau_r)] = T^0(1 - \exp(-k_f \tau_f)) \exp(-k_r \tau_r)$$

$$= T^0 k_f \tau_f (1 - k_r \tau_r) \tag{11}$$

This expression shows that pulse times that are longer will lead to larger shifts in qualitative agreement with the data. Further calculations for distributions of traps are being performed.

Summary

Experimental data confirms that at very short times, reversible charging and discharging can occur leaving no residual threshold voltage shift. The direction of the threshold voltage shift is consistent with the accumulation of positive charge. It is probable that in most experiments performed to date, this term has been ignored because of instrumental limitations. The preliminary calculations support the hypotheses that holes are indeed the positively charged species are involved in the trapping and they originate in the silicon substrate. This conclusion follows from the successful use of the well-known mechanism of hole tunneling to explain the data. The next phase of these experiments will involve measurements as a function of temperature while the theoretical effort will center upon a more quantitative physics mechanisms.

Acknowledgments

This work has been partially supported by the Defense Threat Reduction Agency (DTRA – Dr. L. Palkuti) through a contract with North Carolina State University. Sandia National Laboratories is a multi-program laboratory managed and operated by Sandia Corporation, a wholly owned subsidiary of Lockheed Martin Corporation, for the U.S. Department of Energy's National Nuclear Security Administration under contract DE-AC04-94AL85000.

References

1. M. White and J. B. Bernstein, JPL Publication 08-5 2/08 (2008)
2. G. Chen, K.Y. Chuah, M. F. Li, D. S. H. Chan, C. H. Ang, J. Z. Zheng, Y. Jin and D. L. Kwong, Proc. IRPS 196-202 (2003)
3. D. Heh, C. D. Young and G. Bersuker, IEEE Electron Dev. Lett., **29** 180-182 (2008)
4. S. M. Sze,"Physics of semiconductor devices" (Wiley, N.Y. 1981) Chapt. 8
5. J. K. Mee and R. A. B. Devine, J. Appl. Phys. **107** 024511 (2010)
6. S. C. Sun and J. D. Plummer, IEEE Trans. Electron. Devices **27** 1497 (1980)
7. D. K. Schroder and J. A. Babcock, J. Appl. Phys. **94** 1 (2003)
8. H. P. Hjalmarson, R. L. Pease and R. A. B. Devine, IEEE Trans. Nucl. Sci., **55** 3009 (2008)
9. T.R. Oldham, A. J. Lelis, and F. B. McLean, IEEE Trans. Nucl. Sci., **33** 1203 (1986)

ECS Transactions, 35 (4) 461-479 (2011)
10.1149/1.3572299 ©The Electrochemical Society

Electrical and structural properties of ternary rare-earth oxides on Si and higher mobility substrates and their integration as high-k gate dielectrics in MOSFET devices

J.M.J. Lopes[1,2], E. Durğun Özben[2], M. Schnee[2], R. Luptak[2], A. Nichau[2], A. Tiedemann[2], W. Yu[2], Q.T. Zhao[2], A. Besmehn[3], U. Breuer[3], M. Luysberg[4], St. Lenk[2], J. Schubert[2], and S. Mantl[2]

[1] Paul-Drude-Institute for Solid State Electronics, 10117 Berlin, Germany
[2] Peter Grünberg Institute (PGI 9-IT), and JARAFIT, Research Center Jülich, Jülich, Germany
[3] Central Division of Analytical Chemistry (ZCH), Research Center Jülich, Jülich, Germany
[4] Institute for Solid State Research and Ernst Ruska Center for Microscopy and Spectroscopy with Electrons, Research Center Jülich, Germany

The continuous downscaling in metal-oxide-semiconductor field effect transistors is approaching fundamental limits. Allied to new device architectures, novel materials are needed in order to continue the evolution of complementary metal-oxide-semiconductor technologies. The combination of high dielectric constant (k) oxides with silicon and other semiconductors having a higher charge carrier mobility (ex.: germanium) is currently a fundamental technologic issue that requires extensive investigation on materials science. The search for high-k oxides (with $k > 20$) that can offer stable interfaces combined with a low density of electrically active defects is a topic of major interest. In this contribution, we will review some of our results on the structural and electrical properties of REScO$_3$ (RE = La, Gd, Tb, Sm) and LaLuO$_3$ amorphous films on Si as well as on high mobility substrates, showing their potential as high-k dielectrics for future CMOS applications.

Introduction

Ternary *RE* oxides on Si

The reduction of feature size in silicon-based integrated circuits over the last decades has led to the replacement of the SiO$_2$-based gate dielectric in order to reduce the high leakage current levels in metal-oxide-semiconductor field effect transistors (MOSFETs). The introduction of a hafnium-based gate dielectric has already been realized for the recent technological nodes [1]. However, as predicted by the roadmap for semiconductors [2], further scaling will require gate dielectrics (combined with proper metal gates) capable of scaling down to equivalent oxide thickness (EOT) below 0.9 nm, which may be unfeasible for most of the hafnium-based gate materials due to their relatively low

461

dielectric constants (k) ≤ 20. Therefore, the search for other dielectrics having k values higher than 20 is currently mandatory. Moreover, the simultaneous integration of such gate stacks with new channel materials such as strained-Si (sSi), Ge, or SiGe, allied with innovative device concepts, is essential for the realization of the ultimate complementary metal-oxide-semiconductor (CMOS) scaling [2,3,4].

Among different high-k materials that eventually could replace Hf-based dielectrics in future CMOS devices, ternary RE oxides such as the scandates (REScO$_3$, with RE = La, Gd, Tb, Sm) and LaLuO$_3$, are quoted as promising candidates [5,6]. Amorphous thin films of these materials on Si substrates offer k values ranging from 23 to 32 [7-12], as well as large optical band gaps (>5 eV) and band offsets to silicon (2-2.5 eV) [13,14]. Additionally, due to their high crystallization onset temperatures ranging from 800°C to 1000°C, they show a great potential of integration into a gate first transistor scheme. A high-k dielectric that keeps the amorphous structure after high temperature processes is still desirable in order to prevent the formation of grain-boundary-related defects, which could be responsible for deep states in the band gap and therefore would act as leakage paths in the gate stack [15,16]. In this context, ternary RE oxides offer a figure of merit that exceeds that of some Hf-based dielectrics (ex.: HfSiO$_x$ and HfAlO$_x$), which possess an enhanced thermal stability of the amorphous phase up to temperatures close to 1000°C, however with lower k values (≤20) [17,18]. Table 1 below summarizes some of the properties for amorphous films of scandates and LaLuO$_3$ on Si substrates.

Table 1. Dielectric constant (k), crystallization onset temperatures, optical band gap (E_g), conduction (ΔE_c) and valence (ΔE_v) band offsets to silicon, for amorphous films of the scandates and LaLuO$_3$. The k values were extracted from EOT plots. The crystallization temperatures were determined by X-ray diffraction (XRD) measurements for films annealed at different temperatures. The energy band parameters were obtain by internal photoemission and photoconductivity measurements [11,13,14].

	k- value	Thermal stability (°C)	$E_g \pm 0.1$ [eV]	$\Delta E_c \pm 0.1$ [eV]	$\Delta E_v \pm 0.1$ [eV]
GdScO$_3$	23	≤ 1000	5.6	2.0	2.5
DyScO$_3$	23	≤ 1000	5.6	2.0	2.5
TbScO$_3$*	26	≤ 1000	-	-	-
LaScO$_3$	28	≤ 800	5.6	2.0	2.5
SmScO$_3$*	29	≤ 800	-	-	-
LaLuO$_3$	30-32	≤ 1000	5.2	2.1	2.1

* Energy band parameters not determined for amorphous films.

It is also important to mention that the deposition of high quality scandate and LaLuO$_3$ films by atomic layer deposition (ALD) has also been demonstrated [19-21], showing the feasibility of implementing this class of materials into a large scale industrial process. As a remarkable example, Wang et al. achieved high quality LaLuO$_3$ films on Si without a detectable interfacial layer and offering EOTs down to 0.86 nm, which showed two orders of magnitude lower leakage currents than that of the SiO_2 layers with the same EOT.

Ternary *RE* oxides on high-mobility substrates

Following the integration of high-*k* dielectrics with Si, the next step is to deposit them on other channel materials with intrinsic higher carrier mobilities than Si [22]. As already mentioned before, sSi, Ge, SiGe, and also III-V semiconductors are currently being investigated as potential substitutes to Si for the oncoming generation of devices offering improved properties and lower power consumption. While combining high-*k* oxides with sSi or conventional Si leads to similar chemical characteristics at the interface [23], in the cases of Ge and SiGe (the topic of III-V semiconductors will not be addressed in this contribution), it still remains as a challenge the achievement of a high quality interface between the semiconductor surface and the high-*k* film. Research on alternative high-*k* oxides that could offer intrinsically stable interfaces (chemically and electronically) is of major importance.

The investigation of *RE*-based high-*k* oxide films on Ge shows very promising results. It is reported that amorphous thin films of binary oxides such as La_2O_3 and Gd_2O_3 form interfaces with Ge offering a low density of interface states (D_{it}) [24-26]. Such an excellent characteristic has been attributed to the formation of stable *RE*–O–Ge compounds (germanates) at this region, with the *RE* element having a fourfold coordination within the germanate, and thus offering a state-free interface with Ge [27]. However, such oxides seem restricted to be applied as passivating layers between the Ge substrate and another dielectric with a higher *k* value. This is due to their relatively low dielectric constants and therefore questionable capability for an aggressive scaling of the EOT to values below 1 nm. Their implementation as passivation layers and not as the actual gate dielectric appears as an alternative to other approaches proposed in literature. Thermally grown GeO_2 as well Si have been successfully employed for realizing the electrical passivation of the Ge (and also SiGe) surfaces before the deposition of a gate oxide [28-30]. In the same way as for the binary *RE*-oxides, their use may also restrict the EOT scaling.

In this scenario, the integration of scandates and $LaLuO_3$ with high mobility substrates such as Ge and SiGe looks very promising. The main advantage of using them would be the possibility of having a gate dielectric with a high *k* value (giving the EOT scaling a better perspective) and a robust rare-earth-based interface to the channel material (thus avoiding the use of lower-*k* passivation layers). In spite of considerably less work reported in comparison to binary oxides, exciting results for these materials have recently been shown by different groups [30-35]. For $LaLuO_3$/Ge(100) stacks, Toriumi *et al.* [34] reported well behaving capacitance-voltage curves and a low density of interface states (D_{it}), which were attributed to the high stability of the $LaLuO_3$ film as well as of its interface with Ge, preventing thus formation and desorption of GeO species, responsible for the detrimental of the electrical properties. Radtke *et al.* [35] have shown the effects of post-deposition thermal treatments on the properties of similar stacks. Similarly to the case of La_2O_3 [24], the authors observed that the formation of germanate compounds at the $LaLuO_3$/Ge interface (instead of oxidized Ge species) is correlated with improved electrical properties. Regarding scandates films, results for *p*-MOSFETs having a

GdScO$_3$ film as a gate oxide and strained-SiGe/sSi on insulator substrates show promising results [36]. However, in that case a thin Si capping layer is used to passivate the SiGe surface before high-k deposition, making the interaction between the dielectric and SiGe not really accessible.

In this present contribution, we will review some of our results on the electrical and structural characterization of amorphous films of different scandates and LaLuO$_3$, prepared either on silicon or on a high mobility substrate. Although more emphasis will be put on the materials science, recent results on the integration of these oxides into MOSFETs having sSi or SiGe as channel materials will be briefly discussed.

Experimental procedure

Amorphous scandate and LaLuO$_3$, whose properties are partially listed in table 1, were prepared using three different deposition techniques: pulsed laser deposition (PLD), e-gun evaporation, and molecular beam deposition (MBD). While the PLD and e-gun techniques make use of stoichiometric ceramic targets, the MBD films are prepared by co-evaporating the rare-earth elements in an O$_2$ background atmosphere of about 1×10^{-6} mbar, which allows oxidation of the evaporated metal atoms. More details about the deposition processes can be found elsewhere [9-11, 37-39]. For deposition on silicon, standard RCA cleaned (100) silicon with or without an HF-last surface were used as substrates and the film deposition temperatures ranged from 400 to 600°C. In the case of Ge, prior to the deposition process, Ge pieces (2×2 cm^2) were cleaned using sulfuric acid (H$_2$SO$_4$) and hydrofluoric acid (HF) solutions. After the water rinsing, the samples were immediately loaded into the UHV system, and annealed at 700°C for 30 min for desorbing the remaining GeO$_x$ not totally removed by the HF solution. Finally, after the *in-situ* cooling of the samples down to temperatures between 200 and 350°C, the high-k deposition (either GdScO$_3$ or LaLuO$_3$) was realized. The techniques employed for the microstructural and interfacial characterization of the films were Rutherford backscattering spectrometry (RBS), transmission electron microscopy (TEM), X-ray diffraction (XRD), X-ray reflectometry (XRR), X-ray photoelectron spectroscopy (XPS), and time of flight secondary ion mass spectrometry (TOFSIMS).

For the electrical investigation of the films, Pt and TiN top contacts were either deposited through a shadow mask or lithographically patterned on the high-k surface. The metal depositions were performed at room temperature (R.T.) and *ex-situ.* This is proven to not cause deleterious effects to the electrical properties [40]. Contrary to binary oxides [41], ternary *RE* oxides do not present an intense hygroscopic behavior after exposure to the ambient. Following the metallization, ohmic backside contacts were made by deposition of 100-150 nm Al, followed by forming gas annealing (90% N$_2$ + 10% H$_2$) at temperatures ranging from 350 to 450°C for 10 to 20 min. This treatment improves the back-side contact and also helps on the passivation of dangling bonds at the high-k/semiconductor interface. All the MOS stacks whose results are presented here underwent through this forming gas treatment. Capacitance-voltage (*C-V*), current-voltage (*I-V*), and conductance-voltage (*G-V*) curves were measured using an impedance analyzer HP 4192A. D_{it} values were extracted by considering the peak conductance, which represents the loss due to the exchange of carriers with interface traps, and by applying Terman's method to the *C-V* measurements [42]. The EOT of the films were determined by fitting the experimental C-V curves with a Hauser fit, taking into account quantum-mechanical corrections [43].

Finally, the integration of ternary RE oxides into transistors was realized through gate-last and full-replacement-gate processes. Fully depleted n- and p-MOSFETS were processed on SOI, sSOI (Si on insulator and sSi on insulator, respectively), and SiGe substrates. More details can be found elsewhere [30,44-47].

Results and Discussion

Ternary RE oxides on Si

RBS was used to determine the stoichiometry of the deposited films (Fig. 1 illustrates a typical spectrum for a 30 nm thick LaLuO$_3$ on Si). The results show a general trend for the scandates and LaLuO$_3$. Independent of the deposition technique employed, a ratio varying between 1:0.9 and 1:1.2 for the metallic elements is measured for films deposited at different conditions. However, the oxygen amount strongly depends on the growth parameters.

Films that were deposited at R.T. with the purpose of avoiding the formation of an interfacial layer, always contained an excessive amount of oxygen. The achievement of an stoichiometric content (i.e. ~1.5 oxygen atoms per metal atom) could be obtained by thermally treating them after growth [38,40,48]. In parallel to the oxygen reduction, an improvement on the dielectric properties of the films was observed. By comparing the structural properties of the samples before and after annealing, it is observed that the post-deposition annealing not only reduces the oxygen content but also readjusts the short-range order in the amorphous matrix. For instance, XPS analyses of MBD LaLuO$_3$ films (which after being thermally treated at 800°C for 5 minutes show $k \cong 30$) reveals that the La3d, Lu4f, and O1s peaks become narrower after post-deposition annealing. This indicates a reduction in the degree of amorphicity after annealing [50,51]. In fact, it is observed that the full width at half maximum (FWHM) of the peaks is comparable to those measured for a LaLuO$_3$ crystalline film, which also exhibits a high permittivity around 45 [52]. The low k value (~17) determined for the as-deposited layers (containing an excessive oxygen amount) may be a result of not only a higher degree of disorder in the structure, but also due to a lower quality of the films, perhaps as a result of a high CO$_2$ absorption during the low temperature deposition.

Although XPS analyses show a structural rearrangement of the films after thermal treatment, XRD and high-resolution TEM measurements (see Fig. 2) reveals that the films remains in its initial amorphous state. Also seen in the TEM image in Fig. 2 is the formation of an interfacial layer as a side-effect to the dielectric constant improvement. The growth of such a thick interface (> 3 nm thick in this case) mainly composed of SiO$_2$ is undesired since it prevents further EOT scaling.

The achievement of high dielectric constant values as for the annealed samples, but with a minimization of the interfacial layer growth, could be achieved by depositing the films at higher temperatures and controlling the O$_2$ partial pressure in the growth chamber. Fig. 3 shows a TEM image of a LaLuO$_3$ film deposited by MBD on a RCA cleaned (100) Si substrate at 450°C. As for the sample illustrated in Fig. 2, the film also presents an amorphous structure. However, in this case no interfacial layer is visible between the dielectric and the substrate. Interestingly, the TEM image shows an very abrupt interface, even though the existence of a ~1 nm thick chemical SiO_2 (as a product of the RCA cleaning) should be visible. The absence of a clear contrast between the $LaLuO_3$ film and this SiO_2 interlayer may be related to a reaction taking place in this region, leading to an atomic intermixing of the two oxides. In fact, XPS analyses of

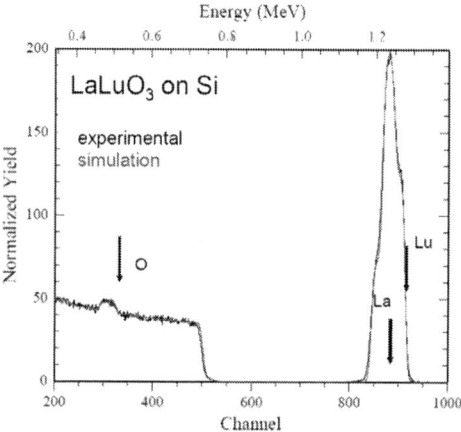

Fig. 1. RBS spectrum of a ~30 nm thick $LaLuO_3$ film on Si. The simulation of the data was performed by using the code RUMP.

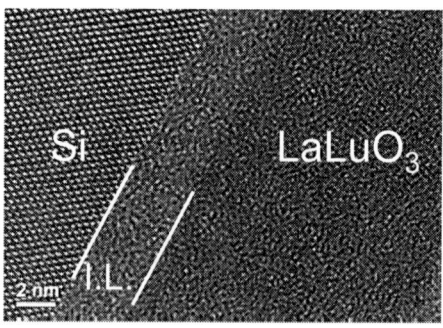

Fig. 2. TEM micrograph of an amorphous $LaLuO_3$ film deposited on a RCA cleaned Si at R.T., and submitted to a post-deposition annealing at 800°C for 5 minutes in N_2 ambient. The k value of the film is 17 before annealing and 30 after it (I.L. stands for interfacial layer).

Fig. 3. TEM micrograph of a LaLuO₃ film deposited on a RCA cleaned Si substrate at a substrate temperature of 450°C. The k value for this film is around 30.

LaLuO₃ films (see Fig. 4) reveal the existence of mainly silicate near the interface with a minor contribution of SiO_2. The Si 2s region was chosen for the analysis in order to avoid any interference with La 4d peaks which is the case for the Si 2p. For a LaScO₃ film (see Fig. 5), the formation of silicate at the interface is predominant and no SiO_2 is visible. Such a SiO_2-free interface was also observed for TbScO₃ films on Si [47].

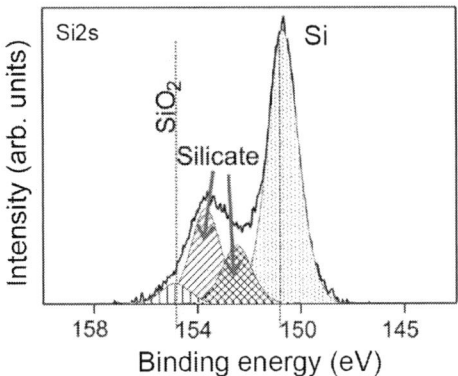

Fig. 4. XPS spectrum for the Si 2s region measured for a ~3.5 nm thick LaLuO₃ film deposited by MBD on a RCA-cleaned silicon substrate. The substrate temperature during deposition was 450°C. The dashed lines found in the figure indicate the binding energies for Si and SiO_2.

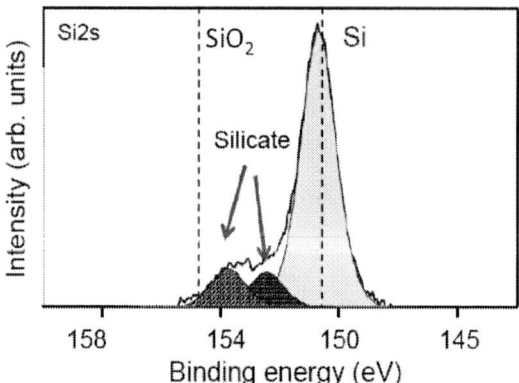

Fig. 5. XPS spectrum for the Si 2s region measured for a ~3 nm thick LaScO₃ film deposited by MBD on a RCA-cleaned silicon substrate. The substrate temperature during deposition was 350°C. The dashed lines found in the figure indicate the binding energies for Si and SiO₂.

The electrical characterization of the films through *C-V* and *I-V* measurements reveal well-behaving MOS stacks. As an example, Fig. 6 shows *C-V* forward and reverse bias sweeps measured at a frequency of 100 kHz for a TiN/LaLuO₃ gate stack prepared on *n*-type Si(100). The estimated EOT for this sample is 1.5 nm. The curve exhibits no

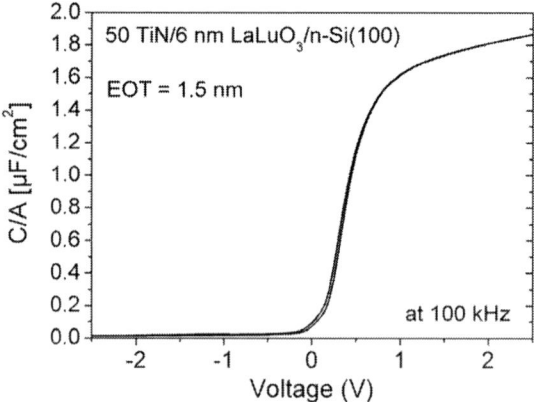

Fig. 6. *C-V* measurement of a TiN/LaLuO₃ gate stack on *n*-type Si(100).

humps or irregularities and is almost free of hysteresis. The D_{it} values for LaLuO$_3$ and also for scandate films on Si are typically in the range of 1 to 5×10^{11} eV^{-1} cm^{-2} [45-47,53].

Low leakage current density levels are observed for this class of materials. Fig. 7 is a plot of the leakage current density for ternary *RE* oxides on Si prepared by different techniques, as a function of their respective EOT. Here only the results for samples prepared with optimized growth conditions and therefore with high *k* values (as listed in table 1) are taken into account. For the sake of comparison, leakage currents for gate stacks containing SiO$_2$, HfO$_2$, and HfO$_2$/Al$_2$O$_3$ as gate dielectrics are also plotted.

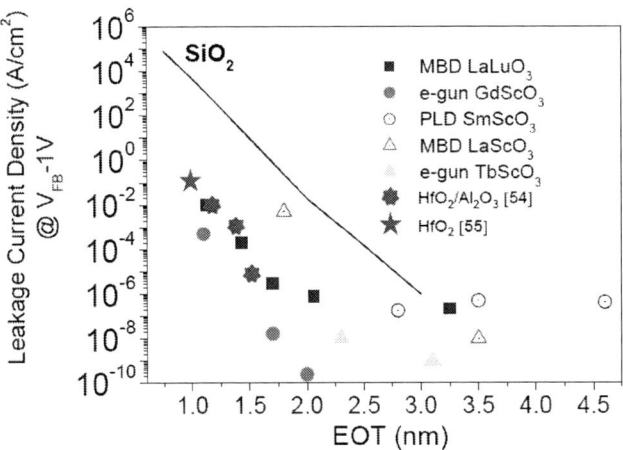

Fig. 7. Leakage current density *vs.* EOT for different ternary *RE* oxides. The plotted values are for $|V_G$-$V_{FB}| = 1$ V, where V_G and V_{FB} are the gate and flat-band voltages, respectively.

Ternary *RE* oxides on high-mobility substrates

The integration of amorphous films of LaLuO$_3$ and GdScO$_3$ with high mobility substrates such as Ge, SiGe, and strained Si (sSi) also shows great potential. Fig. 8 illustrates *C-V* measurements for a ~4.5 nm thick amorphous LaLuO$_3$ film deposited directly on *n*-type Ge(100) substrates without any surface passivation layer. *C-V* curves free of humps and irregularities could be obtained. When comparing curves measured at different frequencies, one can see a dispersion of less than 5% on the accumulation side and a small flatband voltage shift of about 0.2 V between 10 kHz and 1 MHz. Most importantly, an EOT below 1 nm could be achieved with leakage current densities kept below 10^{-1} A/cm^2 and D_{it} levels in the range of 1 to 5×10^{12} eV^{-1}cm^{-2} (not shown here).

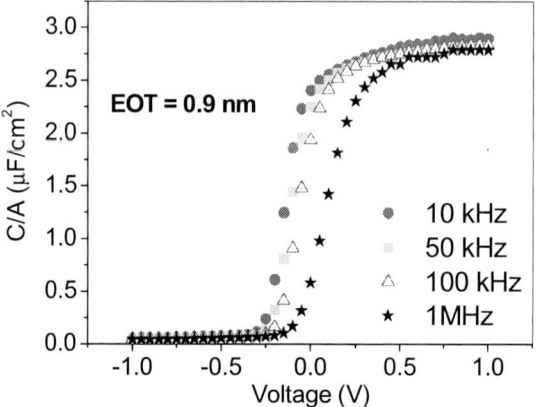

Fig. 8. *C-V* curves measured at different frequencies for a ~4.5 nm thick LaLuO$_3$ film deposited on *n*-type Ge(100) by molecular beam deposition. In this case Pt top contacts were used for the MOS stacks.

Amorphous GdScO$_3$ films of different thicknesses (from 2.7 to 11 nm, as determined by XRR) were deposited on *n*-type Ge(100) substrates by e-gun evaporation. As for the LaLuO$_3$, no passivation layer between the high-*k* and the semiconductor was used. Fig. 9 shows the results for the measured MOS stacks prepared with Pt top electrodes. Similarly to LaLuO$_3$, a small dispersion and flat-band voltage shift is observed for curves measured at different frequencies. For MOS stacks offering EOTs higher than 1 nm, the *C-V* curves also show an inversion capacitance at low frequency as a product of the high intrinsic carrier concentration and the effect of minority carriers, which is stronger in Ge due to its smaller band gap as compared to Si [24]. For a ~2.7 nm thick GdScO$_3$ film offering a low EOT ~0.8 nm, the absence of inversion capacitance and also the appearance of bumps in the depletion region is seen and it is probably due to a higher density of interface states and leakage current.

The D_{it} levels for these films (see Fig. 10) were roughly determined by considering the peak conductance from *G-V* curves (as depicted in Fig. 10(a)) by applying the formula [42]:

$$D_{it} \sim 2.5 G_{max} / q \omega A \qquad (1),$$

where G_{max} is the peak conductance, q is the electronic charge, ω is the angular frequency and A the area of the electrode used for the measurement.

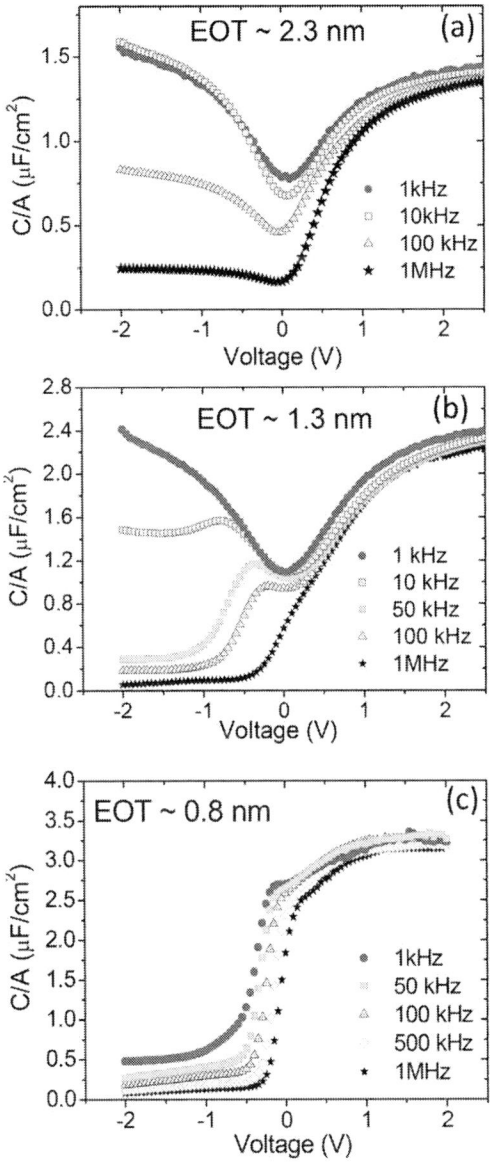

Fig. 9. *C-V* curves measured at different frequencies for Pt/GdScO$_3$/*n*-Ge(100) MOS stacks with a ~11 nm (a), ~4.5 nm (b), and ~2.7 nm (c) thick GdScO$_3$ film.

Unfortunately, it is observed that the EOT scaling brings together an increase in the D_{it} levels, reaching $\sim 10^{13}$ $eV^{-1}cm^{-2}$ for EOTs below 1 nm. More investigation has to be conducted aiming at the proper combination of ultra-low EOTs and D_{it} values lying in the low to medium 10^{11} $eV^{-1}cm^{-2}$ range.

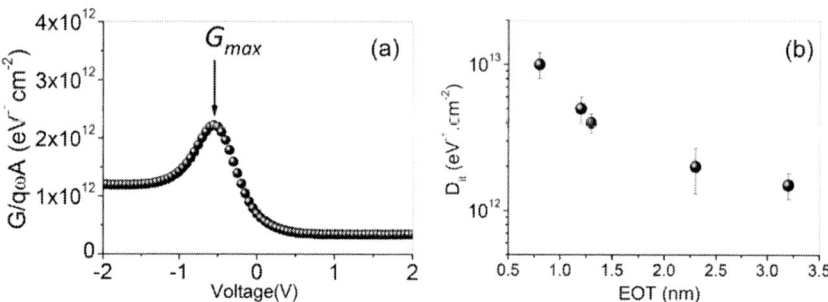

Fig. 10. Conductance curve at 100 kHz illustrating the peak conductance used to extract D_{it} (a). D_{it} value obtained from equation (1) plotted as a function of the EOTs measured for Pt/GdScO$_3$/n-Ge MOS stacks.

The dielectric constant of films prepared on Ge substrates was obtained in the same way as for Si substrates, i.e. by plotting the EOT values versus the physical thickness of the films. Obviously, in this case the EOT determination from a Hauser fit [43] took into account the specific properties of Ge as a substrate. Fig. 11 illustrates the plot for GdScO$_3$. As for films grown on Si (see table 1), the k value is found to be around 23.

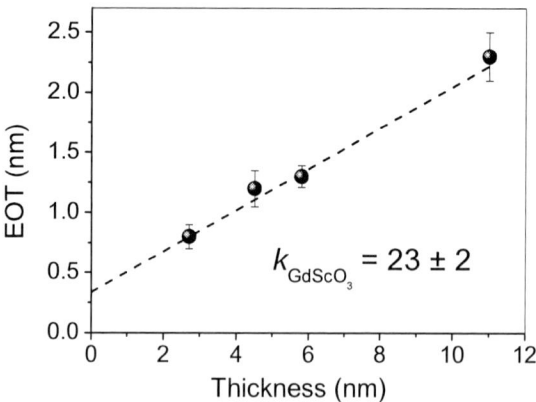

Fig. 11. EOT plot for amorphous GdScO$_3$ films deposited by e-gun evaporation on n-Ge (100) substrates at 300°C.

The intersection of the linear fit with the EOT axis indicates the existence of a very thin (~0.35 nm) lower-k interfacial layer between oxide and semiconductor. This value was obtained from the EOT calculation, which still considered SiO_2 as a reference dielectric ($k = 3.9$). Knowing that a GeO_2-based interface must be formed in this case (and therefore with $k = 5$-6 [24,28]), the actual thickness of the interface is then about 0.6 nm.

Regarding the chemistry of $LaLuO_3$ and $GdScO_3$ on Ge, XPS and TOF-SIMS analyses reveal the formation of germanate compounds taking place at the interface. Fig. 12(a) depicts TOF-SIMS and XPS spectra for a ~4.5 nm thick $GdScO_3$ film deposited directly on Ge (the C-V curves for the same sample are those shown in Fig. 9(b)). XPS shows that intermixing occurs at the interface leading to the formation of germanate-like compounds at this region. Although GeO_2 is not present, the existence of sub-stoichiometric oxidized species can not be completely ruled out. Although the TOF-SIMS profile shows Ge diffusion within the oxide film, depth-profiling analysis of the XPS peaks confirms intermixing (i.e. germanate formation) only close to the interface. This result corroborates the EOT plot findings (see Fig. 11), which showed a very thin lower-k interlayer between film and substrate. Moreover, in case the whole oxide film would have been converted to a RE-based germanate layer, a much lower dielectric constant would be expected [24-26]. Similar results were obtained for $LaLuO_3$ films on Ge and will not be shown here.

In order to investigate the effect of a GeO_2 passivation layer at the scandate/Ge interface, some Ge samples went through an oxidation process before the high-k film deposition. A GeO_2 layer (~1-1.5 nm thick) was thermally grown in 1 mbar O_2 for 60 min using a conventional furnace. Afterwards the oxidized Ge samples were *ex-situ* transferred to the deposition chamber where the $GdScO_3$ film deposition took place (for these samples the annealing procedure at 700°C to desorb remaining GeO_x was obviously not performed). Fig. 12(b) shows similar TOF-SIMS depth profiles as compared to the films deposited directly on Ge, however with a curiously more pronounced Ge diffusion within the film taking place. In addition, segregation of the RE elements in the GeO_2 seems to occur. XPS analysis confirms reaction at the interfacial region combined with the existence of GeO_2. In terms of electrical properties (not shown here), the use of the GeO_2 as a passivation layer reduced the D_{it} by a factor of 5. On the other hand, an increment of the EOT up to 40% is observed, thus preventing the required scaling to values below 1 nm. Finally, as another alternative (not shown here), the use of Si as a surface passivation layer for SiGe substrates also resulted in similar beneficial effects for the electrical properties [30,36] of MOS stacks containing $LaLuO_3$ and $GdScO_3$ as gate dielectrics. As for the case of Ge, the integration of ternary RE oxides with SiGe is now at its early stage and, in spite of the great potential, more research still has to be conducted.

The implementation of ternary RE oxides in MOSFETs having Si and high-mobility channel materials such as sSi or SiGe as channel materials were also studied and will be briefly presented here. More details can be found in recent publications [30,44-47]. As an example, Fig. 13 shows results obtained for n-MOSFETs processed on SOI and sSOI substrates and having either $LaLuO_3$, $LaScO_3$, or $TbScO_3$ as the gate dielectric. The electrical characterization of the devices reveals nearly ideal subthreshold slopes around

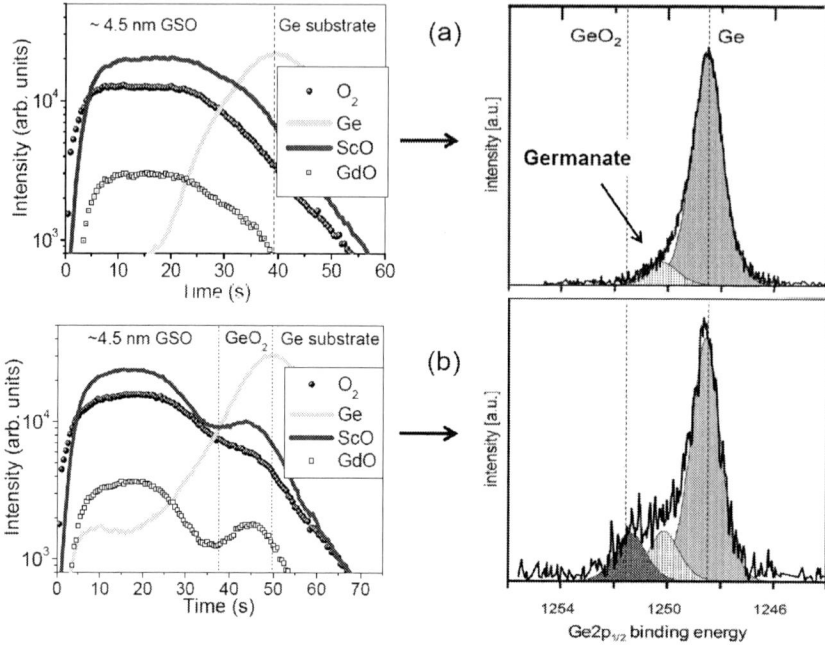

Fig. 12. TOF-SIMS and XPS analyses of a ~4.5 nm thick GdScO₃ film (denoted as GSO) deposited on a clean Ge surface (a) and on a GeO₂-passivated Ge surface (b).

72 mV/dec and high I_{on}/I_{off} ratios up to 10^8 (see Figs. 13(a) and (b)). Besides it, as observed for MOS structures on conventional Si substrates, D_{it} in the low to medium 10^{11} $eV^{-1}cm^{-2}$ range were measured (not shown). For MOSFETs on SOI substrates electron mobilities around 180 cm^2/Vsec were obtained, while for sSOI this value was improved by a factor of ~2. Fig. 13(c) shows the electron mobility curves for ternary *RE* oxides on sSOI *n*-MOSFETS. Although not shown here, comparable values were also determined for GdScO₃ gate oxides on similar substrates [44]. The mobility improvement determined for sSOI is a product of the elastic strain lifting the degeneracy of the six fold conduction band valleys. This increases the occupancy of the lower mass subband and decreases carrier scattering, resulting in higher electron mobility [56].

Aditionally, *p*-MOSFETs having a compressive strained SiGe as the channel and TiN/LaLuO₃ as the gate stack were processed (not shown). The devices yield good transfer characteristics with a subthreshold swing of 92 mV/dec and an I_{on}/I_{off} ratio of 10^5. The hole mobility extracted from split *C-V* measurements is comparable to the reported values of other high-*k* materials on SiGe. In this case a ~3 nm thick Si capping layer was used to passivate the SiGe surface prior the high-*k* deposition. Therefore the existence of a silicate-like interface at the gate/channel interface is expected as for the case of SOI and sSOI substrates.

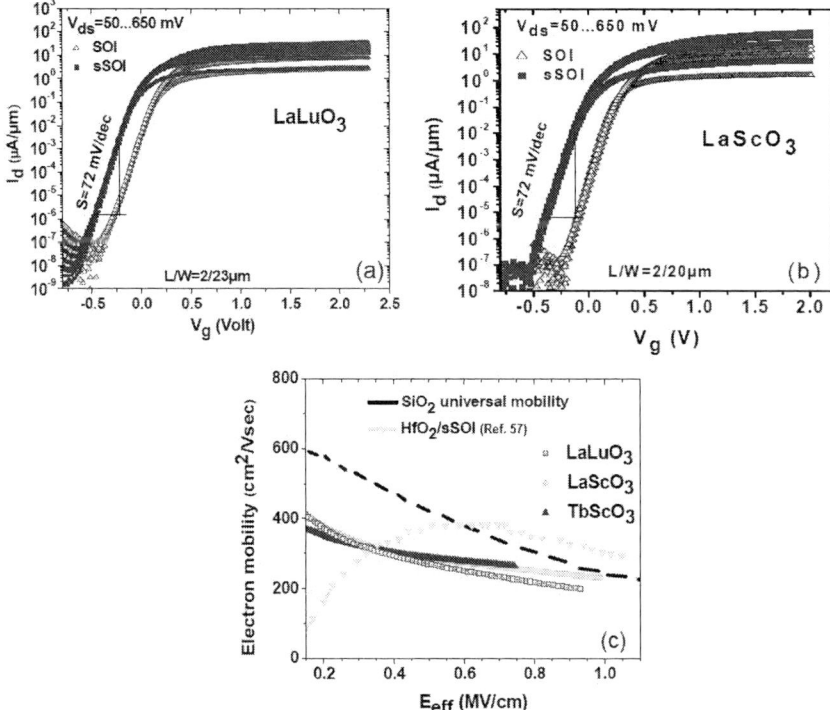

Fig. 13. Transfer characteristics of fully depleted SOI and sSOI and *n*-MOSFETS containing LaLuO$_3$ as a gate oxide (a), and SOI and sSOI *n*-MOSFETS with LaScO$_3$ as gate oxide (b). The EOTs were found to be 1.55 and 2.4 nm for LaLuO$_3$ and LaScO$_3$, respectively. L and W stands for the gate channel length and width of the devices. Electron mobility curves for the *n*-MOSFETS using LaLuO$_3$, LaScO$_3$, and TbScO$_3$ as gate dielectric (c). The mobility curves for devices with SiO$_2$ or HfO$_2$ are plotted in (c) for comparison.

For the ultimate CMOS scaling, the concept of having an interlayer between the gate dielectric and the channel material may have to be reconsidered. No matter how thin it can be intentionally created (ex.: GeO$_2$ or Si), or if it is formed as a product of spontaneous reactions (as the silicate- or germanate-like ones showed in this work), non-abrupt dielectric/semiconductor interfaces might not allow the conjunction of ultra-thin EOTs and low D$_{it}$. In this context, epitaxial high-*k* dielectrics either on Si or high-mobility substrates may play an important role in the future. Recent studies showing heteroepitaxy of LaLuO$_3$ on Si or GaAs substrates reveal the growth of high-quality films with sharp interfaces to the substrates [58-60]. For LaLuO$_3$ on GaAs [60], promising electrical properties such as high *k* and low D$_{it}$ were also realized. The integration of such ternary *RE* crystalline oxides into devices is a very interesting topic that still needs to be explored.

Summary

In summary, the results show that ternary *RE* oxides are very promising as alternative high-*k* dielectrics for future CMOS applications. Thin films of these materials keep their amorphous structure up to 1000°C, offer large optical band gaps (>5 eV) and band offsets to silicon (2-2.5 eV) and, most remarkably, have dielectric constants (*k*) between 23 up to 32. Intermixing at the dielectric/semiconductor interfacial region leading to the formation of silicate-like compounds seems to act favorably towards the electrical properties. MOS stacks with low D_{it} levels and EOT scalability could be realized. In the case of Ge, although the results show great potential, further investigation is still necessary to achieve interfaces with better qualities. Finally, transistors with these materials, although still at their early stage of development, show similar performance as those obtained for Hf-based devices, offering however the advantage of higher dielectric constants and thus potential for improved scaling.

Acknowledgments

This work was supported in part by the Project KZWEI which is funded in line with the technology funding for regional development (ERDF) of the European Union and by funds of the Free State of Saxony, by the German Federal Ministry of Education and Research through the MEDEA+ project DECISIF under Grant 2T104, and by the Nanosil network from the European Community under FP7 Grant 216171.

References

1. M.T. Bohr, R.S. Chau, T. Ghani, and K. Mistry, *IEEE Spectr.* **44**, 29 (2007).
2. 2008 International Technology Roadmap for Semiconductors (Semiconductor Industry Association, San Jose, CA, 2008).
3. J.-P. Colinge, *Silicon-on-Insulator Technology*, 3rd ed., 3rd ed. London, U.K.: Springer-Verlag, 2004.
4. C.K. Maiti, S. Chattopadhyay, and L.K. Bera, *Strained-Si Heterostructure Field Effect Devices*. New York: Taylor & Francis, 2007.
5. D.G. Schlom, S. Guha, S. Datta, *MRS Bulletin* **33**, 1017 (2008).
6. J. Robertson, *J. Appl. Physics* **104**, 124111 (2008).
7. S. Van Elshocht, P. Lehnen, B. Seitzinger, A. Abrutis, C. Adelmann, B. Brijs, M. Caymax, T. Conard, S. De Gendt, A. Franquet, C. Lohe, M. Lukosius, A. Moussa, O. Richard, P. Williams, T. Witters, P. Zimmerman, and M. Heyns, *J. Electrochem. Soc.* **153 (9)**, F219 (2006).
8. R. Thomas, P. Ehrhart, M. Luysberg, M. Boese, R. Waser, M. Roeckerath, E. Rije, J. Schubert, S. Van Elshocht, and M. Caymax, *Appl. Phys. Lett.* **89**, 232902 (2006).

9. M. Wagner, T. Heeg, J. Schubert, St. Lenk, S. Mantl, C. Zhao, M. Caymax, and S. De Gendt, *Appl. Phys. Lett.* **88**, 172901 (2006).
10. E. Durğun Özben, J.M.J. Lopes, M. Roeckerath, St. Lenk, B. Holländer, Y. Jia, D.G. Schlom, J. Schubert, and S. Mantl, *Appl. Phys. Lett.* **93**, 052902 (2008).
11. J.M.J. Lopes, M. Roeckerath, T. Heeg, E. Rije, J. Schubert, and S. Mantl, V.V. Afanas'ev, S. Shamuilia, and A. Stesmans, Y. Jia and D.G. Schlom, *Appl. Phys. Lett.* **89**, 222902 (2006).
12. M. Roeckerath, T. Heeg, J.M.J. Lopes, J. Schubert, S. Mantl, A. Besmehn, P. Myllymäki, L. Niinistö, *Thin Solid Films* **517**, 201 (2008).
13. V.V. Afanas'ev, A. Stesmans, C. Zhao, M. Caymax, T. Heeg, J. Schubert, Y. Jia, D.G. Schlom, and G. Lucovsky, *Appl. Phys. Lett.* **85**, 5917 (2004).
14. V. V. Afanas'ev, S. Shamuilia, M. Badylevich, A. Stesmans, L.F. Edge, W. Tian, D.G. Schlom, J.M.J. Lopes, M. Roeckerath, J. Schubert, *Microelectron. Eng.* **84**, 2278 (2007).
15. E.P. Gusev, V. Narayanan, M.M. Frank, *IBM J. Res. & Dev.* **50**, 387 (2006).
16. B.H. Lee, S.C. Song, R. Choi, and P. Kirsch, *IEEE Trans. Elec. Dev.* **55**, 8 (2008).
17. J. Robertson, *Rep. Prog. Phys.* **69**, 327 (2006).
18. C.R. Essary, K. Ramani, V. Craciun, R.K. Singh, *Appl. Phys. Lett.* **88**, 182902 (2006).
19. K.H. Kim, D.B. Farmer, J.-S.M. Lehn, P.V. Rao, R.G. Gordon, *Appl. Phys. Lett.* **89**, 133512 (2006).
20. P. Myllymäki, M. Roeckerath, J.M.J. Lopes, J. Schubert, K. Mizohata, M. Putkonen, L. Niinistö, *J. Materials Chemistry* **20**, 4207 (2010).
21. H. Wang, J.-J. Wang, R. Gordon, J.-S. M. Lehn, H. Li, D. H, and D. V. Shenai, *Electrochem. Solid-State Lett.* **12**, G13-G15 (2009).
22. M. Heyns and W. Tsai, *MRS Bulletin* **34**, 485 (2009).
23. E. Durğun Özben, J. M. J. Lopes, M. Roeckerath, A. Nichau, R. Luptak, S. Lenk, A. Besmehn, B. Ghyselen, Q.-T. Zhao, J. Schubert and S. Mantl, *Proceedings of the 11th International Conference on ultimate integration on Silicon (ULIS 2010)*, p. 93 (2010).
24. G. Mavrou, S. Galata, P. Tsipas, A. Sotiropoulos, Y. Panayiotatos, A. Dimoulas, E.K. Evangelou, J.W. Seo, and Ch. Dieker, *J. Appl. Phys.* **103**, 014506 (2008); and references herein.
25. J. Song, K. Kakushima, P. Ahmet, K. Tsutsui, N. Sugii, T. Hattori, and H. Iwai, *Microelectron. Eng.* **84**, 2336 (2007).
26. G. Mavrou, P. Tsipas, A. Sotiropoulos, S. Galata, Y. Panayiotatos, A. Dimoulas, C.Marchiori, and J. Fompeyrine, *Appl. Phys. Lett.* **93**, 212904 (2008).
27. M. Houssa, G. Pourtois, M. Caymax, M. Meuris, and M.M. Heyns, *Appl. Phys. Lett.* **92**, 242101 (2008).
28. A. Delabie, F. Bellenger, M. Houssa, T. Conard, S.V. Elshocht, M. Caymax, M. Meyns, and M. Meuris, *Appl. Phys. Lett.* **91**, 82904 (2007).
29. J. Huang, P.D. Kirsch, J. Oh, S.H. Lee, P. Majhi, H.R. Harris, D.C. Gilmer, G. Bersuker, D. Heh, C.S. Park, C. Park, H.-H.Tseng, R. Jammy, *IEEE Elec. Dev. Lett.* **30**, 285 (2009).
30. W. Yu, B. Zhang, Q.T. Zhao, J.-M. Hartmann, D.Buca, A. Nichau, J.M.J. Lopes, J. Schubert, B. Ghyselen, S. Mantl, in *Proceedings of 6th Workshop of the thematic network on Silicon-on-Insulator Technology, Devices and circuits (EUROSOI)*, p. 25 (2010).
31. T. Tabata, C.H. Lee, K. Kita and A. Toriumi, *ECS Trans.* **16**, 479 (2008).

32. P.D. Ye, J.J. Gu, Y.Q. Wu, M. Xu, Y. Xuan, T. Shen, and A. T. Neal, *ECS Trans.* **28**, 51 (2010).
33. J.J. Gu, Y.Q. Liu, M. Xu, G.K. Celler, R.G. Gordon, and P.D. Ye, Appl. Phys. Lett. **97**, 012106 (2010).
34. A. Toriumi, T. Tabata, C.H. Lee, T. Nishimura, K. Kita, K. Nagashio, *Microelectron. Eng.* **86**, 1571 (2009).
35. C. Radtke, C. Krug, G.V. Soares, I.J.R. Baumvol, J.M.J. Lopes, E. Durğun Özben, A. Nichau, J. Schubert, and S. Mantl, *Electrochem. Solid-State Lett.* **13**, G37 (2010).
36. R.A. Minamisawa et al. *manuscript in preparation.*
37. M. Wagner, T. Heeg, J. Schubert, C. Zhao, O. Richard, M. Caymax, V.V. Afanas'ev, S. Mantl, Sol. Stat. Elec. **50**, 58 (2006).
38. J.M.J. Lopes, U. Littmark, M. Roeckerath, St. Lenk, J. Schubert, S. Mantl, A. Besmehn, *J. Appl. Phys.* **101**, 104109 (2007).
39. C. Zhao, T. Witters, B. Brijs, H. Bender, O. Richard, M. Caymax, T. Heeg, J. Schubert, V.V. Afanas'ev, A. Stesmans, and D.G. Schlom, *Appl. Phys. Lett.* **86**, 132903 (2005).
40. J.M.J. Lopes, M. Roeckerath, T. Heeg, J. Schubert, U. Littmark, S. Mantl, A. Besmehn, P. Myllymäki, L. Niinistö, C. Adamo, D.G. Schlom, *ECS Trans.* **14**, 311 (2007).
41. S. Jeon, H. Hwang, *J. Appl. Phys.* **93**, 6393 (2003).
42. E.H. Nicollian and J.R. Brews, *MOS Technology*, John Wiley & Sons, New York (1982).
43. J.R. Hauser and K. Ahmed, *AIP Conf. Proc.*, **449**, 235 (1998).
44. M. Roeckerath, J.M.J. Lopes, E. Durğun Ozben, C. Sandow, S. Lenk, T. Heeg, J.Schubert, S. Mantl, *Appl. Phys. A: Mater. Sci. Process.* **94**, 521 (2009).
45. M. Roeckerath, J.M.J. Lopes, E. Durğun Ozben, C. Urban, J. Schubert, S. Mantl, *Appl. Phys. Lett.* **96**, 013513, (2010).
46. E. Durğun Özben, J.M.J. Lopes, A. Nichau, M. Schnee, S. Lenk, A. Besmehn, K.K. Bourdelle, Q.T. Zhao, J. Schubert, and S. Mantl, *IEEE Elec. Dev. Lett.*, *In press* (available on line: Digital Object Identifier 10.1109/LED.2010.2089423).
47. E. Durğun Özben, J.M.J. Lopes, A. Nichau, R. Lupták, S. Lenk, A. Besmehn, K.K. Bourdelle, Q.T. Zhao, J. Schubert and S. Mantl, *IEEE Trans. Elec. Dev.*, *Accepted for publication.*
48. J.M.J. Lopes, U. Littmark, M. Roeckerath, E. Durğun Özben, S. Lenk, U. Breuer, A. Besmehn, A. Stärk, P.L. Grande, M.A. Sortica, C. Radtke, J. Schubert, S. Mantl, *Appl. Phys. A: Mater. Sci. Process.* **96**, 447 (2009).
49. Y.A. Teterin and A.Y. Teterin, *Russ. Chem. Rev.* **71**, 347 (2002).
50. J.P. Espinos, A.R. Gonzalez-Elipe, J.A. Odriozola, *Appl. Surf. Sci.* **29**, 40 (1987).
51. J. Schubert, O. Trithaveesak, W. Zander, M. Roeckerath, T. Heeg, H.Y. Chen, C.L. Jia, P. Meuffels, Y. Jia, and D.G. Schlom, *Appl. Phys. A: Mater. Sci. Process.* **90**, 577 (2008).
52. J.M.J. Lopes, M. Roeckerath, T. Heeg, U. Littmark, J. Schubert, S. Mantl, Y. Jia, D.G. Schlom, *Microelectron. Eng.* **84**, 1890 (2007).
53. Y.Y. Gomeniuk, Y.V. Gomeniuk, A.N. Nazarov, P.K. Hurley, K. Cherkaoui, S. Monaghan, H.D.B. Gottlob, M. Schmidt, J. Schubert, J.M.J. Lopes, and O. Engström, *ECS trans.* **33**, 221 (2010).

54. R. Lupták, J.M.J. Lopes, St. Lenk, B. Holländer, E. Durğun Özben, A.T. Tiedemann, M. Schnee, J. Schubert, and S. Mantl, *J. Vac. Sci. Tech. B, Accepted for publication.*

55. T. Ando, M.M. Frank, K. Choi, C. Choi, J. Bruley, M. Hopstaken, M. Copel, E. Cartier, A. Kerber, A. Callegari, D. Lacey, S. Brown, Q. Yang, and V. Narayanan, in *IEDM Tech. Dig.*, p. 423 (2009).

56. S.F. Feste, Th. Schäpers, D. Buca, Q.T. Zhao, J. Knoch, M. Bouhassoune, A. Schindlmayr, and S. Mantl, *Appl. Phys. Lett.* **95,** 182101 (2009).

57. F. Andrieu, T. Ernst, O. Faynot, Y. Bogumilowicz, J.-M. Hartmann, J. Eymery, D. Lafond, Y.-M. Levaillant, C. Dupré, R. Powers, F. Fournel, C. Fenouillet-Beranger, A. Vandooren, B. Ghyselen, C. Mazure, N. Kernevez, G. Ghibaudo, and S. Deleonibus, in *Proc. IEEE Int. SOI Conf.*, p. 223 (2005).

58. T. Watahiki, F. Grosse, W. Braun, V. M. Kaganer, A. Proessdorf, A. Trampert, and H. Riechert, *Appl. Phys. Lett.* **97,** 031911 (2010).

59. T. Watahiki, F. Grosse, V.M. Kaganer, A. Proessdorf, and W. Braun, *J. Vac. Sci. Technol. B* **28,** C3A5 (2010).

60. Y. Liu, M. Xu, J. Heo, P.D. Ye, and R. G. Gordon, *Appl. Phys. Lett.* **97,** 162910 (2010).

480

ECS Transactions, 35 (4) 481-495 (2011)
10.1149/1.3572300 ©The Electrochemical Society

High-k Integration and Interface Engineering for III-V MOSFETs

H. J. Oh, Sumarlina A. B. S., and S. J. Lee

Department of Electrical and Computer Engineering, National University of Singapore, Singapore 117576

In this work, we report the comprehensive study of performance enhancement of InGaAs n-MOSFET by plasma PH_3 passivation. The calibrated plasma PH_3 passivation of the InGaAs surface before CVD high-k dielectric deposition significantly improves interface quality, resulting in suppressed frequency dispersion in C-V, increase in drive-current with high electron mobility, and excellent thermal stability.

Introduction

The dominance of silicon technology enabled by geometrical scaling is considered to encounter its downsizing limit in next few generations [1]. Among several emerging technologies and approaches replacing conventional silicon-based planar CMOS devices, III–V compound semiconductor MOSFETs stand out to hold promise as potential device candidates to be integrated onto the silicon platform for enhancing circuit functionality and also for extending Moore's Law [2-4]. The development of surface passivation techniques with proper high-k gate dielectric is one of the most critical requirements for successful implementation of III-V MOSFETs. In this work, we report the plasma PH_3 passivation on InGaAs channel with HfO_2 and HfAlO dielectrics and the performances of n-MOSFETs fabricated by self-aligned gate –first process.

Experiments

The process flow to fabricate InGaAs n-MOSFETs with a conventional self-aligned gate-first scheme is shown in Figure 1. A p-type $In_{0.53}Ga_{0.47}As$ layer (Zn-doped, ~1 x 10^{17} cm^{-3}) grown by MBE on a p^+ InP wafer with a InP buffer layer was used in this study. The $In_{0.53}Ga_{0.47}As$ surface was cleaned by 1% dilute HF for 2 min to remove native oxides and treated in $(NH_4)_2S$ for 5 min at room temperature. By using a UHV multi-chamber CVD system, a plasma-PH_3 treatment (1% PH_3/N_2), MOCVD HfO_2 ($Hf(OC(CH_3)_3)_4$ and O_2) or HfAlO ($HfAl(MMP)_2(OiPr)_5$) deposition, and post deposition annealing were subsequently conducted without breaking vacuum. In the passivation processing, the chamber pressure was stabilized with PH_3 gas diluted at 1% with N_2 for 10 s first, followed by a main surface passivation step with rf (13.56 MHz) plasma. Process parameters such as temperature, pressure, rf plasma power, and time were varied within the range of maximum hardware capacity to examine their effects on $In_{0.53}Ga_{0.47}As$ substrate and find optimal conditions. A sputtered TaN was deposited as a gate electrode and patterned by Cl gas based reactive ion etching. Source/Drain (S/D) was implanted with Si at 50KeV with a dose of $1\times10^{14}cm^{-2}$ and activated at 600°C for 60sec, 700°C 10sec and 750°C 5sec by RTA (Rapid Thermal Annealing). AuGe/Ni/Au and Ti/Pt/Au contacts were deposited for front and back side contacts, respectively.

481

○ MBE growth of $In_{0.53}Ga_{0.47}As$ and $In_{0.7}Ga_{0.3}As$ substrate
○ HCl and $(NH_4)_2S$ surface cleaning
○ In-situ PH_3 passivation
○ MOCVD high-k deposition and in-situ PDA
○ Sputter TaN deposition 150nm
○ Lithography
○ Gate etch and high-k etching
○ Si implantation $(50KeV/1x10^{14}cm^{-2})$
○ S/D activation @ RTA 600 °C 60s, 700 °C 10s, 750°C 5s
○ S/D and backside contacts
○ Metal alloy annealing @ RTA 400 °C 60s

Figure 1. Process flow for fabrication on InGaAs MOSFET with self-aligned gate-first process

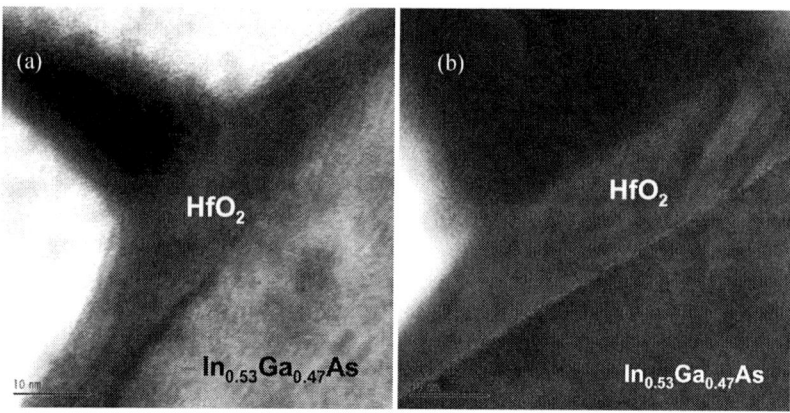

Figure 2. Cross sectional TEM images of TaN/HfO$_2$/InGaAs gate stacks (a) without passivation and (b) with plasma PH$_3$ passivation

Results and discussion

Figure 2 shows the cross sectional TEM (Transmission Electron Microscopy) images of InGaAs/HfO$_2$ gate stacks. As can be seen in Figure 2 (a), the direct deposition of HfO$_2$ on the InGaAs substrate without any passivation layer results in a rough interface and possible inter diffusion. On the other hand, the plasma PH$_3$ passivated InGaAs gate stack shows sharp and smooth interface between InGaAs substrate and HfO$_2$ layer with ultra thin interfacial layer. Electron Dispersion Spectroscopy (EDS) measurement was performed in the vicinity of the transistor channel region. Results show that outdiffusions of Oxygen and Hafnium to the channel are well-controlled. This is important for the achievement of high-quality gate dielectric interface and high electron mobility. Atomic Force Microscope (AFM) is used to examine the change in surface morphology due to passivation. InGaAs surface profiles after HCl/$(NH_4)_2$S pre-gate cleaning shows RMS roughness of 0.68 nm in Figure 3 (a). After plasma PH_3 passivation, its roughness is reduced to be 0.31 nm in Figure 3 (b). Inversion-mode MOSFET uses a surface channel, therefore, a smooth surface reduces interface scattering which degrades the carrier mobility in the channel at a high gate bias.

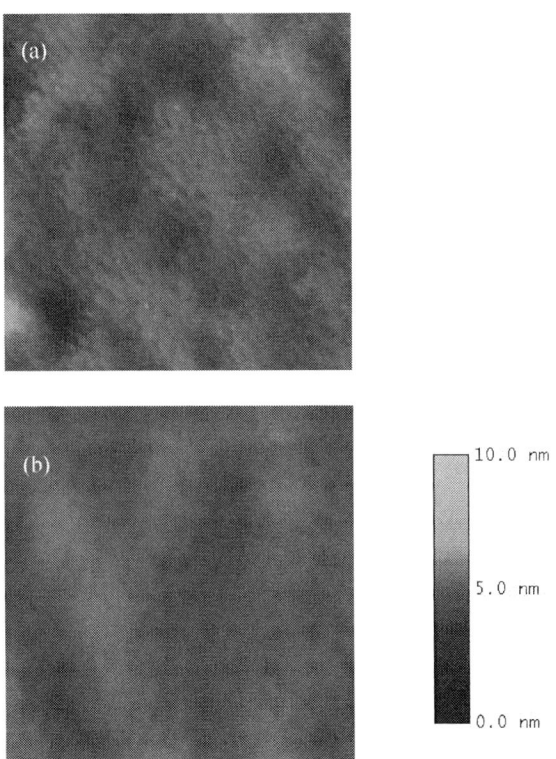

Figure 3. Atomic force microscopy (AFM) image of surface profile. The scanning size is 5x5 μm^2. (a) After HCl and $(NH_4)_2$S pre-gate cleaning; the root mean square (RMS) roughness is about 0.68 nm. (b) After pre-gate cleaning and plasma PH_3 passivation. The RMS roughness is about 0.31 nm. Plasma PH_3 passivation reduces the surface roughness.

The interface trap densities (D_{it}) in the HfO_2/InGaAs gate stacks are extracted by conductance method. Directly deposited samples show D_{it} of 4.3×10^{12} $eV^{-1}cm^{-2}$. D_{it} is reduced in the passivated InGaAs/HfO_2 gate structure to 8.6×10^{11} $eV^{-1}cm^{-2}$.

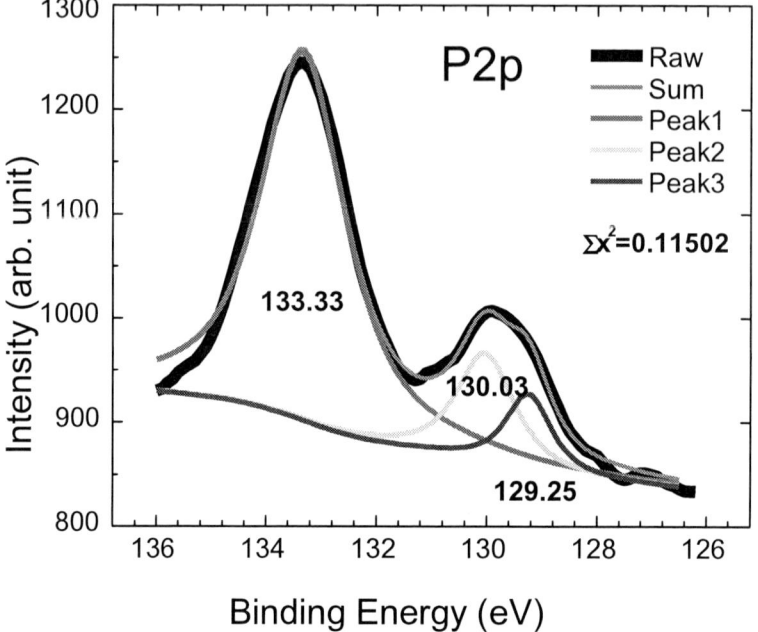

Figure 4. XPS P 2p spectrum of InGaAs surface with plasma PH₃ passivation treatment.

In order to investigate the formation and properties of the passivation layers, a XPS (X-ray Photoelectron Spectroscopy) analysis was carried out. Core level spectra of As 3d, Ga 3d, and In 3d (not shown) shows that the formation of native oxides, which may induce the Fermi-level pinning and degradation of interface quality, was not detected from the plasma PH₃ passivated InGaAs gate stacks. Figure 4 shows the P 2p core level spectrum of the plasma PH₃ passivated InGaAs gate stack. The PH₃ passivated sample exhibits P–metal bond formation with the corresponding P 2p peak at lower binding energies of 129.1–129.6 eV. [5, 6] The additional high energy peaks at ~133.3 eV clearly appear with Gaussian shapes (Full Width Half Maximum - FWHM = 1.8 ~ 1.9 eV) in their P 2p spectra, showing the formation of the P_xN_y compound as the major composition of the phosphorus-incorporated layer. The formation of a P_xN_y passivation layer is also evidenced by the presence of the high energy N 1s peak at ~ 398.2 eV (not shown), which was not observed in the case of non passivated samples. The thickness of the passivation layer was estimated using XPS depth profiles for the P 2p and N 1s spectra. Using an in-situ sputtering with an etching rate of 1 Å/s, it is found that the P-N bond almost disappeared after etching of 2 s, which is equivalent to the film thickness of ~2 Å The Capacitance-Voltage (C-V) characteristic of the HfO₂/InGaAs MOS system is measured at 1 kHz in Figure 5. The measured high frequency C-V curve matches well with the low frequency simulation curve, indicating high quality gate stack. From the curve fitting, the Effective Oxide Thickness (EOT) is extracted to be 3 nm and dielectric constant for gate dielectric is 12.9. The gate leakage densities in directly-deposited and passivated samples are measured in Figure 6. Similar leakage levels are observed.

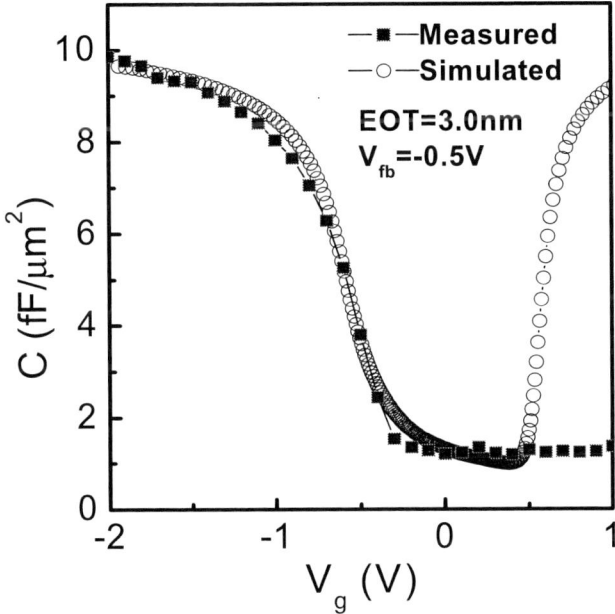

Figure 5. Capacitance-Voltage (C-V) characteristic at 1 kHz of plasma PH_3 passivated MOS capacitor with $TaN/HfO_2/InGaAs$ gate stack (square). Simulation of ideal C-V characteristic is fitted to the flatband voltage and EOT (circle). A close match between the simulated and measured curve indicates a high-quality MOS capacitor.

By grounding the substrate and measuring the capacitance between gate and S/D (standard split C-V measurement for inversion capacitance), the C-V characteristic is measured from low frequency of 500 Hz to high frequency 1 MHz in Fig. 7. After the 600 °C annealing for S/D dopant activation, the MOS gate stack in the transistor maintains stable. It exhibits very low frequency dispersion (< 5%) and small threshold voltage difference (< 0.1 V) from 500 Hz to 1 MHz. We attribute this performance to the interfacial layer formed after passivation to prevent the substrate from oxidizing during the thermal treatment and to reduce the interface state density.

Bi-directional C-V measurement is done at 10 KHz, sweep rate of 160 mV/s and sweep range from −0.5 to 2 V. C-V hysteresis of 0.17 V was observed as shown in Fig. 8. A previous MOS capacitor study on InGaAs and GaAs showed that high Indium concentration in the interface effectively improve trapping behavior of dielectric, thus the hysteresis [7].

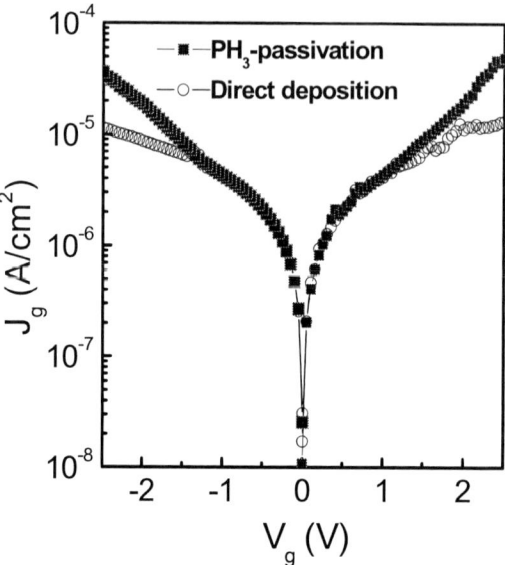

Figure 6. Gage leakage density of direct deposition and passivated MOS capacitors. Similar leakage levels are observed.

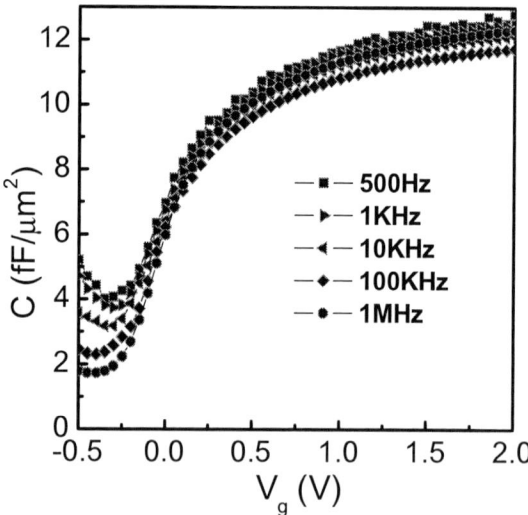

Figure 7. Plasma PH₃ passivated MOSFET inversion capacitance versus applied gate bias from 500 Hz to 1 MHz. It has low frequency dispersion, indicating low interface traps in a wide range of frequencies and successful channel formation.

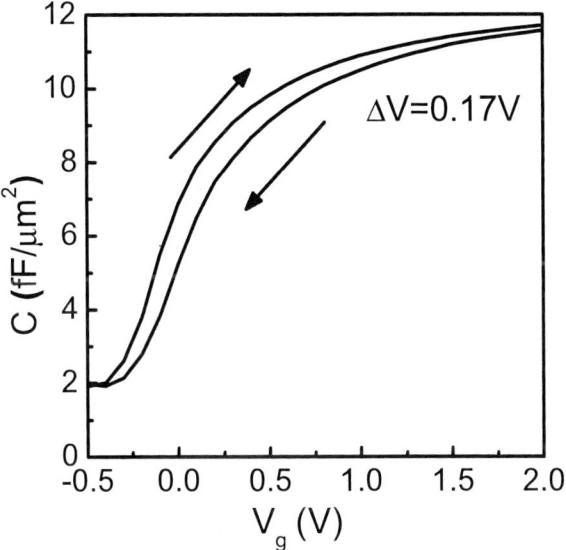

Figure 8. Hysteresis of inversion capacitance measurement at 10 KHz from -0.5 V to 2 V measuring range. Inversion capacitance measures between gate to source and drain with substrate grounded, shown in the inset.

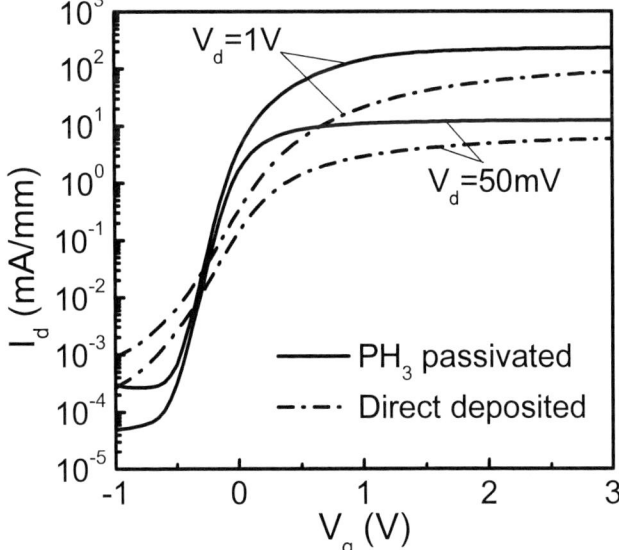

Figure 9. Drain current at the subthreshold region. I_d-V_g of InGaAs n-MOSFET of 4 μm gate length for passivated MOSFET and directly deposited control.

Figure 9 shows the I_d-V_g characteristics for the directly deposited and passivated InGaAs n-MOSFET of 4 μm gate length. By using the plasma PH3 passivation, we achieved mean subthreshold slope of 100 mV/dec, I_{on}/I_{off} ratio from 5 orders to 6 orders, and drain induced barrier lowering (DIBL) from 92 to 18 mV/V. These results indicate that both input transfer and output characteristics can be improved by using the plasma PH3 passivation scheme for InGaAs n-MOSFET.

Figure 10 shows the subthreshold slope of passivated and non-passivated MOSFET at various temperatures. The comparison shows the plasma PH3 passivation results in a significant reduction and smaller variation of subthreshold slope (S.S.) at a specific temperature. As temperature rises, the increase in S.S. of passivated MOSFET is less than directly deposited ones, indicating greatly suppressed interface traps.

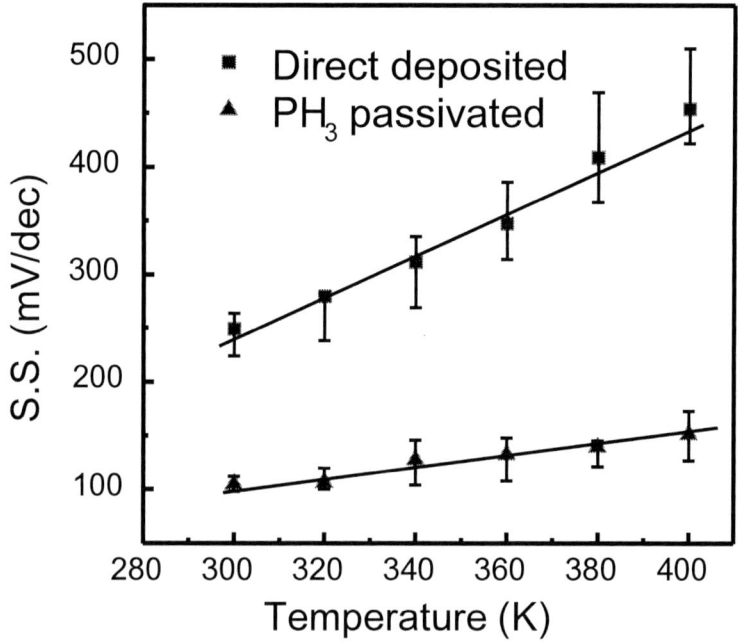

Figure 10. The subthreshold slope comparison of InGaAs n-MOSFET with plasma PH3 passivated and direct deposited dielectrics over a temperature from 300 K to 400 K.

Figure 11 shows the I_d-V_d characteristic of InGaAs n-MOSFET of gate length 4 μm. Well behaved output characteristic is observed. At maximum bias conditions ($V_g = V_d = 3$ V), 400 mA/mm drain current is obtained in passivated n-MOSFET. There is a four-time leap in on current for passivated sample over the directly deposited control sample. Using split C-V method without correction of parasitic series resistance and D_{it}, the peak mobility of 2557 cm^2/Vs is obtained at $E_{eff} = 0.24$ MV/cm.

The linear I_d-V_g characteristic and corresponding transconductance are shown in Figure 12 for the passivated transistor under drain bias of 0.05 V and 1 V. The gate length is 4 μm. The maximum transconductance is 20 mS/mm and 180 mS/mm for the two drain biases, respectively. The threshold voltages are ~ 0 V. The threshold voltage is determined by the effective metal work function, channel doping and Fermi-level pinning in the gate stacks. With a modulation of work function of gate electrode, the threshold voltage can be modulated to fulfill the enhancement-mode n-MOSFET operation. The effective gate work function tuning for TaN and high-k dielectric has been reported in Silicon n-MOSFET [8].

The drop of transconductance in high gate bias ($V_g > 1.6$ V) can be attributed to the S/D series resistance. S/D series resistance, measured by the transmission line method, is about 2.1 Ω.mm. This value is about half of the total resistance ($R_{channel}$ and R_{sd}) at $V_g = 1.6$ V. It becomes a limiting factor for the continuous current increase. Therefore, the reduction of S/D series resistance is imperative for further exploration of the intrinsic performance, such as the inter-valley scattering, of InGaAs n-MOSFET at high gate overdrive. This result implies that new technologies, such as efficient dopants introduction and activation techniques and metallic Schottky S/D contact formation techniques, to reduce the S/D parasitic resistance should be developed in future.

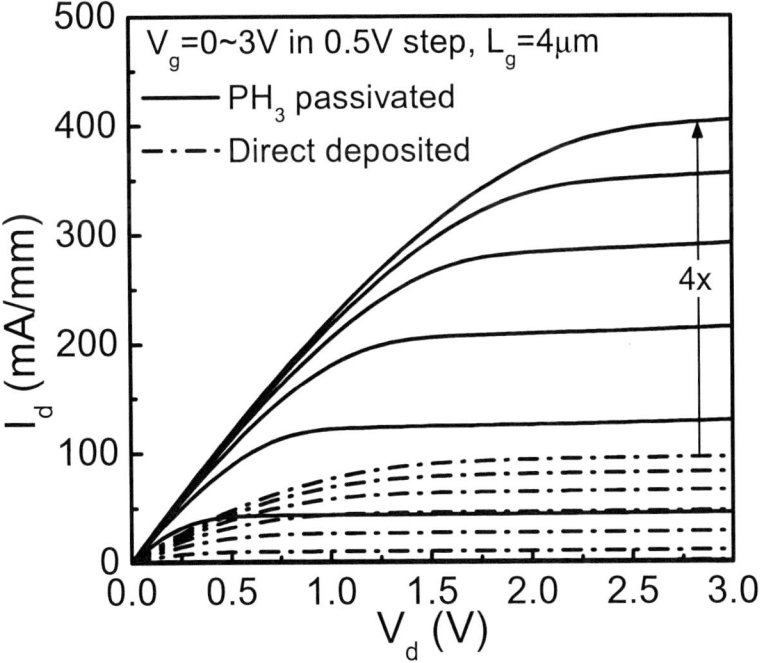

Figure 11. I_d-V_d characteristics of InGaAs n-MOSFET for passivated and directly deposited n-MOSFET.

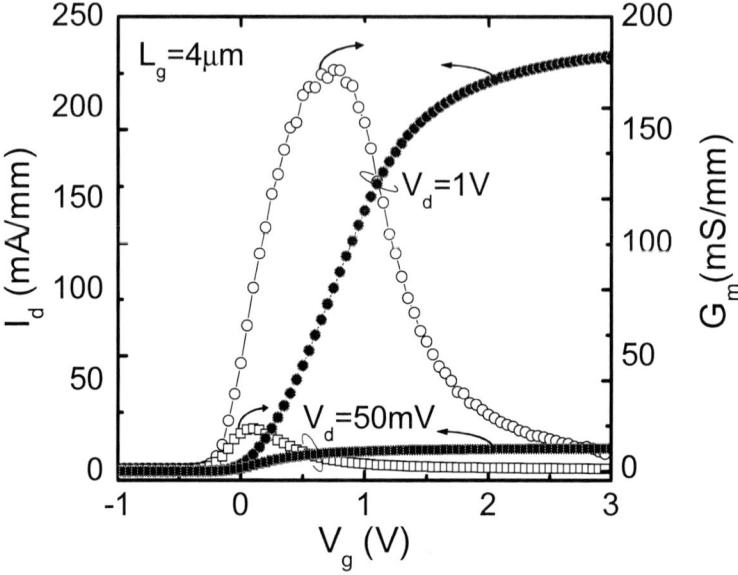

Figure 12. Transconductance for passivated InGaAs n-MOSFET with 4 μm gate length at drain bias of 0.05 V and 1 V.

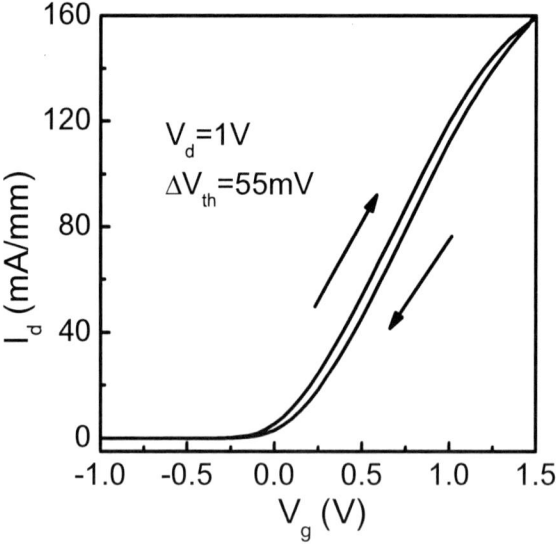

Figure 13. Drain current hysterisis for a bi-directional sweeping of gate voltage from -1 V to 1.5 V at drain bias of 1 V

In Figure 13, the drain current hysteresis is shown. A threshold voltage shift of 55 mV is observed from a sweeping range of -1 V to 1.5 V at V_d = 1 V. It indicates the dielectric has a high quality and is immune to bulk and interface trappings.

Figure 14 compares the on current at V_g = 2 V and V_d = 1 V for the passivated n-MOSFET and directly deposited control. Significant increase of drain current from passivation is observed in the plot of on current as a function of gate length. The electron mobility in the channel is extracted by split C-V method without considering D_{it} and series resistance.

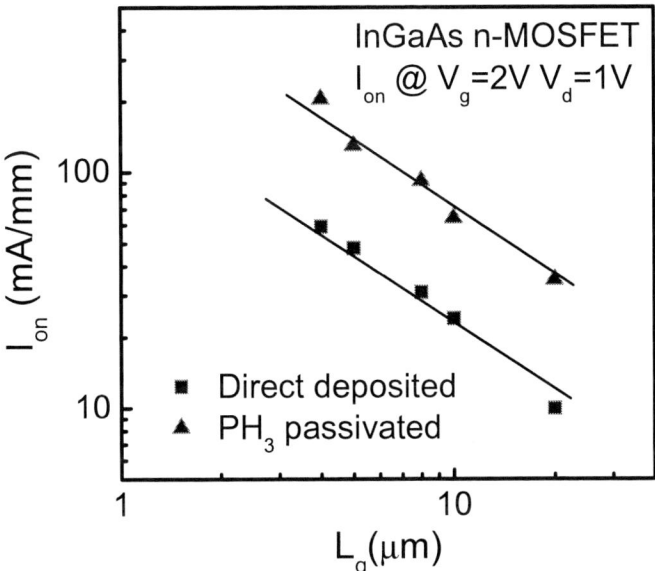

Figure 14. On-current as a function of gate length comparison under similar gate overdrive in n-MOSFET with passivated and directly deposited InGaAs n-MOSFET. A significant increase in drive current is obtained with the interface engineering on HfO_2/InGaAs gate stack.

Thermal stability of plasma-PH_3 passivated HfAlO/ $In_{0.53}Ga_{0.47}As$ gate stack was studied by applying thermal annealing processes after MOS capacitor fabrication, a Forming Gas Anneal (FGA) with 10% H_2/ 90% N_2 concentration at 400°C for 10min, followed by rapid thermal annealing processes at varying temperature steps i.e. 500°C 1min, 600°C 1min, 700°C 1min and 800°C 5s in N_2 ambient. The inversion C-V, gate leakage current and equivalent oxide thickness were monitored after each annealing step.

Fig. 15 shows well behaved inversion C-V characteristics, measured by Split C-V, after annealing the plasma-PH_3 passivated $InGa_{0.53}As_{0.47}$ MOSFETs up to 800°C.Almost identical C-V shifts to the positive direction observed after thermal annealing processes can be attributed to different level of fixed charges compared to the device without anneal. The plasma-PH_3 passivated devices show negligible changes in EOT up to 800°C anneal, with EOT ~ 3.1nm before anneal and EOT ~ 3.0nm after 800°C anneal.

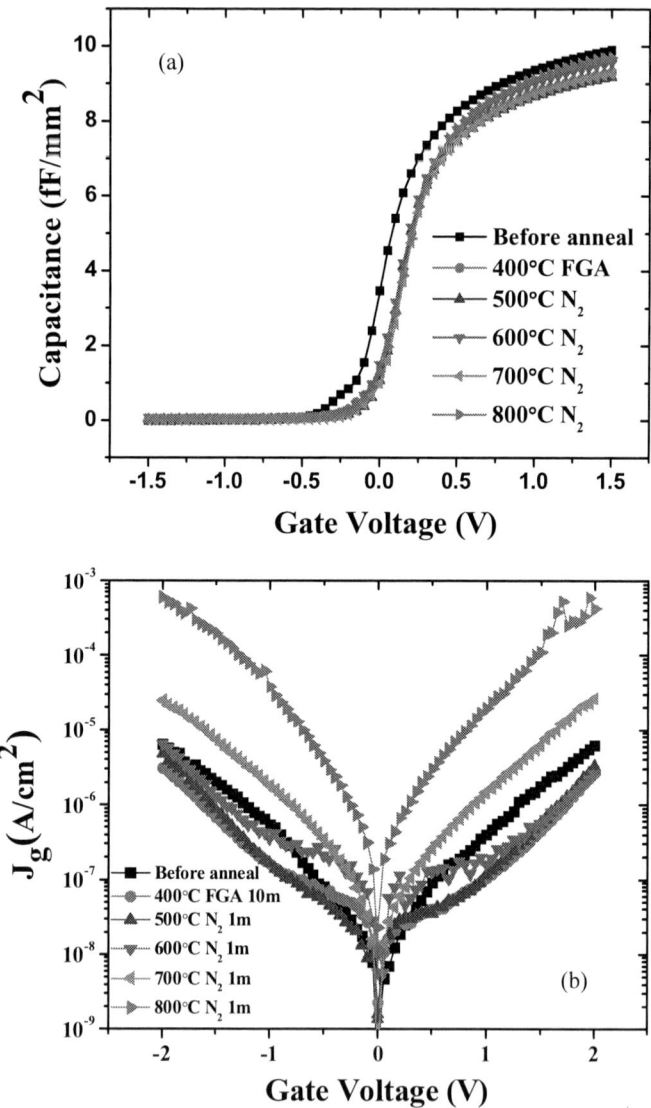

Figure 15. (a) Split C-V and (b) I-V curves of Plasma-PH_3 passivated HfAlO/ $In_{0.53}Ga_{0.47}As$ gate stack measured up to 800°C.

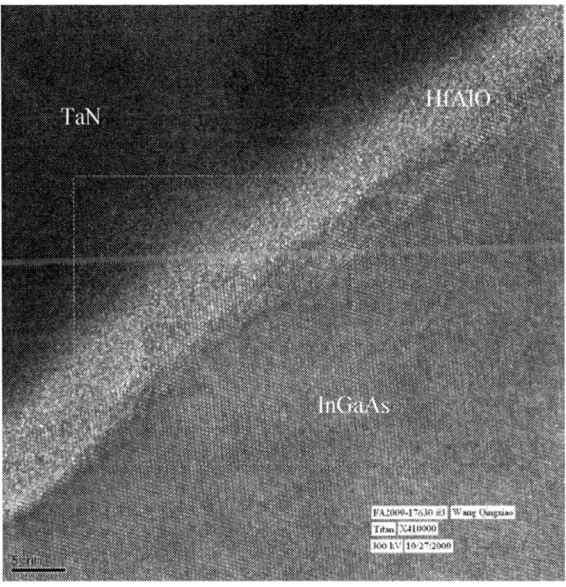

Figure 16. TEM image of TaN/HfAlO/InGa$_{0.53}$As$_{0.47}$ gate stack with FGA 400°C followed by 800°C rapid thermal anneal.

Figure 16 shows the TEM image of HfAlO/plasma-PH$_3$/InGa$_{0.53}$As$_{0.47}$ gate stack with a 400°C FGA followed by a 800°C rapid thermal anneal. It is well known that the crystallization temperature of HfO$_2$ film is increased by alloying with Al [9]. As can be seen in the TEM image, the HfAlO layer remains amorphous up to 800°C anneal. The interface between the InGaAs and the passivation layer is smooth and conformal up to 600°C anneal.

However, the sharp transition from the crystalline In$_{0.53}$Ga$_{0.47}$As substrate to the P$_x$N$_y$ layer can no longer be observed after 800°C anneal, due to the structural changes seen at the localized rough In$_{0.53}$Ga$_{0.47}$As interface caused by the poor thermal stability of In$_{0.53}$Ga$_{0.47}$As that starts to show Ga out-diffusion at 800°C [10], which is also confirmed by energy dispersion X-ray analysis (not shown). A localized thinning of the HfAlO is observed after 800°C anneal and this may be caused by the out-diffusion of Ga/As, the inter-diffusion between HfAlO and the In$_{0.53}$Ga$_{0.47}$As substrate, and the growth of InGaAs into the passivation layer due to the in-diffusion of phosphorus from the InP substrate into the In$_{0.53}$Ga$_{0.47}$As at high temperature anneals beyond 800°C [10] causing outdiffusion of Ga/As into HfAlO. This Ga/As outdiffusion, which is also proven from the energy dispersion X-ray analysis (not shown), and the localized HfAlO thinning could explain the rise in J$_g$ at 800°C observed in Figure 15, suggesting the need to find an optimal condition for plasma PH$_3$ passivation process. Nevertheless, the frequency dispersion ($\Delta C_{10kHz\sim1MHz}/C_{10kHz}$) and hysteresis (difference in flat band voltage for 100 kHz C-V curves obtained from forward and reverse sweeps) in inversion C-V of the plasma-PH_3 passivated devices are not significantly altered, ranging between 2.3% to 2.9% and 48mV to 52mV for frequency dispersion and hysteresis, respectively, with varying annealing temperature before anneal to 800°C. These results imply that no significant degradation of the interface quality of the gate stack is observed, possibly due to no significant defect formation caused by this Ga/As outdiffusion and/or the preservation of the gate stack with the passivation layer up to 800°C anneal.

Summary

In this work, a plasma PH_3 passivation of $In_{0.53}Ga_{0.47}As$ was applied to the fabrication of n-channel MOSFET. A calibrated plasma PH_3 treatment forming a P_xN_y interface layer provides high interface quality and performance improvement for high-k/InGaAs integration. The $In_{0.53}Ga_{0.47}As$ n-MOSFET demonstrated excellent Capacitance-Voltage characteristic with small frequency dispersion (< 5%) and hysteresis. The n-MOSFET showed significant increase in drive current (4 times) and reduction in subthreshold slope (< 100 mV/dec). Peak electron mobility measured by split-capacitance-voltage without any corrections was 2557 cm^2/Vs. In addition, we report the thermal stability of the plasma-PH_3 passivated HfAlO/$In_{0.53}Ga_{0.47}As$ gate stack up to 800°C anneal as proven from the negligible changes in the EOT, frequency dispersion, and hysteresis up to 800°C anneal. A slight increase in J_g exists at 1V and 1.5V, with J_g being 1.95×10^{-5}A/cm^2 and 1.10×10^{-4} A/cm^2 respectively. Studies using TEM revealed the HfAlO oxide still being amorphous at 800°C anneal but with localized thinning of gate dielectric and rough interface caused by the out-diffusion of Ga/As causing the rise in J_g. However, it is found that the HfAlO/InGaAs gate stack maintains good electrical integrity up to 800°C anneal, which enables the successful fabrication of self-aligned gate-first n-MOSFETs.

References

1. International Technology Roadmap for Semiconductors 2009 Edition, ITRS2009
2. S. Takagi and M. Takenaka., VLSI Symposium, 147, (2010).
3. H.-J. Oh, J. Lin, S. A. B. Suleiman, G. Q. Lo, D. L. Kwong, D. Z. Chi and S. J. Lee, Tech. Dig. -Int. Electron Devices Meet., 339, (2009).
4. M. Radosavljevic, G. Dewey, J. M. Fastenau, J. Kavalieros, R. Kotlyar, B. Chu-Kung, W. K. Liu, D. Lubyshev, M. Metz, K. Millard, N. Mukherjee, L. Pan, R. Pillarisetty, W. Rachmady, U. Shah, and Robert Chau, Tech. Dig. -Int. Electron Devices Meet., 126, (2010).
5. M. Losurdo, P. Capezzuto, G. Bruno, G. Leo and E. A. Irene, *J. Vac. Sci. Technol.,* A **17**, 2194, (1999).
6. G. Bruno, *Appl. Surf. Sci.,* **235**, 239, (2004).
7. H.-J. Oh, J. Lin, S.J. Lee, G.K. Dalapati, A. Sridhara, D.Z. Chi, S.J. Chua, G.Q. Lo, and D.L. Kwong, *Appl. Phys. Lett.*, **93**, 062107, (2008).
8. X. P. Wang, H.Y. Yu, M.-F. Li, C.X. Zhu, S. Biesemans, A. Chin, Y.Y. Sun, Y.P. Feng, A. Lim, Y.-C. Yeo, W.Y. Loh, G. Q. Lo, and Dim-Lee Kwong, *IEEE Electron Device Letters,* **28**, 258, (2007).

9. M. S. Joo, B. J. Cho, C. C. Yeo, D. S. H. Chan , S. J. Whoang, S. Mathew, L. K. Bera, N. Balasubramanian, and D. L. Kwong, *IEEE Trans. Electron Devices*, **50**, 2088, (2003).
10. K. Kiziloglu, M. M. Hashemi, L.W. Yin, Y. J. Li, P. M. Petroff, U. K. Mishra and A. S. Brown, *J. Appl .Phys,* **72**, 8, (1992).

496

ECS Transactions, 35 (4) 497-513 (2011)
10.1149/1.3572301 ©The Electrochemical Society

Plasma Enhanced Atomic Layer Deposition of ZrO₂: A Thermodynamic Approach

E. Blanquet[1], D. Monnier[2], I. Nuta[1], F. Volpi[1], B. Doisneau[1], S. Coindeau[1], J. Roy[3], B. Detlefs[3], Y. Mi[3], J. ZegenHagen[3], C. Martinet[4], C. Wyon[5], and M. Gros-Jean[2]

[1]Science et Ingénierie des Matériaux et Procédés (SIMaP),
Grenoble INP/CNRS/UJF, BP 75 38402 Saint Martin d'Hères, France
[2]STMicroelectronics, 850 rue Jean Monnet,
38926 Crolles, France
[3]ESRF 6 rue Jules Horowitz 38000 Grenoble, France.
[4]LPCML 10 rue André-Marie Ampère 69622 Villeurbanne, France
[5]CEA 17 rue des Martyrs 38054 Grenoble, France.

In the pursuit of smaller and faster devices manufacture, integration of new materials exhibiting a high dielectric permittivity is going on to replace silicon oxide SiO_2 in Metal/Insulator/Metal (MIM) capacitors and in Dynamic Random Access Memory (DRAM). Among these materials, zirconium oxide, ZrO_2, in its highest dielectric permittivity phase (the high temperature tetragonal one) is investigated. Atomic Layer Deposition (ALD) of out-of-equilibrium ZrO_2 thin films in 3D architectures is explored using various approaches: evaluation of the zirconium gaseous precursor, influence of operating conditions, and thermal behavior of the deposited films. Thermodynamic models are used to better understand the film growth.

Introduction

As a consequence of the continuous miniaturization of silicon-based electronic devices, intense research and development activities are focused on the introduction of new materials as well as the integration of new processes and structures at the different technology levels. For instance, introduction of new functional insulators exhibiting higher dielectric permittivity (i.e. "high-k") than conventional silicon oxide SiO_2 is needed. High-k materials have been massively studied during the last decade for their introduction as gate transistor isolator. They also allow density improvement for Metal/Insulator/Metal (MIM) capacitors and Dynamic Random Access Memory (DRAM) technologies [1]. The main challenges related to these last technologies are to ensure, in 3D architectures such as trenches, a uniform film deposition with controlled physical and electrical properties, and a controlled interface quality with the metallic layer on which the isolator is deposited. Atomic Layer Deposition (ALD) and Plasma Enhanced Atomic Layer deposition (PEALD) techniques which are based on the sequential self-limiting surface reactions from generally two gaseous precursors provide an ideal way for depositing ultrathin and conformal films in 3D architectures.

Among the promising materials to replace SiO_2, ZrO_2 is investigated [2]. Compared to the value of 3.9 for silica, its dielectric permittivity, ε_r, is much higher. It depends on its crystalline structure and is equal to 17, 37 or 47 for the monoclinic, cubic or tetragonal phases, respectively [3]. The most promising phase to reach a capacitance of 10 fF/μm²,

497

the tetragonal one, is stable at high temperature according to the binary phase diagram (Fig. 1) [4, 5]. To increase the capacitance density, 3D architectures are used, introducing the need for uniform internal surface coverage of cavities (Fig. 2). PVD (Physical Vapor deposition) or even CVD (Chemical Vapor Deposition) techniques are not suited and need to be replaced by Atomic Layer Deposition techniques (ALD). ALD or PEALD presents the advantage of its main characteristics i.e. the growth-controlled process at atomic level [6]. As the process is based on saturating surface reactions between the substrate and each of the reactant needed for the compound to be grown [7], it results in excellent coverage and uniformity of the film in any kind of complex shape substrates. The thickness is controlled at the nanometer scale, even for high aspect ratio structures [8, 9]. The temperature of the reaction chamber is generally lower in ALD compared to conventional CVD processes. Specific growth conditions [10, 11] have to be adjusted in order to obtain the desired phase (tetragonal ZrO_2) at low temperature.

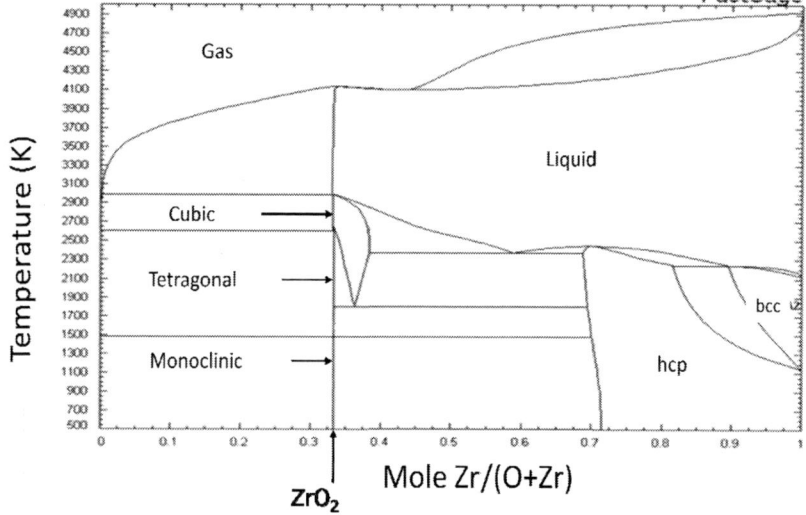

Figure 1. O-Zr binary phase diagram [4, 5].

In that context, this paper describes the development and the optimization of an out-of-equilibrium phase ALD deposition process. It requires the knowledge of all the reactions mechanisms i.e. the gas phase and surface reactions. A methodology based on the correlations among the nature of the precursor molecule, its thermodynamical behavior, the nature of the substrate and the properties of the deposited films is illustrated with the example of the PEALD deposition of ZrO_2 from the zirconium organometallic TEMAZ (TetrakisethylmethylaminoZirconium) precursor and oxygen. First, specific attention has been focused on the selection of the zirconium precursor. A system composed of a Knudsen effusion cell coupled with a mass spectrometer specifically designed for the thermodynamic study of organometallic precursors has been used to identify and quantify the actual species present in the gaseous phase formed during the vaporization of the TEMAZ ALD gaseous precursor [12-14]. Then, the influence of the major process parameters (essentially the process type -ALD vs PEALD- and the nature of the substrate) on the deposited film properties has been analyzed, to identify conditions providing the tetragonal phase. Annealing behavior of the films has been explored during the fabrication of the MIM structures (ALD TiN electrodes deposition at 400°C). Thermodynamic simulations are carried out to explore the substrate/film reactivity and to better understand the film growth.

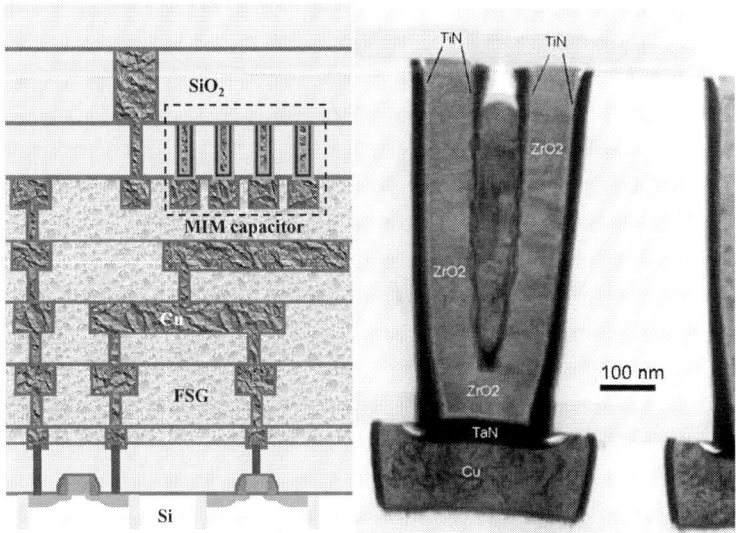

Figure 2. Left: Cross section of an integrated circuit with inserted 3D MIM capacitors Right: 3D MIM capacitor TEM cross section.

Experimental

Experimental study of the zirconium precursor TEMAZ

The gaseous precursors compounds used for ALD and PEALD processes have to meet several physicochemical property requirements including high reactivity vs surface substrate, relatively high volatility, convenient decomposition behavior and thermal stability (decomposition temperature higher than the deposition one, generally below 400°C) [15, 16]. Metal halide precursors like $ZrCl_4$ have been widely used in ALD for zirconium oxide deposition [17-19]. However chloride contamination of ZrO_2 films occurs when deposited with $ZrCl_4$. Metal alkyl amide precursors such as Tetrakis(ethylmethylamino)zirconium (TEMAZ) have been studied to replace them [15, 20, 21]. To evaluate the thermodynamic behavior of TEMAZ and identify the gaseous species formed during its vaporization and transport in the gaseous phase towards the substrate surface in the ALD operating conditions range, a thermodynamical study has been carried out. The specifically designed Knudsen effusion cell coupled with a mass spectrometer, (Fig. 3), has been described elsewhere in more detail [12]. The vaporization cell reproduces conditions as in the bubbler of an ALD system. The thermal cracking cell (upper part of the system) reproduces conditions as in the heated transport lines from the bubbler to the reactor and in the deposition chamber of the ALD set up.

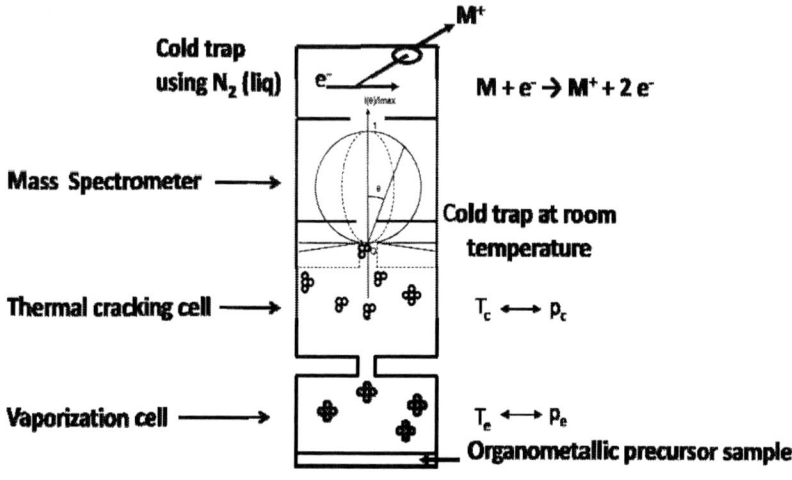

Figure 3. Schematic of the special vaporization/cracking analysis system [12].

Experimental study of the PEALD ZrO₂ deposition from the TEMAZ precursor and O₂

Influence of the choice of ALD vs PEALD has been first investigated. It is known that PEALD process leads to higher film growth rate compared to ALD process [20]. Plasma addition provides highly reactive surface species and consequently increases the process performance.

60 nm thick ZrO_2 layers were deposited on a Si (100)/ ALD TiN (10 nm) either by ALD (TEMAZ and ozone –more reactive than oxygen) or by PEALD (TEMAZ and oxygen radicals generated by an O_2 plasma) in a commercial ALD reactor (Emerald) manufactured by ASM® at a deposition temperature of 250°C.

A schematic layout of the deposition reactor is shown in Fig. 4. The wafer/substrate is located on a heated susceptor. The reactants are introduced from above the wafer through a showerhead system. This showerhead is connected to a RF generator (13.56 MHz), enabling plasma generation above the substrate. The electrical system is designed to force the showerhead to behave as a cathode, thus limiting ion bombardment of the substrate. The RF is turned on/off in the same manner as the pulsing valves.

The PEALD deposition cycle was composed of four steps. First, TEMAZ was delivered into the reactor during 1700ms. An Ar purge of 1750 ms followed the TEMAZ pulse. O_2 was then introduced during 500 ms. The oxygen remote plasma was started

200 ms after the beginning of the oxygen pulse and maintained until the end of the oxygen pulse. An Ar pulse of 1500 ms was set again to purge the oxygen precursor. A direct plasma (13.56 MHz, capacitive) is applied sequentially with a RF power. A low RF power ranging from 75 to 200 W was used to obtain a low plasma density which limits the ion bombardment and UV radiation of the substrate. The pressure within the reactor was maintained at 2 Torr (266 Pa).

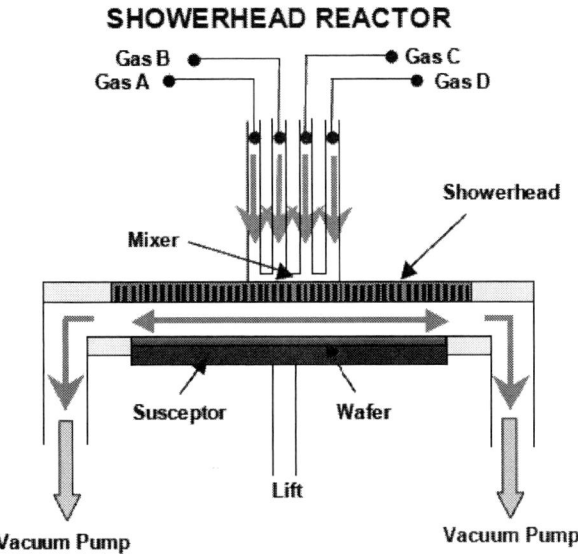

Figure 4. ALD/PEALD reactor schematic.

Influence of the substrates has also been investigated. Films were grown either on TiN/Si(100) or on Si(100) substrates [22]. TiN 45nm-thick layers were grown by sputtering of a Ti target with Ar^+ ions under an N_2 atmosphere in a different reactor.

Interfaces between ZrO_2 films and the different substrates as well as the ZrO_2 film microstructure were observed by High Resolution Transmission Electron Microscopy (HRTEM). Chemical analysis of the ZrO_2 interface with TiN films was carried out by Synchrotron – X-Ray Photoelectron Spectroscopy (SR-XPS). Photoemission experiments were performed on the European Synchrotron Radiation Facility (ESRF- located in Grenoble, France) beam line ID32. An X-Ray source of energy hv = 2595.9 eV was used for excitation with a resolution of 0.5eV. The spectra were taken at two distinct angles of incidence θ: normally to the surface (θ = 0°) and at an angle of 60° (θ = 60°). Raman spectroscopy with a He-Cd laser source emitting at 325nm wavelength has been used to identify and evaluate the presence of each ZrO_2 phase. The tetragonal, cubic and monoclinic phases present six, one and fifteen characteristic Raman bands, respectively [23]. The band characteristics of the tetragonal and monoclinic phases are close except one band (269 cm^{-1}) , which can be used to identify the presence of tetragonal ZrO_2 [23].

Results and discussions

Experimental study of the zirconium precursor TEMAZ

The vaporization of a non electronic grade commercial TEMAZ liquid solution, synthesized by Air Liquid, was analyzed using the special vaporization/cracking analysis system described in Fig. 3, with only the evaporation cell. The precursor contains few metallic impurities (Zn, Pb, Na, Li ...) with very low amounts (<1ppm). At room temperature, TEMAZ, $Zr(NC_2H_5CH_3)_4$, is liquid and colorless. Its structure is composed of a central zirconium atom connected with four ethyl methyl amino ligands, (NEtMe). Nine experiments of TEMAZ effusion were carried out by increasing the temperature of the vaporization cell from 5°C to 60°C [14]. The effusion cell was loaded in a recycled Argon atmosphere glove box with a large amount of liquid sample (5 ml) for all the mass spectrometric studies. The glove box humidity is lower than 1 ppm.

The variation of the TEMAZ vapor pressure vs temperature is shown in Fig. 5. The linear behavior of the experimental data shows that TEMAZ vaporization acts as the vaporization of a pure component or an azeotropic mixture, i.e. that the composition of the liquid phase at the boiling point is identical to that of the vapor in equilibrium with it. Using the second law of thermodynamics, the enthalpy of sublimation of TEMAZ was deduced to be 79.4 ± 2.4 kJ/mol.

The gaseous phase composition was analyzed. After considerations based on the known bond energies (for instance Zr-NEtMe bonds), the measured ionization potentials and the possible ionization processes of TEMAZ or impurities, it was concluded that [14]:
(i) no thermal decomposition of liquid TEMAZ occurred.
(ii) the gaseous phase is a mixture of five species $Zr(NEtMe)_4(g)$, $Zr(NEt_2)_4(g)$, $Zr(NEtMe)_2(NEt_2)_2(g)$, $Zr(NEtMe)_3(NEt_2)(g)$ and $Zr(NEtMe)_3OH(g)$. Thus, the analyzed precursor can be considered as a mixture containing 97% TEMAZ, 2 % of other organometallics compounds (identified as $Zr(NEt_2)_4$, $Zr(NEtMe)_2(NEt_2)_2$ and $Zr(NEtMe)_3(NEt_2))$ and 1% of $Zr(NEtMe)_3OH(g)$. This latter molecule might come either from the formation of TEMAZ or from a reaction between TEMAZ and the glove box atmosphere moisture.
(iii) no metallic impurities Zn, Na, Li... present in the liquid TEMAZ were detected in the gaseous phase.

All these results show that the TEMAZ solution presents a constant vaporization since there is no thermal decomposition and that it is a reproducible and pure zirconium source for ZrO_2 deposition at low temperature (below 333 K). That validates its selection for any low temperature ALD or CVD process.

However, at higher temperatures (above 333 K), the extrapolation of our results (dashed line of Fig. 5) indicates higher values than the ones proposed by the supplier (square points). It suggests that a slight decomposition of the precursor might occur at the entrance of the PEALD reactor. The decomposition products might react with residual oxygen and form particles via homogeneous nucleation. Particles with a size greater than 0.16 μm are observed with an optical detection tool on the surface of the films deposited by PEALD with the TEMAZ precursor (Fig. 6).

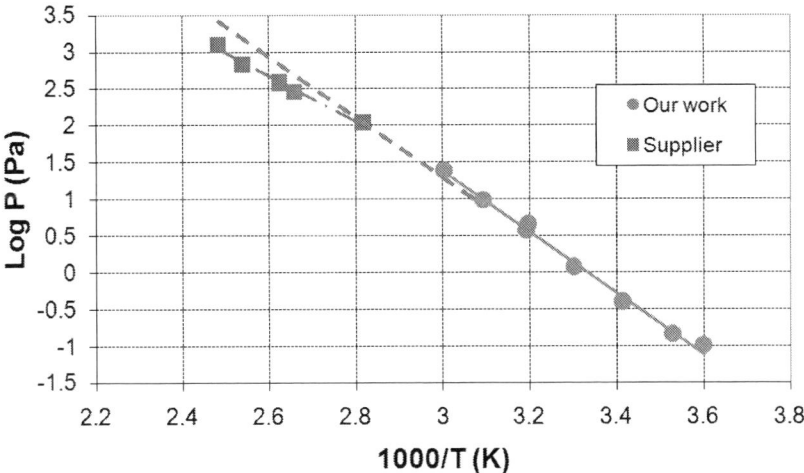

Figure 5. TEMAZ vaporization: total vapor pressure vs temperature.

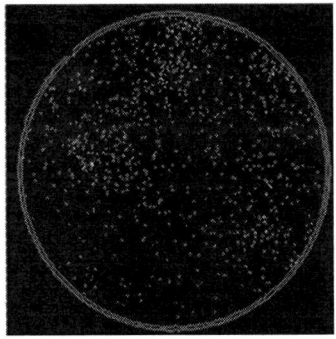

Figure 6. Surface of 60 nm ZrO_2 thick film deposited by PEALD with the TEMAZ precursor. Particles size cartography > 0.16 μm.

Experimental study of the PEALD ZrO_2 deposition from the TEMAZ precursor and O_2

Effect of the deposition techniques and conditions. Raman spectra of ZrO_2 films deposited by ALD and PEALD are shown on Fig. 7. Spectra of Si(100) substrate and Si(100)/ ALD TiN (10 nm) stack are also displayed to break out the Si and TiN contributions in the other spectra. For the Si(100) spectrum, a band is detected at 322cm^{-1} and a very large one at 521 cm^{-1}, typical of the silicon [24]. Compared to the Si(100), no additional peaks corresponding to TiN are detected on the Si(100)/ ALD TiN (10 nm) stack spectrum. This is probably due to the small thickness of the TiN films.

The tetragonal phase characteristic band at 269 cm^{-1} is detected in the spectra of PEALD films deposited with a RF power of 75W and 200W but not for the ALD film. Band intensity is similar for both PEALD films. The other five tetragonal bands, reported by [23], are not detected. The cubic band is also not detected in any spectrum but this band is difficult to measure as it is well-known to present a very low intensity. Monoclinic bands, easily detectable in Raman Spectroscopy, are not visible in any spectrum. According to these results, the microstructures of PEALD films seem to be similar and composed of a mixture of tetragonal and cubic phases, while ALD films are only in the cubic phase.

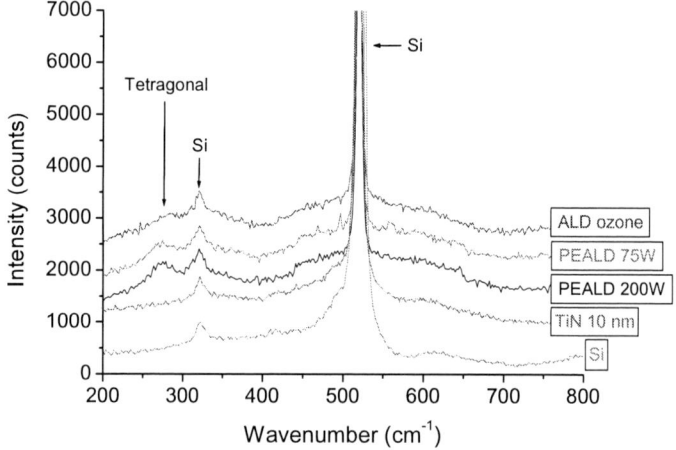

Figure 7. Raman spectra of 60 nm thick ZrO$_2$ films deposited on Si(100)/ ALD TiN (10nm) stack by ALD and PEALD at various RF power.

The PEALD technique appears to be more favorable to the synthesis of films containing the tetragonal phase. The effect of plasma treatment on the formation of the tetragonal phase can be explained. Plasma increases the reactivity of the nucleation sites present on the sample surface (Fig. 8a and b). The nucleation sites are activated by the plasma and nuclei can begin to grow from these sites (Fig. 8c and d). Moreover, plasma treatment allows the formation of more reactive oxygen species. These species are able to react with a larger amount of nucleation sites. The growth of more nuclei (Fig. 8e and f) induces the formation of films with smaller grains (Fig. 8g and h). This microstructure with small grains is favorable to the tetragonal phase formation as its surface energy is lower than the monoclinic one [25]. However, RF power has no influence on the formation of the tetragonal phase since both PEALD films have the same ratio of tetragonal phase.

ECS Transactions, 35 (4) 497-513 (2011)

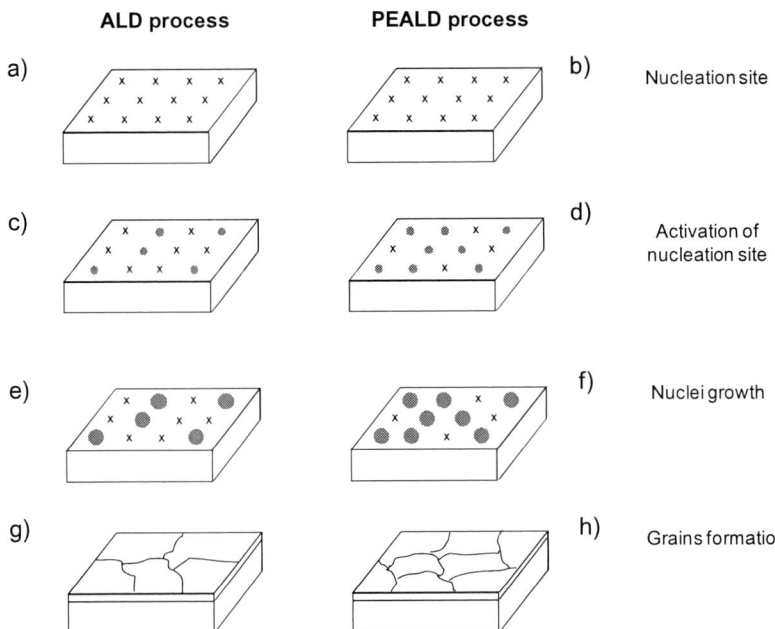

Figure 8. Schematic description of ZrO_2 films growth deposited by ALD or PEALD. Nucleation, activated nucleation sites and nuclei are represented by a cross, a grey dot and a grey round, respectively.

Effect of the nature of the substrate. ZrO_2 films deposited by PEALD with a RF power of 75W at 266 Pa were deposited on three different substrates: Si(100), Si(100)/ ALD TiN (10 nm) and Si(100)/ PVD TiN (45 nm). Fig. 9 shows the Raman spectra recorded on these films. Two bands at 266 and 642 cm^{-1} are visible on the spectra of films deposited on the Si(100) and Si(100)/ ALD TiN substrates. These bands are attributed to the tetragonal phase. They present a lower intensity for the film deposited on the Si(100)/ ALD TiN substrate. For these two spectra, a band at 615 cm^{-1} is also measured which corresponds to the cubic phase. Therefore, these two zirconia layers are probably constituted of a mixture of tetragonal and cubic phases with a lower amount of tetragonal phase in the film produced on the ALD TiN substrate. On the contrary, the band at 642 cm^{-1} is not detected on the spectrum of the layer deposited on the PVD TiN substrate and the band at 266 cm^{-1} presents an even lower intensity than on the spectrum of the layer deposited on the ALD TiN substrate. This film is mainly formed of cubic zirconia with a small amount of tetragonal phase.

These results confirm the influence of the substrate nature on the zirconia microstructure. ZrO_2 films contain more tetragonal phase when they are deposited on Si(100). As mentioned above, the variability on the tetragonal phase formation may be explained by a different nucleation mechanism according to the substrate nature. More nucleation sites should be present on the Si(100) substrate surface compared to ALD and

505

PVD TiN substrates surfaces. This higher number of nucleation sites leads to the formation of a zirconia film with smaller grains which are favorable to the stabilization of the tetragonal phase.

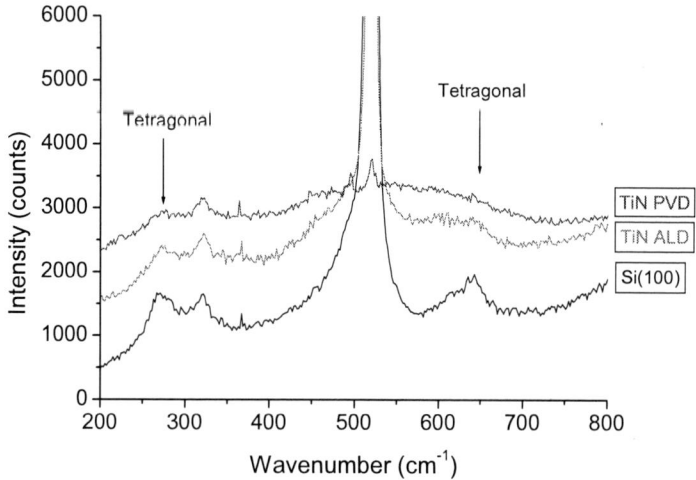

Figure 9. Raman spectra of 60 nm thick ZrO_2 films deposited on various substrates.

The presence of an interfacial layer between the substrate (TiN or Si) and the deposited ZrO_2 film has also been observed as shown in the cross section HRTEM images presented in Fig. 10. This observation has been already reported for both systems [1, 26, 27].

To focus on the interface between ZrO_2 and TiN films, SR-XPS analyses were performed to study this latter interfacial layer [22]. A dedicated 4 nm ultra-thin ZrO_2 film was prepared on TiN PVD/Si substrate. A TiN PVD layer on Si sample was used as reference to evaluate the TiN layer contribution in the XPS spectra of the analyzed sample. XPS spectra of the Zr 4p and Ti 3p region and of the O-1s region recorded at $\theta = 0°$ are displayed on Fig. 11 a and b, respectively for a PEALD ZrO_2 film. An analysis of the spectra recorded at $\theta = 0°$ and $60°$ confirms the presence of a $Zr_xTi_yO_z$ compound at TiN/ZrO_2 interface [22]. This interfacial layer might have also a favourable impact on the formation of tetragonal ZrO_2 phase [11].

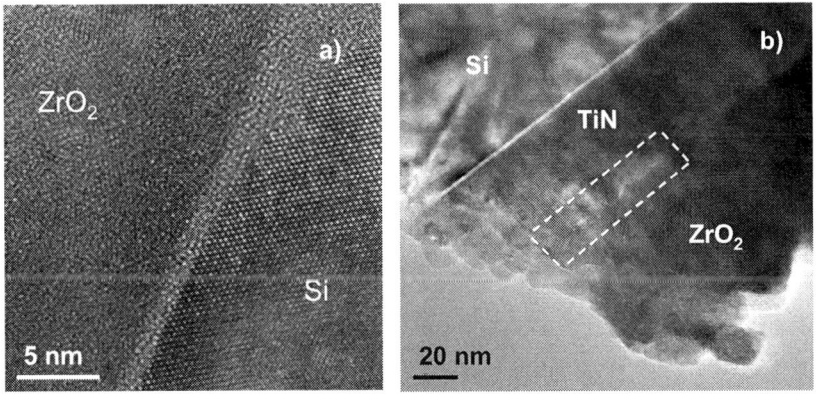

Figure 10. HRTEM image of the 60 nm PEALD ZrO_2 film deposited on: a) Si(100) and b) TiN(PVD)/Si(100) substrate.

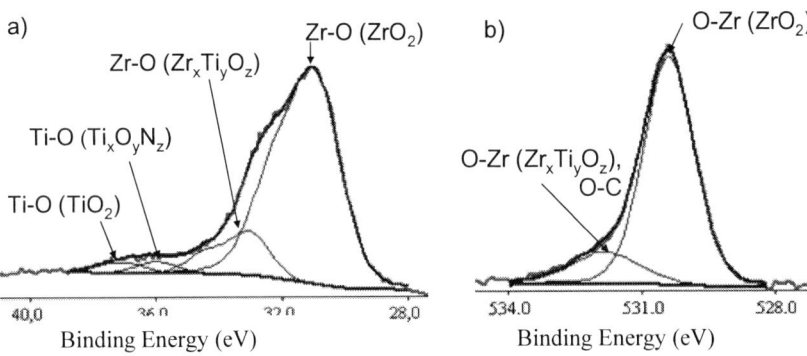

Figure 11. Deconvolution profile of the XPS spectra of the ZrO_2 film in the a) Zr-4p, Ti-3p and b) O-1s region at $\theta = 0°$.

Thermodynamic simulations of the PEALD ZrO_2 deposition from the TEMAZ precursor and O_2 on TiN/Si substrates

In order to confirm this assignment, a thermodynamic study has been performed on the quaternary Ti-N-Zr-O system. The Ti-N-Zr-O quaternary phase diagram calculated at T = 250°C using the FTOXID and SGPS 2004 thermodynamic databases and the Factsage software [4] is given in Fig. 12. ZrO_2 is in equilibrium with TiN. We can see on the Ti-O-N ternary isothermal section that TiN is in equilibrium with TiO_2 and other titanium oxides (TiO, Ti_2O_3, Ti_4O_7). As shown on the Ti-Zr-O diagram, TiO_2 is not in direct equilibrium with ZrO_2 but with an intermediary compound, $ZrTi_2O_6$. This ternary compound, $ZrTi_2O_6$, is itself in equilibrium with ZrO_2.

Thermodynamic calculations have been made to predict the nature of the interfacial layer between ZrO_2 and TiN and to model the chemical reactions on the TiN surface during the PEALD process within the reactor. For each PEALD step (summarized in Fig. 13), the equilibrium between the gaseous reactants and the elements constituting the sample free surface has been determined. Calculations have been made using the Factsage software [4] on the complex equilibrium Ti-Zr-O-N-C-H system using the FTOXID, SGPS 2004 databases at T = 250°C [5].

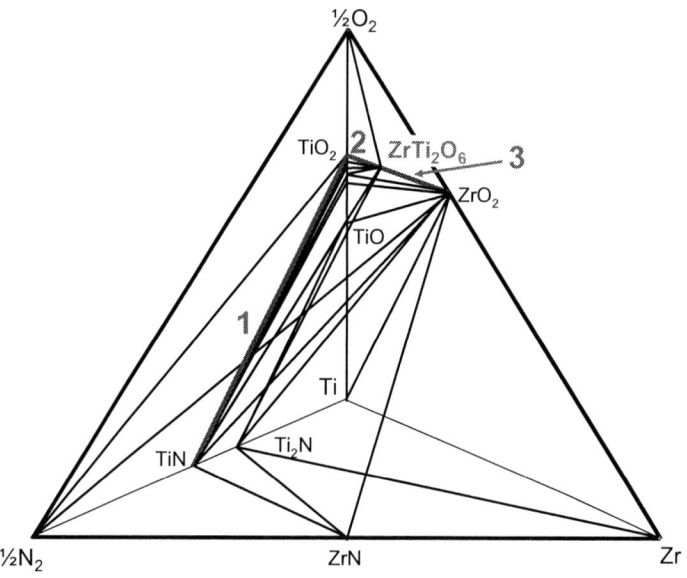

Figure 12. Ti-N-Zr-O quaternary diagram calculated at 250°C [4, 5].

Calculations were made by the method of Gibbs free energy minimization to determine the equilibrium composition. As no thermodynamic data are available on TEMAZ, calculations have been carried out considering complete decomposition of the TEMAZ molecule (1 Zr, 4 N, 12 C and 32 H atoms).

First, we calculated the equilibrium between a TiN layer, TEMAZ and Ar (no oxygen) to simulate the first zirconium precursor pulse of the PEALD cycle (Fig. 13a). The obtained results show that no zirconium was deposited on the TiN layer and no zirconium compound such as ZrN was produced (Fig. 13b). Similar simulations were carried out for a TiN layer in equilibrium with O_2 and Ar (no more TEMAZ) to simulate the oxygen pulse of the PEALD cycle (Fig. 13c). In this case, a thin TiO_2 layer was formed with a N_2 gas production (step 1 in Fig. 12 and Fig. 13d). The following zirconium pulse of the PEALD cycle was modeled by studying the equilibrium between TEMAZ, Ar and the new TiO_2 surface, assuming that TiO_2 layer covers completely the TiN surface (Fig. 13e). We noticed the formation of a new compound $ZrTi_2O_6$ (step 2 in Fig. 12). Therefore, a stack of $ZrTi_2O_6/TiO_2/TiN$ layers is formed (Fig. 13f). Then, simulation of the PEALD cycle oxygen pulse is repeated. The equilibrium between O_2, Ar and the new $ZrTi_2O_6$ surface is calculated (Fig. 13g). We assume again here that the $ZrTi_2O_6$ layer covers completely the TiO_2 surface. ZrO_2 started to grow on the $ZrTi_2O_6$ film (step 3 in Fig. 12). By repeating the PEALD cycle, a thick layer of ZrO_2 was grown on the $ZrTi_2O_6/TiO_2/TiN$ stack (Fig. 13h).

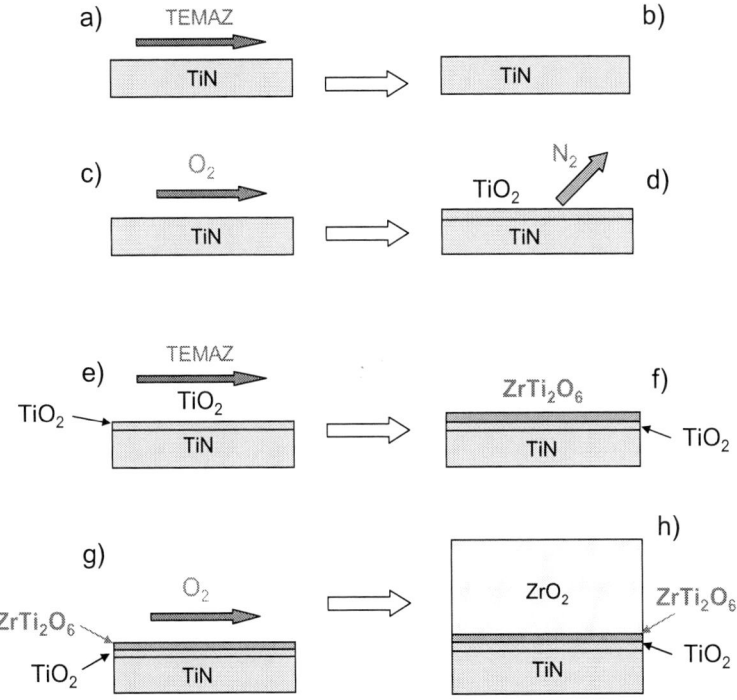

Figure 13. Schematic view of the successive steps of ZrO_2 film growth on TiN substrate.

Effect of thermal annealing on ZrO_2 structural properties

When integrated into MIM or DRAM capacitors, ZrO_2 films are annealed to a temperature of about 400°C, during the ALD TiN electrode deposition. We have thus analyzed the effect of a 400°C anneal of ZrO_2 layers deposited on TiN. Fig. 14 shows the Raman spectra of a non annealed sample and of two samples annealed at 400°C under 266 Pa of N_2 for 30s and 300s. We show here results obtained for an initial layer containing the tetragonal phase. It is clear that the anneal increases the tetragonal phase content in the material since not only the 274 cm^{-1} peak increases, but also the 642 cm^{-1} peak, which is characteristic of the tetragonal phase, appears.

Two hypotheses may explain this result. The first one is related to the nano-crystalline structure of ZrO$_2$. It has been shown by Garvie [28] that the tetragonal phase is more favorable for small crystallites because of the balance between the volume energy and the surface energy. When the crystallites are sufficiently small, surface energy dominates and the tetragonal phase formation leads to a Gibbs free energy gain. This behavior is modulated by the internal strain in the oxide material [29]. We can suppose that thermal treatment may modify internal strain in the ZrO$_2$, including dilatation, thus leading to the tetragonal phase formation. The second hypothesis is related to oxygen vacancies. Annealing the TiN/ZrO$_2$ stack may produce a transfer of oxygen from the oxide towards the electrode, creating oxygen vacancies. Indeed, by analogy, defects produced in HfO$_2$ by the heating of a metal/oxide contact have been evidenced by many authors [30-32]. Moreover TiN can dissolve a large amount of oxygen atoms [33], the solubility limit being about 0.49. Some studies have shown that formation of the ZrO$_2$ tetragonal phase is correlated to the presence of oxygen vacancies [34-37]. With the presence of vacancies, ZrO$_2$ relaxes from a Zr seven-fold coordination of monoclinic to the Zr eight-fold coordination of the cubic or tetragonal phase.

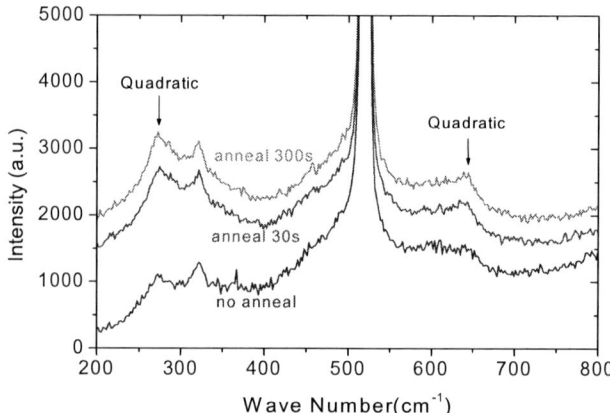

Figure 14. Raman spectra of 60nm thick ZrO$_2$ films deposited on ALD TiN, just after deposition, after 30s anneal or 300s anneal at a temperature of 400°C.

The final structure of ZrO$_2$ is thus controlled by the maximum temperature at which the material is heated, whatever its initial structure is. We can thus suspect that the dielectric constant of ZrO$_2$, after integration in a conventional device fabrication process, will converge at the same value. To verify this hypothesis, we have extracted the dielectric constant of ZrO$_2$ embedded between two ALD TiN electrodes, this MIM structure being fully embedded at the level of the interconnections of a standard 45nm technology. The results are presented in the Fig. 15. The dielectric constant is about 35 irrespective of the structure of ZrO_2 just after deposition, confirming that the thermal budget is one of the main factors controlling the final structure of ZrO_2.

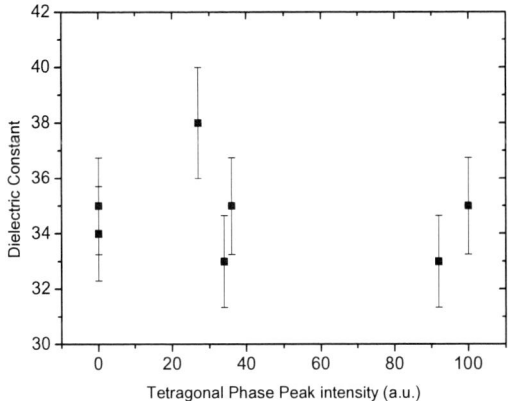

Figure 15. Dielectric constant of a ZrO_2 MIM capacitor as a function of the tetragonal phase peak intensity from the Raman spectra of the as deposited films.

Conclusions

We have studied the Atomic Layer Deposition of out-of-equilibrium tetragonal ZrO_2 phase to be used in MIM (Metal/Insulator/Metal) capacitors. Thermodynamic approaches have been applied to better understand the deposition process and evaluate the influence of deposition parameters on the structure of the deposited film. Together with the experimental deposition study, it has been found that the plasma power and the thermal budget are the main parameters which control the structure of the deposited films. Additionally, the formation of an interfacial layer between TiN and ZrO_2 at the first stages of the deposition may play a role on the growth.

Acknowledgments

The authors acknowledge C. Chatillon, A. Crisci for fruitful discussions and L. Artaud and H. Collas for mass spectrometry experiments.

References

1. J.-H. Kim, V. A. Ignatova, J. Heitmann, and L. Oberbeck, *J. Phys. D: Appl. Phys.*, **41**, 172005 (2008).
2. G. D. Wilk, R. M. Wallace, and J. M. Anthony, *J. Appl. Phys.*, **89**, 5243 (2001).
3. X. Zhao, and D. Vanderbilt, *Phys. Rev. B*, **65**, 075105 (2002).

4. Factsage, in, http://www.factsage.com.
5. S.G.T.E, Scientific Group Thermodata Europe, BP 166, 38402 Saint Martin d' Heres , France, in, http://www.sgte.org/.
6. T. Suntola, and M. Simpson, *Atomic Layer Epitaxy,* Chapman and Hall, Editors, New York (1990).
7. T. Suntola, *Handbook of Crystal Growth,* D. T. J. Hurle, Editor, Part B, Elsevier, Amsterdam, **3**, p. 601 (1994).
8. R. Puurunen, *J. Appl. Phys.*, **97**, 121301 (2005).
9. M. Leskela, and M. Ritala, *Angew. Chem.-Int. Edit.*, **42**, 5548 (2003).
10. W. Qin, C. Nam, H. L. Li, and J. A. Szpunar, *Act. Mater.*, **55**, 1695 (2007).
11. U. Troitzsch, A. G. Christy, and D. J. Ellis, *Phys. Chem. Minerals,* **32**, 504 (2005).
12. P. Violet, I. Nuta, L. Artaud, E. Blanquet, and C. Chatillon, *Rapid Commun. Mass Spectrom.*, **23**, 793 (2009).
13. P. Violet, E. Blanquet, D. Monnier, I. Nuta, and C. Chatillon, *Surf. Coat. Technol.*, **204**, 882 (2009).
14. D. Monnier, I. Nuta, C. Chatillon, M. Gros-Jean, F. Volpi, and E. Blanquet, *J. Electrochem. Soc.*, **156**, 71 (2009).
15. D. M. Hausmann, E. Kim, J. Becker, and R. G. Gordon, *Chem. Mater.*, **14**, 4350 (2002).
16. M. Leskelä, and M. Ritala, *Thin Solid Films*, **409**, 138 (2002).
17. J. Aarik, A. Aidla, H. Mandar, T. Uustare, and V. Sammelselg, *Thin Solid Films*, **408**, 97 (2002).
18. C. M. Perkins, B. B. Triplett, P. C. McIntyre, K. C. Saraswat, S. Haukka, and M. Tuominen, *Appl. Phys. Lett.*, **78**, 2357 (2001).
19. R. Matero, A. Rahtu, M. Ritala, M. Leskelä, and T. Sajavaara, *Thin Solid Films*, **368**, 1 (2000).
20. S. J. Yun, J. W. Lim, and J. H. Lee, *Electrochem. Solid-State Lett.*, **7**, F81 (2004).
21. S. K. Kim, and C. S. Hwang, *Electrochem. Solid-State Lett.*, **11**, G9 (2008).
22. D. Monnier, M. Gros-Jean, E. Deloffre, B. Doisneau, S. Coindeau, A. Crisci, J. Roy, Y. Mi, B. Detlefs, J. Zegenhagen, C. Wyon, C. Martinet, F. Volpi, and E. Blanquet, *ECS Trans.*, **25**(8), 235 (2009).
23. C. Urlacher, and J. Mugnier, *J. Raman Spectrosc.*, **27**, 785 (1996).
24. G. Abramof, N. G. Ferreira, A. F. Beloto, and A. T. Ueta, *J. Non Crystalline Solids*, **139**, 338 (2004).
25. R. Chaim, A. H. Heuer, and D. G. Brandon, *J. Am. Ceram. Soc.*, **69**, 243 (1986).
26. Y. Kim, J. Koo, J. Han, S. Choi, H. Jeon, and C.-G. Park, *J. Appl. Phys.*, **92**, 5443 (2002).
27. C. Y. Ma, and Q. Y. Zhang, *Vacuum*, **82**, 847 (2008).
28. R. C. Garvie, *J Phys. Chem.*, **69**, 1238 (1965).
29. F. F. Lange, *J. Mater. Sci.*, **17**, 225 (1982).
30. K. Shiraishi, K. Yamada, K. Torii, Y. Akasaka, K. Nakajima, M. Konno, T.Chikyow, H.Kitajima, T. Arikado, and Y. Nara, *Thin Solid Films*, 305 (2006).
31. J. Robertson, O. Sharia, and A. A. Demkov, *Appl. Phys. Lett.* , **91**, 132912 (2007).
32. G. Bersuker, C. S. Park, H.-C. Wen, K. Choi, J Price, P Lysaght, H-H Tseng, O. Sharia, A Demkov, J. T. Ryan, and P. Lenahan, *IEEE Trans. Electron Devices*, **57**, 2047 (2010).
33. H. Kim, P. C. McIntyre, C. O. Chui, K. C. Saraswat, and S. Stemmer, *J. Appl. Phys.*, **96**, 3476 (2004).

34. G. Gottardi, N. Laidani, V. Micheli, R. Bartali, and M. Anderle, *Surf. Coat. Technol.*, **202**, 2332 (2008).
35. C. Y. Ma, F. Lapostolle, P. Briosi, and Q. Y. Zhang, *Appl. Surf. Sci.*, **253**, 8718 (2007).
36. P. Scardi, M. Leoni, and L. Bertamini, *Surf. Coat. Technol.*, **76-77**, 106 (1995).
37. G. Dutta, K. P. S. S. Hembram, G. M. Rao, and U. V. Waghmare, *J. Appl. Phys.*, **103**, 016102 (2008).

514

V_T Stability Of High-K/Metal Gate Stacks With Device Scaling
In 30nm FDSOI Technology

X. Garros[a], L. Brunet[a,b], M. Cassé[a], O. Weber[a], F. Andrieu[a], D. Lafond[a], C. Gaumer[b], G. Reimbold[a] and F. Boulanger[a]

[a] CEA-Leti Minatec, 17 rue des Martyrs, Grenoble, France. Email: xavier.garros@cea.fr
[b] STMicroelectronics, 850 rue Jean Monnet, Crolles, France.

The paper investigates the stability of the transistor threshold voltage V_T with scaling in 30nm High-K/Metal Gate technology. For HfO_2 and HfZrO oxides, large V_T instabilities, up to 230mV, are observed when the device width (W) is scaled down to 80nm. It is explained by undesirable lateral oxygen diffusion through the spacers, which mainly modifies the metal workfunction in narrow transistors. HfSiO(N) oxides exhibit a much better immunity to this effect, attributed to a different crystallinity of the High-K layer. Moreover, V_T stability is not modified by Aluminum incorporation in the gate stack but can significantly be degraded by Lanthanum doping.

Introduction

Today, large efforts are made to adjust the V_T of the transistors by tuning the metal gate workfunction ϕ_M [1-4]. However, only little attention is paid to the stability of V_T with device scaling, especially with transistor width W [1], [5-6]. Yet, it is fundamental that V_T remains stable with device geometry for basic circuit cell performance and reliability. In particular, the performance of SRAM cells, which are very sensitive to V_T variability, can dramatically be affected by small changes of V_T with transistor length L and width W. Therefore it becomes crucial to give insight on the mechanisms responsible for the V_T instability with device scaling in order to overcome this issue.

In this study we present a systematic characterization and a comprehensive analysis of the gate stack stability, focusing on the V_T. The first part deals with the V_T variations with L and W for various high-k/metal gate stacks (HfO_2, $HfZrO_2$ vs. HfSiO(N)). A large set of electrical properties, Equivalent Oxide Thickness (EOT), interface states (Nit), Bias Temperature Instability (BTI) reliability, are compared in the different oxides in order to identify the key parameters affecting the V_T stability. The second part proposes a physical model to explain the V_T variations with scaling. It is fully supported by many physical characterizations like High Resolution Transmission Electron Microscopy (HRTEM), Electron Energy Loss Spectroscopy (EELS) and Attenuated Total Reflectance spectroscopy (ATR).

Experimental

Fully Depleted Silicon On Insulator (FDSOI) transistors down to 30nm gate lengths were fabricated in a gate-first integration scheme on 300mm SOI wafers (with a 1050°C activation anneal). A device HRTEM picture is shown in Fig. 1 with an 8nm thin silicon film. The buried SiO_2 Oxide BOX is 145nm thick. The dielectric stack consists of a thin 0.8nm SiO_x interfacial oxide and of a 2-2.5nm High-K layer. Several High-K (HK) oxides were integrated: pure Hafnium oxide HfO_2, Hafnium Zirconate oxides $Hf_{0.5}Zr_{0.5}O$ and nitrided or non nitrided Hafnium silicates HfSiO(N). The gate material is a thin TiN film deposited either by PVD or ALD. More details on the process can be found in [2]. For all the various stacks, the EOT values are around 12Å and the gate leakage currents are 2 orders of magnitude smaller than in SiON/Poly-Si technologies.

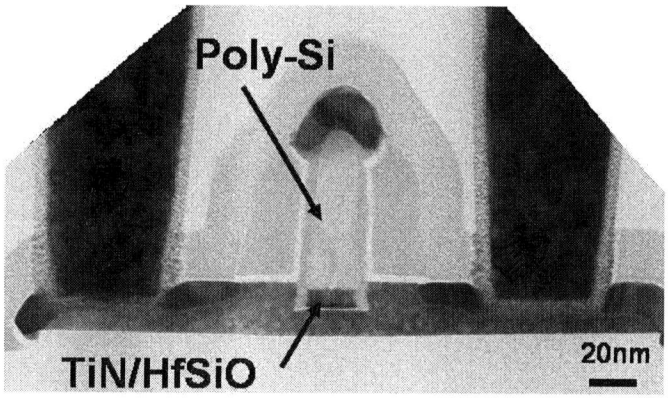

Figure 1. HRTEM picture of a 30nm gate length FDSOI MOSFETs with a 8nm thin Si film and a HK gate stack. The dielectric stack consists of a HfSiO oxide and of a TiN metal electrode.

Results and Discussion

V_T stability in Hf-based dielectrics

Fig. 2 shows the V_T variation as a function of the channel width W for several HfZrO/TiN stacks. We can first notice that V_T strongly depends on the TiN thickness with a 210mV difference between 5nm and 10nm TiN gates at W=10µm. The higher V_T measured on 10nm thick gate devices is actually explained by a higher metal gate workfunction for the 10nm TiN gate compared to the 5nm one. This is clearly demonstrated in Fig. 3 by internal photo emission measurements [7] which reveal a higher barrier height Φ_b between the dielectric and the gate material, for a 10nm TiN gate. Therefore reducing the metal gate thickness can be used to tune V_T as already proposed in [2],[4],[8-9].

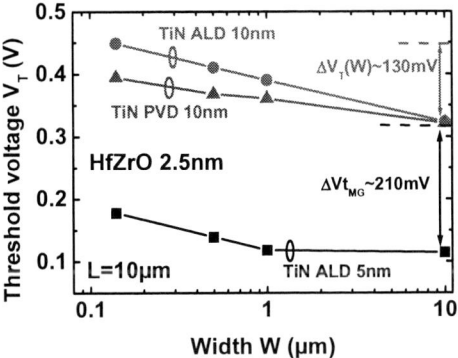

Figure 2. Stability of V_T with channel width W for HfZrO/TiN stacks. V_T increases when W is scaled regardless of the TiN gate.

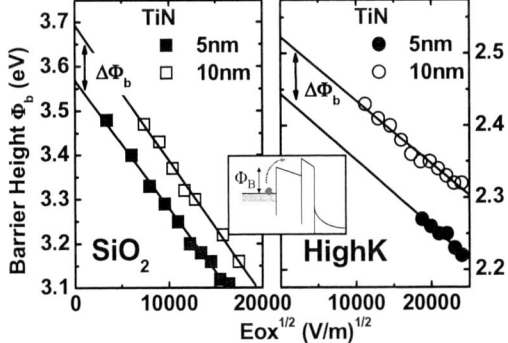

Figure 3. Barrier height Φ_b between the dielectric and the metal gate measured by Internal PhotoEmission technique on both SiO_2 and HK oxides. The barrier height is reduced on 5nm metal gates, regardless of the gate oxide. This means that the TiN workfunction with respect to the vacuum level is smaller in 5nm TiN gates: $WF_{5nm}=4.47eV$ compared to $WF_{10nm}=4.6eV$.

But the most striking feature in Fig.2 is that V_T is clearly highly W-dependent. For the 10nm ALD TiN gate, V_T is increased by 130mV when W is scaled. This effect is enhanced when the channel length L is also reduced (see Fig. 4). For the 5nm ALD TiN gate, the low V_T value of 0.12V measured on large devices, is further increased by 230mV on narrow devices. Such an unexpected and undesirable effect sometimes appears during the development of a technology. It is generally explained by the fact that the device architecture is not well optimized. In particular, a not well controlled etching of the edges of the transistor can create defects responsible for V_T instability with W scaling.

However this W-dependence of V_T observed here is not explained by a non-optimized device architecture. It has a clearly different origin since it is actually very dependent on the HK material. As shown in Fig 5, the variation in ΔV_T between wide (W=10μm) and narrow (W=80nm) devices, is about ~200mV for both HfO_2 and HfZrO oxides whereas it is limited to 30mV for HfSiO(N) oxides. Moreover, this V_T instability appears on both NMOS and PMOS with the same sign and magnitude (see Fig. 6).

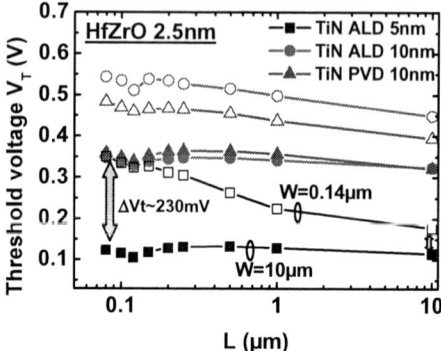

Figure 4. V_T vs. channel length L for large or narrow nMOS. V_T instability with W increases on short channel devices.

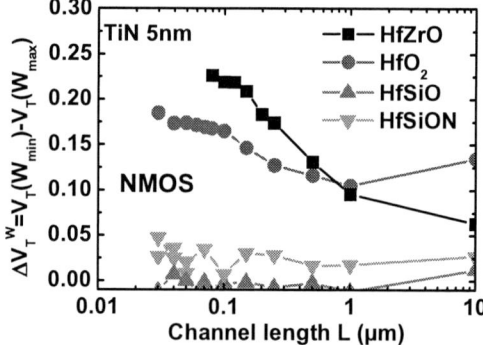

Figure 5. $\Delta V_T = V_T^{W=80nm} - V_T^{W=10\mu m}$ vs. channel length L for various HK oxides. ΔV_T is drastically reduced for HfSiO(N) compared to HfO$_2$ and HfZrO.

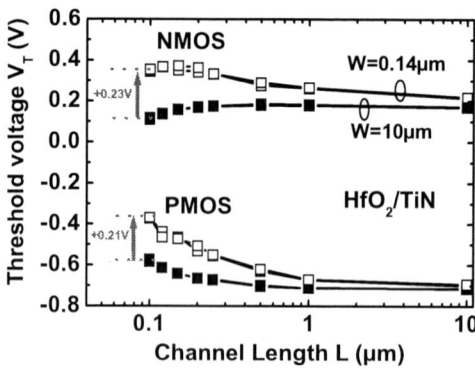

Figure 6. V_T versus L for wide and narrow NMOS and PMOS transistors. A same positive V_T shift of ~+0.2V is observed on narrow devices for both NMOS and PMOS

Origin of the V_T instability with device scaling

To find the origin of this V_T instability in HfZrO and HfO$_2$ oxides, the variation of V_T between wide and narrow transistors at fixed channel length L is recalled in (1)

$$\Delta V_T(W) = V_T^{W\,min} - V_T^{W\,max} \approx (\phi_M^{W\,min} - \phi_M^{W\,max}) - (\frac{Q_{ox}^{W\,min}}{C_{ox}^{W\,min}} - \frac{Q_{ox}^{W\,max}}{C_{ox}^{W\,max}}) \qquad (1)$$

where Wmax and Wmin designate the width of the wide and narrow transistors respectively, ϕ_M the metal workfunction, Q_{ox} the effective density of charge at the interface and Cox the oxide capacitance.

Therefore, by looking more carefully at this expression (see Fig. 7), we can deduce that $\Delta V_T(W)$ can be explained either by (1) a change of the metal work function $\Delta\phi_M = \phi_M^{W\,min} - \phi_M^{W\,max}$, (2a) by the generation of negative fixed charges in the HK layer ΔQ_{HK} or in the bottom oxide ΔQ_{SiO2} and (2b) by a possible bottom oxide SiO$_2$ regrowth ΔT_{SiO2}.

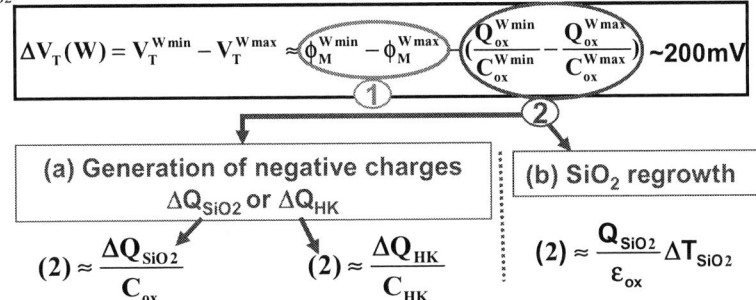

Figure 7. Possible explanations for W dependence of V_T in HfO$_2$ and HfZrO oxides. It can be due to (1) $\Delta\Phi_M$ change (2a) ΔQ_{SiO2} or ΔQ_{HK} generation (2b) SiO2 regrowth. Experimental $\Delta V_T(W)$ is about 200mV

To discriminate the different possible mechanisms, in depth electrical and physical characterizations are performed on transistors with different geometries. Unlike V_T, the other electrical parameters are actually little impacted by device scaling.

Firstly, Fig. 8 shows that the interface state density N_{it} is not significantly modified by W and L shrinking. The maximum additional charge $\Delta Q_{SiO2}=q\Delta Nit$ induced by W scaling is estimated to be ~2.10^{11}/cm^2. This leads to a maximum $\Delta V_T(W)$ of 10mV much lower than the experimental one ~200mV. Therefore, the generation of fixed charges at the Si interface $\Delta Q_{SiO2}=q\Delta Nit$ is not the mechanism responsible for the V_T instability. This result is further supported by Negative Bias Temperature Instabilities (NBTI) measurements shown in Fig. 9. NBTI is known to be very sensitive to the bottom interface degradation [10,11], and hence it can be used to reveal the presence of additional defects in the bottom oxide induced by W scaling. NBTI normalized at same oxide field to account for the V_T shift, is completely independent of W and L. This confirms that the generation of fixed charges, directly at the Si interface or inside the bottom oxide ΔQ_{SiO2} is not at the origin of the V_T instability with device scaling.

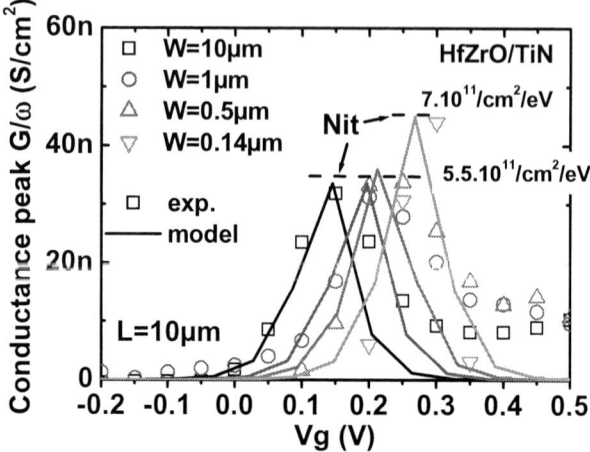

Figure 8. Nit extracted from conductance measurements for different channel width W. Nit hardly changes with W. Max ΔQ_{SiO2} is estimated to $2.10^{11}/cm^2$.

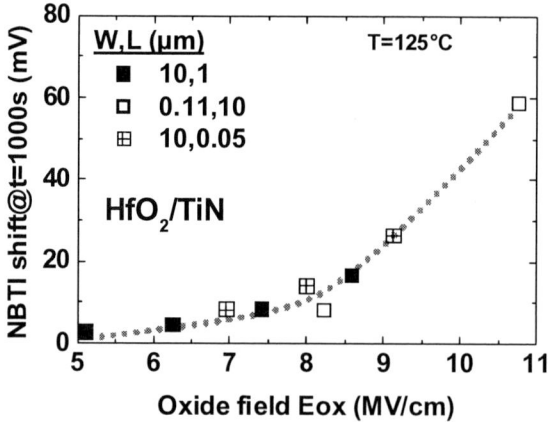

Figure 9. NBTI vs. oxide field Eox~Vg-V_T/EOT. NBTI is independent of W and L. This confirms that no additional defects are generated in the bottom oxide when the device is scaled.

Secondly, dynamic $I_D V_G$ measurements reported in Fig.10 demonstrate that fast electron trapping in HK defects [12-13] is negligible in narrow transistors. It means that the concentration of oxygen vacancies is very low in these oxides, and, above all, does not increase with W and L reduction. These results are further confirmed by Fig.11 which shows that Positive Bias Temperature Instabilities (PBTI) is not enhanced in HfO_2 compared to HfSiON and is independent of device geometry. Therefore ΔQ_{HK} in HfO_2 and HfZrO is negligible, ruling out completely the hypothesis (2a) to explain W dependence of V_T.

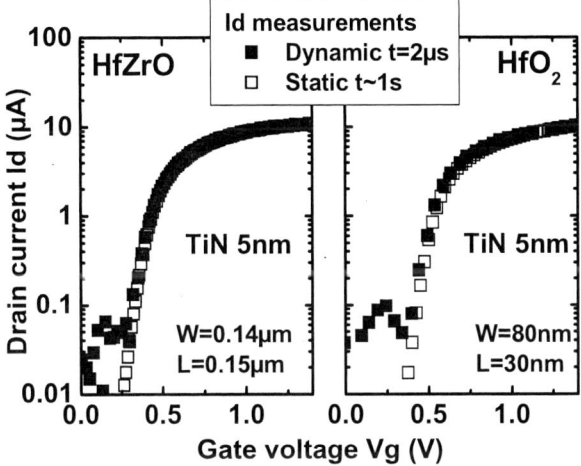

Figure 10. Fast and static Id-Vg measurements on HfZrO and HfO_2 V_{TS} do not change with time. This proves that fast trapping in O vacancies is negligible here, even in narrow and short devices.

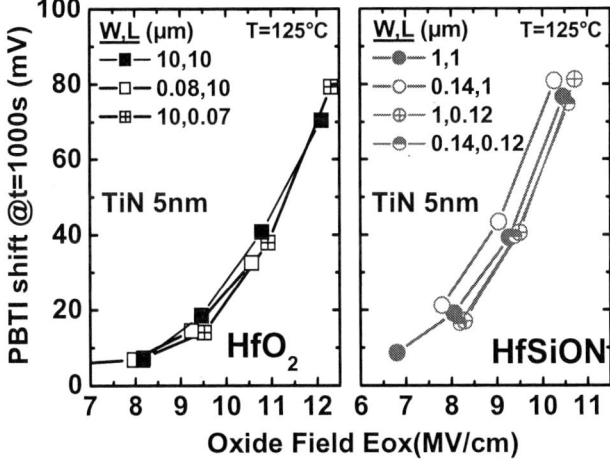

Figure 11. PBTI vs. oxide field Eox. PBTI does not change with W and L and is equivalent for HfO_2 and HfSiON. Hence no additional negative charge ΔQ_{HK} is generated in narrow devices.

Finally, a possible regrowth of the interfacial oxide ΔT_{SiO2} has been investigated through different electrical and physical characterizations.

In HfZrO oxide, the EOT remains stable with W reduction on long channel devices L=10μm (see Fig. 12). This proves that no significant interfacial layer regrowth occurs on these narrow devices with large L=10μm. Moreover, gate oxide leakage current J_g measured on shorter channel lengths L<1μm does not exhibit any variation change with ·

W (see Fig. 13). As the leakage current depends exponentially of the physical thickness of the bottom oxide, a small increase of T_{SiO2} should lead to a significant reduction of leakage current J_g. The fact that J_g is unchanged confirms that SiO_2 regrowth is negligible in narrow devices for short and long device lengths. This result is further evidenced by TEM analysis (see Fig.19) where the SiO_2 interfacial observed at the active/isolation edge is measured to its targeted value of 8 Å.

In HfO_2 oxide, EOT is increased by 3Å when W is reduced. But such a small bottom oxide regrowth leads to a maximum $\Delta Vt = Q_{SiO2} \Delta T_{SiO2}/\epsilon ox$ of ~10mV, considering that $Q_{SiO2} = qNit = 7.10^{11}/cm^2$ (see Fig. 8). This ΔV_T of 10 mV is again largely below the experimental V_T shift of ~200mV. This lead us to conclude again that bottom SiO_2 regrowth in HfO_2 is not the main mechanism responsible for the $V_T(W)$ instability.

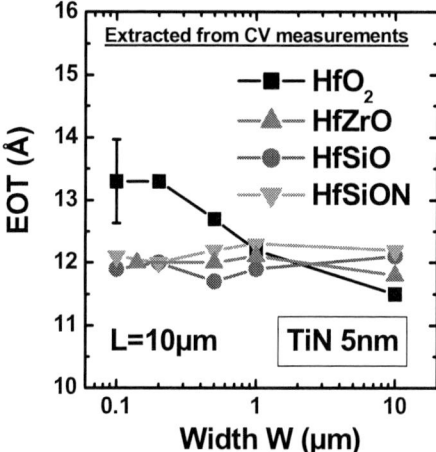

Figure 12. EOT vs transistor width W for various HK oxides. EOT is extracted from CV characteristics. T_{SiO2} regrowth is only visible in HfO_2 and limited to ~3Å.

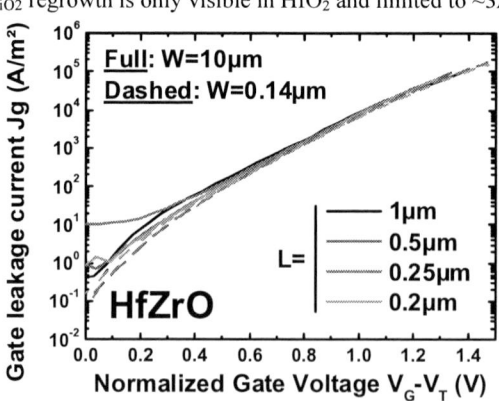

Figure 13. Gate leakage current J_g of HfZrO oxides vs gate overdrive $Vg-V_T$ for different device geometries. J_g is completely independent of the transistor size. This means that no oxide regrowth occurs in HfZrO oxide.

Therefore, although charges and/or dipoles can fix the V_T on large and long devices [14], the $V_T(W)$ instability itself is not explained by additional defects generated in the gate oxide at the edge of the devices. Consistently with the expression (1), this instability can only be explained by a modification of the metal workfunction $\Delta\phi M$ between wide and narrow transistors. In depth physical characterization, correlated to electrical results, is now provided to identify the origin of $\Delta\phi_M$.

Mechanism of lateral oxygen diffusion and gate oxidation

Early works have demonstrated that oxidation of the metal gate through lateral oxygen diffusion after spacer removal [15] or oxygen doping [16-17], is very effective to tune the V_T of PMOS. Consistent with these observations, a schematic model, involving O diffusion, is proposed in Fig.14 to explain $V_T(W)$ instability in (HfZrO) HfSiON oxides, respectively.

In HfZrO or HfO$_2$ oxides, oxygen O can diffuse from the four edges of the gate through the spacers, and locally oxidize the bottom of the TiN gate. Thus, the shorter and narrower the device is, the larger the {TiO$_x$, TiON} region is. The formation of the TiO$_x$ oxide does not increase the global EOT because of its strong k value [18]. It rather modifies the workfunction of the metal gate ϕ_M and, in turn, the V_T. On the contrary, in HfSiO(N), O diffusion is limited and the TiN oxidation is strongly reduced. As a result, V_T remains stable when W and L are scaled. The higher O diffusion in HfO$_2$ and HfZrO is assumed to be related to the different crystallinity of these HK dielectrics compared to HfSiO. This model is completely supported by many physical and electrical characterizations that are described hereafter.

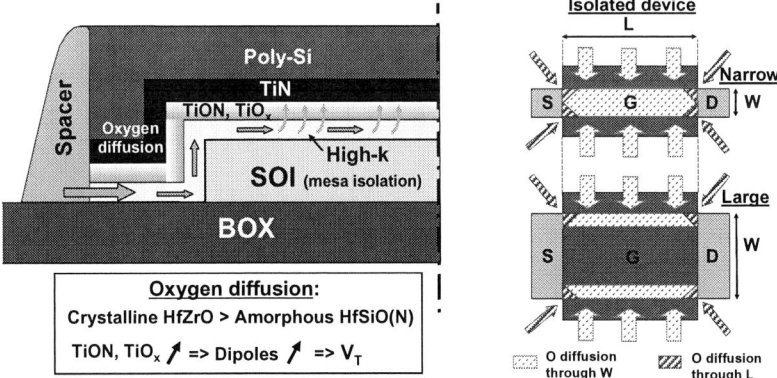

Figure 14. Schematic view of the gate edge effects: oxygen diffusion in crystalline high-k layers enhances TiO$_x$ formation and changes the metal workfunction ϕ_M

Firstly, the O diffusion through the four edges of the gate can be clearly evidenced by comparing V_T vs L curves for isolated devices and for a matrix of 100 parallel devices (W=80nm) with common gate (see Fig.15). For L=10μm, the V_T^{100ch} of the matrix is nearly the same than the V_T of a large (W=10μm) isolated device. It is due to the fact that oxygen diffusion and TiN oxidation only affects the edge transistors as schematically illustrated in Fig.16. As only 2% of the transistors of the matrix have seen their gate workfunction modified by the TiN oxidation, the final V_T^{100ch} of the matrix is closer to the one of a wide isolated transistor W=10μm rather then the one of a narrow transistor W=80nm.

On the contrary, for L=30nm, the V_T^{100ch} is identical to the V_T of a narrow isolated device (W=80nm). It is explained by the fact that O diffusion and gate oxidation impacts, in this case, all the transistors of the matrix (see Fig. 17). As a result, all the workfunctions of the matrix transistors have been now changed by TiN oxidation and the resulting V_T^{100ch} of the matrix is the same as the one of a narrow isolated transistor.

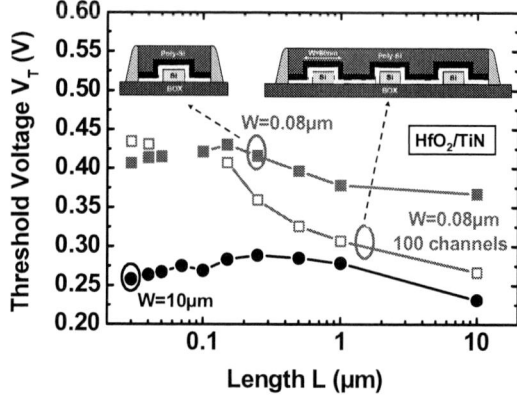

Figure 15. V_T vs L for isolated and for a matrix of 100 parallel transistors. The V_T of the array increases from the value of the isolated W=10μm device to the one of the W=80nm device, when L is scaled. It is consistent with a O diffusion through W and L edges

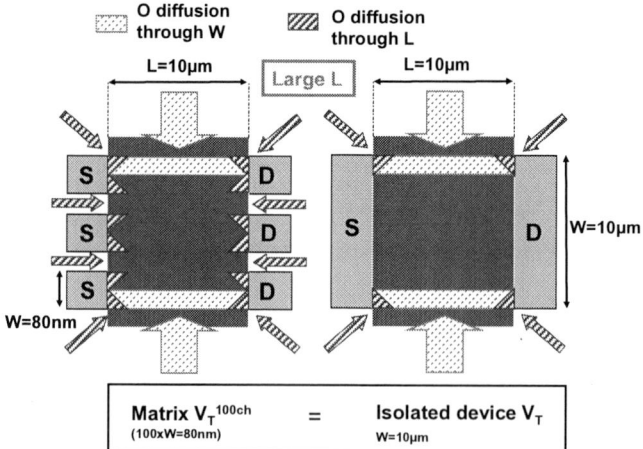

Figure 16. Schematic mechanism of O diffusion and TiN oxidation in large devices L=10μm. O diffusion only affects the workfunction of the edge transistors in red. This explains why V_T^{100ch} of the matrix with W=80nm is very similar to the one of a wide isolated transistor with W=10μm

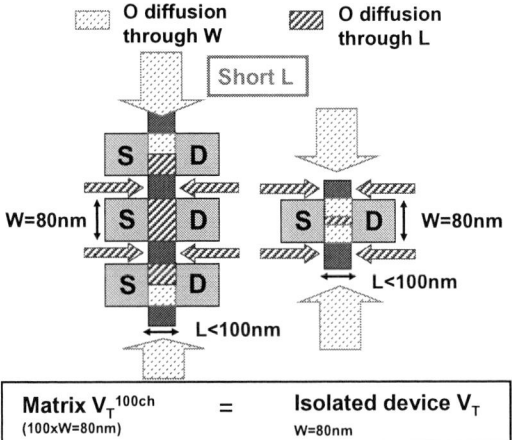

Figure 17. Schematic mechanism of O diffusion and TiN oxidation in short devices L<100nm. O diffusion modifies the workfunction of all the transistors in red and blue. This explains why V_T^{100ch} of the matrix with W=80nm is now very similar to the one of a narrow isolated transistor with W=80μm

Furthermore we assume in our physical model that the V_T instability in HfO_2 and HfZrO results from the modification of the TiN workfunction due the metal gate oxidation. Such an oxidation of the TiN gate for HfO_2 has been confirmed by EELS measurements made on narrow devices (see Fig.18). For HfO_2 oxide, the O peak is clearly superimposed to the Ti and N peaks suggesting a complete oxidation of the whole 5nm TiN layer. For HfSiO oxides, this oxidation of the TiN gate, also seen by EELS, is reduced. The O peak only reaches the Ti level at the the TiN/PolySi interface and not at the interface of interest between HfSiO and TiN. This last result is perfectly consistent with the better stability of V_T with scaling for HfSiO oxides when compared with HfO_2.

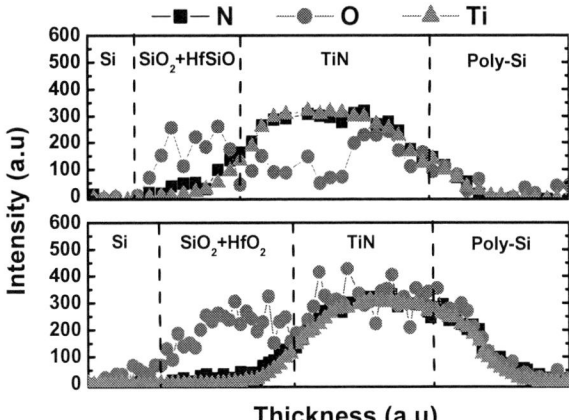

Figure 18. Electron Energy Loss Spectroscopy made on narrow devices W=80nm for HfO_2 and HfSiO oxides. Higher O content is found in the TiN gate for HfO_2

To clearly understand why O diffusion and resulting TiN oxidation are smaller in HfSiO oxides than in HfO$_2$ or HfZrO, TEM and ATR characterizations have been performed on both kinds of oxides. Fig. 19 shows HRTEM images of the isolation edges of a transistor for HfZrO and HfSiO oxides. Clearly, no SiO$_2$ regrowth, which could induce V$_T$ variation, is observed at the transistor edge for both oxides. The thickness of the bottom SiO$_2$ oxide is of 0.8nm which is in perfect agreement with the EOT value of 1.2nm measured previously. In fact, the only difference between HfSiO and HfZrO material arises from their crystallinity. Indeed HfZrO oxide exhibits a crystalline phase clearly identified by fringes in the HRTEM picture whereas HfSiO oxide still remains mainly amorphous, even after the complete CMOS process. This result is further confirmed by ATR measurements shown in Fig. 20. For HfZrO, the apparition of a clear peak in the ATR spectrum at σ=700μm indicates that this oxide has crystallized in a tetragonal phase during the activation anneal at 1050°C [19]. On the other hand, for HfSiO oxides, no clear peak is visible in the ATR spectrum, for the same range of σ. This confirms the amorphous nature of the HfSiO oxide even after high temperature anneal.

Finally all the last results support the idea that the oxygen species uses diffusion paths inside the dielectric itself, to reach and oxidize the TiN metal gate, as depicted in Fig. 14. And the diffusion velocity in the oxide is strongly dependent of the oxide crystallinity. In crystalline High-K dielectrics like HfZrO or HfO$_2$, O atoms can easily and rapidly diffuse through the oxide during the different high temperature anneals and leads to significant TiN oxidation at the origin of the V$_T$ instability with device geometry. In amorphous High-K dielectrics like HfSiO, the O diffusion is limited and the V$_T$ remains stable whatever the device size.

To conclude, the key point to suppress this V$_T$ instability is to prevent oxygen from diffusing from the edge of the transistors towards the High-K/TiN interface. This can be achieved by integrating amorphous High-K dielectrics like HfSiO or maybe by adding blocking barriers to O diffusion at the isolation edges of the transistor.

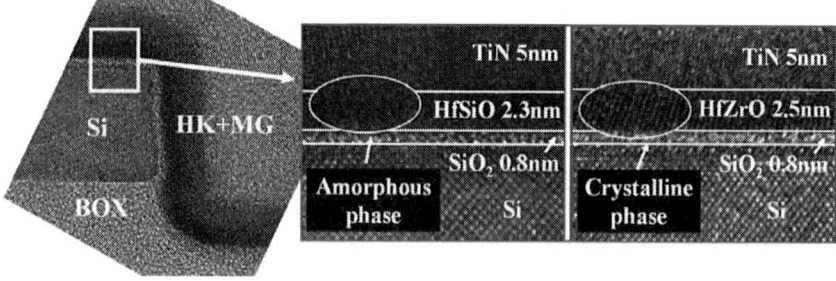

Figure 19. HRTEM images for HfZrO and HfSiO oxides at the mesa edge (gate edge). (1) No SiO$_2$ regrowth is visible for both stacks. (2) Fringes are visible in the HfZrO layer, characteristic of a crystalline phase whereas the HfSiO layer is mainly amorphous.

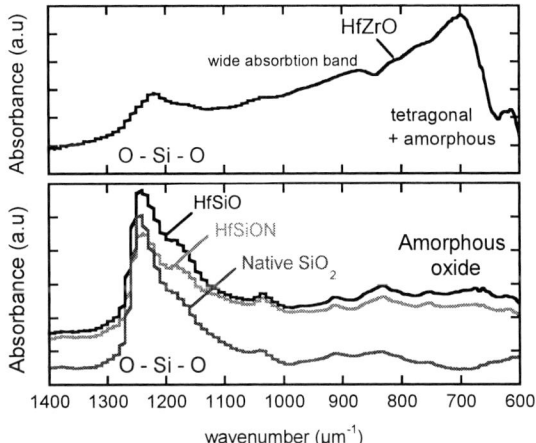

Figure 20. ATR spectra for HfZrO and HfSiO(N) oxides after 5nm ALD TiN + Poly-Si deposition+spike annealing at 1050°C.

Impact of dopants incorporation on V_T (in)stability

Today several dopants are used in High-K/Metal gate technology to modulate the V_T of transistors in a gate first integration scheme. Typically Aluminum and Lanthanum atoms are chosen to reduce the V_T of PMOS and NMOS transistors, respectively [20-22]. Basically these Al and La dopants incorporated in the High-K layer form dipoles at the SiO_2/High-k interface responsible for the V_T adjustment. So far they were shown to be very effective to fix the V_T of long and short channel devices but their impact on the stability of V_T with device geometry was only very little investigated [1,6]. Therefore, in this study, we try to evaluate how Al and La affect the V_T stability with W scaling.

Fig. 21 compares the V_T variation with W scaling for HfO_2 and HfSiON oxides with and without Al doping. For HfSiON, the V_T remains stable even after La doping. For HfO_2, the Al incorporation does not modify the original V_T instability with W scaling observed in standard oxides without Al. As a result, Al dopants used to adjust the V_T of PMOS does not affect its stability with device scaling. On the other hand, a recent study made by Inoue et al. reports severe degradation of the V_T stability with W scaling due to La doping [1]. As shown in Fig. 22, the V_T of La doped HK/MG transistors strongly increases when W is scaled, whereas, the V_T of the references without La doping remains stable. It means that the V_T instability observed here has a clearly different origin from that observed in crystallized oxides like HfO_2. It is completely ascribed to La incorporation. In fact, the authors explained this V_T instability induced by La by a reduced amount of La dopants at the isolation edges of the device. For wide transistor, the lack of La dopants at the STI edges has little impact on the V_T. On the contrary, on narrow transistors, this reduced concentration of La dopants induces a reduced amount of dipoles responsible for V_T adjustment, and the V_T becomes closer to the V_T of reference transistors without La.

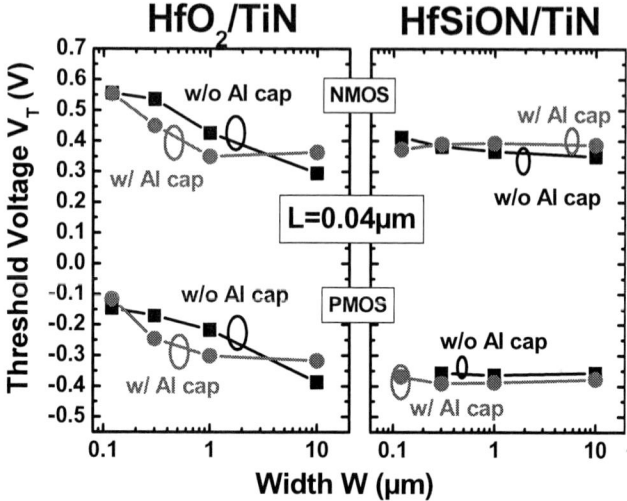

Figure 21. V_T vs. transistor width W for HfO_2 and HfSiON with and without Al capping. Al incorporation only hardly affects V_T instability with W for HfO_2

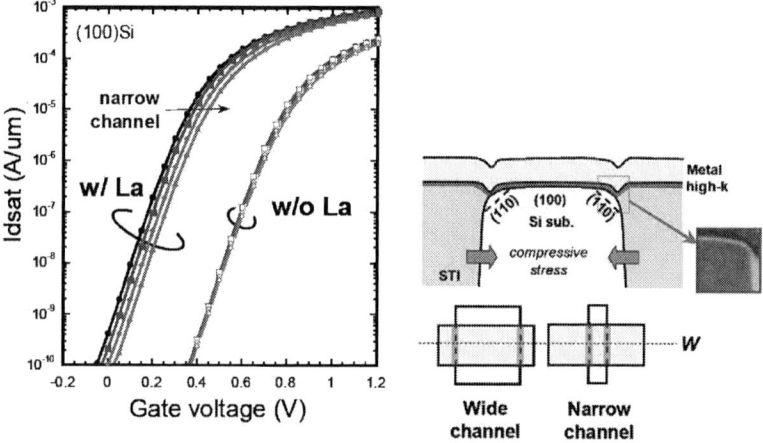

Figure 22. Id-Vg characteristics for different transistor widths W from [1]. La incorporation clearly induces V_T instability of these HK/MG transistors for 32nm node. This V_T instability is mainly explained by a reduced amount of La dopants at the isolation edge of the device.

Conclusion

The V_T stability with device scaling has been largely investigated by electrical and physical measurements in 30nm High-K/Metal Gate technologies. It is demonstrated that crystallized High-K dielectrics like HfO_2 and HfZrO exhibit a systematic and detrimental V_T instability with device scaling W and L. This instability has been undoubtedly related to an uncontrolled oxidation of the TiN metal gate, which modifies its workfunction. This oxidation phenomenon was shown to be limited by the lateral diffusion of oxygen through the High-K layer. HfSiON oxides present a better robustness to this effect. It has been ascribed to the amorphous structure of these oxides, which affects the O diffusion mechanism. Finally, the incorporation of Al dopants in the dielectric stack does not impact this V_T instability. On the other hand, La doping can significantly degrade the V_T stability with device geometry, due to a non-uniformity of La concentration over the device area.

Acknowledgments

This work has been supported by the CEA-LETI and ST Microelectronics joint program and the MEDEA+ DECISIF project

References

1. M. Inoue M. Inoue, Y. Satoh, M. Kadoshima, S. Sakashita, T. Kawahara, M. Anma, R. Nakagawa, H. Umeda, S. Matsuyama, H. Fujimoto and H. Miyatake, *Proc. of the VLSI Tech. symposium*, p.40, (2009)
2. C. Fenouillet-Beranger, P. Perreau, L. Pham-Nguyen, S. Denorme, F. Andrieu, L. Tosti, L. Brevard, O. Weber, S. Barnola, T. Salvetat, X. Garros, M. Cassé, C. Leroux, J.P Noel, O. Thomas, B. Le-Gratiet, F. Baron, M. Gatefait,Y. Campidelli, F. Abbate, C. Perrot, C. de-Buttet, R. Beneyton, L. Pinzelli, F. Leverd, P. Gouraud, M. Gros-Jean,A. Bajolet, C. Mezzomo, C. Leyris, S. Haendler, D. Noblet, R.Pantel, A. Margain, C. Borowiak, E. Josse, N. Planes, D. Delprat, F. Boedt, K. Bourdelle, B.Y. Nguyen, F. Boeuf, O. Faynot and T.Skotnicki, *Int. Elec. Dev. Meeting Tech. Dig.,* p. 667, (2009)
3. H. Takahashi, H. Minakata, Y. Morisaki, S. Xiao, M. Nakabayashi, K. Nishigaya, T. Sakoda, K. Ikeda, H. Morioka, N. Tamura, M. Kase and Y. Nara., *Int. Elec. Dev. Meeting Tech. Dig.,* p. 427, (2009)
4. E. Cartier, M. Hopstaken, and M. Copel., *App. Phys. Lett.*, **95**, p.042901, (2009)
5. L. Brunet, X. Garros, M. Cassé, O. Weber, F. Andrieu, C. Fenouillet-Béranger, P. Perreau, F. Martin, M. Charbonnier, D. Lafond, C. Gaumer, S. Lhostis, V. Vidal, L. Brévard, L. Tosti, S. Denorme, S. Barnola, J.F. Damlencourt, V. Loup, G. Reimbold, F. Boulanger, O. Faynot, A. Bravaix, *Proc. of the VLSI Tech. symposium*, p. 29, (2010)
6. T. Morooka, M. Sato, T. Matsuki, T. Suzuki, K. Shiraishi, A. Uedono, S. Miyazaki, K. Ohmori, K. Yamada, T. Nabatame, T. Chikyow, J. Yugami, K. Ikeda, and Y. Ohji, *Proc. of the VLSI Tech. symposium*, p. 33, (2010)
7. M. Charbonnier, C. Leroux, V. Cosnier, P. Besson, E. Martinez, N. Benedetto, C. Licitra, N. Rochat, C. Gaumer, K. Kaja, G. Ghibaudo, F. Martin, and G. Reimbold, *IEEE Trans. Elec. Dev.*, **57**, p. 1809, (2010)
8. S. K. Han, H-S. Jung, H. Lim, M. J. Kim, C-K. Lee, M. Lee, Y. You, H. S. Baik, Y. S. Chung, E. Lee, J-H. Lee, N. I. Lee and H-K. Kang., *Int. Elec. Dev. Meeting Tech. Dig.,* p. 621, (2006)

9. M. Kadoshima, T. Matsuki, N. Mise, M. Sato, M. Hayashi, T. Aminaka, E. Kurosawa, M. Kitajima, S. Miyazaki, K. Shiraishi, T. Chikyo, K. Yamada, T. Aoyama, Y. Nara and Y. Ohji, *Proc. of the VLSI symposium*, p.48, (2008)

10. J. F. Zhang, M. H. Chang, Z. Ji, L. Lin, I. Ferain, G. Groeseneken, L. Pantisano, S. De Gendt, and M. M. Heyns, *IEEE Electron Dev. Lett.,* **29**, p.1360-1363 (2008)

11. X. Garros,M. Casse, G. Reimbold, M. Rafik, F. Martin, F. Andrieu, V. Cosnier and F. Boulanger, *Microelec. Eng.,* **86**, p. 1609, (2009)

12. A. Kerber, E. Cartier, L. Pantisano, R. Degraeve, T. Kauerauf, Y. Kim. A. Hou, G. Groeseneken, H.E. Maes and U. Schwalke, *Elec. Dev. Lett.,* **24**, 3, p.87, (2003)

13. C. Leroux, J. Mitard, G. Ghibaudo, X. Garros, G. Reimbold, and F. Martin, *Int. Elec. Dev. Meeting Tech. Dig.*, p.737, (2004)

14. K. Kita and A. Toriumi, *Int. Elec. Dev. Meeting Tech. Dig*, p. 29, (2008)

15. E. Cartier, M. Steen, B.P. Linder, T. Ando, R. Iijima, M. Frank, J.S. Newbury, Y.H. Kim, F.R. McFeely, M. Copel, R. Haight, C. Choi, A. Callegari, V.K. Paruchuri and V. Narayanan, *Proc. of the VLSI Tech. symposium*, p.42, (2009)

16. W. Mizubayashi, K. Akiyama, W. Wang, M. Ikeda, K. Iwamoto, Y. Kamimuta, A. Hirano, H. Ota, T. Nabatame and A. Toriumi, *Proc. of the VLSI Tech. symposium*, p.42, (2008)

17. B. Chen, R. Jha, H. Lazar, N. Biswas, J. Lee, B. Lee, L. Wielunski, E. Garfunkel and V. Misr, *Elec. Dev. Lett.*, **27**, 4, p. 228, (2006)

18. J. Robertson, *Solid State Elec.* , **49**, 3 ,p. 283, (2005)

19. S. Lhostis, C. Gaumer, C.Bonafos, S. Schamn, N. Cherkashin, F. Pierre, A. Fanton, C. Morin, F. Ferrieu, M. Casse , X. Garros and C. Leroux, *Trans. Electrochem. Soc.*, **13**, pp. 101, (2008)

20. P. D. Kirsch, M. A. Quevedo-Lopez, S. A. Krishnan, C. Krug, H. AlShareef, C. S. Park, R. Harris, N. Moumen, A. Neugroschel, G. Bersuker, B .H. Lee, J.G. Wang, G. Pant, B. E. Gnade, M. J. Kim, R. M. Wallace, J. S. Jur, D. J. Lichtenwalner, A. I. Kingon and R. Jammy., *Int. Elec. Dev. Meeting Tech. Dig.,* p.1, (2006)

21. P. Sivasubramani, T. S. Böscke, J. Huang, C. D. Young, P. D. Kirsch, S. A. Krishnan, M. A. Quevedo-Lopez, S. Govindarajan, B. S. Ju, H. R. Harris, D. J. Lichtenwalner, J. S. Jur, A. I. Kingon, J. Kim, B. E. Gnade, R. M. Wallace, G. Bersuker, B. H. Lee and R. Jammy., *Proc. of the VLSI Tech. symposium*, p. 68, (2007)

22. S. Guha, V. K. Paruchuri, M. Copel, V. Narayanan, Y. Y. Wang, P. E. Batson, N. A. Bojarczuk, B. Linder and B. Doris, *App. Phys. Lett.*, **90**, 092902, (2007)

ECS Transactions, 35 (4) 531-543 (2011)
10.1149/1.3572303 ©The Electrochemical Society

Investigation of Electron and Hole Charge Trapping in LaLuO₃ Stack MOS Capacitor Using the Three-Pulse CV Technique

N. Sedghi[a], I. Z. Mitrovic[a], J. M. J. Lopes[b], J. Schubert[b], and S. Hall[a]

[a] School of Electrical Engineering and Electronics, University of Liverpool, Liverpool L69 3GJ, UK
[b] Institute of Bio and Nanosystems and JARAFIT, Research Centre Jülich, 52428 Jülich, Germany

> The three-pulse CV measurement technique has been used to study electron and hole charge trapping and de-trapping in $SiO_2/LaLuO_3$ gate stacks on p-type silicon substrate. Variation of flat-band voltage shift due to electron and hole trapping with charging time and pulse amplitude has been investigated. The pulsed CV measurements have been taken at various ramp rates and times to delineate the influence of interface states. The time constant of interface states responses estimated experimentally. The three-pulse CV measurements at different ramp rates are also used to evaluate the reliability of the measurement technique.

Introduction

High-κ materials are used increasingly as the gate dielectric in metal-oxide-semiconductor field effect transistors (MOSFETs). The understanding of charge trapping phenomena and the development of new, fast measurement techniques to reduce the effects of charge tapping during the measurement are of high importance. The concentration of traps in high-κ dielectrics is generally higher than that in conventional SiO_2 [1, 2] and trapping dynamics tend to be much faster. Conventionally, the trapped charge is measured from the shift of flat-band voltage in high frequency (HF) capacitance-voltage (CV) curves or from the shift of threshold voltage on transfer characteristics of MOSFETs. However, these measurements, which take fractions to a few seconds, are generally too slow to probe the traps in high-κ dielectric layers. Several groups have used fast, pulsed current-voltage (IV) and pulsed CV to study trapping and de-trapping in high-κ dielectrics [1-10].

Ternary rare earth oxides are promising candidates for high-κ applications, amongst which is lanthanum lutetium oxide ($LaLuO_3$) [11, 12]. This material has excellent physical and electrical properties, namely a high dielectric constant up to 32, large band gap of 5.2 eV, symmetrical and high band conduction band offset of 2.1 eV, relatively low concentration of bulk electron traps, and high crystallization temperature over 1000 °C [13, 14]. The two-pulse capacitance voltage (CV) measurement technique reported by Puzzilli et al. and later by Zhang et al. on SiO_2/Al_2O_3 gate stack MOS capacitors with n-type silicon substrate was shown to be a useful method to investigate trapped charges whilst maintaining trapping and de-trapping effects to a minimum [6, 7]. However, we have observed evidence for both negative and positive charge trapping in $SiO_2/LaLuO_3$ gate stacks on p-type silicon [15]. The two-pulse technique is not adequate

531

to investigate the effects of each charge type independently. The three-pulse technique has been used to separate the contributions of negative and positive trapped charge in $SiO_2/LaLuO_3$ dielectric stack MOS capacitors [15].

Sample Preparation and Measurement Setup

Sample Preparation

Lanthanum lutetium oxide film with thickness of 20 nm was deposited on 1-1.5 nm thick SiO_x layer on *p*-type silicon using molecular beam deposition. Details of the deposition procedure can be found elsewhere [11]. The capacitors were prepared by electron beam evaporation of 70 nm of platinum metal gate through a shadow mask with device area of 245×245 μm^2. The ohmic, backside contact was made by deposition of 120 nm aluminum. Finally, the sample received a forming gas anneal ($90\% N_2 + 10\% H_2$) at 450 °C for 10 min. This treatment improved the backside aluminum contact as well as the interface between the oxide stack and silicon substrate. Further details concerning device fabrication, together with physical and electrical characterization can be found in [13]. The cross-section of the device is shown in Fig. 1.

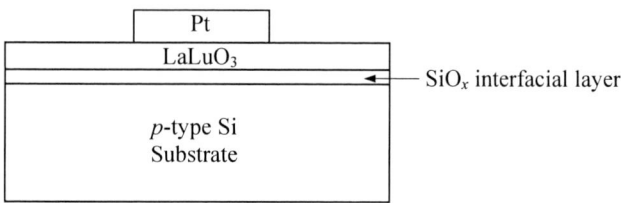

Figure 1. Device cross section.

Measurement Setup

The pulsed CV measurements set-up is shown in Fig. 2. The capacitance is calculated from the displacement current induced by the ramp voltage at the pulse edges. In most measurements, the rise and fall time of the pulses are set at 200 μs and the ramp rate is set at 25 kV/s firstly, to minimize the trapping and de-trapping during the measurement in bothe the oxide bulk and also in interface states; secondly, to achieve displacement current much larger than the oxide leakage current over the range swept. However, measurements at other ramp times and rates have also been performed to study any effect due to interface states and also for further evaluation of this relatively new measurement technique.

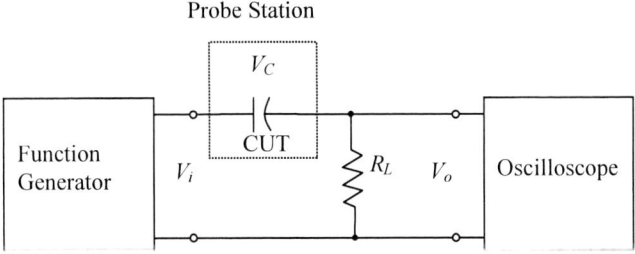

Figure 2. Measurement setup.

Measurement Procedure

Current voltage (IV) measurements were first performed on the devices, to investigate the oxide leakage current and identify the breakdown voltage. The leakage current density, shown in Fig. 3, is about 1×10^{-4} A/cm^2 at an applied voltage of -5 V to the gate and the absolute value of breakdown voltage is between 5 and 6 V corresponding to about 2 MV/cm in the high-κ layer. More detailed investigation of breakdown and any role of the interfacial layer is beyond the scope of the paper. The leakage current at positive applied voltage is much lower since, as is well known, the silicon is forced into deep depletion due to the lack of minority carrier electrons to provide the current. Conventional high frequency (HF) capacitance-voltage (CV) measurements, shown in Fig. 4 (a), were performed to allow estimation of the flat-band and mid-gap voltages and also as comparison with pulsed CV results. A flat-band voltage of about -0.5 V was calculated from the second derivative of $1/(C/C_{ox})^2$, as shown in Fig. 4 (b).

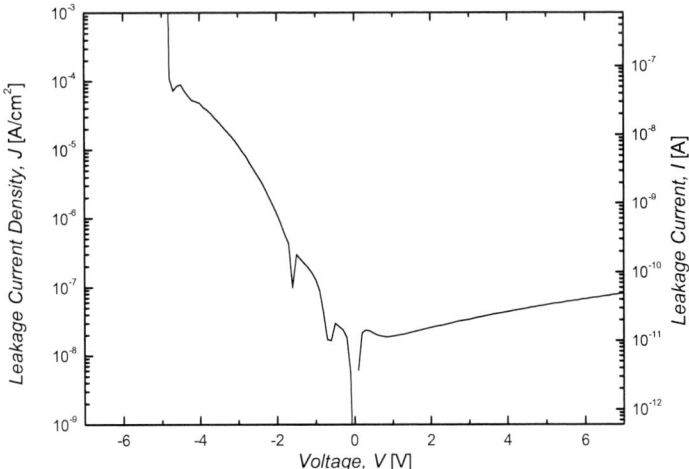

Figure 3. Leakage current.

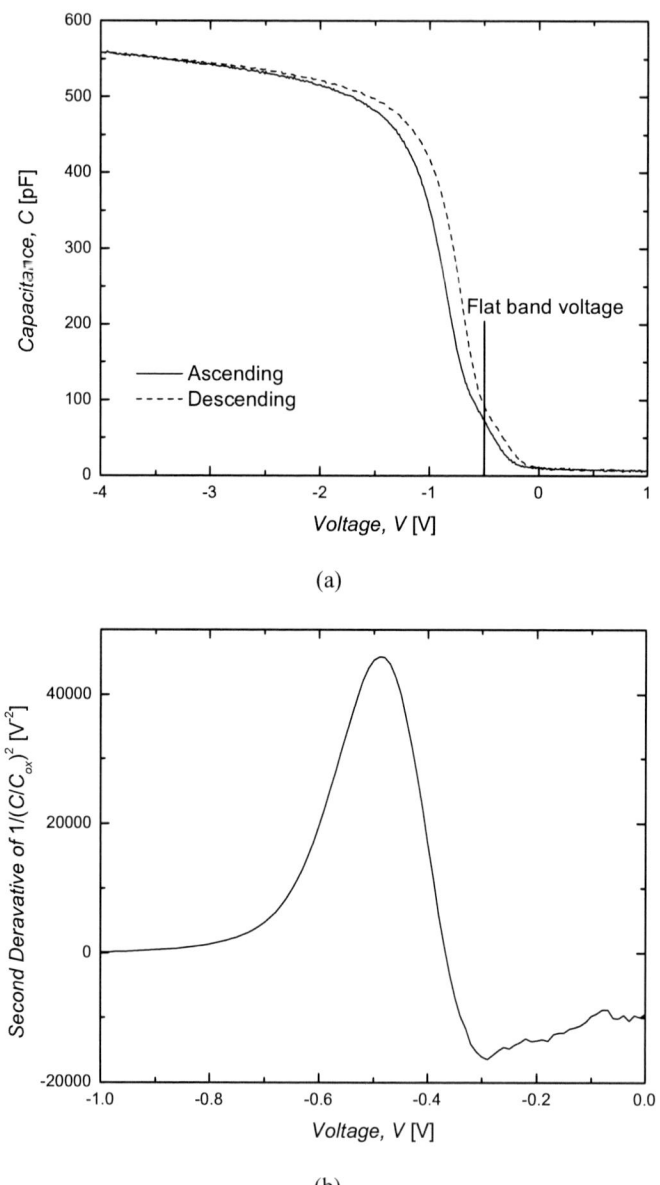

(b)

Figure 4. (a) High frequency CV and (b) plot to find flat band voltage.

The pulse schemes for two-pulse and three-pulse CV measurements are shown in Fig. 5 (a) and (b), respectively. The applied voltage is maintained at voltage V_H for a long time before the start of first pulse, to allow equilibration, with the device in inversion at the start of the measurement. In other measurements V_H was kept at zero volt (device in depletion) to avoid the build-up of the inversion layer. The initial CV measurement on a fresh sample is performed at the ramp-down of the first pulse, here referred to as Ramp 1. The displacement current in the MOS capacitor is proportional to dV_C/dt where V_C is the voltage across the capacitor. The value of load resistor is chosen such that the voltage across it is much smaller than the applied voltage. Therefore,

$$V_O = R_L C \frac{dV_C}{dt} \approx R_L C \frac{dV_i}{dt} \qquad \because \qquad C = \frac{V_O}{R_L \, dV_i/dt}. \qquad [1]$$

The oxide is charged during the negative pulse-width of V_L, for a duration t_{charge} and another CV measurement is then performed at the ramp-up of the first pulse, Ramp 2. In the two-pulse CV method, the device is then kept at voltage V_H for period of $t_{discharge}$ and stored charge is released over this period. A third CV measurement is then taken at the ramp-down of the second pulse, Ramp 3 in Fig. 5 (a).

The three-pulse CV technique uses the same principle as that of the two-pulse method but with the addition of a third pulse which serves to separate positive and negative trapped charge contributions. Typical input and output waveforms of the three-pulse CV technique for $t_{charge} = 5$ ms and $t_{discharge} = 10$ ms are shown in Fig. 5 (b). The initial measurement on a fresh sample is performed at Ramp 1. The CV measurement after charging is performed at Ramp 2, in the same way as for the two-pulse method, although in the present case of p-type material, the shift is affected by the net value of both negative and positive trapped charge. However, in the 3-pulse method, a further CV measurement is performed after a very short time, 1-10 µs, at Ramp 3 to remove the negative charge. The technique is based therefore, on the assumptions that the positive hole discharge during this time is negligible whereas the negative electron charge is released before the end of Ramp 2 when the voltage at gate electrode is positive. Experimental evidence for the assumptions follows from the observation that pulsed-CV plots measured at Ramp 2 and Ramp 4 in Fig. 5 are completely identical, whereas they have a significant shift to the left at Ramp 3. Trapped holes show negligible discharge if this time is much less than charging time.

Experimental Results and Discussion

Two-Pulse CV

It is useful to describe the limitations of the two-pulse technique to illustrate the advantages of the three-pulse technique. Two-pulse CV measurement plots for $t_{charge} = 500$ µs and $t_{discharge} = 1$ ms are shown in Fig. 6. We now describe the sequence of events as the pulse waveform is applied. A positive voltage, V_H is first applied to the gate electrode for a long time, to establish equilibrium. The capacitor therefore starts in the inversion condition. The inversion capacitance can be seen on the initial CV plot at voltages above -0.8 V on Ramp 1, after a transient time related to the RC ramp response of the circuit. As the ramp voltage goes increasingly negative, inversion charge is removed by injection into the substrate, where it quickly recombines. The device then goes into depletion and subsequently, accumulation. The oxide is then charged during

t_{charge} before it is ramped back with the positive-going pulse, designated 'Ramp 2'. For Ramp 2, the CV plot is seen to shift to the right, which is an indication of the introduction of negative charge into the oxide. Furthermore, the CV plot is steeper than that for Ramp 1; this latter effect will be discussed in the next section.

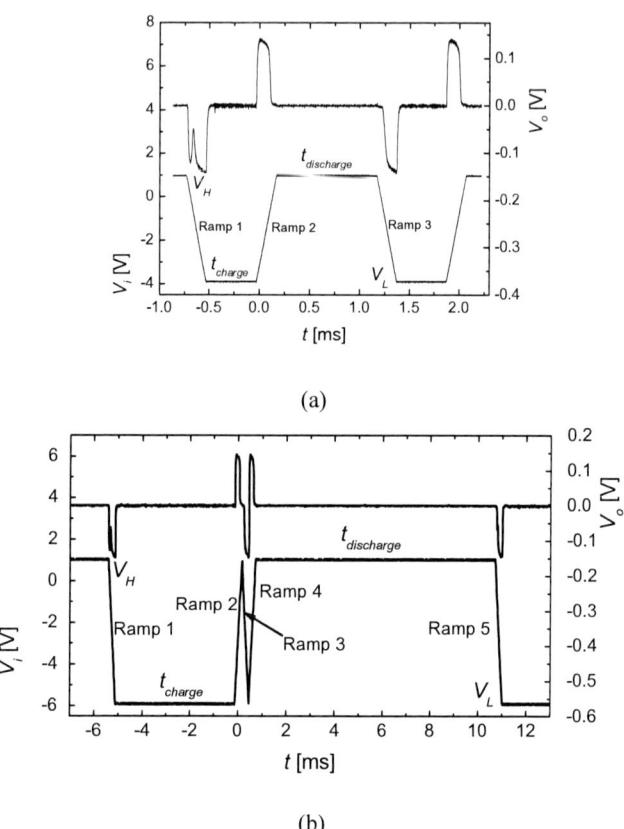

Figure 5. Waveforms for two-pulse and three-pulse CV techniques.

The most obvious source of the overall shift due to negative charge is the consequence of electrons injected from the gate electrode into the oxide. However, there are also possible mechanisms for positive (hole) charge to be injected and stored in the device, which would limit the extent of the shift ascribed to electron trapping. The first possible mechanism is that due to the emptying of electrons from the donor-like fast interface states in the lower half of the silicon band gap (designated hole capture), and the second, the tunneling of holes into oxide bulk traps (slow states), as indicated in the energy band diagram of Fig. 7. These latter traps are likely to be within the interfacial SiO_x layer and its interface with the high-κ layer. When the positive voltage, V_H is applied to the gate during $t_{discharge}$, the negative charge and part of the positive component may be released to the gate and substrate, respectively. The CV plot measured at Ramp 3 is shifted to the left, presumably due to positive charge remaining in the oxide. However, the amount of initial positive charge cannot be measured using the two-pulse CV method and therefore it is not possible to calculate the amount of released charge. This section therefore has served to identify the problem that relates to the inability of the two-pulse technique to allow de-convolution of the contributions of positive and negative charge trapping.

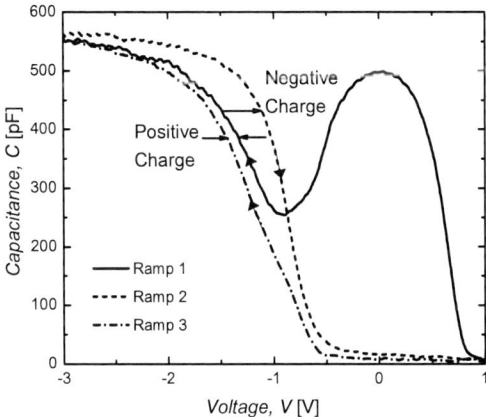

Figure 6. Typical two-pulse CV plots.

We now go on to explain in the next section, how the CV plot shift due to the initial positive charge after t_{charge} can be measured using the three-pulse technique [15]. The essential idea is that a CV plot measured during the edge of a further ramp gives the shift of CV plot due to positive trapped charge and this can be used as a reference to calculate the released charge after $t_{discharge}$. It should be noted that the situation is simpler to interpret for the case of MOS devices on n-type silicon where only electron trapping occurs.

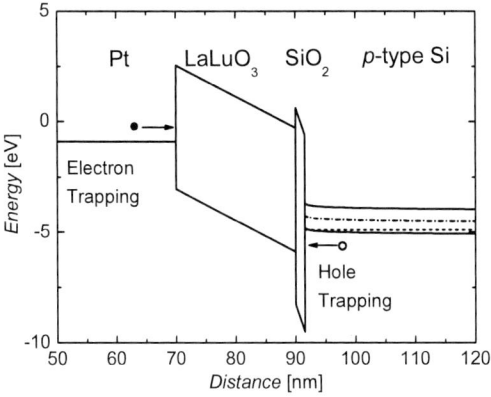

Figure 7. Energy band diagram.

Three-Pulse CV

The technique has been described before [15] but is presented here for completeness. Typical three-pulse CV plots are shown in Fig. 8, measured at negative and positive ramps on various pulse edges with a fixed ramp time of 200 µs and ramp rate of 25 kV/s. The charge and discharge times are both set at 10 ms. The ramp time is chosen to be short enough so that hole discharge during the measurement is negligible but it is much larger than typical time constants for interface states, which are expected to be of the order of microseconds. The effect of interface states will be discussed in a separate section below. The initial measurement is performed at Ramp 1 from positive to negative voltage with the device biased in inversion for a long time to allow equilibration, as for the two-pulse method. Measurements with different circuit time constants have shown that the RC transient affects only the start of the CV plots and so has no effect on the subsequent shifts from which trapping information is extracted. It can be noted that even if the measurement sequence is started from the depletion condition before the first ramp, the ensuing CV characteristics are found to be the same for both cases, namely initial depletion and inversion.

The CV plot measured after trapping at Ramp 2 in Fig. 8 is affected by both negative and positive trapped oxide charge. The negative charge is then released by Ramp 3, and the CV plot is seen to shift to the left due to the dominant positive charge. This positive charge is the result of hole trapping in the interfacial layer. A fraction of the positive charge is released, dependent on the duration of discharge time, and the CV plot measured at Ramp 5 shifts to the right towards the initial CV plot. It needs to be mentioned that the CV plot for Ramp 3 in Fig. 8 cannot be achieved by the two-pulse technique and hence the amount of initial and released positive charge cannot be calculated using that method. Hereafter we refer to positive and negative charge as holes and electrons, respectively.

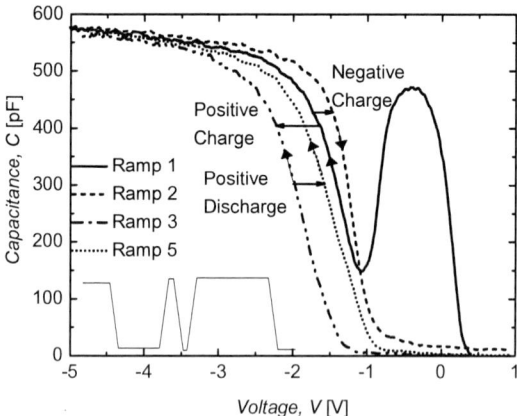

Figure 8. The three-pulse CV plots for $t_{charge} = t_{discharge} = 10$ ms.

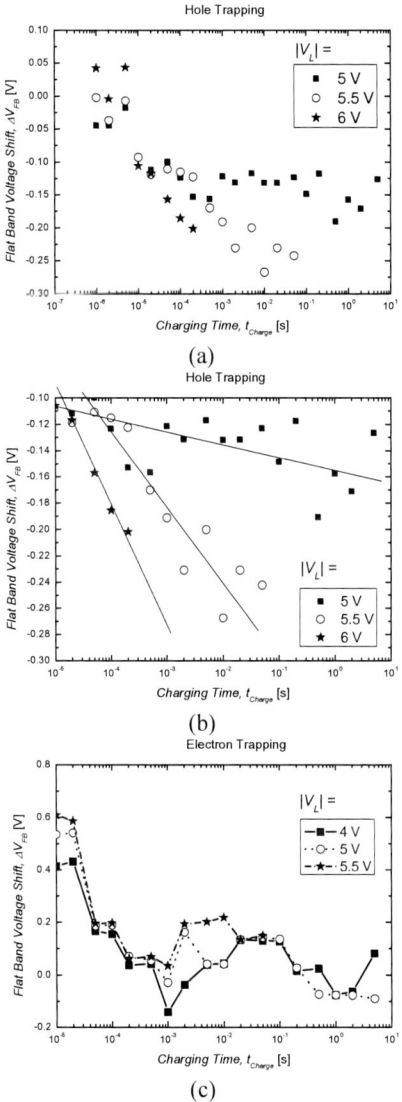

Figure 9. Variation of flat band voltage due to hole (a, b) and electron (c) trapping with charging time.

Effect of Charging Time

Having separated the contributions to CV shifts of electrons and holes, the dynamics of each are now described in turn. Considering first the hole trapping, the variation of flat-band voltage shifts measured at Ramp 3 with charging time at different pulse amplitudes is shown in Fig. 9 (a). The large change at low charging times between 1 μs and 10 μs is possibly due to transient effect of electron trapping and is subject of further investigations. The same plot for ramp times more than 10 μs is shown in Fig. 9 (b). The flat-band voltage shifts follow a logarithmic relationship with charging time. A similar relationship has been reported by Rao *et al.* on trapping in Al_2O_3 on p-type Si using the pulsed CV technique [8]. However, the latter is seen on a much longer time scale than that for $LaLuO_3$.

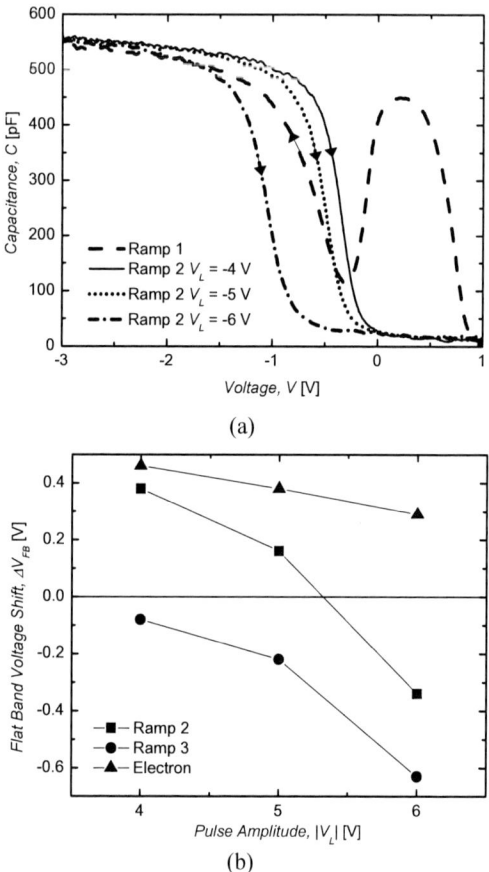

Figure 10. (a) Variation of CV plots measured at Ramp 2 with pulse amplitude and (b) variation of flat band voltage shift with pulse amplitude.

Turning to the case of electron trapping, the variation of flat band voltage with charging time is shown in Fig. 9 (c). The shift was extracted by subtracting the relative shift of CV plots measured at Ramp 3 (affected by hole trapping) from those at Ramp 2 (affected by both electron and hole trapping). The large variation for charging times less than 50 μs is ascribed to the influence of interface states on Ramp 2, since the CV plots were measured at ramp times less than 10 μs to prevent charge trapping and de-trapping during the ramp. For longer charging times, the flat band shift does not show a significant variation and is virtually independent of pulse amplitude, implying a saturation effect in the trap occupancy.

Effect of pulse amplitude

The shift of flat-band voltage due to hole trapping with charging time depends also on the pulse amplitude, as shown in Fig. 10 (b). At low pulse amplitudes, electron trapping is dominant and the variation of flat-band voltage shift due to hole trapping with charging time has a low rate. By increasing the pulse amplitude above 5 V, electron trapping saturates and hence hole trapping becomes dominant. This can be seen in Fig. 10 (b) and also in CV plots measured at Ramp 2 in Fig. 10 (a). Variation of flat-band voltage shift on CV plots measured at Ramp 2 and Ramp 3 with pulse amplitude is shown in Fig. 11 (b) which shows an increase in rate for amplitudes more than 5 V. It should be noted, in particular, that the change in flat-band voltage shift at Ramp 2 is positive at amplitudes less than 5 V, indicative of the dominance of electron trapping, but is negative at higher pulse amplitudes, indicating the dominance of hole trapping. The change in flat-band voltage shift with pulse amplitude for electron trapping is insignificant.

Effect of Interface States

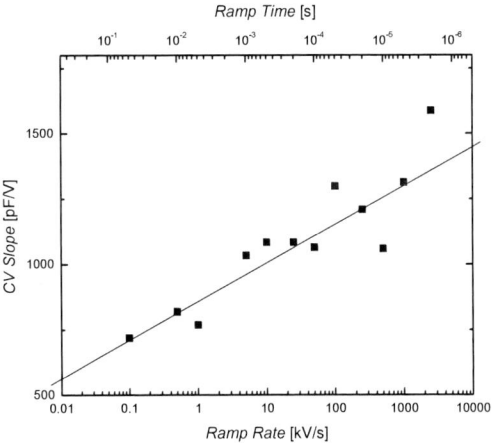

Figure 11. Variation of slope of CV plots at Ramp 2 with ramp rate and ramp time.

The CV plots measured at Ramp 2 in Fig. 6, 8 exhibit a steeper slope than those measured on downward ramps. Figure 11 shows further, the variation of the CV slope of upward (Ramp 2) CV plots as a function of ramp time and rate. This increase in slope for upward ramps and the differing slopes for upward and downward slopes can be ascribed to the influence of interface states. On the upward ramps, from negative to positive voltage, majority carrier holes tend to be emitted from interface states in response to the changing surface potential. However, the occupancy of the states governed by hole emission, cannot follow the fast ramps, resulting in a steeper slope close to that of an ideal CV plot. However, the downward ramp is influenced by hole capture into the states which is relatively faster. The states have time to charge causing the CV plots to be stretched out; that is, the slope of the CV plot decreases. The slope shows a logarithmic relationship with ramp time or ramp rate.

Conclusions

Trapping and de-trapping of positive and negative trapped charges in $SiO_2/LaLuO_3$ gate stacks on p-type silicon substrate was studied using the three pulse CV measurement technique. The positive trapped charge is ascribed to trapping of holes in the SiO_x interfacial layer and at the interface of SiO_x and high-κ dielectric. The negative charge is assumed to be close to the interface of the metal gate and high-κ dielectric. The flat-band voltage shift after hole trapping has a logarithmic relationship with charging time. At small pulse amplitudes, the electron trapping is dominant whereas at pulse amplitudes greater than 5 V, hole trapping dominates. The influence of interface states on the CV plot response has been identified as giving rise to steeper CV slopes at faster ramps due to the inability of the state occupancy to remain in equilibrium with the voltage ramp.

Acknowledgments

The work was funded by the European Union Framework Programme 7, Network of Excellence, NANOSIL.

References

1. G. Groeseneken, L. Pantisano, L.-Å. Ragnarsson, R. Degraeve, M. Houssa, T. Kauerauf, P. Roussel, S. De Gendt, and M. Heyns, *P. IEEE Int. Freq. Cont.*, 147 (2004).
2. C. Z. Zhao, J. F. Zhang, M. B. Zahid, B. Govoreanu, G. Groeseneken, and S. De Gendt, *J. Appl. Phys.*, **100**, 093716 (2006).
3. A. Kerber, E. Cartier, L. Pantisano, M. Rosmeulen, R. Degraeve, T. Kauerauf, G. Groeseneken, H. E. Maes, and U. Schwalke, *Int. Rel. Phy.*, 41 (2003).
4. C. D. Young, R. Choi, J. H. Sim, B. H. Lee, P. Zeitzoff, Y. Zhao, K. Matthews, G. A. Brown, and G. Bersuker, *Int. Rel. Phy.*, 75 (2005).
5. C. D. Young, Y. Zhao, D. Heh, R. Choi, B. H. Lee, and G. Bersuker, *IEEE T. Electron Dev.*, **56**, 1322 (2009).
6. G. Puzzilli, B. Govoreanu, F. Irrera, M. Rosmeulen, and J. Van Houdt, *Microelectron. Reliab.*, **47**, 508 (2007).
7. W. D. Zhang, B. Govoreanu, X. F. Zheng, D. Ruiz Aguado, M. Rosmeulen, P. Blomme, J. F. Zhang, and J. Van Houdt, *IEEE Electron Device Lett.*, **29**, 1043 (2008).
8. R. Rao, P. Lorenzi, G. Ghidini, F. Palma, and F. Irrera, *IEEE T. Electron Dev.*, **57**, 637 (2010).
9. D. Ruiz Aguado, B. Govoreanu, W. Zhang, M. Jurczak, K. De Meyer, and J. Van Houdt, *IEEE T. Electron Dev.*, **57**, 2726 (2010).

10. R. Rao and F. Irrera, *J. Appl. Phys.*, **107**, 103708 (2010).
11. J. M. J. Lopes, E. Durğun-Özben, M. Roeckerath, U. Littmark, R. Lupták, St. Lenk, M. Luysberg, A. Besmehn, U. Breuer, J. Schubert, and S. Mantl, *Microelectron. Eng.*, **86**, 1646 (2009).
12. I. Z. Mitrovic and S. Hall, *Journal of Telecommunications and Information Technology*, **4** ,51 (2009).
13. J. M. J. Lopes, M. Roeckerath, T. Heeg, E. Rije, J. Schubert, S. Mantl, V. V. Afanas'ev, S. Shamuilia, A. Stesmans, Y. Jia, and D. G. Schlom, *Appl. Phys. Lett.*, **89**, 222902 (2006).
14. K. Xiong and J. Robertson, *Appl. Phys. Lett.*, **95**, 022903 (2009).
15. N. Sedghi, I. Z. Mitrovic, S. Hall, J. M. J. Lopes, and J. Schubert, *J. Vac. Sci. Technol. B*, **29**, 01AB03-1 (2011).

544

ECS Transactions, 35 (4) 545-561 (2011)
10.1149/1.3572304 ©The Electrochemical Society

Inelastic Electron Tunneling Spectroscopy (IETS) Study of Ultra-thin Gate Dielectrics for Advanced CMOS Technology

T.P. Ma

Yale University, New Haven, CT, 06520, USA
Email: t.ma@yale.edu

As the thickness of the gate dielectric in a CMOS transistor continues to shrink in each new generation of integrated circuits in order to meet the targeted gains in performance and circuit density, it becomes increasingly difficult for some conventional dielectric characterization tools, such as infrared spectroscopy, Raman spectroscopy, neutron scattering, and Rutherford scattering, to reveal their structural and compositional information. In contrast, the Inelastic Electron Tunneling Spectroscopy (IETS) technique, which relies on tunneling current to probe the ultra-thin gate dielectric in a metal-insulator-semiconductor (MIS) sandwich, becomes more sensitive when the tunneling current increases, which is in the direction of the CMOS scaling trend. IETS can address materials issues related to reactions and intermixing at interfaces, as well as properties related to carrier mobility and reliability, such as phonon modes, impurities, and charge traps, for structures that are difficult to accurately characterize by other techniques. The principle of operation, experimental considerations, and examples will be shown in this paper to illustrate the capabilities and limitations of the IETS technique.

I. Introduction

The basic principle of the Inelastic Electron Tunneling Spectroscopy (IETS) technique [1-10] is illustrated in Fig. 1, where one can see that, without any inelastic interaction, the I-V characteristic is a smooth curve, and its 2nd derivative is zero. When the applied voltage causes the Fermi-level separation to be equal to the characteristic interaction energy of an inelastic energy loss event for the tunneling electron (see the 2nd band diagram in Fig. 1), then an additional conduction channel (due to inelastic tunneling) is established, causing the slop of the I-V characteristic to increase at that voltage, and a peak in its 2nd derivative plot, where the voltage location of the peak corresponds to the characteristic energy (in eV) of the inelastic interaction, and the area under the peak is proportional to the strength of the interaction. In a typical MOS sample, there are numerous inelastic modes, as a wide variety of inelastic interactions may take place, including interactions with phonons, various bonding vibrations, bonding defects, and impurities. Therefore, Fig. 2 is shown to represent a typical IETS spectrum taken on a SiO_2/Si stack, where the various features on the spectrum have been identified and will be discussed in a later section.

545

Inelastic Electron Tunneling Spectroscopy

An Inelastic Tunneling Event at E=eV = hν Causes
(a) *I-V* to increase slope;
(b) a step in *dI/dV*;
(c) a peak in *d²I/dV²*

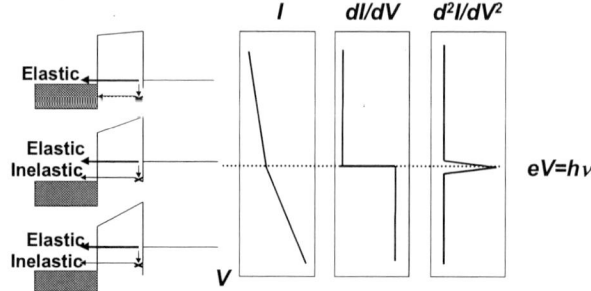

Fig. 1. Inelastic interaction causes a slope increase of I-V, a step in dI/dV, and a peak in 2nd derivative, all occurring at the voltage corresponding to the characteristic

Various Inelastic Modes in the Barrier (Left)
May Be Reflected in IETS (Bottom Right)

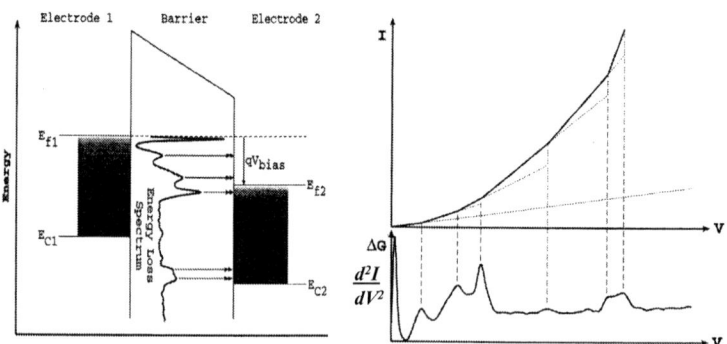

IETS probes phonons, bonding vibrations, impurities, and Traps

Fig. 2. Typical IETS spectrum (lower right) reveals various inelastic modes depicted in the energy diagram on the left.

II. IETS Measurement Technique

As discussed in the previous section, the IETS spectrum is essentially the d^2I/dV^2 vs V plot, where d^2I/dV^2 is the 2nd derivative of the I-V curve, and V is the applied voltage (which equals the Fermi level separation between the two electrodes). Therefore, the simplest way to obtain the IETS spectrum, at least conceptually, is to take the I-V curve and then employ mathematical differentiation of the data to determine its 1st and 2nd derivatives. Unfortunately, however, we have not been successful with this approach despite numerous attempts, due to inadequate signal-to-noise ratios that are required to resolve some of the fine structures that are important to our study. Therefore, an ac modulation/phase-sensitive detection technique is used throughout this study [1]. This technique uses a small sinusoidal signal to modulate the voltage across the sample under test and measure the response of the ac current through the sample to this modulation. Specifically, one uses a lock-in amplifier to measure the 2nd harmonic component of the ac current, and the amplitude of that component is taken as the IETS signal.

Mathematically, the above can be derived by taking a Taylor expansion of the I(V) curve around a voltage bias V_b with $V_m cos\omega t$ as a small perturbation:

$$I \left(V_b + V_m cos\omega t \right) = I(V_b) + dI/dV \mid_{Vb} V_m cos\omega t + \frac{1}{2}\, d^2I/dV^2 \mid_{Vb} Vm^2 cos^2 2\omega t + \ldots\ldots\ldots$$

$$= I(V_b) + G \mid_{Vb} V_m cos\omega t + \frac{1}{4}\, \Delta G \mid_{Vb} Vm^2 (1 + cos2\omega t) + \ldots\ldots \qquad (1)$$

From Equation (1), it is clear that the coefficients of $cos2\omega t$ and $cos2\omega t$ in the Taylor expansion are proportional to the first and second derivative of the I-V characteristic, respectively. As a result, synchronous detection of the first (ω) and second (2ω) harmonic gives a scaled measure of the first and second derivative of the I-V, respectively. Experimentally, the detection of ω and 2ω components is realized by a lock-in amplifier. During the lock-in measurement, a reference signal with the same frequency of the ac modulation is generated. Then the device signal and the reference signal are multiplied by phase-sensitive detector (PSD) of the lock-in. The PSD has the function of singling out the component of the signal at a specific reference frequency and phase. Noise signals at other frequencies are rejected. The phase-sensitive detection process can by mathematically explained as follows. Suppose the device signal is $Vcos(\omega t + \theta)$ and the reference signal is $V_R cos(\omega t + \theta_R)$, the output of the PSD is the product of the two [11]

$$V_{output} = VV_R \cos(\omega t + \theta)\cos(\omega t + \theta_R)$$
$$= \frac{1}{2} VV_R \{\cos[2\omega t + (\theta + \theta_R)] + \cos(\theta - \theta_R)\} \qquad (2)$$

By adjusting the phase difference to zero, one gets:

$$V_{output} = \frac{1}{2} VV_R [\cos(2\omega t + 2\theta) + 1] \qquad (3)$$

Therefore, a filtered DC output of the PSD is proportional to the amplitude of the device signal.

Figure 3 shows a Schematic diagram of the IETS measurement setup used in this study [1].

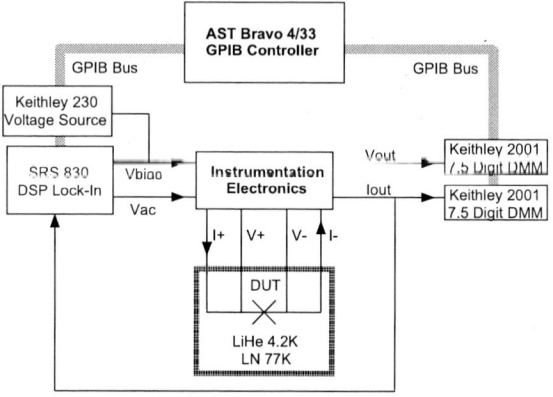

Fig. 3. Schematic diagram of the IETS measurement setup (After W.K.. Lye [1])

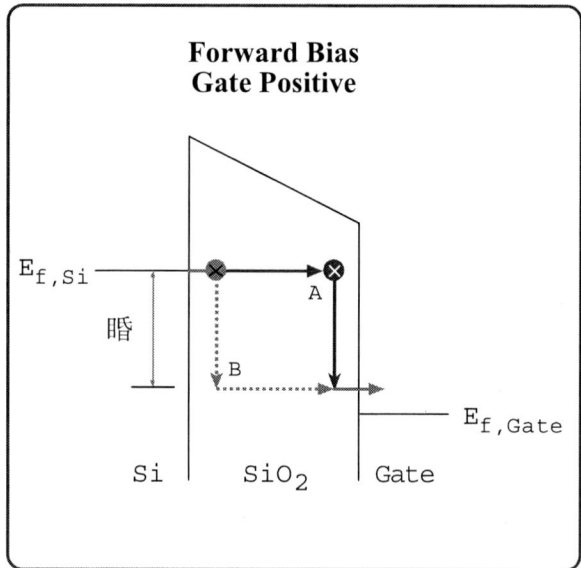

Fig. 4. The tunneling probability of an electron decreases once it loses some energy to an inelastic interaction. As a result, the inelastic feature in an IETS spectrum is more prominent for interactions near the positive biased electrode. In other word, the IETS measurement tends to preferentially reveal the interactions near the anode.

III. Bias Polarity Dependence of Inelastic Interaction

As shown in Fig. 4, the tunneling electron loses energy after an inelastic interaction, and its subsequent tunneling probability will reduce due to the increased effective tunnel barrier. As a consequence, the intensity of the peak in the IETS spectrum due to an inelastic interaction event near the cathode will be smaller than that due to a similar event near the anode interface, simply because the tunneling electron in the former case travels through most of the barrier with a lower electron energy (and thus it sees a higher tunnel barrier) than that in the latter case. A more quantitative theoretical treatment of this phenomenon can be found in [1]. This bias polarity dependence can be conveniently used to distinguish the microstructures and bonding defects between the two different interfaces in a MIS structure.

IV. Results and Discussion

IV.1 SiO_2/Si stacks

In thjis study, the IETS measurement was first applied to the SiO_2/Si system, not only because of its historical importance in CMOS technology, but also because of the vast knowledge base that exists about this system that can be used to calibrate the IETS technique. Figure 5 shows an example of the IETS spectrum for a high quality SiO_2 thermally grown on Si, where the most prominent features are those associated with Si phonons (between 10 mV and 70 mV) and those associated with various SiO_2 modes (between 135 mV and 170 mV).

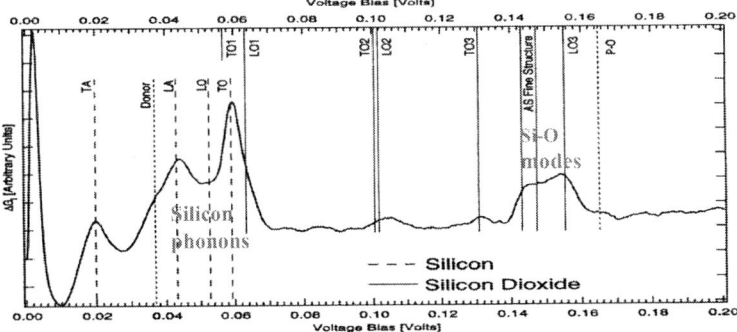

Fig. 5. IETS spectrum of SiO_2 on Si, where the Si phonon modes and SiO_2 vibrational modes are highlighted.

Figure 6 depicts a close-up view of the 10 mV-70 mV portion of the IETS spectrum shown in Fig. 5, and Fig. 7 depicts a close-up view of the 135mV-170mV portion of the IETS spectrum shown in Fig. 5. With the help of existing IR data and neutron spectroscopy published in the literature, our detailed analysis of the IETS data has enabled us to make assignments of various excitations associated with both the oxide barrier and the Si electrodes. For instance, the portion of the spectrum between 10 mV and 70 mV has been correlated with the expected Si phonon modes (both acoustic and optical ones), whereas the features between 135mV and 170mV are consistent with the excitation of several oxide modes. Because of the adequate signal-to-noise ratios and the very high stability and reproducibility of the IETS signals in these samples, we have been able to do fairly precise spectrum analysis by deconvolving overlapping features and quantifying their relative contributions. The deconvolution of the spectra was done by least-squares Gaussian fits (Lorenzians do not converge) utilizing commercial data analysis software.

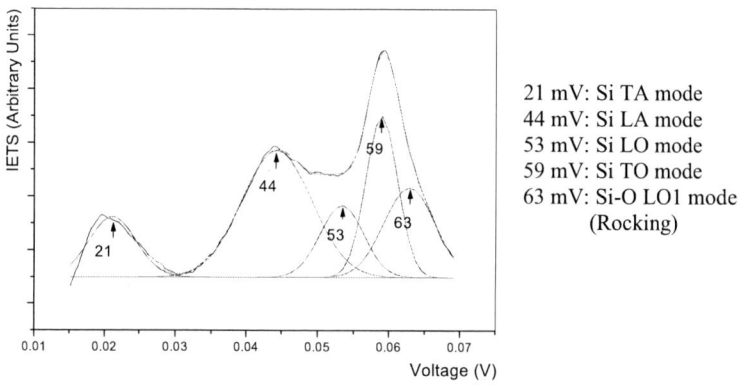

21 mV: Si TA mode
44 mV: Si LA mode
53 mV: Si LO mode
59 mV: Si TO mode
63 mV: Si-O LO1 mode
(Rocking)

Fig. 6. Deconvolution of Si phonon modes in the IETS spectrum. The voltage locations of individual peaks are marked. Note that the composite envelope formed by adding all 5 dashed curves together coincides almost exactly with the original curve.

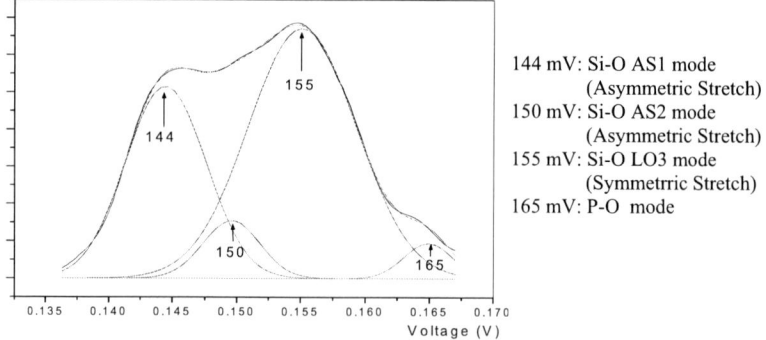

144 mV: Si-O AS1 mode
(Asymmetric Stretch)
150 mV: Si-O AS2 mode
(Asymmetric Stretch)
155 mV: Si-O LO3 mode
(Symmetrric Stretch)
165 mV: P-O mode

Fig. 7. Deconvolution of silicon oxide vibrational modes in the IETS spectrum. The voltage locations of individual peaks are marked. Note that the composite envelope formed by adding all 5 dashed curves together coincides almost exactly with the original curve.

Having demonstrated IETS' ability to probe microscopic bonding structures in the SiO_2/Si system, we decided to use IETS to study the effects of electrical stress on the IETS spectrum. Figure 8 shows that the electrical stress caused significant changes in the IETS spectrum. Specifically, the feature between 130 and 170 mV, which is associated with the bonding structure of the silicon oxide barrier, changes gradually with increasing stress time, and new features between 200 and 320 mV are created as the electrical stress progresses. The changes between 130 and 170 mV are consistent with the reordering of the microscopic bonding configurations in the silicon oxide barrier due to hot-electron induced structural damage, while the newly created features between 200 and 320 mV are associated with electron traps generated by the electric stress, which will be discussed in more detail in Section IV.4.

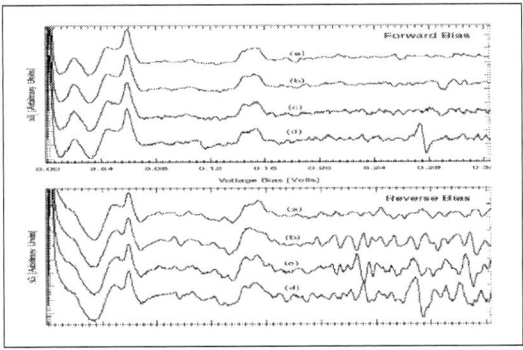

Fig. 8. IETS spectra of electrically stressed sample. Curve (a): before stress; (b) after 100 sec. of stress; (c) after 200 sec. of stress, and (d) after 400 sec. of stress. The stress voltage was –4V with a current density of approximately $7A/cm^2$.

IV.2 High-k/Si stacks

HfO_2 is the high-k gate dielectric of choice for current CMOS technology, which has also received most attention in this study among all high-k gate dielectrics.

Figure 9 shows the IETS spectra of an Al/HfO_2/Si sample, measured at both voltage polarities.   One can see that the reverse-biased (gate negative) spectrum contains many Si-O related features, suggesting a high density of Si-O bonds near the Si/HfO_2 interface, while the forward-biased (gate positive) spectrum shows primarily Hf-O features near the electrode-HfO_2 interface.

It has been shown that the soft optical phonons in the high-k gate dielectric cause degradation of the carrier mobility in MOSFETs made of high-k gate dielectrics [12] This "remote phonon scattering" mechanism is significant in high-k gated MOSFETs because the optical phonon energies in high-k dielectrics are low enough for efficient electron (hole)-phonon interactions, while the corresponding phonons in lower-k gate dielectrics have too high energies for such interactions.

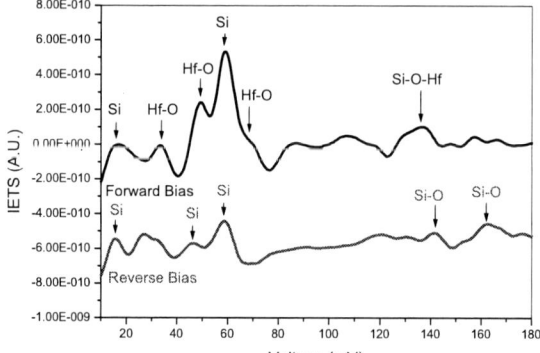

Fig. 9.   Reverse-bias (gate negative) IETS spectrum reveals information near Si/HfO₂ interface, while forward-bias (gate positive) IETS reveals information near electrode/HfO₂ interface.

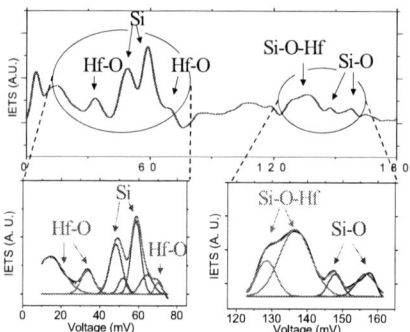

Fig. 10.   IETS spectrum taken on a Al/HfO₂/Si sample reveals Hf-O soft phonon modes, which are believed to contribute to degraded channel mobility in MOSFETs with HfO₂ as gate dielectric.

We believe IETS is ideally suited for probing the aforementioned mechanism, as electron-phonon interactions will appear explicitly as peaks in the IETS spectrum. Figure 10 shows an example of the IETS spectra taken on an Al/HfO₂/Si sample, where one can see clearly the Hf-O vibrational modes around 15, 35, and 70 meV. Note that these modes are very close to the substrate Si phonon energies, and are much lower than Si-O vibrational mode energies. Since the Si phonons are efficient scatters for the channel electrons, one would expect the Hf-O modes to be efficient scatters as well, on account of their very similar energies. In contrast, the energies of the Si-O vibrational modes are too high to cause any significant scattering of the channel carriers.

Figures 11-13 show the effects of various post-deposition processing details on the IETS spectrum of a HfO$_2$/Si stack, where Fig. 11 highlights the effect of post-deposition annealing temperature, Fig. 12 reveals the effect of water-vapor anneal, and Fig.13 compares the difference between furnace anneal and rapid thermal anneal (RTA). One major difference between furnace anneal and RTA in Fig. 13 is the numerous additional features appearing in the RTA curve, which can be attributed to traps, which will be discussed in Section IV.4.

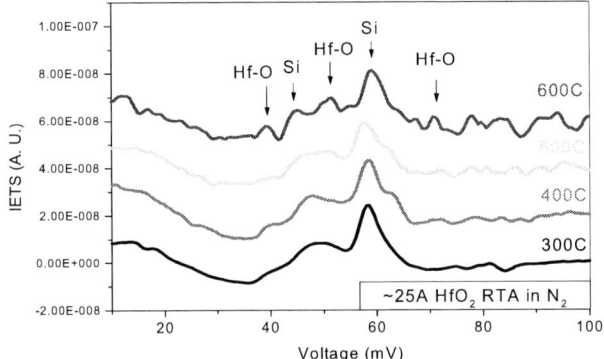

Fig. 11. IETS spectrum of HfO$_2$/Si stack as affected by the post-deposition annealing temperature

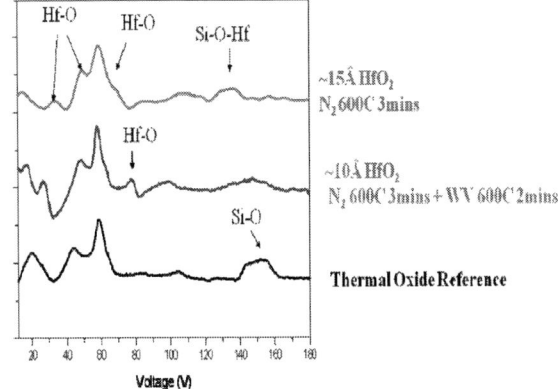

Fig. 12. Example of IETS spectrum of HfO$_2$/Si stack with or without water vapor (WV) anneal at 600C.

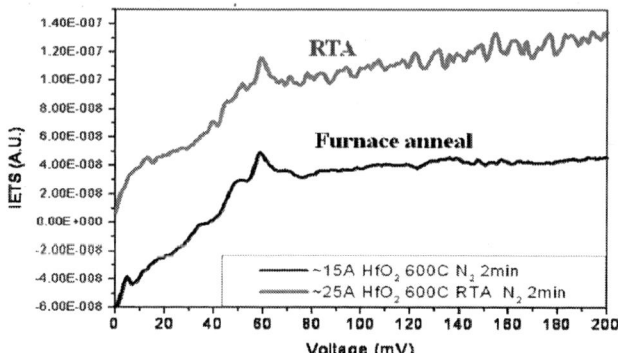

Fig. 13. IETS spectrum of a HfO$_2$/Si stack after furnace anneal (lower curve) and rapid thermal anneal (RTA, upper curve) in N$_2$ at 600C.

In addition to HfO$_2$, we have also studied other high-k gate dielectrics, of which LaAlO$_3$ is of particular interest, because of its possibility of epitaxial growth on Si as well as the formation of LaAlO$_3$ directly on Si without a SiO$_2$-like interfacial layer [13,14].

Figures 14 (a) and (b) show the IETS spectra of an Al/LaAlO/Si measured under positive gate voltage (a) and negative gate voltage (b).

The absence of Si-O bonds at the LaAlO$_3$/Si interface is evident in Fig.15, where the IETS spectra for a crystalline LaAlO$_3$ epitaxially grown on Si were depicted. One can see that no detectable features appear in the range where the vibrational energies for Si-O bonds are distributed.

IV.3 High-k/GaAs stacks

It is widely believed that III-V semiconductors with high electron mobility, such as GaAs, InGaAs, and InAs are likely to succeed Si in a decade or so, once the conventional CMOS scaling trends cannot be continued [15].

Figure 16 shows the IETS spectra of an Al/(TiO$_2$-Al$_2$O$_3$)/p-GaAs capacitor. The upper spectrum was collected with the Al gate positively biased; whereas the bottom one was negatively biased. It can be seen in Fig. 17 that the bottom spectrum in the 0-50 mV range can be fitted with phonon modes associated with GaAs [16], as well as trivalent gallium oxide (Ga$_2$O$_3$) [17]. No signal associated with gallium sub-oxides or with arsenic oxide is detected in this IETS spectrum.

Figure 18 shows a close-up view of the same spectrum in the range of 37 mV and 99 mV. Again, this portion of the spectrum can be fitted by phonon vibration modes associated with stoichiometric Al$_2$O$_3$ [18,19] and TiO$_2$ [20] in the gate dielectric stack, which are consistent with our XPS data.

ECS Transactions, 35 (4) 545-561 (2011)

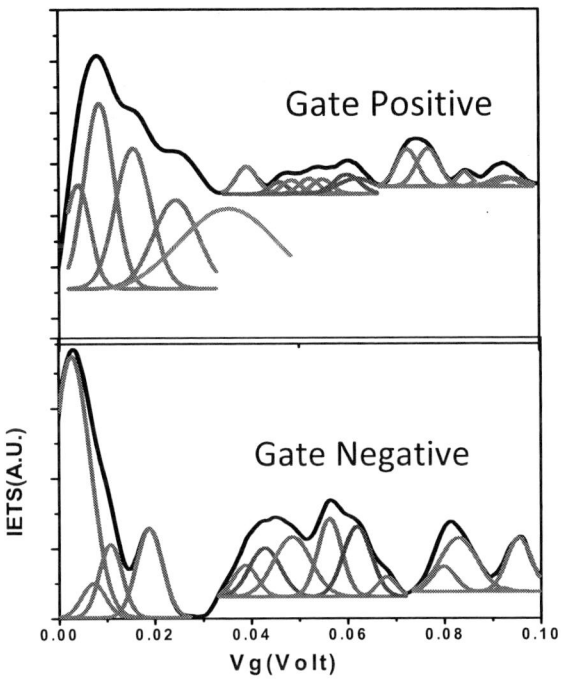

Fig. 14. IETS spectra of a Al/LaAlO₃/Si structure: (1) Gate positively biased; (2) Gate negatively biased. Deconvolution of the IETS spectrum reveals the various vibrational modes of the LaAlO dielectric.

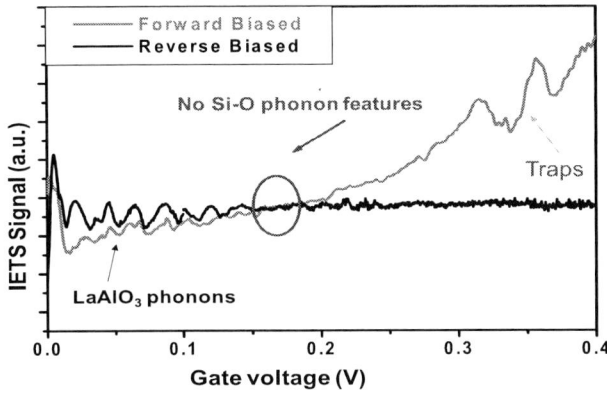

Fig. 15. IETS spectra of LaAlO₃ epitaxially grown on Si, showing no sign of SiO₂ formation at the LaAlO₃/Si interface.

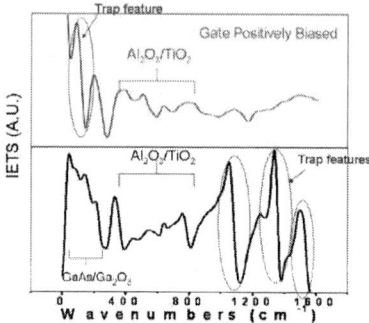

Fig. 16. IETS spectra obtained on an Al/TiO₂-Al₂O₃/GaAs capacitor. Upper curve: gate positively biased; Lower curve: gate negatively biased

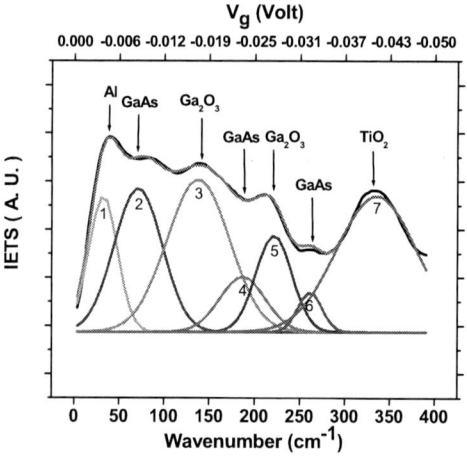

Fig. 17. The bottom curve in Fig. 16 in the voltage range below 50mV can be fitted to Gaussian distributions corresponding to GaAs and Ga_2O_3 phonons

IV.4 Traps

It is well known that traps in the gate dielectric can cause at least two different kinds of adverse effects in a MOSFET: (1) trapping of carriers, which causes shifts in the threshold voltage and reduced carrier density, and (2) trap-assisted conduction, which causes increased gate leakage current. This section will shed some lights on the nature of these two kinds of traps, which are not available from conventional measurements. Particular attention is paid to the electrical stress-induced traps.

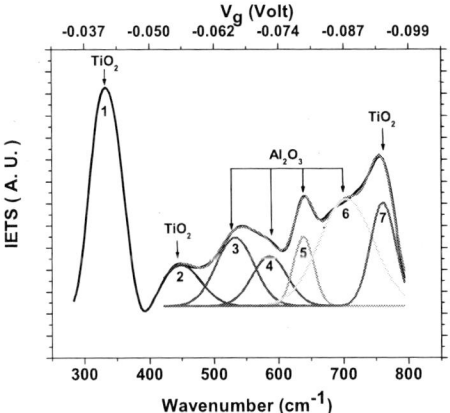

Fig. 18. The bottom curve in Fig.21 in the range between 37 mV and 99 mV is fitted by Gaussian line shapes associated with TiO_2 (TO: #1; LO: #2, #7) and Al_2O_3 (TO: #3, #4, #5, #6) phonons.

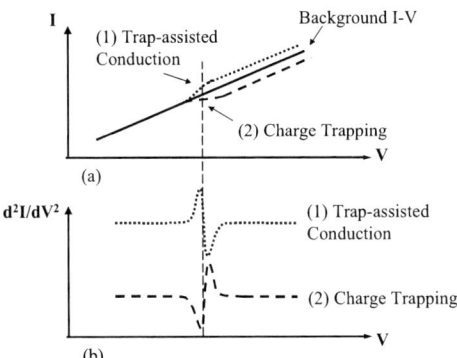

Fig. 19. Illustration of two kinds of trap effects on (a) I-V, and (b) d^2I/dV^2 –V characteristics. In each case, Curve (1) represents trap-assisted conduction, and Curve (2) represents carrier trapping.

Figure 19 illustrates schematically the aforementioned trap effects on the I-V and d^2I/dV^2-V plots. Curve (1) in Figure 19(a) indicates that the I-V curve will show an increase in slope over a small range of voltage when trap-assisted conduction mechanism takes place, where the voltage at which this occurs corresponds to the trap energy (in eV) above the Fermi-level in equilibrium, and the width of this region corresponds to the energy spread of this particular trap, including thermal spread. Curve (2) in Figure 19 (a) shows that charge trapping effect causes a small horizontal plateau, where the width of the plateau is proportional to the trapped charge ($\Delta V = qN_t/C_{ox}$, where q is the electron charge, N_t is the areal trap density, and C_{ox} is the gate dielectric capacitance). Note that this effect is analogous to the well-known C-V shift due to charge trapping. Figure 19 (b) shows the corresponding second derivative of the I-V curve; specifically, the second derivative will show a peak-followed-by-a-valley for the trap-assisted conduction mechanism (Curve 1), while a valley-followed-by-a-peak for the charge trapping mechanism (Curve 2). In contrast, the IETS features due to phonons or other vibrational modes manifest themselves only as peaks, with no valleys.

Figure 20 shows an example of trap generation in gate dielectrics due to electrical stress, where IETS spectra of an Al/HfO$_2$/Si structure were depicted, in which curve (a) was taken before the stress, and the most prominent peak at 58 mV corresponds to Si phonon vibration; other features are relatively weak, but if sufficiently amplified we could identify Hf-O vibration at about 80 mV, Hf-O-Si and Si-O vibration modes from 130mV to 160mV, and possible hydrogen-bonds related vibration modes at about 110mV and 255mV. Curve (b) in Fig.20 was taken after constant-voltage stress at 1.2V for 200 seconds. Many additional features appear between 130 and 300 mV, indicating electrical stress-induced build-up of defects. Curve (c) in Fig.20 was taken after constant-voltage stress for 800 seconds, and one can see a dramatic increase in magnitude of the features between 160 and 200 mV, and between 250 mV and 290 mV. It seems that both kinds of trap effects as depicted in Fig.19 are found in Curve (c), where the feature marked (1) may be attributed to trap-assisted conduction, the feature marked (2) may be associated to charge trapping, while that marked (3) may also be associated with charge trapping. Similarly, the features between 250 mV and 290 mV also appear to contain both kinds of traps.

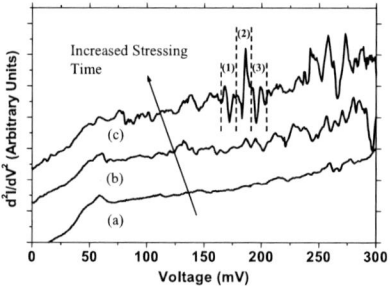

Fig. 20. IETS Spectra for HfO$_2$-gated MOS capacitor (a) before stress, (b) after 200 seconds of constant-voltage stress at 1.2 V, and (c) after 800 seconds of constant-voltage stress at 1.2 V. Three trap-related features are highlighted on Curve (c).

Figure 21 shows the IETS spectrum of an Al/HfO$_2$/Si structure where prominent trap features are revealed in both the forward and reversed-bias spectrum. From a spectrum of this kind (i.e., extending both voltage polarities), the energy level and the physical location of the corresponding trap in the gate dielectric can be estimated, as described below. It should be noted that the fact that the feature in the positive-bias region is much stronger than its negative-bias counterpart is due to the asymmetry of the tunnel barrier, which gives rise to a much higher tunneling probability in the positive-bias region.

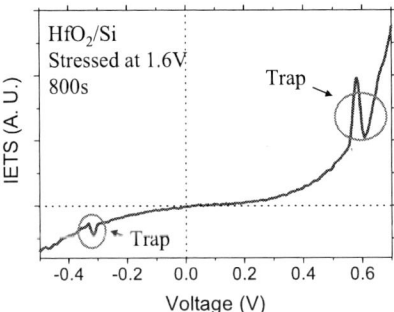

Fig. 21. An IETS spectrum (2nd derivative of the tunneling I-V characteristic) of an Metal-Oxide-Silicon structure with HfO_2 as the dielectric. Two strong trap related features appear at ~580mV in the forward-bias range and ~320mV in the reverse-bias region.

Figure 22 (a) shows the schematic of a tunnel dielectric sandwiched between two electrodes (made of a metal or a semiconductor) with zero bias voltage, where the Fermi energies of electrodes 1 and 2 are at the same level. The total physical thickness of the dielectric is x_0. Assume a non-uniform dielectric constant, $\varepsilon = \varepsilon(x)$, and a trap is located at x_t from the electrode-1 interface, with an energy level at eV_t, defined as the energy above the Fermi levels of the electrodes at zero bias. An applied voltage bias will lower the trap energy with respect to the negatively biased electrode. As the applied voltage reaches V_f, the Fermi level of the negative-biased electrode (electrode-2) reaches the energy level of the trap, as shown in Figure 22 (b), and trap-assisted tunneling will start to take place. Considering the electric field across the dielectric, V_f can be expressed as:

$$V_f = \int_0^{x_0} dx\, D_f / \varepsilon(x)$$

(4)

where D_f is the electric displacement, defines as the electric field E_f times the dielectric constant ε. On the other hand, V_f can be written as:

$$V_f = V_t + \int_0^{x_t} dx\, D_f / \varepsilon(x)$$

(5)

Similarly, for a reverse bias (with electrode-2 positive) of V_r, the Fermi level of the negatively biased electrode (electrod-1) reaches the energy level of the trap, as shown in Figure 22(c), and one can write:

$$V_r = \int_0^{x_0} dx\, D_r / \varepsilon(x)$$

(6)

and

$$V_r = V_t + \int_{x_t}^{x_0} dx\, D_r / \varepsilon(x)$$

(7)

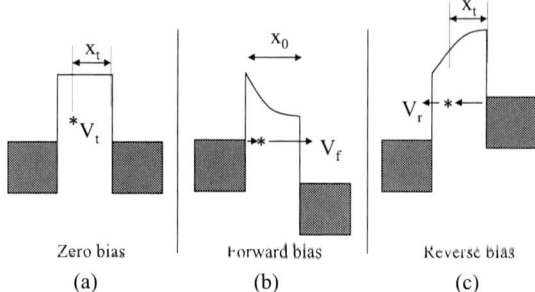

Zero bias   Forward bias   Reverse bias

(a)   (b)   (c)

Fig. 22. Schematic diagrams of a tunnel junction (a) with zero bias voltage, (b) with a forward bias when the trap energy level is reached by the Fermi level of an electrode, and (c) with a reverse bias when the trap energy level is reached by the Fermi level of another electrode.

Solving Eqs. (4) – (7) by cancelling out D_f and D_r, one has the trap energy level expressed as:

$$V_t = V_f V_r / (V_f + V_r)$$

(8)

and the physical location of the trap expressed as:

$$d_t = d_0 V_f / (V_f + V_r)$$

(9)

where

$$d_0 = \int_0^{x_0} dx / \varepsilon(x)$$

is the electrical effective thickness of the dielectric, and

$$d_t = \int_0^{x_t} dx / \varepsilon(x)$$

is the electrical effective distance. Equations (8) and (9) imply that, by recording trap-assisted tunneling features in both bias polarities of the IETS spectrum, one can calculate the energy level and physical location of the trap that causes the related IETS feature.

Using the equations shown above, we estimate the trap energy level associated with the features in Fig. 21 to be about 0.2 eV above the Fermi level at zero bias. The physical location of the traps is estimated to be ~1nm from the silicon substrate, which is in close proximity to the interface between the HfO$_2$ and the interfacial oxide near the silicon substrate.

Acknowledgements

The author would like to thank many of his former and current students whose work made this review article possible, especially Drs. W. K. Lye, Wei He, Miaomiao Wang, and Mr. Zuoguang Liu. He also gratefully acknowledges support from the Semiconductor Research Corporation over a period of 5 years, and partial support from the National Science Foundation under Contract No. MRSEC DMR 0520495.

References

1. W. K. Lye, *Ph.D Thesis*, Yale University, New Haven, CT (1998).
2. W. K. Lye, E. Hasegawa, T. P. Ma, R. C. Barker, Y. Hu, J. Kuehne, and D. Frystak, *Appl. Phys. Lett.*, **71**, 2523 (1997).
3. P. Balk, S. Ewert, S. Schmitz and, A. Steffen, *J. Appl. Phys.*, **69**, 6510 (1991).
4. G. Salace, C. Petit and, D. Vuillaume, *J. Appl. Phys.*, **91**, 5896 (2002).
5. R. C. Jaklevic and J. Lambe, *Phys. Rev. Lett.*, **17**, 1139 (1966).
6. P. K. Hansma, *Tunneling spectroscopy: capabilities, applications, and new techniques*, Plenum Press, New York 1982.
7. W. He and T. P. Ma, *Appl. Phys. Lett.*, **83**, 2605 (2003).
8. W. He and T. P. Ma, *Appl. Phys. Lett.*, **83**, 5461 (2003).
9. M. M. Wang, W. He and, T. P. Ma, *Appl. Phys. Lett.*, **86**, 192113 (2005).
10. M. Wang, W. He, T.P. Ma, L.F. Edge, and D.G. Schlom, *Appl. Phys. Lett.*, **90**, 053502 (2007).
11. *SR830 DSP Lock-in Amplifier Operating Manual and Programming Reference, Stanford Research Systems*, 1993.
12. M. V. Fischetti, D. A. Neumayer, and E. A. Cartier, *J. Appl. Phys.*, **90**, 4587 (2001).
13. J. W. Reiner, A. Posadas, M. Wang, M. Sidorov, Z. Krivokapic, F. J. Walker, T. P. Ma, and C. H. Ahn, *J. Appl. Phys.*, **105**, 124501 (2009).
14. L. F. Edge, D. G. Schlom, R. T. Brewer, Y. J. Chabal, J. R. Williams, S. A. Chambers, C. Hinkle, G. Lucovsky, Y. Yang, S. Stemmer, M. Copel, B. Hollander, and J. Schubert, *Appl. Phys. Lett.*, **84**, 4629 (2004).
15. 2009 ITRS, published by the Semiconductor Industry Association.
16. I. M. Tiginyanu, G. Irmer, J. Monecke, A. Vogt, and H. L. Hartnagel, *Semicond. Sci. Technol.*, **12**, 491 (1997).
17. Yanyan Zhao, Jing Yang, and Ray L. Frost, *J. Raman Spectroscopy*, **39**, 1327 (2008).
18. B. G. Frederick, G. Apai, and T.N. Rhodin, *Phys. Rev. B*, 44, (1991).
19. C. Pecharroman, T. González-Carreño, and Juan E. Iglesias, *J. Mater. Res.*, **11**, 1 (1996).
20. G. Rocker and J. A. Schaefer, *Phys. Rev. B*, **30**, 370 (1984).

562

ECS Transactions, 35 (4) 563-580 (2011)
10.1149/1.3572305 ©The Electrochemical Society

High-*k* Gate Dielectric MOSFETs: Meeting the Challenges of Characterization and Modeling

M. M. De Souza, S. B. F. Sicre, and D. Casterman

Department of Electrical and Electronic Engineering,
University of Sheffield, Sheffield S1 3JD, UK

Invited Paper

Although high-*k*/metal gate MOSFETs have made it into manufacturing at the 45 nm CMOS technology node there still remain many challenges associated with their implementation under debate ranging from unknowns in stack parameters to frequency dispersion during measurement. We elucidate methodologies to address frequency dispersion in depletion and inversion via the conductance technique using typical test case of a poly/TiN and polysilicon gate. Analytic expressions for the extraction of transistor parameters are compared with those from Schrodinger-Poisson simulations. The poly/TiN gate has a significantly reduced EOT and a similar mobility to the polysilicon gate. Using insight from Quantum Mechanical calculations, it is demonstrated that the quantity and location of fixed oxide charge plays a decisive role in the mobility degradation.

Introduction

Current understanding of metal gate/high-*k* dielectric stacks has come a long way since the 90's when the need for a higher dielectric constant material to replace conventional silicon dioxide in mainstream CMOS was first envisaged [1]. To date, at least one technology node has been successfully implemented in production with high-*k* dielectric by several (though not all) semiconductor manufacturers. There are however, gaps in our knowledge of high-*k* gate stacks, related to process control, which if tackled would help design better and more reliable consumer electronic products. One such issue is the control of interactions between the various distinctly disparate layers of the stack when subject to high processing temperatures under ambient gases: silicon, Interfacial Layer (IL), high-*k*, metal and polysilicon [2]. These generate fixed oxide charge at the interface boundaries, which in particular cause a higher than proportional threshold voltage shift (or "V_T roll-off") at scaled Equivalent Oxide Thicknesses (EOTs) [3] especially in the p-MOSFET. Their quantity via trapping/de-trapping can vary depending upon the process flow. The identification of fixed oxide as well as interface states is often inaccessible to conventional electrical characterization techniques either due to the interfaces being buried, the presence of high leakage current or an absence of detectable signal in measured data using conventional methods such as the conductance technique [4]. Moreover aggressively scaled oxides, particularly in alternate materials, often show strong frequency dispersion of electrical characteristics [5] which, given their uncertainties, makes the values of dependent transistor parameters such as the flatband voltage, Equivalent Oxide Thickness or Inversion layer charge debatable.

563

ECS Transactions, 35 (4) 563-580 (2011)

We describe the methodology to address frequency dispersion in ultra-thin high-k gate stacks [6]. In combination with our quantum mechanical coulomb scattering model [7], which allows an arbitrary number of gate stack layers of varying dielectric constant, we demonstrate "closing of" the gap between experimental and predicted mobility. The Coulomb charge required to explain mobility degradation is significantly less than that in [8]. This work demonstrates the impact of a poly/TiN gate on charge traps across the gate stack in comparison with a polysilicon gate under an identical fabrication process. Our approach is generic to materials beyond silicon and will become increasingly indispensable as industry continues its relentless pursuit of scaling.

Experimental Samples

MOS capacitors on p-type silicon substrate with a doping density ~3.5×10^{17} cm^{-3} are used in this work. The gate stacks consist of 1 nm of SiO_2 by thermal oxidation and 3 nm HfO_2 deposited via ALD followed by anneal at 600 °C in N_2 ambient prior to electrode deposition. The two sets of samples have the same dielectric processing and differ only with respect to the type of gate: (i) poly-Si/TiN metal and (ii) poly-Si. Both electrodes are deposited via CVD and activated via 10s RTA in Nitrogen ambient at 1000 °C. The capacitance and conductance characteristics are measured after open/short corrections to take into account parasitic impedances of cables, adaptors and probes using an Agilent E4980A LCR bridge in parallel mode at frequencies ranging from 1 kHz to 2 MHz and for large negative voltages to small positive voltages (-2.0 V to 0.2 V). The static conductance is extracted through the derivative of the gate current measured with a Keithley 4200 current meter. The experimental mobility is characterized using the split CV technique using MOSFETs of size 10x10 μ m^2 for measurement of the drain conductance.

Methodology of the Conductance Technique

The conductance technique used in conjunction with the appropriate equivalent circuit model is a powerful method for the detection of interface states in a MOS structure. The method is based on the response of interface states to a small signal superimposed on the gate, which causes the trap to capture and discharge an electron or a hole as a function of the applied bias and measurement frequency. The capture and emission of charge in an MOS structure causes an energy loss which is represented by an equivalent parallel combination of the conductance G_p and the capacitance C_p [9]. The decomposition of C_p and G_p into components of the generic MOS model in the substrate is shown in Figures 1a-1c.

The equivalent MOS model (see Figure 1c) can be differentiated into the corresponding equivalent models for the accumulation, depletion and inversion modes shown in Figures 1e to 1g, respectively. In the case of accumulation and depletion, the current flow during measurement is through the substrate, since the high of the instrument is connected to the gate and the low to the substrate, while the source and drain terminals are connected to ground. On the other hand, in inversion, the source/drain and substrate are connected together to the low of the instrument, whereas the high is connected to the gate. The supply of carriers from source and drain causes the resistance of the channel (particularly a low mobility channel) to be convolved with the interface states, resulting in a contribution of channel resistance to the measured signal. The corresponding equivalent circuit represented by a transmission line model (see Figure 1g) has been presented in [10, 12].

564

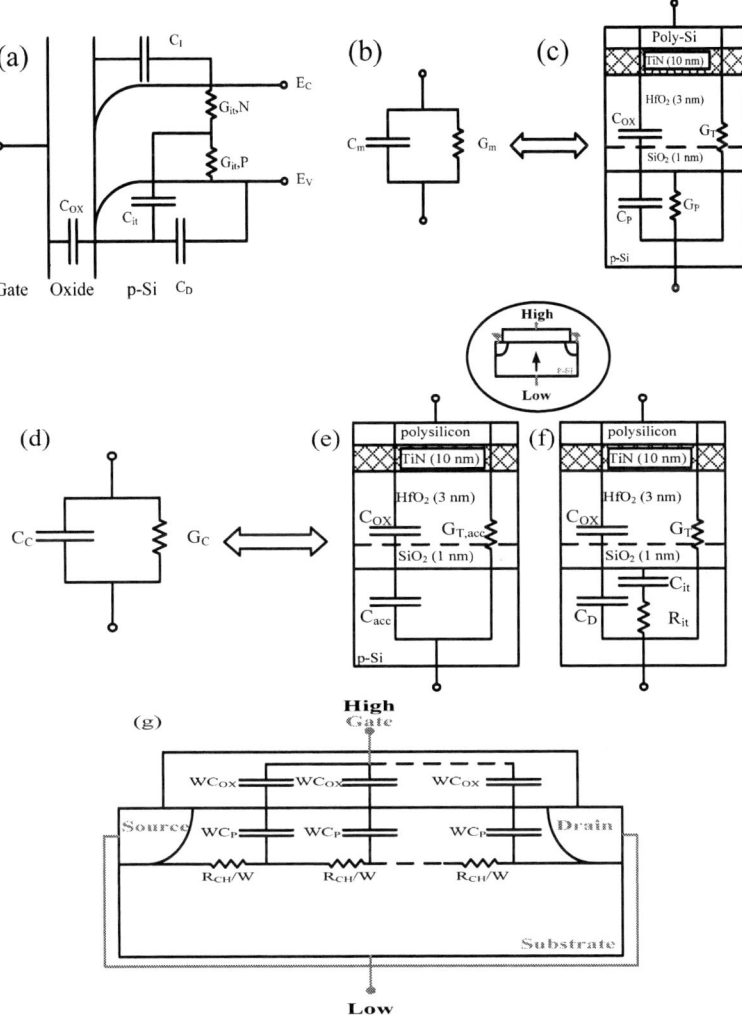

Figure 1. a) Equivalent circuit of the MOS structure in a p-Si substrate, indicating the presence of equivalent capacitances and conductances: C_I (inversion layer capacitance), C_{it} (interface state capacitance), C_D (depletion capacitance), C_{OX}, oxide capacitance. b) C_m and G_m represent the measured capacitance and conductance respectively, which can be modeled into a corresponding parallel C_p and G_p as shown in c). Figure c) includes G_T the tunneling conductance to account for leakage in ultra-thin oxides. Figure d) represents the corrected capacitance C_C and the corrected conductance G_C which can be further expanded for accumulation e), depletion f) and inversion g) modes respectively. In accumulation and depletion, the current path is through the substrate, whereas in inversion, the supply of carriers from source and drain results in a convolution of the resistance of the channel with that of the interface states.

From the equivalent circuit in depletion, G_P is given by [6],

$$G_P = \frac{G_m - G_T}{\left(\dfrac{G_m - G_T}{\omega \cdot C_{OX}}\right)^2 + \left(1 + \dfrac{C_m}{C_{OX}}\right)^2} \tag{1}$$

where C_m and G_m are measured capacitance and conductance respectively and ω is the frequency, G_T the tunneling conductance and C_{OX} the oxide capacitance. In inversion the expressions are [10]:

$$\frac{G_m}{\omega} = -\text{Im}[\ C'\frac{\tanh \lambda}{\lambda}] + (\frac{G_{it}}{C_I \omega})\text{Im}[\frac{C'\tanh \lambda}{\lambda}] + \frac{G_I}{\omega}\text{Im}[\frac{\tanh \lambda}{\lambda}][F/cm^2] \tag{2}$$

where $C' = [C_{ox}C_I]/[C_{ox} + C_I + G_{it}/j\omega]$, $\quad C_I = C_{inv} + C_{it}[F/cm^2]$, $\quad \lambda = \gamma L/2$,

$\gamma^2 = r_1[j\omega C' + \dfrac{C'G_{it}}{C_I} + G_T]$, and $r_1 = \dfrac{W}{L}g_{ds}$.

In the conductance-frequency method the conductance G_P is extracted as a function of the frequency at fixed gate bias as G_P attains its maximum at a frequency corresponding to the time constant of the interface state. The density of interface states is extracted by fitting the peaks of $G_P/\omega(\omega)$ using an appropriate model. Such a model may either be related to the presence of a single time constant, a continuum of states or the influence of the surface potential fluctuations through a statistical model [9]. In the case of a single time constant, the equivalent parallel conductance G_P/ω is equal to:

$$\frac{G_P}{\omega} = \frac{q\omega \cdot D_{it} \cdot \tau_{it}}{1 + \omega^2 \cdot \tau_{it}^2} \tag{3}$$

In the case of a continuum of states G_P/ω can be expressed as [9]:

$$\frac{G_P}{\omega} = \frac{q}{2}\frac{D_{it}}{\omega \cdot \tau_{it}} \cdot \ln\left(1 + (\omega \cdot \tau_{it})^2\right) \tag{4}$$

Finally, the surface potential fluctuations usually broaden the G_P/ω peaks and can be taken into account via a Gaussian distribution [9] by:

$$\frac{G_P}{\omega} = \frac{q}{2}\int_{-\infty}^{+\infty} \frac{D_{it}}{\omega \cdot \tau_{it}} \cdot \ln\left(1 + (\omega \cdot \tau_{it})^2\right) \cdot \frac{1}{\sqrt{2\pi\sigma^2}}\exp\left(-\frac{\Psi_S - \overline{\Psi_S}}{2\sigma^2}\right) \cdot d\Psi_S \tag{5}$$

where σ is the standard deviation of the surface potential fluctuations, q is the electron charge, Ψ_S is the surface potential and $\overline{\Psi}_S$ is the mean surface potential depending on the gate voltage. The response of the above distributions of observed G_P/ω curves is shown in Figure 2.

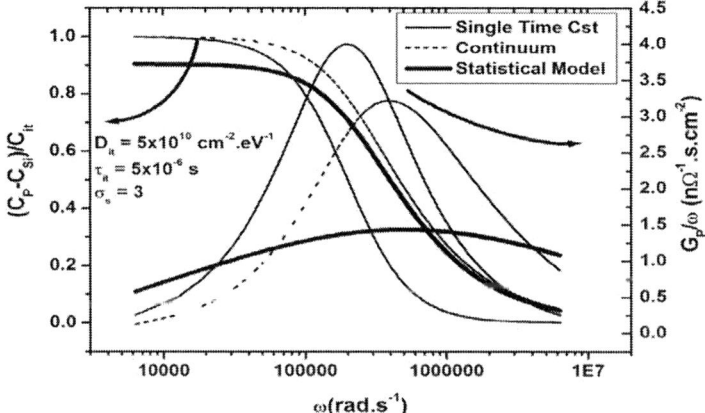

Figure 2. Illustration of types of interface states response observed through equivalent parallel conductance and capacitance measurements.

The position of interface states in the band gap can be determined if the surface potential is known via the Berglund integral [11]:

$$\psi_S = \int_{V_{FB}}^{V} 1 - \frac{C_c}{C_{ox}}.dV_G \qquad (6)$$

where ψ_S is the surface potential and V_{FB} the flat band voltage. However, in high-k metal gates and alternate materials, intrinsic device parameters may not be easily derived. Hence it is worth reviewing the performance of analytic techniques versus quantum mechanical Schrodinger-Poisson simulations in the assessment of their accuracy. For SP simulations, we use an in-house solver SP-SCATTER [12]. The performance and accuracy of SP-SCATTER has been benchmarked with other well-known solvers SCHRED [13] and UTQuant [14]. SP-SCATTER allows for more than two layers of gate stack material, alternate semiconductors and orientations, wave penetration in the calculation of the oxide capacitance, non-parabolicity, exchange correlation, image potential, interface states and fixed oxide charge.

Extraction of Transistor Parameters

Oxide capacitance and Equivalent Oxide Thickness (EOT)

The physical thicknesses of the gate stack layers in the poly and poly/TiN stacks are identical before gate processing and should therefore theoretically yield identical EOT of 2.06 nm for both, assuming $k \sim 22$ for HfO$_2$. Nevertheless, it is apparent from the low frequency Capacitance-Voltage characteristics in Figure 3 that the metal gate has a significantly lower EOT (by 0.5 nm) than the polysilicon gate. The cause of this reduction cannot be ascertained electrically but physical characterization such as TEM can provide evidence of the layer structures post processing and in this case can occur due to a transition region at the gate/high-k interface. The k value of the IL required to fit the experimental CV is 5.7, assuming the k value of the HfO$_2$ layer is 22. The thicknesses of the two layers in the simulations are assumed as deposited.

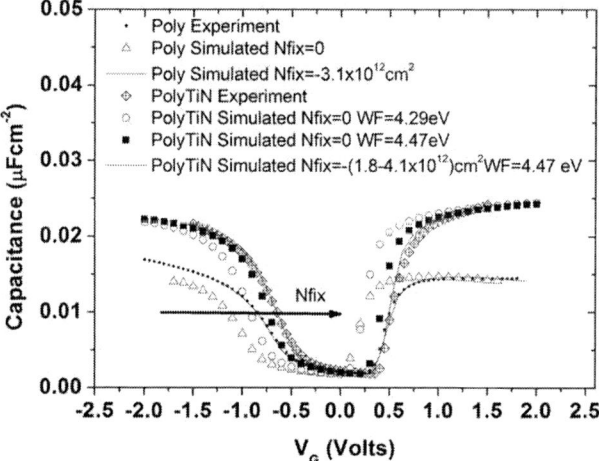

Figure 3. Fit achieved between quantum-mechanical simulations using SP-SCATTER and experimental Capacitance-Voltage characteristics for polysilicon and poly-TiN gates indicating the dependence of fixed oxide charge and work function of the metal gate.

Analytical methods for the extraction of the oxide capacitance have been proposed by Kar [15] and Islam [16]. In [15] an exponential behavior of the silicon capacitance in accumulation is assumed, corresponding to Boltzmann statistics expressed as:

$$C_{Si} = \alpha \exp(\beta.\psi_S) \tag{7}$$

where C_{Si} is the silicon capacitance in accumulation, α and β are constants. From Equations 6 and 7, one can express the total corrected capacitance C_C as a function of the oxide capacitance C_{OX} as,

$$\left| \frac{dC_C^{-2}}{dV_G} \right|^{1/2} = \left(2|\beta| \right)^{1/2} \cdot \left(\frac{1}{C_C} - \frac{1}{C_{OX}} \right) \tag{8}$$

The oxide capacitance is then obtained from the linear intercept of $\left(dC_C^{-2} / dV_G \right)^{1/2}$ versus $1/C_C$. On the other hand, in [16] it is assumed that quantum mechanical effects are better accounted for by assuming that part of the silicon capacitance behaves linearly in strong accumulation as:

$$C_{Si} = \alpha + \beta.\psi_S \tag{9}$$

This linear behavior in strong accumulation has been developed via SP simulations which includes wave penetration in the oxide [16]. Consequently Equations 6 and 9 yield,

$$\left|\frac{dC_C}{dV_G}\right|^{1/3} = \beta^{1/3}\left(1-\frac{C_C}{C_{OX}}\right) \tag{10}$$

In this case, the oxide capacitance is obtained from the linear intercept of $\left(\left|dC_C\right|/\left|dV_G\right|\right)^{1/3}$ versus C_C.

The Kar and Islam techniques give a significant discrepancy of 0.5-0.6 nm for the two gate stacks examined herein as highlighted in Figure 4 and Table I. The discrepancy between the two analytic techniques confirms the underestimation of quantum mechanical effects in Eq. 8 which assumes Boltzmann statistics as previously highlighted by Islam [16]. Assuming a linear behavior of silicon capacitance in strong accumulation gives results closest to those of quantum-mechanical simulations, as expected for metal gates.

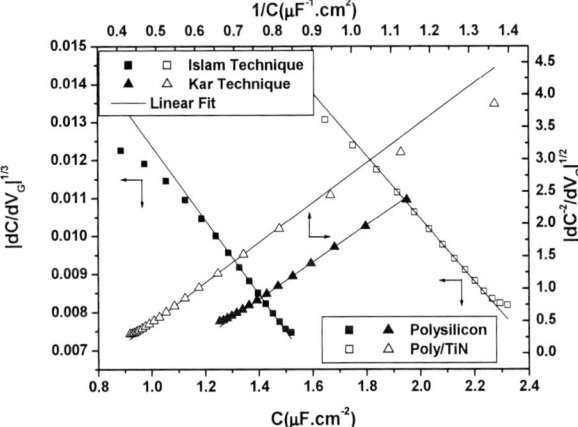

Figure 4. Extraction of the EOT using analytical methods. The EOT is extracted from the linear fit of the expressions (3) and (5).The open symbols are for the poly/TiN gate and the close symbols are for the polysilicon gate.

Estimation of the Fixed Oxide Charge

The density of fixed oxide charge helps determine the flatband voltage of the transistor. The density is extracted assuming the knowledge of the gate work function. In the case of the polysilicon gate, the work function ϕ_m is equal to 4.18 eV for a gate doping of $\sim 1 \times 10^{20}$ cm^{-3}, and the work function of the silicon ϕ_S is equal to 5.17 eV for a substrate doping of 3.5×10^{17} cm^{-3}. Using these parameters the fixed oxide charge density, N$_{FIX}$, is adjusted in the SP solver to match the CV characteristics as shown in

Figure 3. A total negative density of fixed charges is estimated to be around -3.1×10^{12} cm^{-2} in the case of the polysilicon gate.

In the case of TiN/HfO$_2$ interface, it has been shown that Fermi pinning effect is reduced by the introduction of a TiN layer [17] though it may not be completely eliminated or be as relevant for an n+ doped gate [18], since the underlying cause is proposed to be Hf-Si bonds which lie in the upper half of the silicon bandgap. The difficulty with estimation of fixed oxide charge in polyTiN gate arises from the uncertainty in the gate workfunction. Both parameters have a similar impact of shifting the CV characteristics, which requires a reliance on additional experimental techniques such as IPE to determine the work function [20]. Moreover, the workfunction depends upon the annealing temperature, fabrication process and the thickness of the TiN layer with the creation of a dipole at the HfO$_2$/SiO$_2$ interface [20]. In the present situation, we assume a similar level of fixed oxide charge as the polysilicon gate since both are subject to identical process conditions and derive the work function from a fit to the CV to be 4.47. This is consistent with reported values in the range 4.29 – 4.47 eV [19].

Flat Band Voltage And Surface Potential

Determination of the position of the interface states in the band gap via the Berglund integral also requires a knowledge of the flat band voltage V_{FB}, which is an equally important if not the most important parameter for technology development. V_{FB} is extracted based on the work by Maserjian [21] and further developed by [22-23]. In this technique the extraction of the flat band voltage is based on the properties of the following Y function:

$$Y = \frac{1}{C_C{}^3} \frac{\partial C_C}{\partial V_G} = \frac{1}{C_{Si}{}^3} \cdot \frac{\partial C_{Si}}{\partial \psi_S} \tag{11}$$

Expression 11 is obtained from Eq. 6 and the fact that the total capacitance C_C is a series combination of the oxide capacitance C_{OX} and the silicon capacitance C_P. Using the classical approximation of the silicon capacitance in depletion [23] one obtains:

$$\frac{1}{C_C^2} = \frac{1}{C_{OX}^2} + \frac{2}{q \varepsilon_{Si} N_A} (V_G - V_{FB}) \tag{12}$$

Hence the function Y is independent of the EOT (see Figure 5) and has a minimum in depletion given by:

$$Y_{MIN} = \frac{-1}{q \cdot \varepsilon_{si} \cdot N_A} \tag{13}$$

The doping density N_A, using Eq. 13, is found to be $\sim 3.5 \times 10^{17}$ cm^{-3} for both kinds of gate and is identical to that from SP simulations. More generally, it has been demonstrated that at flat band voltage [23], Y is equal to:

$$Y_{FB} = \frac{-1}{3 \cdot q \cdot \varepsilon_{Si}} \cdot \frac{\frac{\partial^2 p}{\partial E_f^{\,2}}}{\left(\frac{\partial p}{\partial E_f}\right)^2} \tag{14}$$

where p is the density of majority carriers and E_f is the Fermi energy. Assuming Boltzmann statistics at flat band yields:

$$\frac{\partial^2 p}{\partial E_f^{\,2}} = \left(\frac{p}{kT}\right)^2 \tag{15}$$

Consequently, Eq. 15 yields:

$$Y_{FB} = \frac{-1}{3 \cdot q \cdot \varepsilon_{Si} \cdot N_A} = \frac{Y_{MIN}}{3} \tag{16}$$

However, as demonstrated by [22], the inclusion of quantum mechanical effects via SP calculations increases the ratio Y_{MIN}/Y_{FB} slightly with increasing doping density thus shifting the extracted values of the flat band voltage. Therefore the flat band voltage can be extracted in two different ways: (i) SP calculations to match CV characteristics obtained in experiment and plotting the surface potential directly from the simulations (ii) the classical ratio as in Eq. 16 using data derived from an SP solver following the knowledge of the doping density from Eq. 13.

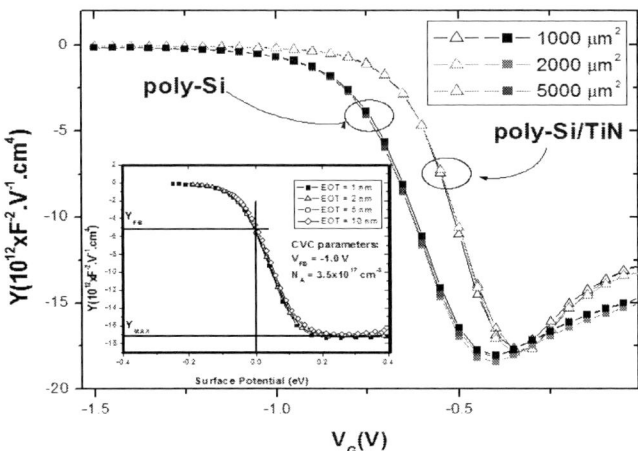

Figure 5. Extraction of the flat band voltage using Eq. 11. The inset shows the simulation of ratio Y_{MIN}/Y_{FB}.

The flat band voltage for the polysilicon gate is obtained as -0.73 V from SP-SCATTER corresponding to a work function difference, ϕ_{ms}, equal to -0.99 eV, whereas in the case of poly/TiN gate, the flat band voltage is equal to -0.62 V with a work function difference varying from -0.7 and -0.88 eV. These values correspond well to those reported in [24].

The flat band voltages differ significantly between the polysilicon and poly-TiN gates due to the different work functions. The difference in flatband voltage predicted by SP simulations and analytic methods is in the range of ~20 meV (see Table I). The extraction of the flat band voltage via SP simulations alone can be expected to be *less* accurate because of its dependence on the EOT, doping density and the work function of the gate whereas in the methodology with the function Y in Eq. 13, only the doping density needs to be estimated.

Finally it is noted that evaluating the Y_{MIN}/Y_{FB} ratio from quantum mechanical simulations does not represent a significant shift with the classical assumption made in Eq. 16 which remains valid for these doping densities. Consequently, for the purpose of a rapid evaluation of the V_{FB}, the use of Eq. 16 which assumes Boltzmann statistics remains reasonable for metal/high-k gate stacks in silicon.

TABLE I. Comparison of the Cox, EOT, N_A and V_{FB} extracted for the poly-Si and polySi/TiN gates.

Method	Parameter	Polysilicon	Poly-Si/TiN
SP-SCATTER	$Cox(\mu F.cm^{-2})$	2.16	2.87
Kar [15]		1.77	2.27
Islam [16]		2.61	3.26
SP-SCATTER	EOT (Angstrom)	15.5	10.5
Kar [15]		19.45	15.21
Islam [16]		13.19	10.56
SP-SCATTER	$N_A(x10^{17}cm^{-3})$	3.5	3.5
Leroux [22]		3.5	3.5
SP_SCATTER	V_{FB}	-0.73	-0.62
Leroux [22]		-0.71	-0.60
SP-SCATTER	Nfix $(x10^{12} cm^{-2})$	-3.0	-1.5-4.6
SP-SCATTER	Workfunction (eV)	4.18	4.29-4.6

The Conductance Technique in the Presence of Frequency Dispersion Arising from Parasitic Effects

In practice, significant frequency dispersion in measured conductance data of ultra-thin high-k dielectrics necessitates further modifications of the standard circuit models in Fig 1 to make them of any practical use. Because the capacitance and conductance measurements converge at the lowest frequency, intrinsic parameters may be extracted at a suitable low frequency in strong accumulation on small area devices using the methods explained above. However, small area devices may not be ideal for conductance measurements which necessitate a resolution of the frequency dispersion.

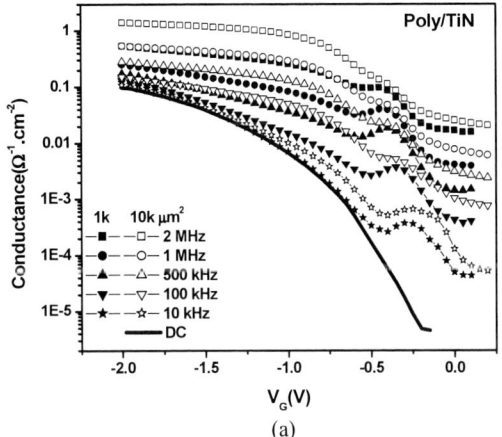

(a)

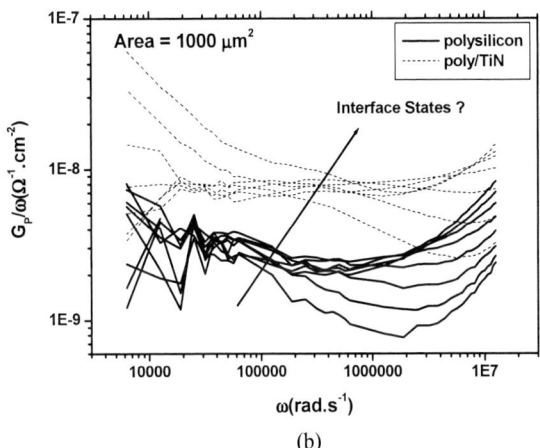

(b)

Figure 6. (a) Conductance-Voltage characteristics for poly/TiN gate. Open (close) symbols represent measurement on a 10 kμm² (1k) gate area. Lower frequency dispersion occurs with smaller area. (b) $G_P / \omega(\omega)$ appears featureless in depletion even on the smallest area for both polysilicon and poly/TiN gates.

Figure 6 shows frequency dispersion, increasing with gate area. This is a signature that confirms parasitic effects as being the cause of dispersion. Despite peaks corresponding to interface states observed in depletion in the GV characteristics ($V_G >$ $V_{Flat-Band} = 0.61$ V) and moving towards flat band voltage with increasing frequencies, a significant presence of parasitic elements affects their correct response to the small ac signal. Furthermore $G_P / \omega(\omega)$ characteristics appear featureless and suggest that a correction for such parasitic effects is required to unveil the proper response of interface states. The frequency dispersion can arise from the resistance of silicon

[25], the resistance and inductance of cables and contacts [26], the capacitance and resistance of the probe chuck [27], as well as the quality of the gate/oxide interface [28]. However, none of the models reported to date can be used to fit the variation of capacitance over a wide range of frequency. In the case of poly/TiN, we have demonstrated a 6-element model (Figure 7a) for accumulation, arising from parasitic effects, which can satisfy the experimental data for capacitance and conductance with a single set of parameters, independent of area and frequency (Figure 7b) [6]. The model consists of the corrected capacitance C_C and corrected conductance G_C, in series with impedances Z_1 and Z_2, which have differing underlying origins. Z_1 for example may be related to the gate because of its larger resistive effects whereas Z_2 has a negligible influence on the frequency dispersion in both kinds of gates. Z_1 and Z_2 are in series with inductive elements L and R respectively as shown.

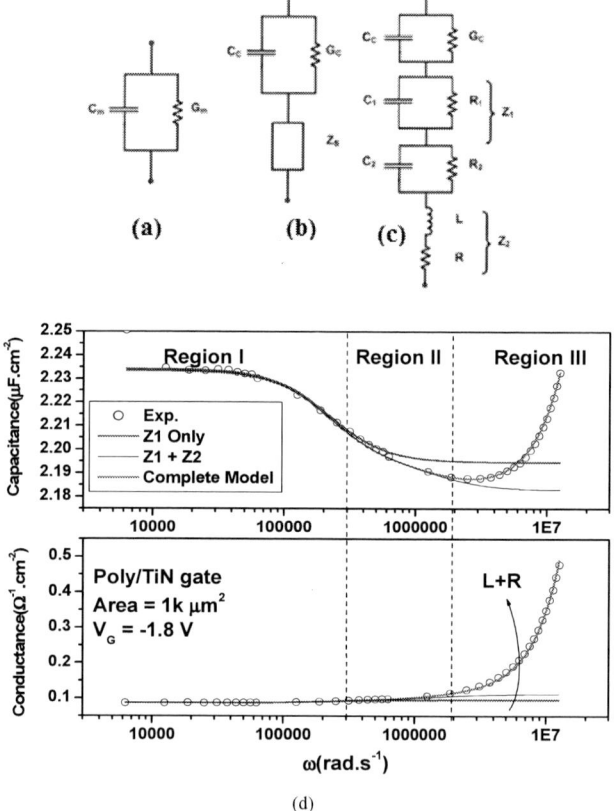

Figure 7. A six element model (a-c) required for resolution of frequency dispersion of parasitic effects in poly and poly/TiN gates over the entire frequency range as shown in (d). The model is resolved via matching in various regions of frequency [6].

The extraction process is briefly highlighted here. The frequency dispersion is first separated into three regions as shown in Fig 7. Z_S is extracted as a function of frequency in accumulation, assuming it is voltage independent. The corrected impedance Z_C can then be determined since,

$$Z_S = Z_{m,ACC} - Z_{C,ACC} \qquad (17)$$

Using the values of extracted Z_S, the corrected capacitance C_C and conductance G_C at all gate voltages can be obtained via the real and imaginary parts of the intrinsic impedance:

$$Z_C(V_G) = Z_m(V_G) - Z_S(V_G) \qquad (18)$$

The model is first applied to fit at best the capacitance towards low frequency in *region I* using only Z_C and Z_1. Once the value of Z_1 is established, a numerical fit is applied using Z_2 to fit at best *region II* until the capacitance begins to increase, identifying the onset of region III. Finally L and R are included in the fit to match both the capacitance and the conductance at higher frequencies in *region III*. The corrected capacitance and conductance from the model are given as:

$$C_C(V_G) = \frac{1}{\omega} \cdot \text{Im} \left(\frac{1}{Z_m(V_G) - \dfrac{1}{1/R_1 + j\omega C_1} - \dfrac{1}{1/R_2 + j\omega C_2} - \dfrac{1}{R} - j\omega L} \right) \qquad (19)$$

$$G_C(V_G) = \text{Re} \left(\frac{1}{Z_m(V_G) - \dfrac{1}{1/R_1 + j\omega C_1} - \dfrac{1}{1/R_2 + j\omega C_2} - \dfrac{1}{R} - j\omega L} \right) \qquad (20)$$

The values thus extracted are summarized in Table II:

TABLE II. Parameters extracted for polysilicon and poly/TiN gates.

Gate	Poly/TiN		Polysilicon	
Area(k μm^2)	1	10	1	10
C_1(nF)	1.02	1.4	1.25	3.2
R_1(kΩ)	4.39	3.22	24.6	4.20
C_2(nF)	4.9	9.1	5.1	12.3
R_2(Ω)	151	75	237	81
L(μH)	5.56	0.44	6.91	0.7
R(Ω)	40	24	70	40

Results

Distribution of Interface States in Polysilicon and Poly/TiN

The density of interface states in the upper half of the bandgap is extracted via Transmission Line modeling whereas that in the lower half is extracted using the 6-element model as explained above. The combined density of interface states in depletion and inversion, unveiled from "corrected" G_P curves is shown in Figure 8. Electron traps are found to be an order of magnitude higher than hole traps. Moreover, in the upper half of the band gap, states are extracted with a single trap model, revealing a capture cross section of (σ_p ~10^{-18}-10^{-19} cm^{-3}) which reinforces their different nature in comparison to hole traps which are derived by a statistical model. The capture cross section of the hold traps correspond well to P_b centers (σ_p ~10^{-16}-10^{-17} cm^{-3}). Integrating over the band gap, the total density of interface states N_{it} for the polysilicon gate is around ~1.9×10^{11} cm^{-2} and ~4.2×10^{11} cm^{-2} with the poly/TiN gate which are not significantly high numbers. The lower density for hole traps appears to suggest a similarity to those in an Si/SiO$_2$ interface whereas those in the upper half of the bandgap are significantly increased and might be related to oxygen vacancies [29]. The difference in D_{it} might be attributed to poorer passivation with TiN due to strain of the metal electrode [30]. The large thickness of the TiN (~10 nm) may be involved in the physical stress applied on the whole gate stack. In the present case the physical stress may be compressive and tends to distort and break chemical bonds in the ultra-thin SiO$_2$ layer [31, 32].

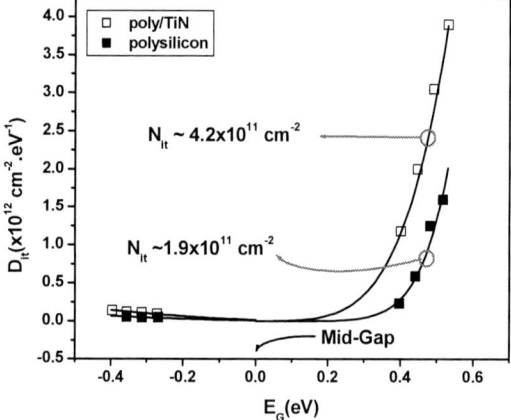

Figure 8. A comparison of the interface states of poly and poly/TiN gates extracted via the conductance technique. A 6 elements model is used for accumulation and a transmission line model is used in inversion.

Mobility

Given the extracted quantities of interface states and fixed oxide charge from experiment, it is worth comparing the theoretical mobility attributable to this charge. For this purpose, we use our Coulomb scattering model [7] implemented in the mobility code SCATTER. For intervalley phonon scattering, Jacobani's parameters [33] and an N_s dependent acoustic deformation potential D_{ac} varying from 8.7 -15 at $N_s = 1 \times 10^{11}$ cm^{-2} to 5×10^{13} cm^{-2} [34] respectively are used. Surface Roughness scattering is implemented using Ando's model with $\Lambda = 1.7$ nm and $\Delta = 0.1$ nm and an exponential auto-covariance function [35]. The screening model employed for Coulomb and Surface Roughness takes the density correlation function [36] into account. Mobility is calculated within the code SCATTER using the relaxation time approximation as [37]:

$$\mu_{xx} = \frac{q\pi\sqrt{\dfrac{m_y}{m_x}}}{\pi^2 k_B T \hbar^2 N_s} \sum_{\mu} g_{\mu} \int_0^{\infty} E(1+\alpha E)(1+2\alpha E)\tau(E) f_0 (E+E_{\mu})\left[1 - f_0 (E+E_{\mu})\right] dE$$

(21)

where k_B is the Boltzmann constant, T is the temperature, $\hbar$ is the reduce Plank's constant, g_{μ} is the valley degeneracy factor, α is the non-parabolic parameter, τ is the sum of relaxation times for bulk phonon, surface roughness and coulomb scattering, f_0 is the Fermi distribution, q the electron charge and $m_{x,y}$ the electron effective masses along x and y axis where z is the axis perpendicular to the Si/SiO$_2$ interface. SP-SCATTER is used to obtain the distribution of the electron wavefunctions, subbands, carrier density and the electrostatic potential. SCATTER has been benchmarked against experimental data for SiO$_2$, using published data for interface states and substrate doping [38, 39].

The mobility is first benchmarked against universal mobility and the same parameters are then used for simulating the high-k dielectric MOSFETs. The results for the two gate stacks are shown in Figures. 9 and 10 assuming a distribution of charge as highlighted in the inset. A Gaussian distribution of interface states centered on the Si-SiO$_2$ interface is assumed, the fixed oxide charge is maintained identical for

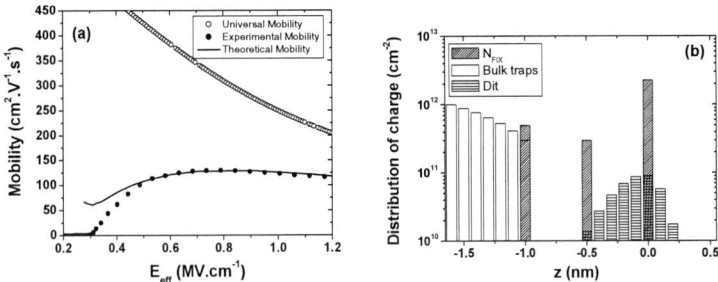

Figure 9. A comparison of the experimental and theoretical mobility for the Poly/TiN gate (a) using the charge distribution as extracted from experiment (b).

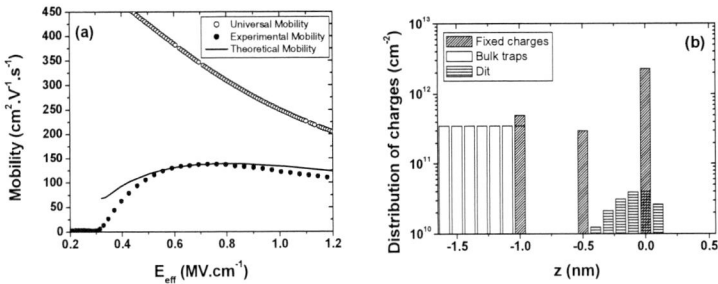

Figure 10. A comparison of the experimental and theoretical mobility for the polysilicon gate (a) using the charge distribution as extracted from experiment (b).

both samples at 3.1×10^{12} cm^{-2} and the corresponding bulk traps are 4.55×10^{12} cm^{-2} and 2.5×10^{12} cm^{-2} for the poly/TiN and polysilicon samples respectively as extracted from CV measurements. In Figures 9 and 10, the primary Coulomb scatterer is the fixed oxide charge distributed at the centre of the IL and the two interfaces HfO$_2$/SiO$_2$, and Si/SiO$_2$.

The "effective" fixed oxide charge at the Si/SiO$_2$ interface in a high-k gate stack represents a projection of the fixed oxide charge within the IL to the Si/SiO$_2$ interface [3]. Evidence of such charge can be found in work function experiments as a function of the IL thickness and the extraction of V_{FB} from terraced oxide structures [40]. A possible origin of this charge is surmised to be oxygen deficiency in the IL [3]. The location of the fixed oxide charge plays a crucial role in achieving a match of experimental mobility via theory. The influence of the charge is diminished if located at the HfO$_2$/SiO$_2$ interface alone, and alternate degradation mechanism within the entire effective field range and/or a significantly higher interface state density, unsubstantiated by experiment then needs to be invoked to explain the mobility degradation.

Conclusion

Extraction methodologies for transistor parameters and interface states have been evaluated using an example of a polysilicon and a poly/TiN high-*k* gate stack. The introduction of a poly/TiN gate increases the density of interface states in comparison to that of a polysilicon gate. However, this has a minimal impact on the mobility, which is similar for the two stacks in the present samples, because of the higher scattering potential in the polysilicon gate. The primary cause of Coulomb degradation (85%) is attributable to the fixed oxide charge at the two interfaces, Si-SiO$_2$ and the SiO$_2$-HfO$_2$ which are considered the same in the two samples.

References

1. J. D. Wilk, R. M. Wallace, and J. M. Anthony, *J. Appl. Phys.*, **89**, 5243 (2001).
2. S. Guha and V. Narayanan, *Annual Rev. Mater. Res.*, **39**, 181 (2009).
3. G. Bersuker, C. S. Park, H.-C. Wen, K. Choi, J. Price, and P. Lysaght, *IEEE Trans. Electron Devices*, **57**, 2047 (2010).
4. E.Vogel, W. K. Henson, C. A. Richter, and J. S. Suehle,, *IEEE Trans. Electron Devices*, **47**, 601 (2000).
5. K. Martens, C. O. Chui, G. Brammertz, B. De Jaeger, D. Kuzum, M. Meuris, M. M. Heyns, T. Krishnamohan, K. Saraswat, H. E. Maes, and G. Groeseneken, *IEEE Trans. Electron Devices*, **55**, 547 (2008).
6. S. Sicre and M. M. De Souza, *IEEE Trans. Electron Devices*, **57**, 1642 (2010).
7. D. Casterman and M. M. De Souza, *J. Appl. Phys.*, **107**, 063706 (2010).
8. M. Cassé, L. Thevenod, B. Guillaumot, L. Tosti, F. Martin, J. Mitard, O. Weber, F. Andrieu, T. Ernst, G. Reimbold, T. Billon, M. Mouis, and F. Boulanger, *IEEE Trans. Electron Devices*, **53**, 759 (2006).
9. E.H. Nicollian and J.R. Brews, *MOS (Metal Oxide Semiconductor) Physics and Technology*, Wiley-Interscience, New York (1982).
10. A. Ali, H. Madan, S. Koveshnikov, S. Oktyabrsky, R. Kambhampati, T. Heeg, D. Schlom, and S. Datta, *IEEE Trans. Electron Devices*, **57**, 742 (2010).
11. C. N. Berglund, *IEEE Trans. Electron Devices*, **13**, 701 (1966).

12. S. Sicre, *"Influence of traps on the mobility of ultra-thin high-k MOSFETs"*, PhD. Thesis, University of Sheffield (2010).
13. D.Vasileska, S.S. Ahmed, M. Mannino, A. Matsudaira, G. Klimeck, and M. Lundstrom, SCHRED 2.2, Purdue University, available at http://www.nanohub.org
14. UT-Quant 2.4 University of Texas, Austin, 2004, National Nanotechnology Infrastructure Network, http://www.nnin.org/nnin_utquant.html
15. S. Kar, *IEEE Trans. Electron Devices*, **50**, 2112 (2003).
16. A. E. Islam and A. Haque, *IEEE Trans. Electron Devices*, **53**, 1364 (2006).
17. A. Kuriyama, O. Faynot, L. Brevard, A. Tozzo, L. Clerc, S. Deleonibus, J. Mitard, V. Vidal, S. Cristoloveanu, and H. Iwai, in Proceedings of the 36th European Solid State Device Research Conference, ESSDERC 2006, p. 109.
18. C. C. Hobbs, L. R. C. Fonseca, A. Knizhnik, V. Dhandapani, S. B. Samavedam, W. J. Taylor, J. M. Grant, L. R. G. Dip, D. H. Triyoso, R. I. Hegde, D. C. Gilmer, R. Garcia, D. Roan, M. L. Lovejoy, R. S. Rai, E. A. Hebert, H.-H. Tseng, S. G. H. Anderson, B. E. White, and P. J. Tobin, *IEEE Trans. Electron Devices*, **51**, 971 (2004).
19. F. Fillot, T. Morel, S. Minoret, I. Matko, S. Maitrejean, B. Guillaumot, B. Chenevier, and T. Billon, *Microelec. Eng.*, **82**, 248 (2005).
20. M. Charbonnier, C. Leroux, V. Cosnier, P. Besson, E. Martinez, N. Benedetto, and C. Licitra, *IEEE Trans. Electron Devices*, **57**, 1809 (2010).
21. J. Maserjian, G. Petersson, and C. Svensson, *Sol. State Elec.*, **17**, 335 (1974).
22. C. Leroux, G. Ghibaudo and G. Reimbold, *IEEE Trans. Electron Devices*, **47**, 660 (2007).
23. G. Ghibaudo, S. Bruyere, T. Devoivre, B. De Salvo, and E. Vincent, *IEEE Trans. Sem. Manuf.*, **13**, 152 (2000).
24. G. Bersuker, C. S. Park, J. Barnett, P. S. Lysaght, P. D. Kirsch, C. D. Young, R. Choi, B-H. Lee, B. Foran, K. van Benthem, S. J. Pennycook, P. M. Lenahan, and J. T. Ryan, *J. Appl. Phys.*, **100**, 094108 (2006).
25. K.J. Yang and C. Hu, *IEEE Trans. Electron Devices*, **46**, 1500 (1999).
26. W. H. Wu, B. Y. Tsui, Y. P. Huang, F. C. Hsieh, M. C. Chen, Y. T. Hou, Y. Jin, H. J. Tao, S. C. Chen, and M. S. Liang, *IEEE Electron Devices Lett.*, **27**, 399 (2006).
27. K.J. Yang and C. Hu, *IEEE Trans. Electron Devices*, **46**, 1500 (1999).
28. P.A. Kraus, K.Z. Ahmed, and J.S. Williamson Jr., *IEEE Trans. Electron Devices*, **51**, 1350 (2004).
29. O. Ghobar, D. Bauza, and B. Guillaumot, *IEEE IIRW Final Report*, 94 (2007).
30. G.S. Lujan, T. Schram, G. Sjoblom, T. Witters, S. Kubicek, S. De Gendt, M. Heyns, and K. De Meyer, in Proceedings of the 34th European Solid State Device Research Conference, ESSDERC 2004, p. 325.
31. S.C. Song, Z. Zhang, C. Huffman, J.H. Sim, S.H. Bae, P.D. Kirsch, P. Mahji, R. Choi, N. Moumen, and B.H. Lee, *IEEE Trans. Electron Devices*, **53**, 979 (2006).
32. T.C. Yang and K.C. Saraswat, *IEEE Trans. Electron Devices*, **47**, 746 (2000).
33. C. Jacoboni and L. Reggiani, *Rev. Mod. Phys.*, **55**, 645 (1983).
34. R. Shah and M. M. De Souza Lecture Notes in Engineering and Computer Science, **1**, 417, Proceedings World Congress on Engineering, London, UK (2009).
35. T. Ando, A. B. Fowler, and F. Stern, *Rev. Mod. Phys.*, **54**, 437 (1982).
36. A. Pirovano, A. L. Lacaita, G. Zandler, and R. Oberhuber, *IEEE Trans. Elec. Dev.*, **47**, 718 (2000).
37. S. Barraud, O. Bonno, and M. Cassé, *J. Appl. Phys.*, **104**, 073725 (2008).

38. J. Koga, S. I. Takagi, and A. Toriumi, *IEDM Tech. Dig.*, 475 (1994).
39. J. Koga, S. Takagi, and A.Toriumi, *IEEE Trans. Electron Devices*, **49**, 1042 (2002).
40. G. Brown, G. Smith, J. Saulters, K. Matthews, H. C. Wen, P. Majhi, and B. H. Lee, Proceedings of SISC Conference, p. 15 (2004).

Universal Set/Reset Characteristics of Metal-Oxide Resistance Switching Memories

Daniele Ielmini

Dipartimento di Elettronica e Informazione and IU.NET, Politecnico di Milano, Italy

Resistance switching occurs in metal oxides after dielectric breakdown, as a result of localized thermally-activated ion migration and chemical reduction/oxidation. These phenomena form the fundamental basis for the resistive switching memory (RRAM), which is a novel memory device with high scaling potential for future non-volatile memories. The device is programmed/erased by the set/reset operations, consisting of the electrically-induced formation and dissolution of a conductive filament (CF) through the metal oxide layer. Understanding and modeling the set and reset processes is essential for developing device simulation tools and predicting the scalability and reliability of RRAM. This work studies set/reset characteristics for unipolar/bipolar RRAM devices. It is shown that the cell resistance after set and the reset current are universal functions of the compliance current, that is the maximum current flowing during the set operation for the formation of the CF. The universal reset characteristic is described in terms of a universal reset voltage, resulting from a weak dependence of the dissolution temperature on material parameters. Finally, the reliability of RRAM is reviewed focussing on data retention and discussing possible methods to improve resistance stability for non-volatile applications.

Introduction

The resistive-switching memory (RRAM) is a novel memory device based on the resistance change in a transition metal oxide, e.g. NiO [1–3], TiO_2 [4–6] or HfO_2 [7, 8]. The memory operation is initiated by a dielectric breakdown, causing the electrical *forming* of a conductive filament (CF) with a relatively low resistance, typically in the range from 0.1 to 10 kΩ. After the forming process, the device resistance can be reversibly changed between the low value of the CF to a higher value, generally higher than 10 kΩ, corresponding to a discontinued or dissolved filament. The disruption of the CF is called the *reset* operation, while the re-formation of a continuous CF is called *set* operation. Electrical set/reset allows the repeatable transition from high to low resistance and vice versa, allowing the storage of the logic bit as the resistance level in the cell. This two-terminal device, generally implemented as a simple metal-insulator-metal (MIM) structure, has a strong scaling capability in terms of both cell size and 3D stacking, thus attracts a wide applicative interest as potential post-Flash technology [9]. Other interesting advantages of the memory concept include a large programming window, enabling multilevel cell (MLC) operation [10], and switching speeds below/around 50 ns (11–13), thus compatible with RAM applications. Finally, extremely low programming currents in the sub-10 μA range have been recently shown [14–16], thus offering a perspective for low-power applications. At the same time, many reliability issues exist, including large switching variability [17], unstable set/reset [18, 19] and stability of the logic states [20–22]. To address these issues, a detailed investigation of set/reset and reliability performance as a function of active material composition/stacks and operation conditions is needed [23].

This work addresses the set and reset characteristics in unipolar and bipolar metal-oxide RRAM. It is shown that the set resistance and the reset current as a function of the compliance current used during the set operation display a universal behavior, since they show only a minor dependence on active material and set/reset polarity. The universal reset characteristic is explained based on the fundamental physical mechanisms of thermally-activated dissolution of the CF. The RRAM reliability is finally addressed, reviewing the available characterization results for the data retention of different metal oxide devices and discussing the ways to improve data stability by material engineering.

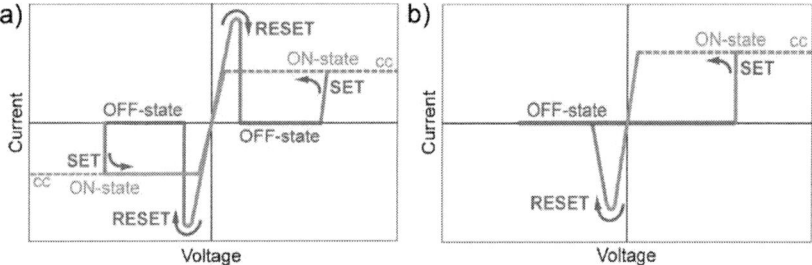

Figure 1. Schematic I-V curves for the set and reset transitions in unipolar (a) and bipolar (b) switching RRAM [24].

Switching Characteristics

Set and reset processes are achieved by electrical voltage pulses or sweeps, with either unipolar or bipolar operation, as summarized in Fig. 1 [24]. In the case of *unipolar* switching (Fig. 1a), the set and reset processes can take place irrespective of the voltage polarity used. The set process is observed at a voltage $\pm V_{set}$, where a sudden increase of current to the set (or ON) state is observed. The current flowing in the device at the set transition is generally limited by a compliance current (CC) mode, allowing a maximum current I_C. This is necessary to avoid disruption or damage of the CF soon after its formation. The subsequent reset process occurs at $\pm V_{reset}$, where the current suddenly drops from the reset current I_{reset} to a lower value, corresponding to the high resistance reset (or OFF) state. Since set and reset processes are independent on the voltage polarity, the switching mechanism is generally explained in terms of thermo-chemical transitions, and the unipolar RRAM is sometimes referred to as thermo-chemical memory TCM [24]. In a TMC, the applied voltage and current are effective in providing localized Joule heating, which accelerates chemical reduction or oxidation during set and reset transitions, respectively. In *bipolar* switching (Fig. 1b), the set and reset operations are similar to the unipolar case, except for the polarity of set and reset transitions: If set is achieved under positive polarity, as shown in Fig. 1b, the reset operation must take place at negative polarity, and vice versa. The strong polarity dependence of switching in bipolar RRAM highlights the role of ion migration for set and reset. For this reason, the memory is sometimes referred to as valence change memory (VCM), where oxygen ion migration, possibly assisted by oxygen vacancies, results in different local valence, hence conductivity [24]. Note that the bipolar switching behavior is very similar to the conductive bridge RAM (CBRAM) operation, although in CBRAM the CF is formed through cation migration from an active electrode (e.g. Ag, Cu) within a solid-state electrolyte, generally a chalcogenide glass or SiO_2 [25, 26].

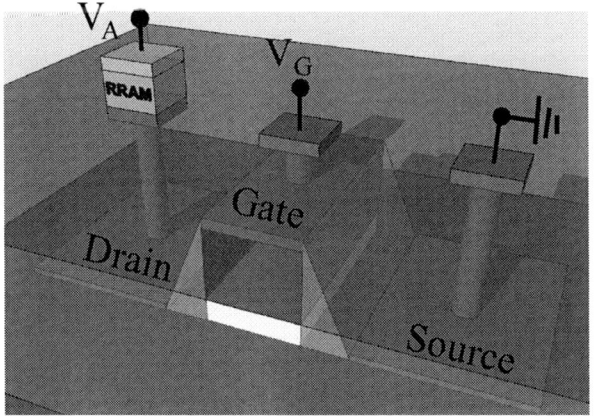

Figure 2. Schematic view of a 1T-1R structure, including a MOSFET and a RRAM element integrated on top of the drain contact [23].

As shown in Fig. 1, a key ingredient in the set transition of both unipolar and bipolar RRAM is a proper control of the compliance current I_C. The most straightforward experimental setup for a variable current control during set and reset is the one transistor – one resistor (1T1R) structure, where the MIM device is connected to a MOS transistor. This is schematically shown in Fig. 2 [23], and corresponds to the three terminal structure which is usual in NOR memory architectures, such as DRAM and Flash, where each device can be selectively accessed by applying a high voltage to the wordline (controlling the gate of the MOSFET) and the bitline (controlling the top electrode of the MIM in Fig. 2). It was shown that integrated 1T1R device structures allow for minimum parasitic capacitance affecting the RRAM cell [27], thus avoiding unwanted current spikes that may affect the transition from high to low resistance [28]. In comparison, current compliance modes of commonly used current meters are affected by a long delay for voltage readjustment after switching, thus cannot effectively limit the current to the desired value. Fig. 3 shows typical I-V curves for an integrated 1T1R structure [23]. The set transition takes place at about 1.7 V, where the current suddenly increases to the saturated drain current I_C of the MOSFET (80 μA for the case considered in the figure). The reset transition, evidenced by a sudden increase of resistance R to the reset state, was instead observed at a reset current of 90 μA, thus slightly higher than I_C.

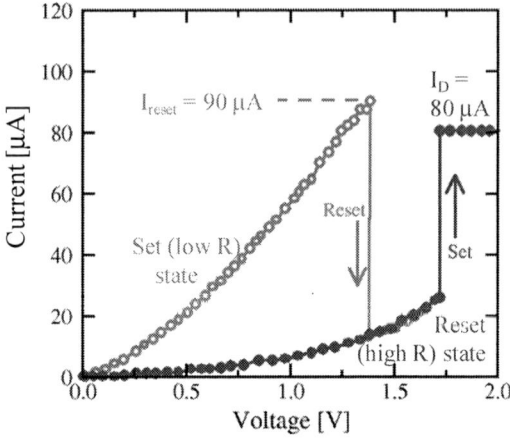

Figure 3. Measured currents for the set (blue) and the reset (red) transition for a 1T1R device. Set was done with a saturated drain current $I_C = 80$ µA, resulting in a reset current of 90 µA, thus close to I_D [23].

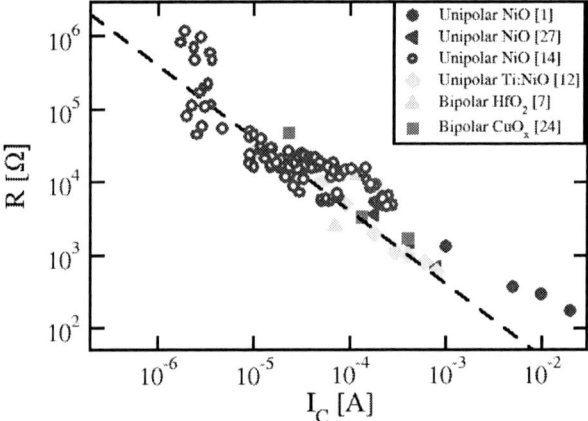

Figure 4. Measured R for the set state as a function of I_C, for different metal oxides and unipolar-bipolar switching modes [23]. All data approximately follow the line $R = V_0/I_C$, with $V_0 = 0.4$ V.

Universal Set and Reset Characteristics

Fig. 4 shows the measured set characteristic, that is the set-state resistance R as a function of the compliance current I_C, for different devices and switching modes [14]. Data were collected for different metal oxides including undoped NiO [1, 14, 27], Ti-doped NiO [12], HfO_2 [7] and CuO_x [29]. The current during set was controlled according to different schemes, including current-compliance mode [1, 7, 29] and control of the gate voltage in integrated 1T1R structures [12, 14, 27]. All data in the figure were obtained through quasi-static voltage sweeps in a relatively long time range (around 1 s). Both unipolar (NiO) and bipolar (HfO$_2$, CuO$_x$) switching modes were considered. All data fall approximately on the same line R = V$_0$/I$_C$, with V$_0$ = 0.4 V, irrespective of the metal oxide material and switching mode. The constant voltage V$_0$ = 0.4 V represents the final voltage across the cell at the end of the set transition. These results suggest that, irrespective of the set voltage V$_{set}$, of the initial resistance, of the metal-oxide composition and of the unipolar/bipolar switching type, the CF resistance is only a function of the compliance current.

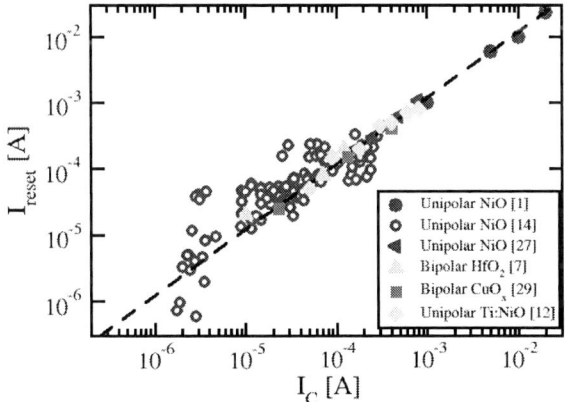

Figure 5. Measured I$_{reset}$ as a function of I$_C$, for different metal oxides and unipolar-bipolar switching modes [23]. All data roughly follow the universal line I$_{reset}$/I$_C$ = 1.2.

Fig. 5 shows the measured reset characteristics, that is the measured reset current I$_{reset}$ as a function of I$_C$ [14]. All data align on a line with slope equal to 1, indicating a linear relationship between I$_{reset}$ and I$_C$, with an approximate ratio of about 1.2. From the results in Figs. 4 and 5, one can derive a relationship between reset current and resistance, namely RI$_{reset}$ = RI$_C$ x I$_{reset}$/I$_C$ = 0.4*1.2 = 0.48 V. Assuming a linear (ohmic) I-V characteristic in the set state, the RI$_{reset}$ product yields the reset voltage V$_{reset}$. Therefore, the results in Figs. 4 and 5 indicate a universal V$_{reset}$ of about 0.5 V.

Universal Reset Model

To understand the universal reset characteristic in Fig. 5, we must refer to the physical reset mechanism responsible for resistance change in the RRAM device. The reset transition is generally attributed to oxidation and/or diffusion of the CF, accelerated by local heating due to the dissipated Joule power. The high conductivity of the CF can be due to segregated metals or to local doping associated to oxygen vacancies or excess metallic elements. This can explain the observed variation of conductive characters, from metallic to semiconductive conductivity for increasing CF resistance. It was shown in fact that the resistance increases with temperature T for low R, thus indicating a metallic conduction behavior [20]. This can be attributed to a degenerate semiconductor structure with the Fermi level in the conduction/valence band as a result of a high doping concentration. For increasing R, the temperature dependence of resistance changes from metallic to semiconductor-like, indicating a transition from high to low doping. During reset, the Joule heating activates gradient-driven diffusion and chemical oxidation of excess metallic elements, thus causing the dissolution of the CF and the increase of resistance [3]. Assuming that diffusion is the limiting step for the reset mechanism, the reset time t_{reset} at constant temperature can be estimated by [23]:

$$t_{reset} = \frac{\phi^2}{D_0} e^{\frac{E_A}{kT}}$$
(1)

where ϕ is the diameter of the CF, assumed of cylindrical shape ($A = \pi\phi^2/4$), D_0 is the pre-exponential constant for diffusivity and E_A is the activation energy for diffusion. From Eq. (1), the temperature T_{reset} needed to complete the dissolution of the CF in a given experimental timescale τ can be obtained as:

$$T_{reset} = \frac{E_A}{k \log \frac{D_0 \tau}{\phi^2}}.$$
(2)

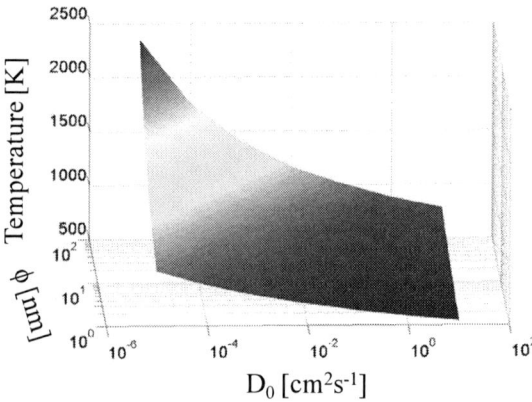

Figure 6. Calculated T_{reset} as a function of ϕ and D_0. An overall variation by a factor of about 5 is seen, despite the wide range of ϕ and D_0 used [23].

Fig. 6 shows T_{reset} calculated by Eq. (2), as a function of parameters D_0 and ϕ. These parameters were changed on a relatively broad range: D_0 was changed by 6 orders of magnitude, between 10^{-5} and 10 cm^{2}s^{-1}, while ϕ was changed by 2 orders of magnitude, between 1 and 100 nm. E_A was kept equal to 1.4 eV, consistently with reset activation energy estimated from the Kissinger plot in [30]. Other experimental results suggest that E_A remains in a comparable range, e.g. activation energies of about 1.3 eV [21] and 1.2 eV [20] were found from temperature-accelerated annealing experiments on MoO-GdO and NiO respectively. An experimental time scale τ of 1 s was used, corresponding to quasi-static reset measurements. The reset temperature decreases for increasing D_0, due to the enhanced diffusivity, and increases for increasing ϕ, as a result of the larger amount of material that needs to be displaced by diffusion. However, due to the logarithmic dependence in Eq. (2), the calculated T_{reset} remains within a relatively narrow range between 500 and 2500 K, despite the extremely large range assumed for D_0 and ϕ. This indicates that the change of D_0 and ϕ due to different materials and CF size results in a minor change of temperature needed for the reset transition.

To describe the results of reset experiments, we need to calculate V_{reset} and I_{reset} needed to cause a local temperature increase to T_{reset} at the CF. Joule heating can be estimated from the Fourier equation for heat conduction. To this purpose, we will assume that the CF is of cylindrical shape and that heat is mostly transferred along the filament toward the top and bottom electrodes, serving as heat sinks. This assumption allows us to use the 1D Fourier equation for heat conduction for the local temperature T:

$$C_p \frac{\partial T}{\partial t} = k_{th} \frac{\partial^2 T}{\partial x^2} + P'''$$

(3)

where C_p is the specific thermal capacitance, k_{th} is the thermal conductivity, t is time, x is the longitudinal coordinate along the CF and P''' is the Joule power, obtained as P''' = JF, where J is the current density and F is the electric field. At steady state ($\partial T/\partial t = 0$) and using boundary conditions $T = T_0$ (the room temperature) at the top and bottom heat sinks, Eq. (3) yields a parabolic dependence of temperature along the x axis, according to:

$$T(x) = T_0 + \frac{P'''}{2k_{th}} \left(\frac{t_{ox}^2}{4} - x^2 \right)$$

(4)

where t_{ox} is the thickness of the metal oxide layer, assumed equal to the length of the CF, and the origin of the x axis is taken in the middle of the CF. The parabolic dependence of temperature is schematically shown in Fig. 7 [3].

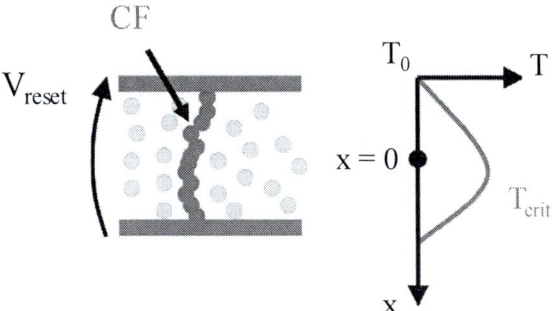

Figure 7. Schematic CF in a metal oxide RRAM and temperature parabolic profile along the CF [3].

Note that heat loss from the surface of the filament through the surrounding metal oxide cannot be neglected for relatively narrow CFs, due to the increased surface/volume ratio [31, 32]. In the narrow CF limit the T profile tends to show a relatively flat profile due to the surface heat-loss contribution [33]. From Eq. (4), the maximum temperature along the CF can be obtained as:

$$T(0) = T_0 + \frac{P'''t_{ox}^2}{8k_{th}} = T_0 + R_{th}VI \tag{5}$$

where the effective thermal resistance $R_{th} = t_{ox}/(8Ak_{th})$ is introduced. From Eq. (5), one can obtain V_{reset} as the voltage needed to raise the maximum temperature along the CF to the critical temperature for reset T_{reset}, thus leading to:

$$V_{reset} = \sqrt{\frac{R}{R_{th}}(T_{reset} - T_0)}. \tag{6}$$

where the ohmic relationship $I = V/R$ was used. For a metallic CF, the ratio between electrical and thermal resistances can be given by the Wiedemann-Franz law for the ratio between thermal conductivity k_{th} and electrical conductivity σ, yielding $R/R_{th} = 8k_{th}/\sigma = 8LT_{reset}$, where $L = 2.48 \times 10^{-8}$ V^2K^{-2} is the Lorenz constant [34].

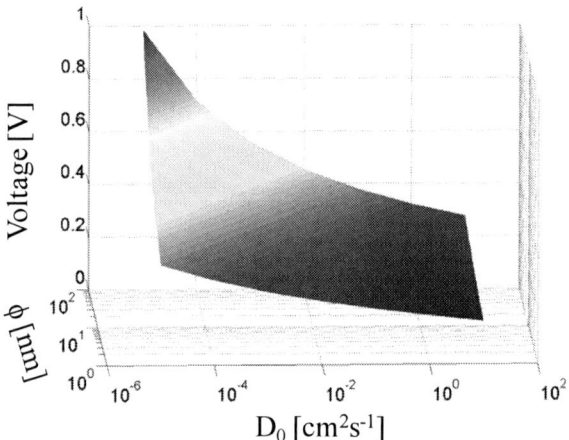

Figure 8. Calculated V_{reset} as a function of ϕ and D_0. An overall variation by a factor of about 5 is seen, despite the wide range of ϕ and D_0 used [23].

Fig. 8 shows V_{reset} calculated by Eq. (6) from the calculated T_{reset} in Fig. 7. The reset voltage is shown as a function of ϕ and D_0 assuming the same ranges of D_0 and ϕ as in Fig. 7. V_{reset} remains within a relatively small range between 0.2 and 1 V, despite the large variation range of D_0 and ϕ. This universal V_{reset} was inherent in the experimental data in Figs. 4 and 5, as discussed in the previous section. In summary, our analytical model for reset accounts for universal reset characteristics due to the logarithmic dependence of reset temperature on diffusion parameters in Eq. (2).

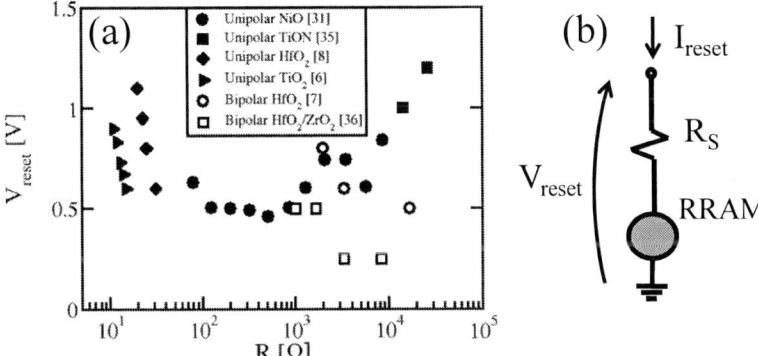

Figure 9. Measured V_{reset} as a function of R for different metal oxide RRAM devices with unipolar or bipolar switching mode (a) and schematic circuit for a RRAM cell with finite series resistance R_S (b). All data in (a) approximately are in a range from 0.25 to 1.2 V.

Fig. 9a shows measured V_{reset} as a function of R, for RRAM devices with different metal oxides including NiO [31], TiON [35], HfO₂ [7, 8], TiO₂ [6] and ZrO₂/HfO₂ [36]. Data are shown for both unipolar [31, 35, 8, 6] and bipolar switching [7, 36]. Data indicate that V_{reset} is generally limited within the range from 0.25 to 1.2 V, consistently with calculations in Fig. 8. Unipolar V_{reset} displays a typical U-shaped dependence on resistance, which can be attributed to series resistance effects and size-dependent CF oxidation and heating [31]. Series resistance effects are summarized in Fig. 9b: the RRAM device is generally accompanied by a series resistance R_S, which includes electrodes and interconnect in the integrated device and in the experimental set up. Usually R_S can be quantified in the 10-100 Ω range. The 'apparent' reset voltage V_{reset} shown in Fig. 9b can thus be written as:

$$V_{reset} = V_{reset}^{*} + R_S I_{reset} = V_{reset}^{*} \frac{R}{R - R_S} \tag{7}$$

where V^{*}_{reset} is the 'true' reset voltage across the RRAM cell and R is the apparent resistance measured across the RRAM device and the series resistance. For large RRAM resistances, R is very large compared to R_S, thus V_{reset} is virtually equal to the true reset voltage. For small RRAM resistances, instead, R becomes almost equal to R_S and the denominator in Eq. (7) approaches zero, thus V_{reset} diverges. This is the reason for the steep increase of V_{reset} for decreasing R in Fig. 9a. Here, different values of R_S, possibly due to different sample layout, electrode material/thickness and experimental setups, result in different R marking the onset of the steep increase.

Size-dependent effects in reset also contribute to dictating the V_{reset} dependence on R in the figure. In addition to series resistance effects discussed above, a V_{reset} increase for small decreasing R is also expected as a result of the decreasing CF cross section, which leads to a lower T_{reset} according to Eq. (2) and Fig. 6. Above approximately 1 kΩ, instead, V_{reset} increases with R as the ratio R/R_{th} in Eq. (6) increases due to the surface heat-loss contribution to R_{th} at small CF size. This is described in Fig. 10, showing the calculated thermal resistance as a function of CF diameter (a) and the calculated T profile along the CF for different CF sizes (b). The effective thermal resistance was obtained from 3D solutions of the Poisson equation for electrical conduction and of the Fourier equation for thermal conduction for a metallic cylindrical CF through a NiO film with thickness $t_{ox} = 160$ nm. The thermal resistance was calculated as the ratio between the maximum temperature increase within the CF and the dissipated Joule power VI, according to Eq. (5). R_{th} decreases for increasing area A, due to the larger area available for heat conduction. Different R_{th} behaviors were found depending on the thermal conductivity of the metal oxide $k_{th,NiO}$. Assuming that NiO behaves as an ideal thermal insulator ($k_{th,NiO.} = 0$), R_{th} is inversely proportional to A, displaying a slope of about -1 on the bilogarithmic plot. For non-zero thermal conductivity of the metal oxide, however, R_{th} changes to a shallower dependence at small A. This is the result of parallel out-of-CF contribution to heat loss, which is increasingly important for larger $k_{th,NiO}$ and smaller ϕ. The effects of the parallel heat loss on the temperature profile can be clearly seen in Fig. 10b: for large ϕ, the T profile is almost perfectly parabolic due to the prevalence of heat conduction along the CF. For smaller ϕ, the T profile is flattened due to the significant surface heat loss contribution, which is almost position independent.

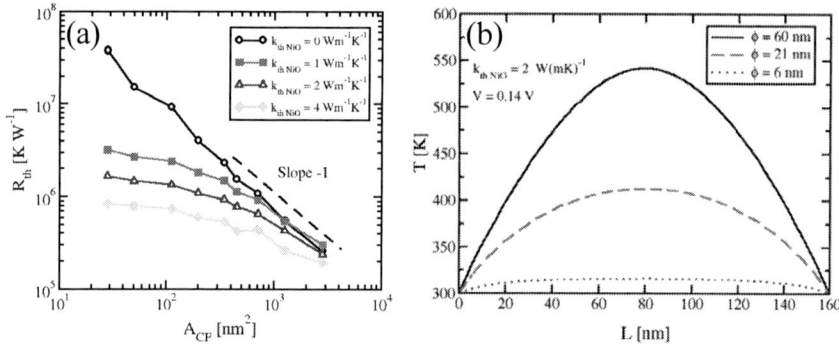

Figure 10. Calculated R_{th} as a function of the CF area in NiO, for different metal-oxide thermal conductivities $k_{th,NiO}$ (a) and calculated T profile along the CF for different CF size (b).

Data Retention

Since RRAM technology is targeting nonvolatile memory application, the reliability study and prediction is essential. Among the main reliability issues for RRAM, cycling endurance [37], switching variability [17], data retention [20–22] and resistance instabilities due to resistance drift [10] and noise [38] are the most relevant. For nonvolatile memory applications, the typical specification for data retention is 10 years stability of the programmed logic bit below 85°C. According to the physical interpretation of the CF as due to a local segregation of metallic elements responsible for high conductivity, data retention may be affected by thermally-activated oxidation of the CF. This is similar to the reset process discussed in the previous section, but at relatively low temperature. Based on these considerations, data retention for NiO RRAM devices was studied by temperature-accelerated experiments starting from the low resistance state.

Elevated-temperature annealing was performed in a chamber with air ambient conditions. The annealing times were long enough (at least 30 minutes) so that the transient heating and cooling times (around few minutes) could be considered negligible. About 50 RRAM cells were first programmed in the low R state, then subjected to a high-T bake condition, finally measured again at room temperature. After each post-annealing readout, the cells were programmed again at the same initial resistance and subjected to a new bake for a different annealing time. The failure time τ_R was defined in correspondence of a 10x increase of resistance with respect to the initial value. Fig. 11 shows the Arrhenius plot of retention time τ_R for various percentiles, from $f = 25\%$ to 90%, and for an initial resistance between 0.2 and 1 kΩ. Data indicate an Arrhenius dependence with an activation energy $E_A = 1.21$ eV at $f = 50\%$. Extrapolations according to the Arrhenius law at 50 % results in a temperature of about 100°C at 10 years, thus higher than 85°C which is the minimum for non-volatile memories. However, note the wide spread of retention times at various percentiles, which is still insufficient to satisfy the necessary statistical reliability in large memory arrays. Also, as already observed for the reset operation, size-dependent oxidation effects may cause a degradation of data retention. To check for the dependence of retention time on CF size, Fig. 12a shows the Arrhenius plot of τ_R, for $f = 90\%$ and for three different ranges of R, namely R < 0.2 kΩ, 0.2 < R < 1 kΩ, and R > 1 kΩ. The Arrhenius dependence of τ_R is confirmed, although with a smaller E_A between 0.64 and 1 eV. No obvious dependence of E_A on initial resistance can be seen. Most importantly, τ_R decreases for increasing R (hence decreasing ϕ), indicating a clear dependence of τ_R on CF size. Note that a similar relationship between τ_R and R can be expected irrespective of the choice of f, as suggested by the similar behaviors for different f in Fig. 11.

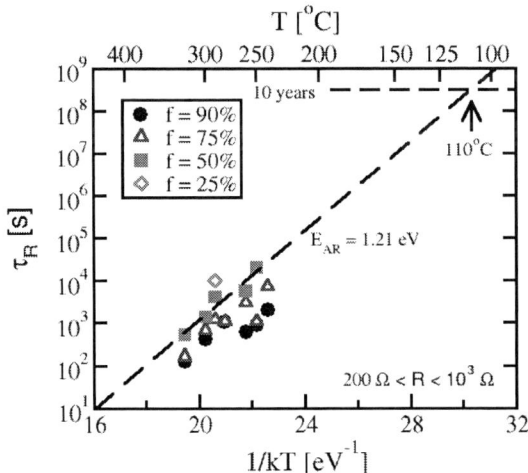

Figure 11. Arrhenius plot of measured retention time τ_R for different percentiles f, for NiO RRAM devices. Data indicate an activation energy of about 1.2 eV [20].

The R dependence of τ_R can be easily understood based on Eq. (2), which shows that the reset temperature decreases for decreasing CF size ϕ. Fig. 12b shows the failure temperature at $\tau_R = 10^3$ s as a function of R, from the best fitting lines in Fig. 12a [20] and from MoO-GdO data [21]. Calculations by Eq. (2) for $E_A = 1.4$ eV, $D_0 = 10^{-6}$ cm²s⁻¹ are also shown. The CF size was estimated based on the following relationship between resistance and CF size:

$$R = \rho_0 \frac{t_{NiO}}{A} e^{\frac{E_{AC}}{kT}}$$

(8)

where the activation energy for conduction was assumed $E_{AC} = 0$, corresponding to a metallic conductivity, a resistivity $\rho_0 = 100$ μΩ cm was used, consistent with metallic conductivity in ultranarrow Ni nanowires [20], and the area of the CF cross section was estimated as $A = \pi\phi^2/4$. The comparison in the figure shows that the improved retention for decreasing resistance can quantitatively be accounted for by a size effect.

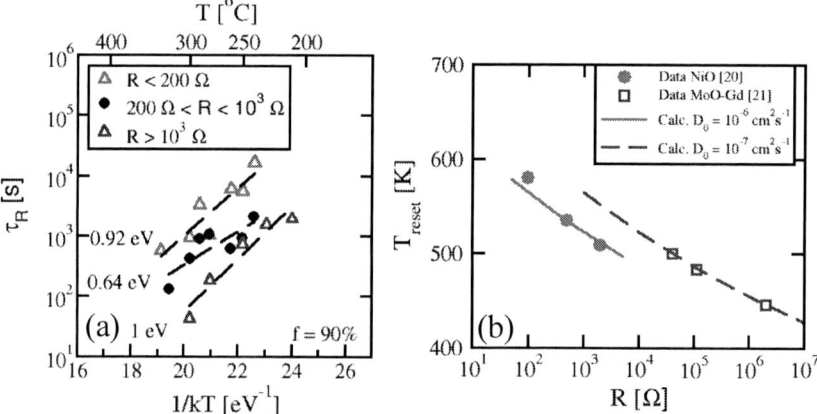

Figure 12. Measured retention times at 90% percentile [20] (a) and measured/calculated retention temperatures as a function of initial resistance [20, 21] (b). Three different initial resistances were used in (a), namely less than 200 Ω, between 200 and 1000 Ω, and above 1000 Ω. The retention times still obey the Arrhenius law, in agreement with data in Fig. 12, however τ_R decreases for increasing R, as a result of size-dependent oxidation [20]. The retention temperature change at 1000 s can be reproduced by the analytical reset model of Eq. (2).

To check for possible universal behaviors of data retention, we collected the Arrhenius parameters of temperature-accelerated data retention for metal-oxide RRAM devices, namely NiO [20], MoO/GdO [21] and TaO$_x$ [22]. In all of the considered studies, the retention time obeys the Arrhenius law:

$$\tau_R = \tau_0 e^{\frac{E_A}{kT}}$$

(9)

where the pre-exponential time is given by $\tau_0 = \phi^2/D_0$, according to Eq. (1). Fig. 13 shows the experimental pre-exponential time τ_0 as a function of the activation energy E_A of data retention for RRAM devices, compared to phase change memory (PCM) devices with

$Ge_2Sb_2Te_5$ active material [39], high-k Flash [40] and nanocrystal Flash [41]. Data are compared to the correlation line corresponding to T = 85°C and τ_R = 10 years, that is the condition for reliability of nonvolatile memories. Nonvolatile memory behavior corresponds to the region at high E_A and high τ_0 with respect to the 10 years correlation line. Arrhenius parameters for metal oxide RRAM are between E_A = 1 eV and 1.5 eV and remain close to the 10 years line, confirming that RRAM may be compatible with nonvolatile memory applications. Note however that resistance-dependent data retention and its large statistical spread need to be understood and solved to enable full nonvolatile reliability of large arrays [20].

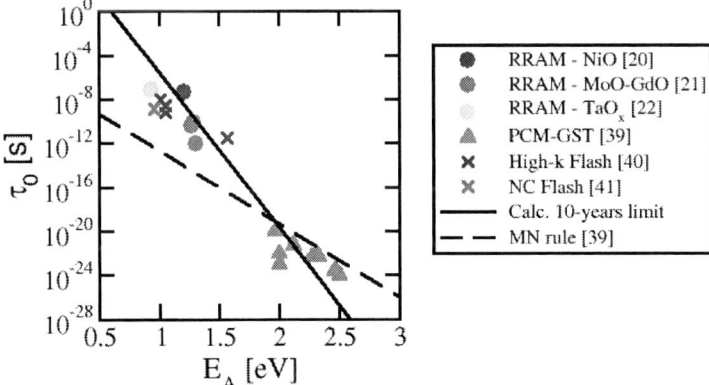

Figure 13. Extracted Arrhenius parameters for data retention of RRAM with different metal oxides (NiO [20], MoO-GdO [21] and TaO_x [22]), PCM [39], high-k Flash [40] and nanocrystal Flash [41]. The calculated 10-year retention line and the MN rule for PCM devices [39] are also shown.

A general way to improve data retention is to enhance the activation energy E_A. In fact, atomic/electronic diffusion and migration generally obey to the Meyer Neldel (MN) rule, given by [39, 42, 43]:

$$\tau_0 = \tau_{00} e^{-\frac{E_A}{kT_{MN}}} \tag{10}$$

where τ_{00} is a pre-exponential constant and T_{MN} is the isokinetic temperature, defining the temperature for which all Arrhenius extrapolations cross each other, hence feature the same kinetic [43]. The MN parameters in Eq. (10) are generally not known, and require detailed characterization of diffusion/migration phenomena for variable sample parameters, such as chemical composition, doping, thickness, and so on. For instance, MN parameters were extracted for PCM devices based on measurements of structural relaxation and crystallization times at elevated temperature, and were estimated to be τ_{00} = $4x10^{-6}$ s and T_{MN} = 760 K [39]. The MN rule for PCM devices is also shown in Fig. 13 for reference. Note that, according to the MN line, one moves from the volatile to the nonvolatile domain for increasing E_A, which thus highlights a possible direction for material selection to improve reliability. Increasing E_A may be achieved by material/electrode engineering and by improving the size of the CF, thus providing a non-defective metallic phase as the starting material for diffusion.

Conclusions

Reset and set parameters, including voltages, currents, kinetics and statistics, play an important role in identifying a feasible technology for RRAM memories. Therefore, comparison of set and reset parameters among different metal oxide materials for unipolar or bipolar switching is essential. In this work, the set and reset characteristics of different RRAM devices have been reviewed, highlighting a universal dependence of set-state resistance and reset current on the compliance current, that is the maximum current during the set operation for the formation of the CF. The universal reset is discussed based on the physical mechanism for reset, that is diffusion and oxidation of metallic elements from the CF. It is shown that, due to the weak dependence of reset temperature on diffusion parameters, such as pre-exponential factor and activation energy, and CF size, the reset voltage remains substantially constant irrespective of the specific metal oxide and programming scheme. The universal reset voltage is confirmed by comparison of data from the literature, showing a V_{reset} generally in the range from 0.2 to 1.5 V. The reset model is then used to discuss data retention, which is attributed to diffusion and oxidation processes. Data retention can be predicted by Arrhenius law, providing a useful extrapolation law from temperature-accelerated experiments to the long term. Arrhenius parameters are compared for different memory technology, discussing the gaps and possible improvement methodologies to achieve 10 years nonvolatile data retention.

Acknowledgments

The author would like to thank C. Cagli, S. Larentis, F. Nardi, and A. L. Lacaita from Politecnico di Milano for several discussions. This work was supported in part by Intel (Project 34523), Fondazione Cariplo (Project 2010-0500) and EU (Emma Project under Grant FP6-033751).

References

1. S. Seo, M. J. Lee, D. H. Seo, E. J. Jeoung, D.-S. Suh, Y. S. Joung, I. K. Yoo, I. R. Hwang, S. H. Kim, I. S. Byun, J.-S. Kim, J. S. Choi, and B. H. Park, *Appl. Phys. Lett.*, **85**, 5655 (2004).
2. I. G. Baek, M. S. Lee, S. Seo, M. J. Lee, D. H. Seo, D.-S. Suh, J. C. Park, S. O. Park, H. S. Kim, I. K. Yoo, U.-In Chung and J. T. Moon, *IEDM Tech. Dig.*, 587 (2004).
3. U. Russo, D. Ielmini, C. Cagli and A. L. Lacaita, *IEEE Trans. Electron Devices*, **56**, 186 (2009).
4. B. J. Choi, D. S. Jeong, S. K. Kim, C. Rohde, S. Choi, J. H. Oh, H. J. Kim, C. S. Hwang, K. Szot, R. Waser, B. Reichenberg and S. Tiedke, *J. Appl. Phys.*, **98**, 033715 (2005).
5. W. Wang, S. Fujita and S. S. Wong, *IEEE Electron Device Lett.*, **30**, 733 (2009).
6. K. M. Kim and C. S. Hwang, *Appl. Phys. Lett.*, **94**, 122109 (2009).

7. H. Y. Lee, P. S. Chen, T. Y. Wu, Y. S. Chen, C. C. Wang, P. J. Tzeng, C. H. Lin, F. Chen, C. H. Lien, M.-J. Tsai, *IEDM Tech. Dig.*, 297 (2008).
8. Y.-M. Kim and J.-S. Lee, *J. Appl. Phys.*, **104**, 114115 (2008).
9. K. Kim, *IEDM Tech. Dig.*, 1 (2010).
10. D. Ielmini, F. Nardi, A. Vigani, E. Cianci and S. Spiga, as discussed at *IEEE Semiconductor Interface Specialist Conference (SISC)*, Arlington, VA, Dec. 3-5 (2009).
11. Y. Hosoi, Y. Tamai, T: Ohnishi, K. Ishihara, T. Shibuya, Y. Inoue, S. Yamazaki, T. Nakano, S. Ohnishi, N. Awaya, I. H. Inoue, H. Shima, H. Akinaga, H. Takagi, H. Akoh and Y. Tokura, *IEDM Tech. Dig.*, 793 (2006).
12. K. Tsunoda, K. Kinoshita, H. Noshiro, Y. Yamazaki, T. Iizuka, Y. Ito, A. Takahashi, A. Okano, Y. Sato, T. Fukano, M. Aoki and Y. Sugiyama, *IEDM Tech. Dig.*, 767 (2007).
13. K. Aratani, K. Ohba, T. Mizuguchi, S. Yasuda, T. Shiimoto, T. Tsushima, T. Sone, K. Endo, A. Kouchiyama, S. Sasaki, A. Maesaka, N. Yamada and H. Narisawa, *IEDM Tech. Dig.*, 783 (2007).
14. F. Nardi, D. Ielmini, C. Cagli, S. Spiga, M. Fanciulli, L. Goux, D. J. Wouters, *2010 IEEE International Memory Workshop (IMW)*, 66 (2010).
15. Y. Wu, B. Lee; H.-S. P. Wong, *IEEE Electron Device Lett.*, **31**, 1449 (2010).
16. C. H. Cheng, A. Chin and F. S. Yeh, *IEEE Electron Device Lett.*, **31**, 1020 (2010).
17. J. Park, M. Jo, J. Lee, S. Jung, S. Kim, W. Lee, J. Shin, and H. Hwang, *IEEE Electron Device Lett.*, **32**, 63 (2011).
18. D. Ielmini, *IEEE Electron Device Lett.*, **31**, 552 (2010).
19. D.-K. Kim, D.-S. Suh, and J. Park, *IEEE Electron Device Lett.*, **31**, 600 (2010).
20. D. Ielmini, F. Nardi, C. Cagli and A. L. Lacaita, *IEEE Electron Device Lett.*, **31**, 353 (2010).
21. J. Park, M. Jo, E. M. Bourim, J. Yoon, D.-J. Seong, J. Lee, W. Lee, and H. Hwang, *IEEE Electron Device Lett.*, **31**, 485 (2010).
22. Z. Wei, Y. Kanzawa, K. Arita, Y. Katoh, K. Kawai, S. Muraoka, S. Mitani, S. Fujii, K. Katayama, M. Iijima, T. Mikawa, T. Ninomiya, R. Miyanaga, Y. Kawashima, K. Tsuji, A. Himeno, T. Okada, R. Azuma, K. Shimakawa, H. Sugaya, T. Takagi, R. Yasuhara, K. Horiba, H. Kumigashira, M. Oshima, *IEDM Tech. Dig.*, 293 (2008).
23. D. Ielmini, F. Nardi, C. Cagli, as discussed at *IEEE Semiconductor Interface Specialist Conference (SISC)*, San Diego, CA, Dec. 2-4, 2010.
24. R. Waser, R. Dittmann, G. Staikov and K. Szot, *Adv. Mat.*, **21**, 2632 (2009).
25. M. N. Kozicki, M. Park, and M. Mitkova, *IEEE Trans. Nanotechnology*, **4**, 331, (2005).
26. C. Schindler, S. C. P. Thermadam, R. Waser, M. N. Kozicki, *IEEE Trans. Electron Devices*, **54**, 2762 (2007).
27. K. Kinoshita, K. Tsunoda, Y. Sato, H. Noshiro, S. Yagaki, M. Aoki and Y. Sugiyama, *Appl. Phys. Lett.*, **93**, 033506 (2008).
28. D. Ielmini, C. Cagli and F. Nardi, *Appl. Phys. Lett.*, **94**, 063511 (2009).
29. T.-N. Fang, S. Kaza, S. Haddad, A. Chen, Y.-C. Wu, Z. Lan, S. Avanzino, D. Liao, C. Gopalan, S. Choi, S. Mahdavi, M. Buynoski, Y. Lin, C. Marrian, C. Bill, M. VanBuskirk and M. Taguchi, *IEDM Tech. Dig.*, 789 (2006).
30. C. Cagli, F. Nardi and D. Ielmini, *IEEE Trans. Electron Devices*, **56**, 1712 (2009).
31. D. Ielmini, C. Cagli, F. Nardi and A. L. Lacaita, *Proc. ISIF*, 211 (2010).
32. D. Ielmini, *ECS Trans.*, **33**, 3, 323 (2010).

33. E. Pop, *Nanotechnology*, **19**, 295202 (2008).
34. G. S. Kumar, G. Prasad, and R. O. Pohl, *J. Mater. Sci.*, **28**, 4261 (1993).
35. Y. H. Tseng, C.-E. Huang, C.-H. Kuo, Y.-D. Chih and C. J. Lin, *IEDM Tech. Dig.*, 109 (2009).
36. J. Lee, J. Shin, D. Lee, W. Lee, S. Jung, M. Jo, J. Park, K. P. Biju , S. Kim, S. Park, H. Hwang, *IEDM Tech. Dig.*, 452 (2010).
37. J. J. Yang, M.-X. Zhang, J. P. Strachan, F. Miao, M. D. Pickett, R. D. Kelley, G. Medeiros-Ribeiro, and R. S. Williams, *Appl. Phys. Lett.*, **97**, 232102 (2010).
38. D. Ielmini, F. Nardi and C. Cagli, *Appl. Phys. Lett.*, **96**, 053503 (2010).
39. D. Ielmini and M. Boniardi, *Appl. Phys. Lett.*, **94**, 091906 (2009).
40. B. Govoreanu, and J. Van Houdt, *IEEE Electron Device Lett.*, **29**, 177 (2008).
41. C. Gerardi, G. Molas, G. Albini, E. Tripiciano, M. Gely, A. Emmi, O. Fiore, E. Nowak, D. Mello, M. Vecchio, L. Masarotto, R. Portoghese, B. De Salvo, S. Deleonibus, A. Maurelli, *IEDM Tech. Dig.*, 821 (2008).
42. R. S. Crandall, *Phys. Rev. B*, **43**, 4057 (1991).
43. A. Yelon, B. Movaghar, and H. M. Branz, *Phys. Rev. B*, **46**, 244 (1992).

ECS Transactions, 35 (4) 597-603 (2011)
10.1149/1.3572307 ©The Electrochemical Society

Resistive Switching Behaviors of ReRAM having W/CeO$_2$/Si/TiN Structures

C. Dou[a], K. Mukai[a], K. Kakushima[b], P. Ahmet[a], K. Tsutsui[b], A. Nishiyama[b], N. Sugii[b], K. Natori[a], T. Hattori[a], and H.Iwai[a]

[a] Frontier Research Center,
[b] Interdisciplinary Graduate School of Science,
Tokyo Institute of Technology, 4259-S2-20, Nagatsuta, Midori-ku, Yokohama, 226-8503, Japan

As one of emerging next-generation nonvolatile memories, Resistive RAM (ReRAM) still calls new material technology to improve its performance. By utilizing the special characteristics of cerium oxides, this paper proposes a new method to improve the performance of CeO$_2$ based ReRAM devices by using Si buffer layer. It is confirmed that the device having W/CeO$_2$/Si/TiN structure shows significant advantage over the device without Si layer for memory application in terms of lower forming voltage, smaller compliance current, larger window and better endurance characteristic. The effect of Si buffer layer is discussed in detail and a model based on filament switching mechanism was also proposed to explain the underlying reasons.

Introduction

Resistive random access memory (ReRAM), which utilizing resistive switching of certain materials, has attracted much attention as non-volatile memory in the near future because of its high speed, low consumption and great potential for scalability [1]. Resistive switching behaviors in various insulating oxides, ranging from simple binary oxides such as NiO$_2$ [2] and TiO$_2$ [3] to complex oxides such as perovskites [4] and chalcogenides [5], has been reported. Untill now, relatively few studies on the resistive switching behaviors of cerium oxides has been reported. Among various insulating oxides, cerium oxides have strong potential for ReRAM application for the following reasons: 1) cerium oxides have high dielectric constant and moderate band gap, 2) cerium atom exhibits both +3 and +4 oxidation states in cerium oxides, which is suitable for valency change switching processes [6], 3) oxygen ions/vacancies, which play a important role in resistive switching [7], produce high conductivity in CeO$_2$ due to its fluorite structure, and 4) cerium oxide is easy to react with Si to form silicate [8]. The last reason provides us a method to modify the structure and concentration of vacancies in cerium oxide film by forming cerium silicate. In this paper, cerium oxides based ReRAM devices with silicon buffer layer were fabricated by utilizing W and TiN as electrodes. The resistive switching behaviors of the devices and the influence of the Si buffer layer on the resistive switching were investigated in detail.

597

Experimental

The structure of the device and the measurement circuit are schematically shown in Fig. 1(a), while fabrication process flow is shown in Fig. 1(b). Highly doped Si wafer was used as substrate for better contact between the bottom electrode and the cathode. A 200- nm-thick SiO_2 insulator layer was grown on the substrate by thermal oxidation to protect the device from leakage current, and contact windows though the SiO_2 layer were patterned for the connection between the bottom electrode (BE) and the substrate. Then, a 15-nm-thick TiN BE layer and a 1-nm- or 2-nm-thick Si buffer layer were successively formed by RF sputtering. Next, a 20-nm-thick CeO_2 layer was deposited by electron beam evaporation as the resistive switching layer, then a 50-nm-thick W top electrode (TE) layer was in-situ deposited by sputtering and was patterned to form electrodes having area of 20 μm square. Finally, Al was evaporated as a back contact followed by rapid thermal annealing (RTA) in N_2 ambient for 30 s at 400°C. Furthermore, the device without Si buffer layer was also fabricated to evaluate the influence of Si buffer layer on switching characteristics.

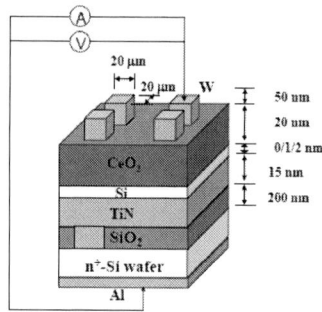

Fig. 1(a). Schematic illustration of the ReRAM device having $W/CeO_2/Si/TiN$ structure

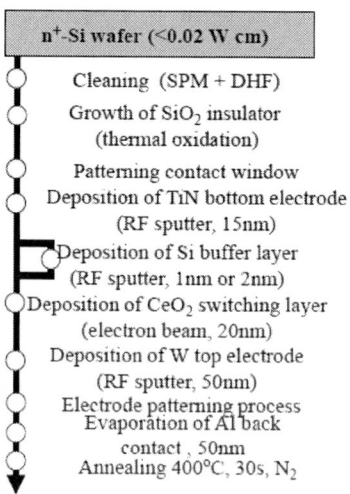

Fig. 1(b). $W/CeO_2/Si/TiN$ device fabrication process flow

Results and Discussion

Resistive Switching Behaviors of the W/CeO₂/Si/TiN Device

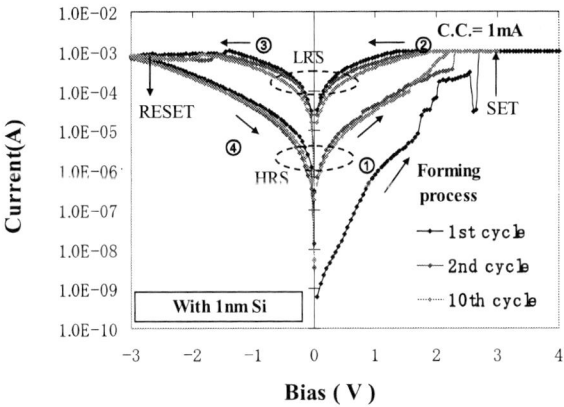

Fig. 2. I-V curve of W/CeO₂(20 nm)/Si(1 nm)/TiN for the following range of bias voltages in each cycle: ①0 V to 3 V, ②3 V to 0 V, ③0 V to -3 V and ④-3 V to 0 V. A forming process is necessary in the first cycle.

The typical current-voltage curve measured at room temperature is shown in Fig. 2. The bias potential was applied to the W TE, and the TiN BE was grounded. A forming process of fresh W/CeO₂/Si/TiN device at the first cycle requires low voltage because Si buffer layer is introduced. The devices can be operated in the range of voltage from -3 V to 3 V with compliance current (CC) of 1 mA. According to this figure, bipolar resistive switching from high resistance state (HRS) to low resistance state (LRS) under positive bias is observed, while it returns to HRS under negative voltage. The endurance characteristic and retention characteristic are illustrated in Fig. 3. The device is able to keep a nearly 30 times big window even after 60 cycles and also keep the window for 10^5 s. The unique feature of this device does not require high voltage in the forming process. As a result, this process not only protects devices from irreversible breakdown, but also simplifies the external circuits of ReRAM.

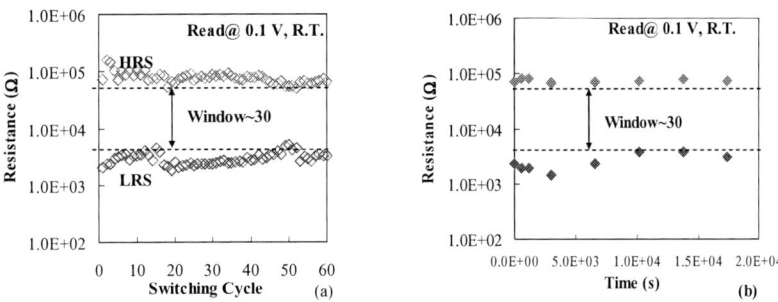

Fig. 3. (a) Endurance and (b) retention characteristics of W/CeO₂(20 nm)/Si(1 nm)/TiN device.

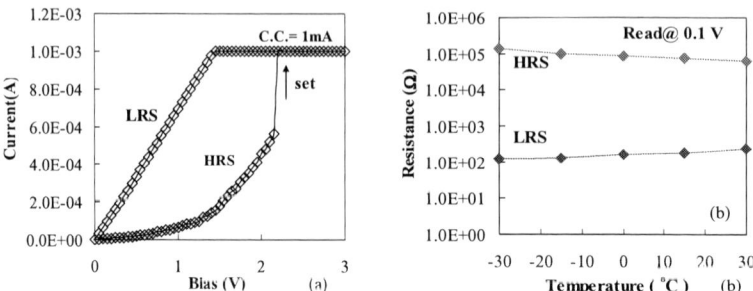

Fig. 4. (a) I-V curve of set process in linear scale, and (b) temperature dependence of HRS and LRS of W/CeO$_2$(20 nm)/Si(1 nm)/TiN device.

Figure 4(a) shows the I-V relationships at HRS and LRS, whose conduction mechanisms are clearly different to each other. Current and voltage follow linear relation, in other words ohmic relation, at LRS and implies the formation of conductive filament (CF). Figure 4(b) shows the temperature dependence of HRS and LRS. With increase in bath temperature, resistance of HRS decreases like a semiconductor. On the other hand, with the increase in the bath temperature resistance of LRS gradually increase like a conductor..

The Effect of the Si Buffer Layer

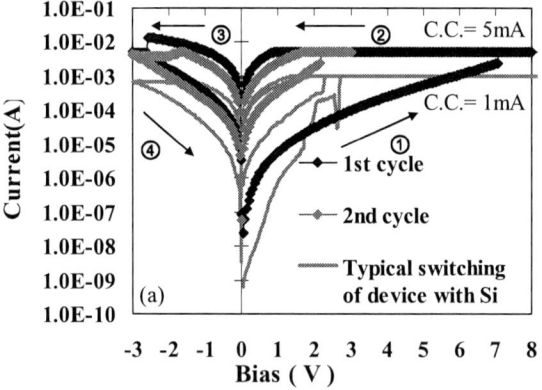

Fig. 5 (a) Typical I-V curve of W/CeO$_2$(20 nm)/TiN device. A large voltage is necessary for the forming process at the first cycle. Typical switching behavior of the device with 1-nm-thick Si layer is also shown by the grey lines.

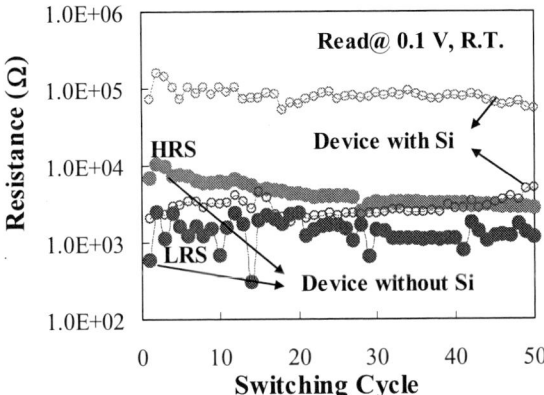

Fig. 5 (b) Endurance characteristic of the W/CeO$_2$(20 nm)/TiN device (filled circle) and that of the device with 1-nm-thick Si layer is also shown (open circle).

The influence of Si buffer layer on the resistive switching of W/CeO$_2$/TiN device, which was treated in same annealing condition, was also investigated. As shown in Fig. 5(a), the device without Si buffer layer requires a high voltage for the forming process. The device without Si buffer layers requires a high CC compared with the device with Si buffer layer. Figure 5(b) shows the endurance characteristic of the devices, and the window is relatively small (about 10) and degrades quickly in accordance with switching is repeated. Furthermore, the resistance at HRS of device without Si buffer layer is far lower than that of the device with Si buffer layer. This implies that in most part of cerium oxide the irreversible breakdown of oxide film occurs in the device without Si buffer layer. By inserting a Si buffer layer, the resistive switching is easier to be triggered at lower voltage with lower CC. This protects the device from hard breakdown. Considering the low forming voltage and the reaction between Si and CeO$_2$, the following model is proposed to explain the influence of the Si buffer layer on the resistive switching of CeO$_2$ film.

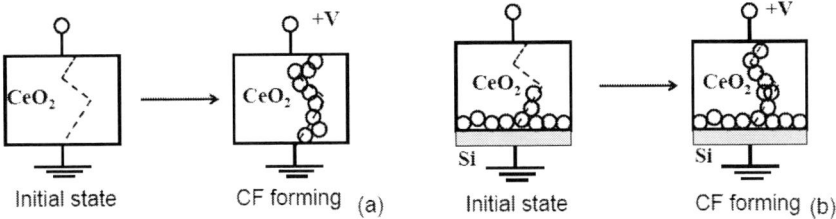

Fig. 6. Schematic illustration of forming process: (a) the device without Si buffer layer, and (b) device with Si buffer layer. The dash line represents the grain boundary in CeO$_2$ film and the open circle represents the oxygen vacancies.

Proposed Model for the Infulence of Si Buffer Layer on Resistive Switching

Although the conductive filaments (CFs) was not directly observed in this work, we apply the model based on the formation of CFs to explain the experimental results by considering ohmic relation in LRS and the forcing process. The model is schematically illustrated in Fig. 6. Figure 6(a) shows the normal forming process of device without Si buffer layer. There is not any pre-existed CFs in the thin film at initial state, and thus the formation of CFs at fresh devices requires a large voltage. After the formation of CFs, the large voltage is not necessary for the set and reset processes are just the partially rapture and formation of the CFs. Fig. 6(b) shows the formation of CFs in the devices with silicon buffer layer. After suitable thermal treatment, for example, in N_2 at $400^\circ C$ for 30 s, oxygen vacancies can be introduced near the interface by the reaction between cerium oxide and silicon ($CeO_2 + Si = CeSiO_4 + 2V_o$¨ or $Ce_2O_3 + Si = Ce_2SiO_5 + 2V_o$¨). As a result, some of oxygen vacancies diffuse along the GBs and thus partial part of CFs have been formed at the initial state, which allows the forming process to be finished at lower voltage. Consequently, hard breakdown of the oxide film caused by high voltage is avoided and the device is able to work at low compliance current.

The Thickness Effect of Si Buffer Layer:

The influence of Si buffer-layer-thickness on the resistive switching of W/CeO_2 (20 nm)/Si(2nm)/TiN device, which was treated by same annealing condition, was also studied, whose typical I-V curve and C-V characteristics and the change in resistance as a function of switching cycles are shown in Figs. 7 and 8, respectively, where the corresponding values of the device with 1 nm thick Si buffer layer are also shown for comparison. As shown in Fig. 8, the window of the device with 2 nm thick Si buffer layer is smaller than those with 1 nm thick Si buffer layer. Furthermore, two kinds of devices have similar LRS and the difference comes from HRS. Using the aforementioned model, the thickness effect of the Si buffer layer can be explained by the following mechanism. Although thicker Si layer results in the introduction of lager amount of oxygen vacancies, the formation of CFs is largely determined by set voltage and CC. Because the devices with 1 nm and 2 nm thick Si buffer layer are operated at close voltage with the same CC,

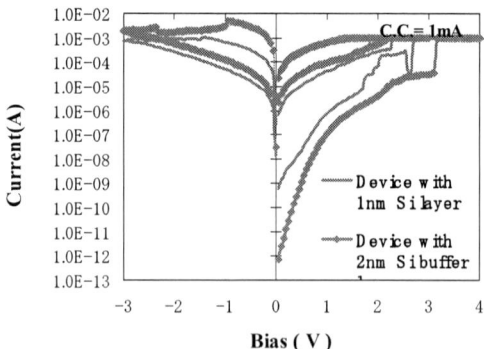

Fig.7. Typical forming process and resistive switching of the device with 1 nm and 2 nm thick Si buffer layer.

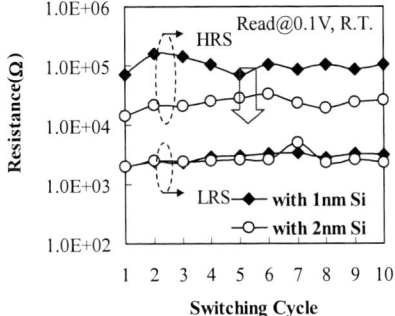

Fig. 8 Resistive switching vs switching cycles of devices with
1 nm (filled square) and 2 nm (open circle) thick Si buffer layer

similar CFs are formed to have similar LRS in these devices. However, lager amount of oxygen vacancies make the rapture of CFs become more difficult. As a result, the device with 2 nm Si layer is not able to reset to a HRS as high as the one of the device with 1nm, which also can be observed in Fig.7. The influence of the Si buffer-layer-thickness on the resistive switching indicates that in order to achieve the best performance of the devices, the thickness of Si buffer layer should be optimized.

Conclusion

In summary, $W/CeO_2/Si/TiN$ ReRAM device, in which the Si buffer layer is used, was found to achieve better memory performance compared with the device without Si layer. The Si buffer layer helps the device finish forming process without requiring high voltage, and thus enlarges the window, improves the endurance characteristic and also lowers power consumption. Furthermore, the influence of the silicon buffer layer on the resistive switching process can be explained by the formation of Ce-silicate, which introduces additional oxygen vacancies to the interface between the Si layer and the CeO_2 layer. Finally, it was found from the study on the influence of the Si buffer-layer-thickness that thickness of Si buffer layer should be optimized to obtain the best performance of the device.

References

1. A. Sawa, *Materials today*, **11**, 5 (2008).
2. Seo et al., *Appl. Phys. Lett.*, **85**, 23 (2004).
3. Yang et al., *Nature nanotechnology*, **3**, p.429 (2008).
4. Szot et al., *Nature materials*, **5**, p. 312 (2006).
5. Maimon J et al., Chalcogenide-based non-volatile memory technology. IEEE Aerospace 2001 (2001).
6. R. Waser et al., IEDM, p.289 (2008).
7. R. Waser et al., *Nature Materials*, **6**, p. 833 (2007).
8. K. Kakushima et al., VLSI Tech. Dig. p.69-70(2010).

604

ECS Transactions, 35 (4) 605-627 (2011)
10.1149/1.3572308 ©The Electrochemical Society

Electrically Detected Magnetic Resonance in Dielectric Semiconductor Systems of Current Interest

P.M. Lenahan[a], C.J. Cochrane[a], J.P. Campbell[b], and J.T Ryan[b]

[a]Pennsylvania State University, University Park, PA16802
Email:pmlesm@engr.psu.edu
[b]The National Institute of Standards and Technology, Gaithersburg, MD 20878

Several electrically detected magnetic resonance techniques provide insight into the physical and chemical structure of technologically significant deep level defects in solid state electronics. Spin dependent recombination is sensitive to deep level defects within semiconductors or at semiconductor dielectric interfaces. Spin dependent trap assisted tunneling can identify defects in dielectric films and , under some circumstances, can provide fairly precise information relating energy levels to physical/ structural information about the defects under observation.

Introduction

The performance of solid state devices is inevitably affected by the presence of point defects with energy levels within semiconductor or insulator band gaps. Although many electrical measurements such as deep level transient spectroscopy, capacitance versus voltage, charge pumping, gated diode measurements… can provide information about defect energy levels and densities, only electron paramagnetic resonance (EPR) offers the analytical power necessary to provide detailed structural and chemical information about the underlying point defect structures [1,2]. Unfortunately, conventional EPR sensitivity is about 10^{10}, a sensitivity too low for measurements in almost any fully processed solid state device. Conventional EPR also has the disadvantage that it is sensitive to all paramagnetic defects within a sample under study, whereas the active volume of a solid state device, for example MOSFETs, is typically a very small fraction of the semiconductor substrate volume. Electrically detected magnetic resonance offers several significant advantages over conventional EPR in device physics studies [3-11]. The sensitivity of EDMR is at least seven orders of magnitude greater than that of conventional EPR [5-11]. EDMR is sensitive only to electrically active defects within the active area of the device under study [5-11]. EDMR can, at least crudely, provide information about both the spatial distribution and the energy levels of the defects under study.

Most EDMR studies to date have utilized spin dependent recombination (SDR). These measurements, like conventional EPR are sensitive to paramagnetic defects.

605

Paramagnetic defects usually have an odd number of electrons, though under certain circumstances, paramagnetic ESR active defects with an even number of electrons may also be observed, for example defects with an electron spin of one. Both conventional ESR measurements and EDMR measurements involve the simultaneous application of a large, slowly varying, magnetic field and a microwave frequency field. Resonance occurs when the microwave photon energy equals the Zeeman splitting energy of the electrons. In the simplest case, $h\nu = g\beta H$. In this expression, h is Planck's constant, ν is the microwave frequency, β is the Bohr magneton, and H is the magnetic field at resonance [1,2]. The g depends upon the relationship between the magnetic field vector and the orientation of the defect under observation. The behavior of g is expressed as a matrix sometimes called the g tensor [1,2]. The magnetic resonance condition becomes more complex when the paramagnetic site involves magnetic nuclei. The nuclear moment generates a local magnetic field which alters the magnetic resonance condition, depending on the nuclear spin quantum number, the magnitude of the nuclear moment, and the electron wave function [1,2]. The effect of a nuclear moment on the resonance condition is referred to as a hyperfine interaction [1,2]. Additional factors can also play a role in determining EPR spectra, for example, when multiple electrons are involved at a paramagnetic site, so called zero-field splitting or fine structure can also affect the EPR spectrum [1].

An analysis of EPR spectra utilizing the well developed understanding of these factors quite frequently allows the physical and structural nature of the defects under study to be identified. The atoms involved as well as electron wave function parameters are frequently elucidated from an analysis of EPR spectra. Since the EDMR spectra are very nearly identical to those of conventional EPR, this analytical power is also available in the EDMR measurement.

Most EDMR measurements have involved SDR. In SDR [3-11], one detects a magnetic resonance spectrum by measuring a change in device recombination current dominated or at least strongly influenced by recombination of electrons and holes at a deep level defect. The SDR effect can be understood in a qualitative way by considering the Shockley-Read-Hall model for recombination through deep level defects [3-11]. The process involves the capture of an electron and then a hole at a deep level center. The sequence can, of course, be reversed. It is difficult to envision such a process without the involvement of a paramagnetic charge carrier capture event at a paramagnetic deep level. The capture of a paramagnetic charge carrier at a paramagnetic defect is spin dependent. Consider a simple "dangling bond." If the dangling bond is occupied by an unpaired electron and then captures a conduction electron, the two electrons must have different spin quantum numbers. The recombination process must involve a singlet. This is so because of the Pauli exclusion principle. If the conduction electron and the deep-level defect electrons have the same spin quantum number, the process is forbidden. However, in the magnetic resonance process, the electron's spin is flipped from one quantum number to the other when the resonance condition is satisfied. Therefore, flipping the deep-level defect spin makes a previously forbidden transition allowed, thereby increasing the recombination current [3-11].

Lepine [3] was first to address potential models for SDR. He envisioned a process in which two spins interact essentially in an instantaneous collision; in his model, the size of the effect is limited by the product of the polarization of the two spin systems; that of a charge carrier and that of a paramagnetic deep level defect. For room temperature measurements involving simple defects, the product of the polarization of two spin systems would be approximately 10^{-6} at the widely utilized X-band frequencies and corresponding fields. This is so because the polarization of a system of very weakly interacting electron spins, with g matrix components all close to 2, is approximately $g\beta HkT$, where β is the Bohr magneton, H is the magnetic field, k is Boltzmann's constant, and T is absolute temperature [1]. In such measurements, all the g matrix components are typically all close to 2, and the magnetic field at resonance is typically about 3400 Gauss. If we take room temperature to be 294 K, these parameters yield a polarization of about 1×10^{-3} for each of the spin systems and thus a product of about 1×10^{-6}. In a Lepine like process then, the maximum possible effect would be a current change in about one part in one million. Kaplan et al. [4] proposed an SDR model in which they envisioned a coupling between a pair of spins for a finite time. The model of Kaplan et al. [4] could be consistent with a much larger effect.

There are only a handful of reports of EDMR observations via SDT in the literature [8,9,10,11] and very little in the way of detailed models; nevertheless, the fundamental concept is similar to that of SDR. The spin dependent tunneling event involves two spins. One might envision a conduction electron tunneling into a deep-level defect or tunneling from one paramagnetic deep-level defect to another paramagnetic deep-level defect. A tunneling event involving two unpaired electrons with the same spin quantum numbers will be forbidden. A tunneling event involving two unpaired electrons with different spin quantum numbers could be allowed. Thus, satisfying the resonance condition of a paramagnetic deep-level defect involved in the tunneling process should also be detectable, if one can measure a device current due to or partially due to trap-assisted tunneling through the defect in question.

These two EDMR techniques have been applied to a number of systems of interest in solid state electronics. This paper reviews several recent EDMR studies involving important MOS reliability problems, the negative bias temperature instability (NBTI) and stress induced leakage currents as well as new materials based MOS technology, specifically SiC based MOS technology. In these studies, EDMR measurements have been carried out in parallel with more conventional electronic measurements such as DCIV and simple tunneling current versus voltage measurements. The review illustrates the power of these EDMR techniques with regard to chemical and structural information as well as energy levels and physical location.

Spin Dependent Recombination and Interface/ Near Interface Traps

 SDR has been applied to several metal oxide semiconductor (MOS) systems [5-11]. A comparison of the SDR response in two of them, illustrates both the analytical power of the technique as well as the (as yet somewhat limited) capability of the technique to provide information about the spatial extent of specific defect centers. In the silicon/silicon dioxide MOS system, interface trapping is dominated by defect centers precisely at the semiconductor/dielectric interface. In the "new materials" silicon carbide/silicon dioxide MOS system, this is not necessarily the case.

The Negative Bias Temperature Instability in Si/SiO₂ Interface Traps

A silicon/silicon dioxide MOS problem of current interest involves an instability in p-metal oxide silicon field effect transistors (p-MOSFETS) which occurs when the devices are subjected to negative bias at moderately elevated temperatures [12]. The so called negative bias temperature instability (NBTI) is arguably the most important reliability problem in present day conventional MOS technology. Recently, Campbell et al. [7,8] have explored the generation of interface trap generation due to NBTI in both conventional silicon dioxide and nitride silicon dioxide based MOS devices [7,8]. Some representative NBTI results are illustrated below for the case of a pure silicon dioxide gate device. Negative bias temperature stress (NBTS) was applied to the 7.5 nm gate device at 140°C for 250,000 s with −5.7 V on the gate contact. Both pre- and post-NBTS gate-controlled diode DCIV measurements on a device are shown in Fig. 1.

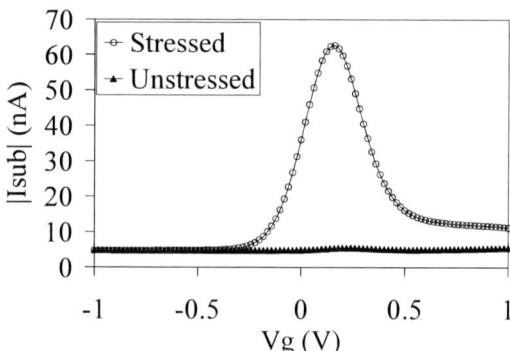

Fig. 1. Gate-controlled diode DCIV measurements on a pMOSFET before and after the application of NBTS (140°C for 250,000 s with −5.7 V on the gate contact).

In the gate-controlled diode DCIV measurement, the source and drain are shorted, the source/drain to substrate diode is slightly forward biased (0.33 V in these measurements) and the source/drain to substrate recombination current is monitored as a function of gate bias. Fitzgerald and Grove [13] found that, if the deep level defects are all or nearly all at the semiconductor dielectric interface, a peak appears in recombination current at the biasing condition which yields equal densities of conduction electrons and hoes at the semiconductor dielectric interface. This peak is characterized by ΔIsub (difference between the peak and the baseline) which scales with the interface state density and can be approximated by

$$\Delta I_{sub} = (1/2)\ qn_i\sigma_S v_{th}D_{it}Aq|V_F|\ \exp(q|V_F|/2kT) \tag{1}$$

where q is the electronic charge, ni is the intrinsic number of carriers, σ_S is the geometric mean of the electron and hole capture cross section, v_{th} is the thermal velocity, A is the effective gate lateral area, V_F is the forward bias applied to the source/drain to substrate junction, k is Boltzmann's constant, and T is the temperature. The change in recombination current (ΔI_{sub}) is proportional to the interface state density within an energy window $1/2\ q|VF|$ [13]. These results illustrate all known observations that NBTS induces a large increase in the interface state density. Following the analysis of Fitzgerald and Grove [13], D_{it} values were extracted for pre-NBTS (9×10^9 cm^{-2}eV^{-1}) and post-NBTS (7×10^{11} cm^{-2}eV^{-1}), taking the mean capture cross section $\sigma_s = 1.4\times10^{-16}$ cm^2.

Fig. 2 illustrates the corresponding pre- and post-NBTS SDR traces with the magnetic field vector perpendicular to the (100) surface. The interface defect density in the unstressed device is below the Campbell et al. SDR detection limit. After NBTS, two strong signals appear at $g = 2.0057 \pm 0.0003$ and at $g = 2.0031 \pm 0.0003$ (Fig. 2). They attribute the $g = 2.0057$ signal to P_{b0} centers and the $g = 2.0031$ signal to P_{b1} centers. P_{b0} and P_{b1} defects are both silicon dangling bond defects in which the central silicon atom is back-bonded to three other silicon atoms precisely at the Si/SiO$_2$ boundary [2,6,14-16].

Figs. 3 and 4 show schematic drawings of P_{b0} and P_{b1} centers. The P_{b0} center structure is well understood; the P_{b1} drawing should be viewed as, at best a provisional sketch. The main differences between the two defects are in the dangling bond axes of symmetry [2,6,14-16] and electronic density of states [17-20]. The P_{b0} dangling-bond orbital points along the [111] directions, while the P_{b1} dangling-bond orbital points along the [211] directions. Both the (111) Si P_b center and (100) Si analog, the P_{b0} center, have a broadly peaked density of states centered about midgap with the (+/ 0) and (0/−) transitions separated by about 0.7 eV [17-20]. The P_{b1} density of states is much narrower and is skewed towards the lower part of the silicon band gap [17].

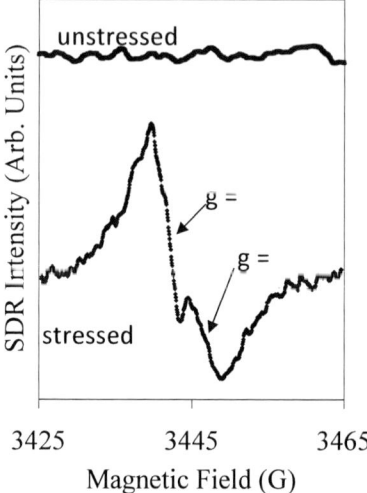

Fig. 2. SDR traces of pMOSFET with the magnetic field vector perpendicular to the (100) surface both before and after the application of NBTS (140°C for 250,000 s with −5.7 V on the gate contact.

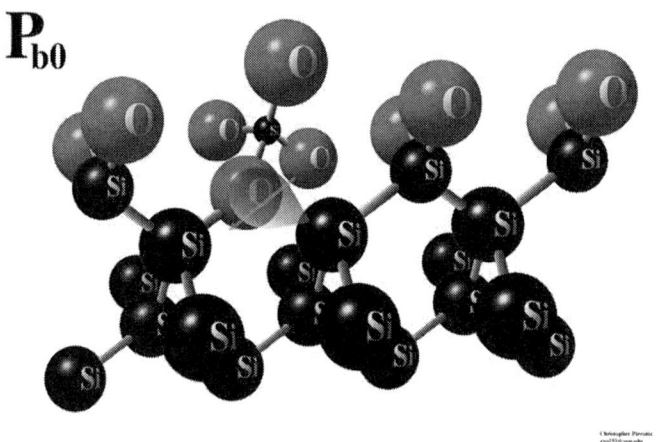

Fig. 3. Schematic drawing of the P_{b0} Si/SiO$_2$ interface defect.

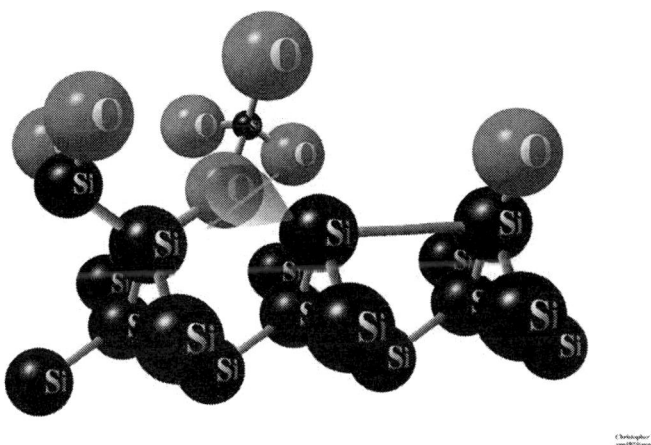

Fig. 4. Schematic drawing of the P_{b1} Si/SiO$_2$ interface defect. It should be noted that this drawing in particular should be viewed as a cartoon which merely indicates that P_{b1} involves a silicon back-bonded to three silicons at the interface.

Fig. 5 illustrates the P_{b0} and P_{b1} SDR signal amplitudes as a function of gate bias for the same pMOSFET that was subjected to 140°C for 250,000 s with −5.7 V on the gate contact. The dashed lines in Fig. 5 are only a guide for the eye. Note that both curves are strongly peaked, indicating that both P_{b0} and P_{b1} centers must be present at a single plane, in this case, that of the silicon/silicon dioxide interface. The fact that these amplitudes are large only when the quasi–Fermi levels are close to symmetrically split about the interface intrinsic energy level is strong evidence for this physical location.

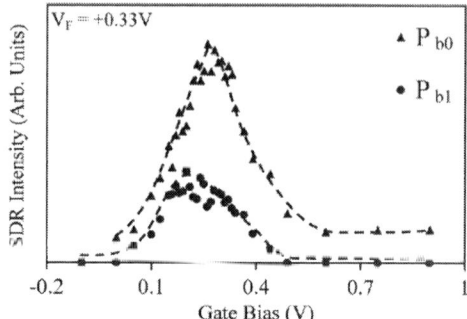

Fig.5 SDR-derived P_{b0} and P_{b1} signal amplitudes as a function of applied gate bias for the pMOSFET which had been subjected to NBTS (140°C for 250,000 s with −5.7 V on the gate contact). The dashed lines are included as merely a guide for the eye. In these measurements, the magnetic field vector is perpendicular to the (100) surface.

These recent SDR results of Campbell et al.[7,8] involve defects which are precisely located at a semiconductor/dielectric boundary. The strongly peaked SDR response corresponding to equal numbers of electrons and holes at the Si/SiO$_2$ boundary is strong evidence for this location (It should be noted that this location may be inferred from other results as well [2]). Recent EDMR results of Cochrane et al. [9] on a "new materials" based MOS system, that of silicon carbide/silicon dioxide, yields a somewhat different conclusion regarding defect location. A comparison of the results of Cochrane et al.[9] and those of Campbell et al. illustrate the (so far limited and qualitative/ semi-quantitative) capability of SDR to provide information about the spatial distribution of electrically active defects.

Interface/Near Interface Traps in SiC/SiO$_2$ MOS Systems

Quite recently, Cochrane et al. [9] reported on a fairly extensive EDMR study of SiC based MOS systems. This study is of general interest for several reasons, among them, the potential of SiC based MOS technology and, more broadly, the materials physics issues of compound semiconductor based MOS technology. There is growing interest in metal oxide semiconductor field effect transistors composed of materials other than the classical Si/SiO$_2$ chemistry. Among the new material systems, SiC/SiO$_2$ is perhaps the most promising. SiC offers great promise for MOSFETs in high-power and high-temperature applications [21]. Unfortunately, these devices are plagued with performance limiting defects which are frequently viewed primarily as interface traps. It is often explicitly or implicitly assumed that these traps exist essentially right at the SiC/SiO$_2$ boundary and some recent studies suggest that the traps are quite similar to those which dominate the classical Si/SiO$_2$ system, semiconductor/insulator interface "dangling bond" centers [22]. Several fundamental questions about these defects have yet to be resolved. (1) What is the physical and chemical nature of the trapping centers? (2) Do these traps have a consistently defined physical location, for example, the semiconductor/ insulator interface? (3) Do these traps have a common origin in all or nearly all SiC/SiO_2 structures? This is so in the Si/SiO_2 case.

Cochrane et al. [9] found a variety of SDR responses, which were extremely dependent upon device processing. Their study involved SiC lateral n-channel MOSFETs with gate areas of 100μm x 100μm. The devices all had essentially the same geometry. All were n-channel lateral MOSFETs with 50 to 70 nm gate dielectrics and 10^4 $μm^2$ gate areas. In the representative results illustrated herein, all transistors were doped during epitaxial growth of the silicon carbide.

Figure 6 illustrates EDMR results on a silicon carbide device with an entirely deposited oxide. Figure 6(a) illustrates a narrow EDMR scan and Figure 6(b) a wider scan. A gate bias of 4 V was applied during the measurement. The traces illustrate a single strong central line accompanied by much weaker side peaks. This line has an anisotropic g, with $g_{\parallel}$ =2.0026 ± 0.0002 and $g_{\perp}$=2.0010 ± 0.0002. This anisotropy is illustrated in the g-map of Fig. 7. The symmetry axis is, within experimental error, the crystalline c-axis, which is very close to the SiC/SiO_2 surface normal.

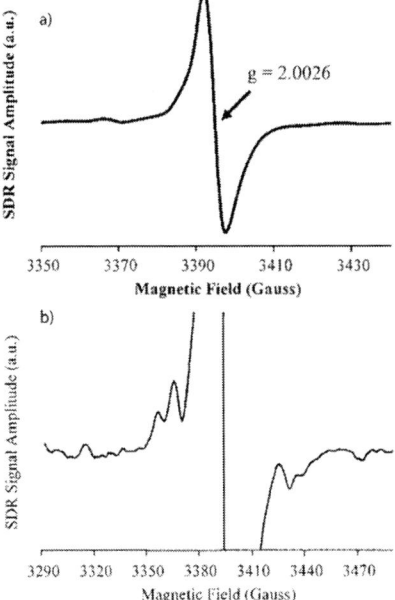

Fig. 6. Narrow (a) and wider (b) scan EDMR traces taken on a deposited gate dielectric 4H–SiC MOSFET configured as a gated diode. The magnetic field is approximately parallel to the crystalline c-axis and also very nearly parallel to the SiC/SiO2 interface normal. The traces show a strong central line; at this orientation $g_{\parallel}$=2.0026 ± 0.0002.

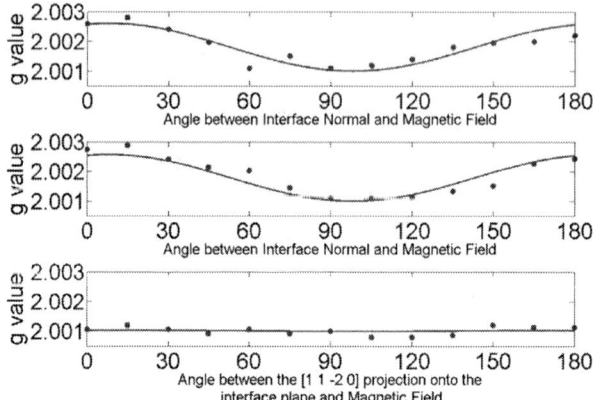

Fig. 7. The g vs. magnetic field orientation with respect to the surface normal rotation about the three perpendicular axes. (a) This axis corresponds approximately to the [112¯0] axis and the surface normal is 8° from the [0001] crystalline axis. (b) The g vs magnetic field orientation with respect to the surface normal rotation about the integrated circuit side edge axis. This axis corresponds approximately to the [11¯00] axis. (c) The g vs magnetic field orientation with respect to the edge axis of the integrated circuit for rotation around the surface normal. Note: the solid lines correspond to calculated g values utilizing the correct crystalline orientation and $g_{\parallel}$=2.0026 and $g_{\perp}$=2.0010.

Elementary magnetic resonance theory [1,2] predicts that simple "dangling bond" defects with an unpaired electron primarily localized on one p-character orbital on an atom with zero nuclear spin would yield a magnetic resonance spectrum consistent with the results of figure 6 and 7. (About 95% of silicon nuclei possess zero nuclear spin.) Simple theory [1,2] predict that the resonance g matrix would be axially symmetric with the axis of symmetry of the high p character orbital. Theory predicts that the g corresponding to this orientation, $g_{\parallel} \approx 2.0023$. The largest deviation from this value would occur when the magnetic field perpendicular to the symmetry axis, this would be $g_{\perp}$ [1,2]. A simple "ball and stick" model at the SiC/SiO$_2$ interface does indicate the possible presence of dangling bond defects with symmetry axes corresponding very nearly to the SiC/SiO$_2$ interface normal. If these spectra came only from conventional EPR, one might erroneously conclude that the observed defects are all SiC/SiO$_2$ interface dangling bonds much like the Si/SiO$_2$ P$_b$ centers. However, as mentioned previously, SDR can provide some information about the physical distribution of defects.

Figure 8 illustrates DCIV and EDMR amplitudes versus gate voltage from the transistor utilized in Figs. 6 and 7, a 4H-SiC device with a deposited oxide/nitride/oxide dielectric 50 nm thick. Fig. 8(a) illustrates the EDMR amplitude as a function of gate bias and Fig. 8(b) illustrates the recombination current, that is the DCIV measurement as a function of gate bias. Both responses are peaked near zero volts, though the DCIV peak is rather weak. Under this biasing condition, the SiC/SiO$_2$ interface region is depleted. As the analysis of Grove and co-workers [13] indicates, at modest junction forward biases, 2.15 V in this case, the current is dominated by recombination in the depletion region. With a gate bias which provides equal numbers of electrons and holes at the location of the highest density of deep levels, the recombination current will peak [13]. This is so because, if nearly all the deep-level defects responsible for recombination are at a specific location, the recombination current will be maximized if equal densities of electrons and holes are present there [13]. In the Si/SiO_2 system, this location is invariably the Si/SiO_2 interface. If the SiC device had a physical distribution of defects similar to that of a silicon device (near perfect bulk semiconductor, deep-level defects predominately at the semiconductor/dielectric boundary) the EDMR amplitude would be expected to exhibit a fairly strong peak but only a very weak peak is evident in the DCIV curve. This indicates that, in these devices, there is not a particularly predominant density of defects at a specific plane within the device (the SiC/SiO_2 interface) but a broader distribution; so the result indicates a moderately high density of deep-level defects at the SiC/SiO_2 boundary (thus the weak peak) but that high densities extend into the "bulk" of the SiC epilayer. The overall similarity between the DCIV and EDMR curves suggests that the defects observed in magnetic resonance are the dominating deep levels in this device

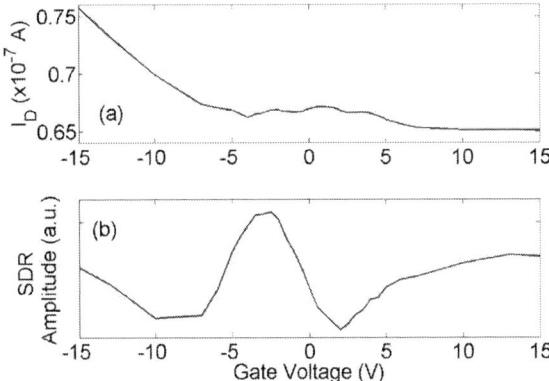

Fig. 8 DCIV (a) and EDMR (b) amplitude vs gate voltage for the 4H-SiC transistor utilized in Figs. 6 and 7

Figure 9 illustrates EDMR results on a MOSFET built on a 6H–SiC substrate. Figure 9(a) illustrates a narrow EDMR scan and Fig. 10(b) illustrates a wider scan. A gate bias of 5 V was applied during the measurement. In this case the strong central line has an isotropic g; g=2.0026 ± 0.0002. In these traces, strong but poorly resolved side peaks appear symmetrically located about the central line separated by about 15 G. Much weaker peaks are also present with a separation of about 60 to 70 G.

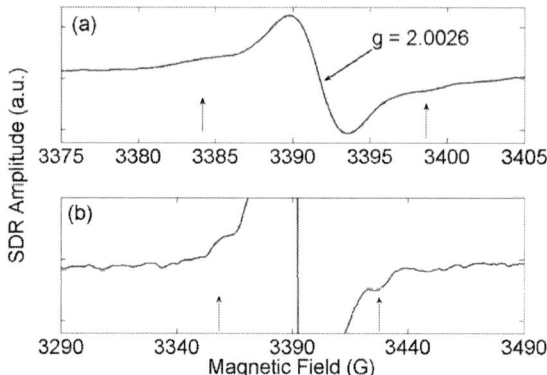

Fig. 9 Narrow (a) and wider (b) scan EDMR traces taken on sample B, a thermal oxide/silicon nitride/thermal oxide gate 6H–SiC MOSFET configured as a gated diode. The magnetic field is approximately parallel to the crystalline c-axis and is also very nearly parallel to the SiC/SiO$_2$ interface normal. The traces show a strong central line with g=2.0026 ± 0.002 and strong but poorly resolved side peaks separated by 14 G and much weaker and more distant side peaks separated by 60 to 70 G.

The results of figures 9 and 10, a strong central line accompanied by a strong (through poorly resolved) side peaks near the center line and much weaker more distant side peaks can be interpreted [9] in terms of a silicon vacancy. As discussed in some detail by Cochrane et al. [9] such a pattern is qualitatively consistent with both an extensive conventional EPR literature on large volume samples [23,24] and a rudimentary analysis of the silicon vacancy defect structure [9]. A cartoon model of a silicon vacancy is provided by figure 10. (It should be noted that the correspondence between the conventional EPR spectrum reported for this defect and the SDR/EDMR results are only qualitative. [9,23,24] It is conceivable that the assignment of Cochrane et al is in error but the fact that the SDR results involve measurements in a highly defective SiC/SiO$_2$ interface region may plausibly account for the differences.)

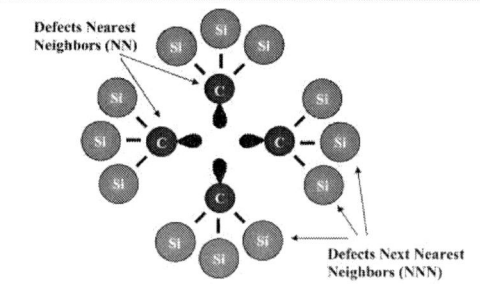

Fig. 10. A cartoon schematic of a silicon vacancy in SiC.

Figure 11 illustrates DCIV and EDMR amplitudes versus gate voltage from the transistors utilized in Fig 9, a device with a 50 nm dielectric with a thermal oxide on 6H–SiC. Figure 10(a) illustrates the EDMR amplitude as a function of gate bias and Figure 10(b) illustrates the recombination current (DCIV) versus gate bias. Note that both the EDMR and DCIV responses are significantly different from the results of Fig. 8. In this case, the EDMR and DCIV amplitudes are quite strongly peaked at modest gate bias; EDMR signals virtually disappear at large positive and negative voltages. These results indicate that, in this device, there is a specific plane (the SiC/SiO_2 interface) at which the deep-level defect density is much higher than in the near-interface SiC. The fairly close similarity between the DCIV and EDMR suggest that, in this device, the defects observed in magnetic resonance are largely responsible for the dominating deep levels.

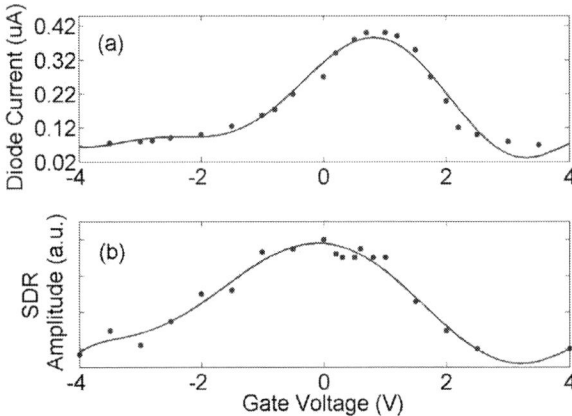

Fig. 11 DCIV (a) and SDR (b) amplitude vs. gate voltage for the 6H SiC sample utilized in Figure 9. Note the qualitative correspondence between the SDR and DCIV responses.

The Utility of EDMR/SDR in Providing Information About Defect Location

The results just discussed show that SDR results in silicon/silicon dioxide systems and silicon carbide silicon dioxide systems are quite significantly different. The differences provide useful information. In MOSFETs based on the silicon/silicon dioxide system, as the representative results of Campbell illustrate, and as other results in the literature not shown here also illustrate [2,5,6] the electrically active defects of interest are truly INTERFACE traps. This is indicated by the strongly peaked SDR response and the strongly peaked DCIV response. The dominating defects must be primarily present at a plane, the semiconductor/ dielectric interface. Strong clues supporting this conclusion are also provided by the magnetic resonance spectrum itself [2]. In MOSFETs based on silicon carbide/ silicon dioxide systems, the SDR and DCIV response does not necessarily have this simple peaked structure, indicating a more complex system of interface/ near interface trapping defects. The magnetic resonance spectra are also somewhat more complex with significant differences appearing in differently processed devices and in devices prepared on different polytypes.

Energy Resolved Spin Dependent Trap Assisted Tunneling in Very Thin Dielectrics

Recently Ryan et al. [10,11] reported on a very simple approach to spin dependent trap assisted tunneling (SDT) which allows for the evaluation of the energy levels of traps in thin dielectric films. The approach exploits advantages provided by extremely thin 1.2 nm effective oxide thickness (EOT) dielectrics. The enormous difference between the very high capacitance of the thin dielectric and the much lower capacitance of the Si depletion layer allows a modest applied voltage to sweep through most of the Si band gap with very little net potential drop across the dielectric. The approach yields direct information about defect energy levels and provides magnetic resonance spectra with excellent sensitivity. The dielectrics utilized in their recent study were silicon oxynitride films quite widely utilized in essentially "state of the art" complementary metal oxide silicon (CMOS) integrated circuits. Deep level defects were generated within these films by subjecting them to high electric fields. The defect generating conditions were chosen because they represent the circumstances under which an important instability in present day CMOS integrated circuits occurs: stress induced leakage current (SILC) [25,26]They showed that SDT can be utilized to extract information about the electronic levels of dominating point defects generated by the stressing of this system. The approach is almost certainly widely applicable to the study of other important defects. The SDT samples utilized in their study were 1.2 nm EOT nitrided SiO_2 p-channel metal-oxide-silicon capacitors with p+ poly-Si gates. The very high p+ doping of the gate effectively pins the gate Fermi energy (E_F) very close to the gate Si valence band edge (V_{BE}). The gate areas were 10^4 μm^2. Deep level defects were generated in the dielectrics by room temperature stressing of 2.2 V for 10^4 s.

Figure 12 illustrates a normalized tunneling current, a ratio of tunneling current measurements taken before and after stressing, which emphasizes the contribution of the (stressing induced) trap assisted tunneling component in the total current. Here, J_0 is the gate current density pre-stress and ΔJ is the gate current density post -stress (J_t) minus J_0. The $\Delta J / J_0$ versus V_G plot illustrates the difference between the I_G-V_G curves before and after stress due to trap assisted tunneling current in the post-stress I_G-V_G measurement. The peak of this curve (around V_G=0.35 V) corresponds to the maximum fractional contribution of trap assisted tunneling current, not the maximum total current. This is so because the total current is the sum of the trap assisted current and all other sources of current. The only other significant source of current in these very thin dielectrics, direct tunneling, is exponentially increasing with voltage. At higher voltages the current is dominated by direct band to band tunneling which overwhelms the trap assisted tunneling current. Values for $\Delta J / J_0$ around V_G=0 V are not included because the amplitude of the currents are below the detection limit of the I_G-V_G measurements.

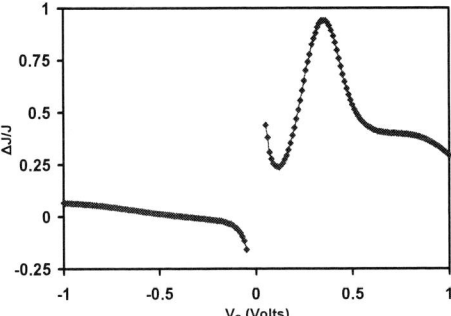

Figure 12: $\Delta J/J$ vs. V_G. The peak in the curve is caused by a trap assisted tunneling current in the stressed I_G-V_G measurement.

Figure 13 illustrates a representative SDT measurement taken with V_G biased to correspond to the peak in the $\Delta J/J_0$ curve in Fig.12, (V_G=0.35 V). In this figure, the measurement was made with the Si/dielectric interface normal parallel to the applied magnetic field (0°). The spectrum is a single line with a g of 2.0030 +/- 0.0002 and a line width of about 15 G. The spectrum does not change when the sample is rotated; this very strongly suggests that the defects are located in an amorphous material. If the defects existed at specific orientations, as they would in a crystalline environment or were precisely at the Si/dielectric interface, the g value would almost certainly change as the sample is rotated in the magnetic field. For example, as discussed earlier in this paper, the g values of the dominating interface defects in conventional Si/SiO$_2$, P$_b$ centers, change considerably as the sample is rotated in the magnetic field. The defects observed in this study do not follow such a pattern, ruling out a direct role for Si/dielectric P$_b$ centers in the spin dependent trap assisted tunneling process. The magnetic field orientation independence, the zero crossing g value of 2.003, and the 15 G linewith of the observed defect spectrum are all consistent with the K center found in Si$_3$N$_4$ and some SiO$_x$N$_y$ films [27,28,29] .

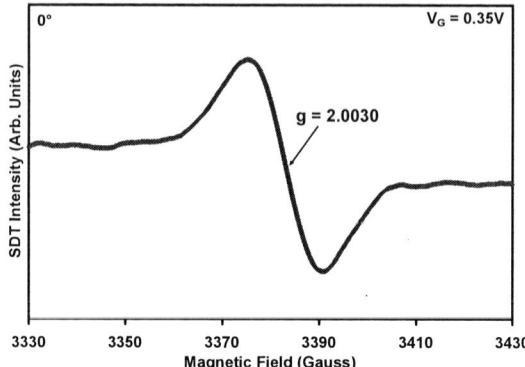

Figure 13: Representative SDT measurement taken with the gate biased to correspond to the peak in the $\Delta J/J$ curve of figure 2. The measurement was taken with the magnetic field parallel to the Si/dielectric interface normal

In Figure 14 we illustrate a cartoon figures of the K center, a silicon back bonded to three nitrogen atoms. Figure 15 illustrates a comparison between the normalized SDT intensities as a function of V_G (a) and the $\Delta J / J_0$ versus V_G (b) plot of Fig. 12. The normalization of Fig.15 (a) is achieved by dividing the spin dependent modification to the tunneling current (I_{SDT}) by the total dc current (I). The $I_{SDT}/$ I response very closely follows the characteristic trap assisted tunneling peak of Fig.15(b), a very strong indication that we are observing spin dependent trap assisted tunneling current due to the defects largely responsible for the tunneling current.

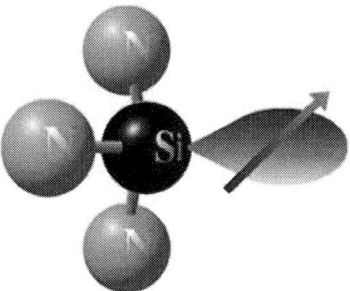

Figure 14: Schematic illustration of the K-center.

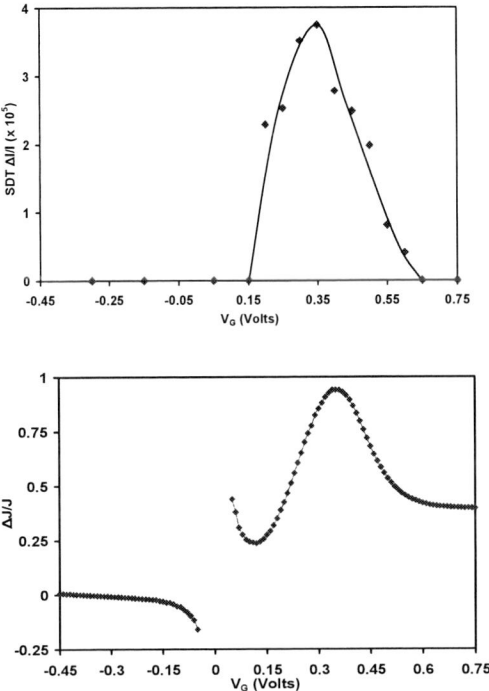

Figure 15: Comparison between SDT signal intensity ($\Delta I/I$) vs. V_G (a) and the $\Delta J/J$ vs. V_G curve (b) of figure 2. The SDT response ($\Delta I/I$) very closely follows the characteristic trap assisted tunneling peak of (b).

In an attempt to delineate between the spin dependent trap assisted tunneling current and the direct tunneling current, Fig.16 shows the spin dependent modification to the tunneling current (I_{SDT}) as a function of V_G. It peaks at about 0.5 V, indicating that, as one would expect, the peak at $V_G=0.35$ V in ($I_{SDT}/$ I) Fig.15(a) is shifted downward because direct tunneling overwhelms the trap assisted tunneling process at higher bias. Since the direct tunneling is not spin dependent, the SDT response is not affected by the large direct tunneling current response which overwhelms the "electrically" measured trap assisted tunneling current at higher bias.

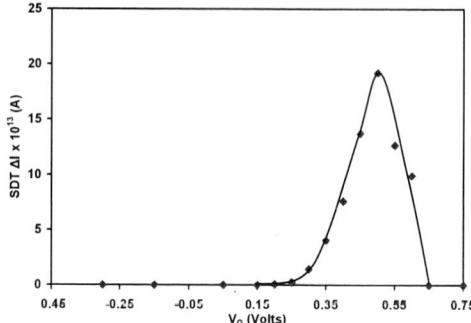

FIG. 16 SDT spin dependent modification to the tunneling current I_{SDT} as a function of V_G. Note that it peaks at about V_G=0.5 V indicating the peak at V_G=0.35 V in the SDT $I_{SDT}/$ I is shifted downward because direct tunneling overwhelms the trap assisted tunneling process at higher voltages

Figure 17 illustrates the poly-Si/SiO$_x$N$_y$/crystalline-Si SOS band diagram for three quite different biasing conditions: V_G=0, 0.55, and 1.0 V. For simplicity of presentation, only two levels of a single dielectric trap are included in diagrams. These band diagrams were calculated using the Boise State Univeristy band diagram program.[30] Note first that there is very little band bending in the dielectric at any of the illustrated biasing levels. The dielectric is so thin that the relationship between the crystalline-Si/dielectric E_F and the defect energy level is nearly independent of the physical position of the defect with respect to the crystalline-Si/dielectric interface. This is so because of the enormous difference between the capacitance of the 1.2 nm EOT dielectric and the much thicker Si depletion region. Nearly all the voltage appears across the Si. Figure 16 shows that the SDT response appears at a V_G of about 0.2 V, peaks at 0.5 V, and has completely disappeared at about 0.65 V. At V_G=0.2 V, where SDT appears, the crystalline-Si/dielectric E_F is 0.26 eV above the V_{BE}. At V_G=0.65 V, where the SDT disappears, the E_F is about 0.68 eV above the Si V_{BE}. This narrow response must reflect a narrow distribution in K center levels.

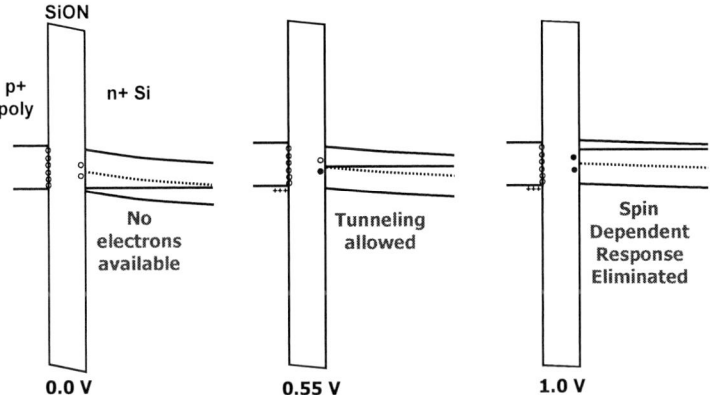

FIG.17. Energy band diagrams for the sample at three different values of V_G. Note that the only plausible explanation for the tunneling current must involve electron tunneling through defects with levels corresponding to the range of the silicon band gap. The simplified sketch illustrates two dielectric defect levels, consistent with experimental results.

An explanation of the response can be gleaned from a brief consideration of the physics of spin. The SDT process, like all EDMR processes, must involve a pair of spins initially separated physically. One of the spin sites is a K center. K centers, especially those nearest the crystalline-Si/dielectric boundary, can act like interface traps in that, as the E_F is advanced from the V_{BE} toward the conduction band edge C_{BE}, the empty dangling bond trap levels (+/0) will accept an electron as the E_F crosses the relevant energy. This process is not spin dependent, whether or not it involves paramagnetism at the K center site, it does not involve paramagnetism from the valence band. However, once the K center is rendered paramagnetic, interactions of the K center site with another paramagnetic site would be spin dependent and thus susceptible to SDT. Should the K center accept an additional electron, it would be rendered diamagnetic again, insensitive to the SDT process.

Consider tunneling of an electron from a paramagnetic K center site to another paramagnetic site in the (highly defective) poly-Si gate. The process would be allowed only if the unpaired electron spins have opposite spin quantum numbers. If the two sites had electron spins with the same spin quantum number, the tunneling process would be forbidden by the Pauli exclusion principle. However, if the K center electron spin were to be "flipped" via EPR ($h\upsilon = g\beta H$) the previously forbidden tunneling event would be allowed. Thus, magnetic resonance could modulate such a tunneling process. The SDT process would thus "turn on" when E_F crosses the energy level corresponding to the first K center electron (+/0) transition which places one electron in the defect's dangling bond orbital. Figure 18(a), a replotting of the results of Fig. 16 in which V_G is replaced by E_F, indicates that the SDT response begins to appear with E_F at about 0.26 eV above the V_{BE}. The process peaks with E_F at about 0.54 eV.

Very crudely speaking, the energy range of 0.26 to 0.54 eV would correspond to the range of energy over which the K centers accept the first electron (+/0 transition). The SDT response drops from 0.54 eV to below our detection limit at 0.68 eV. So, to a rough approximation, the energy range of 0.54–0.68 eV corresponds to the range of energy over which the K centers accept the second electron (0/-) transition.

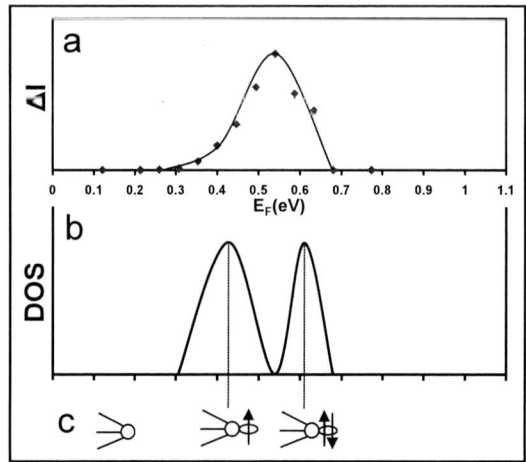

Figure 18: (a) The SDT response as a function of interface E_F, (b) a crude representation of K center density of states, and (c) a cartoon representation of the charge states of the K centers.

To a very crude approximation, we could approximate the collective K center density of states (DOS) by the absolute value of the derivative dI_{SDT}/dE_F. This is illustrated in Fig. 18(b). The cartoons of Fig. 18(c) illustrate the spin states (and charge) of the K centers versus E_F. We can understand how this is so by first considering an array of precisely identical defects which have precisely identical energy levels. This array of defects was schematically illustrated in figure 17. Figure 19 (a) illustrates a more physically reasonable DOS in which each of the levels is broadened to take into account disorder. If the E_F is below the (+/0) level, the defect's unoccupied dangling bond orbital does not have an electron to contribute to the tunneling. The defect is also diamagnetic (no unpaired electron) and cannot take part in magnetic resonance. Thus, with E_F below the (+/0) level, no SDT signal can be observed.

However, if E_F crosses the (+/0) level of some of the K centers, these centers can contribute to the tunneling and are paramagnetic and do take part in magnetic resonance. Therefore, the SDT response begins to turn on as the E_F level crosses the lower (+/0) levels and increases as long as E_F continues to cross these levels. However, as the E_F begins to cross the (0/-) level, the orbitals begin to accept a second electron and become negative. When this happens, the centers lose their paramagnetism, because they are now occupied by two electrons of opposite spin, and can no longer take part in magnetic resonance; thus, the SDT response is reduced. The SDT response drops to zero when all of the K centers accept the second electron. This SDT response is illustrated in Fig.19(b).

Figure 19(c) illustrates the derivative of the SDT amplitude versus energy response of Fig. 19(b). Notice that the maximum on the left side of the trace occurs at the same energy as the (+/0) peak in Fig. 19 (a). This is so because the increase in SDT amplitude versus energy will be greatest at the lower peak of the curve in Fig. 19(a). Analogously, since the rate of decrease in SDT amplitude versus energy will occur at the (0/+) peak, the minimum on the right will occur at that (0/+) energy. Thus, the absolute value of the derivative shown in Fig. 19(d) is a fairly good first order representation of the defect DOS illustrated in Fig.19(a). It is important to point out that this absolute value of the derivative is only a first order representation of the actual DOS. If the (+/0) and (0/-) transition peaks overlap, the absolute value of the derivative will incorrectly indicate a zero in the DOS between the two peaks. Also, the tunneling transmission probability from the K centers to defects in the poly-Si gate will not be precisely constant throughout the energy range (about 0.4 eV) over which the SDT is observed. However, the transmission probability will vary relatively slowly over the energy range. Thus, although the experimental evaluation of the defects density of states is crude, it should still provide a reasonable measure of the (+/0) and (0/-) transition levels and thus the electron-electron correlation energy.

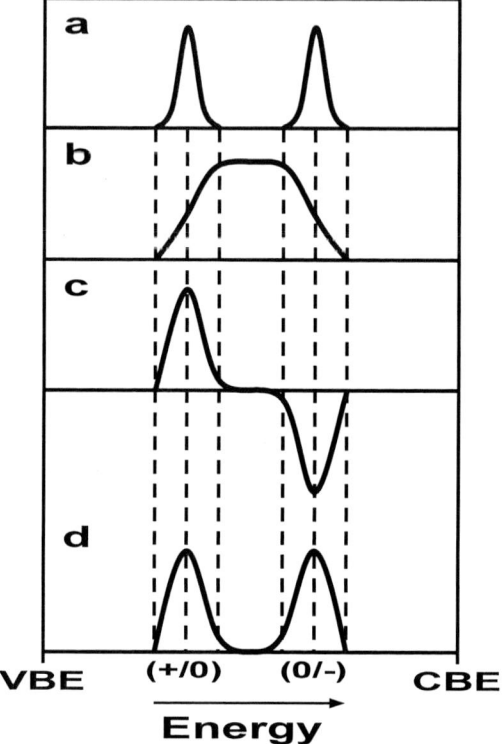

FIG.19. (a) A more physically reasonable DOS in which each of the levels is broadened to take into account disorder. (b) The SDT response from the levels of (a). (c) Schematic illustration of the derivative of the SDT amplitude vs energy response of (b). (d) The absolute value of the derivative (c)

Conclusions

Two electrically detected magnetic resonance techniques, one based upon spin dependent recombination, the other upon spin dependent tunneling, have the sensitivity and analytical power to provide fundamental information about the physical and chemical nature of electrically active defects in fully processed solid state electronic devices of current technological interest. These techniques possess all of the analytical power of conventionally detected magnetic resonance, but with greatly enhanced sensitivity. In addition, the techniques can provide information about the spatial distribution and the energy levels of these defects.

References

1. J. A. Weil, J. R. Bolton, and J. E. Wertz, *Electron Paramagnetic Resonance: Elementary Theory and Practical Applications* (Wiley, New York, 1994.)
2. P. M. Lenahan and J. F. Conley, *J. Vac. Sci. Technol. B* **16**, 2134 (1998)
3. D. Kaplan, I. Solomon, and N. F. Mott, *J. Phys. _Paris_, Lett.* **39**, 51 (1978)
4. D. J. Lepine, *Phys. Rev. B* **6**, 436 (1972)
5. P. M. Lenahan and M. A. Jupina, *Colloids Surf.* **45**, 191 (1990).
6. J. W. Gabrys, P. M. Lenahan, and W. Weber, *Microelectron. Eng.* **22**, 273
7. J. P. Campbell, P. M. Lenahan, A. T. Krishnan, and S. Krishnan, *J. Appl. Phys.* **103**, 044505 (2008)
8. J. P. Campbell, P. M. Lenahan, A. T. Krishnan, and S. Krishnan, *Appl. Phys. Lett.* **91**, 133507 (2007)
9. C.J. Cochrane, P.M. Lenahan, A.J. Lelis, to be published in *Journal of Applied Physics*
10. J.T. Ryan, P.M. Lenahan, At. Krishnana, and S. Krishnan, *Appl. Phys. Lett.* **96**, 223509 (2010)
11. J.T. Ryan, P.M. Lenahan, At. Krishnan, and S. Krishnan, *J. Appl. Phys.* **108**, 064511 (2010)
12. T. Grasser, W. Gos, V. Sverdlov, and B. Kaczer, *Proceedings of the IEEE International Reliability Physics Symposium* (2007) p. 268.
13. D. J. Fitzgerald and A. S. Grove, *Surf. Sci.* **9**, 347 (1968)
14. E.H. Poindexter, P.J. Caplan, B.E. Deal, and R.R. Razouk, *J. Appl. Phys.* 52, 879 (1981)
15. Y. Y. Kim and P. M. Lenahan, *J. Appl. Phys.* **64**, 3551 (1988)
16. A. Stesmans, B. Nouwen, and V.V. Afanas'ev, *Phys. Rev. B* **58**, 15801 (1998)
17. P. Campbell and P. M. Lenahan, *Appl. Phys. Lett.* **80**, 1945 (2002)
18. P. M. Lenahan and P. V. Dressendorfer, *Appl. Phys. Lett.* **41**, 542 (1982)
19. P. M. Lenahan and P. V. Dressendorfer, *J. Appl. Phys.* **55**, 3495 (1984)
20. G. J. Gerardi, E. H. Poindexter, P. J. Caplan, and N. M. Johnson, *Appl. Phys. Lett.* **49**, 348 (1986)
21. J.A. Cooper, *Phys. Status Solidi A* **162**, 305 (1997)
22. J.L. Cantin, H.J. Von Bardelebin, Y.S. Niskin, Y.Re, R. P. Devaley, and W.J. Choyke, *Phys. Rev. Lett.* **92**, 015502 (2004)
23. E. Janzen, A. Gali, P. Carlsson, A. Gallstrom, B. Magnusson, and N.T. Son, *Physica* **B409**, 4354 (2009)
24. M. Bocksledte, A. Gali, A. Mattdusch, O. Pankratov, and J. W. Steeds, *Phys. Stat. Sol. B* **245**, 1281 (2001)
25. L.F. Register, E. Rosenbaum, and K. Yang, *Appl. Phys. Lett* **74**, 457 (1999)
26. E. Rosenbaum and L.F. Register, *IEEE Trans Electron Dev.* **44**, 317 (1997)
27. J. P. Campbell, P. M. Lenahan, A. T. Krishnan, and S. Krishnan, *J. Appl. Phys.* **103**, 044505 (2008)
28. D. T. Krick, P. M. Lenahan, and J. Kanicki, *J. Appl. Phys.* **64**, 3558 (1988)
29. W. L. Warren and P. M. Lenahan, *Phys. Rev. B* **42**, 1773 (1990)
30. R. G. Southwick and W. B. Knowlton, *IEEE Trans. Device Mater. Reliab.* **6**, 136(2006)

628

Synthesis, Pore Morphology, and Dielectric Property of Mesoporous Low-*k* Material PSMSQ using a Reactive High-Temperature Porogen, TEPSS

S.-Y. Chiu, H.L. Hsu, M.L. Che, and J. Leu[*]

Department of Materials Science and Engineering, National Chiao Tung University, Hsinchu, Taiwan
1001 University Road, Hsinchu, Taiwan 30049

A high-temperature reactive porogen, triethoxy(polystyrene)silane (TEPSS) (M_w=3,500 g/mole), suitable for late-porogen removal integration scheme has been synthesized in *p*–xylene via atom transfer radical polymerization. TEPSS was then grafted onto poly(methyl-silsesquioxane) (MSQ) matrix (k=2.9) to circumvent possible phase separation between matrix and porogen in the hybrid approach and porogen aggregation. Our results shows porous low-k MSQ films possess uniform pore size, 24 nm for porosity up to 40%, primarily due to low PDI and reactive porogen, and the dielectric constant is decreased to 2.37 at 40% porosity. In addition, less porogen aggregation was observed at porogen loading ~40 v%.

Introduction

For the application of ultra-low-k dielectrics (k ≤ 2.5) as ILD for 32 nm node and beyond, large porosity is introduced into low-k materials matrix using porogens which is burned out immediately after dielectrics deposition or after completion of a Cu/low-metallization layer in a late-porogen removal scheme. The common method for preparing porous low-k films is to mix porogen into a matrix, and then spin-coat the matrix/porogen solution onto a substrate. For spin-on dielectrics with large porosity, the pore size is relatively large with wide distribution compared to the porous dielectrics deposited by plasma-enhanced chemical vapor deposition (PECVD) method using small molecules as the porogen. The large pores in the porous spin-on dielectrics may be caused by the aggregation of porogen in the solution at room temperature [1] and in the spin-coated film during the curing step. Large pore size and interconnected pores may detrimentally degrade the mechanical strength of ultra-low-k films [2]. However, for a mechanically robust low-k material and better reliability at trench sidewall, it is highly desirable to have small and well-dispersed pore sizes with tight distribution in the low-k films [3,4]. Approaches such as fast heating rate in the cure step or addition of surfactants [5] had been undertaken to retard the porogen aggregation in the solution and casted film to improve the pore size and distribution in porous low-k materials with limited success.

To overcome these issues, Nguyen *et al.* reported the fabrication of the nanoporous MSQ films via a star-shaped poly(ε-caprolactone) porogen in the PMSSQ/PCL hybrids [6]. Pore sizes of nanoporous PMSSQ films were smaller than 20 nm in size. Further, Ree *et al.* demonstrated the use of six-armed poly(ε-caprolactone) (mPCL6) as a reactive porogen in the MSQ matrix [7]. Pore sizes of porous low-k films which used mPCL6 as porogen was observed around 5 nm with 40% porogen loading [8]. However, such porogens cannot be used in the late-porogen removal integration scheme due to their low thermal stability.

In this study, a high-temperature, reactive porogen with siloxane backbone and polymer chains of well-controlled molecular weight is designed and synthesized for the preparation of low-k dielectrics suitable for the late-porogen removal integration scheme. In specific, a reactive high-temperature porogen, triethoxy-(polystyrene)silane (TEPSS) was synthesized in p-xylene by atom transfer radical polymerization (ATRP) [9] to obtain tight molecular weight distribution, then grafted onto the poly(methyl-silsesquioxane) (MSQ) matrix to circumvent possible phase separation between matrix and porogen in the conventional hybrid approach. The chemical structures and thermal characteristics of TEPSS reactive porogen and TEPSS-grafted MSQ (PSMSQ) were examined and verified by ^{1}H-NMR, ^{29}Si-NMR, and thermogravimetric analyzer (TGA), respectively. The morphology and pore size of porous PSMSQ low-k films were investigated by scanning electron microscopy (SEM). The porosity and dielectric property of porous PSMSQ low-k film were measured by x-ray reflectometry (XRR) and C-V dot.

Experimental

Materials

Styrene (Aldrich, 99%) was distilled and stored at 5°C. Copper(I) bromide (CuBr) (Sterm Chemicals, 98%) was sublimated before using and stored at 15°C. Methyltrimethoxysilane (MTMS) (Acros, 97%), N,N,N',N',N'-pentamethyldiethylene-triamine (PMDETA) (Acros, 99$^+$%), hexafluorophosphoric acid (HPF$_6$) (Acros, 60 wt%), and (3-chloropropyl)(triethoxy)silane (CPTES) (TCI, 95%), were used as-received. Tetrahydrofuran (THF) (Echo, 99%) was distilled to keep anhydrous before use. P-xylene (Aldrich, 99%) was used as-received.

Synthesis of Triethoxy(polystyrene)silane (TEPSS)

CuBr(I) (2.0x10^{-3} mole) was added to a 100mL two-necked flask equipped with a magnetic stirring bar and a condenser. After sealing, the flask was degassed for 1.5 hours. Subsequently, CPTES (1.0x10^{-3} mole) as initiator, solvent (4 ml) (p-xylene), and styrene (5.0x10^{-2} mole) were degassed by three freeze–pump–thaw cycles in order to remove the dissolved oxygen, respectively. After pre-treatment, solvent, PMDETA (2.0x10^{-3} mole), and CPTES were transferred into the flask, which was immediately sealed under argon. Styrene was then added into this flask, which was then degassed to remove oxygen by three freeze–pump–thaw cycles. The mixture was placed in an oil bath and stirred at 80°C for 12 hours under argon atmosphere. At the end of the polymerization reaction, the flask was quenched in cold water, subsequently diluted with THF. Afterwards, Cu^{2+} was removed through an alumina column. The reactive porogen TEPSS was obtained by precipitating the solution into alcohol and drying in the vacuum overnight. The molar ratio of styrene/CPTES was chosen at 500:1 to obtain a TEPSS.

Synthesis of MSQ from MTMS using Sol-gel Reaction

Methyltrismethoxysilane (MTMS) (2.4×10^{-2} mole), distilled water (DI water) (1.8×10^{-2} mole), and hexafluorophosphoric acid solution (HPF$_6$) (3.0×10^{-4} mole) were added into an alumina dish, which was then heated in an oven at 80°C for 7.5 minutes. The function of DI water was to hydrolyze MTMS, while the function of HPF$_6$ was to dehydrate MTMS and initiate sol-gel reaction.

Grafting TEPPS onto MSQ through Sol-gel Reaction.

TEPSS (1.6×10^{-6} mole) was first dissolved in 2 ml THF in a 100 ml flask. Upon the completion of the MSQ sol-gel reaction, MSQ (1.6×10^{-3} mole) was immediately added into the flask. Afterwards, 0.06 g of dilute HCl (10%) were added into the flask slowly. This reaction was carried out at 50°C in an oil bath under N$_2$ atmosphere for 14 h. The grafting reaction product was named PSMSQ. The ratio of TEPSS/MSQ was increased from 0.001:1 to 0.02:1 to increase the loading of high-temperature reactive porogen, TEPSS. (*i.e.* the porosity in the low-k MSQ matrix after the burn-out of porogens.)

Preparation of Porous Low-k Film

PSMSQ was dissolved in THF to form a 20 wt% solution. Prior to coating, the solution was filtered through a 0.2 μm PTFE filter (Millipore Inc.). The PSMSQ solutions were spin-coated at 2000 rpm on a silicon wafer. The PSMSQ film was sequentially baked at 100°C for 1 minute, then cured on a hot plate preheated at 200°C for 30 minutes to form a crosslinked MSQ structure. The TEPSS reactive porogens were thermally decomposed at 400°C for 90 minutes to form a porous low-k MSQ films. The nominal thickness of PSMSQ film is 150 nm, unless specified otherwise.

Measurements

The chemical structures of TEPSS and PSMSQ were first validated by ^{1}H-NMR spectra and ^{29}Si-NMR spectra, respectively. ^{1}H-NMR spectra were recorded in solution using a Bruker AC-300P (300 MHz) spectrometer, with the tetramethylsilane (TMS) proton signal as an internal standard. ^{29}Si-NMR spectra for powder samples were carried out by a Bruker DSX-400WB NMR spectrometer.

The number-average (M_n), weight-average (M_w) molecular weight, and the molecular weight distribution (*i.e.* polydispersity, PDI), were determined by gel permeation chromatography (GPC) using a Waters chromatography unit interfaced with a Water 2414 differential refractometer (Waters Corporation, Ashland), and using tetrahydrofuran (THF) as an eluant and polystyrene as the standard. TGA was employed to measure the decomposition temperatures of TEPSS and PSMSQ using TA Q500 (TA Instruments). The temperature was raised from 100°C to 900°C in nitrogen at 10°C/min.

The density of porous MSQ film was measured by XRR. The films were scanned by D8 Discover, Bruker with Cu Kα source (λ=0.154 nm) using ω-2θ method with a scanning angle ranging from 0° to 2° at an increment of 0.002°. Porosity was then calculated by the Eq. 1.

$$\rho = \rho_s (1-p) \tag{1}$$

where ρ is film density (g/cm^3), ρ_s is silica density (1.63 g/cm^3), and "p" refers to porosity.

The pore morphology of porous low-k films was examined by using a dual beam FIB/SEM (Nova Nanolab 2000 system, FEI Company) system. The pore size and distribution were calculated using Image J analysis [10] of >200 pores in the SEM graphs. The film thicknesses of PSMSQ and porous MSQ film were measured using n&k Analyzer 1280 (n&k Technology, Inc.) at wavelengths ranging from 190 to 900 nm. The dielectric constant of porous low-k MSQ film was measured by C-V dots (Keithley 590 C-V Analyzer) with a sweeping frequency of 1 MHz based on Metal/Insulator/Metal (MIM) configuration at room temperature.

Results and Discussion

Synthesis of Triethoxy(polystyrene)silane (TEPSS) and TEPSS-grafted MSQ (PSMSQ)

(a) Synthesis of TEPSS by using ATRP. In order to obtain a siloxane-functionalized polystyrene, TEPSS as a reactive and high-temperature porogen, the synthesis was carried out at 80 °C by adding styrene as monomers, CPTES as initiator, CuBr/PMDETA as catalyst using ATRP method [11] as schematically illustrated in Figure 1. p-xylene was selected as the solvent for synthesizing TEPSS due to its better control of its molecular weight. In specific, TEPSS, with a low molecular weight (3,500 g/mole) and a low PDI (1.12) were used in this study, and the yield of TEPSS was 58% as shown in Table I.

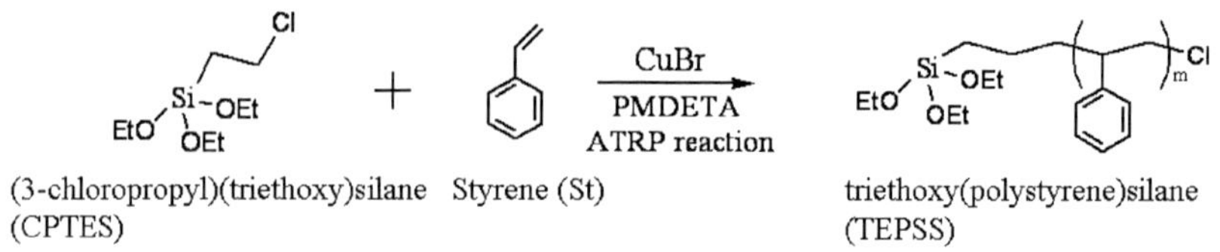

(3-chloropropyl)(triethoxy)silane   Styrene (St)   triethoxy(polystyrene)silane
(CPTES)   (TEPSS)

Figure 1. The synthesis route of TEPSS

Table I. The number-average (M_n) molecular weight, weight-average (M_w) molecular weight, PDI, and yield of TEPSS.

	M_n (g/mole)	M_w (g/mole)	PDI	Yield
TEPSS	3900	3500	1.12	58%

(b) Grafting TEPSS onto MSQ through Sol-gel method. The oligomer MSQ was synthesized and prepared in good yield by sol-gel method as outlined in Figure 2(a). MSQ has siloxane groups on the end caps when MSQ was synthesized from monomer MTMS. TEPSS was grafted onto MSQ with siloxane groups by sol-gel reaction through the reaction between MSQ and TEPSS in dry THF as shown in Figure 2(b). This yields reaction product, PSMSQ.

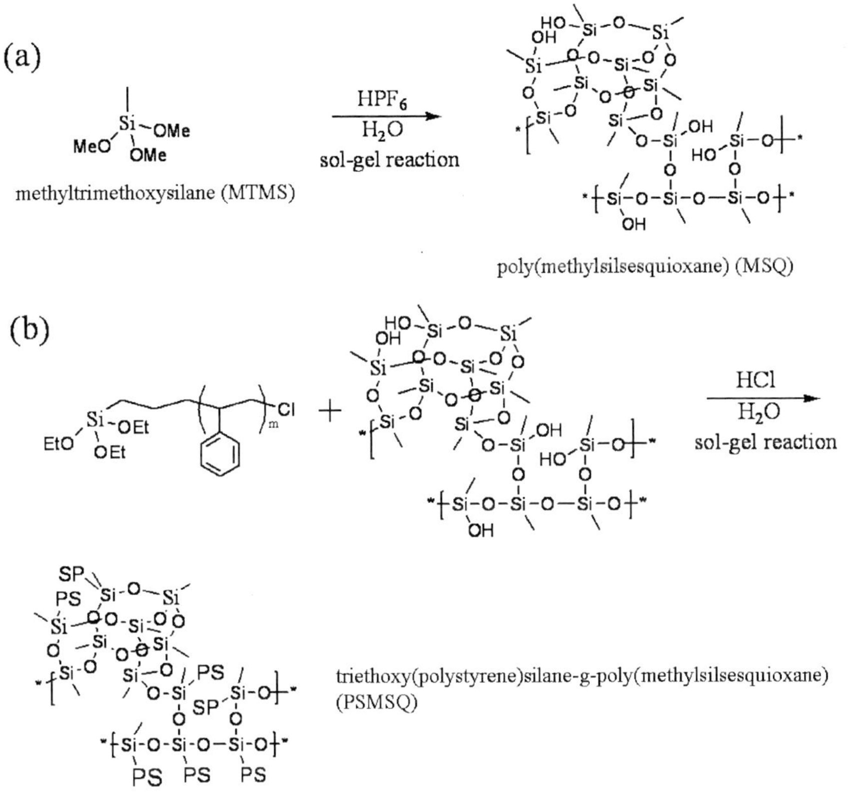

Figure 2. (a) synthesis route of MSQ from MTMS by sol-gel method, and (b) synthesis of PSMSQ by grafting TEPSS onto MSQ.

Chemical Structure and Thermal Properties of TEPSS and PSMSQ

(a) Chemical Structures of TEPSS and PSMSQ. The chemical structure of TEPSS was confirmed by ^{1}H-NMR spectra as shown in Figure 3. From ^{1}H-NMR spectra, two characteristic signals of TEPSS are located: (1) in the 6.45-7.18 ppm range representing the aryl protons of styrene units and (2) in the 1.25-1.84 ppm range assigned to the methylene protons of styrene backbone. Moreover, signals located in the 3.6 ppm range are attributed to -Si-(OCH$_3$)$_3$ end group. From the ^{1}H-NMR result, it is shown that TEPSS has been successfully synthesized by ATRP method. Figure 4 exhibits the ^{29}Si-NMR spectra of (a) MSQ and (b) PSMSQ baked at 60°C. Two types of silicon atoms are located at -54 ppm and -69 ppm, which correspond to T^2 and T^3 species, respectively. T^2 refers to silicon nuclei with hydroxyl group termination, [RSi-(OSi)$_2$OH] or [RSi-(OSi)(OR)OH], and the chemical shift is from -54 ppm to -61 ppm. Beside, chemical shift of [RSi-(OSi)(OR)OH] would close to -54 ppm, and [RSi-(OSi)$_2$OH] would close to -61 ppm. On the other hand, T^3 represents three -O-Si bonds and one alkane group, [RSi-

(OSi)$_3$], and the chemical shift is from -64 ppm to -69 ppm [12-14]. In this study, T^3 peak has been normalized to check the change of T^2 peak on MSQ and PSMSQ. It can be seen that T^2 peak of PSMSQ is closer with -54 ppm than MSQ. It means that PSMSQ has more [RSi-(OSi)(OR)OH] structures than MSQ, due to the triethoxysilyl side group of TEPSS are less reactive than the trimethoxysilyl side group of MSQ [15]. On the other word, in the same sol-gel condition, the trimethoxysilyl side group would become [RSi-(OSi)$_2$OH] structure largely. However, the triethoxysilyl side group maintain [RSi-(OSi)(OR)OH] structure and become to [RSi-(OSi)$_2$OH] structure slowly. Therefore, when TEPSS was grafted onto MSQ, the triethoxysilyl side group would decrease siloxane condensation and cause PSMSQ has more [RSi-(OSi)(OR)OH] structure. Thus, the results of ^{29}Si-NMR confirmed that TEPSS has been grafted onto MSQ successfully.

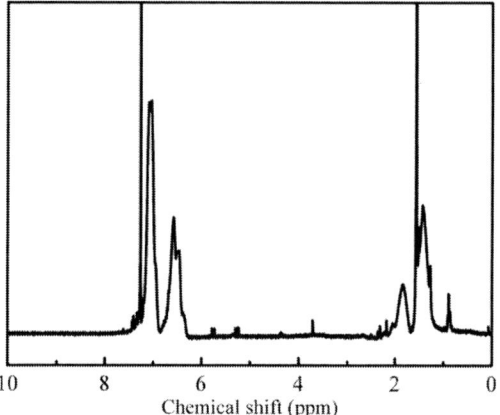

Figure 3. ^{1}H-NMR spectra of TEPSS

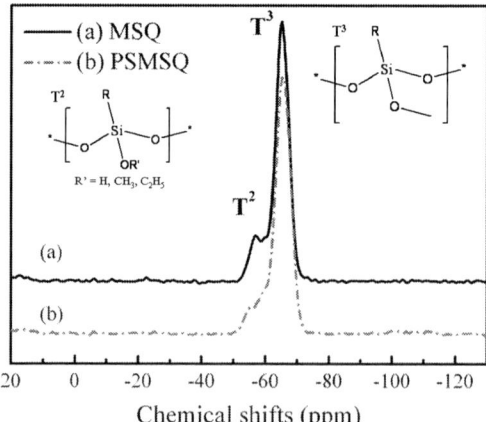

Figure 4. ^{9}Si-NMR spectra of MSQ and PSMSQ

(b) Thermal properties of TEPSS and PSMSQ. In this paper, we can use not only ^{29}Si-NMR but TGA analysis to demonstrate that TEPSS has been grafted onto MSQ. Thermo-gravimetric curve of TEPSS is shown in Figure 5(a) under nitrogen at a ramp rate of 10 °C/min. The decomposition temperature of TEPSS was investigated to determine the maximum processing temperature of the porous low-k film. As illustrated in Figure 3(a), reactive porogen, TEPSS, had a high decomposition temperature (T_d) of 362°C, at which weight loss 5% occurred. Moreover, TEPSS could be completely removed at T > 435°C under N_2 atmosphere in dynamic mode. In addition, TEPSS can be completely removed at 400°C for 1 hour in static mode as shown in Figure 5(b), and the 8% residual was silane skeleton. As a result, TEPSS is confirmed to be an excellent porogen for a porous low-k dielectric in late-porogen removal scheme. Next, the thermal stability of MSQ and PSMSQ in nitrogen using TGA are shown as in Figure 6. MSQ showed a high thermal stability with a mere 2.5% weight losses for temperature ranging from 200°C to 600°C. The TGA of PSMSQ showed 1% weight loss from 376°C to 460°C. The weight loss can be attributed to the decomposition of the long chain polystyrene in TEPSS. The T_d of grafted PS in PSMSQ was calculated to be 383°C, which is 21°C higher than the T_d of TEPSS (362°C). The increased T_d of styrene in PSMSQ is due to the confinement and interactions of styrene long chain by the MSQ matrix structure.

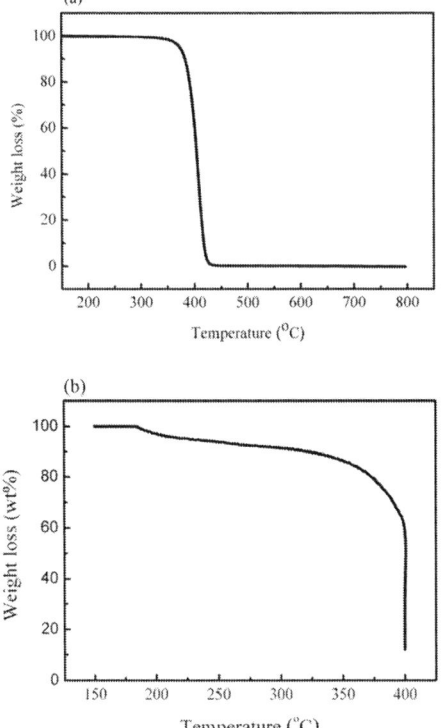

Figure 5. The (a) dynamic, and (b) 400°C isothermal TGA curves of TEPSS

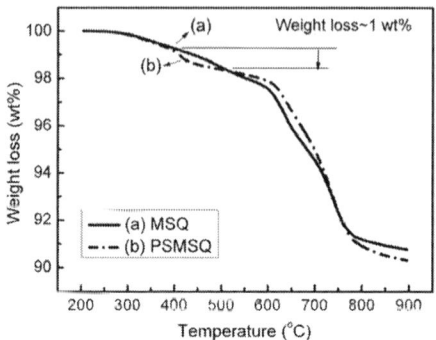

Figure 6. TGA thermograms of MSQ and PSMSQ

Pore Morphology and Pore Size of Porous MSQ Films with Different Porosities

After the chemical structure and thermal properties of PSMSQ are demonstrated, this paper continues to study the changing of porosity in the different porogen loading, and PSMSQ porous films morphology and pore size. The PSMSQ films were prepared by dissolving 20 wt% PSMSQ into THF solution. Moreover, various weight ratios of TEPSS and MSQ were used in this study to change the loading of reactive porogen, and in turn the porosities. The densities and porosities obtained by XRR for various porogen loadings are summarized in Table II. The porosity was increased from 18% to 54%, while the film density decreased from 1.80 to 10.1 g/cm^3 when the porogen loading was raised from 0.4% to 20%. The pore morphology of porous low-k MSQ films at various porosities after the removal of porogen at 400 °C by using scanning electron microscopy are shown in Figures 7(a) through 7(d): (a) 16%, (b) 27%, (c) 40%, and (d) 54%, respectively. From these top-view SEM graphs, porous PSMSQ films were found to be continuous and smooth. The pores were spherical and uniformly distributed without rod-like, interconnected pores. Pore sizes of PSMSQ porous films were estimated to be 18 nm at 16% porosity, 19 nm at 27% porosity, 24 nm at 40% porosity, and 38 nm at 54% porosity. Compared to the porous MSQ prepared from MSQ/PS hybrid, porous low-k film at $\leq$ 40% porosity based on PSMSQ through the grafting of reactive TEPSS porogen onto MSQ can deliver smaller, mesoporous ~18-24 nm, which are in spherical shape and uniformly distributed. However, severe pore aggregation was observed for porosity at 54% porosity presumably due to the aggregation of reactive porogen, TEPSS and poor dispersion, at relatively high loading prior to grafting reaction. A better dispersion of TEPSS at high loading in the solvent is critical for fabricating porous low-k MSQ films with porosity > 40 v%.

Table II. The dielectric constant of porous PSMSQ films with various porosities

Porogen loading	Density g/cm^3	Porosity (%)	Dielectric constant	Pore size (nm)
MSQ	2.20	0%	2.90	-
0.4 %	1.80	18%	2.66	18
1 %	1.60	27%	2.45	19
8 %	1.34	40%	2.37	24
20%	1.01	54%	2.30	38

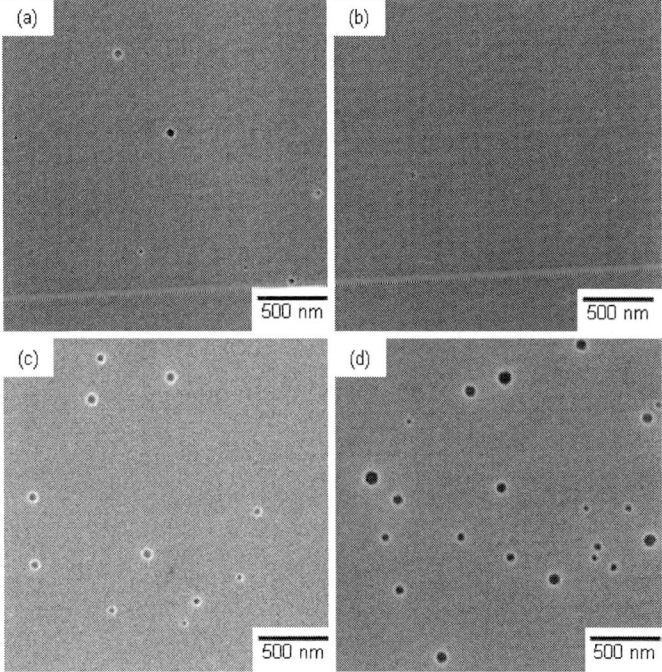

Figure7. SEM viewgraphs of porous low-k MSQ films at various porosities: (a) 16%, (b) 27%, (c) 40%, and (d) 54%

Dielectric Property of Porous Low-k MSQ Films

The dielectric constants of porous low-k MSQ films prepared from PSMSQ films at various porosity are also summarized in Table II. For comparison, the dielectric constant of pure MSQ film was 2.9 (k = 2.9). The dielectric constants of PSMSQ are 2.66, 2.45, 2.37, and 2.30 for porous MSQ films at 18%, 27%, 40%, and 54% porosity, respectively. The dielectric constant decreases with increasing porosity as expected. The result suggests that the dielectric constant of the PSMSQ films decreased with increasing TEPSS loading (i.e. increasing PSMSQ films porosity). However, the experimental k values of porous PSMSQ films were higher than theoretical values presumably due to adsorbed water, whose k value is close to 80 [16]. Moisture absorption may be attributed to the residual silanol groups in the low-k films. Efforts in reducing residual silanol groups in the cure step and cautions in the dielectric measurement with appropriate procedures for removing adsorbed moisture have been undertaken. Results will be reported and discussed.

Conclusion

This study introduced a novel method for preparing mesoporous materials, PSMSQ by using a reactive high-temperature porogen, TEPSS, suitable for the late-porogen removal integration scheme. In specific, a reactive high-temperature porogen, triethoxy-(polystyrene)silane (TEPSS) (M_w=3,500 g/mole) was synthesized in p-xylene by atom transfer radical polymerization (ATRP) to obtain tight molecular weight (MW) distribution, then grafted onto the poly(methyl-silsesquioxane) (MSQ) matrix to circumvent the phase separation between matrix and porogen in the conventional hybrid approach. TEPSS (T_d = 362°C) is confirmed to be an excellent high-temperature porogen as observed from the enhanced T_d of PS in PSMSQ and complete removal of progen at 400°C for 1 hr. In addition, the extremely low polydispersity by using ATRP and controllable reactivity of high-temperature porogen, TEPSS onto MSQ matrix yielded uniformly distributed and spherical pores (~24 nm) with less aggregation at high porosity up to 40%. The density and the electric properties of porous PSMSQ film were measured by XRR and C-V dots, respectively. The dielectric constant of porous low-k MSQ film was decreased from 2.90 to 2.37 when the porosity was raised from 0% to 40%.

Acknowledgments

The authors appreciate the financial support in part by National Science Council of ROC under contracts: NSC97-2221-E009-160 and NSC98-2221-E009-177.

References

1. Y. H. Chen, U.S. Jeng, and J. Leu, *J. Electrochem. Soc.*, **158**, G58 (2011).
2. J. Kovacik, *J. Mate. Sci. Lett.*, **18**, 1007 (1999).
3. J. P. Hsu, S. H. Hung, and W. C. Chen, *Thin Solid Films*, **473**, 185 (2005).
4. Y. Oku, K. Yamada, T. Goto, Y. Seino, **A.** Ishikawa, T. Ogata, K. Kohmura, N. Fujii, N. Hata, R. Ichikawa, T. Yoshino, C. Negoro, **A.** Nakano, Y. Sonoda, S. Takada, H. Miyoshi, S. Oike, H. Tanaka, H. Matsuo, K. Kinoshita, and T. Kikkawa, *2003IEDM*, O3-139 (2003).
5. Y. H. Chen and J. Leu, *2008 MRS Fall Meeting*, 510449 (2008).
6. C. V. Nguyen, K. R. Carter, C. J. Hawker, J. L. Hedrick, R. L. Jaffe, R. D. Miller, J. F. Remenar, H. W. Rhee, P. M. Rice, M. F. Toney, M. Trollss, and D. Y. Yoon, *Chem Mater.*, **11**, 3080 (1999).
7. B. Lee, W. Oh, J. Yoon, Y.Hwang, J. Kim, B. G. Landes, J. P. Quintana, and M. Ree, *Macromolecules*, **38**, 8991 (2005).
8. B. Lee, W. Oh, Y. Hwang, Y. H. Park, J. Yoon, K. S. Jin, K. Heo, J. Kim, K. W. Kim, and M. Ree, *Adv. Mater.*, **17**, 696 (2005).
9. J. F. Lutz and K. Matyjaszewski, *Macromol. Chem. Phys.*, **203**, 1385 (2002).
10. L. Shamir, J. D. Delaney, N. Orlov, D. M. Eckley, and I. G. Goldberg, *PLoS Comput. Biol.*, **6**, e1000974 (2010).
11. M. Degirmenci, O. Izgin, A. Acikses, and N. Genli, *React, Funct. Polym.*, **70**, 28 (2010).
12. L. Bourget, D. Leelereq, and A. Vioux, *J. Sol-Gel Sci. Technol.*, **14**, 137 (1999).
13. K. J. Shea, D. A. Loy, and O. Webster, *J. Am. Chem. Soc.*, **114**, 6700 (1992).
14. H. Jo and F. D. Blum, *Langmuir*, **15**, 2444 (1999).
15. C. Gualandris, F. Babonneau, M. T. Janicke, and B. F. Cjmelka, *J. Sol-Gel Sci. Technol.*, **13**, 75 (1998).
16. D. Shamiryan, T. Abell, F. lacopi, and K. Maex, *Materials Today*, **7**, 34 (2004).

Electrical Characteristics Analysis at "Oxide Flat-band Voltage"
for Al-SiO$_2$-Si Capacitor

Han-Wei Lu, Tzu-Yu Chen, and Jenn-Gwo Hwu*

Graduate Institute of Electronics Engineering/Department of Electronics Engineering,
National Taiwan University
No. 1, Sec. 4, Roosevelt Road, Taipei, 10617 Taiwan
Tel: 886-2-3366-3646, Fax: 886-2-23671909,
*hwu@cc.ee.ntu.edu.tw

In this paper, a new noun named "oxide flat-band voltage" is
proposed meaningfully and further electrical characteristics study
focused on non-uniformity in MOS capacitor will be demonstrated.
Based on the rectangular barrier approximation, the slope of the
ln[Jg@V$_{ox,fb}$] versus oxide thickness plot of -12.18 is much closer
to the theoretical slope of -12.42 than the slope of the ln[Jg@V$_{fb}$]
versus oxide thickness plot of -11.09. It is because that there are
some negative charges existing in the SiO$_2$ dielectrics during the
film-grown anodization system resulting in the non-uniformity
phenomenon. Furthermore, various device areas of MOS capacitor
were designed to do an advanced non-uniformity study. Both in the
analysis of differential J-V curves and the slope of natural log of
gate current density versus oxide thickness plot, it indicates
evidently and identically that the larger the gate electrode area, the
worse uniform the dielectric is.

Introduction

Among semiconductor technology, complementary metal-oxide-semiconductor
(CMOS) plays an important role due to its high noise immunity and low static power
consumption. Power is only drawn while the transistors in the CMOS device are
switching between on and off states. Moreover, CMOS also allows a high density of logic
functions on a chip. CMOS circuits use a combination of p-type and n-type metal-oxide-
semiconductor field-effect transistors (MOSFETs); while, the Metal-Oxide-
Semiconductor (MOS) capacitor structure is the heart of the MOSFET devices.

In today's ultra-large-scale integration (ULSI) industry, silicon is the mainstream
technology because of its abundant amount in the earth. Another reason for the success of
silicon ICs is the fact that the excellent native oxide, SiO$_2$, can be grown directly from the
surface of silicon. This conventional dielectric material, SiO$_2$, has the outstanding
characteristics as an insulator. The oxide can be used as the gate dielectric layer in
MOSFET and can also be used as an insulator, such as field oxide between devices. Most
other semiconductors wouldn't form this kind of native oxides like SiO$_2$ that has
excellent quality in device fabrication.

According to Moore's Law in 1975 [1], the number of transistor per unit area will double and the feature length of each technology node will multiple by 0.7 for every 18 months. In addition, Moore's law has been proved accurately for more than 40 years. Ultra-thin gate dielectric layer of MOSFET is of considerable interest in modern microelectronics. Nevertheless materials encounter the intrinsic or physical limit and have a hard time. The International Technology Roadmap for Semiconductors (ITRS) predicts that the equivalent oxide thickness will down to 7.5Å and the gate current density will reach a value of 1.0×10^3 A/cm^2 in 2012. It indicates that there are no known solutions existing beyond 2012, due to the unacceptable high gate leakage current [2]. Another effect following by the device scaling is that when the oxide thickness is shrinking, the effect of quantum mechanical becomes significant. At that time, the gate oxide of MOS cannot be treated as good insulator and the direct tunneling current occurs. Therefore, the conventional equations of MOSFET cannot be used and the power consumption becomes higher than before. As the device scale shrinks, and the thickness of SiO$_2$ decreases less than 2 nm, the problem of leakage current should be concerned seriously. It is not allowable for such a high leakage current density anymore so the issue of reducing the gate tunneling current density is of importance.

As the oxide thickness is below 2.5 nm, direct tunneling current becomes a dominant influence in the leakage current mechanism. One of the most important factors affecting the tunneling current is the tunneling probability. A reasonable approximation of the transmission current density for a special situation will be illustrated in our work. At the same time, traps existing in the oxide would increase the leakage current directly and result in non-uniformity of dielectric layer. This phenomenon cannot be disregarded anymore especially for the ultra-thin gate oxide device. The influence of the oxide thickness non-uniformity on the tunneling current has been theoretically considered in some works [3]-[7]. Continuous trends to reduce the oxide thickness causes an increasing significance of investigating the tunneling mechanism in MOS devices, and therefore results in the extreme importance of the needs of the uniformity of oxide thickness and the good quality of the Si-SiO$_2$ interface. Oxide thickness non-uniformity becomes a critical parameter to the tunneling current value.

At previous studies, the criterion to determine the quality of oxide at a certain thickness is almost using the gate leakage current density at V_{fb}-1. And, V_{fb} means the gate voltage at which the Si is in flat-band condition; that is, the situation of Ψ_s=0. It is said that the lower the gate leakage current density at V_{fb}-1, the better the oxide quality is. However, the inconsistent electrical field across different oxide thicknesses will be brought out because no matter what thickness it is, the method always use V_{fb}"-1" as the criterion. It will result in an inaccurate estimation of oxide quality. In our study, it is the first time to present a novel way to determine the quality of oxide dielectrics by approximating the oxide as a rectangular barrier. At first, we faced a difficulty to find out at which voltage, the oxide acts most like the rectangular barrier. By observing the basic J-V curves, it is clear to see a transition point. We extracted the voltage at this point and put it into the rectangular barrier approximation formula; interestingly, the result of this transition point is much more accurate to the theoretical one than the results of V_{fb} or even V_{fb}-1. Here, we assert the transition point in J-V curve is the voltage of oxide acting as a rectangular barrier, means the change of the oxide band bending direction and which is named after "oxide flat-band voltage ($V_{ox,fb}$)".

In this work, we focus on the electrical characteristics analysis of dielectric layer for MOS devices. Thanks to accurate technological control, Al-SiO$_2$-Si structure became an excellent bench for non-uniformity testing. The tunneling current behavior in metal-oxide-semiconductor structure is presented here. The oxide thickness non-uniformity is a critical parameter to the tunneling current. The aim of our work is to try to develop a method as mentioned above to quantify the degree of non-uniformity by analyzing the tunneling electrical characteristics of MOS capacitor. Moreover, the comparison between three different gate electrode areas will be demonstrated to prove the feasibility of the method we brought up. With this quite practical and assistant technique, we can determine rapidly about the technology of the non-uniformity effects in solid state devices.

Experimental

The whole experimental process flow is indicated briefly in Fig. 1(a). The substrate of Al/SiO$_2$/Si capacitor which was studied in this paper was a 3-inch, boron-doped p-type (100)-oriented silicon wafer with a resistivity of 1-10 Ωcm. After standard RCA clean process, oxidation was carried out by the method called anodization. It was first proposed by P. F. Schmidt and W. Michel [8]. From then on, many studies concerning about anodization methods had been conducted [9]-[14]. There are many advantages of the anodization process such as cost-effective, grown at the room-temperature, superior performance mentioned in the previous literature [15]-[17], earned different oxide thickness on one wafer simultaneously, etc. The schematic diagram of anodization setup system is shown in Fig. 1(b). Deionized water was used as the electrolyte and the Si wafer was served as anode and the platinum plate as cathode. Owing to the tilted angle of the cathode plate, the MOS capacitor with the same experimental condition but various SiO$_2$ thicknesses would be obtained on the same wafer. According to Ghandhi [18], when a dc voltage was applied, the chemical reaction equations occurred in the electrolyte and in the anode are listed in Eq. (1)-(5).

In electrolyte,

$$2H_2O \leftrightarrow 2H^+ + 2(OH)^- \qquad (1)$$

At anode,

$$Si + 2h^+ \rightarrow Si^{2+} \qquad (2)$$

$$Si^{2+} + 2(OH)^- \rightarrow Si(OH)_2 \qquad (3)$$

$$Si(OH)_2 \rightarrow SiO_2 + H_2 \qquad (4)$$

$$Si + 2h^+ + 2H_2O \rightarrow SiO_2 + 2H^+ + H_2 \qquad (5)$$

It is evident that the oxide thickness is related to the growth time, the distance between anode and cathode, and quantity of the applied voltage. From the equations mentioned above, it is believed that the anodic oxide was formed by the reaction between OH⁻ and Si. The pre-trapped negative charges in anodic oxide are possibly caused by the incorporation of OH⁻ groups during oxide preparation. Post oxidation annealing (POA) was implemented in the rapid thermal chamber in N_2 ambient at 950℃ for 15 seconds in order to reduce interface trap and densify the oxide. Subsequently, the aluminum gate film was evaporated and then patterned by conventional photolithography with three different defined areas of $1 \times 2.25 \times 10^{-4}$, $4 \times 2.25 \times 10^{-4}$, $16 \times 2.25 \times 10^{-4}$ cm^2 for our further non-uniformity study. At last, back native oxide was removed by buffered oxide etchant (BOE) and Al film was then evaporated again as the back contact of our MOS capacitor.

In the measurement, the capacitance-voltage (C-V) and current-voltage (I-V) curves were measured by an HP4284 precision LCR meter and an HP4140B picoampere meter, respectively. Oxide thickness d_{ox} was extracted by fitting the corrected C-V curves of 100 kHz and 1 MHz under the consideration of quantum mechanical effect [19]-[21].

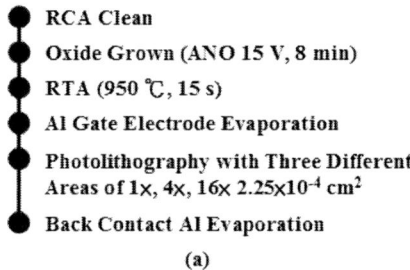

(a)

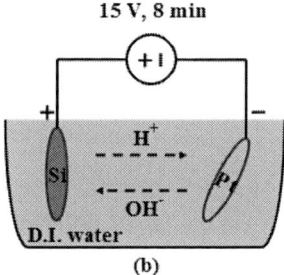

(b)

Fig. 1 (a) A process flow of the Al-SiO₂-Si Capacitor. (b) The schematic diagram of tilted cathode anodization system by applying a dc voltage of 15 V for 8 minutes.

Results and Discussion

Based on the tilted cathode anodization system, it is known that different distances between the anode Si wafer and the cathode platinum plate will result in different electric fields. Hence, oxides with various thicknesses on the same wafer would be acquired at one time avoiding unnecessary repeatedly experimental steps. Fig. 2 is the electrical characteristics of the basic J-V curves measured from 2V to -2.5V with dielectric thicknesses varying from 1.8 to 2.4 nm. For p-type MOS capacitor, the right side of Fig. 2 is the saturation behavior of inversion tunneling current density and it is clear that the thicker the oxide, the increasing saturated current density would be got [22]. On the other hand, the left side of Fig. 2 represents the leakage current density of our MOS capacitor. As the SiO_2 film continues to scale down, it shows obviously that the leakage current density would increase tremendously due to the raise of the leakage path.

Moreover, we find out an interesting phenomenon in the left side of Fig. 2 just as the square pointed out. When voltage sweeps into negative bias, the increasing rate of gate current density will slow down first and then speed up continuously in every J-V curve of different oxide thickness. The current density-voltage characteristics of forward biased $Al-SiO_2-Si(p)$ tunnel diodes having insulator thicknesses in the 1.8-2.4 nm range are investigated in this work. To study this characteristic, at the beginning, we try to draw out this turning point around this region.

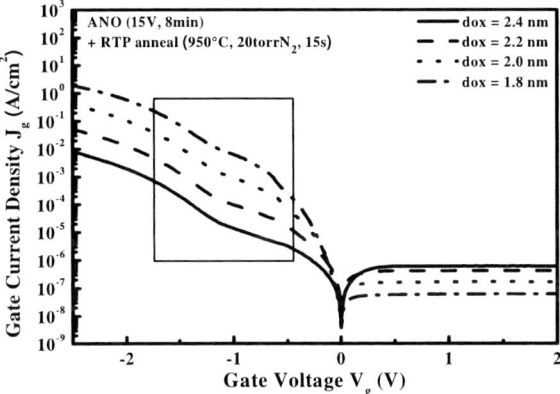

Fig. 2 The J-V curves of thin SiO_2 films with thicknesses varying from 1.8 to 2.4 nm.

By differentiating the gate current density versus gate voltage for advanced observation as shown in Fig. 3, it is intentional to get a minimum derivative around this increasing rate of current density switching region. In this work, we define the minimum slope of J-V curve as the oxide flat-band voltage ($V_{ox,fb}$). It is supposed to acquire the best ability to block the leakage current point. Here, in this 1.9 nm SiO_2 dielectrics, the corresponding oxide flat-band voltage ($V_{ox,fb}$) is obtained easily as -1 V.

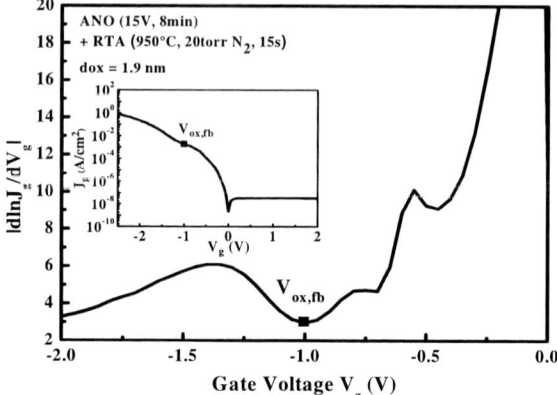

Fig. 3 The method to extract oxide flat-band voltage by the first order derivative of gate current density versus gate voltage for 1.9 nm SiO_2 film. The inset is the J-V curve of 1.9 nm SiO_2 film.

Based on the definition of oxide flat-band voltage in this work, $V_{ox,fb}$ of different oxide thicknesses were drew out in Fig. 4; meanwhile, the conventional flat-band voltage (V_{fb}) of different oxide thicknesses were also shown. The flat-band voltage is defined as the applied gate voltage such that there is no band bending in the semiconductor. For an ideal MOS capacitor model, flat-band voltage is equal to work function difference between silicon and metal gate (φ_{ms}) and in our Al/ SiO_2/Si case, the ideal flat-band voltage is equal to -0.9 V as indicated in Eq. (6).

$$V_{fb} = \varphi_{ms} = -0.9V \qquad (6)$$

However, in real case, the charges in SiO_2 (Q_{eff}) and the traps near Si/SiO_2 interface (Q_{it}) may cause the non-ideal effect in MOS structure. Therefore, we have to correct Eq. (6) to the following.

$$V_{fb} = \varphi_{ms} - \frac{Q_{eff}}{C_{ox}} - \frac{Q_{it}(\psi_s = 0)}{C_{ox}} \qquad (7)$$

In Fig. 4, the flat-band voltages (V_{fb}) extracted by conventional C_{fb} method would shift rightward compared to the ideal value of -0.9 V. That is owing to the SiO_2 grown system we adopted in this work is anodization. From Eq. (3) and Fig. 1(b), the anode Si wafer would attract OH^- ion in D. I. water so the grown dielectrics would have more negative charges than positive charges. On the other words, the effective charges (Q_{eff}) are negative. Take it into Eq. (7), it is evident that flat-band voltage would shift from -0.9 V to much more positive as shown in the figure. Consequently, due to the existing negative charges in oxide, the oxide flat-band voltage ($V_{ox,fb}$) extracted by derivative J-V curves would shift to much more negative.

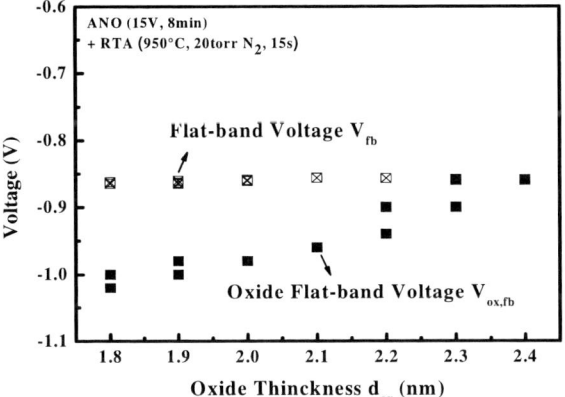

Fig. 4 Extraction of flat-band voltage (V_{fb}) and oxide flat-band voltage ($V_{ox,fb}$) of oxides on the same wafer with various thicknesses.

Subsequently, the reason why we named the transition point as the oxide flat-band voltage will be introduced below. For one thing, the minimum increasing rate of gate current density supposes to be the best condition to block the leakage current. We guess it's the change bending direction voltage of the oxide and at that time, the oxide is just like a rectangular barrier. A reasonable approximation of the transmission current density for a rectangular barrier can be written as [23]

$$J_{tunneling} \propto \exp(-2kd_{ox}) \qquad (8)$$

where

$$k = \frac{\sqrt{2m*(U-E)}}{\hbar} \qquad (9)$$

Take the energy barrier height ($q\phi_B$) as 3.2 eV [24] and the effective mass ($m*$) as $0.46m_0$ [25], a theoretical slope of $\ln[J_{tunneling}]$ versus oxide thickness (d_{ox}) of -12.42 is therefore obtained.

Comparing to the theoretical analysis, the experimental slopes of $\ln[J_g@V_{fb}\text{-}1, V_{fb}, V_{ox,fb}]$ versus oxide thickness are indicated in Fig. 5. It concludes that the slope of -12.18 gotten by oxide flat-band voltage is the closest one to the theoretical slope -12.42. It is because that $V_{ox,fb}$ is exact at "oxide flat-band" condition just as the rectangular barrier theory we adopted in this work contrast to V_{fb} and $V_{fb}\text{-}1$ illustrated in the inset. As mentioned in Fig. 4, the flat-band voltages shift to more positive because there are some negative charges remained in the dielectrics for the sake of the grown anodization system. This phenomenon results in the non-uniformity of the SiO_2 dielectric layer and affects the flat-band condition of our MOS capacitor directly. It is clear that once the charges exist in oxide, the flat-band voltage applied in the gate electrode could only let the Si substrate band bending in the flat-band condition but the oxide couldn't band bend in the flat-band condition simultaneously. For this reason, the analyzed slope of -11.09 of $\ln[J_g@V_{fb}]$ versus oxide thickness would be a little bit diverging from the theoretical slope of -12.42 compared with the analyzed slope of -12.18 of $\ln[J_g@V_{ox,fb}]$ versus oxide thickness due to the non-uniformity phenomenon in SiO_2 dielectrics.

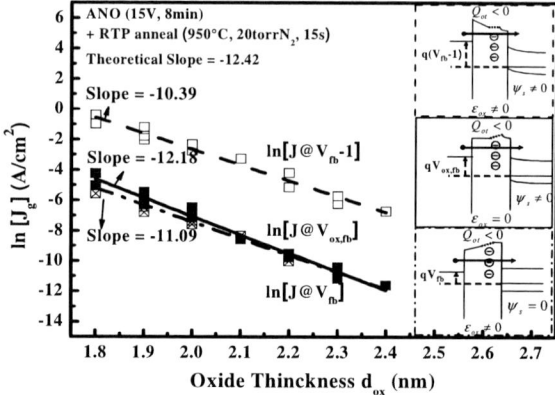

Fig. 5 The experimental data of natural log of gate current density at V_{fb}, V_{fb}-1, and $V_{ox,fb}$ versus oxide thickness compared to the theoretical one.

Following is the non-uniformity exploration by different device areas depending on our new methodology and we demonstrate and realize it from experiment and analysis of MOS capacitor.

Fig. 6 depicts the derivatives of J-V curves with three different gate electrode areas of $1\times2.25\times10^{-4}$, $4\times2.25\times10^{-4}$, and $16\times2.25\times10^{-4}$ cm² for SiO_2 dielectrics of 1.9 nm. By observation, we can find out some essential differences between these three different area devices pointed out by a square shown in the figure. Under the same oxide thickness condition, each device has its own specific non-uniformity situation that affects the derivatives of the J-V curves. The larger the device area, the more notable change of the slope will be carried out. It is supposed that large area device would cover more traps in oxide dielectrics and consequently the non-ideal phenomenon would become severe. That is to say, as to different area conditions, the 1 time area device is much more uniform than the 4 and 16 times area devices. Here, we can make an assertion by the analysis of different gate electrode areas that the derivative of J-V curve versus gate voltage is sensitive to the oxide uniformity and in the further study, this method can be a reference to determine the uniformity of any dielectric layer.

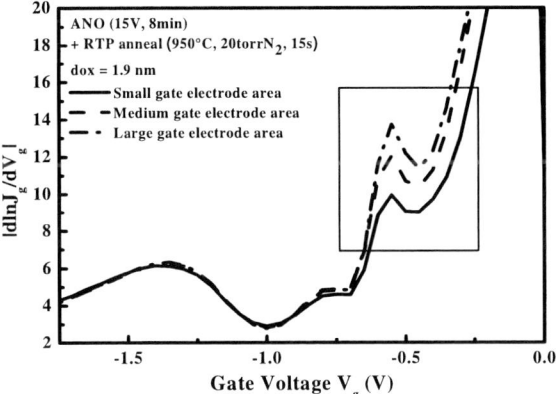

Fig. 6 The derivatives of J-V curves versus gate voltage with three different 1:4:16 device areas, and the unit area is 2.25×10^{-4} cm^2.

After analyzing the characteristics of the derivatives of J-V curves plot; similarly, we try to draw out all minimum differential values; that is, the oxide flat-band voltage ($V_{ox,fb}$) of different oxide thicknesses with different gate electrode areas as shown in Fig. 7. We could obtain the same oxide flat-band voltage tendency as the Fig. 4 indicated. $V_{ox,fb}$ extracted by derivative J-V curves would shift to much more negative when oxide thickness decreases despite different gate electrode areas.

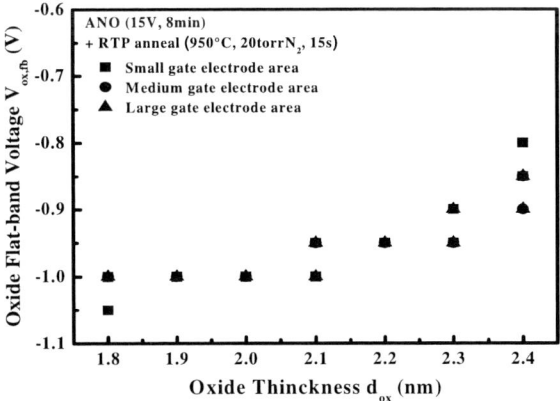

Fig. 7 The oxide flat-band voltage ($V_{ox,fb}$) of different SiO$_2$ dielectrics thicknesses with three different gate electrode areas of 1×, 4×, and 16× 2.25×10^{-4} cm^2.

Moreover, we try to apply the methodology referred in Fig. 5 to quantify the degree of non-uniformity of different gate electrode areas by the analysis of natural log of gate current density versus oxide thickness plot. Fig. 8 is the ln[Jg] versus oxide thickness plot of three different gate electrode areas of 1×, 4×, 16× 2.25×10^{-4} cm^2 and the corresponding slopes are carried out. As to different area conditions, the non-uniformity factor for 4 and 16 times area devices are more remarkable than the 1 time area device due to the experimental slope diverging from the theoretical one. It is suggested that the device with larger area would be easier to cover more defects and there are more charges in the oxide; hence, the real oxide would not represent that flat-band like a rectangular barrier, the theory adopted in our methodology. For this reason, the slopes of ln[Jg] versus oxide thickness are far from the theoretical slope when the gate electrode areas become larger. Simply speaking, we analyze and compare the characteristics of SiO$_2$ film in different device areas with two methods: one is the derivative J-V curves in Fig. 6, and the other is the slope of the ln[Jg] versus oxide thickness plot in Fig. 8. Both of them prove that the device with smaller gate electrode area exhibited better oxide performance, quality and uniformity.

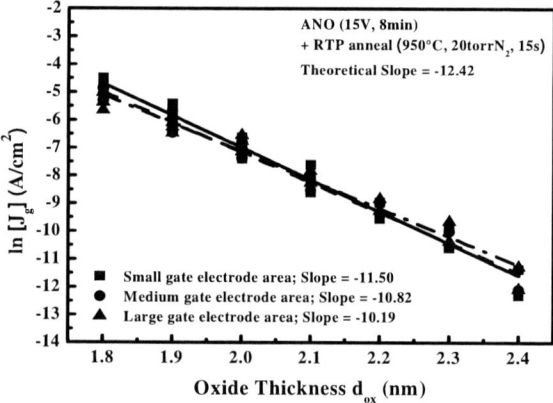

Fig. 8 The comparison of slopes of different device areas.

Conclusion

In this paper, we focus on essential and fundamental electrical characteristics especially the non-uniformity phenomenon in MOS capacitor. At first, the definition of oxide flat-band voltage ($V_{ox,fb}$) and the method to obtain the oxide flat-band voltage from basic J-V curve was depicted. Subsequently, a simple theory adopted in this work to prove that the slope of the ln[Jg@$V_{ox,fb}$] versus oxide thickness plot is much closer to the theoretical slope of -12.42 than the slope of the ln[Jg@V_{fb}] versus oxide thickness plot.

And the following, we compare the characteristics between different gate electrode areas of MOS capacitor in order to know more about the development of the technology of examining non-uniformity effects in solid state devices. Based on the analysis mentioned in the paper, both the derivatives of J-V curves and the slope of ln[Jg] versus oxide thickness plot could be reference index of uniformity for MOS devices. Depends on these two respects, the same conclusion that the smaller the device area, the more uniform the oxide is would be carried out.

Acknowledgments

This work is supported by the National Science Council, R.O.C., under Contract No. NSC96-2628-E-002-246-MY3.

References

1. G. E. Moore, *IEDM Tech. Dig.*, 11-13 (1975)
2. International Technology Roadmap for Semiconductors (ITRS), *Semiconductor Industry Association (SIA)* (2009)
3. J. Sune, Y. Placencia, F. Campabadal, and X. Aymerich, *Surf. Sci.*, **189/190**, 346 (1987)
4. Z. Hurych, *Solid-State Electron*, **9**, 967 (1966)
5. B. Majkusiak, and A.Strojwas, *J. Appl. Phys.*, **74**, 9 (1993)
6. C. K. Chow, *J. Appl. Phys.*, **34**, 2599 (1963)
7. J. Pochobradsky, *Solid-State Electron*, **10**, 973 (1967)
8. P. F. Schmidt and W. Michel, *Journal of the Electrochemical Society*, **104**, 230 (1957)
9. P. F. Schmidt, T. W. O'Keffe, J. Oroshnik, and A. E. Owen, *Journal of the Electrochemical Society*, **112**, 800 (1965)
10. G. C. Jain, A. Prasad and B. C. Chakravarty, *Journal of the Electrochemical Society*, **126**, 89 (1979)
11. S. K. Sharma, B. C. Chakravarty, S. N. Singh, B. K. Das, D. C. Parashar, J. Rai and P. K. Gupta, *J. Phys. Chem. Solids*, **50**, 679 (1989)
12. J. A. Bardwell, N. Draper, and P. Schmuki, *J. Appl. Phys.*, **79**, 8761 (1996)
13. Vitali Parkhutik, *Electrochimical Acta*, **45**, 3249 (2000)
14. M. Grecea, C. Rotaru, N. Nastase, and G. Craciun, *Journal of Molecular Structure*, 607 (1999)
15. M. Grecea, C. Rotaru, N. Nastase, and G. Craciun, *J. Mol. Struct.*, **480-481**, 607 (1999)
16. G. C. Jain, A. Prasad, and B. C. Chakravarty, *J. Electrochem. Soc.*, **126**, No. 1, 89 (1979)
17. C. C. Ting, Y. H. Shih and J. G. Hwu, *IEEE Trans. Electron Devices*, **49**, No. 1, 179 (2002)
18. Sorab K. Ghandhi, *VLSI Fabrication Principles, 2nd ed.*, Wiley-Interscience, p. 487 (1994)
19. K. Yang, Y. C. King, and C. Hu, *VLSI Symp. Tech. Dig.*, 77 (1999)
20. K. J. Yang, and C. Hu, *IEEE Trans. Electron Devices*, **46**, 1500 (1999)
21. S. –H. Lo, D. A. Buchanan, Y. Taur, and W. Wang, *IEEE Electron Device Letter*, **18**, 209 (1997)
22. C. H. Chen, K. C. Chuang, and J. G. Hwu, *IEEE Trans. Electron Devices,* **56**, No. 6, 1262 (2009)

23. A. Beiser, *Concepts of modern physics*, 6th edition, McGraw-Hill, p.185 (2003)
24. T. Hori, *Gate dielectrics and MOS ULSIs*, Springer, p. 37
25. N M Ravindra and Jin Zhao, *Smart Mater. Struct.*, **1**, 197 (1992)

Novel Hardmask For sub-20nm Copper/Low k Backend Dual Damascene Integration

Li-Qun Xia*, David Cui, Mihaela Balseanu, Victor Nguyen, Kevin Zhou, Jeremiah Pender, and Mehul Naik

Applied Materials, Inc.,
3225 Oakmead Village Dr. M/S 1290, Santa Clara, CA 95012;
*Email: li-qun_xia@amat.com

A new boron-based hardmask material was developed using a conventional CVD approach to address the integration challenges associated with the use of metal hardmask for low k dielectric patterning. Its low and tunable stress eliminates any patterning concerns due to line bending for device nodes below 20nm. Defectivity is also reduced because by-products from its F-based etch (BF_x) are volatile unlike TiF_x based defects, which dramatically widens the manufacturing process window. Extensive study concludes that boron content can be optimized to achieve good etch selectivity to the porous low k oxide. Preliminary electrical test using 45nm node 2-metal level structure showed good yield and 9% RC reduction compared to the conventional tri-layer integration scheme

Introduction

Dual damascene concept was introduced for manufacturing by IBM in the 90s for integrated circuits (IC) back-end-of-line (BEOL) integration using copper as metal interconnect [1]. The insulation between the metal lines has also shifted to C-doped oxide films since the 90nm device node to reduce the dielectric constant (k <3.0) from conventional oxide. Via-first scheme was the mainstream copper/oxide BEOL integration scheme. Photo-resist (PR) "poisoning" often occurs during line imaging lithography after vias are etched into the C-doped oxide. The source of the poison is reported to be the interaction between the photo-resist and the liberation of the amine compounds from insulators, formed either during deposition or post integration steps. The situation becomes much worse with the increasing permeability of the low k insulators, as well as the adoption of 193nm resist. To overcome this patterning issue, a tri-layer scheme was adopted [2], Figure 1, which uses an organic planarizing layer (OPL) after via etch, followed by a thin hermetic oxide layer to protect the resist from low k under-layers. Photo-resist was then used to pattern the oxide and transferred to the OPL. After line (trench) etch, OPL is stripped to form the dual damascene pattern.

An alternative approach using dual hardmask scheme was also proposed [3]. There is no low k dielectric etch prior to patterning, as shown in Figure 2. Trench pattern was first transferred to the top hardmask, followed by via patterning then partial etch and resist strip. Both trench and via were then etched simultaneously to form the dual damascene structure. In this case the hardmask layers protect the resist from poisoning. TiN metal hardmask was selected as the top hardmask mainly due to its superior etch selectivity to low k dielectric, since it has to survive through multiple etch steps. After etch, the remaining TiN stays through copper metallization and finally removed by the chemical-mechanical polishing (CMP) step.

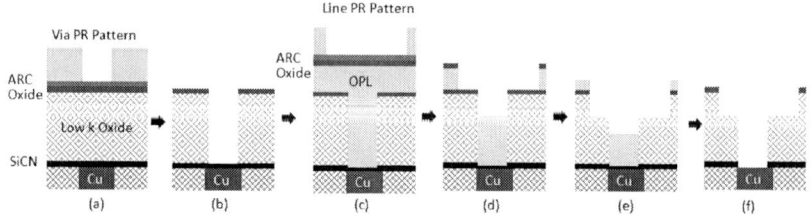

Figure 1: Low k dual damascene patterning, tri-layer scheme: (a) Via patterning of low k stack, (b) Via etch and resist strip; c) Line patterning; (d) Pattern transfer to oxide; (e) Line etch; and (f) OPL strip and SiCN open

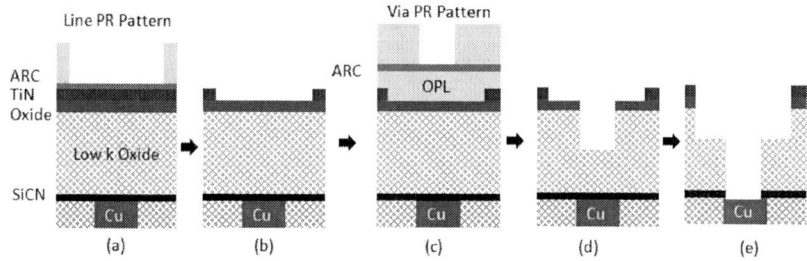

Figure 2: Low k dual damascene patterning, dual hardmask scheme: (a) Line patterning of low k stack, (b) Pattern transfer to TiN and resist strip; (c) Via patterning; (d) Partial via etch and resist strip; (e) Line and full via etch

Both schemes have been used adopted in high volume manufacturing since the 90nm technology node. As the industry continues to advance, porous low k materials were introduced at 45nm with k <2.5 to further reduce the RC. At sub-20nm technology, the industry is exploring even lower k dielectric materials, which present significant integration challenges due to potential film damage resulting from carbon loss and moisture uptake. For the tri-layer patterning scheme, the photoresist/OPL strip step following the dielectric etch induces significant trench sidewall and bottom damage leading to effective k increase. An example of the side-wall damage can be shown in Figure 3 for a simple 3-layer film stack using SiCN as an etch stop, bulk film is a porous low k film and capped with a dense oxide layer. After line litho, the structure was etched using CHFx based chemistry to form trenches and followed by resist strip using O_2 plasma. A wet dip using 100:1 DHF was used to highlight the damage between the oxide cap and the underlying low k film. Increasing concerns for low k damage during plasma etch and strip motivated the migration to a metal hardmask patterning schemes. The use of a metal hardmask enables the shifting of the resist strip process after via etch. Any dielectric damage layer, mostly on the sidewall, will be removed during trench etch.

Despite its technical advantages, conventional TiN hardmask faces many challenges ranging from technical extendibility to its manufacturing robustness. The baseline TiN is formed using physical vapor deposition (PVD). PVD's high plasma density creates a dense TiN film beneficial for etch selectivity, but also makes its stress highly compressive. As device geometries shrink and narrower lines are printed, resist and line bending becomes an issue, so controlling film stresses becomes critical. Additionally, manufacturing concerns with defectivity and shelf-life control shrink the overall process window [4-6]. One source of defectivity comes from the interaction between moisture and TiFx, a by-product formed between TiN and F-based dielectric etch chemistry. In this paper we present the development of an alternative hardmask (HM) which can be integrated using the "metal hardmask flow" but does not present the challenges encountered when using the TiN film.

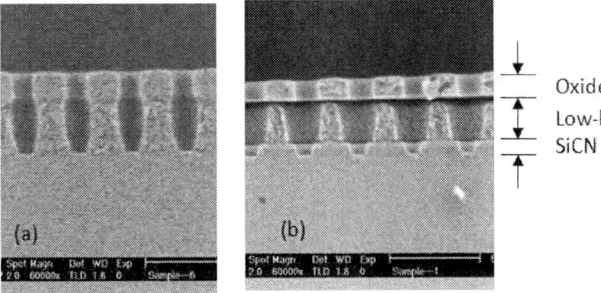

Figure 3: Low k damage through etch and resist strip. Film stack is SiCN as etch stop, bulk porous low k and oxide cap. (a) Post trench etch and resist strip; and (b) Post 100:1 DHF dip

Experimental

The novel hardmask film was developed in a plasma-enhanced chemical vapor deposition (PECVD) reactor on an Applied Materials Producer GT^{TM} platform using 300mm silicon substrates. A simple schematic of the reaction chamber design is shown in Figure 4, depicting the parallel plate plasma system equipped with a ceramic heater to control the wafer temperature.

A 300mm wafer was delivered under vacuum into the chamber by a mechanical robot and placed on the heater. Both boron and nitrogen containing gases were pre-mixed and fed through the faceplate for uniform distribution. Chamber pressure can be modulated by a throttle valve. Radio frequency (RF) power can be applied to the faceplate to generate the plasma between the heater and the faceplate. The plasma density can be controlled by the spacing between these two electrodes, as well as chamber pressure. A secondary low frequency plasma can also be added to enhance the ion bombardment. Diborane (B_2H_6) gas was the main source for boron. Nitrogen comes from NH_3 or N_2 dissociation and various carbon and silicon containing molecules were also tested and compared. The film composition can be simply modulated by the gas ratio. To optimize the film properties, a variety of process conditions were explored to deposit the hardmask film, including RF power, chamber pressure and gas flows.

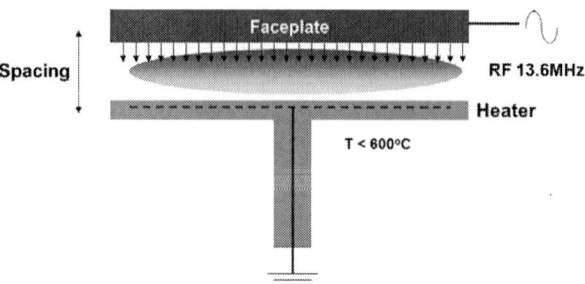

Figure 4: Schematic for Applied Materials PECVD reactor using 13.6MHz RF plasma and a ceramic heater to control wafer temperature below 600°C. Heater is grounded and the RF power is added to the faceplate, which also acts as gas distribution plate for the reactants to achieve good within wafer uniformity.

Initial film characterization was performed on blanket silicon substrates to determine the stress, density, hardness, modulus and wet etch resistance. Film thickness and some other basic optical properties (refractive index and extinction coefficient) were measured using ellipsometry. Stress was calculated from the wafer bow induced by the film deposition, whereas film density was monitored using X-ray Reflection (XRR) technique. Mechanical properties, such as hardness and Young's modulus were measured using a nanoindentor. Wet etch rates were measured and compared with the standard silicon oxide and nitride films in multiple solutions typically used in the BEOL integration flow: diluted hydrofluoric acid (100:1 DHF), hot phosphoric acid, Piranah SPM (H_2SO_4 : H_2O_2) solution, etc. Blanket dry etch test was also carried out using conventional CHFx chemistry to compare and screen the compositional impact from Si, C, B and N content in the film.

Dual damascene etch tests were performed using 45nm node, 2-level via/trench structure. A porous C-doped oxide film with k~2.5 was used as the main insulation material, capped with silicon oxide and the new B-based material as the dual hardmask layers. The film stack built to evaluate and optimize the etch selectivity of the boron-based new hardmask to the porous low k oxide is shown in Figure 5. The dielectric etch study was conducted using Applied Materials' Enabler[TM] etch system to understand the film interaction with etch chemistry.

Integration of these materials was completed using 32nm node (90nm pitch) single level metal process flow (line/space of 45nm) and 45nm node 2-level metal structure for electrical parametric testing. Test conditions are applied after M2 Cu CMP. Measurements include RC and line-to-line I-V. Similar structures were used for determining etch selectivity after inspection with SEM. EELS and TEM were used to verify the complete removal of the hardmask after the CMP step.

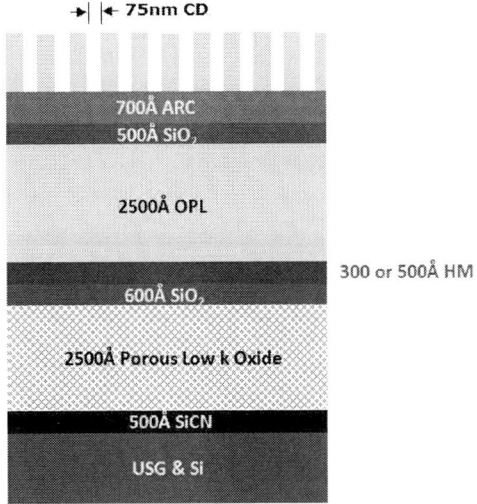

Figure 5: Film stack for dual hardmask etch study using the B-based film as etch hardmask above porous low k material. The thickness of the HM varies from 300Å to 500Å to match the TiN baseline.

Results and Discussion

Blanket Film Characterization

The drive to reduce the low k dielectric damage during patterning led to the introduction of metal hardmasks which can result in defectivity due to TiFx residue [4-6] and stress induced line bending. The TiN etch byproduct TiFx reacts with air and moisture to form a metallic salt. Conventional metal fluorides, e.g. TiF_4 (284°C), TaF_5 (229.5°C), are much less volatile than non-metal molecules, such as BF_3 (-100.3°C) and SiF_4 (-86°C), and can lead to defectivity issues. The non-volatile post-etch residue inside the via and trench can cause yield loss. Similarly, residue can buildup inside the etch chamber resulting in particles and poor manufacturability performance. On the other hand, non-metal based hardmask materials often suffer from poor etch selectivity during low k etch using F chemistry and can't survive multiple etch steps required to form dual damascene structure. An ideal hardmask material will match the etch selectivity of the TiN without the manufacturability issues introduced by non-volatile etch byproducts and high film stress. In addition, advanced hardmask materials need to show excellent oxidation resistance and integrability with the current process flow developed for TiN patterning to insure easier adoption.

Boron-based materials have been used for multiple applications outside the realm of IC devices. The most typical applications are in areas requiring superior mechanical properties. Boron-based materials are excellent choices for abrasives, cutting tool coatings and parts for high temperature equipment due to their extreme hardness (comparable to diamond) and chemical and thermal resistance [7-8]. One drawback of using any of these boron-based materials is the processing required to deposit them: high temperature (>>800°C) and pressures (>5GPa) are not compatible with the film used in IC processing. Recent studies showed the potential use of boron-based materials deposited by conventional chemical vapor deposition (CVD) technique in the realm of nanotechnology as nanotubes (hexagonal boron nitride) [9-10] or IC processing as Copper low k barrier films (boron nitride BN, boron carbon nitride, BCN) [11-12] at deposition temperature below 500°C.

In this paper we discuss the development of boron-based films deposited by conventional CVD as dual damascene patterning hardmask. Wafer temperature during deposition was controlled at ≤400°C for low k BEOL integration. Requisite properties for a new hardmask are low stress and good etch selectivity to low k dielectric. From previous literature study, the composition and structure of the boron compounds can be varied widely by the process conditions [9-12]. In our studies, the boron content was modulated by the precursor flow ratio in the CVD reactor. Hardness, modulus, film stress and refractive index were monitored as a function of the boron content to screen the candidates for hardmask applications. As shown in Figure 6, the refractive index increases and film becomes more tensile with increasing B content. Hardness and modulus also improve.

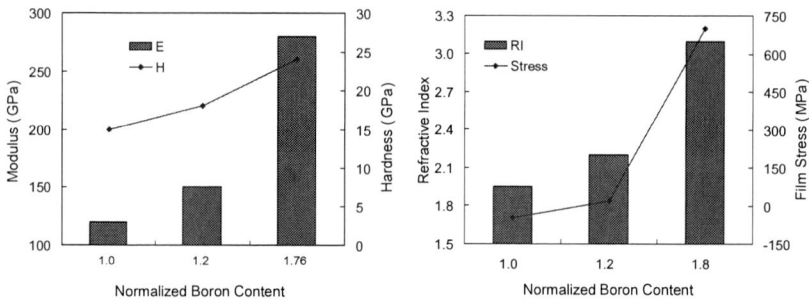

Figure 6: Boron-based material properties function of Boron content

By subjecting these films to wet etch solutions typically used for semiconductor manufacturing, we found that the boron-based film etch behavior is also a very strong function of the boron content. Figure 7 shows the wet etch rate variations for these films. For reference, thermal oxide and thermal nitride samples were etched at the same time in all solutions. In most cases, these B-based hardmask materials showed lower removal rates, especially the B-rich films.

In 100:1 diluted HF, where oxide typically etches faster than nitride, most of the B-based films show very little etch, lower than nitride. This indicates that B films are more resistant to F etch. Hot H_3PO_4 (150°C) solution is mainly used to remove nitride over oxide, where all the new hardmask films showed almost no etch. SPM solution composed of peroxide and sulfuric acid, is a strong oxidizer used to remover organic residue. Good oxidation resistance is critical for hardmask material to avoid CD increase during litho re-work steps or litho-etch-litho-etch (LELE) process. As seen from the graph, the SPM etch rate was low, but it is worth noting that decreasing boron content will however degrade the oxidation resistance. Therefore we have concluded that a B-rich film is more suitable for hardmask applications.

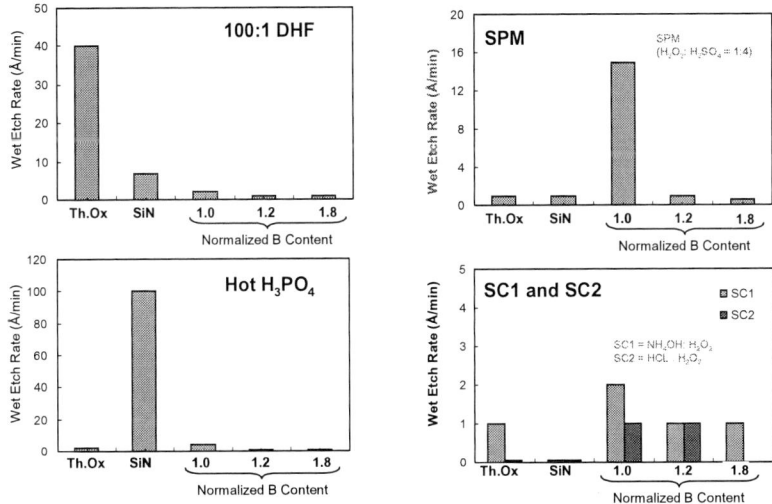

Figure 7. Wet etch rates as a function of B content in the new hardmask films

Figure 8 shows the O profile comparison for the B-rich hardmask film to TiN baseline using either H_2 or O_2 plasma based resist strip process in the etch chamber. The B-rich film clearly shows more resistance to oxidation.

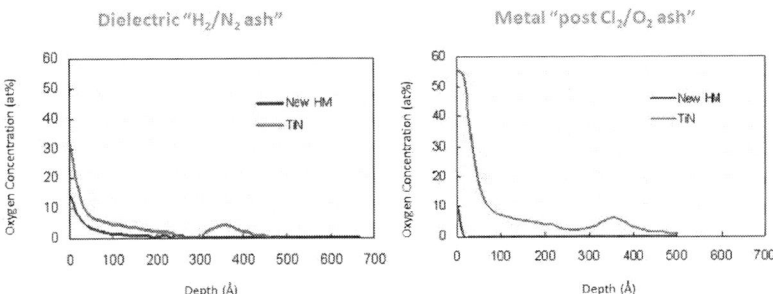

Figure 8 Oxidation comparison for TiN and B-rich hardmask films during conventional resist strip processes: 1) H_2 and 2) O_2 plasma. In both cases, B-rich film shows much less O into the film

Blanket dry etch using a CHF_3/CF_4 chemistry showed similar trend as wet etch rates, i.e. higher boron content yields lower etch rate irrespective of the etch condition (Figure 9).

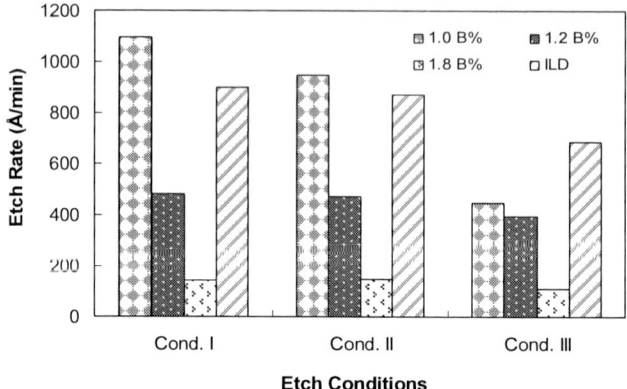

Figure 9. Dry etch rate performance using a CHF_3/CF_4 chemistry

Once we converged on the optimum film composition for good etch selectivity, the next task was to modulate the film stress. As observed in Figure 6, the film stress increases with boron content. The film stress was modulated by the plasma density during deposition. Figure 10 shows the stress range of the hardmask films under different plasma conditions. The baseline stress is 700MPa tensile and we can tune the film stress from tensile to neutral, or compressive to achieve the best performance of the resist line bending. TiN normally has over 1GPa compressive stress, and based on patterning studies, low tensile film seems more beneficial to reduce line bending. Since film composition is mainly controlled through gas flow ratio, we can maintain the same composition while varying the stress level of these films. So the etch selectivity will not degrade as we tune to the desired film stress.

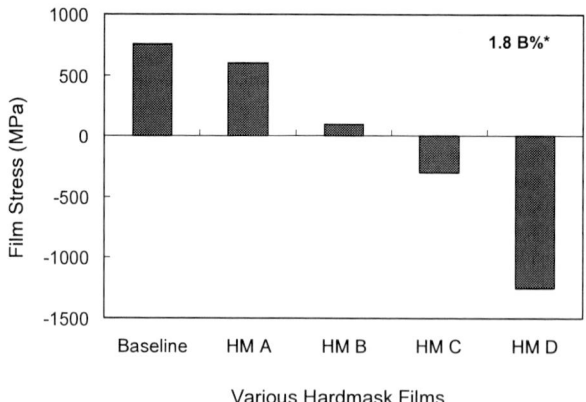

Figure 10: Boron-based hardmask has wider stress tuning window without composition change

Figure 11 shows both blanket dry etch and CMP rates for B-rich hardmask films with similar composition but different stress level. CMP test was performed using Applied Materials Reflection LK™ CMP system. While CMP rates are almost the same among these films, dry etch rate showed ~20% further improvement for the highly compressive film. XRR measurement reveals that the high compressive film has 0.1g/cm³ higher density, possibly due to stronger RF bombardment.

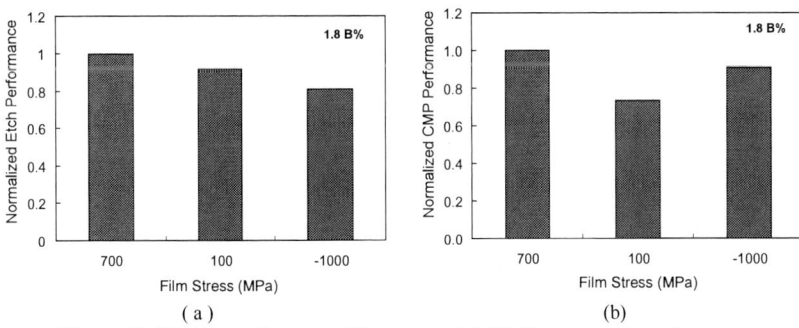

(a)   (b)

Figure 11 (a). Dry etch rate vs film stress; (b) CMP rate vs. film stress

Dual Damascene Structural Etch

As seen from the blanket test results, etch selectivity is tunable mainly by varying the film composition. Hardmask open etch selectivity was verified on patterned structures using the test stack shown in Figure 5. Using 500Å B-based hardmask on-top of 600Å oxide, we first investigated the hardmask loss after mask open step, as shown in Figure 12. Films with different boron content were used for this testing. Figure 12a shows the remaining film stack and the post etch HM thickness is marked in Figure 12b. As expected, the film with much higher boron content shows the best etch selectivity as defined by the ratio between the hardmask thickness loss and depth of the ILD etch. There is minimal thickness loss for the film with 1.8 normalized %B content. The film retains its original shape while etching into low k films, a key requirement to match TiN performance. Similar test was done with 300Å HM thickness and the result is summarized in Table I. This data demonstrates that even with thin hardmask thickness, B-based films can meet the etch selectivity requirements.

Table I: Remaining Hardmask thickness after HM open and trench etch with various boron concentration. Minimal film loss for films with much higher B content.

Pre-etch Thickness	Post Etch Thicknesses	
	1.2 Norm. %B	1.8 Norm %B
50nm	41.6 nm	49.6 nm
30nm	23.8 nm	29.8 nm

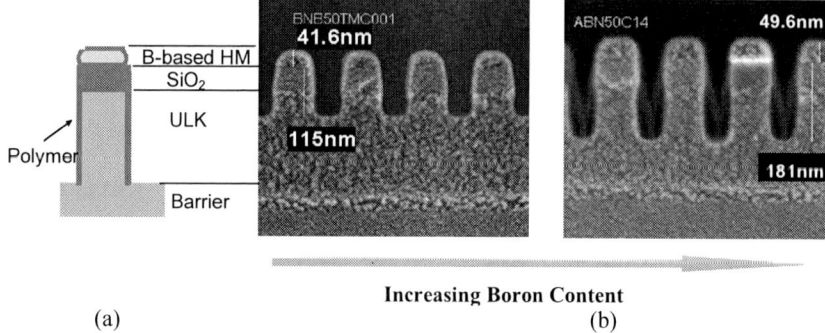

Increasing Boron Content

(a)   (b)

Figure 12: Hardmask Open Etch Selectivity. (a) Remaining film stack after mask open, (b) Post mask open and trench etch stack thickness for the remaining boron-based hardmask layer

The etch selectivity of the film with optimized boron content was demonstrated using a 2-level metal structure with 300Å HM thickness. The film used for this evaluation had tensile stress of 700MPa. The post etch profile is shown in Figure 13 for the dual damascene M2 trench and via chains using a 45nm node structure, where the line width and line spacing are both 75nm. Selectivity on the order of 20:1 (vs. porous low k oxide, ULK) was realized after HM etch process tuning. The post-etch profile did not exhibit undercut or bowing with controlled taper at the trench/via interface. From these results, it was estimated that a HM thickness of ≤250Å would be sufficient to fabricate 22nm/16nm node structures.

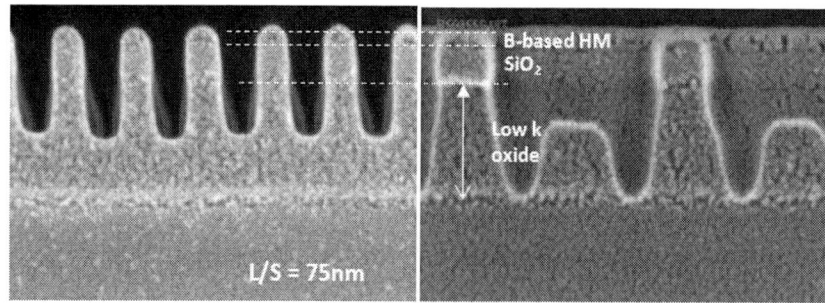

Figure 13. 45nm node dual damascene trench and via chains profile

To demonstrate the film stress benefit, we also tested the extendibility of this hardmask from 75nm down to 35nm line/spacing, shown in Figure 14. All structures patterned using the 700MPa tensile boron-based hardmask showed straight line profile and good line edge roughness (LER) which is one of the challenges with high compressive stress TiN film. This indicates the extendibility of the new hardmask film beyond 20nm. An example of the top view using 45nm node L/S can also be seen in Figure 15, for both dense array and open area.

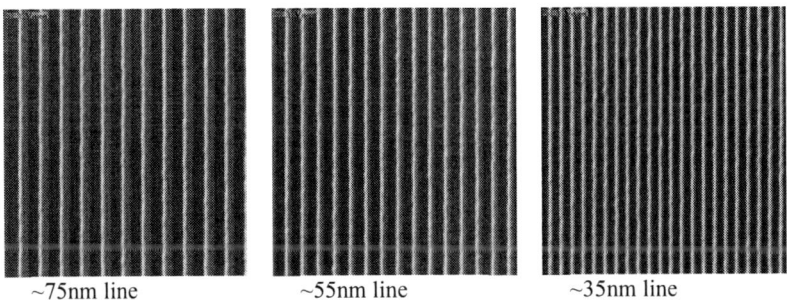

| ~75nm line | ~55nm line | ~35nm line |

Figure 14. No line bending or breaks seen in 75nm to 35nm dielectric lines

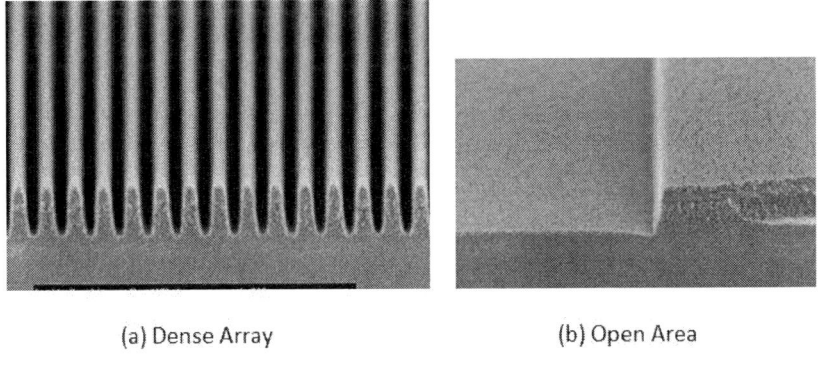

(a) Dense Array   (b) Open Area

Figure 15. 45nm Line/Spacing M1 post etch profile for both dense and open area

As earlier mentioned, defect management in a high volume manufacturing environment is another challenge for TiN, as the etch by-product TiFx interacts with moisture [4]. The shelf time control requirement is eliminated with the new boron-based hardmask, simply because of the volatility of the etch by-product (BF_x) during F etch: the boiling point of BF_3 is -100.3°C vs. TiF_4 which is a solid at room temperature (b.p. 284°C). As shown in Figure 15 for both HM open and trench etch, no defect growth is observed after over 1 week of air exposure. The same defect-free performance is observed on both dense and field areas in Figure 16.

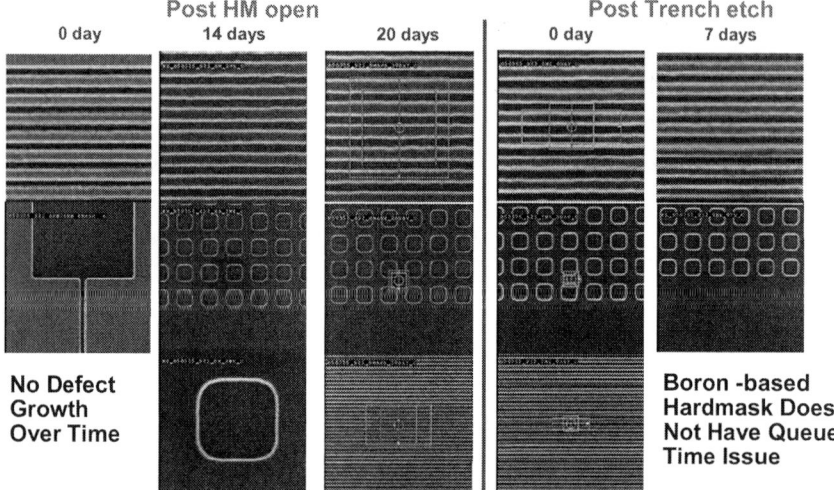

Figure 16. Surface residue inspection using CDSEM on structures of post HM trench open and via etch, with queue time of 7 days.

Integration and Electrical Test

The test vehicles used in this study for integration and electrical performance evaluation include 32nm node (45nm half-pitch) single level metal and 45nm node 2-level metal. Electron-Energy-Loss Spectroscopy (EELS) and TEM were used to verify complete removal of the hard-mask after CMP. Electrical test characterization included RC and Line to line breakdown voltage measurements.

The TEM of the 2-level metal structure in Figure 17 shows complete hardmask removal by CMP on both dense and isolated lines.

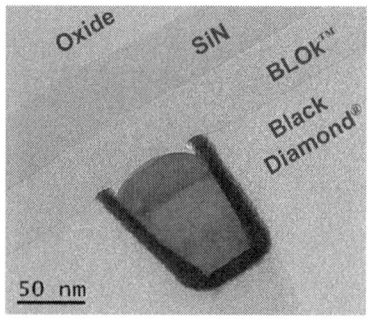

Figure 17: 2Level Metal M2 and M1 TEM profile, no boron residue on top of ULK or trench sidewall

To verify the integration of the boron-based films in the exiting patterning flow, the hardmask was removed using the same CMP baseline process used for the TiN scheme. Figure 18 shows the EELS line scan on both the top and bottom of the ILD structure.

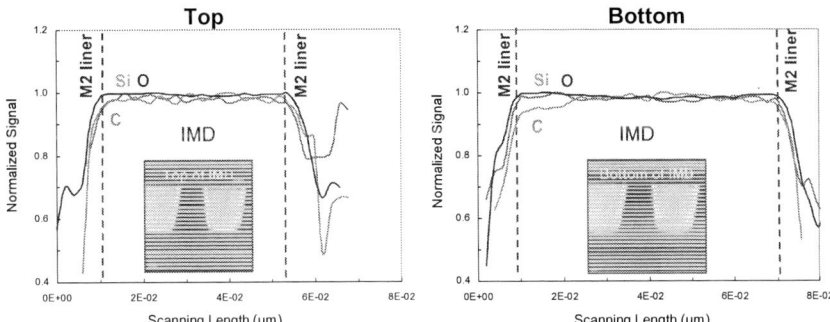

Figure 18: Dielectric composition profile for the 2-level metal (M2) integrated structure by EELS, comparing surface layer after CMP and film at the bottom in between metal lines - no boron residue on top of ULK or trench sidewall detected

The EELS analysis at the top of the structure was performed to confirm the complete removal of the B-based hardmask during the CMP step. An additional EELS analysis was performed lower in the structure across the ILD layer to determine if any boron out-diffusion of the hardmask layer occurred before it was removed. As seen in Figure 18, no boron is detected in the ILD layer demonstrating complete removal and no boron out-diffusion during the integration process.

B diffusion into ULK or incomplete HM removal will cause leakage and low breakdown voltage. As seen in Fig 19, the breakdown field for the integrated structure was >5MV/cm. This electrically verifies the TEM-EELS results and confirms complete removal of hardmask during CMP and absence of B within the dielectric.

Boron-based HM flow also needs to have comparable or better electrical performance than the baseline TiN or tri-layer integration process flow. In order to achieve good yield (>95%) and tight die-to-die distribution, a post-etch treatment (PET) was used to eliminate the etch polymer residue, followed by a wet clean using ST250 solution. RC improvement (vs. conventional tri-layer scheme) was ~9%, confirming the effectiveness of the boron-based HM flow to reduce low k damage. These results indicate that the boron-based hardmask can be integrated into sub 20nm BEOL structures.

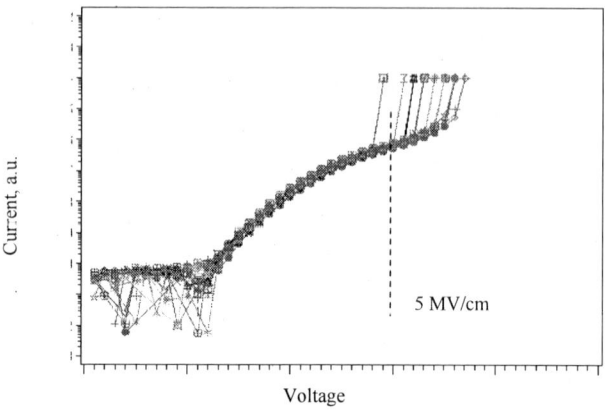

Figure 19: IV characteristic for 2ML structures with different L/S

Summary

The use of boron-based hardmask material was investigated as an alternative to TiN for use in the dual-hardmask process flow for low k patterning. The film characterization showed that etch selectivity is improved by increasing the boron content while the stress can be modulated over a wide range: from -1GPa compressive to 700MPa tensile. The performance of the boron-based hardmask for the low k patterning was investigated using a 2-metal level process flow with structures down to 45nm L/S. No line bending or residue formation was observed at any step in the patterning process. To evaluate the process defectivity, the wafers were monitored for up to 20 days after mask open and after trench etch. No residue was detected in either cases demonstrating that the boron-based hardmask does not suffer from queue-time issues.

The electrical characteristics of the devices built using boron-based hardmask for the low k ILD patterning show similar performance to the TiN baseline. Those results indicate that the boron-based film can be integrated into sub 20nm BEOL process flow with no impact on device performance while addressing manufacturability issues encountered with conventional hardmasks.

Acknowledgement

The authors sincerely wish to thank all collaborators for their contributions: Applied Materials CMP group for their support with hardmask removal; Defect and Thin Film Characterization Laboratory (DTCL) for the TEM analysis.

References

1. D. Edelstein et al., *Electron Devices Meeting, 1997. IEDM Technical Digest. IEEE International*, p.773 (1997).
2. W. Cote, et al., *AMC 2006 Proc. San Diego*, p.43 (2006).
3. O. Hinsinger et al., *Electron Devices Meeting, 2004. IEDM Technical Digest. IEEE International*, p.317 (2004).
4. MC Lin, et al, *Mater. Res. Soc. Symp. Proc.*, p. 287, 914 (2006).
5. H. Cui; S. J. Kirk, D. Maloney, *Advanced Semiconductor Manufacturing Conference, 2007. ASMC 2007. IEEE/SEMI* , p.366, 11-12 June 2007.
6. N. Posseme, et al, *IITC*, pp. 240, 2009.
7. W.J. Croft, N.C. Tombs and J.F. Fitzgerald, *Mater. Res. Bull,* p.489, 5 (1970).
8. J. Hunag, Y.T. Zhu, *Defect and Diffusion Forum*, p.1, 186-187 (2000).
9. L. Song et al., *Nano Letters*, p.3209, 10 (8), (2010).
10. S. Saha, et al., *Chemical Physics Letters*, p.86, 421 (2006).
11. Y. Chen et al., *IITC*, p.1, 6-9 June 2010.
12. T. Sugino, T. Tai, Y. Etou, *Diamond and Related Materials*, p.1375, 10 (2001).

666

Study of porous SiOCH patterning using metallic hard mask:
challenges and solutions

N.Posseme [a], T.David [a], T.Chevolleau [b], M.Darnon [b],
F.Bailly [c], R.Bouyssou [c], J.Ducote [c], H. Chaabouni [b], M. El Kodadi [b], C. Licitra [a]
C.Verove [c] and O.Joubert [b]

[a]CEA-LETI, Minatec campus, 38054 Grenoble, France
[b]LTM-CNRS/UJF/INP, 38054 Grenoble, France
[c]STMicroelectronics, 38926 Crolles cedex, France
Contact : nicolas.posseme@cea.fr

The choice of copper/low-k interconnect architectures is instrumental in achieving high device performances. Today, the implementation of porous low-k materials becomes mandatory in order to compensate metal resistance increase upon RC product. However, their introduction, which was initially planned for the 65nm technological node, was delayed to 45nm node due to integration issues. Using an intégration strategy which combines porous SiOCH materials and metal hard masks, the difficulties and possible solutions are presented in this paper with emphasis on plasma etching.

I.   Introduction

One of the most serious challenges in semiconductor manufacturing is to produce low cost integrated circuits as well as to develop high performance devices. On a technical point of view, a serious problem is that electric signal propagation through metal interconnects is delayed by the resistance (R) of the metal lines and the capacitance (C) between adjacent metal lines. Nowadays, the chosen way to decrease the runtime delays and thus the RC product is to use a less resistive metal (ie. copper) and decrease the power consumption and cross-talk (given by the capacitance) by integrating lower dielectric constant insulators [1]. To achieve this, many low-k materials have been investigated. There are two categories of materials: silicon containing (silica-based and silsesquioxane) or non silicon containing [2-5] materials presenting different advantages and drawbacks in term of chemical and physical properties, thermal stability and expansion, etc.. Finally, from the 90 nm technological node, the semiconductor industry selected $Si-CH_3$ containing organosilicate materials close to the well known SiO_2 [6]. These materials present k value of about 3 and can be reduced further by introducing porosity leading in the meantime to a weakening of its mechanical properties [2]. The introduction of porous SiOCH material was initially planned for the 65nm technological node, but the complexity of their introduction delayed their introduction for the 45nm interconnect technology node. For instance, previous study performed on blanket wafers, has shown that porous SiOCH materials are sensitive to fluorocarbon based plasma. These results obtained on spin-on p-SiOCH films presenting a porosity variation between 40% and 50% demonstrated the film degradation (methyl group depletion and moisture uptake) directly scales with porosity in the material [7].

In this paper, we propose to analyse the issues revealed during the integration of a porous low-k material (deposited by plasma enhanced chemical vapor deposition, with dielectric constant 2.35), and the associated solutions from an etching point of view. P-SiOCH line patterning can be achieved using different hard mask strategies (metallic or organic hard masks) [8-10], both presenting advantages and drawbacks in terms of etching performance etc. [11]. In this paper we will only focus on the metallic hard mask approach to discuss the etching challenges.

II.   Experimental

The presentation of p-SiOCH integration difficulties with a metallic hard mask approach will be discussed and characterized using several techniques briefly described below.

II.1   Porous SiOCH materials

In this study, porous SiOCH material presenting a porosity of 28% with dielectric constant 2.35 is investigated. These materials are deposited by plasma enhanced chemical vapor deposition (PECVD) in a Producer^{TM} mainframe from Applied Materials. Using a co-injection gas feed, two precursors are simultaneously vaporized (Diethoxymethylsilane ($C_5H_{14}O_2Si$, DEMS) as a matrix precursor and a norbornadiene (NBD) component as a porogen precursor and injected in a capacitively coupled plasma chamber operated at 13.56 MHz. After depositing the hybrid film (SiOCH matrix plus the porogen), a thermal curing assisted by ultraviolet radiations is performed at 400 °C under He environment in a curing chamber to facilitate the porogen removal and generate the porous film [12, 13].

II.2   Porous SiOCH integration with a metallic hard mask

In this work, the patterning of porous SiOCH trenches is performed using a metallic hard mask with the following stack on silicon substrate: porous carbon-doped silicon-oxide (PECVD p-SiOCH, with 28% porosity) / PECVD silicon dioxide (SiO_2) / titanium nitride deposited by physical vapor deposition (PVD TiN) and BARC/193 nm resist patterns.

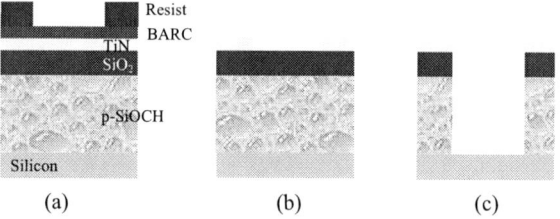

(a)   (b)   (c)

Fig. 1: Metallic hard mask (MHM) description: (a) Line lithography with organic BARC, (b) HM etching and resist stripping, (c) Line etching.

In the metallic hard mask (MHM) approach, a thin metallic hard mask layer is deposited on the top of a dense dielectric layer (SiO_2) which encapsulates the underlying p-SiOCH material. The trench photolithography is performed using a positive tone photoresist coupled with a bottom antireflective layer (BARC) (Fig.1(a)).

The BARC and titanium nitride layers are etched in a chlorine-based chemistry (Cl_2/BCl_3) using an inductively coupled plasma (ICP), followed by resist stripping in an O_2 micowave plasma (Fig.1(b)). Then the SiO_2 capping and porous SiOCH layers are patterned in fluorocarbon (FC) based plasmas ($C_4F_8/N_2/Ar/O_2$) using a multi-frequency capacitive coupled plasma (CCP) etcher (Fig.1(c)).

II.3   Characterization techniques of p-SiOCH modification
In this study, the film modification will be presented through experiments performed on blanket wafers (to represent bottom p-SiOCH film modification) and on patterned wafers to analyse the sidewall film modifications. Different complementary techniques like X-ray photoelectron spectroscopy (XPS), infrared spectroscopy, decoration method and ellipsometric porosimetry (EP) will be used.

II.3.1 X-ray Photoelectrons spectroscopy (XPS)
The surface composition of the p-SiOCH after FC based plasma exposure has been analyzed on 300mm blanket and patterned wafers using ex-situ XPS analyses.
XPS analyses spectra are performed on blanket and patterned wafers using a Thermo Fisher Scientific Theta 300 spectrometer operating with a monochromatic Al Kα x-ray source (hυ = 1486.6 eV) and an electron energy analyzer operating in a constant pass energy mode of 20eV. Chemical compositions are derived from the areas of the different XPS spectra. Spectral decomposition is performed to extract the Si2p, O1s, C1s, N1s and F1s peak intensities. Individual line shapes are simulated with the combination of Lorentzian and Gaussian functions. The background subtraction is performed by using a Shirley function. After XPS analysis, the integrated intensities are divided by the theoretical Scofield (S) cross section (C1s:1; O1s:2.93; Si2p:0.82; F1s:4.43; N1s:1.8). The sum of the concentration of the different elements present on the analysed surface is equal to 100%. The hydrogen content is not taken into account in this calculation since hydrogen cannot be detected by XPS. More details of the experimental characterization conditions can be found elsewhere [7, 14].
The determination of the p-SiOCH sidewalls film composition is achieved thanks to the specific die where different zones exhibiting regular arrays of lines, blanket substrate and unpatterned mask material are present. From the concentrations of the different elements present on a blanket area (covered with the metal hard mask) and by recording XPS spectra in the perpendicular mode in dense features, we can obtain the chemical composition of the patterns sidewalls. However, the contributions originating from the SiO_2 sidewalls and those originating from the p-SiOCH sidewalls cannot be separated. The experimental protocol is described in more details elsewhere [15]. To overcome this issue we developed an improvement of the experimental procedure by using the subtraction between two analysed patterned areas (Fig.2).
Two XPS acquisitions are performed in the perpendicular mode with contributions originating from the top (TiN), SiO_2 and p-SiOCH sidewalls of the features in a large pitch (LP) (Fig.2(a)) is performed first. A second XPS acquisition is then performed in a smaller pitch (SP) where contributions originate from the top (TiN) and SiO_2 sidewalls of the features only (Fig.2(b)).
Assuming the compositions on top of metallic hard mask and sidewall of SiO_2 trenches are similar between LP and SP, the substraction of the signal (LP-SP) corrected to the ration P2/P1 provides the contribution of the p-SiOCH sidewalls only. For these specific measurements the TiN and SiO_2 thickness are 15 nm and 125 nm, respectively while the p-SiOCH thickness is set at 670 nm.

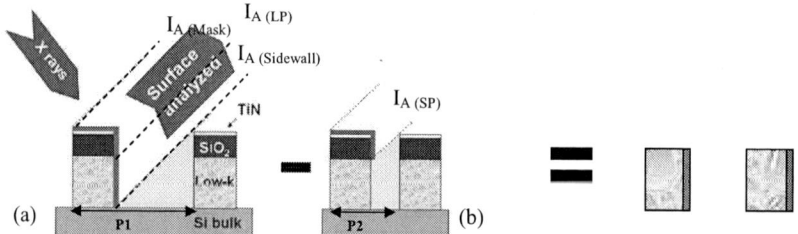

Fig 2: Desciption of XPS analyses on patterned wafers in large pitch (P1 – 800nm, feature dimension= 200nm) (a) and small pitch (P2= 340nm, feature dimension = 200nm) (b) areas to obtain p-SiOCH sidewall film composition.

Using this experimental protocol we can determine the p-SiOCH sidewall composition as follows:

- The concentrations of Ti, O, N, F, Si, and C atoms are extracted from the Ti2p, O1s, N1s, F1s, Si2p and C1s core-level energy regions, respectively.

- The background subtraction is performed using a Shirley function calculated from a numerical iterative method.

- The intensity of photoelectrons coming from hard mask and SiO_2 ($I_{A(Mask)}$) in large area is calculated as following:

$$I_A(Mask) = I_A(SP) * \frac{P2}{P1}$$

- The sidewall porous SiOCH intensity is calculated for each element (extracted from the Ti2p, O1s, N1s, F1s, Si2p and C1s core-level energy regions).

For a given element A :

$$I_A(Sidewall) = I_A(LP) - I_A(Mask)$$

$$I_A(Sidewall) = I_A(LP) - I_A(SP) * \frac{P_2}{P_1}$$

- Then the concentration of the given element A is quantified with respect to the other elements detected as follows :

$$[A] = \frac{\dfrac{I_A}{S_A}}{\sum_k \dfrac{I_k}{S_k}}$$

This method gives a rough estimate of the p-SiOCH composition on the feature sidewalls.

II.3.2 Infrared Spectroscopy

Infrared Spectroscopy analyses in transmission (T-FTIR) and in multi internal reflection (MIR-FTIR) are performed using a Bruker IFS-55 FTIR spectrometer. For the T-FTIR analyses, the IR beam passes trough the wafer and is focused on a DTGS detector. The T-FTIR mode covers a spectral range from 400 to 4000 cm^{-1}. For the MIR-FTIR analyses, the sample is coupled to the s-polarised IR beam with two silicon prisms whose spacing is adjusted to Z=6 cm. A s-polarized IR beam coming from FTIR spectrometer is directed on the coupling area of the input prism, which ensures optical tunneling inside the wafer. After being internally reflected about 100 times in our conditions (with an incidence angle of 34° and a wafer thickness of about 600 µm), the IR beam is coupled out of the wafer by the second prism and focused onto a liquid-N_2-cooled HgCdTe detector [16, 17]. The MIR technique is 100 times more sensitive than the standard transmission mode (T-FTIR). The IR beam propagates inside the wafer so that the samples have to be double-side polished. The MIR-FTIR mode covers a spectral range from 2600 to 4000 cm^{-1}. The spectral range is lower than this with the T-FTIR mode due to absorption of the IR beam at higher wave numbers into the silicon wafer during the internal reflection. Spectra are recorded with a 2 cm^{-1} spectral resolution and an average of 32 scans and 200 scans with the T-FTIR and MIR-FTIR modes, respectively. Before the spectrum acquisition, 5 min nitrogen purge is performed in the spectrometer to remove H_2O and CO_2 vapor. The baseline of IR spectra is removed with the software "Resolution Pro" using a spline curve. For T-FTIR acquisition, a reference spectrum is measured using a virgin silicon wafer in order to remove the spectral contribution of the silicon wafer and the optical bench.

II.3.3 Spectroscopic Ellipsometry (SE)

The refractive index, extinction coefficient and the thickness of the porous SiOCH films were measured using a UV-visible spectroscopic ellipsometry from Jobin Yvon. To determine those values, the ellipsometry parameters were fitted with an absorbent Cauchy model for the porous SiOCH.

II.3.4 HF decoration method

We can determine the thickness of the modified p-SiOCH film thanks to spectroscopic ellipsometry (SE) using the decoration technique. This technique relies on the principle that the pristine porous SiOCH film is not consumed during hydrofluoric acid (HF) dip while the modified layer (with moisture uptake and carbon depletion) is removed [18-20]. The thickness of the remaining p-SiOCH film is determined by spectroscopic ellipsometry (SE) before and after HF dip allowing the thickness of the modified layer to be determined. On patterned wafers, the wafers are cleaved and dipped into a 1% diluted HF solution during 15 s after p-SiOCH etching. Using this protocol, we can estimate the thickness of the damaged layer on the trench sidewalls by using scanning electron microscopy (SEM). Before SEM observations, the trenches are filled with resist in order to prevent potential feature collapse during SEM exposure.

II.3.4 Ellipsometric porosimetry (EP)

The Ellipsometric porosimetry technique has been used to characterize the impact of FC based plasma exposure on the mean pore radius, the pore size distribution and the open porosity of p-SiOCH. Ellipsometry porosimetric measurements are performed in the visible range on an EP12 ellipsometric porosimeter from SOPRALAB [21]. It consists of a rotating polarizer spectroscopic ellipsometer coupled with a vacuum chamber which is operated in a pressure range between 10^{-3} Torr and the saturation vapor pressure (Ps) of the adsorptive. For a solvent partial pressure P/Ps, the adsorptive penetrates into the low-k and condenses into the pores leading to an increase of the refractive index. Indeed the vapour of the adsorptive condenses in the pores for a vapor pressure (P_a) lower than a flat surface liquid (P_s) and the value of partial pressure P_a/P_s depends on the pore size, surface tension and molar volume of the solvent.

Two solvents are used as adsorptive: methanol (P_s~105 Torrs, n_{met}=1.329 at 633 nm) and water (P_s~20 Torrs, n_{Water}=1.33 at 633 nm). The methanol solvent penetration kinetics over time through the damaged p-SiOCH surface gives information on the sealing of the surface [22]. The degree of hydrophilization associated to the plasma damage is determined with water vapors thanks to a dedicated bi-layer analysis. The spectra are saved with a CCD detector between 1.5 and 4 eV at an angle of incidence set at 60.25°. Thickness and refractive index of the layers are calculated from the ellipsometric data as a function of the relative pressure ($P_{rel}=P/P_s$, with P the chamber pressure) using a Cauchy law in adjunction with a Lorentz oscillator to take into account the absorption band in the ultraviolet region. On the basis of the Lorentz-Lorenz effective medium approximation, the volume fraction of solvent adsorbed in the pores is calculated as described in a previous paper [21].

The p-SiOCH sidewall modification of patterned structures differs from at the bottom of the trenches. An innovative non-destructive method has recently been developed to measure the porous properties of p-SiOCH patterned layers as integrated into circuit called Scatterometric Porosimetry (SP) [23, 24]. It mainly consists in the use of an EP tool (previously described) to record the scatterometric response of periodic structures made of porous material as a function of the relative pressure of the solvent. The patterned structures are chosen to have a critical dimension which is equivalent to the interconnect line size except that they typically consist of parallel periodic lines. The porous properties are subsequently extracted from the SP measurement with the use of a specially-developed scatterometric modeling [23]. For such analyses, patterning of the porous SiOCH trenches is performed using a metallic hard mask (Figure 1) with the following film thickness: a titanium nitride layer (15 nm thick), an oxide capping layer (125 nm thick) and the porous dielectric material (670 nm thick and 28% of porosity).

II.3.5 Capacitance measurement: Mercury Probe
The dielectric constant of blanket films has been calculated before and after the different plasma exposures using capacitance-voltage (C-V) measurements at 0.1 MHz. C-V measurements have been performed using a mercury probe capacitance measurement (C-V) system (model SSM495).

II.5.1 Bottom line roughness
Atomic force microscopy (AFM) measurements are performed at ambient atmosphere with a Nanoscope III multimode microscope from Digital Instruments. The RMS roughness is determined by tapping mode AFM with the measurement of the standard deviation of the height distribution (RMS). More detail of the measurement can be found elsewhere [25].

III. P-SiOCH film modification induced by fluorocarbon based plasma

A low-k material must withstand different integration steps, in particular the etching step. However, p-SiOCH materials are much more sensitive than SiO_2 to plasma exposure (chemical and physical impact of the plasma may induce p-SiOCH damage which in turn alters the dielectric film properties) [2]. A large contribution of the sensitivity comes from porosity introduction [7]. In this section, we propose to evaluate the sensitivity of p-SiOCH (PECVD p-SiOCH with 28% porosity) material after trench etching under FC based plasmas exposure. The bottom (estimated thanks to experiment performed on blanket wafers) and sidewalls p-SiOCH film modifications are compared and the potential solutions are discussed.

III.1 Bottom p-SiOCH film modification after partial etching

In this part, experiments have been performed on blanket p-SiOCH films to estimate bottom p-SiOCH film modifications induced by fluorocarbon plasmas. First, the surface and structural composition of the pristine p-SiOCH are investigated. Then to mimic material consumption for 45 nm technology node, 120 nm of p-SiOCH has been etched in CF_4 /C_4F_8/ N_2/Ar used for p-SiOCH trench etching. P-SiOCH film has been analyzed on blanket wafers by XPS, T-FTIR and MIR-FTIR before and after partial etching.

III.1.1 Surface analyses

The XPS survey spectrum of as-deposited p-SiOCH (not shown), indicates the presence of silicon, oxygen and carbon (hydrogen is not detected by XPS). The surface composition, reported in Table 2, shows that p-SiOCH is composed of Si (36%), O (40%) and C (24%).

Table 2: XPS surface composition of p-SiOCH (porosity of 28% and k=2.35) on blanket and patterned wafers after FC etching

Material	Si(%)	O(%)	C(%)	F(%)	N(%)	Ti (%)
p-SiOCH as deposited (blanket wafer)	36	40	24	0	0	0
p-SiOCH after FC etching (blanket wafer)	25	30	34	8	3	-
p-SiOCH after FC etching (sidewall)	17	19	38	23	1	2

After p-SiOCH partial etching in CF_4 /C_4F_8/ N_2/Ar plasma, XPS analyses reveal that the surface is composed of 25%Si, 30%O, 34% C, 8%F and 3%N (See Table 2).

III.1.2 Volume analyses

T-FTIR analyses

Fig.3 shows a typical T-FTIR spectrum of the as deposited porous SiOCH. In the 1000 to 1500 cm^{-1} spectral range, T-FTIR spectrum exhibits three main absorption bands at 1022, 1065 and 1140 cm^{-1} corresponding to C-Si-O and Si-O-Si stretching vibration bands [21, 26]. Two additional peaks are also observed at 1275 cm^{-1} and 2960 cm^{-1}, which are assigned to the Si-CH$_3$ and C-H$_3$ vibration bands, respectively. After p-SiOCH exposure to FC based plasma, the FTIR spectrum (see Fig.3) exhibits the same vibration modes as those detected in the pristine material.

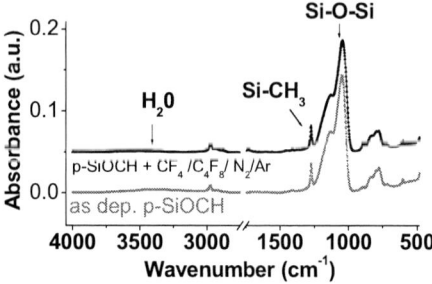

Fig. 3: T-FTIR spectra of p-SiOCH before and after partial etching in FC based plasma

Based on these spectra, the methyl content in the film can be monitored by calculating the peak area ratio from the T-FTIR spectra:

$$R_{SiCH3} = \frac{A_{Si-CH_3}}{A_{Si-CH_3} + A_{Si-O-si}}$$

where R_{Si-CH3} represents the methyl ratio, A_{Si-CH3} the area of the Si-CH$_3$ vibration bands and $A_{Si-O-Si}$ the area of the Si-O-Si vibration bands[14, 34]. In these conditions, after partial etching, the methyl depletion is estimated at about 25% compared to the pristine material.

MIR-FTIR analyses
In the 2700 to 4000 cm^{-1} spectral range, the MIR-FTIR spectrum of porous SiOCH after partial etching shows vibration bands which are assigned to OH and CH$_x$ bonds (see Fig.4) [16].

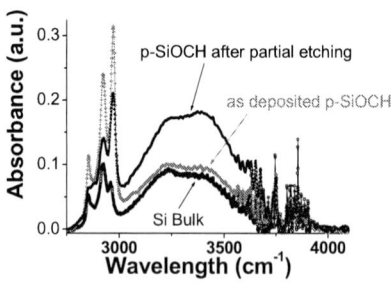

Fig. 4: MIR-FTIR spectra of p-SiOCH before and after partial etching in FC based plasma

CH$_x$ vibration bands are assigned to four peaks: CH$_2$ at 2860 and 2930 cm^{-1}, CH$_3$ at 2880 and 2965 cm^{-1}. The large OH vibration band into the modified porous SiOCH can be assigned to four peaks: two from Si-OH (isolated and linked) and two from water H$_2$O (free and bonded) at 3672, 3551, 3425 and 3225 cm^{-1}, respectively [17, 20, 27]. This broad OH vibration bands exhibits the same intensity as the OH vibration band originating from the Si bulk spectrum. Therefore the OH vibration is mainly attributed to the Si bulk coming from the native silicon oxide and it is then difficult to prove the presence of OH bonds in the as deposited materials. It has to be noticed that the OH vibration bands are hardly detected on the T-FTIR spectra due to the lower sensitivity compared to MIR-FTIR technique.

After partial etching, the absorption of the OH vibration bands is much higher than in the as deposited porous SiOCH. This result shows an increase in Si-OH bonds formation due to water uptake with free and bonded H$_2$O. It has to be noticed that no OH vibration bands have been observed with FTIR measurement due to the lack of sensitivity. After partial etching, the CH$_x$ peaks (CH$_3$ and CH$_2$) are slightly lower than those of the as deposited porous SiOCH indicating lower methyl depletion in the remaining porous SiOCH (which is in good agreement with the FTIR analyses).

III.1.3 Depth modification of the p-SiOCH film
Using decoration method
We have then determined the thickness of this modified p-SiOCH film thanks to spectroscopic ellipsometry (SE) using the decoration technique previously described. Based on this protocol, after the etching process, the damaged layer is estimated to be about 30 nm thick. We can notice that after the HF dip which removes the modified layer, no methyl depletion is evidenced after partial etching (not shown) indicating that the methyl depletion is mainly localized in the modified layer.

Using EP
The hydrophilic properties of the modified layer have been also determined with EP measurement using water as solvent. Based on the protocol previously described, the thickness of the modified layers is estimated after FC based plasma exposure to be 29nm.

Both techniques show similar modified layer thickness confirming the correct estimation of modified layer thickness on blanket wafers.

III.1.4 Impact of the modified layer on pore sealing
The impact of the thick hydrophilic modified layer on open porosity and permeability has been investigated thanks to EP analyses with methanol as solvents (Fig.5). The degree of porosity is followed as function of time at a fixed pressure (0.8 of vapour pressure) to determine the total porosity and permeability of the porous SiOCH modified layer (as described in the previous section). The open porosity which is measured at the plateau remains the same as the pristine material (of about 27%) with or without partial etching. This result demonstrates no significant change in pore density and pore size in the remaining porous SiOCH films. This also indicates that the porosity of the modified layer is similar to the pristine material.

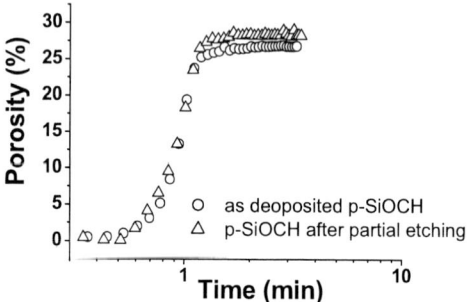

Fig. 5 : Evolution of porosity as a function of time at a fixed relative pressure (0.8Ps) with methanol thanks to EP measurement

These experiments performed on blanket p-SiOCH show that the film is altered under FC based plasma exposure leading to the formation of a hydrophilic perturbed layer depleted in methyl groups. This perturbed layer induces an increase of the dielectric constant from 2.35 to 2.7 after partial etching. These results are similar than those observed for a spin-on p-SiOCH material [7]. In this previous study, we demonstrated that the film modification could be assigned to fluorine reactive species diffusion and to the impact of the ion bombardment. In the first case, the diffusion of fluorine atoms into the porous material generates an important degradation of the material since fluorine is very reactive toward carbon and silicon. While in the second case, the ion bombardment can break Si–O, Si–CH$_3$, and C–H bonds [7].

The study of the film modification performed on blanket wafer gives a good picture of the p-SiOCH film modification at the bottom of trenches. However, can we draw similar conclusion on the p-SiOCH sidewalls modification? These results are now compared with the film modification analyses performed on patterned structure.

III.2   p-SiOCH film sidewall modification after partial etching
After patterning of p-SiOCH trenches in CF$_4$ /C$_4$F$_8$/ N$_2$/Ar plasma, no sidewall modification is clearly evidenced using the decoration technique. This can be attributed to of a lack of sensitivity of this technique for thin damage layer (Fig.6).

Fig. 6: Estimation of the sidewall modification of p-SiOCH trenches after CF$_4$ /C$_4$F$_8$/ N$_2$/Ar etching using the decoration technique

Indeed the hydrophilic properties of the modified layer determined with SP measurement using water as solvent for EP analyses show different behaviour. The thickness of the damaged layer after etching is estimated in this case to 10 nm. This result tends to indicate the SP can be a more accurate technique to estimate the sidewall p-SiOCH film modification compared to decoration method.

Sidewall composition analyses performed using chemical topography analyses by XPS is summarized in Table 2. After etching, the presence of fluorocarbon species are detected on the p-SiOCH trench sidewalls while Ti is hardly detected. The presence of this small amount of Ti can be attributed either to the sputtering of the hard mask during etching or to the element concentration measurement error induced by the experimental protocol.

The impact of this FC layer formation on pore sealing was determined thanks to EP analyses using methanol as solvent. The methanol diffuses through the p-SiOCH sidewalls (not shown) indicating that p-SiOCH sidewalls are not pore sealed by the presence of the fluorocarbon layer. These results indicate that a CF_x layer grows on the p-SiOCH sidewalls (porosity of 28% and k = 2.35) during FC based plasma and that no pore sealing is induced by this CFx layer.

These observations are similar to previous observations on blanket wafers. Therefore, we can propose similar mechanisms to explain the p-SiOCH sidewall film modification which can be assigned to fluorine reactive species diffusion on the sidewalls leading to the creation of a hydrophilic layer. The effect of the ion bombardment being lowered since the ions deflected on the hard mask have lower energy than ions directly striking the p-SiOCH blanket wafers.

The whole results show that p-SiOCH (bottom and sidewall) is very sensitive to FC based plasma leading to the formation of a hydrophilic layer. This modified layer cannot be neglected since it alters the dielectric film properties. Indeed, k value measurement performed on blanket and patterned wafers shows that the perturbed layer leads to an increase of the dielectric constant. The k degradation is more pronounced as the thickness of porous SiOCH decreased [28]. Furthermore, the modified layer can also degrade the reliability performance of the dielectric layer. Therefore solution must be found to limit this p-SiOCH film modification or to improve the hydrophobic properties of the dielectric film after etching.

III.3 Solutions to limit p-SiOCH film modification

As suggested, p-SiOCH film modification (bottom and sidewalls) can be induced by fluorine, ion bombardment. Based on these results, different solutions were proposed in the literature to limit the p-SiOCH film modification.

One approach is to reduce fluorine reactive species diffusion in the film and the ion bombardment impact. This can be achieved by using high polymerising chemistries [7, 29]. Fig.7 shows that higher CH_2F_2 concentration in the CF_4/Ar gas mixture enables a reduction of the p-SiOCH (spin-on with 45% porosity) sidewall film modification. We have shown that such polymerizing etch chemistries lead to the formation of thick fluorocarbon layers on the p-SiOCH sidewalls which act as protective layers against fluorine diffusion inside the porous SiOCH material. Less sidewall modification are observed in high polymerizing chemistries such as $CF_4/Ar/CH_2F_2$ which produce thicker FC layers on the sidewalls than in low polymerizing chemistries such as CF_4/Ar [29].

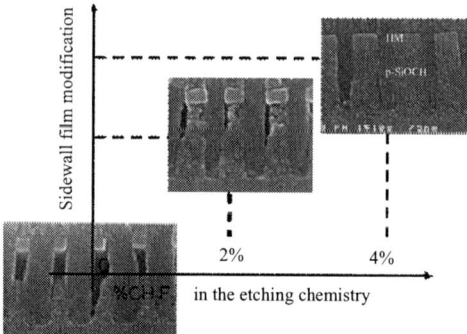

Fig. 7: Sidewall film modification of porous SiOCH (containing 45% porosity) after etching as a function of the CH_2F_2 concentration in CF_4/Ar gas mixture. The sidewall modification is determined using the decoration method.

Finally an emerging solution, as proposed in the literature, is to restore the hydrophobic properties of the modified layer by using a curing process: a silylation based process [28]. For instance, for the 45nm interconnect technology node, *Chaabouni et al* [28] show that after p-SiOCH (porosity of 25% and k=2.5) integration using a metallic hard mask approach, the integrated k value is about 3 (see Fig.8). After a post etching, restoration process using hexamethyldisilazane (HMDS: $(CH_3)_3–Si–NH–Si–(CH_3)_3$) enables to decrease the integrated k value to 2.6 close to the dielectric constant of the as-deposited p-SiOCH [28]. HMDS molecules selectively react with the silanol species and convert them into hydrophobic trimethyl-siloxy ($–O–Si–(CH_3)_3$) groups leading to an increase of the carbon content that prevents water uptake.

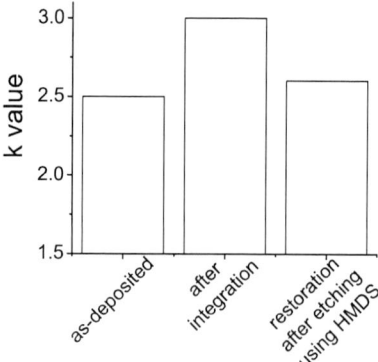

Fig. 8: Effect of sylilation process after etching on integrated k value [40]

III.4 Summary

Based on these results, we have shown that the damages of porous SiOCH sidewalls induced by FC plasma exposure are inevitable. They can alter dielectric film properties and generate an increase of dielectric constant and/or reliability degradation. The film modification due to reactive fluorine species diffusion and ion bombardment impact can be limited by using high polymerizing FC chemistries. In addition, a complementary solution is the restoration of the hydrophobic properties of the modified layer using silylation process.

We have shown the porosity introduction into SiOCH material induces higher film sensitivity to FC based plasma. But porosity introduction also leads to new issue like roughening of bottom line. In this case, are the solutions proposed to limit p-SiOCH damage compatible with the reduction of bottom line roughness?

IV. Bottom line roughness

As illustrated in Fig.9, increasing porosity of PECVD SiOCH material also leads to an increase of the bottom line roughness after line etching using a metal hard mask.

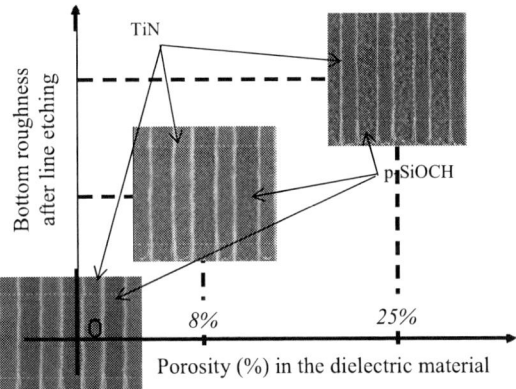

Fig. 9: Bottom line roughness evolution after line etching using a metal hard mask (with similar etching conditions) as a function of the p-SiOCH film porosity.

The impact of such a roughness on reliability performances has been evaluated by comparing two etching processes inducing either a low (Fig.10(a)) or a high bottom line roughness (Fig.10(b)). Whatever the etching chemistries, Fig.10(c) shows that the electromigration trend is not impacted while in the meantime, the time dependent dielectric breakdown (TDDB) lifetime is degraded (Fig.10(d)) when the bottom line roughness increase.

These results demonstrate that bottom line roughness strongly impact the TDDB. Therefore it's mandatory to limit the bottom line roughness in order to prevent TDDB degradation.

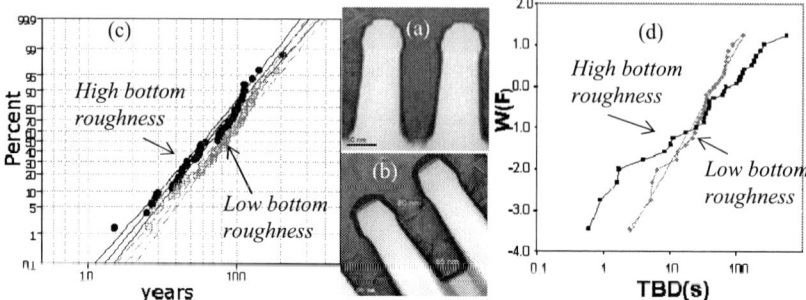

Fig. 10: Impact of the bottom line roughness high (a) and low (b) on electromigration (EMG) (c) and Time Dependent Dielectric Breakdown (TDDB) (d)

IV.1 Mechanism of bottom line roughness formation

The evolution of the roughness during plasma etch processes has already been studied [30 – 39]. Plasma processes usually smooth the etched surfaces. The isotropic component of the etch process, [33, 34] as well as the shadowing effect [40, 41] and the preferential sputtering of the material for off-normal ions [33, 42, 43] contribute to this effect during plasma etch processes. Nevertheless, several groups have reported that etched surfaces can also be roughened when exposed to plasma etch processes. [39, 44–47]. Several explanations such as preferential etching by re-emitted species [42, 43] or ion induced defects [46] have been proposed. However, recent studies [39, 47] tend to prove that etch induced roughness rather originates from micromasking phenomena induced by particles sputtered form the chamber walls during the etch process. Etch induced roughness has also been observed with p-SiOCH materials [30-33, 48]. These papers claim that, the fluorocarbon species deposited on the porous low-k surface induce p-SiOCH roughness. Tatsumi and Urata [49] suggest that low-k materials are roughened when a thin fluorocarbon layer (4 nm) is deposited on the material surface, while no fluorocarbon or thicker fluorocarbon layers do not lead to roughening. Yin et al. [31, 48] attribute the etching induced roughness to fluorocarbon species micromasking the material surface. They attribute the non uniform coverage of the surface to the presence of pores in the material. Lazzeri et al. [30] propose that the larger surface of porous materials leads to an incomplete coverage of the surface of the dielectric by the fluorocarbon layer, compared to dense dielectric layers. Due to this incomplete coverage, plasma species can directly interact with the low-k material, leading to harsher etch conditions and roughness. In both cases, the models assume that the roughness is created by an incomplete or non homogeneous coverage of the porous SiOCH surface by fluorocarbon species.

However in our experimental conditions close to the industrial ones, we demonstrated the roughness of porous SiOCH also occurs under fluorocarbon free plasma exposure (SF_6 plasma) suggesting complementary mechanism [25]. We proposed that: (1) during the first few seconds of the etch process, the surface of porous SiOCH materials gets denser. (2) Cracks are formed, leading to the formation of deep and narrow pits. (3) Plasma radicals diffuse through those pits and the pore network and modify the porous material at the bottom of the pits. (4) The difference in material density and composition between the surface and the bottom of the pits leads to a difference in etch rate and an amplification of the roughness [25].

IV.2 Solutions

Previous studies performed on blanket wafers indicate that the solution to smooth p-SiOCH during etching is to use highly polymerising FC chemistries [25]. The impact of the etch chemistry on porous SiOCH (p-SiOCH with 25% porosity) roughening is illustrated in Fig.11 by comparing the RMS roughness on blanket wafers induced by two different fluorocarbon based etching chemistries (different degrees of polymerizing). CH_2F_2 addition to CF_4/Ar (more polymerizing chemistry) induces a decrease in roughness.

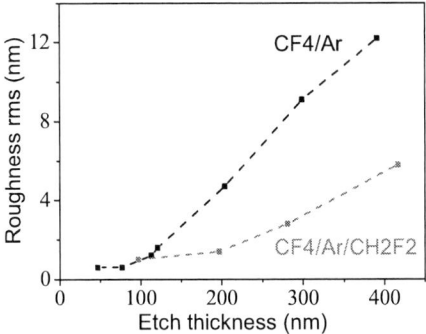

Fig. 11: Roughness evolution as a function of the etched depth for PECVD porous SiCOH (25% porosity) exposed to either CF_4 / Ar or CF_4 /Ar/CH_2F_2 plasmas. RMS value of the roughness measured by AFM.

However, this solution does not take into account the impact of the metallic hard mask on the p-SiOCH roughness formation. Indeed, on patterned wafers, in addition to the roughening of the p-SiOCH induced by fluorocarbon plasmas, additional roughness also originates from the micromasking induced by TiFx compounds coming from the TiN mask sputtering [50, 51]. The two key parameters to limit this metallic hard mask impact on roughness formation are the substrate temperature and the polymerising rate of the chemistry as illustrated below.

IV.2.1 Impact of Substrate temperature on p-SiOCH roughness formation

The impact of the substrate temperature on porous SiOCH roughness is illustrated Fig.12 by comparing the etching of p-SiOCH patterned structure using CF_4 based chemistry with two different substrate temperatures (higher or lower than 50°C). Increasing substrate temperature induces a lower bottom p-SiOCH roughness and straight profiles as evidenced by the SEM cross section in Fig.12(b). This trend is correlated with a lower titanium (5%) concentration at the bottom of trenches (determined by XPS), compared to lower substrate temperature where 12% titanium is measured by XPS (Fig.12(a)).

Therefore increasing substrate temperature above 50°C favours the formation of more volatile TiFx etch products and therefore decrease the micromasking induced effect of Ti based compounds deposited on p-SiOCH surfaces.

 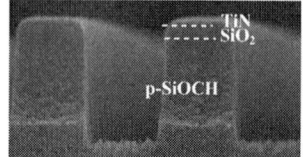

Fig. 12: Illustration of p-SiOCH etching in a metallic hard mask environment with a CF_4 based etching chemistry and a substrate temperature lower (a) or higher (b) than 50°C, respectively.

Another way to limit this metallic hard mask sputtering impact on p-SiOCH roughness formation is to protect it during etching. This can be achieved by using polymerizing chemistry.

IV.2.2 Impact of the etching chemistry on the p-SiOCH roughness

For substrate temperatures above 50°C, the impact of the etch chemistry on the porous SiOCH (p-SiOCH with 25% porosity) roughening is now presented in Fig.13 by comparing two CF_4 based etching chemistries with and without C_4F_8 addition.

With C_4F_8 addition to CF_4, a lower bottom p-SiOCH roughness is clearly observed on top SEM-CD pictures (Fig.13(b)) compared to CF_4 only (Fig.13(a)). XPS analyses reveal that C_4F_8 addition to CF_4 leads to the formation of a thick fluorocarbon layer on top of the metal hard mask (25% C and 13% F) decreasing its sputtering rate (5% titanium is detected at the bottom of trenches). Without polymerizing gas addition, the presence of higher titanium species concentration (10%) at the bottom of trenches correlated with a thinner fluorocarbon layer is detected on top of the metal hard mask (15% C and 10% F).

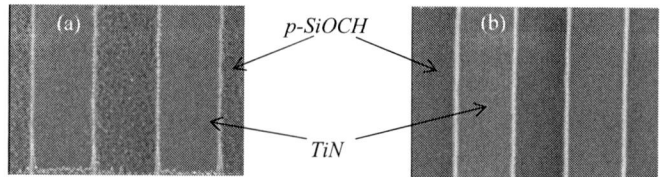

Fig. 13: Illustration of p-SiOCH etching in a metallic hard mask environment with either a CF_4 (a) or C_4F_8 (b) based etching chemistry for a substrate temperature higher than 50°C.

These results show that the use of highly polymerizing chemistries protect the metal hard mask, and limit its sputtering rate. Consequently, Ti based micromasking is reduced limiting the development of rough p-SiOCH surfaces.

IV.2.3 Summary

In summary, we have shown that the bottom surface roughness is directly related to the degree of SiOCH porosity and such a roughness can lead to the degradation of the TDDB while the electromigration is not altered.

The roughness formation is explained by different mechanisms such as the intrinsic p-SiOCH roughness developed under plasma exposure or micromasking coming from reactor wall conditioning, FC species or hard mask environment.

With a metal hard mask approach, the solution to limit p-SiOCH formation is to develop etching processes at substrate temperature above 50°C (to limit sputtered metal species condensation leading to micromasking) and the use of highly polymerising chemistry (to protect MHM with FC layer and also playing the role of p-SiOCH smoothing).

Conclusions

The major difficulties of p-SiOCH material integration using a metallic hard mask integration have been presented from the etching point of view. The first problem encountered with porosity introduction into SiOCH material is the high sensitivity of p-SiOCH film during line etching with fluorocarbon based plasma. In this case, fluorine reactive species diffusion and ion bombardment leads to the formation of a hydrophilic layer (bottom and sidewalls), not sealing the pores, and altering the dielectric film properties. While the second problem presented with bottom line roughness formation and explained by intrinsic roughness or micromasking (coming from reactor wall conditioning or FC species and increased by the low volatile compounds (TiFx) coming from the sputtered hard mask) strongly impacts reliability performances like TDDB.

The common solution to limit these issues is to develop high polymerising chemistry at high substrate temperature above 50°C. Developing etching processes at substrate temperature above 50°C allows limiting sputtered metal species condensation leading to micromasking. While the use of high polymerising chemistry leads to the formation of thick fluorocarbon layers which act as protective layers against fluorine diffusion inside the porous SiOCH material and ion bombardment impact but also protect the metallic hard mask and smooth the p-SiOCH film. In addition a complementary solution to limit the impact of the film modification is the restoration of the hydrophobic properties of the modified layer using silylation process.

References

1. M. R. Baklanov and K. Maex, 206. Phil. Trans. R. Soc. A (2006)
2. D. Shamiryan, T. Abell, F. Iacopi and K. Maex, Mat. Today, N1, 34 (2004)
3. D. Fuard, O. Joubert, L. Vallier and M. Bonvalot, J. Vac. Sci. Technol. B, **19**, 447, (2001)
4. M. Fayolle, G. Passemard, O. Louveau, F. Fusalba, J. Cluzel, journal Microelectronic Eng., **70**, 255, (2003).
5. Willi Volksen, Robert D. Miller and G. Dubois, Chemical Reviews, 110, 56, (2010).
6. A. Grill, D. Edelstein, D. Restaino, M. Lane, S. Gates, E. Liniger, T. Shaw, X.-H. Liu, D. Klaus, V. Patel, S. Cohen, E. Simonyi, N. Klymko1, S. Lane, K. Ida, S. Vogt, T. Van Kleeck, C. Davis, M. Ono, T. Nogami, T. Ivers, in: International Interconnect Technology Conference , 54 (2004).
7. N. Posseme, T. Chevolleau, O. Joubert, L. Vallier, N. Rochat. J. Vac. Sci. Technol. B, **22**, 2772, (2004).
8. D. L. Keil, B. A. Hemer and S. Lassig, J. Vac. Sci. Technol. B, **21**, 1969 (2003).

9. N. Posseme, C. Maurice, Ph. Brun, E. Ollier, M. Guillermet, C. Verove, T. Berger, R. Fox, and O. Hinsinger, International interconnect technology conference (IITC) proceedings, 36 (2006).

10. H. Struyf, D. Hendrickx, J. Van Olmen, F. Iacopi, O. Richard, Y. Travaly, M. Van Hove, W. Boullart and S. Vanhaelemeersch, International interconnect technology conference (IITC) proceedings, 30 (2005).

11. M. Darnon, T. Chevolleau, D. Eon, R. Bouyssou, B. Pelissier, L. Vallier, O. Joubert, N. Posseme, T. David, F. Bailly, J. Torres, Microelectronic Eng. **85**, 2226, (2008).

12. A. Zenasni, B. Remiat, C. Waldfried, C. Le Cornec, V. Jousseaume, and G. Passemard, Thin Solid Films, **516**, 1097, (2008).

13. V. Jousseaume, L. Favennec, A. Zenasni, and O. Gourhant, Surf. Coat. Technol., **201**, 9248 (2007).

14. N. Posseme, T. Chevolleau, O. Joubert, L. Vallier, P. Mangiagalli., J. Vac. Sci. Technol. B, **21**, 2432, (2003).

15. Pargon, E. and O. Joubert, J. Vac. Sci. Technol. B, **22**, 1869, (2004).

16. N.Rochat, M.Olivier, A.Chabli, F.Conne, G.Lefeuvre, C.Boll-Burdet, Applied Physics Lett. **77**, 2249, (2000).

17. N. Rochat, A. Troussier, A. Hoang and F. VinetMaterials Science and engineering, **23**, 99, (2003).

18. Q. T. Le, M. R. Baklanov, E. Kesters, A. Azioune, H. Struyf, W. Boullart, J. -J. Pireaux, S. Vanhaelemeersch., Electrochem. Solid-State Lett. **8**, F21, (2005).

19. O. Louveau, C. Bourlot, A. Marfoure, I. Kalinovski, J. Su, G. H. Hills and D. Louis, Microelectronic Eng., **73**,351, (2004).

20. N. Posseme, T. Chevolleau, T. David, M. Darnon, O. Louveau and O. Joubert, J. Vac. Sci. Technol. B. **25** , 1928, (2007).

21. C. Licitra, F. Bertin, M. Darnon, T. Chevolleau, C. Guedj, S. Cetre, H. Fontaine, A. Zenasni, L. L. Chapelon, Physica Status Solidi, **5**, 1278, (2008).

22. W. Puyrenier, V. Rouessac, L. Broussous, D. Rébiscoul, A. Ayral, Microporous Mesoporous Mat. 106, 40, (2007).

23. R. Bouyssou, M. El Kodadi, C. Licitra, T. Chevolleau, M. Besacier, N. Posseme, O. Joubert, P. Schiavone, J. Vac. Sci. Technol. B. **28**,31, (2010)

24. A. Bourgeois, Y. Turcant, C. Walsh, and C. Defranoux, Adsorption-Journal of the International Adsorption Society **14**, 457, (2008).

25. F.Bailly, T. David, T. Chevolleau, M. Darnon, N. Posseme, R. Bouyssou, J. Ducote, O. Joubert, and C. Cardinaud, J. Appl. Phys. **108**, 014906 (2010).

26. Theil, J. A. Tsu, D. V. Watkins, M. W. Kim, S. S. Lucovsky, G. J. Vac. Sci. Technol. A **8**, 1374 (1990).

27. A. Goullet, C. Vallee, A. Granier, G. Turban, J. Vac. Sci. Technol. A **18**, 2452 (2000).

28. H. Chaabouni, L.L. Chapelon, M. Aimadeddine, J. Vitiello, A. Farcy, R. Delsol, P. Brun, D. Fossati, V. Arnal, T. Chevolleau, O. Joubert, J. Torres, Microelectronic Eng. **84**, 2595, (2007).

29. M. Darnon, T. Chevolleau, T. David, J. Ducote, N. Posseme, R. Bouyssou, F. Bailly, D. Perret, and O. Joubert. J. Vac. Sci. Technol. B **28**, 149 (2010).

30. P. Lazzeri, X. Hua, G. S. Oehrlein, M. Barozzi, E. Iacob, and M. Anderle, J. Vac. Sci. Technol. B **23**, 1491 (2005).

31. Y. Yin and H. H. Sawin, J. Vac. Sci. Technol. A **25**, 802 (2007).

32. Y. Yin and H. H. Sawin, J. Vac. Sci. Technol. A **26**, 151 (2008).

33. G. M. Gallatin and C. B. Zarowin, J. Appl. Phys. **65**, 5078, (1989).
34. W. W. Mullins, J. Appl. Phys. **30**, 77 (1959).
35. R. P. U. Karunasiri, R. Bruinsma, and J. Rudnick, Phys. Rev. Lett. **62**, 788 (1989).
36. J. H. Yao and H. Guo, Phys. Rev. E 47, 1007 (1993).
37. C. Roland and H. Guo, Phys. Rev. Lett. **66**, 2104 (1991).
38. G. S. Bales and A. Zangwill, Phys. Rev. Lett. **63**, 692 (1989).
39. M. Martin and G. Cunge, J. Vac. Sci. Technol. B **26**, 1281 (2008).
40. J. T. Drotar, Y.-P. Zhao, T.-M. Lu, and G.-C. Wang, Phys. Rev. B **61**, 3012 (2000).
41. E. Zakka, V. Constantoudis, and E. Gogolides, IEEE Trans. Plasma Sci. 35, 1359 (2007).
42. M. Schaepkens, G. S. Oehrlein, C. Hedlund, L. B. Jonsson, and H.-O. Blom, J. Vac. Sci. Technol. A **16**, 3281 (1998).
43. D. Flamm and D. Manos, Plasma Etching: An Introduction (Academic, New York, 1989).
44. Y.-P. Zhao, J. T. Drotar, G.-C. Wang, and T.-M. Lu, Phys. Rev. Lett. **82**, 4882 (1999).
45. G. S. Hwang, C. M. Anderson, M. J. Gordon, T. A. Moore, T. K. Minton, and K. P. Giapis, Phys. Rev. Lett. **77**, 3049 (1996).
46. R. Pétri, P. Brault, O. Vatel, D. Henry, E. André, P. Dumas, and F. Salvan, J. Appl. Phys. **75**, 7498 (1994).
47. E. Gogolides, C. Boukouras, G. Kokkoris, O. Brani, A. Tserepi, and V.Constantoudis, Microelectron. Eng. 73–74, 312 (2004).
48. Y. Yin, S. Rasgon, and H. H. Sawin, J. Vac. Sci. Technol. A, **24**, 2360 (2006).
49. T. Tatsumi and K. Urata, J. Vac. Sci. Technol. A **23**, 938 (2005).
50. N. Posseme, T. David, M. Darnon, T. Chevolleau, and O. Joubert, 6th International Conference on Microelectronics and Interfaces, 2005
51. M. Darnon, T. Chevolleau, D. Eon, R. Bouyssou, B. Pelissier, L. Vallier, O. Joubert, N. Posseme, T. David, F. Bailly, and J. Torres, Microelectron. Eng. **85**, 2226 (2008).

686

ECS Transactions, 35 (4) 687-699 (2011)
10.1149/1.3572313 ©The Electrochemical Society

Process Challenges for Integration of Copper Interconnects with Low-k Dielectrics

J.P. Gambino

IBM Microelectronics, 1000 River Street, Essex Junction, VT 05452
Email: gambinoj@us.ibm.com

Copper interconnects have gained wide acceptance in the microelectronics industry due to improved resistivity and reliability compared to Al interconnects. Initially SiO_2 was used as the interlevel dielectric. To reduce interconnect capacitance, C-doped SiO_2 or SiCOH was introduced at the 90 nm node, and porous SiCOH was introduced at the 45 nm node, to achieve a dielectric constant of 2.5 or less. However, there are many process problems with the integration of Cu interconnects and porous low-k dielectrics, including patterning, liner coverage, chemical mechanical polishing (CMP), and packaging. In this paper, some of the key integration challenges are discussed.

Introduction

Copper interconnects have gained wide acceptance in the microelectronics industry due to improved resistivity and reliability compared to Al interconnects [1]. Copper cannot be easily patterned by reactive ion etching (RIE), due to the low volatility of Cu chlorides and Cu fluorides. Hence, Cu interconnects are formed using the "dual damascene" process (Fig. 1) [1]. After processing of M1, the V1/M2 dielectric is

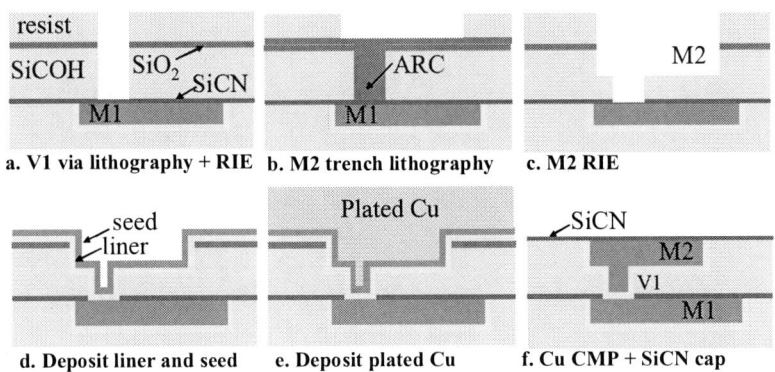

Figure 1. Schematic of process flow for dual damascene copper.

deposited (SiCOH, for example) and V1 vias are patterned (Fig. 1a), stopping on the SiCN layer that protects the Cu from oxidation. Next, the M2 trenches are patterned (Fig. 1b and 1c), the final step being the removal of the SiCN etch stop from the bottom of the

687

via. The first part of the metallization is sputter deposition of a TaN/Ta barrier layer (which prevents Cu from diffusing into the dielectric and a Cu seed layer (Fig. 1d). The vias and trenches are then filled with Cu by electroplating (Fig. 1e). The excess metal over the field regions is removed by chemical mechanical polishing (CMP). The final step is deposition of an SiCN capping layer, that protects the Cu from oxidation (Fig. 1f). In addition, the Cu must be capped with hermetic barrier layers (SiN or SiCN) to protect it from oxidation during processing or during device operation. These materials have much higher dielectric constants than that of the interlevel dielectric; for SiN, $k \sim 7$, and for SiCN, k ranges from ~ 4 to 5, depending on the processing [2]. Hence, the effective dielectric constant is typically 10% higher than that of the interlevel dielectric. These steps are repeated for each metal level. After the last metal layer is fabricated, thick dielectric passivation layers are deposited and vias are opened to the bond pads.

Low-k Dielectrics

Initially, SiO_2 was used as the interlevel dielectric surrounding the Cu wires (Fig. 2). For process integration, SiO_2 has many good properties [3,4]. It is thermally and chemically stable, and therefore does not degrade during processing. It is mechanically rigid (i.e. high elastic modulus) and is relatively impermeable to moisture (at least at the operating temperature of integrated circuits), which simplifies packaging. In addition, high quality films can be deposited by plasma enhanced chemical vapor deposition (PECVD). Of course, the disadvantage of using SiO_2 is that the dielectric constant is higher than desired.

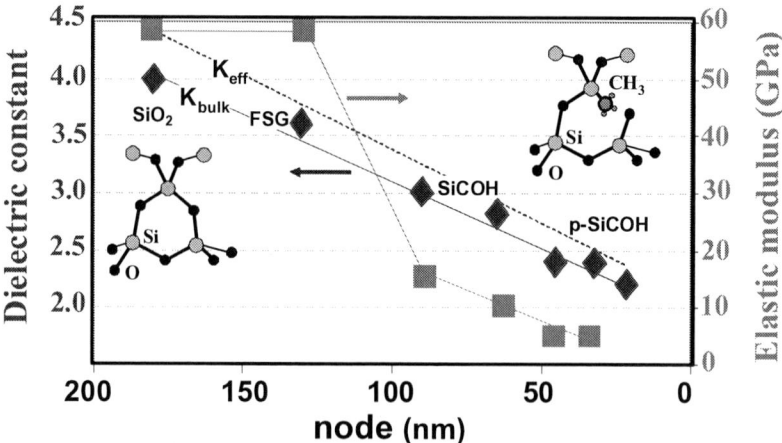

Figure 2. Trend for low-k dielectric scaling [4].

The dielectric constant of SiO_2 can be reduced by using carbon doping [5,6]. Bridging Si-O bonds are replaced by non-bridging $Si-CH_3$ bonds (Fig. 1), resulting in a lower density, and hence a lower dielectric constant. In addition, the Si-C bonds have lower polarizability than Si-O bonds. The C-doped SiO_2 is often called SiCOH, which corresponds to the chemical components in the film. The dielectric constant of non-porous SiCOH is typically 2.7 to 3.0. However, even lower dielectric constants (2.2 or less) are possible by adding pores to the SiCOH. Because of the improved performance associated with the lower dielectric constant, non-porous SiCOH is used at the 90 nm and 65 nm technology nodes, and porous SiCOH is used at the 45 nm node and below. However, the integration of Cu interconnects in SiCOH dielectrics requires many process changes, both during wafer processing and during packaging, especially for porous SiCOH. The modulus of SiCOH dielectrics is much lower than that of SiO_2 (Fig. 1) and the films are brittle. In addition, SiCOH has lower chemical stability compared to SiO_2. Hence, the design and process must be optimized when using SiCOH instead of SiO_2, to ensure yield and reliability. These will be addressed in the following sections.

SiCOH films used in manufacturing are deposited by plasma enhanced chemical vapor deposition (PECVD). [5,6]. It is essential to have small (< 2 nm diameter) isolated pores for a number of reasons [5-9]. Isolated pores are desirable to prevent water and other contaminants from diffusing into the dielectric step during wet clean steps prior to metallization and during CMP [7,8]. If pore connectivity is too high (Fig. 3), water may be absorbed in the dielectric, resulting in higher dielectric constant and/or degraded reliability for dielectric breakdown [7,8].

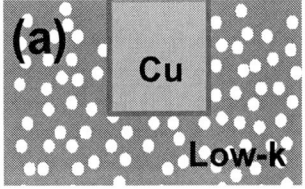 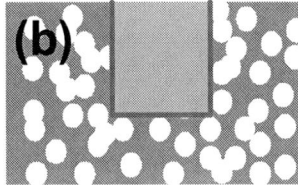

Figure 3. Schematic of porous low-k material with (a) closed pore and (b) open pore structure.

Dielectric Film Patterning

One of most difficult challenges of patterning the SiCOH dielectric is to minimize damage from the reactive ion etch and resist strip processes. The resist strip processes are especially damaging, because ions and radicals in the resist strip process remove methyl groups from the surface of the SiCOH (Fig. 4) [10]. The surface becomes hydrophilic, resulting in water absorption and an increase in the dielectric constant of the material. A number of processes must be optimized to minimize the damage from resist strip, including the C content and bonding in the SiCOH [11-13], the resist strip chemistry [14,15], and use of silylation to repair damage [10].

In oxygen-containing plasmas, damage to the sidewalls of trenches and vias occurs due to oxygen ions and radicals that diffuse into the structure and react with $Si-CH_3$ bonds [13]. SiCOH films with higher C concentration, and in particular, films with higher order hydrocarbons in the side chains, are less susceptible to damage, because the oxygen species react with the hydrocarbon chains rather than the $Si-CH_3$ bonds (Fig. 5). Small pore size is also important, to minimize diffusion of oxygen species into the film.

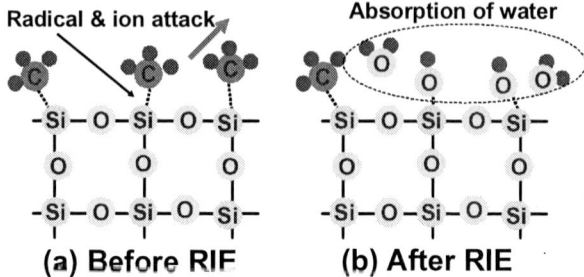

Figure 4. Resist strip damages SiCOH dielectric by removing methyl groups from the surface [10].

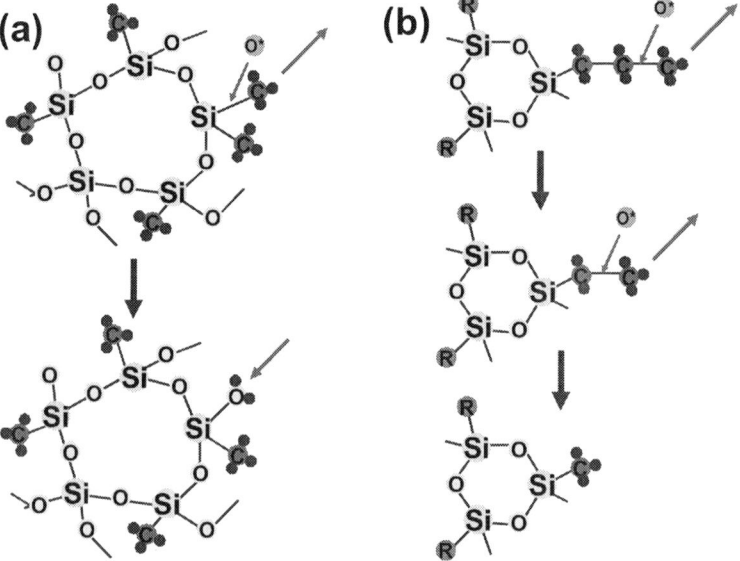

Figure 5. Schematic of surface degradation during exposure to an O_2 plasma for (a) oxygen-rich SiCOH and (b) carbon-rich SiCOH [13].

Conventional resist strips in oxygen plasmas can damage the low-k material, even if the C concentration and pore size have been optimized. Hence, the process integration and resist strip conditions must be chosen carefully to minimize etch damage [14-18]. There are two basic integration approaches for dual damascene patterning at the 32 nm node and below; the metal hardmask method [15,16,19] and the multilayer resist method [15,20] (Fig. 6 and 7). These complicated methods are needed for patterning small features because the resist thickness must be reduced as feature size decreases to ensure an adequate process window for lithography [21]. In the metal hardmask approach, the resist is stripped prior to the trench etch and via etch into the SiCOH, so there is minimal resist strip damage [19]. However, there are a number of problems with the metal hardmask approach [15,16]. Polymer can form on the sidewalls of the trenches during the trench etch; this polymer must be removed without damaging the low-k material to ensure high yield [16]. Metal residues can form on the etched surfaces and block etching of the low-k material [16]. Finally, stress in the metal layer must be minimized to avoid pattern deformation after the etch [15]. The multilayer resist approach avoids the metal residue and metal stress problems associated with the metal hardmask approach [15,20]. However, the low-k material is fully exposed to the resist strips. Hence, resist strips with low damage must be used with the multilayer resist approach.

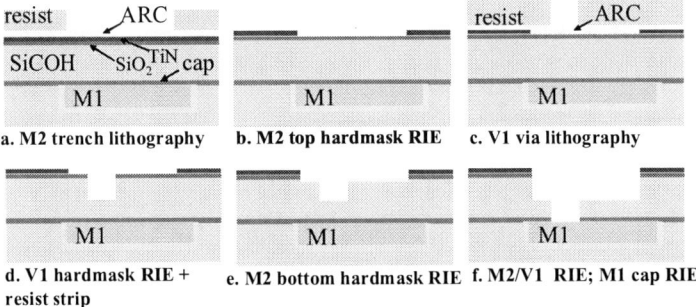

Figure 6. Dual damascene patterning with a metal hardmask [19].

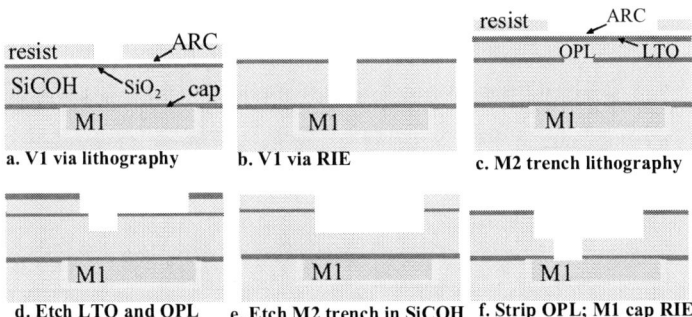

Figure 7. Dual damascene patterning with multilayer resist [20].

There are two different approaches for resist strip for low-k materials; downstream H_2 chemistry [18,22-24] and CO- or CO_2-based reactive ion etching (RIE) [14,25,26]. Direct exposure of the low-k materials to O_2, N_2, or H_2 plasmas (i.e., with ion bombardment) causes significant damage [22], with more damage for porous materials compared to non-porous materials. With downstream plasma exposure (i.e., no ion bombardment), the damage to the low-k material is significant for O_2 plasmas, but greatly reduced for N_2 plasmas, and there is no measurable damage for H_2 plasmas. Acceptable resist etch rates (> 100 nm/min) can be achieved by using a high temperature (260°C) down-stream H_2 plasma. It has been reported that residues are left after a downstream H_2 plasma strip, which must be removed with a wet clean [23]. The downstream H_2 plasma only reacts with H in the low-k film, in a replacement reaction, without altering the stoichiometry of the film [24]. Hence, there is no change in the film thickness or dielectric constant.

The CO or CO_2-based resist strip approaches are generally run in RIE tools [14,25,26]. Argon is often added to the strip chemistry and a bias is applied to the wafer to enable a high removal rate of resist [26]. During patterning, the top surface of the low-k dielectric is typically capped with a hard dielectric such as SiO_2 (i.e. in the multilayer resist approach, Fig. 7). So the regions at risk for damage are the sidewalls of vias and trenches and the bottoms of trenches. The sidewalls are exposed to very little ion bombardment and primarily react with neutral species in the plasma. The low damage associated with CO_2-based resist strips is at least partly due to the lower amount of atomic oxygen present in the plasma compared to O_2 resist strips [26]. Another possible reason for low damage with CO- and CO-based strips is formation of a C-rich passivation layer on pores and sidewalls of the low-k material [14].

Even if plasma damage is minimized, the removal of methyl groups from the surface of the SiCOH is likely to occur. With the loss of methyl groups, the surface becomes hydrophilic and absorbs water [10,27,28]. The absorbed water can cause problems with reliability, such as stress-induced voiding [7]. Hence, it may be necessary to restore the hydrophobic surface of the patterned SiCOH material prior to metallization. A number of silylation methods have been reported, consisting of high temperature exposure (150 to 350°C) of the etched surfaces to a silylating agent such as hexamethyldisilazane (HMDS), trimethysilyl-dimethylamine (TMSDMA), or tetramethylcyclotetrasiloxane (TMCATS) [28].

Metal Deposition and Chemical Mechanical Polishing (CMP)

The metallization and CMP processing are dependent on the pore size distribution in the low-k dielectric. Deposition of thin, continuous metal barrier layers (such as TaN/Ta) is more difficult as pore size increases; incomplete barrier coverage can result in Cu diffusion into the dielectric [9]. The target for barrier layer thickness at the 22 nm node is ~ 3nm. Hence, even for well designed porous low-k materials (i.e., with isolated pores less than 2 nm in diameter) the metal thickness is approaching the pore size. To ensure reliability, it may be necessary to seal the pores prior to metallization, using plasma treatments or conformal dielectric deposition [9,29,30]. A number of dielectrics have been examined as pore sealing materials including SiCH, SiOC, SiO_2 [30] and divinyl-siloxane benzocyclobutene polymer (p-BCB) [29]. The drawback to using an additional pore sealing layer is that the RC delay will be increased [31].

An additional problem is that moisture trapped in the porous dielectric can oxidize the TaN liner (Fig. 8a-c) [32-34]. Copper adheres poorly to oxidized Ta. Hence, the

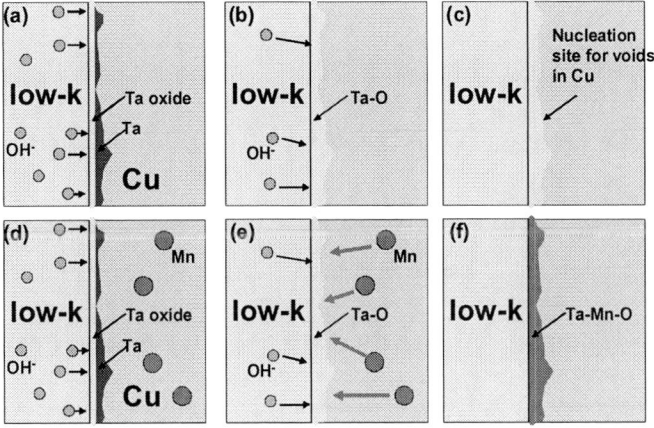

Figure 8. Schematic of barrier oxidation (a-c) and Cu-Mn seed layer barrier restoration (d-f). [34]

oxidized TaN/Ta barrier can cause poor yield and degrade reliability. A number of approaches have been used to minimize problems associated with barrier oxidation. One method is to increase the nitrogen content of the TaN; the oxidation is greatly reduced when TaN stoichiometry is changed from 4:1 to 2:1 (i.e. higher nitrogen content) [33]. Another approach is to use a Mn-doped alloy seed layer to restore the barrier (Fig. 8d-f) [34]. This process begins with sputter deposition of a Ta-based barrier, like the conventional process. The CuMn alloy seed layer is deposited, followed by plating and annealing. If the thin Ta barrier is oxidized, the Mn will segregate at the interface and form a Ta-Mn-O phase, thereby enhancing the barrier (Fig. 8f).

There are a number of problems with Cu CMP in a porous low-k structure, including Cu dishing and insulator erosion [35], cracking and adhesion loss in the dielectric stack [35], and scratching or contamination of the low-k material by components or the slurry or reaction by-products [36-39]. The problems with dishing/erosion and cracking / adhesion loss can be minimized by reducing the downforce during CMP and improving the adhesion between layers in the stack [35].

There are two basic integration schemes for Cu CMP with porous low-k structures; the permanent polish stop method (Fig. 9a-c) [8,35,36] and the direct CMP method (Fig. 9d-f) [8]. In the permanent polish stop approach, a relatively dense material, such as SiO$_2$ [35] or non-porous SiCOH [8,36], is used on top of the porous low-k material. The advantage of this approach is that the porous low-k material is protected from CMP-related scratches and contamination. The disadvantage is that the effective dielectric constant of the stack increases. Hence, there is much research on minimizing damage and contamination when the polish stops directly on the porous low-k dielectric [37,40-42].

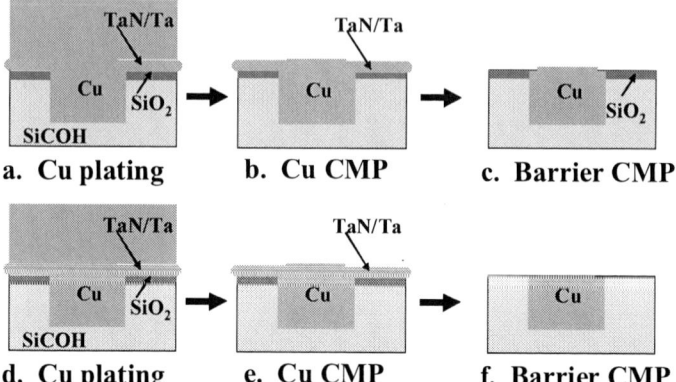

a. **Cu plating**   b. **Cu CMP**   c. **Barrier CMP**

d. **Cu plating**   e. **Cu CMP**   f. **Barrier CMP**

Figure 9. Schematic of CMP options; (a-c) polish stop method and (d-f) direct CMP method [40].

For direct CMP on the low-k dielectric, the first requirement is a low removal rate for the porous low-k material during the polish. Organic compounds such as surfactants are used to lower the polish rate of the low-k material with respect to the metal layer [40-43]. The surfactants selectivity segregate to the surface of the SiCOH and thereby reduce the polish rate with respect to the metals (Fig. 10a) [43]. However, it is often observed that the presence of surfactants in the slurry increases the dielectric constant of the porous SiCOH [40-41]. Additional $-CH_2$ and C-H bonds are observed in the bulk of the porous SiCOH after exposure to surfactants in the slurry, which are responsible for the increase in dielectric constant [41]. Significant diffusion of both linear and branched surfactants into porous low-k materials has been observed at room temperature (Fig. 10b), consistent with this model [42].

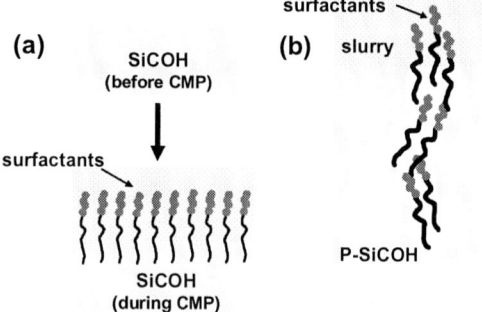

Figure 10. Schematic of (a) surfactant coverage on surface of the SiCOH [43] and (b) surfactant diffusion into pores [42].

There are a number of approaches to minimize the change in dielectric constant. One approach is to optimize the slurry to prevent residues from forming in the pores [40,41]. Another approach is to do a post-CMP anneal at 350°C to restore the dielectric constant

[40,41]. A third approach is to use a bilayer porous SiCOH film, with the near surface region (that is exposed to CMP) having a lower porosity than the bulk of the film [8].

Packaging

Packaging is also more difficult with low-k dielectrics, especially for porous dielectrics, due to the lower modulus of these materials compared to SiO_2 [44,45]. Stress occurs in the die during packaging due to the mismatch in coefficient of thermal expansion (CTE) between the die and the package materials. For wirebond packages, damage can occur under the bond pad, due to force associated with the wirebond process, and at the chip corners, due to the epoxy coating. The low-k dielectric under the bond pads must be protected, by using SiO_2 as the dielectric for the last wiring layers, and by using vias to strengthen the structure [45,49]. For flip chip packages, the stress occurs underneath the solder bumps, (typically during chip joining) that can cause cracks in the low-k dielectrics. The problem becomes even more difficult with Pb-free solder, due to the high modulus and higher melting point of Pb-free solder compared to Pb-based solder [46].

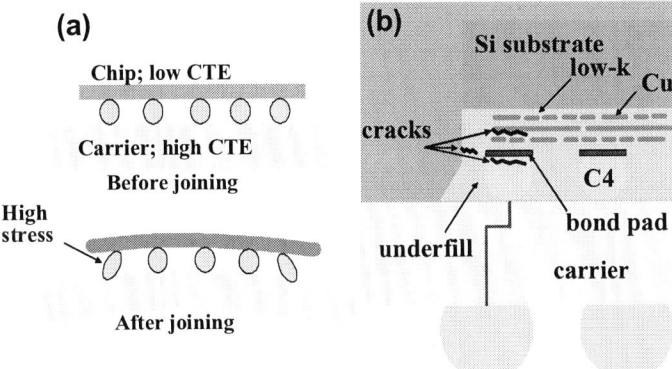

Figure 11. Schematic of cracks that can form in die or in solder due to stress from flip-chip attach.

Design and process changes must also be made to allow reliable packaging of these die. Design solutions include improved layout of the crack stop and edge seal [47,48], and the bond pads [46,50]. Packaging process changes include optimizing the dicing process (two-step dicing or laser dicing, Fig. 12) [50-52] , the underfill (lower modulus) [53], the molding compound (lower coefficient of thermal expansion, CTE) [54], the solder composition, the solder reflow process, and the adhesion of the dielectrics and barrier layers in the low-k stack [46].

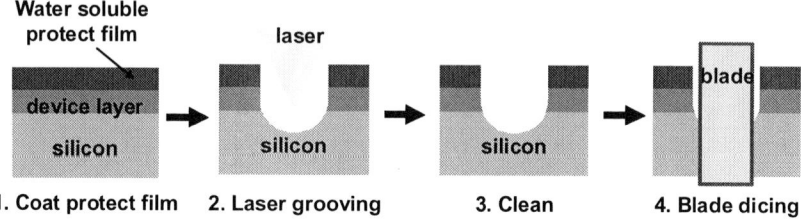

1. Coat protect film   2. Laser grooving   3. Clean   4. Blade dicing

Figure 12. Schematic of two-step dicing using laser [52].

Conclusions

Copper interconnects have gained wide acceptance in the microelectronics industry due to improved resistivity and reliability compared to Al interconnects. To reduce interconnect capacitance, porous SiCOH has been introduced at the 45 nm node, to achieve a dielectric constant of 2.5 or less. There are many process problems with the integration of Cu interconnects and porous low-k dielectrics, including patterning, liner coverage, chemical mechanical polishing (CMP), and packaging. These processes must be modified when using porous low-k dielectrics to ensure adequate yield and reliability.

References

1. D. Edelstein, J. Heidenreich, R. Goldblatt, W. Cote, C. Uzoh, N. Lustig, P. Roper, T. McDevitt, W. Motsiff, A. Simon, J. Dukovic, R. Wachnik, H. Rathore, R. Schulz, L. Su, S. Luce, J. Slattery, *IEEE Int. Electron Devices Meeting Proc.*, 1997, p. 773.
2. L.M. Matz, T. Tsui, E.R Engbrecht, K. Taylor, G. Haase, S. Ajmera, R. Kuan, A. Griffin, R. Kraft, A.J. McKerrow, in *Proc. Advanced Metallization Conf. 2005*, S.H. Brongersma, T.C. Taylor, M. Tsujimura, K. Masu, eds., (Materials Research Society, Pittsburgh, PA, 2006), p. 437.
3. J. Gambino, A. Stamper, T. McDevitt, V. McGahay, S. Luce, T. Pricer, B. Porth, C. Senowitz, R. Kontra, M. Gibson, H. Wildman, A. Piper, C. Benson, T. Standaert, P. Biolsi, E. Cooney, E. Webster, R. Wistrom, A. Winslow, E. White., *Proc. IEEE Int. Symp. on the Physical & Failure Analysis of Integrated Circuits*, 2002, p. 111.
4. J. Gambino, *IEEE Custom Integrated Circuits Conf. Proc.*, 2009, p. 141
5. A. Grill, in *Dielectric Films for Advanced Microelectronics*, M. Baklonov, M. Green, K. Maex, eds. (John Wiley & Sons, 2007), p. 1.
6. A. Grill, *Ann. Rev. Mat. Sci.*, **39**, 49 (2009).
7. F. Ito, T. Takeuchi, H. Yamamoto, T. Ohdaira, R. Suzuki, Y. Hayashi, in *Proc. Advanced Metallization Conf. 2007*, (Materials Research Society, Pittsburgh, PA, 2008), p. 465.
8. T. Seo, Y. Oka, K. Seo, K. Goto, H. Chibahara, H. Korogi, S. Suzuki, M. Hamada, N. Suzumura, K. Tsukamoto, A. Ueki, T. Furuhashi, D. Kodama, S. Kido, J. Izumitani, K. Tomita, E. Kobori, A. Ikeda, Y. Kawano, T. Ueda, *IEEE Int. Interconnect Technology Conf. Proc.*, 2010, p. 5.5.

9. S. Chikaki, K. Kinoshita, T. Nakayama, K. Kohmura, H. Tanaka, M. Hirakawa, E. Soda, Y. Seino, N. Hata, T. Kikkawa, S. Saito, *IEEE Int. Electron Devices Meeting Proc.*, 2007, p. 969.

10. A. Kojima, N. Nakamura, N. Matsunaga, H. Hayashi, K. Kubota, R. Asako, K. Maekawa, H. Shibata, T. Yoda, T. Ohiwa, in *Proc. Advanced Metallization Conf. 2006*, S.W Russell, M.E. Mills, A. Osaki, T. Yoda, eds., (Materials Research Society, Pittsburgh, PA, 2007), p. 301.

11. S.M. Gates, A. Grill, C. Dimitrakopoulos, V. Patel, S.T. Chen, T. Spooner, E.T. Ryan, S.A. Cohen, E. Simonyi, E. Liniger, Y. Ostrovski, R. Bhatia, in *Proc. Advanced Metallization Conf. 2008*, (Materials Research Society, Pittsburgh, PA, 2009), p. 531.

12. N. Inoue, N. Furutake, F. Ito, H. Yamamoto, T. Takeuchi, Y. Hayashi, *Jap. J. Appl. Phys.*, **47**, 2468 (2008).

13. Y. Hayashi, H. Ohtake, J. Kawahara, M. Tada, S. Saito, N. Inoue, F. Ito, M. Tagami, M. Ueki, N. Furutake, T. Takeuchi, H. Yamamoto, M. Abe, *IEEE Trans. Semiconductor Manufacturing*, **21**, 469 (2008).

14. H. Shi, H. Huang, J. Im, P.S. Ho, Y. Zhou, J.T. Pender, M. Armacost, D. Kyser, *IEEE Int. Interconnect Technology Conf. Proc.*, 2010, p. 8.12.

15. T. Chevolleau, N. Posseme, T. David, R. Bouyssou, J. Ducote, F. Bailly, M. Darnon, M. El Kodadi, M. Besacier, C. Licitra, M. Guillermet, A. Ostrovsky, C. Vervore, O. Joubert, *IEEE Int. Interconnect Technology Conf. Proc.*, 2010, p. 5.1.

16. V. Travaly, J. Van Aelst, V. Truffert, P. Verdonck, T. Dupont, E. Camerotto, O. Richard, H. Bender, C. Kroes, D. De Roest, G. Vereecke, M. Claes, Q.T. Le, E. Kesters, M. Van Cauwenberghe, J. Beynet, S. Kaneko, H. Struyf, M. Baklanov, K. Matsushita, N. Kobayashi, H. Sprey, G. Beyer, *IEEE Int. Interconnect Technology Conf. Proc.*, 2008, p. 52.

17. O.V. Braginsky, A.S. Kovalev, D.V. Lopaev, Y.A. Mankelevich, E.M. Malykhin, O.V. Proshina, T.V. Rakhimova, A.T. Rakhimova, A.N. Vasilieva, D.G. Voloshin, S.M. Zyryanov, M.R. Baklanov, in *Materials, Processes and Reliability for Advanced Interconnects for Micro- and Nanoelectronics - 2009*, M. Gall, A. Grill, F. Iacopi, J. Koike, T. Usui, eds., vol. **1156**, Mat. Res. Soc., Pittsburgh, PA, 2009, p. D01-06.

18. M.R. Baklanov, A. Urbanowicz, G. Mannaert, S. Vanhaelemeersch, *Proc. 8th Int. Conf. Solid-State Integrated Circuits Technology*, 2006, p. 291.

19. O. Hinsinger, R. Fox, E. Sabouret, C. Goldberg, C. Verove, W. Besling, P. Brun, E. Josse, C. Monget, O. Belmont, J. Van Hassel, B.G. Sharma, J. P. Jacquemin, P. Vannier, A. Humbert, D. Bunel, R. Gonella, E. Mastromatteo, D. Reber, A. Farcy, J. Mueller, P. Christie, V.H. Nguyen, C. Cregut, T. Berger, *IEEE Int. Electron Devices Meeting Proc.*, 2004, p. 317.

20. W.Cote, D. Edelstein, C. Bunke, P. Biolsi, W. Wille, H. Baks, R. Conti, T. Dalton, T. Houghton, W.-K. Li, Y.-H. Lin, S. Moskowitz, D. Restaino, T. Van Kleeck, S. Vogt, T. Ivers, in *Proc. Advanced Metallization Conf. 2006*, S.W Russell, M.E. Mills, A. Osaki, T. Yoda, eds., (Materials Research Society, Pittsburgh, PA, 2007), p. 289.

21. *International Technology Roadmap for Semiconductors, Interconnect*, 2009. http://www.itrs.net/

22. X. Hua, M. Kuo, G.S. Oehrlein, P. Lazzeri, E. Iacob, M. Anderle, C.K. Inoki, T.S. Kuan, P. Jiang, W. Wu, *J. Vac. Sci. Technol.*, **B24**, 1238 (2006).

23. O. Louveau, C. Bourlot, A. Marfoure, I. Kalinovski, J. Su, G. Hills, D. Louis, *Microelectronic Eng.*, **73-74**, 351 (2004).
24. P. Lazzeri, G.S. Oehrlein, G.J. Stueber, R. McGowan, E. Busch, S. Pederzoli, C. Jeynes, M. Bersani, M. Anderle, *Thin Solid Films*, **516**, 3697 (2008).
25. J. Lee, W.-J. Park, D.-H. Kim, J. Choi, K. Shin, I. Chung, *Thin Solid Films*, **517**, 3847 (2009).
26. M.-S. Kuo, A.R. Pal, G.S. Oehrlein, P. Lazzeri, M. Anderle, *J. Vac. Sci. Technol.*, **B28**, 952 (2010).
27. S.V. Nitta, S. Purushothaman, N. Chakrapani, O. Rodriguez, N. Kymko, E.T. Ryan, G. Bonilla, S. Cohen, S. Molis, K. McCullough, in *Proc. Advanced Metallization Conf. 2005, S H. Brongersma, T.C. Taylor, M. Tsujimura, K. Masu, eds., (Materials Research Society, Pittsburgh, PA, 2006), p. 325.
28. K. Kinoshita, K. Kinoshita, S. Chikaki, E. Soda, K. Tomioka, H. Tanaka, K. Kohmura, T. Nakayama, T. Kikkawa, S. Saito, A. Kojima, , in *Proc. Advanced Metallization Conf. 2007*, (Materials Research Society, Pittsburgh, PA, 2008), p. 513.
29. M. Tada, T. Tamura, F. Ito, H. Ohtake, M. Narihiro, M. Tagami, M. Ueki, K. Hijioka, M. Abe, N. Inoue, T. Takeuchi, S. Saito, T. Onodera, N. Furutake, K. Arai, M. Sekine, M. Suzuki, Y. Hayashi, *IEEE Trans. Elec. Dev.*, **53**, 1169 (2006).
30. A. Furuya, K. Yoneda, E. Soda, T. Yoshie, H. Okamura, M. Shimada, N. Ohtsuka, S. Ogawa, *J. Vac. Sci. Technol.*, **B23**, 2522 (2005).
31. M. Gallitre, L.G. Gosset, A. Farcy, B. Blampey, R. Gras, C. Bermond, B. Flechet, J. Torres, *IEEE Int. Interconnect Technology Conf. Proc.*, 2007, p. 132.
32. M. Hamada, K. Ohmori, K. Mori, E. Kobori, N. Suzumura, R. Etou, K. Maekawa, M. Fujisawa, H. Miyatake, A. Ikeda, *IEEE Int. Interconnect Technology Conf. Proc.*, 2010, p. 13.4.
33. A.H. Simon, F. Baumann, T. Bolom, J.G. Park, C. Child, B. Kim, P. DeHaven, R. Davis, O. Ogunsola, M. Angal, in *Advanced Interconnects and Chemical Mechanical Planarization for Micro- and Nanoelectronics*, J.W. Bartha, C.L. Borst, D. DeNardis, H. Kim, A. Naeemi, A. Nelson, S.S. Papa Rao, H.W. Ro, D. Toma, eds., vol. **1249**, Mat. Res. Soc., Pittsburgh, PA, 2009, p. F01-02.
34. A. Haneda, T. Tabira, H. Sakai, H. Kudo, M. Sunayama, Ohtsuka, A. Tsukune, N. Shimizu, in *Proc. Advanced Metallization Conf. 2007*, (Materials Research Society, Pittsburgh, PA, 2008), p. 59.
35. S. Kondo, B.U. Yoon, S. Tokitoh, K. Misawa, S. Sone, H.J. Shin, N. Ohashi, N. Kobayashi, *IEEE Int. Electron Devices Meeting Proc.*, 2004, p. 151.
36. L.L. Chapelon, H. Chaabouni, G. Imbert, P. Brun, M. Mellier, K. Hamioud, M. Vilmay, A. Farcy, J. Torres, *Microelectronic Eng.*, **73-74**, 351 (2004).
37. N. Heylen, E. Camerotto, H. Volders, Y. Travaly, G. Vereecke, G.P. Beyer, Z. Tokei, *IEEE Int. Interconnect Technology Conf. Proc.*, 2010, p. 17.
38. D. Oshida, T. Takewaki, M. Iguchi, T. Taiji, T. Morita, Y. Tsuchiya, S. Yokogawa, H. Kunishima, H. Aizama, N. Okada, *IEEE Int. Interconnect Technology Conf. Proc.*, 2008, p. 222.
39. M. Ueki, T. Onodera, A. Ishikawa, S. Hoshino, Y. Hayashi, *Jap. J. Appl. Phys.*, **49**, 04C029 (2010).
40. S. Gall, C. Euvard, S. Shhun, S. Maitrejean, M. Assous, P.-H. Haumesser, M. Rivoire, in *Proc. Advanced Metallization Conf. 2007*, (Materials Research Society, Pittsburgh, PA, 2008), p. 115

41. M. Kodera, T. Takahashi, G. Mimamihaba, *Jap. J. Appl. Phys.,* **49**, 04DB07 (2010).
42. T.-S. Kim, T. Konno, T. Yamanaka, R.H. Dauskardt, *IEEE Int. Interconnect Technology Conf. Proc.*, 2008, p. 171.
43. J. Bian, in *Advances and Challenges in Chemical Mechanical*, G. Zwicker, C. Borst, L. Economikos, A. Philipossian, eds., vol. **991**, Mat. Res. Soc., Pittsburgh, PA, 2009, p. C09-03.
44. T. Furusawa, K. Goto, J. Izumitani, M. Matsuura, M. Fujisawa, N. Kawanabe, T. Hirose, E. Hayashi, S. Baba, Y. Asano, T. Ichiki, Y. Takata, *IEEE Int. Interconnect Technology Conf. Proc.*, 2010, p. 9.2.
45. S.Y. Hou, C.W. Shih, W.C. Wu, C.H. Hsieh, A.J. Su, C.H. Tung, S.P. Jeng, M.J. Li, D.C.H. Yu, *IEEE Int. Interconnect Technology Conf. Proc.*, 2010, p. 9.1.
46. R.A. Susko, T.H. Daubenspeck, T.A. Wassick, T.D. Sullivan, W. Sauter, J. Cincotta, ECS Transactions, **16**, 51 (2009).
47. W. Landers, D. Edelstein, L. Clevenger, S. Das, C.-C. Yang, T. Aoki, F. Beaulieu, J. Casey, A. Cowley, M. Cullinan, T. Daubenspeck, C. Davis, J. Demarest, E. Duchesne, L. Guerin, D. Hawkin, T. Ivers, M. Lane, X. Liu, T. Lombardi, C. McCarthy, C. Muzzy, J. Nadeau-Filteau, D. Questad, W. Sauter, T. Shaw, J. Wright, *IITC Proc.*, 2004, p. 108.
48. T.C. Huang, C.T. Peng, C.H. Yao, C.H. Huang, S.Y. Li, M.S. Liang, Y.C. Wang, W.K. Wan, K.C. Lin, C.C. Hsia, M.-S. Liang, *IITC Proc.*, 2006, p. 92.
49. M. Saran, R. Cox, C. Martin, G. Ryan, T. Kudoh, M. Kanasugi, J. Hortaleza, M. Ibnabdeljalil, M. Murtuza, D. Capistrano, R. Roderos, R. Macaraeg, *IRPS Proc.*, 1998, pp. 225.
50. W. ZhiJie, S. Wang, J.H. Wang, S. Lee, Y. SuYing, R. Han, Y.Q. Su, *Proc. IEEE Conf. Electronic Packaging Tech.*, 2005, pp. 262.
51. J. Li, H. Hwang, E.-C. Ahn, Q. Chen, P. Kim, T. Lee, M. Chung, T. Chung, *Proc. ECTC*, 2007, p. 761.
52. S.M. Sullivan, *ECS Trans.*, **18**, 2009, p. 745.
53. P.-H. Tsao, C. Huang, M.-J. Lii, B. Su, N.-S. Tsai, *Proc. ECTC*, 2004, p. 767.
54. M. Tagami, H. Ohtake, M. Abe, F. Ito, T. Takeuchi, K. Ohto, T. Usami, M. Suzuki, T. Suzuki, N. Sashida, Y. Hayashi, *IITC Proc.*, 2005, p. 12.
55. C. Goldberg, S. Downey, V. Fiori, R. Fox, K. Hess, O. Hinsinger, A. Humbert, J.-P. Jacquemin, S. Lee, J.-B. Lhuillier, S. Orain, S. Pozder, L. Proenca, F. Querica, E. Sabouret, T.A. Tran, T. Uehling, , *IITC Proc.*, 2005, p. 3.

700

Patterning with Amorphous Carbon Thin Films

G. A. Antonelli[a], S. Reddy[b], P. Subramonium[b], J. Henri[b], J. Sims[b], J. O'loughlin[b], N. Shamma[c], D. Schlosser[c], T. Mountsier[c], W. Guo[d], and H. Sawin[d]

[a] External Research & Development, Novellus Systems, Albany, New York 12203, USA
[b] PECVD Business Unit, Novellus Systems, Tualatin, Oregon 97062, USA
[c] Customer Integration Center, Novellus Systems, San Jose, California 95136, USA
[d] Department of Chemical Engineering, Massachusetts Institute of Technology, Cambridge, Massachusetts 02139, USA

Amorphous carbon hard mask films grown with plasma enhanced chemical vapor deposition are an enabling technology for advanced front-end-of-line patterning technologies. These films must have a low etch rate and be weakly roughened in dielectric etch chemistries, high transparency at lithography alignment wavelengths, and the mechanical properties to mitigate elastic instabilities such as line bending. The deposition process affects all of these parameters through the resulting structure and composition. Highly graphitic films deposited at 550°C are common; however, other process spaces relying on ion bombardment rather than temperature can create less graphitic films with improved film properties like transparency, hardness, and etch selectivity.

Introduction

Amorphous carbon films grown by plasma enhanced chemical vapor deposition (PECVD) have gained wide acceptance among memory manufacturers employing 193nm lithography [1]. 193nm photoresists (PR) have lower selectivity and a mechanical strength which is poor when compared to 248nm PR [2]. This lower selectivity demands a thicker layer of PR to be used for successful pattern transfer. However, pattern collapse arising from the poor mechanical strength and lower depth of focus of 193nm light limits the PR thickness [3]. For this reason, a high selectivity hard mask layer is often used in this lithography scheme, as shown in Figure 1. The pattern is first transferred using a thin PR layer to the hard mask layer which in turn is used to transfer the pattern to the film of interest. Once the pattern transfer to film of interest is complete, the hard mask layer can be removed with an oxygen-containing plasma ash operation. Amorphous carbon is an example of an ashable hard mask (AHM).

In addition to high selectivity, an ideal AHM should be transparent with an extinction coefficient < 0.10 at the mask alignment wavelength, i.e., 582 or 633 nm. It must have good step coverage to protect the alignment marks from being attacked during etching. The step coverage must not have a reentrant profile to ensure the anti-reflective layer (ARL) can protect the alignment mark, enabling the possibility of PR rework. When used as a hard mask in gate patterning process, it is desirable for AHM to have small line edge roughness and good line profile integrity without bending. The phenomenon of line bending is mechanical buckling and arises due to a combination of the line geometry as well as the thicknesses and mechanical properties of the AHM and ARL films. In this paper, we will further explore the basic material properties of these AHM films as well as their effect on the pattern transfer process.

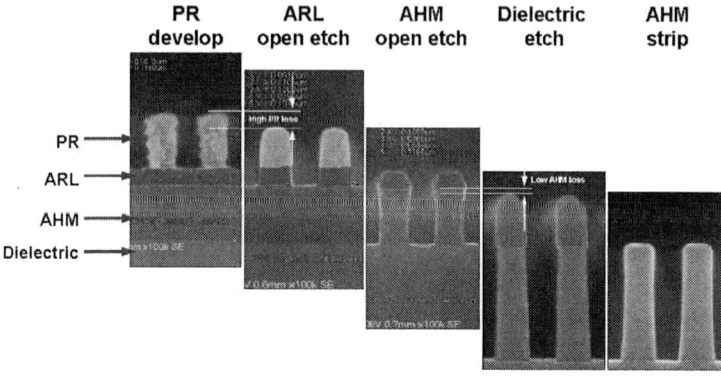

Figure 1. 193 nm lithography scheme with AHM.

Synthesis

In 1911, Bolton reported the deposition of seed crystals of "diamond" from the decomposition of C_2H_2 in the presence of Hg vapor [4]. In 1955, Schmellenmeier studied plasma deposited carbon films and discovered the presence of diamond-like carbon (DLC) [2]. In the early 1980s, DLC films based on carbon and hydrogen received considerable attention due to extreme hardness, high transparency in the infrared, high thermal conductivity, good electrical insulation, and resistance to chemical attack. Amorphous carbon films can be deposited by many methods including CVD, PECVD, and ion beam deposition [5, 6]. Typically, these films are composites of crystalline diamond which has sp^3 hybridized orbitals, amorphous carbon and graphite which has sp^2 hybridized orbitals. Depending on the processing technology and process conditions, the relative concentration of the respective phases differs.

All AHM materials in this study were grown using a hydrocarbon precursor such as CH_4, C_2H_2 , C_2H_6, etc. with a combination of carrier gases such as He, H_2, and N_2, at pressures on the order of a Torr by plasma enhanced chemical vapor deposition (PECVD) in the Novellus VECTOR® Express system on 300 mm [100] Si wafers. The reactor was capacitively coupled and excited by two radio frequency sources operating at 13.5 MHz and 0.4 MHz. The temperature of the wafer and the ion energy/flux in the plasma were varied.

Deposition temperature is often used as the primary parameter to meet the etch selectivity target [7]. PECVD AHM films grown at high temperature (>400°C) tend to be dense but also have a high sp^2 bond content, i.e., graphitic. These films will have high absorption coefficients implying they are less transparent. The etch selectivity and transparency requirements are thus in conflict. That conflict can be avoided if films are grown at lower temperatures, < 300°C, but in a higher ion energy environment, creating a material with a greater sp^3 bond content which can be both dense and highly transparent.

Structural Characterization

Table I summarizes the material properties of representative AHM films. The carbon and hydrogen were measured by Rutherford backscattering and hydrogen forward scattering, respectively. The n, refractive index, and k, the absorption coefficient, were measured with a Woollam spectroscopic ellipsometer, the density by x-ray reflectivity (XRR) on Bruker D8 Discover diffractometer, and hardness with an MTS Nano-Indentor™. Type 1 is a low density film deposited at 400°C with low ion energy. Type 2 was deposited at >500°C with low ion energy. This film is high density and carbon rich. Type 3 was deposited at <300°C with high ion energy. Type 3 films have high selectivity, good transparency, and mechanical strength.

TABLE I. AHM Material Properties.

	% C by atom	% H by atom	n @ 633 nm	k @ 633 nm	Density (g/cm³)	Hardness (GPa)
Type 1	56	42	1.81	0.05	1.23	2.7
Type 2	70	27	2.00	0.38	1.44	1.8
Type 3	60	40	1.99	<0.10	1.48	9.5

The differences in bonding structures for these materials were characterized with Fourier transform infrared spectroscopy (FTIR). The data was collected on films with a thickness in the range of 200 to 500 nm deposited on a 300 mm [100] silicon substrate at multiple sites in transmission mode using a Nanometrics QS1200. The complete spectra for the films in Table I are shown in Figure 2, while Figure 3 focuses on the absorbance in the CH_x and C=C regions. All spectra have had their backgrounds removed.

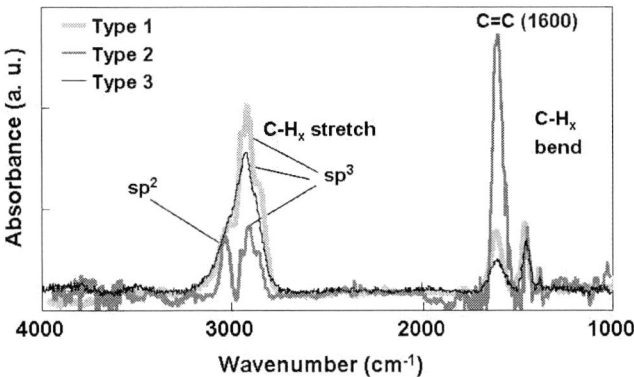

Figure 2. Infrared absorption spectra of Table I materials.

Type 2 is significantly more sp^2 (graphitic) dominated compared to Type 1 and 3. CH_x stretching vibrations of the Type 2 network have more sp^2 CH_2 (2950 cm^{-1}) and sp^2

CH (3050 cm^{-1}) bonds compared to sp^3 bonds. Temperatures >500°C used for deposition of Type 2 favor sp^2 rich amorphous carbon network, as the high temperature enables easy breakage of C-H bonds and subsequent diffusion of hydrogen. Thus, with the aid of thermal energy, dense carbon rich films can be deposited in a low ion energy process space.

Type 1 and 3 materials deposited at lower temperature (≤400°C) have similar or better transparency than high temperature Type 2. Type 3 has less sp^2 bonding compared to Type 2 film and are dominated by sp^3 CH$_2$ peaks. The presence of sp^3 CH$_2$ bonds increases the cross linking compared to terminating sp^3 CH$_3$ (2955 cm^{-1}) bonds which are in much greater number in the Type 1 material. The highly cross linked sp^3 structures in Type 3 films are created by operating at low temperatures and employing high ion energy. These high ion energies lead to compact packing which increases film density thereby improving selectivity. The good transparency and low absorption coefficient of these films are attributed to the sp^3 dominated network.

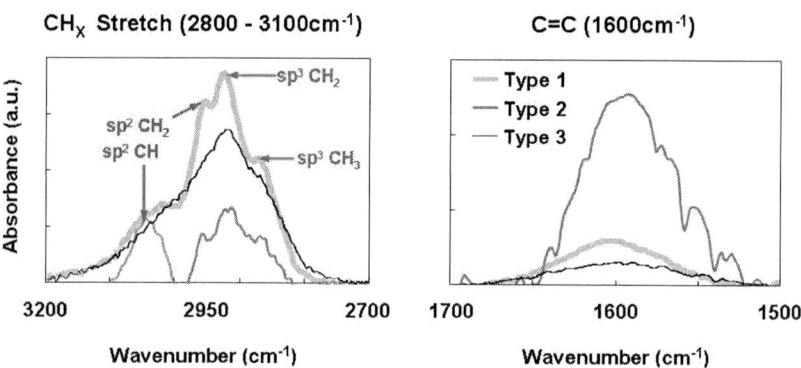

Figure 3. Infrared absorption spectra of Table I materials in CH$_x$ and C=C regions.

Further information regarding the structure of these materials can be gleaned from an analysis of the Raman spectra. Examples of such spectra are shown in Figure 4 for Type 2 and 3 materials. This data was collected in the Stokes region using a micro-Raman setup constructed on an optical bench operating in backscattering mode with a spot size of a few microns. The light source was a solid state diode laser with a wavelength of 532 nm. The films used in these measurements were 500 nm thick and deposited on a thick sputtered copper film on a [100] silicon substrate to eliminate any Raman signal from the silicon.

Robertson gives an excellent summary to the analysis of Raman spectra for these amorphous carbon materials, but we will briefly summarize the basic tenants [6]. The Raman response can be grossly described in terms of two modes labeled G and D. G is a Raman active mode observed in single crystal graphite at 1580 cm^{-1} and relates to the stretching vibration of any pair of sp^2 sites in C=C or an aromatic ring [8]. The Raman active mode, D, which stands for disorder, not diamond, occurs in amorphous carbon materials at approximately 1350 cm^{-1} and is the breathing mode of sp^2 sites only in rings

[9]. The type 2 material has both strong G and D peaks while the G peak is stronger than the D peak in type 3. The Type 2 material has greater fundamental sp^2 content when compared to Type 3. Indeed, the Type 2 material falls between the categories of nano-crystalline graphite (nc-G) and sp^2 amorphous carbon, while Type 3 is between sp^2 and sp^3 amorphous carbon. Although both of these materials should technically remain in the category of sp^2 amorphous carbon, these spectra imply that the Type 2 should have much larger domains of graphitic material as compared to the more disordered Type 3. That information on the short range order of these materials cannot be obtained from infrared absorption measurements and is an important factor to be considered in any detailed study of the etch mechanism of these materials.

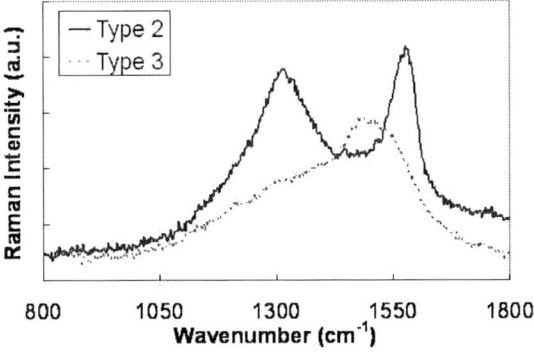

Figure 4. Raman spectra of Type 2 and 3 materials.

The densities of the Type 2 and 3 in Table I are similar; however, there is a dramatic difference in hardness. This difference is ultimately due to the bonding in the Type 3 material. Raman results show lower sp^2 content and a shift away from nano-crystalline graphite for Type 3 while the infrared absorption exhibits much greater sp^3 bond content. The high energy environment used to create the Type 3 material has fostered not only this higher density but also a shift towards a more sp^3 or perhaps more correctly, a less sp^2 matrix, and an increased network connectivity. This will enable use of thinner hard masks in applications where maintaining lower hard mask aspect ratio is critical for successful pattern transfer.

Etch Rate & Roughening Mechanisms

Blanket etch rates for these films in a $C_xF_y/O_2/Ar$ etch chemistry are shown in Figure 5. The wafers were processed in a commercial 300 mm etch system. The etch rate was determined by multiplying the inverse of the etch time by the difference in pre and post etch thickness. The thickness was measured by spectroscopic ellipsometry on a KLA-Tencor Aset F5. In this plot, refractive index is used as a proxy for density. As can be confirmed by Table I, the density and refractive index strongly correlate.

The Type 1 material has a much higher etch rate as compared to Type 2. From the data in Table I, Type 1 has a lower density which tends to scale directly with etch rate.

The structural data obtained by FTIR leads to the same conclusion. Although both Type 1 and 2 are dominated by sp^2 bonds, the Type 1 material exhibits a much greater quantity of terminating CH_3 bonds. The lower etch rate of Type 2 results from this cross linked sp^2 network which increases the carbon content and density of these films. However, due to the lower optical band gap of sp^2 carbon, these films are not transparent, leading to mask alignment and overlay issues and hence making them unsuitable for thick hard mask applications, i.e., >200 nm. The Type 3 process space combines advantageous aspects of both Type 1 and 2. The Type 3 example given in Table I has an etch rate similar to the Type 2, but with very high film transparency; further, substantially lower etch rates are clearly possible with the Type 3 films.

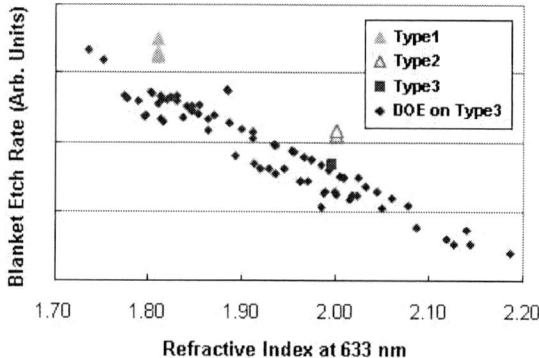

Figure 5. Blanket AHM film etch rate in $C_xF_y/O_2/Ar$ chemistry.

To better understand these results, Type 2 and Type 3 materials were processed in the inductively-coupled plasma chamber shown in Figure 6. The plasma is generated in the upper part of chamber using a three-turn copper coil to power the discharge. A typical RF matching network with L configuration is used to maximize power transfer, with the power density high enough to insure inductive coupling of the RF power. The gas flow rate was 2-10 sccm controlled by mass flow controllers. A ceramic/quartz liner is used to isolate the plasma from the wall, which enables the plasma to be biased up to 400 V using a metal electrode placed at the interior of the beam source chamber. A plasma beam consisting of both ions and neutrals is extracted through a mesh toward the lower chamber, in which the pressure is maintained at 10^{-5}–10^{-6} Torr by turbo pump and cryo pump under typical operating conditions. The ion beam is charge compensated by electron emission from a tungsten filament to maintain the directionality of the bombardment. The neutrals pass through the grid producing a cosine distribution of neutrals from the grid to the sample. The sample stage is located below the neutralizing filament and can be rotated around its axis to vary the ion and neutral incidence angle to the substrate surface. Sidewall roughening is simulated by etching blanket films at glancing angles.

The films were etched in pure Ar, a low polymerizing fluorocarbon chemistry (0.2 sccm C_4F_8 / 2.6 sccm Ar, or 7% C_4F_8/Ar), or an oxidized fluorocarbon chemistry (0.2

sccm O_2 / 0.2 sccm C_4F_8 / 2.6 sccm Ar, or 7% O_2 / 7% C_4F_8/Ar), all of which are characteristic of chemistries used in the dielectric etch process. In all chemistries the plasma source power was fixed at 400 W with a DC bias of 350 V. The surface etching rates were measured by determining the amount of material removed for a given ion dose. The ion dose was 3×10^{17} ions/cm^2 for ion incidence angle less than 82° and 1.5×10^{17} ions/cm^2 for ion incidence angle at 82°.

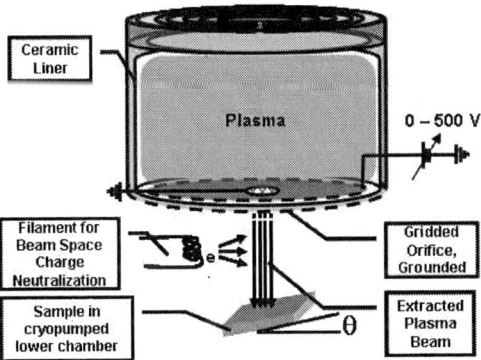

Figure 6. Schematic of a newly designed beam chamber system. The beam source locates at the upper part of the main chamber and the plasma is inductively coupled. This beam system has the flexibility to control the plasma chemistry, ion bombardment energy, and incident angle independently.

A Tencor P-10 profilometer was used to measure the thickness of the carbon films. A razor blade was used to scratch a fine line on top of the etched sample and a diamond stylus is moved laterally across the line to measure the depth of the film. An atomic force microscope (AFM) (ex situ Digital Instruments 3100) in tapping mode with standard silicon tips was used to measure the root mean square (RMS) roughness of the carbon films. The sampling region in AFM scan was 1 μm^2. Captured topographic profiles were subject to a zero-order plane fit before calculating the RMS roughness with standard DI software.

Angular Etching Yield

The angular etching yield is strongly dependent upon substrate composition and etching chemistry [10]. The angular etching yield of the Type 2 and 3 materials in the three plasma chemistries are plotted in Figure 7.

At normal ion incidence, 0° off-normal angle, the etching yield of the Type 3 material increases in the order of Ar, C_4F_8/Ar, and O_2/C_4F_8/Ar, given the same operating conditions. The Type 2 material has a similar etching yield for Ar and C_4F_8/Ar but has a substantially higher value for O_2/C_4F_8/Ar. C_4F_8 dissociates into atomic fluorine and CF_x species. Oxygen accelerates the dissociation by forming volatile species with carbon. This mechanism is related to the well-known theory of Coburn and Winters, who qualitatively plotted the etching versus deposition characteristics of fluorocarbon gases as a function of the fluorine to carbon (F/C) ratio of the chemically active plasma species [11]. They found that a transition from polymer deposition to etching occurs as the F/C ratio increases. The F/C ratio could also be influenced by the addition of other gases. For example, oxygen addition forms CO or CO_2, consuming carbon-containing etching species and thereby increasing the F/C ratio. For this reason, the etching yield of the Type 2 and 3 materials in $O_2/C_4F_8/Ar$ is almost double that in C_4F_8/Ar at normal ion incidence.

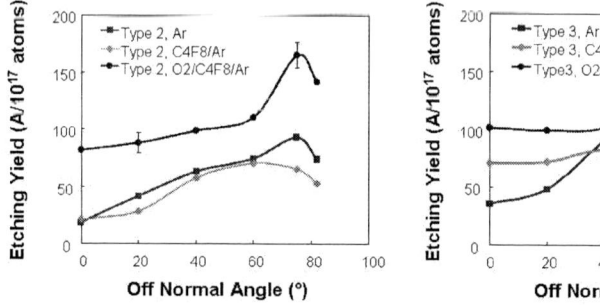

Figure 7. Angular etching yield Type 2 and 3 AHM in three plasma chemistries: (1) Ar, C_4F_8/Ar, and $O_2/C_4F_8/Ar$. In all cases the plasma source power is 400 W, dc bias is 350 V, beam source pressure level is 4mTorr.

It is interesting to compare the angular etching yield of the Type 2 and 3 materials in pure Ar, where sputtering is the only etch mechanism. At normal ion incidence, the Type 2 material etches much more slowly than Type 3. That is mainly attributed to the film composition and the subsequent configuration. The Type 3 material as shown in Table I contains 40% hydrogen, which bonds only to carbon atoms, terminating the carbon bond and thus weakening the matrix of the film. Furthermore, with terminal carbon, carbon is often sputtered in conjunction with hydrogen. These effects facilitate sputtering of the hydrogen-rich carbon film. In contrast, the Type 2 material contains only 27% hydrogen and is thus harder to sputter. The momentum transfer, which occurs during sputtering, is another factor to consider. High-energy ions penetrate deeper along the closed-packed directions [12, 13]. Consequently, a densely packed film sputters more easily. These two materials have similar density, but the sp^3 bonding in the Type 3 material further contributes to a close-packed configuration. Hence, the Type 3 material should sputter faster than the Type 2.

TABLE II. AHM Material Properties.

| | α = Etch Yield $_{MAX}$/ Etch Yield $_{Normal}$ | | |
	Ar	C_4F_8/Ar	$O_2/C_4F_8/Ar$
Type 2	5.0	3.0	2.0
Type 3	3.1	1.3	1.3

In order to further understand the variation of angular etching yield with chemistry, α, (defined as the ratio of the maximum etching yield--about 70° off-normal angle-- to the etching yield at normal ion incidence), calculations were run on all curves shown in Figure 7, and summarized in Table II. As the chemistry transitions from pure Ar to

$O_2/C_4F_8/Ar$, α drops off gradually from 5.0 (3.1) to 2.0 (1.3) for the Type 2 (3) material. The decrease in α indicates the overall etching mechanism shifts from physical sputtering to chemical etching. The amount of ion-enhanced etching compared to physical sputtering increases dramatically as the discharge becomes more chemistry-driven. Ion-enhanced etching may take place more rapidly due to polymer proliferation and fluorination on the surface. And yet it is unclear whether the physical sputtering is suppressed or not. For both C_4F_8/Ar and $O_2/C_4F_8/Ar$ chemistry, α is identical for the Type 3 material. That result suggests that oxygen addition did not appreciably alter the amount of sputtering relative to ion-enhanced etching, although it may accelerate the overall etching by forming CO_x species.

The Type 2 material always has higher α compared to Type 3 in the same chemistry. Keeping all other conditions constant, the more carbon-rich, graphitic Type 2 material has an angular dependence more like physical sputtering, where the Type 3 has a more fundamentally chemical etch process. A similar effect of film composition on the angular dependence has been observed previously by other researchers. Yin, et al., found that low-k dielectrics exhibit more sputtering-like angular dependence than oxide at identical operating conditions [10]. The ratio presented in this paper was about 2.0 for low-k dielectrics and 1.75 for thermal oxide, both being etched in 10% C_4F_8/Ar plasma. The carbon present in the film may moderate the ratio between the ion-enhanced etching over physical sputtering. It could occur in multiple fashions, strengthening one type of reaction and/or weakening the other. For example, carbon in the film network slows down the desorption of ion-enhanced etching products, and/or carbon is more rapidly sputtered away from the substrate than other substances

Surface Roughening

For most dielectric materials, surface roughening varies drastically with ion incidence angle. For thermal oxide and Novellus Coral™ low-k dielectric films, the substrate was observed to remain smooth at normal or grazing ion incidence angles [14]. At intermediate angles (40°-75°), the roughness varied both topographically and quantitatively. At normal ion incidence, the facets on the surface are smoothed out by the higher etching yield at high off-normal angles, according to the angular dependence curve of physical sputtering. At intermediate ion angles, the surface coarsening results from the competition between coarsening initiated by curvature dependence and smoothing due to diffusion, according to Bradley and Harper's model [15]. At grazing incidence angles (≥80°), ion scattering dominates leading to channeling effects, and etching is not as significant. As a result, the surface is entirely deposited with polymers, leading to a smooth plane.

The results in this study were consistent with this model. AFM measurements of the RMS roughness of the Type 1, 2, and 3 materials prior to etching were 0.37 nm, 0.37 nm, and 0.25 nm, respectively. At near normal ion (0°-60°) and grazing incidence angles (≥80°), the surface remains relatively smooth, with the RMS level comparable to the initial roughness present on the film. In this work, similar trends were also observed, with roughening taking place mostly at intermediate angles; therefore, the following discussion will focus on a 75° off-normal angle in the three plasma chemistries. Figure 8 contains AFM images of the Type 2 and Type 3 materials after etching in pure Ar, C_4F_8/Ar, and $O_2/C_4F_8/Ar$ at 75°. The roughness at this high angle is also indicative of the roughness that would be present on the sidewall of a patterned structure. Thus, these measurements can quantify the potential impact on line edge roughness for these two materials, which impacts the fidelity of pattern transfer in hard mask based patterning schemes.

The roughening pattern on the Type 1 material in pure Ar is particularly interesting. As the ion flux impinges on the surface of the sample, striations build up on the surface perpendicular to the ion beam. The edge aligned with the ion beam has narrow and sharp peaks while the edge perpendicular to ion beam has broad and smooth features. These patterns proved the existence of transverse striations on the Type 1 material in pure Ar at 75°. However, this result seems to be distinctive from what has been observed on other materials including silicon, silicon dioxide and low-k. More specifically, it has been accepted that in pure Ar transverse striations appear only at intermediate ion angles around 40° for the dielectric films mentioned above [15-17]. At high off-normal angles such as 75°, either striations turn to be parallel to the ion beam (various low-k dielectrics), or the surface is isotropically roughened (silicon dioxide). As for the Type 3 material, the surface is relatively flat due to its mechanical hardness. The Type 2 material carbon has isotropic, hemispherical blocks.

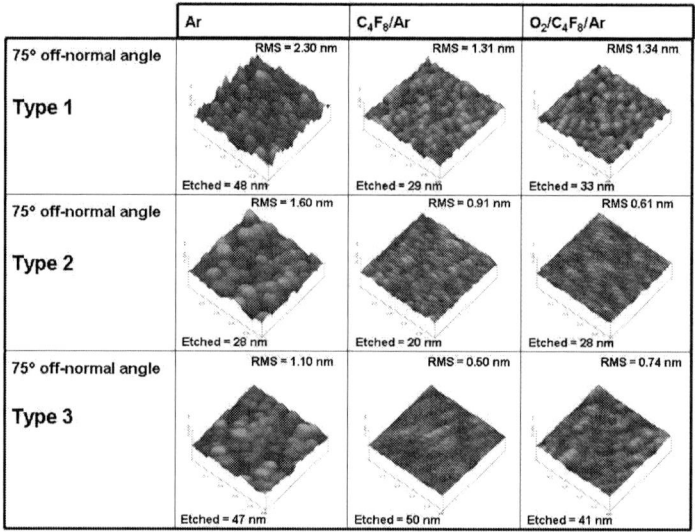

Figure 8. Surface AFM images Type 1, 2 and 3 materials after etching at 75° off-normal angle three plasma chemistries: Ar, C_4F_8/Ar, and O_2/C_4F_8/Ar. Ions reach the surface from the upright direction. The plasma source power is 400 W, dc bias is 350 V, and beam source pressure level is 4mTorr. Ion dose is 3×10^{17} ions/cm^2. The vertical scale is 15 nm and both of the images represent 1 μm^2 of the sample surface.

Quantitatively, the roughness increases from Type 3 to Type 2 to Type 1 materials. The incoming roughness is one factor to consider. The Type 3 material showed a slightly smaller initial roughness compared to the other two materials, and in most but not all cases, the roughness of this film is lower. Yet, Type 1 and 2 had the same incoming roughness but varied substantially in roughness post etch. Hardness in Table I does not perfectly correlate with this trend. Type 3 has a much higher hardness than either Type 1 or Type 2 and is indeed the smoothest; however, type 1 has a greater hardness than Type 2 with Type 2 smoother than Type 1. Density is the difference. Taking incoming roughness, hardness, and density into account, the above trend in roughness is clearer. However, there remain some small differences which are likely due to the difference in composition and structure of these materials.

The three films exhibit distinctively different post-etch topography in the C_4F_8/Ar chemistry. The Type 1 material is roughened in an isotropic manner, Type 3 develops mild and broad grooves parallel to the ion beam on the surface, and fine striations perpendicular to the ion beam form on Type 2. Similar to the analysis in pure Ar, the morphological transitions are dependent upon the amount of etching and the etching rate of a specific substrate. Typically transverse corrugations appear first and evolve toward grooves parallel to ion beam at intermediate ion angle. Therefore, the lower etch rate Type 2 material has a transverse striation, whereas the higher etch rate Type 3 film has a parallel corrugation. As for the roughness level, the Type 1 and 2 materials are roughened substantially more than the Type 3. For the same ion dose $3x10^{17}$ ions/cm^2, the amounts etched for three films are closer compared to those etched in pure Ar. The fluorocarbon discharge provides reactive neutral species, especially carbonaceous species, increasing the ion-enhanced reaction rates for all films regardless of their initial substrate compositions. In this case, sputtering only accounts for a small fraction of overall etching. Comparing these roughness values to those in pure Ar, it is evident that the surface becomes much smoother with the addition of C_4F_8.

The O_2/C_4F_8/Ar results are similar to those with C_4F_8/Ar. The Type 1 material exhibits isotropic bumps regardless of ion beam direction, Type 3 pose mild parallel striations relative to ion beam, and Type 2 has finer structures transverse to ion beam. In summary, adding oxygen to the processing gases did not significantly alter the roughness pattern for all films, whether they are parallel, transverse or isotropic. Again the Type 1 material is the most roughened of the three films given the same amount of ion flux (and similar amount of thickness removed), suggesting film density plays a major role in the coarsening of the film.

Comparing the roughness across the plasma chemistries, the fluorocarbon discharges appear to smooth the surface more than pure Ar, although the material etched away by fluorocarbon gas can be more than that in pure Ar. Oxygen addition does not significantly roughen or alter the roughness pattern for the amorphous carbon films. This supported the polymerizing effect induced by fluorocarbon gas, helping to prevent roughening from growing out of control.

Examples of Feature Patterning

The pattern etch performance of these films was evaluated using one of the AHM materials discussed, along with an ARL and PR stack as shown in Figure 1. Cross-sectional SEM images of dielectric substrate/AHM lines after main etch prior to AHM strip for Type 1 and 3 materials are shown in Figure 9. Although post AHM critical dimensions (CD) were matched, after main etch the Type 1 film showed a significant CD loss compared to Type 3. This loss is attributed to the higher density of Type 3 films which reduces the lateral erosion rate. SEM images after main etch for via patterning for Type 1, 2, and 3 materials are shown in Figure 10.

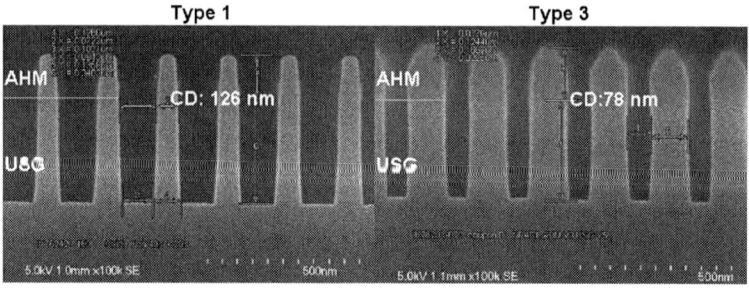

Figure 9. Cross-sectional scanning electron micrograph of line patterning with Type 1 and 3 AHM materials.

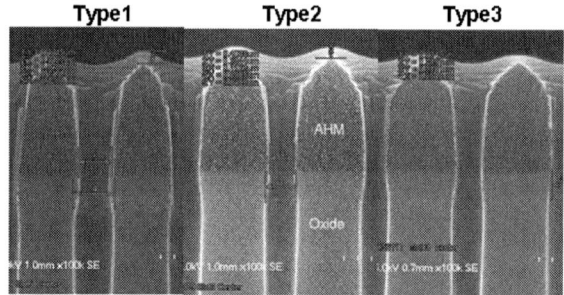

Figure 10. Cross-sectional scanning electron micrograph of via patterning with Type 1, 2, and 3 AHM materials.

Type 2 and 3 materials offer better CD control for via patterning. However, Type 2 films are not transparent which limits the thickness (< 200 nm) that can be used. In addition, 200 nm is not adequate for good CD control making highly selective and transparent Type 3 films the preferred choice for thick, 300 – 800 nm hard mask applications.

Line Integrity

The CD of the patterned AHM line is not the only physical property of importance. The morphology of the line can have a substantial impact on the resulting pattern transferred to the desired dielectric substrate. Two of the most relevant concerns are line edge roughness and line bending.

Line Edge Roughness

In the pattern transfer process, roughness induced on the sidewall of the patterned feature during the lithography step will also transfer to the final pattern to create an artificially roughened final feature. This roughness is termed line edge roughness (LER). Simply moving from a PR/ARL stack to a PR/ARL/AHM stack allows a substantial decrease in LER, as illustrated in Figure 11. This decrease in LER could be understood through increased etch selectivity of the AHM to underlying layers. The increased selectivity of hard mask material is thought to decrease line edge roughness through two mechanisms: (a) during the hard-mask open etch, a tougher mask will enable smoothening of the high frequency modulations (LER) of layers above AHM through lateral etch process and (b) during the final layer etch (e.g. Oxide or Poly), the high selectivity of AHM film reduces thickness marginality related to the transfer of striations. However, improvement beyond the inclusion of the AHM material is also possible. As observed in one of the previous sections, sidewall roughness of the AHM material itself can by modulated by altering the hardness, density, and composition of the AHM. The Type 3 material was shown to be substantially smoother compared to the Type 1 and 2 materials under a variety of etch chemistries.

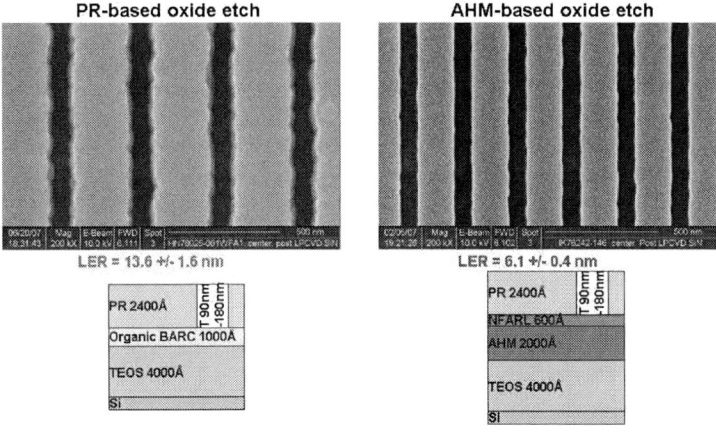

Figure 11. Top-down scanning electron micrograph showing line edge roughness in patterned TEOS-based SiO$_2$ lines created with PR/BARC and PR/ARL/AHM film stacks.

Line Bending

As the CD of linear features shrink, the aspect ratio (AR) of AHM lines increases. As this AR increases, the AHM lines start to bend as shown in Figure 12. Maintaining the integrity of high aspect ratio lines is critical for successful pattern transfer during gate formation process. Clearly, this line bending phenomenon must be eliminated.

The mechanics behind line bending is simple mechanical buckling, and can be observed even in blanket films. [18-19] The AHM materials tend to have a compressive stress. Beam mechanics teaches that all beams have a critical compressive stress, σ_{cr}, above which they will buckle, given by:

$$\sigma_{cr} = \pi^2/12 * \bar{E}/AR^2, \qquad [1]$$

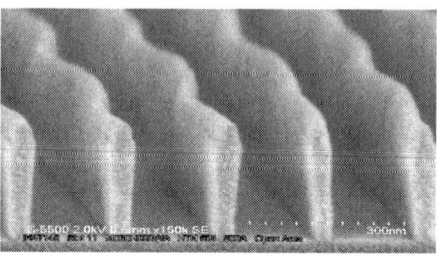

Figure 12. Cross-sectional scanning electron micrograph of line bending in AHM lines.

where $\bar{E}$ is plane strain modulus ($E/(1-v^2)$), E is Young's modulus, v is Poisson's ratio, and AR is the aspect ratio of the beam (height divided by width). From this equation it is clear that as the AR of the line increases, the critical stress will decrease. Buckling can be mitigated by increasing the elastic modulus or by decreasing the intrinsic stress of the AHM material. The effect of Young's modulus on line bending is readily demonstrated by Figure 13, in which lines with the same CD and pitch were etched into two AHM materials with the same nominal intrinsic compressive stress, but one of the AHM materials has a lower value of Young's modulus and one with a much higher value. The AHM with the higher elastic modulus exhibits much less line bending than the basic beam bending model would predict.

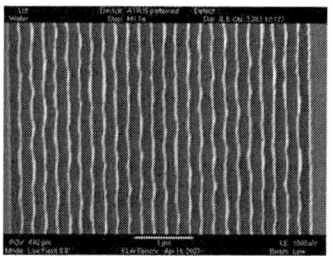 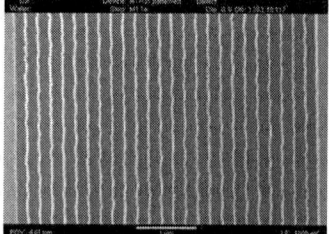

Film-1 : E = 19 GPa   **Film-2 : E = 68 GPa**

Figure 13. Top down scanning electron micrograph of 50 nm AHM lines created using films that differ in elastic modulus.

However, this simple picture is not completely accurate as it ignores the impact of the ARL. In the top of Figure 14, top-down SEM images are shown for a single as deposited PR/ARL/AHM stack as a function of line CD and AR. Interestingly, the features show an oscillation in the line bending with AR. The reason for this effect is clear when one observes the bottom cross-sectional SEM images in the bottom of Figure 14. As the CD decreases, the amount of remaining ARL on top of the AHM line is changing. The ARL film also has an intrinsic stress which it will impart to the line as well. Returning to our simple model, the beam is now a bilayer. In this case, the geometric and mechanical properties of both the ARL and AHM films are relevant and must be controlled.

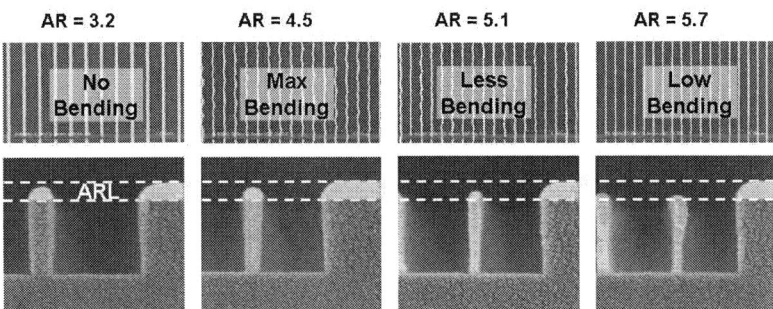

Figure 14. Top-down and cross-sectional scanning electron micrographs showing line bending as a function of aspect ratio on an ARL/AHM stack.

This same behavior is observed when highly compressive TiN films are used as hard masks in the patterning of ultra-low-k (ULK) dielectric materials in the back-end-of-line. [20] The line bending in that case can be much more severe due to the very low elastic modulus of the ULK dielectric compared to the TiN. Interestingly, one solution posed to this problem is to replace the TiN with an AHM material [21].

Conclusions

PECVD amorphous carbon thin film materials are widely used for advanced hard mask applications due to their high etch selectivity, transparency, and ashability. In this work, we demonstrated the ability to tune and control the bonding characteristics of these films by varying their deposition conditions like temperature and plasma conditions. Through depositing films at low temperature, but in a higher ion energy environment, we were able to create films with more sp^3 matrix, increased network connectivity, and yielding traditionally elusive combinations of advantageous film properties such as high selectivity with high lithographic alignment transparency. The sp^2 to sp^3 bonding ratio of the matrix, which were varied widely through deposition conditions, were characterized by Raman spectroscopy, and the short range order of these films was shown to play a significant role in impacting the fundamental etching nature of these films. Films with a highly graphitic sp^2 configuration were seen to etch through an angular dependence like physical sputtering, while the more disordered materials with a greater sp^3 character were seen to be etching through chemical aspects. The bonding characteristics of hard masks were also shown to play a significant impact on the pattern etch performance of key features. For example, by controlling the network connectivity of films, as reflected by FTIR spectroscopy, we were able to create different classes of materials that have control over the CD dimensions and LER. In the same way, the ability to tailor the mechanical properties was found to be advantageous as it allowed the elimination of the line bending phenomenon. Measurements of hardness, density, composition, configuration, etching, and others taken together enabled a much deeper understanding of these materials and hence how they might be used. Fundamentally, solutions in materials science lie at the intersection between material characterization and a thorough understanding of the application.

Acknowledgments

The authors would like to thank Semiconductor Research Corporation for research fund support. We want to thank MIT Center for Materials Science and Engineering (CMSE) for AFM and profilometry usage

References

1. K.A. Pears, M. Stavrev, A. Scire, R. Koepe, M. Markert, U. Egger, and L. Donohue, *Microelectron. Eng.*, **81**, 156 (2005).
2. K. Ronse, *Microelectron. Eng.*, **67-68**, 300 (2003).
3. H.B. Cao, W.D. Domke, and P.F. Nealey, *J. Vac. Sci. Technol. B*, **18**, 3303 (2000).
4. W. Von Bolton, *Electrochem.*, **17**, 971 (1911).
5. H. Schmellenmeier, *Phys. Chem.*, **35**, 17 (1955).
6. C.V. Deshpandey and R. F. Bunshah, *J. Vac. Sci. Technol. A*, **7**, 2294 (1989).
7. J. Robertson, *Mat. Sci. Eng. R.*, **37**, 129 (2002).
8. A.C. Ferrari and J. Robertson, *Phys. Rev. B*, **61**, 14095 (2000).
9. F. Tuinstra and J.L Koenig, *J. Chem. Phys.*, **53**, 1126 (1970).
10. Y. Yin and H.H. Sawin, *J. Vac. Sci. Technol. A*, **25**, 802 (2007).
11. J.W. Coburn and H.F. Winters, *J. Vac. Sci. Technol.*, **16**, 391 (1979).
12. A.E. Andrew, E.H. Hasseltine, N.T. Olson, and H.P. Smith, *J. Appl. Phys.*, **37**, 3344 (1966).
13. H.F. Winters, J.W. Coburn, and T.J. Chuang, *J. Vac. Sci. Technol. B*, **1**, 469 (1983).
14. S. Rasgon, PhD thesis, MIT (2004).
15. R.M. Bradley and J.M.E. Harper, *J. Vac. Sci. Technol. A*, **6**, 2390 (1988).
16. E.O. Yewande, A.K. Hartmann, and R. Kree, *Phys. Rev. B*, **71**, 195405 (2005).
17. S. Rusponi, G. Costantini, C. Boragno, and U. Valbusa, *Phys. Rev. Lett.*, **81**, 2735 (1998).
18. L.B Freund and S. Suresh, *Thin Films Materials: Stress, Defect Formation and Surface Evolution*, p. 312, Cambridge University Press, Cambridge (2003).
19. X.D. Zhu, K. Narumi and H. Naramoto, *J. Phys.: Condens. Matter*, **19**, 236227 (2007).
20. M. Darnon, T. Chevolleau, O. Joubert, S. Maitrejean, J.C. Barbe, and J. Torres, *Appl. Phys. Lett.*, **91**, 194103 (2007).
21. T. Chevolleau, N. Posseme, T. David, R. Bouyssou, J. Ducote, F. Bailly, M. Darnon, M. El Kodadi, M. Besacier, C. Licitra, M. Guillermet, A. Ostrovsky, C. Verove, and O' Joubert, *Proc. IITC 2010*, 1 (2010).

Ultra Low Dielectric Constant Materials for 22 nm Technology Node and Beyond

Mikhail R. Baklanov, Evgeny A. Smirnov, Larry Zhao*

IMEC, Leuven, Belgium
INTEL assignee at IMEC, Leuven, Belgium

Metrology and approaches necessary for selection of ultra low-k dielectric materials for future generation of IC devices are discussed. It is shown that porogen residue formed during the UV curing of porogen based PECVD low-k materials increases the leakage current density and decreases the breakdown field. The amount of porogen residue increases with porosity because of larger amount of co-deposited porogen. Electrical properties of these films are significantly worse in comparison with low-k films prepared by using of self-assembling approaches without porogen. Therefore, utilization of porogen based technologies for preparation of ultra low-k films is limited. New metrologies allowing evaluation of intrinsic properties of low-k materials are discussed. It is shown that UV spectroscopy and a unique planar capacitor test vehicle are extremely important for evaluation of intrinsic properties of ultra low-k films.

Introduction

For almost 50 years since the 1960's, the density of transistors in Integrated Circuit (IC) chips has been doubling every 1.5 years. This progression of circuit fabrication is known as Moore's law, after Gordon Moore, one of the early IC pioneers and founders of Intel Corporation. Throughout the 1980s, the semiconductor industry primarily focused on improvements related to the speed of individual transistors and enhanced performance through scaling - by squeezing more transistors into a single device. But by the early 1990s, the distances between device components were becoming incredibly small, and the relative effect of the interconnect delay became a greater portion of the overall signal propagation delay and could no longer be neglected. As a result, both line widths and the spacing between the conduction lines also shrank. This scaling meant that the total length of the interconnect wires was increasing by power law and that each IC chip soon contained several miles of conduction lines. The total resistance (R) of the interconnect structure became a significant factor affecting chip performance. At the same time, the capacitance (C) between the wires was increasing proportionally to the decreasing spacing between the wires. Both these factors significantly increased the RC delay of IC circuits [1]. Even as the speed of transistors increased dramatically, the interconnect delay became a limiting factor. The microelectronics community realized the need to improve the interconnect delay by making changes to the materials used for wires and the materials used to insulate the wires.

The 1994 National Technology Roadmap for Semiconductors (NTRS) – the US industry's technology strategy document at that time - stated that materials with a lower dielectric constant (i.e. "low-k materials") would be needed for wires insulation as the feature sizes of IC devices became smaller. The NTRS projected that within 10 years the industry should be able to achieve a standard dielectric constant of less than 1.5 in their production interconnect material. However, the real situation has been much more complicated. After several revisions, the latest edition of the International Technology Roadmap for Semiconductors (ITRS) indicates that low-k material with only k–2.5 are considered to be integrated by 2012 (Figure 1). The reasons for such a delay are the huge challenges of integration of porous materials. Porous low-k materials are generally soft, mechanically weak, and do not adhere well to silicon or metal wires. Further, porous low-k materials do not withstand conventional processing (i.e., they degrade during the plasma and chemical processing, crack or delaminate) [2].

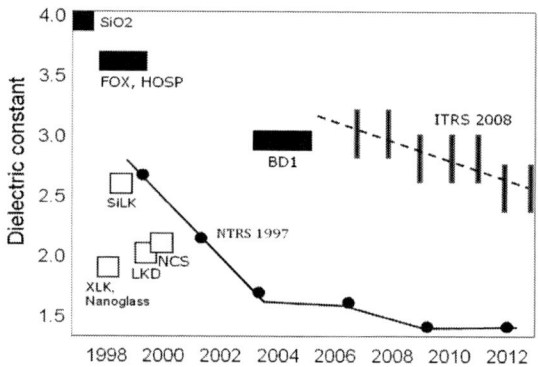

Figure 1. Predictions by NTRS'1997 and the latest corrected version (ITRS'2008). The delay and corrections are related to the difficulty of integrating porous dielectric materials. Only dielectric materials with low porosity FOX (fluorinated oxide), HOSP (dense MSQ) and BD1 (organosilicate glass without artificial porosity) were successfully integrated into IC devices. Meanwhile, chemical companies developed ultra low-k materials (SiLK, XLK, Nanoglass, LKD, NCS etc.) [1] more than 10 years ago, but their integration is still extremely challenging.

However, by 1997, chip manufacturers started integrating insulating materials with dielectric constants smaller than the value typical for traditional SiO_2 (k ≈ 4.0). Fluorosilicate glasses (FSG, FOX), which were created by adding fluorine to silicon dioxide, had a k value of around 3.6 and required very little change in the production process for semiconductor manufacturers. Thus, FSG was quickly and widely adopted by the industry. However, very little progress had been made towards identifying suitable materials with a dielectric constant below 2.7, which would likely require less dense (i.e., more porous) materials.

Several different types of low-k materials were considered as candidates for further reduction of dielectric constant. Low-k dielectrics based on organic polymers were able to provide the lowest k–value without requiring the introduction of porosity. However, aliphatic C–C, C–H and C–N bonds generally become unstable at temperatures greater than 300–400°C and, in some cases, at even lower temperatures. Only materials composed of non-aliphatic C–C, C–O, C–N and C–S bonds, aromatic structures and cross-linked or ladder structures can withstand the temperatures necessary for interconnect technology (450-500°C). Most of the organic low-k films with sufficient thermal stability have dielectric constants close to 2.6–2.8. However, significant efforts to integrate these materials into IC circuits were not sufficiently successful. In addition to poor mechanical and thermal properties, the most important problems were related to relatively low coefficient of thermal expansion (CTE) in comparison with other components of integrated circuits.

Further reduction of dielectric constant required decreasing the density by the introduction of pores. The relative dielectric constant of porous materials, k_r, depends on the porosity and dielectric constant of the film skeleton (k_s) [1]:

$$\frac{k_r - 1}{k_r + 2} = (1 - P) \cdot \frac{(k_s - 1)}{(k_s + 2)}$$

Materials with relatively small k_s values provide smaller k_r values at lower porosity and this is the reason why different types of matrix materials were considered for preparation of porous low-k films. The most important ones were organic polymers and hybrid materials: inorganic silica based materials (Silica xerogel, aerogels) with incorporated organic hydrophobic agents and hydrid silsesquioxanes (SSQ). Porous organic polymers were able to provide k = 2.2 but their integration was challenging. Low CTE still was one of the most critical factors of porous organic polymers. The advantage of silica and SSQ based materials is that their chemical properties are similar to traditional SiO_2 and that makes possible the use of traditional technologies and chemistries during the integration.

The principal difference between silica–based and SSQ materials is the structure of their elementary units. Pure silica has a tetrahedral elementary unit. To reduce the k value and make them hydrophobic, part of the oxygen atoms are replaced with F or alkyl groups CH_x. The addition of CH_x not only introduces less polar bonds, but also creates free volume. Such organosilicate glasses (OSG) are normally deposited by CVD and they are constitutively porous. The carbon concentration in the most SiOCH materials varies between 10 and 30 %. The carbon concentration must be sufficiently large to provide their hydrophobic properties but small enough to deliver good mechanical properties.

In SSQ materials, Si and O atoms are arranged in a form of cage or ladder. The cage structure creates free volume, decreasing the material's density and, therefore, its k value. The cages in polymerized SSQ are connected to each other through oxygen or -CH_2- groups, while other cage corners are terminated by hydrogen (HSQ), methyl (MSQ) or other aliphatic groups. MSQ matrix materials have a lower dielectric constant as compared to HSQ because of the larger size of the CH_3 group and lower polarizability of the Si-CH_3 . SSQ cages are metastable and tend to break down to silica tetrahedral, especially during the curing at elevated temperature. Because of low temperature stability of the cage, SSQ based materials are prepared only by SOG technology.

The skeleton dielectric constant of both PECVD SiCOH and SOG SSQ materials are defined by polarisability of Si-O bonds and free volume that depends on CH_x

concentration. However, deposition of ultra low-k materials needs the introduction of artificial porosity in both SOG and PECVD technologies.

Deposition of porous materials by spin-on glass (SOG) technology.

In the spin-on deposition, the film coating is performed by dispensing a liquid precursor at the center of the substrate, which is placed on a spinner. Rotation of the substrate creates centrifugal forces that ensure a uniform distribution of material on the surface. The thickness of the coating is a result of balance between centrifugal forces and viscous forces, determined by the viscosity of the solution. Normally, the spinning step is followed by heating or "soft bake" at temperatures typically below 250°C, for removal of the solvents. The latter step can also initiate cross-linking of the film. Finally, a sintering at temperatures varying from 350°C to 600°C ('cure') is required to obtain a stable film. This "cure" step induces the final cross-linking of the polymer chains and results in a mechanically stable film structure.

Numerous methods of introducing subtractive porosity into spin-on deposited materials exist, but they can be divided into two main categories. The first category groups all materials where the porosity is introduced exclusively through Sol-Gel processes, while the second group includes the materials where the porosity is formed through the use of sacrificial particles (porogens) that are desorbed during film cure.

Subtractive porosity by sol-gel based techniques. There exist two main approaches based on Sol-Gel techniques to the formation of subtractive porosity: the first takes advantage of aging processes and the second relies on a hierarchical organization of the primary particles in the sol (self-assembly).

The formation of a more or less rigid skeleton structure before extraction of the liquid from a wet gel is a key point in the formation of high porosity materials. Even if the gel-point is reached after material spinning, a long time is still required before the hydrolysis and condensation reactions are complete. For this purpose, an additional step (aging) before drying the wet gel is introduced. The aim of this step is to accelerate the sol-gel reactions, typically by relying on the pH and the water content in the ambient. Once the network structure is strengthened, extraction of the solvent can take place without collapse of the network backbone. The level of residual porosity is generally tuned through the ratio of solvent to solid content in the sol.

In sol-gel science numerous studies have been performed in the synthesis of *self-assembled* materials. Hierarchical ordering of the aggregates by preferential solvent evaporation during spin coating is reported for a solution of surfactants, swelling agent and soluble silica. By this method, ordered materials with dielectric constant as low as 1.3 have been synthesized. In this case, the final film porosity and pore structure is related to the way in which the primary particles are assembled and ordered.

Subtractive porosity by macromolecular porogens. This technique is based on the addition in the dielectric precursor of molecular or supramolecular particles ('porogens') with tailored thermal stability. The stability of these particles is such that they are not affected by the drying step, and they are removed by pyrolysis during final film sintering or cure (typically in the range from 300 to 400°C). Their volume distribution in the film at the moment of desorption represents the template for the residual pores in the layer. In the ideal case, the film's final porous fraction is directly related to the amount of porogen as a function of the total solid part in the precursor solution, and the size of the sacrificial particles is directly related to the final pore size.

There are two ways in which the sacrificial porogens are brought into the precursor solution. One method is dispersion of porogens in the solution. The second is chemically linking sacrificial particles (grafted) to the network polymers. This second method grants an inherent control of the volume distribution of porogens in the dielectric film.

Plasma Enhanced Chemical vapor deposition (PECVD).

The semiconductor industry has long relied on insulating films of SiO_2 deposited from gas phase (plasma) by oxidation of silane (SiH_4) and derivatives. Therefore, most of the attempts at producing low-k materials by different versions of CVD have been with the doped versions of SiO_2. The main dopants in the beginning were fluorine and carbon. Their introduction is done by replacing standard silane by fluoro- and alkylsilanes like $Si_2H_2F_2$ and $(CH_3)_xSiH_y$ with (x+y)=4. Doping a film with alkyl groups terminates some of the silicon bonds within the oxide lattice and lowers the electronic polarizability of the film. The relatively large molecular volume of the alkyl groups decreases the film density and provides the k- value 2.8-3.0.

Various techniques have been employed to produce PECVD SiOCH films with subtractive porosity and k value smaller than 2.8. One method utilizes a multiphase deposition. The SiOCH precursor (alkylsilane derivatives and/or organosilicates) is mixed with a thermally unstable hydrocarbon phase (porogen) during deposition. This unstable phase is thermally decomposed and removed from the film during subsequent anneal leaving behind pores in the organosilicate matrix. The resulting porosity depends on the matrix/porogen precursors ratio. Efficiency of porogen removal can be enhanced by UV light and/or electron beam.

This approach has the benefit of allowing manufacturers to use the same CVD (PECVD) equipment already operating in fab's, modified only by additional gas delivery lines. Thus, adoption of such a process requires a relatively low integration cost and gives the industry a level of confidence in the new process. This is the reason why during the last 5 years PECVD materials have been most popular and considered as the most possible candidates for integration. However, although the lowest k of a known PECVD films has been reported as 2.2 [3,4]. only low-k materials with k > 2.4-2.5 have shown acceptably good electrical characteristics and reliability after integration.

Therefore, it is very important to characterize critical properties and make selection of ultra low-k dielectrics for future generations of IC devices requesting the k-value smaller than 2.3. Generally, it is difficult to identify suitable materials if the final conclusion is made after complete integration because the materials' properties changed significantly during the technological processing. The low-k films are strongly modified and degraded during the various technological treatments such as plasma exposure (etch, strip), CMP, barrier deposition etc. As a result of such degradation the final dielectric constant can become significantly larger than the target value and reliability performance could become worse than what is required. Therefore, the final quality is not only related to the intrinsic properties of low-k materials but also to process issues which might be fixed by further process optimization. For this reason special attention must be spent to evaluation of intrinsic materials properties as a first step of selection.

The main purpose of this work is to analyze the most critical issues related to the low-k materials selection for further integration. The most important attention will be paid to their intrinsic properties because the challenges related to their degradation during the processing is a secondary factor. Intrinsically good materials can provide good final results but intrinsically bad materials cannot be improved during the integration.

To obtain reliable results related to intrinsic properties and to analyze fundamental limitations of low-k materials, we evaluated different types of PECVD and SOG materials with k values below than 2.5. More attention was paid to materials with the k values approaching to 2.0. In addition to traditional instrumentation giving information related to chemical composition, structure, mechanical and chemical properties, relatively new characterization techniques have also been used. Ultraviolet ellipsometry has been used as a UV spectrometer to characterize optical characteristics and formation of porogen residue, and electron spin resonance has been used for characterization of specific defects that can be electrically active [5,6]. Finally, new electrical test structure recently developed at IMEC allowed characterization of intrinsic electrical characteristics and intrinsic reliability of low-k materials [7,8]. This test structure also allows separate study of certain integration steps (for instance, to study barrier and low-k interactions) and extremely helpful in understanding some of the key challenges in low-k scaling.

Results and Discussion

Materials of research.

Table 1 shows the most important characteristics of low-k materials discussed in this paper. Refractive index, porosity and pore size were determined by spectroscopic ellipsometry and ellipsometric porosimetry [9], the k-value was measured by using metal dots [10], and mechanical properties were evaluated by Nanoidentation [11].

CVD1-CVD3 films were deposited using PECVD by mixing an organosilicate matrix precursor with organic porogen precursor. The CVD1 film was deposited in a different chamber than CVD2 and CVD3. The CVD4 film was deposited from an organosilicate precursor without using a sacrificial porogen. Different porosities and k-values were achieved by changing the porogen loading. After deposition, the samples were UV-cured at temperatures of 430-450°C in a nitrogen ambient, which removes porogen to form a porous structure and provide cross-linkage of the film skeleton. Different light sources were used for the curing: lamps emitting nearly monochromatic light with wavelength, $\lambda \approx 172$ nm to form the films CVD2 and CVD3 (lamp A), and broadband source (lamp B) with $\lambda > 200$ nm to form the film CVD1. ALKB is also PECVD material prepared by a technology originally patented by IMEC. It was deposited in conditions similar to CVD3 but before UV curing, the film was exposed in downstream He/H$_2$ plasma. All porogen was removed during this stage and the films were cured afterwards by a broadband UV (B) light at 430°C. The final curing can also be done by lamp A to get larger Young Modulus (ALKA) but in this paper we are only analyzing ALKB. More detailed description of this technology and films can be found elsewhere [12].

Only thermal curing (T) was used for organic polymer SumiV, SOG3 and CVD4. The films SOG1 and SOG3 were deposited using self-assembling approach without sacrificial porogen, while SOG2 was deposited with porogen and cured by monochromatic 172 nm light. Therefore, comparison of these materials allows analyzing effects of UV wavelength during the curing and effect of scaling on materials and electrical characteristics.

Table 1. The most important properties of characterized low-k materials.

Material's Label	Curing	Porosity (%)	Pore radius (R)	RI	K value (100 kHz)	E (GPa)	H (GPa)
CVD1	B	24	0.8	1.35	2.5	6.5	-
CVD2	A	24	0.8	1.36	2.5	7.5	-
CVD3	A	28	0.75	1.37	2.3	4.48	0.39
CVD4	T	(±5)	n/a	1.45	3.0	12.5	2.0
ALKB	B	46	1.6	1.23	2.0	3.48	0.36
SumiM	T	34	0.5	1.5	2.3	6.09	0.48
SOG1	B	40	1.8	1.24	2.0	4.77	0.54
SOG2	A	50	1.8	1.23	2.0	4.47	0.38
SOG3	T	35	1.5	1.28	2.3	6.38	0.36

The principal difference between SumiM and previous generations of organic low-k materials (SiLK and Flare) is the relatively low CTE (30 ppm/°C) that makes it more suitable for integration. Another important feature of SumiM is it's stability in different cleaning solutions and plasmas [13]. In addition, this material has good adhesion to barrier materials compared with other organic materials. The adhesion energies of SumiM with conductive and dielectric barriers were comparative with hybrid OSG materials [14]. Integration of SumiM has been successfully demonstrated with integrated k values close to the pristine value [13]. Therefore, this material can be quite promising for IC devices and technology generations requiring k values of 2.2 – 2.3. However, further scaling of dielectric characteristics of this film might be challenging.

The films SOG1 and SOG3 were thermally cured at the temperature range 400-450°. It is necessary to mention that the mechanical properties of these films can be improved by additional UV curing. So after broadband curing (B), the Young Modulus of SOG1 increased up to 6 GPa without any degradation of dielectric constant. UV light from lamp A increased the Young Modulus up to 10 GPa but the degradation of the k-value was already quite significant.

FTIR spectra of all hybrid materials are similar and quite typical. The major part of these materials is the silica matrix (1000-1200 cm^{-1}) with some oxygen atoms in the matrix being replaced by CH_3 groups (1260-1290 cm^{-1}) (Figure 2). The most important difference was related to UV curing. Figure 2 shows SOG3 film that after thermal curing was additionally exposed in UV light. The additional curing was able to further reduce the amount of remaining hydrocarbons (2900-3000 cm^{-1}). There was almost no change in film composition after exposure by lamp B while lamp A reduces the concentration of residual hydrocarbon (2900-3000 cm^{-1}) and Si-CH$_3$ groups (1260-1290 cm^{-1}) and forms Si-H bonds ($\approx$ 2250 cm^{-1}). For this reason, pristine CVD2, CVD3 and SOG2 films cured by 172 nm light also contained certain amount of Si-H bonds. The mechanism of reduction of Si-CH$_3$ group concentrations and formation of Si-H groups during UV curing was discussed by Prager et al. [15]. Quantum-chemical calculations on the model substances octamethyl-and tetramethylcyclotetrasiloxane as well as on hexamethyl-and tetramethyldisiloxane resulted in threshold wavelengths for the excitation of the molecule into the first excited singlet state of 190 and 198 nm as well as of 189 and 192 nm, respectively. After excitation and intersystem crossing in an excited triplet state the scission of the Si–CH$_3$ bond may occur gaining an energy benefit of around 50 kcal mol^{-1}. These findings reveal the presumption that only photons with $\lambda > 200$ nm can generate Si centered radicals which subsequently attract protons from neighboring methyl groups. It is believed that formation of Si-H groups has a negative impact in chemical resistance of low-k films. Therefore, application of 172 nm light for UV curing is becoming limited despite of significant benefit in improvement of mechanical properties.

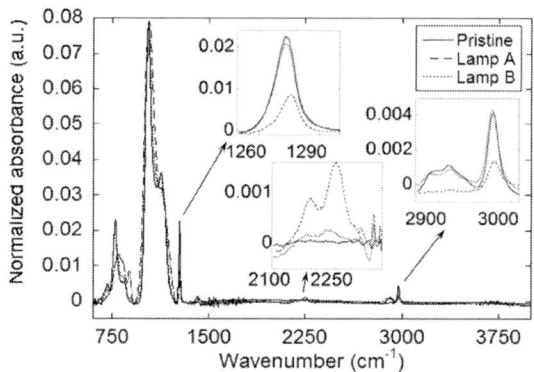

Figure 2. FTIR spectra of thermally cured SOG3 film after additional curing by lamps A and B.

Traditional analytical methods used for the materials selection include evaluation of chemical composition (FTIR, XPS etc.), porosity and pore size (EP, PALS, SAXS), mechanical properties (Nanoidentor, acoustic waves) and dielectric properties (Hg probe, MIS/MIM capacitors) [1]. Although these studies provide information extremely important for further integration, often one can only realize that some important information was missing during the evaluation of intrinsic properties.

FTIR spectrometry is one of the most important tools for analysis of chemical composition. However, it is not sensitive to non polar bonds like sp^2 carbon that can be formed as a result of degradation of porogen molecules during the UV curing. This is the reason why so far researches didn't pay much attention to formation of porogen residues. Meanwhile detection of such groups in pristine low-k materials is extremely important because sp^2 carbon based residues (conjugated carbon polymers and amorphous carbon) are conductive and therefore they can be the reasons of high leakage current and low breakdown voltage in the integrated materials. Generally such residues can be detected by Raman spectroscopy [16] but application of these measurements for thin low-k films is quite challenging. Eslava and Marsik developed a method for detection of porogen residue from UV absorption spectra. The measurements can be done by ultraviolet ellipsometry which is well suited for thin films [5,17]. Marsik et al. carried out detailed study of porogen decomposition in CVD2 film [17]. It was found that the absorption bands of as deposited porogen in the ultraviolet range has maximum at 6.5 eV. UV curing decreases this band but this peak is not completely disappearing even after long curing time. Moreover, new maximum at 4.5 eV (as a shoulder) forms during the curing. The peak at 4.5 eV suggests formation of porogen residues in the low-k material. The incomplete removal is also observed for deposited porogen-only layers treated with the 172 nm UV-cure process. It was concluded that these residues were mainly sp^2 bonded (amorphous or aromatic) carbon because of the appearing absorption of C=C bond vibration in the infrared and the π–π* electronic transitions between sp^2 carbon orbitals in the ultraviolet. The intensity of this peak is smaller when lamp B was used for UV curing.

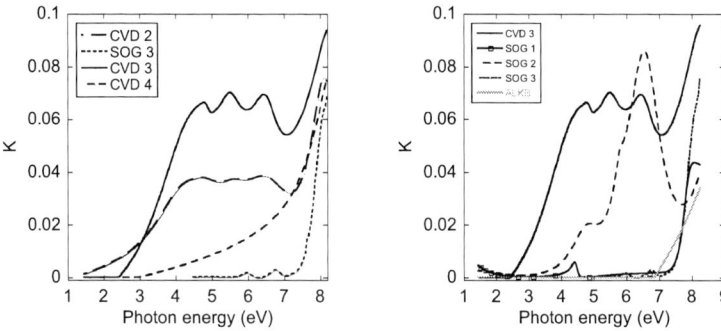

Figure 3. UV absorption spectra of low-k films prepared by different technologies.

Figure 3a shows the absorption spectra of UV cured PECVD low-k films with different porosity. As already mentioned, to deposit materials with lower dielectric constant, it is necessary to increase the ratio of porogen precursor to matrix precursors. This figure clearly shows that the deposition of highly porous films needs more deposited porogen, which then generates more residue during the UV curing. CVD3 is the film with 28 % porosity and k-value 2.3. CVD2 has porosity 24% and k=2.5 and less porogen was co-deposited together with matrix. CVD4 is a film deposited without porogen and with k-value equal to 3.0. Therefore, the formation of more residue in highly porous films is obvious from this graph. Both Figures 3a and 3b also shows that SOG films prepared using self-assembling chemistry without porogen do not show any absorption bands below 7.5 eV. SOG2, which is a spin-on film deposited with porogen and cured with 172 nm light has absorption band at 6.5 eV, which is different than in the UV cured PECVD films (4.5 eV). In principal, this fact demonstrates that the type of porogen residue depends on type of porogen. Electron Spin Resonance spectroscopy (ESR) showed that the carbon clusters related to porogen residues are the major sources of paramagnetic defects in low-k insulators. While the defect density increases with increasing porosity in the PECVD films, all three SOG films are found to be less defective and much more resistant to the ion sputtering damage [6].

Another important question is related to possibility to deposit residue free films using PECVD processes. Urbanowicz et al. [12] developed a process allowing one to form porogen free low-k films using a PECVD process. The matrix material and porogen were co-deposited at 300°C. This stage was completely identical to deposition of CVD3 film. Assuming, that this temperature is sufficient for formation of a more or less rigid skeleton and necessary agglomeration of porogen molecules, the porogen is completely removed before UV curing using He/H_2 downstream plasma. The plasma system was excluded the influence of energetic ions and VUV photons from the plasma to the film. Then the low-k films are UV cured to finalize the formation of cross-linked and rigid skeleton. The proposed method allows one to obtain porogen residue free low-k films with variable thicknesses. The obtained films demonstrate Young Modulus of 5–9.5 GPa after curing by lamp A and 3.5-4 GPa after the lamp B. The open porosity of these films were in the range of 41%–46%, the k-value of 1.8–2.6. The presented method shows the potential for fabrication of PECVD ultra low-k dielectric films for further microelectronic technology nodes.

Intrinsic electrical characteristics of various low-k materials

In the previous section we demonstrated that the properties of ultra low-k materials strongly depend on deposition approach. Even if the materials have sufficiently low dielectric constant and good intrinsic electrical characteristics, the integrated structures might show non satisfactory properties because of degradation during the integration. Therefore, the final results depend on both intrinsic properties and quality of integration processes and the selection of the most promising low-k materials will be much more efficient if to separate evaluation of intrinsic properties and post integration properties.

As mentioned above, the presence of porogen residue (sp^2 carbon) is one of the key factors affecting the intrinsic electrical characteristics of low-k dielectrics. To verify this assumption we evaluated electrical characteristics of the materials shown in the Table 1 using a unique test vehicle based on a planar capacitor design [18].

Figure 4 shows the intrinsic electrical characteristics of several low-k films shown in Table 1. The strong increase of the leakage current density with increasing porosity (Figure 4a) supports our previous assumption that highly porous ultra low-k PECVD materials deposited with higher amount of porogen form more porogen residues during the UV curing (Figure 3) and results in high leakage current density and low breakdown field. Figure 4b compares CVD3 film and SOG3 (SOG_inorg). As mentioned above SOG3 film was prepared using self-assembling chemistry without any porogen and shows no absorption bands at 4.5 eV (Figure 3). Therefore, these results also prove the importance of the development of porogen residue free low-k materials. Also, it is clear from the results presented in Figure 4a that the problem of porogen residue is especially important for ultra low-k materials. It is interesting that organic low-k film (SOG_org) has lower leakage than OSG film with porogen residue. It suggests that mostly sp^2 based residue is responsible for deterioration of electrical characteristics.

Electrical properties of other low-k materials presented in the Table 1 were also characterized [19]. These results completely support our conclusions. Materials without porogen residue (SOG1, SOG2 and ALKB) showed extremely low leakage current density and sufficiently high breakdown field. SOG2 film that has different type of porogen residue (not sp2) also had low leakage current density. This fact supports our conclusion that only formation of sp2 carbon is responsible for deterioration of electrical characteristics.

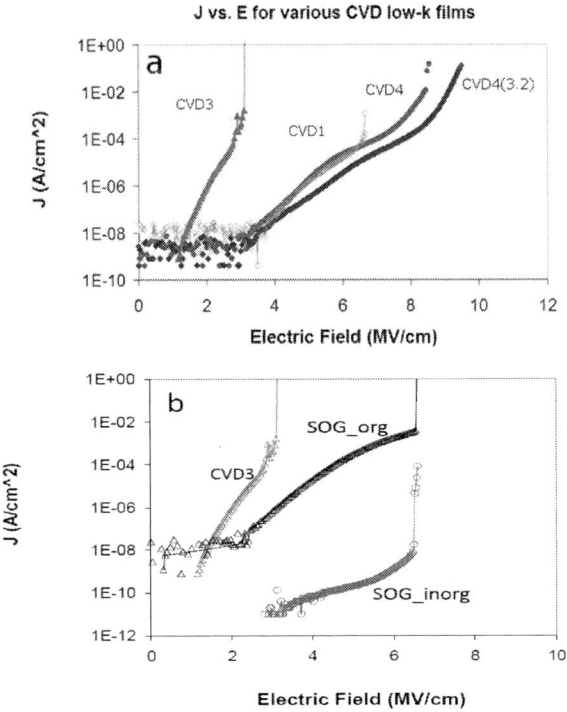

Figure 4. Leakage current density versus electric field for different low-k films. CVD4(3.2) is low-k film prepared similar as CVD4 but with k-value 3.2.

Conclusions

Selection of ultra low-k materials for future generations of IC devices is becoming more and more important. However, it is difficult to identify suitable materials based on integration results if the integration is not optimized and causes damage to the low-k material. Therefore, development and application of new metrology allowing evaluation of intrinsic properties is extremely important. It is demonstrated that application of UV spectroscopy allows detection of sp^2 carbon (porogen residue) that strongly affects electrical characteristics of low-k materials.

Detailed study of different low-k materials showed that further scaling of low-k materials needs development of new approaches. Utilization of porogen based technology for fabrication of ultra low-k films without fundamental modification of the curing technology is problematic. From this point of view, application of non-porogen based materials prepared with self-assembling technology (both organic and hybrid) looks more promising.

References

1. K. Maex, M. R. Baklanov, D. Shamiryan, F. Iacopi, S. Brongersma, Z. Sh. Yanovitskaya. *J. Appl. Phys.*, **93**, 8793 (2003).
2. R. J. O. M. Hoffman, G. J. A. M. Verheijden, J. Michelon, F. Iacopi, Y. Travaly, M. R. Baklanov, Zs. Tokei, G. P. Beyer. *Microelectron. Eng.*, **80**, 337 (2005).
3. A. Grill. *Annu. Rev. Mater. Sci.*, **39**, 49 (2009).
4. N. Kemeling K. Matsushita, N. Tsuji, K. Kagami, M. Kato, S. Kaneko, H. Sprey, D. de Roest, and N. Kobayashi,. *Microelectron. Eng.*, **84** (11), 2575 (2007).
5. S. Eslava, G. Eymery, P. Marsik, F. Iacopi, C. E. A. Kirschhok, K. Maex, J. A. Martens, M. R. Baklanov. *J. Electrochem. Soc.*, **155**, G115 (2008).
6. V. Afanasyev, K. Keunen, M. Jivanescu, A. Stesmans, Zs. Tokei, M. R. Baklanov, G. P. Beyer. *Spring MRS 2011*.
7. L. Zhao Tokei, Z.; Gianni, G.; Pantouvaki, M.; Croes, K. and Beyer, G. P. *Proc. IRPS 2009*, p. 848 (2009).
8. L.Zhao, Z. Tokei, G. Gischia, H. Volders, G. Beyer, *Proc. IITC 2009*, p. 206 (2009).
9. M. R. Baklanov, K. P. Mogilnikov, V. G. Polovinkin, and F. N. Dultsev. *J. Vac. Sci. Technol., B* **18**, 1385 (2000).
10. I. Ciofi, M. R. Baklanov, Zs. Tokei, G. P. Beyer. *Microelectron. Eng.*, 87, 2391 (2010).
11. K. Vanstreels and A. M. Urbanowicz, *J. Vac. Sci. Technol.*, B **28**, 173 (2010).
12. A. M. Urbanowicz, K. Vanstreels, P. Verdonck, D. Shamiryan, S. De Gendt, M. R. Baklanov. *J. Appl. Phys.*, 107, 104122 (2010).
13. M. Pantouvaki. C. Huffman, L. Zhao, N. Heylen, Y. Ono, M. Nakajima, K. Nakatani, G. P. Beyer and M. R. Baklanov. *Jap. J. Appl. Phys.* In press.
14. K. Vanstreels, M. Pantouvaki, A. Ferchichi, P. Verdonck, T. Conard,

    M. R. Baklanov. *J. Appl. Phys.* In press.
15. L. Prager, P. Marsik, L. Wennrich, M. R. Baklanov, S. Naumov, L. Pistol, D. Schneider, J. W. Gerlach, P. Verdonck, M. R. Buchmeiser. *Microelectron. Eng.*, **85**, 2094 (2008).
16. M. Tada, H. Yamamoto, T. Takeuchi, N. Furutake, F. Ito, and Y. Hayashi. *Journal of The Electrochem. Soc.*, **154** (7) D354 (2007).
17. P. Marsik, P. Verdonck, D. De Roest, and M. R. Baklanov, *Thin Solid Films* **518**, 4266 (2010).
18. M. R. Baklanov, L. Zhao, E.Van Besien, M. Pantouvaki. *Microelectron. Eng.*, in press.
19. I. Ciofi, E. Van Besien. *IMEC's PTW reports*, Leuven, 2010.

ECS Transactions, 35 (4) 729-746 (2011)
10.1149/1.3572316 ©The Electrochemical Society

Development of Porosimetry Techniques for the Characterization of Plasma-Treated Porous Ultra Low-*k* Materials

C. Licitra[a], T. Chevolleau[b], R. Bouyssou[b], M. El Kodadi[b], G. Haberfehlner[a], J. Hazart[a], L. Virot[a], M. Besacier[b], N. Posseme[a], M. Darnon[b], R. Hurand[b], P. Schiavone[b], and F. Bertin[a]

[a]CEA-LETI, Minatec Campus, 17 rue des Martyrs, 38054 Grenoble, France
[b]LTM, UJF-Grenoble1/Grenoble-INP/CNRS/CEA, Minatec Campus, 17 rue des Martyrs, 38054 Grenoble, France

> For the sub-32 nm node, porous SiCOH dielectrics (p-SiCOH) are integrated using dual damascene patterning by etching trenches and vias into the porous material. One challenge is to control the process conditions to minimize the plasma-induced damage of p-SiCOH materials at the bottom and at the sidewall of the trenches. Ellipsometric Porosimetry has been adapted to characterize the plasma-treated materials for both horizontal and vertical surfaces. Since surface modifications can cause adsorption and desorption delays and hydrophobicity loss, porosimetry measurements operated with multi-solvent and kinetic protocols are required. Quantitative measurements of vertically patterned materials are demonstrated using periodic structures of porous material and a Scatterometric Porosimetry analysis. Results show a good sensitivity of the measurement to the different process conditions but also highlight a different impact of the plasma processes on patterned materials compared with blanket films.

Introduction

Down-scaling of complementary metal oxide semiconductor (CMOS) devices requires the integration of copper/porous ultra low-*k* (ULK) materials to reduce the interconnect resistance-capacitance delay. For the sub-32 nm node, porous SiCOH dielectrics (p-SiCOH) are integrated using dual damascene patterning by etching trenches and vias into the porous material (Figure 1). Since the pore structure leads to higher sensitivity of the material to environmental and process conditions, controlling the profiles of the etched structures and minimizing the plasma-induced damage of p-SiCOH materials are the two main patterning challenges. The sensitive areas are the bottom and the sidewall of the trenches where surface modifications, post-etch residues, or p-SiCOH roughening can occur (1, 2, 3). Characterization techniques with low spatial resolution including X-ray Photoelectron Spectroscopy, X-ray Reflectivity, Time of Flight - Secondary Ion Mass Spectrometry, Contact Angle, Mercury Probe Capacitance, Infrared Spectroscopy, or Ellipsometric Porosimetry can be easily set up on blanket films to replicate the bottom of the trench (1). Unfortunately the porous properties of the sidewall region differ from the bottom of the trench. Few studies have been performed on patterned structures to determine the sidewall modification while this information is critical for device performance (4). In this study, Ellipsometric Porosimetry (EP) has been adapted to characterize the plasma-treated materials for both horizontal and vertical geometries. EP

729

is a non destructive technique (5) based on the acquisition of ellipsometric spectra of a layer during the adsorption and desorption cycles of an adsorptive (solvent). However, it can suffer from measurement artifacts when used with plasma-treated films because of surface modifications. We have therefore evaluated different experimental configurations. Results show that using solvents with different molecular diameters, e.g. toluene and methanol, allows surface densification to be detected. This plasma-sealing effect can also be quantified by studying the solvent penetration kinetics over time through the damaged surface. Finally for hydrophobic materials, the plasma-induced degree of hydrophilization of the surface can also be determined through EP with water vapors. On the other hand, the properties of vertical sidewalls can be determined using the recently developed Scatterometric Porosimetry (SP) technique (6, 7). It mainly consists in the use of the EP tool to record the scatterometric response of periodic structures made of porous material as a function of the relative pressure of the solvent. The patterned structures are chosen to have a critical dimension (CD) which is equivalent to the dielectric line size of the interconnect except that they typically consist of parallel periodic lines. The porous properties are subsequently extracted from the SP measurement with the use of a specially-developed scatterometric modeling. We show a side-by-side comparison between EP and SP on different plasma-treated samples. A different effect of the plasma processes is observed on patterned materials compared with blanket films. These results illustrate that the porous material modifications strongly depend on the sample geometry and highlight the interest of Scatterometric Porosimetry to characterize sidewall damage after each step of the etch process.

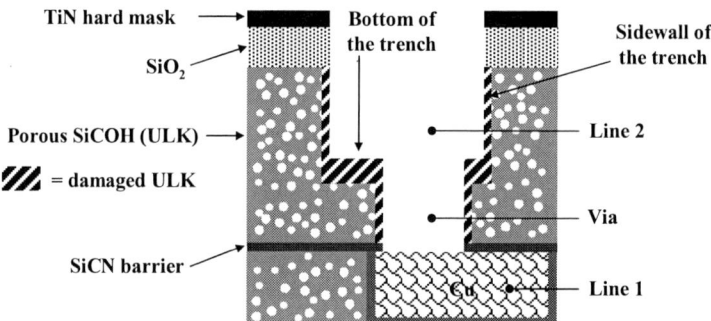

Figure 1. Dual damascene structure fabricated using a metallic hard mask.

Experimental Details

Integration process

Porous SiCOH films with a thickness around 120 nm were deposited by Plasma Enhanced Chemical Vapor Deposition on blanket 300 mm silicon wafers. The material is the BD2.35^{TM} from Applied Materials with k=2.35, 28 % porous, 1.2 nm mean pore radius and hydrophobic by design. Porosity is achieved by co-depositing two precursors to make a SiCOH skeleton containing organic species which are subsequently removed using an ultraviolet assisted thermal cure at 400°C. In order to simulate the integration process, porous blanket films were etched in conventional fluorocarbon based plasmas (CF_4 / C_4F_8 / N_2 / Ar) and after partial etching the remaining films was exposed to post-etching treatments such as NH_3, H_2, CH_4 or O_2 based plasmas (8). The etching and post-etching plasma treatments were performed in a dual frequency capacitive discharge Flex45DD^{TM} from Lam Research.

On the other hand, the patterning of porous ULK lines was performed using a dual metallic hard mask strategy. This architecture is mainly used to minimize the porous dielectric exposure to environmental contamination and to stripping plasmas. A simplified stack was used to replicate the dual damascene vertical structures. It consists of a 15 nm titanium nitride layer, a 125 nm oxide capping layer, and the BD2.35^{TM} p-SiCOH dielectric material ~600 nm thick. In the visible range, the 15 nm titanium nitride layer is transparent allowing optical measurements through the layer. After lithography, the TiN hard mask patterning step was performed in inductively coupled plasma using chlorine-based chemistries while the SiO$_2$ capping and the p-SiCOH layers were etched in fluorocarbon-based plasmas using a capacitive coupled plasma (Flex45DD^{TM}). After patterning, porous structures were also exposed to NH$_3$, H$_2$, CH$_4$, or O$_2$ based plasmas. A Scanning Electron Microscopy (SEM) image of such a patterned structure is shown in Figure 2 after the etching process steps.

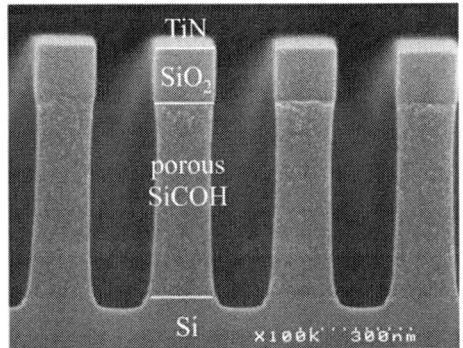

Figure 2.   Test structure for Scatterometric Porosimetry with critical dimension CD=180 nm and pitch=340 nm.

Characterization tool

EP and SP measurements are performed in the visible range on an ellipsometric porosimeter from SOPRALAB (SEMILAB). It consists of a rotating polarizer spectroscopic ellipsometer coupled with a vacuum chamber which is operated at a pressure ranging from 0.133 Pa to the saturation vapor pressure (P_s) of the adsorptive (Figure 3) to allow capillary condensation into the open pores. Three solvents were separately used as adsorptive: toluene (P_s~3333 Pa at room temperature, refractive index at 633 nm n_{tol}=1.492), methanol (P_s~14000 Pa at room temperature, refractive index at 633 nm n_{met}=1.329) and water (P_s~2666 Pa, n_{water}=1.333 at 633 nm). Pressure values are plotted using a relative scale P_{rel}=P/P_s, with P the chamber pressure. The standard porosimetry cycle consists of an adsorption sequence (P_{rel}=0 to 1) followed by a desorption sequence (P_{rel}=1 to 0) which are called isotherms, but measurements can also be done as a function of time at fixed pressures. In our experiments the dwell time for each pressure step is less than 1 min for measurements over pressure and time interval is set to 5 s for measurements over time. For each step, ellipsometric spectra are recorded at an incidence angle of 60.15°, in a typical wavelength range between 1.5 and 4 eV using a multichannel detector. More details about the experimental setup and EP technique can be found elsewhere (9, 10).

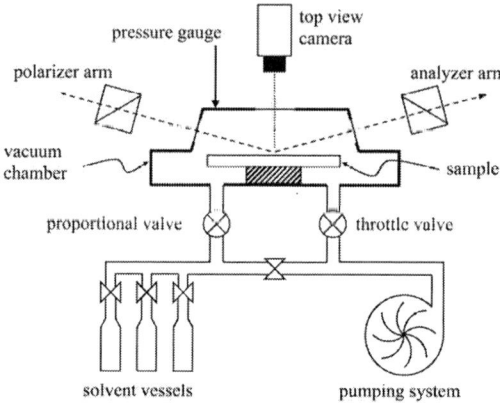

Figure 3. Ellipsometric porosimeter setup.

Data modeling

The experimental data acquired with the SOPRALAB system are the ellipsometric coefficients of the sample: tan(Ψ) and cos(Δ) or α and β as a function of photon energy and relative pressure. These coefficients can be used with an ellipsometric modeling to determine the thickness and the optical properties of the porous film in the form of the complex refractive index $N=n+ik$ at each pressure step. If the sample is a patterned structure instead of a thin film, scatterometry must be used to model α and β coefficients. Scatterometry is a non-destructive optical method that allows the geometry of a periodic structure such as a grating to be determined by modeling its optical diffracted response (11). It is usually applied to materials with fixed refractive index. However we have adapted the scatterometric analysis to extract the properties of porous gratings by modeling the optical index variations of the porous material during the adsorption and desorption sequences. Since a real circuit cannot be properly modeled directly, specific periodic structures made of porous materials were used to replicate the interconnect levels that need to be characterized. These gratings can be chosen to have a critical dimension which is equivalent to the interconnect line size except that they typically consist of parallel periodic lines. This kind of structures can usually be found into the service areas of the photomasks or can be prepared specifically using the same process as for the interconnect levels manufacturing. α and β coefficients of the grating depend on its geometry including the CD, height, side wall angle; the periodicity of the lines; and on the optical indices of all component materials. During the optical measurement we used the standard scatterometry configuration where the periodic lines are placed perpendicular to the plane of incidence of the light beam. An example of EP and SP measurements is given in Figure 4. Because of refractive index variations, we observe a shift of the tan(Ψ) and cos(Δ) spectra while the p-SiCOH is being filled with methanol.

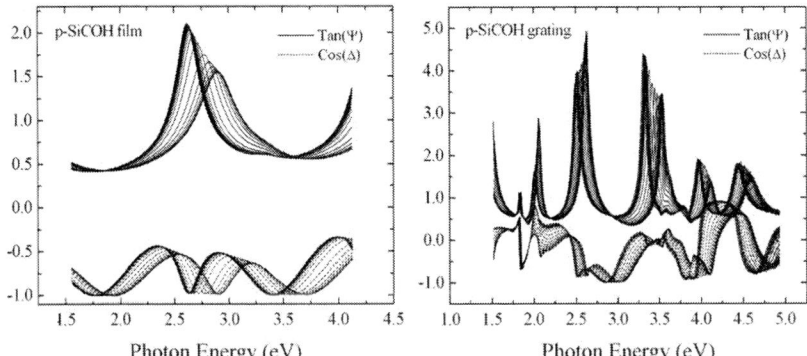

Figure 4. tan(Ψ) and cos(Δ) spectra during methanol adsorption. Left: porous film, right: porous grating.

During the porosimetry cycle the effective refractive index of the porous material partially filled with solvent must be modeled at each relative pressure. Absorbance measurements were performed using a Cary 5 spectrophotometer from Varian through 1 cm of liquid toluene, methanol and deionized water to ensure that no absorption peak from the solvent should be included in the porous material refractive index (Figure 5, left). A 130 nm p-SiCOH film was also measured after deposition using a broad range M2000 spectroscopic ellipsometer from Woollam. This type of dielectric material shows several absorption peaks in the ultraviolet region as depicted in Figure 5, right. However within the restricted EP spectral range the p-SiCOH refractive index can be modeled using classical dispersion laws such as Tauc-Lorentz, Forouhi-Bloomer or Cauchy law with a Lorentz oscillator. Each of these laws has 5 variable parameters. In this study we chose a Cauchy law with a Lorentz oscillator to easily model the p-SiCOH refractive index by adjusting only 2 Cauchy parameters during the condensation. Indeed the solvents extinction coefficients calculated from the absorbance spectra are lower than 10^{-7} thus the Lorentz oscillator which takes into account the porous material absorption can remain constant during the adsorption and desorption sequences.

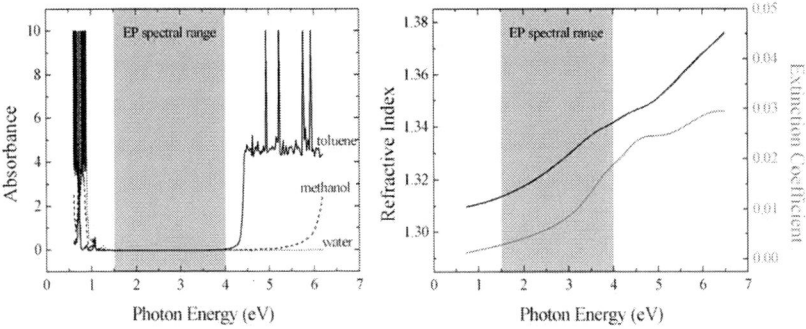

Figure 5. Left: absorbance spectra through 1 cm of liquid toluene, methanol and deionized water. Right: refractive index and extinction coefficient of an as-deposited porous SiCOH film after curing.

EP or SP porosimetry modeling then consists in fitting all the spectra recorded as a function of photon energy and relative pressure. On one hand, the ellipsometric simulated signatures of blanket films can be easily derived from the reflection coefficients of the p- and s-polarized waves known as Fresnel equations (12). On the other hand, the simulated signatures of a grating are obtained using a rigorous electromagnetic solver based on the Modal Method by Fourier Expansion (MMFE) detailed by Li et al. (13). The MMFE method uses a modal decomposition of Maxwell's equations with appropriate boundary conditions at each interface of the geometric structure to solve the electric field in each region of the sample. The simulation of the optical signatures is known as the direct problem. In order to model the experimental data versus photon energy, a minimization algorithm such as the least-squares algorithm can be used. This optimization is commonly employed for ellipsometry and for EP data analysis. The least-squares can be used for SP modeling but it is also possible to apply a faster library-based method to model the optical response of the grating. First a library of optical responses is built by mapping the optical indices of the porous material and the variable geometric parameters of the sample including CD, height, side wall angle or other required parameters. Then the experimental signatures are compared to the library to quickly find the unknown parameters. This method is widely used for in-line scatterometry with fixed refractive indices and for real-time scatterometry (14). Both the least-squares and the library-based minimization techniques were tried to model SP measurements and gave similar results. The modeling of the optical responses versus pressure can be done in different manners. We have tested the 3 following modeling strategies.

Sequential strategy. This technique is commonly used for EP data analysis with commercially available software. A first ellipsometric or scatterometric modeling step is used to calculate the optical indices and the geometric parameters: thicknesses in the case of a blanket stack or geometry of the grating including CD, height, side wall angle or other required parameters. This calculation is done on the first measurement corresponding to the vacuum state when no pores are filled. The optical indices of each material should preferably be determined on blanket wafers to get a robust initial model. A sequential modeling step with less variable parameters is then applied to calculate the p-SiCOH refractive index variations during the adsorption or desorption of the adsorptive. During this step the geometry can be kept constant or also fitted if necessary. As the solvent doesn't absorb light in the visible range only the two Cauchy parameters are sequentially fitted. The optical indices of each non porous material are kept constant.

Reversed sequential strategy. This modeling is a sequential strategy but the refractive indices and the geometric parameters are determined at relative pressure $P_{rel}=1$ instead of $P_{rel}=0$ and the sequential fit is done in the reversed order from $P_{rel}=1$ to 0.

Combined strategy. The sequential strategies can lack of precision when some parameters are correlated. We have therefore adapted our minimization algorithm to the specificity of porosimetry experiments. Indeed EP and SP measurements contain redundant information because multiple measurements are done at the same location. In some cases this information can be used to reduce the modeling errors because several model parameters are constant with pressure. Usually the optical indices of each non porous material, the Lorentz peak, and potentially some geometric parameters are fixed especially for SP. In the combined strategy these fixed parameters are no longer determined under vacuum only but by fitting measurements at several relative pressures simultaneously (usually up to 5 pressures at once). This guarantees that these parameters are fitted to achieve a low modeling error over the whole pressure range. After this first combined fit the parameters varying with pressure that is to say the 2 Cauchy parameters and some varying geometric parameters are then fitted sequentially at each pressure.

For both EP and SP the fraction of solvent adsorbed in the pores is finally calculated at each relative pressure with the Lorentz-Lorenz effective medium approximation (15) and by knowing the effective refractive index of the material and the refractive index of condensed solvent at 633 nm (9). The open porosity is given by the solvent volume fraction when the relative pressure is equal to 1 and all open pores are filled with solvent. Finally the pore size distribution (PSD) of mesopores, 1 nm < radius < 25 nm, can be calculated using the Kelvin equation of capillary condensation (16, 17) and by knowing the thickness of the layer of solvent condensed on the pore walls before capillary condensation occurs.

An example of different EP modeling strategies is given in Figure 6 (left) for methanol adsorption in a 130 nm blanket porous SiCOH film. In the case of the combined strategy only the Lorentz oscillator was simultaneously fitted at different relative pressures distributed over the range. The mean square error (Chi^2) extracted from the modeling is shown as a function of the relative pressure for the different strategies. As expected we observe that Chi^2 is lower at $P_{rel}=0$ for the sequential fit and lower at $P_{rel}=1$ for the reversed sequential fit. However the combined fit shows a relatively low average Chi^2 over the whole pressure range. The corresponding solvent volume fraction is shown in Figure 6 (right). The open porosity varies from 25.3% to 26.1% depending on the modeling. This can be attributed to parameter correlation between the thickness and the refractive index.

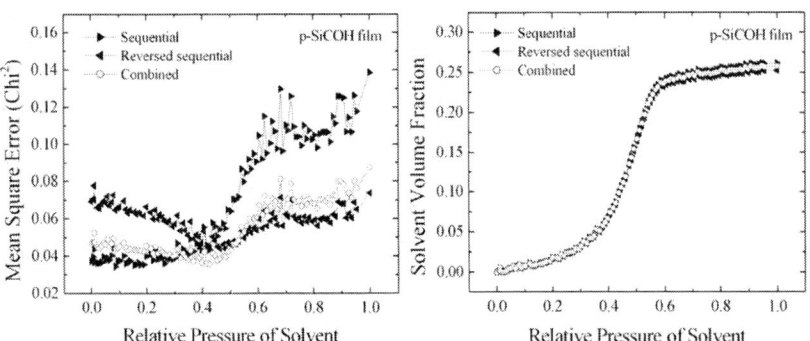

Figure 6. Left: mean square error for different modeling strategies of a 130 nm blanket porous SiCOH film. Right: corresponding solvent volume fractions for the different modeling strategies.

The same modeling strategies were applied to the porous grating shown in Figure 2. In the case of the combined strategy, the Lorentz oscillator and the geometric parameters were all simultaneously fitted at different relative pressures distributed over the range. For the 2 sequential strategies, the geometric parameters were fixed to lower the correlations between parameters. Mean square errors and solvent volume fractions are shown in Figure 7. As the geometry was fixed at P_{ref}=0 the 2 sequential strategies results are very close meaning that the Lorentz parameters were found identical at P_{ref}=0 and P_{ref}=1. However the Chi^2 is high at saturation pressure. On the contrary, the combined modeling allows the Chi^2 to be lowered over the whole pressure range. As scatterometry modeling has more parameters than ellipsometry, this strategy is therefore interesting to lower the correlations between the parameters especially when fitting the geometric parameters of the structure.

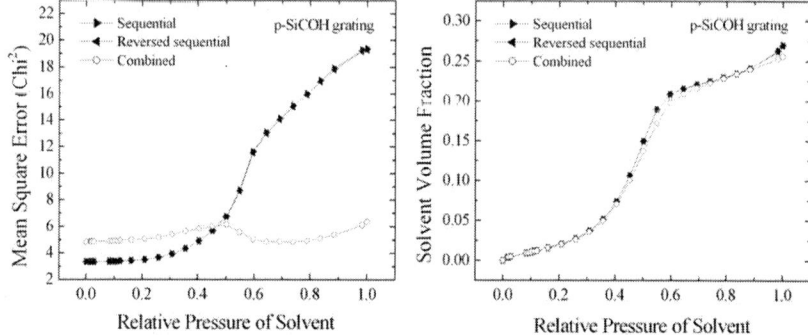

Figure 7. Left: mean square error for different modeling strategies of a porous grating. Right: corresponding solvent volume fractions for the different modeling strategies.

Results and Discussion

Effect of plasma processes on blanket p-SiCOH films

After patterning, the p-SiCOH can be exposed to NH_3, H_2, CH_4, or O_2 based plasmas. Such reducing and oxidizing treatments can be used as i) ashing processes (1), ii) cleaning processes after etching, and/or iii) plasma processes to prevent the metallic barrier diffusion (18). Unfortunately they can also lead to surface modifications that can cause the formation of a damaged surface layer (1, 18, 19, 20) and increase the ULK effective dielectric constant. Indeed the damaged surface may present a barrier-like effect because of the densification of the material. The p-SiCOH which is hydrophobic by design can also become partly hydrophilic after plasma treatments because of a methyl group depletion converting the surface layer into a hydrophilic material. Figure 8 shows the water isotherms obtained with EP of an as-deposited p-SiCOH film compared to a film after exposure to O_2 downstream plasma (k=2.5 and 25% porosity in that case). The O_2 downstream plasma is known to remove all the methyl groups of the ULK leading to a fully hydrophilic material (1). Indeed no water adsorption is detected for the as-deposited film whereas 24.7% of the material is hydrophilic after the O_2 downstream plasma which is almost the total initial open porosity.

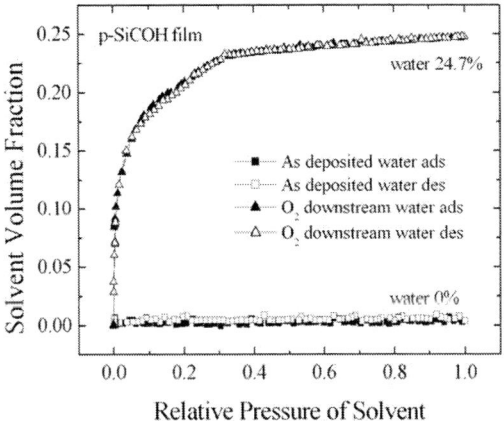

Figure 8. Water isotherms of as-deposited and O_2 downstream post-treated p-SiCOH films.

In contrast to the O_2 downstream plasma, the p-SiCOH film is partially modified after exposure to the different post-etching treatments. The modified layers can be etched in HF solution whereas the pristine hydrophobic material cannot (21). The thicknesses of the modified layers were therefore estimated by measuring the samples before and after HF etching using spectroscopic ellipsometry (Table I). After exposure to the different plasmas, dielectric constant measurements were performed on blanket wafers using a mercury probe capacitance measurement system. As expected, all the treatments lead to an increase of the dielectric constant (Table I) which is linked to the modification of the surface layer.

TABLE I. Dielectric constant and surface properties of p-SiCOH blanket samples.

Porous films	Dielectric constant	Modified layer thickness (nm)
ULK (pristine)	2.35±0.05	0
ULK+etch	2.76±0.05	28
ULK+etch+NH3	2.98±0.05	28
ULK+etch+H2	2.92±0.05	>40
ULK+etch+CH4	2.91±0.05	18
ULK+etch+O2	2.69±0.05	32

Previous workers have employed EP to characterize plasma-treated ULK films by using either methanol (18), or toluene and water (22), or by studying the solvent penetration kinetics (23). However the quantification of plasma-sealing strongly depends on the size of the probe molecule and the choice of the solvent is critical to determine the porosity of a plasma-treated film. The degree of plasma damage is also hard to determine because water vapors only partially condense into the modified part of the layer. In order to better understand the plasma impact, an EP analysis using a multi-solvent protocol has been systematically performed. As the pores can be modified during the treatment, methanol and toluene were used to compare the adsorption of molecules with a different size, and water was used to check the degree of plasma damage which is correlated to the hydrophilic volume properties of the layer (22). The ULK film after partial etching and post-etching plasmas can be considered as a two-layer system with separate optical constants: a plasma-modified surface layer on top of a bulk undamaged ULK layer. However the small refractive index difference (less than 0.05) between the modified layer and the unmodified ULK cannot allow performing a reliable ellipsometric fit. Indeed, in the case of toluene and methanol, the condensation occurs into the modified layer as well as in the bulk unmodified ULK. Toluene and methanol measurements were therefore analyzed using a single layer model giving the whole porosity of the remaining film. On the contrary, the water condensation only occurs into the modified top layer due to its hydrophilicity, which leads to a significant change in refractive index. In that case, the EP measurements can be analyzed with a two-layer model if we assume the thickness and the optical properties of the undamaged bulk layer. Thus we can extract the properties of the hydrophilic pores in the modified layer.

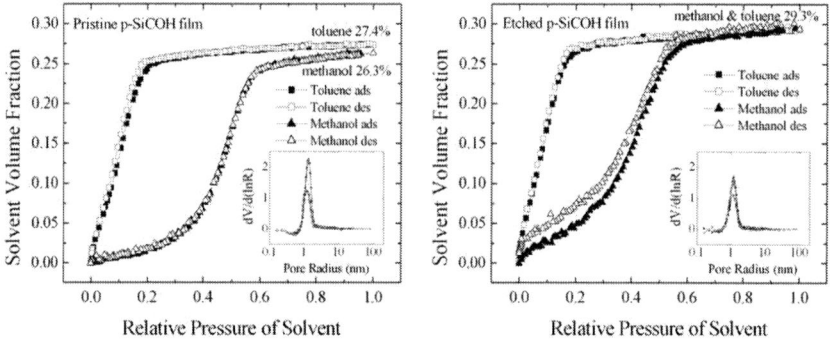

Figure 9. Toluene and methanol isotherms of pristine and etched p-SiCOH blanket films, inset: corresponding pore size distribution.

Figure 9 (left) shows, for the pristine ULK, the solvent volume fractions and pore size distributions obtained with toluene and methanol during the adsorption (ads) and desorption (des) sequences. The open porosity of the unmodified film is 27.4 % with toluene and 26.3 % with methanol, which values are close to the nominal 28 %. Pore size distributions for toluene and methanol give an average pore radius around 1.2 nm. EP results after partial etching are also presented in Figure 9 (right). The isotherms of toluene and methanol are quite similar to that of the pristine p-SiCOH and a porosity of 29.3 % is achieved considering a single layer model. The PSD calculation with both solvents gives the same values compared to the pristine p-SiCOH. The EP measurements with water, summarized in Table II, show that after etching, the water condenses partially into the modified layer (16% of porosity) indicating that only a few pores are hydrophilic. On the contrary, as the pristine material is hydrophobic, no porosity is detected with water. At last the etching plasma does not induce significant changes of porosity and pore size distribution of the remaining ULK film but involves a change of the hydrophobic properties of a part of the pores.

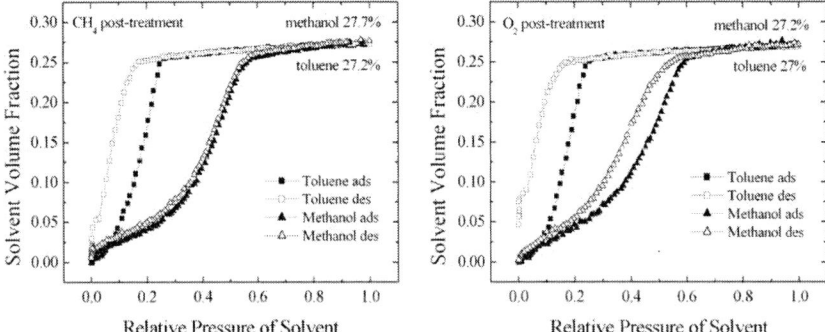

Figure 10. Toluene and methanol isotherms of CH_4 and O_2 post-treated p-SiCOH films.

EP results after the CH_4 and O_2 post-treatments are presented in Figure 10. In both cases the isotherm curves look similar to the unmodified sample with a porosity around 27 %. However we can notice the apparition of a small hysteresis in the case of CH_4 with toluene and in the case of O_2 with both toluene and methanol. As the pristine ULK measurement did not show this type of hysteresis, we can assume that the treatment slightly modifies the surface properties of the layer which changes the adsorption and desorption equilibrium conditions. In that case, the isotherm curves must be carefully interpreted and the corresponding PSD cannot be calculated. In addition, CH_4 and O_2 plasmas also lead to water condensation (Table II) into a fraction of the pores of the modified layer (12.3 % and 19.2 % respectively).

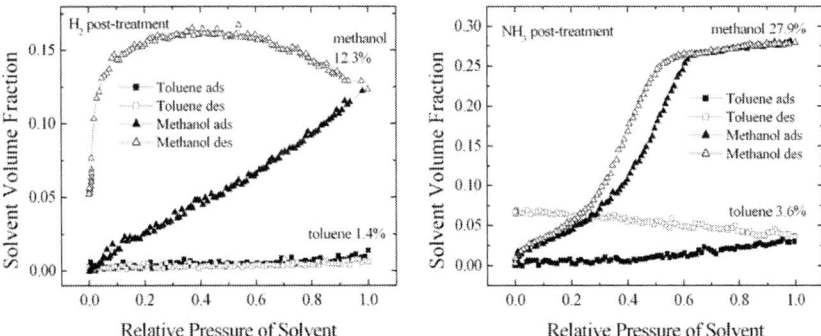

Figure 11. Toluene and methanol isotherms of H_2 and NH_3 post-treated p-SiCOH films.

EP results after H_2 post-treatment are presented in Figure 11 and Table II. As expected, this plasma treatment also induces p-SiCOH damage detected with water condensation in the modified layer (up to 16.7 %). However, the methanol adsorption curve does not have the typical plateau near the saturation vapor pressure indicating the total filling of the pores, and the desorption curve shows that the layer is still adsorbing methanol during the desorption. At the end of the desorption cycle, there is still around 5 % of methanol in the layer. In contrast to methanol, toluene (a bigger molecule) is no longer adsorbed after plasma (only 1.4 %). Those behaviors indicate a slow solvent condensation kinetic. This is due to the densification of the surface which acts as a membrane. The filling time of the pores will depend on the molecule size and the solvent polarity and viscosity. In our experiments the dwell time between each pressure which is lower than one minute is not sufficient to reach the equilibrium condensation conditions that guarantee the complete filling of the pores. The dwell time should be adapted to each plasma-treated material but it is not predictable and in some cases it can become too long to make comfortable EP measurements. To a lesser extent, a similar behavior is observed after the NH_3 post-treatment. The modified layer is also partially hydrophilic (water content up to 10.2 %) and almost acts as a membrane for the toluene with a slow adsorption and desorption condensation kinetic. Methanol adsorption and desorption isotherms are still reversible but present a hysteresis loop and they reach the nominal open porosity of 27 9 %. In the end, all these measurements give useful information when integrating ULK materials about the surface modification but the results must be carefully interpreted because the isotherms are not acquired in equilibrium conditions. For example the PSD cannot be accurately calculated in these conditions and the porosity may be wrong in some cases because of a delayed solvent adsorption.

TABLE II. Porosimetry results of p-SiCOH blanket samples.

Porous films	Toluene	Methanol	Water uptake (%) in the modified layer
ULK (pristine)	Correct isotherms	Correct isotherms	0
ULK+etch	Correct isotherms	Correct isotherms	16
ULK+etch+NH_3	Slow adsorption	Delayed adsorption	10.2
ULK+etch+H_2	Slow adsorption	Slow adsorption	16.7
ULK+etch+CH_4	Delayed adsorption	Delayed adsorption	12.3
ULK+etch+O_2	Delayed adsorption	Delayed adsorption	19.2

In order to compare more precisely the membrane properties of the plasma-induced modified layers, we have performed EP kinetic measurements (23). Instead of monitoring the change of refractive index versus the relative pressure, the adsorptive relative pressure is quickly increased from the residual vacuum to 80% of the saturation pressure and the measurements are done versus time. At such a relative pressure, the measurements done on the pristine ULK film (Figure 9) show that the capillary condensation has already occurred and all the pores are filled. The kinetic measurements have been performed on blanket films with methanol and toluene (Figure 12). As expected the CH_4, O_2 and NH_3 post-treatments show a methanol condensation kinetic nearly similar to the partially etched material which is well correlated with the solvent introduction in the chamber. At this relative pressure all the pores are quasi instantly filled due to the capillary condensation. However a small adsorption delay up to 2 min is measured between the partially etched and the post-treated films. This can be attributed to a densification of the surface that prevents the solvent adsorption. After the H_2 plasma treatment, the layer is filled after 240 min indicating a strong densification of the top surface. With the toluene, a higher delay is observed after the NH_3 and H_2 plasma treatments, respectively 250 and 1300 min, which indicates that the modified layer acts as a membrane. The whole results show that 1) after O_2 and CH_4 plasmas, the modified layer does not prevent the solvents diffusion, 2) after NH_3 plasma, the modified layer can be considered as a membrane and 3) the H_2 plasma treatment leads to a quasi pore sealing. The time to reach the plateau (equilibrium state of the adsorptive) is always higher with the toluene than with the methanol which indicates a slower diffusion kinetic mainly due to a bigger molecular diameter. We also noticed that after all the post-treatments the value of the porosity is about 28 % which is the porosity of the pristine ULK material. This value is not always achieved with the EP measurements versus relative pressure because of an insufficient stabilization time. Those results show that only the surface is modified and there is no significant change of porosity in the remaining ULK material compared to the pristine ULK regardless of the plasma treatments.

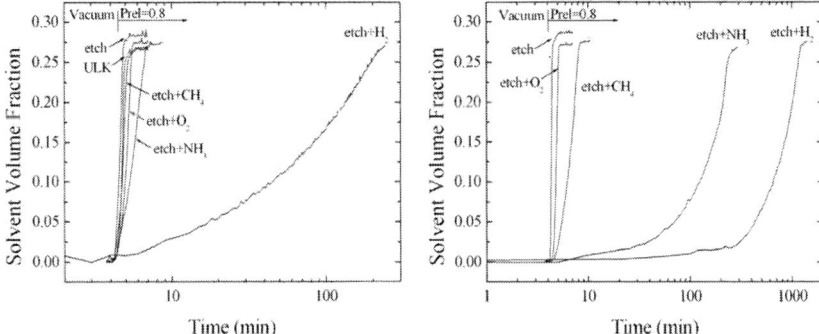

Figure 12. Methanol (left) and toluene (right) penetration kinetics of etched and post-treated p-SiCOH films.

Effect of plasma processes versus sample geometry

In order to compare EP and SP we have exposed a blanket porous film and a masked porous stack to similar etching plasma. After partial etching, the remaining blanket film has been analyzed with EP (Figure 9). After etching of the masked porous stack, a patterned structure is obtained (Figure 2). It consists of periodic lines of titanium nitride on top of an oxide capping layer and the p-SiCOH dielectric material. This grating has been analyzed by means of SP with the geometry described in Figure 13. The corresponding solvent volume fraction and pore size distribution obtained with methanol are shown in Figure 13. The pore size distribution is also centered on 1.2 nm and the open porosity is 27.8% which is a little smaller than the blanket film. However the two p-SiCOH were deposited on separate wafers with different thicknesses which may cause small discrepancies. In addition isotherms are consistent with those obtained on the blanket film with EP. Nevertheless SP isotherms show less hysteresis effects indicating a limited surface modification of the sidewalls during the etching step. These first SP results show that the technique is able to measure the porous properties of patterned structures and also to detect small differences between blanket and patterned structures.

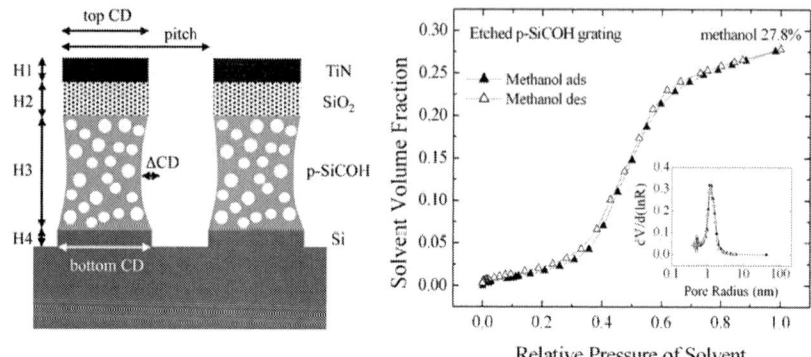

Figure 13. Left: definition of the parameterized geometrical parameters for the p-SiCOH grating. Right: methanol isotherms and PSD of the etched p-SiCOH grating.

In order to compare the sensitivity of horizontal and vertical porous surfaces to plasma treatments we have performed SP kinetic measurements with methanol. We have exposed patterned structures to the same NH_3, H_2, CH_4, or O_2 based plasma treatments. The SP kinetic measurements were done on different gratings with the same geometry CD=180 nm and pitch=340 nm. Solvent volume fraction results are presented in Figure 14. Unlike previous thin films results, all the post-treatments present a fast methanol condensation kinetic (less than 30 s) indicating a reduced diffusion barrier. This can be explained by the vertical geometry of the sidewalls which are less exposed to ion bombardment during the plasma processes. These differences between blanket and patterned films highlight the interest of the technique for real interconnect process monitoring. EP measurements on blanket films may not be fairly representative of plasma-induced modifications on vertical structures.

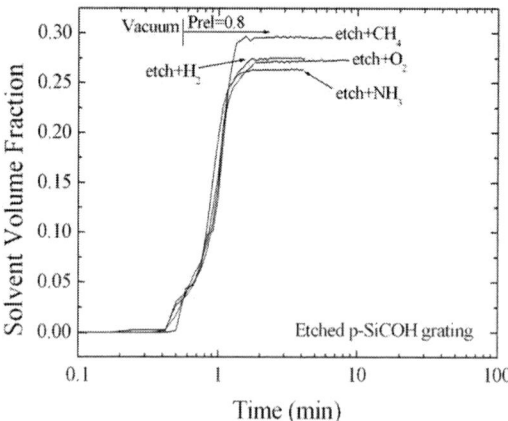

Figure 14. Methanol penetration kinetics of etched and post-treated p-SiCOH gratings.

Finally SP measurements were also performed using water as solvent to detect any change of the ULK hydrophobic properties. Two gratings were prepared using the standard etching process and post-treated with either capacitive O_2 plasma (FLEX45DD^{TM}) or O_2 downstream plasma. SP results considering a scatterometric model with a uniform porous material are shown in Figure 15. As expected water condensation clearly occurs after the O_2 downstream plasma because of the full methyl group depletion. The measured porosity is close to the pristine film value indicating that almost all the pores are hydrophilic. On the contrary the O_2 post-treatment allows a small amount of water to be condensed (7.6%). In that case, water only condensates in the pores located near the sidewall surface which is modified by the plasma. The small hysteresis observed in that case is linked to surface modifications which prevent water to be adsorbed in the same conditions than in uniform hydrophilic samples. Through SP with water vapors, it is therefore possible to detect the plasma-induced damage of porous materials as integrated in a structure.

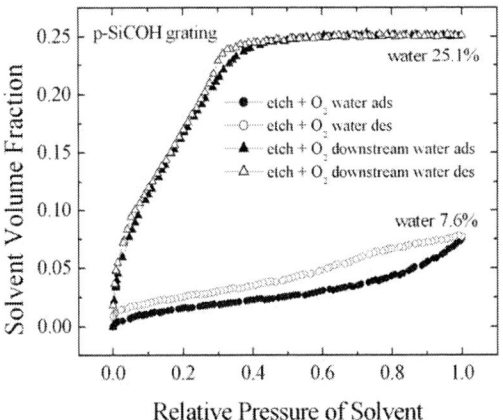

Figure 15. Water isotherms of O_2 and O_2 downstream post-treated p-SiCOH gratings.

The quantification of plasma-induced damage could also be possible through SP. However it adds more complexity to the model and therefore more correlation issues. Indeed the refractive index of modified sidewalls when the pores are empty is close to the bulk p-SiCOH then it is not possible to define directly a geometric model with modified sidewalls. However as we know the refractive index of the hydrophilic material filled with water from previous measurements after O_2 downstream plasma, it is possible to model a structure by varying the sidewall thickness and keeping constant the refractive index of the fully hydrophilic material. The thickness value found at saturation pressure will therefore be equivalent to the modified sidewall thickness with the assumption that all the pores of the modified layer are hydrophilic. We have compared this complex geometric modeling to the conventional decoration method. In this method cleaved patterned samples are exposed to 1% diluted HF for 15 s. Indeed the damaged layer is quickly removed by the HF solution, while the unmodified material is hardly consumed. The trenches are filled with resist prior to the HF dip to prevent potential collapse of features during SEM exposure (Figure 16). The comparison of the modified layer thickness estimated with both techniques is shown in Figure 17. Similar thicknesses are observed with both techniques except after the etching plasma but the SEM measurement has a worse accuracy than scatterometry. We found that the thickness of the modified sidewall depends on the post-treatment plasma. In conclusion plasma modifications were observed for both horizontal and vertical geometries regardless of the plasma treatments. In addition to plasma modifications, surface densification was also observed in the bottom of the trench which is more exposed to ion bombardment.

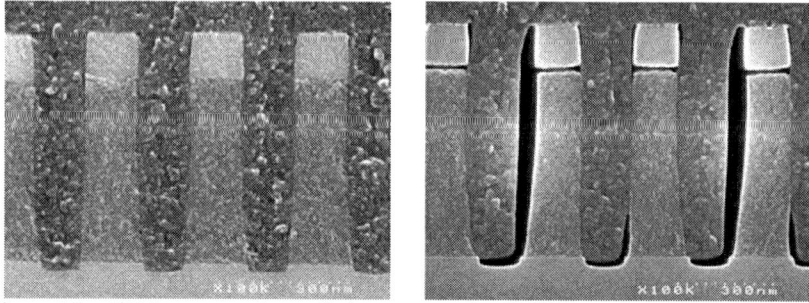

Figure 16. SEM cross-sections of plasma-treated p-SiCOH gratings before (left) and after (right) HF dip.

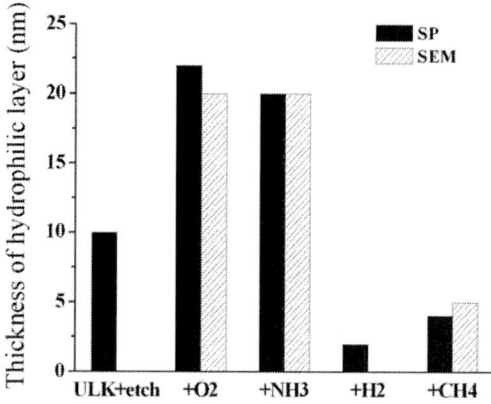

Figure 17. Estimation of the plasma-modified thickness of a porous grating with SP and decoration technique.

Conclusion

The properties of porous materials need to be assessed during their integration in CMOS interconnect levels. In the dual damascene architecture, plasma-induced modifications of porous materials are likely to occur in the bottom of the trench and in the sidewall regions, i.e. respectively in horizontal and vertical geometry. We have used Ellipsometric Porosimetry with a multi-solvent protocol and a kinetic measurement mode on plasma-treated blanket films to replicate the bottom of the trenches. No significant changes of porosity and pore size were evidenced in the remaining ULK film regardless of the plasma. However surface modifications such as densification and hydrophilization are observed. The modified layers induced by the etching plasma, the O_2, and the CH_4 based plasmas do not prevent the solvent condensation. On the contrary the modified layers induced by the NH_3,and the H_2 based plasmas act as a membrane which slows down the condensation kinetic and leads to a quasi pore sealing. We have used the recently developed Scatterometric Porosimetry technique to characterize the sidewall of patterned structures. One advantage of SP is that it can be implemented with an already existing EP tool using a specific modeling procedure. However SP measurements must be done on periodic structures which have to be equivalent to the real circuit in terms of dimensions and stack. We have performed a side-by-side comparison between EP and SP on different plasma-treated samples using the same multi-solvent and kinetic protocols. The results showed that the effect of plasma processes is different on patterned structures compared with blanket films. In particular, different surface densification is observed highlighting that porous material modifications strongly depend on the sample geometry. Such a characterization technique is expected to be useful for microelectronic applications on patterned wafers as it is a potential technique to characterize sidewall damage after each step of the etch process. It also appears as a good complementary technique to EP which only provides quantitative measurements on continuous layers.

References

1. N. Posseme, T. Chevolleau, T. David, M. Darnon, O. Louveau, O. Joubert, *J. Vac. Sci. Technol. B*, **25**, 1928 (2007).
2. N. Posseme, T. Chevolleau, R. Bouyssou, T. David, V. Arnal, J. P. Barnes, C. Verove, and O. Joubert, *J. Vac. Sci. Technol. B*, **28**, 809 (2010).
3. F. Bailly, T. David, T. Chevolleau, M. Darnon, N. Posseme, R. Bouyssou, J. Ducote, O. Joubert, and C. Cardinaud, *J. Appl. Phys.*, **108**, 014906 (2010).
4. M. Darnon, T. Chevolleau, D. Eon, R. Bouyssou, B. Pelissier, L. Vallier, O. Joubert, N. Posseme, T. David, F. Bailly, and J. Torres, *Microelectron. Eng.*, **85**, 2226 (2008).
5. M.R. Baklanov, K.P. Mogilnikov, V.G. Polovinkin, F.N. Dultsev, *J. Vac. Sci. Technol. B*, **18**, 1385 (2000).
6. Patent pending No. FR 09 55027.
7. R. Bouyssou, M. El Kodadi, C. Licitra, T. Chevolleau, M. Besacier, N. Posseme, O. Joubert, and P. Schiavone, *J. Vac. Sci. Technol. B*, **28**, L31 (2010).
8. N. Posseme, T. Chevolleau and R. Bouyssou, T. David, V. Arnal, M. Darnon, Ph. Brun, C. Verove, O. Joubert, *J. Vac. Sci. Technol. B*, **29**, in press (2010).
9. C. Licitra, R. Bouyssou, T. Chevolleau, F. Bertin, *Thin Solid Films*, **518**, 5140 (2010).
10. A. Bourgeois, Y. Turcant, C. Walsh, C. Defranoux, *J.-J. Int. Adsorp. Soc.*, **14**, 457 (2008).
11. H. Kleinknecht, H. Meier, Appl. Opt., **19**, 525 (1980).
12. Fujiwara, H., Principles of Optics, in *Spectroscopic Ellipsometry: Principles and Applications*, John Wiley & Sons, Ltd, Chichester, UK (2007).
13. L. Li, C. W. Haggans, *J. Opt. Soc. Am. A*, **10**, 1184 (1993).
14. M. El Kodadi, S. Soulan, M. Besacier, and P. Schiavone, *J. Vac. Sci. Technol. B*, **27**, 3232 (2009).

15. Fujiwara, H., Data Analysis, in *Spectroscopic Ellipsometry: Principles and Applications*, John Wiley & Sons, Ltd, Chichester, UK (2007).
16. A.W. Adamson, A.P. Gast, *Physical Chemistry of Surfaces*, John Wiley & Sons, New York (1997).
17. P. Revol, D. Perret, F. Bertin, F. Fusalba, V. Rouessac, A. Chabli, G. Passemard, *J. Porous Mater.*, **12**, 113 (2005).
18. N. Posseme, T. Chevolleau, T. David, M. Darnon, J.P. Barnes, O. Louveau, C. Licitra, D. Jalabert, H. Feldis, M. Fayolle, O. Joubert, *Microelectron. Eng.*, **85**, 1842 (2008).
19. K. Maex, M.R. Baklanov, D. Shamiryan, F. Iacopi, S.H. Brongersma, Z.S. Yanovitskaya, *J. Appl. Phys.*, **93**, 8793 (2003).
20. A. Grill, V. Sternhagen, D. Neumayer, V. Patel, *J. Appl. Phys.*, **98**, 074502 (2005).
21. Q.T. Le, M.R. Baklanov, E. Kesters, A. Azioune, H. Struyf, W. Boullart, J.J. Pireaux, S. Vanhaelemeersch, *Electrochem. Solid-State Lett.*, **8**, F21 (2005).
22. M.R. Baklanov, K.P. Mogilnikov, Q.T. Le, *Microelectron. Eng.*, **83**, 2287 (2006).
23. W. Puyrenier, V. Rouessac, L. Broussous, D. Rébiscoul, A. Ayral, *Microporous Mesoporous Mater.*, **106**, 40 (2007).

An Electron Paramagnetic Resonance Study of Defects in Interlayer Dielectrics

B. C. Bittel[a], T. A. Pomorski[a], P. M. Lenahan[a], and S. W. King[b]

[a]The Pennsylvania State University, University Park, Pennsylvania 16802, USA
[b]Intel Corporation, Hillsboro, Oregon 97125, USA

The electronic properties of thin film low-κ interlayer dielectric (ILD) and etch stop layers (ESL) are important issues in present day ULSI development. Low-κ ILD and ESLs with dielectric constants significantly less than those of SiO_2 and SiN are utilized to reduce capacitance induced RC delays in ULSI circuits. Leakage currents, time dependent dielectric breakdown (TDDM) and stress induced leakage currents (SILC) are critical problems that are not yet well understood in ILD. A topic of current interest is ultraviolet light (UV curing) of low-k materials. We have made electron spin resonance (ESR) and current density versus voltage measurements on a moderately extensive set of dielectric/silicon structures involving materials of importance to low-k interconnect systems. Most of the dielectrics studied involve various compositions of SiOC:H. In addition we have also made measurements on other dielectrics including SiO_2, SiCN:H and SiN:H.

Introduction

There is interest in finding new materials for use as inter-layer dielectrics (ILDs) and etch stop layers (ESLs) for use in ULSI. The reliability of these novel ILDs and ESLs are of particular concern [1-6]. TDDM and SILC are of particular interest due to the relatively low breakdown strength of these films [1-6].

There is a vast literature dealing with electron spin resonance (ESR) in Si based dielectric thin films as well as large volume samples of silicon based amorphous insulators [7-15]. The wealth of knowledge provided in the literature on previous ESR studies of Si based dielectrics, bulk Si, and glass offers a foundation for understanding these reliability related defects in ILDs and ESLs. In this work, we compare ESR and electronic measurements to provide insight into performance limiting defects associated with reliability concerns.

In our study we have made ESR and current density versus voltage measurements both before and after exposing the dielectrics to UV light, and films that have experienced an industrial UV curing process. We observe extremely gross differences in the ESR spectra and leakage cur-rent versus voltage response of these low-k films. We find that UV exposure consistently increases both the density of paramagnetic defects and the leakage current density at a given field. Paramagnetic point defects observed in these films include, E' centers, silicon dangling bond defects in which the silicon is back bonded to oxygen, possibly silicon and carbon dangling bond centers and likely organic radicals. We have also made electrically detected magnetic resonance (EDMR) spin dependent trap assisted tunneling measurements on some ILD films. The close correspondence between the ESR and SDT result establishes a direct link between the defects observed in ESR and the defects responsible for the increased tunneling currents. Our preliminary results suggest the UV curing process creates paramagnetic centers which take part in trap assisted tunneling. This tunneling increases dielectric leakage current. Our preliminary results indicate quite clearly that the processing parameters have extremely gross effects upon defect densities within these films. These apparently modest changes in composition result in large changes in defect density, which corresponds to a large change in leakage currents between the films. The films (without UV irradiation or UV curing) have defect densities that differ by a factor of about 15. The UV cured sample results also suggest that our UV irradiation has a similar effect on the films as the post deposition UV curing method.

Experimental

All thin film materials investigated in this study were deposited on 300mm (100) Si wafers using commercially available plasma enhanced chemical vapor deposition (PECVD) tools and various combinations of silane, organosilanes, hydrogen, helium, oxidizers, and porogens. Deposition temperatures were on the order of 250-400°C and film thicknesses ranged from 500-2,000 nm. UV irradiation was accomplished by a 30 minute exposure to 254 nm photons at an intensity of approximately 1×10^{16} photons/cm^{2}s. ESR measurements utilized a Bruker X-band spectrometer with a 300 series bridge, a TE_{104} double cavity and a weak pitch standard. Bias was applied across the dielectric structures with a corona discharge apparatus [16]. Surface potential was measured with a Kelvin probe electrostatic voltmeter [16].

Results

Figure 1 shows the ESR results taken on a k = 6.5 ESL SiN:H film pre and post UV irradiation. We observe a large increase in paramagnetic defect density from 0.3 x 10^{14}cm^{-2} to 4 x 10^{14}cm^{-2} with UV irradiation. The ESR response is dominated by a spectrum with a zero crossing g = 2.0027 and a peak to peak line width of 15 Gauss (1.5 mill-Tesla.) In the simplest cases, the ESR magnetic resonance condition is given by equation 1.

$$h\nu = g\beta H \qquad (1)$$

Where h is Planck's constant, β is the Bohr magneton, ν is the frequency of the microwave radiation and H is the magnetic field at resonance. The g typically depends on the relationship between magnetic field vector and the orientation of the defect under observation. We assign this spectrum to that of the well known K-center since the zero crossing g-value and line width match precisely with literature values [9-10]. The K-center is shown schematically in figure 3.

The leakage current of the k = 6.5 SiN:H film is also greatly increased by exposure to UV irradiation as figure 2 demonstrates. This result suggests that leakage current likely involves the K-centers via trap assisted tunneling.

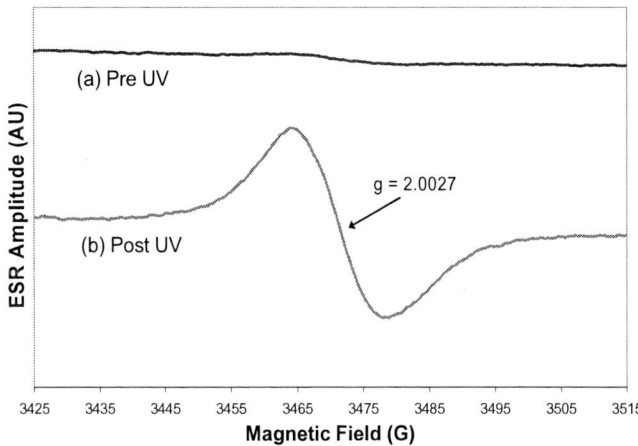

Figure 1. ESR results on a k = 6.5 SiN:H film (a) pre and (b) post UV irradiation. Note that post UV ESR trace is much larger and has a g-value of 2.0027 and a peak to peak line width of 15 Gauss, consistent with a K-center spectrum.

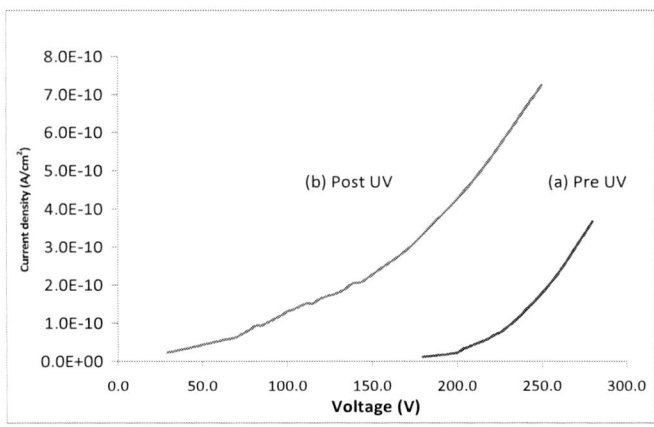

Figure 2. Leakage current results on a k = 6.5 SiN:H film (a) pre and (b) post UV irradiation. Note that UV irradiation greatly increase the amount of leakage.

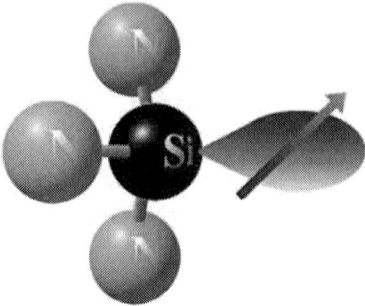

Figure 3. Schematic representation of a K-center, a silicon vacancy back bonded to three nitrogen atoms.

Figure 4 shows ESR results from a k = 4.4 ILD SiO_2 film pre and post UV irradiation. We observe a large increase in the amplitude of the zero crossing 2 Gauss wide g = 2.0007 spectrum from 0.3 x $10^{14}cm^{-2}$ to 1.3 x $10^{14}cm^{-2}$. We assign this spectrum to E' centers, since the observed spectrum is very similar in zero crossing g-value and line width to that of the E' centers reported in the literature [8]. The most prominent E' variant has a zero crossing g= 2.0005+/- 0.0003 and a 2.5 Gauss line width and is shown schematically in figure 5. Figure 6 shows a pre- UV irradiated sample with gain increased to illustrate additional side structure in this SiO_2 film. This additional structure is likely due to the presence of organic contaminants and probably non-bridging oxygen hole or peroxy centers (13). (Note: This side structure vanishes after the UV exposure.) Figure 7 shows a post-UV SiO_2 film with the gain increased compared to figure 4 to illustrate a very different post UV side structure. Note that a pair of side-peaks symmetric with respect to the E' center with a separation of 74 Gauss is observable. This structure is almost certainly due to a hydrogen complexed E' center called the 74 Gauss doublet as the separation (74 Gauss) and the width of the two side-peaks (~3 Gauss) are all consistent with the literature for this center [14-15]. The 74 Gauss doublet is shown schematically in figure 8. The leakage current of the k = 4.4 SiO_2 film is also greatly increased by exposure to UV irradiation as figure 9 demonstrates. This result suggests that leakage current is likely correlated to the existence of the paramagnetic centers detected in the film. Since E' centers dominate the ESR spectrum and have energy levels in the SiO_2 band-gap appropriate for trap assisted tunneling our results suggest that they are largely responsible for the increase leakage in these films [14-15].

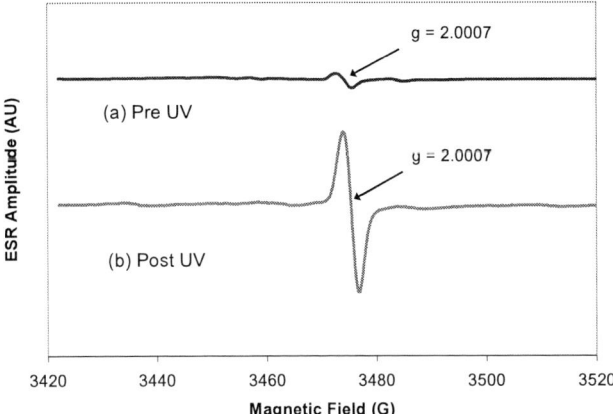

Figure 4. ESR results on a k = 4.4 SiO_2 film (a) pre and (b) post UV irradiation. Note that post UV ESR trace is significantly larger and both spectra have a g-value of 2.0007 and narrow line-width which is consistent with an E' spectra.

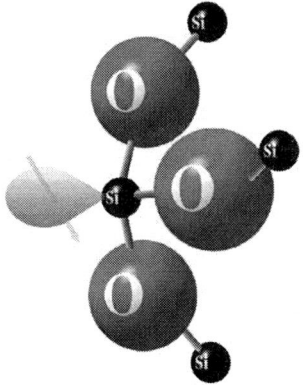

Figure 5. Schematic illustration of an E' center, a silicon vacancy back bonded to three oxygen atoms.

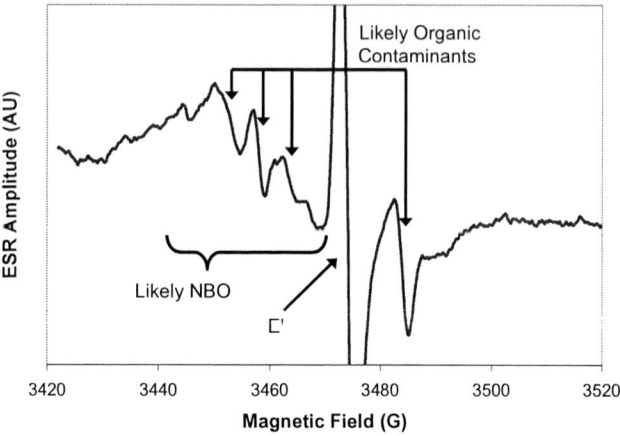

Figure 6. ESR results on a k = 4.4 SiO_2 pre UV film with gain increased to showcase additional structure. This structure is likely due to organic contaminants and possibly due to the nonbridging oxygen hole center (NBO) or peroxy centers.

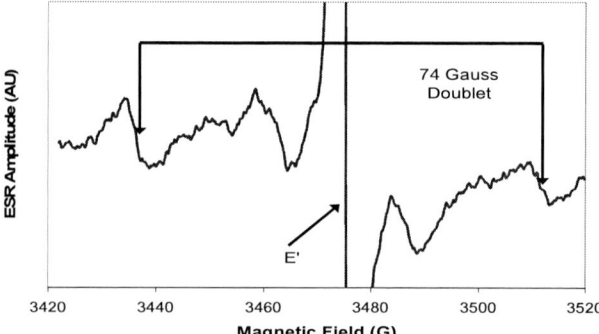

Figure 7. ESR results on a k = 4.4 SiO_2 post UV film with gain increased to showcase additional structure. Distant side peaks are almost certainly due to the 74 Gauss doublet.

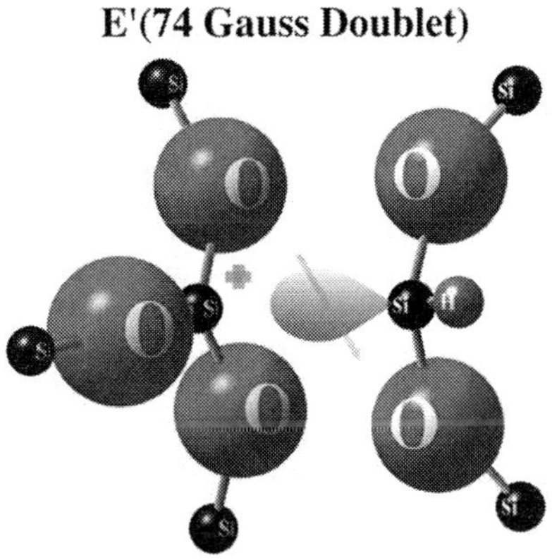

E'(74 Gauss Doublet)

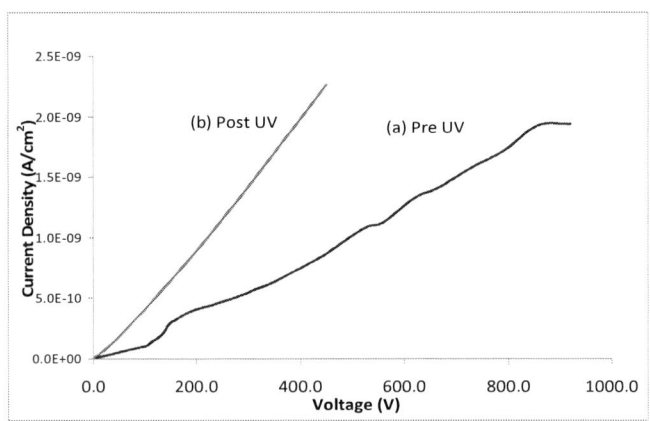

Figure 8. Schematic representation of the 74 Gauss Doublet, an E' center complexed to a hydrogen atom.

Figure 9. Leakage current results on an k = 4.4 SiO2 film (a) pre and (b) post UV irradiation. Note that UV irradiation greatly increase the amount of leakage.

Figure 10 shows ESR results from a k = 3.0 ILD SiOC:H film pre and post UV irradiation. We observe a large increase in defect density of the g = 2.0026 spectrum. We tentatively assign this spectrum to that of silicon or carbon dangling bond. The 2.006 center is a P_{b0} center (P_{b0} centers are silicon dangling bonds located precisely at the silicon dielectric interface) and is thus likely irrelevant to the ILD performance. The P_{b0} center is shown schematically in figure 11. The leakage current of the k = 3.0 SiOC:H film is also greatly increased by exposure to UV irradiation as figure 12 demonstrates.

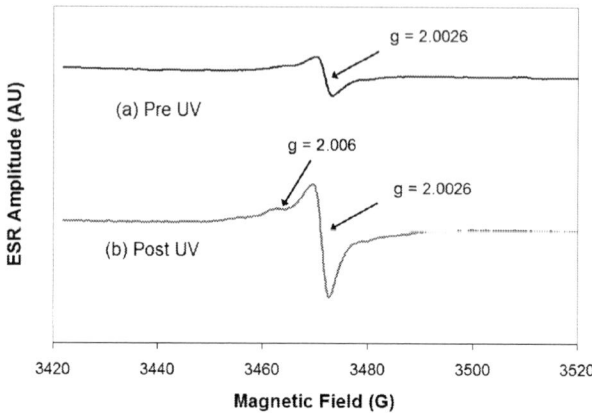

Figure 10. Leakage current results on an k = 3.0 SiOC:H film (a) pre and (b) post UV irradiation. Note that UV irradiation greatly increases the amount of leakage.

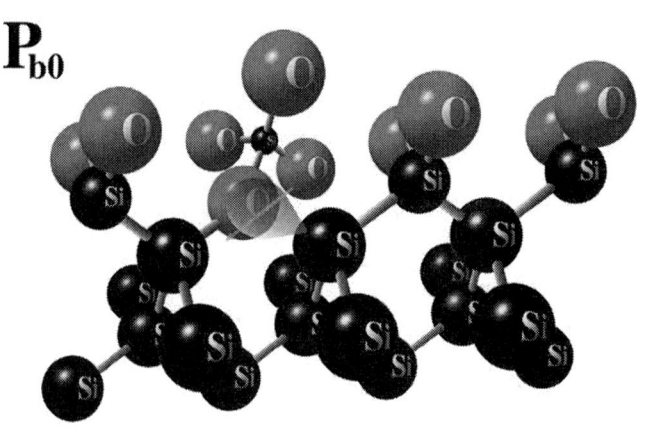

Figure 11. Schematic illustration of the P_{b0} center, which is a silicon dangling bond located precisely at the silicon / dielectric interface and is thus likely not relevant to interlayer performance.

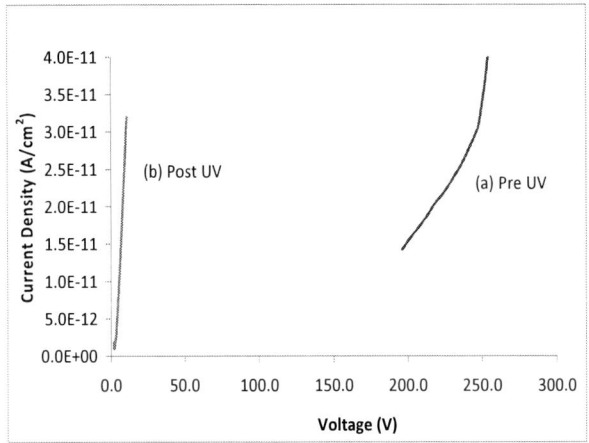

Figure 12. Leakage current results on an k = 3.0 SiOC:H film (a) pre and (b) post UV irradiation. Note that UV irradiation greatly increase the amount of leakage.

Table 1 is a comparison of seven other SiOC:H films that we have examined with ESR. It is interesting to note the large differences in spin densities observed with ESR corresponding to small changes in film composition. The dominating defects in this material system most likely involve unpaired spins on silicon or carbon atoms.

Table I. Comparison of seven representative SiOC:H films.

Film	composition	k-value	10^{17} spins cm^{-3}	width (G)	g-value
SiOC:H	43% Si, 35% C, 22% O	4.8	4.09	4.6	2.0026
SiOC:H	44% Si, 32% O, 8 %C	4.3	5.88	5.9	2.003
SiOC:H	47% Si, 39% O, 14% C	3.2	1.83	10.7	2.0029
SiOC:H	40% Si, 51% O, 9%C	3.7	1.34	10.1	2.0042
SiOC:H	33% Si, 29% C, 39% O	3.05	1.60	3.2	2.0025
SiOC:H	%C > 15%	2.8	2.31	5.4	2.0027
SiOC:H	32% Si, 31% C, 37% O	2.8	0.03	3.1	2.0025

Conclusions

Paramagnetic point defects observed in these films include, E' centers, silicon dangling bond defects in which the silicon is back bonded to oxygen, the 74 Gauss doublet which is E' center complexed to a hydrogen atom, the K-center which are silicon vacancies back bonded to three nitrogens, P_{b0} centers which are silicon dangling bonds which exist precisely at the Si/dielectric interface (and are thus likely not important to ILDs), and possibly silicon and carbon dangling bond centers and likely organic radicals. Leakage current phenomena are likely closely correlated to the existence of these performance limiting defects. Our preliminary results suggest the UV curing process creates paramagnetic centers which take part in trap assisted tunneling. Our results indicate quite clearly that processing parameters, such as film composition, have extremely gross effects upon defect densities within these films.

Acknowledgments

Work at The Pennsylvania State University was supported by Intel Corporation.

References

1. F. Chen, K. Chanda, J. Grill, et al., *Proc. Forty Third Int. Rel. Phys. Sym.*, p. 501 (2008).
2. Y. Ou, P. Wang, M. He, et al., *J. Electrochem. Soc.*, **155**, pp. 283-286 (2008).
3. J. Michelon and R. J. O.M. Hoofman, *IEEE Trans on Dev. and Mcr. Rel.*, **6**, p. 169 (2006).
4. C. Y. Kim, R. Navamathan, H. J. Lee, C.K. Chio, *Surface and Coatings Technology*, **202**, p. 5688 (2008).
5. S. Eslava, G. Eymery, P. Marsik, et al., J. *Electrochem. Soc.* **155**, p. 115 (2008).
6. E. Marhrez, N. Rochet, C. Guedj, et al., *J. Appl. Phys.*, **100**, p. 124106-1 (2006).
7. Y. Nishi, K. Tanaka, A. Ohwada, *J.J. of Appl. Phys.*, **11**, p. 85 (1972).
8. P. M. Lenahan and P. V. Dressendorfer, *J. Appl. Phys.*, **55**, p. 3495 (1984).
9. D. T. Krick, P. M. Lenahan, and J. Kanicki, *J. Appl. Phys.*, **64**, p. 3558 (1988).
10. P. M. Lenahan and S. E. Curry, *Appl. Phys. Lett.*, **56**, p157 (1990).
11. P. J. Caplan, E. H. Poindexter, B. E. Deal, and R. R. Razouk, *J. Appl. Phys.*, **50**, pp. 5847 (1979).
12. D.L. Griscom and E.J. Friebele, *Phys. Rev. B.*, **24**, p. 4896 (1981).
13. J. Vitko, *J. Appl. Phys.*, **49**, p. 5530 (1978).
14. J.F. Conley and P.M. Lenahan, *Appl. Phys. Lett.*, **62**, p. 40 (1992).
15. Z.A. Weinberg, W.C. Johnson, and M.A. Lampert, *J. Appl. Phys.*, **47**, p. 248 (1976).

ECS Transactions, 35 (4) 757-771 (2011)
10.1149/1.3572318 ©The Electrochemical Society

Development of Voltammetry-based Techniques for Characterization of Porous Low-k/Cu Interconnect Integration Reliability

Choong-Un Kim[a], L. S. Chen[a], N. Michael[b],
W. H. Bang[a], Young-Joon Park[c], E. Todd. Ryan[d], and S. King[e]

[a]Mat. Sci. & Eng., University of Texas at Arlington, TX 76019;
[b] Mech. & Aerospace Eng., University of Texas at Arlington, TX 76019;
[c]Texas Instruments Inc., 13121 TI Boulevard, MS 366, Dallas Texas 75243
[d]GLOBALFOUNDRIES, 255 Fuller Rd, Albany, NY 12203
[e]Logic Technology Development, Intel Corporation, Hillsboro, OR 97124

contact email: choongun@uta.edu

This paper concerns the new method of detecting the integration failures in porous low-k (PLK)/Cu interconnects using simple voltammetry-based techniques. In essence, the technique takes advantage of the fact that pores in PLK allow permeation of liquid, including electrolyte, into interconnect structures. The infiltration of electrolyte allows the formation of a micro-cell, consisting of two mating Cu interconnect electrodes and the electrolyte in PLK, where simple linear voltammetry can examine various integration reliability issues pertinent to PLK/Cu interconnects. Specifically, the technique is proven to be effective in detection of 1) failure in Ta barrier, 2) cracks in the capping layer, and 3) trapped impurity in pores in PLK. The working principle of the voltammetry technique and demonstration of its effectiveness is introduced in this paper.

Introduction

Continuing miniaturization of microelectronic devices makes it necessary for back-end-of-line (BEOL) interconnects to incorporate radically different materials and process techniques. Among many planned material and process changes, introduction of porous low-k (PLK) dielectrics has proven to be highly challenging in their integration into Cu interconnects, partly because of newly found reliability failure threats [1-4]. One example is the interconnect failure instigated by a defective barrier which exposes underlying Cu to the dielectric. In a conventional BEOL interconnect, where dense dielectrics are employed, barrier failure and its impact on interconnect reliability is not as serious a concern. Material redundancy both in the barrier and dielectric protects the barrier from defect generation during deposition and later processes. However, a robust barrier does not exist in the case of PLK integrated interconnects; rather, barrier failure is a central reliability concern. This is because the etched surface of PLK dielectric is bound to be irregular, resulting in incomplete coverage by the barrier layer. The problem is further exacerbated by the fact that there exists very little material redundancy in the barrier layer as it has to be just a few nanometers thick. Thermal and mechanical load on the barrier, common during interconnect processing, can easily compromise the integrity of the

757

barrier layer, and the resulting barrier layer failure can instigate various physical failures that may slowly develop in the later processes or device operation [5-7]. A striking example of interconnect failure instigated by the barrier failure is the massive out-migration of Cu into PLK, leaving a large void behind in the Cu, exhibited by the dummy metallization in Fig 1. It is found that Cu out-migration occurs via defects in 20nm thick Ta barriers when the interconnect is exposed to oxidation potential. Another serious reliability concern unique to PLK/Cu interconnects stems from the fact that PLK dielectrics are known to have an open pore structure when the porosity exceeds about 25%. With an open pore structure, process gases and liquids can infiltrate the interconnects and may react with components including the barrier. Furthermore, there is a possibility that the infiltrated gas or liquid may not be entirely removed from the pores, resulting in trapped impurities in the PLK dielectrics. Such impurities may become the source of slowly developing failure mechanisms.

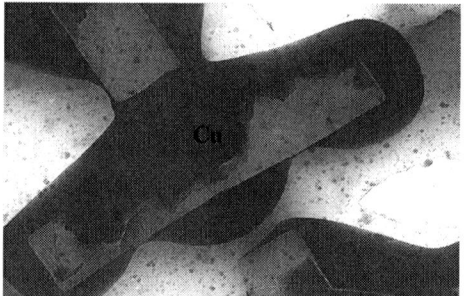

Figure 1. Transmission Electron Microscopy (TEM) micrograph showing an example of Cu out-diffusion by oxidation potential in PLK/Cu interconnects. Notice that a significant fraction of the dummy metallization lost Cu to surroundings.

The newly identified reliability concerns, including barrier failure and trapped impurities, are becoming central issues in PLK technology as they pose significant threat to the success of PLK technology. While extensive developmental efforts have been made to negate the problems, the progress is seriously hindered by the lack of quick, yet accurate, detection techniques for the conditions which lead to the new reliability threats. Standard measurement techniques, ritualized over the years, for characterization of integration success, including microscopic inspection of interconnect structure and series of long-term reliability testing, cannot be effective in revealing the newly problematic features. For instance, microscopic examination of random barrier interface can visualize only a small fraction of the entire interface, but a single tiny failure in any place can be fatal in PLK/Cu interconnects. Moreover, common analytical techniques utilized for chemical inspection of interconnects do not have adequate spatial resolution and sensitivity to detect trapped impurities that may be of small amount and share the same elements with dielectrics. On the other hand, the long-term reliability testing, including electromigration (EM), stress-migration (SM), and dielectric breakdown (DB), may reveal the impact of the concerned defects through exaggerated conditions. However, it not only takes a long time to gain meaningful results, but also often requires extensive additional analysis to identify the failure source. Therefore, the availability of a simple and effective technique that can detect the targeted reliability threats in PLK/Cu interconnects in an as-processed condition is critically important for the desired progress in implementation of PLK technology.

This paper introduces the voltammetry-based interconnect characterization (VBIC) that has been under development in our research for the past few years [8]. The technique is being developed with an aim of accurately detecting component failures in PLK/Cu interconnects without having to rely on complex analysis and sophisticated instrumentation. Specifically, it is proven to be highly effective in detecting failures in the barrier layer, capping layer and the presence of trapped impurities. In this paper, we detail the technique and its working principles, and give examples of how voltammetry is able to detect the defects in advanced interconnect structures provided by various industrial sources.

Background and Experiment

The VBIC methods developed in our research is based on the active utilization of four simple facts. The first is the open-pore property of the PLK dielectrics. With an open-pore structure, PLK dielectrics allow intentional infiltration of any gas or liquid, including an electrolyte solution. This means that an electrolyte containing selected ions can be injected and used as tracers for characterizing failures in interconnect structures. Secondly, the interconnect patterns common in the test and even in the actual devices can be used for voltammetry. For example, common interconnect test wafers contain an inter-digitized comb structure where two Cu interconnect lines are placed in parallel with a dielectric layer in-between. The original intention of such a pattern is for the electrical characterization of PLK dielectrics, such as the measurement of leakage current and DB. However, it also offers an ideal test structure for the application of VBIC. When the PLK layer in such a structure is infiltrated with an electrolyte, the pattern becomes essentially a two-electrode electrochemical cell consisting of two Cu/barrier electrodes and an electrolyte (in PLK). The two-electrode electrochemical cell concept formed in a comb structure is schematically shown in Figure 2. Third is that the motion of ions in electrolyte and electrochemical reaction at interconnects can be controlled and monitored by the application of external potential and measurement of the cell current. Because the pattern offers very large interface area, the resulting electrical current is sizable and can be measured without sophisticated electronics. The fourth idea that makes VBIC applicable to Cu PLK interconnects is that the structures are designed to isolate Cu completely from contact with PLK by barrier or capping layers. Thus, any infiltrated liquid in the electrochemical cell as described above should not find contact with Cu. The application of voltage to the cell creates the conditions for oxidation/reduction of any Cu exposed to the electrolyte. The first two VBIC methods used here have in common the search for this reaction current as an indication that barrier or capping layers are inadequate to keep Cu isolated from the PLK. The third technique expands the concept to find impurities trapped in PLK, rather than inspect for Cu exposure.

The instrumentation required for VBIC is simple and consists essentially of a function generator to modulate the potential of the cell and pico-amp meter to track the resulting current. Figure 3 is a schematic representation of the essential instrumentation needed for the voltammetry system. The functional generator can be of any type as long as it can generate a saw-tooth wave signal at variable rates, but a computer controlled function generator offers many advantages. In our research, both the function generator and pico-amp meter are interfaced with a computer for the convenience of the signal control and data acquisition. One special requirement is an EMI shielding box to reduce the environmental noise during voltammetry measurement.

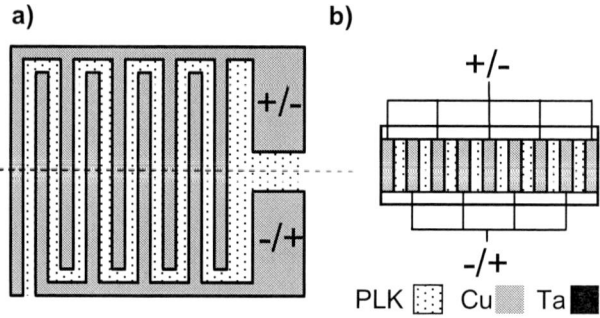

Figure 2. Schematic representation of an interdigitaed comb structure (a) and the formation of two-electrode chemical cell when PLK layer is filled with an electrolyte (b).

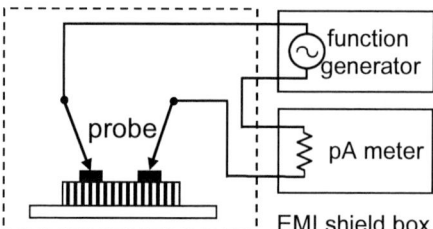

Figure 3. Schematic representation of measurement apparatus used for the Voltammetry.

One of the essential steps for VBIC techniques is the infiltration of the test solution (electrolyte or pure water) into the interconnect structure. In normal circumstances, a guard-rail within the chip surrounds the PLK/Cu test patterns, and liquid infiltration into the interconnect structure is not possible. In order to allow infiltration, therefore, it is necessary to open holes near the test pattern (within the guard rail) or cut through the guard-rail to expose the test structure to the ambient. Our research finds that the latter method works better because it is relatively simple and can be done safely without damaging the test structure. Once the cut is made, the infiltration is done by simply immersing the chip into the test solution. The time to complete infiltration depends highly on the PLK porosity, PLK type (MSSQ vs. SiCOH), test structure, and ambient temperature. For a common comb structure, complete infiltration at room temperature takes place within 2 hours for highly porous PLK (k=2.2-2.4) while it can reach up to several days for less porous PLK. It should be noted that the infiltration is possible even if the dielectric layer does not have an open pore structure. For instance, our research finds that electrolyte infiltration into FSG (fluorosilicate glass) having less than 10% porosity is possible when the chip is immersed into solution for a week. In this sense, the VBIC introduced in this paper are not limited to highly porous low-k interconnects, but can be applied to almost all types of low-k dielectrics. Inducing infiltration at elevated temperature can significantly speed up the process, but the temperature should be below 70°C for protection of the test chip. Once infiltration is completed, the chip should be subjected to the voltammetry immediately to minimize the loss of liquid through vaporization. While voltammetry can be conducted at any ambient temperature, room temperature testing is found to be the adequate for most purposes.

Results

1. Detection of Barrier Failure

As mentioned above, the detection of failed barrier is extremely difficult by any conventional means. Cyclic voltammetry with 1-2% KCl electrolyte solution offers an unconventional resolution to this challenge. The principle idea behind the technique is the fact that Cu makes contact with electrolyte solution only through defects in the barrier and is electrochemically active. Importantly, the normal barrier material (Ta) is electrochemically inactive with an immediate passivation of its surface upon contact with ambient. It is therefore possible to induce reduction or oxidation (redox) reaction of Cu using an external potential and detect the presence of Cu (and thus presence of barrier defects) from the reaction current.

The best condition for the detection of barrier defects is found to be cyclic linear voltammetry, where the external voltage applied to the cell swings linearly between negative and positive values. The ideal voltage range that can be safely employed is +/- 1V with ramp rate ranging from 1-200mV/s. This condition works the best because the cycling bias causes the Cu redox reaction to repeat with each cycle and the resulting current appears as peaks, ideally at the redox potential of Cu, in an IV diagram. In order to better understand the voltammetry behavior of the cell, it may be necessary to consider various barrier conditions and the resulting voltammetry signals. There are three possible variations in the barrier condition: 1) no defect in barrier at both electrodes; 2) equal density of barrier defect at both electrodes (symmetrical defect density); 3) one interface contains more defects than the other (asymmetric defect density). The voltammetry signal varies uniquely with each barrier condition as detailed below.

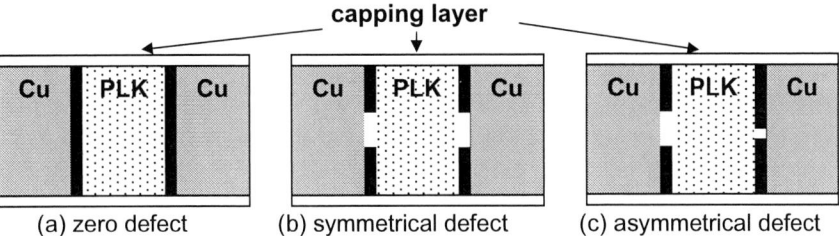

(a) zero defect   (b) symmetrical defect   (c) asymmetrical defect

Figure 4. Schematic representation of three different barrier conditions in PLK/Cu interconnects.

The first to consider is the case of no barrier defect at both electrodes (Fig.4-a). In this case, the cell behaves like a capacitor and it shows a simple and symmetrical IV hysteresis exemplified in Fig. 5. This figure shows the IV hysteresis of the cell formed by comb pattern Cu integrated with k=2.7 FSG dielectric measured under two different voltage ramp rates. It can be seen that the hysteresis quickly becomes steady state (no change with cycles) and that the hysteresis becomes larger under faster ramp rate. The voltammetry data taken from the comb pattern with PLK dielectrics (k=2.2~2.5), regardless of PLK type, shows the same hysteresis pattern when the barrier is not defective. This type of hysteresis forms simply because the cell current is dictated only by the ionic current. In the given voltammetry situation, there are two current components and they are the current by ion migration (K and Cl) and by any reaction that injects additional charges to the cell. When the barrier is defect-free and thus no electrochemical reaction occurs at the Ta interface, the cell current comes only from ion migration (or polarization of electrolyte). In this case, the internal field due to ion migration and the external field work together in determining the total cell current, resulting in the behavior of current saturation with applied bias in each half cycle. The hysteresis becomes larger with increasing ramp rate because the influence of the internal field becomes weaker, as in the case of an electric capacitor. Regardless of the exact mechanism, the formation of a simple hysteresis without peak current indicates the absence of ion injection/drainage at the barrier interface, indicative of intact barrier.

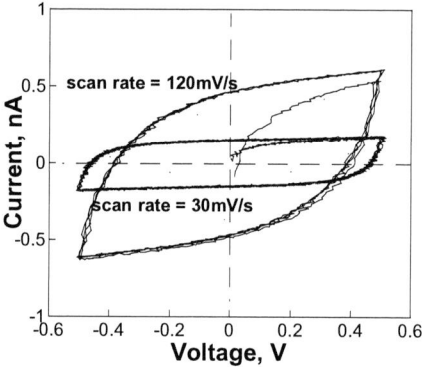

Figure 5. Example of voltammetry singal under two different ramp rates, showing a simple and symmetrical IV hysteresis. This data is taken from Cu interconnect comb structure integrated with FSG (k~2.9) dielectric layer. The width of dielectric layer is 0.15μm, and the thickness of Ta barrier is 20nm.

The second type of voltammetry signal is from interconnects with equal density of barrier defects on the mating interfaces (Fig.4-b). According to our testing on various samples from a few industrial sources, this is the most common type of barrier failure in PLK/Cu interconnects. In this case, the electrolyte can reach the underlying Cu electrode through the holes and cracks in the barrier, making it possible for Cu to react with electrolyte under bias. With an equal area of Cu exposure to the electrolyte on both electrodes, the cell configuration essentially becomes similar to electroplating. In one electrode, Cu dissolves into the electrolyte (oxidation), while Cu plating occurs at the opposite electrode (reduction). The dissolution and deposition of Cu introduces new ions to the electrolyte, not only resulting in an increase in the overall current magnitude, but also a peak in the voltammetry signal because the reactions should occur at a specific redox potential. An example of the resulting voltammetry signal is shown in Fig. 6, and is one of numerous examples we have collected from interconnects of varying PLK types, porosity, pattern density, and barrier thickness. This particular data is taken from Cu interconnects integrated with MSSQ type PLK having 40% porosity and 10nm thick Ta as a barrier layer. The electrolyte used for this measurement is 2% KCl solution. It can be seen that the voltammetry shows markedly different characteristics from the one shown in Fig. 4. The first difference to notice is the overall level of cell current, which is more than an order of magnitude higher. The second is the presence of current peaks at each half cycle of IV hysteresis (marked with an arrow). Testing on various samples reveals that the peak generally appears at +/-0.3-0.4V. Since these two features would not appear without reaction at an electrode and the only known component that can react with the electrolyte in PLK/Cu interconnect is Cu, it is not unreasonable to conclude that the barrier under testing is defective and Cu electrode is exposed for reaction. It is worth mentioning here that we fail to visualize defects in the Ta barrier using microscopic inspection of failed samples, using both scanning electron microscopy (SEM) and TEM. Nonetheless, the presence of defects is confirmed after high temperature baking experiments in which Cu out-migration (as in Fig.1) is observed from the interconnects that voltammetry indicates have defective barriers. These results evidence the near impossibility of detecting barrier defects using microscopic inspection and the effectiveness of the VBIC technique.

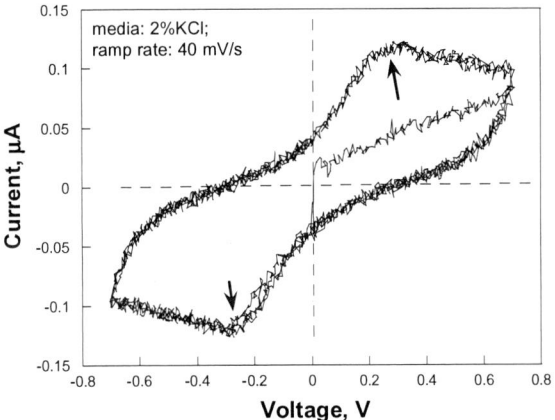

Figure 6. Example of voltammetry singal showing a symmetrical IV hysteresis with reaction peaks at each half cycle. The data is taken from Cu interconnect comb structure integrated with MSSQ (k~2.2) having 40% porosity and 10nm Ta barrier. Note the presence of the current peaks near at +/-0.3V.

The fact that the current peak occurs near at +/-0.3-0.4V appears to suggest that the reaction at the electrode involves a reduction/oxidation of Cu into cuprous ions because:

$$Cu \rightarrow Cu^{++} + 2e \quad E° = 0.34V \tag{1}$$

where E° represents the standard oxidation potential measured against a hydrogen electrode [9]. Our investigation indicates that it is indeed the case [10]. In the given cell configuration, oxidation reaction occurs at the anode and reduction reaction occurs at the cathode. Among these two reactions, it is found that the anode reaction, that is the oxidation reaction, is the rate controlling reaction. This occurs because the oxidation (dissolution) requires external potential in order for it to proceed, while the reduction reaction (deposition) occurs spontaneously. This is indicated by the positive sign of the standard half-cell potential shown in eq.(1). In the two-electrode configuration, the true potential at each electrode is impossible to measure accurately, however, since reduction occurs without impedance, it is likely that the potential drop at the cathode electrode is near zero. This causes the external potential to be applied mostly to the anode electrode, resulting in the peak potential being close to the half-cell potential of Cu oxidation. It is not yet clear why the redox reaction in the PLK voltammetry does not involve the usual two-step reaction path of Cu, involving cupric (Cu^+) ion formation, namely .

$$Cu \rightarrow Cu^+ + e \quad E° = 0.52V \qquad [2]$$
$$Cu^+ \rightarrow Cu^{++} + e \quad E° = 0.15V \qquad [3]$$

We speculate that this is related to the sluggish motion of ions in PLK structure and the unstable nature of cupric ions, but it remains to be investigated.

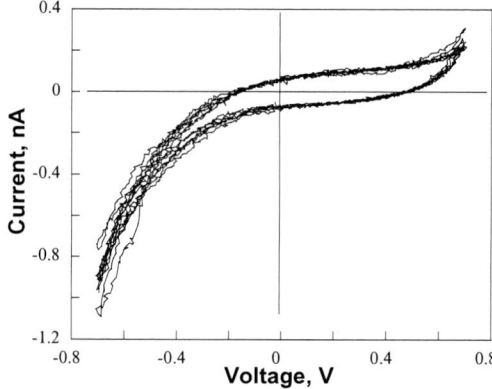

Figure 7. Example of voltammetry signal at the quasi-steady state (after 10 cycles) showing an asymmetrical IV hysteresis resulting from an extreme difference in barrier defect density at two mating electrodes. This data is taken from a comb structure made of Cu interconects with 20nm barrier and FSG low-k dielectrics.

The third type of voltammetry signal, which is found to be rather rare in real interconnects, occurs when the mating electrodes have dissimilar density of barrier defects (Fig. 4-d). In this case, which includes the situation where one side of barrier is completely intact and the other side is defective, the amount of cuprous ions injected by oxidation and drained by reduction is no longer balanced. This leads to development of asymmetrical IV hysteresis as a reflection of the unbalanced reactions at two electrodes. Our investigation using a simulation cell, which is made of two Cu plates with different coverage of Ta, reveals numerous insights helpful in understanding the formation of IV hysteresis under dissimilar defect density. It is found that the level of peak current at each half-cycle is no longer the same when the difference in defect density is small. Also, peak potential starts at the oxidation potential of Cu during early cycles, but drifts away from it. This is attributed to the development of sizable voltage drop at the cathode electrode due to unbalanced cuprous ion drainage compared to its injection rate. When the defect density is excessively dissimilar, the current peak can disappear completely. Rather, the IV hysteresis shows the extreme asymmetric behavior displayed in Fig. 7. This data is obtained from FSG (k=2.9)/Cu interconnects with 10nm Ta barrier after infiltration of 2% KCl. It can be seen that the hysteresis lacks the peak but develops distinctive asymmetricity. Another noteworthy behavior of the asymmetrical hysteresis is the fact that, unlike in the first two cases, it does not reach steady state within the first few cycles. It often takes a few tens of cycles before it reaches the quasi-steady state. We believe that this behavior is result of excessive cuprous ions injected in the first few cycles that cannot be effectively drained. Such behavior is also visible in the hysteresis shown in Fig. 7. Note that the long-tail end of hysteresis varies with cycle.

Many variations in voltammetry signal can exist with variation in interconnect structure, PLK types, and barrier condition in mating electrodes. With such variations, the detection of barrier defect may appear to require complex analysis of the IV signal. However, it turns out that detection is rather simple. Any interconnect with IV hysteresis that deviates from Fig. 5, that is, a simple and symmetrical hysteresis, can be declared to contain defective barrier in one or both of the mating electrodes if the interconnect test structure has symmetrical interface area (a comb structure, for example). When the interface area is not the same, further analysis may be necessary to include the influence of the different electrode area on the hysteresis shape. However, it is found that the number of cycles to reach the steady state hysteresis can be used as an alternative criterion. For an intact barrier, the hysteresis reaches the steady state within the first few cycles even though the interface area is not the same. On the other hand, when the barrier is defective, the voltammetry signal resembles that seen in Fig.7.

2. Detection of Capping Layer Failure

One of the newly arisen concerns of PLK technology, which is as equally vexing as barrier failure, is the possibility of developing cracks in the capping layer. The capping layer refers to the dielectric layer that covers the top of the interconnects of one layer for the purpose of providing electrical isolation from other layers in multi-level structures. The capping layer is subjected to a significant level of mechanical stress during CMP (chemical mechanical polishing) and thus prone to failure by cracking. Similar to the case of barrier failure, cracks in the capping layer can not be characterized by any normal microscopic means because they can be randomly located and they do not produce sufficient contrast to be detectable. Nevertheless, because such crack can allow permeation of process gases and liquids, especially CMP solution, they need full characterization before qualification of PLK integration processes.

The voltammetry technique has proven to be highly effective in detecting cracks in the capping layer. The characterization method is simple and shares many similarities with the technique used in barrier characterization. It starts with the infiltration of an electrolyte solution, typically 2% KCl, into multi-layer interconnect test chip. Then, two interconnect patterns facing across the capping layer are chosen for application of the cyclic bias. Figure 8-(a) shows one of many electrode configurations possible for capping layer integrity characterization. In this particular case, the comb structure at M1 and a large dummy Cu electrode at M2 are chosen. However, any two sets of electrode, one from the bottom and the other from top, can be chosen.

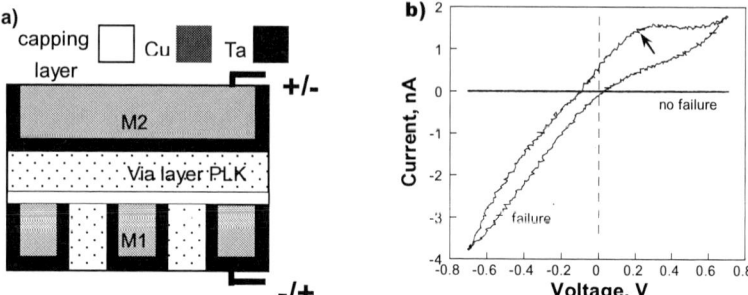

Figure 8. (a) Schematic diamgram showing one possible electrode configuration ideal for capping layer voltammetry characterization in multi-level interconnects; (b) the resulting IV hysteresis conducted on Cu/PLK (SiCOH: k~2.6, ~35% porosity) with 40nm SiCN capping layer using the electrode configuration of (a). Two different cases, indicating intact and failed capping layer, are compared in this plot.

After infiltration, PLKs both at M1 and via layer (and also M2 layer) are filled with the electrolyte. When the capping layer is intact, ion migration across the capping layer is physically prohibited. This means that the conduction circuit by ion migration between M1 and M2 does not exist, leading to open circuit behavior in the IV hysteresis. Specifically, no current flows under any external bias during voltammetry. However, when the capping layer contains physical cracks, ion migration can take place across the capping layer, making the M1 and M2 electrode form a two-electrode electrochemical cell. The exact IV hysteresis can vary significantly with the choice of electrodes, crack density, and location of the crack. However, any hysteresis in current other than zero is an indication of the failure in the capping layer. Figure 8-(b), where voltammetry results are shown from Cu/PLK (SiCOH: k~2.6) interconnects with 40nm SiCN capping layer, presents an example IV hysteresis indicating the failure in the capping layer. Because the electrode configuration in Fig. 8-(a), in which the electrode interface area is asymmetrical, is used for this measurement, the IV hysteresis deviates from a simple and symmetrical hysteresis. However, the simple fact that a sizable amount of current flows with hysteresis formation evidences the conduction by ion migration across the capping layer. Also noteworthy is the fact that the hysteresis contains current peak at the right hand side, marked with an arrow. We believe that the peak originates from a Cu oxidation reaction which occurs because the capping layer crack formed directly on top of the M1 Cu interconnects. Therefore, it may be possible to identify the crack location when the detailed analysis on the hysteresis shape becomes available.

3. Detection of Trapped Impurities

The possibility of trapped impurity is another persistent but less well addressed concern in PLK/Cu interconnection technology. As previously mentioned, the pores in PLK dielectrics are interconnected, increasing the likelihood of infiltration of process chemicals through the pore network and leaving chemical residues on the pore surface. Aqueous chemicals, such as solutions for resist cleans, electroplating, or CMP (chemical-mechanical-polishing), are particularly troublesome, as there are no practical methods of removing the solution without leaving ionic residues. Equally troublesome is the lack of an effective method of detecting it. Common analytical techniques, such as EPMA (electron probe micro analyzer) and SIMS (secondary ion mass spectrometry), are often inadequate for characterization of such ionic residues. Both the limitations in detection resolution and their inability to detect ionic states of elements make them ineffective for impurity detection, especially when contaminants have the same elements as the interconnect components themselves (Cu, barrier, PLK). Our research finds that a simple extension of this VBIC technique can provide an effective resolution to this characterization challenge.

The second VBIC technique utilizes water as a testing media instead of an electrolyte. Since ionic contaminants on pore surfaces of the low-k dielectric can react with the infiltrated water to become mobile ions, the water can become electrically conductive in proportion to the mobile ion concentration. Therefore, the presence of ionic chemicals on pore surfaces can be determined by measuring the level of electrical conduction across any two sets of mating Cu interconnects like the one used for the barrier characterization. The detection works best when the measurement is combined with the cyclic voltammetry technique. With an electrochemically inert Ta barrier, the motion of ions is confined between the two electrodes. Cyclic bias forces ions to move back and forth between the two electrodes, and thus, allows more reliable measurement of conductivity with less interference from the capacitive behavior of the circuit.

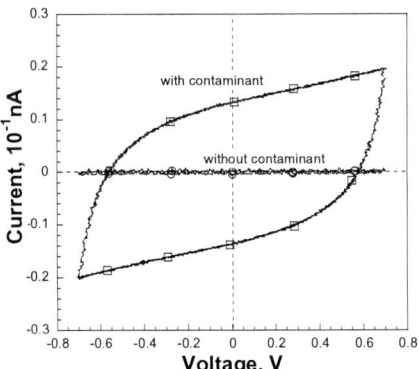

Figure 9. A plot comparing two different I-V voltammetry responses of comb structure taken from the same wafer. The samples used for this characterization is taken from Cu/PLK (SiCOH: k~2.6) comb structure.

The data shown in Figure 9 exemplify the effectiveness of the linear cyclic voltammetry technique in detecting pore contamination when it is conducted on a comb structure after water penetration. The samples used for this test are taken from a standard 2-level interconnect test wafer having PECVD SiCOH as the interlayer dielectric. Two contrasting voltammetry responses from two identical comb patterns from the same wafer are compared in the figure. The first response is the near zero current under linearly cycling voltage. Since de-ionized and degassed water is highly insulating (>10MΩ-cm), its penetration into contaminant-free pores does not impact the electrical conductance of the dielectric, and the cell current remains unchanged by water penetration. Contrasting with this is the second response, which exhibits a simple IV hysteresis with large current, resembling the one in Fig. 5. Within a few cycles, the IV hysteresis reaches a steady state and does not change with further cycling. This is the characteristic voltammetry behavior of an electrolyte cell with non-reacting electrodes.

In our developmental research, a large number of industrial samples at the various stages of integration are examined to determine the effectiveness of water-based VBIC in characterizing the trapped impurities. The technique has proven very successful because it offers extreme sensitivity to the trapped impurity with unprecedented speed. For instance, after infiltration, contamination characterization of test chips in 8" wafer can be completed within an hour even without using automated sample scanning apparatus. Such scanning often showed that the contamination was not universal but located at particular places within a wafer, which may be important information for process engineers. The most time consuming part is the infiltration, as it involves the creation of cut to allow infiltration and the time to complete infiltration. Further development in instrumentation and test process will certainly improve the speed.

The technique needs substantial improvements in two critical areas. The first is a technique for identification of the contaminants detected. In order to prevent the contaminant from entering the PLK, it is important for process engineers to know where those contaminants originate. Our attempts to identify the detected impurity shown in Fig.8 using SIMS (secondary ion mass spectrometry) and EELS (electron energy loss spectroscopy) fail to produce convincing results. Although both characterizations do find near background level of fluorine, which may originate from CH_4 etch gas residue, as a foreign element in PLK, the result is not highly convincing as the spectroscopy signal is too close to the resolution limit. The extreme detection resolution of the voltammetry is possible because it detects the presence of impurities through electrical measurement where the signal can be amplified. Hence, it is desirable to develop an identification method based on a similar approach. One possible route is to combine titration and voltammetry. When contamination is detected, a series of known ions can be added to water and injected to the contaminated pattern. When the injected ions react with impurity ions and produce precipitates, the number of mobile ions decreases, and so does voltammetry signal. Although this approach may require extensive developmental efforts before implementation, it may be the most sensitive method for the contaminant identification.

The second area that requires further development is the quantification of the contaminant. Voltammetry can reveal only the presence or absence of contaminant. The result as given in Fig. 8 cannot provide any indication as to the amount of contaminant. It is, of course, possible to gain a comparative quantification from the strength of voltammetry signal. The chip producing the larger IV hysteresis contains more contaminants when measured at the identical ramp rate and voltage range. However, as knowledge of the exact amount is often desired, it is important to develop a technique for quantification. Among many routes explored in our research, the voltammetry signal itself appears to offer the best route. The voltammetry hysteresis is determined by the ion mobility and quantity. It is therefore possible to determine the ion quantity if the mobility is known. However, if the identity of the ion and its diffusivity in the given PLK media are unknown, the ion mobility is an unknown parameter. Thus, it is necessary to extract both the ion mobility and the quantity from the voltammetry result. In an ideal case, such determination can be done by way of simulating the ion motion and the resulting signal under voltammetry condition. This simulation requires solving the Nernst-Planck equation for a given geometry of PLK/Cu and voltammetry conditions [11-12]. At the present moment, with a lack of suitable analytical solution, the quantification is not yet possible, and further efforts appear to be necessary.

Discussion

As is introduced in this paper, VBIC offers effective resolution to the characterization challenges ahead in PLK/Cu interconnection technology. The methods are extremely simple to apply yet the result is highly accurate. Such accuracy is possible because the technique is specifically tailored to characterize a targeted defect. This eliminates characterization ambiguities and background interference common in other microscopic and spectroscopic techniques. It should be also noted that the technique is probably the only technique that can carry out such characterization for the as-processed condition of the interconnects. While the usefulness of VBIC has been demonstrated in the case of defects in barrier, capping layer, and PLK (contamination), it can be easily extended to other characterization targets. Two examples can be mentioned here.

The first is the characterization of the pore-seal layer that may be deposited prior to the barrier layer with the aim of improving the integrity of the barrier and preventing PLK contamination. One of the critical requirements of the pore-seal layer, created by polymer deposition, is the complete coverage of the PLK surface without pin-holes or cracks. Similar to the case of the barrier and capping layer, flaws in the seal layer are impossible to detect by any microscopic techniques. However, the pore-seal layer can be inspected with the electrolyte based voltammetry technique. The idea behind the characterization is again simple. When the pore-seal is flawless, an electrolyte cannot make contact with the underlying Ta or Cu layer, making the cell an electrically open circuit. No hysteresis formation in voltammetry signal is expected. On the other hand, in case of flaws, the cell becomes electrically conductive because the ions in the electrolyte can reach the Ta/Cu electrode. The voltammetry should yield IV hysteresis with the signal strength reflecting the density of the flaws. Our trials with a few samples from the early developmental stage of pore-seal layer, although limited, do find that such characterization is indeed possible.

The second possible extension of the VBIC is the characterization of the pore structure (size and pore density) in PLK dielectrics, not necessarily in a blank film but at as-processed condition. This capability is important because the pore structure may vary significantly even within a chip, for example, with variation in the pattern density. An extension of the VBIC has the potential to be useful for this purpose as the ions in an electrolyte can be used as a mobility tracer. In this case, another mode of voltammetry can be used, such as step-mode voltammetry after infiltration of 2% KCl as an electrolyte. In the step-mode, the infiltrated comb interconnect pattern with known PLK dimension is subjected to a constant bias until complete polarization of ions by electric field takes place. When the polarization is complete, the cell current becomes zero due to a balanced potential between the external field and the internal field. When the bias is removed at this point, the force to restore chemical equilibrium induces diffusion of polarized ions until attainment of uniform ion composition across the cell. The restoration process produces exponentially decaying current in the opposite direction that can be measured using an ammeter. Since the kinetics of the decaying current are not affected by external field (no external field exists) and governed only by internal diffusion, diffusivity can be computed based on Fickian diffusion model. Then, the pore structure can be indirectly characterized from the ion diffusivity using its sensitivity to pore size and porosity, as is presented elsewhere [13].

The final note on the potential extensions of VBICs is the necessity for more fundamental understanding on voltammetry behavior. As is demonstrated, VBIC fails to yield quantitative data on the characterization target, such as the defect density in the barrier and capping layer. This limitation stems from the difficulty in developing a suitable electrochemistry model because the majority of electrochemical theories are developed for the three-electrode configuration, that is, the cell with working, counter, and reference electrodes. Because the three electrode configuration allows the tracking of the chemical potential at the working electrode, quantitative model relating the measured current to the reaction or ion polarization is possible. However, the VBIC techniques are based on two-electrode configuration where each electrode serves as a working as well as a counter electrode for the other electrode. This is rarely considered in the field of electrochemistry, but our research appears to suggest that two-electrode electrochemical cells can yield useful information and that the theory needs attention.

Summary

This paper introduces the working principles of VBIC for detecting the presence and absence of defects in PLK/Cu interconnect structure. Specifically, the technique of detecting 1) the defects in the barrier, 2) the defects in capping layer, and 3) contaminants trapped in PLK, is shown. As is demonstrated, VBIC allows simple yet effective characterization of component failures in PLK/Cu interconnects without having to rely on complex and sophisticated analysis, yet it needs further development in order to reach its full potential as a characterization metrology. Presenly, developed VBIC methods are incapable of providing quantitative data on the characterization target such as the defect density in the barrier and impurity concentration. Quantitiatve characterization may become possible when suitable models relating electrochemical reaction and electrolyte motion to the resulting voltammetry signal in two-electrode cell are developed.

Acknowledgements

This research is supported by SRC (Semiconductor Research Corporation) under contract #1292.052 and 2071.009.

References

1. K. Maex, M. R. Baklanov, D. Shamiryan, F. Iacopi, S. H. Brongersma and Z. S. Yanovitskaya, *J. Appl. Phys.*, **93**, 8793 (2003).
2. M. Morgen, E. T. Ryan, J.-H. Zhao, C. Hu, T. Cho and P. S. Ho, *Annu. Rev. Mater. Sci.*, **30**, 645 (2000).
3. International Technology Roadmap for Semiconductors, 2006 ed. (Semiconductor Industry Association, San Jose, CA, 2006) (http://public.itrs.net).
4. M. Fayolle, G. Passemard, O. Louveau, F. Fusalba and J. Cluzel, *Microelectron. Eng.*, **70**, 255 (2003).
5. N.L. Michael, C.-U. Kim, P. Gillespie, and R. Augur, "Mechanism of reliability failure in Cu interconnects with ultralow-k materials", *Appl. Phys. Lett.*, **83**, 1959 (2003).
6. N.L. Michael, D.M. Meng, C.-U. Kim, S.H. Kang, and Y.J. Park, Proceedings of Advanced Metallization Conference, (MRS, 2004), pp.269-273.
7. L.S. Chen, W. H. Bang, Choong-Un Kim, Young-Joon Park, "Observation of space charge limited current by Cu ion drift in porous low-k/Cu interconnects", Appl. Phys. Lett. 96, 091903 (2010).
8. C.-U. Kim, D.M. Meng, N. Michael, Y.-J. Park, and S. Satyanarayana, Proceedings of Advanced Metallization Conference, (MRS, 2004), pp.679-685.
9. See for example, *Standard Potentials in Aqueous Solutions*, A. Bard, R. Parsons, eds., (Marcel Dekker, New York, 1985), p. 292.
10. D. M. Meng, N. L. Michael, Y.-J. Park and C.-U. Kim. *J. Electron. Mater.*, **37**, 429 (2008).
11. Malcolm Smyth and Johannes G. Vos, eds., *Analytical Voltammetry*, (Elsevier, Amsterdam, 1992).
12. David K. Gosser, Jr., *Cyclic Voltammetry,* (Wiley-VCH, New York, 1993).
13. D. M. Meng, N. L. Michael, C.-U. Kim and Y.-J. Park, *Appl. Phys. Lett.*, **88**, 261911 (2006).

772

ECS Transactions, 35 (4) 773-804 (2011)
10.1149/1.3572319 ©The Electrochemical Society

Recent Findings In Electrical Behavior Of CMOS
High-k Dielectric/Metal Gate Stacks

G. Ghibaudo[a], J. Coignus[a,b], M. Charbonnier[b], J. Mitard[a,c], C. Leroux[b], X. Garros[b],
R. Clerc[a] and G. Reimbold[b]

[a] IMEP-LAHC, MINATEC-INPG, 3 Parvis Louis Néel, 38016 Grenoble, France.
Email: ghibaudo@minatec.inpg.fr
[b] CEA-LETI, MINATEC, 17 rue des Martyrs, 38054 Grenoble, France.
[c] IMEC, Kapeldreef 75, B-3001 Leuven, Belgium.

We first review the kinetics of the trapping/detrapping process
involved in hysteresis phenomena of HfO_2 films on a large
temperature range (20K-500K). We show that, up to 400K, trapping
and detrapping processes are temperature independent after
correction of threshold voltage variation with temperature. In most
cases, the capture and emission into or from available traps are
mainly controlled by tunneling. Then, we address the variations of
the effective metal work function and its relation with the different
fabrication processes. Based on C-V and internal photoemission
measurements, we show the existence of an interfacial voltage drop
(i.e. dipole) at the high-k SiO_2 interface, as well as of variations in
metal work function with metal gate process. Finally, we present a
complete study of the transport mechanisms throughout the
SiO_2/HfO_2 gate stacks combining C-V, I-V and Transmission
Electron Microscopy measurements, on several nMOS transistors
with different interfacial layer and HfO_2 thicknesses, over a large
temperature range (80K to 400K).

Introduction

The miniaturization of CMOS technologies requires the use of high-k
dielectrics/metal gate (HK/MG) stacks in order to suppress excessive gate leakage and
gate polysilicon depletion. During the past years, promising electrical performances have
been achieved with HfO_2-based dielectric materials in terms of equivalent oxide
thickness (EOT) and gate leakage current. However, there are still important issues
related to trapping and reliability concerns, understanding and optimization of threshold
voltage control via HK/MG work function process dependence as well as detailed
comprehension of transport and conduction mechanisms throughout the HK/MG stacks.

In this paper, we first review the kinetics of the trapping/detrapping process
involved in hysteresis phenomena of HfO_2 films on a large temperature range (20K-
500K) [1,2]. We show that, up to 400K, trapping and detrapping processes are
temperature independent after correction of threshold voltage variation with temperature.
However, at higher temperature (above 400K), a thermally and field activated
discharging process is found to limit the trapping efficiency. In all cases, the trapping and

773

detrapping kinetics are almost temperature independent, suggesting that the capture and emission into or from available traps are mainly controlled by tunnelling.

Then, we address the variations of the effective metal work function (WF) and its relation with the different fabrication processes (gate thickness, High-κ material and anneals) [3]. Based on C-V and internal photoemission (IPE) measurements, we show the existence of an interfacial voltage drop (dipole) at the High-κ/SiO$_2$ interface, as well as of variations in metal work function with metal gate process. We find that the dipole voltage is mostly influenced by the choice of the High-κ material in contact with the SiO$_2$. It induces an increase of 200mV with HfO$_2$ to about 900mV with Al$_2$O$_3$. Moreover, we confirm that WF strongly depends on the gate process parameters, increasing with the gate thickness and decreasing with a high temperature anneal.

Finally, we present a complete study of the transport mechanisms throughout the SiO$_2$/HfO$_2$ gate stacks combining C-V, I-V and Transmission Electron Microscopy (TEM) measurements, on several nMOS transistors with different interfacial layer and HfO$_2$ thicknesses, over a large temperature range (80K to 400K) [4]. The gate leakage currents have been found to be very weakly temperature dependent when plotted versus electric field, except on thicker stacks, in inversion regime, between 300 and 400K. These experimental data have been compared with direct tunneling current quantum simulations in inversion regime, using as few as possible fitting parameters. The modeling results show that the oxide interfacial layer differs from pure SiO$_2$ only below 1 nm thickness.

Trapping And Detrapping Mechanisms In HfO$_2$ Films

In order to fulfill the International Technology Roadmap for Semiconductors (ITRS) requirements for equivalent oxide thickness and gate leakage current, the conventional gate oxide dielectric has to be replaced by higher dielectric materials [5]. Hafnium-based dielectrics are widely investigated as being very good candidates for the silicon nitride replacement [6]. However, before their final use in industry, significant issues remain to be understood such as mobility degradation, metal gate compatibility, threshold voltage instability. To address the latter issue, especially the hysteresis phenomena [7-9], it is necessary to analyze the trapping/detrapping mechanisms in pre-existing defects. In this work, we have studied the trapping and detrapping kinetics over a large range of temperature from 20K to 500K with an appropriate methodology. From the experiment, we report the importance of direct tunneling mechanism especially Shockley-Read-Hall (SRH-like) mechanism for all operation conditions.

The impact of temperature on trapping and detrapping kinetics has been measured on a 4.5nm HfO$_2$ oxide. This oxide has been deposited by ALD on a 0.7 nm native oxide, followed by a post-deposition annealing under nitrogen ambient at 600°C. The gate electrode is a CVD TiN metal. Details about MOS transistors fabrication can be found elsewhere [6].

In order to characterize the hysteresis phenomena, we have proposed a novel time resolved measurement technique allowing the characterization of V_t shift transients at short times [9]. In literature, other methods [10,11] have been proposed to extract the V_t shift using the transconductance g_m or its approximation via $(V_g-V_t)/I_d$. However, both techniques underestimate the V_t shift (up to 50% in the worst cases), because they do not take into account the increase of g_m as Id is reduced, or even underestimate gm. So, we have proposed to probe directly the shift of the I_d-V_g curve (i.e. V_t shift), by comparing the transient drain current to the reference I_d-V_g characteristic measured at very short sweep time (here 5 µs) where no trapping is taking place (Fig. 1).

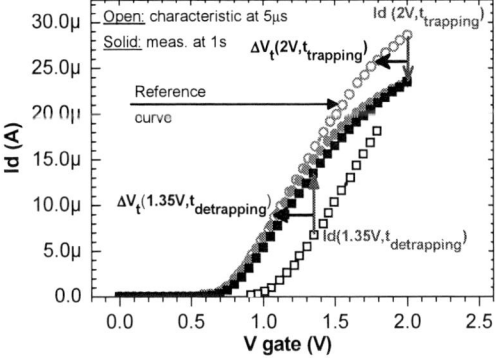

Fig. 1. Procedure used for V_t shift extraction from a 5µs reference curve (after Ref. [1]).

Moreover, we can notice that the up and down curves at different stress times are only shifted along the V_g axis. This means that the mobility variations due to the trapped charges can be neglected in our V_t extraction.

The temperature dependence of the charging effects was first studied with the classical methodology originally described in [7].

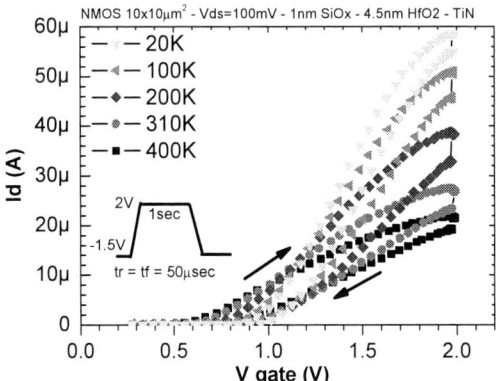

Fig. 2. Dynamic I_d-V_g characteristics as obtained for various temperatures (after Ref. [1]).

As shown in Fig. 2, an increase of V_t-shift is observed with temperature when V_g is swept from -1.5V to a constant bias (here 2V) and back to -1.5V. In Fig. 3, we report the V_t-shift after one second at maximum bias in the 20K-400K range.

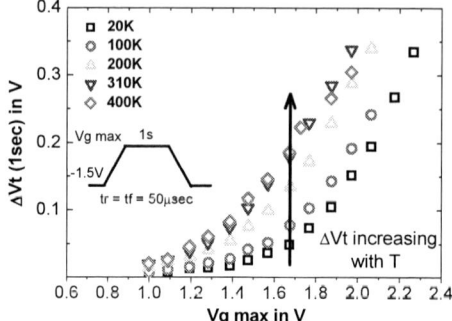

Fig. 3. Evolution of hysteresis in V_t at various temperatures (after Ref. [1]).

The increase of trapping with temperature is related to the decrease of the threshold voltage with temperature. As the trapping depends on the electric field, we should rather plot the trapping level versus V_g-V_t (Fig.4). Thus, we can see that the dependence of trapped charge level with electric field remains nearly the same from 20K up to 400K.

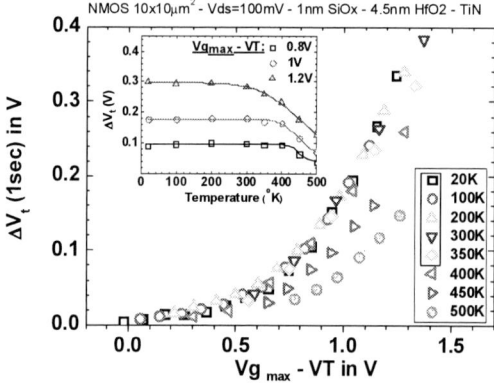

Fig .4. Evolution of hysteresis at various temperatures after V_t correction (after Ref. [1]).

This feature clearly indicates that, for usual conditions (weak V_g-V_t and T up to 400K), tunneling SRH-like trapping mechanism is prevailing. At higher temperatures, there is the onset of an electric field and temperature activated mechanism (Poole Frenkel) limiting the trapped charge amplitude.

In Fig. 5, we compare the trapping kinetics at low temperatures. All the transients well superimpose indicating that the capture process is governed by SRH statistics with a trapping rate $\sigma_n.v_{th}.n_s.T_n$ independent of temperature. The latter result means that the capture cross section does not vary significantly with temperature as already shown in the literature for other types of traps [12]. Moreover, we used in [9] for trapping modeling, a capture trap cross section of the order of $10^{-18}cm^2$, which corresponds to neutral traps and weakly temperature dependent [13]. However, the temperature dependence of σ can differ with the nature of traps. The V_t shift transients have been obtained from drain current transients using the methodology described in the experimental section of Ref [9].

At higher temperatures (see Fig. 6), the transients show a reduction of the trapped charge already illustrated in Fig. 4. However, the trapping kinetics remains almost unchanged, indicating that the trapping mechanism is nearly the same, but with a reduced number of traps.

The detrapping kinetics are reported in Fig. 7 for various temperatures and a fixed $(V_g - V_t)$ value. The reduction of the trapping level with temperature is again demonstrated at short times. The normalized detrapping kinetics (Fig. 8), defined as $[\Delta V_t(0) - \Delta V_t(t)]/[\Delta V_t(0) - \Delta V t(1s)]$, appears to be rather temperature independent above 300K suggesting that the involved detrapping process is mainly by direct tunnelling.

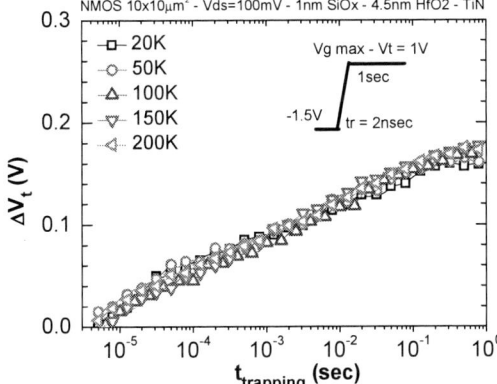

Fig. 5. Trapping kinetics of V_t shift at low temperatures for $V_{gmax} - V_t = 1V$ (after Ref. [1]).

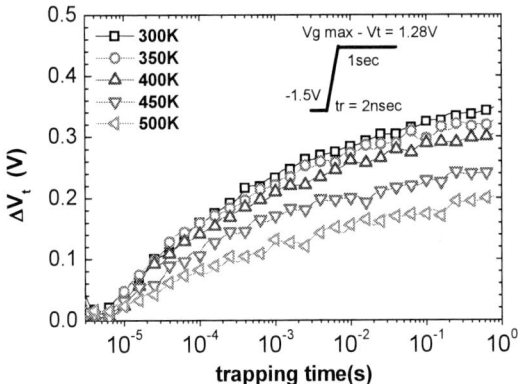

Fig. 6. Trapping kinetics above 300K at $V_{gmax} - V_t = 1.3V$ (after Ref. [1]).

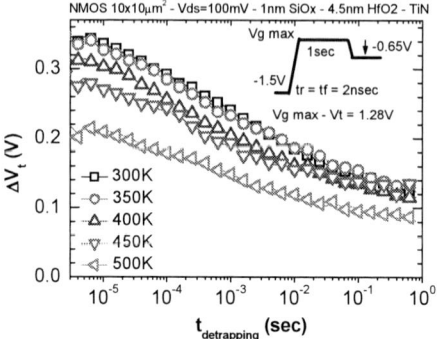

Fig. 7. Detrapping kinetics above 300K at a fixed ($V_{gmax} - V_t$) (after Ref. [1]).

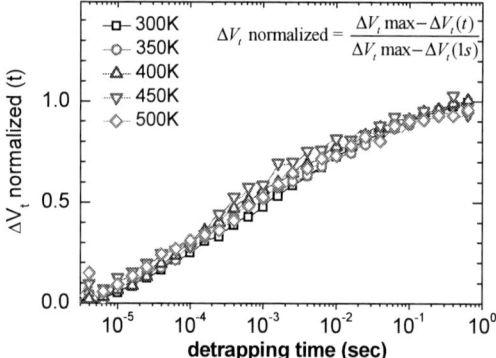

Fig. 8. Normalized detrapping V_t shift (from Fig. 7) showing the same kinetics (after Ref. [1]).

Effective Metal Gate Work Function In Metal/High-k Gate Stacks

The scaling down of MOS transistors leads to thinner gate dielectric, which cannot be obtained with standard SiO_2/Poly-Si stacks without a huge increase of the gate leakage current. The use of new materials in the gate stack, such as high-k instead of SiO_2 and Metal instead of Poly-Si, is now a good solution to reduce both the EOT and the gate leakage current. However, the control of the threshold voltage (V_t) in gate first high-k/Metal gate devices remains a problem. Indeed, V_t is determined by the metal work function (WF_M) but it depends strongly on the process, resulting in an effective metal gate work function (WF_{Meff}) different from WF_M. This phenomenon has been widely investigated during the past years, concluding that one or several potential drops, within the stack, induce a WF_{Meff} variation. However, the origins of these potential drops as well as their locations within the stack are still under discussion in the literature. The Fermi level pinning explains this shift due to the presence of metal-induced gap states at the metal/high-k interface [14,15]. Oxygen vacancies have also been advanced as they could lead to positive charges in the dielectric or to dipoles if the charges are located near the metal/high-h interface [16,17]. Recently, Iwamoto found a relationship between WF_{Meff} and the high-k in contact with the SiO_2 interfacial layer, due to the presence of energy offsets at the high-k/SiO_2 interface [18]. However, the nature of the high-k is not the only parameter involved in the WF_{Meff} shift. Indeed, WF_{Meff} also shows a drastic decrease for interfacial SiO_2 thickness less than 2nm (V_{fb} Roll-Off) [19]. Other process such as gate deposition and annealing step might have also an impact on WF_{Meff} [20].

The aim of this work is to understand the origin of these WF_{Meff} shifts to allow the optimization of metal/high-k MOS stacks. We first describe the studied devices, which have been designed especially to allow both internal photoemission and capacitance versus voltage measurements. Then, we measure the WF_{Meff} of these devices for different high-k dielectrics based on C-V curve analysis. Internal photoemission (IPE) will allow separating the different effects influencing WF_{Meff}. The roll-off phenomenon will also be studied and it will be clearly shown which part of the stack is involved in this phenomenon. Finally, the WF_{Meff} variations due to various processes and materials will be discussed.

Our study relies on the use of two complementary electrical characterization methods, namely capacitance versus voltage measurements and internal photoemission data (IPE). Each method requires the design of specific samples. A reliable extraction of WF_{Meff} using C-V analysis needs varying dielectric layer thickness [21], whereas IPE measurement requires relatively thick dielectrics layers, in order to reduce the parasitic leakage current. It also requires a semi-transparent gate to allow photons reaching the dielectric's interfaces [22]. One could fabricate two sets of samples, one for each measurement. However, the different processes could lead to dissimilar devices and maybe different WF_{Meff}. For this reason, we choose to perform both measurements on the same samples, which must conform to both C-V and IPE requirements.

The studied samples are n-MOS capacitors fabricated on a 200mm p-type Si wafer. The dielectric is a bi-layer material composed of a beveled SiO_2 layer with a high-k dielectric film above it. To perform the SiO_2 bevel, we first grow a 20nm thick SiO_2 layer on the wafers. Then, the wafers are dropped into a HF bath and progressively pulled upwards. The latter leads to a SiO_2 beveled layer with a thickness ranging continuously from 6Å at one side of the wafer to 60Å at the other side. The dielectric stack is completed with an HfO_2 layer of various thicknesses, from 1 monolayer to 10nm, depending on the wafer. In some cases, a thin Al_2O_3 film (5, 10 or 15 ALD cycles, with a growth rate of about 1Å/cycle) has been inserted between the HfO_2 and the SiO_2 bevel. The metal gate consists of a 5 or 10nm thick TiN or TaN layer deposited using either PVD or CVD process. The metal is finally covered with a 200nm polycrystalline Si film and annealed at different temperatures (750°C, 1050°C or no anneal). Such fabricated structures with the SiO_2 bevel and the different HfO_2 thicknesses meet the requirement for the extraction of WF_M by C-V analysis. To get the transparency gate required by IPE, the Poly-Si capping film has been removed on one half of some wafers using a Tetramethylammonium hydroxide (TMAH) wet etching. Such a cold process does not alter the stack properties as shown by additional C-V measurements (not shown here).

The analysis of the C-V response of MOS devices is widely used for its capability to extract the equivalent oxide thickness (EOT) and the flat band voltage (V_{fb}) [23]. The analysis of these parameters allows the comparison of processes and materials properties for a further use in production [24-26]. The flat band condition is realized when there is no charge in the substrate. In this case, the gate bias is equal to V_{fb}. Equation 1 shows that V_{fb} depends on the Silicon work function WF_{Si}, metal gate work function WF_M and internal potential drop in the insulator V_{ins}.

$$V_{fb} = WF_M - V_{ins} - WF_{Si} = WF_{Meff} - WF_{Si} \tag{1}$$

For example, we illustrate in Fig. 9a the effect on WF_{Meff} of a sheet of charges at the Si/SiO$_2$ interface. Any charge in the dielectric would induce a potential drop in the gate stack depending on the charge level and its distance to the metal gate. V_{ins} can also be a combination of fixed charges and interfacial energy drops (see Fig. 9b). However, WF_{Si} is the only known parameter (related to the doping level of silicon), therefore one can only extract the sum of WF_M and V_{ins}, called WF_{Meff}. In order to determine more accurately the value of WF_M, we use a method, which removes the effect of fixed interfacial charges [22]. Indeed, Eq. 2 shows the dependence of V_{ins} in case of fixed charges ($Q_{Si/SiO2}$ and $Q_{HfO2/SiO2}$). V_{ins} is actually a function of the different insulating layer thicknesses (EOT and EOT_{HfO2}) and of the voltage drop δV_1 and δV_2 at the different interfaces.

$$V_{ins} = Q_{Si/SiO2} \times EOT/\varepsilon_{SiO2} + Q_{SiO2/HfO2} \times EOT_{HfO2}/\varepsilon_{SiO2} + \delta V_1 + \delta V_2 \tag{2}$$

So, a linear dependence of WF_{Meff} with both dielectric thicknesses indicates that fixed charges are only present at the interfaces and not in the bulk of the dielectrics. In this case, the linear extrapolation of WF_{Meff} at EOT = 0 and $EOT_{HfO2} = 0$ ($WF_{Meff}(0,0)$) allows to evaluate the work function only influenced by the interfacial voltage drops (δV_1 and δV_2) called WF_{MCV}. This analysis necessitates that EOT and EOT_{HfO2} both sweep a large range of values, as it is the case for the studied devices.

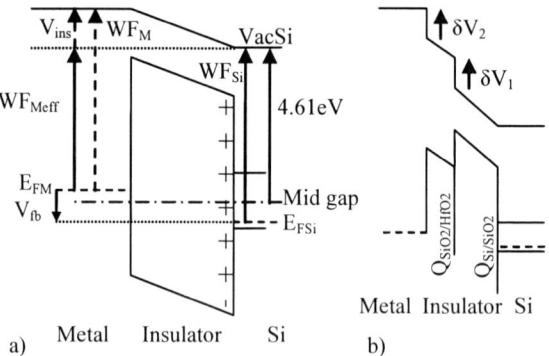

Fig. 9. a) Band diagram of MOS device including a sheet of fixed charges at the Silicon/Insulator interface. b) Part of the band diagram when interfacial energy drops occur (δV_1 and δV_2) within the insulator (after Ref. [3]).

To simplify the analysis, we calculate ΔWF_{Meff}, the difference between WF_{Meff} and the energy from the vacuum of silicon to its mid-gap (4.61eV).The sign of ΔWF_{Meff} can directly be used to identify the type of metal gate: ΔWF_{Meff} positive for P-type material and ΔWF_{Meff} negative for N-type material. Figure 10 shows ΔWF_{Meff} as a function of EOT for devices with a SiO_2 bevel and various HfO_2 thicknesses. Devices without HfO_2 are also shown for comparison. All samples have a 10 nm thick TiN gate and have been annealed at 750˚C. As can be seen, for the same HfO_2 thickness, almost all measurements follow the same straight line, in agreement with Eq. 2. This implies that only a positive sheet charge $Q_{SiO2/Si}$ is present in the SiO_2 dielectric. For SiO_2 only devices, we obtain $WF_{MCV} \approx 4.65eV$. For devices with HfO_2, we observe that all the lines have the same intercept, 200mV higher than for SiO_2 only devices. This fixed intercept value with the EOT_{HfO2} strongly suggests that no fixed charges are present at the SiO_2/HfO_2 interface. It could also be interpreted by a variation of the charge density at the HfO_2/SiO_2 interface due to charge trapping [27]. Nevertheless, to obtain the observed 200mV shift, the charge density should be proportional to $1/EOT_{HfO2}$, corresponding to a very high and unrealistic amount of charges ($> 5.10^{13}$ cm^{-2}) for a 1ML thick HfO_2 layer. Hysteresis trapping analyses did not confirm such density of traps at thin EOT [28]. It should be noted that, in such a case, the amount of charges would be determined by the Fermi level in the substrate. So, the doping level in the substrate should influence WF_{MCV}. However, measurements performed on N- and P-type substrate have demonstrated no WF_{MCV} dependence. This feature leads us to conclude that traps in the dielectrics cannot explain the constancy of WF_{MCV} with EOT_{HfO2}. Moreover, the WF_{MCV} shift observed with HfO_2 cannot be due to fixed charges since it appears on very short distances (1ML). Therefore, the origin of the 200mV positive shift observed with HfO_2 seems likely associated to an interfacial voltage drop at one of the HfO_2 interfaces.

We have fabricated devices with a thin Al_2O_3 layer (several ALD cycles) between the SiO_2 and the HfO_2, since, as reported in the literature, Al inclusion into the gate stack can increase WF_{Meff} [29,30]. Indeed, the presence of the Al_2O_3 layer induces a 800mV increase of WF_{MCV} as compared to the SiO_2 value (see Fig. 11). This shift is higher than values usually reported in the literature (570mV [29]). Moreover, it appears for very thin Al_2O_3 layers and is relatively independent of the HfO_2 thickness (3nm to 10nm). So, it cannot be justified by fixed charges or interaction between the metal gate and Al diffused from the Al_2O_3 layer. This feature suggests that an energy drop exists at the $SiO_2/High$-κ interface controlling WF_{MCV}.

As shown before, the C-V analysis performed on devices with different dielectric thicknesses, allows an assessment of the metal work function without the effect of fixed charges. However, it cannot clearly localize the position of the interfacial voltage drops in the stack. Therefore, we further investigate the various dielectric interfaces through IPE measurement.

In MOS devices, the contact of the insulator with the metal electrode and the silicon electrode forms two energy barriers (respectively Φ_{bM} and Φ_{bSi}) for electrons (see Fig. 12a). Φ_{bM} (Φ_{bSi}) represents the energy difference between the metal Fermi level (silicon valence band) and the dielectric conduction band. The aim of IPE measurement is to determine these energy barriers. To this end, a monochromatic source of a selectable wavelength is used to excite electrons at both electrodes near the insulator (1) (see

Fig. 12b).

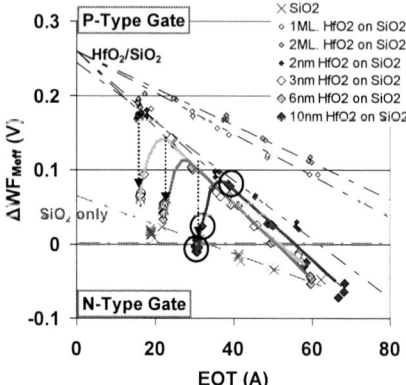

Fig. 10. ΔWF_{Meff} versus EOT for SiO$_2$ only dielectric and different thicknesses of HfO$_2$ (1ML to 10nm) (after Ref. [3]).

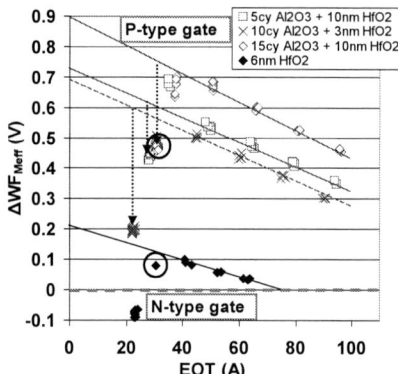

Fig. 11. ΔWF_{Meff} versus EOT for devices including a thin Al$_2$O$_3$ layer between the HfO$_2$ and SiO$_2$ (after Ref. [3]).

If the photons energy, hv, is sufficiently large, photoelectrons can enter the insulator's conduction band (2). Then, the electric field in the insulator E_{ins} drives the photoelectrons from one electrode to the other (3). It is then possible to determine Φ_b at the emitting electrode by analyzing the photocurrent, which is a function of hv. Both Φ_{bM} and Φ_{bSi} can be determined by changing the direction of E_{ins}. Since IPE requires that the photoelectrons reach both sides of the dielectric layer, they must be generated near the dielectric interfaces. Therefore, the gate must be thin enough in order to avoid the absorption of photons. Here, we use thin TiN (10nm or less) layers as gate metal without any capping (poly-Si removed).

The quantum yield Y represents the ratio between the number of electrons crossing the insulator and the number of incoming photons. Y is expected to vary with the energy according to the following function $(\Phi - h\nu)^n$. Φ is the energy at which electrons can enter the dielectric conduction band and n an integer depending on the band structure of the emitting electrode (2 for a metal and 3 for a semiconductor) [31]. As shown in Fig. 13, the linear extrapolation of $Y^{1/n}(h\nu)$ yields the value of Φ. Figure 13 reveals a dependence of Φ with E_{ins}, in reason of the image force effect (IF on Fig. 12b), which induces a reduction of Φ with the square root of E_{ins}. So, Φ_b can be extracted from the linear extrapolation of Φ versus $E_{ins}^{1/2}$ (namely Schottky's plot) at zero field [32]. The determination of both Φ_{bM} and Φ_{bSi} yields another evaluation of WF_M, namely WF_{MIPE} (see Eq. 3 and Fig. 12a).

$$\Delta WF_{MIPE} = \Phi_{bM} - \Phi_{bSi} + E_g/2 \tag{3}$$

Both WF_{MCV} and WF_{MIPE} give a measure of the Fermi level difference between the electrodes at the flat band conditions and they should be equal.

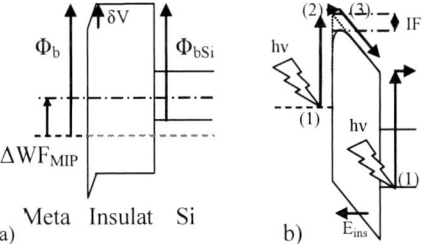

Fig. 12. a) Relation between ΔWF_{MIPE} and the different barrier height. b) The three steps in the internal photoemission phenomenon (after Ref. [3]).

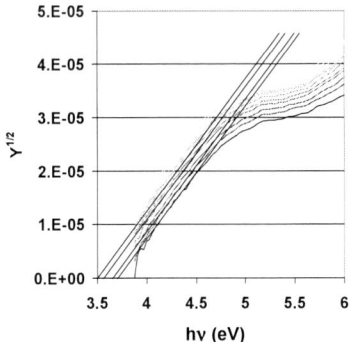

Fig. 13. $Y^{1/2}$ (injection from the metal gate) versus hν for increasing gate bias measured on a TiN/thick SiO$_2$ device (after Ref. [3]).

However, IPE measurements cannot be carried out on every device used in the C-V analysis due to two main constraints. First, the IPE current is very weak (often less than

10^{-5}A/m²). Secondly, it is impossible to measure Φ_b on a bilayer dielectric if the second one (referred to as emitting electrode) has a higher conduction band than the first one. As a result, to be measurable, a device might be composed of a very thin HfO$_2$ layer deposited on a thick SiO$_2$ layer or a very thick HfO$_2$ layer deposited on a thin SiO$_2$ layer, such that photoelectrons can tunnel through the bottom SiO$_2$ layer.

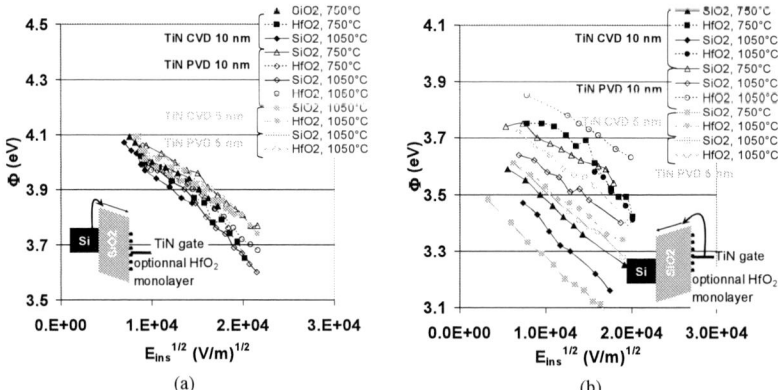

(a)   (b)

Fig. 14. a) Φ versus $E_{ins}^{1/2}$ for devices with and without 1ML HfO$_2$, varying TiN thickness (5 and 10nm) and different anneals (750˚C and 1050˚C). a) Injection from the substrate. b) Injection from the gate (after Ref. [3]).

Φ_{bSi} (the barrier height from the substrate side) is found almost the same for all devices at 4.3eV (see Fig. 14a and b), in agreement with values reported in the literature [33,34]. On the other hand, we observe large variations of Φ_{bM} (the barrier height from the metal side), which varies over 500mV when changing the TiN thickness, anneal temperature or adding 1ML of HfO$_2$. Having Φ_{bM} and Φ_{bSi}, we can extract WF$_{MIPE}$ using Eq. 3 and compare it to WF$_{MCV}$ as obtained previously on the same devices (see Fig. 15). As can be seen, we obtain a good agreement between W$_{MCV}$ and W$_{MIPE}$ confirming that the IPE analysis allows valuable extraction of WF$_M$. The advantage of the IPE is that it permits to decompose WF$_{MCV}$ in two parameters, allowing a detailed analysis of its variations. Here, we clearly demonstrate that all Φ_{bSi} are insensitive to the process and that the observed variations of WF$_{MCV}$ with the process can be explained by a voltage drop at one of the HfO$_2$ interface. However, at this stage, we cannot determine the HfO$_2$ interface responsible for these voltage drops. Another study has already identified a 0.3eV voltage drop at the HfO$_2$/SiO$_2$ interface [35]. This conclusion is based on the assumption that the voltage drop cannot be at the metal/HfO$_2$ interface. However, other works suggest that voltage drops can occur at the metal/high-k interface [36,37]. The aim of the following analysis is to clearly find the interface responsible for the WF$_{Meff}$ variations.

Using IPE, we have followed the evolution of the barrier height at the metal/HfO$_2$ interface when a strong variation of WF$_{Meff}$ occurs (600mV increase), as obtained by the insertion of a thin Al$_2$O$_3$ layer at the HfO$_2$/SiO$_2$ interface. To permit the measurement of Φ_{bM}, we choose devices with a thick HfO$_2$ layer deposited on a thin SiO$_2$ layer. We found that Φ_{bM} remains constant around 2.55eV with and without the thin interfacial Al_2O_3, (see Fig. 16), whereas large variations of WF_{Meff} were observed with C-V measurement on these devices (circles in Fig. 11). This clearly indicates that the metal/HfO_2 interface is not involved in the WF_{Meff} variation.

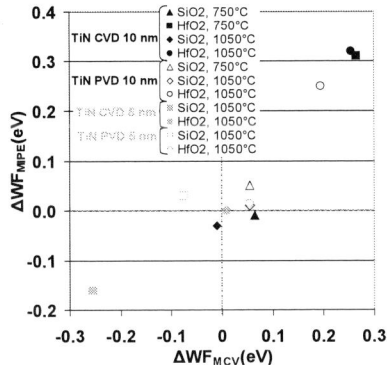

Fig. 15. Comparison between C-V and IPE measurements for devices with different TiN thicknesses (5 or 10nm), dielectrics (SiO_2 or SiO_2 with 1ML HfO_2) and anneals (750°C and 1050°C) (after Ref. [3]).

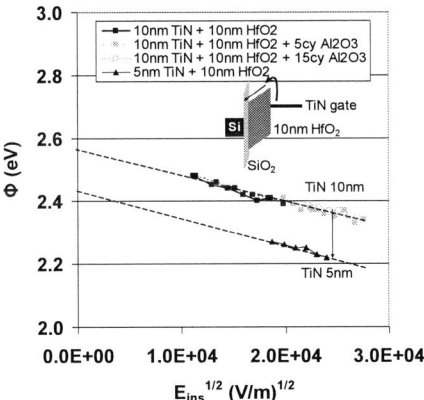

Fig. 16. Φ_{bM} versus $E_{ins}^{1/2}$ for devices with a thick HfO_2 layer on a thin SiO_2 with two different TiN gate thicknesses (5 or 10nm) and an optional Al_2O_3 layer between the SiO_2 and the HfO_2 (after Ref. [3]).

To infer the metal/dielectric interface, we analyze the evolution of the voltage drop at the metal/dielectric interface ($\delta V2$ in Fig. 9) for SiO_2 and HfO_2 dielectrics. Adding the affinity X_{SiO2} (0.90 to 0.95eV) to Φ_{bM} obtained for a SiO_2 layer in contact with a 10nm thick TiN layer, one gets for $WF_{Mdie} = 4.6eV$, which is a measurement of $WF_M + \delta V_2$. Therefore, we observe that WF_{Mdie} is equal to WF_{MCV} obtained for the same devices. This means that WF_{MCV} is controlled by WF_M and by a voltage drop at the metal/SiO$_2$ interface. To compare WF_{Mdie} with WF_{MCV} for devices containing HfO$_2$, we have determined the electronic affinity of the HfO$_2$ layer (X_{HfO2}) by Ultraviolet photoemission spectroscopy (UPS) and ultraviolet (UV) ellipsometry analysis [38]. With UPS measurement, we could probe the energy from the HfO$_2$ valence band to the vacuum. From UV ellipsometry, we could determine the gap of the HfO$_2$. It should be noted that, when HfO$_2$ is crystalline, defects generates band tails yielding two values for the gap Eg and Egd. The latter accounts for band tails and should be used in our case [39]. We get 1.85eV for X_{HfO2} [38], which is slightly smaller than other results of literature 2eV [40,41]. This provides a WF_{Mdie} of 4.40eV (4.55eV with a 2eV electron affinity). We notice that WF_{Mdie} obtained on HfO$_2$ is equal or smaller than WF_{Mdie} obtained on SiO$_2$ (depending on X_{HfO2}). This result contradicts the C-V measurement where we observe a 200mV higher WF_{Meff} for devices with an HfO$_2$ layer. As a result, a voltage drop at this interface cannot be responsible for the variations observed in WF_{Meff} with high-k. Therefore, only a voltage drop at the HfO$_2$/SiO$_2$ interface could explain the variations in WF_{Meff} deduced from C-V measurements. It could be noted that, taking for the affinity $X_{HfO2}=2eV$, it could not be necessary to consider any voltage drop at the metal/high-k interface.

While extracting WF_{MCV}, we assumed that the interfacial charge densities as well as the interfacial voltage drops are independent of EOT. The validity of this assumption can be verified by the linearity of WF_{Meff} versus EOT (Fig. 10 and Fig. 11). However, some devices show a huge reduction of WF_{Meff} (roll-off) when the SiO$_2$ layer is thin (dotted arrows in Fig. 10 and Fig. 11) [20]. This feature is an important issue for the achievement of P-MOS transistors with both a P-type metal gate and a low EOT. The interpretation of this phenomenon is still unclear and several mechanisms have been invoked. Recent studies suggest that there exists a relationship between the roll-off and the creation of oxygen vacancies (Vo) within the SiO$_2$ layer. However, they disagree on the electrostatic effect of these Vo, which are considered, in one case, as positive charges in the SiO$_2$ layer [25,42] or as being at the origin of dipoles at the SiO$_2$/HfO$_2$ interface [43]. The aim of the following analysis is to clearly determine the electrostatic phenomenon responsible for such a roll-off.

To study the evolution of the roll-off amplitude, we define the quantity ΔRoll-Off, which is the difference between WF_{Meff} measured at a given EOT and the linear trends obtained from larger EOT. Figure 17 represents the variations of ΔRoll-Off with the SiO$_2$ layer thickness (EOT$_{SiO2}$) for devices with a large set of HfO$_2$ thicknesses (2nm to 10nm), two different anneals (750°C and 1050°C) and an optional Al$_2$O$_3$ layer between HfO$_2$ and SiO$_2$.

At a first glance, the roll-off phenomenon appears as to be independent of the HfO$_2$ thickness. Indeed, it rapidly increases as the SiO$_2$ thickness is reduced but in the same range from 3nm to 10nm thick HfO$_2$ layers (circles Fig. 17). This clearly indicates that the roll-off is independent of the HfO$_2$ thickness. The two lines emphasize the influence of final activation anneals: 750°C (dotted line) and 1050°C (full line). It shows that the roll-off is thermally activated, which agrees with most of the studies on this phenomenon [25,44].

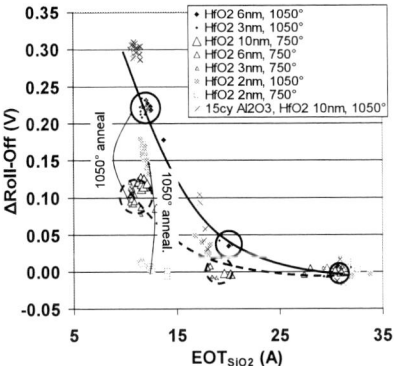

Fig. 17. Roll-off magnitude versus SiO_2 thickness for different HfO_2 thicknesses (2 to 10nm) and anneals (750 and 1050°C) (after Ref. [3]).

The devices with a 2nm thick HfO_2 layer have a special behavior as compared to devices with thicker HfO_2 layers. The former shows no roll-off at a 750°C anneal and it only arises after a 1050°C anneal. This suggests the existence of an HfO_2 threshold thickness, between 2 and 3nm, below which the roll-off is reduced. It is worth noting that HfO_2 crystalline phase could also appear in this range of thickness [45]. This has been inferred by ATR spectroscopy carried out on 2 and 3nm thick HfO_2 layers with and without anneal. As a matter of fact, both 3nm thick HfO_2 layers (annealed at 1050°C and not annealed) exhibit two absorption peaks between wave numbers 600cm^{-1} and 800cm^{-1}. These peaks are representative of a crystalline HfO_2 phase. For 2nm thick HfO_2 films, we have seen no peaks if the layer is not annealed, meaning that the HfO_2 layer is amorphous. However, after a 1050°C anneal, the peaks arise, indicating that the HfO_2 layer is starting to crystallize. This clearly suggests that the roll-off phenomenon is related to the crystalline phase of the HfO_2 layer.

For EOT thicker than 2nm, the independence of the roll-off with the HfO_2 thickness reveals that it cannot be justified by fixed charges, which should induce a linear dependence with the HfO_2 thickness. Besides, IPE measurements were performed for different SiO_2 layer thicknesses, corresponding to the formation of the roll-off (circles Fig. 10). The same metal/HfO_2 barrier height was found independently of the SiO_2 thickness (see Fig. 18), indicating that the roll-off is not related to this interface. Consequently, the roll-off can only be attributed to a voltage drop at the SiO_2/HfO_2 interface.

Therefore, we have clearly established the role of the high-k/SiO_2 interface in the effective metal work function as well as its connection with the roll-off effect. We have shown that the voltage drops at this interface depends on the type of high-k material and on its crystallinity. We will now analyze the influence of the different gate processes on WF_{Meff}.

In Fig. 15, IPE and CV measurements show a clear dependence of WF_{Meff} with the metal gate thickness. This effect arises even with SiO_2 only dielectrics, indicating that it depends on the TiN thickness. IPE measurements were carried out on both 5 and 10nm thick TiN layer deposited on a 10nm thick HfO_2 layer (see Fig. 16). They also show a 150mV reduction of Φ_{bM} for 5nm thick TiN layers, revealing a dependence of the WF_M with the gate thickness.

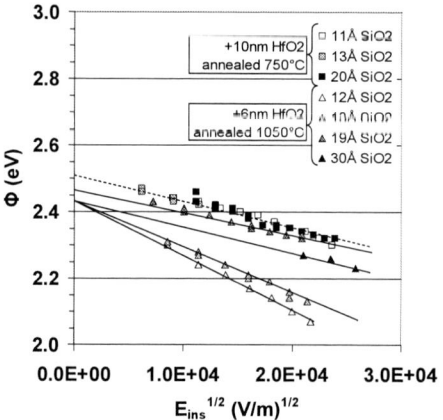

Fig. 18. Φ_{bM} versus $E_{ins}^{1/2}$ at different SiO_2 thicknesses where the roll-off effect occurs (after Ref. [3]).

CVD gates deposited on SiO_2 show a weak dependence (100mV in Fig. 15) of WF_M after high temperature anneal. Therefore, this effect should depend on the deposition process since no influence is observed for PVD gates. C-V measurements have been carried out on a 10nm thick TiN CVD layer deposited on a thick HfO_2 layer. The devices have been annealed at various temperatures: 1050°C, 750°C and not annealed. The results (see Fig. 19) show that there is a reduction of about 100mV in ΔWF_{MCV}, when the temperature decreases and up to 200mV reduction without anneal. IPE measurements made on the devices annealed at 750°C and 1050°C also show a 100mV reduction of Φ_{bM} when increasing the annealing temperature (see Fig. 18). All these experimental results clearly demonstrate that WF_M can be influenced by the thickness, the thermal budget and the deposition process of the metal gate.

In this study, we have demonstrated the major role of interfacial voltage drops in the control of WF_{Meff} (see Fig. 20). The voltage drop at the HfO_2/SiO_2 interface (δV) has a huge influence on WF_{Meff} since it can increase WF_{Meff} of up to 900mV. However, this interface effect is very unstable, since most of the interfacial voltage drop at the high-k/SiO_2 interface can disappear for thin SiO_2 thicknesses. We have also found a reduction of the metal/HfO_2 barrier height, when reducing the TiN thickness or increasing the anneal temperature. Nonetheless, the effects at this interface are smaller than at the high-k/SiO_2 interface, but they could lead to a reduction of 200mV of WF_{Meff}. We have shown that this influence due to the gate process could be related to a diminution of the metal work function.

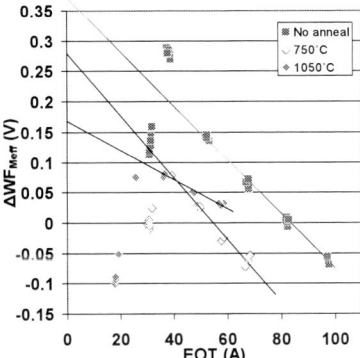

Fig. 19. ΔWF_{Meff} vs EOT for 10nm thick TiN deposited on 10nm thick HfO_2. Different annealed have been performed (no anneal, 750°C and 1050°C) (after Ref. [3]).

In this study, we have demonstrated the major role of interfacial voltage drops in the control of WF_{Meff} (see Fig. 20). The voltage drop at the HfO_2/SiO_2 interface (δV) has a huge influence on WF_{Meff} since it can increase WF_{Meff} of up to 900mV. However, this interface effect is very unstable, since most of the interfacial voltage drop at the high-k/SiO_2 interface can disappear for thin SiO_2 thicknesses. We have also found a reduction of the metal/HfO_2 barrier height, when reducing the TiN thickness or increasing the anneal temperature. Nonetheless, the effects at this interface are smaller than at the high-k/SiO_2 interface, but they could lead to a reduction of 200mV of WF_{Meff}. We have shown that this influence due to the gate process could be related to a diminution of the metal work function.

Therefore, the gate process seems to have no big impact on the voltage drop at the high-k/SiO_2 interface. This is asserted by different results in the literature, which suggest that the interfacial voltage drop at the HfO_2/SiO_2 interface (often considered as a dipole [29,43]) only depends on the choice of the materials [19,29]. This suggests that such a conclusion should be temperate.

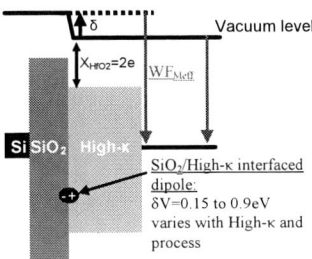

Fig. 20. Band diagram of the two parameters δV and WF_M involved in WF_{Meff} variations (after Ref. [3]).

Indeed, we have used C-V measurements to follow the increase of WF_{Meff} with the TiN thickness (5 and 10nm) for three different dielectrics: SiO_2 only dielectrics, SiO_2 with a 1ML HfO_2 layer and SiO_2 with 10cy Al_2O_3 layer and 3nm HfO_2 on top (see Fig. 21). The results reveal a 150mV raise of WF_{MCV} from 5nm thick TiN layer to 10nm for both SiO_2 and 1ML HfO_2 dielectrics, which is consistent with IPE measurement (see Fig. 16) and C-V measurements (not shown) made on a 10nm thick HfO_2/SiO_2 bevel stack. This means that this effect is independent of the dielectric. However, when introducing a thin Al_2O_3 layer between the SiO_2 layer and HfO_2 layer, the effect is enhanced and WF_{MCV} increases of about 350mV. For this dielectric, we have also observed a strong increase of the roll-off when the gate thickness is augmented from 5nm to 10nm. These two observations entail that, in this case, the effect of the gate thickness is not independent of the dielectric in contact with the SiO_2. Therefore, the effect of one process on WF_{Meff} must not be considered separately from the other ones.

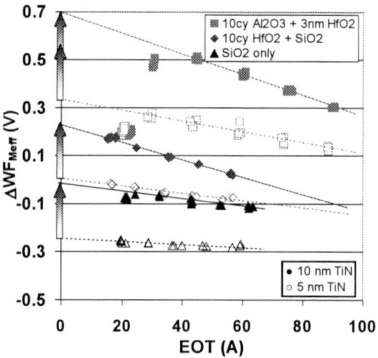

Fig. 21. ΔWF_{Meff} vs EOT for different TiN thicknesses (5nm: open symbols and 10nm: filled symbols) and dielectrics (No High-κ, HfO_2 and $HfO_2+Al_2O_3$) (after Ref. [3]).

In summary, we have found that the voltage drop δV varies from 0.2eV with HfO_2 to about 0.7eV with Al_2O_3. The dependence of WF_M with the final annealing temperature and with its thickness is evident: reducing the thickness from 10nm to 5nm induces a 150mV decrease in WF_M. In the case of CVD gates deposited on a high-k material, we have clearly shown an influence of the gate thickness on δV, depending on the high-k in contact with SiO_2 (the variation reaches 50mV for HfO_2 and 200mV for Al_2O_3). It infers that the choice of the high-k and the metal gate can have a mutual effect on WF_{Meff}. Moreover, our results indicate that the knowledge of the materials present in the stack is not sufficient to predict WF_{MCV}. As a matter of fact, for the same materials (TiN CVD/HfO_2/SiO_2), a variation of about 300mV in WF_{MCV} can be observed.

Transport Processes In HfO_2-Based Gate Stacks

In order to reduce gate leakage in future CMOS devices, the replacement of conventional SiO_2 and SiON gate dielectrics by high-κ materials, and especially HfO_2 and its silicates, has received a great consideration in the past years [44,45]. Already introduced in the 45 nm technology by INTEL [46], high-κ materials are being integrated in both High-Performance and Low-Power 32nm CMOS technologies and beyond. In this context, the understanding of gate leakage processes in HfO_2-based gate stacks appears as a major concern to predict the leakages and study the future high-k materials needed for the 22 nm and 16 nm nodes [47].

Even though this topic has already been discussed in previous works [48-56], an agreement on the nature of conduction has not been reached yet. Many studies have assumed or concluded that leakage currents were due to pure direct tunneling (DT) [45, 47-49], and can be modeled by effective mass DT approximations, as for ultra-thin SiO_2 dielectrics [57]. However, this conclusion is still controversial. In fact, in these works, the extraction method of dimensions and parameters of both HfO_2 and interfacial layers (tunnel barrier height φ, tunneling effective masses, thicknesses and permittivities) is usually incomplete, making doubtful any conclusion on the nature of conduction in such dielectric stacks. In addition, the applicability of effective mass DT approach to HfO_2 gate stacks is not straight forward. Indeed, previously reported extraction of HfO_2 tunneling parameters exhibited a large dispersion (see Fig. 22), advanced tight binding simulations have discussed the physical validity of these extracted parameters and of conventional DT model [51,52], and, it has been shown that ultra thin interfacial SiO_2 layer (below 8Å) differs from thicker SiO_2, for both fundamental and process reasons [53,58]. Moreover, other studies have also indicated that pure DT model cannot totally explain the temperature dependence of gate leakage currents, suggesting the contribution of thermally assisted additional processes [54-56]. This absence of consensus can be partly attributed to the lack a full set of experimental data, enabling to reduce the number of arbitrary assumptions necessary to compare experiment and simulation results.

This work aims at filling this gap by combining capacitance-voltage (C-V), current-Voltage (I-V) and Transmission Electron Microscopy (TEM) measurements performed on various samples featuring different Interfacial Layers (IL) and/or HfO_2 thicknesses. Moreover, the temperature dependence of gate leakage current has been measured from 80K to 400K in order to discriminate the transport mechanisms. Finally, the experimental results have been compared with DT simulations, in order to obtain the most efficient methodology to get a consistent picture of the HfO_2 tunneling parameters.

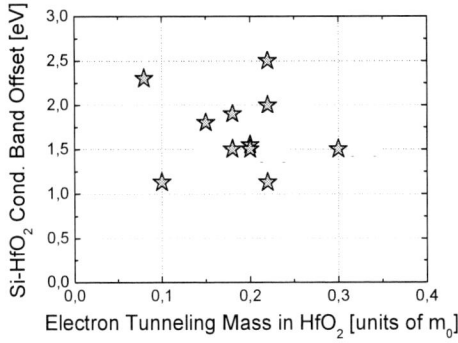

Fig. 22. HfO_2 tunneling parameters dispersion in literature data (after Ref. [4]).

Long channel nMOS (10x10µm and 100x100µm) transistors have been fabricated on (100) p-doped substrates with different gate stacks. First, reference samples have been processed with a standard n$^+$ polysilicon gate, deposited on top of an ultra thin SiO$_2$ obtained by Rapid Thermal Oxidation (RTO). In addition, several SiO$_2$/HfO$_2$/TiN metal gate transistors have been realized with various thicknesses. A first series of devices features SiO$_2$ interfacial layers grown by RTO with different thicknesses (from 1.2 to 2 nm) with a constant 3 nm HfO$_2$ layer. A second series of samples has been fabricated with various HfO$_2$ thicknesses (from 2 to 4.5 nm), with a constant interfacial layer thickness ($\approx$8Å, grown by chemical wet cleaning). In both cases, a TiN metal gate (10nm thick) has been deposited by Chemical Vapor Deposition, and the HfO$_2$ has been deposited by Atomic Layer Deposition (ALD) and annealed at low temperature (600°C). Main sample characteristics are summarized in Table I.

Table I. Sample characteristics used in the experiment (after Ref. [4]).

Samples	t_{IL} (nm)	t_{HfO2} (nm)	Gate	Na (m^{-3})	IL process conditions
i	1.5	-	n+ poly	1.10^{24}	RTO
ii	2.5	-	n+ poly	1.10^{24}	RTO
iii	1.2	3.0	TiN	2.10^{23}	RTO
iv	1.5	3.0	TiN	2.10^{23}	RTO
v	2.0	3.0	TiN	2.10^{23}	RTO
vi	0.8	2.0	TiN	2.10^{23}	Chem. ox.
vii	0.8	3.0	TiN	2.10^{23}	Chem. ox.
viii	0.8	4.5	TiN	2.10^{23}	Chem. ox.

Both capacitance versus voltage and gate current versus voltage measurements have been carried out in all samples, at four different temperatures: 80, 150, 300 and 400K. C-V measurement frequencies and gate area have been carefully chosen in order to prevent any distortion due to either gate leakage current (dissipation factor D < 1) [59,60], too long channel length [60] or series resistance parasitic effects.

The Equivalent Oxide Thickness (EOT) and flat band voltage (V$_{FB}$) have been extracted from C-V measurements for different temperatures, following the procedure described in more details in [61]. It mainly relies on fitting experimental curves by self-consistent Poisson-Schrodinger (PS) simulations, taking into account oxide wave function penetration. Typical experimental and simulated C-V curves are shown in Figures 23 and 24. As expected, the extracted EOT values do not exhibit any significant variation with temperature. Moreover, the experimental temperature dependence of the flat band voltage has been found in good agreement with modelling results, taking into account the substrate Fermi level variation with temperature [62]. Finally, it should be noted that Transmission Electron Microscopy observations have been performed on a selected set of samples.

The direct tunneling current depends on temperature mostly via the dependence of semiconductor charge with temperature (essentially due to flat band and threshold voltage variations). Instead, the temperature-assisted processes may reveal additional temperature dependence. In order to distinguish these two effects (charge variation versus temperature from temperature-assisted mechanism), gate currents have been plotted as a function of total charge (extracted from C-V measurements), and not versus gate voltage as is usually done (Figs 25, 26 and 27).

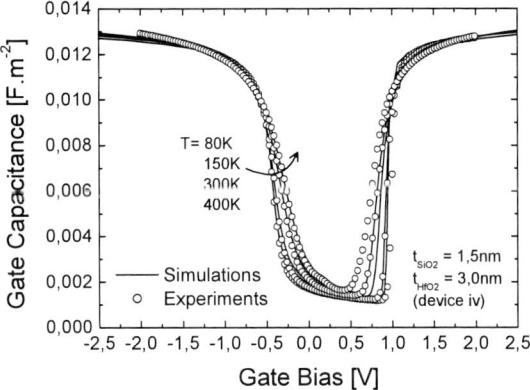

Fig. 22. Gate capacitance versus gate bias for a nominal 1.5nm SiO_2 / 3nm HfO_2 / TiN gate stack (sample v). Measurements and simulations have been performed at different temperatures. V_{FB}= -0.466 / -0.429 / -0.311 / -0.239V and EOT= 2.30 / 2.30 / 2.29 / 2.28 nm have been extracted at 80, 150, 300 and 400K respectively (after Ref. [4]).

As expected, gate leakage currents through SiO_2–Polysilicon gate stacks (Figure 25) do not show any significant temperature variation in the 1.5 - 2.5nm thickness range, for both accumulation and inversion regimes. In this case, and as already reported in previous works [57,63], the absence of any significant thermal activation due to trap-assisted transport mechanisms can be attributed to the good quality and thermal stability of such Si /SiO_2 gate stacks realized by RTO process.

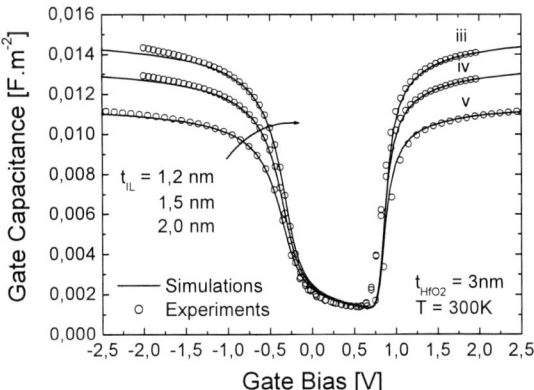

Fig. 24. Gate capacitance versus gate bias for various SiO_2 / HfO_2 / TiN gate stacks, with different t_{IL} (samples iii, iv, v and vi) (after Ref. [4]).

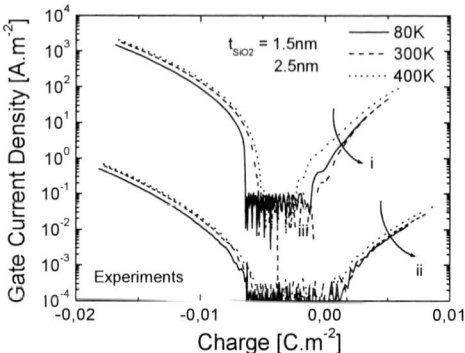

Fig. 25. Gate current density versus substrate total charge for reference samples with different SiO_2 thicknesses (samples i and ii, the indicated thicknesses are nominal values) and polysilicon gate (after Ref. [4]).

For high-k-based gate stacks, gate leakage currents activation with temperature can really be observed in some conditions (see Figs 26 and 27). While the current in accumulation only shows a small variation with temperature, the current in inversion (negative charge) can exhibit a significant increase with temperature, but only at high temperature (300 and 400K). The temperature activation is only important on thick HfO_2 stack (typically 4.5 nm) and decreases with t_{HfO2}, leading to almost no activation for ultra-thin HfO_2 layer (2 nm case).

To quantify this temperature activation, the excess of leakage current with respect to the low temperature measurement (80K) has been plotted in Fig. 28 as a function of 1000/T, for different total charges. As can be seen, the excess current qualitatively follows an Arrhenius law.

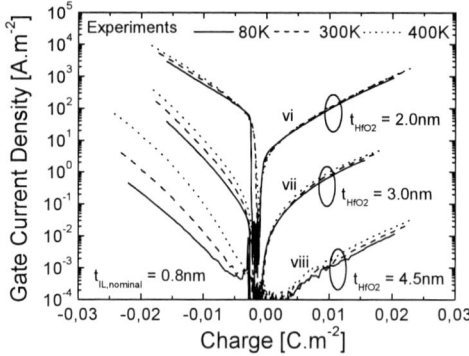

Fig. 26. Gate current density versus substrate total charge for samples with different HfO_2 thicknesses (samples vi, vii and viii, the indicated thicknesses are nominal values) (after Ref. [4]).

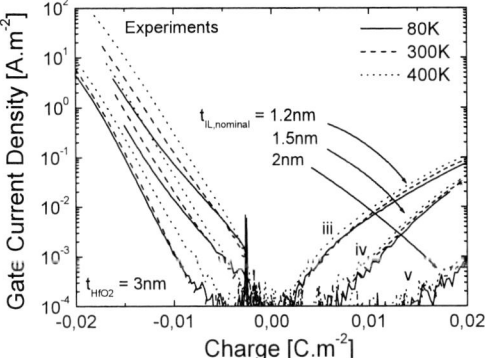

Fig. 27. Gate current density versus substrate total charge for samples with different Interfacial Layer (IL) thicknesses. (samples iii, iv and v, the indicated thicknesses are nominal values) (after Ref. [4]).

Finally, the amplitude of the temperature variation has been found to be relatively independent of the interface layer thickness (see Fig. 27). All these observations clearly suggest the presence of additional transport mechanism in HfO$_2$, likely volume limited and thermally assisted. These findings are in good agreement with other results presented in [54], in which the thermal activation of gate leakage current has been ascribed to traps located below the HfO$_2$ conduction band.

The experimental data have been compared with direct tunneling simulations in inversion regime (electrons emitted from Silicon conduction band). The modeling first consists in solving self-consistent Poisson and Schrodinger equations to calculate the subband structure of the inversion layer. The results are then used to compute the tunnel current using a transparency-based model described elsewhere [50]. This approach takes into account all relevant subbands, wave function penetration and double abrupt barriers. In this modeling, the barrier between Si and SiO$_2$ has been considered as abrupt. The validity of this approximation will be discussed below. This DT model is then applied to analyze gate tunnel leakage at low temperature and for extraction of tunneling parameters.

For aggressively scaled gate stacks, previous works have underlined the key role of structural changes occurring at the interface between Silicon and the SiO$_2$-like interfacial layer. Ab-initio simulations [53] have shown that energy band profile, permittivity and tunneling mass transitions over a few Angstroms at Si-SiO$_2$ interface have a critical impact on the quantization and gate leakage current, especially for Interfacial Layer (IL) thickness reaching sub-nanometer values. Therefore, a specific Poisson-Schrodinger solver has been developed accounting for the continuous transition of potential energy, effective mass and permittivity in the IL region of typical thickness 3-4 Å. In this "SiO$_x$" region, and following the approach proposed by Markov et al. [53], material parameters like permittivity, carrier masses, band profiles, are linearly interpolated from Silicon to SiO$_2$ values, as shown for instance in the inset of Figure 29.

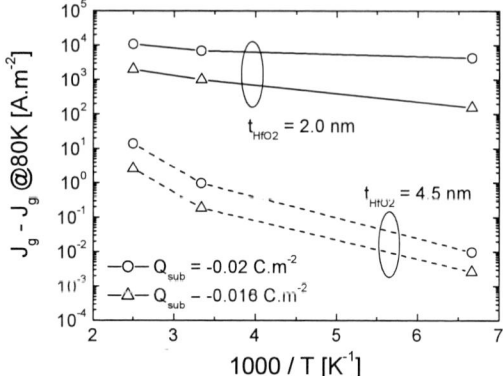

Fig. 28. Arrhenius plot of excess gate current density, for different HfO₂ thicknesses (samples vi and viii) and substrate charges. Additional temperature activation is clearly shown for thick HfO₂ layer (4.5nm) in a large temperature range (after Ref. [4]).

As previously reported, the non-abrupt realistic tunnel barrier profile model leads to an increase of both its capacitance and carriers transmission probability at constant physical thickness. As a result, the tunnel current in the non-abrupt model (curve A) is higher than its counterpart, curve C, in the abrupt model (see Fig. 29). Parameters used for stacks A and C are reported in Table II. However, when studying gate leakage current in advanced devices, it is more appropriate to compare the two approaches at the same EOT (see Fig. 29, curves B (realistic) and C (abrupt)). In this case, it can be seen that abrupt and realistic models do not show any significant differences. Indeed, permittivity transition (which tends to decrease the EOT) and realistic energy band profiles have opposite impact on gate leakage current.

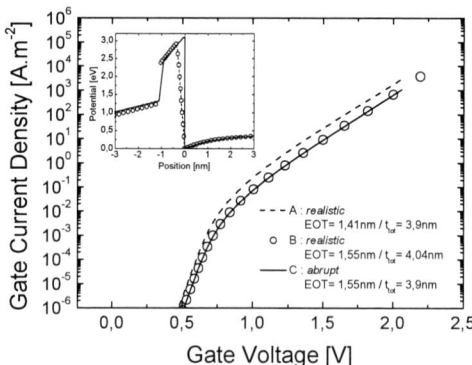

Fig. 29. Gate current simulations versus gate voltage through three different gate stacks differing by their EOT and/or their total physical thickness (see Table II for simulated samples description) (after Ref. [4]).

Table II. Parameters used for simulation of various devices (after Ref. [4]).

Device	$t_{transition}$ (nm)	$t_{pure\,SiO2}$ (nm)	t_{HfO2} (nm)	EOT (nm)	Total physical thickness (nm)
A	0.3	0.6	3	1.41	3.9
B	0.3	0.74	3	1.55	4.04
C	0	0.9	3	1.55	3.9

Since the non-abrupt approach does not significantly modify the tunnel current when the comparison is made at the same EOT, and as it requires one extra fitting parameter (the transition region thickness), in the following, the simulations have been conducted assuming an abrupt interface. First, the SiO$_2$ - Polysilicon reference samples have been compared with the direct tunneling model in inversion region and accounting for gate polydepletion. As can be seen in Fig. 30, a good agreement between model and experiments has been achieved, for all temperatures, using the usual SiO$_2$ tunneling parameters (m_{SiO2}=0.5m$_0$, φ_{SiO2}=3.1eV, ε_{SiO2}=3.9ε_0.) and EOT (i.e. SiO$_2$ physical thickness) extracted from C-V measurements.

The comparison between model and experiment is not so straightforward in the case of HfO$_2$ stacks, because the parameters (including the respective thicknesses) of both the interfacial layer and HfO$_2$ are not accurately known. The permittivity values are very important, since they relate the main gate stack measurable parameter (i.e. EOT) and the DT model inputs, i.e. physical thicknesses of both dielectric layers, according to:

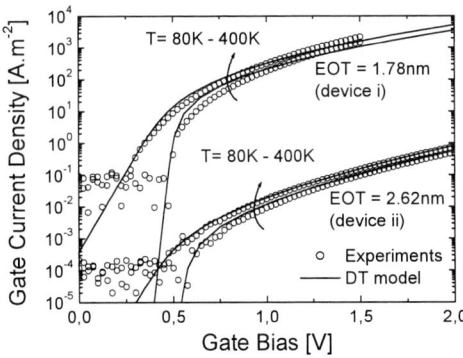

Fig. 30. Experimental and simulated gate current density plotted versus gate bias for samples i and ii (SiO$_2$ – Polysilicon gate stacks) at low and high temperature (80K and 400K). (m_{SiO2}=0.5m$_0$, φ_{SiO2}=3.1eV, ε_{SiO2}=3.9ε_0) (after Ref. [4]).

$$EOT = \varepsilon_{SiO2} \cdot \left(\frac{t_{IL}}{\varepsilon_{IL}} + \frac{t_{HfO2}}{\varepsilon_{HfO2}} \right) \qquad (4)$$

In the following, a methodology to better compare simulation and experimental results is discussed, based on the two main sets of experiments, featuring different HfO$_2$ thicknesses t$_{HfO2}$ at same IL thickness, or different IL thicknesses t$_{IL}$ at constant HfO$_2$ layer thickness. Additional assumptions have been made in order to reduce the number of fitting parameters: i) only experiments at 80K have been considered, in order to suppress the thermally assisted contribution in thicker stacks, ii) the tunneling parameters of HfO$_2$ have been assumed identical in all samples, since the HfO$_2$ layers have been deposited using the same process, and, iii) the HfO$_2$ thickness t$_{HfO2}$ has been assumed equal to its nominal values, since the ALD process allows a good control of the deposited HfO$_2$ thickness (confirmed by TEM observations).

Once extracted the EOT by C-V and t$_{HfO2}$ by TEM, the dielectric constants of both layers are needed to deduce t$_{IL}$. The dielectric constant of HfO$_2$ (ε_{HfO2}) has then been extracted using Eq. 4 from the slope of the EOT versus t$_{HfO2}$ curve(see Fig. 31). It has been found equal to 17ε_0. This extracted value is considered as relatively reliable and will be used in the following.

The application of such a method to extract ε_{IL} from devices having different t$_{IL}$ is more difficult. In fact, once knowing EOT, and, high-k thickness and permittivity, it appears that the extraction of ε_{IL} is highly sensitive to the IL thickness input. This uncertainty on the extracted IL permittivity is increased by the lack of accuracy on thin interfacial layers thickness values, even in samples where SiO$_2$ was realized by accurate and well-controlled RTO process. Previous works [58,64] have indeed underlined the influence of high-k deposition and post-annealing on IL regrowth.

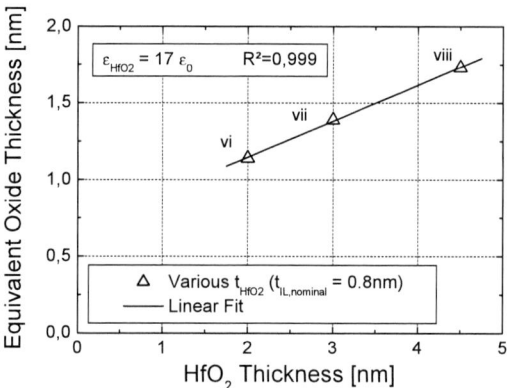

Fig. 31. Equivalent Oxide Thickness (EOT) extracted from C-V measurements plotted versus HfO$_2$ thickness (after Ref. [4]).

Applied to our data, as mentioned in our previous work [65], the extraction of ε_{IL} leads to the value of 4.7ε_0 when taking t$_{IL}$ = t$_{RTO-SiO2, deposited}$. This value is not very realistic for several reasons. First, because it considerably differs from pure SiO$_2$ value of

3.9 ε_0, which is surprising for interfacial layer thicknesses t_{IL} larger than 1 nm. Moreover, this ε_{IL} value is not consistent with what can be deduced from C-V and TEM measurements. In fact, TEM pictures have been taken on devices iii and v, in which the expected value of t_{IL} is 1.2 and 2 nm, respectively. TEM pictures clearly reveal an IL regrowth from 1.2 up to 1.35 nm in device iii, while device v does not show any IL thickness increase. The absence of any regrowth on the thicker IL is reasonable, since in this last case, the Si-SiO$_2$ interface is quite far away from the oxygen source (located at SiO$_2$-HfO$_2$ interface). Finally, using the two new values of t_{IL} measured by TEM, the extracted ε_{IL} has been found around 3.8 ε_0, which close to the pure SiO$_2$ permittivity 3.9ε_0.

The previous conclusions show that, with thicknesses in the range of 1.2 - 2nm, the RTO interfacial layer behaves more like a pure SiO$_2$ layer, indicating that the same tunnel parameters extracted from reference Polysilicon - SiO$_2$ experiments of Figure 30 (m_{SiO2}=0.5m_0, φ_{SiO2}=3.1eV) should be used in order to simulate gate currents through devices iii, iv and v gate stacks.

Once knowing the IL parameters (i.e. permittivity, barrier height and electron tunneling mass), the HfO$_2$ tunnel parameters have been extracted by fitting the experiments of Figure 32, having different t_{IL} at same t_{HfO2} (devices iv, v and vi), yielding m_{HfO2}=0.165m_0 and φ_{HfO2}=1.9eV. These values are in good agreement with previously reported data (see Figure 22).

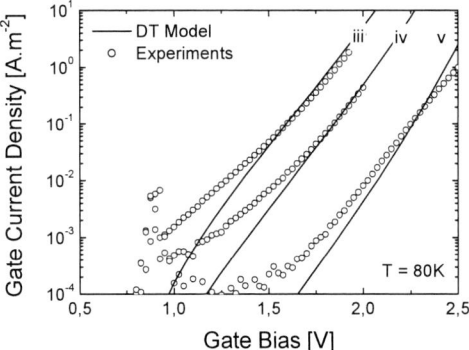

Fig. 32. Experimental and simulated gate current density plotted versus gate bias for samples iii, iv, v and vi (various RTO IL thicknesses with a constant 3nm HfO$_2$ layer) at low temperature (80K). Simulations have been done using the following parameters: m_{IL}=0.5m_0, φ_{IL}=3.1eV, ε_{IL}=3.9ε_0, m_{HfO2}=0.165m_0, φ_{HfO2}=1.9eV, ε_{HfO2}=17ε_0 (after Ref. [4]).

Once knowing ALD-deposited HfO$_2$ parameters, the DT model has then been compared to I-V measurements performed on the second series of samples, i.e. devices vi, vii and viii, having different t_{HfO2} (using same process conditions as before) and a constant DDC-SiO$_2$ interfacial layer with expected thickness of 0.8 nm. The interfacial layer has been evaluated from TEM images, leading to the value of t_{IL}=1nm. In this

ECS Transactions, 35 (4) 773-804 (2011)

second set of samples with DDC-SiO$_2$ interfacial layer, the permittivity value has thus be found equal to ε_{IL}= 5 ε_0. Finally, as shown in Figure 33, using the previously extracted permittivity value the I-V curves can be nicely reproduced without any additional fitting adjustment, i.e. using the same tunneling parameters for the interfacial layer (m$_{IL}$=0.5m$_0$, φ_{IL}=3.1eV) and for the HfO$_2$ (m$_{HfO2}$=0.165m$_0$, φ_{HfO2}=1.9eV).

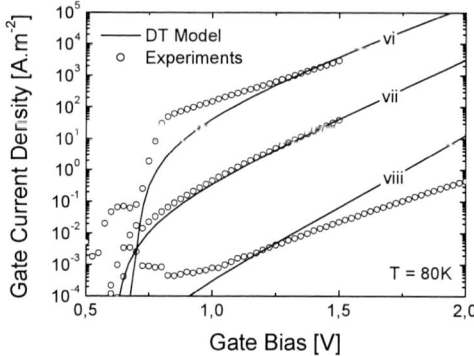

Fig. 33. Experimental and simulated gate current density plotted versus gate bias for samples vii, viii and ix (various HfO$_2$ thicknesses with the same IL, grown by chemical cleaning) at low temperature (80K). Simulations have been done using the following parameters: m$_{IL}$=0.5m$_0$, φ_{IL}=3.1eV, ε_{IL}=5ε_0, m$_{HfO2}$=0.165m$_0$, φ_{HfO2}=1.9eV, ε_{HfO2}=17ε_0 (after Ref. [4]).

The higher permittivity value and very close to SiO$_2$'s tunneling parameters indicate that the interfacial layer in this case is similar to SiO$_x$ (sub-stoechiometric SiO$_2$). These final results suggest that, provided a correct extraction of the interfacial layer thickness (i.e. by TEM measurement), the experimental I-V curves can be well fitted by effective mass direct tunneling model, using a unique set of tunneling parameters. This conclusion emphasizes the critical role of parameter extraction (and especially interfacial layer thickness) when analyzing transport conduction through HfO$_2$ based gate stacks.

Conclusions

In the first part, the kinetics of the trapping/detrapping process involved in hysteresis phenomena of HfO$_2$ films have first been investigated on a large temperature range (20K-500K). Up to 400K, trapping and detrapping processes are temperature independent when they are corrected from threshold voltage variation with temperature. This result strongly suggests that the analysis of trapping and detrapping mechanisms at different temperatures must be compared at the same dielectric field to be relevant. At high temperatures (over 400K for our samples), a thermally and field assisted discharging process (Poole-Frenkel or similar) limits the trapping efficiency. However, in all cases, the trapping and detrapping kinetics are almost temperature independent, indicating that the capture and emission into or from available traps are mainly controlled by direct

800

tunnelling. It should be mentioned that, for sub-2 nm ALCVD HfO$_2$, the trapping amplitude becomes negligible [66].

In the second part, we have measured the variations of the effective metal work function and studied its relation with the different processes (gate thickness, high-k material and anneals) involved in the fabrication of our devices. Using both capacitance versus voltage and internal photoemission measurements, we have evidenced the presence of an interfacial voltage drop at the high-k/SiO$_2$ interface, δV, as well as variations in the metal work function, WF$_M$, with the metal gate process. We have shown that both δV and WF$_M$ variations are sufficient to interpret the evolutions of WF$_{Meff}$, widely observed when including a high-k in the metal gate stack. Nevertheless, δV has a most important contribution in these variations. On one hand, we showed that δV is mostly influenced by the nature of the high-k material in contact with the SiO$_2$. It induces an increase of δV from 200mV with HfO$_2$ to about 900mV with Al$_2$O$_3$. However, our study of the roll-off effect reveals a strong reduction of δV, stemming from a combined effect of the HfO$_2$ crystallinity and the SiO$_2$ layer thickness. Finally, δV is also found very sensitive to the gate process parameters such as the gate thickness. On the other hand, we have revealed that WF$_M$ depends on the gate process parameters. Indeed, WF$_M$ increases with the gate thickness and decreases with a high temperature anneal. As a result, to achieve a P-type gate, one should carefully select its materials, but also take into account the process dependence of WF$_{Meff}$ and especially the roll-off phenomenon, which seems to be difficult to overcome.

Finally, in the third part, capacitance and current measurements have been performed from low to high temperatures, on several MOS structures, observed by TEM, and having various interfacial layers (IL) or HfO$_2$ thicknesses. Those data provide a unique and complete set of experiments, necessary to carefully study the transport mechanism through HfO$_2$ stacks. By plotting the gate current as a function of the total charge enables us to discriminating the charge dependence with temperature from other thermally assisted transport mechanism. We have found that: i) leakage currents are almost identical at 80 and 150K in all experiments, ii) a significant thermal activation has been observed at 300 and 400K only on stacks having a HfO$_2$ thicknesses of 3 and 4.5nm, and under electron injection from the substrate. Finally, low temperature inversion currents have been compared with direct tunneling simulation. Taking advantage of experiments performed on samples having both HfO$_2$ and IL thicknesses variations, it has been possible to show that all experiments can be reproduced by simulation, using the same set of tunneling parameters for both HfO$_2$ and interfacial layer. To this aim, it has been necessary to extract the IL thickness by TEM measurements, in order to avoid any t_{IL} thickness uncertainty, due to IL regrowth during high-k deposition and post-annealing process. Based on an important set of tested devices and following an appropriate extraction methodology, $m_{HfO2}= 0.165m_0$ and $\varphi_{HfO2}= 1.9eV$ have been extracted, providing a better consensus on HfO$_2$ tunneling parameters. Moreover, the value of the extracted barrier height is in good agreement with IPE results [67,68]. Using a large set of experiments at low temperature, completed with physical characterization, this work emphasizes the difficulty to carefully extract HfO$_2$ tunnel parameters. Such a difficulty is certainly one of the reasons explaining the surprising wide diversity of HfO$_2$ tunneling parameters reported in the literature.

Acknowledgments

This work has been partially supported by UTTERMOST Catrene European project.

References

1. J. Mitard, C.Leroux, G.Ghibaudo, G. Reimbold, X.Garros, B.Guillaumot and, F.Boulanger, *Microelectronic Engineering*, **80**, 362 (2005).
2. J. Mitard, X. Garros, L.P. Nguyen, C. Leroux, G. Ghibaudo, F. Martin and, G. Reimbold, Proc Int Rel. Phys. Symp., 2006, p. 174.
3. M. Charbonnier, C. Leroux, V. Cosnier, P. Besson, E. Martinez, N. Benedetto, C. Licitra, N. Rochat, C. Gaumer, K. Kaja, G. Ghibaudo, F. Martin and, G. Reimbold, *IEEE ED.*, **57**, 1809 (2010)
4. J. Coignus, C. Leroux, R. Clerc, R. Truche, G. Ghibaudo, G. Reimbold and, F. Boulanger, *Sol. State Electron.*, **54**, 972 (2010).
5. G. D Wilk, R.M Wallace and, J.M. Anthony, *J. Appl. Phys.*, **89**, 5243 (2001).
6. B. Guillaumot, X. Garros, F. Lime, K. Oshima, B. Tavel, J. Chroboczek, P. Masson, R. Truche, A. M. Papon, F. Martin, J.F. Damlencourt, S. Maitrejean, M. Rivoire, C. Leroux, S. Cristoloveanu, G. Ghibaudo J.L. Autran, T. Skotnicki and, S. Deleonibus, IEDM Tech. Dig., 2002, p. 355.
7. A. Kerber, E. Cartier E, L. Pantisano, M. Rosmeulen, R. Degraeve, T. Kauerauf, G. Groeseneken, H.E. Maes and, U. Schwalke, Proc Int. Rel. Phys. Symp. 2003, p 41.
8. S. Zafar, A. Callegari, E. Gusev and, M.V. Fischetti, IEDM Tech. Dig., 2002, p.517.
9. C. Leroux, J. Mitard, G. Ghibaudo, X. Garros, G. Reimbold, B. Guillemot and, F. Martin, Proc IEDM Tech. Dig., 2004, p. 737.
10. G. Ribes, M. Muller, S. Bruyere, D. Roy, M. Denais, V. Huard, T. Skotnicki and, G. Ghibaudo, Proc ESSDERC 2004, p. 89.
11. G. Bersuker, J.H. Sim, C.D. Young, R. Choi, P.M. Zeitzoff, G.A. Brown, B.H. Lee and, R.W. Murto, *Microelectronics Reliability*, **44**, 1509 (2004).
12. G. Van den Bosch, G. Groeseneken and, H. E Maes, *IEEE ED*, **38**, 1820, (1991).
13. G. Barbottin and A. Vapaille, "Instabilities in Silicon devices", silicon passivation and related instabilities, volume 2, North-Holland collection, Elsevier Science Publishers, 1989 (quoted page: p23 and p24).
14. J. Robertson, *Eur. Phys. J. Appl. Phys.*, **28**, 265 (2004).
15. C. Hobbs, L. Fonseca, A. Knizhnik, V. Dhandapani, S. Samavedam, W.J. Taylor, J.M. Grant, L.G Dip, D.H Triyoso, R.I Hegde, D.C Gilmer, R. Garcia, D. Roan, M.L. Lovejoy, R.S. Rai, E.A. Hebert, Hsing-Huang Tseng, S.H. Anderson, B.E White and, P.J. Tobin., *IEEE EDL*, **51**, 971, (2004).
16. K. Shiraishi, Y. Akasaka, S. Miyazaki, T. Nakayama, T. Nakaoka, G. Nakamura, K. Torii, H. Furutou, A. Ohta, P. Ahmet, K. Ohmori, H. Watanabe, T. Chikyow, M.L. Green, Y. Nara and, K. Yamada, Tech. Digest. IEDM 2005, p. 39.
17. E. Cartier, M. Steen, B.P. Linder, T. Ando, R. Iijima, M. Frank, J.S. Newbury, Y.H. Kim, F.R. McFeely, M. Copel, R. Haight, C. Choi, A. Callegari, V.K. Paruchuri and, V. Narayanan, Tech. Digest. VLSI 2005, p. 230.
18. K. Iwamoto, A. Ogawa, Y. Kamimuta, Y. Watanabe, W. Mizubayashi, S. Migita and, Y. Morita, Tech. Digest. VLSI 2007, p. 70.

19. B. J. O'Sullivan, V. S. Kaushik, L.-Å. Ragnarsson, B. Onsia, N. Van Hoornick, E. Rohr, S. DeGendt and, M. Heyns, *IEEE Electron Dev. Letters*, **27**, (2006).
20. M. Charbonnier, J. Mitard J, C. Leroux and, G. Ghibaudo, Proc ESSDERC 2007, p. 275.
21. R. Jha, J. Gurganos, Y.H. Kim, R. Choi, J. Lee and, V. Misra, *IEEE EDL*, **25**, 420 (2004).
22. A. M. Goodman, *Phys. Rev.*, **144**, 588 (1966).
23. M. Charbonnier, C. Leroux, F. Allain, A. Toffoli, G. Ghibaudo, F. Martin, H.Grampeix, F.Boulanger and, G. Reimbold, Proc ISAGST 2008, p. 56.
24. C. S. Park, S. C. Song, C. Burham, H. B. Park, H. Niimi, B. S. Ju, J. Barnett,C. Y. Kang, P. Lysaght, G. Bersuker, R. Choi, H. K. Park, H. Hwang, B. H. Park, S. Kim, P. Kirsch, B. H. Lee and, R. Jammy, Proc SSDM 2007, p. 14.
25. S. C. Song, C. S. Park, J. Price1, C. Burham, R. Choi, H. C. Wen, K. Choi, H. H. Tseng, B. H. Lee and, R. Jammy, Tech. Digest IEDM 2007, p. 337.
26. W. Wang, Proc. SSDM, 2007, p. 840.
27. J. K. Schaeffer, D. C. Gilmer, S. Samavedam, M. Raymond, A. Haggag, S. Kalpat, B. Steimle, C. Capasso and, B. E. White *J. Appl. Phys.*, **102**, 074511 (2007).
28. C. Leroux, J. Mitard, G. Ghibaudo, X. Garros, G. Reimbold, B. Guillaumot and, F. Martin, Tech. Digest. IEDM 2004, p. 737.
29. Y. Kamimuta, K. Iwamoto, Y. Nunoshige, A. Hirano, W. Mizubayashi, Y. Watanabe, S. Migita, A. Ogawa, H. Ota, T. Nabatame and, A. Toriumi, Tech. Digest. IEDM 2007, p. 341.
30. M. Kadoshima, Y. Sugita, K. Shiraishi, H. Watanabe, A. Ohta, S. Miyazaki, K. Nakajima, T. Chikyow, K. Yamada, T. Aminaka, E. Kurosawa, T. Matsuki, T. Aoyama, Y. Nara and, Y. Ohji, Tech. Digest. VLSI 2007, p. 66.
31. C. N. Berglund and, R. J. Powell. , *J. Appl. Phys.*, **42**, 573 (1971).
32. V. K. Adamchuk and, V. V. Afanasev, *Prog. Surf. Science*, **41**, 111 (1992).
33. J. L. Alay and, M. Hirose, *J. Appl. Phys.*, **81**, 1606 (1996)
34. J. C. Brewer, R. J. Walters, L. D. Bell, D. B. Farmer, R. G. Gordon and, H. A. Atwater, *Appl. Phys. Lett.*, **85**, 4133 (2004)
35. J. Widiez, K. Kita, T. Nishimura and, A. Toriumi, Proc SSDM 2007, p. 97.
36. M. Kadoshima, A. Ogawa, H. Ota, M. Ikeda, M. Takahashi, H. Satake, T. Nabatame and, A. Toriumi, Tech. Digest. VLSI (2006).
37. Y. Akasaka, G. Nakamura, K. Shiraishi, N. Umezawa, K. Yamabe, O. Ogawa, M. Lee, T. Amiaka, T. Kasuya, H. Watanabe, T. Chikyow, F. Ootsuka, Y. Nara and, K. Nakamura, *Jap. J. Appl. Phys.*, **45**, L1289 (2006).
38. E. Martinez, C. Leroux, N. Benedetto, C. Gaumer, M. Charbonnier, C. Licitra, C. Guedj, F. Fillot and, S. Lhostis, *J. Electrochem. Soc.*, **156**, G120 (2009).
39. G. Lucovsky, Y.Zhang, J. Luning, V. Afanase'v, A. Stesmans, S. Zollner, D. Triyoso, B.R. Rogers and, J.L. Whitten, *Microelectonic Engineering*, **80**, 110 (2005).
40. C. C. Fulton, G. Lucovsky and, R. J. Nemanich., *J. Appl. Phys.*, **99**, 063708 (2006).
41. H. Jeon, as discussed at the IEEE SISC (2007).
42. G. Bersuker, C. S. Park, H. C. Wen, K. Choi, O. Sharia and, A. Demkov, Proc. ESSDERC 2008, p. 134.
43. K. Akiyama, W.Wang, W. Mizubayashi, M. Ikeda, H. Ota, T. Nabatame and, A. Toriumi, Tech. Digest. VLSI 2008, p. 80.

44. H. Iwai, S. Ohmi, S. Akama, C. Ohshima, A. Kikuchi, I. Kashiwagi, J. Taguchi, H. Yamamoto, J. Tonotani, Y. Kim, I. Ueda, A. Kuriyama and, Y. Yoshihara, IEDM 2002 Tech. Digest, p. 625.

45. A. Toriumi, K. Kita, K. Tomida, Y. Zhao, J. Widiez, T. Nabatame, H. Ota and, M. Hirose, IEDM 2007 Tech. Digest, p. 54.

46. C. Auth, A. Cappellani, J.-S. Chun, A. Dalis, A. Davis, T. Ghani, G. Glass, T. Glassman, M. Harper, M. Hattendorf, P. Hentges, S. Jaloviar, S. Joshi, J. Klaus, K. Kuhn, D. Lavric, M. Lu, H. Mariappan, K. Mistry, B. Norris, N. Rahhal-orabi, P. Ranade, J. Sandford, L. Shifren, V. Souw, K. Tone, F. Tambwe, A. Thompson, D. Towner, T. Troeger, P. Vandervoorn, C. Wallace, J. Wiedemer and, C. Wiegand, Proc. VLSI 2007 Symposium, p. 128.

47. Y.C. Yeo, T.-J. King and, C. Hu, *IEEE ED*, **50**, 1027 (2003).

48. B. Govoreanu, P. Blomme, K. Henson, J. V. Houdt and, K. D. Meyer, *Solid-State Electronics*, **48**, 617 (2004).

49. F. Li, S. P. Mudanai, Y.-Y. Fan, L. F. Register and, S. K. Banerjee, *IEEE ED*, **53**,1096 (2006).

50. J. Coignus, R. Clerc, C. Leroux, G. Reimbold, G. Ghibaudo and, F. Boulanger, *J. Vac. Sci. Technol. B*, **27**, 338 (2009).

51. M. Stadele, B. R. Tuttle and, K. Hess,, *J. Appl. Phys.*, **89**, 348, (2001).

52. F. Sacconi, J. Jancu, M. Povolotskyi and, A. Di Carlo, *IEEE ED*, **54**, 3168 (2007).

53. S. Markov, P. V. Sushko, S. Roy, C. Fiegna, E. Sangiorgi, A. L. Shluger and, A. Asenov, *Phys. Stat. Sol. (a)*, **205**, 1290 (2008).

54. A. Campera, G. Iannaccone and, F. Crupi, *IEEE EDL*, **54**, 83 (2007).

55. I. Z. Mitrovic, Y. Lu, O. Buiu and, S. Hall, *Micro. Engineering*, **84**, 2306 (2007).

56. Z. Xu, M. Houssa, S. De Gendt, and M. Heyns, *Appl. Phys. Lett.*, **80**, 1975 (2002).

57. S. H. Lo, D.A. Buchanan, Y. Taur and, W. Wang, *IEEE EDL*, **18**, 209 (1997).

58. J.-F. Damlencourt, O. Renault, D. Samour, A. M. Papon, C. Leroux, F. Martin, S. Marthon, M. N. Séméria and, X. Garros, *Solid-State Electronics*, **47**, 1613 (2003).

59. R. Clerc, A. S. Spinelli, G. Ghibaudo, C. Leroux and, G. Pananakakis, *Micro. Reliability*, **41**, 1027 (2001).

60. E. Vogel, K.Z. Ahmed, B. Hornung, W.K. Henson, P. McLarty, G. Lucovsky, J.R. Hauser and, J.J. Wortman, *IEEE ED*, **45**, 1350 (1998).

61. C. Leroux, F. Allain, A. Toffoli, G. Ghibaudo and, G. Reimbold, M*icro. Engineering*, **84**, 2408 (2007).

62. M. A. Green, *J. Appl. Phys.*, **67**, 2944 (1989).

63. J. Cai and C. T. Sah, *J. Appl. Phys.*, **89**, 2272 (2001).

64. V. Cosnier, P. Besson, V. Loup, L. Vandroux, S. Minoret, M. Casse M, X. Garros X, J.M. Pedini, S. Lhostis, K. Dabertrand, C. Morin, C. Wiemer, M. Perego and, M. Fanciulli, *Micro. Engineering,* **84**, 1886 (2007).

65. J. Coignus, C. Leroux, R. Clerc, G. Ghibaudo, G. Reimbold and, F. Boulanger, Proc. ESSDERC 2009, p. 169.

66. J. F. Zhang, C. Z. Zhao, M. B. Zahid, G. Groeseneken, R. Degraeve, and S. De Gendt, *IEEE EDL*, **27**, 817 (2006).

67. M. Charbonnier, C. Leroux, F. Allain, A. Toffoli, G. Ghibaudo and, G. Reimbold, *Micro. Engineering*, **86**, 1740 (2009).

68. J. Widiez, K. Kita, K. Tomida, T. Nishimura and, A. Toriumi, *Jap. J. Appl. Phys.*, **47**, 2410 (2008).

ECS Transactions, 35 (4) 805-813 (2011)
10.1149/1.3572320 ©The Electrochemical Society

Flatband Voltage Tuning of HfSiON-based Gate Stacks: Impact of High Temperature Activation Annealing and LaO$_x$ Capping Layers

R. Boujamaa[a], S. Baudot[a], E. Martinez[b], O. Renault[b], B. Detlefs[d], J. Zegenhagen[d], V. Loup[b], F. Martin[b], M. Gros-Jean[a], F. Bertin[b], C. Dubourdieu[c]

[a] STMicroelectronics, 850, rue Jean Monnet, 38926 Crolles, France
[b] CEA-LETI, MINATEC Campus, F38054 Grenoble, France
[c] LMGP, CNRS, Grenoble INP, 3 parvis L. Néel, BP 257, 38016 Grenoble, France
[d] European Synchrotron Radiation Facility, 6 rue Jules Horowitz, F-38000 Grenoble, France

> We have investigated high temperature annealing effects on the flatband voltage (V_{fb}) modulation of TiN/LaOx/HfSiON/SiON/Si gate stacks with different LaO$_x$ thicknesses. Using hard X-ray photoelectron spectroscopy with synchrotron radiation and capacitance versus voltage measurements, we show band alignments shifts of the gate stacks attributed to La diffusion into the bottom HfSiON/SiON interface and the subsequent creation of an interfacial dipole. From electrical measurements on beveled oxide capacitors, we demonstrate that the effective work function of the metal can be increased to the desired value for transistor operation, which is a key point for optimizing these advanced devices.

Introduction

The aggressive scaling of metal-oxide-semiconductor field-effects transistors (MOSFETs) faces the challenge of metal gate (MG) and high-k (HK) dielectric integration to reduce power consumption [1]. Hf-based oxides and silicates, such as HfSiON, are considered as the most promising candidates for next-generation gate dielectrics, owing to their high permittivity, with a sufficiently wide band gap and a good thermal stability [2]. However, it remains critical to develop metal gate electrodes that control the effective work functions (EWFs) over a sufficiently wide range to keep suitable threshold voltage (V_{th}) for n-type and p-type FETs in a gate first integration scheme [3]. To overcome this problem, capping layers are incorporated into the gate stack in order to modulate the V_{th} transistor [4]. LaO$_x$ capping layers have been reported to provide a negative V_{th} shift, yielding the necessary decrease of the EWF of the gate for nFET devices [5, 6], while keeping the mobility sufficiently high [7, 8]. The mechanism of this voltage shift is attributed to La-induced dipoles at the HK/Si interface [9]. For this reason, the location of LaOx capping layer within the gate stack is a key factor for optimizing the transistor V_{th}. In our case, capping layers are deposited between the HK dielectric and the MG. Previous works from our group have shown that a high temperature annealing of typically 1065°C used for dopant activation induces the lanthanum diffusion at the HK/Si interface with the formation of a La-silicate. It is therefore crucial to understand and investigate the impact of this La diffusion on the

805

electronic structure and band discontinuity of the gate stack after the high-temperature activation anneal. This will help further optimization of the gate EWF for the advanced n-MOSFET devices. In this paper, we have investigated the impact of high temperature thermal annealing and LaOx capping layer on the gate EWF modulation of TiN/LaOx (0, 0.4, 1.0 nm) /HfSiON/SiON/Si gate stacks by coupling synchrotron based hard X-ray photoelectron spectroscopy (HAXPES) with capacitance versus voltage (C-V) measurements. Combining these techniques is particularly useful for providing information on the electronic structure, such as band alignment.

Experimental

The studied samples were prepared in a gate first integration process on 300 mm p-type Si(100) wafers. For electrical analysis, we performed C-V measurements on metal–insulator–semiconductor (MIS) capacitors across the wafers by using bevel structures in order to determinate the metal gate EWF accurately from equivalent oxide thickness (EOT)-V_{fb} plots. C-V characteristics of MIS capacitors were acquired at a frequency of 100 kHz using a HP4156 semiconductor parameter analyser with a HP4284A LCR Meter. Knowing V_{fb} and the substrate doping level, we calculate the gate EWF using Eq.1

$$V_{fb} = EWF - \Phi_s - \frac{EOT \cdot Q_{ox}}{\varepsilon_{ox}}$$ (1)

$$EWF = \Phi_{m,vac} + \Delta V$$

where EWF is the gate stack work function in vacuum ($\Phi_{m, vac}$) plus the contribution from interfacial dipoles (ΔV), Φ_s is the silicon work function, EOT is the equivalent oxide thickness, ε_{ox} is the permittivity of SiO$_2$ and Q_{ox} is the equivalent oxide charge density assumed to be located at the dielectric/silicon interface. Therefore, as follows from [1] and using test structures in which the high-k film thickness is fixed and varying the thickness of the underlying interfacial layer (IL), we extract EWF from the Y-axis intercept of the linear fit of the V_{fb}-EOT plots. Figure 1 shows the steps involved in the devices fabrication.

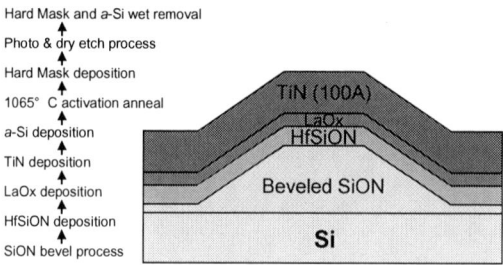

Figure 1. Process and schematic for beveled oxide capacitor structures. The SiON bevel (thickness: 1.5 to 9 nm) allows extraction of the metal EWF.

After a HF-SC1 clean, a 90 nm thick SiO_2 oxide layer is formed by oxidation of the silicon substrate with in-situ steam generation (ISSG) at 1025°C. The SiO_2 films were then etched down by a HF solution followed by decoupled plasma nitridation (DPN) and post nitridation anneal (PNA) to obtain a SiON beveled layer with a thickness varying from 1.5 to 90 nm across the wafer radius. Then, 1.7 nm HfSiON films were deposited by metal organic chemical vapor deposition (MOCVD) of HfSiO followed by a DPN and PNA. LaO_x-based capping layers were formed by physical vapor deposition (PVD) and two thicknesses were investigated: 0.4 and 1.0 nm. The metal gate stack was composed of a 6.5 nm TiN film deposited by PVD followed by a 60 nm amorphous Si (*a*-Si) film deposited using chemical vapor deposition (CVD) with a maximum thermal budget of 600°C. Reference samples without LaO_x capping layers were also prepared. To simulate the transistor fabrication process, high temperature activation annealing was performed at 1065°C for 1.5 s on some wafers. Finally, after patterning, we remove the silicon top layer by a wet etching in order to perform C-V measurements directly on TiN metal electrodes. In this way, we were able to characterize the impact of the high temperature activation annealing on the V_{fb} tuning by comparing C-V plots of MIS capacitor structures with and without the standard 1065°C anneal. The beveled SiON oxide structure provides a series of thicknesses on the same wafer, enabling a reliable and precise extraction of the TiN work function from numerous EOT-V_{fb} plots.

For HAXPES analysis, "full sheet" samples with 1.5 nm-thin SiON interfacial layers were used. The HAXPES measurements were performed at the European Synchrotron Radiation Facility (ESRF, Grenoble, France) on the ID32 beam line using a photon energy of 3.81 keV. The use of high energy x-rays (hard x-rays) enables to successfully probe, in a non-destructive way, the nanoscale buried layers and interfaces, without removing the top TiN metal gate, which is of utmost importance to insure the integrity of the full stack. Photoelectrons were detected at a take-off angle of 80° with respect to the sample surface (i.e., normal emission). The overall energy resolution was 0.3 eV thanks to a beam spectral width optimized with a high resolution Si (311) double crystal post-monochromator. In these conditions, we were able to perform a non-destructive high-energy and high-resolution analysis of chemical and electronic properties of extremely thin buried layers and interfaces typically present in CMOS structures.

Results and Discussion

HAXPES Analysis

The analysis of core level shifts by HAXPES gives detailed information on the electronic structures such as band alignments of the gate stacks.

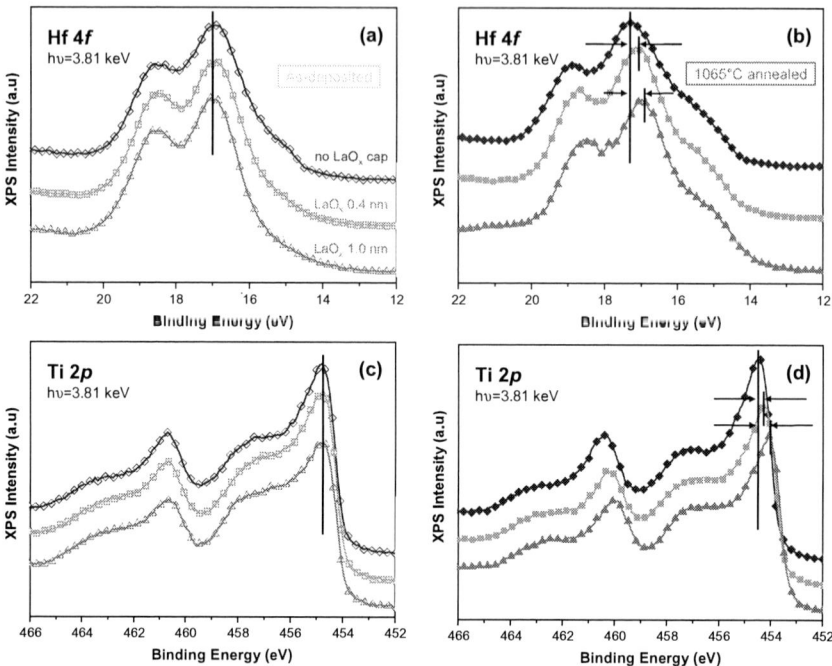

Figure 2. Evolution of the HAXPES spectra as a function of LaOx thickness. Hf 4f core level of as-deposited (a) and annealed (b) samples. The Ti 2p core-level spectra obtained for as-deposited (c) and annealed samples (d).

Figures. 2(a) and 2(b) show the evolution of Hf 4f core-level emission for as-deposited and annealed samples, respectively, as a function of LaOx thickness. Binding energies (BE) were referenced to the Fermi level of the sample. Before annealing, no significant chemical shift is observed with the increase of LaOx thickness inserted between the HfSiON and the TiN. For the others core-levels analysed (not shown), the same result is observed. Therefore, no band offset is expected when the LaOx capping layer is inserted between the MG and the HK dielectric. In contrast, after annealing, we observe a distinct chemical shift of the Hf 4f core level towards lower binding energy from 0.2 to 0.4 eV as the LaOx thickness increases from 0.4 to 1.0 nm, indicating the presence of interfacial dipoles or fixed charges in the stack. Consequently, we infer that the core level energy shift observed in the present gates stacks is strongly related to the thermally-induced La diffusion into the HK/IL interface. The increased La atoms diffusion (thicker starting LaOx film), which is associated to an increased La-silicate formation results in a stronger shift. The analysis of Ti2p core level emission for the as-deposited and annealed samples shown in Figs. 2(c) and 2(d), respectively, reinforces this assumption. After a 1065°C spike anneal, we observe the same gradual decrease in Ti 2p core-level binding energy from 0.18 to 0.4 eV with the increase of the LaOx thickness compare to the samples without anneal, where no chemical shift is observed. We have therefore direct evidence that the high-k band alignment is affected by the La diffusion process to the bottom HfSiON/SiON interface. Figure 3 shows a schematic picture of band discontinuity at the TiN/HfSiON/SiON/Si stack structures.

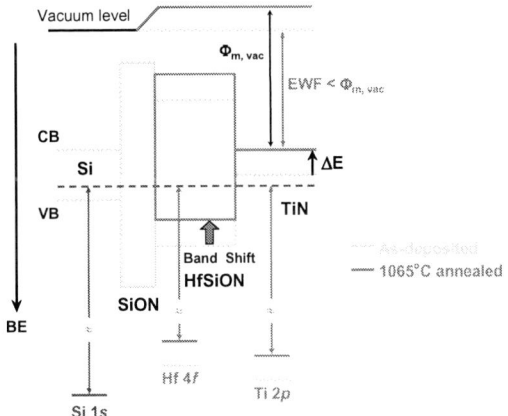

Figure 3. Schematic picture of the band discontinuity for TiN/LaOx/HfSiON/SiON/Si stack structures for the as-grown and annealed samples, Si band bending is not represented.

Upon annealing, La atoms diffuse from the LaOx capping layers to the HfSiON/SiON interface and the band discontinuity is affected. This result confirms that the location of La atoms in the gate stacks is crucial for optimizing the electrical properties of the advanced n-MOSFET devices. The incorporation of La atoms into the HK/IL interface plays the dominant role for the V_{fb} tuning compared to the MG/HK interface, which is in agreement with previous studies [10-12]. Thus, our photoemission results are qualitatively consistent with V_{fb} shift based on interface dipole and/or a fixed charge effect at HK/IL.

Electrical Analysis

In order to investigate the flatband voltage shifts, capacitance-voltage (C-V) measurements were performed. Figure 4 reproduces the C-V characteristics of TiN/HfSiON/SiON/Si gate stack with the beveled oxide structure.

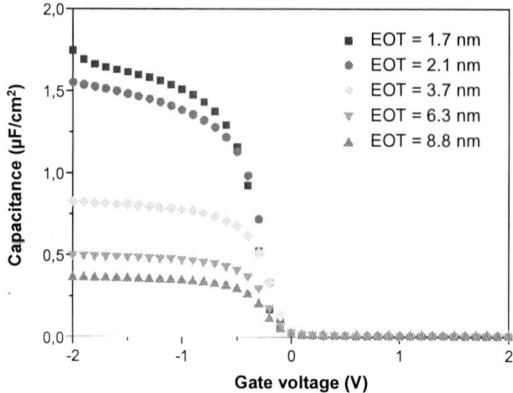

Figure 4. C-V characteristics of TiN/LaO$_x$/HfSiON/SiON/Si gate stack with the beveled SiON oxide structure. The C-V curves were measured from -2 V to +2 V at 100 kHz.

This figure exhibits various values of capacitance on a single wafer without any degradation of electrical characteristics, indicating that the bevel oxide technique is definitely effective for the extraction of the gate EWF, and also has an excellent compatibility with the high-k materials such as HfSiON.

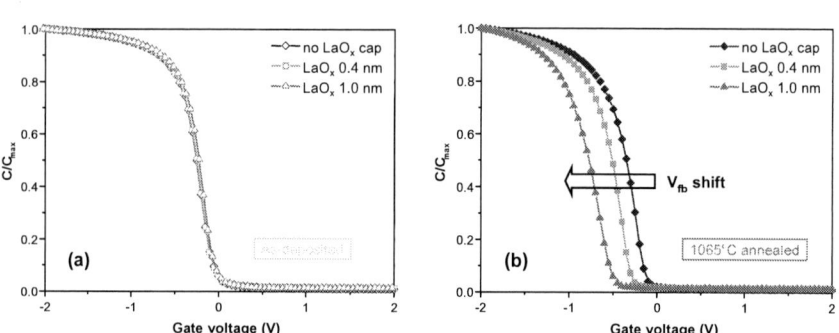

Figure 5. Comparison of capacitance versus voltage (CV) characteristics for as-deposited (a) and annealed (b) gate stacks as a function of LaOx thickness.

Figures. 5(a) and 5(b) show the CV curves of as-deposited and annealed gate stacks, respectively, as a function of LaO$_x$ thickness. It should be noted that the electrical measurements before and after anneal were performed on the same MIS capacitors, i.e. at the same positions of the SiON bevel. Before annealing, the CV curves indicate no shift in the flatband voltage of the devices with increasing LaOx content In contrast, after annealing, the CV curves show a negative shift in V$_{fb}$ by nearly -0.2V with the initial addition of 0.4 nm of LaOx capping layer and by -0.4V with the addition of 1.0 nm of capping layer. These results are similar to previous results obtained on HfO$_2$/SiO$_2$ stacks [9, 13, 14]. The V$_{fb}$ modulation is conditioned by the La diffusion process into the

HK/IL interface, which is, in our case, thermally activated through the 1065°C spike anneal. The main origin for the V_{fb} shift is at the HK/IL interface and the contribution of dipoles and/or fixed charge effects at the MG/HK interface can be neglected.

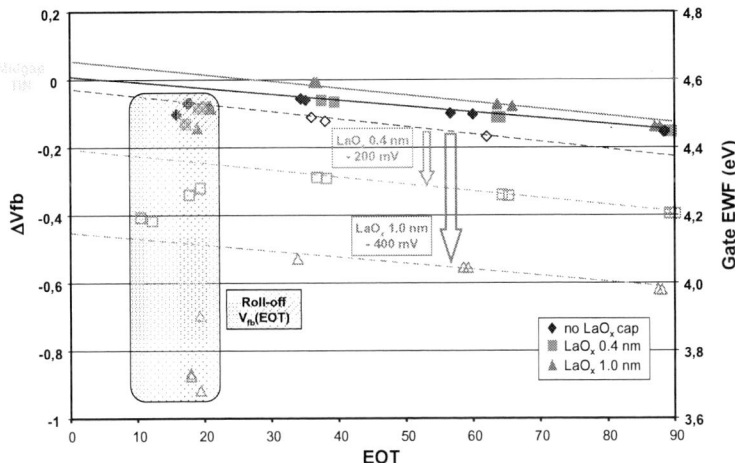

Figure 6. ΔV_{fb} versus EOT for different LaOx thicknesses (0, 0.4 and 1.0 nm) as a function of annealing.

Figure 6 represents the flatband voltage (V_{fb})-EOT plots obtained on the beveled capacitor structures for extracting the gate EWF. This figure clearly depicts a linear relationship between EOT and V_{fb}, indicating the absence of bulk charges in the SiON. Using Eq. 1, we extract an EWF value of 4.62eV for the non-annealed reference stack, revealing that the TiN acts as a mid-gap material. By adding LaOx capping layers from 0.4 to 1.0 nm above the HfSiON dielectric, the EWF values extracted are almost the same without annealing: 4.62 and 4.64 eV respectively. However, after high temperature processing, we observe a gradual decrease of the gate EWF of -40 meV for the reference sample and from -200 to 400 meV as the inserted LaOx capping layer thickness increases from 0.4 to 1.0 nm respectively. The EWF shift toward n-type like values is achieved upon high annealing temperatures, indicating that EWF tuning is correlated by a diffusion-related mechanism. Note that the slopes of the curves are almost parallel, suggesting that oxide charges inside the dielectric due to La atoms are negligible, even after a 1065°C dopant activation anneal. The results also reveal that the sensitivity of EWF to the gate stack process is especially highlighted with thin IL thicknesses. It results in a V_{fb} drop when the IL becomes thinner. This V_{fb} behaviour is referred to as "V_{fb} roll-off" phenomenon. The roll-off remains almost constant before annealing even with the increase of the LaOx thickness inserted between the HfSiON and the TiN. However, upon annealing, we observe that the roll-off amplitude becomes larger. This indicates that the roll-off mechanism is thermally activated, which is consistent with most of the studies performed on this phenomenon [15-17]. More importantly, this additive roll-off effect increases after annealing with increasing initial amount of LaOx in the gate stack after annealing. We previously reported that, after high temperature annealing. Therefore, the roll-off mechanism must be strongly related to the presence and amount of La atoms at the HK/IL interface. Consequently, the EWF at low EOT (<20Å) can be interpreted as the result of La-induced dipoles at the HK/IL and of V_{fb} roll-off effect. The mechanism of this latter effect is beyond the scope of this paper and will be discuss elsewhere.

Conclusion

By coupling hard X-ray photoelectron spectroscopy using synchrotron radiation and capacitance versus voltage measurements, we have analysed the impact of high temperature annealing and LaOx capping layer on electronic structure and band alignment for TiN/LaOx/HfSiON/SiON/Si gate stacks. Based on the core level shifts, we show that the V_{fb} shift is related to the thermally-induced La diffusion at the HK/IL interface. The increase of the initial LaOx thickness results in an increased La-silicate layer formation and in a larger V_{fb} shift. The electrical measurements on beveled oxide capacitors demonstrate that the metal EWF can be tuned toward n-type like values by modulating the amounts of inserted LaOx in the gate stacks, which is a key step for optimizing the advanced n-MOSFET devices.

Acknowledgments

This work was partially supported by the European EUREKA/CATRENE program in the frame of the CT206 UTTERMOST project.

References

1. G. D. Wilk, R. M. Wallace and J. M. Anthony, *J. Appl. Phys.*, **89**, 5243 (2001).
2. M. R. Visokay, J. J. Chambers, A. L. P. Rotondaro, A. Shanware and L. Colombo, *Appl. Phys. Lett.*, **80**, 3183 (2002).
3. C. C. Hobbs, L. R. C. Fonseca, A. Knizhnik, V. Dhandapani, S. B. Samavedam, W. J. Taylor, J. M. Grant, L. G. Dip, D. H. Triyoso, R. I. Hegde, D. C. Gilmer, R. Garcia, D. Roan, M. L. Lovejoy, R. S. Rai, E. A. Hebert, H. H. Tseng, S. G. H. Anderson, B. E. White and P. J. Tobin, *IEEE Trans. Electron Devices*, **51**, 971 (2004).
4. V. Narayanan, V. K. Paruchuri, N. A. Bojarczuk, B. P. Linder, B. Doris, Y. H. Kim, S. Zafar, J. Stathis, S. Brown, J. Arnold, M. Copel, M. Steen, E. Cartier, A. Callegari, P. Jamison, J. P. Locquet, D. L. Lacey, Y. Wang, P. E. Batson, P. Ronsheim, R. Jammy, M. P. Chudzik, M. Ieong, S. Guha, G. Shahidi and T. C. Chen, in *VLSI Technology Digest of Technical Papers*, p. 178 (2006).
5. K. Kita and A. Toriumi, *Appl. Phys. Lett.*, **94**, 132902 (2009).
6. H. N. Alshareef, M. Quevedo-Lopez, H. C. Wen, R. Harris, P. Kirsch, P. Majhi, B. H. Lee, R. Jammy, D. J. Lichtenwalner, J. S. Jur and A. I. Kingon, *Appl. Phys. Lett.*, **89**, 232103 (2006).
7. M. A. Quevedo-Lopez, S. A. Krishnan, P. D. Kirsch, G. Pant, B. E. Gnade and R. M. Wallace, *Appl. Phys. Lett.*, **87**, 262902 (2005).
8. T. Ando, M. Copel, J. Bruley, M. M. Frank, H. Watanabe and V. Narayanan, *Appl. Phys. Lett.*, **96**, 132904 (2010).

9. P. D. Kirsch, P. Sivasubramani, J. Huang, C. D. Young, M. A. Quevedo-Lopez, H. C. Wen, H. Alshareef, K. Choi, C. S. Park, K. Freeman, M. M. Hussain, G. Bersuker, H. R. Harris, P. Majhi, R. Choi, P. Lysaght, B. H. Lee, H. H. Tseng, R. Jammy, T. S. Boscke, D. J. Lichtenwalner, J. S. Jur and A. I. Kingon, *Appl. Phys. Lett.*, **92**, 092901 (2008).

10. K. Kakushima, K. Okamoto, M. Adachi, K. Tachi, P. Ahmet, K. Tsutsui, N. Sugii, T. Hattori and H. Iwai, *Solid-State Electron.*, **52**, 1280 (2008).

11. Y. Yamamoto, K. Kita, K. Kyuno and A. Toriumi, *Jpn. J. Appl. Phys. Part 1 - Regul. Pap. Brief Commun. Rev. Pap.*, **46**, 7251 (2007).

12. W. J. Maeng, W. H. Kim and H. Kim, *J. Appl. Phys.*, **107**, 074109 (2010).

13. M. Di, E. Bersch, R. D. Clark, S. Consiglio, G. J. Leusink and A. C. Diebold, *Journal of Applied Physics*, **108**, 114107 (2010).

14. S. Guha, V. K. Paruchuri, M. Copel, V. Narayanan, Y. Y. Wang, P. E. Batson, N. A. Bojarczuk, B. Linder and B. Doris, *Appl. Phys. Lett.*, **90**, 092902 (2007).

15. K. Akiyama, W. Wang, W. Mizubayashi, M. Ikeda, H. Ota, T. Nabatame and A. Toriumi, *VLSI Technology Digest of Technical Papers*, 72 (2007).

16. H. H. Tseng, P. Kirsch, C. S. Park, G. Bersuker, P. Majhi, M. Hussain and R. Jammy, *Microelectron. Eng.*, **86**, 1722 (2009).

17. K. Akiyama, W. Wang, W. Mizubayashi, K. M. A. Salam, M. Ikeda, H. Ota, T. Nabatame and A. Toriumi, *Extended Abstracts of the IWDTF*, **63** (2007).

814

ECS Transactions, 35 (4) 815-834 (2011)
10.1149/1.3572321 ©The Electrochemical Society

Physical and Electrical Effects of the Dep-Anneal-Dep-Anneal (DADA) Process for HfO$_2$ in High K/Metal Gate Stacks

R. D. Clark[a], S. Aoyama[b], S. Consiglio[a], G. Nakamura[a] and G. J. Leusink[a]

[a]TEL Technology Center, America, 255 Fuller Rd, Albany, NY 12203, USA
[b]Tokyo Electron AT, 650 Mitsuzawa, Hosaka-cho, Niraski, Yamanashi, 407-0192, Japan

In this work, we present physical and electrical characterization of HfO$_2$ films deposited using the Dep-Anneal-Dep-Anneal (DADA) deposition scheme. Electrical results from MOSCAP devices fabricated using a low temperature (Gate Last-like) integration flow are presented. In addition, we report detailed physical analyses of the films and changes in the films versus as-deposited ALD HfO$_2$ and films undergoing a single post-deposition anneal (PDA). This comparison shows the correlation between observed physical changes in the film and electrical results. Observed physical changes using high resolution Rutherford backscattering spectroscopy (HR-RBS), high resolution transmission electron microscopy (HR-TEM), secondary ion mass spectrometry (SIMS), X-ray photoelectron spectroscopy (XPS) and X-ray reflectivity (XRR) include crystallization, densification, Si intermixing, reduction of in-film carbon and improved etch resistance leading to improved leakage vs. EOT and electrical non-uniformity. Dependence of these changes on the underlying interface layer (e.g. SiO$_2$ vs. SiON) is also described.

Introduction

The use of anneals interspersed during hafnium oxide deposition to optimize the electrical properties of the resulting films has recently been reported (1-4). In particular, Professor Toriumi's group has reported anneals after every cycle (layer-by-layer deposition and anneal) to optimize atomic layer deposition (ALD) hafnium oxide films (1,3). Other groups have reported electrical improvements in films deposited by chemical vapor deposition (CVD) when the depositions are interrupted by anneals.(2,4) We have extended this approach by performing multiple ALD cycles before each anneal to produce optimized hafnium oxide films in a process that is demonstrable and manufacturable on 300 mm wafers for advanced high K/metal gate stacks, which we have termed Dep-Anneal-Dep-Anneal (DADA). Specifically, the DADA process includes the following steps: 1) Depositing 5-30 cycles of ALD HfO$_2$, 2) Annealing at 700-950 °C, 3) repeating steps 1-2 at least one time. The electrical improvements observed in highly scaled gate first devices using this process have recently been reported and attributed to a reduction in leakage current associated with trap assisted tunneling through the film, Figure 1 (5,6). In this transaction we report electrical results in MOSCAP devices including DADA HfO$_2$ that were fabricated using a low thermal budget (gate last-like) flow in which the thermal processing after HfO$_2$ deposition never exceeds 540 °C, the deposition temperature of the oxide hard mask used to pattern the MOCSCAP structures

815

used in this study. Because this temperature is significantly lower than the anneal temperatures we use in the DADA process, we believe our results confirm that the improvements previously reported using DADA HfO_2 are a result of a true improvement in the HfO_2 film from the DADA process and not contributed to significantly by thermal effects in the gate first integration schemes previously utilized.

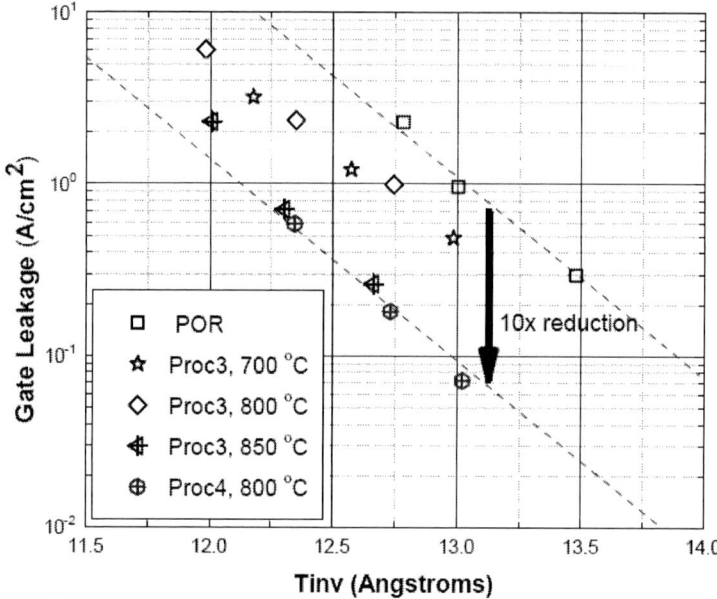

Figure 1. Gate Leakage (measured at 1V) vs. T_{inv} of gate stack with conventional and optimized HfO_2 films in a gate first integration scheme, reprinted from (5). Temperatures noted are the temperature of the anneal in the DADA process used to optimize the HfO_2 films.

We have further undertaken to increase our understanding of the processes underlying the observed electrical improvements from the DADA process. To that end, we have conducted additional physical analysis of DADA HfO_2 films. These analyses are primarily performed on films without a metal gate, but we believe the physical changes within the HfO_2 film we observe correlate to the effects found in our MOSCAP devices due to the low thermal budget of our MOSCAP flow, as discussed above. Using these results we propose mechanisms for the physical changes that correlate to the electrical improvements observed.

Sample Preparation

MOSCAP Fabrication

The fabrication scheme for the MOSCAP devices presented in this study is outlined below, Figure 2. All devices were fabricated on 300mm wafers using equipment located in the Albany Nanotech cleanroom facility located on the campus of the University at Albany, Albany, NY. After initial cleaning, a sacrificial oxide layer was grown and then removed during pre-gate clean. The pre-gate clean included a rinse with ozone/deionized water that resulted in an SiO_2 interface layer on the order of 0.6-0.8 nm in thickness. In most cases the SiO_2 layer was then nitrided by a remote plasma Radical Flow (RF) nitridation process that slightly thickens the interface and results in an approximate 0.7-0.9 nm SiON interface layer. Following interface layer formation, the HfO_2 deposition was performed using tetrakis(ethylmethylamido)hafnium and water in an ALD process that has been described previously (7,8). Splits were performed at this step by varying the anneal time and temperature, and cycle counts during DADA deposition, and in some cases by depositing films without an anneal or with only a single post-deposition anneal (PDA). After HfO_2 deposition, the metal gate was formed by 50nm of CVD TiN. The MOSCAP devices were then formed by depositing a hardmask, patterning and etching the MOSCAPs and cleaning and stripping the hard mask. The devices then underwent a 30 minute H_2 sinter prior to C-V and I-V testing on an in-line Auto-tester. Equivalent oxide thickness (EOT) of the devices tested was calculated from the C-V curves using the N.C. State CVC program (9). Gate leakage current (Jg) is reported per unit area using leakage measured at CVC calculated flatband voltage minus one volt (Vfb − 1V). The mask set used in this study provides a range of capacitor sizes within each die from 100-10000 μm^2 with 93 die per wafer. Most data reported in this paper come from 1600 μm^2 capacitors, which are the smallest devices that can be measured reliably in-line using this device flow. In some cases 900 or 2500 μm^2 capacitors are used instead to overcome measurement issues (e.g. a wafer with bad probe alignment). In all cases, MOSCAP data

Low Temperature MOSCAP Flow

- Pre-clean for Sacrificial Oxide
- Sacrificial Oxide : Thermal SiO_2 10nm
- Pre-clean : DHF then Ozone based chemical oxide
- Optional RF Nitridation to form SiON interface
- HfO_2 Deposition/Anneal (DADA, PDA or as deposited)
- CVD TiN 50 nm (500 °C)
- Hardmask: SiN (480 °C) & SiO_2 (540 °C)
- Gate Litho
- Gate Etch & Ash
- HM Removal
- H_2 Sinter : 400 °C, 4% H_2, 30min
- E-test : TEL P12 Auto-tester (I-V, C-V)

Figure 2. Outline of low temperature MOSCAP flow used to produce devices for this study including thermal budget of high temperature processes.

is reported within a lot from devices of the same area. We believe lot to lot comparisons are possible even in the case when devices measured have slightly different areas because all device data is normalized by area.

Samples for Physical Analysis

For some analysis techniques, for instance high resolution TEM imaging (HR-TEM), it is possible to perform physical analysis directly on the MOSCAP devices that were measured. However, for some analyses it was necessary to prepare separate samples for physical analysis. Samples for physical analysis were prepared by an abbreviated version of the above fabrication scheme. Bare Si wafers were first cleaned with the gate pre-clean used above and then RF nitridation was optionally performed followed by HfO_2 deposition and anneal. The same equipment was used to deposit films in most cases to that used for films deposited for MOSCAP devices. Except where noted, the system used to deposit HfO_2 films for this study consists of a traveling wave-type geometry counter-flow ALD system with horizontal flows of precursor and oxidant across the wafer which has been described previously (7, 8).

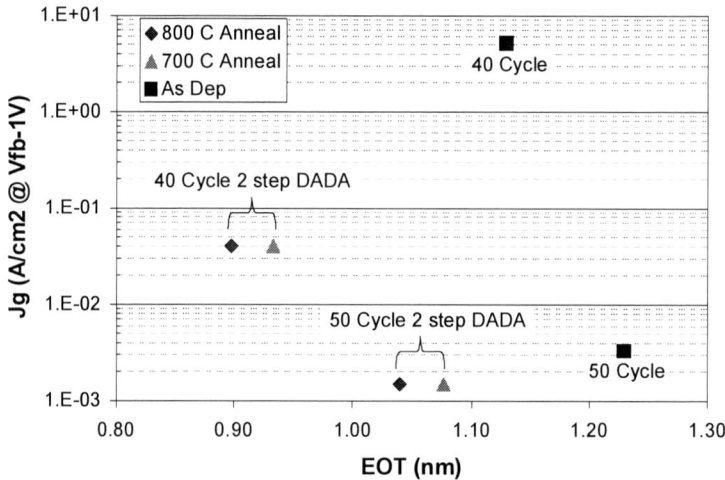

Figure 3. Comparison of results using 2 step DADA films on SiON interface with 40 second anneals at 700 °C and 800 °C with as deposited films with equivalent cycle counts. Each point represents an average of 9 devices measured across a 300 mm wafer. Anneal temperature is noted in the legend and cycle counts are labeled on the graph. The 40 cycle as deposited sample exhibited a high leakage current level and is therefore likely to have an actual EOT lower than that reported due to severe C-V roll off.

Results

Annealing Temperature and Time

As reported previously, the DADA process results in crystalline HfO_2 films (5). The temperature effect of the anneal on results with the DADA process is shown in Figure 3, which compares MOSCAP results between 2-step DADA films with 700 °C and 800 °C anneals with the result from as deposited (unannealed) HfO_2 with equivalent cycle counts. As can be seen in the graph the DADA process results in significantly better scaling in terms of EOT at equivalent or lower gate leakage levels than unannealed films. We find a striking difference in morphology between the as deposited films and the DADA films measured in this study, Figures 4 and 5. As can be seen in the HR-TEM images, the films undergoing the DADA process are nano-crystalline while the as deposited film remains amorphous. In addition, a greater degree of crystallinity and ordering, and larger grain size seems apparent in the film using 800 °C anneals, versus the film utilizing 700 °C anneals.

As shown in Figure 6, the effect of anneal time on electrical performance is less well defined than the effect of anneal temperature. It is clear, however, that a longer dwell time during the DADA anneals can provide an additional benefit in terms of lower gate leakage in the measured devices. HR-TEM images, Figure 7, of devices from the wafers used in Figure 6, show that increasing the DADA anneal time results in increased crystallinity and ordering within the film.

One effect of increased crystallinity can be observed when the gate edge of the device is imaged, Figure 8. While we find that the gate edges in devices with the as deposited films are undercut, it is clear that DADA films result in HfO_2 footing using the

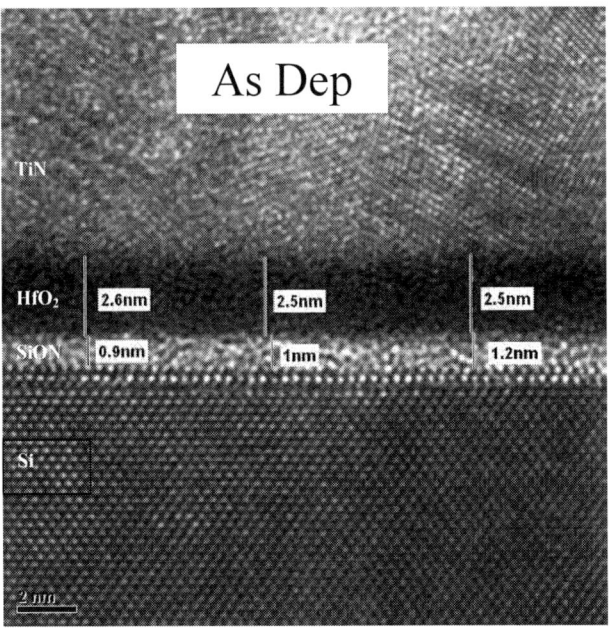

Figure 4. HR-TEM cross-section of a MOSCAP device incorporating a 40 cycle (~2.5 nm) as deposited (unannealed) HfO_2 film. This device is from the wafer measured for the 40 cycle as deposited data point in shown in Figure 3.

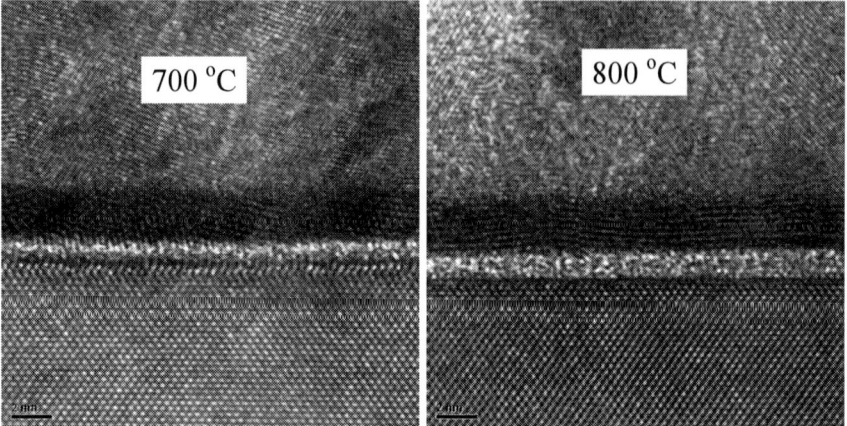

Figure 5. HR-TEM cross-sections of MOSCAP devices incorporating 2-step DADA HfO$_2$ films with a total of 40 cycles in each film, anneal temperature is indicated in the insets. These devices are from the wafers measured for the 40 cycle DADA data points shown in Figure 3.

Effect of DADA anneal Time

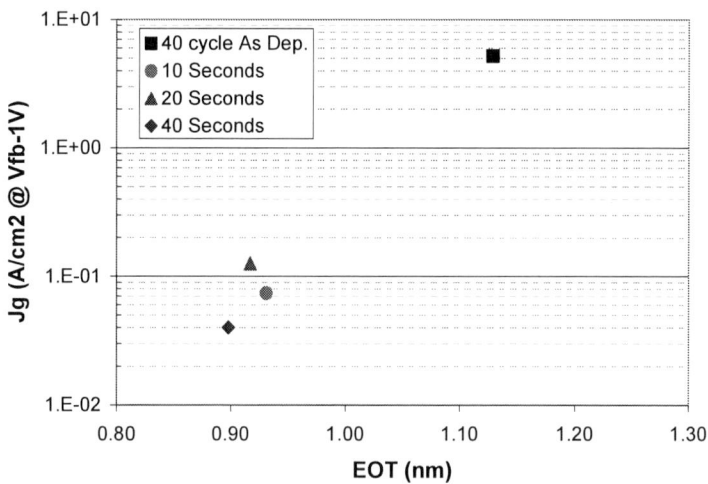

Figure 6. Comparison of results using 2 step 40 cycle DADA films on SiON interface with anneals at 800 °C and varying anneal time with as deposited films with equivalent cycle counts. Each point represents an average of 9 devices measured across a 300 mm wafer. Anneal dwell time is noted in the legend. The 40 cycle as deposited sample exhibited a high leakage current level and is therefore likely to have an actual EOT lower than that reported due to severe C-V roll off.

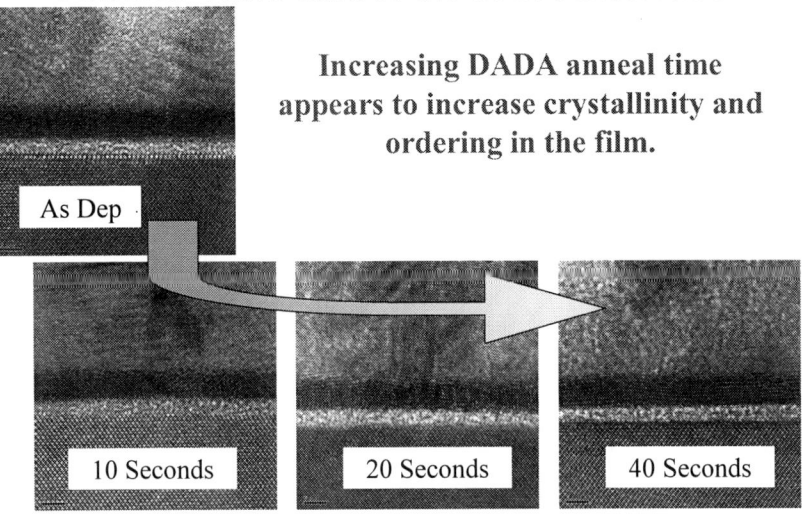

Figure 7. HR-TEM cross-sections of MOSCAP devices using 2 step 40 cycle DADA films on SiON interface with anneals at 800 °C and varying anneal time compared with an as deposited film with equivalent cycle counts. These devices are from the wafers measured for Figure 6.

same integration scheme. We believe that this results from a difference in etch resistance to the hard mask removal etch used in our device flow. While the as deposited films are etched more quickly than the TiN gate metal and thus result in undercut, the DADA films are more etch resistant than the TiN, and thus show significant footing at the gate edge.

Overall, we observe that the more crystalline and ordered films resulting from DADA provide superior electrical performance in our MOSCAP devices. In general, the trends and magnitude of improvement we see in this study match those previously reported for gate first MOSFET devices and this suggests that the mechanism for the improvement is similar in this study to those previously reported (5,6).

Film Densification

Thickness and density of DADA films were compared with as deposited films by measuring X-ray reflectivity (XRR) on samples prepared for physical analysis, Figure 9. As reported previously (5), DADA films are found to be consistently thinner with higher density at low total cycle counts (e.g. 40 total cycles or less). For higher total cycle counts, 2-step DADA films have similar thickness and density to as deposited films. This result suggests that there is a critical limit to the number of ALD cycles that can be included in a DA cycle during the DADA process. For the 2-step processes shown, we have used half of the total cycles in each DA cycle, thus from this result we can infer that the critical limit to cycles per DA cycle is between 20 and 40 cycles in order to realize the benefits of the DADA process.

Gate Edge Comparison

As Deposited   2 step DADA – 800 °C, 40 sec.

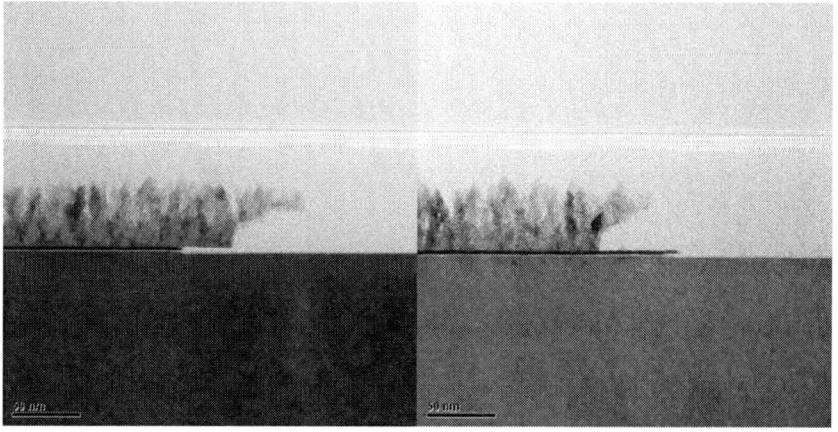

Figure 8. HR-TEM cross-sections of MOSCAP devices with as deposited HfO$_2$ versus 2 step 40 cycle DADA films on SiON interface showing the gate edge.

Interface Dependence

We have found that DADA films on chemical oxide without nitridation seem to have a different critical limit for ALD cycles in a DA cycle, Figure 10. We find in this case that the 50 cycle DADA films actually show degraded performance compared with the as deposited film. This behavior is quite different from that observed on an SiON interface layer, shown in figure 3, and is observed for 50 cycle processes using 700 and 800 °C anneal, and with short and long anneal times. A further example of the different behavior of DADA films on chemical oxide versus SiON can be seen using high resolution Rutherford backscattering (HR-RBS). Depth profiles of DADA on chemical oxide show significant Si intermixing throughout the film, however, on SiON, Si intermixing with the HfO$_2$ is limited to the interface.

Further evidence for interface modification during the DADA process on chemical oxide is found from EDX-EELS profiling, Figure 12, as well as in XPS, Figure 13. In each case the observed data is consistent with increased interface intermixing. Another possibility for the observed data is interface layer growth. Interface growth is not observed, however, by high resolution RBS.

Interface Thickness

As mentioned above, some measurements showing evidence of interface modification could be explained by interface growth. Of course, the HR-RBS data presented above

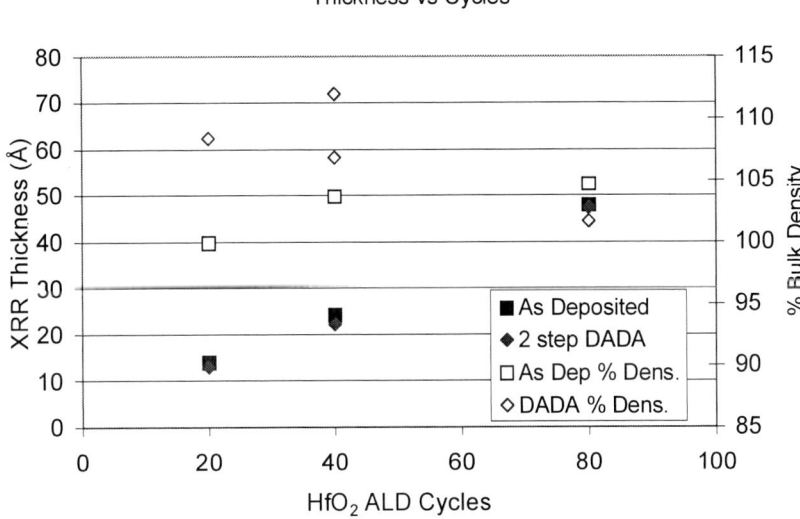

Figure 9. Comparison of XRR Thickness and % Bulk Density measured for as deposited versus 2 step DADA at equivalent total cycles of ALD HfO_2. Bulk density was assumed to be 10.5 g/cm^3 for HfO_2.

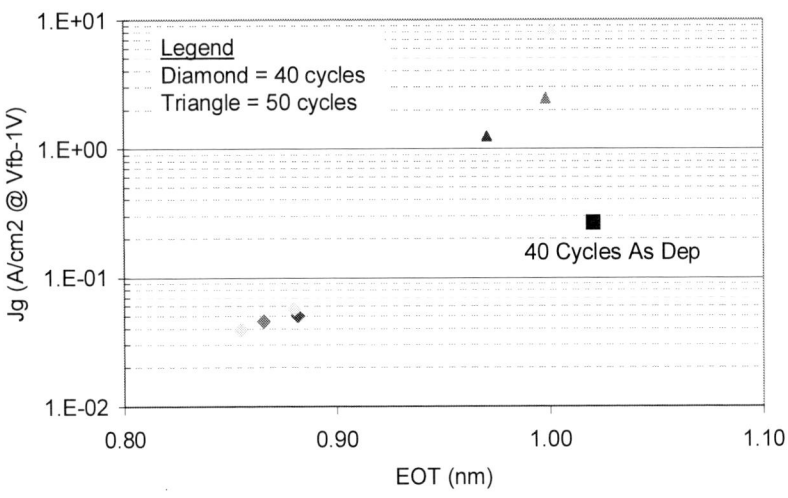

Figure 10. Comparison of electrical results of MOSCAP devices incorporating 2-step DADA films with 40 and 50 total ALD cycles on non-nitrided chemical oxide. A point showing performance of 40 cycles as deposited on chemical oxide is included from a separate lot for comparison.

High Resolution RBS depth profile

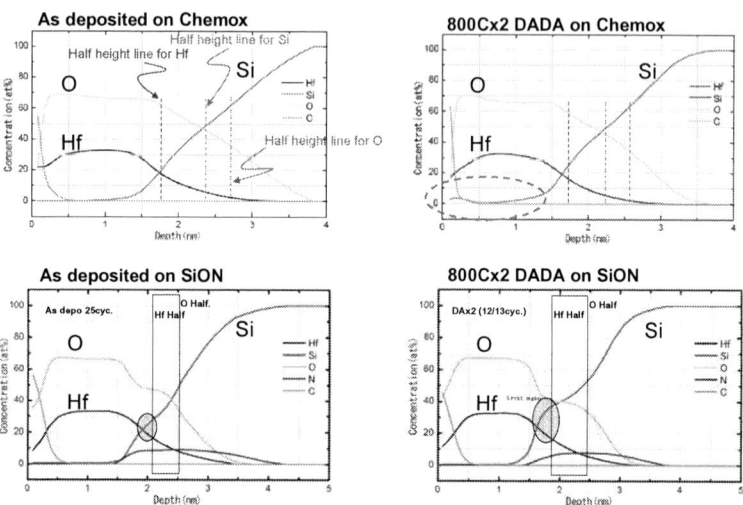

Figure 11. HR-RBS Depth profiles of as deposited and 2-step DADA films on chemical oxide and on SiON.

suggests interface mixing with the HfO_2 is taking place- which can also explain the XPS and EDX-EELS results, though it does not unambiguously show that interface growth is not occurring. While it is possible to extract interface thicknesses from HR-TEM, we do not consider the current HR-TEM micrographs are useful for interface measurement since interfacial SiO_2 layers in High K gate stacks have previously been reported to grow during HR-TEM measurements (10). Further, we consider that changes less than 2 Å in thickness should be within the measurement error inherent in the TEM measurement. Based on the electrical data presented here and elsewhere (5,6) we propose it is not possible for interface SiO_2 to increase significantly during the DADA process and simultaneously provide 1-2 Å lower EOT. We did, however, seek to measure the interface on a series of samples for physical analysis by using a combination of spectroscopic ellipsometry (SE) with XRR (11). In this measurement we treat the HfO_2 and the interface layer as a single dielectric stack, and allow the ellipsometer to fit the thickness of the dielectric stack and the refractive index over the spectral range using a standard HfO_2 model for the layer. We use XRR to measure the HfO_2 thickness because of the density difference between HfO_2 and SiO_2. For this measurement, we use only the center point of each measurement on the wafer and subtract the HfO_2 thickness (XRR) from the dielectric stack thickness (SE). As can be seen in Figure 14, this measure is consistent with the electrical and HR-RBS results suggesting interface layer thinning, presumably due to interface/HfO_2 intermixing.

EDX/EELS: POR as deposited vs. DADA 2-step

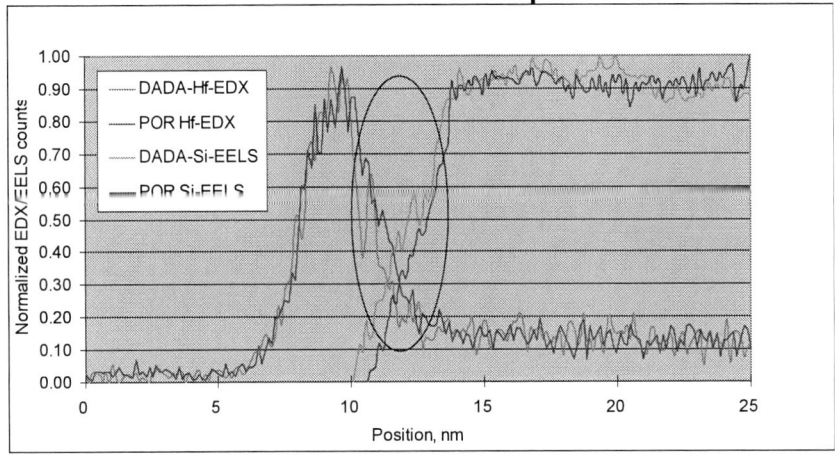

Figure 12. EDX (Hf) and EELS (Si) Depth profiles of as deposited and 2-step DADA films on chemical oxide in MOSCAP devices.

Si2p core level comparison – 25 cycle ALD DADA vs. As Deposited

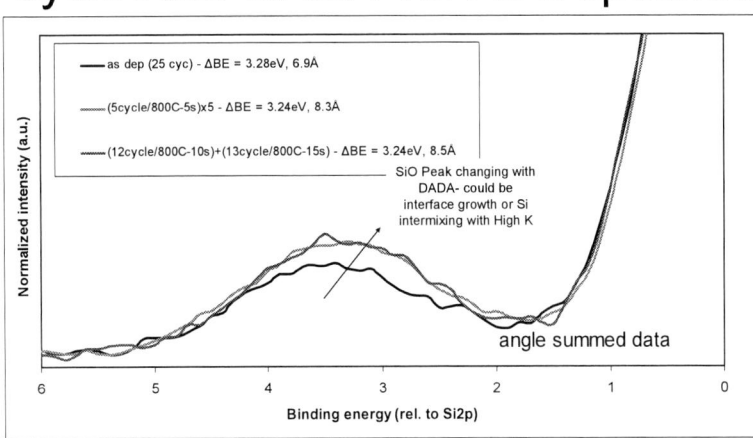

Figure 13. XPS of as deposited and DADA films on chemical oxide showing the Si 2p peak from Si bonded to O normalized to the Si substrate peak.

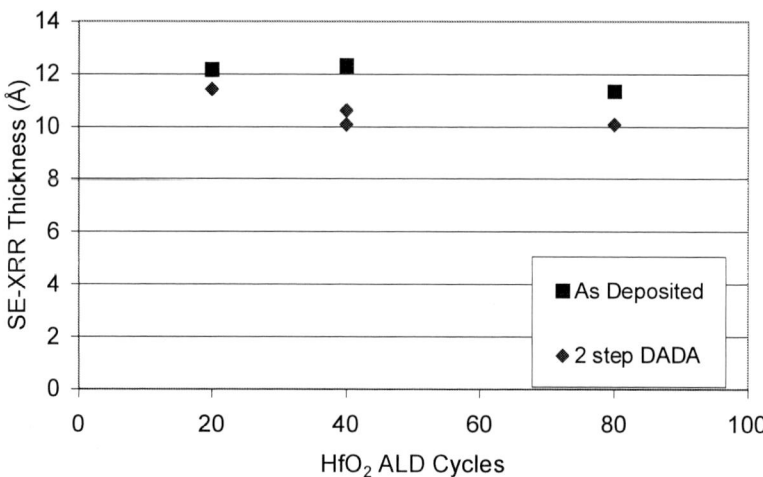

Figure 14. Measured interface layer thickness at equivalent total ALD cycles based on subtracting HfO_2 measured thickness by XRR from dielectric stack thickness measured by spectroscopic ellipsometer (SE).

Discussion

We have presented evidence above that the DADA process results in significantly improved scaling versus as deposited films in MOSCAP devices fabricated using a low temperature device flow. In addition we have found significant physical differences in films produced by the DADA process versus as deposited films. We have also shown that the DADA process can be optimized by changing the anneal temperature and time used in the process. In Figure 15, we present data collected from DADA films deposited using optimized hardware for HfO_2 deposition on an SiON interface. This hardware provides faster cycle times during the deposition process and results in an additional electrical improvement suggesting that the improvements from the DADA process are additive even when the HfO_2 deposition has been significantly optimized. In addition, this data illustrates the performance advantage of the DADA process versus simple PDA that has been previously reported (6).

Mechanism of Si Intermixing

In this report we have shown that the DADA process results in significant Si intermixing with HfO_2 along with interface thinning, especially in the case of a non-nitrided interface layer. Comparing DADA films with as deposited films and films undergoing a single PDA on chemical oxide it is apparent that PDA alone is not sufficient to drive Si into the film. These results suggest that Si can travel through only thin layers of HfO_2, and thus we propose the mechanism shown in Figure 16 to explain the data.

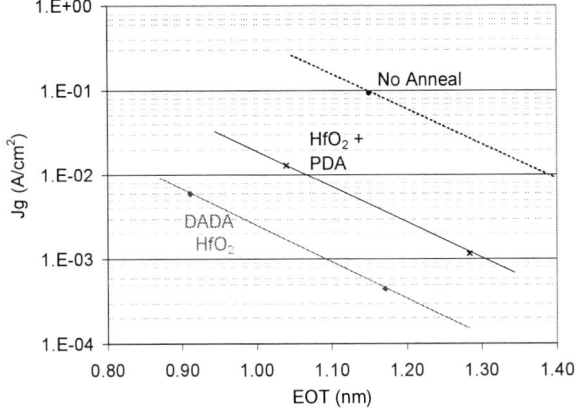

Figure 15. Scaling trends observed in low temperature MOSCAPs incorporated using optimized HfO₂ films including as deposited, DADA and films undergoing a single post-deposition anneal (PDA) matching the temperature and total dwell time of the DADA anneals. Lines are a guide to the eye showing expected scaling performance based on the data collected to date.

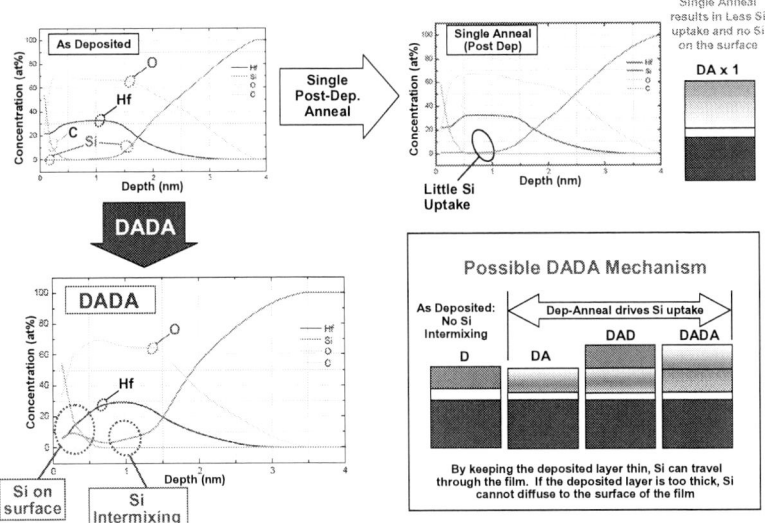

Figure 16. Mechanism of Si uptake proposed to account for Si intermixing observed in DADA films on chemical oxide and comparison with as deposited and PDA films by HR-RBS depth profile.

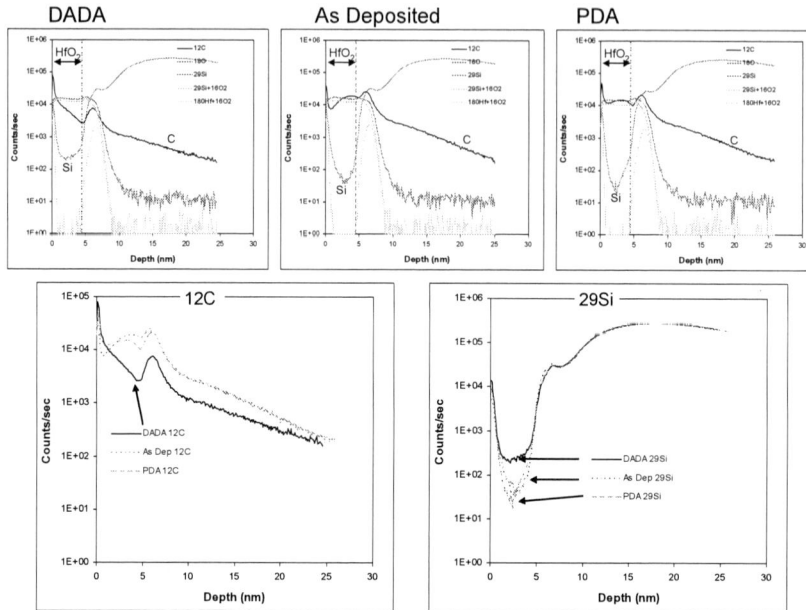

Figure 17. SIMS Depth profiles comparing 100 cycle DADA HfO$_2$ with As Deposited HfO$_2$ and HfO$_2$ after PDA. Comparison of C and Si profiles are shown in the bottom graphs.

We investigated in-film Si and C in thick (100 cycle) films by SIMS and found that DADA HfO$_2$ has both lower C and higher Si than as deposited or PDA samples, Figure 17. A similar reduction in C has previously been observed in HfO$_2$ processed with DADA-type processes (1, 5, 6). Since it has been previously reported that C removal from ultra-thin HfAlO becomes less efficient as the film thickness increases above about 1.8 nm (1), our observation of a limited thickness for Si uptake on the same order suggests that Si uptake and C removal could be related. We therefore investigated the effect of annealing on Si uptake in ultra-thin (5 ALD cycles) HfO$_2$ using HR-RBS and found clear evidence of Si uptake in ultra-thin HfO$_2$ after anneal, Figure 18. This data suggests that Si uptake and C removal take place concurrently during anneal, and results in a film that is both lower in C and richer in oxygen. We therefore propose that C, in a reduced form, is being displaced by SiO diffusing up from the interface during anneal. Such a mechanism could explain both C removal and Si uptake in ultra-thin HfO$_2$. In addition, this mechanism accounts for the additional oxygen content found in the annealed film. Presumably, additional oxygen in the film would result in fewer oxygen vacancies and therefore a positive flatband voltage (Vfb) shift versus PDA or as deposited films (12). Indeed we find that the Vfb for DADA films is shifted to more positive values versus the Vfb observed for as deposited films and those undergoing a single PDA, Table I. We thus conclude that the DADA process results in a film with lower C and fewer oxygen vacancies than as deposited films or films undergoing a single PDA. In the case of chemical oxide interface layers, we find Si incorporation throughout the DADA film, whereas in the case of DADA on SiON, we find that Si intermixing is limited to the lower portion of the HfO₂ film.

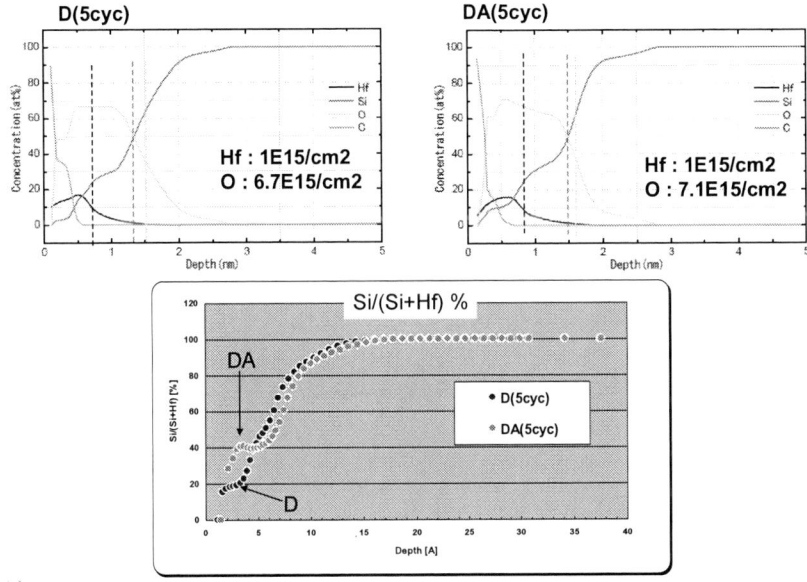

Figure 18. HR-RBS of Ultra-thin HfO₂ with and without anneal. Inset shows Si% (as [Si]/[Si+Hf]) in as deposited 5 cycle HfO₂ (D) versus 5 cycle HfO₂ undergoing anneal for clarity.

TABLE I. EOT and flatband voltage (Vfb) for DADA vs. As Deposited and PDA. Average values of 9 points across a 300mm wafer. Data is from the wafers used for Figure 14.

Process	Total ALD Cycles	EOT (nm)	Vfb (v)
As Dep	40	1.15	-0.28
PDA	40	1.04	-0.26
PDA	50	1.28	-0.32
DADA	40	0.91	-0.20
DADA	50	1.17	-0.22

Crystallinity

We have observed that films deposited with the DADA process are nano-crystalline, as shown above. We have further found that by increasing the dwell time of anneals in the DADA process it is possible to increase the apparent ordering in the DADA films with a corresponding improvement in electrical properties. And we have found that DADA films with 20 cycles or less in a DA cycle are thinner and denser than as deposited films deposited with the same total number of cycles as measured by XRR. Though we observe Si uptake in DADA films and low level Si incorporation has been reported to enhance the formation of higher K tetragonal or cubic phases of HfO_2 (13), our data to date (not shown) indicates that the DADA processes described herein result in monoclinic or possibly orthorhombic HfO_2 phases. We are continuing to investigate the crystal phase resulting from the DADA process, but currently we are confident that the electrical results we report herein are not indicative of the formation of a higher K tetragonal/cubic phase. Rather, we believe that film densification occurs during the DADA process because the amorphous as deposited film is simply less dense than any of the crystalline forms of HfO_2. Presumably the amorphous as deposited film is also more defective than a well ordered crystalline HfO_2 film. Thus we attribute the EOT reduction we observe in DADA films to film densification, and we attribute the observed leakage current reduction to reduced trap assisted tunneling current (5, 6) through the film as a result of a reduction in oxygen vacancies, carbon levels, and overall defectivity in the film. Of course this does not explain fully why DADA films would outperform films crystallized with a single PDA.

It has previously been shown that the DADA process does not result in any significant improvement in breakdown resistance of the HfO_2 film (5, 6). We believe this is a natural result based on a mechanism analogous to that now proposed to account for breakdown or forming within this type of film when used in a resistive memory (14). Prior to breakdown, leakage current is due primarily to trap assisted tunneling along grain boundaries in a polycrystalline HfO_2 film. It therefore follows that a reduction in the number of grain boundaries within the area of the device should result in a reduction in the total leakage current associated with trap assisted tunneling through the HfO_2. When the film undergoes electrical breakdown, a filament has been proposed to form along a single grain boundary path within the device. Essentially, breakdown occurs at the weakest point within the device which happens to be along a crystalline grain boundary. So, as long as there are multiple grain boundaries contained within the area of the device breakdown will occur at approximately the same time in a device with a given number of grain boundaries (x) as in a device with even twice as many grain boundaries (2x). In essence unless the grain boundaries are completely eliminated, breakdown should still occur at approximately the same stress level.

One potential explanation that has been proposed for the electrical improvements observed in films undergoing multi-step deposition and anneal is that there are multiple layers formed within the film and that each layer is polycrystalline and crystallized in such a way that the grain boundaries and pin holes are misaligned with each other, Figure 19 (2). We have been unable to find any evidence for such a structure forming in our devices by HR-TEM. On the contrary, crystal grains in our devices generally appear to extend from the interface layer through the thickness of the film to the metal gate and would be best represented by the picture presented in (a) in Figure 18. Further, if the mechanism proposed in Figure 19 were correct for devices with DADA films, we would expect a significant improvement in breakdown resistance, which is not observed (5, 6).

We therefore need to explain the improved performance of DADA versus PDA of the same thermal budget by another mechanism. The mechanism we propose is shown in Figure 20, and includes nucleation of a thin layer followed by crystallization in a bottom up crystallization. The epitaxial crystallization of HfO_2 has recently been reported by rapid thermal crystallization of an amorphous ALD HfO_2 layer directly in contact with Si (15). The critical temperature for crystallization of HfO_2 reported was in the range of 400 to 500 °C, so it could be expected that ALD HfO_2 films undergoing a 540 °C thermal budget in contact with crystalline TiN would crystallize as well. We believe that the

(a)

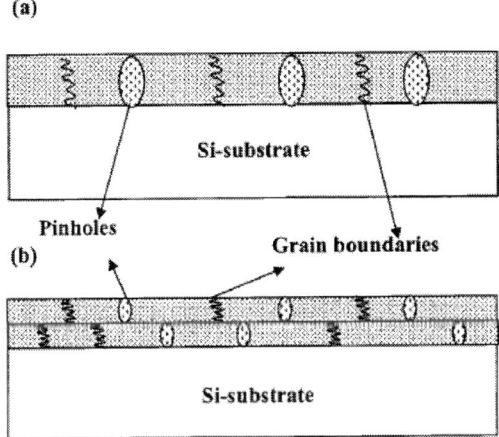

(b)

Figure 19. Schematic drawing of the idea on how multistep deposition can improve leakage current after annealing. (a) Single-step deposition and (b) offset of leakage current path by double-step deposition of HfO$_2$. From (2).

reason our as deposited films remain amorphous is that neither the top interface (with TiN) nor the bottom interface (with SiO$_2$) provide a good seed layer for crystallization, and therefore a higher temperature is required to form nucleation sites suitable for crystallization within the film. The critical temperature for ultra-thin HfO$_2$ to form nucleation sites seems to be between 540 °C and 700 °C based on our results, and we expect that it may depend on film thickness in the initial DA cycle. When the film is ultra-thin there is a strong driving force to minimize the ratio of the film surface area to film volume, which pushes the film to become smoother and align with the surface of the substrate, essentially "wetting" the wafer surface in a process akin to sintering, this type of process has previously been reported for ALD HfO$_2$ on ultra-thin interface layers (16). After the initial DA cycle, crystallization will continue in a bottom up fashion because the initial film provides a good nucleation layer for further crystallization. This is analogous to the mechanism of epitaxial HfO$_2$ growth recently reported. The resulting film is smoother with a larger grain size than a single film crystallized randomly by PDA.

Although our proposed mechanism is at best speculative, we believe it provides a better model for the data we have collected on the DADA process than the previously proposed model based on grain boundary misalignment. It is worth noting that the interface dependence of the DADA process can be explained within our model if we consider that the driving force for HfO$_2$ wetting of the interface is increased for SiON versus SiO$_2$. Further, we have found that AFM measurements of DADA films compared with amorphous and PDA show less roughness for DADA versus PDA, in fact roughness of the DADA films approaches that of the amorphous as deposited films, Figure 21.

Finally, we would like to address the possibility that a crystalline high K film might result in increased variability versus an amorphous high K film. This assumption is reasonable from the perspective that the addition of anneals will add an additional source of variation, and that variable crystal orientation might result in variability in device parameters. While the results in this study should not be considered equivalent to the results one might obtain in a well controlled manufacturing line, we nonetheless sought to

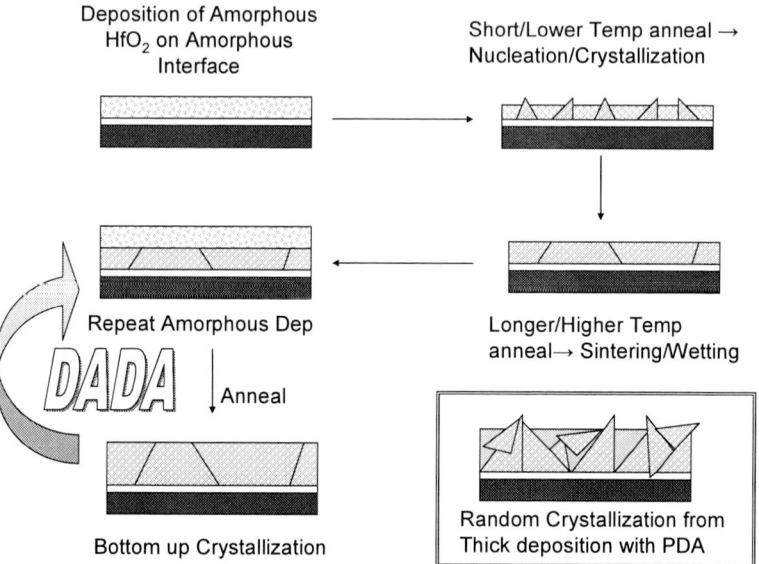

Figure 20. Proposed mechanism for DADA crystallization and comparison with random crystallization from PDA.

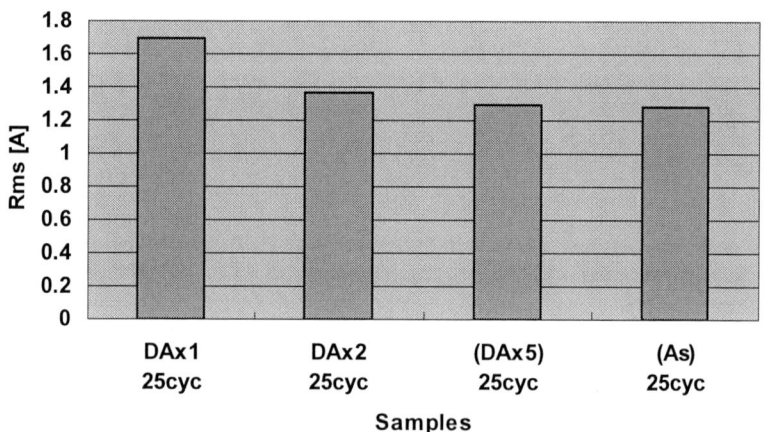

Figure 21. Comparison of AFM roughness measured on films undergoing PDA at 800 C versus 2-step (DAx2) and 5-step (DAx5) DADA films versus as deposited (As).

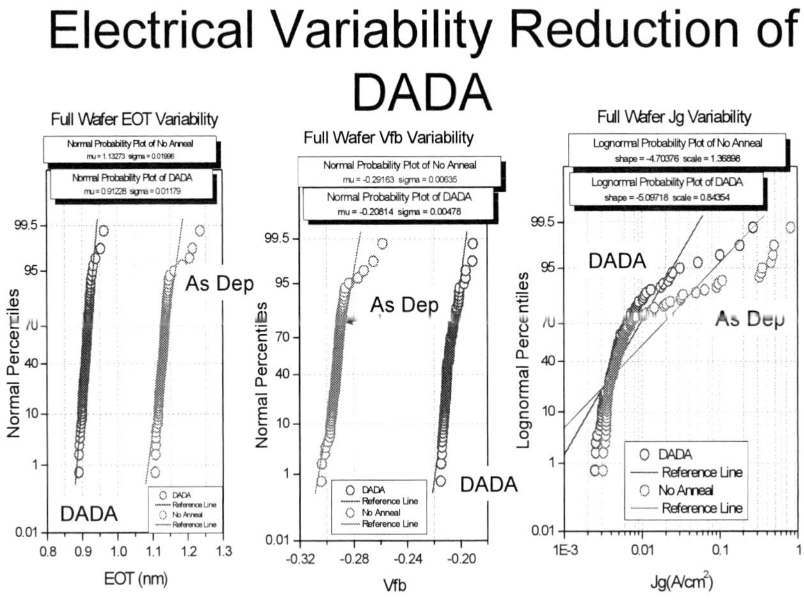

Electrical Variability Reduction of DADA

Figure 22. 93 point data measuring all devices on wafers in the same lot shows DADA improves electrical variability across the wafer while scaling EOT. Comparison is for 40 total cycle HfO$_2$ depositions using an optimized deposition chamber. Data from the same wafers was previously presented in Figure 15 and Table I.

investigate the effect the DADA process might have on variability within a wafer. In order to check variability, we measured identical MOSCAP devices in all 93 die in two different 300 mm MOSCAP wafers - one wafer with DADA HfO$_2$ and the other with as deposited HfO$_2$, Figure 22. Both wafers were from the same lot and represent the 40 total cycle DADA and as deposited samples for which data was presented in Table I, and Figure 14. As can be seen, we observe a significant reduction in variability for devices containing DADA films versus as deposited HfO$_2$. It's possible that this variability reduction is integration related, but we believe it may also be evidence that DADA can help to reduce high K variability effects by providing a more uniformly oriented and crystallized HfO$_2$ film.

Conclusion

We have presented a series of results on the DADA process for HfO$_2$ deposition. MOSCAP devices with DADA films show significant electrical improvement over as deposited HfO$_2$, including improved EOT scaling at equivalent or lower gate leakage current. The cause of the observed electrical improvements in the DADA films is proposed to include a combination of compositional and morphological changes as a result of the DADA process. The compositional changes observed include a reduction in the amount of C and the number of oxygen vacancies within the film as well as increased intermixing at the High K/Si interface. The morphological changes observed include increased crystallinity and long range order in the HfO$_2$ film. Mechanisms for both the compositional and morphological changes have been proposed herein.

Acknowledgments

The authors wish to acknowledge the helpful contributions of the operations and engineering staff at TEL Technology Center America, and at TEL-AT Ltd. We would also like to thank Tat Ngai of Sematech for assistance in obtaining HR-TEM and EDX-EELS data presented in this paper.

References

1. T. Nabatame, K. Iwamoto, H. Ota, H. Hisamatsu, T. Yasuda, K. Yamamoto, W. Mizubayashi, Y. Morita, N. Yasuda, M. Ohno, T. Horikawa, and A. Toriumi, *Symp. VLSI Tech.*, p. 25 (2003).
2. C. C. Yeo, B. J. Cho, M. S. Joo, S. J. Whoang, D. L. Kwong, L. K. Bera, S. Mathew, and N. Balasubramanian, *Electrochem. Solid State Lett.*, **6** (11) F42 (2003).
3. A. Ogawa, K. Iwamoto, H. Ota, M. Takahashi, A. Hirano, T. Nabatame, and A. Toriumi, *Microlectron. Eng.*, **84**, 1861 (2007)
4. D. Ishikawa, S. Kamiyama, E. Kurosawa, T. Aoyama, and Y. Nara, *Jpn. J. Appl. Phys.*, **48**, 04C004 (2009).
5. H. Jagannathan, R. D. Clark, S. Consiglio, P. Jamison, B.P. Linder, M. Hopstaken, G. J. Leusink, V. K. Paruchuri, and V. Narayanan, *ECS Trans.*, **33** (3) 157-164 (2010).
6. M. Trentzsch *International Symposium on Advanced Gate Stack Technolgy* Troy, NY (2010).
7. R. D. Clark, C. S. Wajda, G. J. Leusink, L. F. Edge, J. Faltermeier, P. Jamison, B. P. Linder, M. Copel, V. Narayanan, M. Gribelyuk, R. Loesing, and R. Murphy, *ECS Trans.,* **11** (3), 55 (2007).
8. R. D. Clark, S. Consiglio, C. S. Wajda, G. J. Leusink, T. Sugawara, H. Nakabayashi, H. Jagannathan, L. F. Edge, P. Jamison, V. K. Paruchuri, R. Iijima, M. Takayanagi, B. P. Linder, J. Bruley, M. Copel, and V. Narayanan, *ECS Trans.*, **16** (4) 291-305 (2008).
9. J. R. Hauser and K. Ahmed, "Characterization of Ultrathin Oxides Using Electrical CV and I-V Measurements," presented at Characterization and Metrology for ULSI Technology: 1998 International Conference, 1998.
10. R. S. Rai and S. Subramanian *Progress in Crystal Growth and Characterization of Materials,* **55**, 63-97 (2009).
11. J. J. Gallegos, D. H. Triyoso, and M. Raymond *Microelectronic Engineering,* **85**, 49–53 (2008).
12. E. Cartier *ECS Trans.*, **33** (3) 83-94 (2010).
13. K. Tomida, K. Kita, and A. Toriumi *App. Phys. Lett.* **89**, 114902 (2006).
14. G. Bersuker, D. C. Gilmer, D. Veksler, J. Yum, H. Park, S. Lian, L. Vandelli1, A. Padovani1, L. Larcher, K. McKenna, A. Shluger, V. Iglesias, M. Porti, M. Nafría, W. Taylor, P. D. Kirsch, and R. Jammy *IEDM Technical Digest* 19.6.1 (2010)
15. S. Migita, Y. Morita, W. Mizubayashi, and H. Ota *IEDM Technical Digest* 11.5.1 (2010)
16. F. Bohra, B. Jiang, and J.-M. Zuoa *APP. PHYS. LETT.*, **90**, 161917 (2007).

ECS Transactions, 35 (4) 835-845 (2011)
10.1149/1.3572322 ©The Electrochemical Society

Interface Structure and Charge Trapping in Hf-incorporated Y_2O_3 Gate Dielectrics on Germanium

C. Mahata[*], S. Mallik, T. Das, M . K. Hota, C.K. Maiti

Dept. of Electronics and ECE, Indian Institute of Technology, Kharagpur, India
e-mail: chandreswar@gmail.com

Y_2O_3 and $HfYO_x$ (~12nm) high dielectric constant films, with RF sputtering, have been prepared and studied with respect to their chemical bonding and electrical properties. The rapid thermal annealing after high-k dielectric deposition in N_2 ambient showed major impacts on crystalinity properties. The incorporation of Hf into Y_2O_3 causes reduction in generation of positive oxide charge and leakage current density. The mechanism of leakage current reduction is considered to be due to Hf-induced compensation of existing oxygen vacancies. Moreover, $HfYO_x$ films exhibit superior electrical performances in terms of charge trapping induced flatband voltage instability and gate leakage current compared to directly deposited Y_2O_3 high-k dielectric films on Germanium. The density of interface traps (D_{it}) and border traps (N_{bt}) estimated from high-frequency CV curves (1 MHz) indicates that incorporation of Hafnium can improve the interfacial defects.The breakdown field in MIS capacitor with Y_2O_3 gate dielectric was significantly enhanced by incorporating Hf.

Introduction

As the aggressive scaling of Si-based complementary metal-oxide semiconductors (CMOS) approaches its fundamental physical limitations, various attempts to improve the device performance have been investigated by enhancing carrier mobility in the channel region and by modifying channel structure. Major breakthrough may be achieved if the conventional Si substrate is replaced by alternative semiconductor materials, such as Ge, which have high intrinsic carrier mobility. Because of its higher low-field carrier mobility and smaller mobility band gap for supply voltage scaling, there have been many attempts to use Ge as a channel material in high-speed field-effect transistors [1, 2]. Recently, Ge-based MOS capacitors and transistors incorporating high-k gate dielectrics have been demonstrated with respectable electrical properties [3, 4]. In particular, HfO_2 is an interesting system because of its good band offsets and thermal stability and also HfO_2/Ge stacks with equivalent oxide thickness (EOT) down to 1.2-0.5 nm and with low leakage current have been already reported [5-9]. However, it has also been demonstrated that surface pretreatments are a necessity [10]. In particular, surface nitridation (Ge-oxynitride passivation) of Ge has been investigated and has provided improvements in the electrical properties of high-k/Ge gate stacks and was found to improve the electrical performance of the gate stacks, yielding a leakage current density about two orders of magnitude lower compared to the non-nitrided gate stacks [10-13].

In recent works Kita *et al* have investigated the impact of high-k material selection on the electrical characteristics of high-k/Ge MIS capacitors. They found that Y_2O_3/Ge structure maintained good electrical characteristics even after annealing at high temperature, while HfO_2/Ge seriously deteriorated. From crystallinity they have concluded that Y_2O_3 film was amorphized at the interface by forming Y-germanate, which would work as a good interfacial layer up to 600°C, while HfO_2 films lost the amorphous GeO_x interfacial layer by annealing [14]. The results indicate that the HfO_2/Ge interface is detorieted more severely than the Y_2O_3/Ge one, while the bulk Ge quality is almost the same for both cases and it can be concluded that Y_2O_3 is better than HfO_2 as a high-k material on Ge in terms of its thermal robustness during the device fabrication process.

Recently it has also been investigated by many groups that composite high-k with different compositional range of Y_2O_3 and HfO_2 is more stable than pure dielectric. A pure cubic phase of HfO_2 can be stabilized for certain percentage of Y. The composite films show higher permittivity than pure high-k dielectric [15-17].

In this work, Y_2O_3 and $HfYO_x$ dielectric films are deposited by RF sputtering. Effect of Hf incorporation in Y_2O_3 has been extensively investigated through physical and electrical characteristics. The chemical nature was analyzed by X-ray photo electron spectroscopy (XPS) and X-ray diffraction (XRD) analysis. Electrical characteristics was monitored by capacitance-voltage characteristics and the impact of electrical stressing under different constant voltage and current on flatband voltage (V_{fb}) of Ge MIS capacitors have been analyzed and the trap generation mechanisms due to the electrical stressing have been discussed.

Experimental

Y_2O_3 and $HfYO_x$ films were obtained with RF sputtering using a Y_2O_3 and HfO_2 targets (99.99% purity), followed by a post deposition annealing (PDA). Prior to RF sputter deposition, p-type Ge (100) substrates with resistivity of 18-20 Ω-cm were given a HF-cleaning and plasma nitridation with NO. During deposition, the chamber pressure was kept at 2.5 mTorr and the Ge substrates were held at a temperature of 50°C with 100 W RF power in Ar ambient. To compare the thermal stability in terms of the crystallization temperature post deposition annealing was performed with temperature varied from 450-750°C in N_2 ambient for both types of sample. Al was evaporated and patterned as gate electrode with an area of $4x10^{-3}cm^{-2}$. Finally, a post metal anneal was performed at 300°C for 10 min to achieve better electrode contacts. High-frequency (1 MHz) capacitance-voltage (C-V) characteristics were measured at room temperature using AgilentE4980A precision LCR meter. Gate-leakage current was measured by Agilent4156C precision semiconductor parameter analyzer. The thickness and the refractive index, as measured by ellipsometer (α-SE, J. A. Woollam) at a wavelength of $\lambda = 632.8$ nm were d_{high-k} ~8 (for XPS) and 12 nm and n = ~1.8, respectively. High-field stress, with the capacitors biased in accumulation and constant current stress ($\pm 25mA/cm^2$) by Agilent4156C precision semiconductor parameter analyzer, was used to examine device reliability in terms of gate leakage increase and flatband voltage changes. All electrical measurements were carried out under a light-tight and electrically-shielded condition.

Results and Discussion

Physical Characterization

X-ray Photo Electron Spectroscopy analysis . The surface chemical state of Y_2O_3 and $HfYO_x$ films were analyzed using high-resolution XPS (model: ESCA-2000 Multilab apparatus, VG Microtech) using a non-monochromatic M_gK_α (hv= 1253.6 eV) excitation source radiation at an angle of 30° between the analyzer axis and the sample normal with pass energy of 50 eV, at a residual gas pressure in the range of 10^{-10} Torr with an instrumental resolution of 0.6 eV. Effects on chemical structure of Hf-incorporation in Y_2O_3 have been studied using XPS. Figure 1 show the photoelectron spectra of Y3d signal initially located at 156.6eVand 158.5eV due to $Y3d_{5/2}$ to $Y3d_{3/2}$, respectively; 0.2eV shifts towards higher binding energy (BE) in accordance with the changing atomic configuration of $HfYO_x$ film. However, BE values corresponding to $Hf4f_{7/2}$ and $4f_{5/2}$ main peaks in the Hf4f spectra (see inset of Figure 1), located at 17.1eV and 18.7 eV, respectively confirms that hafnium is in +4 oxidation state. The behavior of Ge at the Ge/High-k interface is of great interest because the electronic properties of these interface states, which play crucial role in the device operation, are mainly determined by the chemical formations there. The incorporation of Ge with Y_2O_3 can be found by XPS because a considerable amount of Ge is included in the Y_2O_3 film only in the vicinity of the Ge substrate. In the Ge 3d spectrum of Figure 2, a broad shoulder is seen between 30.5 eV and 32.5 eV which is at lower binding energy than GeO_2. Such a chemical shift can be attributed to the $YGeO_x$ formation. The peak of the Y 3d spectrum, which appears at a higher binding energy than that of pure Y_2O_3, also suggests the $YGeO_x$ formation. Though we found $YGeO_x$ formation is lower in case of $HfYO_x$. It seems that Ge atoms close to the Y_2O_3 and $HfYO_x$ interface are oxidized during the deposition. It is likely that Ge atoms extract O atoms from the underlying high-k film. These indicate formation of a thin layer of GeO_x composed of Ge_2O, GeO, and Ge_2O_3 at the interface which we found in higher binding energy.

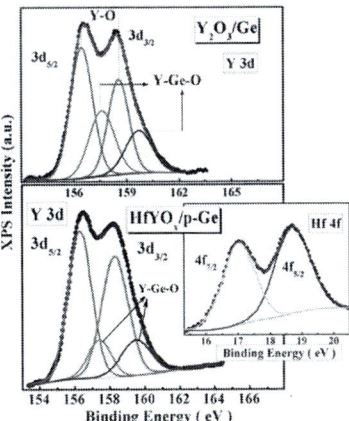

Fig. 1. XPS of Y3d for Y_2O_3 and $HfYO_x$ high-k gate dielectric on p-Ge.

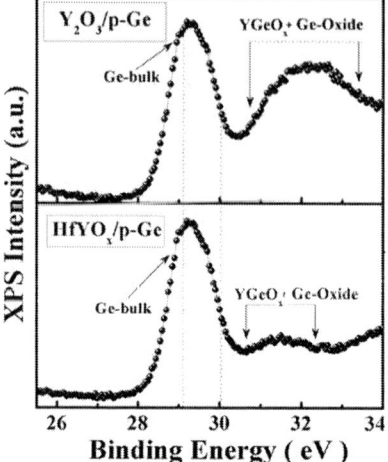

Fig. 2. XPS of Ge3d from Y_2O_3 and $HfYO_x$ high-k gate dielectric on p-Ge.

O1s photoelectron energy loss spectra of Y_2O_3 and $HfYO_x$ thin film is shown in Figure 3(a) and (b). In the spectra, the onset of band-to-band exitation which corresponds to band gap was defined as an intercept of the linear extrapolation of the leading edge to the background level [18]. Band gap of Y_2O_3 and $HfYO_x$ were measured to be ~4.7eV and 4.85eV respectively after annealing at 500°C. Bandgap (E_g) was not changed significantly after incorporation of Hafnium.

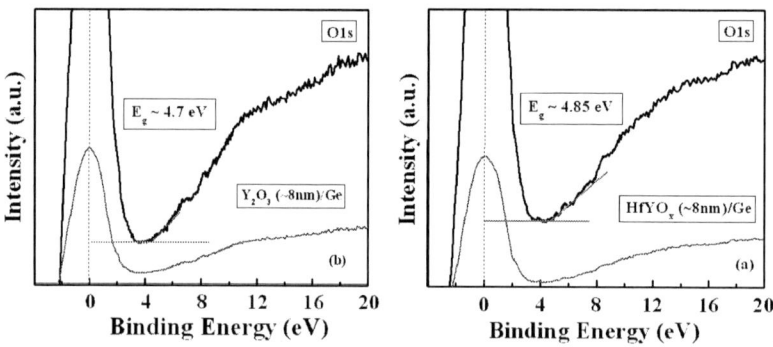

Fig. 3. O1s photoelectron spectrum measured from Y_2O_3/Ge and $HfYO_x$/Ge structures. Band gap of Y_2O_3 and $HfYO_x$ were measured to be ~4.7eV and 4.85eV respectively.

X-ray Diffraction analysis

The structural analysis of as-deposited and RTA samples of thickness ~30nm was carried out by grazing incidence X-ray diffraction (GIXRD) with a Philips PW 1710 diffractometer using CoK_α radiation. Data were recorded at 20mA in the 2θ scanning range of 20° to 80° using a constant step width of 0.05°. Figure 4(a) and 4(b) shows typical XRD patterns of the ~30nm thick Y_2O_3 and $HfYO_x$ films, respectively, as a function of annealing temperature. It showed that in Figure 4(a) the refractive peaks of 2θ were observed at 20.6, 43.6, and 57.8 for annealed Y_2O_3 film from 450 to 750°C. These peaks were corresponding to the (2 1 1), (3 3 2), and (6 2 2) planes, which can be indexed to the pure body-centered cubic Y_2O_3. No peak of any other phase was detected. The present result is in agreement with that of a recent investigation by Chien *et al* [19]. As seen in Figure 4(b), the annealed $HfYO_x$ film shows a featureless diffraction, characteristics of amorphous state due to the scattering of X-rays by the short range order in the amorphous phase. However, at higher annealing temperature the film exhibits only a very weak crystallization peak corresponding to the HfO_2 monoclinic (011) phase. With increasing the annealed temperature, the peak seems more pronounced indicating higher order of HfO_2 crystal structure. This result indicates that Hf acts as a crystallization inhibitor or stabilizer of the amorphous phase and causes an increase in the crystallization temperature of yttrium-based gate dielectrics. Compared to Y_2O_3 film with the same thickness, it is obvious that the crystallization temperature of Hf-incorporated Y_2O_3 films has been increased.

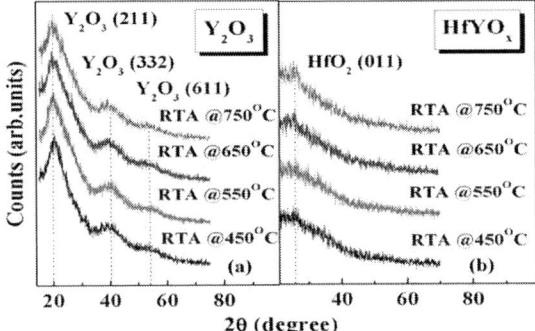

Fig. 4. XRD characteristics of Y_2O_3 and $HfYO_x$ high-k gate dielectric under different RTA temperature.

Electrical Characterization

High-k/Ge interface analysis. Electrically measured capacitance-voltage and leakage current-voltage characteristics are given in Figure 5(a) and 5(b) respectively. The cyclic C-V curves obtained when the gate voltage was first swept from accumulation to inversion (forward sweep) and then from inversion to accumulation (reverse sweep) for the MIS structures with Y_2O_3 high-k dielectric with and without Hf-incorporation. However this hysteresis can be interpreted by the following hole trapping mechanism at the Ge/high-k interface. In the forward sweep, when the gate bias is varied from -2.5V to 2.5V so that the Ge interface goes from accumulation to inversion, the Ge interface is under high accumulation state at the beginning. In Y_2O_3 and $HfYO_x$, there could be a high density of hole trapping sites at the Ge/IL/high-k structure and these trapping sites could be filled with majority carrier holes at the accumulation state. The excess positive charge thus trapped at the gate stack will give rise to the negative V_{fb} shift of the forward C-V curve. Consequently, an injection type hysteresis is observed. When the gate voltage is brought to the inversion region (2.5V) de-trapping of the trapped holes will occur by charge exchange with the underlying Ge substrate. Reasonably good electrical property is being demonstrated for $HfYO_x$ films over that of Y_2O_3 films with lower hysteresis voltage (~200mV in compare to ~320mV for Y_2O_3). The effective border trap density N_{bt} was calculated by integrating the absolute value of the difference between the forward and reverse hysteresis curves using the following equation [20,21]:

$$\Delta N_{bt} \approx (1/qA) \int \left| \left(C_{inverse} - C_{forward} \right) \right| dV \tag{1}$$

Y_2O_3 sample has the higher value of border trap density, when compared with Hf-incorporated Y_2O_3. This higher value of border trap density is thus speculated to be due to the presence of a higher dangling bond centre in bulk dielectrics stack for the Y_2O_3 sample [22]. The border traps density, N_{bt} values were estimated to be in the range of ~4×10^{11} and ~1.5×10^{11} for Y_2O_3 and $HfYO_x$ respectively. Y_2O_3 sample show a gradual shift of C-V characteristics to more negative voltages which should be attributed to a net positive charge trapping in the gate stack. Higher negative flat band voltage shift (V_{fb}) before Hf-incorporation indicates a high density of net positive defect charge (~6-8×10^{12}/cm^2) in the Y_2O_3 film compared to $HfYO_x$ film. It is expected that Hf-incorporation can repair the oxygen deficiency induced defects in bulk high-k dielectric. But, it is noted that Y_2O_3/Ge shows a little better C-V characteristics than $HfYO_x$/Ge in terms of D_{it} formation (less stretching out behavior). This is probably caused by the presence of interfacial states induced by the lattice mismatch between $HfYO_x$ and Ge. The complex bonding of mixed oxides may also be attributed to the little higher amount of D_{it}. $HfYO_x$ sample shows an opposite trend of hysteresis. Significant dispersion in inversion region at same frequency indicates the presence of slow negative interface states. Interface trap-assisted minority carrier generation can also produce a deviation in high frequency C-V characteristics in the depletion/inversion regime.

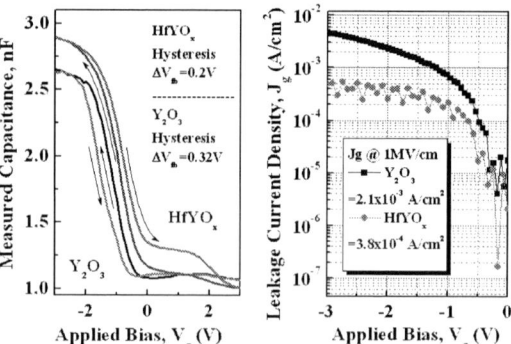

Fig. 5. (a) High-frequency (1MHz) hysteresis and (b) leakage current density-voltage characteristics of Y_2O_3 and $HfYO_x$ high-k gate dielectric.

The J-V curves for all the samples are not symmetric under negative (accumulation mode) and positive (inversion mode) biases. One of the main reasons for that is the difference in the barrier heights at the two interfaces [23]. We have taken the accumulation part of I-V characteristics. The leakage current density for $HfYO_x$ measured at gate electric field of 1 MV/cm is $3.8 \times 10^{-4} A/cm^2$, about one order of magnitude lower than the value of $Y_2O_3 (2.1 \times 10^{-3})$ layer with similar thickness. The incorporation Hf into Y_2O_3 has reduced the number of dangling bond defects (O_2 defects), thus notably lowering the electrical leakage current.

Charge trapping under constant current stressing

The deterioration of the high-k/Ge interface is a major reliability concern for the long-term operation of the device. An enhanced trap generation rate would lead to the electrical degradation of the dielectric and the subsequent failure of the oxide. Leakage current density-gate voltage (J-V) curve has been recorded to detect stress-induced leakage current (SILC) as an indication for the device degradation. Also flat-band voltage shifts (ΔV_{fb}) were investigated using high-frequency capacitance-voltage (HFCV) measurements under constant current stressing ($\pm 25 mA/cm^2$). To measure the SILC induced by CCS, the stress was periodically stopped after 100sec to measure the leakage current at, $E_{ox} = 1MV/cm$ of $Al/Y_2O_3/Ge$ and $Al/HfYO_x/Ge$ structures shown in Figure 6(a) is plotted as a function of stress time. Under substrate hole injection (negative bias at gate) ($-25mA/cm^2$) SILC was gradually increase without showing a clear saturation which is due to generated defects during the stress [24]. SILC generation was found lower in case of $HfYO_x/Ge$. The observed lower value of SILC is attributed to the pre-existing lower trap generation in this gate stack due to the incorporation of Hf into Y_2O_3. Under substrate electron injection (positive bias at gate), SILC behavior was found lower in both structures which could be due to a less number of trap generation in the interface. Figure 6(b) shows the flatband voltage shift (ΔV_{fb}) after both positive and negative fluence at gate. For $HfYO_x$ dielectric both hole (or new electron trap generation) and electron trapping was observed after negative and positive fluence respectively whereas for Y_2O_3 only electron trapping was monitored.

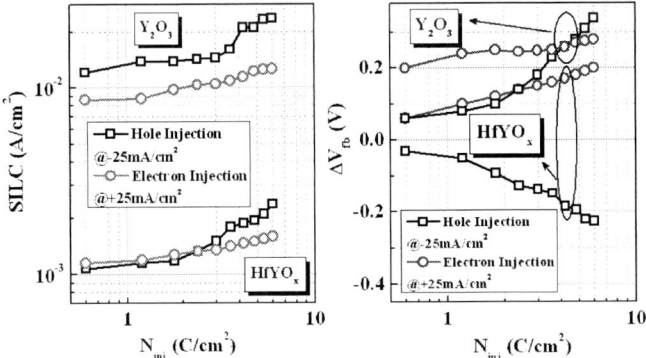

Fig. 6. (a) SILC variation and (b) change in flatband voltage after CCS at both $\pm 25 mA/cm^2$.

Saturation behavior was further confirmed from flatband shift under positive fluence for both dielectrics. This is further substantiated below from a detailed study of the generation of interface and oxide traps during constant voltage stressing (CVS).

Charge trapping under constant voltage stressing

Figure 7 shows the change in current density ($\Delta J_s = J_{stress} - J_{without\ stress}$) for Y_2O_3 and $HfYO_x$ gate stack under constant voltage stressing range from -2V to -6V. Figure 7(a) shows that under substrate hole injection for $HfYO_x$/Ge, ΔJ_s shifts towards negative direction upto -5.2V CVS. These features imply filling existing hole traps in the $HfYO_x$ high-k gate stacks. After -5.2 to -6V as shown in Figure 7(a) negative charge trapping dominates at higher constant voltage. This phenomenon is thought to be due to the presence of additional interface traps (mainly electron traps) in case of $HfYO_x$/IL/Ge gate stack.However, this electron trapping is absent before Hf incorporation. For Y_2O_3 gate stack net positive charge trapping dominates over the entire stress range. As is seen only gradual change of the current occurs for all stressed capacitors (no abrupt increase of the current) which corresponds to SILC generation. The extent of current increase depends on the level of stress. The current increase is suggested to appear from the trapping and release of carriers in the stress-generated traps which act as stepping sites for tunneling. Time dependent dielectric soft breakdown, TDDSBD as shown in Figure 7(b) has also been observed for Y_2O_3 samples without any hard breakdown (HBD) which might have resulted from a weak localized percolation path between the gate electrode and the substrate depending on the size and sphere of influence of defects present in the oxide. After Hf-incorporation no such SBD was observed indicates that Hf-incorporation further increase the breakdown field minimizing the trap generation probality.

The generated interface trap density ΔD_{it} and border trap density ΔN_{bt} for the MIS capacitor with Y_2O_3 and $HfYO_x$ stacked gate dielectric during constant voltage stress (CVS) at V_g from -2 to -6V are shown in Figure 8, where ΔD_{it} was extracted by Hill's method [Hill] and ΔN_{bt} was calculated from the difference between flatband voltage of the forward and reverse C-V characteristics[25]. The increment of D_{it} and N_{bt} (Figure 8(a) and (b)) for the capacitor with Y_2O_3 is much higher than that for the capacitor with $HfYO_x$ dielectric. In addition, from Figure 8(a) trap generation of $HfYO_x$/Ge structure during CVS was preferred to be within the bulk dielectric rather than at the interface (very small change in D_{it}), implying that the dielectric reliabilities at low stress voltage would be dominated by the bulk properties of the gate stacks. Since both interface traps and bulk-oxide traps contribute to charge centroid [26], absence of any contribution of interface traps on ΔV_{fb} implies that bulk-oxide traps are dominant. No considerable variation was observed in interface state density (D_{it}) of $HfYO_x$ devices as a function of stress voltage. On the other hand, the interface-trap density was found rapidly increased (ΔD_{it}) for Y_2O_3 gate dielectric MOS capacitor. Superior electrical characteristics and dielectric reliabilities were therefore obtained for the Y_2O_3 dielectric with incorporation of Hf.

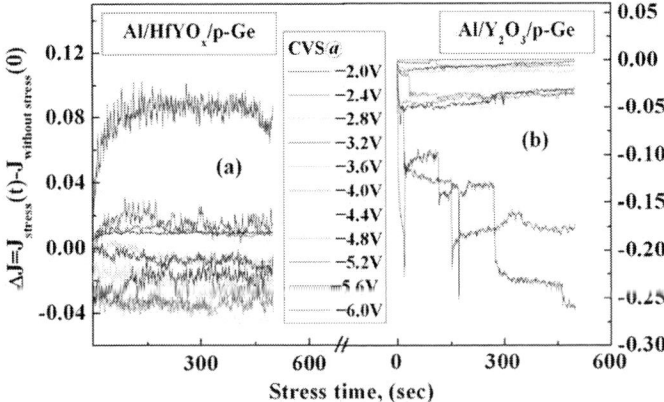

Fig. 7. Change in current density after CVS at different stress biases ranging from -2V to -6V.

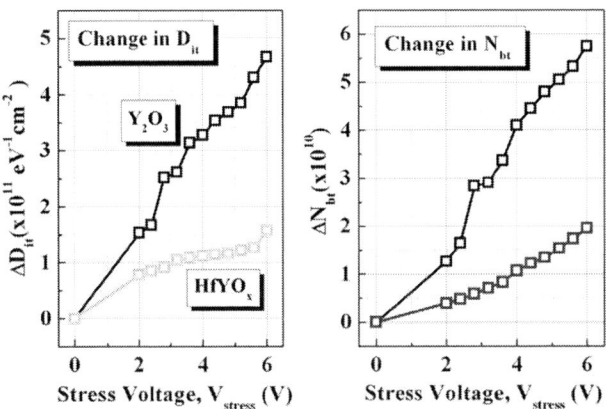

Fig. 8. Change in D_{it} and N_{bt} after CVS at several stress biases between -2V to -6V.

Conclusions

Effect of Hf incorporation in Y_2O_3 has been studied through XPS, XRD and electrical characteristics. From XPS analysis of Y 3d and Ge 3d spectrum formation of $YGeO_x$ was found in both samples. Though we found $YGeO_x$ formation is lower in case of $HfYO_x$. Bandgap was not significantly altered after Hf incorporation. Hf-incorporated Y_2O_3 films shows higher crystallization temperature in XRD characteristics compared to Y_2O_3. The leakage current density for $HfYO_x$ measured at gate electric field of 1 MV/cm is 3.8×10^{-4} A/cm^2, about one order of magnitude lower than the value of Y_2O_3(2.1×10^{-3}) layer with similar thickness. For $HfYO_x$ dielectric both hole and electron trapping was observed whereas for Y_2O_3 only electron trapping was monitored under CVS/CCS. Hf-incorporation further increases the breakdown field minimizing the trap generation probability under constant voltage stressing.

References

1. K. C. Saraswat, C. O. Chui, T. Krishnamohan, A. Nayfeh and P. McIntyre, Microelectron. Engg., **80**, 15-21 (2005).
2. H. Kim, P. McIntyre, J. Korean Phys. Soc., **48**, 5 (2006).
3. J. W.Seo, C. Dieker, J.-P. Locquet, G. Mavrou, A. Dimoula$_2$s, Appl. Phys. Lett., **87**, 221906 (2005).
4. W. P. Bai, N. Lu, J. Liu, A. Ramirez, D. L. Kwong, D. Wristers, A. Ritenour, L. Lee, and D. Antoniadis, Symposium on VLSI Technology, Kyoto, Japan, IEEE, New York (2003) 121.
5. V. V. Afanasev and S. Stesmans, Appl. Phys. Lett., **84**, 2319 (2004).
6. N. Lu, W. Bai, A. Ramirez, C. Mouli, A. Ritenour, M. L. Lee, D. Antoniadis, and D. L. Kwong, Appl. Phys. Lett., **87**, 051922 (2005).
7. N. Wu, Q. Zhang, C. Zhu, C. C. Yeo, S. J. Whang, D. S. H. Chan, M. F. Li, B. J. Cho, A. Chin, D.-L. Kwong, A. Y. Du, C. H. Tung, and N. Balasubramanian, Appl. Phys. Lett., **84**, 3741 (2004).
8. X. Chen, S. Joshi, J. Chen, T. Ngai, and S. Banerjee, IEEE Trans. Electron Devices, **51**, 153 (2004).
9. A. Dimoulas, G. Mavrou, G. Vellianitis, E. Evangelou, N. Boukos, M. Houssa, and M. Caymax, Appl. Phys. Lett., **86**, 032908 (2005).
10. E. P. Gusev, H. Shang, M. Copel, M. Gribelyuk, C. D'Emic, P. Kozlowski, and T. Zabel, Appl. Phys. Lett., **85**, 2334 (2004).
11. C. Mahata, M. K. Bera, P.K. Bose and C. K. Maiti, Semicond. Sci.Technol., **24**, 025026 (2009).
12. D. W. Wang, Q. Wang, A. Javey, R. Tu, H. J. Dai, H. Kim, P. C. McIntyre, T. Krishnamohan, and K. C. Saraswat, Appl. Phys. Lett., **83**, 2432 (2003).
13. J. J. H. Chen, N. A. Bojarczuk, H. L. Shang, M. Copel, J. B. Hannon, J. Karasinski, E. Preisler, S. K. Banerjee, and S. Guha, IEEE Trans. Electron Devices, **51**, 1441(2004).
14. K. Kita, T. Nishimura, K. Nagashio, and A. Toriumi, International Conference on Solid State Devices and Materials (SSDM), pp.8-9, (2008) Tsukuba.
15. K. Kita, K. Kyuno, and A. Toriumi, Appl. Phys. Lett., **86**, 102906 (2005).
16. E. Rauwel, C. Dubourdieua, B. Holländer,N. Rochat,F. Ducroquet,M. D. Rossell and G. Van Tendeloo,B. Pelissier, Appl. Phy. Lett., **89**, 012902 (2006).
17. J. Y. Dai, P. F. Lee, K. H. Wong, H. L. W. Chan, and C. L. Choy, J. Appl. Phys., **94**, 912 (2003).
18. S. Miyazaki, Applied Surface Science, **190**, 66 (2002).
19. W. C. Chien, J. Cryst. Growth, **290**, 554 (2006).
20. D. K. Chen, R. D. Schrimpf, D. M. Fleetwood, K. F. Galloway, S. T. Pantelides, A. Dimoulas, G. Mavrou, A. Sotiropoulos, and Y. Panayiotatos, IEEE Trans. Nucl. Sci., **54**, 971 (2007).

21. D. M. Fleetwood, N. S. Saks, J. Appl. Phys., **79**, 1583 (1996).
22. D. M. Fleetwood, IEEE Trans. Nucl. Sci., 39, 269 (1992).
23. A. Paskaleva, E. Atanassova and N. Novkovski, J. Phys. D: Appl. Phys., **42**, 025105 (2009).
24. R. Rodríguez, E. Miranda, R. Pau, J. Suñé, M. Nafría, X. Aymerich, Microelectron. Reliab., **40**, 707 (2000).
25. W. A. Hill and C. C. Coleman, Solid-State Electron., **23**, 987 (1980).
26. N. A. Chowdhary, R. Garg, and D. Misra, Appl. Phys. Lett., **85**, 3289 (2004).

846

CHAPTER 4

POSTER SESSION

848

Schottky Barrier Height at Dielectric Barrier/Cu Interface in low-k/Cu Interconnects

S.W. King[a], M. French[a], M. Jaehnig[a], M. Kuhn[a], B. French[b]

[a] Logic Technolog Development, Intel Corporation, Hillsboro, Oregon 97124, USA
[b] Octotillo Materials Laboratory, Intel Corporation, Chandler, Arizona 85248, USA

In order to understand the various possible leakage mechanisms in low-k/Cu interconnects, a knowledge of the basic band alignment between Cu and low-k dielectric materials is needed but has gone largely unreported. In this regard, we have utilized X-ray Photoelectron Spectroscopy (XPS) to measure the Schottky Barrier at interfaces of importance to Cu/low-k interconnects. Specifically, we have utilized XPS to determine the Schottky Barrier at the interface between Cu and low-k SiCN capping layers deposited on Cu via Plasma Enhanced Chemical Vapor Deposition (PECVD). We have also utilized XPS to investigate the impact of various plasma surface treatments on the band alignment at these interfaces. The cumulative results indicate that electron transport along the SiCN:H/SiOC:H interface may represent the lowest energy barrier path for line-line Schottky emission based leakage.

Introduction

Electrical leakage in low-k/Cu interconnect structures is becoming a growing vital concern as the nano-electronics industry moves to increasingly tighter metal spacing for sub 22 nm technology nodes and continues capacitance scaling by replacing low density / non-porous "low-k" SiOC:H interlayer dielectric (ILD) materials with porous SiOC:H dielectrics (1). In order to understand the various possible leakage mechanisms in low-k/Cu interconnects, knowledge of the basic electronic band alignment between Cu and the various dielectric and metal cladding materials is needed but has gone largely unreported (2). Prior studies have focused primarily on investigating the energy level alignment between the Cu via/trench barrier (typically Ta or TaN) and the low-k ILD. Internal photoemission (IPE) measurements by Shamuilia have indicated the barrier at this interface is fairly large at 4.5 eV and independent of the metal (Ta, TaN_x, TiN_x, Al, Au) and the particular low-k ILD material (k = 3.0, 2.3, and 2.0) (3). These measurements have been partially confirmed by the independent IPE measurements of Atkin which determined a barrier of 4.0±0.5 for a Au/SiOC:H (k=2.40) interface (4). In contrast though, the XPS measurements of Martinez have indicated a much smaller barrier height of 0.8 eV for a $TaN_{0.1}$/a-SiOC:H (25% porous) interface (2).

Interpretation of the discrepancies between the Shamuilia and Martinez results for the TaN/SiOC:H interface are complicated by the differences in surface preparation and TaN_x nitrogen content in each study, both of which could affect the band alignment via differences in density of interface states and TaN work function respectively. These differences also complicate correlations to real Cu/low-k interconnect structures where the low-k ILD surface is exposed to vastly different wet/dry plasma etches and cleans prior to Ta or TaN_x deposition. A further complication concerns the horizontal orientation of the low-k ILD surface in the Shamuilia and Martinez studies versus the predominantly vertical orientation of TaN/ILD interfaces in Cu/low-k interconnect structures. This latter point can significantly modulate the type and extent of damage to the low-k ILD surface imparted by the wet and dry plasma etches utilized to pattern the low-k ILD. These details underscore the need for continued investigation of the band alignment at this interface.

Electron transport across the ILD/Ta/Cu interface, however, represents only one possible leakage path in low-k/Cu interconnects. Another possible and pertinent leakage path is electron injection across the top Cu surface/dielectric capping layer interface and propagation along the ILD/Cu capping dielectric interface (see Fig. 1). The dielectric Cu capping layer serves multiple purposes including passivating the top Cu surface to minimize electromigration, preventing Cu diffusion into the low-k ILD, serving as a plasma etch stop for low-k ILD patterning, and protecting the Cu from plasma etch and wet chemical cleans (5). Accordingly, the Cu/dielectric cap interface is deserving of further investigation for numerous reasons. In this regard, we have utilized X-ray Photoelectron Spectroscopy (XPS) to measure the Schottky Barrier (Φ_B) present at the interface between Cu and various dielectric capping layers.

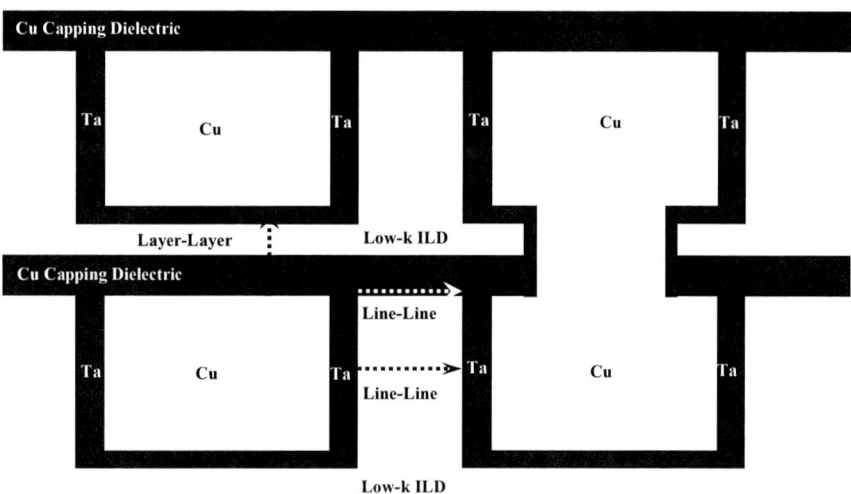

Figure 1. Schematic of a typical low-k/Cu interconnect structure and possible electrical leakage paths.

XPS has been used extensively for determining the band alignment of numerous semiconductors to other semiconductors, metals, and dielectrics (6). In this report, we demonstrate that XPS can also be utilized to determine the band alignment at interfaces between amorphous dielectrics and metals of interest to the low-k/Cu interconnects industry. Specifically, we have utilized XPS to determine the Schottky Barrier present at the interface between Cu and low-k a-SiC(N):H thin films formed via the Plasma Enhanced Chemical Vapor Deposition of a-SiC(N):H on polished Cu surfaces. **The composition of the a-SiC(N):H films included a-SiC:H, a-SiC$_{0.6}$N$_{0.5}$:H, and a-SiN:H and represents thin films commonly utilized in the semiconductor industry as Cu capping layers. (5). Additionally,** the planar orientation of the Cu surfaces in both our study and those of real low-k/Cu interconnects allows us to both more realistically mimic the Cu surface preparation prior to dielectric capping layer deposition and more directly apply our results to Cu/dielectric capping layer interfaces of interest to the semiconductor industry

Experimental

The Cu thin films utilized for these experiments consisted of electrochemically plated (ECP) Cu that had been chemically mechanically polished (CMP) using a Cu ECP and CMP process optimized for 45nm interconnect technologies. The ECP Cu was plated on a Cu seed and TaN adhesion layer sputter deposited on 300mm (100) Si substrates on which 100nm of thermal oxide had been previously grown. The a-SiC(N):H thin films were deposited on the Cu thin films by PECVD at temperatures on the order of 400°C using a standard commercially available 300 mm parallel plate capacitance PECVD tool. Process gases included various silane and methlysilane like sources diluted in gases such as N$_2$, NH$_3$, H$_2$ and He (7,8). Table I summarizes some of the key material properties for the a-SiC:H, a-SiC$_{0.6}$N$_{0.5}$:H, and a-SiN:H dielectric Cu capping layers investigated in this study including dielectric constant, refractive index, and intrinsic film stress. Prior to a-SiC(N):H deposition, an H$_2$ plasma pre-treatment was performed *in-situ* to remove Cu corrosion inhibitors left behind by the Cu CMP process and to reduce Cu surface oxides formed by ambient exposure (9). Such plasma pre-treatments are commonly performed prior to a-SiC(N):H deposition on Cu interconnects in order to improve the electromigration performance of the a-SiC(N):H/Cu interface (10). To test the impact of plasma pre-treatments on the Schottky barrier, the SiC$_{0.6}$N$_{0.5}$:H film was deposited with and without the H$_2$ plasma treatment. Additional NH$_3$ and He *in-situ* plasma pre-treatments reported to improve SiN/Cu electromigration were also investigated (11,12). After PECVD deposition, the SiC(N):H/Cu samples were transferred *ex-situ* to a VG Theta 300 XPS system equipped with a hemispherical analyzer and a monochromated Al anode x-ray source (1486.6 eV). The emitted photoelectrons were detected using an angle resolved detector and a pass energy of 80 eV that generated a full width half maximum of < 0.55 eV for the Ag 3d$_{5/2}$ core level from a Ag reference sample. For the a-SiC(N):H/Cu samples, high resolution scans of the Cu3p, Si2p, C1s, O1s and N1s core levels were acquired. Removal of surface contamination and oxidation from ex-situ transfer was achieved using a 2 keV Ar$^+$ ion sputtering beam. Ion beam sputtering at these energies has been previously shown to remove surface oxides from a-SiN$_x$ while maintaining the bulk composition at the surface (i.e. no preferential sputtering) (13).

The method of Grant and Waldrop was utilized to determine the Schottky barrier at the a-SiC(N):H/Cu interface (14). This method relies on referencing the core levels of the dielectric to the valence band maximum and then measuring how the position of the core levels change with the addition of the metal. Specifically:

$$\Phi_B = E_g - (E_{CL})_{int} + (E_{CL}-E_v)_{bulk} \qquad [1]$$

Where Φ_B is the Schottky barrier at the dielectric/metal interface, E_g is the band gap of the dielectric, $(E_{CL})_{int}$ is the dielectric core level energy at the interface (relative to the Fermi level), and $(E_{CL}-E_v)_{bulk}$ is the position of the dielectric core level relative to the valence band maximum measured at least 10nm away from the interface. Typically this experiment consists of determining the position of the dielectric valence band maximum relative to a core level from a clean dielectric surface and then measuring the changes in position of the core level as thin metal layers are deposited (10). In our case, we have performed the experiment in reverse by depositing the dielectric directly on the metal surface. Specifically, we deposited a thin 3-5nm layer of SiC(N):H directly on Cu to measure $(E_{CL})_{int}$, and then deposited thicker (> 25nm) SiC(N):H layers on separate Cu samples to determine $(E_{CL}-E_v)_{bulk}$. This method has the advantage of mimicking the process flow in low-k/Cu interconnect fabrication and also in eliminating potentially spurious surface photovoltage effects (15).

Table I. Summary of properties for dielectric Cu capping layers utilized in this study.

Film	k	RI	Stress (MPa)	E_g (eV)
SiC:H	6.5	2.24	-138	2.8
$SiC_{0.6}N_{0.5}$:H	5.85	2.03	-350	2.94
SiN:H	6.5	1.99	-180	3.14

Equation 1 also requires knowledge of E_g for the dielectric material. In our case, we determined E_g for the a-SiC(N):H films by reflective electron energy loss spectroscopy (REELS) using the method of Miyazaki (see Table I).[16] The REELS spectra were collected using a separate VG 350 Auger Electron Spectroscopy system equipped with a 0.5-25 keV electron gun and a hemispherical electron energy analyzer with a pass energy of 12eV.

Results

Fig. 2 presents an example EELS spectrum collected using a 500 eV electron beam from a 25nm a-SiC:H/Cu sample. The band to band excitation was defined by a linear extrapolation to the leading edge of the background noise level. As shown in Fig. 2, this indicates an E_g of 2.8±0.1 eV for the a-SiC:H sample. This value is consistent with prior reports of E_g = 2.6 – 3.2 eV determined by other techniques for PECVD a-SiC:H thin films of similar composition (17,18). The values of E_g similarly determined for the a-$SiC_{0.6}N_{0.5}$:H and a-SiN:H samples are summarized in Table I and are also consistent with other reports from similar samples (19-21).

In Fig. 3, we present the XPS valence band spectra acquired from 25nm a-SiC:H, a-SiCN:H, and a-SiN:H films deposited on Cu. A prominent feature at approximately 9.8

eV exists in the spectra for all three films. Although the valence band of SiC and Si_3N_4 materials are predominantly comprised of C and N 2p and 2s states (22-28), the peak at 9.8 eV is actually attributed to Si3s states due to the significantly higher photoemission cross section for this state relative to C2p and N2p states at the photon energies employed in this study (26). Another prominent feature exists in the valence band spectra for both the a-SiN:H and a-$SiC_{0.6}N_{0.5}$:H samples at 19.5 eV and is attributed to the N2s state (21,26). For the a-SiN:H sample, two additional peaks are also clearly observed at 7.5 and 12.25 eV and have been attributed to Si-H and N-H bonding states respectively as well as a mix of N2p-Si3p and N2p-Si3s bonding states (24). Overall, our a-SiN:H XPS valence band spectrum is qualitatively similar to those obtained from similarly deposited a-SiN_x:H films by Iqbal (21), Karcher (26) and Kubler (29).

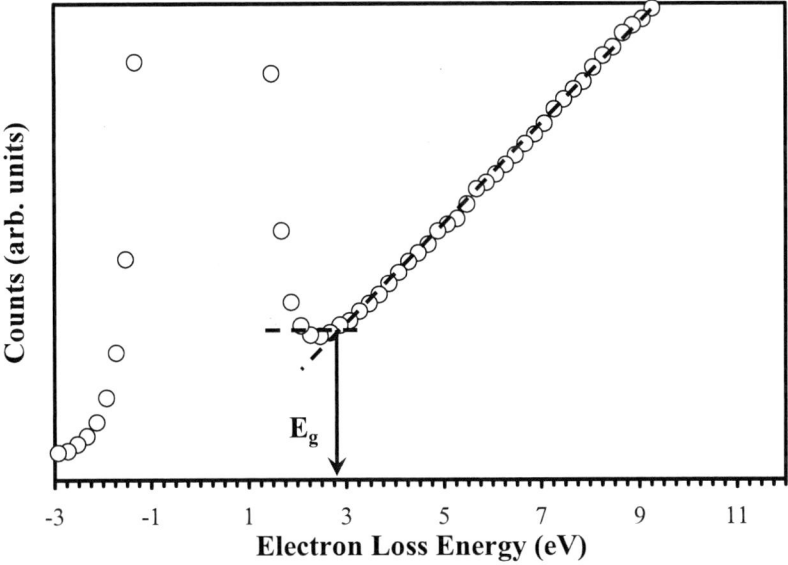

Figure 2. EELS spectrum for 25nm a-SiC:H on Cu.

For a-SiC:H, a striking resemblance is observed between our valence band spectra and that obtained by Porte (30), Parrill (31), and King (32) from single crystal α and β-SiC samples. In addition to the Si3s state at 9.8 eV, a broad C2s state at 16 eV is observed in the valence band spectra for both the a-SiC:H and α-SiC samples (see Figure 3c). Similarities also exist between our a-SiC:H valence band spectra and those obtained from other PECVD a-SiC:H films by Fang (33), Evangelisti (34), and Katayama (35) using lower energy photon sources (23-40.8 eV). As perhaps expected, the XPS valence band spectra from the a-$SiC_{0.6}N_{0.5}$:H sample (see Figure 3b) shows similarities to the spectra from both the a-SiC:H and a-SiN:H samples. For this sample, both the N-H/N2p-Si3p and N2s states at 12.25 and 19.5 eV are observed as well as the C2s state at 16 eV.

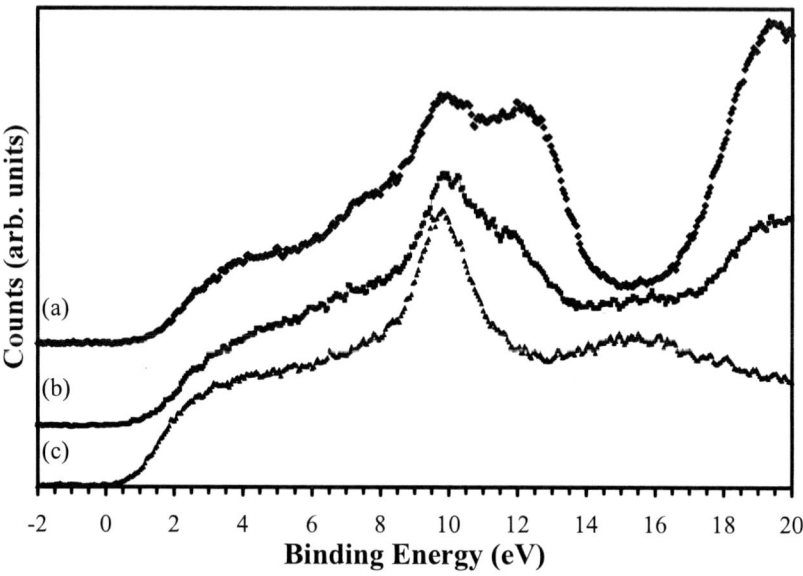

Figure 3. XPS valence band spectrum for 25nm (a) a-SiN:H, (b) a-SiC$_{0.6}$N$_{0.5}$:H, and (c) a-SiC:H deposited on Cu.

In Fig. 4, we present a close up of the XPS valence band spectra in the vicinity of the valence band maximum for a 25nm a-SiC:H on Cu sample. Using a linear extrapolation of the leading edge of the valence band emission, we determine the valence band maximum to be located at 0.75±0.05 eV. The peak position for the Si2p core level from the same a-SiC:H film was determined to be 100.4±0.02 eV. From these two measurements, we therefore determine the position of the Si2p relative to the VBM to be 99.65±0.1 eV. This value is in excellent agreement with other reported Si2p-VBM values, the majority of which range from 98.9 to 99.98 eV for both a-SiC:H and single crystalline SiC materials (see Table II) (33-41). The agreement in Si2p-VBM between a-SiC:H and c-SiC is consistent with the similarities between the valence band spectra of a-SiC:H, a-SiC, 3C-SiC, and 2H-SiC observed by others (30). Using the same methods, we determine Si2p-VBM to be 100.3±0.1 and 100.8±0.1 for the a-SiC$_{0.6}$N$_{0.5}$:H and a-SiN:H thin films respectively. As shown in Table II, these values are also consistent with those reported in the literature for PECVD a-SiN$_x$ and a-SiN$_x$:H materials by Iqbal and Karcher (21,26).

For 3-5nm layers of a-SiC:H, a-SiC$_{0.6}$N$_{0.5}$:H, and a-SiN:H deposited on Cu with an *in-situ* H$_2$ plasma pre-treatment, the position of the Si2p core level was determined to be 100.8±0.02, 101.35±0.02, and 102.15±0.02 eV respectively. Using these values and the E$_g$ and Si2p-VBM values previously determined, we calculate Φ_B to be 1.65±0.1, 2.0±0.1, and 1.8±0.1 eV for a-SiC:H, a-SiC$_{0.6}$N$_{0.5}$:H, and a-SiN:H/Cu interfaces respectively (see Figure 5 for schematic of band alignment).

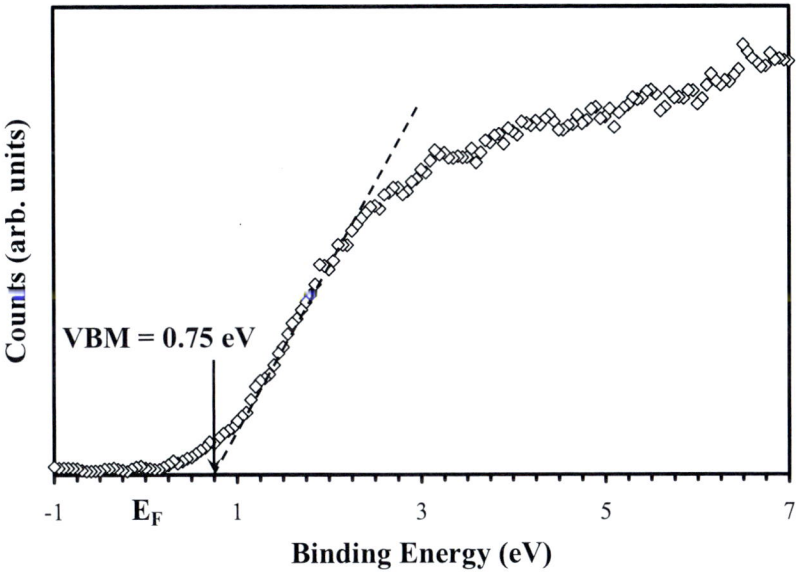

Figure 4. XPS valence band spectrum of 25nm a-SiC:H on Cu.

Table II. Summary of Si2p-VBM values for SiC, SiCN, and SiN materials in this study and other reports.

Material	Si2p-VBM (eV)	References
(0001) 2H-SiC	99.5	33
(0001) 4H-SiC	99.34	36
(0001) 4H-SiC	98.98±0.13	37
(0001) 6H-SiC	99.7	38
(0001) 6H-SiC	99.1	32
(111) 3C-SiC	99.3	39
(001) 3C-SiC	99.1±0.2	40
a-SiC:H	99.9	41
a-SiC:H	98.9	33
a-SiC:H	99.65±0.05	This Work
a-SiC$_{0.6}$N$_{0.5}$:H	100.3±0.05	This Work
a-SiN$_{1.4}$:H	100.2	21
a-SiN$_{1.5}$	100.6	26
a-SiN	100.3	26
a-SiN$_{0.8}$	100.4	26
a-SiN:H	100.8±0.05	This Work

Table III. Summary of Cu/SiC(N):H Φ_B from this study and other reports. Note: RT = Room Temperature, T_{int} = Interface formation temperature.

Interface	Φ_B (eV)	Method	T_{int}	References
Cu/n-6H-SiC(0001)	1.18	IPE	400°C	42
Cu/n-6H-SiC(0001)	1.12	IV	400°C	42
Cu/n-6H-SiC(0001)	1.23	CV	400°C	42
Cu/p-6H-SiC(0001)	1.53	IPE	400°C	42
Cu/p-6H-SiC(0001)	1.6	IV	400°C	42
Cu/p-6H-SiC(0001)	1.53	CV	400°C	42
Cu/n-6H SiC(0001)	1.22	IPE	RT	43
Cu/n-6H SiC(0001)	1.36	IPE	300°C	43
Cu/n-6H SiC(0001)	1.45	IPE	500°C	43
Cu/n-6H SiC(0001)	1.3	IPE	700°C	43
Cu/n-($000\bar{1}$)$_C$ 6H-SiC	1.5±0.1	XPS	RT	44
Cu/n-($000\bar{1}$)$_C$ 6H-SiC	1.6±0.1	XPS	150°C	44
Cu/n-($000\bar{1}$)$_C$ 6H-SiC	1.3±0.1	XPS	700°C	44
a-SiC:H/Cu	1.65±0.1	XPS	400°C	This Work
a-SiCN:H/Cu	2.0±0.1	XPS	400°C	This Work
a-SiN:H/Cu	1.8±0.1	XPS	400°C	This Work

Discussion

The value of 1.65±0.1 is on the high end but consistent with previous investigations of the Cu/(0001) 6H-SiC interface for which Φ_B was determined to range from 1.1 – 1.6 eV (see Table III) (42-44). These values were determined by a number of different techniques including: XPS, Internal Photoemission (IPE), current voltage (IV), and capacitance voltage (CV) measurements. The IPE, IV, and CV measurements by Aboelfotoh indicated a Φ_B of 1.2 eV for Cu deposited on n type (0001) 6H-SiC surfaces and annealed at 400°C, while higher values of 1.5-1.6 eV were observed for Cu deposited on similarly prepared p-type (0001) 6H-SiC surfaces (42). The IPE measurements by Suezaki observed a similar Φ_B of 1.22 eV for room temperature deposited Cu on (0001) 6H-SiC that increased to 1.45 eV after annealing at 500°C and then decreased to 1.3 eV after annealing at 700°C (43). In contrast, XPS measurements by Dontas found Φ_B for room temperature deposited Cu on the ($000\bar{1}$)$_C$ n-type 6H-SiC to be 1.5±0.1 eV (44). After annealing at 150 and 700°C degrees, Φ_B first increased to 1.6±0.1 eV and then decreased to 1.3±0.1 eV. In both cases, the decrease in Φ_B on annealing at temperatures > 500°C appears to be the result of $CuSi_x$ formation at the Cu/SiC interface (43,45).

Overall, the best agreement between our a-SiC:H/Cu Φ_B results and those for the Cu/c-SiC interfaces appears to be for p-type and carbon terminated crystalline SiC surfaces. This is consistent with our observation of a slight p-type character for our a-SiC:H films in that E_f-VBM is only 0.75±0.05 eV. It is also consistent with the hydrophobic character typically observed for a-SiC:H films and which has been attributed to CH_x surface termination for these and other low-k dielectric materials (46,47). Lastly, comparison between the a-SiC:H/Cu and Cu/c-SiC Φ_B results argues against the formation of significant levels of $CuSi_x$ at our a-SiC:H/Cu interfaces. In particular, we note that $CuSi_x$ formation at Cu/c-SiC interfaces was only observed at temperatures in excess of the a-SiC:H deposition temperature (44,45).

For the a-$SiC_{0.6}N_{0.5}$:H and a-SiN/Cu interfaces, we are not aware of any prior reports for Φ_B. However some useful comparisons can be made to Φ_B reported for other noble metal/SiN interfaces. Using IPE and XPS measurements, DiMaria (48) and Weinberg (49) were able to determine the hole barrier (Φ_{Bh}) at the Au/Si_3N_4 interface to be 1.9 and 1.5 eV respectively. Taking $\Phi_{Bh} = E_g - \Phi_B$, we determine Φ_{Bh} for SiN:H/Cu to be 1.35±0.1 eV which is in reasonable agreement with the Weinberg results. In contrast though, the IV measurements of Sze indicate a much smaller Φ_B of 1.3±0.2 eV for Au/Si_3N_4 interfaces (50). The lower value could be attributed to subtle differences between a-SiN_x.H/Cu and Au/Si_3N_4 interfaces, or to the observation of Frenkel Poole charge transport in the Sze measurements. Frenkel-Poole processes differ from Schottky emission in that charge transport occurs through charge hopping between bulk traps/defects. In this case, Φ_B represents the trap depth instead of the charge barrier at the interface.

Similar but contrasting behavior has been observed for other noble metals on SiC. In the XPS, CV, and IV measurements by Waldrop, the Au/(0001) 6H-SiC Φ_B was determined to be 1.4±0.05 eV for both ex-situ cleaned n and p-type surfaces. However for Au on ($000\bar{1}$)$_C$, Φ_B was determined to be 1.2±0.08 and 1.53 eV for n-type and p-type surfaces respectively (51-52). For Au/6H and 15R-SiC interfaces prepared by in-situ vacuum cleaving prior to Au evaporation, the CV and IPE measurements by Hagen indicated Φ_B = 1.39-1.695 eV (53). Lastly, the IV, CV, and IPE measurements by Aboelfotoh showed Au/(0001) 6H-SiC Φ_B = 1.43±0.02 for both p and n-type surfaces (42). Collectively, these values are also consistent with our results for Cu/a-SiC:H.

The above variation in Φ_B is likely due to differences in Fermi level pinning at the metal/dielectric interface that is typically attributed to interface gap states (51). The density, type, and energy level of these interface states can be largely dependent on surface preparation and interface chemistry (55). The observed decrease in Φ_B for Cu/(0001) 6H-SiC with some interfacial $CuSi_x$ formation is one example (43,44). To investigate what impact surface preparation might have on Φ_B for a-SiC(N):H/Cu interfaces, we additionally determined Φ_B for $SiC_{0.6}N_{0.5}$:H/Cu interfaces prepared with in-situ He, NH_3, and no plasma pre-treatments. The Φ_B for H_2, He, and NH_3 pre-treated interfaces were observed to all be within ±0.1 eV which is within the statistical error of these measurements. However when no plasma pre-treatment was performed, a large variation in Φ_B was observed with values as low as 1.5 eV and as high as 2.2 eV being observed. This large variability is probably due to differences in the uncontrolled amount of corrosion inhibitor and surface oxides present on the Cu surface prior to SiCN:H deposition. These results perhaps underscore the importance of the plasma pre-treatments in improving both electromigration and providing reliable/consistent electrical performance (56).

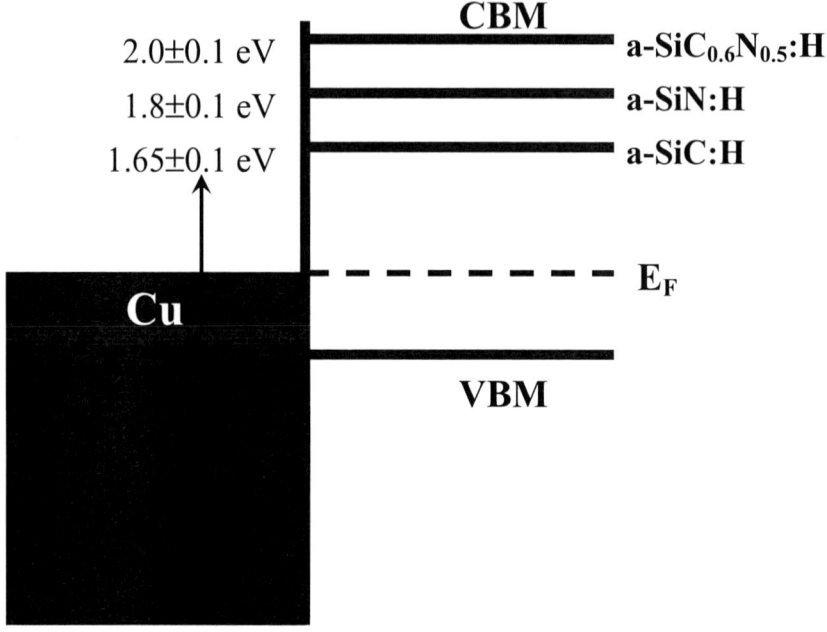

Figure 4. Schottky Barrier at Cu/Dielectric Interface for various a-SiC(N):H dielectric capping films.

Conclusions

In summary, we have utilized XPS to investigate the Schottky Barrier present at the interface between Cu and various SiC(N):H dielectric capping layers utilized in low-k Cu interconnect structures. The results highlight the importance of performing an *in-situ* plasma pre-treatment prior to a-SiC(N):H deposition in obtaining a consistent Schottky barrier at the SiC(N):H interface. The Schottky barrier Φ_B was found to be 1.8±0.2 eV across the full SiC-SiN composition window

Acknowledgments

The authors would like to acknowledge Drs. Bruce Tufts, Boyan Boyanov, and Jose Maiz of Intel for their thoughtful discussion and support in preparing this manuscript.

References

1. G. Gischia, K. Croes, G. Groeseneken, Z. Tokei, V. Afanas'ev, and L. Zhao, *Proc. 2010 IEEE Int. Rel. Phys. Symp.*, p. 549.

2. E. Martinez, C. Geudj, D. Mariolle, C. Licitra, O. Renault, F. Bertin, A. Chabli, G. Imbert, and R. Delsol, *J. Appl. Phys.* **104**, 073708 (2008).
3. S. Shamuilia, V. Afanas'ev, P. Somers, A. Stesmans, Y. Li, Z. Tokei, G. Groeseneken, and K. Maex, *Appl. Phys. Lett.* **89**, 202909 (2006).
4. J. Atkin, D. Song, T. Shaw, E. Cartier, R. Laibowitz, and T. Heinz, *J. Appl. Phys.* **103**, 94104 (2008).
5. C. Chiang, Z. Wu, W. Wu, M. Chen, C. Ko, H. Chen, S. Jang, C. Yu, and M. Liang, *Jap. J. Appl. Phys.* **42**, 4489 (2003).
6. G. Margaritondo and P. Perfetti, in *Heterojunction Band Discontinuities: Physics and Device Applications*, F. Capasso and G. Margaritondo, Editors, pp. 59-113, Elsevier, New York, (1987).
7. S. King, R. Chu, G. Xu, and J. Huening, *Thin Solid Films* **518**, 4898 (2010).
8. G. Stan, S. King, and R. Cook, *J. Mater. Res.* **24**, 2960 (2009).
9. A. Nishi, M. Sado, T. Niki, and Y. Fukui, *Appl. Surf. Sci.* **203-204**, 470 (2003).
10. T. Usui, H. Miyajima, H. Masuda, K. Tabuchi, K. Watanabe, T. Hasegawa, and H. Shibata, *Jap. J. Appl. Phys.* **45**, 1570 (2006).
11. A. Urbanowicz, M. Baklanov, J. Heijlen, Y. Travaly, and A. Cockburn, *Electrochem. Sol. Stat. Lett.* **10**, G76 (2007).
12. W. Qin, Z. Mo, L. Tang, B. Yu, S. Wang, and J. Xie, *J. Vac. Sci. Technol. B* **19**, 1942 (2001).
13. G. Ingo, N. Zacchetti, D. Sala, and C. Coluzza, *J. Vac. Sci. Technol. A* **7**, 3048 (1989).
14. J. Waldrop, R. Grant, Y. Wang, and R. Davis, *J. Appl. Phys.* **72**, 4757 (1992).
15. M. Hecht, *Phys. Rev. B* **41**, 7918 (1990).
16. S. Miyazaki, *Appl. Surf. Sci.* **190**, 66 (2002).
17. K. Mui, D. Basa, F. Smith, and R. Corderman, *Phys. Rev. B* **35** 8089 (1987).
18. T. Rajagopalan, X. Wang, B. Lahlouh, C. Ramkumar, P. Dutta, and S. Gangopadhyay, *J. Appl. Phys.* **94** 5252 (2003).
19. K. Kamata, Y. Maeda, and M. Moriyama, *J. Mater. Sci.* **5**, 1051 (1986).
20. S. Hasegawa, M. Matsuda, and Y. Kurata, *Appl. Phys. Lett.* **58**, 741 (1991).
21. A. Iqbal, W. Jackson, C. Tsai, J. Allen, and C. Bates, *J. Appl. Phys.* **61**, 2947 (1987).
22. S. Ren, and W. Ching, *Phys. Rev. B* **23**, 5454 (1981).
23. E. Ferreira, and C. da Silva, *Phys. Rev. B* **32**, 8332 (1985).
24. G. Pacchioni, and D. Erbetta, *Phys. Rev. B* **60**, 12617 (1999).
25. J. Justo, F. Mota, and A. Fazzio, *Phys. Rev. B* **65**, 732202 (2002).
26. R. Karcher, L. Ley, and R.L. Johnson, *Phys. Rev. B* **30**, 1896 (1984).
27. J. Roberston, *Philos. Mag. B* **66**, 615 (1992).
28. V. Ivashchenko, P. Turchi, V. Shevchenko, L. Ivashchenko, and G. Rusakov, *J. Phys. Condens. Matter* **15**, 4119 (2003).
29. L. Kubler, R. Haug, E. Hill, D. Bolmont, and G. Gewinner, *J. Vac. Sci. Technol. A* **4**, 2323 (1986).
30. L. Porte, *J. Appl. Phys.* **60**, 635 (1986).
31. T. Parrill, and V. Bermudez, *Sol. Stat. Comm.* **63**, 231 (1987).
32. S. King, M. Benjamin, R. Nemanich, R. Davis, and W. Lambrecht, *MRS Symp. Proc.* **395**, 375 (1996).
33. R. Fang, and L. Ley, *Phys. Rev. B* **40**, 3818 (1989).
34. E. Evangelisti, P. Fiorini, C. Giovannella, F. Patella, P. Perfetti, C. Quaresima, and M. Capozi, *Appl. Phys. Lett.* **44**, 764 (1984).

35. Y. Katayama, T. Shimada, T. Uda, and K. Kobayashi, *J. Non-Cryst. Sol.* **59**, 561 (1983).
36. J. Kohlscheen, Y. Emirov, M. Beerbom, J. Wolan, S. Saddow, G. Chung, M. MacMillan, and R. Schlaf, *J. Appl. Phys.* **94**, 3931 (2003).
37. B. Zhang, G. Sun, Y. Guo, P. Zhang, R. Zhang, H. Fan, X. Liu, S. Yang, Q. Zhu, and Z. Wang, *Appl. Phys. Lett.* **93**, 242107 (2008).
38. M. O'Brien, C. Koitzsch, and R. Nemanich, *J. Vac. Sci. Technol. B* **18**, 1776 (2000).
39. S. King, R. Davis, C. Ronning, and R. Nemanich, *J. Elect. Mater.* **28**, L34 (1999).
40. V. Bermudez, *J. Appl. Phys.* **63**, 4951 (1988).
41. P. Perfetti, *Surf. Sci.* **168**, 507 (1986).
42. M. Aboelfotoh, C. Frojdh, and C. Petersson, *Phys. Rev. B* **67**, 75312 (2003).
43. T. Suezaki, K. Kawahito, T. Hatayama, Y. Uraoka, and T. Fuyuki, *Jap. J. Appl. Phys.* **40**, L43 (2001).
44. I. Dontas, S. Ladas, and S. Kennou, *Diam. Rel. Mater.* **12**, 1209 (2003).
45. Z. An, M. Hirai, M. Kusaka, T. Saitoh, and M. Iwami, *Jap. J. Appl. Phys.* **40**, 1927 (2001).
46. A. Flannery, N. Mourlas, C. Storment, S. Tsai, S. Tan, J. Heck, D. Monk, T. Kim, B. Gogoi, and G. Kovacs, *Sens. Act. A* **70**, 48 (1998).
47. M. Baklanov, K. Mogilnikov, and Q. Le, *Microelectron. Eng.* **83**, 2287 (2006).
48. D. DiMaria, and P. Arnett, *Appl. Phys. Lett.* **26**, 1975 (1975).
49. Z. Weinberg, and R. Pollak, *Appl. Phys. Lett.* **27**, 254 (1975).
50. S. Sze, *J. Appl. Phys.* **38**, 2951 (1967).
51. J. Waldrop, R. Grant, Y. Wang, and R. Davis, *J. Appl. Phys.* **72**, 4757 (1992).
52. J. Waldrop, *J. Appl. Phys.* **75**, 4548 (1994).
53. S. Hagen, *J. Appl. Phys.* **39**, 1458 (1968).
54. R. Tung, *Phys. Rev. Lett.* **84**, 6078 (2000).
55. K. Tracy, P. Hartlieb, S. Einfeldt, R. Davis, E. Hurt, and R. Nemanich, *J. Appl. Phys.* **84**, 3939 (2003).
56. P. Liu, T. Chang, Y. Yang, Y. Cheng, and S. Sze, *IEEE Trans. Elec. Dev.* **47**, 1733 (2000).

ECS Transactions, 35 (4) 861-871 (2011)
10.1149/1.3572324 ©The Electrochemical Society

Global and Local Stress Characterization of SiN/Si(100) Wafers using Optical Surface Profilometer and Multiwavelength Raman Spectroscopy

Woo Sik Yoo, Junya Kajiwara, Takeshi Ueda, Toshikazu Ishigaki, and Kitaek Kang

WaferMasters, Inc., 246 East Gish Road, San Jose, CA 95112, USA

Thin SiN films with various stress levels, from tensile +1.7 GPa to compressive -3.5 GPa, (as estimated by average curvature after SiN film deposition using Stoney's equation) were deposited on blanket Si(100) wafers by plasma enhanced chemical vapor deposition (PECVD). Global and local distortion and stress of SiN/Si(100) wafers were characterized using an optical surface profilometer and multiwavelength Raman spectroscopy. While wafers with tensile stress and compressive SiN films showed significant changes in wafer bowing (direction and curvature), negligible changes in Raman shift in the Si below the SiN film of all wafers were observed, as indicated by applying various excitation wavelengths. Periodic Raman shift fluctuations were observed from all SiN/Si(100) wafers, suggesting a self stress relaxation mechanism at the lattice level. The importance of global (wafer level), local (wafer level) and lattice level stress characterization and its contribution to proper understanding of the mechanisms involved in wafer bowing and stress build up is discussed.

Introduction

External stresses in Si are known to affect band structure and carrier mobility. Stress engineering has become a very popular technique for enhancing complementary metal oxide semiconductor (CMOS) device performance beyond 90 nm technology nodes, in addition to conventional device miniaturization [1-3]. Compressive stress and tensile stress have contrasting effects on device performance. Stress is generated by many different techniques depending on device design strategies. Dual stress liners (DSLs), embedded silicon germanium (eSiGe) or carbon (C) doping in the source and drain (S/D), and stress memorization, are strategically used to obtain the desired level of channel stress to overcome the unintentional stress from shallow trench isolation (STI). Stress controlled SiN plasma enhanced chemical vapor deposition (PECVD) films are frequently used as stress liners to transfer compressive or tensile stresses to the underlying Si channel.

The stress value of a SiN PECVD film is typically determined from Stoney's formula [4] by comparing the change in radius of curvature of the blanket Si wafer, before and after film deposition. The wafer curvature is determined by measuring the angle of deflection of a laser beam off the Si surface or SiN/Si interface of a Si wafer. However, it uses the unrealistic assumptions of perfectly uniform film/wafer properties and stress/curvature states over the entire film/wafer system [4, 5]. Localized distortion of the wafer, non-uniform film/wafer thickness and inhomogeneous material properties are completely ignored. Simple interpretation of blanket wafer curvature measurement data

861

cannot represent Si stress in device wafers with uneven and non-continuous SiN/Si interfaces. Appropriate global stress, local stress and lattice stress characterization methodologies should be introduced for accurate measurement of meaningful physical parameters.

In this paper, we have introduced two types of new, non-contact, in-line process and/or material property monitoring methods which use various forms of interactions (reflection, diffraction, interference, elastic and inelastic scattering) between semiconductor wafers and a laser beam. The laser-based, optical surface profilometry (WaferMasters OSP-300) system, which utilizes laser beam reflection and diffraction from a wafer surface, can map wafer-level global and local curvature, warpage and distortion of blanket and device wafers. A multi-wavelength Raman spectroscopy (WaferMasters MRS-300) system characterizes lattice level stress/strain and crystallinity. A number of 300 mm Si(100) wafers with various CVD deposited SiN films are characterized using the OSP-300 and MRS-300 systems and compared with results from a conventional, laser deflection-based wafer curvature measurement tool.

Experiment

Optical Surface Profilometry (OSP-300) System

Laser-based optical wafer surface profiling techniques, using laser beam reflection from the blanket wafer surface, have been widely used in the semiconductor industry for rough inspection of wafer flatness, wafer bow and process induced stress after film deposition. However their application was limited to rough surface profile measurements of blanket wafers only. Due to the optical sensing, wafer holding and wafer rotation/translation mechanisms used in conventional wafer flatness and profile inspection systems, fine measurement wafer inspection, in the sub-micron range, is not possible. In these systems, interaction of the laser beam with a patterned wafer leads to diffraction of the incident laser beam and limits the accuracy of the surface characterization using conventional optical wafer surface profiling techniques.

For very high magnification surface characterization, such as surface roughness measurements, an atomic force microscope (AFM) is often used. In an AFM, a constant force is maintained between the probe and sample by measuring the force with a "cantilever" (or light lever) sensor and using feedback control electronic circuits to control the position of the Z-axis piezoelectric positioning device. The motion of the probe over the surface is generated and controlled by piezoelectric positioning devices that move the probe and force the sensor across the surface in the X and Y directions [6]. The probe is typically raster scanned across the surface of a very limited area (typically, a few mm^2). By monitoring the motion of the probe as it is scanned across the surface, a three dimensional (3D) image of the surface is constructed.

To overcome the difficulties of conventional optical reflection techniques (in sub-micron level fine measurement and patterned wafer inspection) and the difficulties of the AFM (in large area surface characterization), a new optical surface profilometry system (OSP-300) has been developed [7-9]. Figures 1 (a) and (b) illustrate the primary components of the OSP-300 system and an AFM system, respectively. As seen in Figs. 1 (a) and (b), there are many similarities between the OSP-300 system and AFM.

The OSP-300 system irradiates a wafer with a laser beam at a fixed incident angle and captures optical (reflected, diffracted and scattered) images projected to the screen from the wafer (either blanket or patterned) to characterize the wafer surface profile and pattern distortions. The system generates wafer maps of vector plots, grid plots, intensity, height contours, distortion and 3D surface profiles. It also generates the height profile and estimated curvature along the major crystal orientations, as well as a histogram of wafer surface tilt (or slope) from the wafer stage. Process induced surface profiles can be estimated and traced by comparing wafer surface profiles before and after a process step or a series of process steps in both blanket and patterned production wafers. Wafer surface profile maps of the accumulated impact of these process steps can be generated for process control and diagnosis. The OSP-300 system has achieved both sub-micron spatial resolution and sub-nanometer height resolution (from blanket and patterned wafers) which is very useful for advanced semiconductor device research and development. The OSP-300 system differentiates between reflected and diffracted beams and thereby avoids complications from patterned wafers.

The OSP-300 is designed to characterize the surface profiles of full size wafers (up to 300mm in diameter) with very high magnification (sub-nanometer height resolution) and high lateral (sub-micron) resolution, while the AFM focuses on very high magnification (sub-nanometer) surface characterization over a very small area using the light lever (cantilever). The operating principles of the OSP-300 and AFM are quite similar. The sub-nanometer height resolution and sub-micron lateral resolution of the OSP-300 system were achieved by using appropriate optical magnification and lateral resolution (stage resolution: 0.5μm/step). The OSP-300 has resolution comparable to the AFM but with the ability to characterize the surface detail of the entire wafer.

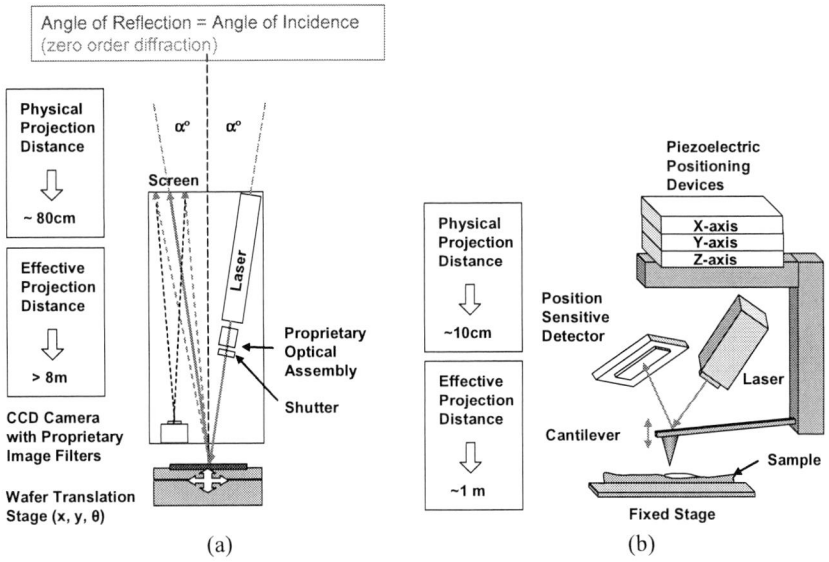

Figure 1. Schematic illustration of the primary components of optical surface profilometry (OSP-300) system (a) and an atomic force microscopy (AFM) system (b).

Graphical representations of surface profiles from blanket wafers, measured by the OSP-300 system, are shown in Fig. 2. Grid plots, vector plots, intensity plots, height maps, beam position distribution plots, curvature plots along major crystal axes and 3D plots can be obtained from a single series of measurements. It also generates the height profile and estimated curvature along the major crystal axes, as well as a histogram of wafer surface tilt (or slope) from the wafer stage, allowing easy statistical process control (SPC) and data analysis. By comparing wafer surface profiles before and after a certain process step or a series of process steps, process induced surface profiles, and the accumulated impact of these process steps, can be estimated and displayed as wafer maps.

The grid plots, vector plots and 3D contour maps provide visual images of global and local surface distortions at the time of measurement. The intensity plots provide an indication of the direction of distortion (either concave or convex). The height profile, beam position distribution, with histogram, curvature and height range along major crystalline axes offer very powerful tools for SPC, as well as valuable hints for diagnosis of process and/or equipment related problems.

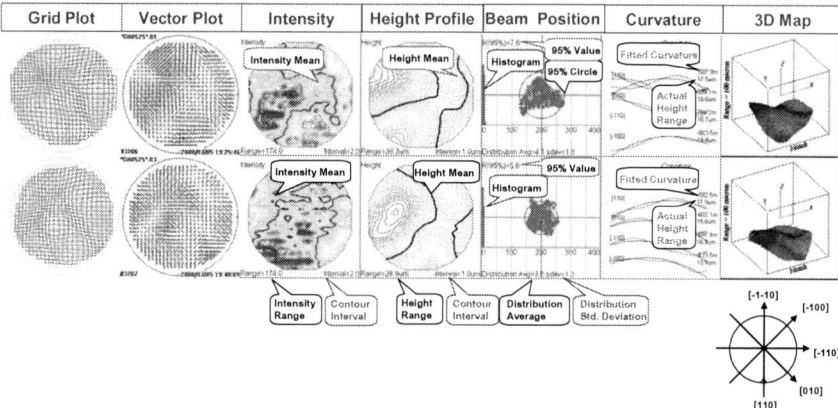

Figure 2. Graphical representation of surface profiles measured from two blanket Si wafers by the OSP-300 system

Multiwavelength Raman Spectroscopy (MRS-300) System

For in-line monitoring purposes, a very high spectral resolution Raman system with high measurement repeatability and stability is required. A multi-wavelength excitation capability is also extremely valuable as it allows depth profiling of the basic parameters. Conventional Raman spectroscopy systems are limited in their effectiveness by repeatability and stability issues, as well as distortion, resolution and an inability to make measurements at exactly the same spot with depth profiling.

To overcome these common issues with conventional Raman systems using a single excitation wavelength, we have designed a Polychromator-based, multi-wavelength Raman spectroscopy system (MRS-300) with non-contact and non-destructive in-line characterization capabilities [10]. This enables analysis of basic material parameters such as Si stress/strain, Ge content, B content, crystallinity, and thickness of $Si_{1-x}Ge_x$ epitaxial layer [11-18]. To improve the accuracy, consistency and usefulness of the MRS-300, it was designed with no moving parts and a minimum of optical components for improved stability and repeatability, and a longer focal length for enhanced spectral resolution [10]. The system has three thermoelectrically-cooled charge coupled device (CCD) cameras for measuring Raman peaks from the same measurement point, under three different excitation wavelengths, and without any disruption (i.e., without scanning the monochromator, switching the excitation laser or calibration) (Fig. 3). Three major spectral lines (457.9 nm, 488.0 nm and 514.5 nm) from a multi-wavelength Ar ion laser are used as the excitation source. Intensity, shift, and full width at half maximum (FWHM) characteristics of the Raman signals are measured at each frequency at the same wafer location. Preliminary results on the performance of the newly designed Raman system (MRS-300) and Raman characterization of mechanically stressed Si, Si stress around through-silicon-vias (TSVs) and $Si_{1-x}Ge_x/Si$ were reported previously [15-16].

Measurement repeatability of the MRS-300 system was evaluated using a prime Si (100) wafer for a test period of three months. Range of variations in Raman shift and FWHM at all three excitation wavelengths was less than 0.05cm^{-1}. Very reliable, high resolution Raman studies and in-line monitoring of stress, strain, crystallinity characterization of Si and $Si_{1-x}Ge_x$ epitaxial layers on Si wafers are now possible.

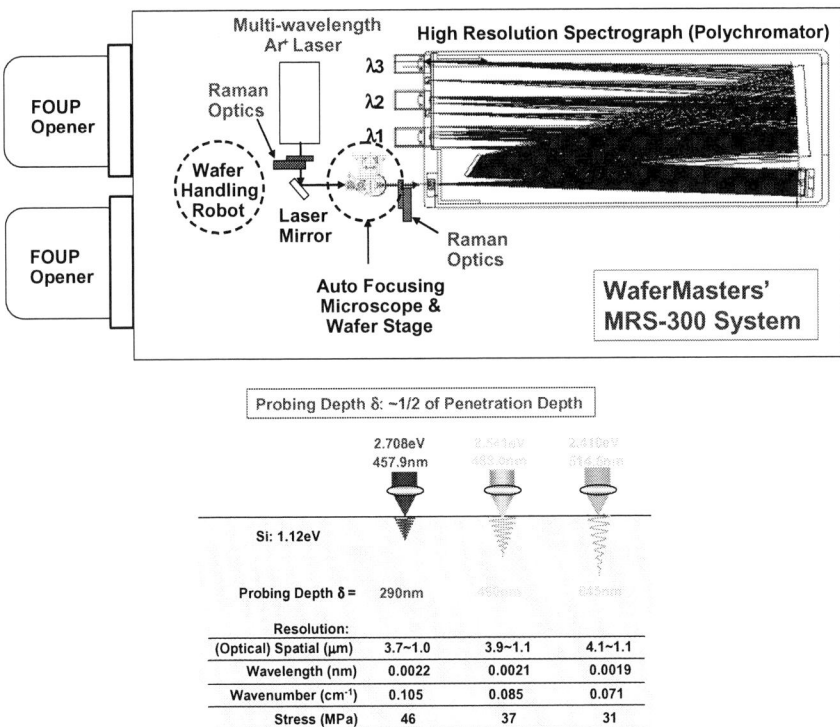

Figure 3. Schematic illustration of multi-wavelength Raman spectroscopy (MRS-300) system and summary of measurement capability.

PECVD Deposited SiN/Si(100) Wafers

A number of Si(100) wafers, with PECVD deposited SiN thin films, were prepared at ~400°C under capacitively coupled 13.56MHz plasma for comparative study. The SiN film thickness was fixed at ~30 nm. The film thickness was measured by ellipsometry. The targeted stress of the SiN films was in the range of +2.0 GPa (tensile) ~ -3.5 GPa (compressive). The stress of the SiN film was adjusted by changing PECVD conditions such as source gas flow rate, rf power and process pressure. The film stress was measured without additional thermal cycle. The stress values of the PECVD SiN films were in the range of tensile +1.7 GPa to compressive -3.5 GPa, as estimated by a conventional stress estimation method using the laser deflection, but ignoring localized wafer distortion. The stress values are calculated from the average curvature using the Stoney's formula, assuming perfectly uniform film/wafer properties and stress/curvature states over the entire film/wafer system.

Results and Discussions

All Si(100) wafers with PECVD SiN thin films were characterized and compared using the OSP-300 and MRS-300 systems. Blanket Si(100) wafers were also characterized for reference. Correlation between SiN thin film stress, as estimated by the conventional laser deflection measurement results, and Si stress under PECVD SiN films, as measured by multiwavelength Raman spectroscopy, were investigated to understand the effect of the stress of SiN capping layer on underlying Si.

Wafer Level Global and Local Flatness/Curvature Characterization by OSP-300

Figure 4 shows selected images (vector plots, curvature along major crystal axes and 3D wafer maps) of wafer surface profile measurements performed using the OSP-300 system for: (a) a reference blanket Si wafer, (b) SiN/Si (100) wafers with +1.7 GPa tensile film stress, (c) similar wafers with +3.1 GPa compressive film stress and (d) similar wafers with +3.5 GPa compressive film stress. The (reference) blanket Si wafer showed a reasonably flat surface with localized distortions. All Si(100) wafers with PECVD SiN thin films showed curved surface profiles. The SiN/Si(100) wafer with +1.7 GPa tensile film stress showed slight convex curvature along [110] and [-100] axis while the other two major crystal axes of [010] and [-110] showed slight concave curvature. Localized distortion of the wafer was observed along all major crystal axes. The other two Si(100) wafers, with compressive SiN film stress (+3.1 GPa and +3.5 GPa), showed convex curvature in all four major crystal axes. As the compressive stress increases, the wafer bow increases and radius of curvature becomes smaller. The general trends of wafer curvature, including the direction of curvature as a function of SiN film stress, were in good agreement with the conventional laser deflection measurement results.

All wafers showed some degree of localized distortion due to a combination of thickness non-uniformity, material property inhomogeneity and process induced stress. It clearly indicates the potential problem of casual interpretation of SiN film stress values using the average curvature of very limited line scans from conventional laser deflection-based measurements. Asymmetrical vector plots, grid plots, curvature along major crystal axes, contour maps and 3D plots, all generated by the OSP-300 system, clearly indicate the degree of localized distortions, stress variations and/or process induced deformations. This type of within wafer curvature variation is totally ignored by conventional curvature measurement and can be a significant source of device performance variations in advanced devices, which utilize performance enhancement by stress engineering.

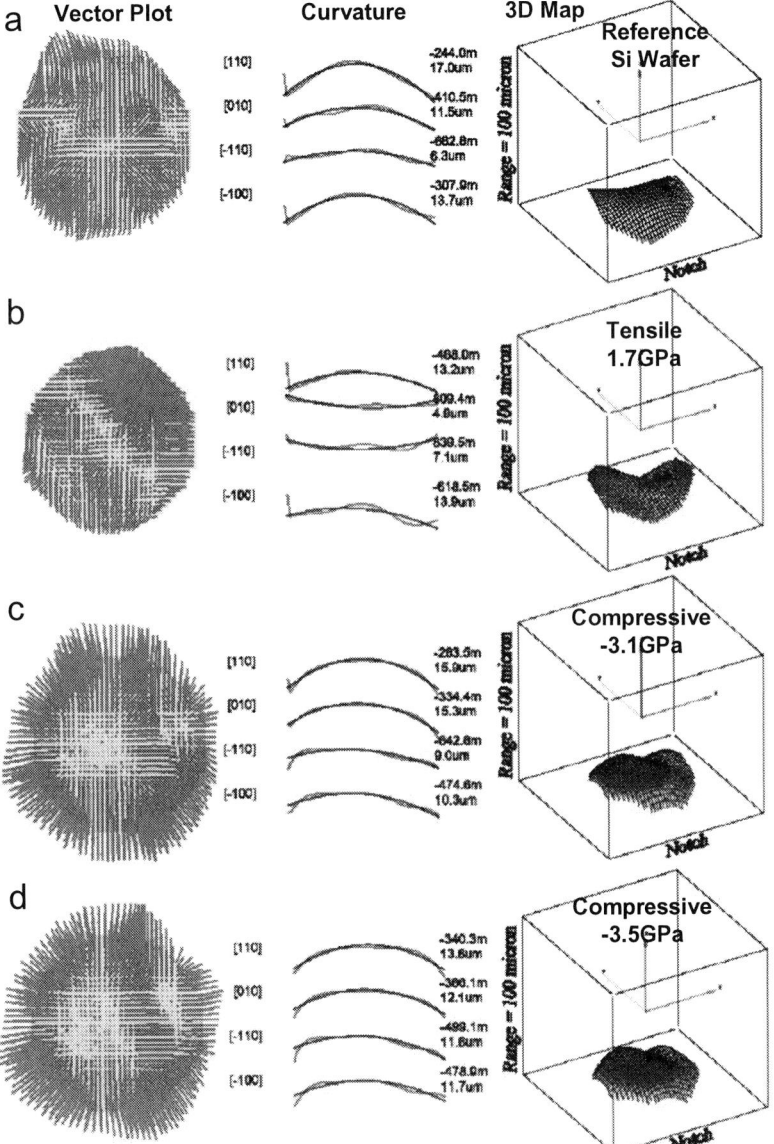

Figure 4. Vector plots, curvature along major crystal axes and 3D wafer maps generated from optical surface profile measurements of a blanket reference Si wafer (a) and SiN/Si (100) wafers with (b) +1.7 GPa tensile film stress, (c) +3.1 GPa compressive film stress and (d) +3.5 GPa compressive film stress.

Lattice Level Stress Characterization by MRS-300

To investigate the effect of SiN thin films with various stress levels on the lattice level stress of the underlying Si wafer, the in depth direction, multiwavelength Raman measurements were performed. A narrow and symmetrical Raman signal from a stress free Si wafer is observed at 520.3 cm^{-1} [10, 15-19]. The position and FWHM of a Raman shifted peak are very sensitive to crystalline stress and crystallinity of Si. The FWHM (shape) of the Raman signal is determined by combinations of crystallinity of the Si, coherency of the excitation laser beam and resolution capability of the spectrograph. Compressive stress and tensile stress in Si results in Raman peak shifts towards the higher wavenumber and lower wavenumber side, respectively. A Raman peak shift of 1 cm^{-1} is equivalent to ~450 MPa of biaxial mechanical stress ($\sigma_{xx} + \sigma_{yy}$) [19-23].

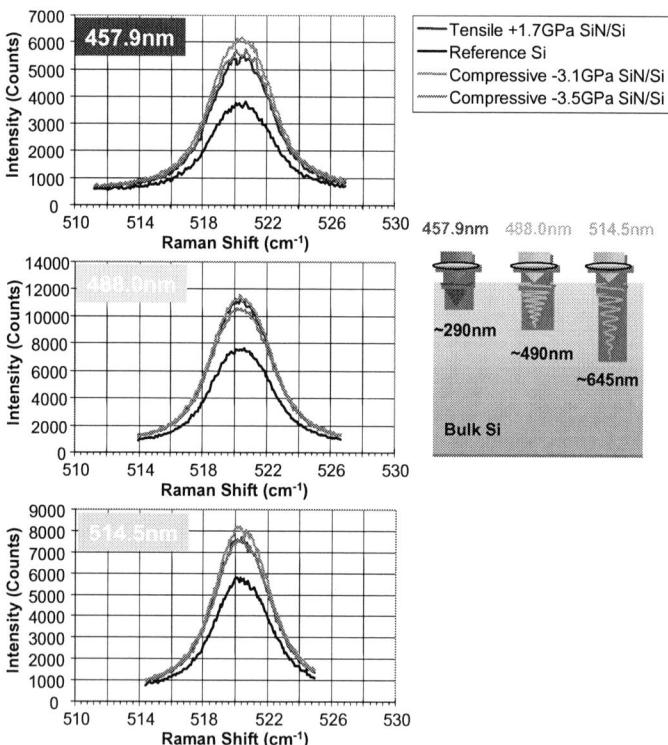

Figure 5. Very high resolution, multiwavelength Raman spectra from a reference blanket Si (100) wafer and SiN/Si (100) wafers with tensile and compressive stresses. Penetration depth at excitation frequencies is depicted in the insert.

Figure 5 shows Raman signals from reference Si and three SiN/Si wafers with different SiN thin film stresses, under three different excitation wavelengths (457.9 nm, 488.0 nm and 514.5 nm). The probing depth of the three excitation wavelengths in Si is also illustrated. All wafers showed almost identical Raman signals, suggesting negligible impact of the presence of the SiN capping thin films, with different stress levels, on the Si lattice of blanket Si wafers. It is quite surprising that the impact of blanket SiN thin films with various film stress levels on the stress of the underlying Si lattices is almost negligible, compared to stress induced by STI structures [19]. UV Raman measurement under 363.8 nm with a probing depth of ~10 nm from the SiN/Si interface also showed a negligible lattice stress in all samples.

Correlation between SiN Film Stress and Si Stress

While wafers with tensile SiN film stress and compressive SiN film stress showed significant changes in both wafer bowing direction and curvature (Fig. 4), Raman signals from the base Si, beneath the SiN film, showed negligible crystalline stress as indicated by the induced shift of the Si Raman peak under all three excitation wavelengths (Fig. 5) for all wafers. Correlation between SiN film stress measured by wafer curvature from the laser deflection method, and relative shift of the Raman peak from that of the blanket reference Si wafer is plotted in Fig. 6. SiN film stress and Si stress directions and values are also noted.

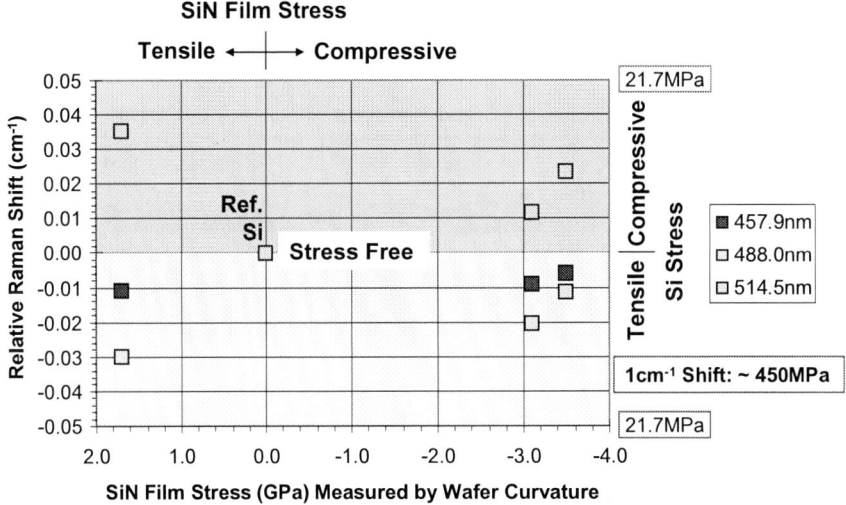

Figure 6. Correlation between SiN film stress measured by wafer curvature and Si stress under SiN measured by multiwavelength Raman spectroscopy under 457.9, 488.0 and 514.5 nm excitation.

To investigate micrometer scale stress variations within wafers, Raman line scan and area mapping measurements were done in 1 μm increments in x- and y-directions. We have observed periodic Raman shift fluctuations of 0.1 cm^{-1} in every 10 ~ 20 μm interval depending on the SiN thin film stress. We believe that this is the stress relaxation mechanism of SiN/Si at the lattice level. For Si wafers with sub-micron isolated features, the stress accumulation and relaxation mechanism will be significantly different from that in blanket wafers.

Summary

Global stress, local stress and lattice level stress of SiN/Si(100) wafers with various SiN film stress levels are characterized using the optical surface profilometer (OSP-300) and multiwavelength Raman spectroscopy (MRS-300) systems. Thin (~30 nm) SiN films were deposited on blanket Si(100) wafers by PECVD. The SiN film stress was varied from tensile +1.7 GPa to compressive -3.5 GPa (as estimated by average curvature after SiN film deposition using Stoney's equation). Wafers with tensile and compressive SiN film stresses showed significant change in both wafer bowing direction and curvature. However, for all wafers, negligible changes in Raman shift from the Si below the SiN film were measured under various excitation wavelengths with probing depths ranging from 290 nm ~ 645 nm. Stress of the SiN film has very limited effect on stress in blanket Si wafers at the lattice level. The periodic Raman shift fluctuations of Raman line scans and area mapping, in micrometer scale, suggests the existence of self stress relaxation mechanisms at the lattice level in all SiN/Si(100) wafers. Global (wafer level), local (wafer level) and lattice level stress characterization is increasingly important for proper understanding of wafer bowing and stress build up mechanisms.

References

1. T. Ghani, T., M. Armstrong, C. Auth, M. Bost, P. Charvat, G. Glass, T. Hoffmann, K. Johnson, C. Kenyon, J. Klaus, B. McIntyre, K. Mistry, A. Murthy, J. Sandford, M. Silberstein, S. Sivakumar, P. Smith, K. Zawadzki, S. Thompson and M. Bohr, *IEDM Tech. Digest*, 978 (2003).
2. V. Moroz, X. Xu, F. Nouri and Z. Krivokapic, *Solid State Technology*, **47**, 49 (2004).
3. R. Borges, V. Moroz and X. Xu, *Semiconductor International*, **Nov.**, 1 (2007).
4. G. G. Stoney, *Proc. R. Soc. London, Ser. A*, **82**, 172 (1909).
5. X. Feng, Y. Huang and A. J. Rosakis, *Trans. ASME*, **74**, 1276 (2007).
6. AFM/SPM Principles, www.afmuniversity.org.
7. W. S. Yoo, T. Ueda, J. Kajiwara, T. Ishigaki and K. Kang, *ECS Trans.*, **13** (1), 359 (2008).
8. W. S. Yoo, T. Ueda, T. Ishigaki and K. Kang, *ECS Trans.*, **19** (1), 315 (2009).
9. W. S. Yoo, T. Ueda, T. Ishigaki and K. Kang, *ECS Trans.*, **28** (1), 261 (2010).
10. W. S. Yoo, K. Kang, T. Ueda and T. Ishigaki, *Appl. Phys. Exp.*, **2**, 116502 (2009).
11. W. S. Yoo, T. Ueda, T. Ishigaki and K. Kang, *ECS Trans.*, **28**(1), 168 (2010).
12. W. S. Yoo, T. Ueda, T. Ishigaki and K. Kang, *ECS Trans.*, **28** (1), 253 (2010).
13. W. S. Yoo, T. Ueda, T. Ishigaki and K. Kang, *ECS Trans.*, **33** (6), 1003 (2010).
14. W. S. Yoo, T. Ueda, T. Ishigaki and K. Kang, *ECS Trans.*, **33** (6), 877 (2010).
15. A. D. Trigg, L. H. Yu, C. K. Cheng, R. Kumar, D. L. Kwong, T. Ueda, T. Ishigaki, K. Kang and W. S. Yoo, *Int. Conf. on Solid State Devices and Materials*, (2010) P-2-3.
16. A. D. Trigg, L. H. Yu, C. K. Cheng, R. Kumar, D. L. Kwong, T. Ueda, T. Ishigaki, K. Kang and W. S. Yoo, *Appl. Phys. Exp.*, **3**, 086601 (2010).

17. Y. F. Tzeng, S. Ku, S. Chang, C. M. Yang, C. S. Chern, J. Lin, N. Hasuike, H. Harima, T. Ueda, T. Ishigaki, K. Kang and W. S. Yoo, *Appl. Phys. Exp.*, **3**, 106601 (2010).
18. Y. F. Tzeng, S. Ku, S. Chang, C. M. Yang, C. S. Chern, J. Lin, N. Hasuike, H. Harima, T. Ueda, T. Ishigaki, K. Kang and W. S. Yoo, *J. Mater. Res.*, in press.
19. S. Nishibe, T. Sasaki, H. Harima, T. Isshiki, M. Yoshimoto, K. Kisoda, T. Yamazaki and W.S. Yoo, *Proc. of 14th Int. Conf. on Advanced Thermal Processing of Semiconductors* (RTP 2006, Kyoto), 211 (2006).
20. I. De Wolf, J. Vanhellemont, A. Romano-Rodrígues, H. Norström, H. E. Maes, and S. K. Jones, *J. Appl. Phys.*, **71**, 898 (1992).
21. I. De Wolf, *Semicond. Sci. Technol.*, **11**, 139 (1996).
22. I. De Wolf, H. E. Maes, and S. K. Jones, *J. Appl. Phys.*, **79**, 7148 (1996).
23. I. De Wolf, *Spectroscopy Europe*, **15**/2, 6 (2003).

872

ECS Transactions, 35 (4) 873-887 (2011)
10.1149/1.3572325 ©The Electrochemical Society

Mechanisms of Difficulty to Correlate the Leakage Current of High-k Capacitor Structures with Defect States Detected Spectroscopically by the Thermally Stimulated Current Technique

W.S. Lau

School of EEE, Nanyang Technological University, Singapore 639798, Singapore

> Historically, it has been difficult to correlate the leakage current of capacitor structures involving high-k dielectric materials and defect states detected spectroscopically by the thermally stimulated current (TSC) technique. Four mechanisms are proposed and solutions are explained with tantalum oxide as an example. One of the mechanisms is the limitation of the TSC technique itself because of the presence of a parasitic current due to the bias voltage used. This can be solved by migrating to more advanced versions of TSC like zero-bias TSC.

Introduction

Intuitively it can be imagined that the leakage current of high-k capacitor structures may arise from some defect states in the high-k dielectric material, which can be detected by thermally stimulated current (TSC) spectroscopy. However, it has been difficult to make a successful correlation between the leakage current and defect states detected by TSC. For example, Dr. Y. Nishioka, who was a pioneer working on high-k dielectric material, pointed out to the author that no correlation can be seen between leakage current and defect states detected by TSC in tantalum oxide through a private communication [1]. However, the author strongly believes that there is a correlation. Since Dr. Nishioka and his co-workers never publish their work on defect states detected by TSC in tantalum oxide, this problem has become some sort of mystery. In this paper, the author would like to point out that the correlation of the leakage current with defect states detected by TSC can really be difficult but this problem may not be really so mysterious. Besides TSC, other techniques have been tried. For example, Alers et al. used photoluminescence (PL) for the detection of defect states in tantalum oxide [2]; they pointed out that the defect band detected by PL could be significantly reduced by a nitrogen/oxygen plasma annealing process. However, photoluminescence is not convenient for failure analysis of practical capacitor structures which involve both a top metal electrode and a bottom metal electrode. Furthermore, there is another family of techniques based on the shift of the I-V characteristics because of charge trapping by defect states. For example, Zhao et al. [3]-[4] applied this approach to study ultrathin HfO_2/SiO_2 structure; they pointed out a shortcoming of this technique is that if the defect state causes leakage, then the defect state is not efficient to cause charge trapping and vice versa. The problem of this approach is to use the I-V characteristics to detect defect states and then correlate the detected defect states with the I-V characteristics. Another variant is to use the shift of the C-V characteristics due to charge trapping by defect states [5]. This approach can also suffer from the problem that the defect states responsible for leakage may not be able to cause significant charge trapping and vice versa. The author's approach is to use TSC to

873

detect defect states and then correlate with the I-V characteristics and so there is no such problem.

Theory

Mechanism A

In the 1910's, Poole reported his experimental observations on mica insulators [6]. In 1938, Frenkel followed up on Poole's work and proposed his theory that an electric field can enhance the ionization of defect states in a semiconductor [7]. According to standard solid state theory, there is no basic difference between a semiconductor and an insulator except that the bandgap energy is larger for an insulator compared to a semiconductor. Hence, the Poole-Frenkel (P-F) effect can be observed in both semiconductors and insulators. For the P-F mechanism, the leakage current through an insulator is given by

$$J_{PF} = BE\exp\{[\phi_B - ((qE)/(\pi\varepsilon oK))^{1/2}]/(kT/q)\} \qquad (1)$$

In equation (1), B is a constant while E , ϕ_B , k, T, q, εo and K are the electric field, barrier height of defect state, Boltzmann constant, absolute temperature, electronic charge, vacuum permittivity and dielectric constant.

Beside the P-F effect, leakage current can also be due to Schottky emission. For the Schottky emission mechanism, the leakage current through an insulator is given by

$$J_{SK} = A^{**}T^2\exp\{[\phi_B - ((qE)/(4\pi\varepsilon oK))^{1/2}]/(kT/q)\} \qquad (2)$$

In equation (2), A** is Richardson constant while ϕ_B is the barrier height at the metal-insulator interface.

In general, it is not easy to distinguish between the P-F mechanism and the Schottky emission mechanism because for both cases the logarithm of leakage current plotted against the square root of voltage is a straight line. In addition, there is also a controversy whether the dielectric constant in eq. (1) and eq. (2) is the dielectric constant at low frequency or that at optical frequency [8]-[9]. The author would like to point out that this problem arises because equation (1) and equation (2) are used to fit the same I-V characteristics, resulting in two different values of K.; then these two different values of K will be compared with the known value of K in order to see whether equation (1) or equation (2) fits better. Then there is a controversy whether the measured value of K is the known DC dielectric constant or the dielectric constant at optical frequencies. In this paper, the author would like to point out a different approach: equation (1) and equation (2) can be used to fit two different portions of the same I-V characteristics. For example, the I-V characteristics may show a forward characteristics with current rising faster as a function of voltage compared to the reverse characteristics; then equation (1) can be used to fit the forward I-V characteristics while equation (2) can be used to fit the reverse I-V characteristics.

As discussed above, there is a controversy regarding the leakage current versus voltage relationship is governed by the Schottky mechanism or by the Poole-Frenkel mechanism for several decades. The Schottky mechanism does not involve defect states in the bulk of the high-k dielectric; however, the Poole-Frenkel mechanism involves defect states in the bulk of the high-k dielectric. In this paper, the author points out that these two mechanisms actually can happen simultaneously and a unified Schottky-Poole-Frenkel model, as shown in Fig. 1, can be used to explain observed experimental data.

The unified Schottky-Poole-Frenkel model shown in Fig. 1 is actually similar to a model involving two back-to-back Schottky diodes suggested by Lai and Lee in 1999 [10]. The only difference is the author's addition of a non-linear resistor RNL, which can

Fig. 1 A capacitor structure involving a high-k dielectric can be thought as two back-to-back Schottky diodes D1 and D2 with a non-linear resistor RNL in between. The high-k dielectric is usually a metallic oxide with oxygen vacancy type of defect states. An oxygen vacancy is a deep double donor; a high-k dielectric can be considered as a very weakly n-type large bandgap semiconductor such that D1 and D2 are drawn for metal to n-type semiconductor Schottky diodes. D1 and D2 actually represent the two interfacial regions of the capacitor and RNL represents the bulk region of the high-k dielectric.

be used to represent the Poole-Frenkel effect. Lai and Lee analyzed capacitor structures involving an as-deposited tantalum oxide film which is very leaky [10]; according to the model shown in Fig. 1, RNL is approximately zero for their work. If a high-k dielectric has oxygen vacancies, which are deep double donors, as the dominant type of donors, the high-k dielectric can be considered a very slightly n-type large bandgap semiconductor. The model shown in Fig. 1 is drawn assuming that the high-k dielectric behaves like a very slightly n-type large bandgap semiconductor. The physical origin of the model shown in Fig. 1 is the author's hypothesis that the distribution of oxygen vacancies in a high-k dielectric is non-uniform; for an MIM capacitor, it can be easily imagined that the concentration of oxygen vacancies is greatest at the two metal/high-k interfaces because the high-k dielectric, which is usually a metallic oxide, is chemically reduced by the metal. As discussed above, oxygen vacancies are deep double donors. It is well known that when the concentration of donors is very large, the donors will appear as much shallower donors compared to the situation when the concentration of donors is small according to Pearson and Bardeen [11]; the same situation is true for acceptors. Thus near the interface between the metal and the high-k dielectric, the high-k dielectric can behave like an n-type large bandgap semiconductor.

As discussed above, one basic difficulty to correlate the leakage current with defect states (Mechanism A) is the difficulty to distinguish whether the leakage current follows the Schottky mechanism which does not depend on bulk defect states or the Poole-Frenkel mechanism which depends on bulk defect states. This problem can be solved in the following manner. The analysis of the structure shown in Fig. 1 can be very greatly simplified if one of the two Schottky diodes D1 and D2 has a significantly lower barrier height than the other one such that it can be considered an Ohmic contact. According to the theoretical analysis by Robertson [12], this is the case if the metal is n^+-Si and the high-k dielectric is tantalum oxide (Ta_2O_5) or titanium oxide (TiO_2). As shown in Table I, the calculated conduction band (CB) offset on Si for Ta_2O_5 or TiO_2 is quite small and so the Schottky barrier height of n^+-Si on Ta_2O_5 or TiO_2 is quite small. As shown in Table I, the CS offset for Ta_2O_5 on Si is 0.36 eV according to theoretical calculation; experimentally, Miyazaki [13] reported a value of 0.28 eV, which is quite close. Similarly, as shown in Table I, the CS offset for TiO_2 on Si is 0 eV according to theoretical calculation; experimentally, Perego et al. [14] reported a negative value. As shown in Table I, the bandgap of TiO_2 is significantly smaller than that of Ta_2O_5 and CB offset on silicon is so low such that the leakage current of TiO_2 capacitors on silicon is expected to be much higher such that quite frequently Ta_2O_5 may be more suitable for microelectronics applications. Thus, in this paper, we will concentrate on $M/Ta_2O_5/n^+$-Si capacitors where M stands for "metal". If D2 is the Schottky diode with the metal n^+-Si and the high-k dielectric Ta_2O_5, then D2 is like an Ohmic contact and so it can be ignored such that the model is Fig. 1 has only D1 and RNL left. When the metal M is positively biased, D1 is forward biased and RNL is likely to dominate over D1 such that the Poole-Frenkel mechanism dominates over the Schottky mechanism. Conversely, when the metal M is negatively biased, D1 is reverse biased and D1 is likely to dominate over RNL such that the Schottky mechanism dominates over the Poole-Frenkel mechanism. (Note: According to basic MOS theory, applying positive bias to $M/Ta_2O_5/n^+$-Si capacitors will lead to "accumulation"; applying negative bias to $M/Ta_2O_5/n^+$-Si capacitors may lead to "depletion" or "inversion", resulting in the formation of a depletion region or an inversion layer such that the model shown in Fig. 1 has to be modified to include the effect of a depletion region or an inversion layer. Thus, in order to see the Schottky mechanism dominating over the Poole-Frenkel mechanism when M is negatively biased, the magnitude of the negative bias voltage cannot be too big. Alternatively, the doping concentration of the n^+-Si can be made larger such that it is not so easily depleted or inverted.) Experimental evidence to support the above theory can be found in the work by Matsuhashi and Nishikawa for Ta_2O_5 [15]. Similarly, experimental evidence to support the above theory can be found in the work by Sun and Chen for TiO_2 [16]. Thus it is more likely to have a correct correlation of the leakage current with bulk defect states if the $M/Ta_2O_5/n^+$-Si capacitors are biased positively on the metal M than when the capacitors are biased negatively on the metal M.

The above discussion shows that $M/Ta_2O_5/n^+$-Si capacitors with positive voltage applied to M can be more easily understood because of the small CB offset of Ta_2O_5 on Si. The same theory with some modification can be applied to $M/Ta_2O_5/p^+$-Si capacitors with positive voltage applied to M. It appears to the author that there may be some positive charge in tantalum oxide due to ionized donors such that $M/Ta_2O_5/p^+$-Si capacitors are quite frequently inverted. (Note: In the absence of a large positive charge in Ta_2O_5, $M/Ta_2O_5/p^+$-Si capacitors can also be inverted because of a large enough positive voltage applied to M.) In this way, sometimes, an $M/Ta_2O_5/p^+$-Si capacitor can be weakly inverted or even strongly inverted; it can behave like $M/Ta_2O_5/n^-$-Si/p^+-Si or even like $M/Ta_2O_5/n^+$-Si/p^+-Si. In other words, an $M/Ta_2O_5/p^+$-Si capacitor can behave like an $M/Ta_2O_5/n$-Si capacitor in series with an "induced" n-Si/p^+-Si p-n junction or even an "induced" n^+-Si/p^+-Si tunnel diode. There is some voltage drop across the n-Si/p^+-Si p-n junction or n^+-Si/p^+-Si tunnel diode such that usually $M/Ta_2O_5/p^+$-Si capacitors are less leaky than $M/Ta_2O_5/n^+$-Si capacitors for the same positive voltage applied to M. For the author, the significance of the $M/Ta_2O_5/p^+$-Si capacitor structure is that he only managed to perform ZBTSC on an $M/Ta_2O_5/p^+$-Si capacitor structure for as deposited CVD Ta_2O_5, which is very leaky; this is because even ZBTSC has some sort of very small parasitic voltage (probably of the mV level) applied to the sample [17]. It was observed that ZBTSC on an $M/Ta_2O_5/p^+$-Si capacitor structure for as deposited CVD

Ta_2O_5 sometimes can be significantly easier than ZBTSC on an $M/Ta_2O_5/n^+$-Si capacitor structure for as deposited CVD Ta_2O_5. In order to make a study of the defect states after high temperature annealing compared to the defect states for an as deposited sample using ZBTSC, the $M/Ta_2O_5/p^+$-Si capacitor structure sometimes can be the better choice.

Table I   Calculated conduction band (CB) offset on Si for Ta_2O_5 and TiO_2

Insulator	Eg (eV)	Electron affinity (eV)	Calculated CB offset on Si (eV)
SiO_2	9	0.9	3.5 (experimental)
Si_3N_4	5.3	2.1	2.4 (experimental)
Ta_2O_5	4.4	3.2	0.36
TiO_2	3.05	3.9	0

Note: The values of the experimental conduction band offset on Si for SiO_2 and Si_3N_4 are shown in Table I for the purpose of comparison. Data in Table I originate from Robertson [12].

Mechanism B

Another difficulty to correlate the leakage current with defect states arises from a basic shortcoming of the TSC technique (Mechanism B): for an insulating film with finite thickness, the physical location of the defect states detected by TSC cannot be distinguished. Naturally, the greatest quantity of defect states occur at the two interfacial regions of the high-k dielectric capacitor but RNL is only controlled by the defect states in the bulk. Thus the defect states detected by TSC represent both the bulk and the interfacial region. Careful interpretation of TSC spectrum is therefore necessary to handle this issue.

Mechanism C

There is a third mechanism (Mechanism C). TSC may have a limited range such that only some of the defect states can be detected while some important defect states may be out of range. This situation is particularly serious if the high-k dielectric is in the form of an ultrathin film. This problem can be partially solved by using a novel zero-bias thermally stimulated current (ZBTSC) spectroscopy technique [17]-[19]. The purpose of zero bias is to get rid of a parasitic leakage current that can interfere with the measurement. Experience shows that even ZBTSC may not be sufficient such that a more sophisticated technique like zero-temperature gradient ZBTSC has to be used [20]-[21], as shown in Fig. 2.

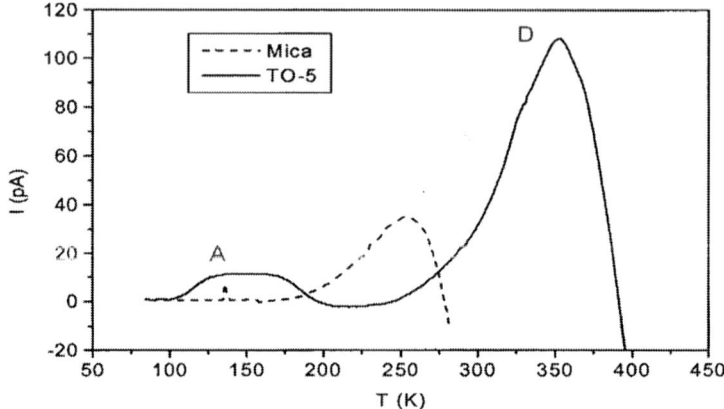

Fig. 2 Defect D (first ionization state of the oxygen vacancy) detected by a novel zero-temperature-gradient ZBTSC technique in an ultrathin Ta_2O_5 capacitor structure. A TO-5 metal can was used to house the sample when zero-temperature-gradient ZBTSC was performed.

Sometimes the high-k dielectric thin film can be quite leaky. For example, as deposited CVD Ta_2O_5 thin film is quite leaky. For such a situation, even ZBTSC may run into problem. One way to solve this problem is to modify the sample structure. It turns out that $M/Ta_2O_5/p^+$-Si tends to be much less leaky than $M/Ta_2O_5/n^+$-Si. For as deposited CVD, usually, the $M/Ta_2O_5/p^+$-Si structure is used instead of the $M/Ta_2O_5/n^+$-Si structure. This case will be discussed in more detail subsequently in this paper.

Mechanism D

There is a fourth mechanism (Mechanism D). The TSC family of techniques involves two steps in general: Step 1) filling up of defect states in the sample by carriers (electrons or holes) usually at a relatively low temperature by electrical injection or optical injection and Step 2) heating up the sample from the filling temperature to a higher temperature. During Step 2, the current due to carriers released from the defect states is recorded as a function of temperature, resulting in a spectrum. There is a possibility that it may be difficult to fill up the defect states during Step 1. Previously, the author pointed out the first ionization level of the oxygen vacancy double donor (Defect D) has an electron repulsive energy barrier such that Defect D is quite difficult to be filled by electrons at low temperature [21]. Another possibility is that the carrier lifetime may be small. For example, in as deposited Ta_2O_5, the carrier lifetime may be quite small such that the filling of defect states by optical injection is quite inefficient. Shining light may generate "electrons" and "holes"; shining light can also excite electrons captured by the defect states back into the conduction band. Thus it is possible that the defect states measured may be significantly lower than the actual amount of defect states present.

After a high temperature annealing, the carrier lifetime may be significantly improved such that the filling of defect states by optical injection becomes significantly more efficient. Because of this, there may be difficulty to compare the defect states spectrum of as deposited Ta_2O_5 with that of annealed Ta_2O_5.

Experimental

Ta_2O_5 was deposited onto (100) n^+-Si or p^+-Si wafers by low-pressure metal-organic chemical vapor deposition (LP-MOCVD), as discussed before [17]-[22]. The precursor used was tantalum ethoxide with the chemical formula of $Ta(OC_2H_5)_5$. As deposited Ta_2O_5 film is amorphous and very leaky. A post-deposition annealing in an oxidizing environment is necessary to lower the leakage current. Then Al dots with a diameter of 1 mm were evaporated through a shadow mask onto the front side of the wafer to form an $Al/Ta_2O_5/n^+$-Si or $Al/Ta_2O_5/p^+$-Si capacitor structure. The film deposited on the backside was removed by chemical etching and then Al was evaporated to form a backside contact.

ZBTSC (zero-bias thermally stimulated current) measurements were performed at a ramp rate of 0.5 K/s as before [17]-[22]. Conventional TSC technique suffers from a serious parasitic current problem because of the need to apply a bias voltage to the sample. The purpose of "zero bias" is to solve this parasitic current problem. The energy level of the defect was estimated using $E_T = 23kT_m$, where T_m is the peak temperature and k the Boltzmann constant. The validity of this equation was checked for Ta_2O_5 by comparison with the initial rise method, as shown in Table II.

Table II. A comparison between the $23kT_m$ method and the initial rise method

Defect level	E_T by the $23kT_m$ method (eV)	E_T by the initial rise method (eV)
B'	0.373	0.332
C'	0.522	0.499

Defect states in tantalum oxide

Various types of defect states have been detected by the ZBTSC technique as shown in Table III. People have been speculating that the oxygen vacancy is an important type of defect in tantalum oxide. As discussed above, the precursor was $Ta(OC_2H_5)_5$ and naturally CVD tantalum oxide will be contaminated by carbon and hydrogen. In addition, as discussed above, the substrate used for deposition was silicon and so it is not too difficult to imagine that tantalum oxide may be contaminated by silicon from the silicon substrate by thermal diffusion [23] or by recombination enhanced diffusion [24]. The author believes that hydrogen contamination can be eliminated relatively easily by heating but carbon contamination is relatively difficult to remove. This can be seen by the SIMS (Secondary Ion Mass Spectrometry) data presented by Shinriki and Nakata in their Fig. 14 in 1991 [25]; annealing in an oxidizing ambient quite frequently only removes the carbon contamination near the surface for a thick tantalum oxide film while hydrogen can be removed relatively easily throughout the whole film. Thus the author believes that tantalum oxide is usually only contaminated by carbon and silicon, which can be expected to substitute for tantalum, resulting in acceptors in tantalum oxide. In the presence of a large quantity of oxygen vacancies, which are deep double donors, it is not too difficult to imagine that carbon and silicon can complex with oxygen vacancies. As shown in Table III, the author believes that the defect states detected by the TSC family of techniques are silicon oxygen vacancy complex, carbon oxygen vacancy complex and the first ionization level of the oxygen vacancy. Please note the defect states discussed above are all donor states. However, silicon oxygen vacancy complex and carbon oxygen vacancy complex are relatively shallow single donors while the oxygen vacancy is a relatively deep double donor.

Table III. Various defect states in Ta_2O_5 identified by our previous ZBTSC study

Defect state	Peak temperature (K)	Estimated energy level (eV)	Remark
A, B, B'	~100,150,200	~0.2, 0.3, 0.4	Tentatively identified as the Si/O vacancy complex shallow single donor in slightly different configurations
C", C', C	~250,270,290	~0.49,0.53,0.57	Tentatively identified as the C/O vacancy complex shallow single donor in slightly different configurations
H	~250	~0.49	C" or C' may be a H_2O related defect H
D	~380	~0.8	Tentatively identified as the first ionization level of the O vacancy deep double donor.

Results and Discussion

As explained above, there is some sort of mystery regarding how to correlate the leakage current of tantalum oxide with defect states detected by TSC. The author would like to point out that once the nature of the defect states is understood this sort of mystery can be solved as follows. Furthermore, as discussed above, a basic shortcoming of the TSC technique (Mechanism B) is that for an insulating film with finite thickness, the physical location of the defect states detected by TSC cannot be distinguished. It is not too difficult to imagine that the correlation between leakage current and defect states detected by TSC is easier when the insulator film is very thin. However, as discussed above, TSC may have a limited range (Mechanism C) and this problem becomes more serious when the insulator film is very thin; this problem is partially solved by using ZBTSC. ZBTSC also suffers from Mechanism B but it suffers less from Mechanism C.

Relatively thick tantalum oxide capacitor on silicon

Fig. 3 shows the ZBTSC spectra of as deposited CVD Ta_2O_5 on p^+-Si. Fig. 4 shows the ZBTSC spectra of Ta_2O_5 film on p^+-Si after annealing in O_2 at 800°C for 1 hour. It is interesting to note that 15 min. UV excitation produces a stronger signal than 30 min. UV excitation. The thickness of the Ta_2O_5 film was 98.6 nm thick.

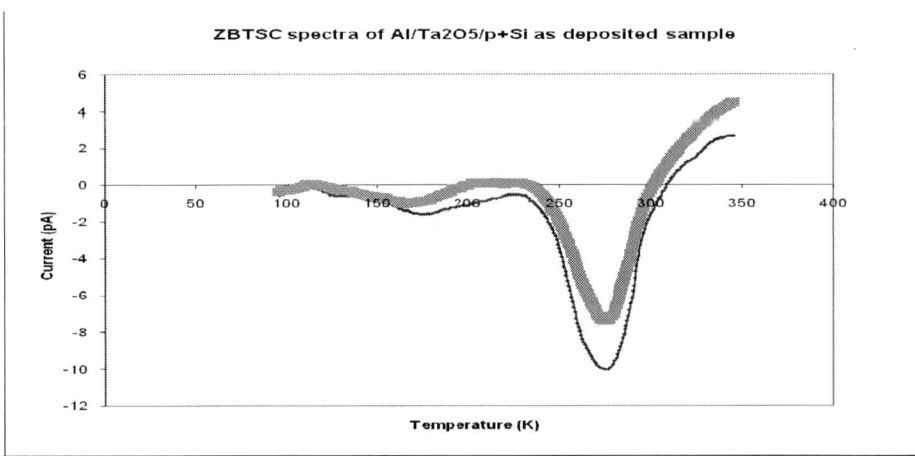

Fig. 3 ZBTSC spectra of as deposited CVD Ta_2O_5 film on p^+-Si with 15 min. UV excitation (thin line) and 30 min. UV excitation (thick line).

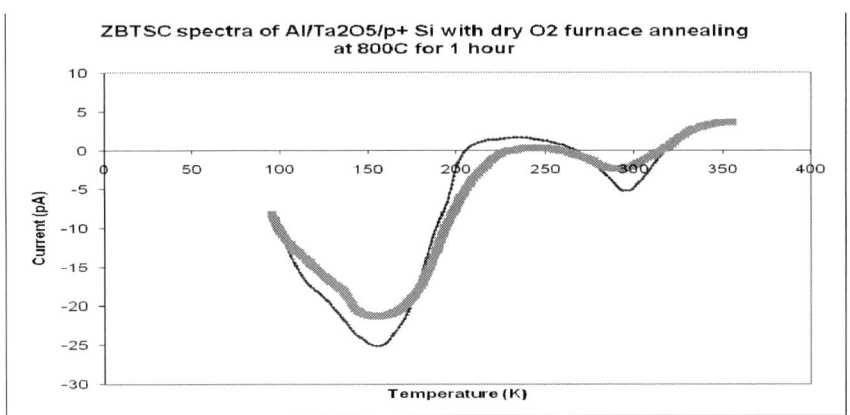

Fig. 4 ZBTSC spectra of CVD Ta_2O_5 on p^+-Si after annealing in O_2 at 800°C for 1 hour with 15 min. UV excitation (thin line) and 30 min. UV excitation thick line.

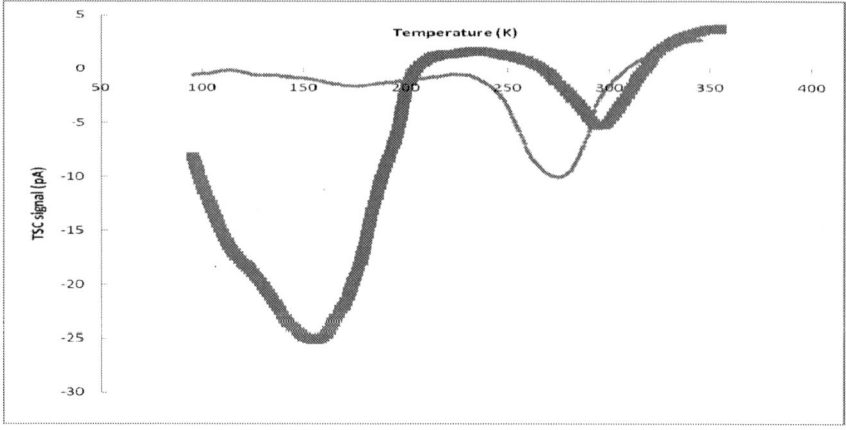

Fig. 5 ZBTSC spectra of CVD Ta_2O_5 film on p^+-Si (a) as deposited (thin line) and (b) after annealing in O_2 at 800°C for 1 hour (thick line) with 15 min. UV. The thickness of the Ta_2O_5 film was 98.6 nm thick.

Fig. 5 shows the comparison of ZBTSC spectra for as deposited sample and annealed sample. By examining Fig. 5 with the help of Table III, it can be easily seen that the carbon oxygen vacancy complex single donor is slightly reduced by annealing but the the silicon oxygen vacancy complex single donor is strongly enhanced by annealing. In terms of leakage current, it is significantly reduced by the annealing, as seen in Fig. 6 and Fig. 7. As explained above, as deposited CVD Ta_2O_5 film on silicon suffers from strong carbon contamination but not much silicon contamination. Annealing in an oxidizing ambient at high temperature results in a reduction of the carbon contamination near the exposed Ta_2O_5 surface and silicon can diffuse from the Ta_2O_5/Si interface into the Ta_2O_5 film. A 2-zone model can be proposed as follows. Zone I is a region near the exposed Ta_2O_5 surface with carbon contamination depleted by annealing in an oxidizing ambient at high temperature. Zone II is a region between Zone I and the Ta_2O_5/Si interface with strong carbon contamination not removed by annealing in an oxidizing ambient at high temperature. For a thick CVD Ta_2O_5 film, Zone I can be much smaller than Zone II. If Zone II is large enough such that most of the silicon contamination coming from the silicon substrate resides in Zone II. It can be easily imagined that Zone II has a lot of carbon oxygen vacancy complex and silicon oxygen vacancy complex single donors such that it is highly leaky whereas Zone I does not have much carbon oxygen vacancy complex and silicon oxygen vacancy complex single donors such that it is highly insulating such that the leakage current is controlled by Zone I instead of by Zone II. However, our experience shows that annealing in an oxidizing ambient at high temperature only slightly reduces the ZBTSC signal from the carbon oxygen vacancy complex but strongly increases the ZBTSC signal from the silicon oxygen vacancy complex while the leakage current is strongly reduced. This may appear difficult to understand without the 2-zone model. The strongly reduced leakage current is actually correlated with the strongly reduced carbon oxygen vacancy complex in Zone I. However, the TSC family technique measures essentially the total quantity of defect states regardless of physical location such that the ZBTSC spectrum only shows a small reduction of carbon oxygen vacancy complex.

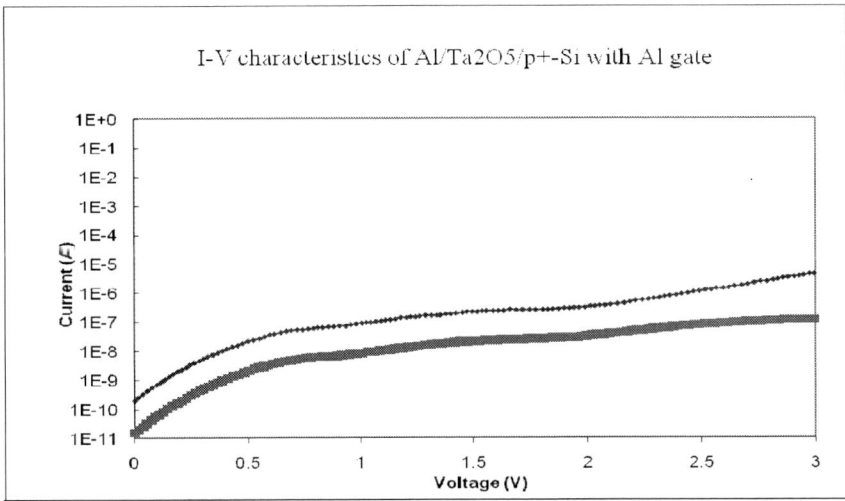

Fig. 6 I-V characteristics of CVD Ta_2O_5 film on p^+-Si (a) as deposited (thin line) and (b) after annealing in O_2 at 800°C for 1 hour (thick line) with 15 min. UV. Al metal was biased positively.

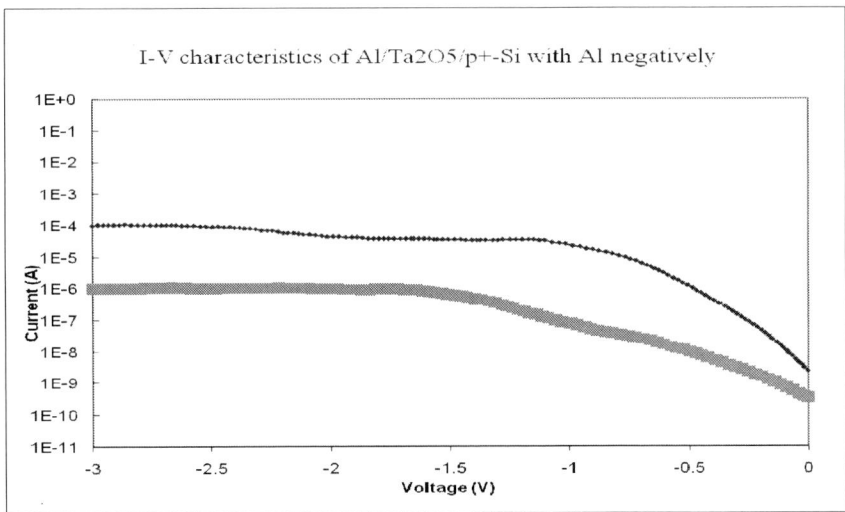

Fig. 7 I-V characteristics of CVD Ta_2O_5 film on p^+-Si (a) as deposited (thin line) and (b) after annealing in O_2 at 800°C for 1 hour (thick line) with 15 min. UV. Al metal was biased negatively.

As discussed above, it can be difficult to distinguish between the Schottky mechanism, which does not involve bulk defect states, or by the Poole-Frenkel mechanism, which involves bulk defect states (Mechanism A). This problem can be solved as explained above by using the I-V characteristics of $M/Ta_2O_5/n^+$-Si capacitors with M positively biased. As discussed above, for $M/Ta_2O_5/p^+$-Si capacitors in the inverted state, they can be considered similar to $M/Ta_2O_5/n^+$-Si capacitors with some voltage drop in the silicon substrate.

<u>Ultrathin tantalum oxide capacitor on silicon</u>

It is not too difficult to imagine that an ultrathin Ta_2O_5 film is quite different from a thick Ta_2O_5 film. Experimental observation shows that carbon contamination can be relatively easily suppressed throughout an ultrathin Ta_2O_5 film by a short annealing in an oxidizing ambient at high temperature [26] such that carbon oxygen vacancy complex simply disappears and the 2-zone model discussed above is not necessary. As shown in Fig. 2, there are only two kinds of defect states left: silicon oxygen vacancy complex and simple oxygen vacancies. Thus interpretation of experimental data can become much more straightforward. As shown in Fig. 8, the leakage current is obviously reduced because of RTN (rapid thermal nitridation) of silicon substrate before deposition. This can be correlated with the suppression of the silicon oxygen vacancy complex seen in ZBTSC spectra. RTN helps to prevent silicon diffusing into the tantalum oxide film because silicon nitride is a diffusion barrier and so suppress oxygen vacancy formation by the reaction between silicon and tantalum oxide.

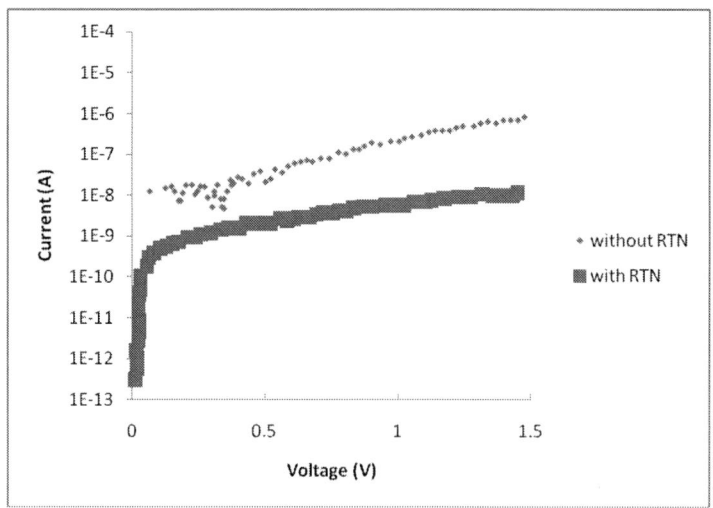

(a)

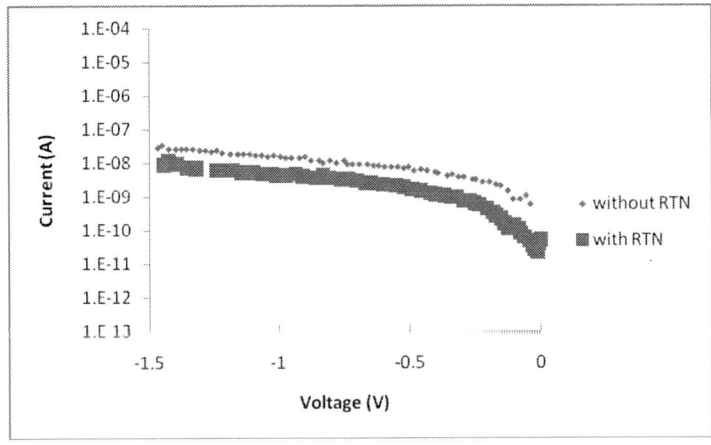

(b)

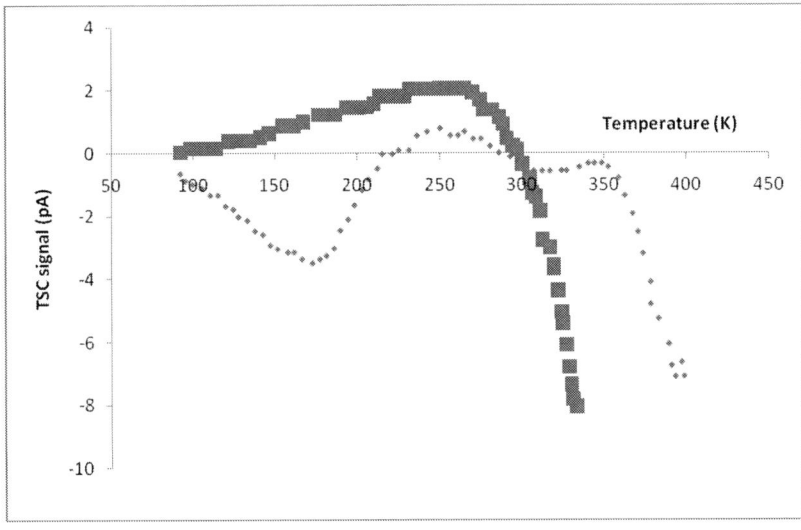

(c)

Fig. 8 I-V characteristics of Al/Ta$_2$O$_5$/n$^+$-Si capacitors with O$_2$ RTA at 800°C for 30 s with RTN (thick line) and without RTN (thin line). The thickness of Ta$_2$O$_5$ film was 8.4 nm. Diameter of metal contact = 1 mm. **a)** Al was biased positively **b)** Al was biased negatively. **c)** ZBTSC spectra of Al/Ta$_2$O$_5$/n$^+$-Si capacitors with O$_2$ RTA at 800°C for 30 s with RTN (thick line) and without RTN (thin line). The thickness of Ta$_2$O$_5$ film was 8.4 nm. RTN = rapid thermal nitridation.

Conclusion

Four mechanisms regarding why it is difficult to correlate the leakage current of high-k dielectric capacitor structures with defect states detected by the TSC family of techniques have been discussed. Methods to handle the three types of problems have been discussed above. Mechanism A is that it is possible that only part of the I-V characteristics is related to the presence of bulk defect states. Therefore it is necessary to identify which portion of the I-V characteristics is sensitive to the variation of bulk defect states. Mechanism B is that the defect states may not have a uniform distribution in the high-k dielectric. This problem can only be handles on a case-to-case basis. There is no general outline for solving this problem. Mechanism C is that the TSC technique may have a limited range because of the presence of a parasitic current. This problem can be solved by migrating to ZBTSC or even ZTGZBTSC. Mechanism D is that it may be difficult to fill some defect states at low temperature. This problem can be handled by a two-scan TSC technique: one scan is used for defect states with this problem while the other scan is used for defect states without this problem. In conclusion, it is possible to make a successful correlation between the leakage current of high-k dielectric capacitor structures with defect states detected by the TSC family of techniques even though it is difficult in general.

Acknowledgments

The author would like to acknowledge the help of his previous graduate students and colleagues. In addition, he would like to acknowledge the support of Dr. Taejoon Han and Mr. Neal Sandler.

References

1. Y. Nishioka, private communication.
2. G.B. Alers, R.M. Fleming, Y.H. Wong, B. Dennis, A. Pinczuk, G. Redinbo, R. Urdahl, E. Ong and Z. Gasan, *Appl. Phys. Lett.*, **72**, 1308 (1998).
3. C.Z. Zhao, M.B. Zahid, J.F. Zhang, G. Groeseneken, R. Degraeve, and S. De Gendt, *Microelectronic Engineering*, 80, 366 (2005).
4. C.Z. Zhao, J.F. Zhang, M.H. Chang, A.R. Peaker, S. Hall, G. Groeseneken, L. Pantisano, S. De Gendt and M. Heyns, *IEEE Trans. Electron Dev.*, 55, 1647 (2008).
5. W.D. Zhang, B. Govoreanu, X.F. Zheng, D. Ruiz Aguado, M. Rosmeulen, P. Blomme, J.F. Zhang and J. Van Houdt, *IEEE Electron Dev. Lett.*, 29, 1043 (2008).
6. H.H. Poole, *Philosophical Magazine*, **32**, 112 (1916).
7. J. Frenkel, *Phys. Rev.*, **54**, 647 (1938).
8. J.R. Yeargan and H.R. Taylor, *J. Appl. Phys.*, **39**, 5600 (1968).
9. D.S. Jeong, H.B. Park and C.S. Hwang, *Appl. Phys. Lett.*, **86**, 072903 (2005).
10. B.C. Lai and J. Y. Lee, *J. Electrochem. Soc.*, **146**, 266 (1999).
11. G.L. Pearson and J. Bardeen, *Phys. Rev.*, **75**, 865 (1949).
12. J. Robertson, *J. Vac. Sci. Technol. B*, **18**, 1785 (2000).
13. S. Miyazaki, *J. Vac. Sci. Technol. B*, **19**, 2212 (2001).

14. M. Perego, G. Seguini, G. Scarel, M. Fanciulli and F. Wallrapp, *J. Appl. Phys.*, **103**, 043509 (2008).
15. H. Matsuhashi and S. Nishikawa, *Jpn. J. Appl. Phys.*, **33**, 1293 (1994).
16. S.C. Sun and T.F. Chen, *Jpn. J. Appl. Phys.*, **36**, 1346 (1997).
17. W.S. Lau, T.S. Tan, N.P. Sandler and B.S. Page, *Jpn. J. Appl. Phys.*, **34**, 757 (1995).
18. W.S. Lau, L. Zhong, A. Lee, C.H. See, T. Han, N.P. Sandler and T.C. Chong, *Appl. Phys. Lett.*, **71**, 500 (1997).
19. W.S. Lau, T. Han, G.Y. Zhang, P.W. Qian, L.L. Leong, S.T. Che and P. Wong, *ECS Transactions*, **1(5)**, 577 (2006).
20. W.S. Lau, K.F. Wong, T. Han and N.P. Sandler, *Appl. Phys. Lett.*, **88**, 172906 (2006).
21. W.S. Lau, *Appl. Phys. Lett.*, **90**, 22904 (2007).
22. K.A. McKinley and N.P. Sandler, *Thin Solid Films*, **290-291**, 440 (1996).
23. W.S. Lau, K.K. Khaw, P.W. Qian, N.P. Sandler and P.K. Chu, *J. Appl. Phys.*, **79**, 8841 (1996).
24. W.S. Lau, M.T.C. Perera and P.K. Chu, *Appl. Phys. Lett.*, **96**, 083501 (2010).
25. H. Shinriki and M. Nakata, *IEEE Trans. Electron Dev.*, **38**, 455 (1991).
26. W.S. Lau, G. Zhang, L.L. Leong, P.W. Qian, T. Han, J. Das, N.P. Sandler and P.K. Chu, *MRS Symp. Proc.*, **864** , E4.22.1 (2005).

888

ECS Transactions, 35 (4) 889-900 (2011)
10.1149/1.3572326 ©The Electrochemical Society

The Degradation of MILC P-Channel Poly-Si TFTs under Dynamic Hot-Carrier Stress Using a Novel Test Structure

Cheng-I Lin, Wen-Chiang Hong,Tin-Fu Lin, Horng-Chih Lin* and Tiao-Yuan Huang
Department of Electronics Engineering and Institute of Electronics, National Chiao Tung University

1001 Ta-Hsueh Road, Hsinchu, Taiwan 300, ROC
*Phone:+886-35712121 ext. 54193, Fax:+886-3-5724361, E-mail: hclin@faculty.nctu.edu.tw

In this study, dynamic hot carrier effect in the MILC p-channel TFT device has been characterized by the unique struture. This novel structure is capable of spatially resolving the hot carrier effect and is highly sensitive to detect the defect-rich region. The dynamic hot carrier stress has been focused on the impacts of the frequency, the rise time and the fall time. In varied frequency stress condition, the degradation in the drain-sided monitor transistor (DMT) increases monotonically with increasing frequency, infering that more defects are generated by extra dynamic stress contribution in the drain side and degrade the characteristic of device. Under varied fall time stress condition, the on-current degradtion is severe with decreasing fall time due to the extra voltage drop during voltage switch. The final part is effect of rise time. While device switches, the large voltage drop exists in the junction between the channel and the drain, which resulted in anothor hot carrier degradation.

Introduction

In the past several decades, the amorphous silicon thin film transistor (TFT) was the main building block for manufacturing active-matrix liquid crystal display (AMLCD). However, the lower field-effect mobility prohibits the integration of amorphous silicon TFTs in the peripheral driver circuit. In this regard, the high-mobility polycrystalline silicon (poly-Si) TFTs prepared by excimer laser anneal (ELA) and metal-induced lateral crystallization (MILC) are desirable for integration of system on the panel to enhance device and circuit performance [1]. For MILC, low cost and low process temperature are its unique advantages for practical production [2]. While the mobility has improved with the advances in crystallization method, the reliability issues become one of the major concerns during device operation[3]. Recnetly, more studies has focused on the reliability of the n-channel poly-Si TFTs. In this work, we employed a novel test structure previously proposed by our group [4][5] to analyze the hot-carrier degradation in the MILC p-channel poly-Si TFTs.

Experimental

First, a 100 nm thermal oxide was grown on the Si wafer. Next, a 100 nm amorphous silicon layer was deposited by LPCVD at 550°C, followed by the deposition of a 100 nm PECVD oxide as the blocking layer. After formation of the seeding windows,

889

a thin 5 nm Ni layer was deposited by E-gun. And then, the amorphous silicon layer was crystallized into polycrystalline silicon by annealing in a furnace at 540°C for 24 hours in N_2 ambient. Next, an isolation region was patterned after removing the blocking oxide and the Ni layer. Afterwards, a 30 nm-thick silicon oxide layer was deposited by PECVD, followed by the deposition of 150 nm-thick poly-Si layer for gate material. Then, a self-aligned BF_2^+ implantation was performed with a dose of 5×10^{15} /cm^2, and the dopant was activated at 600°C for 12 hours in N_2 ambient by furnace. Next, a PECVD silicon oxide layer of 200 nm thickness was deposited which served as the passivation layer to prevent the penetration of humidity and impurities. After opening the contact hole in the passivation layer, metallization was performed, followed by a sintering treatment at 400°C for 30 minutes in H_2 ambient.

Fig.1 shows the structure of the tester, consisting of one lateral transistor, denoted as TT, and three monitor transistors, denoted as SMT, CMT, and DMT. The dynamic stress is exerted on the TT, and the hot carrier effect is analyzed by the monitor transistor.

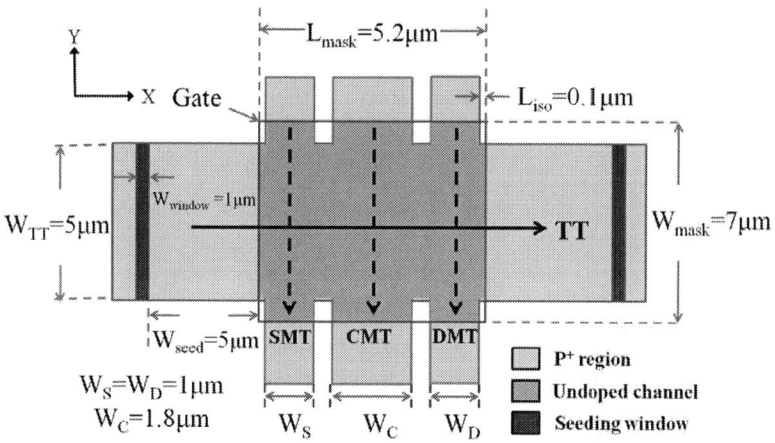

Fig.1 Top view of the test structure

Fig.2 shows the waveform of the AC signal applied during stress. The standard AC stress condition used in the experiment is set with V_G from -7.5 V (V_{G_low}) to 0 V (V_{G_high}) and set with V_D at -20 V for 1000s. The frequency is set at 500 kHz with fall time (ft) and rise time (rt) is equal to 100 ns, and the duty ratio is 50%. Fall time is the time that voltage signal falls from 90% to 10% of the amplitude ($V_{G\_high} - V_{G\_low}$), and vice versa for rise time. Subsequently, the information that should be required in the above parameters is varied to study their impact on the device degradation. The frequency is varied from 100 kHz to 1MHz, and both rise time and fall time are varied from 100 ns to 10 ns.

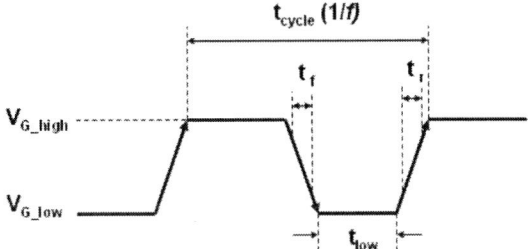

Fig.2 The waveform of AC stress signal

Results and Discussion

<u>Frequency</u>
Fig. 3(a) shows electrical characteristics of the TT before and after AC stress. The stress condition is V_G = -7.5 V and V_D = -20 V with rise time and fall time equal to 100ns at 100kHz for 1000s. With such stress condition, less hot-carrier induced degradation is observed by the TT, because the grain size in the MILC device is large, implying less grain boundary in the channel. Once the device is exerted hot carrier stress, less defects generate in the grain boundary. The monitor transistors embedded in the same test structure can be employed to help address the issue, and the results observed by CMT and DMT are shown in Fig. 3(b) and (c), respectively. The typical cahracteristic of SMT(not shown) is the same as that of CMT, which exhibits negligible degradation in the typical characteristic curves after AC stress. Apparently, Fig. 3(c) shows significant shift in subthreshold characteristics of DMT. Device degradations in terms of on-current degradation and threshold voltage shift are observed obviously after AC stress. This indicates that defects, like interface states and traps in the grain-boundaries are generated and form a defect-rich region. Moreover, electrons are trapped in the gate oxide near the drain side of TT after the AC stress.

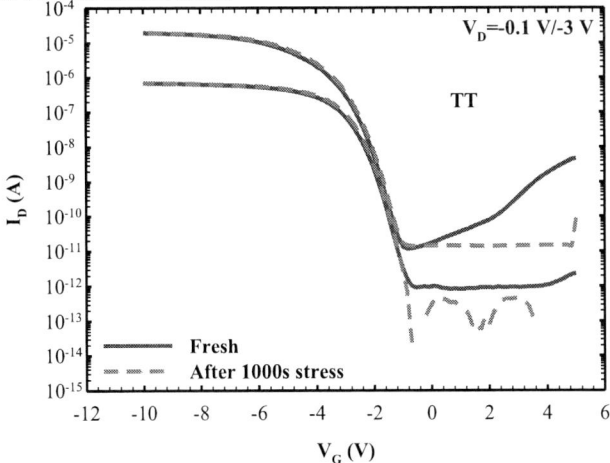

Fig.3 (a) The electrical characteristics of the TT before and after AC stress with V_G = -7.5 V and V_D = -20 V at 100kHz for 1000s.

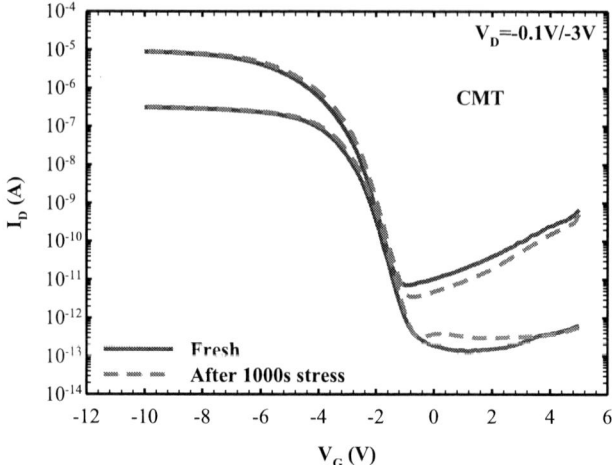

Fig.3 (b) The electrical characteristics of the CMT before and after AC stress with V_G = -7.5 V and V_D = -20 V at 100kHz for 1000s.

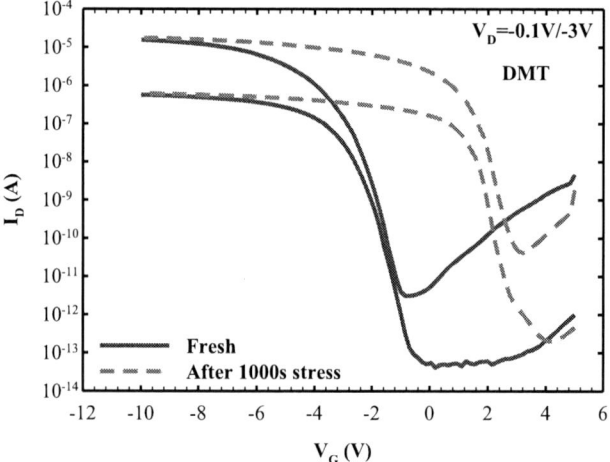

Fig.3 (c) The electrical characteristics of the DMT before and after AC stress with V_G = -7.5 V and V_D = -20 V at 100kHz for 1000s.

Fig. 4 and Fig. 5 show the on-current degradation and threshold voltage shift of TT and MTs after 1000 sec AC stress as a function of frequency. The results indicate that the damage induced in SMT and CMT is negligible and almost independent of the frequency. The on-current degradation is significant for the DMT, and it increases with increasing frequency, inferring that additional damage is caused near the drain side as the frequency increases. This inference is the same as the trend verified by the unique structure as shown in Fig. 5, where the DMT shows significant shift. These results clearly demonstrate that the MTs of test structure can definitely resolve the non-uniform damage location and their excellent sensitivity for detecting the frequency-dependent degradation.

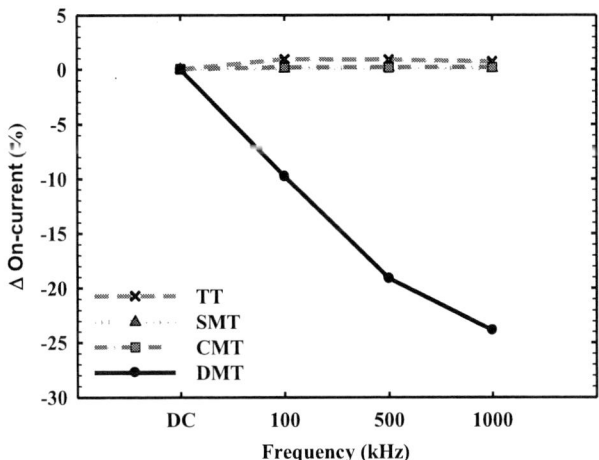

Fig.4 The on-current degradation of TT and MT after 1000 sec AC stress as a function of frequency

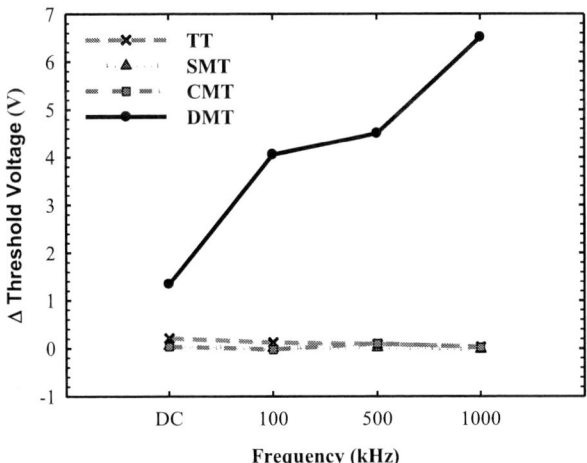

Fig.5 The threshold voltage shift of TT and MT after 1000 sec AC stress as a function of frequency

Rise Time

In this section, we investigate the effect of rise time. The AC stress condition is the same as the standard one except the rise time. Fig. 6 and Fig. 7 show subthreshold characteristics of DMT before and after AC stress with the rise time of 100 ns and 10 ns, respectively. The results show that the device performance degrades more under a faster rise time of AC stress. Fig. 8 and Fig. 9 show the on-current degradation and threshold voltage shift, respectively, under AC stress with fall time = 100 ns but with varying rise time. Still, the DMT exhibits the most serious degradation. Moreover, the damage becomes even more severe as the rise time is shortened.

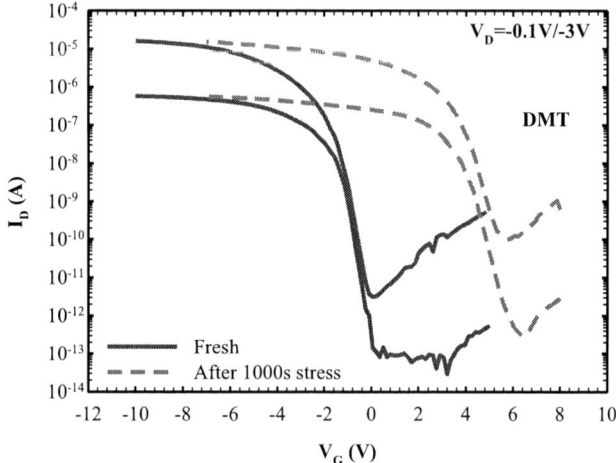

Fig.6 The electrical characteristics of the DMT before and after AC stress with the rise time of 10 ns

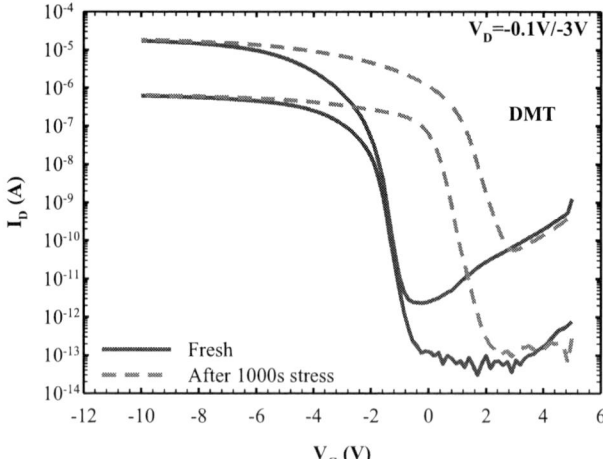

Fig.7 The electrical characteristics of the DMT before and after AC stress with the rise time of 100 ns

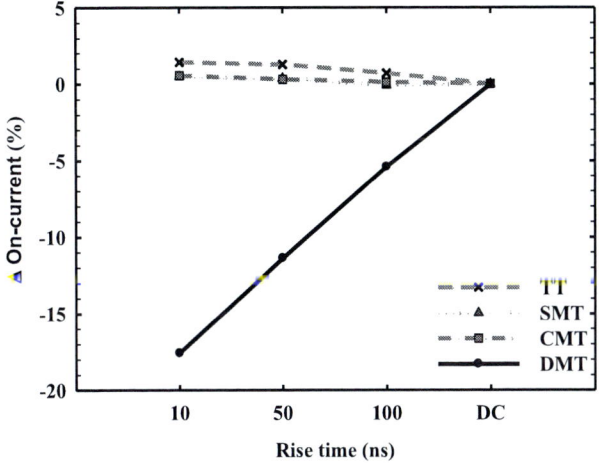

Fig.8 The on-current degradation of TT and MT after 1000 sec AC stress as a function of rise time

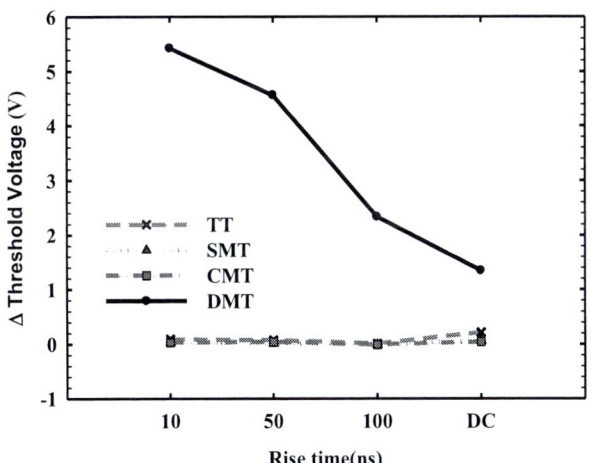

Fig.9 The threshold voltage shift of TT and MT after 1000 sec AC stress as a function of rise time

The degradation mechanism in the transient stage from VG_low to VG_high under AC stress is illustrated in Fig. 10[6]. As shown in the illustration, a high absolute pulse voltage is applied to the gate while the source is grounded and a high absolute bias is applied to the drain. For gate pulse at VG_low of -7.5 V, a sheet of holes is induced in the channel and the defects are attributed to the impact ionization occurring near the drain, as shown in Fig. 10(a). Defect generation results in on-current reduction. Some portions of electrons induced by impact ionization may trap in the gate oxide and resulted in shift of threshold voltage. During the transient period (VG_low rise to VG_high), the inversion holes remained in the channel are mainly attracted by the negative drain bias and accelerated toward the drain, resulting in additional damages and more electrons trapped in the gate oxide, as shown in Fig. 10(b). In the case of a slow rise time, most of the holes have enough time to relax through collisions. Thus, the hot-carrier issue is also relaxed. On the other hand, in the case of a fast rise time, the voltage drop across the drain junction is increased by ΔE (Fig. 10) in a short time, and more hot holes are expected to be generated, causing more damage in the regions of channel near the drain of the TT. For this reason, the on-current degradation of DMT, and the threshold voltage shift of DMT are related to rise time during AC stress.

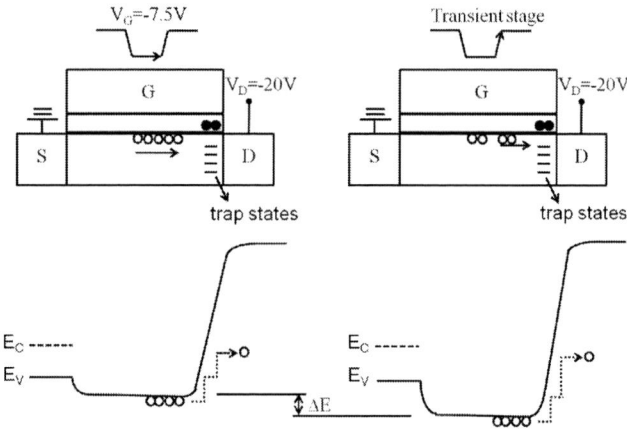

Fig.10 Degradation mechanism of variable rise time under AC stress

Fall Time

Here, the effect of fall time is discussed. The AC stress condition is the same as the standard one except the fall time. Fig. 11 and Fig. 12 show typical characteristics of DMT before and after AC stress with fall time of 100 ns and 10 ns, respectively. The results indicate that the degradation of device under the AC stress is higher as the fall time becomes shorter. Figure 13 and Fig. 14 show the on-current degradation and threshold voltage shift, respectively, under AC stress with fixed rise time = 100 ns as function of fall time. Comparing the on-current degradation and threshold voltage shift of the three MTs under AC stress, only those of the DMT increase dramatically with decreasing fall time, indicating that additional defect generation and electron trapping in gate oxide occur as the fall time is shortened.

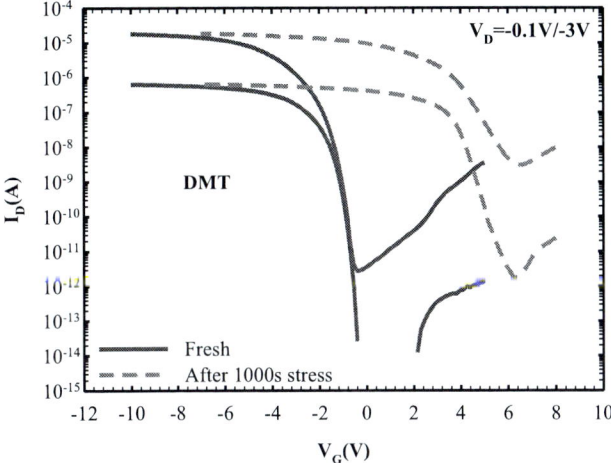

Fig.11 The electrical characteristics of the DMT before and after AC stress with the fall time of 10 ns

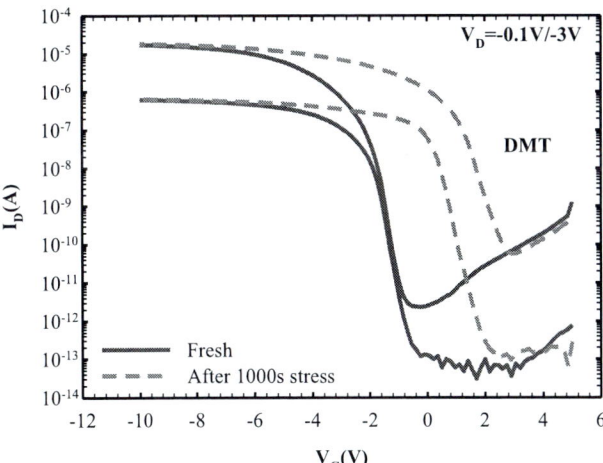

Fig.12 The electrical characteristics of the DMT before and after AC stress with the fall time of 100 ns

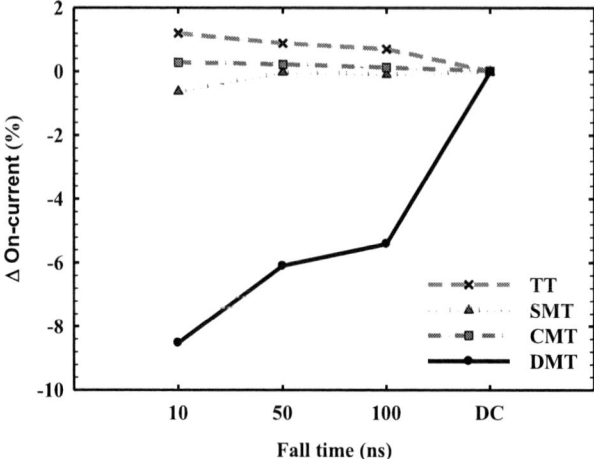

Fig.13 The threshold voltage shift of TT and MT after 1000 sec AC stress as a function of fall time

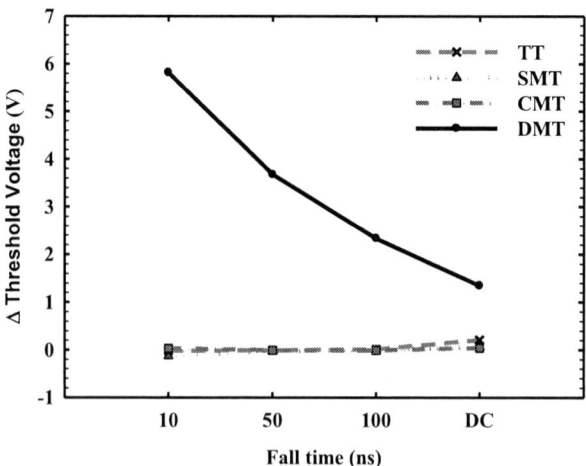

Fig.14 The threshold voltage shift of TT and MT after 1000 sec AC stress as a function of fall time

Fig. 15 shows a proposed scenario for explaining the effect of fall time under AC stress. When a VG_high = 0 V is applied to the gate, high gate-induced drain leakage by band to band tunneling (BTBT) dominates the conduction due to the high voltage difference between the gate and the drain, as shown in Fig. 15(a). Under the situation, the channel field is more or less uniform, and the electrons appear at the tunneling junction (i.e., drain junction of the TT) would drift toward the source by the field. When the gate voltage is switched from VG_high to VG_low, a high voltage drop is developed at the channel/drain junction in a short time due to the formation of the inversion hole layer, as shown in Fig. 15(b). Portions of the electrons coming from the gate-induced-drain-leakage (GIDL) current remained at the original tunneling junction would be accelerated by the suddenly presenting field. This leads to the generation of hot electrons, which create defects in the channel and trapped into the oxide (near the drain of the TT). Such phenomenon becomes more significant as the fall time is reduced. This explains why the on-current degradation of DMT shown in Fig. 13, and threshold voltage shift of DMT shown in Fig. 14.

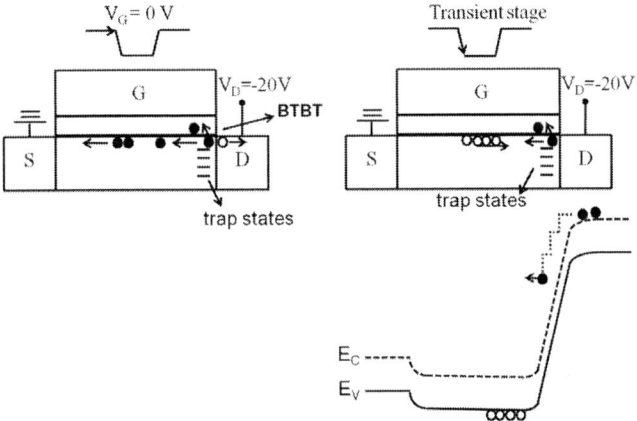

Fig.15 Degradation mechanism of variable fall time under AC stress

Conclusions

A new approach for characterizing and resolving the hot-carrier degradation in thin-film transistors is proposed and sucessfully demonstrated using a novel test structure. With AC stress, we can identify the relationship between the frequency, the fall time, and rise time for the AC hot carriers. In this work, the major damage induced near the drain side of the TT and the trapping of electrons in the gate oxide can be accutely resolved by means of the DMT embedded in the test structure. The new test structure is highly sensitive for detecting degradations in the device.

Acknowledgment

The authors would like to thank the National Nano Device Laboratories (NDL) for assistance in device fabrication. This work was supported in part by the National Science Council under contract No. NSC99-2221-E-009-172

References

[1]Z. Meng, M. Wang, M. Wong, *IEEE Trans. Electron Devices,* **47**, 404 (2000).
[2]S.W. Lee, S. K. Joo, *IEEE Electron Device lett.,* **17**, 162 (1996).
[3]Y. Toyota, M. Matsumura, M. Hatano, Takeo Shiba, Makoto Ohkura, *IEEE Trans. Electron Devices,* **57**, 429 (2010).
[4]H.C. Lin, M. H. Lee, K. H. Chang, *IEEE Electron Device lett.* **27**, 561 (2006).
[5]M.H. Lee, K. H. Chang, H.C. Lin, *J. Appl. Phys.,***101**, 54518 (2007).
[6]Y. Uraoka, T. Hatayama, T. Fuyuki, T. Kawamura, Y. Tsuchihashi, *IEEE Trans. Electron Devices,* **48**, 2370 (2001).

ECS Transactions, 35 (4) 901-908 (2011)
10.1149/1.3572327 ©The Electrochemical Society

Solution Processed High-*k* Lanthanide Oxides for Low Voltage Driven Transparent Oxide Semiconductor Thin Film Transistors

S. Choi [a], B-Y. Park [a], M. Jang [a,b], S. Jeong [a], J-Y. Lee [a], B-H. Ryu [a], T.-Y. Seong [b], and H-K. Jung [a]

[a] Advanced Materials Division, Korea Research Institute of Chemical Technology, P.O.Box 107, Daejeon 305-600, Korea.
Email: shochoi@krict.re.kr

[b] Department of Materials Science and Engineering, Korea University, Anam-dong, Seoul 136-701, Korea.

ZnO-based low power operating thin film transistors (TFTs) were fabricated by a simple and robust solution process. Combined the amorphous oxide semiconductors with the high capacitance lanthanide oxides thin film, Y_2O_3 and Gd_2O_3, as a gate insulator, the resultant device exhibits an enhanced device performance; lower threshold voltage, increased carrier mobility, and smaller subthreshold slope with exceptionally low operating voltage than the typical TFT devices using a SiO_2 gate insulator.

Introduction

Thin film transistor (TFT) gate dielectric must be smooth, dense free from pinholes and charge-trapping defects, and exhibit low-leakage current under applied bias. Certainly the gate dielectric with a high dielectric constant (high-*k*) is needed for the low operating voltage and high carrier mobility semiconductor TFTs applicable to the display device backplanes [1,2]. Reducing the threshold voltage and the subthreshold swing is essential for operating TFTs at low-voltage levels. To ensure the insulating properties as well as those enhanced device performances, dielectric layer with high-*k* materials should be considered. Furthermore, developing an operational solution-based procedure is beneficial to device fabrications since the solution process is an alternative thin film deposition method which is a simple and cost effective and enables large area coating and high throughput. [3,4].

Recently the lanthanide oxides have attracted much attention to replace SiO_2 as the gate dielectrics [5,6]. They offer many advantages such as high capacitance with low leakage current, and good thermodynamic stability with respect to various substrates.

In the present work, we have investigated the microstructure and dielectric properties. Dielectric constant, leakage current and breakdown voltage characteristics of a solution-processed lanthanide oxides, Y_2O_3 and Gd_2O_3 thin films, were obtained. Moreover the switching property of a bottom-gate structured TFTs combined with the amorphous ZnO-based active layer was also examined to confirm the expected performance of high-*k* gate insulators.. Recently, amorphous oxide semiconductors have attracted a great deal of attention as the active layer in transparent TFTs. In particular, Zn-Sn-O (ZTO) thin films have been used as a channel layer with good carrier mobility [7].

Experimental

Both the semiconductor and insulator precursor solutions were prepared by dissolving metal salts in appropriate solvent. Precursor solution for the insulator layer (P1) were synthesized using gadolinium nitrate hexahydrate ($Gd(NO_3)_3 \cdot 6H_2O$, 99.99 %) and polyvinylpyrrolidone (($C_6H_9NO)_n$, PVP) dissolved in a mixture of 2-methoxyethanol ($C_3H_8O_2$, 99 %) and distilled water. The volume ratio of the two solvents was 4:1. The concentration of Gd ion was 0.5M and same to the PVP addition. The as-prepared precursor was aged for 24 h at room temperature prior to a spin coating.

The precursor solution for semiconductor (P2) was prepared using zinc chloride ($ZnCl_2$, 99.99 %) and tin chloride dihydrate ($SnCl_2 \cdot 2H_2O$, 99.99 %). Stoichiometric zinc and tin chlorides were dissolved in 2-methoxyethanol. The overall metal cation concentration was 0.2M and the composition was Zn:Sn = 12:44. Ethanolamine (C_2H_7NO, 99 %) was added to stabilize a precursor solution with the concentration of 1.2 M. The as-prepared precursor solution was stirred for 3 h in air prior to spin coating.

For Metal-Insulator-Metal structure (MIM), P1 was filtered through a 0.2 μm syringe filter and then dropped onto the pre-cleaned ITO/glass substrate followed by spin-coating at 4500 rpm for 30 s. After spin casting, the film was dried at 180 °C for 3 min on the hot plate. The overall procedure was repeated several times to achieve the desired film thickness. Finally, the film was then annealed in air at various annealing temperature in the range of 350-500 °C. The dot-patterned Al upper electrodes were evaporated on top of the resultant film layer through a shadow mask.

Fabrication of TFT is similar to that of the MIM samples. The precursor solution, P2, was filtered using a syringe filter and spin coated (at 3500 rpm for 30 sec) on the as-prepared Gd_2O_3/ITO/glass substrate, followed by annealing at 400 °C in air. For top-contact structured TFTs, Al source/drain electrodes were thermally evaporated on semiconductor layer through a shadow mask. The channel length and width were 100 and 1000 μm, respectively. To clarify the role of high-k Gd_2O_3 gate insulator, the transistor based on a low-k SiO_2 gate insulator was fabricated by depositing a ZTO precursor solution on heavily-doped silicon wafer with 100 nm-thick SiO_2 (DASOM RMS Inc.). The Al source/drain electrode architecture was identical to that of transistor based on a Gd_2O_3 gate insulator.

Transmission electron microscopy (TEM) and thermogravimetric (TG) analysis were conducted to verify the phase formation. Film transmittance in the visible wavelengh region was measured with the Shimadzu UV-2501PC UV-VIS recording spectrophotometer. Capacitance-frequency behaviour was measured using an Agilent Technologies 4294A semiconductor parameter analyzer. Also, the I-V characteristics of the transistors and insulator were measured in air using an HP 4145B semiconductor parameter analyzer to assess the electrical properties.

Results and Discussion

TG analysis for prapared lanthanides precursor solution showed that, when heated in air its thermal decomposition was completed around 600 °C for a Y_2O_3 film and 325 °C for a Gd_2O_3 film, respectively (Figure 1). These results suggest that a relatively low annealing temperature (~350 °C) would be enough to form lanthanide compounds using the corresponding solution precursors, although the formation of Y_2O_3 is not sufficient. We have used various water soluble lanthnides salts, like acetate, sulfate, and chloride, but only the nitrate-based precursor solution is suitable for growing smooth and uniform thin films without any particulate residues during spin coating.

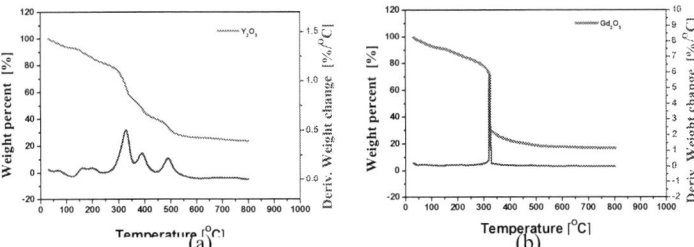

Fig. 1. TG analysis of the Y_2O_3 ((a)) and Gd_2O_3 ((b)) precursor solutions.

The resultant films are semi-crystalline, homogeneous, smooth and highly transparent under visible wavelength (400~700 nm). Such a good transparency is also confirmed by the readability of some characters through the film sample in the inset image in Fig. 2 (a). The TEM images of the solution processed Gd_2O_3 and Y_2O_3 film annealed at 400 °C are presented in the Fig. 2 (b) and (c). The film has smooth and dense microstructure. It can be seen that the overall film thickness is about 100 nm within our synthetic conditions. The electron diffraction patterns is represented on inset image. The rings correspond to the lattice spacings of the cubic phase Gd_2O_3 (JCPDS #12-0797) and Y_2O_3 (JCPDS #5-0574). Interestingly, we can realize that the formation of Y_2O_3 phase is enough even at lower annealing temperature, 400 °C, than that of the thermal decomposition as expected from the TG analysis.

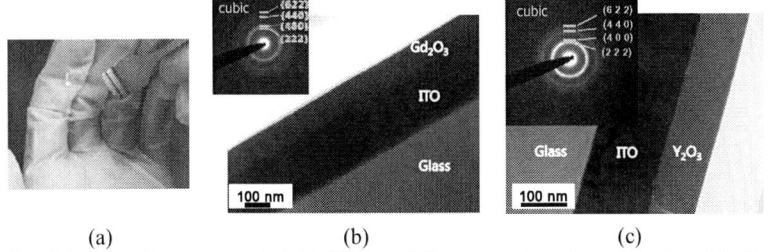

(a)   (b)   (c)

Fig. 2. (a) Image of transparent Gd_2O_3 film on ITO-coated glass. Cross-sectional TEM image and corresponding selected area diffraction pattern (inset) of Gd_2O_3 ((b)) and Y_2O_3 ((c)) films.

Initially, MIM capacitor structures were evaluated to analyze the relative dielectric constant, ε_r. Analysis of the behavior of the small-signal ac voltage capacitance measured between the two output electrodes provides valuable insight into the electrical behavior of the MIM system.

Frequency variable dielectric constant (ε_r-f) characteristic were carried out on MIM with as-prepared Y_2O_3 and Gd_2O_3 films annealed at 350 and 500 °C. The calculated dielectric constant of Gd_2O_3 (Y_2O_3) was in the range of 9-15 (8-15), while that of the typical silicon oxide film is 4, and this value was comparable to the previously reported lanthanide thin films grown by the e-beam evaporations (ε_r=14-18) [8]. Increase in dielectric constant with annealing temperature can be understood in view of the proportionally increased volume fraction of the phase pure Y_2O_3 and Gd_2O_3. It was confirmed by the X-ray diffraction results as shown in our previous work [9]. A detectable crystalline phase forms above 400 °C, which is the temperature interval over which nitrate begins to decompose. Compared to the thermally grown SiO_2 film, the capacitance of sol-gel prepared Gd_2O_3 film gradually decreases with frequency as expected owing to the remnant organic residues such as carbonyl groups induced by the PVP, which limit the polarization response time.

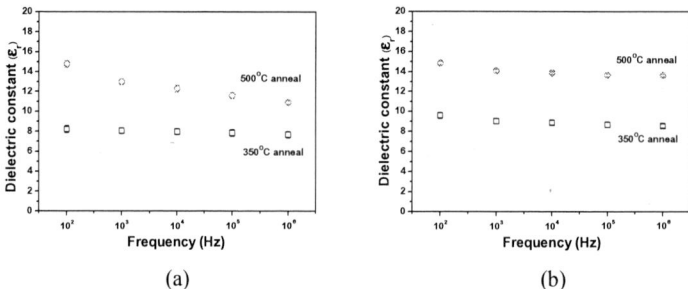

(a)   (b)

Fig. 3. Calculated dielectric constant of Y_2O_3 ((a)) and Gd_2O_3 ((b)) with varying the applied frequency. Annealing temperatures were also denoted.

In order to validate the usefulness of high-k lanthanides film as a gate insulator for thin-film transistor, the devices with top-contact electrode architecture were fabricated. The solution-processable ZTO precursor solution was spin-coated on lanthanide/ITO/glass stack, followed by annealing. The annealing temperature for ZTO layer was from 350 to 400 °C, which depended on the undercoated lanthanide film layer. In addition, for comparison, the transistor with identical device structure was fabricated using a 100 nm-thick SiO_2 as a gate insulator.

The transfer characteristics of transistors are shown in Figure 4; The electrical performance parameters of devices are summarized in Table 1. Interestingly, by replacing a low-k SiO_2 with a high-k lanthanide film, the operation voltage was drastically decreased to ≤ 10 V and the field-effect mobility was significantly improved by a factor of 3.75. For transistors based on a Gd_2O_3 gate insulator annealed at 500 °C, the mobility was 2.5 cm²V⁻¹s⁻¹, whereas the mobility was 0.8 cm²V⁻¹s⁻¹ for transistor comprising a

SiO$_2$ gate insulator. In devices based on the high-k gate insulators, the enhancement of field-effect mobility is attributed to a high capacitance of gate insulator. The mobility of oxide semiconductor film depends on the carrier concentration, which is determined by capacitance of gate dielectric at a given gate voltage, since the carrier transport is governed by percolation conduction over trap states and is enhanced at high carrier concentrations by filling the trap states [10,11]. The improvement of mobility with increasing annealing temperature from 350 to 500 °C is also associated with the increased capacitance.

Hydroxyl groups act as a trap site for electrons [12,13], a charge carrier for n-type semiconductor, which deteriorates the subthreshold characteristic at the interface between semiconductor and gate insulator. As per dielectric results (Fig. 3), hydroxyl groups are present inside and at the surface of lanthanides film at 350 °C and might be completely vanished above 500 °C. To ensure these results, the corresponding Fourier Transform Infrared (FTIR) spectra of Gd$_2$O$_3$ films prepared with different annealing temperatures were investigated. The broad absorption band around 3440 cm^{-1} due to residual hydroxyl groups gradually decreased with annealing temperature. The band around 545 cm^{-1} is assigned to the Gd-O vibration of cubic Gd$_2$O$_3$. The existence of a characteristic Gd-O bond with diminished –OH species annealed above 350 °C confirmed that the precursor was fully converted to Gd$_2$O$_3$ phase, as per the TG results. (data not shown in here) Thus, for TFT based on high-k lanthanides gate insulator annealed at 500 °C, the improvement of subthreshold characteristic arises from the high capacitance and hydroxyl-free interface.

Optimized Y$_2$O$_3$ and Gd$_2$O$_3$ films were successfully integrated as gate dielectrics in ZTO TFTs verifying the exceptional performance such as enhanced field-effect mobility and subthreshold characteristic under low operation voltage. It is believed that a device quality lanthanides are considered to be the most promising materials for high-k gate insulator applications.

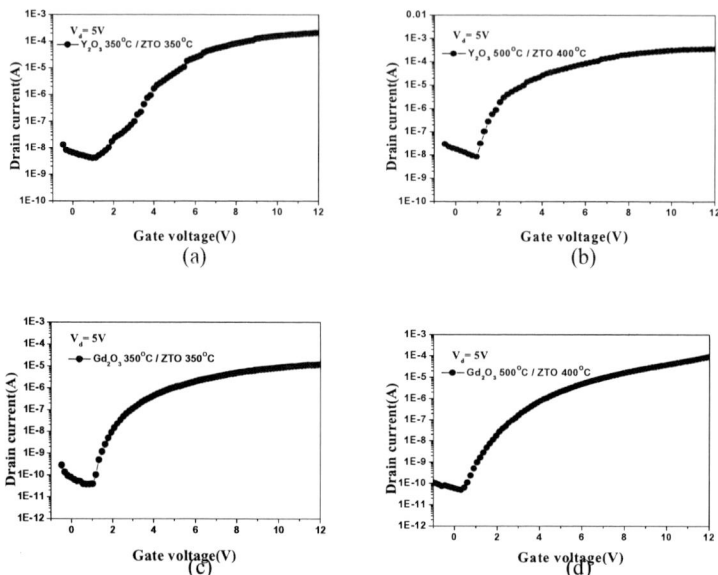

Fig. 4. Transfer characteristics of ZTO-TFTs with the solution-processed lanthanides gate insulator annealed at either 350 or 500 °C; (a) and (b) for Y_2O_3, (C) and (d) for Gd_2O_3.

Table 1 The electrical parameters of ZTO-TFTs based on either Y_2O_3 or Gd_2O_3 gate insulator. Data for using the thermally grown 100nm-thick SiO_2 are also available

Samples		μ [cm^2/V·s]	V_{th} [V]	SS[#] [V/dec]
Insulator (annealing temp.)	Semiconductor (annealing temp.)			
Gd_2O_3 (350°C)	ZTO (350°C)	0.3	1.8	0.53
Gd_2O_3 (500°C)	ZTO (400°C)	2.5	3.4	0.84
Y_2O_3 (350°C)	ZTO (350°C)	3.0	3.4	1.28
Y_2O_3 (500°C)	ZTO (400°C)	2.9	1.8	0.55
SiO_2	ZTO (400°C)	0.8	9.2	2.31

subthreshold swing

Conclusions

Using the solution-processed lanthanides film, we have investigated both the dielectric property and the charge transport in oxide-TFT devices. The overall trends showed that device with high-k oxides, especially on Y_2O_3 and Gd_2O_3, were properly working within low power consumption. The dielectric constant of Gd_2O_3 (Y_2O_3) was in the range of 9-15 (8-15). The high output current, $\sim 10^{-4}$ A, at less than 10 V for this device is particularly attractive for devices needing high drive current at low voltage. Both the low voltage-driven switching property and the increased carrier mobility with high-k dielectrics were presumably induced by the increased effective mobile carriers in semiconductor/insulator interface. Device performance based on Y_2O_3 and Gd_2O_3 high-k materials can lead to marked improvements in solution-based amorphous oxide TFTs for on-going flat panel display backblane such as organic light emitting diodes and liquid crystal displays.

Acknowledgements

This research was supported by a grant (F0004073-2010-33) from the Information Display R&D Center, one of the Knowledge Economy Frontier R&D programs funded by the Ministry of Knowledge Economy of the Korean government.

References

1. M. Zirkl, A. Haase, A. Fian, H. Schon, C. Sommer, G. Jakopic, G. Leising, B. Stadlober, I. Graz, N. Gaar, R. Schwodiauer, S. Bauer-Gogonea, and S. Bauer, *Adv. Mater.*, **19**, 2241-2245 (2007).
2. M. Kim, H.S. Kim, Y. Ha, J. He, M.G. Kanatzidis, A. Facchetti, and T.J. Marks, *J. Am. Chem. Soc.*, **132**, 10352-10364 (2010).
3. J.T. Anderson, C.L.Munsee, C.M. Hung, T.M. Phung, G.S. Herman, D.C. Johnson, J.F. Wager, and D.A. Keszler, *Adv. Func. Mater.*, **17**, 2117-2124 (2007).
4. B.N. Pal, B.M. Dhar, K.C. See, and H.E. Katz, *Nature Materials.*, **8**, 898-903 (2009).
5. A.P. Milanov, K. Xu, A. Laha, E. Bugiel, R. Ranjith, D. Schwendt, H.J. Osten, H. Parala, R.A. Fischer, and A. Devi, *J. Am. Chem. Soc.*, **132**, 36-37 (2010).
6. P.K. Hurley, K. Cherkaoui, E.O'Connor, M.C. Lemme, H.D.B. Gottlob, M. Schmidt, S. Hall, Y. Lu, O. Buiu, B. Raeissi, J. Piscator, O. Engstrom, and S.B. Newcomb, *J. Electrochem. Soc.*, **155**, G13-G20 (2008).
7. S-J. Seo, C.G. Choi, Y.H. Hwang, and B-S. Bae, *J. Phys. D: Appl. Phys.*, **42**, 035106 (2009).
8. J. Kwo, M. Hong, A. R. Kortan, K. T. Queeney, Y. J. Chabal, J. P. Mannaerts, T. Boone, J. J. Krajewski, A. M. Sergent, and J. M. Rosamilia, *J. Appl. Phys. Lett.*, **77**, 130 (2000).
9. S. Choi, B-Y. Park, T. Ahn, J.Y. Kim, C.S. Hong, M.H. Yi, and H-K. Jung, *Thin Solid Films*, in-press (2011).
10. K. Nomura, H. Ohta, A. Takagi, T. Kamiya, M. Hirano, and H. Hosono, *Nature*, **432**, 488 (2004).
11. H. Hosono, *J. Non-Cryst. Solids*, **352**, 851 (2006).

12. E. H. Nicollian, C. N. Berglund, P. F. Schmidt, and J. M. Andrews, *J. Appl. Phys.*, **42**, 5654 (1971).
13. L. Chua, J. Zaumseil, J. Chang, E. W. Ou, P. H. Ho, H. Sirringhaus, and R. Friend, *Nature*, **434**, 194 (2005).

ECS Transactions, 35 (4) 909-921 (2011)
10.1149/1.3572328 ©The Electrochemical Society

Reliability Properties and Current Conduction Mechanisms of HfO$_2$ MIS Capacitor with Dual Plasma Treatment

Kow-Ming Chang[a,b], Ting-Chia Chang[a,*], Shou-Hsien Chen[a], and I-Chung Deng[c]

[a] Department of Electronics Engineering & Institute of Electronics, National Chiao Tung University, 1001 Ta Hsueh Road, Hsinchu, Taiwan 30010, R.O.C.
[b] College of Electrical and Information Engineering, I-Shou University, Kaohsiung, Taiwan 84001, R.O.C.
[c] Department of Electronic Engineering, Technology and Science Institute of Northern Taiwan

The incorporation of nitrogen in HfO$_2$ gate dielectrics has been reported to be beneficial for electrical performance. The improvement in the electrical characteristics of HfO$_2$ thin film with plasma nitridation process or plasma fluorination process has also been examined. In this study, dual plasma, CF$_4$ pre-treatment and nitrogen post-treatment, treatments were performed on HfO$_2$ MIS capacitor for further improvement on reliability characteristic. We examine the reliability properties and the current conduction mechanism of HfO$_2$ thin films. The frequency dispersion and constant voltage stress (CVS) characteristics of the samples were analyzed to estimate the improvement. According to the present study, dual plasma treatment could be better than single plasma treatment and would be an effective approach for HfO$_2$ dielectric improvement.

Introduction

The rapid progress of complementary metal oxide semiconductor (CMOS) integrated circuit technology has met several serious technological challenges over the past few years. According to the prediction of the International Technology Roadmap for Semiconductor (ITRS), the conventional gate dielectric layer will reach its physical limits [1]. Gate dielectric scaling of CMOS will increase the speed and the packing density of modern circuits. However, the aggressive shrinking of the gate length and gate dielectric thickness accompanies excessive leakage current and reliability problems. To solve these problems, a major solution is to replace the traditional SiO$_2$ or SiON by other higher dielectric constant material. Using high dielectric constant material for gate dielectric could have larger physical thickness and maintain smaller equivalent oxide thickness (EOT). As a result, high-dielectric-constant (high-κ) thin films have been considered as suitable gate dielectric for modern CMOS technology. There are various high-κ thin film has been investigated [2-4]. Among these high-κ materials, HfO$_2$ is considered as the most promising candidate because of high dielectric constant (~25), wide band gap (~5.7 eV), and large band offset with Si conduction band (~1.5 eV) [2, 5]. Nevertheless, there are still some issues which need to be considered, such as the reliability and thermal stability of the dielectrics [6, 7].

909

It has been reported that nitrogen incorporated into HfO_2 gate dielectrics has beneficial effect on performance [8]. Nitrogen could be incorporated into dielectric layer by ICP plasma nitridation process at lower temperature [9, 10]. As reported in previous study: nitrogen incorporation can suppress crystallization during high temperature treatment, reduce dopant penetration, increase dielectric constant, and reduce leakage current by about 3-4 orders of magnitude [11, 12]. Umezawa et al [13] noted that nitrogen could deactivate the oxygen vacancy related states within HfO_2 band gap. The absence of gap states leads to the removal of electron leakage path.

In addition, the incorporation of fluorine in HfO_2 gate dielectrics also has been reported to be beneficial for electrical and reliability performance [14-16]. The quality of interfacial layer (IL) becomes more and more important due to gate dielectric scaling. Wong et al [17] noted that the applied electric field would be largely distributed in the low-κ region for high-κ/low-κ stack layer because of Gauss's law. The first breakdown happened in the low-κ layer [17]. Recently, several studies have used fluorine incorporation to improve IL quality at HfO_2/Si interface because Si-F bond (5.73 eV) is stronger than Si-H bond (3.18eV) [18, 19]. Moreover, pre-CF_4 plasma treatment has been shown to effectively suppress the IL formation [20].

In this study, we propose to combine two kinds of plasma treatment (denoted as dual plasma treatment), CF_4 pre-treatment and nitrogen post-treatment, in order to achieve further improvement. We have examined the reliability properties and the current conduction mechanism of HfO_2 MIS capacitor structure. First of all, the capacitance-voltage (C-V) characteristics and current-voltage (J-V) characteristics will be briefly described. Second, the frequency dispersion and constant voltage stress (CVS) characteristics of the samples will be analyzed to estimate the improvement. Finally, current conduction mechanisms, such as Schottky emission, Frenkel-Poole (F-P) emission, and Fowler-Nordheim (F-N) tunneling will be discussed. Schottky barrier height, F-P barrier height, and F-N barrier height will be extracted.

Experimental

After standard RCA cleaning, the samples were treated in CF_4 plasma (denoted as CF_4 pre-treatment) for various times. The substrate temperature in a plasma enhanced chemical vapor deposition (PECVD) system was set at 300 °C. The process pressure and the CF_4 flow rate were 500 mTorr and 100 sccm, respectively. The RF power was set at 20 W and the exposure times were varied in the range of 10-40 sec. After CF_4 pre-treatment, HfO_2 thin film was deposited on the samples by the metal organic chemical vapor deposition (MOCVD) system. Then post deposition annealing (PDA) was performed at 600 °C for 30 sec in a rapid temperature annealing (RTA) system. After PDA, samples were nitrided in nitrogen plasma (denoted as nitrogen post-treatment) by PECVD. The flow rate of nitrogen gas, which is N_2 or NH_3, was set at 100 sccm. Post nitridation annealing (PNA) was performed with RTA equipment at 600 °C for 30 sec in order to reduce plasma damage. Thereafter, 40 nm-Ti films and 400 nm Al films were deposited by e-beam evaporation system. The top electrodes were defined lithographically and etched to define a gate area of 5000 μm^2. Finally, backside Al electrodes were deposited by the thermal evaporation to form the Ohmic contact. The capacitance-voltage (C-V) and current-voltage (I-V) characteristics of MIS structure were measured by using a C-V measurement (Hewlett-Packard 4284) and Hewlett-Packard

4156C semiconductor parameter analyzer, respectively. The condition of frequency dispersion measurement was set as from 1 kHz to 100 kHz. The stress condition of CVS measurement was set as a constant voltage -3V for 0 to 500 sec. Furthermore, the I-V characteristics were measured at elevated temperatures from 25 to 125 °C, 25 °C per step, in order to analyze current conduction mechanisms.

Results and Discussion

<u>C-V and J-V Characteristics</u>

Figure 1 shows the C-V characteristics of the HfO_2 thin films treated in CF_4 plasma for different process durations and N_2 plasma for 120 sec. The frequency used in the high frequency C-V measurement was set at 50 kHz. The sample was treated only in N_2 plasma for 120 sec (RF power = 50 W) shows much higher capacitance density than the sample with no treatment. The higher capacitance could be attributed to the PDA process [21-23] and the nitrogen incorporation in the HfO_2 thin film. The nitrogen incorporation could enhance the electronic polarization as well as the ionic polarization, which result in the increase of dielectric constant [12, 24]. On the other hand, the capacitance density and interface characteristics show further improvement with the combination of CF_4 pre-treatment for 10 sec and N_2 post-treatment for 120 sec. With CF_4 pre-treatment, fluorine atoms would pile up at the HfO_2/Si interface, improve the quality of interface [19], and suppress the IL formation [20]. Besides, for CF_4 pre-treatment times longer than 10 sec, the plasma damage caused the degradation of HfO_2/Si interface and the degradation of capacitance density.

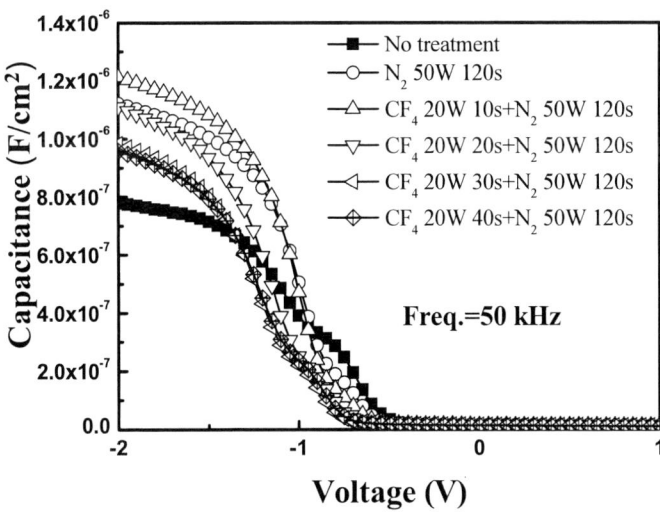

Figure 1. The C-V characteristics of the HfO_2 thin films treated in CF_4 plasma for different process durations and N_2 plasma for 120 sec.

Figure 2 shows the J-V characteristics of the HfO_2 thin films treated in CF_4 plasma for different process durations and N_2 plasma for 120 sec. Compared with the sample with no treatment, the gate leakage current decreased by about 4 orders of magnitude for the sample with CF_4 pre-treatment for 10 sec and N_2 post-treatment for 120 sec. The reduction of the gate leakage could be attributed to defect passivation. Oxygen vacancy related states and interface states could be passivated by nitrogen and fluorine atoms [13, 16]. On the other hand, for CF_4 pre-treatment times longer than 10 sec, the plasma damage caused the degradation of interface and the increase of gate leakage current. The best condition of dual plasma treatment is as follows: CF_4 pre-treatment (time=10s, RF Power=20W) and N_2 post-treatment (time=120s, RF Power=50W). The gate leakage of the sample with best condition is 1.05×10^{-5} A/cm^2 at V_g = -1.5 V.

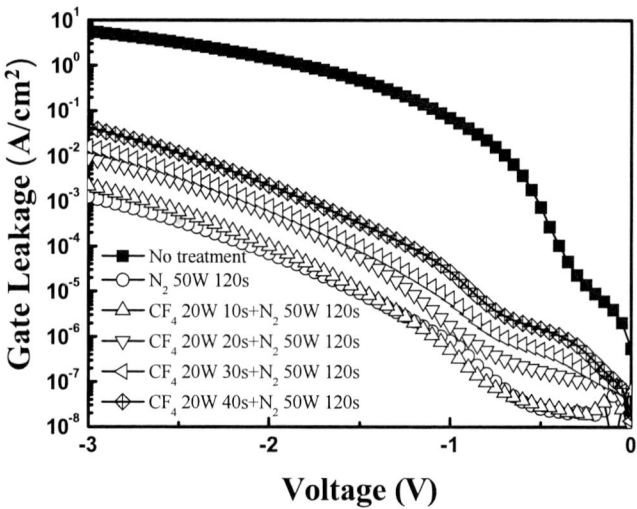

Figure 2. The J-V characteristics of the HfO_2 thin films treated in CF_4 plasma for different process durations and N_2 plasma for 120 sec.

In Figure 3 and Figure 4, the C-V and the J-V characteristics of the HfO_2 gate dielectrics, treated in CF_4 plasma for different process durations and NH_3 plasma for 120 sec, are presented. The RF power of NH_3 post-treatment was set at 40 W. As mentioned before, the reason of the improvement in the NH_3 plasma nitridation process could be the same as the one in the N_2 plasma nitridation process. From the similar analysis, the best condition of dual plasma treatment is as follows: CF_4 pre-treatment (time=10s, RF Power=20W) and NH_3 post-treatment (time=120s, RF Power=40W). The gate leakage of the sample with best condition is 1.62×10^{-5} A/cm^2 at V_g = -1.5 V. In summary, C-V and I-V Characteristics could be further improved by dual plasma treatment.

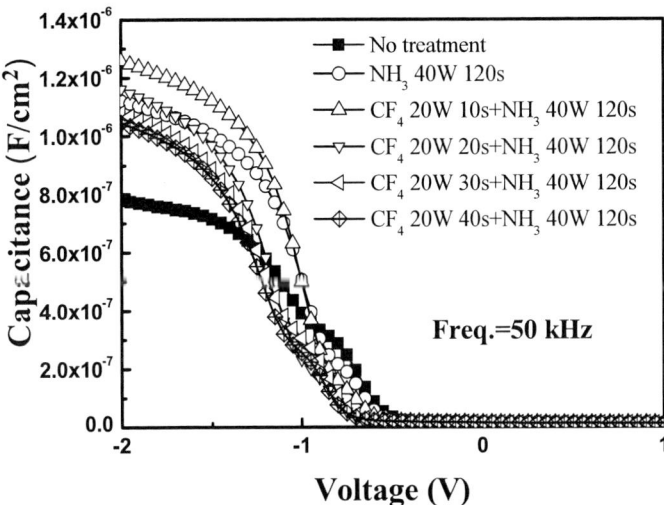

Figure 3. The C-V characteristics of the HfO$_2$ thin films treated in CF$_4$ plasma for different process durations and NH$_3$ plasma for 120 sec.

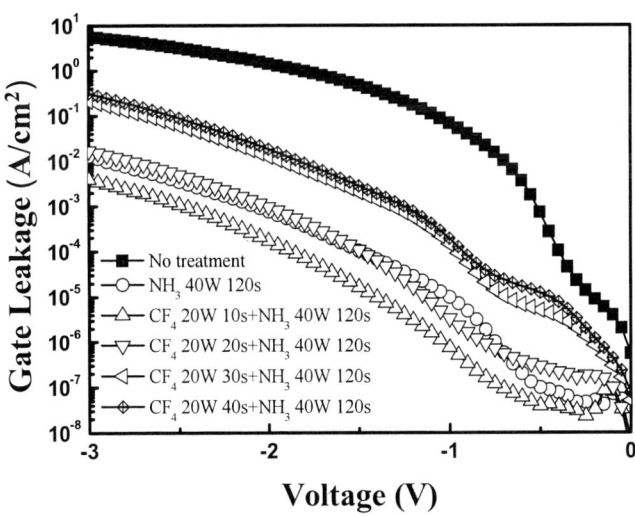

Figure 4. The J-V characteristics of the HfO$_2$ thin films treated in CF$_4$ plasma for different process durations and NH$_3$ plasma for 120 sec.

Frequency Dispersion Characteristics

The C-V characteristics of the HfO_2 thin films, treated with CF_4 plasma for different process durations and N_2 plasma for 120 sec, have been measured as a function of frequency as shown in Figure 5. The measurements were made in the frequency range of 1-100 kHz (1, 10, and 100 kHz). Frequency dispersion could be observed because of the response of trap charges to signal frequency. At low frequencies, interface traps generated the additional capacitance because some of traps could follow the change of gate voltage [25]. The frequency dispersion in the accumulation region and the hump in the depletion region are significant for the sample with no treatment. The sample treated by N_2 plasma showed relatively smaller frequency dispersion and smaller hump than the sample with no treatment. On the other hand, it was obvious that the sample treated by CF_4 pre-treatment for 10 sec and N_2 post-treatment for 120 sec exhibited nearly no dispersion in the accumulation region and nearly no hump in the depletion because interface states could be improved effectively [25-28] by dual plasma treatment. However, the frequency dispersion and the hump became severe again when the CF_4 pre-treatment time is longer than 10 sec owing to plasma damage at interface.

Figure 6 displays the C-V frequency dependence of the HfO_2 thin films treated in CF_4 plasma for different process durations and NH_3 plasma for 120 sec. The sample treated by CF_4 pre-treatment for 10 sec and NH_3 post-treatment for 120 sec exhibited nearly no dispersion and nearly no hump during C-V measurement. This indicated that dual plasma treatment could effectively eliminate interface states and greatly enhance IL quality than single plasma treatment.

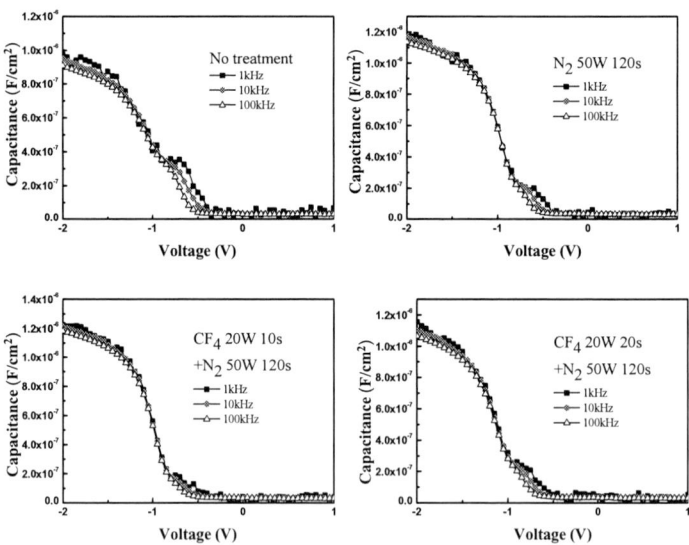

Figure 5. C-V frequency dependence of the HfO_2 thin films treated in CF_4 plasma for different process durations and N_2 plasma for 120 sec.

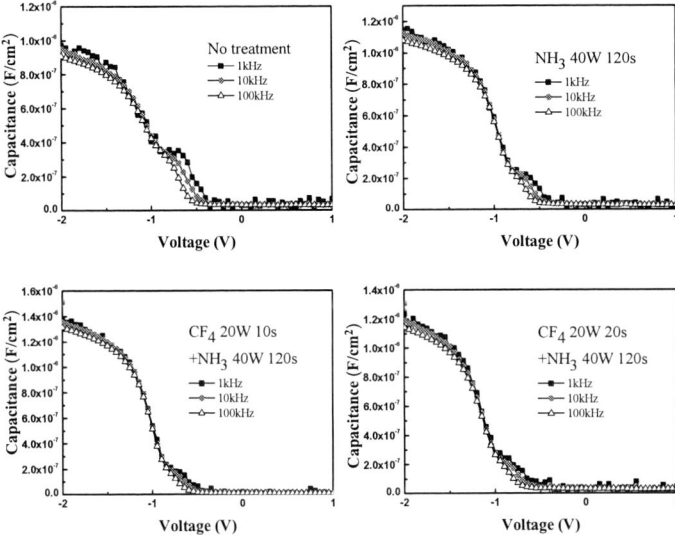

Figure 6. C-V frequency dependence of the HfO_2 thin films treated in CF_4 plasma for different process durations and NH_3 plasma for 120 sec.

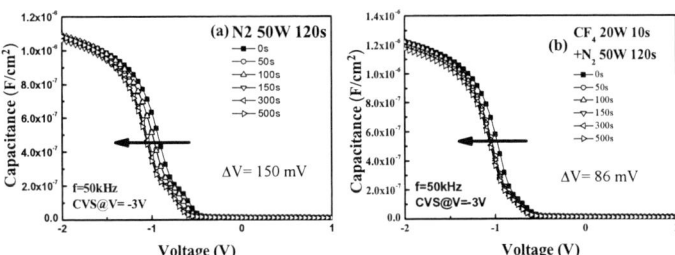

Figure 7. The C-V curves before and after CVS characteristics of the HfO_2 thin films treated by (a) single plasma treatment and (b) dual plasma treatment.

Constant Voltage Stress Characteristics

Figure 7 displays the C-V curves before and after CVS testing of the HfO_2 thin films treated by N_2 plasma post-treatment (as shown in Figure 7(a)) and the combination of CF_4 plasma pre-treatment and N_2 plasma post-treatment (as shown in Figure 7(b)), respectively. The stress voltage was set at -3 V. The stress times were made in a range from 0 to 500 sec. All the C-V curves shift to left as stress time increase indicated that there were positive charges trapped in the high-κ dielectric layer. Trapping of positive charges could be explained by Anode hole injection model [29]. During constant negative bias stress at a fixed gate voltage, the injected electrons traveled through dielectric and arrived at the interface. These electrons gained energy to liberate the hydrogen at the interface, leading to the generation of Si dangling bond. The liberated hydrogen diffused into the dielectric through the oxide field, trapped in the dielectric, leading to the creation of positively charged centers [30, 31]. The C-V curves had smaller V_{fb} shift and less distortion for samples with dual plasma treatment, indicating that samples with dual plasma treatment had less interface trap charges generated at the dielectric/Si interface and had better reliability properties than samples with single plasma treatment.

Schottky Emission

Figure 8(a) depicts the J-E plots for the sample with no treatment at different temperatures from 298 K to 398 K (25 K per step). The inset of Figure 8(a) is the band diagram demonstrating Schottky emission. The electric field E is an "effective" electric field (E = V/CET), while capacitance effective thickness (CET) is extracted from the capacitance density in the accumulation region [32]. The standard Schottky emission could be expressed as

$$J_{SE} = A^* T^2 \exp\left[\frac{-q\left(\phi_B - \sqrt{qE/4\pi\varepsilon_r\varepsilon_0}\right)}{kT}\right], \quad A^* = 120\frac{m^*}{m_0}\left(\frac{A}{cm^2 K^2}\right), \tag{1}$$

where J_{SE} is the current density, A^* is the effective Richardson constant, E is the effective electric field, T is the absolute temperature, q is the electron charge, $q\phi_B$ is the Schottky barrier height, k is Boltzmann's constant, ε_0 is the permittivity of free space, ε_r is the dynamic dielectric constant, m^* is the electron effective mass in HfO_2, m_0 is the free electron mass. The electron effective mass is 0.1 m_0 [33, 34].

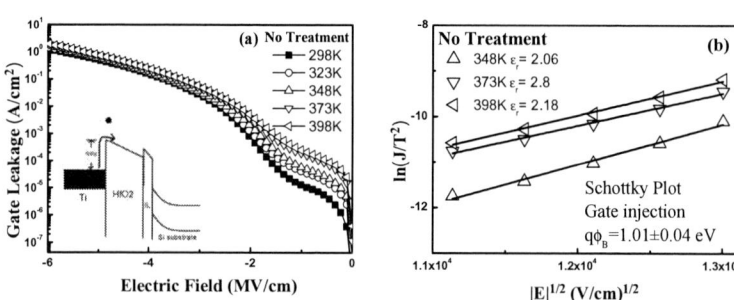

Figure 8. (a) The J-E curves at various temperatures and (b) Schottky emission plots, $\ln(J/T^2)$ versus $E^{1/2}$, for the HfO_2 thin film with no treatment.

For the standard Schottky emission, a plot of $\ln(J/T^2)$ versus $E^{1/2}$ should be a straight line, as shown in Figure 8(b). It was found that Schottky emission is the dominate conduction mechanism in the region of low to medium electric fields (1.7 - 3.0 MV/cm) [35]. Eq. (2) expresses the intercept of the Schottky emission plot with the vertical axis. The barrier height can be extracted from Eq. (2).

$$Intercept = \ln\left(A^*\right) - \frac{q\phi_B}{kT}, \quad A^* = 120\frac{m^*}{m_0}\ (\frac{A}{cm^2 K^2}). \tag{2}$$

Because of effective electric field (E = V/CET), the extracted barrier heights in this study are effective barrier heights. The extracted Schottky barrier heights for the samples with no treatment, single plasma treatment, and dual plasma treatment are listed in Table I. It is clear that the samples had larger barrier height than other samples after CF$_4$ pre-treatment for 10 sec and nitrogen post-treatment for 120 sec.

TABLE I. Schottky Barrier Height Extracted for The Samples with No Treatment, Single Plasma Treatment, and Dual Plasma Treatment.

Barrier height	No treatment	N$_2$, 120s	CF$_4$, 10s + N$_2$, 120s	CF$_4$, 20s + N$_2$, 120s
$q\phi_B(eV)$	1.01±0.04	1.23±0.09	1.32±0.1	1.19±0.09

Barrier height	No treatment	NH$_3$, 120s	CF$_4$, 10s + NH$_3$, 120s	CF$_4$, 10s + NH$_3$, 120s
$q\phi_B(eV)$	1.01±0.04	1.21±0.04	1.32±0.07	1.21±0.07

Frenkel-Poole (F-P) Emission

When gate under negative bias, electrons will inject from gate into HfO$_2$ dielectric layer and will be trapped into shallow trap levels. Thereafter, the electrons transported through the dielectric layer by hopping between these trap levels, leading to leakage current, called Frenkel-Poole (F-P) emission. The standard F-P emission could be expressed as [35]

$$J_{FP} = C_t E \exp\left[\frac{-q\left(\phi_t - \sqrt{qE/\pi\varepsilon_r\varepsilon_0}\right)}{kT}\right], \tag{3}$$

where J$_{FP}$ is the current density, E is effective electric field, C$_t$ is a constant proportional to the density of bulk oxide traps, $q\phi_t$ is the trap energy in HfO$_2$, and other parameters are as defined earlier. For the standard Frenkel-Poole emission, a plot of ln(J$_{FP}$/E) versus $E^{1/2}$ should be linear. It was found that F-P emission is the dominate conduction mechanism in the region of medium to high electric fields (4.0 - 6.0 MV/cm). The trap energy in dielectric layer could be extracted form good F-P fitting, as shown in Figure 9.

Table II lists the trap energy levels of the samples with no treatment, single plasma treatment, and dual plasma treatment. The extracted trap levels for the sample with and without N$_2$ post-treatment were 1.12 eV and 0.72 eV, respectively. In contrast, the trap level for the sample treated by CF$_4$ pre-treatment for 10 sec and N$_2$ post-treatment for 120 sec was 1.14 eV. The deeper trap level means that most of shallow trap levels in HfO$_2$ thin film can be eliminated [32] by using dual plasma treatment. The elimination of shallow trap levels resulted in the reduction of F-P conduction current. Similarly, the sample treated in CF$_4$ pre-treatment for 10 sec and NH$_3$ post-treatment for 120 sec also exhibited larger trap levels than other samples. In short, dual plasma treatment could eliminate the shallow trap levels and greatly reduce the gate leakage current.

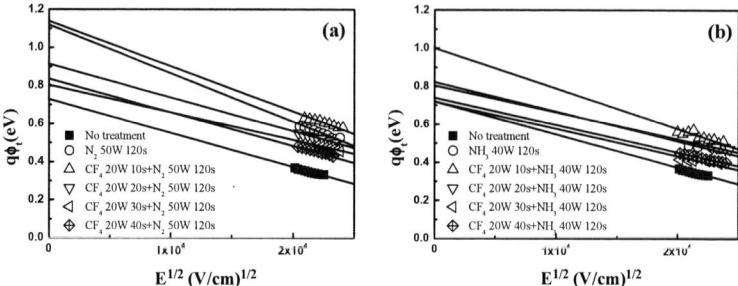

Figure 9. Trapping energy levels extracted from F-P fitting for the samples with dual plasma treatment (a) CF_4 plasma and N_2 plasma (b) CF_4 plasma and NH_3 plasma.

TABLE II. F-P Trapping Level Extracted for The Samples with No Treatment, Single Plasma Treatment, and Dual Plasma Treatment.

Barrier height	No treatment	N_2, 120s	CF_4, 10s $+ N_2$, 120s	CF_4, 20s $+ N_2$, 120s
$q\phi_t(eV)$	0.72	1.12	1.14	0.91

Barrier height	No treatment	NH_3, 120s	CF_4, 10s $+ NH_3$, 120s	CF_4, 10s $+ NH_3$, 120s
$q\phi_t(eV)$	0.72	0.82	1.10	0.80

<u>Fowler-Nordheim (F-N) Tunneling</u>

In higher electric field, the Fowler-Nordheim (F-N) tunneling dominated the conduction mechanism [36]. The standard F-P emission could be expressed as

$$J_{FN} = AE^2 \exp\left[\frac{-8\pi\sqrt{2m^*}\left(q\phi_f\right)^{3/2}}{3qhE}\right], \qquad (4)$$

where J_{FN} is the current density, h is the Plunk's constant, $q\phi_f$ is the potential barrier height, m^* is the electron effective mass in HfO_2, and the other notations were as same as mentioned before. The electron effective mass here is 0.1 m_0 [33, 34]. If the leakage current is dominated by the F-N mechanism, a plot of $\ln(J/E^2)$ versus $1/E$ should be linear. Linear characteristic could be observed at high electric field (> 7 MV/cm), as shown in Figure 10. The slope of each curve in Figure 10 and Eq. (5) could obtain the F-N barrier height, which were listed in Table III.

$$\phi_f = \left(\frac{9h^2}{128\pi^2 m^* q}\right) * (slope)^{2/3}. \qquad (5)$$

Table III listed the F-N barrier height of the samples with no treatment, single plasma treatment, and dual plasma treatment. It could be observed that samples treated in CF_4 pre-treatment for 10 sec and nitrogen post-treatment for 120sec had larger value than other samples. The injection of electrons from the gate entered the conduction band of HfO_2 by tunneling through a triangular potential barrier. The injected electrons interacted with lattice or transferred its energy at anode [37], which resulted in the degradation of the dielectric layer. As a result, the sample with proper dual plasma treatment had bigger F-N barrier height and better reliability properties.

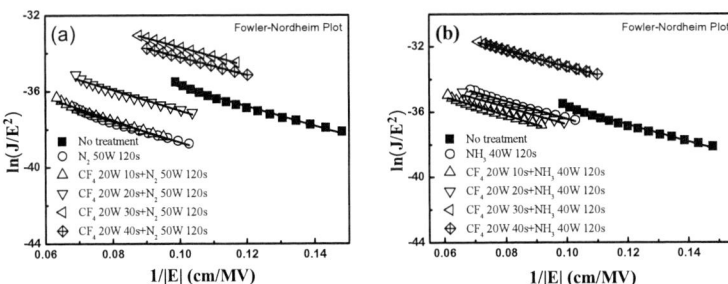

Figure 10. F-N tunneling characteristic, $\ln(J/E^2)$ vs. $1/E$, for the samples with dual plasma treatment (a) CF_4 plasma and N_2 plasma (b) CF_4 plasma and NH_3 plasma

TABLE III. F-N Barrier Height Extracted for The Samples with No Treatment, Single Plasma Treatment, and Dual Plasma Treatment.

Barrier height	No treatment	N_2, 120s	CF_4, 10s + N_2, 120s	CF_4, 20s + N_2, 120s
$q\phi_f(eV)$	1.78	1.86	2.01	1.85

Barrier height	No treatment	NH_3, 120s	CF_4, 10s + NH_3, 120s	CF_4, 10s + NH_3, 120s
$q\phi_f(eV)$	1.78	1.86	1.92	1.83

Conclusion

In conclusion, the reliability properties and current conduction mechanisms of HfO_2 gate dielectric films as a function of dual plasma treatment (the combination of CF_4 pre-treatment and nitrogen post-treatment) have been investigated. First, the best conditions which decided form C-V and J-V characteristics were the samples treated by CF_4 plasma for 10 sec and N_2 (NH_3) plasma for 120 sec. According to the current conduction analysis, the dominant current conduction mechanism was Schottky emission type in the region of low to medium electric fields (1.7 – 3.0 MV/cm); Frenkel-Poole (F-P) emission operated in the region of medium to high fields (4.0 – 6.0 MV/cm); Fowler-

Nordheim (F-N) tunneling was dominant at high fields (> 7 MV/cm). Dual plasma treatment was effective in improving interface quality, eliminating shallow trap levels, and enhancing reliability properties. In summary, the effect of dual plasma treatment could be better than single plasma treatment and dual plasma treatment would be an effective technology to improve the reliability of HfO_2 thin films.

Acknowledgments

The authors gratefully acknowledge the National Nano Device Laboratories (NDL) and the Nano Facility Center of the National Chiao Tung University.

References

1. International Technology Roadmap for Semiconductors, presented at public.itrs.net (2009).
2. G. D. Wilk, R. M. Wallace, and J. M. Anthony, *J. Appl. Phys.*, **89**, 5243 (2001).
3. Y. Shimamoto, J. Yugami, M. Inoue, M. Mizutani, T. Hayashi, K. Shiga, F. Fujita, M. Yoneda, and H. Matsuoka, *Symposium on VLSI Technology Digest of Technical Papers*, IEEE, p. 132, (2005).
4. C. Hobbs, L. Fonseca, V. Dhandapani, S. Samavedam, B. Taylor, J. Grant, L. Dip, D. Triyoso, R. Hegde, D. Gilmer, R. Garcia, D. Roan, L. Lovejoy, R. Rai, L. Hebert, H. Tseng, B. White, and P. Tobin, *Symposium on VLSI Technology Digest of Technical Papers*, IEEE, p. 9, (2003).
5. J. Robertson, *J. Vac. Sci. Technol. B*, **18**, 1785 (2000).
6. S. Yamaguchi, K. Tai, T. Hirano, T. Ando, S. Hiyama, J. Wang, Y. Hagimoto, Y. Nagahama, T. Kato, K. Nagano, M. Yamanaka, S. Terauchi, S. Kanda, R. Yamamoto, Y. Tateshita, Y. Tagawa, H. Iwamoto, M. Saito, N. Nagashima, and S. Kadomura, *Symposium on VLSI Technology Digest of Technical Papers*, IEEE, p. 192, (2006).
7. J.-P. Han, E. M. Vogel, E. P. Gusev, C. D'Emic, C. A. Richter, D. W. Heh, and J. S. Suehle, *IEEE Electron Device Lett.*, **25**, 126 (2004).
8. G. Shang, P.W. Peacock, and J. Robertson, *Appl. Phys. Lett.*, **84**, 106 (2004).
9. K.M. Chang, B.N. Chen, and S.M. Huang, *Applied Surface Science*, **254**, 6116 (2008).
10. K.M. Chang, B.N. Chen, and C.K. Tang, *ECS Trans.*, **19** (2), 773 (2009).
11. C. H. Choi, S. J. Rhee, T. S. Jeon, N. Lu, J. H. Sim, R. Clark, M. Niwa, and D. L. Kwong, *Tech. Dig. - Int. Electron Devices Meet.*, **2002**, 857 (2002).
12. M. Koyama, A. Kaneko, T. Ino, M. Koike, Y. Kamata, R. Iijima, Y. Kamimuta, A. Takashima, M. Suzuki, C. Hongo, S. Inumiya, M. Takayanagi, and A. Nishiyama, *Tech. Dig. Int. Electron. Device Meet.*, p. 849, (2002).
13. N. Umezawa, K. Shiraishi, T. Ohno, H. Watanabe, T. Chikyow, K. Torii, K. Yamabe, K. Yamada, H. Kitajima, and T. Arikado, *Appl. Phys. Lett.*, **86**, 143507 (2005).
14. C.S. Lai, W.C. Wu, J.C. Wang, and T.S. Chao, *Appl. Phys. Lett.* **86**, 222905 (2005).
15. W.C. Wu, C.S. Lai, J.C. Wang, J.H. Chen, M.W. Ma, and T.S. Chao, *J. Electrochem. Soc.*, **154** H561 (2007).

16. K. Tse and J. Robertson, *Appl. Phys. Lett.* **89**, 142914 (2006).
17. H. Wong, B. Sen, V. Filip, and M. C. Poon, *Thin Solid Films*, **504**, 192 (2006).
18. K.I. Seo, R. Sreenivasan, P.C. McIntyre, and K.C. Saraswat, *Tech. Dig. - Int. Electron Devices Meet*, pp. 17.2.1–17.2.4., (2005).
19. C.S. Lai, W.C. Wu, K.M. Fan, J.C. Wang, and S.J. Lin, *Jpn. J. Appl. Phys.*, **44**, pp. 2307-2310, (2005)
20. C.S. Lai, W.C. Wu, T.S. Chao, J.H. Chen, J.C. Wang, L.L. Tay, and N. Rowell, *Appl. Phys. Lett.* **89**, 072904 (2006).
21. G. D. Wilk, M. L. Green, M.-Y. Hot, B. W. Busch, T. W. Sorsch, F. P. Klemens, B. Brijs', R. B. van Dover, A. Komblit, T. Gustafsson, E. Garfunkel, S. Hillenius, D. Monroe, P. Kalavade, and J.M. Hergenrother, *Symposium on VLSI Technology Digest of Technical Papers, IEEE,* p. 88, (2002).
22. J. Molina, K. Tachi, K. Kakushima, P. Ahmet, K. Tsutsui, N. Sugii, T. Hattori, and H. Iwai, *J. Electrochem. Soc.*, **154**, 110 (2007).
23. C.C. Cheng, C.H. Chien, C.W. Chen, S.L. Hsu, C.H. Yang, and C.Y. Changa, *J. Electrochem. Soc.*, **153**, 160 (2006).
24. M. R. Visokay, J. J. Chambers, A. L. P. Rotondaro, A. Shanware, and L. Colombo, *Appl. Phys. Lett.*, **80**, 3183 (2002).
25. H. X. Xu, J. P. Xu, C. X. Li, and P. T. Lai, *Appl. Phys. Lett.*, **97**, 022903 (2010).
26. T.M. Pan, C.S. Liao, H.H. Hsu, C.L. Chen, J.D. Lee, K.T. Wang, and J.C. Wang, *Appl. Phys. Lett.*, **87**, 262908 (2005).
27. H. Kim, C. O. Chui, K. C. Saraswat, and P. C. McIntyre, *Appl. Phys. Lett.*, **83**, 2647 (2003).
28. H. Harris, K. Choi, N. Mehta, A. Chandolu, N. Biswas, G. Kipshidze, and S. Nikishin, *Appl. Phys. Lett.*, **81**, 1065 (2002).
29. D. J. DiMaria, E. Cartier, and D. A. Buchanan, *J. Appl. Phys.*, **80**, 304 (1996).
30. M. Houssa, J. L. Autran, A. Stesmans, and M. M. Heyns, *Appl. Phys. Lett.*, **81**, 709 (2002).
31. E. Efthymiou, S. Bernardini, J.F. Zhang, S.N. Volkos, B. Hamilton, and A.R. Peaker, *Thin Solid Film*, **517**, 207 (2008).
32. W.C. Wu, C.S. Lai, T.M. Wang, J.C. Wang, C.W. Hsu, M.W. Ma, W.C. Lo, T.S. Chao, *IEEE Trans. Electron Devices*, 55, no. 7, pp. 1639–1646 (2008).
33. W.J. Zhu, T. P. Ma, T. Tamagawa, J. Kim, and Y. Di, *IEEE Electron Device Lett.*, **23**, 97 (2002).
34. F.C. Chiu, *J. Appl. Phys.*, **100**, 114102 (2006).
35. C.H. Liu, H.W. Chen, S.Y. Chen, H.S. Huang, and L.W. Cheng, *Appl. Phys. Lett.*, **95**, 012103 (2009).
36. F. El Kamel, P. Gonon, C. Vallée, and C. Jorel, J. Appl. Phys., **106**, 064508 (2009).
37. K.F. Schuegraf, and C. Hu, *J. Appl. Phys.*, **76**, 3695 (1994).

922

A MIM diode with ultra abrupt switching process
and high on/off current ratio

Lijie Zhang and Ru Huang

Key Laboratory of Microelectronic Devices and Circuits
Institute of Microelectronics, Peking University
Beijing 100871, China
Email: ruhuang@ pku.edu.cn

In this work, a TaO_x-based Metal-Insulator-Metal (MIM) diode is demonstrated. This $Cu/TaO_x/W$ diode is fabricated with sputtering method without any high temperature process (~ 400 °C), which shows promising compatabiltiy with CMOS back-end process. The fabricated diode exhibits abrupt switching process (~3 mV/dec) with small switching voltage (~ 0.7 V), high on/off current ratio more than 10^6, low leakage current (~ pA). Current is limited when the diode is switched to the on-state to avoid the hard breakdown caused by the thermal run-away. On/off current ratio can be further improved to 10^8 by the limited current increase at the precondition of ensuring the safty of the diode. This MIM diode can be easily integrated into high density PCM or RRAM cross-point arrays due to its simple fabrication process. Its nearly ideal diode behavior with ultra low on-resistance can also be used as an ESD protection device or as a rectifying element in circuit.

Introduction

A Metal-Insulator-Metal device is a hot research topic in many applications such as MIM capacitor, MIM diode, and MIM RRAM. MIM capacitor, which is usually used in the analog circuit, has been investigated for many years to improve its capacitance, linearity, breakdown voltage, and so on. [1-5]. MIM RRAM, which attracts many attentions in recent years, is used as a nonvolatile memory cell for high density storage, low power, and embedded applications [6-11]. MIM diode is very attractive for detection and mixing in the antenna system due to its extremely fast response time [12]. When compared with the complex fabrication process of thin film transistor, the fabrication of MIM diode is more cost-effective, and thus MIM diode are often used as a switch in the system of liquid crystal displays [13]. Furthermore, since the MIM diode has similar structures to RRAM, MIM diode has a great potential to act as a selection element in the cross-point memory array for high density, 3D-stacking integration. It is necessary to investigate the MIM diode, therefore.

In this work, a MIM diode fabricated without any high temperature process is demonstrated. The characterization of the MIM diode is given including the reliability issue and the temperature impact. The switching mechanism of this diode is explained with the trap or defect level alignment under the electric field.

Experimental

The fabrication process of the $Cu/TaO_x/Pt$ planer MIM diode is similar to our previous work except some modifications of the process condition [14]. The Ti adhesive layer was first deposited on the silicon substrate with the PVD method, and then 200 nm Pt was deposited by PVD method as bottom electrode (BE) to avoid electrode oxidation during the following deposition of the tantalum oxide film. The size of BE was defined by the lift-off process. TaO_x film was then deposited by RF reactive magnetron sputtering from a Ta target in the mixture of oxygen and argon ambient. To elevate the oxygen concentration of the TaO_x film, an annealing process in the oxygen atmosphere at 400°C was used. Vias for the bottom electrode connection are etched by RIE approach. Finally, Cu was deposited on the tantalum oxide film with sputtering method at room temperature and the size of the top electrode (TE) was patterned by lift-off process.

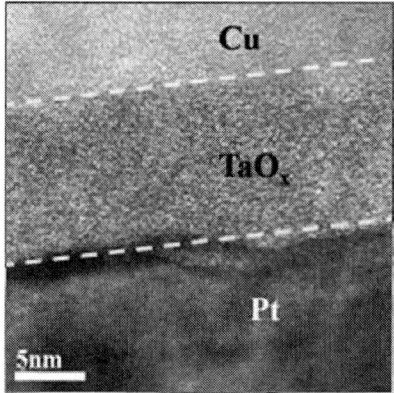

Fig. 1. Cross-sectional TEM image of $Cu/TaO_x/Pt$ diode

Estimated by the cross-sectional high resolution TEM image of the metal-insulator-metal structure in Fig. 1, the thickness of the TaO_x layer is about 10 nm. No obvious crystal lattice can be found in the TaO_x layer, indicating amorphous state of the TaO_x film fabricated by the PVD method.

Depth distribution of elements in the diode was analyzed by the EDX spectrum. Fig. 2 shows that the concentration of Cu decrease sharply with depth, indicating that little diffusion or chemical reaction happens between TE and the TaO_x film, in agreement with the theory of Gibbs free energy [15].

In this work, all of the current-voltage characteristics of the TaO_x-based diode were measured by Agilent B1500A semiconductor parameter analyzer. Voltage was applied on the top electrode with bottom electrode grounded.

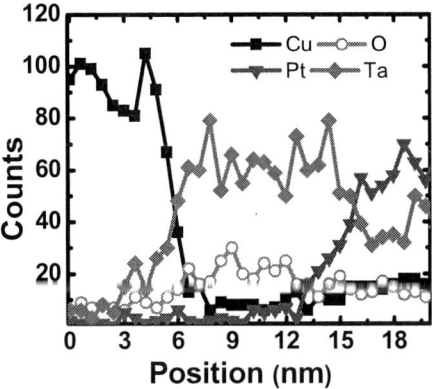

Fig. 2. EDX spectrum analysis across the Cu/TaO$_x$/Pt diode

Results and Discussion

The typical I-V curve of MIM diode with the size of 2500 μm^2 at room temperature is shown in Fig. 3. Compared with silicon PN diode and oxide-based diodes [16, 17], the device exhibits ultra abrupt turn-on switching process (the slope: ~ 3 mV/dec), which is similar to the set process of RRAM or the breakdown process of dielectric [6-11].

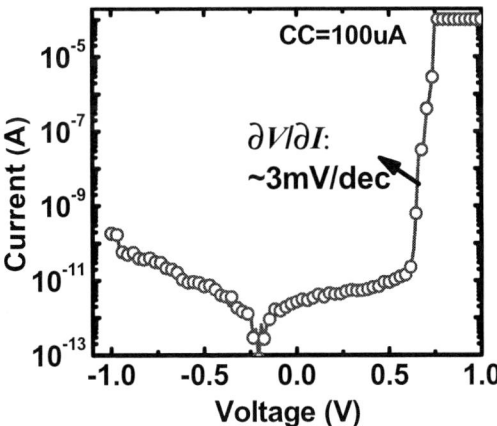

Fig. 3. Typical I-V curve of MIM diode at room temperature, current was limited to 100 μA.

Large on/off current ratio more than 10^6 can also be observed in Fig. 3. When the voltage applied to the device is below 0.7 V, the leakage current of the diode is very small ($\sim 10^{-11}$ A), while above 0.7 V, the device exhibits high electrical conductivity with current compliance of 100 μA. The device can be automatically switched back to high resistance state with no stable memory effect when the applied voltage or current is removed, which is quite different from RRAM. The low leakage current of this diode is possibly caused by the polarization of TaO_x dielectric, which is a common phenomenon in MIM capacitors.[18] It is also found that this device exhibits asymmetric switching behavior in both positive and negative voltage bias region. This device can not be turned on with negative voltage bias except hard breakdown at about -1.5 V (not shown here). This asymmetric switching behavior is possibly caused by the different metals renpectively uned on top and bottom electrode.

The current of the diode is limited to avoid the hard breakdown of the oxide caused by the thermal runaway when the diode is switched to the on-state. The impact of the compliance current on on/off-resistance is investigated with statistic method to rule out some possible variations of the resistance in different switching cycles. As shown in Fig. 4. On-state current increases with compliance current while leakage current changes little. On/off current ratio can be improved with increase of compliance current before the hard breakdown of the diode. The upper limit of the current that is allowed to conduct through the diode is still under investigation. The variations of the off-state get much larger when the compliance current increases to 500 μA, which should be taken into account when it is used in the circuit. Furthermore, it is founded that the compliance current will impact the recovery time of the diode to the off-state when the applied voltage or the current is removed. The exact recovery time of the diode with various compliance currents will be investigated in the future. Large compliance current will cause the increase of the recovery time possibly due to large number of defects or traps involved in the conduction. Therefore, considering the application of this diode, there is a trade off between the on/off current ratio and switching speed.

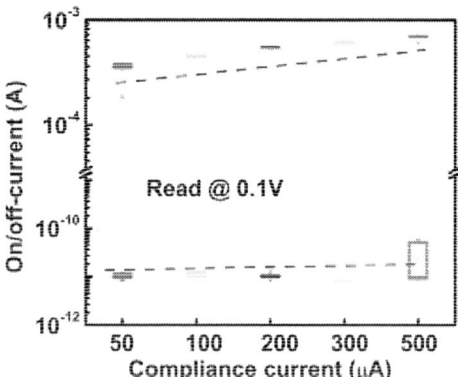

Fig. 4. Statistics of current compliance impact on on/off-state of MIM diode.

The conduction and switching mechanism of this diode is discussed next. There are some traps or defects, e.g. oxygen vacancies in the nonstoichiometric TaO_x film, and these traps are distributed in the oxide possibly with discrete energy levels. When the bias applied to the device is small, leakage is possibly dominated by polarization current, which will be discussed in the following paragraph. With the voltage increasing, the energy band bending of the TaO_x increases. When the band bends to a certain extent, the energy levels of discrete traps or defects are aligned and electron tunneling results in the switching of the diode to the on-state. When the voltage applied on the diode is removed, the bended band will recover and the aligned defect or traps levels will get discrete again. As a result, the diode is turned off.

The switching behaviors of the diode at different temperatures are investigated. The diode is switched under DC sweeping mode at various temperatures.

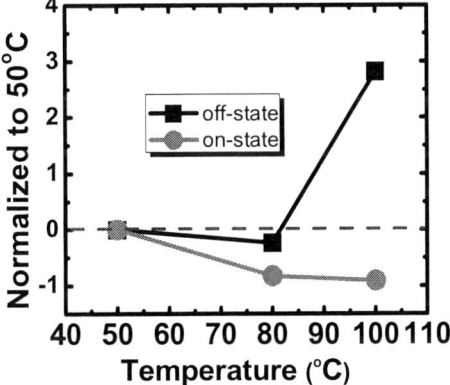

Fig. 5. Temperature dependence of on/off-state resistance of MIM diode. Both the on and off resistance are normalized to 50°C with the equation $(R_T-R_{50°C})$ / $R_{50°C}$. $R_{50°C}$ denotes the resistance at 50°C, R_T denotes the resistance at high temperatures of 80°C, 100°C.

The on-state resistance of the device is immediately read after the turn-on process with a low voltage of 0.1 V. All the measured current values are normalized to values measured at temperature of 50 °C. The leakage current exhibits little temperature dependence as shown in Fig. 5, which is consistent with the phenomenon of the polarization current at various temperatures [18]. The decrease of the on-resistance with the temperature is possibly due to the longer recovery time needed at high temperatures, which needs further investigation. Since this diode can be automatically returned to the off-state in a short time when the applied voltage is removed, the on-state resistance is expected much smaller than that measured considering the testing time. To monitor the current response of this diode under voltage pulse mode is necessary to assess the properties of the diode such as on/off-resistance, speed and so on. This investigation with pulse operation is underway and will be reported in the future.

Based on the above discussion, the switching process of the diode is possibly related with defects or traps in the TaO_x film, so that an investigation into the repeatability of switching is necessary. The reliability of this diode is measured with successive DC sweeping mode. Both the leakage current and the switching voltage exhibit little degradation after successive sweeping cycles as shown in Fig. 6, indicating good reversible switching behavior of the device.

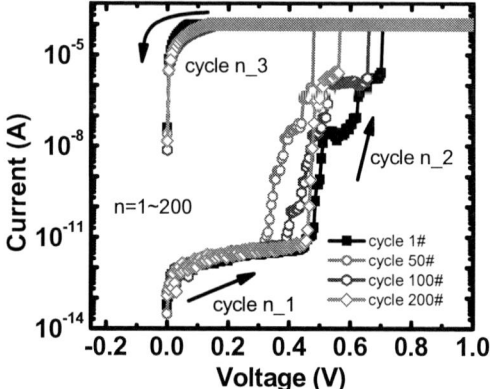

Fig. 6. I-V curves of MIM diode during hundreds of successive sweeping cycles. Arrows indicate the voltage sweeping direction. In contrast with RRAM, this diode can be automatically switched to high resistance state when applied voltage is removed.

Conclusions

In this work, a novel MIM TaO_x-based diode is successfully fabricated. This diode exhibits abrupt switching behaviour with large on/off current ratio, low switching voltage, good reliability property. The switching process of the diode is explained by the trap or defect alignment under the electric field. Due to its similar process to oxide-based RRAM, this diode can be easily integrated with RRAM devices as a memory selector in the crossbar array. Also its excellent switching behavior with low leakage current and abrupt turn-on process can be used as a good rectifier in the circuit.

Acknowledgements

This work was supported by the National Natural Science Foundation of China under Grant 60625403 and 90207004, by 973 Projects under Contract 2006CB302701, by the NCET Program, and by NSFC 61006062.

References

1. C.-M. Hung, Y.-C. Ho, I.-C. Wu, and K. O, in *Proc. IEEE MTT-S Int. Microw. Symp. Dig.* 1998, p. 505
2. J. A. Babcock, S. G. Balster, A. Pinto, C. Dirnecker, P. Steinmann, R. Jumpertz, and B. El-Kareh, *IEEE Electron Device Lett.*, **22**, p. 230, (2001)
3. T. Ishikawa, D. Kodama, Y. Matsui, M. Hiratani, T. Furusawa, and D. Hisamoto, in *IEDM Tech. Dig.*, 2002, p. 940
4. S. B. Chen, J. H. Lai, K. T. Chan, A. Chin, J. C. Hsieh, and J. Liu, *IEEE Electron Device Lett.*, **23**, p. 203, (2002)
5. X. Yu, C. Zhu, H. Hu, A. Chin, M. F. Li, B. J. Cho, D.-L. Kwong, P. D. Foo, and M. B. Yu, *IEEE Electron Device Lett.*, **24**, p. 63, (2003)
6. S. Seo, M. J. Lee, D. H. Seo, E. J. Jeoung, D.-S. Suh, Y. S. Joung, I. K. Yoo, I. R. Hwang, S. H. Kim, I. S. Byun, J.-S. Kim, J. S. Choi, and B. H. Park, *Appl. Phys. Lett.*, **85**, p. 5655, (2004)
7. H. Y. Lee, P. S. Chen, T. Y. Wu, Y. S. Chen, C. C. Wang, P. J. Tzeng, C. H. Lin, F. Chen, C. H. Lien, and M.-J. Tsai, in *IEDM Tech. Dig.*, 2008, p. 297
8. Z. Wei, Y. Kanzawa, K. Arita, Y. Katoh, K. Kawai, S. Muraoka, S. Mitani, S. Fujii, K. Katayama, M. Iijima, T. Mikawa, T. Ninomiya, R. Miyanaga, Y. Kawashima, K. Tsuji, A. Himeno, T. Okada, R. Azuma, K. Shimakawa, H. Sugaya, T. Takagi, R. Yasuhara, K. Horiba, H. Kumigashira, and M. Oshima, in *IEDM Tech. Dig.* 2008, p. 293
9. J. J. Yang, M. D. Pickett, X. Li, D. A. A. Ohlberg, D. R. Stewart, and R. S.Williams, *Nat. Nanotechnol.*, **3**, p. 429, (2008)
10. R. Waser and M. Aono, *Nat. Mater.*, **6**, p. 834, (2007)
11. R. Waser, R. Dittmann, G. Staikov, and K. Szot, *Adv. Mater.*, **21**, p. 2632, (2009)
12. C. Fumeaux, W. Herrmann, F. K. Kneubühl, and H. Rothuizen, *Infrared Phys. Technol.* **39**, p. 123, (1998)
13. D. Baraff, et al., J. R. Long, B. K. MacLaurin, C. J. Miner, R. W. Streater, *IEEE Trans. Electron Devices*, **28**, p. 736, (1981)
14. L. Zhang, R. Huang, M. H. Zhu, S. Q. Qin, Y. B. Kuang, D. J. Gao, C. Y. Shi, and Y.Y.Wang, *IEEE Electron Device Lett.*, **31**, p. 966, (2010)
15. M. W. Chase, Jr., C. A. Davies, J. R. Downey, Jr., D. J. Frurip, R. A. McDonald, and A. N. Syverud, J. Phys. Chem. Ref. Data, 3rd ed. **14**, (1985).
16. K. N. Choi, J. W. Park, H. S. Lee, K. S. Chung, in Mater. Res. Soc. Symp. Proc., 2009, **1108**, 1108-A09-11
17. M.-J. Lee, Park Y. S., B.Soo Kang, Seung-Eon Ahn, C. Lee, Kihwan Kim, X. Wen, Y.Xian, G. Stefanovich, J.-H. Lee, S.-J.Chung, Kim.Y.-H., C.-S. Lee, Park. J.-B. Park, Y. In-Kyeong, in *IEDM Tech. Dig.*, 2007, p. 771
18. Bing Miao, Rajat Mahapatra, Richard Jenkins, Jon Silvie, Nicholas G. Wright, and Alton B. Horsfall, *IEEE Trans. Nuclear Science*, **56**, p. 2916, (2009)

930

Author Index

Ahmet, P.	597	Chang, Y.	39
Ai, C.	39	Chang-Liao, K.	39
Alam, M.	145	Charbonnier, M.	773
Ambacher, O.	205	Che, M.	629
Andrieu, F.	515	Chen, L.	757
Ang, D.	125	Chen, S.	909
Antonelli, G. A.	701	Chen, T.	639
Aoyama, S.	815	Cheng, S.	383
		Chevolleau, T.	667, 729
Bailly, F.	667	Chikyow, T.	403
Baklanov, M. R.	717	Chiu, S.	629
Balseanu, M.	651	Choi, S.	901
Bang, W.	757	Clark, R. D.	815
Baudot, S.	805	Clerc, R.	773
Bauza, D.	95	Cochrane, C.	605
Benkhelifa, F.	205	Coignus, J.	773
Bertin, F.	729, 805	Coindeau, S.	497
Besacier, M.	729	Consiglio, S.	815
Besmehn, A.	461	Cui, Z.	651
Bittel, B. C.	747		
Blanquet, E.	497	Darnon, M.	667, 729
Boogaard, A.	259	Das, T.	835
Boujamaa, R.	805	David, T.	667
Boulanger, F.	515	Deng, I.	909
Bouyssou, R.	667, 729	Deora, S.	145
Breuer, U.	461	De Souza, M. M.	563
Brunet, L.	515	Detlefs, B.	497, 805
		Devine, R.	447
Campbell, J.	605	Dey, K. R.	273
Cao, W.	383	Dhar, S.	369
Cao, Y.	353	Dingemans, G.	191
Cardin, J.	273	Doisneau, B.	497
Cassé, M.	515	Dou, C.	597
Casterman, D.	563	Driad, R.	205
Chaabouni, H.	667	Droopad, R.	175
Chaji, G.	73	Dubourdieu, C.	805
Chang, K.	909	Ducote, J.	667
Chang, T.	909	Dufour, C.	273

Durğun Özben, E.	461	Huang, T.	889
		Hurand, R.	729
El Kodadi, M.	667, 729	Hurley, P.	175
Engström, O.	19	Hwu, J.	639
Enichlmair, H.	321		
Esmaeili-Rad, M.	73	Ielmini, D.	581
		Imai, S.	217
Fleetwood, D.	369	Ishigaki, H.	55
French, B.	849	Ishigaki, T.	861
French, M.	849	Islam, A. E.	145
Fu, C.	39	Iwai, H.	597
Fujieda, S.	55		
Fujii, S.	417	Jaehnig, M.	849
Fujiki, J.	417	Jang, M.	901
Fukutani, K.	55	Jeong, S.	901
		Ji, Z.	81
Gambino, J.	687	Joshi, P.	241
Garros, X.	515, 773	Joubert, O.	667
Gaumer, C.	515	Jung, H.	901
Ghibaudo, G.	773	Jungemann, C.	321
Gourbilleau, F.	273		
Grasser, T.	321	Kajiwara, J.	861
Gros-Jean, M.	497, 805	Kakushima, K.	597
Gu, C.	125	Kambour, K.	447
Guo, W.	701	Kang, K.	861
		Kent, T.	175
Haberfehlner, G.	729	Kessels, W. M.	191
Hall, S.	531	Kim, C.	757
Hartzell, J.	241	King, S.	747, 757, 849
Hattori, T.	115	Kinoshita, T.	115
Hattori, T.	597	Kirste, L.	205
Hazart, J.	729	Kobayashi, H.	217
Heh, D.	39	Kovalgin, A. Y.	259
Henri, J.	701	Kubota, Y.	217
Hjalmarson, H.	447	Kuhn, M.	849
Hong, W.	889	Kumagai, Y.	115
Hota, M.	835	Kummel, A. C.	175
Hou, F.	39	Kusai, H.	417
Hsu, H.	629		
Hsu, Y.	39	Lafond, D.	515
Hsu, Y.	39	Lau, W. S.	873
Huang, D.	383	Lee, J.	901
Huang, R.	923	Lee, S.	481

Lenahan, P. M.	605, 747	Mori, Y.	3
Lenk, S.	461	Morimura, Y.	3
Leroux, C.	773	Mountsier, T.	701
Leu, J.	629	Mukai, K.	597
Leusink, G.	815	Muro, T.	115
Li, F.	73		
Li, M.	383	Nabatame, T.	403
Li, Y.	39	Naik, M.	651
Licitra, C.	667, 729	Nakamura, G.	815
Lin, C.	889	Nalini, R.	273
Lin, H.	889	Nathan, A.	73
Lin, L.	81	Natori, K.	597
Lin, T.	889	Nguyen, V.	651
Liu, Z.	55	Nichau, A.	461
Lopes, J.	461	Nishiyama, A.	597
Lopes, J.	531	Nuta, I.	497
Lösch, R.	205		
Loup, V.	805	O'loughlin, J.	701
Lu, C.	39	Oh, H.	481
Lu, H.	639	Ohi, A.	403
Luptak, R.	461	Ohmi, T.	115
Luysberg, M.	461	Olubuyide, O. O.	287
Ma, T.	545	Pantelides, S. T.	369
Mahapatra, S.	145	Park, B.	901
Mahata, C.	835	Park, J.	321
Maheta, V. D.	145	Park, Y.	757
Maiti, C.	835	Pender, J.	651
Mallik, S.	835	Pomorski, T.	747
Mantl, S.	461	Portier, X.	273
Martin, F.	805	Posseme, N.	667, 729
Martinet, C.	497		
Martinez, E.	805	Reddy, S.	701
Matsumoto, T.	217	Reimbold, G.	515, 773
Mee, J.	447	Renault, O.	805
Melitz, W.	175	Roy, J.	497
Mi, Y.	497	Ryan, J.	605
Michael, N.	757	Ryan, T.	757
Mikulla, M.	205	Ryu, B.	901
Mitard, J.	773	Ryu, S.	369
Mitrovic, I.	531		
Monnier, D.	497	Sawin, H.	701
Moradi, M.	73	Sazonov, A.	73

Schiavone, P.	729	Weber, O.	515
Schlosser, D.	701	Wilde, M.	55
Schnee, M.	461	Wolters, R.	259
Schrimpf, R.	369	Wyon, C.	497
Schubert, J.	461, 531		
Sedghi, N.	531	Xia, L.	651
Seong, T.	901		
Shamma, N.	701	Yao, C.	383
Shen, C.	383	Yasuda, N.	417
Shen, J.	175	Ye, Y.	353
Shen, X.	369	Yoo, W.	861
Sicre, S.	563	Yota, J.	229
Sims, J.	701	Yu, H.	383
Smirnov, E. A.	717	Yu, W.	461
Starkov, I.	321		
Subramonium, P.	701	Zegenhagen, J.	497, 805
Sugii, N.	597	Zhang, C.	369
Sumarlina, A.B.	481	Zhang, E.	369
Suwa, T.	115	Zhang, J.	81
		Zhang, L.	923
Teo, Z.	125	Zhang, W.	81
Teramoto, A.	115	Zhao, L.	717
Tiedemann, A.	461	Zhao, Q.	461
Tsai, W.	39	Zheng, R.	353
Tsao, C.	39	Zhou, K.	651
Tsuchiya, T.	3		
Tsutsui, K.	597		
Tyaginov, S. E.	321		
Ueda, T.	861		
Van de Sanden, M.	191		
Van Helvoirt, C.	191		
Velamala, J.	353		
Verove, C.	667		
Virot, L.	729		
Volpi, F.	497		
Voutsas, A.	241		
Wang, C.	353		
Wang, T.	39		
Watanabe, H.	303		